Handbook of organic materials for optical and (opto)electronic devices

Related titles:

Printed films: Materials science and applications in sensors, electronics and photonics
(ISBN 978-1-84569-988-8)

Organic light-emitting diodes (OLEDs): Materials, devices and applications
(ISBN 978-0-85709-425-4)

Carbon nanotubes and graphene for photonic applications
(ISBN 978-0-85709-417-9)

Details of these books and a complete list of titles from Woodhead Publishing can be obtained by:

- visiting our web site at www.woodheadpublishing.com
- contacting Customer Services (e-mail: sales@woodheadpublishing.com; fax: +44 (0) 1223 832819; tel.: +44 (0) 1223 499140 ext. 130; address: Woodhead Publishing Limited, 80, High Street, Sawston, Cambridge CB22 3HJ, UK)
- in North America, contacting our US office (e-mail: usmarketing@woodheadpublishing.com; tel.: (215) 928 9112; address: Woodhead Publishing, 1518 Walnut Street, Suite 1100, Philadelphia, PA 19102-3406, USA)

If you would like e-versions of our content, please visit our online platform: www.woodheadpublishingonline.com. Please recommend it to your librarian so that everyone in your institution can benefit from the wealth of content on the site.

We are always happy to receive suggestions for new books from potential editors. To enquire about contributing to our Electronic and Optical Materials series, please send your name, contact address and details of the topic/s you are interested in to laura.pugh@woodheadpublishing.com. We look forward to hearing from you.

The team responsible for publishing this book:
Commissioning Editor: Laura Pugh
Publications Coordinator: Adam Davies
Project Editor: Ginny Mills
Editorial and Production Manager: Mary Campbell
Production Editor: Richard Fairclough
Cover Designer: Terry Callanan

Woodhead Publishing Series in Electronic and Optical Materials:
Number 39

Handbook of organic materials for optical and (opto)electronic devices

Properties and applications

Edited by
Oksana Ostroverkhova

WOODHEAD PUBLISHING

Oxford Cambridge Philadelphia New Delhi

Published by Woodhead Publishing Limited,
80 High Street, Sawston, Cambridge CB22 3HJ, UK
www.woodheadpublishing.com
www.woodheadpublishingonline.com

Woodhead Publishing, 1518 Walnut Street, Suite 1100, Philadelphia,
PA 19102-3406, USA

Woodhead Publishing India Private Limited, 303, Vardaan House, 7/28 Ansari Road,
Daryaganj, New Delhi – 110002, India
www.woodheadpublishingindia.com

First published 2013, Woodhead Publishing Limited

British Library Cataloguing in Publication Data
A catalogue record for this book is available from the British Library.

Library of Congress Control Number: 2013941717

ISBN 978-0-85709-265-6 (print)
ISBN 978-0-85709-876-4 (online)
ISSN 2050-1501Woodhead Publishing Series in Electronic and Optical Materials (print)
ISSN 2050-151X Woodhead Publishing Series in Electronic and Optical Materials (online)

The publisher's policy is to use permanent paper from mills that operate a sustainable forestry policy, and which has been manufactured from pulp which is processed using acid-free and elemental chlorine-free practices. Furthermore, the publisher ensures that the text paper and cover board used have met acceptable environmental accreditation standards.

Typeset by Toppan Best-set Premedia Limited, Hong Kong
Printed by Lightning Source

Contents

Contributor contact details

(* = main contact)

Editor

O. Ostroverkhova
Oregon State University
USA

E-mail: oksana@science.oregonstate.edu

Chapter 1

Y. Shirota*
Department of Environmental and Biological Chemistry
Fukui University of Technology
3-6-1 Gakuen
Fukui City
Fukui 910-0850
Japan

E-mail: shirota@fukui-ut.ac.jp, shirota@chem.eng.osaka-u.ac.jp

H. Kageyama
Department of Electrical and Electronics Engineering
Faculty of Engineering
University of the Ryukyus
1 Senbaru
Nishihara
Okinawa 903-0213
Japan

Chapter 2

S. R. Puniredd, W. Pisula* and K. Müllen
Max Planck Institute for Polymer Research
Ackermannweg 10
55124 Mainz
Germany

E-mail: pisula@mpip-mainz.mpg.de; muellen@mpip-mainz.mpg.de

Chapter 3

K. Leo* and M. Hummert
Institut für Angewandte Photophysik
Technische Universität Dresden
George-Bähr-Straße 1
01062 Dresden
Germany

E-mail: leo@iapp.de

Chapter 4

H. Kishida*
Department of Applied Physics
Nagoya University
Furo-cho
Chikusa-ku
Nagoya 464-8603
Japan

E-mail: kishida@nuap.nagoya-u.ac.jp

T. Yamamoto
Chemical Resources Laboratory
Tokyo Institute of Technology
4259 Nagatsuta
Midori-ku
Yokohama 226-8503
Japan

E-mail: tyamamot@res.titech.ac.jp

Chapters 5 and 12

I. Biaggio
Department of Physics
Lehigh University
16 Memorial Drive East
Bethlehem, PA 18015
USA

E-mail: biaggio@lehigh.edu

Chapter 6

M. Jazbinsek* and P. Günter
Nonlinear Optics Laboratory
ETH Zurich
Switzerland
and
Rainbow Photonics AG
Farbhofstrasse 21
CH-8048 Zurich
Switzerland

E-mail: mojca@rainbowphotonics.com; gunter@phys.ethz.ch

Chapter 7

J. C. Sancho-Garcia
Departamento de Quimica-Fisica
Universidad de Alicante
E-03080 Alicante
Spain

E-mail: jc.sancho@ua.es

Chapter 8

J. Godlewski* and M. Obarowska
Gdansk University of Technology
G. Narutowicza 11/12
80-233 Gdansk
Poland

E-mail: jago@mif.pg.gda.pl

Chapter 9

J. Perez-Moreno
Department of Physics
Skidmore College
815 North Broadway
Saratoga Springs, NY 12866
USA
and
Department of Physics and Astronomy
Washington State University
Pullman, WA 99164-2814
USA

E-mail: jperezmo@skidmore.edu; xpmworks@gmail.com

Chapter 10

Z. V. Vardeny*
Department of Physics and Astronomy
University of Utah
Salt Lake City, UT 84112
USA

E-mail: val@physics.utah.edu

C.-X. Sheng
School of Electronic and Optical Engineering
Nanjing University of Science and Technology
Nanjing
Jiangsu
210094
China

Chapter 11

H. Diesinger, E. A. Chan, J. Yin and C. Soci*
Centre for Disruptive Photonic Technologies and Division of Physics and Applied Physics
21 Nanyang Link
Nanyang Technological University
Singapore
637371

E-mail: csoci@ntu.edu.sg

Chapter 13

O. D. Jurchescu
Wake Forest University
Department of Physics
1834 Wake Forest Road
Winston-Salem, NC 27109-7507
USA

E-mail: oana@wfu.edu

Chapter 14

A. Pivrikas
School of Chemistry and Molecular Biosciences
Centre for Organic Photonics and Electronics (COPE)
The University of Queensland
Brisbane
Australia

E-mail: almantas.pivrikas@uq.edu.au

Chapter 15

K. Willets*
The University of Texas at Austin
Department of Chemistry
Welch Hall 2.204
105 E. 24th St. Stop A5300
Austin, TX 78712-1224
USA

K. Mayer
Tufts University
Department of Chemistry
62 Talbot Avenue
Medford, MA 02155
USA

E-mail: kwillets@cm.utexas.edu

Chapter 16

K. D. Singer and Y. Wu*
Case Western Reserve University
Department of Physics
Cleveland, OH 44106-7079
USA

E-mail: kds4@cwru.edu

Chapter 17

M. Hösel, D. Angmo and F. C. Krebs*
Technical University of Denmark
Department of Energy Conversion and Storage
Frederiksborgvej 399
Roskilde
DK-4000
Denmark

E-mail: frkr@dtu.dk

Chapter 18

T. Schwab,* B. Lüssem, M. Furno, M. C. Gather, and K. Leo
Institut für Angewandte Photophysik
TU Dresden
01069 Dresden
Germany

E-mail: tobias.schwab@iapp.de; leo@iapp.de

Chapter 19

Z. V. Vardeny* and T. D. Nguyen
Physics and Astronomy Department
University of Utah
Salt Lake City, UT 84112
USA

E-mail: val@physics.utah.edu

E. Ehrenfreund
Physics Department
Technion–Israel Institute of Technology
Haifa 32000
Israel

Chapter 20

H. E. Katz* and T. J. Dawidczyk
Department of Materials Science and Engineering
Johns Hopkins University
206 Maryland Hall
3400 North Charles Street
Baltimore, MD 21218
USA

E-mail: hekatz@jhu.edu

H. Kong
Korea Research Institute of Chemical Technology (KRICT)
Research Center for Green Fine Chemicals 895-4
Yugok-dong
Jung-gu
Ulsan
681-310
Republic of Korea

E-mail: hkong2@krict.re.kr; lovealkein@gmail.com

Chapter 21

G. Tarabella, N. Coppedè and S. Iannotta
IMEM-CNR
Institute of Materials for Electronics and Magnetism
Viale Usberti 37/A
Parco Area delle Scienze
43100
Parma
Italy

F. Cicoira,* P. Kumar and C. Santato
Département de Génie Chimique
École Polytechnique de Montréal
2500 chemin de Polytechnique
Montréal
Québec
H3T 1J4
Canada

E-mail: fabio.cicoira@polymtl.ca

Chapter 22

M. C. Petty
School of Engineering and Computing Sciences and Centre for Molecular and Nanoscale Electronics
Durham University
Durham DH1 3LE
UK

E-mail: m.c.petty@durham.ac.uk

Chapter 23

M. Oszajca*
Jagiellonian University
Faculty of Chemistry
ul. Ingardena 3
30-060 Kraków
Poland

A. Podborska
AGH University of Science and Technology
Faculty of Non-Ferrous Metals
al. Mickiewicza 30
30-059 Kraków
Poland

K. Szaciłowski
Jagiellonian University
Faculty of Chemistry
ul. Ingardena 3
30-060 Kraków
Poland

and

AGH University of Science and Technology
Faculty of Non-Ferrous Metals
al. Mickiewicza 30
30-059 Kraków
Poland

E-mail: szacilow@agh.edu.pl; szacilow@chemia.uj.edu.pl

Chapter 24

J. Thomas
NanoScience Technology Center
CREOL, College of Optics and Photonics and
Department of Material Science and Engineering
University of Central Florida
Orlando, FL 32826
USA

E-mail: jayan.thomas@ucf.edu

Chapter 25

R. A. Norwood*
College of Optical Sciences
University of Arizona
1630 E. University Blvd
Tucson, AZ 85721
USA

E-mail: rnorwood@optics.arizona.edu

C. T. DeRose
Sandia National Laboratories
Applied Photonics Microsystems
P.O. Box 5800, MS 1082
Albuquerque, NM 87185
USA

E-mail: cderose@sandia.gov

C. Greenlee
Lightwave Logic Corporation
Suite 100
1 Innovation Way
Newark, DE 19711
USA

E-mail: chuck.g@lightwavelogic.com

A. Yeniay
Photon-X LLC
283 Great Valley Parkway
Malvern, PA 19355
USA

E-mail: ayeniay@photon-x.net

Woodhead Publishing Series in Electronic and Optical Materials

1 **Circuit analysis**
J. E. Whitehouse

2 **Signal processing in electronic communications: For engineers and mathematicians**
M. J. Chapman, D. P. Goodall and N. C. Steele

3 **Pattern recognition and image processing**
D. Luo

4 **Digital filters and signal processing in electronic engineering: Theory, applications, architecture, code**
S. M. Bozic and R. J. Chance

5 **Cable engineering for local area networks**
B. J. Elliott

6 **Designing a structured cabling system to ISO 11801: Cross-referenced to European CENELEC and American Standards**
Second edition
B. J. Elliott

7 **Microscopy techniques for materials science**
A. Clarke and C. Eberhardt

8 **Materials for energy conversion devices**
Edited by C. C. Sorrell, J. Nowotny and S. Sugihara

9 **Digital image processing: Mathematical and computational methods**
Second edition
J. M. Blackledge

10 **Nanolithography and patterning techniques in microelectronics**
Edited by D. Bucknall

11 **Digital signal processing: Mathematical and computational methods, software development and applications**
Second edition
J. M. Blackledge

12 **Handbook of advanced dielectric, piezoelectric and ferroelectric materials: Synthesis, properties and applications**
Edited by Z.-G. Ye

13 **Materials for fuel cells**
Edited by M. Gasik

14 **Solid-state hydrogen storage: Materials and chemistry**
Edited by G. Walker

15 **Laser cooling of solids**
S. V. Petrushkin and V. V. Samartsev

16 **Polymer electrolytes: Fundamentals and applications**
Edited by C. A. C. Sequeira and D. A. F. Santos

17 **Advanced piezoelectric materials: Science and technology**
Edited by K. Uchino

18 **Optical switches: Materials and design**
Edited by S. J. Chua and B. Li

19 **Advanced adhesives in electronics: Materials, properties and applications**
Edited by M. O. Alam and C. Bailey

20 **Thin film growth: Physics, materials science and applications**
Edited by Z. Cao

21 **Electromigration in thin films and electronic devices: Materials and reliability**
Edited by C.-U. Kim

22 ***In situ* characterization of thin film growth**
Edited by G. Koster and G. Rijnders

23 **Silicon-germanium (SiGe) nanostructures: Production, properties and applications in electronics**
Edited by Y. Shiraki and N. Usami

24 **High-temperature superconductors**
Edited by X. G. Qiu

25 **Introduction to the physics of nanoelectronics**
S. G. Tan and M. B. A. Jalil

26 **Printed films: Materials science and applications in sensors, electronics and photonics**
Edited by M. Prudenziati and J. Hormadaly

27 **Laser growth and processing of photonic devices**
Edited by N. A. Vainos

28 **Quantum optics with semiconductor nanostructures**
Edited by F. Jahnke

29 **Ultrasonic transducers: Materials and design for sensors, actuators and medical applications**
Edited by K. Nakamura

30 **Waste electrical and electronic equipment (WEEE) handbook**
Edited by V. Goodship and A. Stevels

31 **Applications of ATILA FEM software to smart materials: Case studies in designing devices**
Edited by K. Uchino and J.-C. Debus

32 **MEMS for automotive and aerospace applications**
Edited by M. Kraft and N. M. White

33 **Semiconductor lasers: Fundamentals and applications**
Edited by A. Baranov and E. Tournie

34 **Handbook of terahertz technology for imaging, sensing and communications**
Edited by D. Saeedkia

35 **Handbook of solid-state lasers: Materials, systems and applications**
Edited by B. Denker and E. Shklovsky

36 **Organic light-emitting diodes: Materials, devices and applications**
Edited by A. Buckley

37 **Lasers for medical applications: Diagnostics, therapy and surgery**
Edited by H. Jelínková

38 **Semiconductor gas sensors**
Edited by R. Jaaniso and O. K. Tan

39 **Handbook of organic materials for optical and optoelectronic devices: Properties and applications**
Edited by O. Ostroverkhova

40 **Metallic films for electronic, optical and magnetic applications: Structure, processing and properties**
Edited by K. Barmak and K. Coffey

41 **Handbook of laser welding technologies**
Edited by S. Katayama

42 **Nanolithography: The art of fabricating nanoelectronics, nanophotonics and nanobiology devices and systems**
Edited by M. Feldman

43 **Laser spectroscopy for sensing: Fundamentals, techniques and applications**
Edited by M. Baudelet

44 **Chalcogenide glasses: Preparation, properties and applications**
Edited by J.-L. Adam and X. Zhang

45 **Handbook of MEMS for wireless and mobile applications**
Edited by D. Uttamchandani

46 **Subsea optics and imaging**
Edited by J. Watson and O. Zielinski

47 **Carbon nanotubes and graphene for photonic applications**
Edited by S. Yamashita, Y. Saito and J. H. Choi

48 **Optical biomimetics: Materials and applications**
Edited by M. Large

49 **Optical thin films and coatings**
Edited by A. Piegari and F. Flory

50 **Computer design of diffractive optics**
Edited by V. A. Soifer

51 **Smart sensors and MEMS: Intelligent devices and microsystems for industrial applications**
Edited by S. Nihtianov and A. Luque

52 **Fundamentals of femtosecond optics**
V. Kozlov and V. Samartsev

Preface

Organic (opto)electronic and nonlinear optical (NLO) materials have attracted attention due to their low cost, easy fabrication, and tunable properties. This has spurred the rapid development of the field of organic (opto)electronics and nonlinear optics over the past 15–20 years: a variety of novel materials have been synthesized and characterized, novel experimental techniques that enabled better understanding of physics behind optoelectronic and NLO properties have been established, and various exciting applications have been demonstrated. This book aims to provide an overview of recent developments in all three areas (materials development, characterization techniques, and applications). Although in the edited book it is difficult to keep the level of material coverage uniform from chapter to chapter, the overall level should be sufficiently accessible for researchers who are entering the field of organic (opto)electronics and nonlinear optics (graduate students, postdoctoral researchers, and professionals switching field). Yet, it should be advanced enough to serve as a useful desk reference for both academic and industrial researchers, with a background in physics, chemistry, materials science, and engineering, already working with organic materials and their applications. To help the reader deepen their knowledge of a particular topic, each chapter contains an extensive list of references which provide further information on the subject matter discussed in the chapter.

The book is organized as follows. Part I overviews novel optoelectronic (Ch. 1–3) and NLO (Ch. 4–6) organic materials and examples of structure–property relationships. New directions highlighted in this part include molecular design of high-performance organic semiconductors (Ch. 1), organic thin film processing methods (Ch. 2), tuning of conductive and luminescent properties of organic semiconductors using doping (Ch. 3), molecular assembly design strategies for maximized NLO response (Ch. 4 and 5), and development of organic NLO materials for terahertz (THz) wave generation and detection (Ch. 6).

Part II of the book starts with a general theoretical description of charge generation and transport (Ch. 7), optical and luminescent properties (Ch. 8), and NLO properties (Ch. 9) of organic materials. This are followed by descriptions of experimental methods widely used in the characterization of organic materials, which are illustrated by examples of physical studies (Ch. 10–16). In particular, ultrafast optical, THz, and microwave spectroscopy as a means of studying the dynamics of photoexcited neutral and charged states are reviewed in Ch. 10 and 11. Additional time-resolved techniques for probing exciton and charge carrier dynamics with sub-nanosecond resolution are discussed in Ch. 11 and 12. Measurements of charge carrier mobility and other parameters pertaining to charge transport are detailed in Ch. 13 and 14. The surface enhanced Raman scattering method, applied to studies of metal-organic interactions, which are important for understanding contribution of contacts at the metal-organic interfaces to (opto)electronic properties, is introduced in Ch. 15. Finally, Ch. 16 reviews methods of studying NLO properties of organic materials and highlights applications of these methods to various fields.

Part III of the book is dedicated to selected applications of organic (opto) electronic and NLO materials (Ch. 17–25). In order to facilitate a quick reference to device performance, many chapters in Part III contain tables that summarize device characteristics of the best currently available (as of 2012) organic devices such as organic solar cells (Ch. 17; also see Tables 1.1 and 1.2 in Ch. 1), light-emitting diodes (Ch. 18), chemical and biological sensors (Ch. 20 and 21), and electronic memory and logic devices (Ch. 22 and 23). The relatively young field of organic spintronics is reviewed in Ch.19. Physics and applications of organic NLO materials are illustrated in Ch. 24 (photorefractive materials) and Ch. 25 (electro-optic materials).

As no single book can provide a complete coverage of the field, some important topics have had to be omitted. For information on these topics, the reader is referred to recent comprehensive review papers. Examples of such topics include charge carrier photogeneration in organic materials (see, for example, *Chemical Reviews* **110**(11), 6736 (2010), *Advanced Materials* **22**(37), 4097(2010)); charge injection at metal–organic interfaces (*Advanced Materials* **24**(11), 1357 (2012)); fabrication, characterization, and performance of organic field-effect transistors (*Chemical Reviews* **112**(4), 2208 (2012), *Chemical Reviews* **111**(8), 4833 (2011), *Chemical Reviews* **111**(5), 3358 (2011), *Chemical Reviews* **110**(1), 3 (2010)), photoconductors (*Chemical Reviews* **110**(1), 479 (2010)), and lasers (*Polymer International* **61**(3), 390 (2012)); and many others.

I would like to express my gratitude to all contributing authors for investing their time and effort in writing as accurate, complete, and up-to-date chapters as possible and for keeping on schedule as much as possible. It has also been a great pleasure to work with Adam Davies and Laura Pugh at

Woodhead Publishing – many thanks to them and to the Woodhead Publishing production team for excellent professionalism in managing the project, ensuring timely communication with authors, and getting things done promptly and with high quality. This book could not have happened without a dedicated effort of all the authors, the editor, and the publisher working as a team.

Oksana Ostroverkhova
Corvallis, Oregon, USA

Part I

Materials for organic (opto)electronics and nonlinear optics: structure–property relations

1

Small molecular weight materials for (opto)electronic applications: overview

Y. SHIROTA, Fukui University of Technology, Japan and
H. KAGEYAMA, University of the Ryukyus, Japan

DOI: 10.1533/9780857098764.1.3

Abstract: This chapter gives an overview of organic materials, focusing on molecular materials for use in optoelectronic devices such as organic photoreceptors, organic photovoltaic devices (OPVs), and organic light-emitting diodes (OLEDs). A description of the historical development of organic optoelectronics is provided. This is followed by a review of the principles and operation processes involved in such optoelectronic devices and the molecular materials for use in such devices. Finally, device structures and performance are discussed.

Key words: organic optoelectronics, organic light-emitting diode (OLED), organic photovoltaic device (OPV), organic semiconductor, molecular material.

1.1 Introduction

Electronic, optoelectronic, and photonic devices using inorganic semiconductors, e.g., silicon transistor-based computers, light-emitting diodes (LEDs), devices based on semiconductor lasers, and optical communication devices such as waveguides, have greatly contributed to the arrival of the present information-oriented society. In recent years, new fields of organic electronics, organic optoelectronics, and organic photonics using organic materials have opened up. These newly emerging fields of science and technology are mainly concerned with novel thin-film, flexible devices using organic materials. Devices using organic materials have the following advantages over inorganic semiconductor-based devices; lightweight, flexibility, and potentially low cost. Organic photovoltaic devices (OPVs), organic light-emitting diodes (OLEDs), and organic field-effect transistors (OFETs) have been the central subjects of current research and development in the fields of organic electronics and optoelectronics.

Electrophotography has long been applied to practical copying machines and printers, making significant contributions to the development of information-oriented society. OPVs have been receiving a great deal of attention as promising candidates for next-generation solar cells that contribute to solving the present global issues of the environment and energy. OLEDs have already found successful practical application as portable, flat-panel displays for various purposes and have been expected to be a candidate for next-generation solid-state lighting. OFETs are expected to be key devices for driving various devices in the printed electronics, which aim to fabricate the whole part of thin-film, flexible organic devices by use of printing techniques.

Organic electronics, optoelectronics, and photonics constitute interdisciplinary fields that cover physics, chemistry, biology, and materials science. That is, the science and technology of these fields, which can be termed organic functional materials science or organic device science, include wide areas from the molecular design and synthesis of photo and electroactive organic materials to the elucidation of their physical and chemical properties as well as their structures, fabrication of devices using synthesized organic materials, evaluation of their performance, and creation of new knowledge underlying the operation of organic devices. As photo and electroactive organic materials, including both small molecules and polymers, are expected to be key materials for organic electronics, optoelectronics, and photonics, and for 21st-century industries related to information and energy, the creation of new photo and electroactive organic materials is a challenging subject.

This chapter deals with an overview of organic materials, focusing on molecular materials for use in optoelectronic devices such as organic photoreceptors in electrophotography, OPVs, and OLEDs. Following the historical development in organic optoelectronics, principles of and operation processes involved in optoelectronic devices, molecular materials for use in such devices, and device structures and performance are discussed. Book and review articles on optoelectronic devices such as electrophotography, OPVs, and OLEDs are available (Mort and Pai, 1976; Borsenberger and Weiss, 1993; Miyata and Nalwa, 1997; Thelakkat, 2002; Forrest, 2004; Sun and Sariciftci, 2005; Kafafi, 2005; Müllen and Scherf, 2006; Shirota and Kageyama, 2007; Roncali *et al.*, 2007; Mishra *et al.*, 2009).

1.2 Historical development in organic (opto)electronics: devices and materials

It was in the 1960s that electrophotographic copying machines using amorphous selenium as a photoreceptor material were put into practical use. In the early 1970s, an organic photoreceptor using a single layer of a poly(*N*-vinylcarbazole) (PVCz) / 2,4,7-trinitrofluorenone (TNF) charge-transfer complex was developed (Schaffert, 1971) and used. This can be regarded as the beginning of organic optoelectronics. Photoreceptors with two-layer structures consisting of a charge-carrier generation layer (CGL) and a charge-carrier transport layer (CTL) have soon become the main focus in the development of photoreceptors (Regensburger, 1968; Mort, 1972; Mort and Nielsen, 1972; Melz *et al.*, 1977). The electrophotographic process has later expanded its application to printers for a computer terminal. A number of organic materials have been developed for use in the CGL and CTL in photoreceptors.

Related to applications of organic photoconductors for use in photoreceptors in electrophotography, photoconductivity of organic materials has been extensively studied (Mort and Pai, 1976; Borsenberger and Weiss, 1993). The electric-field dependence of the photogeneration of charge carriers has been analyzed in terms of the Onsager theory (Onsager, 1938) for anthracene single crystals (Batt *et al.*, 1968; Chance and Braun, 1976; Lyons and Milne, 1976), PVCz (Hughes, 1971; Pfister and Williams, 1974), PVCz/TNF system (Melz, 1972), phthalocyanines (Popovic and Menzel, 1979; Popovic, 1982; Popovic, 1984), and molecularly doped polymer systems (Borsenberger and Hoesterey, 1980; Borsenberger *et al.*, 1978).

Significant progress has also been made toward understanding charge transport. The Scher–Montroll theory (Scher and Montroll, 1975) enabled the analysis of dispersive transient photocurrent curves usually observed for disordered systems such as PVCz (Pfister and Griffiths, 1978) and molecularly doped polymer systems (Pfister, 1977), allowing the determination of the transit time and, hence, charge-carrier mobilities. The empirical equation (Gill, 1972) based on the Poole–Frenkel model and the disorder formalism (Bässler, 1981, 1984, 1993) were presented to explain the temperature and electric-field dependencies of charge-carrier drift mobilities. The disorder formalism assumes that charge transport in disordered systems, e.g., polymers and molecularly doped polymer systems, takes place by hopping through a manifold of localized states subject to the fluctuations of both the hopping site energy and the intermolecular π-electron overlap. A large amount of experimental data has been analyzed in terms of these two models. The small-polaron theory was also presented, which assumes that charge transport takes place via the hopping of charge carriers accompanied by the lattice deformation, i.e., small polarons, and that the hopping of charge carriers is assisted by phonons (Schein and Mack, 1988). The hopping-site distance dependence of the activation energy for charge transport in molecularly doped polymer systems has been explained in terms of the small-polaron model (Mack *et al.*, 1989; Schein *et al.*, 1990; Nomura and Shirota, 1997).

Research into the utilization of solar energy, which began in the 1970s, included conversion of solar energy into chemical and electrical energies. With regard to the conversion of solar energy into electrical energy, studies were made of wet photocells using inorganic semiconductor electrodes and redox reactants such as polysulfide (Ellis *et al.*, 1976a,b) and I_3^-/I^-

(Tsubomura *et al.*, 1976) in solution. Recently developed dye-sensitized solar cells (DSCs) using a nanocrystalline TiO_2 electrode, on which organic dye molecules are coated, and I_3^-/I^- redox species in solution have received a great deal of attention because of their high power conversion efficiencies (PCEs) of 7.9–10% for the simulated solar light (O'Regan and Grätzel, 1991; Nazeeruddin *et al.*, 1993). Studies on organic thin-film OPVs also began in the middle of the 1970s in view of their application to organic solar cells. Schottky-type devices using chlorophyll (Tang and Albrecht, 1975a,b) and PVCz/TNF (Reucroft *et al.*, 1974, 1975) fabricated in the early days gave very low PCEs of ca. 10^{-3} to 10^{-2}%. A much higher conversion efficiency of ca. 1.0% was then reported for OPVs using merocyanine dyes (Morel, 1979). Polyacetylene and doped polyacetylene were also used as photoactive materials (Ozaki *et al.*, 1979, 1980; Weinberger *et al.*, 1981; Tsukamoto and Ohigashi, 1982). From the standpoint of reducing the electrical resistivity of organic materials, electrochemically doped pendant polymers such as PVCz (Shirota *et al.*, 1985), a polymer containing a pendant triphenylamine moiety (Shirota *et al.*, 1987), and poly(3-methylthiophene) (Glenis *et al.*, 1986) were used as p-type organic semiconductors in Schottky-type OPVs. It was reported in 1986 that a two-layer pn-heterojunction OPV using vacuum-deposited thin films of crystalline copper phthalocyanine as a p-type organic semiconductor (electron donor) and a perylene dye as an n-type organic semiconductor (electron acceptor) exhibited a PCE of ~1% under simulated AM2 illumination (Tang, 1986). This report has driven extensive studies on pn-heterojunction OPVs (Wöhrle and Meissner, 1991), leading to remarkable improvement in the PCE. At present, PCEs up to 6–10% under simulated sunlight illumination have been attained for OPVs (J. Y. Kim *et al.*, 2007; K. Kim *et al.*, 2007; Chen *et al.*, 2009; Liang *et al.*, 2010; He *et al.*, 2011; Green *et al.*, 2012). The remarkable improvement in the PCE is due to the development of new materials, e.g., poly(3-hexylthiophene) as a p-type polymer semiconductor, fullerenes as n-type organic semiconductors, the implementation of new device structures e.g., planar pn-heterojunction structures with the insertion of an exciton-blocking layer (Peumans *et al.*, 2000; Peumans and Forrest, 2001) or optical spacers (Hayakawa *et al.*, 2007; Park *et al.*, 2009), tandem structures (Hiramoto *et al.*, 1990; Yakimov and Forrest, 2002), p-i-n (Hiramoto *et al.*, 1991; Hiramoto *et al.*, 1992; Xue *et al.*, 2005) and bulk-heterojunction structures (Sariciftci *et al.*, 1992; Yu *et al.*, 1995; Halls *et al.*, 1995), and a deeper understanding of device physics (Brédas *et al.*, 2009).

Organic electroluminescence (EL) had been studied in the early 1960s using single crystals of anthracene with a thickness of ca. 10–20 μm. However, it was necessary to apply a high voltage of ca. 400 V between the anode and the cathode, resulting in low brightness (Pope *et al.*, 1963;

Helfrich and Schneider, 1965). Studies of electro-chemiluminescence resulting from the recombination of cation and anion radicals of molecules produced by the electrolysis of organic molecules, e.g., anthracene and its derivatives, in solution has also been studied since the 1960s (Visco and Chandross, 1964).

An epoch-making report appeared in 1987 that a two-layer device consisting of a vacuum-deposited hole-transport layer of an aromatic diamine and an emissive layer of tris(8-quinolinolato)aluminum (Alq_3) with electron-transporting properties, sandwiched between the ITO anode and the Mg/Ag cathode, emitted green EL resulting from the fluorescence of Alq_3 with a brightness of ca. 1000 cd m^{-2} by the application of ca. 10 V (Tang and VanSlyke, 1987). A yellow-emitting polymer light-emitting diode using poly(*p*-phenylenevinylene) (PPV) was then reported (Burroughes *et al.*, 1990). Various π-conjugated polymers have been applied as materials for polymer LEDs (Braun and Heeger, 1991; Ohmori *et al.*, 1991; Braun *et al.*, 1992; Grem *et al.*, 1992).

In order to achieve high luminous and quantum efficiencies for OLEDs, multi-layer device structures consisting of hole-injection, hole-transport, hole blocking, and electron-transport layers as well as the emitting layer have been employed. Accordingly, materials functioning in each layer have been required. Independently ongoing research into the creation of photo and electroactive small organic molecules that readily form stable amorphous glasses (Shirota *et al.*, 1989; Shirota, 2000), which are referred to as amorphous molecular materials, merged with the research and development of OLEDs, and a number of high-performance amorphous molecular materials suitable for each layer in OLEDs have been developed. As a result, EL external quantum efficiencies (EQEs) close to or over 5%, which is assumed to be a theoretical limit when the light outcoupling efficiency is assumed to be ca. 20%, have been achieved for the fluorescence-emitting OLEDs (Murata *et al.*, 2002; Wu *et al.*, 2004).

In addition to the EL resulting from the fluorescence of emitting materials, the utilization of phosphorescence emission for EL by use of room-temperature phosphorescent materials has revolutionized the improvement in luminous and quantum efficiencies of OLEDs. Metal complexes, e.g., osmium complexes (Ma *et al.*, 1998), platinum complexes (Baldo *et al.*, 1998), and iridium complexes (Baldo *et al.*, 1999), were reported as phosphorescent emissive dopants for OLEDs. The EQEs of 19% together with an internal quantum efficiency of 100% were reported for green-emitting phosphorescent OLEDs (Adachi *et al.*, 2001c; Ikai *et al.*, 2001).

Both fluorescent and phosphorescent OLEDs have been commercialized. Research and development toward putting OLEDs to practical use as solid-state lighting have also been moving ahead. A white-light emitting OLED with a luminous efficiency (power efficiency) of ca. 90 lm W^{-1} at

1000 cd m^{-2}, which exceeds that of the conventional fluorescent lamp, has been reported (Reineke *et al.*, 2009).

1.3 Photo and electroactive organic materials: organic π-electron systems

1.3.1 Organic π-electron systems: organic semiconductors

Organic materials for use in optoelectronic devices are often referred to as photo and electroactive organic materials as they exhibit such phenomena as light absorption and emission, photogeneration and transport of charge carriers, injection of charge carriers from the electrode, when they are used in devices. Generally, organic π-electron systems are characterized by the capability of absorbing and emitting light in the wavelength region from ultraviolet to near infrared, and generating and transporting charge carriers, and by excellent nonlinear optical properties. Photo and electroactive organic materials for use in electronic and optoelectronic devices such as OFETs, OPVs and OLEDs are often termed organic semiconductors. Photo and electroactive organic materials based on π-electron systems include small molecules, oligomers with well-defined structures, and polymers.

Small organic molecules are crystalline in nature. The representative classes of molecular materials are metal and metal-free phthalocyanines (Leznoff and Lever, 1992; Hanack *et al.*, 2002; Claessens *et al.*, 2008), porphyrines, perylenebisdiimides, e.g., anthra[2″,1″,9″:4,5,6:6″5″10″:4′,5′,6′] diisoquino[2,1-a:2′,1′-a′]dibenzimidazole-10,21-dione (PTCBI), poly-condensed aromatic hydrocarbons, e.g., anthracene, pentacene, and fullerenes (Fig. 1.1). Basic studies on electrical, optical, and magnetic properties of molecular crystals have been made extensively (Wright, 1995). The phenomenon of electrical conduction in organic materials was discovered for phthalocyanine (Eley, 1948), violanthrene (Akamatu and Inokuchi, 1950), and a perylene/bromine charge-transfer complex (Akamatu *et al.*, 1954). In addition to crystalline molecular materials, various amorphous molecular materials based on π-electron systems have been created (Shirota *et al.*, 1989; Shirota, 2000, 2005), as described in Section 1.3.4.

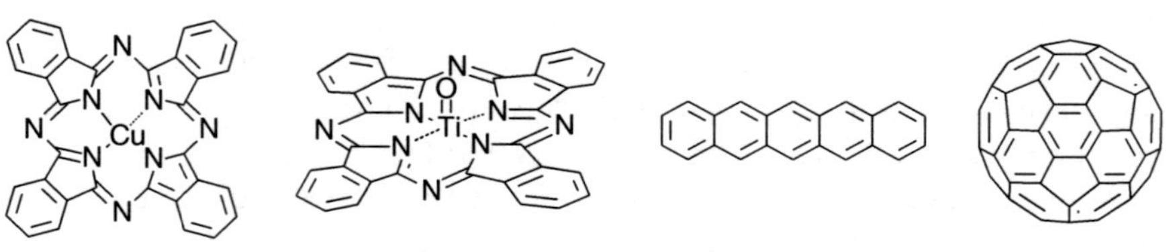

1 (CuPc) **2** (TiOPc) **3** (pentacene) **4** (C_{60})

1.1 Examples of representative crystalline molecular materials.

π-Conjugated polymers have received attention not only as electrically conducting polymers but also as various functional materials, e.g., semiconductors for use in OPVs, OFETs and OLEDs, and nonlinear optical materials. Following the development of freestanding polyacetylene film (Ito *et al.*, 1974), its high electrical conductivity resulting from charge-transfer doping was discovered (Shirakawa *et al.*, 1977; Chiang *et al.*, 1978). Since the appearance of these papers, electrical conductivities of various π-conjugated polymers resulting from charge-transfer doping have been extensively studied (Skotheim, 1998). They include poly(*p*-phenylene)s, poly(*p*-phenylene vinylene)s, poly(9,9-dialkylfluorene)s, polythiophenes, polypyrroles, and polyanilines. Electrically conducting non-conjugated polymers containing pendant π-electron systems, which were produced by electrochemical doping, were also developed (Kanega *et al*, 1984; Shirota *et al.*, 1984; Shirota *et al.*, 1991).

π-Conjugated oligomers with well-defined structures have recently received a great deal of attention not only as models for the corresponding π-conjugated polymers but also as a new family of organic π-electron systems, the properties and functions of which can be controlled by varying the π-conjugation length (Fig. 1.2). Extensive studies have been made of the synthesis, properties, and functions of oligothiophenes with well-defined structures (Beny *et al.*, 1982; Kagan and Arora, 1983; Wynberg and Metselaar, 1984; Nakayama *et al.*, 1988; Martinez *et al.*, 1989; Merz and Ellinger, 1991; Bäuerle *et al.*, 1993, 1995; Noda and Shirota, 1998; Noda *et al.*, 1997a,b, 1999a; Nakanishi *et al.*, 1998, 1999; Malenfant *et al.*, 1998; Malenfant and Fréchet, 1998). Non-conjugated polymers containing pendant oligothiophenes with varying conjugation lengths have also been synthesized and characterized (Nawa *et al.*, 1993a,b; Imae *et al.*, 1997).

1.3.2 Control of structures and morphologies

Control of structures and morphologies of organic materials, small molecules, oligomers with well-defined structures, and polymers is of crucial importance in their applications for devices. In order to make π-conjugated polymers soluble in organic solvents, alkyl groups are introduced as substituents. Highly regioregular poly(3-alkylthiophene)s have been synthesized, and the effect of the regioregularity on their properties and device performance has been discussed (Assadi *et al.*, 1988; Berggren *et al.*, 1994; Chen and Rieke, 1992, 1993; Chen *et al.*, 1995, 1996; Guillerez and Bidan,

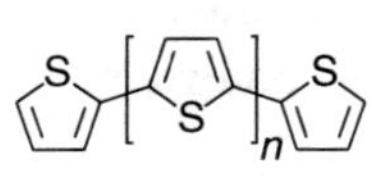

5

1.2 Oligothiophenes with well-defined structures.

1998; McCullough and Lowe, 1992; McCullough *et al.*, 1993a,b, 1997; Marsella and Swager, 1993; Sirringhaus *et al.*, 1998). The correlation between the conjugation length of oligothiophenes and their electrical conductivities has been examined (Shirota, 2000; Noma *et al.*, 1996; Funaoka *et al.*, 1999). Control of morphologies to provide the electrical conduction path in OPVs using copper phthalocyanine as a p-type organic semiconductor has been reported (Peumans *et al.*, 2003; Yang *et al.*, 2005a,b).

1.3.3 Crystalline molecular materials: polymorphism

Small organic molecules usually exist as crystals below their melting temperatures. Oligothiophenes with well-defined structures are also crystalline in nature.

It is noteworthy that certain kinds of small molecules, e.g., copper phthalocyanine (**1**, CuPc) (Ebert and Gottlieb, 1952; Kanemitsu *et al.*, 1991), titanyl phthalocyanie (**2**, TiOPc) (Hiller *et al.*, 1982; Enokida *et al.*, 1990; Saito *et al.*, 1993), and perylene pigments (Del Cano *et al.*, 2002), exhibit polymorphism, taking a few different crystal structures. The crystal structures of materials greatly affect their properties and device performance, as reported for, e.g., titanyl phthalocyanine (Enokida *et al.*, 1990; Saito *et al.*, 1993; Tsuzuki *et al.*, 1996).

1.3.4 Amorphous molecular materials

Studies aimed to create amorphous molecular materials based on π-electron systems, namely, photo and electroactive small organic molecules that readily form stable amorphous glasses above room temperature, and their functions have been performed since the late 1980s (Shirota *et al.*, 1989), and have become very active since the middle of the 1990s in view of their application as materials for OLEDs. It has now been widely accepted that like polymers, small organic molecules also form stable amorphous glasses with definite glass-transition temperatures (T_gs) above room temperature, if their molecular structures are properly designed. In addition, new concepts for photo and electroactive organic materials have been proposed, which include photochromic amorphous molecular materials (Shirota *et al.*, 1998), electrically conducting amorphous molecular materials (Higuchi *et al.*, 1991), amorphous molecular resists (Yoshiiwa *et al.*, 1996), and amorphous molecular materials for OLEDs, OPVs, and photorefractive devices (Shirota, 2000, 2005). Like crystals and liquid crystals, amorphous molecular materials have constituted a new class of versatile materials for various applications.

Amorphous molecular materials have the following characteristic features (Shirota, 2000, 2005):

(1) They are in a thermodynamically nonequilibrium state, and hence, they tend to undergo structural relaxation, exhibiting well-defined T_gs.
(2) They take different organization states, namely, crystal, supercooled liquid, and isotropic liquid as well as amorphous glass.
(3) The crystalline state of a number of amorphous molecular materials exhibits polymorphism, taking a few different crystal structures (Ishikawa *et al.*, 1991, 1993a; Ueta *et al.*, 1994).
(4) They are characterized by the presence of free volume.
(5) They show isotropic and homogeneous properties.
(6) Like polymers, they readily form uniform amorphous thin films by themselves either by vacuum deposition or by spin coating from solution.

Several molecular design concepts for amorphous molecular materials, e.g., nonplanar molecular structures and the presence of multiple conformations, have been presented (Shirota *et al.*, 1989; Ishikawa *et al.*, 1991; Ueta *et al.*, 1994). A variety of amorphous molecular materials have been created (Fig. 1.3). The representative classes of amorphous molecular materials include π-electron starburst molecules such as 4,4′,4″-tris(diphenylamino) triphenylamine (TDATA) family (**6**) (Shirota *et al.*, 1989; Higuchi *et al.*, 1991), 1,3,5-tris(diphenylamino)benzene (TDAB) family (**7**) (Ishikawa *et al.*, 1991), and 1,3,5-tris(diphenylamino)triphenylbenzene (TDATB) family (**8**) (Inada and Shirota, 1993). Other representative classes include diarylaminobenzaldehyde arylhydrazones (Nishimura *et al.*, 1991), tris(oligoarylenyl)amine family (**9**) (Higuchi *et al.*, 1992), triarylborane family (**10**) (Kinoshita *et al.*, 2002), families of π-electron systems end-capped with triarylamine (**11**) (Noda *et al.*, 1997a,b) or a dimesitylboryl group (**12**) (Noda and Shirota, 1998), 1,3,5-triarylbenzene family (Plazek and Magill, 1966; Okumoto and Shirota, 2001a), a family of spiro-linked molecules (**13**) (Salbeck *et al.*, 1997a), tetraphenylmethane derivatives family (**14**) (Oldham *et al.*, 1998), a family of silicone-containing compounds, e.g., α, ω-dimethyloligo(diphenylsilylene)s (Imae and Kawakami, 2005) and tetraarylsilanes (**15**) (Duan *et al.*, 2005), a family of macrocyclic compounds based on π-electron systems (**16**) (Ottmar *et al.*, 2001).

There have been further studies on amorphous molecular materials to date. The principles of crystal engineering can be used in a contrary way for designing molecules that resist crystallization. While hydrogen bonding provides strong directional interactions and normally favors crystallization, examples of amorphous molecular materials by use of hydrogen bonding have been reported (Lebel *et al.*, 2006). The introduction of methylpentaphenyl groups into the triarylamine moiety has been shown to be effective for making amorphous molecular materials (Gagnon *et al.*, 2010). Whether or not any microscopic ordering exists in amorphous molecular materials

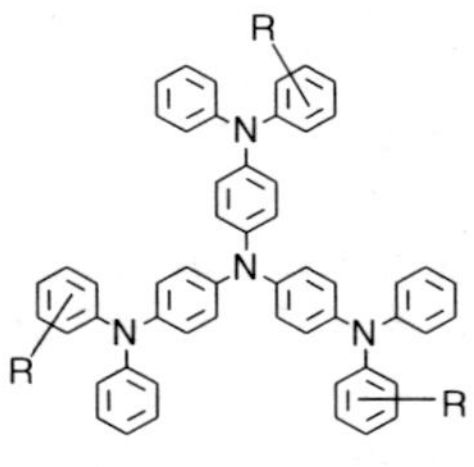

6 (TDATA family)

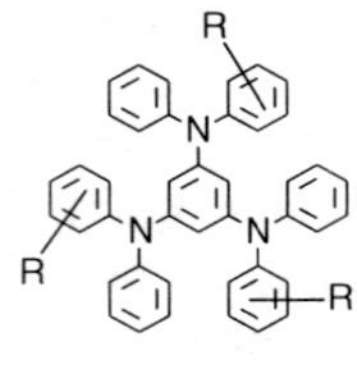

7 (TDAB family)

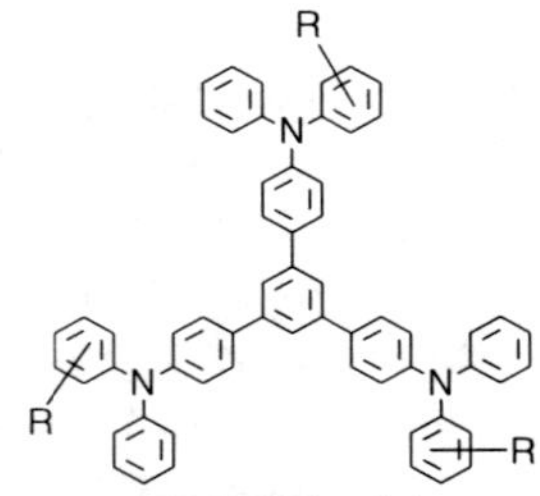

8 (TDAPB family)

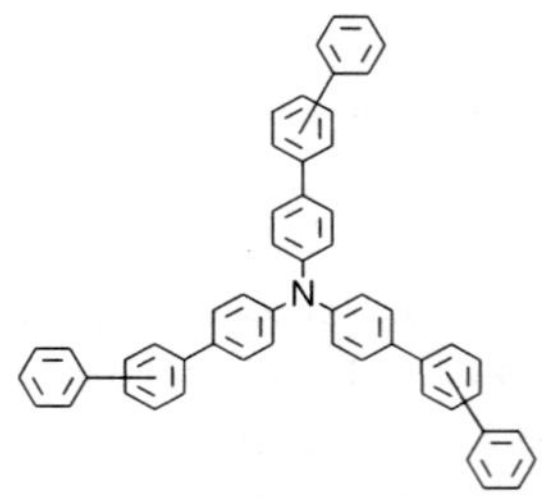

9 (tris(oligoarylenyl)amine family)

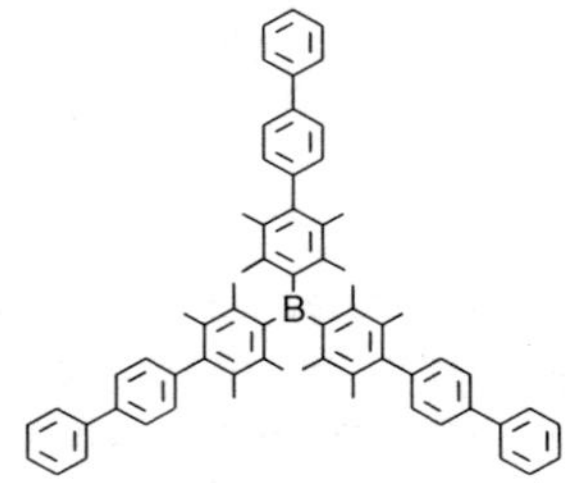

10 (triarylborane family)

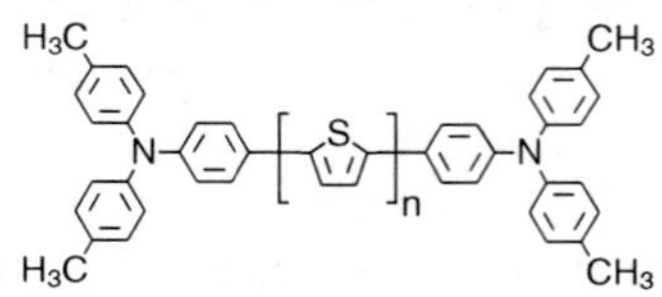

11 (family of π-electron systems end-capped with triarylamine moiety)

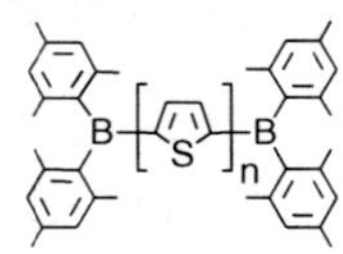

12 (family of π-electron systems end-capped with dimesitylboryl group)

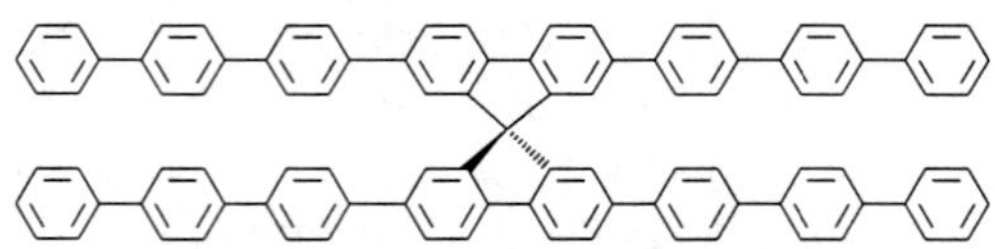

13 (family of spiro-linked molecules)

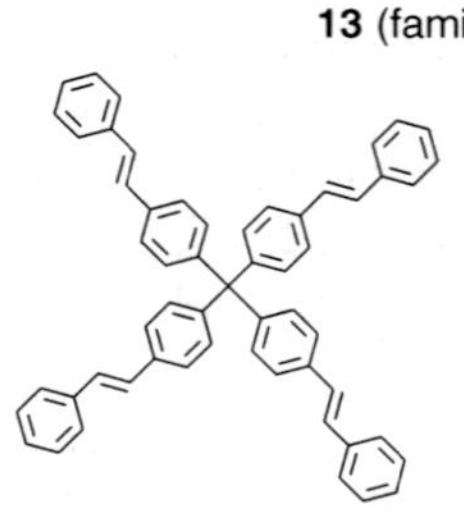

14 (tetraarylmethane family)

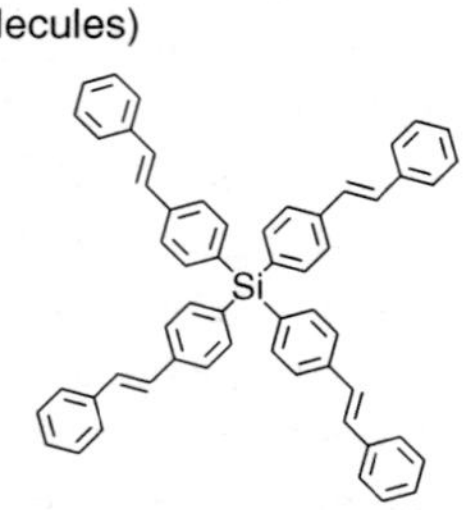

15 (tetraarylsilane family)

1.3 Examples of representative classes of amorphous molecular materials.

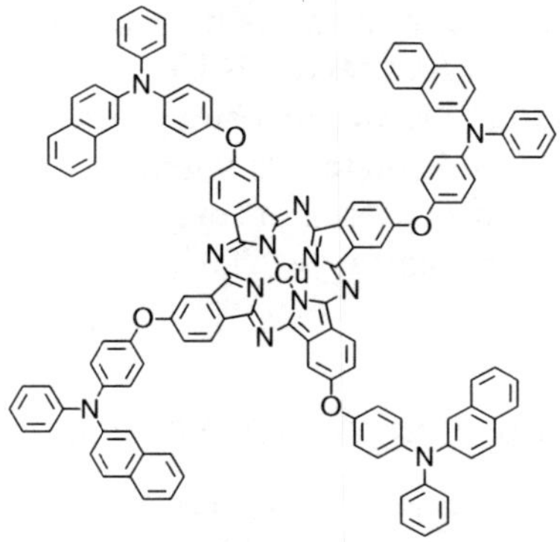

16 (family of macrocyclic compounds based on π-electron systems)

1.3 Continued.

is a subject of interest. Recent studies on amorphous molecular materials have revealed that vacuum-deposited amorphous molecular materials with long or planar molecular structures, such as *N*,*N*,*N*′,*N*′-tetra(biphenyl-4-yl)-[1,1′-biphenyl]-4,4′-diamine, horizontally orient along with the substrate surface (Lin *et al.*, 2004; Yokoyama *et al.*, 2008; Yokoyama, 2011).

1.4 Organic (opto)electronic devices: principles and operation processes

There are distinct differences between inorganic and organic semiconductors. That is, while inorganic semiconductors form energy bands, i.e., valence and conduction bands, organic semiconductors do not necessarily form energy bands. While inorganic semiconductors, which are mostly extrinsic semiconductors having impurities as dopants, are electrically conductive, organic semiconductors used in devices are mostly electrical insulators as they are used without any intentional doping. What are called p-type and n-type organic semiconductors in OPVs stand for electron donor and acceptor molecules, respectively.

The operation processes involved in organic optoelectronic devices such as OPVs and OLEDs should, therefore, be understood in terms of the molecular level instead of the energy band model that are applied for inorganic semiconductors. While photoabsorption by inorganic semiconductors takes place by the band-to-band transition of electrons, photoabsorption by organic semiconductors generates the electronically excited state of a molecule. Charge carriers, i.e., holes and electrons, for organic semiconductors correspond to the cation and anion radicals of molecules.

Photogeneration of charge carriers is involved in photoreceptors in electrophotography and OPVs, whereas charge carrier injection takes place in

OLEDs and OFETs. Charge transport is involved as a process common to photoreceptors, OPVs, OFETs and OLEDs. The operation processes involved in OPVs and OLEDs are opposite to each other.

Charge transport in organic materials together with charge-transporting organic materials has been reviewed in several references (Strohriegl and Grazulevicius, 2002; Coropceanu *et al.*, 2007; Shirota and Kageyama, 2007; Walzer *et al*, 2007).

1.4.1 Organic photoreceptors in electrophotography

The electrophotographic process is also known as the xerographic process and is employed for coping machines and computer terminal printers. It consists of the following sequential processes:

(1) electrostatic charge by means of corona charging;
(2) image exposure to form a latent image on the surface of a photoreceptor;
(3) toner development to form a visible image on the surface of the photoreceptor;
(4) transfer of the visible image formed on the photoreceptor surface onto plain paper;
(5) removal of the remaining charge on the photoreceptor by photoirradiation; and
(6) cleaning of the photoreceptor surface to remove remaining toners.

The toner image transferred onto the plain paper undergoes the process of fixation by means of both heat and pressure using rubber cylinders.

The structures of photoreceptors are classified into two, single-layer and double-layer structures. The latter consists of a CGL and a CTL (Section 1.2). The CGL absorbs visible light to generate the electronically excited state of molecule, leading to the generation of charge carriers either in the CGL or at the interface between the CGL and the CTL. One of the charge carriers generated in the CGL, either holes or electrons, is injected into the CTL and transported across the CTL to neutralize the surface charge (Regensburger, 1968; Mort, 1972; Mort and Nielsen, 1972; Melz *et al.*, 1977, Borsenberger and Weiss, 1993).

1.4.2 OPVs

OPVs can be applied for organic solar cells and photodetectors. Thin-film organic solar cells have been attracting a great deal of attention as potential candidates for next-generation solar cells. A typical structure of OPVs is based on planar pn-heterojunction which consists of the two thin layers of a p-type organic semiconductor (electron donor) and an n-type organic semiconductor (electron acceptor) sandwiched between two metal electrodes that ideally provide ohmic contacts with the organic layers. In such a device, the interface between the electron donor (D) and the electron acceptor (A) plays a vital role in the generation of charge carriers. The photocurrent generation processes involved in OPVs are as follows (Fig. 1.4):

(1) light absorption by molecules to form excitons, i.e., the electronically excited states of molecules;
(2) exciton diffusion to the D/A interface;
(3) exciton dissociation into free carriers, which consists of the two processes, namely, electron transfer from D to A in the electronically excited state of either D or A to generate geminate hole–electron pairs at the D/A interface and then charge separation into free holes and electrons in competition with the recombination of the hole–electron pairs;
(4) charge transport in the organic layers, namely, hole transport across the donor layer and electron transport across the acceptor layer; and
(5) collection of holes and electrons at the anode and the cathode.

The equivalent circuit of OPVs is shown in Fig. 1.5. As eq 1.1 shows, the observed current density (J_{obs}) that flows through the outer circuit is the difference between the photocurrent density J_{ph} and the dark current

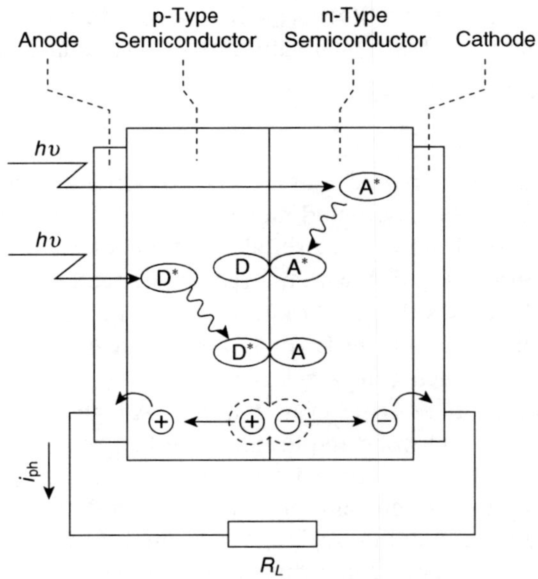

1.4 Generation of photocurrent in OPVs.

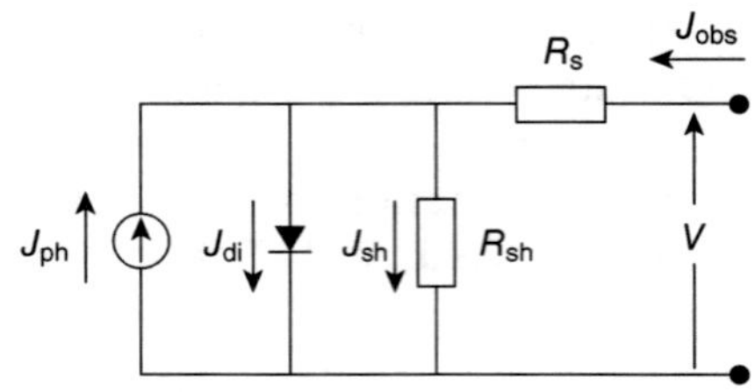

1.5 Equivalent circuit of OPV devices.

density J_d (= J_{di} + J_{sh}, here, J_{di} and J_{sh} represent the current densities that flow through the diode and the shunt resistor, respectively) and is given by eq 1.2, where J_0 is the reverse saturated dark current density, R_s and R_{sh} represent series and shunt resistances, e is the elementary electric charge, V is the cell voltage, n is the diode ideal factor, k is the Boltzmann constant, and T is the absolute temperature.

$$J_{obs} = J_{in} + J_{sh} - J_{ph} \quad [1.1]$$

$$J_{obs} = J_0 \left\{ \exp\left[\frac{e(V - J_{obs} R_s)}{nkT} \right] - 1 \right\} + \frac{V - J_{obs} R_s}{R_{sh}} - J_{ph} \quad [1.2]$$

The PCE for the incident solar power (P_{in}) is defined as eq 1.3, where V_{OC}, J_{SC}, and FF stand for the open circuit voltage, the short-circuit current density, and the fill factor, respectively. V_{OC} is the voltage obtained by substituting 0 for J_{obs} in eq 1.2. Likewise, J_{SC} is the J_{obs} obtained by substituting 0 for V in eq 1.2. FF is defined as $(V \times J)_{max}$ / $(V_{OC} \times J_{SC})$. In order to attain high PCE, it is necessary to obtain high values of V_{OC}, J_{SC}, and FF.

$$\text{PCE} = \frac{V_{OC} \times J_{SC} \times \text{FF}}{P_{in}} \quad [1.3]$$

It has been generally understood that V_{OC} corresponds to the energy difference between the highest occupied molecular orbital (HOMO) of D and the lowest unoccupied molecular orbital (LUMO) of A. To obtain a high V_{OC}, combination of D and A with a large difference in the HOMO level of D and the LUMO level of A should be selected on condition that photoinduced electron transfer from D to A is energetically favorable. In order to obtain high J_{SC}, reducing the series resistance (R_s) of materials and increasing the photocurrent density (J_{ph}) are required. A high J_{ph} is obtained when materials meet the following requirements: effective absorption of sunlight from the visible to near infrared wavelength region, large exciton diffusion length during its lifetime, and effective exciton dissociation into free charge carriers. The probability of charge separation from the ion pair into free ions has been understood in terms of the Onsager theory. Recent studies have pointed out that the charge carrier mobilities of D and A molecules affect the charge separation probability. That is, a Monte Carlo model of charge carrier separation and recombination predicts that an increase in the carrier mobility of D and A facilitates the charge separation process, resulting in an increase in J_{ph}. The equation describing J_{ph} generated from OPVs is not available; however, the equation describing the steady-state photocurrent density (J_{ph}) for organic photoconductors helps to understand the J_{ph} generated from OPVs. The steady-state photocurrent density (J_{ph}) for organic photoconductors is expressed as eq 1.4, where L is the thickness of the sample, I_0 is the total number of photons arriving at the unit surface area of the cell per second, α is the absorption coefficient, η is the photogeneration efficiency of charge carriers, τ is the carrier lifetime, μ is the charge carrier mobility, and E is the electric field. High J_{obs} together with its weak forward voltage dependence and infinitely large shunt resistance (R_{sh}) are the requirements for obtaining high FF.

$$J_{ph} = \frac{1}{L} I_0 [1 - \exp(-\alpha L)] \eta \tau e \mu E \qquad [1.4]$$

1.4.3 OLEDs

OLEDs have been applied for various flat-panel, mobile displays and televisions owing to their characteristic features such as planar emission; high-quality image due to high brightness, wide viewing angle, and good contrast resulting from the self-emission; portability; suitability for moving-image display owing to fast response; and capability of full-color emission. OLEDs are also expected as a promising candidate for next-generation solid-state lighting.

The fluorescence or phosphorescence emitted from the electronically excited singlet or triplet state of a molecule generated by photoabsorption is termed photoluminescence. On the other hand, when the electronically excited singlet or triplet state of a molecule is generated by the recombination between hole and electron charge carriers injected from the electrodes by the application of external voltage, the resulting fluorescent or phosphorescent emission is called EL. OLEDs are devices that emit EL, namely, either fluorescence or phosphorescence from the electronically excited state of a molecule generated by the recombination of hole and electron charge carriers injected from the two electrodes.

OLEDs are composed of organic thin films of ca. 100 nm sandwiched between two dissimilar electrodes, usually the ITO anode and the metal cathode with small work functions, e.g., Al, MgAg alloy, or Ca. Structures consisting of multiple layers of organic thin films, e.g., hole-injection, hole-transport, emitting, hole-blocking, and electron-transport layers, have been employed to attain high performance.

The operation processes of OLEDs involve:

(1) injection of charge carriers, holes and electrons, from the anode and the cathode, respectively, by the application of external voltages;
(2) transport of injected holes and electrons through the hole-transport and electron-transport layers, respectively;
(3) injection of holes and electrons into the emitting layer;
(4) recombination of holes and electrons in the emitting layer to generate the electronically excited states of emitting molecules, and
(5) emission of either fluorescence or phosphorescence from the electronically excited singlet or triplet state of molecule, respectively (Fig. 1.6).

It should be noted that there are distinct differences in the operation processes between LEDs using inorganic semiconductors and OLEDs. In LEDs, minority charge carriers are injected into respective p-type and n-type semiconductors at the pn-junction by the application of external voltages, and emission takes place by the recombination of electrons and holes through the band-to-band transition. In OLEDs, on the other hand, charge carriers, i.e., holes and electrons, are injected from the respective electrodes by the application of external voltages, and the emission takes place from the electronically excited-state molecules produced by the recombination of holes and electrons in the emitting layer.

The luminous efficiency (power efficiency) (η_p) is defined as eq 1.5 when the light emission pattern is assumed to be Lambertian. Here, L is the luminance observed at the normal direction to the device. The EQE for EL is given by eq 1.6, where K_m is the maximum luminous efficacy, $y(\lambda)$ is the normalized photopic spectral response function, $F(\lambda)$ is the EL emission

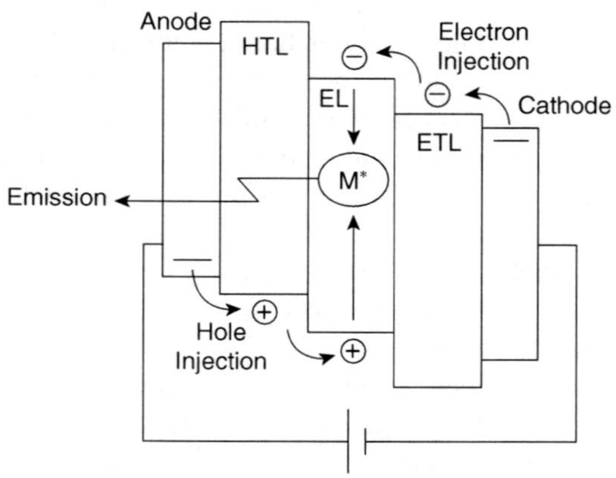

1.6 Generation of EL in OLEDs (HTL, hole-transport layer; EL, emitting layer; and ETL, electron-transport layer).

spectrum for the device, λ is the wavelength, h is the Planck constant, and e is the elementary electric charge (O'Brien *et al.*, 1997).

$$\eta_p = \frac{\pi L}{JV} \quad [1.5]$$

$$\mathrm{EQE} = \frac{\pi \cdot L \int_0^{\infty} \frac{F(\lambda)}{h \cdot c} \mathrm{d}\lambda}{K_m \int_{380}^{780} \frac{F(\lambda) \cdot y(\lambda)}{\lambda} \mathrm{d}\lambda} \div \frac{J}{e} \quad [1.6]$$

There are several factors that determine the EQE for EL. As eq 1.7 shows, they are:

(1) charge balance between holes and electrons injected from the anode and the cathode (Φ_{re});
(2) spin multiplicity of the electronically excited state of emitting molecules, either singlet or triplet state, produced by the recombination of charge carriers (Φ_{spin});
(3) quantum efficiency for emission (Φ_{em}); and
(4) light outcoupling efficiency (α), which is assumed to be ca. 20%.

$$\mathrm{EQE} = \alpha \times \Phi_{re} \times \Phi_{spin} \times \Phi_{em} \quad [1.7]$$

The probability of the generation of the electronically excited singlet state of an emitting molecule by the recombination of charge carriers is 25%. Therefore, when the EL originates from the fluorescence, the maximum EQE for EL becomes 5% in the absence of any other routes. On the other hand, the probability of the generation of the electronically excited triplet state of an emitting molecule by the recombination of charge carriers is 75%. Accordingly, when the EL originates from the phosphorescence, the maximum EQE for EL becomes 15%. In case the intersystem crossing from the singlet state to the triplet state is effectively utilized, the internal quantum efficiency for EL becomes 100%, and the EQE for EL reaches to 20%.

1.5 Molecular materials for organic (opto)electronic devices

Both crystalline and amorphous molecular materials have been used depending upon the kind of devices. Generally polycrystalline materials might prevent smooth, uniform thin-film formation, causing any damage to devices. On the other hand, grain-boundary-free amorphous molecular materials allow smooth, uniform amorphous thin-film formation. Polycrystalline materials have been widely studied for use in OPVs and OFETs because of their higher charge carrier mobilities than those of amorphous materials. Amorphous molecular materials are suitable for solution processes and have also been used for solution-processed bulk heterojunction OPVs. Amorphous molecular materials have found successful application as materials for OLEDs.

1.5.1 Organic photoconductors for photoreceptors

Molecularly doped polymer systems, where photo and electroactive molecular materials are dispersed in inert polymer binders, have been used in organic photoreceptors. The properties required of photoreceptor materials are as follows. They should be electrical insulators in the dark so as to accept and retain electrostatic charge on the surface of photoreceptors. However, they should exhibit photoconductivity. That is, they should absorb visible light, generate charge carriers, and then transport charge carriers to neutralize the surface static charge. The CTL and CGL in photoreceptors take part of these required properties with each other. That is, CGL materials absorb visible light to generate charge carriers in the CGL or at the interface between the CGL and the CTL. CTL materials play the roles of accepting and retaining electrostatic charge on their surface and transporting one of the charge carriers injected from the CGL to neutralize the surface charge.

A number of CGL and CTL materials for photoreceptors were developed and put into practical use one after another. Usually, the surface of the photoreceptor is negatively charged as CTL materials developed mostly transport hole carriers (Fig. 1.7). Such CTL materials include 1-phenyl-3-diethylaminostyryl-5-diethylamino-phenyl-1,3-pyrazoline (**17**), hydrazones (**18**), *e.g.*, diethylaminobenzaldehyde diphenylaminohydrazone (Melz *et al.*, 1977), 2,5-bis(diethylaminophenyl)-1,3,4-oxadiazole (**19**), 1,1-diphenyl-4,4-bis(diethylaminophenyl)-1,3-butadiene (Enokida *et al.*, 1990), and *N*,*N*′-bis(3-methylphenyl)-*N*,*N*′-diphenyl-[1,1′-biphenyl]-4,4′-diamine (TPD, **20**) (Stolka *et al.*, 1984), etc. These CTL materials are used for dispersion in a binder polymer, usually, bisphenol A polycarbonate. Copper phthalocyanine (**1**, CuPc), titanyl phthalocyanine with various polymorphs (**2**), and bisazo compounds, e.g., chlorodiane blue (**23**), have been used as CGL materials (Melz *et al.*, 1977; Enokida *et al.*, 1990) (Fig. 1.8).

1.5.2 Organic semiconductors for photovoltaic devices

p-Type organic semiconductors (electron donors (D)) and n-type organic semiconductors (electron acceptors (A)) for use in OPV devices should desirably meet the following requirements:

(1) Materials should absorb as much of the sunlight as possible, and hence, should have broad spectral sensitivity from the visible to near infrared wavelength region together with large absorption coefficients.

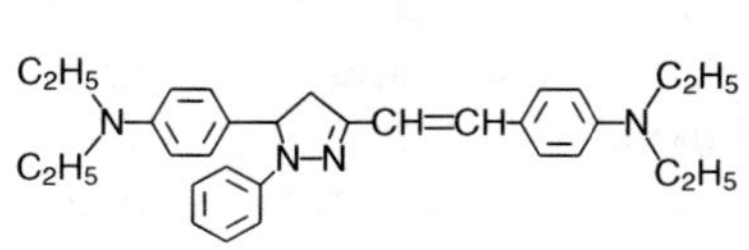

17 (pyrazoline)

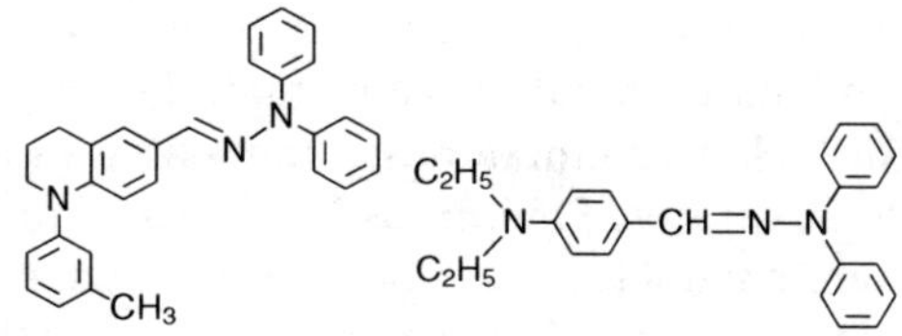

18 (hydrazone)

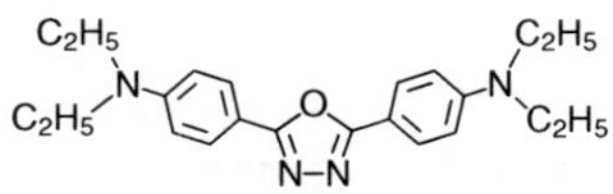

19 (oxadiazole)

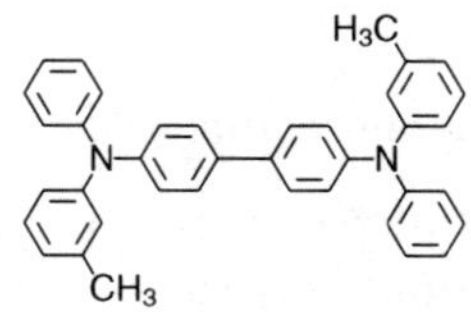

20 (TPD)

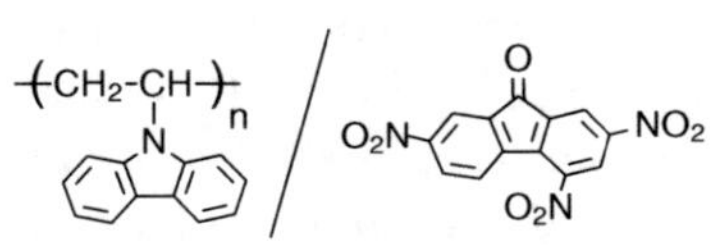

21 (PVCz) / **22** (TNF) CT complex

1.7 Materials for CTL in photoreceptors.

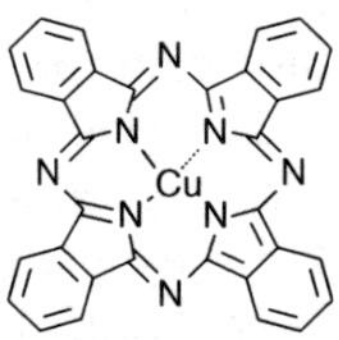

1 (CuPc)

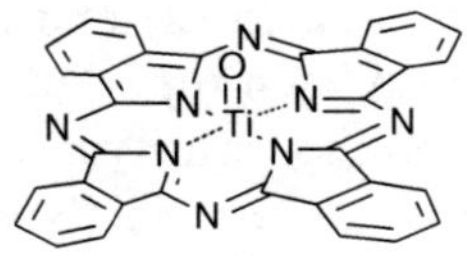

2 (TiOPc)

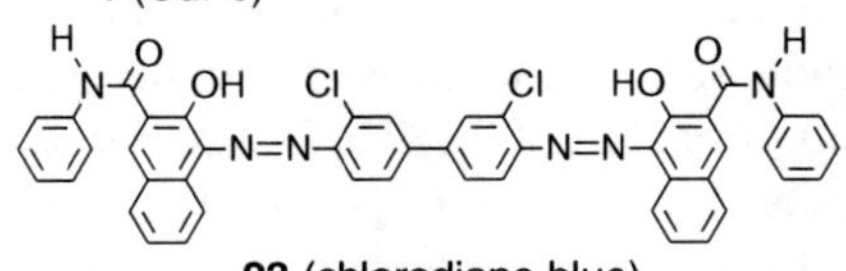

23 (chlorodiane blue)

1.8 Materials for CGL in photoreceptors.

(2) Materials with suitable HOMO and LUMO energy levels should be chosen as D and A, respectively, for efficient interfacial electron transfer from D to A in the electronically excited state of either D or A. It should also be noted that the HOMO level of D and the LUMO level of A affect the open circuit voltage for OPVs.
(3) Materials should have high charge carrier mobilities for efficient charge separation and transport, and for reducing series resistance.
(4) Materials should form uniform thin films with both morphological and thermal stability.
(5) Materials should be durable under photoirradiation.

In addition, it is necessary to consider how to overcome the problems encountered with organic materials, high bulk resistance and short exciton diffusion length, in designing materials and devices.

A variety of materials for OPVs including both p-type and n-type organic semiconductors have been developed. A through review of molecular materials for organic solar cells has recently appeared (Mishra and Bäuerle, 2012).

p-Type organic semiconductors

Representative classes of p-type organic semiconductors include metal and metal-free phthalocyanines, e.g., CuPc (**1**) (Tang, 1986; Xue *et al.*, 2005) and pentacene (**3**) (Yoo *et al.*, 2004; Mayer *et al.*, 2004; Pandey *et al.*, 2006). Molecules with a triphenylamine moiety as a central core have been shown to function well as p-type organic semiconductors for OPVs. They include tris(oligoarylenyl)amines (**24**, **25**) (Kageyama *et al.*, 2009a,b), donor-acceptor type molecules with a triarylamine central core and an electron-accepting dicyanovinyl group or benzothiadiazole moiety, e.g., **26** (Cravino *et al.*, 2006; Roquet *et al.*, 2006), **27** (Chiu *et al.*, 2012), **28** (C. He *et al.*, 2008), **29** (J. Zhang *et al.*, 2009), **30** (Stylianakis *et al.*, 2009), and **31** (Shang *et al.*, 2011). Organic dyes such as a squaraine derivative (**32**) (Wang *et al.*, 2009; Wei *et al.*, 2011; Chen *et al.*, 2012) and a rhodanine-based molecule (**33**) (Li *et al.*, 2012) have also been reported to give high PCEs of 3.2–6.1% in combination with fullerenes as n-type organic semiconductors. Oligothiophene and its derivatives, e.g., α-8T (Noma *et al.*, 1995), **34** (Holzhey *et al.*, 2008), **35** (Schulze *et al.*, 2006; Fitzner *et al.*, 2011), **36** (Wynands *et al.*, 2010), **37** (Steinberger *et al.*, 2011), **38** (Liu *et al.*, 2011), **39** (Ma *et al.*, 2008), have been reported to function as p-type organic semiconductors. Others include pyrrolopyrrole derivatives, e.g., **40** (Walker *et al.*, 2009), **41** (Tamayo *et al.*, 2009), **42** (Wu *et al.*, 2011), and **43** (Loser *et al.*, 2011), oligothiophenes or other π-conjugated moieties end-capped with the triarylamine moiety, e.g., **44** (Q. He *et al.*, 2008), **45** (Kwon *et al.*, 2010), **46** (Lu *et al.*, 2006), **47** (Sun *et al.*, 2007), and **48** (Li *et al.*, 2009), anthradithiophene (**49**) (Lloyd *et al.*, 2007), molecules containing a dithienosilole moiety, **50** (Sun *et al.*, 2012) and **51** (Zhou *et al.*, 2011). Among these molecular materials for OPVs, the materials **1**, **3**, **27**, **32**, **33**, **34**, **35**, **36**, **37**, **38**, **39**, **40**, **41**, **43**, **49**, **50**, and **51** are crystalline, and the materials **24**, **25**, **26**, **28**, **29**, **30**, **31**, **42**, **44**, **45**, **46**, **47**, and **48** are amorphous molecular materials. The chemical structures of these materials are shown in Fig. 1.9.

n-Type organic semiconductors

Fullerenes, e.g., C_{60} (**4**) (Sariciftci *et al.*, 1992, 1993) and C_{70} (**52**), and their derivatives, e.g., $PC_{61}BM$ (**53**) (Yu *et al.*, 1995), $PC_{71}BM$ (**54**) (Wienk *et al.*,

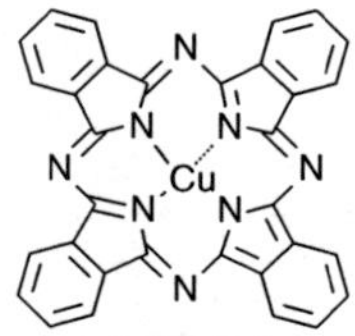

1 (CuPc)

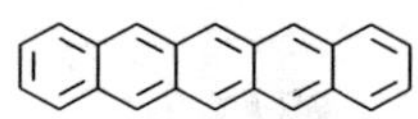

3 (pentacene)

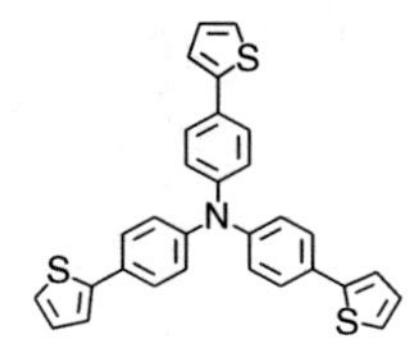

24

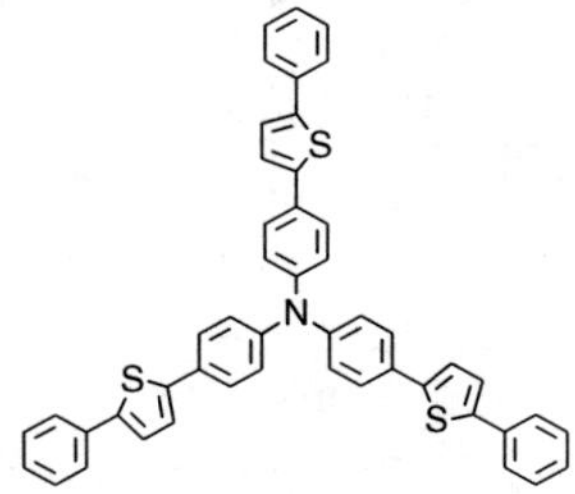

25

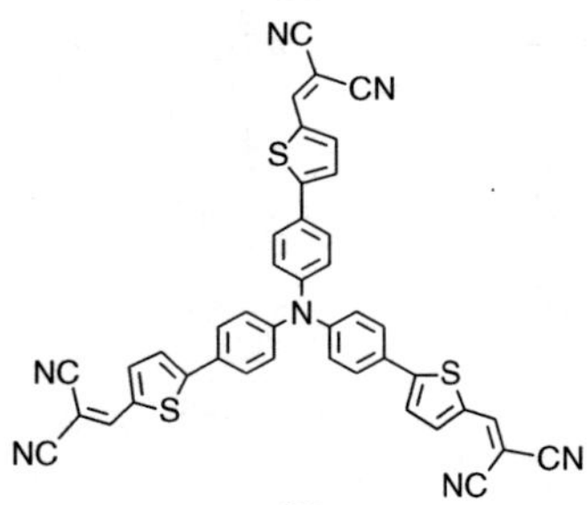

26

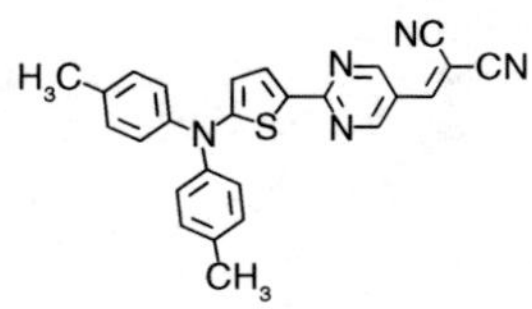

27

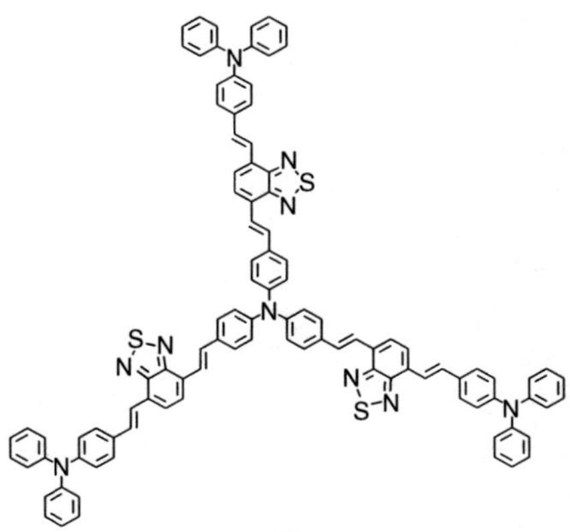

28

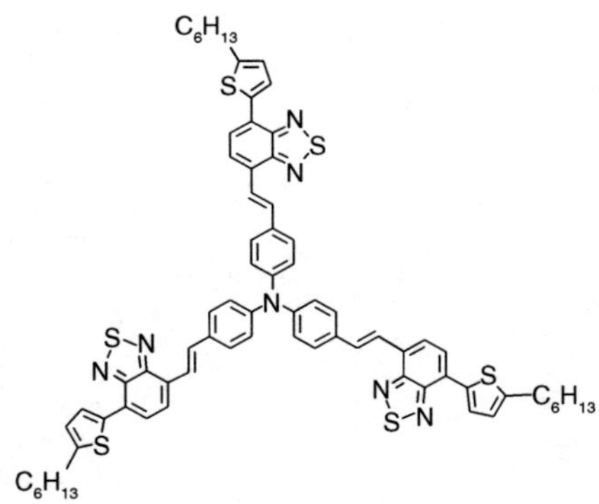

29

1.9 Examples of p-type organic semiconductors (D) for OPVs.

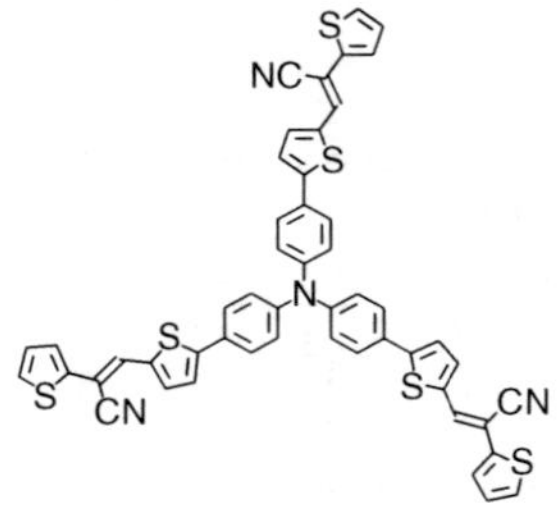

30

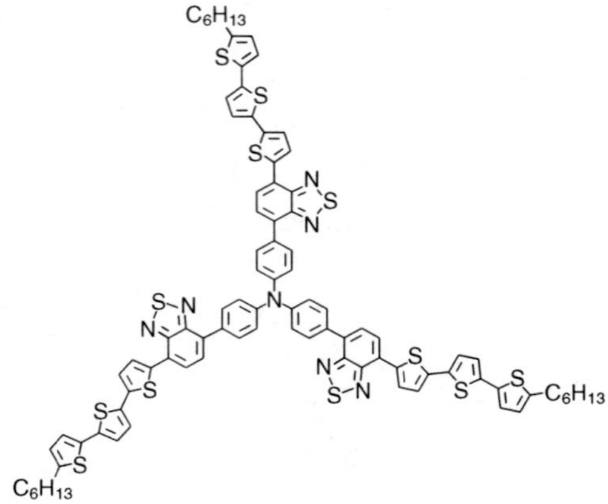

31

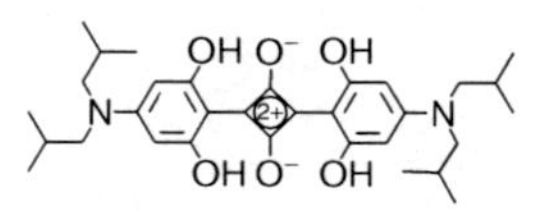

32

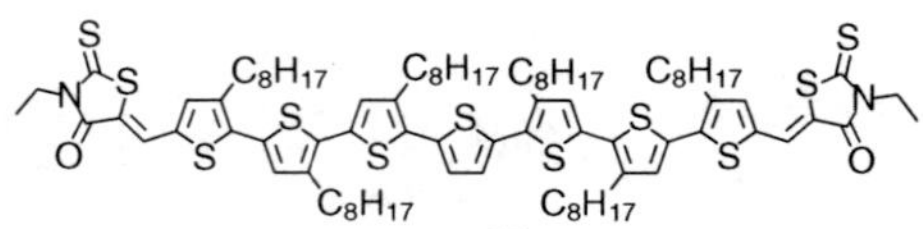

33

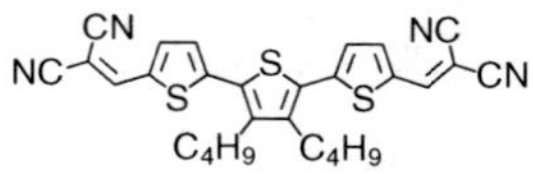

34

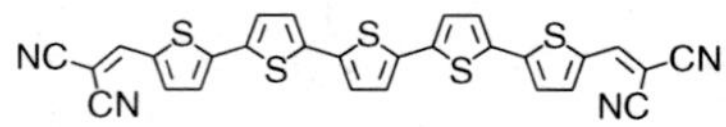

35

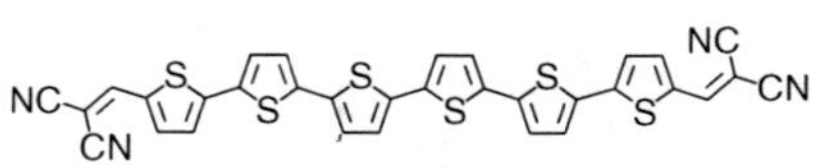

36

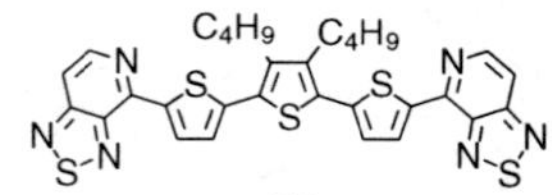

37

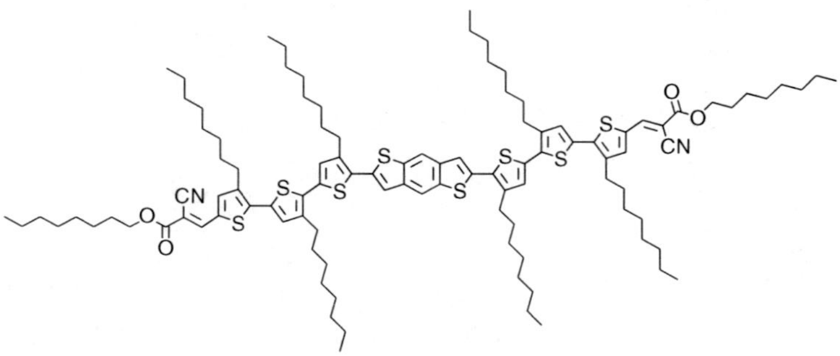

38

1.9 Continued.

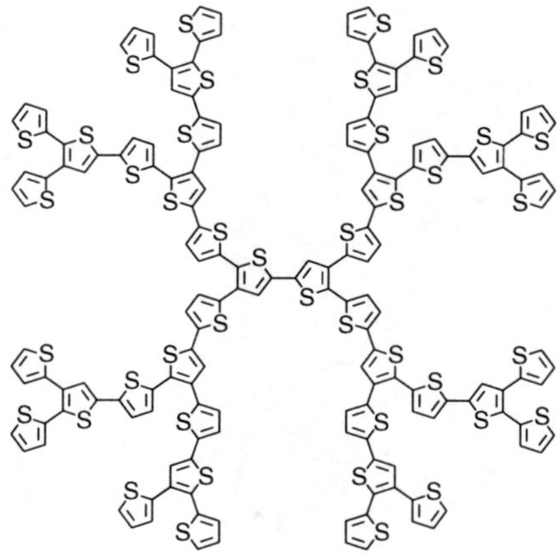

39

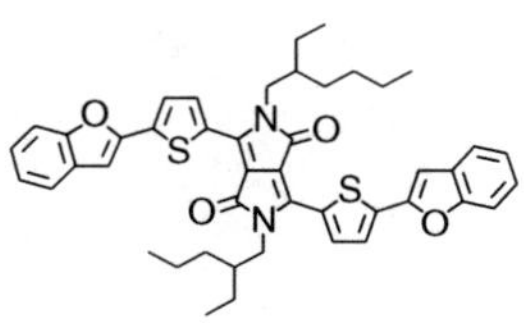

40

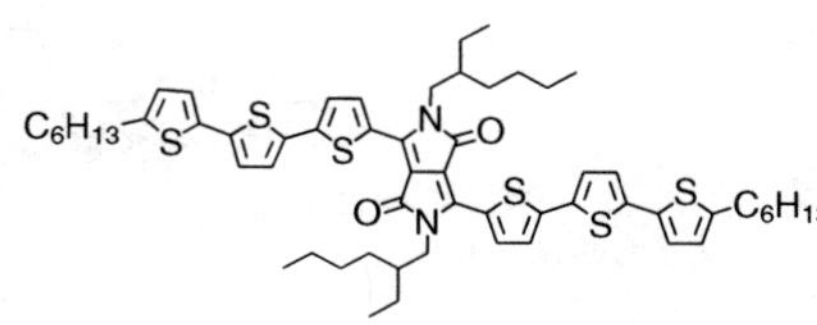

41

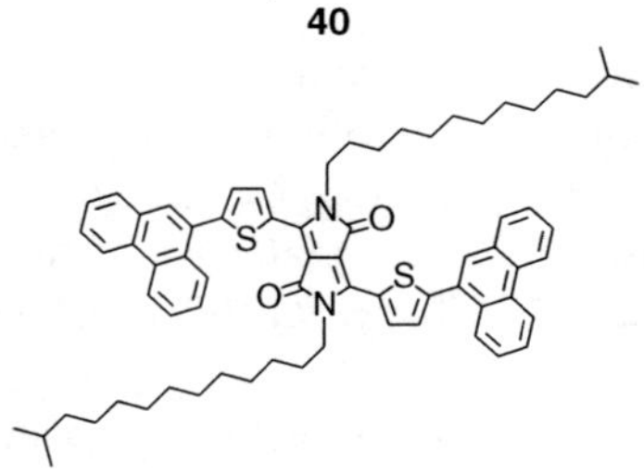

42

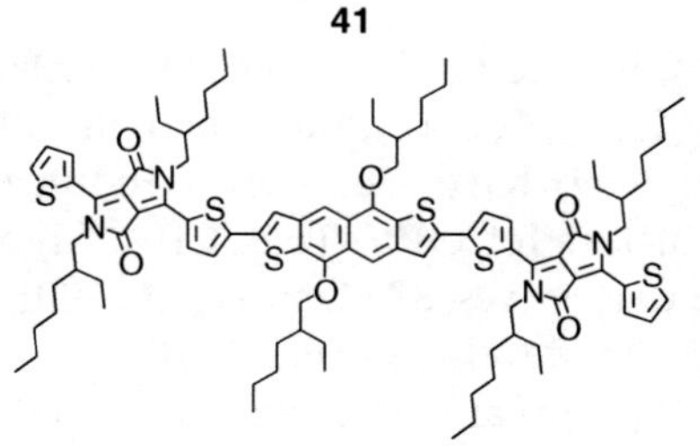

43

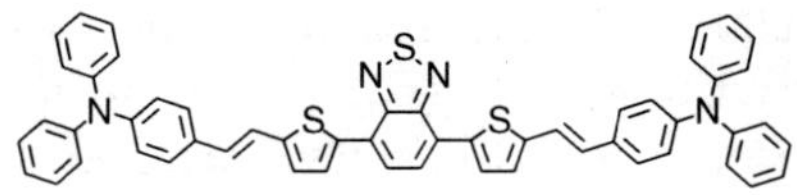

44

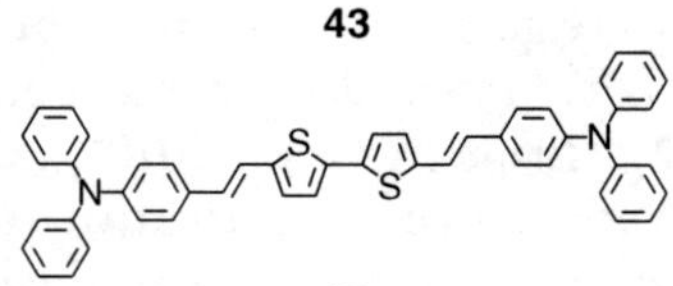

45

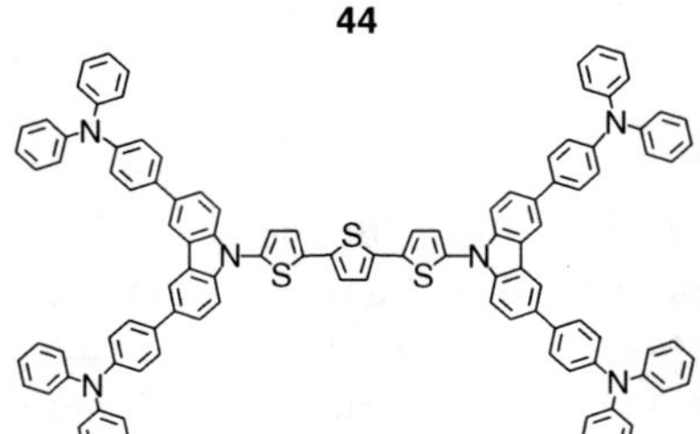

46

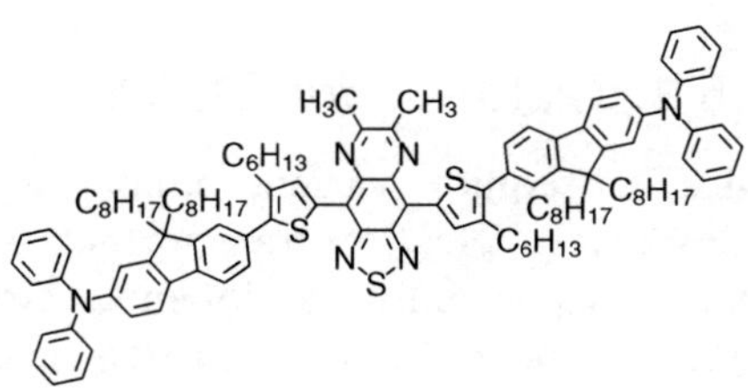

47

1.9 Continued.

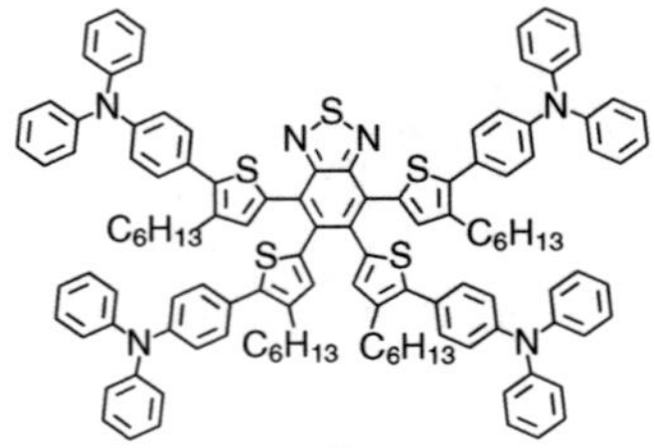

48

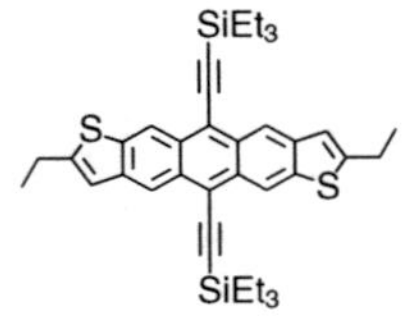

49

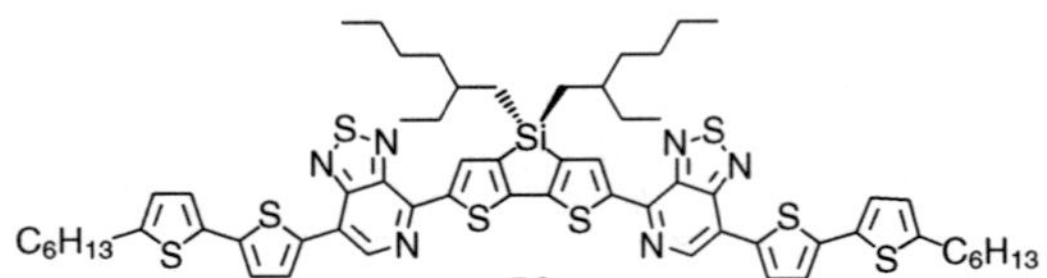

50

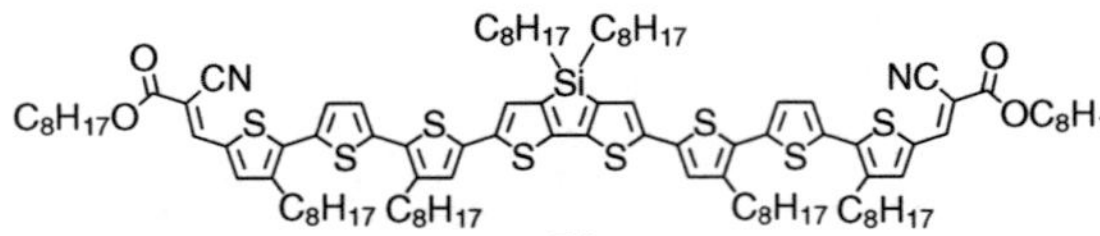

51

1.9 Continued.

2003), indene-C_{60} bisadduct (**55**) (Zhao *et al.*, 2010), indene-C_{70} bisadduct (**56**) (He *et al.*, 2010), and trimetallic nitride endohedral fullerene (**57**) (Ross *et al.*, 2009), have been proven to function as excellent n-type organic semiconductors for OPVs (Fig. 1.10). Amorphous molecular materials of fullerene derivatives, **58** (Ohno *et al.*, 2001) and **59** (X. Zhang *et al.*, 2009), have also been developed (Fig. 1.10).

n-Type organic semiconductors other than fullerenes include perylene bisimides, e.g., perylene–pyrene bisimide (**60**) (Sharma *et al.*, 2010) and PTCBI (**61**) (Tang, 1986; Peumans *et al.*, 2000). 9,9′-Bifluorenylidene derivative (**62**) (Brunetti *et al.*, 2010), diketopyrrolopyrrole (**63**) (Sonar *et al.*, 2010), and pentacene derivative (**64**) (Shu *et al.*, 2011) have also been reported to function as n-type organic semiconductors (Fig. 1.11).

1.5.3 Materials for OLEDs

OLEDs require various kinds of materials that function as hole-transporters, electron-transporters, charge blockers, light emitters, host materials for emissive dopants. Amorphous molecular materials have generally been used in OLEDs except for emissive dopants. General requirements for materials used in OLEDs are: having suitable ionization potentials and electron affinities, ability to form homogeneous, pinhole-less thin films, morphological stability, namely, stable amorphous glass formation, and thermal stability with high T_gs. High-performance amorphous

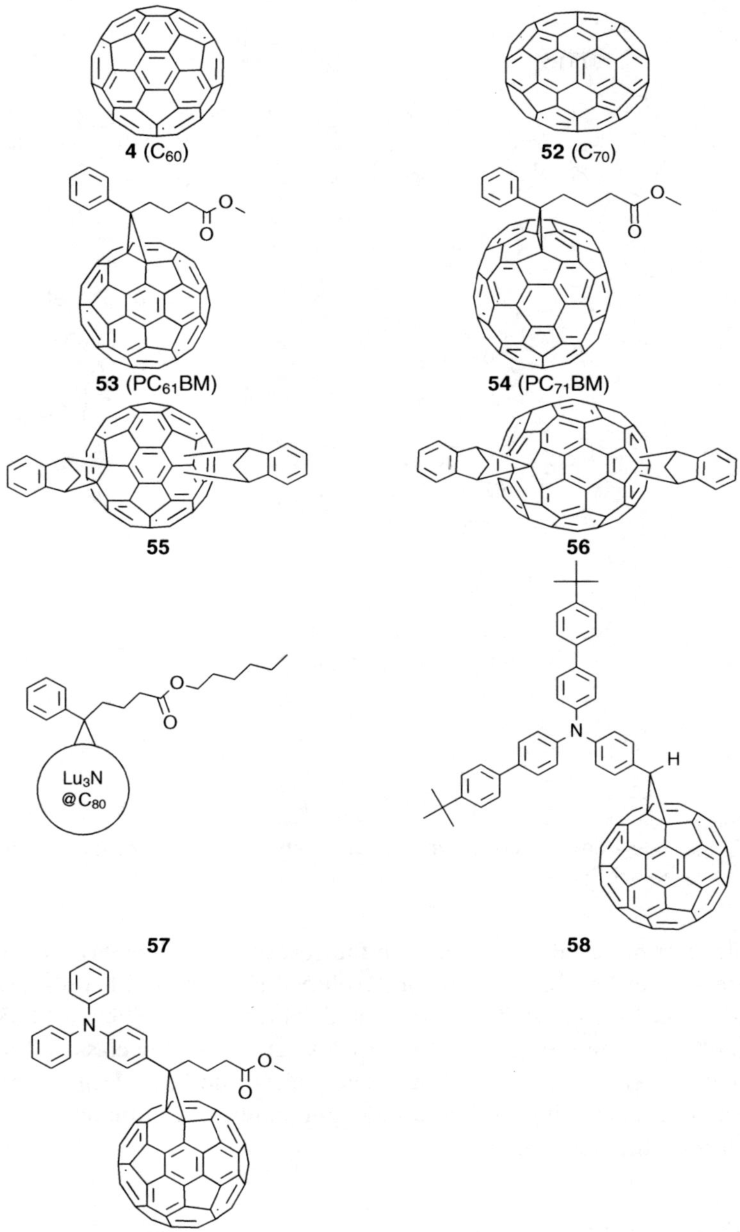

1.10 Examples of fullerene derivatives as n-type organic semiconductors for OPVs.

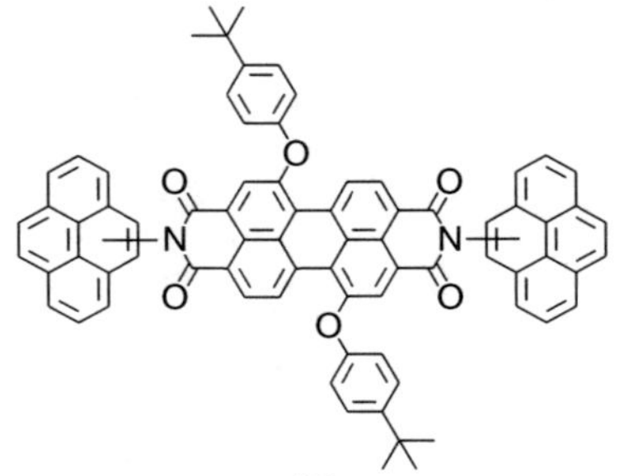

60

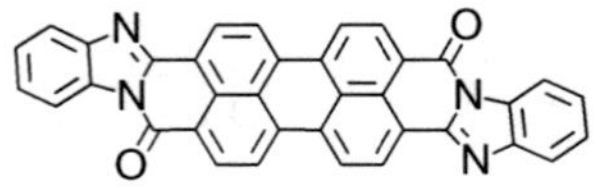

61 (PTCBI)

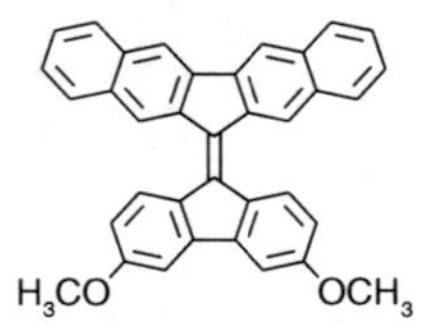

62

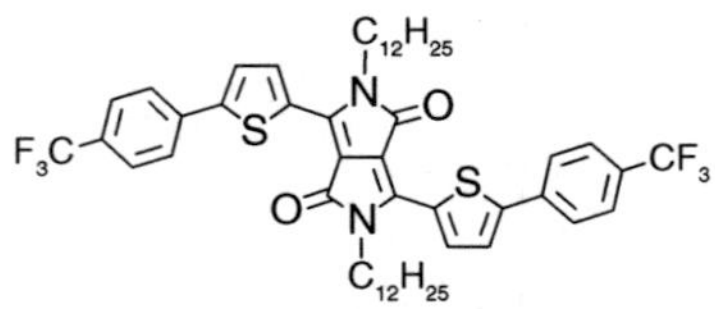

63

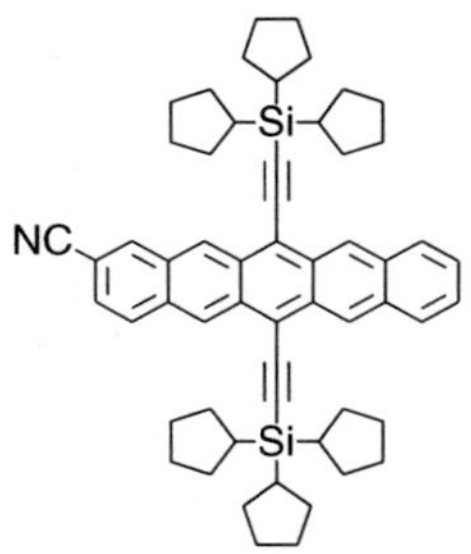

64

1.11 Examples of n-type organic semiconductors other than fullerenes for OPVs.

molecular materials together with fluorescent and phosphorescent dopant materials for OLEDs have been developed (Miyata and Nalwa 1997; Kraft *et al.*, 1998; Shirota, 2000; Mitschke and Bäuerle, 2000; Shirota, 2005; Kafafi, 2005; Müllen and Scherf, 2006). OLEDs can be largely classified into fluorescence-based and phosphorescence-based devices. Hole-transporting, electron-transporting, and hole-blocking materials can be used in common with both the devices.

Hole-transporting materials

The requirements for hole-injection and hole-transporting materials are electron-donating properties with low ionization potentials, capability of transporting hole carriers, and the stability of hole carriers, i.e., cation radicals of molecules. A variety of hole-transporting amorphous molecular

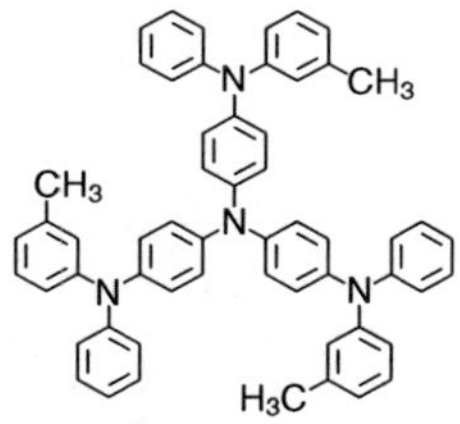

65 (m-MTDATA)

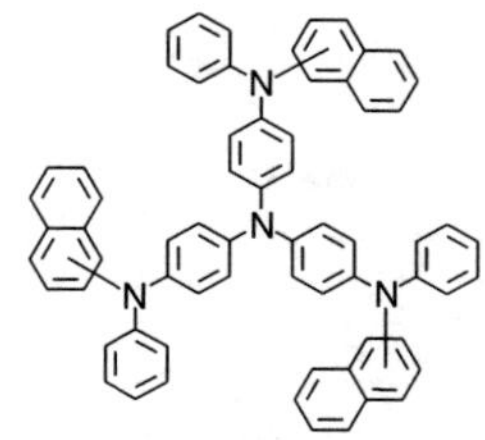

66 (1-, 2-TNATA)

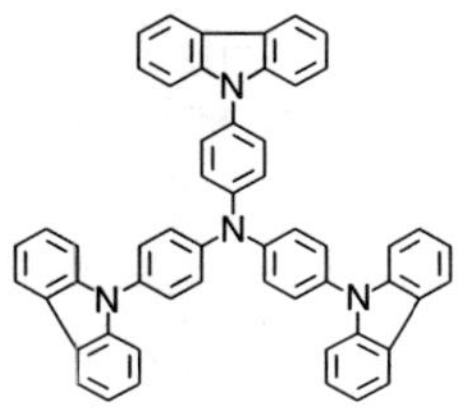

67 (TCTA)

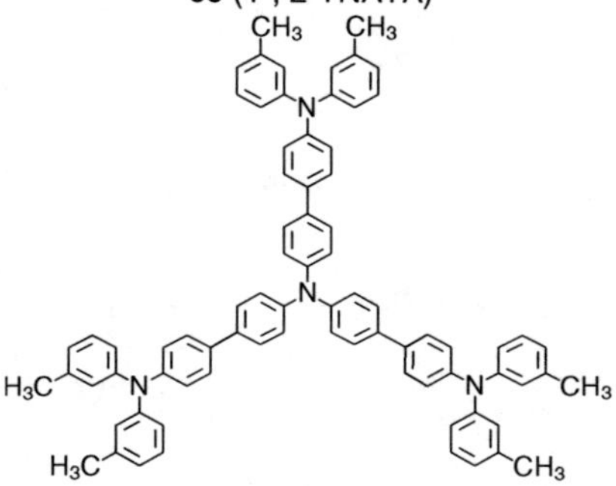

68

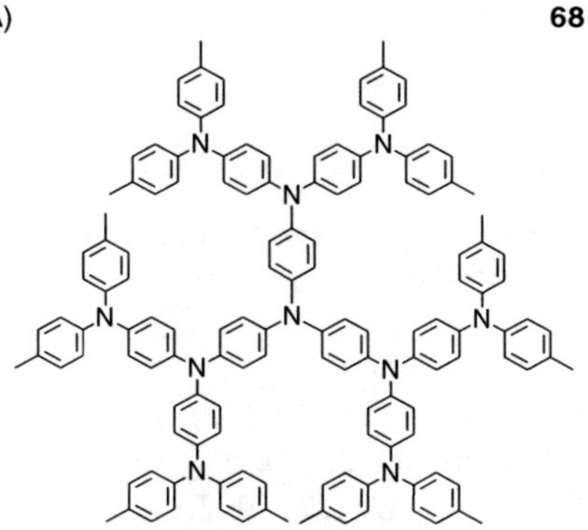

69

1.12 Examples of hole-transporting amorphous molecular materials for OLEDs – TDATA family.

materials have been developed. They are classified into the following classes: TDATA family (Fig. 1.12), e.g., 4,4′,4″-tris(3-methylphenyl(phenyl)amino) triphenylamine (m-MTDATA, **65**) (Shirota *et al.*, 1989), 4,4′,4″-tris(1- or 2-naphthylphenylamino)triphenylamine (1-, 2-TNATA, **66**) (Shirota *et al.*, 1997), 4,4′,4″-tri-N-carbazolyltriphenylamine (TCTA, **67**) (Kuwabara *et al.*, 1994), **68** (Tokito *et al.*, 1997), and **69** (Louie *et al.*, 1997), TDAB family (Fig. 1.13), e.g., p-DPA-TDAB (**70**) (Ishikawa *et al.*, 1993b), MTBDAB (**71**) (Katsuma and Shirota, 1998), **72** (Thelakkat and Schmidt, 1998), and **73** (Thelakkat, 2002), TDAPB family (Fig. 1.14), *o*-, *m*- and *p*-MTDAPB (**74**)

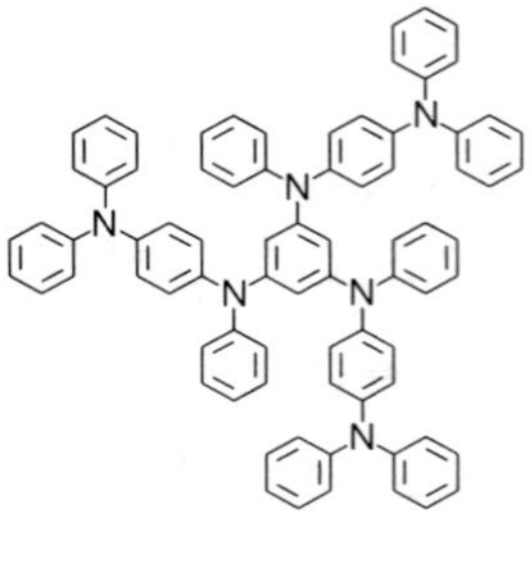

70 (p-DPA-TDAB)

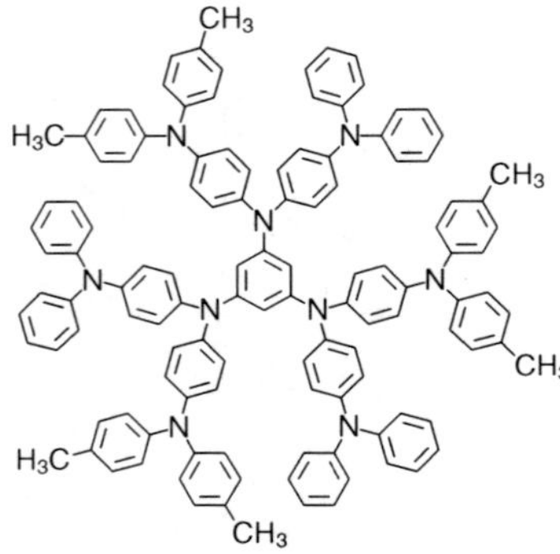

71 (MTBDAB)

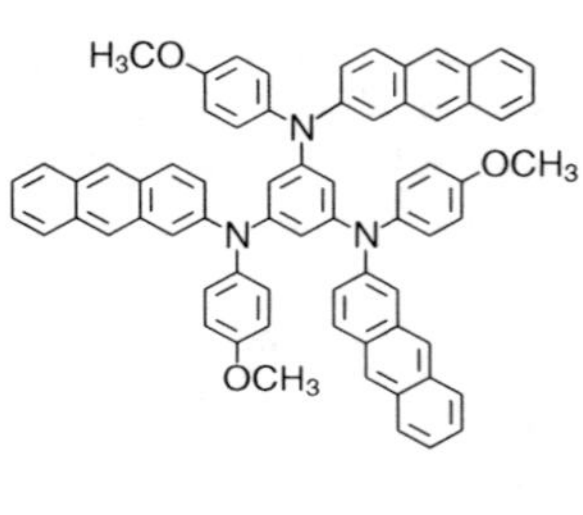

72

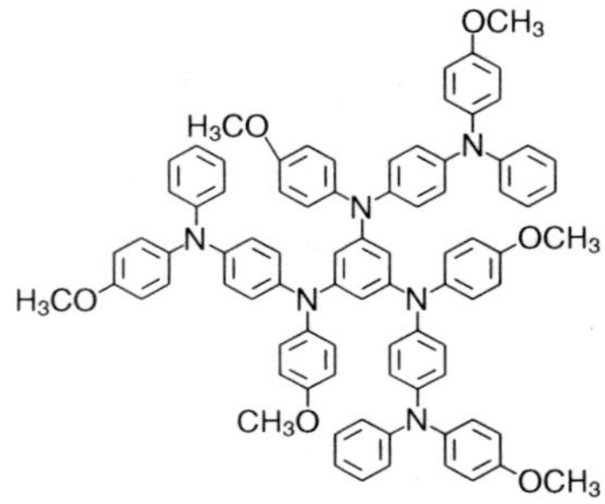

73

1.13 Examples of hole-transporting amorphous molecular materials for OLEDs – TDAB family.

(Inada and Shirota, 1993), MTFAPB (**75**), TBFAPB (**76**) (Okumoto *et al.*, 2002), and TFATr (**77**) (Shirota, 2005), tris(oligoarylenyl)amine family (Fig. 1.15), e.g., p-TTA (**78**) (Higuchi *et al.*, 1992; Ogawa *et al.*, 1997) and **79** (Chou *et al.*, 2004), TPD family (Fig. 1.16), e.g., TPD (**20**), α-NPD (**80**), *o*-, *m*- and *p*-BPD (**81**) (Okumoto *et al.*, 2000), CBP (**82**) (Koene *et al.*, 1998), PPD (**83**) (O'Brien *et al.*, 1998), FFD (**84**) (Okumoto and Shirota, 2000), **85** (Loy *et al.*, 2002), PFFA (**86**) (Okumoto and Shirota, 2001b), **87** (Kundu *et al.*, 2003), and **88** (Tanaka *et al.*, 1996), 1,4-bis(diarylamino)arylene family (Fig. 1.17), e.g., **89** (Koene *et al.*, 1998), NNP (**90**) (Koene *et al.*, 1998), 1,4-bis(carbazolyl)benzenes (**91**) (Koene *et al.*, 1998), **92** (Zhang *et al.*, 2004), **93** (Shan *et al.*, 2005), and diarylaminoanthracene (**94**) (Yu *et al.*, 2002), carbazole-containing arylamine family (Fig. 1.18), e.g., 3,6-bis (*N*, *N*′-[3-methylphenyl(phenyl)amino]-9-phenylcarbazole (**95**) (Grigalevicius *et al.*, 2002), **96** (J. Thomas *et al.*, 2000), and **97** (Kundu *et al.*, 2003), spiro-linked tetrakis(diarylamine) family (Fig. 1.19), e.g., **98** (Kimura *et al.*, 2000), **99** (Salbeck *et al.*, 1997a), **100** (Bach *et al.*, 1998), and **101** (Salbeck *et al.*, 1997b).

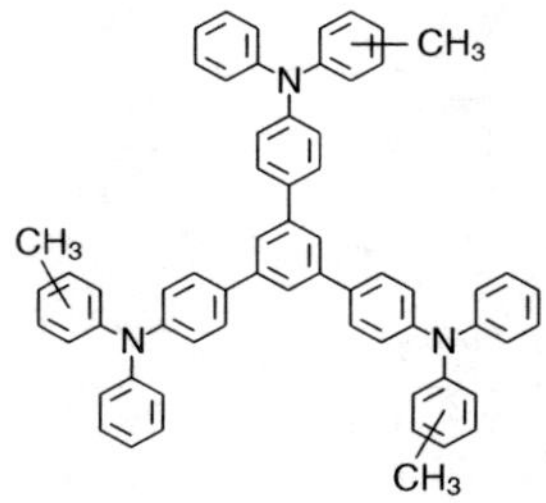

74 (*o*-, *m*- and *p*-MTDAPB)

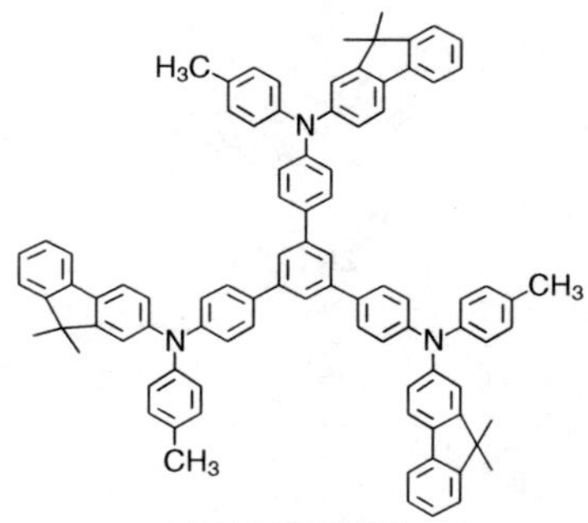

75 (MTFAPB)

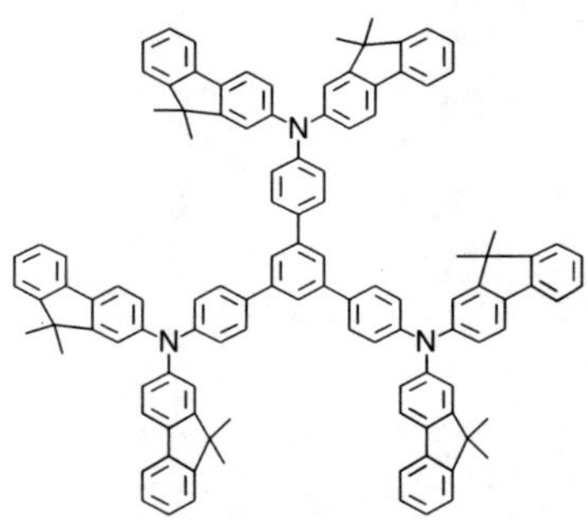

76 (TBFAPB)

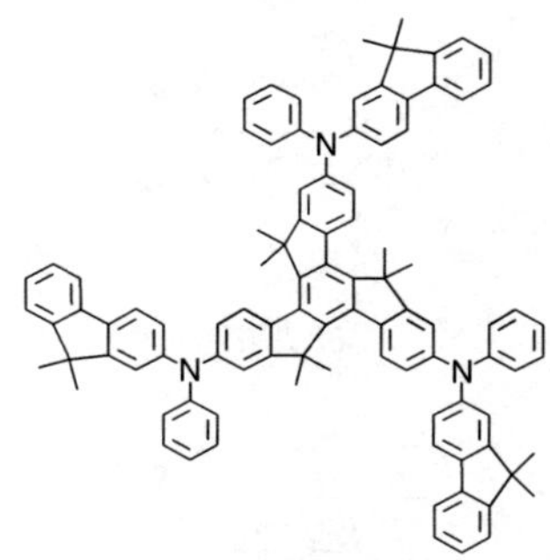

77 (TFATr)

1.14 Examples of hole-transporting amorphous molecular materials for OLEDs – TDAPB family.

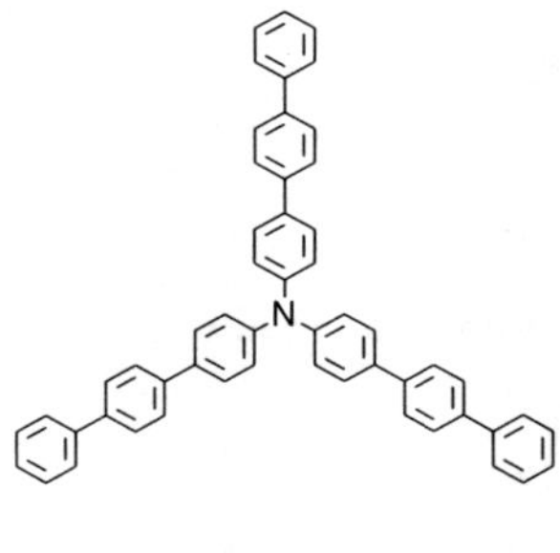

78 (*p*-TTA)

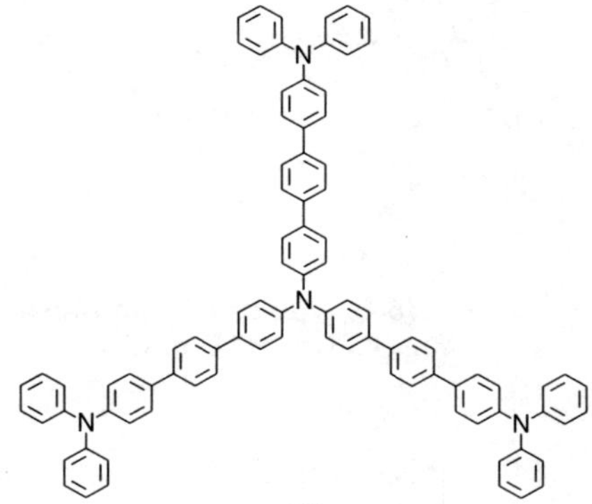

79

1.15 Examples of hole-transporting amorphous molecular materials for OLEDs – tris(oligoarylenyl)amine family.

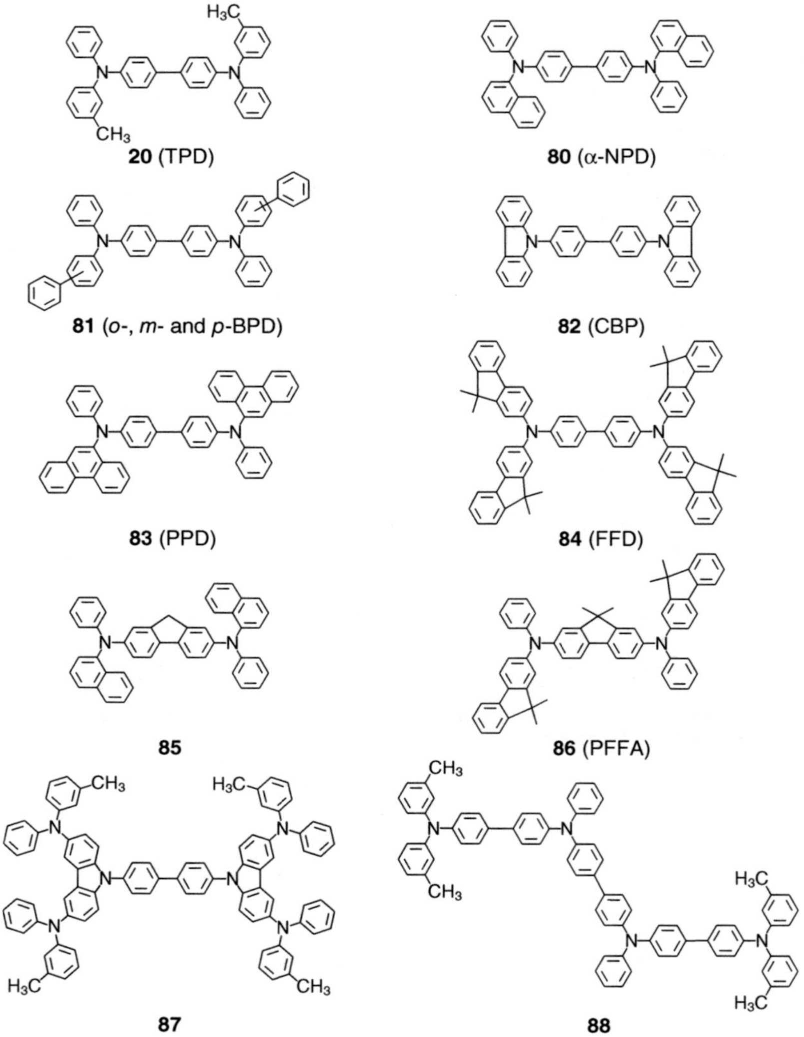

1.16 **Examples of hole-transporting amorphous molecular materials for OLEDs – TPD family.**

Materials of the TDATA family except for TCTA (**67**) are characterized by very low solid-state ionization potentials of ca. 5.10 eV and are used as hole-injection materials (Shirota *et al.*, 1994). Others include π-electron systems endcapped with diarylamino group, 2,2′-diaryl-1,1′-biphenyl derivative (**102**) (Agata *et al.*, 2007), diphenylamino-terminated s-indacenodithiophene (**103**) (Ku *et al.*, 2009), and diphenylamino-terminated indenofluorene (**104**) (Ku *et al.*, 2009) (Fig. 1.20).

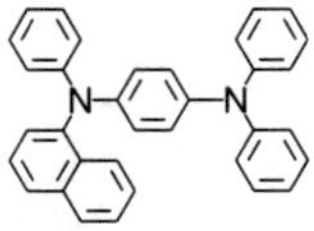

89

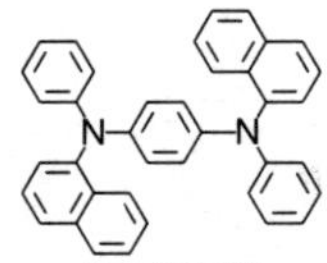

90 (NNP)

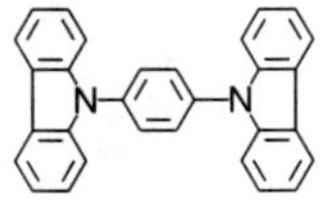

91

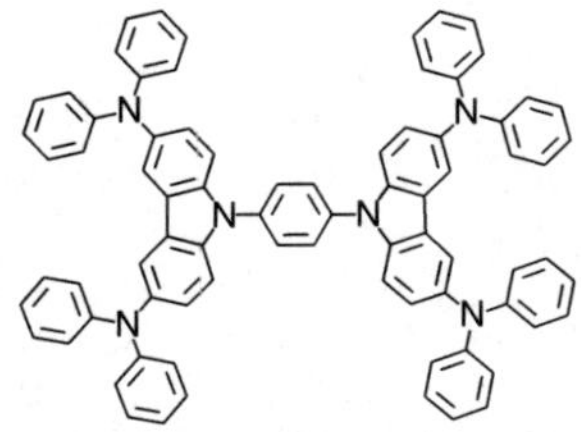

92

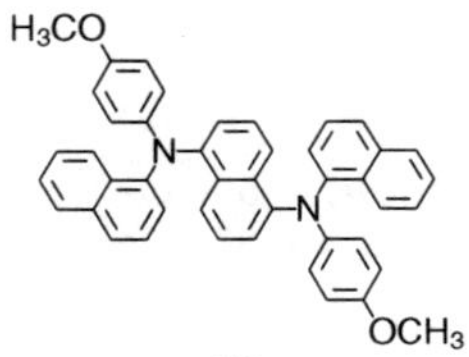

93

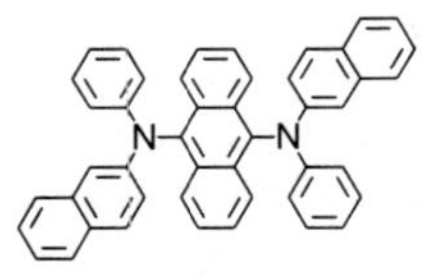

94

1.17 Examples of hole-transporting amorphous molecular materials for OLEDs – 1,4-bis(diarylamino)arylene family.

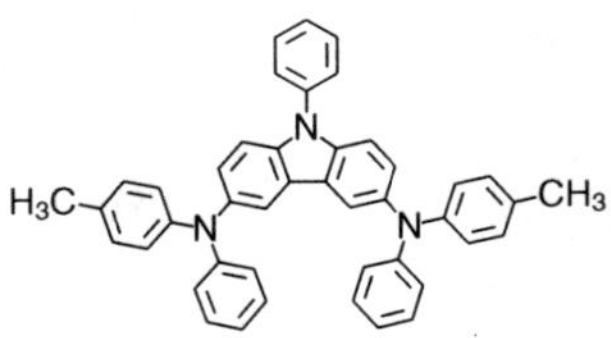

95

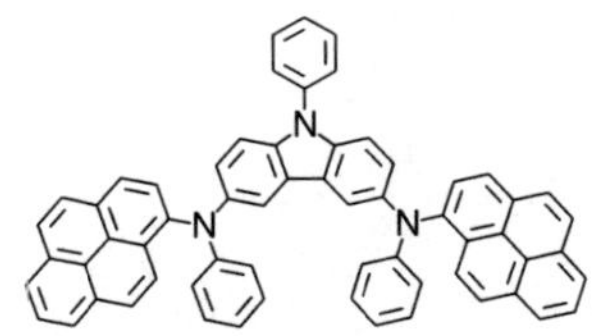

96

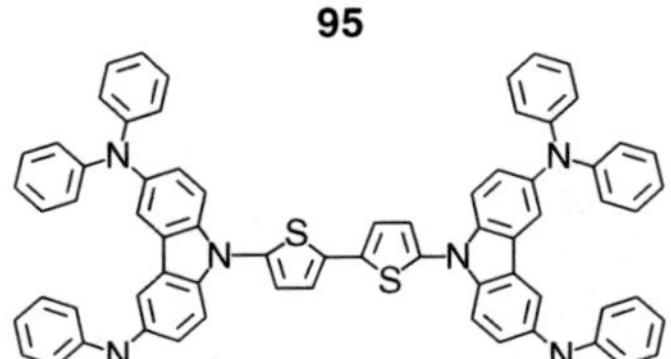

97

1.18 Examples of hole-transporting amorphous molecular materials for OLEDs – a family of carbazole-containing compounds.

Electron-transporting materials

Electron-transporting materials should possess the following properties: suitable electron affinities to accept electrons, ability to transport electron carriers, and the stability of the anion radicals of molecules. Various kinds

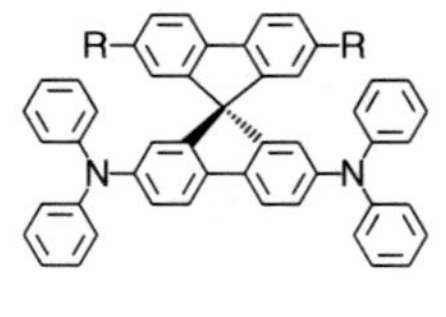

98

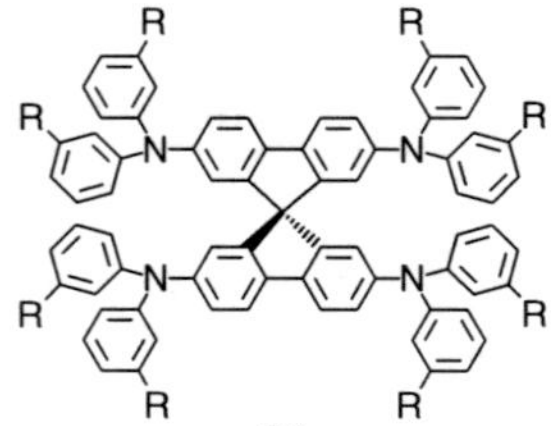

99

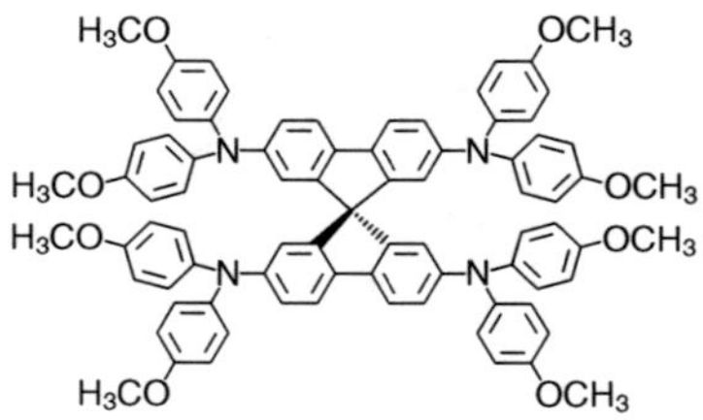

100

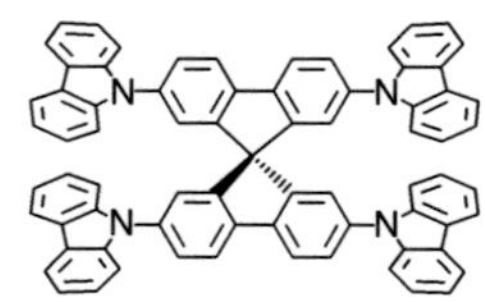

101

1.19 Examples of hole-transporting amorphous molecular materials for OLEDs – spiro-linked tetrakis(diarylamine) family.

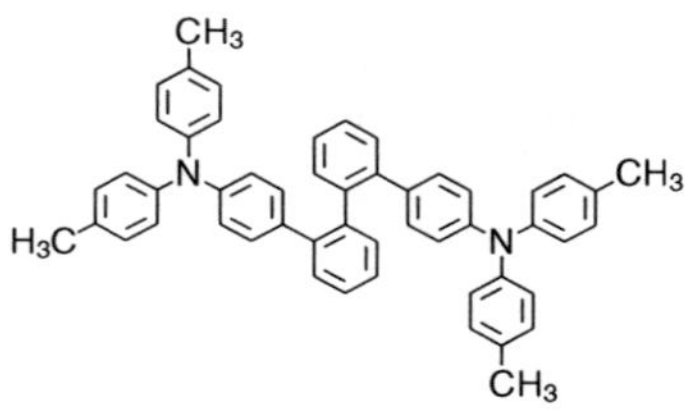

102

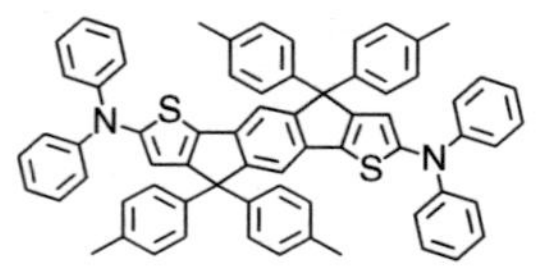

103

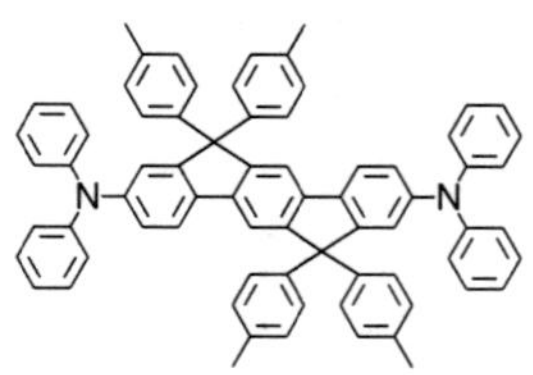

104

1.20 Examples of hole-transporting amorphous molecular materials for OLEDs – π-electron systems endcapped with diarylamino group.

of electron-transporting amorphous molecular materials have been developed (Kulkarni *et al.*, 2004; Hughes and Bryce, 2005; Sasabe and Kido, 2011).

Electron-transporting amorphous molecular materials possess central cores such as benzene, 1,3,5-triphenylbenzene, 1,3,5-triazine, tetraphenylmethane, to which electron-accepting moieties are attached. Electron-accepting moieties include oxadiazole, pyridine, triazine, silole, dimesitylboryl or triarylborane moieties (Fig. 1.21). Here, electron-transporting materials

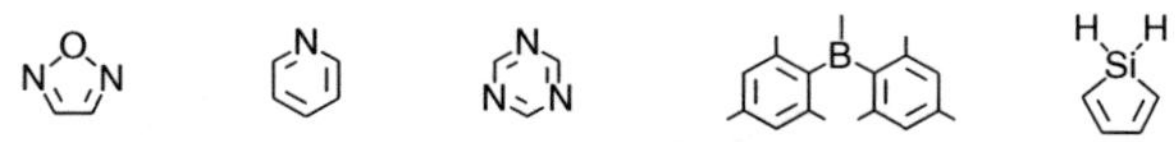

Oxadiazole Pyridine Triazine Dimesitylboryl group Silole

1.21 Examples of representative electron-accepting groups.

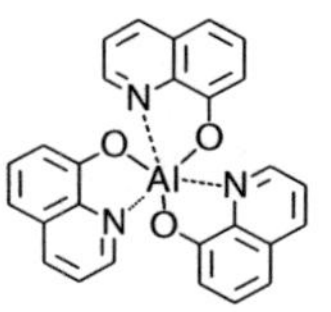

105 (Alq_3)

1.22 Examples of electron-transporting materials for OLEDs – Alq_3.

are classified mainly according to the electron-accepting moieties for the sake of convenience.

Green-emitting Alq_3 (**105**) has been extensively used as an excellent electron-transporting material in OLEDs (Fig. 1.22). Electron-transporting materials containing an oxadiazole or triazole moiety include t-Bu-PBD (**106**) (Adachi *et al.*, 1989), TPOB (**107**) (Bettenhausen and Strohriegl, 1996, Shirota *et al.*, 1997, Ogawa *et al.*, 1998), **108** (Yeh *et al.*, 2001), TAZ (**109**) (Kido *et al.*, 1993) (Fig. 1.23). Electron-transporting materials containing phenylbenzimidazole or phenylquinoxaline moiety are TPBI (**110**) (Gao *et al.*, 1999) and **111** (Jandke *et al.*, 1998) (Fig. 1.23). The examples of electron-transporting materials containing a pyridine moiety are given by **112, 113** (Pang *et al.*, 2002), bis-4,6-(3,5-di-3-pyridylphenyl)-2-methylpyrimidine (**114**) (Tanaka *et al.*, 2007a), 1,3,5-tri(p-pyrid-3-ylphenyl)benzene (**115**) (Su *et al.*, 2008), **116** (Pang *et al.*, 2002), and diphenyl-bis[4-(pyridin-3-yl) phenyl]silane (**117**) (Xiao *et al.*, 2009) (Fig. 1.24). 2,4,6-Triaryl substituted 1,3,5-triazines include **118** (Inomata *et al.*, 2004), **119** (Su *et al.*, 2010), **120** and **121** (Inomata *et al.*, 2004), and **122** (Fink *et al.*, 1998) as well as **113** and **116** (Fig. 1.25).

Silole derivatives have been shown to function as excellent electron-transporting materials for OLEDs, which include PySPy (**123**) (Tamao *et al.*, 1996), PPSPP (**124**) (Palilis *et al.*, 2003a,b), PPySPyP (**125**) (Uchida *et al.*, 2001), PyPSPPy (**126**) (Uchida *et al.*, 2001), PyPySPyPy (**127**) (Palilis *et al.*, 2003a,b), and ASP (**128**) (Palilis *et al.*, 2003a) (Fig. 1.26). The electron mobility of PyPySPyPy (**127**) is 2×10^{-4} cm^2 V^{-1} s^{-1} at 6.4×10^5 V cm^{-1}, which is more than two orders of magnitude higher than that of Alq_3 (Murata *et al.*, 2001).

Amorphous molecular materials containing a dimesitylboryl moiety, BMB-nT (n = 2, 3) (**129**) (Noda and Shirota, 1998; Mäkinen *et al.*, 2001), have been developed and found to be promising candidates for

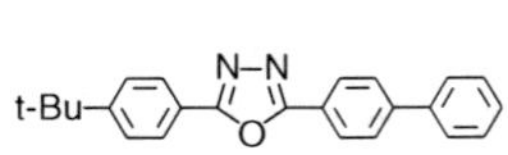

106 (t-Bu-PBD)

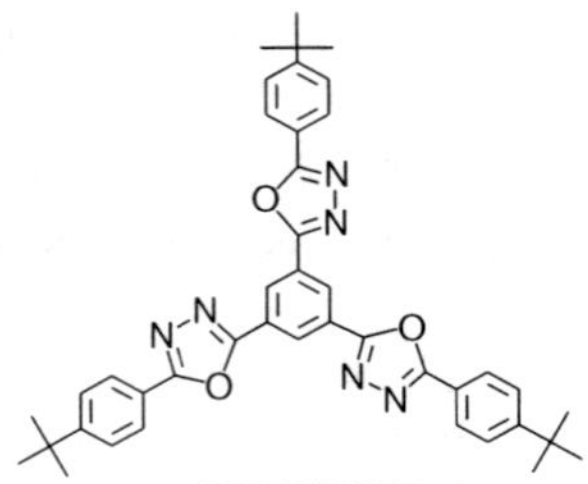

107 (TPOB)

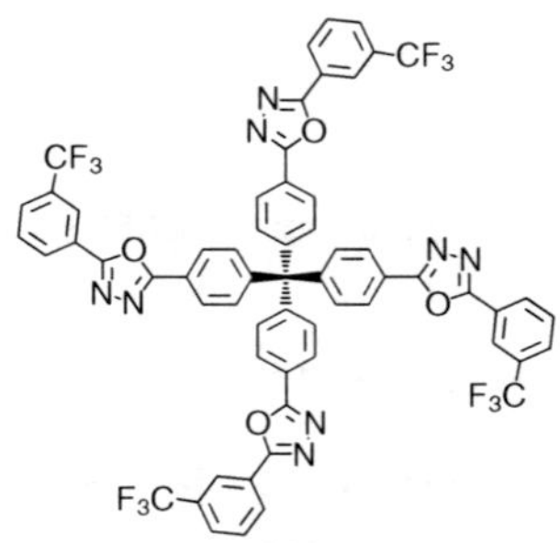

108

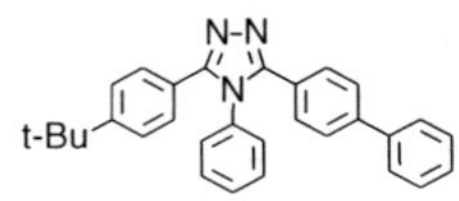

109 (TAZ)

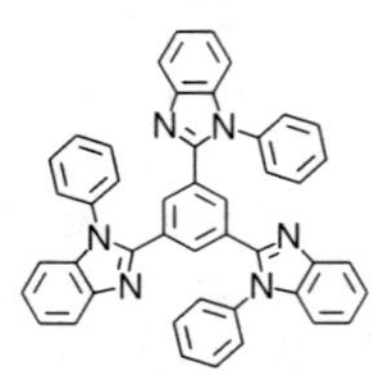

110 (TPBI)

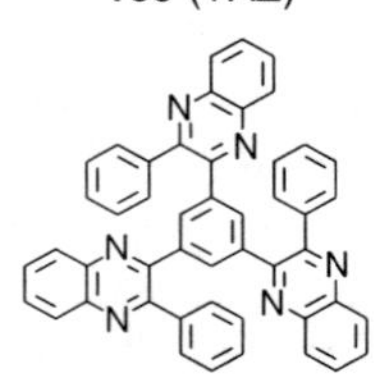

111

1.23 Examples of electron-transporting materials for OLEDs – oxadiazole, phenylbenzimidazole and phenylquinoxaline derivatives.

electron-transporting materials owing to the vacant 2p-orbital of the boron atom. Following this report, other electron-transporting amorphous molecular materials containing a dimesitylboryl group, TMB-TB (**130**) (Kinoshita and Shirota, 2001), B3T (**131**) (Sun *et al.*, 2011), and **132** (Li *et al.*, 2007) have been reported. B3T (**131**) is the molecule where the central core of benzene in TMB-TB (**130**) is replaced by the triazine central core. A number of triarylboranes have also been developed, which include **133** and **134** (Yamaguchi *et al.*, 2000a), **135** (Yamaguchi *et al.*, 2000b), TTPhB (**10**), TBPhB (**136**), and TPhPhB (**137**) (Kinoshita *et al.*, 2002), and **138** (Tanaka *et al.*, 2007b) (Fig. 1.27). Among these electron-transporting materials, materials with weak electron-accepting properties, e.g., **10**, **136** and **137**, have been used as hole-blocking materials.

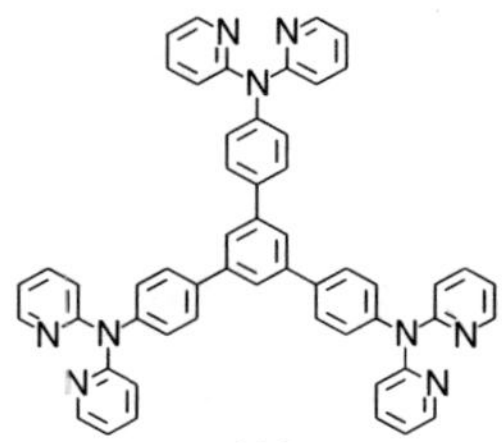

112

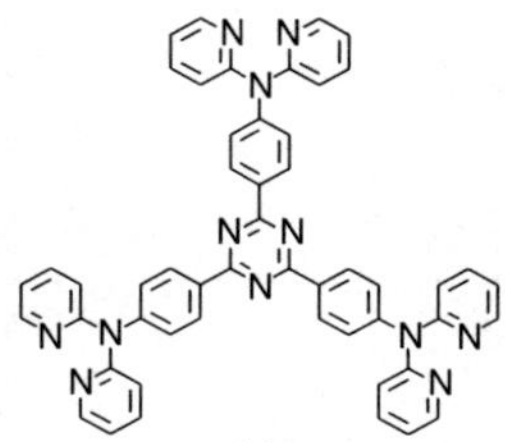

113

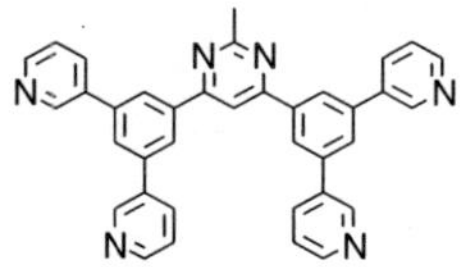

114

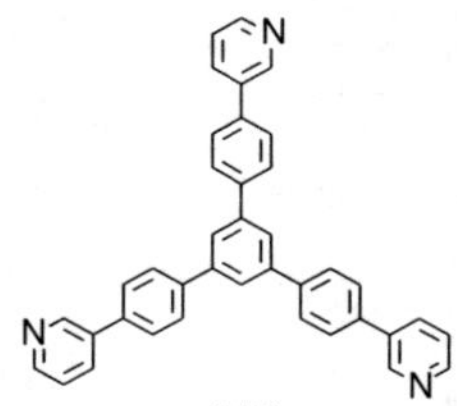

115

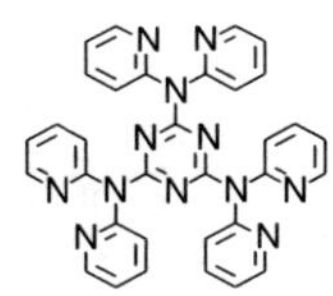

116

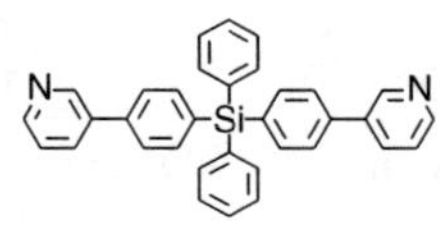

117

1.24 Examples of electron-transporting materials for OLEDs – compounds containing pyridine moiety with a 1,3,5-triphenylbenzene or triazine central core.

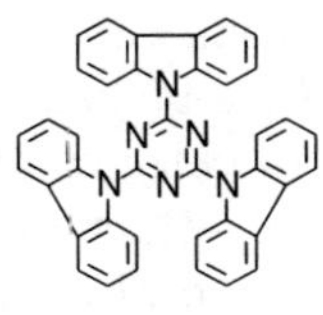

118

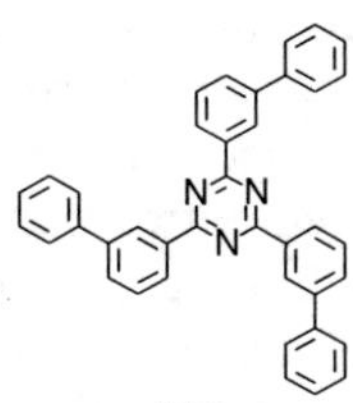

119

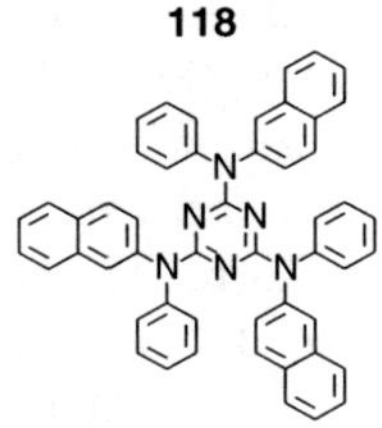

120

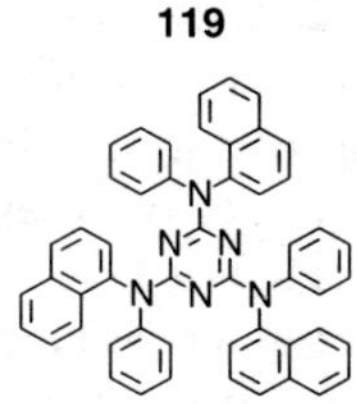

121

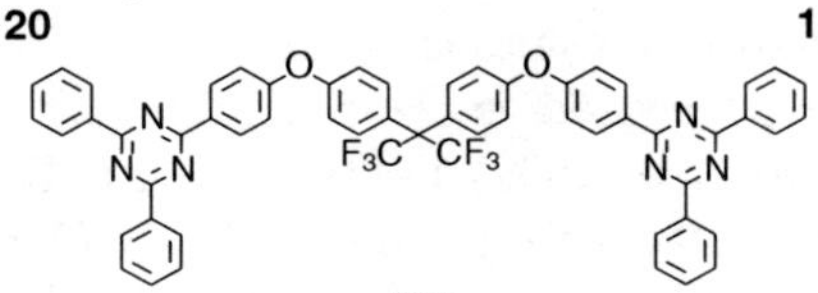

122

1.25 Examples of electron-transporting materials for OLEDs – materials with triazine central core.

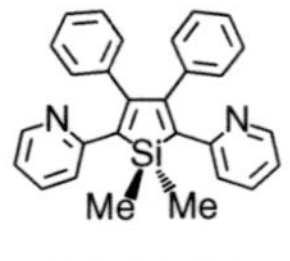

123 (PySPy)

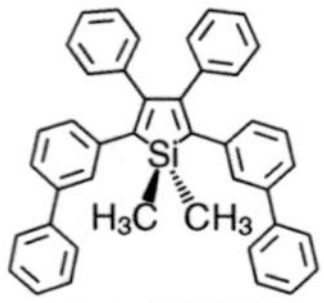

124 (PPSPP)

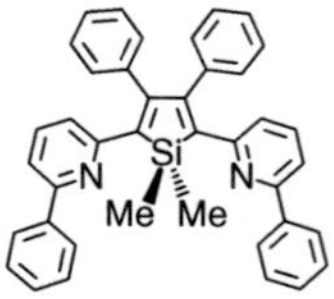

125 (PPySPyP)

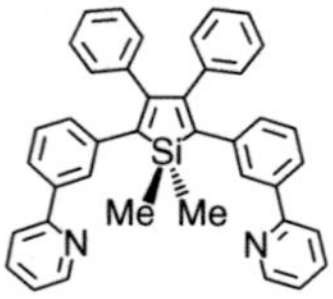

126 (PyPSPPy)

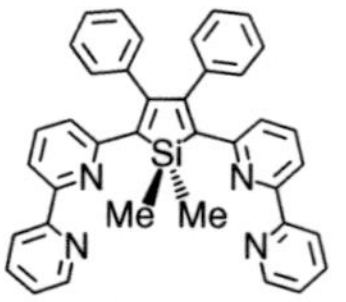

127 (PyPySPyPy)

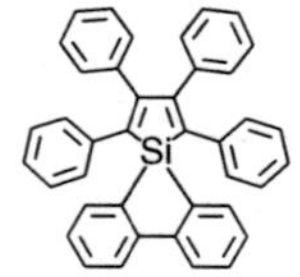

128 (ASP)

1.26 Examples of electron-transporting materials for OLEDs – siloles.

Hole-blocking materials

Hole-transporting and electron-transporting materials have electron-blocking and hole-blocking properties, respectively. When emitting materials with hole-transporting properties are used, a hole-blocking layer is frequently inserted between the emitting layer and the electron-transport layer so as to confine electron and hole carriers in the emitting layer, blocking hole carriers escaping into the electron-transport layer. Electron carriers are injected in a stepwise process from the cathode into the emitting layer via the electron-transport and hole-blocking layers (Fig. 1.28).

Hole-blocking materials should have large ionization potentials to prevent the acceptance of holes from the emitting layer and weak electron-accepting properties so as to accept electrons from the electron-transporting layer and pass electrons to the emitting layer. In addition, their anion radicals should be stable, and they should not form any exciplexes with the emitting materials (Kinoshita *et al.*, 2002).

Materials which have been used as the hole-blocking layer in OLEDs include BCP (**139**) (Kijima *et al.*, 1999), F-TBB (**140**), TFB (**141**) and TFPB (**142**) (Okumoto and Shirota, 2001a; 2003), TBPhB (**136**), TTPhB (**10**), TTPhPhB (**137**) (Kinoshita *et al.*, 2002), X-branched oligophenylenes (**143**) (Li *et al.*, 2006), and tetra(β-naphthyl)silane (**144**) (Duan *et al.*, 2005) (Fig. 1.29).

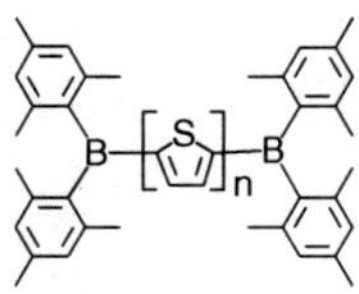

129 (BMB-nT)

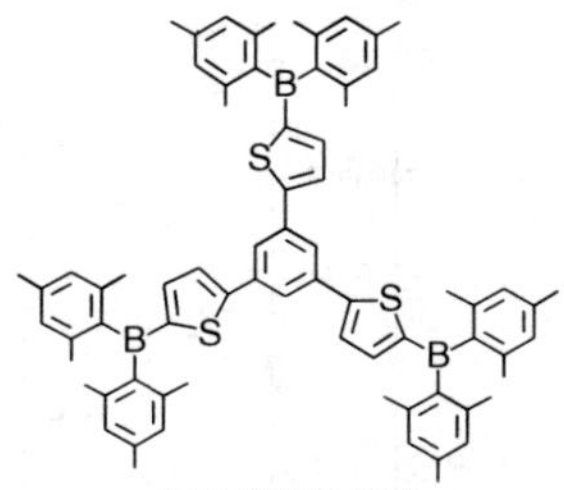

130 (TMB-TB)

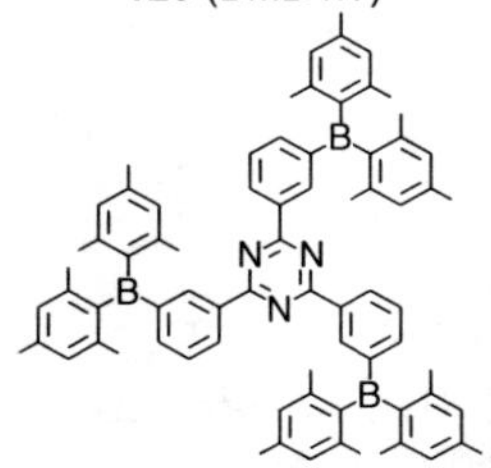

131 (B3T)

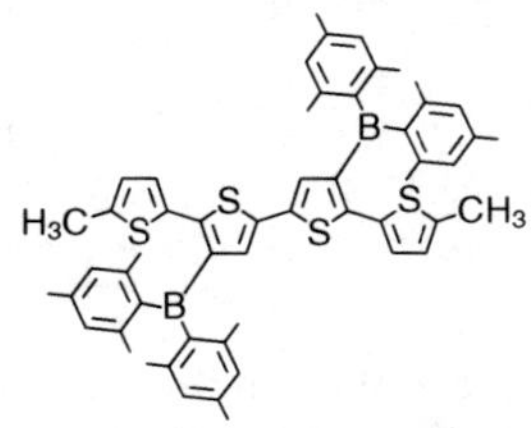

132

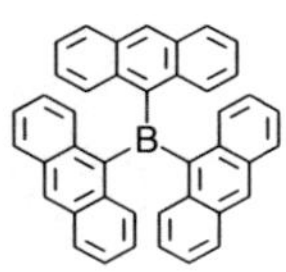

133

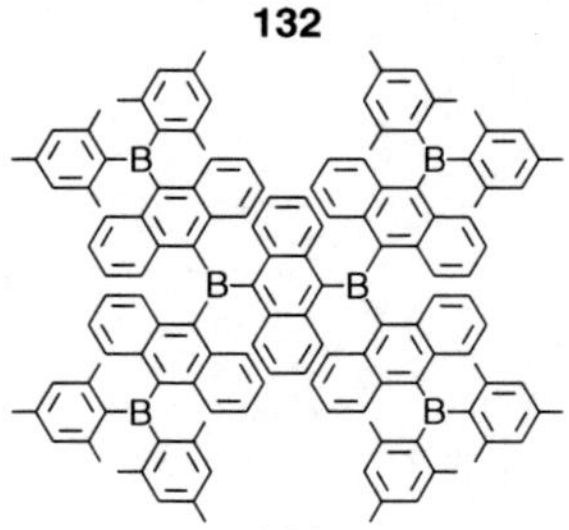

134

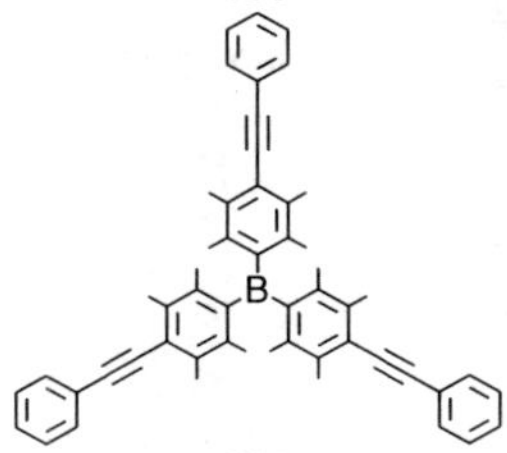

135

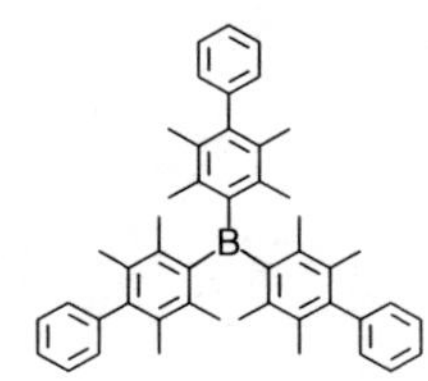

136 (TBPhB)

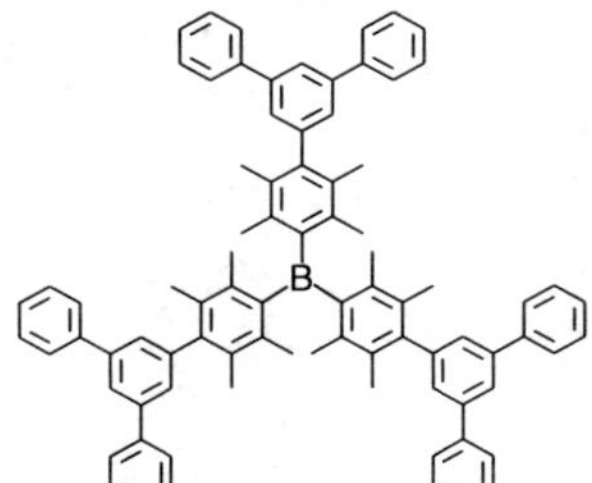

137 (TTPhPhB)

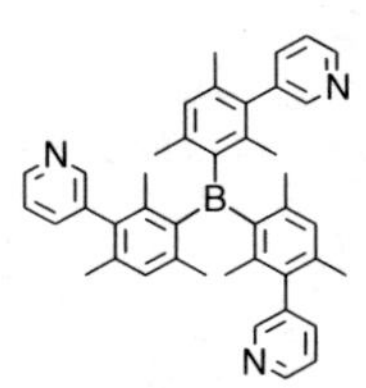

138

1.27 Examples of electron-transporting materials for OLEDs – compounds containing dimesitylboryl or triarylborane moiety.

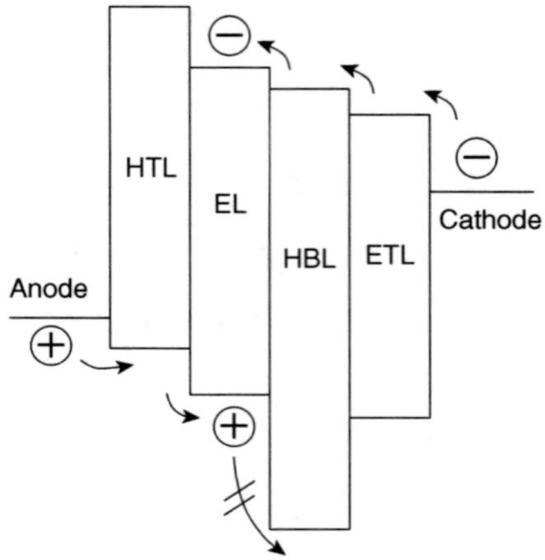

1.28 Role of hole-blocking layer (HBL). HTL, hole-transport layer; EL, emitting layer; and ETL, electron-transport layer.

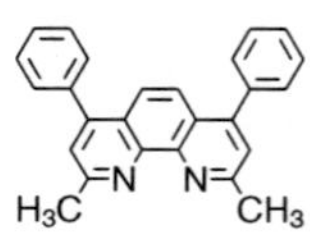

139 (BCP)

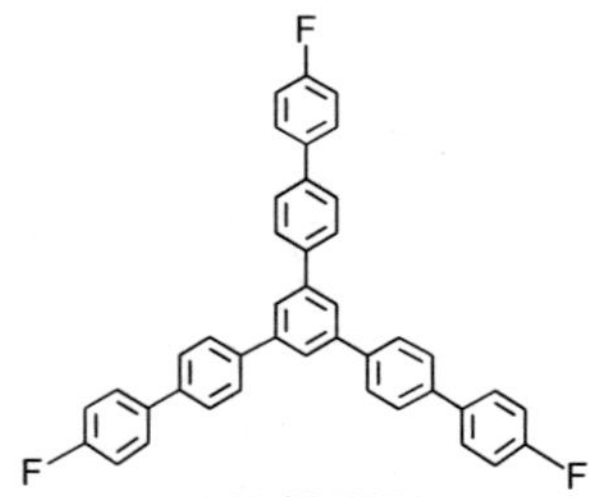

140 (F-TBB)

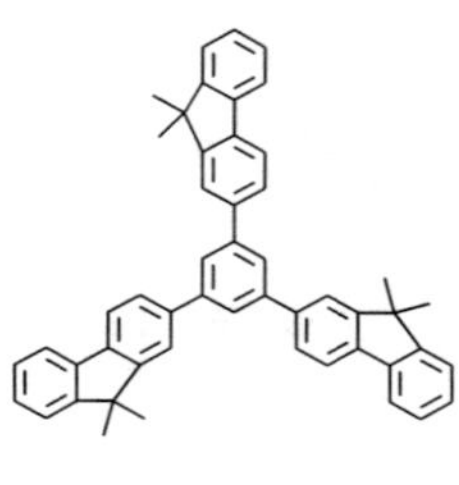

141 (TFB)

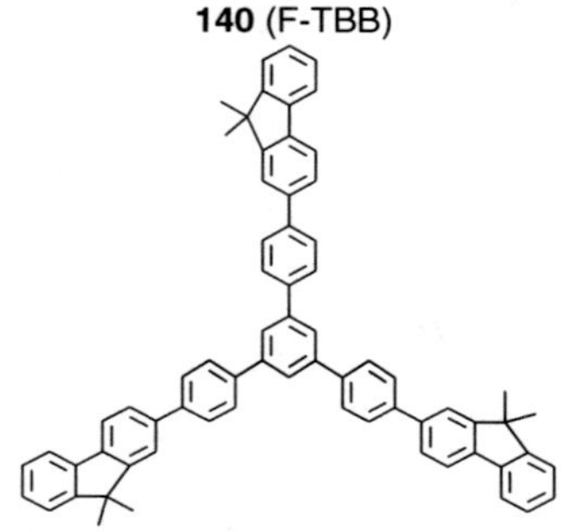

142 (TFPB)

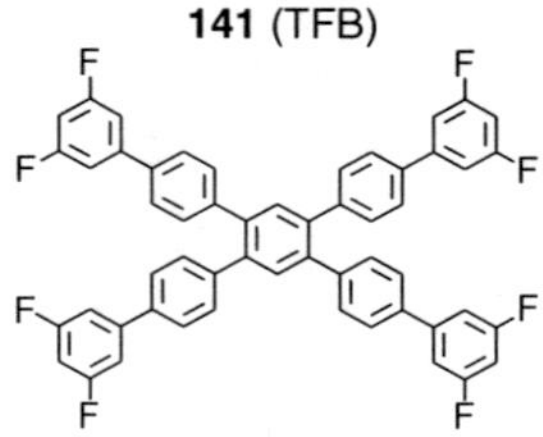

143

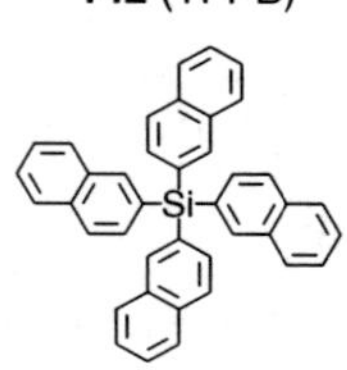

144

1.29 Examples of hole-blocking materials for OLEDs.

Emitting materials

Emitting materials should possess suitable HOMO/LUMO energy levels so that they accept both holes and electrons, confining them in the emitting layer and high quantum efficiency for either fluorescent or phosphorescent emission in addition to good film-forming properties and thermal and morphological stability.

Electron-transporting materials such as BMB-2T (**145**) (Noda *et al.*, 1999b), **146** (Mazzeo *et al.*, 2005), and silole derivatives, e.g., PyPySPyPy (**127**) and 2PSP (**147**) (Murata *et al.*, 2002; Palilis *et al.*, 2003b), have been shown to function as emitting materials with high fluorescence quantum efficiency (Fig. 1.30). In particular, the quantum efficiency for fluorescence of 2PSP (**147**) as vacuum-deposited solid films has been reported to be 100% (Murata *et al.*, 2002).

A number of molecules containing a spirofluorene moiety, **148** (Wu *et al.*, 2004; Tang *et al.*, 2008), **149** (Tang *et al.*, 2008), **150** (Wong *et al.*, 2005), **151** (Shen *et al.*, 2005), **152** (Chen *et al.*, 2004), **153** (Oyston *et al.*, 2005), **154** (Chuen *et al.*, 2004), and **155** (Wei and Chen, 2007), function mostly as blue-emitting materials (Fig. 1.31). Other blue-emitting materials include **156** (Danel *et al.*, 2002), **157** (Chiechi *et al.*, 2006), molecule with an indenopyrazine core (**158**) (Park *et al.*, 2010), dipyrenylbenzenes (**159**) (K.-C. Wu *et al.*, 2008), anthracene-triphenylamine derivative (**160**) (Tao *et al.*, 2008), and **161** (Zhen *et al.*, 2011) (Fig. 1.32).

Emitting materials should accept both holes and electrons and, hence, they should have ambipolar properties as well as high quantum efficiencies for either fluorescence or phosphorescence. That is, they should undergo reversible anodic oxidation and cathodic reduction to yield stable cation and anion radicals. In addition, emitting materials should form uniform amorphous thin films and have high thermal stability (Shirota *et al.*, 2000). Based on this concept, bipolar emitting amorphous molecular materials, BMA-nT (**162**) (Noda *et al.*, 1997a,b, 1999a) and FlAMB-nT (**163**) (Shirota *et al.*, 2000; Doi *et al.*, 2003) have been created (Fig. 1.33). FlAMB-nT (**163**)

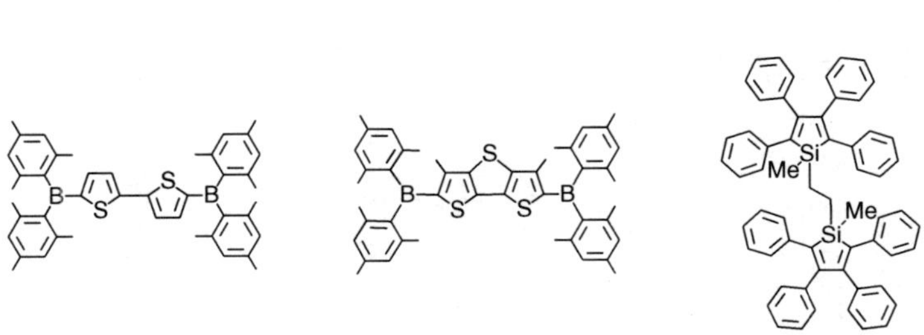

145 (BMB-2T) **146** **147** (2PSP)

1.30 Emitting amorphous molecular materials with electron-transporting property.

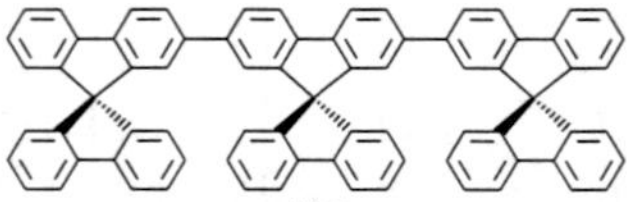

148

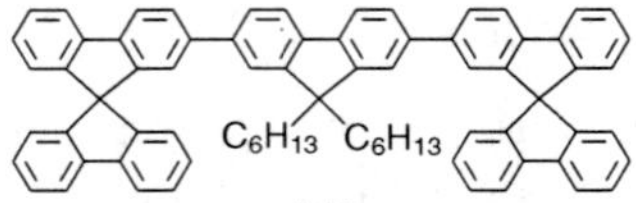

149

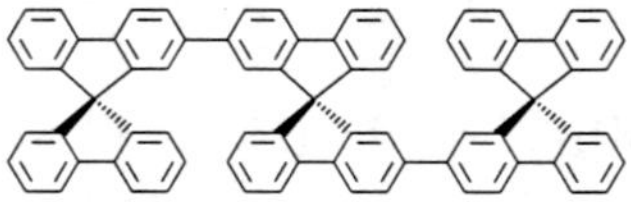

150

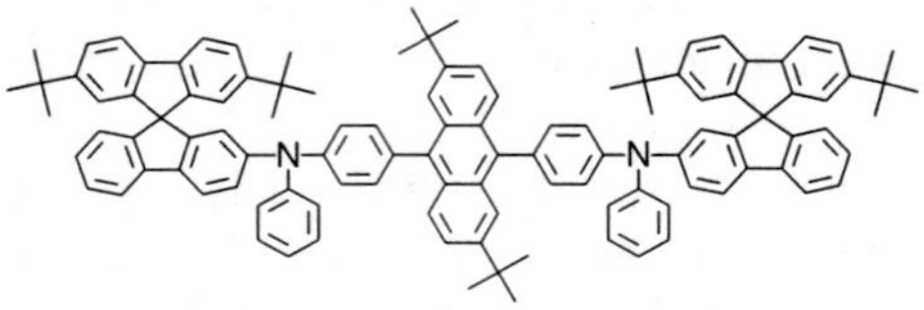

151

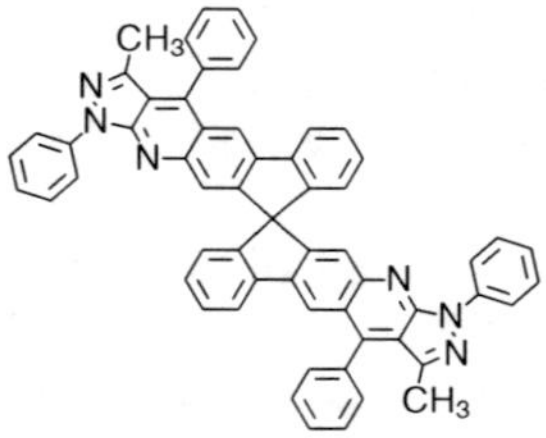

152

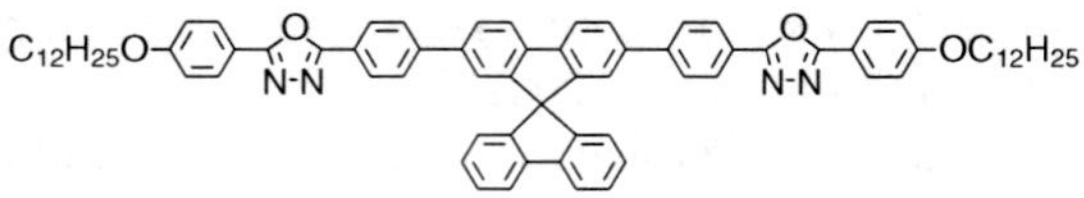

153

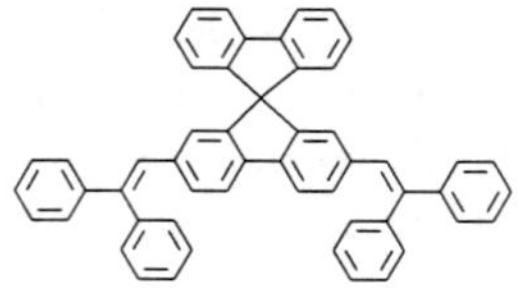

154

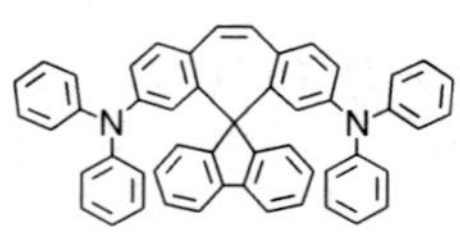

155

1.31 Emitting materials containing spirofluorene moiety.

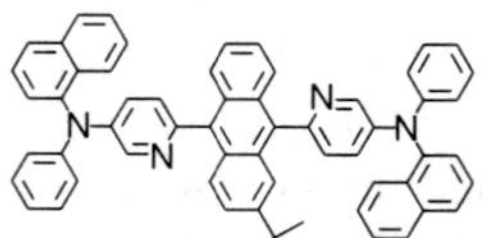

156

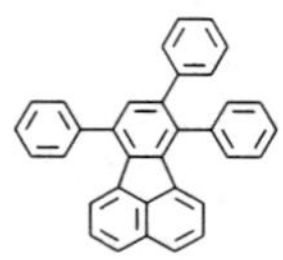

157

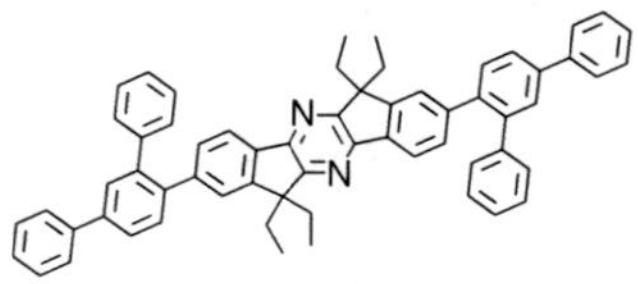

158

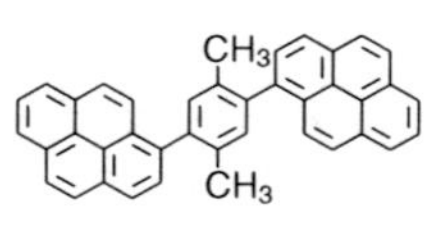

159

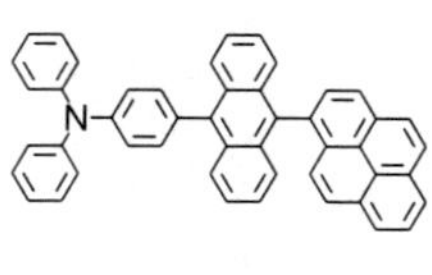

160

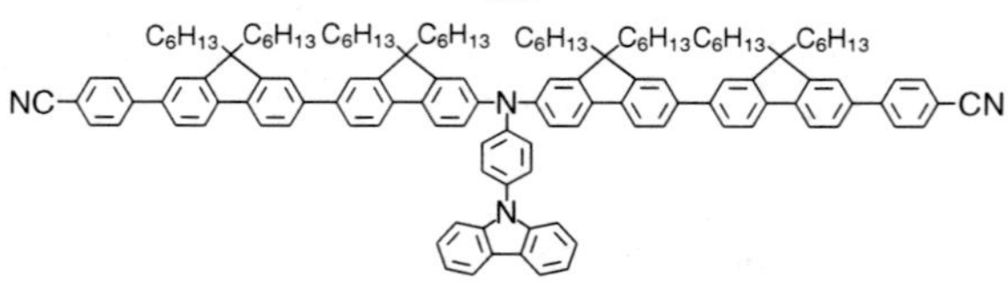

161

1.32 Examples of blue-emitting materials.

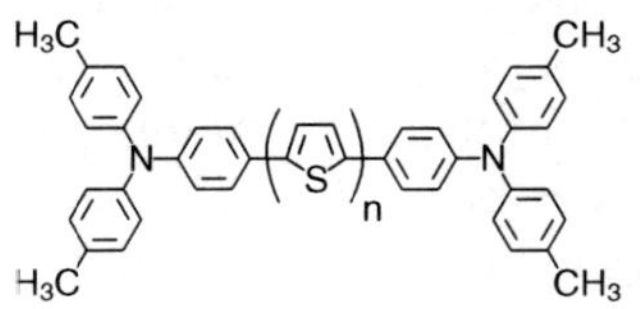

162 (BMA-nT)

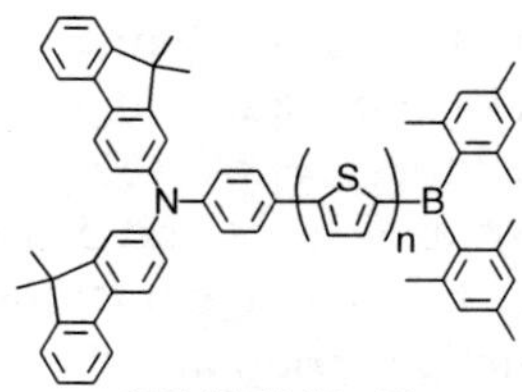

163 (FlAMB-nT)

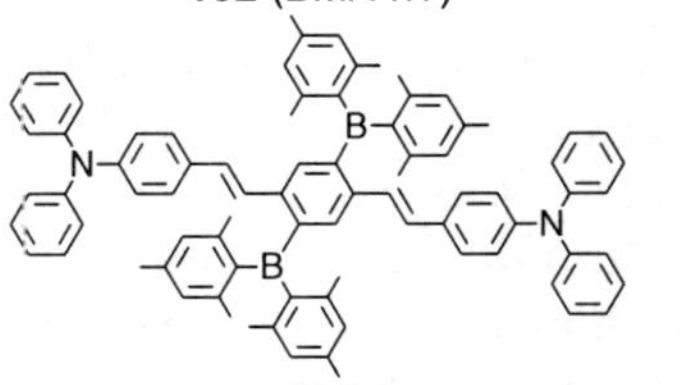

164

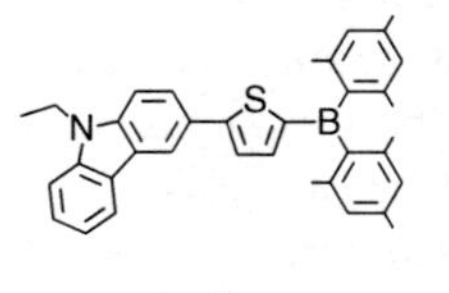

165

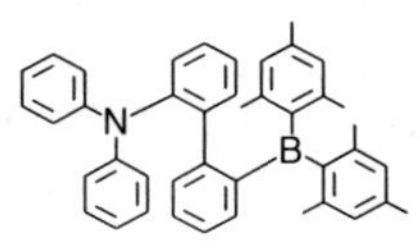

166

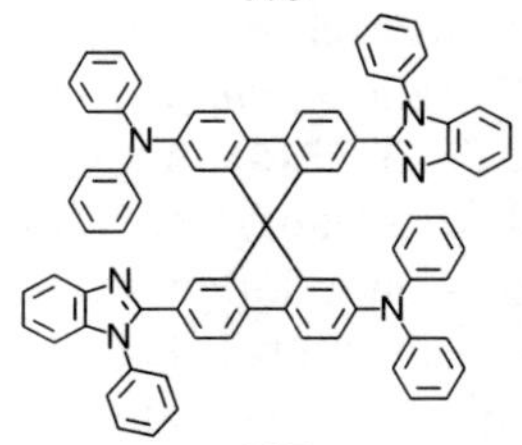

167

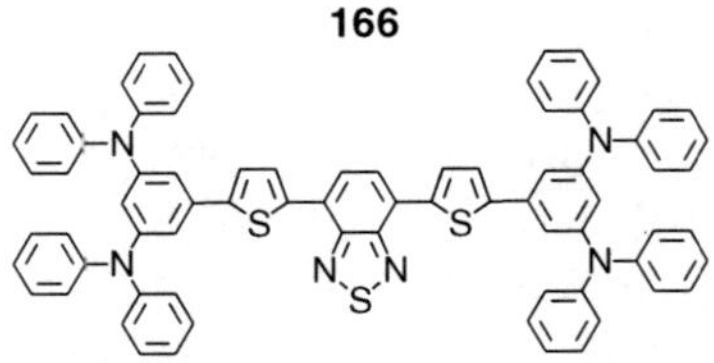

168 (btza)

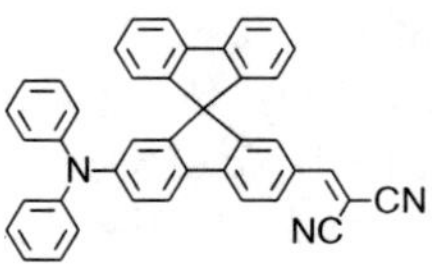

169

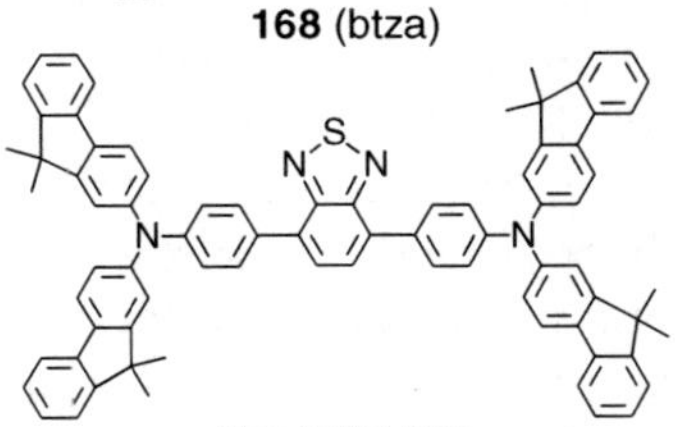

170 (BFA-BT)

1.33 Examples of emitting materials with ambipolar properties.

contains both electron-donating diarylamino and electron-accepting dimesitylboryl groups in a molecule and has intramolecular charge transfer character. The emission color is varied according to the conjugation length of the oligothiophene in BMA-nT and FlAMB-nT. Following these reports, other emitting materials with ambipolar properties have been developed, which include **164** (Zhao *et al.*, 2006), **165** (Lin *et al.*, 2008), biphenyls containing dimesitylboryl and arylamino groups (**166**), where the intramolecular charge transfer takes place by through space (Pan *et al.*, 2011), and a spirobifluorene derivative containing diaryamino and electron-accepting groups (**167**) (Liao *et al.*, 2007). Molecules containing benzo[1,2,5] thiadiazole core and diarylamine moieties, btza (**168**) (Thomas *et al.*, 2004), and molecules containing di(4-tolyl)amino or diphenylamino electron-donating group and dicyanovinyl electron-accepting group, PhSPCV (**169**) (Chiang *et al.*, 2005), function as bipolar red-emitting materials. BFA-BT containing both diarylamino and benzothiadiazole moieties (**170**) function as an orange-color emitting material (Shirota, 2005) (Fig. 1.33).

When molecules with high quantum efficiencies for either fluorescence or phosphorescence lack good film-forming ability, they can be used as emitting dopants for dispersion in the host amorphous molecular materials. The following compounds with high fluorescence quantum efficiencies have been used as emissive dopants: DCM2 (**171**) (Tang *et al.*, 1989), Nile Red (**172**) (Kido *et al.*, 1995a), coumarin 6 (**173**), DPP (**174**) (Picciolo *et al.*, 2001), and ACY (**175**) (Yu and Shirota, 2002) as red dopants, rubrene (**176**) as a yellow dopant, DMQA (**177**) as a green dopant, BCzVBi (**178**) and perylene as blue dopants (Fig. 1.34).

Room-temperature phosphorescent materials hitherto developed are transition metal complexes and molecular crystals (Yuan *et al.*, 2010). They include platinum complexes, e.g., PtOEP (**179**) (Baldo *et al.*, 1998), iridium complexes, e.g., $Ir(ppy)_3$ (**180**) (King *et al.*, 1985; Baldo *et al.*, 1999), $btp_2Ir(acac)$ (**181**) (Adachi *et al.*, 2001a), $(ppy)_2Ir(acac)$ (**182**) (Adachi *et al.*, 2001c), FIrpic (**183**) (Adachi *et al.*, 2001b; Holmes *et al.*, 2003a), and osmium complexes, e.g., **184** (Ma *et al.*, 1998), $[Os(bpy)_2L]^{2+}(PF_6^-)_2$ (**185**) (Bernhard *et al.*, 2002), OsAsNPP (**186**) (Jiang *et al.*, 2002), $OsCF_3$ (**187**) (Kim *et al.*, 2003), and $Os(fptz)_2(PPh_2Me)_2$ (**188**) (Tung *et al.*, 2005; C.-H. Wu *et al.*, 2008; Chien *et al.*, 2009) (Fig. 1.35). While PtOEP and osmium complexes are red emitters, iridium complexes exhibit red, green, and blue emissions depending on the structures of the ligands. $Ir(ppy)_3$ and $(ppy)_2Ir(acac)$ are green emitters, and $btp_2Ir(acac)$ and FIrpic are red and blue emitters, respectively. Since these transition metal complexes have no film-forming ability, they are dispersed in a host material with good film-forming ability.

The selection of host materials for the emitting layer is of crucial importance since charge recombination usually takes place on the host material and the electronic excitation energy transfer, i.e., the singlet–singlet and triplet–triplet energy transfer, takes place from the host material to emissive dopants. The excited triplet energy level of the host material should be higher than that of the emissive dopant molecule for the exothermic energy transfer to take place efficiently. Therefore, host materials for blue phosphorescent dopants are required to possess high triplet energy levels. Host materials such as CBP (**82**) (Koene *et al.*, 1998; O'Brien *et al.*, 1999), TAZ (**109**) (Kido *et al.*, 1993), TCTA (**67**) (Kuwabara *et al.*, 1994), and TDAPB

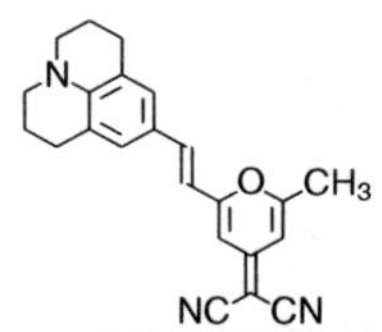

171 (DCM2)

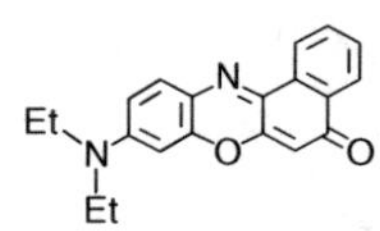

172 (Nile Red)

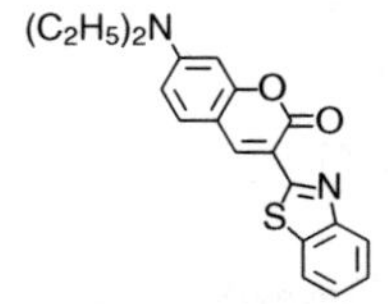

173 (coumarin 6)

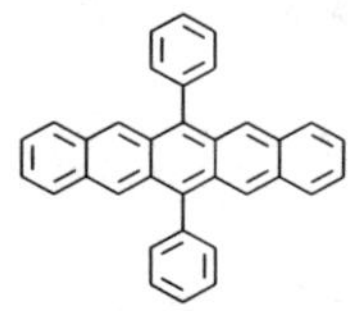

174 (DPP)

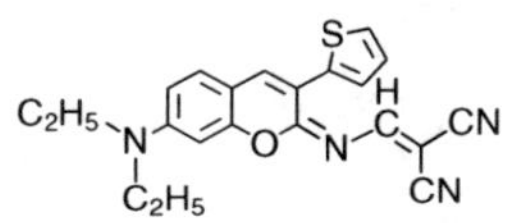

175 (ACY)

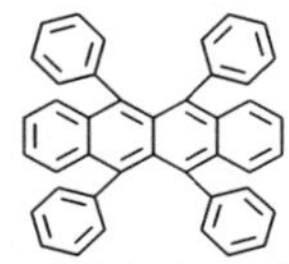

176 (rubrene)

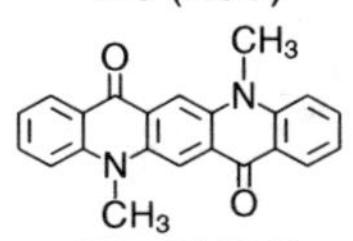

177 (DMQA)

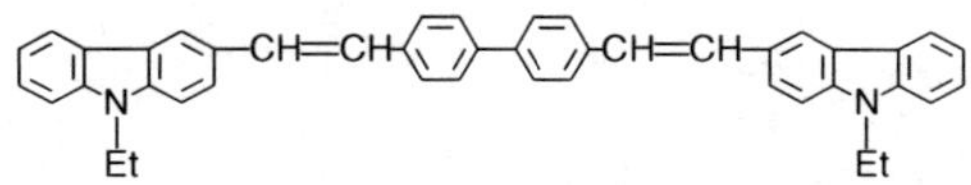

178 (BczVBi)

1.34 Examples of fluorescent dopants for OLEDs.

(**189**) (Hino *et al.*, 2004) for green-emitting dopants, CDBP (**190**) (Tokito *et al.*, 2003), mCB (**191**) (Adamovich *et al.*, 2002), SimCP (**192**) (Yeh *et al.*, 2005), and UGH2 (**193**) (Holmes *et al.*, 2003b) for blue-emitting dopants, and Alq_3 (**105**) (Baldo *et al.*, 1998), TFTPA (**194**) (Shih *et al.*, 2007; C.-H. Wu *et al.*, 2008), and POAPF (**195**) (Chien *et al.*, 2009) for red-emitting dopants have been developed (Fig. 1.36).

1.6 Structures and performance of organic (opto)electronic devices

1.6.1 Photovoltaic devices

Intensive studies have been made on pn-heterojunction type OPVs to improve the PCE on the basis of both the creation of new materials and the design of new device structures. The use of perylenediimides as electron acceptors (A) in planar pn-heterojunction OPVs gave relatively low PCEs (~1.0%). A two-junction tandem structure with a very thin layer of gold or silver metal has been shown to double the V_{OC} for devices using H_2Pc or CuPc as an electron donor (D) and perylenediimides as A (Hiramoto *et al.*, 1990; Yakimov and Forrest, 2002). The use of fullerenes, C_{60} or C_{70}, in place

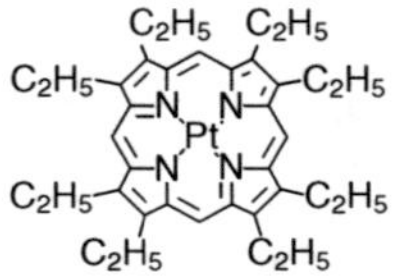

179 (PtOEP)

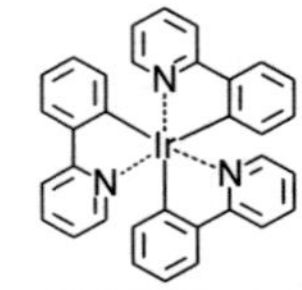

180 ($Ir(ppy)_3$)

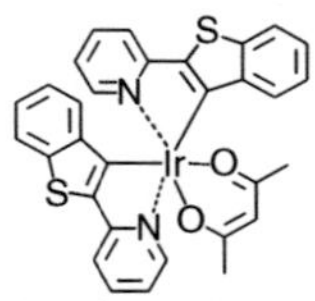

181 ($btp_2Ir(acac)$)

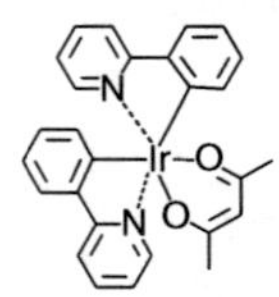

182 ($(ppy)_2Ir(acac)$)

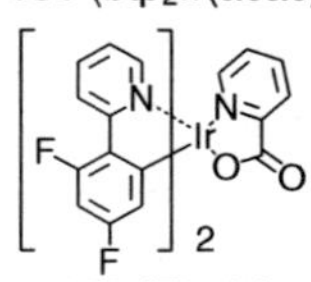

183 (FIrpic)

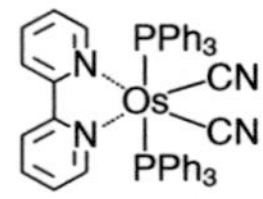

184

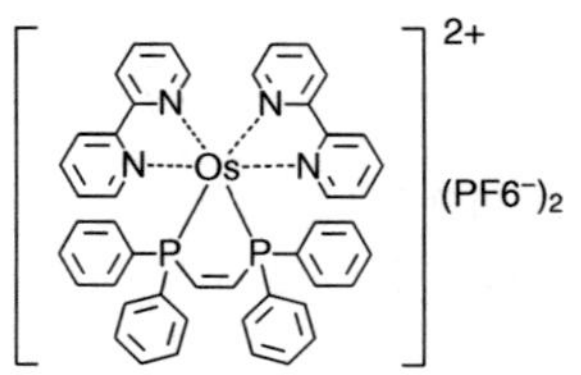

185 ($[Os(bpy)_2L)]^{2+}(PF_6^-)_2$)

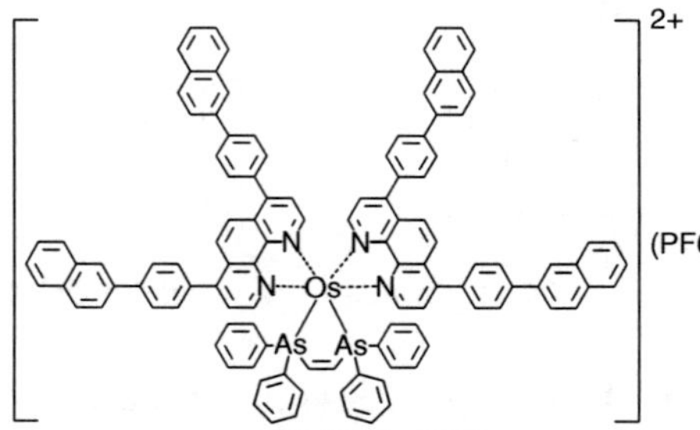

186 (OsAsNPP)

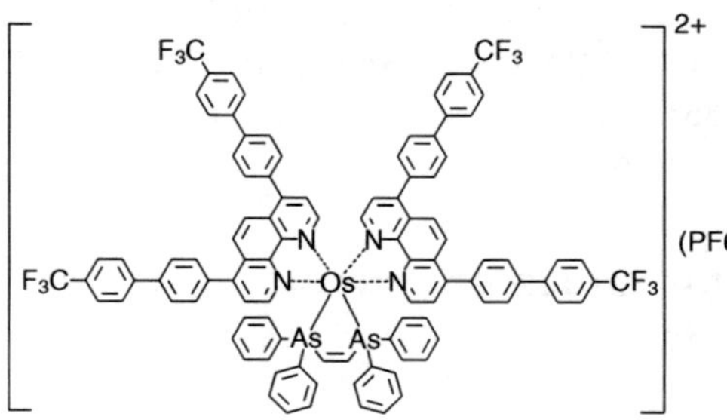

187 ($OsCF_3$)

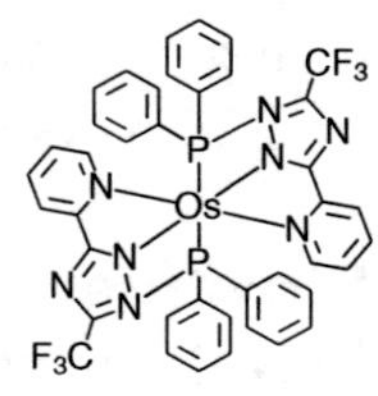

188 ($Os(fptz)_2(PPh_2Me)_2$)

1.35 Examples of phosphorescent dopants for OLEDs.

of perylene diimides as A has brought a jump of PCEs. An OPV using CuPc and C_{60} with the incorporation of a thin layer of bathocuproine (BCP) as an exciton-blocking layer between C_{60} and the Al cathode exhibited a high PCE of 3.6% under AM 1.5 illumination at 150 mW cm^{-2} in that time (Peumans and Forrest, 2001). The p-i-n structure with the insertion of a thin

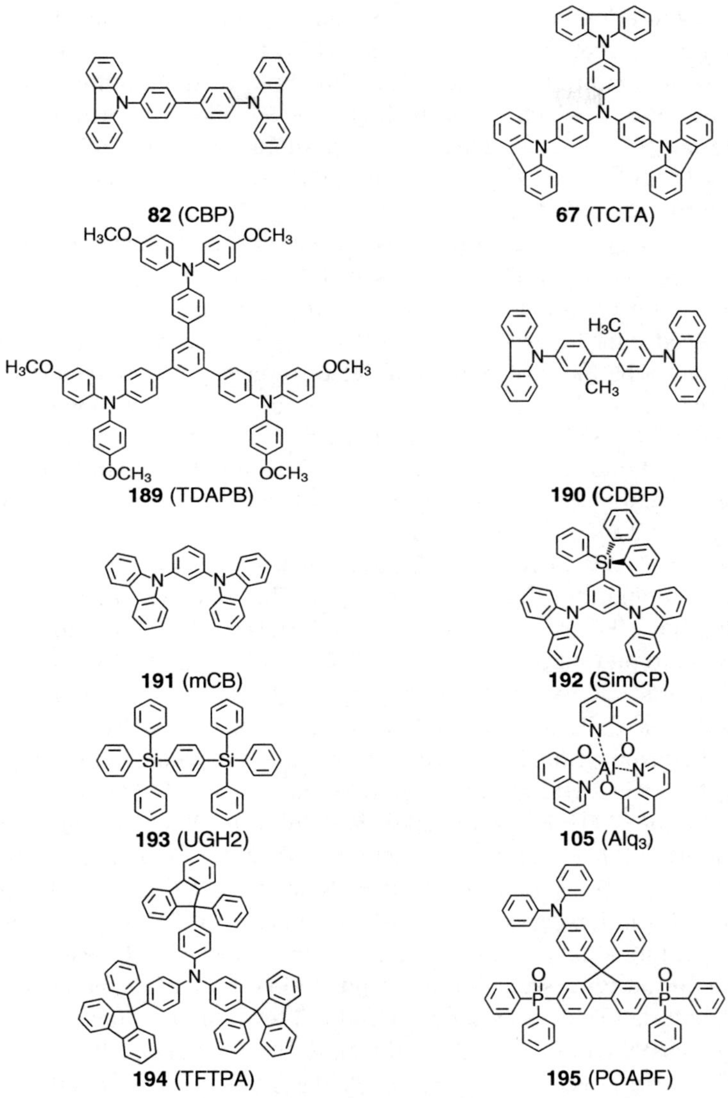

1.36 Examples of host materials for phosphorescent dopants.

mixed layer (10 nm) of CuPc (D) and C_{60} (A) between the two layers gave a higher PCE of 5.0% (Xue *et al.*, 2005).

Various kinds of molecular materials of p-type organic semiconductors (D) for OPVs have been developed (Fig. 1.9). Planar pn-heterojunction OPVs using the newly developed molecular materials such as **26**, **32**, **34**, **35**, and **37** as D and C_{60} or C_{70} as A have been reported to exhibit PCEs of

1.85–3.4% (Cravino *et al.*, 2006; Schulze *et al.*, 2006; Holzhey *et al.*, 2008; Wang *et al.*, 2009; Steinberger *et al.*, 2011; Chen *et al.*, 2012). Amorphous molecular materials **24** and **25** exhibit hole drift mobilities of ca. 1.0×10^{-2} cm^2 V^{-1} s^{-1}, which are of the highest level among amorphous molecular materials (Ohishi *et al.*, 2004). Planar pn-heterojunction OPVs using **24** and **25** as D and C_{60} or C_{70} as A gave PCEs of 1.5–2.2% (Kageyama *et al.*, 2009a,b).

Recently, growing attention has been paid to bulk-heterojunction OPVs using the blend of D and A, which have been fabricated by the co-evaporation of D and A or by the solution process. Studies on solution-processed bulk heterojunction OPVs have started by use of the blend of electron-donating polymers and electron-accepting polymers (Yu and Heeger, 1995; Halls *et al.*, 1995). Fullerenes soluble in common organic solvents, $PC_{61}BM$, $PC_{71}BM$, etc., have been developed (Hummelen *et al.*, 1995), and much higher PCEs have recently been attained for bulk-heterojunction OPVs. A solution-processed bulk-heterojunction OPVs using poly(2-methoxy-5-(2′-ethylhexyloxy)-1,4-phenylenevinylene) (MEH-PPV) as D and $PC_{61}BM$ as A gave 2.9% under 430 nm monochromatic light illumination (Yu *et al.*, 1995). A bulk-heterojunction OPV with additional layers of doped materials (2,3,5,6-tetrafluoro-7,7,8,8-tetracyanoquinodimethane (F_4-TCNQ) doped m-MTDATA (**65**) and Rhodamine B doped dimethyl-perylene-tetracarboxylic-diimide (MPP)) exhibited a PCE of ~3.37% (Gebeyehu *et al.*, 2003). Solution-processed bulk-heterojunction OPVs using crystalline molecular materials as D and $PC_{61}BM$ or $PC_{71}BM$ as A exhibited PCEs of 0.5–6.7% (Cremer *et al.*, 2005; Lloyd *et al.*, 2007; Ma *et al.*, 2008; Walker *et al.*, 2009; Tamayo *et al.*, 2009; Liu *et al.*, 2011; Wu *et al.*, 2011; Zhou *et al.*, 2011; Li *et al.*, 2012; Sun *et al.*, 2012). Amorphous molecular materials are suitable materials for the fabrication of solution-processed bulk-heterojunction OPVs owing to their good film-forming ability by the solution process as well as the vacuum process. The solution-processed bulk-heterojunction OPVs using amorphous molecular materials such as **28**, **29**, **30**, **42**, **44**, **45**, **46**, **47**, and **48** have been reported to exhibit PCEs of 0.26–2.39% (Lu *et al.*, 2006; Sun *et al.*, 2007; C. He *et al.*, 2008; Q. He *et al.*, 2008; Li *et al.*, 2009a; Stylianakis *et al.*, 2009; J. Zhang *et al.*, 2009; Kwon *et al.*, 2010; Wu *et al.*, 2011). Very recently, PCEs of 4.3% and 4.1% have been attained for solution-processed bulk heterojunction OPVs using **31** (Shang *et al.*, 2011) and **25** (Kageyama and Shirota, 2011). Higher PCEs have been achieved for solution-processed bulk-heterojunction OPVs using polymers as D and soluble fullerene derivatives as A (K. Kim *et al.*, 2007; J.Y. Kim *et al.*, 2007; Chen *et al.*, 2009; Liang *et al.*, 2010; He *et al.*, 2011; Green *et al.*, 2012). Higher PCEs achieved for bulk-heterojunction OPVs than those for planar pn-heterojunction OPVs are ascribed to the much higher J_{SC} which stems from the enhanced generation of the hole–electron pair as a result of the larger probability of the photo-generated excitons reaching the D/A interface within their lifetimes. Tables 1.1 and 1.2 list the performance of OPVs using molecular materials and polymers, respectively, as D.

1.6.2 OLEDs

Since the appearance of the report on the bright EL from a vacuum-deposited two-layer device at relatively low driving voltages (Tang and VanSlyke, 1987), extensive research and development on OLEDs have been conducted toward increasing higher luminous and quantum efficiencies for EL, improving longer operational durability, and achieving the realization of full-color and white-color emissions. Our common understanding of the operational processes, e.g., charge injection from electrodes into the organic layers, charge transport in organic materials, recombination of holes and electrons to produce the electronically excited singlet and triplet states, excitation energy transfer from host materials to emissive dopants, and exciplex formation at the solid/solid interface, has greatly deepened, and high-performance materials that meet the requirements of each layer in multi-layered devices have been designed and synthesized. The products of OLEDs came on the marketing in the late 1990s.

OLEDs can be classified into fluorescence and phosphorescence-based devices for convenience. Hole-transporting, electron-transporting, and charge-blocking materials can be used in common with both the devices.

Hole-transporting amorphous molecular materials, m-MTDATA (**65**), 1- and 2-TNATA (**66**), which are characterized by very low solid-state ionization potentials, have been widely used as the hole-injection layer in OLEDs (Shirota *et al.*, 1994; Murata *et al.*, 1999). CuPc (**1**) has also been used as a hole-injection material (VanSlyke *et al.*, 1996). It has been shown that charge-transfer doping of amorphous molecular materials with very low ionization potentials, e.g., m-MTDATA (**65**) and 4,4′,4″-tris(diphenylamino)triphenylamine (TDATA), with electron acceptors such as 2,3,5,6-tetrafluoro-7,7,8,8-tetracyanoquinodimethane (F_4TCNQ) or iodine significantly reduces the operating voltage, leading to the enhancement of hole injection from the anode (Zhou *et al.*, 2001a,b; Endo *et al.*, 2002; Ishihara *et al.*, 2006). Like hole-transporting materials with very low ionization potentials, which are used as the hole-injection layer, electron-transporting materials with strong electron-accepting properties, e.g., BMB-2T, BMB-3T (**129**, n = 2, 3), and PySPy (**123**), have been shown to function as materials for the electron-injection layer that facilitates electron injection from the cathode (Noda and Shirota, 1998; Tamao *et al.*, 1996).

The use of amorphous molecular materials with high T_gs has permitted the fabrication of thermally stable OLEDs. Thermally stable OLEDs up to

Table 1.1 Performance of OPVs using molecular materials

Device structure	J_{SC}/mA cm^{-2}	V_{OC}/V	FF	PCE/%	Light source	Reference
ITO/CuPc (30 nm)/PTCBI (50 nm)/Ag	2.3	0.45	0.65	0.95	AM2 (75 mW cm^{-2})	Tang (1986)
ITO/CuPc (15 nm)/CuPc : C_{60} (1:1, 10 nm)/C_{60} (35 nm)/BCP (10 nm)/Ag (100 nm)	1.8	0.54	0.61	5.0	AM1.5 (120 mW cm^{-2})	Xue *et al.* (2005)
ITO/pentacene (45 nm)/C_{60} (50 nm)/BCP (10 nm)/Al	15	0.363	0.50	2.7	AM1.5 (100 mW cm^{-2})	Yoo *et al.* (2004)
ITO/PEDOT:PSS (~80 nm)/pentacene (20 nm)/C_{60} (40 nm)/CsF (1 nm)/Al (40 nm) (annealed at 200 °C)	6.37	0.46	0.5	1.07	Halogen (115 mW cm^{-2})	Mayer *et al.* (2004)
ITO/PEDOT:PSS (40 nm)/pentacene : PTCDI-C_{13} (1:3, 100 nm)/BCP (8 nm)/Ag (60 nm)	8.6	0.441	0.64	2.0	AM1.5 (80.6 mW cm^{-2})	Pandey *et al.* (2006)
ITO/PEDOT:PSS (ca.30 nm)/**24** (30 nm)/C_{60} (40 nm)/LiF (0.1 nm)/Al (150 nm)	2.5	0.95	0.62	1.5	AM1.5 (100 mW cm^{-2})	Kageyama *et al.* (2009a)
ITO/PEDOT:PSS (ca.30 nm)/**25** (30 nm)/C_{60} (40 nm)/LiF (0.1 nm)/Al (150 nm)	2.6	0.92	0.71	1.7	AM1.5 (100 mW cm^{-2})	Kageyama *et al.* (2009a)
ITO/PEDOT:PSS (ca.30 nm)/**25** (30 nm)/C_{70} (40 nm)/LiF (0.1 nm)/Al (150 nm)	3.6	0.90	0.66	2.2	AM1.5 (100 mW cm^{-2})	Kageyama *et al.* (2009b)
ITO/PEDOT:PSS (80 nm)/**26** (20 nm)/C_{60} (25 nm)/LiF (0.6 nm)/Al (60–100 nm)	4.59	1.15	0.28	1.85	AM1.5 (80 mW cm^{-2})	Cravino *et al.* (2006)
ITO/MoO_3 (20 nm)/**27** (7 nm)/**27** : C_{70} (1:1vol%, 40 nm)/C_{70} (6 nm)/BCP (10 nm)/Ag	12.1	0.95	0.56	6.4	AM1.5 (100 mW cm^{-2})	Chiu *et al.* (2012)
ITO/PEDOT:PSS/**28** : $PC_{61}BM$ (1:3 wt)/Ba (10 nm)/Al (100 nm)	4.18	0.81	0.39	1.33	AM1.5 (100 mW cm^{-2})	C. He *et al.* (2008a)
ITO/PEDOT:PSS/**29** : $PC_{71}BM$ (1:3 wt, ca.60 nm)/Mg/Al	8.58	0.85	0.327	2.39	AM1.5 (100 mW cm^{-2})	Zhang *et al.* (2009a)
ITO/PEDOT:PSS (~50 nm)/**30** : $PC_{61}BM$ (1:1 wt%, ~70 nm)/LiF (0.6 nm)/Al (80 nm)	2.5	0.56	0.29	0.41	AM1.5 (100 mW cm^{-2})	Stylianakis *et al.* (2009)
ITO/PEDOT:PSS/**31** : $PC_{71}BM$ (1:2 wt%)/Ca/Al	9.51	0.87	0.52	4.3	AM1.5 (100 mW cm^{-2})	Shang *et al.* (2011)

ITO/**32** (6.5 nm)/C_{60} (40 nm)/BCP (10 nm)/Al (100 nm)	7.13	0.75	0.60	3.2	AM1.5 (100 mW cm^{-2})	Wang *et al.* (2009)
ITO/MoO_3 (8 nm)/**32** : $PC_{71}BM$ (1:6, 78 nm)/C_{60} (4 nm)/BCP (1 nm)/LiF (0.8 nm)/Al (100 nm) (solvent vapor annealing ; DCM 10 min)	12	0.5	0.92	5.5	AM1.5 (100 mW cm^{-2})	Wei *et al.* (2011)
ITO/MoO_3 (4 nm)/**32** : C_{70} (1:5, 60 nm)/BCP (10 nm)/Al (100 nm)	13.6	0.86	0.52	6.1	AM1.5 (100 mW cm^{-2})	Chen *et al.* (2012)
ITO/PEDOT:PSS (ca.40 nm)/**33** : $PC_{61}BM$ (1:0.5 wt%, ca.80 nm)/LiF (0.8 nm)/Al (60 nm)	13.98	0.92	0.474	6.1	AM1.5 (100 mW cm^{-2})	Li *et al.* (2012)
ITO/PTCBI (40 nm)/α-8T (30 nm)/Au	2.9	0.42	0.50	0.59	Xenon (105 mW cm^{-2})	Noma *et al.* (1995)
ITO/**34** (17 nm)/ZnPc (10 nm)/p-doped TNATA (37 nm)/p-doped ZnPc (10 nm)/Al (100 nm)	5.28	0.68	0.516	1.85	AM1.5 (127 mW cm^{-2})	Holzhey *et al.* (2008)
ITO/Au (1 nm)/highly-doped Di-NPB (30 nm)/Di-NPB (5 nm)/**35** (12 nm)/C_{60} (40 nm)/BPhen (6 nm)/Al	10.6	0.98	0.49	3.4	AM1.5 (118 mW cm^{-2})	Schulze *et al.* (2006)
ITO/C_{60} : NDN1 (2%, 5 nm)/C_{60} (15 nm)/**35** : C_{60} (40 nm)/BPAPF (5 nm)/BPAPF:NDP9 (10%, 10 nm)/Spiro-NPB:NDP9 (10%, 30 nm)/NDP9 (1 nm)/Al (50 nm)	11.1	0.97	0.49	5.2	AM1.5 (102 mW cm^{-2})	Fitzner *et al.* (2011)
ITO/n-doped C_{60} (5 nm)/C_{60} (15 nm)/**36** : C_{60} (2:1, 40 nm)/BPAPF (5 nm)/p-doped BPAPF (10 nm)/p-doped Di-NPD (30 nm)/NDP9 (1 nm)/Al (100 nm)	11.6	0.88	0.570	4.9	AM1.5 (117.7 mW cm^{-2})	Wynands *et al.* (2010)
ITO/C_{60} (15 nm)/**37** (10 nm)/Di-NPB (5 nm)/Di-NPB:NDP9 (5 wt%, 53 nm)/NDP9 (1 nm)/Au (50 nm)	4.7	1.00	0.67	3.15	AM1.5 (100 mW cm^{-2})	Steinberger *et al.* (2011)
ITO/PEDOT:PSS (ca.40 nm)/**38** : $PC_{61}BM$ (1:0.5) (ca.110 nm)/LiF (0.8 nm)/Al (60 nm)	9.77	0.93	0.599	5.44	AM1.5 (100 mW cm^{-2})	Liu *et al.* (2011)
39 : $PC_{61}BM$ (1:2 wt)	4.19	0.97	0.42	1.72	AM1.5 (100 mW cm^{-2})	Ma *et al.* (2008)
ITO/PEDOT:PSS (50 nm)/**40** : $PC_{71}BM$ (60:40, 95 nm)/Al (80 nm) (annealed at 110 °C)	10	0.92	0.48	4.4	AM1.5 (100 mW cm^{-2})	Walker *et al.* (2009)
ITO/PEDOT:PSS (40 nm)/**41** : $PC_{71}BM$ (50:50, ca.80 nm)/Al	9.2	0.75	0.44	3.0	AM1.5 (100 mW cm^{-2})	Tamayo *et al.* (2009)

Table 1.1 Continued

Device structure	J_{SC}/mA cm^{-2}	V_{OC}/V	FF	PCE/%	Light source	Reference
ITO/PEDOT:PSS (30 nm)/**42** : $PC_{71}BM$ (1:3) (ca.60 nm)/LiF (0.7 nm)/Al (100 nm) (annealed at 100 °C for 10 min)	2.79	0.92	0.27	0.69	AM1.5 (100 mW cm^{-2})	Wu *et al.* (2011)
ITO/PEDOT:PSS/**43** : $PC_{61}BM$ (1.5:1.0 w/w, ~75 nm)/LiF (1 nm)/Al (100 nm) (annealed at 110 °C for 10 min)	11.27	0.84	0.42	4.06	AM1.5 (100 mW cm^{-2})	Loser *et al.* (2011)
ITO/PEDOT:PSS/**44** : $PC_{61}BM$ (1:1 wt%, 100 nm)/Ba/Al	0.86	0.76	0.33	0.26	AM1.5 (85 mW cm^{-2})	Q. He *et al.* (2008)
ITO/PEDOT:PSS (30 nm)/**45** : $PC_{61}BM$ = 1:4 (45 nm)/ Bphen or BCP (5 nm)/Al (100 nm)	1.88	0.51	0.34	0.34	AM1.5 (100 mW cm^{-2})	Kwon *et al.* (2010)
ITO/PEDOT:PSS (50 nm)/**46** : $PC_{61}BM$ (1:1.5) (110 nm)/ LiF (1 nm)/Al (100 nm) (annealed at 150 °C for 15 min)	1.011	0.82	0.35	0.29	AM1.5 (100 mW cm^{-2})	Lu *et al.* (2006)
ITO/PEDOT:PSS (50 nm)/**47** : $PC_{61}BM$ (1:4, 70–80 nm)/Ba/ Al (200 nm)	1.95	0.75	0.344	0.75	AM1.5 (100 mW cm^{-2})	Sun *et al.* (2007)
ITO/PEDOT:PSS/**48** : $PC_{71}BM$ (1:3 wt, 75 nm)/LiF (0.6 nm)/Al (80 nm)	4.9	0.92	0.41	1.8	AM1.5 (100 mW cm^{-2})	Li *et al.* (2009)
ITO/PEDOT:PSS (60 nm)/**49** : $PC_{61}BM$ (7:3)/CsF (1 nm)/Al (40 nm)	2.96	0.840	0.40	1	Halogen (100 mW cm^{-2})	Lloyd *et al.* (2007)
ITO/MoO_x (9 nm)/**50** : $PC_{71}BM$ (70:30 wt%, 180 nm)/Al (solvent : chlorobenzene with 0.25 wt% diiodeoctane, annealed at 70 °C for 10 min)	14.4	0.78	0.593	6.70	AM1.5 (100 mW cm^{-2})	Sun *et al.* (2012)
ITO/PEDOT:PSS (40 nm)/**51** : $PC_{61}BM$ (1:0.8 (w/w), 130 nm)/LiF (0.8 nm)/Al (50 nm)	11.51	0.80	0.64	5.84	AM1.5 (100 mW cm^{-2})	Zhou *et al.* (2011)
ITO/PEDOT:PSS (30 nm)/α-NPD (10 nm)/C_{60} (48 nm)/ MgAg (170 nm)	–	0.85	–	1	AM1.5 (97 mW cm^{-2})	Kushto *et al.* (2005)
ITO/PEDOT:PSS (30 nm)/TPD (10 nm)/C_{60} (60 nm)/Ag (40 nm), annealed at 140 °C for 15 min	3.04	0.67	0.53	1.08	AM1.5 (100 mW cm^{-2})	Osasa *et al.* (2007)

Table 1.2 Performance of OPVs using polymers

Device structure	J_{SC}/mA cm^{-2}	V_{OC}/V	FF	PCE/%	Light source	Reference
ITO/MEH-PPV/C_{60}/Au	2.08×10^{-3}	0.44	0.48	0.04	514.5 nm (1 mW cm^{-2})	Sariciftci *et al.* (1993)
ITO/MEH-PPV : $PC_{61}BM$ (1:4)/Ca	–	–	–	2.9	430 nm (20 mW cm^{-2})	Yu *et al.* (1995)
ITO/PEDOT:PSS/$PC_{71}BM$: MDMO-PPV (4.6:1)/LiF (1 nm)/Al (100 nm)	7.6	0.77	0.51	3.0	AM1.5 (100 mW cm^{-2})	Wienk *et al.* (2003)
ITO/PEDOT : PSS/P3HT : $PC_{61}BM$ (1:0.6 wt)/LiF (0.3–0.4 nm)/Al (100 nm) (annealed at 158 °C for 8 min)	–	–	–	6.1	AM1.5 (100 mW cm^{-2})	K. Kim *et al.* (2007)
ITO/PEDOT:PSS (40 nm)/PCPDTBT : $PC_{61}BM$ (0.7:2.5 wt)/TiO_x (20 nm)/PEDOT:PSS (40 nm)/P3HT : $PC_{71}BM$ (1:0.7 wt)/TiO_x (20 nm)/Al (100 nm)	7.8	1.24	0.67	6.5	AM1.5 (100 mW cm^{-2})	J. Y. Kim *et al.* (2007)
ITO/PEDOT:PSS/PBDTTT-CF : $PC_{71}BM$ (1:1.5 wt)/Ca/Al (spincoated from chlorobenzene/1,8-diiodooctane (3 vol%))	15.2	0.76	0.669	7.73	AM1.5 (100 mW cm^{-2})	Chen *et al.* (2009)
ITO/PEDOT:PSS/PTB7 : $PC_{71}BM$ (1:1.5 w/w, 90 nm)/ PFN (5 nm)/Ca/Al	15.75	0.7563	0.7015	8.370	AM1.5 (100 mW cm^{-2})	He *et al.* (2011)

Table 1.2 Continued

Device structure	J_{SC}/mA cm^{-2}	V_{OC}/V	FF	PCE/%	Light source	Reference
ITO/PEDOT:PSS/P3HT : **55** (1:1 wt%)/Ca/Al (annealed at 150 °C for 10 min)	10.61	0.84	0.727	6.48	AM1.5 (100 mW cm^{-2})	Zhao *et al.* (2010)
ITO/PEDOT:PSS (ca.30 nm)/P3HT : **56** (1:1 wt%, 180–220 nm)/Ca (20 nm)/Al (100 nm)	11.34	0.81	0.63	5.79	AM1.5 (100 mW cm^{-2})	He *et al.* (2010)
P3HT/**57**	8.64	0.810	0.61	4.2	AM1.5 (100 mW cm^{-2})	Ross *et al.* (2009)
ITO/ZnO (~50 nm)/C_{60}-SAM/P3HT : **59** (1:0.7 wt, ~200 nm)/Ag (100 nm) (annealed at 150 °C for 10 min)	9.9	0.65	0.62	4.0	AM1.5 (100 mW cm^{-2})	Y. Zhang *et al.* (2009)
ITO/PEDOT:PSS/T : **60** (1:3.5 wt%)/ZnO/Al (annealed at 100 °C for 5 min)	6.3	0.95	0.53	3.17	Tungsten–halogen (100 mW cm^{-2})	Sharma *et al.* (2010)
ITO/PEDOT:PSS/P3HT : **62**/Ba/Al	–	1.10	0.40	1.7	AM1.5 (100 mW cm^{-2})	Brunetti *et al.* (2010)
ITO/PEDOT:PSS (40 nm)/P3HT : **63** (1:2 wt%)/Ca (10 nm)/Al (100 nm)	2.36	0.81	0.52	1.00	AM1.5 (100 mW cm^{-2})	Sonar *et al.* (2010)
ITO/PEDOT:PSS/P3HT : **64** (1:1 wt%)/Cs (0.4 nm)/Al (40 nm)	3.56	0.84	0.42	1.27	AM1.5 (100 mW cm^{-2})	Shu *et al.* (2011)

150 °C and 180 °C were developed by use of hole-transporting materials with high T_gs, e.g., TCTA (T_g = 151 °C) (**67**), TFATA (T_g = 131 °C) (**196**), TFATr (T_g = 208 °C) (**77**), and Alq_3 or BFA-1T (T_g = 158 °C) (**197**) (Kuwabara *et al.*, 1994; Okumoto *et al.*, 2001; Shirota, 2005) as emitting materials.

The insertion of a hole-blocking layer using F-TBB (**140**), TTPhPhB (**137**), and TNS (**144**) between the emitting layer and the electron-transport layer enabled the fluorescent emission from emitting materials with electron-donating properties, e.g., TPD (**20**) and α-NPD (**80**) (Okumoto and Shirota, 2001a; Kinoshita *et al.*, 2002; Yu *et al.*, 2005).

High-performance fluorescence-based OLEDs

Luminous and quantum efficiencies for OLEDs have been steadily improved with times. Fluorescence-based, green-emitting OLEDs with multi-layer structures, ITO/m-MTDATA (**65**)/FlAMB-1T (**163**, *n* = 1)/BMB-2T (**129**, *n* = 2)/Alq_3 (**105**)/LiF/Al (Shirota *et al.*, 2000) and ITO/m-MTDATA (**65**)/β-NPA (**94**)/TPBI (**110**)/MgAg (Yu *et al.*, 2002), showed high performance with a turn-on voltage of 2.5 V, a power efficiency of 3.1 lm W^{-1}, and an EQE of 2.0%, and with a turn-on voltage of 3.7 V, a power efficiency of 7.76 lm W^{-1}, and an EQE of 3.68%, respectively. A blue-green OLED with a device structure of ITO/TPD (**20**)/2PSP (**147**)/PyPySPyPy (**127**)/MgAg gave an EQE of 4.8% for EL (Murata *et al.*, 2002). A green-emitting OLED using an emitter with hole-transporting properties, 4,4′,4″-tris[8-(7,19-diphenylfluoranthenyl)]phenylamine (TDPFPA) (**198**), exhibited a power efficiency of 10.6 lm W^{-1} (Tong *et al.*, 2007). Fluorescence-based blue-emitting OLEDs with structures, ITO/PEDOT:PSS/**199**/**167**/**200**/LiF/Al (Liao *et al.*, 2007) and ITO/TCTA (**67**)/THPF (**201**)/BCP (**139**)/Alq_3 (**105**)/LiF/Al (Cheng *et al.*, 2005), showed high performance with a turn-on voltage of 2.5 V, a power efficiency of 1.55 lm W^{-1}, and EQEs of 1.57% and 1.8%, respectively. A blue OLED, ITO/PEDOT:PSS/TCTA (**67**)/TDAF1 (**202**)/TPBI (**110**)/LiF/Al, showed an EQE of 5.3% with a turn-on voltage of 2.5 V (Wu *et al.*, 2004). A red-emitting OLED, ITO/α-NPD (**80**)/PhSPDCV (**169**)/BCP (**139**)/Alq_3 (**105**)/MgAg, showed an EQE of 3.6% (Chiang *et al.*, 2005). A white-light EL emission was reported for the devices, ITO/TPD (**20**)/p-EtTAZ (**203**)/Alq_3 (**105**)/Nile Red (**172**)-doped Alq_3/Alq_3/LiF/Al (Kido *et al.*, 1995b) and ITO/TPD (**20**)/$Zn(BTZ)_2$ (**204**)/LiF/Al (Hamada *et al.*, 1996). The examples of other white-light emitting OLEDs are given by ITO/m-MTDATA (**65**)/FlAMB-0T (**163**, *n* = 0)/FlAMB-3T (**163**, *n* = 3)/F-TBB (**140**)/Alq_3 (**105**)/LiF/Al (Doi *et al.*, 2003) and ITO/PEDOT:PSS/TCTA (**67**)/rubrene (**176**)/T4 (**149**)/Alq_3 (**105**)/LiF/Al (Tang *et al.*, 2008).

The EQE values of 4.8% and 5.3% achieved for the blue-green OLED (Murata *et al.*, 2002) and for the blue OLED (Wu *et al.*, 2004), respectively, are close to the theoretical limit of 5% for fluorescence-based OLEDs. Much higher EQEs can be expected if the up-conversion from the electronically excited triplet state to the excited singlet state (Endo *et al.*, 2011) or the triplet–triplet annihilation to produce the electronically excited singlet state can be utilized. A green fluorescent OLED exhibiting an EQE of 19.3% has been reported (Uoyama *et al.*, 2012).

High-performance phosphorescence-based OLEDs

Generally, phosphorescence-based OLEDs are expected to give much higher EQEs than the fluorescence-based OLEDs since the statistical probability ratio of the generation of the excited triplet state to the excited singlet state by the recombination of holes and electrons is 3:1. However, the EQEs for the former devices tend to decrease with the increasing injected currents or brightness because of the occurrence of the triplet–triplet annihilation resulting from the longer lifetimes of phosphorescence relative to those of fluorescence (Baldo *et al.*, 1998). Therefore, the lifetimes of phosphorescent dopants should desirably be short. Among the transition metal complexes, the Ir complexes have been thought to be promising candidates for the phosphorescent dopants since their phosphorescent lifetimes (0.5–1.3 μs for $Ir(ppy)_3$ in CBP (Baldo *et al.*, 1999; Endo *et al.*, 2008) and 1.2 μs for FIrpic in mCB (Endo *et al.*, 2008)) are shorter relative to those of the Pt complexes (91 μs for PtOEP in polystrylene (Papkovsky, 1995)). The triplet state of the Ir complexes is understood in terms of metal–ligand charge transfer. The structures of ligands significantly affect the triplet-state level, and hence, emission color can be varied by the structures of ligands (Lamansky *et al.*, 2001). The triplet state of dopants is generally produced by the exothermic triplet–triplet excitation energy transfer from the excited triplet state of the host material to the dopant. It is therefore necessary that the excited triplet state of the host material should be at higher energy level than that of the phosphorescent dopant. That is why host materials with wide optical bandgap energies are needed for blue-emitting dopants.

A green-emitting phosphorescent OLED using $Ir(ppy)_3$ as an emissive dopant, ITO/α-NPD (**80**)/$Ir(ppy)_3$ doped in CBP (**82**)/BCP (**139**)/Alq_3 (**105**)/MgAg, gave an EQE of 8% (Baldo *et al.*, 1999). Following this report, EQEs of ca. 19% (power efficiencies: 60–70 lm W^{-1}), which exceed the theoretical limit of 15% in the absence of the excitation energy transfer from the singlet state to the triplet state, were reported for the green-emitting phosphorescent OLEDs using cyclometalated iridium complexes (Adachi *et al.*, 2001c; Ikai *et al.*, 2001). Red-emitting phosphorescent OLEDs, ITO/α-NPD (**80**)/$btp_2Ir(acac)$ (**181**) doped in CBP (**82**)/BCP (**139**)/Alq_3 (**105**)/MgAg/Al (Adachi *et al.*, 2001a) and ITO/α-NPD (**80**)/$Ir(piq)_3$ (**205**) doped in CBP (**82**)/Bphen (**206**)/KF/Al (Tsuboyama *et al.*, 2003) gave EQEs of 7% (power efficiency: 4.6 lm W^{-1}) and 10.3% (power efficiency: 8.0 lm W^{-1}), respectively. Blue-emitting phosphorescence-based OLEDs, ITO/CuPc (**1**)/α-NPD (**80**)/CBP (**82**)/FIrpiq (**183**) doped in CBP/BAlq (**207**)/LiF/Al (Adachi *et al.*, 2001b) and ITO/PEDOT:PSS/α-NPD (**80**)/FIrpiq (**183**) doped in CDBP (**190**)/BAlq (**207**)/Li/Al (Tokito *et al.*, 2003) exhibited EQEs of 5.7% (power efficiency: 6.3 lm W^{-1}) and 10.4% (power efficiency: 10.5 lm W^{-1}), respectively. Other blue phosphorescent OLEDs showed EQEs of 9–10% (power efficiency: 11–14 lm W^{-1}) (Holmes *et al.*, 2003b) and 10.8% (Sun and Forrest, 2007). White phosphorescence-based OLEDs, ITO/PEDOT:PSS/α-NPD (**80**)/CBP + FIrpiq (**183**)/CBP + $btp_2Ir(acac)$ (**181**)/CBP + $bt_2Ir(acac)$ (**208**)/BCP (**139**)/LiF/Al and ITO/α-NPD (**80**)/TCTA (**67**)/UGH2 (**193**) + PQIr (**209**) + $Ir(ppy)_3$ (**180**) + FIr6 (**210**)/TPBI (**110**)/LiF/Al, showed EQEs of 5.2% (power efficiency: 6.4 lm W^{-1}) (D'Andrade *et al.*, 2002) and 12% (power efficiency : 42 lm W^{-1}) (D'Andrade *et al.*, 2004), respectively.

Further significant improvements in EQEs have been achieved for phosphorescence-based OLEDs. A green OLED using $Ir(ppy)_3$ (**180**) doped in CBP (**82**) as the emissive layer and bis-4,6-(3,5-di-3-pyridylphenyl)-2-methylpyrimidine (B3PYMPM, **211**) as the electron-transport layer has been reported to exhibit high EQEs of 29% at 100 cd m^{-2} (power efficiency: 133 lm W^{-1}) and 26% at 1000 cd m^{-2} (power efficiency: 107 lm W^{-1}) (Tanaka *et al.*, 2007a). OLEDs using osmium(II) complexes have been reported to emit red light and exhibit high performance. A red OLED with a structure of ITO/BPAPF (**212**)/CBP (**82**) + $Os(fptz)_2(PPh_2Me)_2$ (**188**)/BCP (**139**)/Alq_3 (**105**)/LiF/Al showed an EQE of 15.3% (power efficiency: 6.3 lm W^{-1}) (Tung *et al.*, 2005). A red-light emitting OLED using $Os(fptz)_2(PPh_2Me)_2$ (**188**) doped in a host material with bipolar functionalities, 2,7-bis (diphenylphosphoryl)-9-[4-(*N*-diphenylamino)phenyl]-9-phenylfluorene (POAPF, **195**) gave an EQE of 19.9% (power efficiency: 34.5 lm W^{-1}) (Chien *et al.*, 2009).

The phosphorescence-based white-emitting OLED which gave a power efficiency of 90 lm W^{-1} and an EQE of 34% consists of multiple layers, ITO/α-NPD (**80**)/TCTA (**67**) + $Ir(MDQ)_2(acac)$ (**213**)/TCTA (**67**)/TPBI (**110**) + FIrpic (**183**)/TPBI (**110**)/TPBI (**110**) + $Ir(ppy)_3$ (**180**)/TPBI (**110**)/Ag (Reineke *et al.*, 2009) (Fig. 1.37).

Exciplex emissions from OLEDs

In multiple layer OLEDs, charge-transfer interactions often takes place to form exciplexes at the solid interface between the hole-transport layer with electron-donating properties and the emitting layer with electron-accepting properties, and the resulting exciplexes emit light (Itano *et al.*, 1998; Ogawa

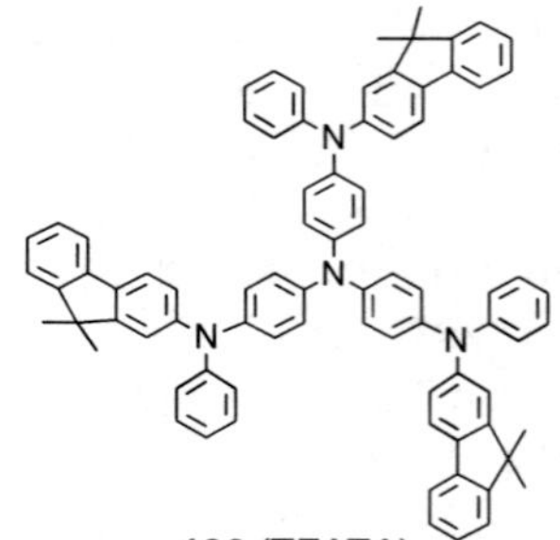

196 (TFATA)

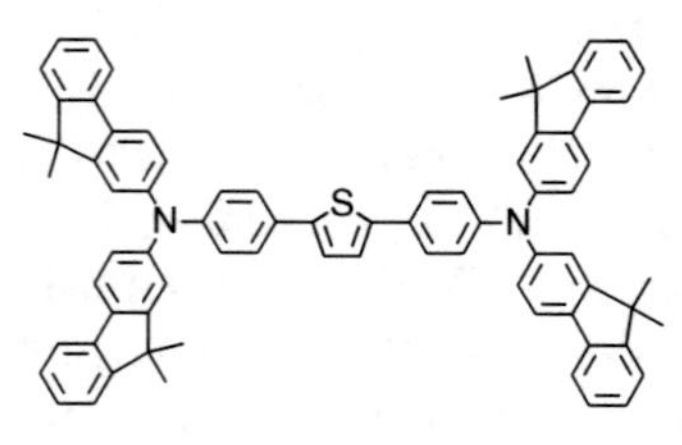

197 (BFA-1T)

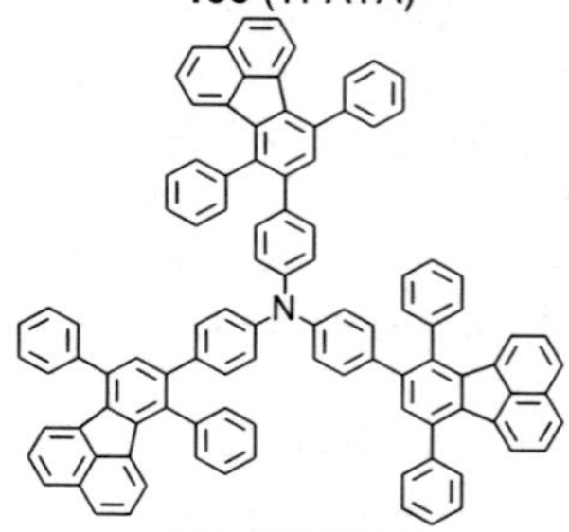

198 (TDPFPA)

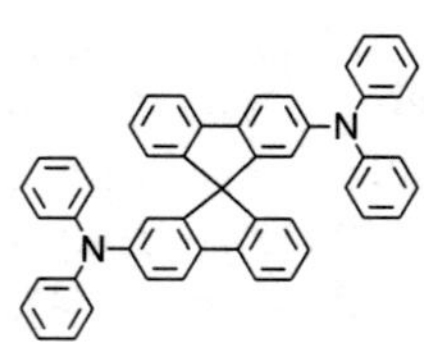

199

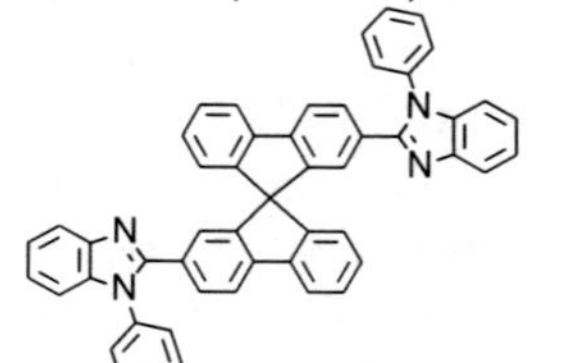

200

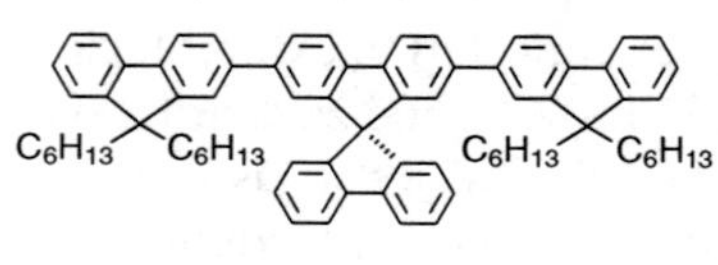

201 (THPF)

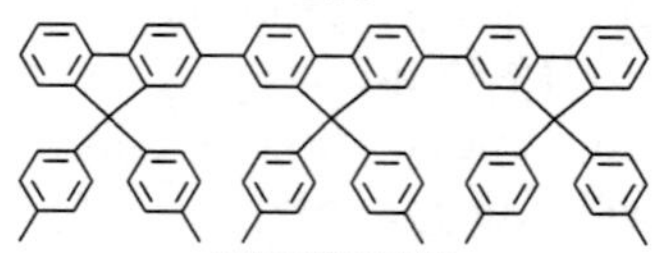

202 (TDAF1)

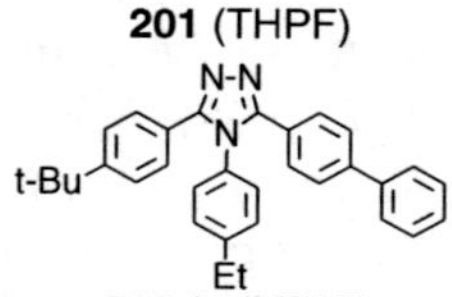

203 (p-EtTAZ)

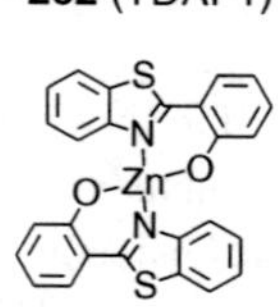

204 ($Zn(BTZ)_2$)

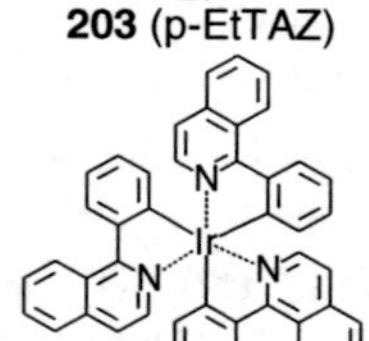

205 ($Ir(pic)_3$)

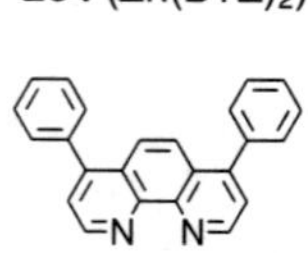

206 (BPhen)

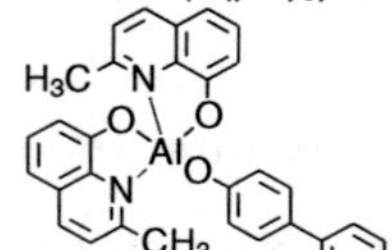

207 (BAlq)

1.37 Materials for OLEDs described in Section 1.6.2.

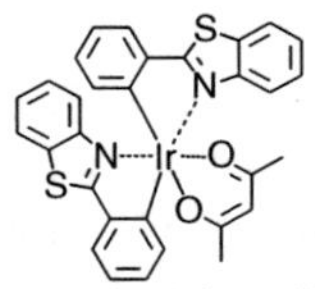

208 ($bt_2Ir(acac)$)

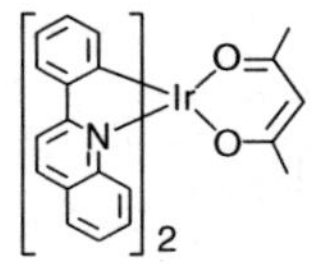

209 (PQIr)

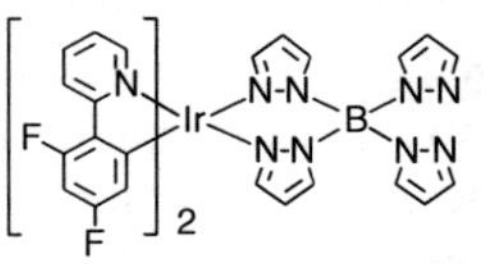

210 (FIr6)

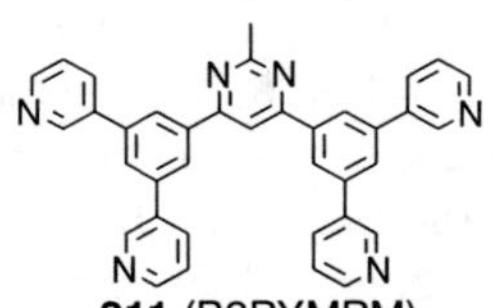

211 (B3PYMPM)

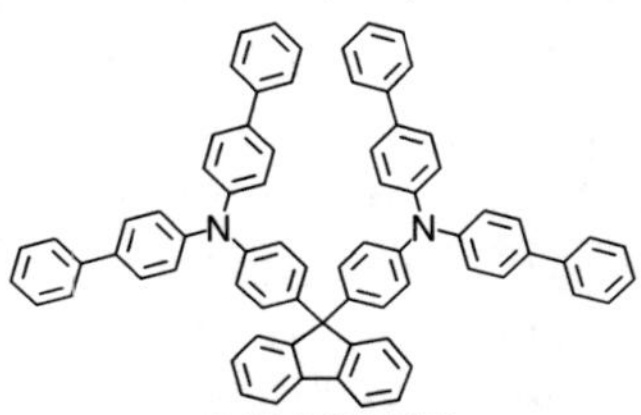

212 (BPAPF)

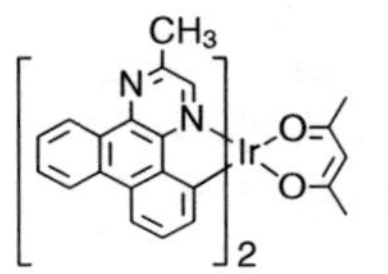

213 ($Ir(MDQ)_2(acac)$)

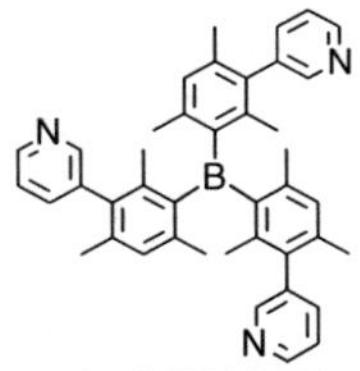

214 (3TPYMB)

1.37 Continued.

et al., 1998; Wang *et al.*, 1998; Cocchi *et al.*, 2002; Palilis *et al.*, 2003b). The emission from the exciplex should be prevented from the standpoint of attaining pure emission color. The emission from the exciplex can be prevented by the insertion of a suitable layer (Noda *et al.*, 1999b). On the other hand, the emission from the exciplex can be utilized for tuning the emission color by varying the applied voltage (Itano *et al.*, 1998) and for obtaining white-light emission (Chao and Chen, 1998; Feng *et al.*, 2001). An OLED using the exciplex-forming system of m-MTDATA (**65**) and 3TPYMB (**214**) has been reported to show an EQE of 5.4% owing to the up-conversion from the excited triplet state to the excited singlet state (Goushi *et al.*, 2012).

1.7 Conclusion and future trends

This chapter is concerned with the survey of a recently emerged new field of organic optoelectronics, highlighting molecular materials for use in optoelectronic devices such as OPVs and OLEDs, and structures and performance of these devices. Historical development in organic optoelectronics and fundamental principles of and operation processes involved in such organic optoelectronic devices are also described. Molecular materials are discussed, taking into account the molecular organization states, crystals, and amorphous glasses.

New fields of organic materials chemistry that deal with functional molecular glasses and organic device physics have been opened up. Steady efforts have been continued up to this time with the aim to improve the PCE for OPVs and the EQE for OLEDs, and significant improvements of PCEs and EQEs have been achieved with times, laying the foundations for the subsequent progress, but problems still remain. Further improvement of PCEs for OPVs is required for realizing their practical use as solar cells. The operational durability of blue-emitting phosphorescent dopants must be improved for white-light emitting OLEDs to be put into practical use for lighting. The developments of devices using organic photorefractive materials and organic photonic devices, which are not taken up in this chapter, are also expected to move ahead. Developments of quantum-dot OPVs, integrated devices, and organic lasers by current injection are also challenging subjects. Research and development on printed electronics will be accelerated in the near future in view of its contribution to solving the environment and energy issues of the 21st century.

1.8 References

Adachi C, Tsutsui T and Saito S (1989), 'Organic electroluminescent device having a hole conductor as an emitting layer', *Appl Phys Lett*, **55**, 1489–1491.

Adachi C, Baldo M A, Forrest S R, Lamansky S, Thompson M E and Kwong R C (2001a), 'High-efficiency red electrophosphorescence devices', *Appl Phys Lett*, **78**, 1622–1624.

Adachi C, Kwong R C, Djurovich P, Adamovich V, Baldo M A, Thompson M E and Forrest S R (2001b), 'Endothermic energy transfer: A mechanism for generating very efficient high-energy phosphorescent emission in organic materials', *Appl Phys Lett*, **79**, 2082–2084.

Adachi C, Baldo M A, Thompson M E and Forrest S R (2001c), 'Nearly 100% internal phosphorescence efficiency in an organic light emitting device', *J Appl Phys*, **90**, 5048–5051.

Adamovich V, Brooks J, Tamayo A, Alexander A M, Djurovich P I, D'Andrade B W, Adachi C, Forrest S R and Thompson M E (2002), 'High efficiency single dopant white electrophosphorescent light emitting diodes', *New J Chem*, **26**, 1171–1178.

Agata Y, Shimizu H and Kido J (2007), 'Syntheses and properties of novel quarterphenylene-based materials for blue organic light-emitting devices', *Chem Lett*, **36**, 316–317.

Akamatu H and Inokuchi H (1950), 'On the electrical conductivity of violanthrone, iso-violanthrone, and pyranthrone', *J Chem Phys*, **18**, 810–811.

Akamatu H, Inokuchi H and Matsunaga Y (1954), 'Electrical conductivity of the perylene-bromine complex', *Nature*, **173**, 168–169.

Assadi A, Svensson C, Willander M and Inganäs O (1988), 'Field-effect mobility of poly(3-hexylthiophene)', *Appl Phys Lett*, **53**, 195.

Bach U, Lupo D, Comte P, Moser J E, Weissörtel F, Salbeck J, Spreitzer H and Grätzel M (1998), 'Solid-state dye-sensitized mesoporous TiO_2 solar cells with high photon-to-electron conversion efficiencies', *Nature*, **395**, 583–585.

Baldo M A, O'Brien D F, You Y, Shoustikov A, Sibley S, Thompson M E and Forrest S R (1998), 'Highly efficient phosphorescent emission from organic electroluminescent devices', *Nature*, **395**, 151–154.

Baldo M A, Lamansky S, Thompson P E and Forrest S R (1999), 'Very high-efficiency green organic light-emitting devices based on electrophosphorescence', *Appl Phys Lett*, **75**, 4–6.

Bässler H (1981), 'Localized states and electronic transport in single component organic solids with diagonal disorder', *Phys Status Solidi B*, **107**, 9–53.

Bässler H (1984), 'Charge transport in molecularly doped polymers', *Philos Mag B*, **50**, 347–362.

Bässler H (1993), 'Charge transport in disordered organic photoconductors', *Phys Status Solidi B*, **175**, 15–56.

Batt R H, Braun C L and Hornig J F (1968), 'Electric-field and temperature dependence of photoconductivity', *J Chem Phys*, **49**, 1967–1968.

Bäuerle P, Pfau F, Schlupp H, Würthner F, Gaudl K-U, Caro M B and Fischer P (1993), 'Synthesis and structural characterization of alkyl oligothiophenes-the first isomerically pure dialkylsexithiophene', *J Chem Soc Perkin Trans*, **2**, 489–494.

Bäuerle P, Fischer T, Bidlingmeier B, Stabel A and Rabe J P (1995), 'Oligothiophenes – yet longer? Synthesis, characterization, and scannnig tunneling microscopy images of homologous, isomerically pure olig(alkylthiophene)s', *Angew Chem Int Ed Engl*, **34**, 303–307.

Beny J-P, Dhawan S N, Kagan J and Sundlass S (1982), 'Synthesis of 3,2′:5′,3″-terthiophene and other terthiophenes by the thiophenecarboxaldehyde → ethynylthiophene → dithienylbutadiyne route', *J Org Chem*, **47**, 2201–2204.

Berggren M, Inganäs O, Gustafsson G, Rasmusson J, Andersson M R, Hjertberg T and Wennerström O (1994), 'Light-emitting diodes with variable colours from polymer blends', *Nature*, **372**, 444–446.

Bernhard S, Gao X, Malliaras G G and Abruña H D (2002), 'Efficient electroluminescent devices based on a chelated osmium(II) complex', *Adv Mater*, **14**, 433–436.

Bettenhausen J and Strohriegl P (1996), 'Efficient synthesis of starburst oxadiazole compounds', *Adv Mater*, **8**, 507–510.

Borsenberger P M and Hoesterey D C (1980), 'Hole photogeneration in aggregate photoreceptors', *J Appl Phys*, **51**, 4248–4251.

Borsenberger P M and Weiss D S (1993), *Organic photoreceptors for imaging systems*, New York, Marcel Dekker.

Borsenberger P M, Contois L E and Hoesterey D C (1978), 'Hole photogeneration in binary solid solutions of triphenylamine and bisphenol-A-polycarbonate', *J Chem Phys*, **68**, 637–641.

Braun D and Heeger A J (1991), 'Visible light emission from semiconducting polymer diodes', *Appl Phys Lett*, **58**, 1982–1984.

Braun D, Gustafsson G, McBranch D and Heeger A J (1992), 'Electroluminescence and electrical transport in poly(3-octylthiophene) diodes', *J Appl Phys*, **72**, 564–568.

Brédas J-L, Norton J E, Cornil J and Coropceanu V (2009), 'Molecular understanding of organic solar cells: the challenges', *Acc Chem Res*, **42**, 1691–1699.

Brunetti F G, Gong X, Tong M, Heeger A J and Wudl F (2010), 'Strain and Hückel aromaticity: driving forces for a promising new generation of electron acceptors in organic electronics', *Angew Chem Int Ed Engl*, **49**, 532–536.

Burroughes J H, Bradley D D C, Brown A R, Marks R N, Mackay K, Friend R H, Burns P L and Holmes A B (1990), 'Light-emitting diodes based on conjugated polymers', *Nature*, **347**, 539–541.

Chance R R and Braun C L (1976), 'Temperature dependence of intrinsic carrier generation in anthracene single crystals', *J Chem Phys*, **64**, 3573–3581.

Chao C-I and Chen S-A (1998), 'White light emission from exciplex in a bilayer device with two blue light-emitting polymers', *Appl Phys Lett*, **73**, 426–428.

Chen C-H, Wu F-I, Shu C-F, Chien C-H and Tao Y-T (2004), 'Spirobifluorene-based pyrazoloquinolines: efficient blue electroluminescent materials', *J Mater Chem*, **14**, 1585–1589.

Chen F, Mehta P G, Takiff L and McCullough R D (1996), 'Improved electroluminescence performance of poly(3-alkylthiophene) having a high head-to-tail (HT) ratio', *J Mater Chem*, **6**, 1763–1766.

Chen G, Sasabe H, Wang Z, Wang X-F, Hong Z, Yang Y and Kido J (2012), 'Co-evaporated bulk heterojunction solar cells with >6.0% efficiency', *Adv Mater*, **24**, 2768–2773.

Cheng G, Zhang Y, Zhao Y, Liu S, Tang S and Ma Y (2005), 'Efficient pure blue electroluminescence from ter(9,9,9″,9″-bihexyl-9′,9′-diphenyl)-fluorenes', *Appl Phys Lett*, **87**, 151905/1–3.

Chen H-Y, Hou J, Zhang S, Liang Y, Yang G, Yang Y, Yu L, Wu Y and Li G (2009), 'Polymer solar cells with enhanced open-circuit voltage and efficiency', *Nature Photon*, **3**, 649–653.

Chen T-A and Rieke R D (1992), 'The first regioregular head-to-tail poly(3-hexylthiophene-2,5-diyl) and a regiorandom isopolymer: Ni vs Pd catalysis of 2(5)-bromo-5(2)-(bromozincio)-3-hexylthiophene polymerization', *J Am Chem Soc*, **114**, 10087–10088.

Chen T-A and Rieke R D (1993), 'Polyalkylthiophenes with the smallest bandgap and the highest intrinsic conductivity', *Synth Met*, **60**, 175–177.

Chen T-A, Wu X and Rieke R D (1995), 'Regiocontrolled synthesis of poly(3-alkylthiophenes) mediated by Rieke zinc: their characterization and solid-state properties', *J Am Chem Soc*, **117**, 233–244.

Chiang C K, Park Y W, Heeger A J, Shirakawa H, Louis E J and MacDiarmid A G (1978), 'Conducting polymers: halogen doped polyacetylene', *J Chem Phys*, **69**, 5098–5104.

Chiang C-L, Wu M-F, Dai D-C, Wen Y-S, Wang J-K and Chen C-T (2005), 'Red-emitting fluorenes as efficient emitting hosts for non-doped, organic red-light-emitting diodes', *Adv Funct Mater*, **15**, 231–238.

Chiechi R C, Tseng R J, Marchioni F, Yang Y and Wudl F (2006), 'Efficient blue-light-emitting electroluminescent devices with a robust fluorophore: 7,8,10-triphenylfluoranthene', *Adv Mater*, **18**, 325–328.

Chien C-H, Hsu F-M, Shu C-F and Chi Y (2009), 'Efficient red electrophosphorescence from a fluorene-based bipolar host material', *Org Electronics*, **10**, 871–876.

Chiu S-W, Lin L-Y, Lin H-W, Chen Y-H, Huang Z-Y, Lin Y-T, Lin F, Liu Y-H and Wong K-T (2012), 'A donor–acceptor–acceptor molecule for vacuum-processed organic solar cells with a power conversion efficiency of 6.4%', *Chem Commun*, **48**, 1857–1859.

Chou M-Y, Leung M-k, Su Y O, Chiang C L, Lin C-C, Liu J-H, Kuo C-K and Mou C-Y (2004), 'Electropolymerization of starburst triarylamines and their application to electrochromism and electroluminescence', *Chem Mater*, **16**, 654–661.

Chuen C H, Tao Y T, Wu F I and Shu C F (2004), 'White organic light-emitting diodes based on 2,7-bis(2,2-diphenylvinyl)-9,9′-spirobifluorene: improvement in operational lifetime', *Appl Phys Lett*, **85**, 4609–4611.

Claessens C G, Hahn U, and Torres T (2008), 'Phthalocyanines: from outstanding electronic properties to emerging applications', *Chem Rec*, **8**, 75–97.

Cocchi M, Virgili D, Giro G, Fattori V, Marco P D, Kalinowski J and Shirota Y (2002), 'Efficient exciplex emitting organic electroluminescent devices', *Appl Phys Lett*, **80**, 2401–2403.

Coropceanu V, Cornil J, da Silva Filho D A, Olivier Y, Silbey R and Brédas J-L (2007), 'Charge transport in organic semiconductors', *Chem Rev*, **107**, 926–952.

Cravino A, Leriche P, Alévêque O, Roquet S and Roncali J (2006), 'Light-emitting organic solar cells based on a 3D conjugated system with internal charge transfer', *Adv Mater*, **18**, 3033–3037.

Cremer J, Mena-Osteritz E, Pschierer N G, Müllen K and Bäuerle P (2005), 'Dye-functionalized head-to-tail coupled oligo(3-hexylthiophenes) – perylene-oligothiophene dyads for photovoltaic applications', *Org Biomol Chem*, **3**, 985–995.

D'Andrade B W, Thompson M E and Forrest S R (2002), 'Controlling excitation diffusion in multilayer white phosphorescent organic light emitting devices', *Adv Mater*, **14**, 147–151.

D'Andrade B W, Holmes R J and Forrest S R (2004), 'Efficient organic electrophosphorescent whtie-light-emitting devices with a triple doped emissive layer', *Adv Mater*, **16**, 624–628.

Danel K, Huang T-H, Lin J T, Tao Y-T and Chuen C-H (2002), 'Blue-emitting anthracenes with end-capping diarylamines', *Chem Mater*, **14**, 3860–3865.

Del Cano T, Duff J and Aroca R (2002), 'Molecular spectra and molecular organization in thin solid films of bis(neopentylimido) perylene', *Appl Spectrosc*, **56**, 744–750.

Doi H, Kinoshita M, Okumoto K and Shirota Y (2003), 'A novel class of emitting amorphous molecular materials with bipolar character for electroluminescence', *Chem Mater*, **15**, 1080–1089.

Duan X, Jiang Z, Yu G, Lu P, Liu Y, Xu X and Zhu D (2005), 'Blue organic electroluminescent device with tetra(β-naphthyl)silane as hole blocking materials', *Thin Solid Films*, **478**, 121–124.

Ebert Jr. A A and Gottlieb H B (1952), 'Infrared spectra of organic compounds exhibiting polymorphism', *J Am Chem Soc*, **74**, 2806–2810.

Eley D D (1948), 'Phthalocyanines as semiconductors', *Nature*, **162**, 819–820.

Ellis A B, Kaiser S W and Wrighton M S (1976a), 'Visible light to electrical energy conversion. Stable cadmium sulfide and cadmium selenide photoelectrodes in aqueous electrolytes', *J Am Chem Soc*, **98**, 1635–1637.

Ellis A B, Kaiser S W and Wrighton M S (1976b), 'Optical to electrical energy conversion. Characterization of cadmium sulfide and cadmium selenide based photoelectrochemical cells', *J Am Chem Soc*, **98**, 6855–6866.

Endo J, Matsumoto T and Kido J (2002), 'Organic electroluminescent devices with a vacuum-deposited Lewis-acid-doped hole-injecting layer', *Jpn J Appl Phys*, **41**, L358–L360.

Endo A, Suzuki K, Yoshihara T, Tobita S, Yahiro M and Adachi C (2008), 'Measurement of photoluminescence efficiency of Ir(III) phenylpyridine derivatives in solution and solid-state films', *Chem Phys Lett*, **460**, 155–157.

Endo A, Sato K, Yoshimura K, Kai T, Kawada A, Miyazaki H and Adachi C (2011), 'Efficient up-conversion of triplet excitons into a singlet state and its application for organic light emitting diodes', *Appl Phys Lett*, **98**, 083302/1–3.

Enokida T, Hirohashi R and Nakamura T (1990), 'Polymorphs of oxotitanium phthalocyanine and their applications for photoreceptors', *J Imaging Sci*, **34**, 234–242.

Feng J, Li F, Gao W, Liu S, Liu Y and Wang Y (2001), 'White light emission from exciplex using tris-(8-hydroxyquinoline)aluminum as chromaticity-tuning layer', *Appl Phys Lett*, **78**, 3947–3949.

Fink R, Heischkel Y, Thelakkat M and Schmidt H-W (1998), 'Synthesis and application of dimeric 1,3,5-triazine ethers as hole-blocking materials in electroluminescent devices', *Chem Mater*, **10**, 3620–3625.

Fitzner R, Reinold E, Mishra A, Mena-Osteritz E, Ziehlke H, Körner C, Leo K, Riede M, Weil M, Tsaryova O, Weiß A, Uhrich C, Pfeiffer M and Bäuerle P (2011), 'Dicyanovinyl-substituted oligothiophenes: structure–property relationships and application in vacuum-processed small molecule organic solar cells', *Adv Funct Mater*, **21**, 897–910.

Forrest S R (2004), 'The path to ubiquitous and low-cost organic electronic appliances on plastic', *Nature*, **428**, 911–918.

Funaoka S, Imae I, Noma N and Shirota Y (1999), 'Structure and electrical conductivity of a solid salt obtained by eletrochemical oxidation of a thiophene dodecamer', *Synth Met*, **101**, 600–601.

Gagnon E, Maris T and Wuest J D (2010), 'Triarylamines designed to form molecular glasses. derivatives of tris(p-terphenyl-4-yl)amine with multiple contiguous phenyl substituents', *Org Lett*, **12**, 404–407.

Gao Z, Lee C S, Bello I, Lee S T, Chen R-M, Luh T-Y, Shi J and Tang C W (1999), 'Bright-blue electroluminescence from a silyl-substituted ter-(phenylene-vinylene) derivative', *Appl Phys Lett*, **74**, 865–867.

Gebeyehu D, Maennig B, Drechsel J, Leo K and Pfeiffer M (2003), 'Bulk-heterojunction photovoltaic devices based on donor–acceptor organic small molecule blends', *Solar Energy Mater Solar Cells*, **79**, 81–92.

Gill W D (1972), 'Drift mobilities in amorphous charge-transfer complexes of trinitrofluorenone and poly-*n*-vinylcarbazole', *J Appl Phys*, **43**, 5033–5040.

Glenis S, Tourillon G and Garnier F (1986), 'Influence of the doping on the photovoltaic properties of thin films of poly-3-methylthiophene', *Thin Solid Films*, **139**, 221–231.

Goushi K, Yoshida K, Sato K and Adachi C (2012), 'Organic light-emitting diodes employing efficient reverse intersystem crossing for triplet-to-singlet state conversion', *Nature Photon*, **6**, 253–258.

Green M A, Emery K, Hishikawa Y, Warta W and Dunlop E D (2012), 'Solar cell efficiency tables (version 39)', *Prog Photovolt*, **20**, 12–20.

Grem G, Leditzky G, Ullrich B and Leising G (1992), 'Blue electroluminescent device based on a conjugated polymer', *Synth Met*, **51**, 383–389.

Grigalevicius S, Blazys G, Ostrauskaite J, Grazulevicius J V, Gaidelis V, Jankauskas V and Montrimas E (2002), '3,6-Di(*N*-diphenylamino)-9-phenylcarbazole and its methyl-substituted derivative as novel hole-transporting amorphous molecular materials', *Synth Met*, **128**, 127–131.

Guillerez S and Bidan G (1998), 'New convenient synthesis of highly regioregular poly(3-octylthiophene) based on the Suzuki coupling reaction', *Synth Met*, **93**, 123–126.

Halls J J M, Walsh C A, Greenham N C, Marseglla E A, Friend R H, Moratti S C and Holmes A B (1995), 'Efficient photodiodes from interpenetrating polymer networks', *Nature*, **376**, 498–500.

Hamada Y, Sano T, Fujii H, Nishio Y, Takahashi H and Shibata K (1996), 'White-light-emitting material for organic electroluminescent devices', *Jpn J Appl Phys*, **35**, L1339–L1341.

Hanack M, Dini D, Barthel M and Vagin S (2002), (2002), 'Conjugated macrocycles as active materials in nonlinear optical processes: optical limiting effect with phthalocyanines and related compounds', *Chem Rec*, **2**, 129–148.

Hayakawa A, Yoshikawa O, Fujieda T, Uehara K and Yoshikawa S (2007), 'High performance polythiophene/fullerene bulk-heterojunction solar cell with a TiO_x hole blocking layer', *Appl Phys Lett*, **90**, 163517/1–3.

He C, He Q, Yi Y, Wu G, Bai F, Shuai Z and Li Y (2008), 'Improving the efficiency of solution processable organic photovoltaic devices by a star-shaped molecular geometry', *J Mater Chem*, **18**, 4085–4090.

He Q, He C, Sun Y, Wu H, Li Y and Bai F (2008), 'Amorphous molecular material containing bisthiophenyl-benzothiadiazole and triphenylamine with bipolar and low-bandgap characteristics for solar cells', *Thin Solid Films*, **516**, 5935–5940.

He Y, Zhao G, Peng B and Li Y (2010), 'High-yield synthesis and electrochemical and photovoltaic properties of indene-C_{70} bisadduct', *Adv Funct Mater*, **20**, 3383–3389.

He Z, Zhong C, Huang X, Wong W-Y, Wu H, Chen L, Su S and Cao Y (2011), 'Simultaneous enhancement of open-circuit voltage, short-circuit current density, and fill factor in polymer solar cells', *Adv Mater*, **23**, 4636–4643.

Helfrich W and Schneider W G (1965), 'Reconbination radiation in antharacene crystals', *Phys Rev Lett*, **14**, 229–232.

Higuchi A, Inada H, Kobata T and Shirota Y (1991), 'Amorphous molecular materials: Synthesis and properties of a novel starburst molecule, 4,4′,4″-tri(*N*-phenothiazinyl)triphenylamine', *Adv Mater*, **3**, 549–550.

Higuchi A, Ohnishi K, Nomura S, Inada H and Shirota Y (1992), 'Tri(biphenyl-4-yl) amine and tri(p-terphenyl-4-yl)amine as a novel class of molecules for amorphous molecular materials', *J Mater Chem*, **2**, 1109–1110.

Hiller W, Strahle J, Kobel W and Hanack M (1982), 'Polymorphie, leitfahigheit und kristallstrukturen von oxo-phthalocyaninato-titan (IV)', *Z Kristallogr*, **159**, 173–183.

Hino Y, Kajii H and Ohmori Y (2004), 'Efficient low-molecule phosphorescent organic light-emitting diodes fabricated by wet-processing', *Org Electronics*, **5**, 265–270.

Hiramoto M, Suezaki M and Yokoyama M (1990), 'Effect of thin gold interstitial-layer on the photovoltaic properties of tandem organic solar cell', *Chem Lett*, 327–330.

Hiramoto M, Fujiwara H and Yokoyama M (1991), 'Three-layered organic solar cell with a photoactive interlayer of codeposited pigments', *Appl Phys Lett*, **58**, 1062–1064.

Hiramoto M, Fujiwara H and Yokoyama M (1992), 'p-i-n Like behavior in three-layered organic solar cells having a co-deposited interlayer of pigments', *J Appl Phys*, **72**, 3781–3787.

Holmes R J, Forrest S R, Tung Y-J, Kwong R C, Brown J J, Garon S and Thompson M E (2003a), 'Blue organic electrophosphorescence using exothermic host-guest energy transfer', *Appl Phys Lett*, **82**, 2422–2424.

Holmes R J, D'Andrade B W, Forrest S R, Ren X, Li J and Thompson M E (2003b), 'Efficient, deep-blue organic electrophosphorescence by guest charge trapping', *Appl Phys Lett*, **83**, 3818–3820.

Holzhey A, Uhrich C, Brier E, Reinhold E, Bäuerle P, Leo K and Hoffmann M (2008), 'Exciton diffusion and energy transfer in organic solar cells based on dicyanovinyl-terthiophene', *J Appl Phys*, **104**, 064510/1–8.

Hughes G and Bryce M R (2005), 'Electron-transporting materials for organic electroluminescent and electrophosphorescent devices', *J Mater Chem*, **15**, 94–107.

Hughes R C (1971), 'Geminate recombination of X-ray excited electron-hole pairs in anthracene', *J Chem Phys*, **55**, 5442–5447.

Hummelen J C, Knight B W, LePeq F, Wudl F, Yao J and Wilkins C L (1995), 'Preparation and characterization of fulleroid and methanofullerene derivatives', *J Org Chem*, **60**, 532–538.

Ikai M, Tokito S, Sakamoto Y, Suzuki T and Taga Y (2001), 'Highly efficient phosphorescence from organic light-emitting devices with an exciton-block layer', *Appl Phys Lett*, **79**, 156.

Imae I and Kawakami Y (2005), 'Novel family of molecular glasses based on silicon-containing compounds', *Chem Lett*, **34**, 290–291.

Imae I, Nawa K, Ohsedo Y, Noma N and Shirota Y (1997), 'Synthesis of a novel family of electrochemically-doped vinyl polymers containing pendant oligothiophenes and their electrical and electrochromic properties', *Macromolecules*, **30**, 380–386.

Inada H and Shirota Y (1993), '1,3,5-Tris[4-(diphenylamino)phenyl]benzene and its methyl-substituted derivatives as a novel class of amorphous molecular materials', *J Mater Chem*, **3**, 319–320.

Inomata H, Goushi K, Masuko T, Konno T, Imai T, Sasabe H, Brown J J and Adachi C (2004), 'High-efficiency organic electrophosphorescent diodes using 1,3,5-triazine electron transport materials', *Chem Mater*, **16**, 1285–1291.

Ishihara M, Okumoto K, Tsuzuki T, Kageyama H, Nakano H and Shirota Y (2006), 'Electrically conducting amorphous molecular material: Iodine doped m-MTDATA as a hole injection buffer layer in organic electroluminescent devices', *Mol Cryst Liq Cryst*, **455**, 259–266.

Ishikawa W, Inada H, Nakano H and Shirota Y (1991), 'Methyl-substituted derivatives of 1,3,5-tris(diphenylamino)benzene as a novel class of amorphous molecular materials', *Chem Lett*, 1731–1734.

Ishikawa W, Inada H, Nakano H and Shirota Y (1993a), 'Polymorphism of starburst molecules: methyl-substituted derivatives of 1,3,5-tris(diphenylamino)benzene', *J Phys D*, **26**, B94-B99.

Ishikawa W, Noguchi K, Kuwabara Y and Shirota Y (1993b), 'Novel amorphous molecular materials: the starburst molecule 1,3,5-tris[N-(4-diphenylaminophenyl) phenylamino]benzene', *Adv Mater*, **5**, 559–561.

Itano K, Ogawa H and Shirota Y (1998), 'Exciplex formation at the organic solid-state interface: yellow emission in organic light-emitting diodes using green-fluorescent tris(8-quinolinolato)aluminum and hole-transporting molecular materials with low ionization potentials', *Appl Phys Lett*, **72**, 636–638.

Ito T, Shirakawa H and Ikeda S (1974), 'Simultaneous polymerization and formation of polyacetylene film on the surface of concentrated soluble Ziegler-type catalyst solution', *J Polym Sci: Polym Chem Ed*, **12**, 11–20.

Jandke M, Strohriegl P, Berleb S, Werner E and Brutting W (1998), 'Phenylquinoxaline polymers and low molar mass glasses as electron-transprot materials in organic light-emitting diodes', *Macromolecules*, **31**, 6434–6443.

Jeon I-R, Noma N, Claridge R F C and Shirota Y (1992), 'Electrochemical doping of poly(3-vinylperylene) and electrical properties of doped polymers', *Polym J*, **24**, 273–279.

Jiang X, Jen A K Y, Carlson B and Dalton L R (2002), 'Red-emitting electroluminescent devices based on osmium-complexes doped blend of poly(vinylnaphthalene) and 1,3,4-oxadiazole derivative', *Appl Phys Lett*, **81**, 3125–3127.

Kafafi Z H (Ed.) (2005), *Organic electroluminescence*, New York, Taylor & Francis.

Kagan J and Arora S K (1983), 'The synthesis of alpha-thiophene oligomers by oxidative coupling of 2-lithiothiophenes', *Heterocycles*, **20**, 1937–1940.

Kageyama H and Shirota Y (2011), 'Development of amorphous molecular materials with high hole drift mobility and their application to organic solar cells', *Proceedings of the International Symposium on Materials Science and Innovation for Sustainable Society (ECO-MATES 2011)*, November 28–30, Osaka, Japan, p. 133.

Kageyama H, Ohishi H, Tanaka M, Ohmori Y and Shirota Y (2009a), 'High performance organic photovoltaic devices using amorphous molecular materials with high charge-carrier drift mobilities', *Appl Phys Lett*, **94**, 063304/1–3.

Kageyama H, Ohishi H, Tanaka M, Ohmori Y and Shirota Y (2009b), 'High-performance organic photovoltaic devices using a new amorphous molecular material with high hole drift mobility, tris[4-(5-phenylthiophen-2-yl)phenyl] amine', *Adv Funct Mater*, **19**, 3948–3955.

Kanega H, Shirota Y and Mikawa H (1984), 'Electrochemical doping of poly(*N*-vinylcarbazole) and poly (3,6-dibromo-*N*-vinylcarbazole) and their electrical conductivities', *J Chem Soc Chem Commun*, 158–159.

Kanemitsu Y, Yamamoto A, Funada H and Masumoto Y (1991), 'Photocarrier generation in metal-free phthalocyanines: effect of the stacking habit of molecules on the photogeneration efficiency', *J Appl Phys*, **69**, 7333–7335.

Katsuma K and Shirota Y (1998), 'A novel π-electron dendrimers for thermally and morphologically stable amorphous molecular materials', *Adv Mater*, **10**, 223–226.

Kido J, Ohtaki C, Hongawa K, Okuyama K and Nagai K (1993), '1,2,4-Triazole derivative as an electron transport layer in organic electroluminescent devices', *Jpn J Appl Phys*, **32**, L917–L920.

Kido J, Shinoyama H and Nagai K (1995a), 'Single-layer white light-emitting organic electroluminescent devices bases on dye-dispersed poly(*N*-vinylcarbazole)', *Appl Phys Lett*, **67**, 2281–2283.

Kido J, Kimura M and Nagai K (1995b), 'Multilayer white light-emitting organic electroluminescent device', *Science*, **267**, 1332–1334.

Kijima Y, Asai N and Tamura S (1999), 'A blue organic light emiting diode', *Jpn J Appl Phys*, **38**, 5274–5277.

Kim J H, Liu M S, Jen A K-Y, Carlson B, Dalton L R, Shu C-F and Dodda R (2003), 'Bright red-emitting electrophosphorescent device using osmium complex as a triplet emitter', *Appl Phys Lett*, **83**, 776–778.

Kim J Y, Lee K, Coates N E, Moses D, Nguyen T-Q, Dante M and Heeger A J (2007), 'Efficient tandem polymer solar cells fabricated by all-solution processing', *Science*, **317**, 222–225.

Kim K, Liu J, Namboothiry M A G and Carroll D L (2007), 'Roles of donor and acceptor nanodomains in 6% efficient thermally annealed polymer photovoltaics', *Appl Phys Lett*, **90**, 163511/1–3.

Kimura M, Inoue S, Shimada K, Tokito S, Noda K, Taga Y and Sawaki Y (2000), 'Spirocycle-incorporated triphenylamine derivatives as an advanced organic electroluminescent material', *Chem Lett*, 192–193.

King K A, Spellane P J and Watts R J (1985), 'Excited-state properties of a triply ortho-metalated iridium(III) complex', *J Am Chem Soc*, **107**, 1431–1432.

Kinoshita M and Shirota Y (2001), '1,5-Bis[5-(dimesitylboryl)thiophen-2-yl]benzene and 1,3,5-tris[5-(dimesitylboryl)thiophen-2-yl]benzene as a novel family of electron-transporting hole blockers for organic electroluminescent devices', *Chem Lett*, 614.

Kinoshita M, Kita H and Shirota Y (2002), 'A novel family of boron-containing hole-blocking amorphous molecular materials for blue-and blue-violet-emitting organic electroluminescent devices', *Adv Funct Mater*, **12**, 780–786.

Koene B E, Loy D E and Thompson M E (1998), 'Asymmetric triaryldiamines as thermally stable hole transporting layers for organic light-emitting devices', *Chem Mater*, **10**, 2235–2250.

Kraft A, Grimsdale A C and Holmes A B (1998), 'Electroluminescent conjugated polymers-seeing polymers in a new light', *Angew Chem Int Ed Engl*, **37**, 402–428.

Ku S-Y, Chi L-C, Hung W-Y, Yang S-W, Tsai T-C, Wong K-T, Chen Y-H and Wu C-I (2009), 'High-luminescence non-doped green OLEDs based on a 9,9-diarylfluorene-terminated 2,1,3-benzothiadiazole derivative', *J Mater Chem*, **19**, 773–780.

Kulkarni A P, Tonzola C J, Babel A and Jenekhe S A (2004), 'Electron transport materials for organic light-emitting diodes', *Chem Mater*, **16**, 4556–4573.

Kundu P, Justin Thomas K R, Lin J T, Tao Y-T and Chien C-H (2003), 'High-T_g carbazole derivatives as blue-emitting hole-transporting materials for electroluminescent devices', *Adv Funct Mater*, **13**, 445–1112.

Kushto G P, Kim W and Kafafi Z H (2005), 'Flexible organic photovoltaics using conducting polymer electrodes', *Appl Phys Lett*, **86**, 093502/1–3.

Kuwabara Y, Ogawa H, Inada H, Noma N and Shirota Y (1994), 'Thermally stable multilayered organic electroluminescent devices using novel starburst molecules, 4,4′,4″-tri(*N*-carbazolyl)triphenylamine (TCTA) and 4,4′,4″-tris(3-methylphenylphenyl-amino)triphenylamine (m-MTDATA), as hole-transport materials', *Adv Mater*, **6**, 677–679.

Kwon J, Lee W, Kim J-y, Noh S, Lee C and Hong J-I (2010), 'Solution processable donor materials based on thiophene and triphenylamine for bulk heterojunction solar cells', *New J Chem*, **34**, 744–749.

Lamansky S, Djurovich P, Murphy D, Abdel-Razzaq F, Lee H-E, Adachi C, Burrows P E, Forrest S R and Thompson M E (2001), 'Highly phosphorescent bis-cyclometalated iridium complexes: synthesis, photophysical characterization, and use in organic light emitting diodes', *J Am Chem Soc*, **123**, 4304–4312.

Lebel O, Maris T, Perron M-È, Demers E and Wuest J D (2006), 'The dark side of crystal engineering: creating glasses from small symmetric molecules that form multiple hydrogen bonds', *J Am Chem Soc*, **128**, 10372–10373.

Leznoff C C and Lever A B P (Eds.) (1992), *Phthalocyanines. properties and applications*, New York, VCH.

Li H, Sundararaman A, Venkatasubbaiah K and Jäkle F (2007), 'Organoborane acceptor-substituted polythiophene via side-group borylation', *J Am Chem Soc*, **129**, 5792–5793.

Li W, Du C, Li F, Zhou Y, Fahlman M, Bo Z and Zhang F (2009), 'Benzothiadiazole-based linear and star molecules: design, synthesis, and their application in bulk heterojunction organic solar cells', *Chem Mater*, **21**, 5327–5334.

Li Z, He G, Wan X, Liu Y, Zhou J, Long G, Zuo Y, Zhang M and Chen Y (2012), 'Solution processable rhodanine-based small molecule organic photovoltaic cells with a power conversion efficiency of 6.1%', *Adv Energy Mater*, **2**, 74–77.

Li Z H, Tong K L, Wong M S and So S K (2006), 'Novel fluorine-containing X-branched oligophenylenes: structure-hole blocking property relationships', *J Mater Chem*, **16**, 765–772.

Liang Y, Xu Z, Xia J, Tsai S-T, Wu Y, Li G, Ray C and Yu L (2010), 'For the bright future - bulk heterojunction polymer solar cells with power conversion efficiency of 7.4%', *Adv Mater*, **22**, E135–E138.

Liao Y-L, Lin C-Y, Wong K-T, Hou T-H and Hung W-Y (2007), 'A novel ambipolar spirobifluorene derivative that behaves as an efficient blue-light emitter in organic light-emitting diodes', *Org Lett*, **9**, 4511–4514.

Lin H-W, Lin C-L, Chang H-H, Lin Y-T, Wu C-C, Chen Y-M, Chen R-T, Chien Y-Y and Wong K-T (2004), 'Anisotropic optical properties and molecular orientation in vacuum-deposited ter(9,9-diarylfluorene)s thin films using spectroscopic ellipsometry', *J Appl Phys*, **95**, 881–886.

Lin S-L, Chan L-H, Lee R-H, Yen M-Y, Kuo W-J, Chen C-T and Jeng R-J (2008), 'Highly efficient carbazole-π-dimesitylborane bipolar fluorophores for nondoped blue organic light-emitting diodes', *Adv Mater*, **20**, 3947–3952.

Liu Y, Wan X, Wang F, Zhou J, Long G, Tian J and Chen Y (2011), 'High-performance solar cells using a solution-processed small molecule containing benzodithiophene unit', *Adv Mater*, **23**, 5387–5391.

Lloyd M T, Mayer A C, Subramanian S, Mourey D A, Herman D J, Bapat A V, Anthony J E and Malliaras G G (2007), 'Efficient solution-processed photovoltaic cells based on an anthradithiophene/fullerene blend', *J Am Chem Soc*, **129**, 9144–9149.

Loser S, Bruns C J, Miyauchi H, Ortiz R P, Facchetti A, Stupp S I and Marks T J (2011), 'A naphthodithiophene-diketopyrrolopyrrole donor molecule for efficient solution-processed solar cells', *J Am Chem Soc*, **133**, 8142–8145.

Louie J, Hartwig J F and Fry A J (1997), 'Discrete high molecular weight triarylamine dendrimers prepared by palladium-catalyzed amination', *J Am Chem Soc*, **119**, 11695–11696.

Loy D E, Koene B E and Thompson M E (2002), 'Thermally stable hole-transporting materials based upon a fluorene core', *Adv Funct Mater*, **12**, 245–249.

Lu J, Xia P F, Lo P K, Tao Y and Wong M S (2006), 'Synthesis and properties of multi-triarylamine-substituted carbazole-based dendrimers with an oligothiophene core for potential applications in organic solar cells and light-emitting diodes', *Chem Mater*, **18**, 6194–6203.

Lyons L E and Milne K A (1976), 'One-photon intrinsic photogeneration in anthracene crystals', *J Chem Phys*, **65**, 1474–1484.

Ma C-Q, Fonrodona M, Schikora M C, Wienk M M, Janssen R A J and Bäuerle P (2008), 'Solution-processed bulk-heterojunction solar cells based on monodisperse dendritic oligothiophenes', *Adv Funct Mater*, **18**, 3323–3331.

Ma Y, Zhang H, Shen J and Che C (1998), 'Electroluminescence from triplet metal-ligand charge-transfer excited state of transition metal complexes', *Synth Met*, **94**, 245–248.

Mack J X, Schin L B and Peled A (1989), 'Hole mobilities in hydrazone-polycarbonate dispersions', *Phys Rev B*, **39**, 7500–7508.

Mäkinen A J, Hill I G, Noda T, Shirota Y and Kafafi Z H (2001), 'Photoemission study of frontier orbital alignment as a metal–organic interface as a function of conjugation length of oligothiophene derivatives', *Appl Phys Lett*, **78**, 670–672.

Malenfant P R L and Fréchet J M J (1998), 'The first solid-phase synthesis of oligothiophenes', *Chem Commun*, 2657–2658.

Malenfant P R L, Groenendaal L and Fréchet J M J (1998), 'Well-defined triblock hybrid dendrimers based on lengthly oligothiophene cores and poly(benzyl ether) dendrimers', *J Am Chem Soc*, **120**, 10990–10991.

Marsella M J and Swager T M (1993), 'Designing conducting polymer-based sensors: selective ionochromic response in crown ether-containing polythiophenes', *J Am Chem Soc*, **115**, 12214–12215.

Martinez F, Voelkel R, Naegele D and Naarmann H (1989), 'Thiophene oligomers: synthesis and characterization', *Mol Cryst Liq Cryst*, **167**, 227–232.

Mayer A C, Lloyd M T, Herman D J, Kasen T G and Malliaras G G (2004), 'Post-fabrication annealing of pentacene-based photovoltaic cells', *Appl Phys Lett*, **85**, 6272–6274.

Mazzeo M, Vitale V, Sala F D, Anni M, Barbarella G, Favaretto L, Sotgiu G, Cingolani R and Gigli G (2005), 'Bright white organic light-emitting devices from a single active molecular material', *Adv Mater*, **17**, 34–39.

McCullough R D and Lowe R D (1992), 'Enhanced electrical conductivity in regioselectively synthesized poly(3-alkylthiophenes)', *J Chem Soc Chem Commun*, 70–72.

McCullough R D, Lowe R D, Jayaraman M and Anderson D L (1993a), 'Design, synthesis, and control of conducting polymer architectures: structurally homogeneous poly(3-alkylthiophenes)', *J Org Chem*, **58**, 904–912.

McCullough R D, Tristram-Nagle S, Williams S P, Lowe R D and Jayaraman M (1993b), 'Self-orienting head-to-tail poly(3-alkylthiophenes): new insights on structure–property relationships in conducting polymers', *J Am Chem Soc*, **115**, 4910–4911.

McCullough R D, Ewbank P C and Loewe R S (1997), 'Self-assembly and disassembly of regioregular, water soluble polythiophenes: chemoselective ionchromatic sensing in water', *J Am Chem Soc*, **119**, 633–634.

Melz P J (1972), 'Photogeneration in trinitrofluorenone-poly(*N*-vinylcarbazole)', *J Chem Phys*, **57**, 1694–1699.

Melz P J, Champ R B, Chang L S, Chiou C, Keller G S, Liclican L C, Neiman R R, Shuttuck M D and Weiche W J (1977), 'Use of pyrazoline-based carrier transport layers in layered photoconductor systems for electrophotography', *Photogr Sci Eng*, **21**, 73–78.

Merz A and Ellinger F (1991), 'Convenient synthesis of α-terthienyl and α-quinquethienyl via a Friedel–Crafts route', *Synthesis*, 462–464.

Mishra A and Bäuerle P (2012), 'Small molecule organic semiconductors on the move: promises for future solar energy technology', *Angew Chem Int Ed Engl*, **51**, 2020–2067.

Mishra A, Fischer M K R and Bäuerle P (2009), 'Metal-free organic dyes for dye-sensitized solar cells: from structure : property relationships to design rules', *Angew Chem Int Ed Engl*, **48**, 2474–2499.

Mitschke U and Bäuerle P (2000), 'The electroluminescence of organic materials', *J Mater Chem*, **10**, 1471–1507.

Miyata S and Nalwa H S (Eds.) (1997), *Organic electroluminescent materials and devices*, New York, Gordon and Breach.

Morel D L (1979), 'Some aspects of the role of solid state chemistry in the performance of organic solar cells', *Mol Cryst Liq Cryst*, **50**, 127–137.

Mort J (1972), 'Transient photoinjection of holes from amorphous Se into poly(*N*-vinyl carbazole)', *Phys Rev B*, **5**, 3329–3336.

Mort J and Nielsen P (1972), 'Transient accumulation of interfacial charge in photoconductor-dielectric systems', *Phys Rev B*, **5**, 3336–3342.

Mort J and Pai D M (Eds.) (1976), *Photoconductivity and related phenomena*, New York, Elsevier.

Müllen K and Scherf U (Eds.) (2006), *Organic light-emitting devices, synthesis, properties, and applications*, Weinheim, Wiley-VCH.

Murata H, Merritt C D, Inada H, Shirota Y and Kafafi Z H (1999), 'Molecular organic light-emitting diodes with temperature-independent quantum efficiency and improved thermal durability', *Appl Phys Lett*, **75**, 3252–3254.

Murata H, Malliaras G G, Uchida M, Shen Y and Kafafi Z H (2001), 'Non-dispersive and air-stable electron transport in an amorphous organic semiconductor', *Chem Phys Lett*, **339**, 161–166.

Murata H, Kafafi Z H and Uchida M (2002), 'Efficient organic light-emitting diodes with undoped active layers based on silole derivatives', *Appl Phys Lett*, **80**, 189–191.

Nakanishi H, Sumi N, Aso Y and Otsubo T (1998), 'Synthesis and properties of the longest oligothiophenes: the icosamer and heptacosamer', *J Org Chem*, **63**, 8632–8633.

Nakanishi H, Aso Y and Otsubo T (1999), 'The longest class of oligothiophenes', *Synth Met*, **101**, 604–605.

Nakayama J, Konishi T and Hoshino M (1988), 'Preparation of thiophene oligomers', *Heterocycles*, **27**, 1731–1755.

Nawa K, Miyawaki K, Imae I, Noma N and Shirota Y (1993a), 'Polymers containing pendant oligothiophenes as a novel class of electrochromic materials', *J Mater Chem*, **3**, 113–114.

Nawa K, Miyawaki K, Imae I, Noma N and Shirota Y (1993b), 'A novel type of semiconducting polymers: synthesis and properties of electrochemically-doped polymers containing pendant oligothiophenes', *Synth Met*, **55–57**, 1176–1181.

Nazeeruddin M K, Kay A, Rodicio I, Humphry-Baker R, Müller E, Liska P, Vlachopoulos N and Grätzel M (1993), 'Conversion of light to electricity by *c*is-X_2Bis(2,2′-bipyricyl-4,4′-dicarboxylate)ruthenium(II) charge-transfer sensitizers ($X = Cl^-$, Br^-, I^-, CN^- and SCN^-) on nanocrystalline TiO_2 electrodes', *J Am Chem Soc*, **115**, 6382–6390.

Nishimura K, Kobata T, Inada H and Shirota Y (1991), 'Arylaldehyde and arylketone hydrazones as a new class of amorphous molecular materials', *J Mater Chem*, **1**, 897–898.

Noda T and Shirota Y (1998), '5,5′-Bis(dimesitylboryl)-2,2′-bithiophene and 5,5″-bis(dimesitylboryl)-2,2′:5′,2″-terthiophene as a novel family of electron-transporting amorphous molecular materials', *J Am Chem Soc*, **120**, 9714–9715.

Noda T, Imae I, Noma N and Shirota Y (1997a), '5,5″-Bis{4-[bis(4-methylphenyl) amino]phenyl}-2,2′:5′,2″-terthiophene and 5,5‴-bis{4-[bis(4-methylphenyl)amino] phenyl}-2,2′:5′,2″:5″,2‴-quaterthiophene as a novel family of amorphous molecular materials', *Adv Mater*, **9**, 239–241.

Noda T, Ogawa H, Noma N and Shirota Y (1997b), 'A novel family of amorphous molecular materials containing an oligothiophene moiety as color-tunable emitting materials for organic electroluminescent devices', *Adv Mater*, **9**, 720–722.

Noda T, Ogawa H, Noma N and Shirota Y (1999a), 'Organic light-emitting diodes using a novel family of amorphous molecular materials containing an oligothiophene moiety as colour-tunable emitting materials', *J Mater Chem*, **9**, 2177–2181.

Noda T, Ogawa H and Shirota Y (1999b), 'A blue-emitting organic electroluminescent device using a novel emitting amorphous molecular material, 5,5′-bis(dimesitylboryl)-2,2′-bithiophene', *Adv Mater*, **11**, 283–285.

Noma N, Tsuzuki T and Shirota Y (1995), 'α-Thiophene octamer as a new class of photoactive material for photoelectrical conversion', *Adv Mater*, **7**, 647–648.

Noma N, Kawaguchi K, Imae I, Nakano H and Shirota Y (1996), 'Electrochemical doping of α-ethyl-disubstituted oligothiophenes and electrical conductivities of the resulting radical-cation salts', *J Mater Chem*, **6**, 117–118.

Nomura S and Shirota Y (1997), 'Concentration dependence of the activation energy for the hole drift mobility of 9-ethylcarbazole-3-carbaldehyde diphenylhydrazone dispersed in polycarbonate', *Chem Phys Lett*, **268**, 461–464.

O'Brien D, Bleyer A, Lidzey D G, Bradley D D C and Tsutsui T (1997), 'Efficient multilayer electroluminescence devices with poly(*m*-phenylenevinylene-co-2,5-dioctyloxy-*p*-phenylenevinylene) as the emissive layer', *J Appl Phys*, **82**, 2662–2670.

O'Brien D F, Burrows P E, Forrest S R, Koene B E, Loy D E and Thompson M E (1998), 'Hole transporting materials with high glass transition temperatures for use in organic light-emitting devices', *Adv Mater*, **10**, 1108–1112.

O'Brien D F, Baldo M A, Thompson M E and Forrest S R (1999), 'Improved energy transfer in electrophosphorescent devices', *Appl Phys Lett*, **74**, 442–444.

Ogawa H, Ohnishi K and Shirota Y (1997), 'Tri(p-terphenyl-4-yl)amine as a novel blue-emitting material for organic electroluminescent devices', *Synth Met*, **91**, 243–245.

Ogawa H, Okuda R and Shirota Y (1998), 'Tuning of the emission color of organic electroluminescent devices by exciplex formation at the organic solid interface', *Appl Phys A: Mater Sci Process*, **67**, 599–602.

Ohishi H, Tanaka M, Kageyama H and Shirota Y (2004), 'Amorphous molecular materials with high carrier mobilities: thiophene- and selenophene-containing tris(oligoarylenyl)amines', *Chem Lett*, **33**, 1266–1267.

Ohmori Y, Uchida M, Muro K and Yoshino K (1991), 'Blue electroluminescent diodes utilizing poly(alkylfluorene)', *Jpn J Appl Phys*, **30**, L1941–L1943.

Ohno T, Moriwaki K and Miyata T (2001), 'Intramolecular charge-transfer interaction in a new dyad based on C_{60} and bis(4′-tert-butylbiphenyl-4-yl)aniline (BBA) donor', *J Org Chem*, **66**, 3397–3401.

Okumoto K and Shirota Y (2000), 'New hole-transporting amorphous molecular materials with high glass-transition temperatures for organic light-emitting diodes', *Chem Lett*, 1034–1035.

Okumoto K and Shirota Y (2001a), 'Development of high-performance blue-violet-emitting organic electroluminescent devices', *Appl Phys Lett*, **79**, 1231–1233.

Okumoto K and Shirota Y (2001b), 'Development of new hole-transporting amorphous molecular materials for organic electroluminescent devices and their charge-transport properties', *Mater Sci Eng B*, **85**, 135–139.

Okumoto K and Shirota Y (2003), 'New class of hole-blocking amorphous molecular materials and their application in blue-violet-emitting fluorescent and green-emitting phosphorescent organic electroluminescent devices', *Chem Mater*, **15**, 699–707.

Okumoto K, Wayaku K, Noda T, Kageyama H and Shirota Y (2000), 'Amorphous molecular materials: charge transport in the glassy state of *N*,*N*′-diphenyl-[1,1′-biphenyl]-4,4′-diamines', *Synth Met*, **111–112**, 473–476.

Okumoto K, Ohara T, Noda T and Shirota Y (2001), 'A thermally stable greenish blue organic electroluminescent device using a novel emitting amorphous molecular material', *Synth Met*, **121**, 1655–1656.

Okumoto K, Doi H and Shirota Y (2002), 'Development of new hole-transporting amorphous molecular materials with high glass-transition temperatures and their application in thermally stable organic electroluminescent devices', *J Photopolym Sci Technol*, **15**, 239–241.

Oldham Jr. W J, Lachicotte R J and Bazan G C (1998), 'Synthesis, spectroscopy, and morphology of tetrastilbenoidmethanes', *J Am Chem Soc*, **120**, 2987–2988.

Onsager L (1938), 'Initial recombination of ions', *Phys Rev*, **54**, 554–557.

O'Regan B and Grätzel M (1991), 'A low-cost, high-efficiency solar cell based on dye-sensitized colloidal TiO_2 films', *Nature*, **353**, 737–740.

Osasa T, Yamamoto S and Matsumura M (2007), 'Organic solar cells by annealing stacked amorphous and microcrystalline layers', *Adv Funct Mater*, **17**, 2937–2942.

Ottmar M, Ichisaka T, Subramanian L R, Hanack M and Shirota Y (2001), '2,3-Tetrakis[4-(*N*-2-naphthyl-*N*-phenylamino)phenoxy]-substituted metal-free and metal phthalocyanines as a novel class of amorphous molecular materials', *Chem Lett*, 788–789.

Oyston S, Wang C, Hughes G, Batsanov A S, Perepichka I F, Bryce M R, Ahn J H, Pearson C and Petty M C (2005), 'New 2,5-diaryl-1,3,4-oxadiazole-fluorene hybrids as electron transporting materials for blended-layer organic light emitting diodes', *J Mater Chem*, **15**, 194–203.

Ozaki M, Peebles D L, Weinberger B R, Chiang C K, Gau S C, Heeger A J and MacDiarmid A G (1979), 'Junction formation with pure and doped polyacetylene', *Appl Phys Lett*, **35**, 83–85.

Ozaki M, Peebles D, Weinberger B R, Heeger A J and MacDiarmid A G (1980), 'Semiconductor properties of polyacetylene p-(CH)x : n-CdS heterojunctions', *J Appl Phys*, **51**, 4252–4256.

Palilis L C, Murata H, Uchida M and Kafafi Z H (2003a), 'High efficiency molecular organic light-emitting diodes based on silole derivatives and their exciplexes', *Org Electronics*, **4**, 113–121.

Palilis L C, Mäkinen A J, Uchida M and Kafafi Z H (2003b), 'Highly efficient molecular organic light-emitting diodes based on exciplex emission', *Appl Phys Lett*, **82**, 2209–2211.

Pan H, Fu G-L, Zhao Y-H and Zhao C-H (2011), 'Through-space charge-transfer emitting biphenyls containing a boryl and an amino group at the *o,o′*-positions', *Org Lett*, **13**, 4830–4833.

Pandey A K, Dabos-Seignon S and Nunzi J-M (2006), 'Pentacene: PTCDI-$C_{13}H_{27}$ molecular blends efficiently harvest light for solar cell applications', *Appl Phys Lett*, **89**, 113506/1–3.

Pang J, Tao Y, Freiberg S, Yang X-P, D'Iorio M and Wang S (2002), 'Syntheses, structures, and electroluminescence of new blue luminescent star-shaped compounds based on 1,3,5-triazine and 1,3,5-trisubstituted benzene', *J Mater Chem*, **12**, 206–212.

Papkovsky D B (1995), 'New oxygen sensors and their application to biosensing', *Sens Actuators B*, **29**, 213–218.

Park S H, Roy A, Beaupré S, Cho S, Coates N, Moon J S, Moses D, Leclerc M, Lee K and Heeger A J (2009), 'Bulk heterojunction solar cells with internal quantum efficiency approaching 100%', *Nature Photon*, **3**, 297–303.

Park Y, Kim S, Lee J-H, Jung D H, Wu C-C and Park J (2010), 'New blue-violet emitters based on an indenopyrazine core for OLEDs: effects of the position of m-terphenyl side group substitution on optical and electroluminescence properties', *Org Electronics*, **11**, 864–871.

Peumans P and Forrest S R (2001), 'Very-high-efficiency double-heterostructure copper phthalocyanine/C_{60} photovoltaic cells', *Appl Phys Lett*, **79**, 126–128.

Peumans P, Bulović V and Forrest S R (2000), 'Efficient photon harvesting at high optical intensities in ultrathin organic double-heterostructure photovoltaic diodes', *Appl Phys Lett*, **76**, 2650–2652.

Peumans P, Uchida S and Forrest S R (2003), 'Efficient bulk heterojunction photovoltaic cells using small-molecular-weight organic thin films', *Nature*, **425**, 158–162.

Pfister G (1977), 'Hopping transport in a molecularly doped organic polymer', *Phys Rev B*, **16**, 3676–3687.

Pfister G and Williams D J (1974), 'Photogeneration processes in poly (*N*-vinylcarbazole)', *J Chem Phys*, **61**, 2416–2426.

Pfister G and Griffiths C H (1978), 'Temperature dependence of transient hole hopping transport in disordered organic solids: carbazole polymers', *Phys Rev Lett*, **40**, 659–662.

Picciolo L C, Murata H and Kafafi Z H (2001), 'Organic light-emitting devices with saturated red emission using 6,13-diphenylpentacene', *Appl Phys Lett*, **78**, 2378–2380.

Plazek D J and Magill J H (1966), 'Physical properties of aromatic hydrocarbons. I. viscous and viscoelastic behavior of 1:3:5-tri-α-naphtyl benzene', *J Chem Phys*, **45**, 3038–3050.

Pope M, Kallmann H P and Magnante P (1963), 'Electroluminescence in organic crystals', *J Chem Phys*, **38**, 2042–2043.

Popovic Z D (1982), 'Electric field induced fluorescence quenching and photocarrier generation in x-metal-free phthalocyanine', *J Chem Phys*, **76**, 2714–2719.

Popovic Z D (1984), 'A study of carrier generation in β-metal-free phthalocyanine', *Chem Phys*, **86**, 311–321.

Popovic Z D and Menzel E R (1979), 'Electric field-induced fluorescence quenching in organic photoconductors', *J Chem Phys*, **71**, 5090–5096.

Regensburger P J (1968), 'Optical sensitization of charge carrier transport in poly(*N*-vinyl carbazole)', *Photochem Photobiol*, **8**, 429–440.

Reineke S, Lindner F, Schwartz G, Seidler N, Walzer K, Lüssem B and Leo K (2009), 'White organic light-emitting diodes with fluorescent tube efficiency', *Nature*, **459**, 234–239.

Reucroft P J, Takahashi K and Ullal H (1974), 'Theoretical efficiency in an organic photovoltaic energy conversion system', *Appl Phys Lett*, **25**, 664–666.

Reucroft P J, Takahashi K and Ullal H (1975), 'Theoretical and experimental photovoltaic energy conversion in an organic film system', *J Appl Phys*, **46**, 5218–5225.

Roncali J, Leriche P and Cravino A (2007), 'From one- to three-dimensional organic semiconductors: in search of the organic silicon?', *Adv Mater*, **19**, 2045–2060.

Roquet S, Cravino A, Leriche P, Alévêque O, Frère P and Roncali J (2006), 'Triphenylamine-thienylenevinylene hybrid systems with internal charge transfer as donor materials for heterojunction solar cells', *J Am Chem Soc*, **128**, 3459–3466.

Ross R B, Cardona C M, Guldi D M, Sankaranarayanan S G, Reese M O, Kopidakis N, Peet J, Walker B, Bazan G C, Keuren E V, Holloway B C and Drees M (2009), 'Endohedral fullerenes for organic photovoltaic devices', *Nature Mater*, **8**, 208–212.

Saito T, Sisk W, Kobayashi T, Suzuki S and Iwayanagi T (1993), 'Photocarrier generation process of phthalocyanines studied by photocurrent and electroabsorption measurements', *J Phys Chem*, **97**, 8026–8031.

Salbeck J, Weissörtel F and Bauer J (1997a), 'Spiro linked compounds for use as active materials in organic light emitting diodes', *Macromol Symp*, **125**, 121–132.

Salbeck J, Yu N, Bauer J, Weissörtel F and Bestgen H (1997b), 'Low molecular organic glasses for blue electroluminescence', *Synth Met*, **91**, 209–215.

Sariciftci N S, Smilowitz L, Heeger A J and Wudl F (1992), 'Photoinduced electron transfer from a conducting polymer to buckminsterfullerene', *Science*, **258**, 1474–1476.

Sariciftci N S, Braun D, Zhang C, Srdano V I, Heeger A J, Stucky G and Wudl F (1993), 'Semiconducting polymer-buckminsterfullerene heterojunctions: diodes, photodiodes, and photovoltaic cells', *Appl Phys Lett*, **62**, 585.

Sasabe H and Kido J (2011), 'Multifunctional materials in high-performance OLEDs: challenges for solid-state lighting', *Chem Mater*, **23**, 621–630.

Schaffert R M (1971), 'A new high-sensitivity organic photoconductor for electrophotography', *IBM J Res and Dev*, **15**, 75–89.

Schein L B and Mack J X (1988), 'Adiabatic and non-adiabatic small polaron hopping in molecularly doped polymers', *Chem Phys Lett*, **149**, 109–112.

Schein L B, Glatz D and Scott J C (1990), 'Observation of the transition from adiabatic to nonadiabatic small polaron hopping in a molecularly doped polymer', *Phys Rev Lett*, **65**, 472–475.

Scher H and Montroll E W (1975), 'Anomalous transit-time dispersion in amorphous solids', *Phys Rev B*, **12**, 2455–2477.

Schulze K, Uhrich C, Schüppel R, Leo K, Pfeiffer M, Brier E, Reinold E and Bäuerle P (2006), 'Efficient vacuum-deposited organic solar cells based on a new low-bandgap oligothiophene and fullerene C_{60}', *Adv Mater*, **18**, 2872–2875.

Shan J, Yap G P A and Richeson D S (2005), 'Synthesis, thermal, electrochemical, and photophysical characterization of 1,5-bis(diarylamino)naphthalene derivatives as potential hole transport OLED materials', *Can J Chem*, **83**, 958–968.

Shang H, Fan H, Liu Y, Hu W, Li Y and Zhan X (2011), 'A solution-processable star-shaped molecule for high-performance organic solar cells', *Adv Mater*, **23**, 1554–1557.

Sharma G D, Suresh P, Mikroyannidis J A and Stylianakis M M (2010), 'Efficient bulk heterojunction devices based on phenylenevinylene small molecule and perylene–pyrene bisimide', *J Mater Chem*, **20**, 561–567.

Shen J Y, Lee C Y, Huang T-H, Lin J T, Tao Y-T, Chien C-H and Tsai C (2005), 'High T_g blue emitting materials for electroluminescent devices', *J Mater Chem*, **15**, 2455–2463.

Shih P-I, Chien C-H, Wu F-I and Shu C-F (2007), 'A novel fluorene-triphenylamine hybrid that is a highly efficient host material for blue-, green-, and red-light-emitting electrophosphorescent devices', *Adv Funct Mater*, **17**, 3514–3520.

Shirakawa H, Louis E J, MacDiarmid A G, Chiang C K and Heeger A J (1977), 'Synthesis of electrically conducting organic polymers: hologen derivatives of polyacetylene, $(CH)_x$', *J Chem Soc Chem Commun*, 578–580.

Shirota Y (2000), 'Organic materials for electronic and optoelectronic devices', *J Mater Chem*, **10**, 1–25.

Shirota Y (2005), 'Photo- and electroactive amorphous molecular materials – molecular design, synthesis, reactions, properties, and applications', *J Mater Chem*, **15**, 75–93.

Shirota Y and Kageyama H (2007), 'Charge carrier transporting molecular materials and their applications in devices', *Chem Rev*, **107**, 953–1010.

Shirota Y, Noma N, Kanega H and Mikawa H (1984), 'Preparation of an electrically conducting polymer by the electrolytic polymerization of *N*-vinylcarbazole', *J Chem Soc Chem Commun*, 470–471.

Shirota Y, Kakuta T, Kanega H and Mikawa H (1985), 'Rectification and photovoltaic properties of a Schottky barrier cell using electrochemically-doped poly(*N*-vinylcarbazole)', *J Chem Soc Chem Commun*, 1201–1202.

Shirota Y, Noma N and Mikawa H (1987), 'Electrochemical doping of poly[4-(*N,N*-diphenylamino)phenylmethyl methacrylate], and rectification and photovoltaic properties', *Synth Met*, **18**, 399–404.

Shirota Y, Kobata T and Noma N (1989), 'Starburst molecules for amorphous molecular materials. 4,4′,4″-Tris(*N,N*-diphenylamino)triphenylamine and 4,4′,4″-tris[*N*-(3-methylphenyl)-*N*-phenylamino]triphenylamine', *Chem Lett*, 1145–1148.

Shirota Y, Noma N, Shimizu Y, Kanega H, Jeon I-R, Nawa K, Kakuta T, Yasui H and Namba K (1991), 'Preparation of electrically conducting polymers containing pendant π-electron systems by electrochemical doping, and properties and applications of doped polymers', *Synth Met*, **41–43**, 3031–3036.

Shirota Y, Kuwabara Y, Inada H, Wakimoto T, Nakada H, Yonemoto Y, Kawami S and Imai K (1994), 'Multilayered organic electroluminescent device using a novel starburst molecule, 4,4′,4″-tris(3-methylphenylphenylamino)triphenylamine, as a hole transport material', *Appl Phys Lett*, **65**, 807–809.

Shirota Y, Kuwabara Y, Okuda D, Okuda R, Ogawa H, Inada H, Wakimoto T, Nakada H, Yonemoto Y, Kawami S and Imai K (1997), 'Starburst molecules based on π-electron systems as materials for organic electroluminescent devices', *J Lumin*, **72–74**, 985–991.

Shirota Y, Moriwaki K, Yoshikawa S, Ujike T and Nakano H (1998), '4-[Di(biphenyl-4-yl)amino]azobenzene and 4,4′-bis[bis(4′-tert-butylbiphenyl-4-yl)amino]azobenzene as a novel family of photochromic amorphous molecular materials', *J Mater Chem*, **8**, 2579–2581.

Shirota Y, Kinoshita M, Noda T, Okumoto K and Ohara T (2000), 'A novel class of emitting amorphous molecular materials as bipolar radical formants: 2-{4-[bis(4-methylphenyl)amino]phenyl}-5-(dimesitylboryl)thiophene and 2-{4-[bis(9,9-dimethylfluorenyl)amino]phenyl}-5-(dimesitylboryl)thiophene', *J Am Chem Soc*, **122**, 11021–11022.

Shu Y, Lim Y-F, Li Z, Purushothaman B, Hallani R, Kim J E, Parkin S R, Malliaras G G and Anthony J E (2011), 'A survey of electron-deficient pentacenes as acceptors in polymer bulk heterojunction solar cells', *Chem Sci*, **2**, 363–368.

Sirringhaus H, Tessler N and Friend R H (1998), 'Integrated optoelectronic devices based on conjutated polymers', *Science*, **280**, 1741.

Skotheim T A (1998), *Handbook of conducting polymers*, Third Edition, New York, Marcel Dekker.

Sonar P, Ng G-M, Lin T T, Dodabalapur A and Chen Z-K (2010), 'Solution processable low bandgap diketopyrrolopyrrole (DPP) based derivatives: novel acceptors for organic solar cells', *J Mater Chem*, **20**, 3626–3636.

Steinberger S, Mishra A, Reinold E, Levichkov J, Uhrich C, Pfeiffer M and Bäuerle P (2011), 'Vacuum-processed small molecule solar cells based on terminal acceptor-substituted low-band gap oligothiophenes', *Chem Commun*, **47**, 1982–1984.

Stolka M, Yanus J F and Pai D M (1984), 'Hole transport in solid solutions of a diamine in polycarbonate', *J Phys Chem*, **88**, 4707–4714.

Strohriegl P and Grazulevicius J V (2002), 'Charge-transporting molecular glasses', *Adv Mater*, **14**, 1439–1452.

Stylianakis M M, Mikroyannidis J A, Dong Q, Pei J, Liu Z and Tian W (2009), 'Synthesis, photophysical and photovoltaic properties of star-shaped molecules with triphenylamine as core and phenylethenylthiophene or dithienylethylene as arms', *Solar Energy Mater Solar Cells*, **93**, 1952–1958.

Su S-J, Chiba T, Takeda T and Kido J (2008), 'Pyridine-containing triphenylbenzene derivatives with high electron mobility for highly efficient phosphorescent OLEDs', *Adv Mater*, **20**, 2125–2130.

Su S-J, Sasabe H, Pu Y-J, Nakayama K and Kido J (2010), 'Tuning energy levels of electron-transport materials by nitrogen orientation for electrophosphorescent devices with an 'ideal' operating voltage', *Adv Mater*, **22**, 3311–3316.

Sun C, Hudson Z M, Helander M G, Lu Z-H and Wang S (2011), 'A polyboryl-functionalized triazine as an electron transport material for OLEDs', *Organometalics*, **30**, 5552–5555.

Sun M, Wang L, Zhu X, Du B, Liu R, Yang W and Cao Y (2007), 'Near-infrared response photovoltaic device based on novel narrow band gap small molecule and PCBM fabricated by solution processing', *Solar Energy Mater Solar Cells*, **91**, 1681–1687.

Sun S S and Sariciftci N S (Eds.) (2005), *Organic photovoltaics, mechanisms, materials and devices*, New York, CRC Press.

Sun Y and Forrest S R (2007), 'High-efficiency white organic light emitting devices with three separate phosphorescent emission layers', *Appl Phys Lett*, **91**, 263503/1–3.

Sun Y, Welch G C, Leong W L, Takacs C J, Bazan G C and Heeger A J (2012), 'Solution-processed small-molecule solar cells with 6.7% efficiency', *Nature Mater*, **11**, 44–48.

Tamao K, Uchida M, Izumizawa T, Furukawa K and Yamaguchi S (1996), 'Silole derivatives as efficient electron transporting materials', *J Am Chem Soc*, **118**, 11974–11975.

Tamayo A B, Dang X-D, Walker B, Seo J, Kent T and Nguyen T-Q (2009), 'A low band gap, solution processable oligothiophene with a dialkylated diketopyrrolopyrrole chromophore for use in bulk heterojunction solar cells', *Appl Phys Lett*, **94**, 103301/1–3.

Tanaka D, Sasabe H, Li Y-J, Su S-J, Takeda T and Kido J (2007a), 'Ultra high efficiency green organic light-emitting devices', *Jpn J Appl Phys*, **46**, L10–L12.

Tanaka D, Takeda T, Chiba T, Watanabe S and Kido J (2007b), 'Novel electron-transport material containing boron atom with a high triplet excited energy level', *Chem Lett*, **36**, 262–263.

Tanaka H, Tokito S, Taga Y and Okada A (1996), 'Novel hole-transporting materials based on triphenylamine for organic electroluminescent devices', *Chem Commun*, 2175–2176.

Tang C W (1986), 'Two-layer organic photovoltaic cell', *Appl Phys Lett*, **48**, 183–185.

Tang C W and Albrecht A C (1975a), 'Chlorophyll-a photovoltaic cells', *Nature*, **254**, 507–509.

Tang C W and Albrecht A C (1975b), 'Photovoltaic effects of metal–chlorophyll-a–metal sandwich cells', *J Chem Phys*, **62**, 2139–2149.

Tang C W and VanSlyke S A (1987), 'Organic electroluminescent diodes', *Appl Phys Lett*, **51**, 913–915.

Tang C W, VanSlyke S A and Chen C H (1989), 'Electroluminescence of doped organic thin films', *J Appl Phys*, **65**, 3610–3616.

Tang S, Liu M, Lu P, Cheng G, Zeng M, Xie Z, Xu H, Wang H, Yang B, Ma Y and Yan D (2008), 'Fluorene trimers with various 9,9′-substituents: the synthesis, characteristics, condensed state structures, and electroluminescence properties', *Org Electronics*, **9**, 241–252.

Tao S, Zhou Y, Lee C-S, Lee S-T, Huang D and Zhang X (2008), 'Highly efficient nondoped blue organic light-emitting diodes based on anthracene-triphenylamine derivatives', *J Phys Chem C*, **112**, 14603–14606.

Thelakkat M (2002), 'Star-shaped, dendrimeric and polymeric triarylamines as photoconductors and hole transport materials for electro-optical applications', *Macromol Mater Eng*, **287**, 442–461.

Thelakkat M and Schmidt H-W (1998), 'Synthesis and properties of novel derivatives of 1,3,5-tris(diarylamino)benzenes for electroluminescent devices', *Adv Mater*, **10**, 219–223.

Thomas K R J, Lin J T, Tao Y-T and Ko C-W (2000), 'Novel green light-emitting carbazole derivatives: potential electroluminescent materials', *Adv Mater*, **12**, 1949–1951.

Thomas K R J, Lin J T, Velusamy M, Tao Y-T and Chuen C-H (2004), 'Color tuining in benzo[1,2,5]thiaziazole-based small molecules by amino conjugation/deconjugation: bright red-light-emitting diodes', *Adv Funct Mater*, **14**, 83–90.

Tokito S, Tanaka H, Noda K, Okada A and Taga Y (1997), 'Thermal stability in oligomeric triphenylamine/tris(8-quinolinolato)aluminum electroluminescent devices', *Appl Phys Lett*, **70**, 1929–1931.

Tokito S, Iijima T, Suzuri Y, Kita H, Tsuzuki T and Sato F (2003), 'Confinement of triplet energy on phosphorescent molecules for highly-efficient organic blue-light-emitting devices', *Appl Phys Lett*, **83**, 569–571.

Tong Q-X, Lai S-L, Chan M-Y, Lai K-H, Tang J-X, Kwong H-L, Lee C-S and Lee S-T (2007), 'Efficient green organic light-Emitting devices with a nondoped dual-functional electroluminescent material', *Appl Phys Lett*, **91**, 153504/1–3.

Tsubomura H, Matsumura M, Nomura Y and Amamiya T (1976), 'Dye sensitised zinc oxide: aqueous electrolyte: platinum photocell', *Nature*, **261**, 402–403.

Tsuboyama A, Iwawaki H, Furugori M, Mukaide T, Kamatani J, Igawa S, Moriyama T, Miura S, Takiguchi T, Okada S, Hoshino M and Ueno K (2003), 'Homoleptic cyclometalated iridium complexes with highly efficient red phosphorescence and application to organic light-emitting diode', *J Am Chem Soc*, **125**, 12971–12979.

Tsukamoto J and Ohigashi H (1982), 'Characteristics of Schottky barrier solar cells using polyacetylene, $(CH)_x$', *Synth Met*, **4**, 177–186.

Tsuzuki T, Kuwabara Y, Noma N, Shirota Y and Willis M R (1996), 'Effect of morphology on photovoltaic properties of titanyl phthalocyanine', *Jpn J Appl Phys*, **35**, L447–L450.

Tung Y-L, Lee S-W, Chi Y, Tao Y-T, Chien C-H, Cheng Y-M, Chou P-T, Peng S-M and Liu C-S (2005), 'Organic light-emitting diodes based on charge-neutral Os(II) emitters: generation of saturated red emission with very high external quantum efficiency', *J Mater Chem*, **15**, 460–464.

Uchida M, Izumizawa T, Nakano T, Yamaguchi S, Tamao K and Furukawa K (2001), 'Structural optimization of 2,5-diarylsiloles as excellent electron-transporting materials for organic electroluminescent devices', *Chem Mater*, **13**, 2680–2683.

Ueta E, Nakano H and Shirota Y (1994), 'Glass formation and phase transition of novel π-electron starburst molecules, 1,3,5-tris(phenyl-2-thienylamino)benzene and 1,3,5-tris(phenyl-3-thienylamino)benzene', *Chem Lett*, 2397–2400.

Uoyama H, Goushi K, Shizu K, Nomura H and Adachi C (2012), 'Highly efficient organic light-emitting diodes from delayed fluorescence', *Nature*, 492, 234–238.

Van Slyke S A, Chen C H and Tang C W (1996), 'Organic electroluminescent devices with improved stability', *Appl Phys Lett*, **69**, 2160–2162.

Visco R E and Chandross E A (1964), 'Electroluminescence in solutions of aromatic hydrocarbons', *J Am Chem Soc*, **86**, 5350–5351.

Walker B, Tamayo A B, Dang X-D, Zalar P, Seo J H, Garcia A, Tantiwiwat M and Nguyen T-Q (2009), 'Nanoscale phase separation and high photovoltaic efficiency in solution-processed, small-molecule bulk heterojunction solar cells', *Adv Funct Mater*, **19**, 3063–3069.

Walzer K, Maennig B, Pfeiffer M and Leo K (2007), 'Highly efficient organic devices based on electrically doped transport layers', *Chem Rev*, **107**, 1233–1271.

Wang J-F, Kawabe Y, Shaheen S E, Morrell M M, Jabbour G E, Lee P A, Anderson J, Armstrong N R, Kippelen B, Mash E A and Peyghambarian N (1998), 'Exciplex electroluminescence from organic bilayer devices composed of triphenyldiamine and quinoxaline derivatives', *Adv Mater*, **10**, 230–233.

Wang S, Mayo E I, Perez M D, Griffe L, Wei G, Djurovich P I, Forrest S R and Thompson M E (2009), 'High efficiency organic photovoltaic cells based on a vapor deposited squaraine donor', *Appl Phys Lett*, **94**, 233304/1–3.

Wei G, Wang S, Sun K, Thompson M E and Forrest S R (2011), 'Solvent-annealed crystalline squaraine: $PC_{70}BM$ (1:6) solar cells', *Adv Energy Mater*, **1**, 184–187.

Wei Y and Chen C-T (2007), 'Doubly ortho-linked *cis*-4,4′-bis(diarylamino)stilbene/fluorene hybrids as efficient nondoped, sky-blue fluorescent materials for optoelectronic applications', *J Am Chem Soc*, **129**, 7478–7479.

Weinberger B R, Gau S C and Kiss Z (1981), 'A polyacetylene:aluminum photodiode', *Appl Phys Lett*, **38**, 555–557.

Wienk M M, Kroon J M, Verhees W J H, Knol J, Hummelen J C, van Hal P A and Janssen R A J (2003), 'Efficient methano[70]fullerene/MDMO-PPV bulk heterojunction photovoltaic cells', *Angew Chem Int Ed Engl*, **42**, 3371–3375.

Wong K-T, Liao Y-L, Lin Y-T, Su H-C and Wu C-C (2005), 'Spiro-configured bifluorenes: highly efficient emitter for UV organic light-emitting device and host material for red electrophosphorescence', *Org Lett*, **7**, 5131–5134.

Wright J D (1995), *Molecular crystals*, 2nd edition, Cambridge, Cambridge University Press.

Wu C-C, Lin Y-T, Wong K-T, Chen R-T and Chien Y-Y (2004), 'Efficient organic blue-light-emitting devices with double confinement on terfluorenes with ambipolar carrier transport properties', *Adv Mater*, **16**, 61.

Wu C-H, Shih P-I, Shu C-F and Chi Y (2008), 'Highly efficient red organic light-emitting devices based on a fluorene-triphenylamine host doped with an Os(II) phosphor', *Appl Phys Lett*, **92**, 233303/1–3.

Wu K-C, Ku P-J, Lin C-S, Shih H-T, Wu F-I, Huang M-J, Lin J-J, Chen I-C and Cheng C-H (2008), 'The photophysical properties of dipyrenylbenzenes and their application as exceedingly efficient blue emitters for electroluminescent devices', *Adv Funct Mater*, **18**, 67–75.

Wu Z, Li A, Fan B, Xue F, Adachi C and Ouyang J (2011), 'Phenanthrene-functionalized 3,6-dithiophen-2-yl-2,5- dihydropyrrolo[3,4–c]pyrrole-1,4-diones as donor molecules for solution-processed organic photovoltaic cells', *Solar Energy Mater Solar Cells*, **95**, 2516–2523.

Wynands D, Levichkova M, Leo K, Uhrich C, Schwartz G, Hildebrandt D, Pfeiffer M and Riede M (2010), 'Increase in internal quantum efficiency in small molecular oligothiophene: C_{60} mixed heterojunction solar cells by substrate heating', *Appl Phys Lett*, **97**, 073503/1–3.

Wynberg H and Metselaar J (1984), 'A convenient route to polythiophenes', *Synth Commun*, **14**, 1.

Wöhrle D and Meissner D (1991), 'Organic solar cells', *Adv Mater*, **3**, 129–138.

Xiao L, Su S-J, Agata Y, Lan H and Kido J (2009), 'Nearly 100% internal quantum efficiency in an organic blue-light electrophosphorescent device using a weak electron transporting material with a wide energy gap', *Adv Mater*, **21**, 1271–1274.

Xue J, Rand B P, Uchida S and Forrest S R (2005), 'A hybrid planar-mixed molecular heterojunction photovoltaic cell', *Adv Mater*, **17**, 66–71.

Yakimov A and Forrest S R (2002), 'High photovoltage multiple-heterojunction organic solar cells incorporating interfacial metallic nanoclusters', *Appl Phys Lett*, **80**, 1667–1669.

Yamaguchi S, Akiyama S and Tamao K (2000a), 'Tri-9-anthrylborane and its derivatives: new boron-containing π-electron systems with divergently extended π-conjugation through boron', *J Am Chem Soc*, **122**, 6335–6336.

Yamaguchi S, Shirasaka T and Tamao K (2000b), 'Tridurylboranes extened by three arylethynyl groups as a new family of boron-based π-electron systems', *Org Lett*, **2**, 4129–4132.

Yang F, Shtein M and Forrest S R (2005a), 'Morphology control and material mixing by high-temperature organic vapor-phase deposition and its application to thin-film solar cells', *J Appl Phys*, **98**, 014906/1–10.

Yang F, Shtein M and Forrest S R (2005b), 'Controlled growth of a molecular bulk heterojunction photovoltaic cell', *Nature Mater*, **4**, 37–41.

Yeh H-C, Lee R-H, Chan L-H, Lin T-Y J, Chen C-T, Balasubramaniam E and Tao Y-T (2001), 'Synthesis, properties, and applications of tetraphenylmethane-based molecular materials for light-emitting devices', *Chem Mater*, **13**, 2788–2796.

Yeh S-J, Wu M-F, Chen C-T, Song Y-H, Chi Y, Ho M-H, Hsu S-F and Chen C H (2005), 'New dopant and host materials for blue-light-emitting phosphorescent organic electroluminescent devices', *Adv Mater*, **17**, 285–289.

Yokoyama D (2011), 'Molecular orientation in small-molecule organic light-emitting diodes', *J Mater Chem*, **21**, 19187–19202.

Yokoyama D, Sakaguchi A, Suzuki M and Adachi C (2008), 'Horizontal molecular orientation in vacuum-deposited organic amorphous films of hole and electron transport materials', *Appl Phys Lett*, **93**, 173302.

Yoo S, Domercq B and Kippelen B (2004), 'Efficient thin-film organic solar cells based on pentacene/C_{60} heterojunctions', *Appl Phys Lett*, **85**, 5427–5429.

Yoshiiwa M, Kageyama H, Shirota Y, Wakaya F, Gamo K and Takai M (1996), 'Novel class of low molecular-weight organic resists for nanometer lithography', *Appl Phys Lett*, **69**, 2605–2607.

Yu G and Heeger A J (1995), 'Charge separation and photovoltaic conversion in polymer composites with internal donor/acceptor heterojunctions', *J Appl Phys*, **78**, 4510–4515.

Yu G, Gao J, Hummelen J C, Wudl F and Heeger A J (1995), 'Polymer photovoltaic cells: enhanced efficiencies via a network of internal donor-acceptor heterojunctions', *Science*, **270**, 1789–1791.

Yu G, Xu X, Liu Y, Jiang Z, Yin S, Shuai Z, Zhu D, Duan X and Lu P (2005), 'Efficient blue electroluminescent device using tetra(β-naphthyl)silane as a hole-blocking material', *Appl Phys Lett*, **87**, 222115/1–3.

Yu J and Shirota Y (2002), 'A new class of high-performance red-fluorescent dyes for organic electroluminescent devices, [7-diethylamino-3-(2-thienyl)chromen-2-ylidene]-2,2-dicyanovinylamine and {10-(2-thienyl)-2,3,6,7-tetrahydro-1h,5h-chromeno[8,7,6-ij]quinolizine-11-ylidene}-2,2-dicyanovinylamine', *Chem Lett*, 984–985.

Yu M-X, Duan J-P, Lin C-H, Cheng C-H and Tao Y-T (2002), 'Diaminoanthracene derivatives as high-performance green host electroluminescent materials', *Chem Mater*, **14**, 3958–3963.

Yuan W Z, Shen X Y, Zhao H, Lam J W Y, Tang L, Lu P, Wang C, Liu Y, Wang Z, Zheng Q, Sun J Z, Ma Y and Tang B Z (2010), 'Crystallization-induced phosphorescence of pure organic luminogens at room temperature', *J Phys Chem C*, **114**, 6090–6099.

Zhang J, Yang Y, He C, He Y, Zhao G and Li Y (2009), 'Solution-processable star-shaped photovoltaic organic molecule with triphenylamine core and benzothiadiazole–thiophene arms', *Macromolecules*, **42**, 7619–7622.

Zhang Q, Chen J, Cheng Y, Wang L, Ma D, Jing X and Wang F (2004), 'Novel hole-transporting materials based on 1,4-bis(carbazolyl)benzene for organic light-emitting devices', *J Mater Chem*, **14**, 895–900.

Zhang Y, Yip H-L, Acton O, Hau S K, Huang F and Jen A K-Y (2009), 'A simple and effective way of achieving highly efficient and thermally stable bulk-heterojunction polymer solar cells using amorphous fullerene derivatives as electron acceptor', *Chem Mater*, **21**, 2598–2600.

Zhao C-H, Wakamiya A, Inukai Y and Yamaguchi S (2006), 'Highly emissive organic solids containing 2,5-diboryl-1,4-phenylene unit', *J Am Chem Soc*, **128**, 15934–15935.

Zhao G, He Y and Li Y (2010), '6.5% Efficiency of polymer solar cells based on poly(3-hexylthiophene) and indene-C_{60} bisadduct by device optimization', *Adv Mater*, **22**, 4355–4358.

Zhen C-G, Dai Y-F, Zeng W-J, Ma Z, Chen Z-K and Kieffer J (2011), 'Achieving highly efficient fluorescent blue organic light-emitting diodes through optimizing molecular structures and device configuration', *Adv Funct Mater*, **21**, 699–707.

Zhou J, Wan X, Liu Y, Long G, Wang F, Li Z, Zuo Y, Li C and Chen Y (2011), 'A planar small molecule with dithienosilole core for high efficiency solution-processed organic photovoltaic cells', *Chem Mater*, **23**, 4666–4668.

Zhou X, Blochwitz J, Pfeiffer M, Nollau A, Fritz T and Leo K (2001a), 'Enhanced hole injection into amorphous hole-transport layers of organic light-emitting diodes using controlled p-type doping', *Adv Funct Mater*, **11**, 310–314.

Zhou X, Pfeiffer M, Blochwitz J, Werner A, Nollau A, Fritz T and Leo K (2001b), 'Very-low-operating-voltage organic light-emitting diodes using a p-doped amorphous hole injection layer', *Appl Phys Lett*, **78**, 410–412.

2

Influence of film morphology on optical and electronic properties of organic materials

S. R. PUNIREDD, W. PISULA and K. MÜLLEN,
Max Planck Institute for Polymer Research, Germany

DOI: 10.1533/9780857098764.1.83

Abstract: During recent years, the development of processing techniques from simple spin coating to more sophisticated methods has resulted in a more advanced control over the macroscopic order and molecular alignment on the surface of organic semiconductors. As a consequence, the charge carrier transport in the thin surface layer and thus the performance of electronic devices, especially field-effect transistors, has been significantly enhanced. Moreover, comprehensive understanding of the relation between molecular design, processing parameters, molecular order on the surface and charge carrier transport has been obtained. The discussed approaches are potential alternatives to traditional processing methods such as inkjet or screen printing which lower the charge carrier transport in the fabricated device.

Key words: self-assembly, solution processing, field-effect transistor (FET), long-range orientation, organic semiconductors.

2.1 Introduction

Organic electronics have found their way from basic academic research to industrial applications over recent years in a quickly growing market. This market covers especially application fields in light-emitting diodes (LEDs),[1,2] field-effect transistors (FETs)[3,4] and solar cells[5] which open the window for a novel type of technologies. The main element are the semiconducting, conjugated organic materials, which can be processed into devices from solution, allowing large area and low-cost fabrication.[6,7] Moreover, many organic semiconductors are mechanically flexible and therefore applicable to bendable electronic elements.[8] The mass fabrication of the devices occurs by high-speed and inexpensive methods at low temperatures and is realized by well-known continuous in-line roll-to-roll technologies, such as inkjet, offset, rotary screen printing and others, which are adapted to the specific requirements of the organic materials.[9] Due to the low cost processing, one-way applications of electronic elements are realizable such as radio-frequency identification (RFID) tags and sensors in which FETs play a major role.[10,11] Thereby, the performance of a transistor is mainly determined by the speed of the charge carrier transport from one electrode to the other, which in turn depends on various factors such as molecular design, supramolecular organization, thin film microstructure and charge carrier transport.[12,13]

Organic semiconductors based on polymers or small molecules consist of π-conjugated bonds, which lead to delocalized filled and empty π-orbitals that affect the electrical and optical behavior of the material.[14] In the solid state, the molecules weakly interact, in comparison to inorganics, by π-stacking, van der Waals and dipole–dipole forces, which are responsible for the material's properties.[15] Therefore, the device operation can be controlled by chemical design of the molecules.[16–18] Substituents such as aliphatic side chains, which are attached to the aromatic part, are one tool to tune the material's properties such as solubility which is essential for the processing of the molecules from solution.[19,20] Substituents also lead to the formation of a distinct superstructure driven by the local phase separation between the flexible aliphatic part and rigid aromatic fraction.[21,22] Additionally, the molecular packing is strongly influenced by using such molecular units. For instance, the steric demand of the substituents can modulate the packing distance of neighboring molecules and the molecular arrangement of the building blocks to each other.[23] This in turn has consequences on the π-orbital overlap and thus on the transport of charge carriers from one molecule to another. An increased π-stacking distance reduces the hopping rate of charge carriers. Changes in the crystal unit cell due to chemical modifications of small molecules have a great impact on the device performance,[24] while in most cases polymeric backbones with alkyl side chains with so-called lamellae superstructures are less sensitive.[25] But the modification of the macromolecular structure has a strong effect on the microstructure of polymers, which includes the organization of the molecules into domains within a range from few tens of nanometers to few micrometers.[26]

Most small molecules processed from solution form only highly crystalline thin layers,[27] while the microstructure of polymers is more complex and in many cases dependent on the polymer design, such as alkyl position and molecular weight. The most prominent example in this context is poly(3-hexylthiophene) (P3HT) for which the polymer regularity determines the thin film topography and order.[28,29] This is an important point since the charge carrier transport is a function of density of structural defects and, domain boundaries, and also of the molecular orientation. Defects and boundaries are trapping sites for charge carriers, while molecules oriented in the transport direction promote the charge carrier mobility.[30] All three factors can be tailored by the processing conditions, such as concentration, type of solvent, temperature and surface energy.[31] Interestingly, few conjugated polymers form only low ordered thin surface layers, but reveal at the same time FET mobilities far above 0.1 $cm^2/V\,s$.[32,33] Such materials are of

Table 2.1 Overview of the discussed materials, fabrication methods and transistor performance

Process	Material	Device	Mobility (cm^2/V s)/ (on–off ratio)	Reference
Drop casting	**1**	BGBC	0.30 (4×10^8)	34
Spin coating	Regioregular P3HT	BGTC	0.1 (10^6)	28
Pre-alignment layer	**2**	BGTC	10^{-3} (10^4)	53
Solvent vapor annealing	**3**	BGBC	0.1 (2×10^3)	64
Langmuir–Blodgett	Polt(3-hexylthiophene)/ arachidic acid and quinquethiophene/ arachidic acid	BGBC	10^{-7}–10^{-4}	69
	Regioregular P3HT	BGTC	2×10^{-2}	89
Dip coating	**7**	BGTC	1.7 (10^7)	80
Zone casting	**8**	BGTC	1.4 (10^4–10^5)	81
	9	BGTC	0.1 (10^4)	84
	10	BGTC	0.7 (10^6–10^7)	85
	11	BGTC	5×10^{-3} (10^4)	86

(Note. BGBC = Bottom Gate Bottom Contact and BGTC = Bottom Gate Top Contact).

great interest for mass production. Rapid roll-to-roll techniques generally yield poorly ordered films on the substrate surface and lead, for most organic semiconductors, to mobilities at least two orders of magnitude lower than those obtained in the laboratory. Polymers with less order-dependent device performance would overcome this problem. To ensure an identical device performance after mass production, new processing approaches are required which can induce a higher molecular order on the meso- and macroscopic scale, or alternatively orient the molecules in the desired direction of charge carrier transport. This chapter discusses various processing methods of prominent organic semiconductors, small crystalline molecules and polymers, for FET applications which are separated in two groups: continuous and discontinuous techniques (Table 2.1).

2.2 Discontinuous processing

2.2.1 Drop casting

Drop casting is the simplest way to produce a thin film and is important also for ink jet printing which can be used to fabricate the whole circuit on a flexible substrate.[34] Drop casting in a saturated environment has been shown to give good results; good film ordering over very large areas has been demonstrated. In this coating method a horizontal surface is required on which the solution of the organic semiconductor is dropped or poured. After the solvent has evaporated, the material forms a thin film on the surface. This method provides no major control with respect to film thickness, whereby typically thicker layers of more than a few hundred nanometers are obtained. The main drawback of this technique is the inhomogeneous film thickness and high surface roughness which make the deposition of top contacts on the active layer for FET device fabrication impossible. As an advantage, the drying conditions, mainly the drying temperature and surface energy,[35] can be controlled precisely, providing additional parameters to tune the molecular arrangement and microstructure. As an interesting effect, the deposited material mainly orients in the direction of the solvent evaporation. Theoretically, at optimized conditions a radial alignment of the superstructure of the organic semiconductor should be formed by simple drop casting a homogeneous film. The alignment would then coincide with the direction of the evaporated solvent. During evaporation a concentration gradient is established, which is the driving force for the molecular alignment on the surface. However, this is quite difficult, since the concentration within the solution drop changes with ongoing evaporation of the solvent and material precipitation. Therefore, it is not possible to achieve stationary conditions and a homogeneous film morphology by drop casting.

On close inspection of the drop-cast layer, one can notice that most of the material is deposited at the rim of the precipitated drop, whereas the film interior is poor of material. This phenomenon is the so-called coffee-stain effect and can be explained by the thermodynamic factors and kinetics in solution.[36] Inhomogeneous material distribution in the resulting film is based on a capillary flow of the solvent together with the solutes within the fixed drop from the center to the peripheral contact line to compensate for the evaporation loss. The application of a mixed-solvent system induces convective and Marangoni flows during drying in the drop that improve the thin layer homogeneity.[37] This approach was adapted to a solution processable pentacene derivative **1** (Fig. 2.1(a)) which was deposited from bi-solvent mixtures to form well-ordered crystals on circular electrodes in an FET (Fig. 2.1(b)). The self-directed grown layer yielded mobilities up to 0.30 cm^2/V s and current on/off ratio of 4×10^8.[38,39]

2.2.2 Spin coating

Among solution processable thin film preparation techniques, spin coating is the preferred method for small-scale laboratory studies, yielding thin, uniform films on flat substrates. Nevertheless, spin coating is applicable for

(a)

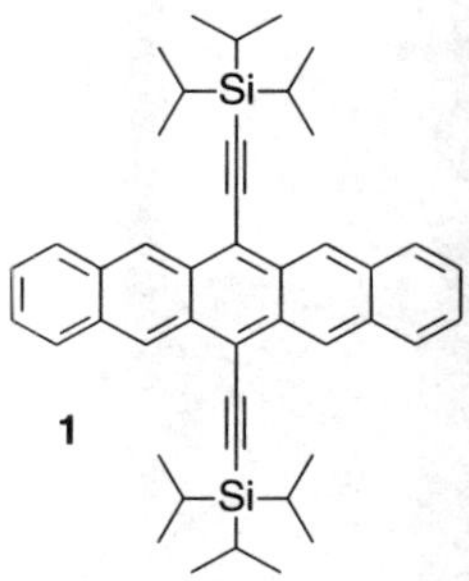

1

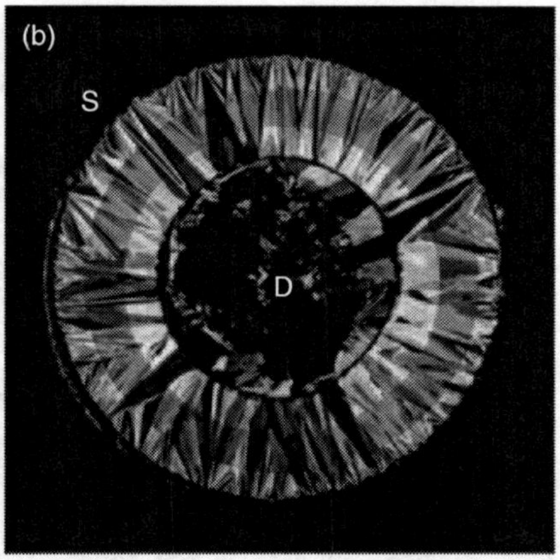

2.1 (a) Chemical structure of 6,13-bis(triisopropylsilylethynyl) TIPS pentacene (**1**), and (b) polarized optical micrograph of a TIPS pentacene crystal transistor deposited by drop casting. Reprinted with permission from ref. 38, Copyright 2009, Wiley-VCH Verlag GmbH.

large-scale production as shown by the example of compact disc production. The spin-coating process is very simple. An excess amount of an organic semiconductor solution is placed on the substrate. During solution deposition the disc, together with the substrate, can either be static or rotating at a low angular velocity, followed by rapid acceleration of the angular velocity. The film thickness is controlled in a straightforward way by changing the concentration, viscosity of the solution or substrate rotation speed. Higher concentrations or viscosities result in thicker films, while higher angular velocities lead to thinner layers. The choice of solvent is important. Higher solvent volatilities lead to thicker films at a given initial concentration and initial viscosity. The solvent can also be highly volatile so that chilling effects become dominant which result in non-uniformities. Furthermore, greater solvent/semiconductor compatibility is important for the film uniformity, i.e. less topographical variation in the film surface. The undesirable skinning process, which can cause meso- to macroscopic defects in the film, is prevented or reduced by partially saturating the atmosphere above the rotating disc by solvent vapor, or by using a multi-component solvent. The outcome of the spin-coating process is not only influenced by the angular velocity and the physical properties of the solution, but in addition is highly sensitive to parameters such as temperature, airflow velocity, relative humidity and thermal surroundings for the evaporating solvent (heat transfer). Thus, in order to perform systematic studies or to fabricate reproducible films of organic semiconductors, it is necessary to obtain a fixed set of operational conditions for a given spin-coating apparatus.[40–42]

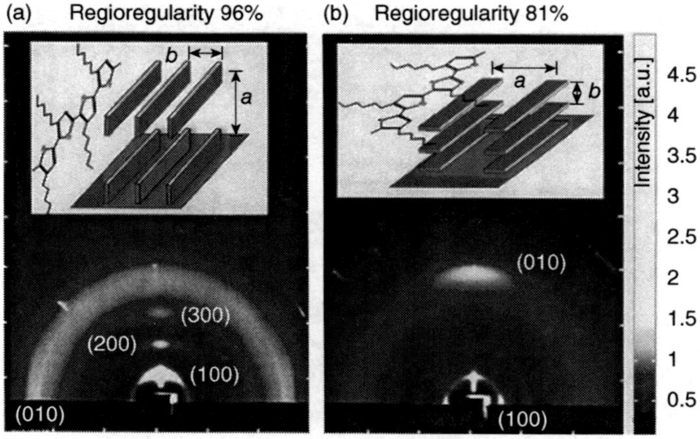

2.2 Two-dimensional grazing incidence X-ray scattering patterns for two different orientations of ordered spin-coated P3HT domains with respect to the substrate: P3HT with regioregularity of (a) 96% and (b) 81%. Reprinted with permission from ref. 28, Copyright 1999, Nature Publishing Group.

To control the aggregation in spin-coated semiconductor films, a systematic study of the spin-coating conditions on the film morphology and the electronic properties of poly(2-methoxy-5-(2′-ethyl-hexyloxy)-1,4-phenylene vinylene) (MEH-PPV) was performed.[43] The change in properties was related to formation of aggregates in solution prior to deposition. In this case, aggregation is caused by short-range interchain attraction forces, which are negligible for diluted solutions, i.e. when polymer chains are isolated. For increasing concentration the interchain forces become significant and the polymer coils start to entangle and to form so-called 'loose aggregates' that turn into even more entangled, so-called 'strong aggregates', at higher concentrations. The morphology of such spin-coated thin film is expected to depend on angular velocity when the centrifugal force is comparable to the cohesive force of the solution, i.e. the viscosity of the polymer solution should, to some degree, reflect the intermolecular forces. It has also been demonstrated that the conjugated plane of P3HT varies from an edge-on to a plane-on orientation depending on the rotation speed during spin coating and also on the regioregularity (Fig. 2.2).[28]

Other processing variables that affect the rate of solvent evaporation during spin coating, such as the air flow conditions and the solvent saturation of the spin environment, are also likely to affect the microstructure and the performance of FETs. Because most additive patterning processes envisioned for solution-processable semiconductors will include a drying step, control over the drying rate should be carefully considered in any process design.

2.2.3 Pre-alignment layers

This technique consists of mechanically depositing a thin, oriented layer of poly(tetrafluoroethylene) (PTFE) or polyimide onto a smooth counterface such as glass and silicon, by dragging the polymer at controlled temperature, pressure and speed against the substrate. Subsequently, the species of interest are deposited onto these PTFE films from the melt or vapor phase or from solution to yield surprisingly well-oriented materials.[44] Electron diffraction of the orientation-inducing films revealed that the molecules are oriented parallel along the dragging direction. It is postulated that the PTFE layers were not molecularly smooth, and that their orienting effect derived, in part, from the multitude of ridges in the film, which presumably provides effective nucleation sites that induced the oriented growth of a great number of materials deposited onto the PTFE films. Today such pre-aligned layers are commercially used for orientation of cholesteric liquid crystal cells in display panels for consumer electronics.[45]

Ordered thin films are required for efficient charge transport and it has been a challenge to establish optimal structural organization using small conjugated molecules and polymers. For example, when P3HT was solvent-cast on a PTFE surface, its polymer-chain axis was observed to align along the PTFE friction-transfer direction, and higher mobility along this direction was observed. The anisotropic properties of this film are explained by the intrinsic property as well as microstructure of the P3HT film.[46] More prominent examples for conjugated polymers oriented by a pre-alignment layer are based on fluorene homopolymers and copolymers which show a nematic liquid crystalline state at elevated temperatures.[47,48] Typically, the polymer chains are processed by alignment of the material on the rubbed substrate and subsequent quenching from the liquid crystalline state into a monodomain glass.[49] Such highly oriented layers produce polarized light emission in organic light-emitting diode (OLED) devices. In FETs the maximum charge carrier mobility is observed along the conjugated chain orientation with values reaching 0.02 cm^2/V s for this type of fluorene-based polymers.[50]

Besides polymers, small molecules such as pentacene were also observed to orient uniaxially on a PTFE layer. From electron diffraction and polarized absorption measurements, it was concluded that pentacene molecules adsorb in an edge-on geometry on the PTFE layer, with molecular columns aligned normal to the sliding direction. The FET mobility of pentacene along the molecular columns (transverse of the direction of friction) was 4.2×10^{-1} cm^2/V s, which was about 100 times higher than that along the friction direction.[51]

It has been reported that the solution processed discotic hexa-*peri*-hexbenzocoronenes (HBCs) **2** form films with supramolecularly ordered

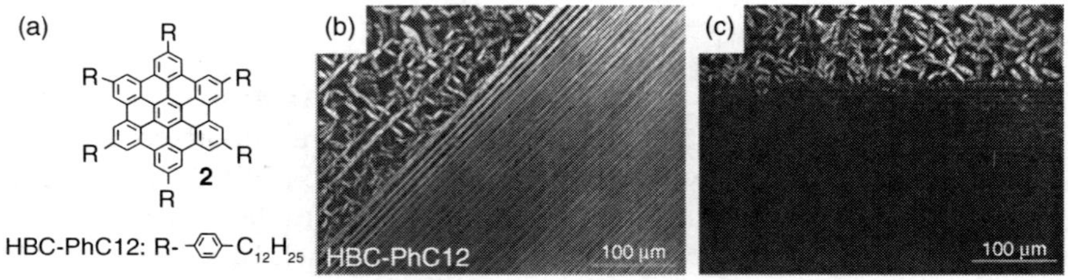

2.3 (a) Molecular structure of **2**, optical micrographs between cross-polarizers of PTFE aligned layers in two different arrangements (b) 45° and (c) parallel to the polarizer. Reprinted with permission from ref. 53, Copyright 2003, Wiley-VCH Verlag GmbH.

columns which lie parallel to the substrate and are oriented uniaxially onto the PTFE layer (Fig. 2.3). The alignment of the HBC derivative occurs via epitaxial growth on the oriented PTFE over several hundred micrometer dimensions.[52] The high interfacial order achieved by solution processing under ambient conditions is as required for FETs. Charge carrier mobilities for p-type accumulation mode FET devices up to 10^{-3} cm^2/V s have been found on implementing these aligned semiconducting films of **2**. The combination of solution processing and high interfacial order together with useful semiconductor properties shows considerable scope for practical applications of discotic liquid crystalline materials.[53]

2.2.4 Solvent vapor annealing

Solvent vapor annealing (SVA) is an industrially important technique since it causes a rapid morphological equilibration of films at room temperature without thermal damage to the material or other complications which are disadvantages of high-temperature annealing.[54–56] It has to be emphasized that this process is a post-treatment method of solution deposited surface layers. It is especially suitable for organic electronics, where sensitive materials with strong intermolecular interactions are used. It is performed by simply placing any cast sample into a vessel where a vapor–liquid equilibrium is attainable for the chosen solvent.[57] This choice requires consideration of all interactions playing a role on the molecular organization from solution on a surface such as molecule–solvent, molecule–substrate, molecule–molecule, and solvent–substrate interactions.[58] In the case of conjugated macromolecules, SVA increases the polymer-chain mobility, giving the system sufficient time to modify its morphology towards its thermodynamic equilibrium. This provides a path for the polymer to evolve towards an advantageous microstructure. Once a desired morphology is obtained, the system is 'frozen' to prevent further changes. SVA involves the exposure of the film to a solvent atmosphere under controlled conditions. The film swells by solvent take-up, causing an increase in polymer-chain mobility similar to temperature annealing above the melting or glass transition temperature.

An interesting example concerns perylene derivatives which form nanoscopic needles after spin-coating on Si/SiO*x* surfaces.[59] When these needles are subjected to SVA in tetrahydrofuran (THF), the material on the surface reorganizes to a large extent, leading to the *in situ* formation of millimeter-long fibers that exhibit aspect ratios exceeding 10^3.[60,61] Their formation has been studied through real-time optical microscopy imaging, and the fibers have been found to grow through a nucleation-governed, Avrami-type mechanism. If successive SVA steps are carried out by employing THF or $CHCl_3$ alternately, which is a better solvent for perylene derivatives, the fibers undergo assembly/disassembly cycles, always involving a remarkable mass transport with rearrangement of the deposited material occurring over hundreds of micrometers. Self-assembly of very long fibers of perylene dyes is also obtained in solution, using solvent-induced precipitation or similar techniques, though SVA promotes controlled and reversible growth directly on the substrate of interest.[62,63]

The one-step, solvent vapor-annealing process can also significantly improve the electrical properties of a solution processed FET as shown on the example of the p-type triethylsilylethynyl anthradithiophene (**3**, Fig. 2.4(a)).[64] The as-deposited films of **3** are featureless and amorphous over large areas as revealed by optical microscopy (Fig. 2.4(b)) and X-ray diffraction. For these layers low saturation mobilities of 0.002 cm^2/V s and severe current hysteresis were determined in FET devices. Exposing the same devices to dichloroethane vapor changed the morphology dramatically leading to macroscopically large grains with high birefringence contrast in the optical microscopy (Fig. 2.4c). These solvent-vapor annealed FETs showed significantly increased saturation mobilities of 0.1 cm^2/V s

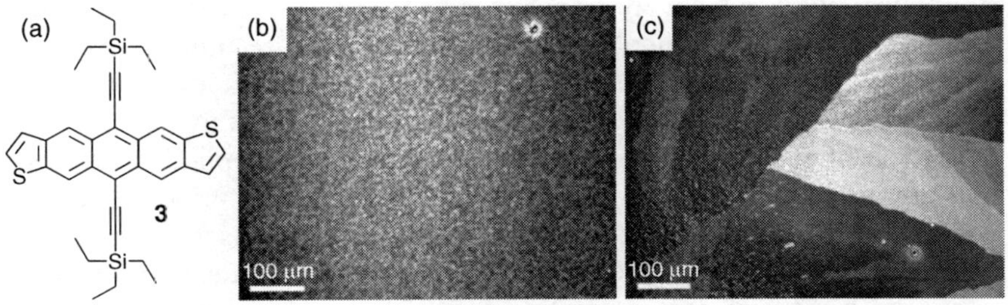

2.4 (a) Chemical structure of triethylsilylethynyl anthradithiophene (**3**), optical microscopy images of (b) an as-spun film and (c) after dichloroethane solvent-vapor annealing. Reprinted with permission from ref. 64, Copyright 2006, Wiley-VCH Verlag GmbH.

in air proving that this post-treatment procedure is efficient for device improvement.

2.3 Continuous processing

2.3.1 Langmuir–Blodgett (LB)

Conjugated polymers and small molecular semiconductors can also be aligned through the Langmuir–Blodgett (LB). In the LB technique, the semiconductor solution is spread onto an air–water interface to form a surface film. The film is compressed immediately to result in ordered structures, and then transferred to the substrate which is moved out by a constant velocity of the water. The advantage of the LB technique is that it provides a high degree of order and precision in thickness. For instance, polythiophenes have been explored for this assembly method. Attempts to assemble pure poly(alkylthiophenes) generally fail because the polymers do not arrange in a stable monolayer at the air–water interface due to the formation of aggregates in spite of the hairy decoration.[65,66] To prevent this aggregation phenomena, long-chain fatty acids such as stearic acid have been mixed with poly(alkylthiophenes), successfully generating a stable monolayer.[67,68] These FET devices showed relatively low charge-carrier mobility, typically in the range of 10^{-7}–10^{-4} cm^2/V s, due to the irregularity of the alkyl side chains in regiorandom P3HT, and the presence of nonconducting fatty acids.[69,70] However, the later reported LB processed FET using pure regioregular P3HT revealed mobilities as high as 2×10^{-2} cm^2/V s. For regioregular alternating amphiphilic polythiophenes even a better control of molecular orientation at the air–water interface is available (Fig. 2.5).[71,72] The amphiphilic nature generated from alternating hydrophobic and hydrophilic side-chains along the polymer backbone induced the polythiophenes to orient their well-packed π-plane perpendicular to the air–water interface in the 2-dimensional polycrystalline monolayers (Fig. 2.5). These monolayers were selectively transferred onto a gold-coated silicon wafer with prepatterned hydrophobic and hydrophilic areas, resulting in a charge carrier mobility orders of magnitude higher in the direction normal to the π-plane as compared to the direction parallel to the π-plane.[73]

The LB alignment was one of the first techniques applied to produce well-organized thin films of discotic liquid crystals. The principle for the orientation of discotic molecules is identical to that one used for polymers. After deposition of the molecules at the air–water interface a monolayer is formed by lateral compression (Fig. 2.6(a)). The transfer onto a solid support is realized by either vertical transfer or a horizontal lift of the substrate (Fig. 2.6(b)). For FET applications a molecular edge-on arrangement on the substrate is required. Therefore, hydrophilic side-groups are attached

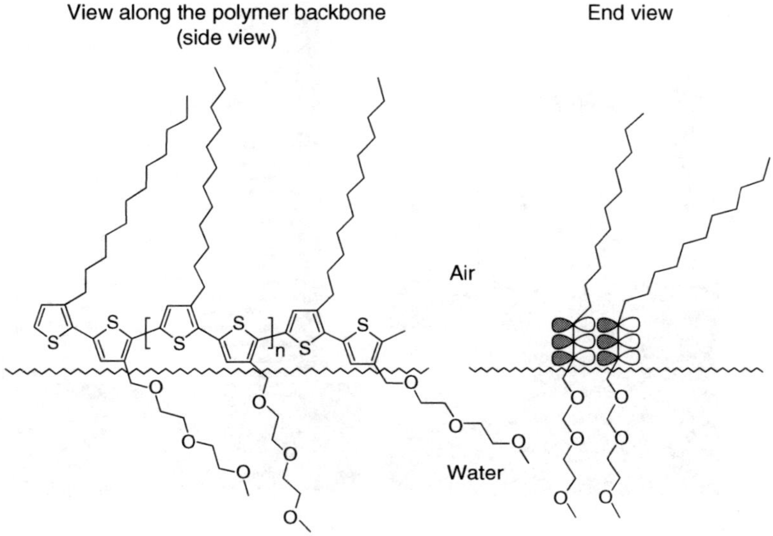

2.5 Single chain of regioregular amphiphilic polythiophene at the air–water interface seen from the side and end view. Reprinted with permission from ref. 72, Copyright 2000, American Chemical Society.

to the aromatic core, leading to the corresponding organization of the discs already at the air–water interface. The molecule is arranged perpendicular to the interface with two or more side-chains submerged in the water. This configuration maximizes the π-stacking interactions between the conjugated cores within the columnar stacks. Typical examples for LB aligned amphiphilic discotics are phthalocyanines (**4**),[74] triphenylenes (**5**)[75,76] and hexa-*peri*-hexabenzocoronenes (**6**)[77] (Fig. 2.6(c)).

2.3.2 Dip coating

Dip coating consists of the dipping of the substrate in a solution followed by the vertical withdrawal of the sample at a controlled speed. The dip-coating procedure was used for the first time in industrial applications for the production of an antireflection layer on glass panes. The dip-coated film thickness is a function of withdrawal speed, solvent evaporation rate and concentration of the solution. By taking the substrate out of the solution at a specific rate, solvent evaporation and material crystallisation are controlled, potentially resulting in alignment of the molecules in the dip-coating direction (Fig. 2.7(a)). In addition, the reservoir provides replenishment of the material used up for layer generation by the meniscus, in this way allowing the creation of continuous films. Therefore, there are mainly two parameters responsible for film formation: boiling point of the solvent

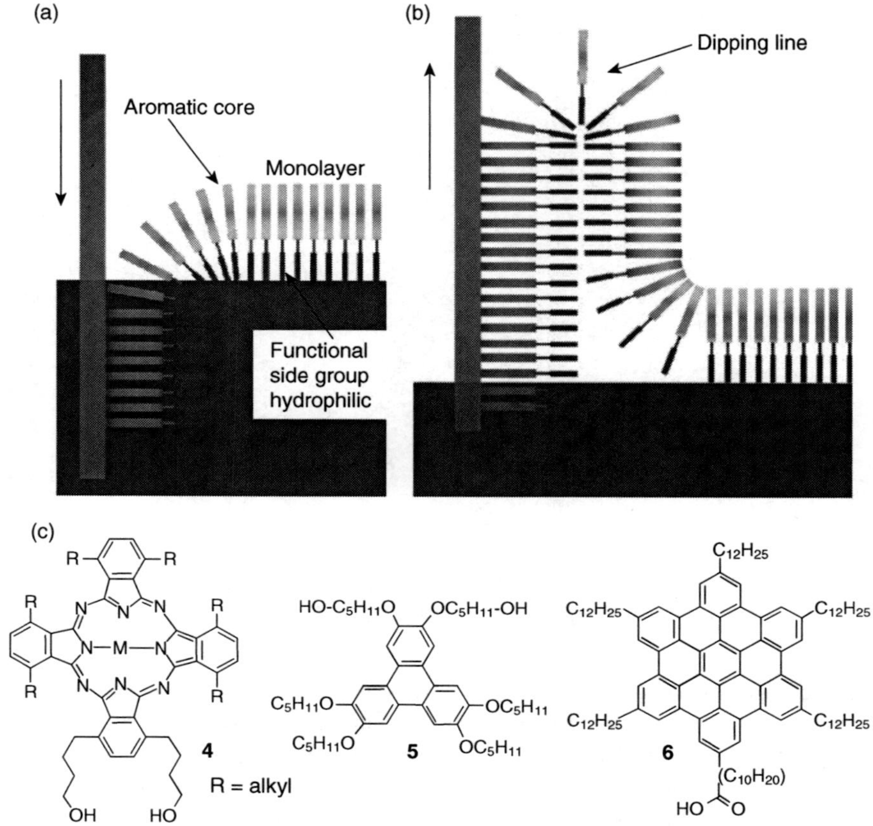

2.6 (a) and (b) schematic illustration of the deposition of discotic molecules onto a substrate with the edge-on arrangement using the Langmuir–Blodgett technique, (c) examples for different amphiphilic discotic liquid crystals. Reprinted with permission from *Chem. Rev.* 2007, **107**, 718; Copyright 2007, American Chemical Society.

and dipping rate. The optimization of these two factors determines the directional alignment of the molecules.

Processing by dip-coating allows slower operation, which better adapts to the kinetic parameters for the assembly of heteroacenes with fused thiophene units, which are highly attractive organic semiconductors for FET applications.[78,79] The compound **7** (Fig. 2.7(b)) with four symmetrically fused thiophene-ring units shows a FET mobility of 10^{-2} cm^2/V s after spin-coating.[80] Based on this good transistor performance, the device characteristics has been improved by dip coating the material in order to enlarge the size of crystalline domains to minimize the grain boundaries, and to orient the charge carrier pathways between the drain and source

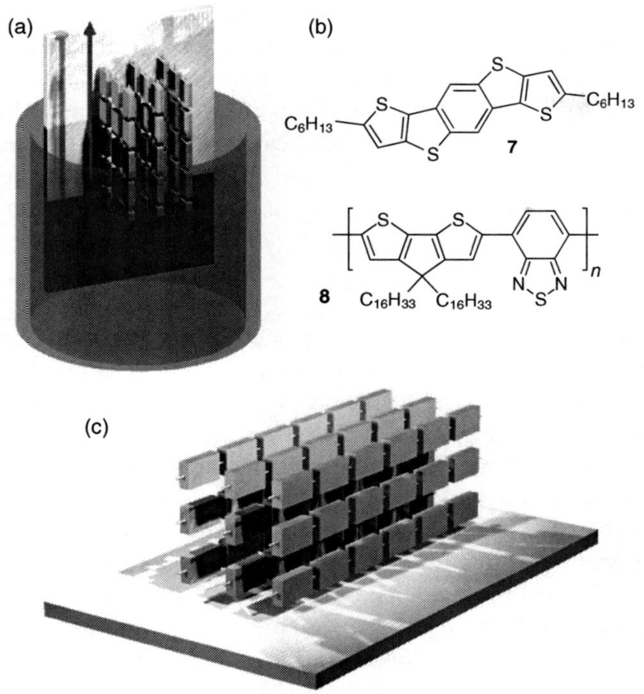

2.7 (a) Schematic illustration of the dip coating process, (b) chemical structures of **7** and **8**, edge-on arrangement and uniaxial orientation of the polymer chains of **8** on the substrate after the process. Reprinted with permission from ref. 81, Copyright 2009, Wiley-VCH Verlag GmbH.

electrodes. Under optimized solvent evaporation and dipping rates, highly crystalline domains ranging over several square millimeters have been obtained. The structural analysis has revealed a parallel orientation of the π-stacking axis with respect to the substrate. Such orientation is ideal for transistor applications and leads to devices featuring excellent hole mobilities of 1.7 cm^2/V s.

This strategy has been also applied successfully to the improvement of polymer-based transistor characteristics. After spin-coating, the cyclopentadithiophene-benzothiadiazole copolymer (Fig. 2.7(b), CDT-BTZ **8**, average molecular weight M_n = 50 kg mol^{-1}) exhibits a FET mobility of 0.67 cm^2/V s, while orientation of the polymer chains by dip coating along the direction of the charge transport significantly enhances this property to a maximum value of 1.4 cm^2/V s.[81] Processing by dip coating has induced an edge-on arrangement towards the surface (Fig. 2.7(c)), with highly ordered structures and long-range orientation of small oligomers and polymers. The improved organization has resulted in an excellent device performance with significantly larger mobility being achieved for small molecules by alignment. This can be attributed firstly to a large reduction of grain boundaries, secondly to the alignment of the conjugated polymer backbones along which the fastest charge carrier transport takes place, and thirdly to the quasi-perfect arrangement of the crystal structure across the electrodes, which favors charge migration in the dip-coated layers.[82,83]

2.3.3 Zone casting technique

The zone casting technique is a one-step method for preparing large area oriented films of solution-processible materials without a preoriented substrate. A solution is spread out by means of a nozzle onto a moving support such as glass or silicon (Fig. 2.8(a)). Thereby, a meniscus is formed between the nozzle and the support. During the solvent evaporation, a concentration gradient adjusts within the meniscus. When the critical concentration is attained, the material begins to precipitate or to nucleate from the solution and then crystallizes directionally onto the moving support, in this way forming the aligned thin layer. The film morphology is controlled by the processing parameters which are, for example, the evaporation temperature and polarity of the solvent, concentration, temperatures of the heating blocks, solvent flow and substrate velocity. The optimized zone casting conditions strongly depend on the solution behavior of the material. Therefore, prior to the processing it is important to obtain a fundamental understanding about the processes such as self-aggregation taking place in solution and during the self-organization in drop-cast films on surfaces. These findings are essential for the application of the appropriate processing parameters. Generally, the size or length of the surface layer prepared by the zone-casting processing is restricted to the limited volume of the applied syringe. Theoretically, it is possible to zone-cast films of unlimited length, as long as the steady-state conditions are not changed during the process.

So far zone casting has been employed only for small molecular, crystalline organic semiconductors such as tetrathiafulvalene derivatives (**9**) and pentacene (**10**), yielding high order in the films with the stacking direction of the molecules parallel to the substrate (Fig. 2.8(b)). The mobilities for these examples were 0.1 $cm^2/V\,s$ for **9** and 0.7 $cm^2/V\,s$ for **10**.[84,85] A particularly high degree of order was observed for hexa-dodecyl hexa-*peri*-hexabenzocoronene (**11**) where long-range aligned columnar structures were achieved (Fig. 2.8(c)).[86] When simply drop-cast, **9** forms randomly distributed fibers consisting of columns. However, after zone casting, highly oriented columns are present, clearly highlighting the effectiveness of zone casting in obtaining directionally arranged supramolecular structures. Even though this technique has so far been successfully applied only for

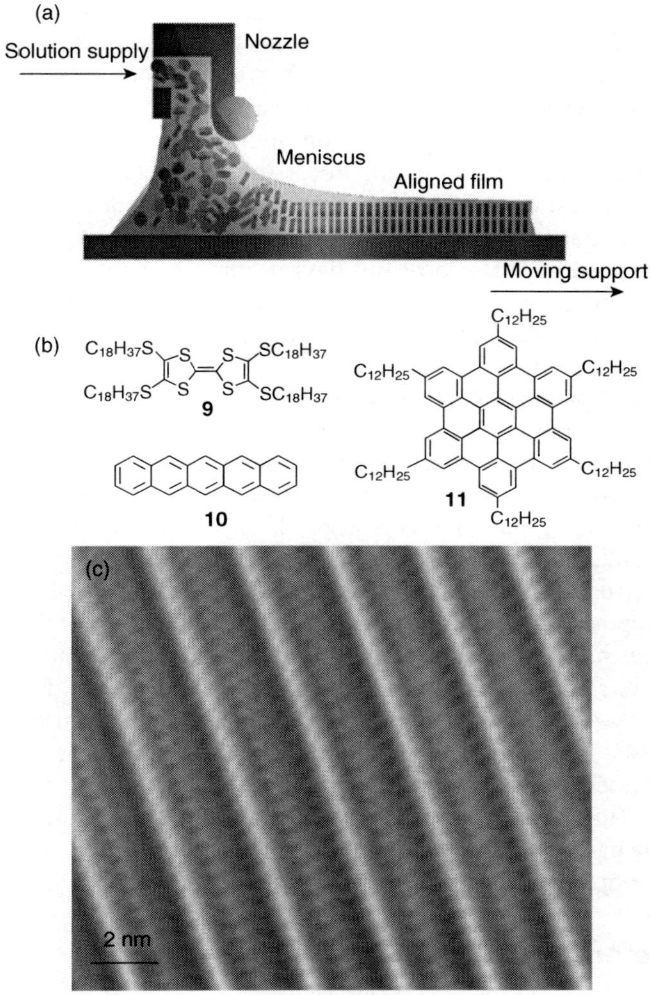

2.8 (a) Small molecules aligned by zone-casting into FETs, (b) schematic illustration of the zone-casting processing of HBCs and (c) uniaxially oriented columnar HBC structures (Fourier-transform image). Reprinted with permission from ref. 86, Copyright 2005, Wiley-VCH Verlag GmbH.

small molecules,[87,88] it is also expected to work equally well for conjugated polymers.

2.4 Conclusion

During recent years, the development of processing techniques from simple spin coating to more sophisticated methods has resulted in more advanced control over the macroscopic order and molecular alignment on the surface of organic semiconductors. As a consequence, the charge carrier transport in the thin surface layer and thus the performance of the electronic devices, especially field-effect transistors, have been significantly enhanced. Over the last few years comprehensive understanding of the relation between molecular design, processing parameters, molecular order on the surface and charge carrier transport has been obtained. Therefore, it is believed that further progress in the field of processing technologies will give an additional boost in device improvement. The discussed approaches are potential alternatives to traditional processing methods such as inkjet or screen printing which lower the charge carrier transport in the fabricated device.

The focus of processing development should not be limited to just FETs, but should be extended to other electronic applications like solar cell and OLEDs for which, molecular ordering is only partly important. Furthermore, a deeper insight into the complex fundamental relation between different processing parameters is needed, since the self-assembly mechanism of organic semiconductors during solution deposition depends on the molecular structure. This includes the initial aggregation stage in solution, solvent evaporation, solidification and microstructure formation on the surface. Feasibility proofs concerning the efficiency, yield and quality are required while transferring from the lab scale to pilot scale, for which an interdisciplinary engineering input is necessary between chemists and material scientists. In addition to this, questions concerning patterning have to be addressed while discussing more complex circuits for active-matrix flat-panels. This clearly indicates that this research area has progressed significantly in the past few years, but it remains a major topic for the development of organic electronics.

2.5 References

1. Y. Ohmori, *Laser Photonics Rev.*, **4** (2), 2010, 300–310.
2. B. Geffroy, P. le Roy, C. Prat, *Polym. Int.*, **55** (6), 2006, 572–582.
3. R. A. Street, *Adv. Mater.*, **21**, 2009, 2007–2022.
4. K. A. Arpin, A. Mihi, H. T. Johnson, A. J. Baca, J. A. Rogers, J. A. Lewis, P. V. Braun, *Adv. Mater.*, **22** (10), 2010, 1084–1101.
5. C. J. Brabec, N. S. Sariciftci, J. C. Hummelen, *Adv. Funct. Mater.*, **11** (1), 2001, 15–26.
6. S. Allard, M. Forster, B. Souharce, H. Thiem, U. Scherf, *Angew. Chem. Int. Ed.*, **47** (22), 2008, 4070–4098.
7. Z. Bao, *Adv. Mater.*, **12** (3), 2000, 227–230.
8. T. Sekitani, U. Zschieschang, H. Klauk, T. Someya, *Nature Materials*, **9**, 2010, 1015–1022.
9. A. C. Arias, J. D. MacKenzie, I. McCulloch, J. Rivnay, A. Salleo, *Chem. Rev.*, **110**, 2010, 3–24.

10. D. J. Gundlach, *Nature Materials*, **6**, 2007, 173–174.
11. A. N. Sokolov, M. E. Roberts, Z. Bao, *Materials Today*, **12**, 2009, 12–20.
12. P. M. Beaujuge, W. Pisula, H. N. Tsao, S. Ellinger, K. Müllen, J. R. Reynolds, *J. Am. Chem. Soc.*, **131** (22), 2009, 7514–7515.
13. X. Guo, R. P. Ortiz, Y. Zheng, Y. Hu, Y.-Y. Noh, K.-J. Baeg, A. Facchetti, T. J. Marks, *J. Am. Chem. Soc.*, **133** (5), 2011, 1405–1418.
14. A. Facchetti, *Materials Today*, **10** (3), 2007, 28–37.
15. L. Brunsveld, B. J. B. Folmer, E. W. Meijer, R. P. Sijbesma, *Chem. Rev.*, **101** (12), 2001, 4071–4098.
16. A. Mishra, C.-Q. Ma, P. Bäuerle, *Chem. Rev.*, **109**, 2009, 1141–1278.
17. T. Lei, Y. Cao, Y. Fan, C.-J. Liu, S.-C. Yuan, J. Pei, *J. Am. Chem. Soc.*, **133**, 2011, 6099–6101.
18. I. McCulloch, C. Bailey, M. Giles, M. Heeney, I. Love, M. Shkunov, D. Sparrowe, S. Tierney, *Chem. Mater.*, **17**, 2005, 1381–1385.
19. Z. Chen, A. Lohr, C. R. Saha-Möller, F. Würthner, *Chem. Soc. Rev.*, **38**, 2009, 564–584.
20. S. A. Ponomarenko, E. A. Tatarinova, A. M. Muzafarov, S. Kirchmeyer, L. Brassat, A. Mourran, M. Moeller, S. Setayesh, D. de Leeuw, *Chem. Mater.*, **18**, 2006, 4101–4108.
21. M. Hasegawaa, M. Iyoda, *Chem. Soc. Rev.*, **39**, 2010, 2420–2427.
22. T. F. A. De Greef, M. M. J. Smulders, M. Wolffs, A. P. H. J. Schenning, R. P. Sijbesma, E. W. Meijer, *Chem. Rev.*, **109** (11), 2009, 5687–5754.
23. W. Pisula, M. Kastler, D. Wasserfallen, M. Mondeshki, J. Piris, I. Schnell, K. Müllen, *Chem. Mater.*, **18** (16), 2006, 3634–3640.
24. M. Mas-Torrent, C. Rovira, *Chem. Rev.*, **37**, 2008, 827–838.
25. A. Facchetti, *Chem. Mater.*, **23**, 2011, 733–758.
26. A. Salleo, *Materials Today*, **10** (3), 2007, 38–45.
27. W. Wu, Y. Liu, D. Zhu, *Chem. Soc. Rev.*, **39**, 2010, 1489–1502.
28. H. Sirringhaus, P. J. Brown, R. H. Friend, M. M. Nielsen, K. Bechgaard, B. M. W. Langeveld-Voss, A. J. H. Spiering, R. A. J. Janssen, E. W. Meijer, P. Herwig, D. M. de Leeuw, *Nature*, **401**, 1999, 685–688.
29. J.-M. Verilhac, G. LeBlevennec, D. Djurado, F. Rieutord, M. Chouiki, J.-P. Travers, A. Pron, *Synth. Met.*, **156**, 2006, 815–823.
30. S. Liu, W. M. Wang, A. L. Briseno, S. C. B. Mannsfeld, Z. Bao, *Adv. Mater.*, **21**, 2009, 1–16.
31. A. A. Virkar, S. Mannsfeld, Z. Bao, N. Stingelin, *Adv. Mater.*, **22**, 2010, 3857–3875.
32. H. Yan, Z. Chen, Y. Zheng, C. Newman, J. R. Quinn, F. Dötz, M. Kastler, A. Facchetti, *Nature*, **457**, 2009, 679–686.
33. H. N. Tsao, D. M. Cho, I. Park, M. R. Hansen, A. Mavrinskiy, D. Y. Yoon, R. Graf, W. Pisula, H. W. Spiess, Klaus Müllen, *J. Am. Chem. Soc.*, **133** (8), 2011, 2605–2612.
34. D. Kim, S. H. Lee, S. Jeong, J. Moon, *Electrochem. Solid-State Lett*, **12** (6), 2009, H195–H197.
35. J. A. Lim, W. H. Lee, D. Kwak, Kilwon Cho, *Langmuir*, **25** (9), 2009, 5404–5410.
36. R. D. Deegan, O. Bakajin, T. F. Dupont, G. Huber, S. R. Nagel, T. A. Witten, *Nature*, **389**, 1997, 827.
37. J. Park, J. Moon, *Langmuir*, **22** (8), 2006, 3506–3513.

38. J.-P. Hong, S. Lee, *Angew. Chem. Int. Ed.*, **48**, 2009, 3096–3098.
39. J. A. Lim, W. H. Lee, H. S. Lee, J. H. Lee, Y. D. Park, K. Cho, *Adv. Funct. Mater.*, **18**, 2008, 229–234.
40. D. M. DeLongchamp, B. M. Vogel, Y. Jung, M. C. Gurau, C. A. Richter, O. A. Kirillov, J. O. D. A. Fischer, S. Sambasivan, L. J. Richter, E. K. Lin, *Chem. Mater.*, **17**, 2005, 5610–5612.
41. H. Yang, S. W. LeFevre, C. Y. Ryu Zhenan Bao, *App. Phys. Lett.*, **90**, 2007, 172116.
42. K. Norrman, A. Ghanbari-Siahkali, N. B. Larsen, *Annu. Rep. Prog. Chem., Sect. C*, **101**, 2005, 174–201.
43. Y. Shi, J. Liu, Y. Yang, *J. Appl. Phys.*, **87**, 2000, 4254.
44. J. C. Wittmann, P. Smith, *Nature*, **352**, 1991, 414.
45. O. Bunk, M. M. Nielsen, T. I. S<o/>lling, A. M. van de Craats, N. Stutzmann, *J. Am. Chem. Soc.*, **125**, 2003, 2252–2258.
46. K. R. Amundson, B. J. Sapjeta, A. J. Lovinger, Z. N. Bao, *Thin Solid Films*, **414**, 2002, 143.
47. D. Neher, *Macromol. Rapid Commun.*, **22**, 2001, 1365.
48. M. Grelf, D. D. C. Bradley, M. Inhasekarun, E. P. Woo, *Adv. Mater.*, **9**, 1997, 798.
49. M. Grell, W. Knoll, D. Lupo, A. Meisel, T. Miteva, D. Neher, H. G. Nothofer, U. Scherf, A. Yasuda, *Adv. Mater.*, **11**, 1999, 671.
50. H. Sirringhaus, R. J. Wilson, R. H. Friend, M. Inbasekaran, W. Wu, E. P. Woo, M. Grell, D. D. C. Bradley, *Appl. Phys. Lett.*, **77**, 2000, 406.
51. H. Kihara, Y. Ueda, A. Unno, T. Hirai, *Mol. Cryst. Liq. Cryst.*, **424**, 2004, 195.
52. O. Bunk, M. M. Nielsen, T. I. S<o/>lling, A. M. van de Craats, N. Stutzmann, *J. Am. Chem. Soc.*, **125** (8), 2003, 2252–2258.
53. A. M. van de Craats, N. Stutzmann, O. Bunk, M. M. Nielsen, M. Watson, K. Müllen, H. D. Chanzy, H. Sirringhaus, R. H. Friend, *Adv. Mater.*, **15**, 2003, 495.
54. S. Huttner, M. Sommer, A. Chiche, G. Krausch, U. Steiner, M. Thelakkat, *Soft Matter*, **5**, 2009, 4206–4211.
55. J. C. Conboy, E. J. C. Olson, D. M. Adams, J. Kerimo, A. Zaban, B. A. Gregg, P. F. Barbara, *J. Phys. Chem. B*, **102**, 1998, 4516.
56. B. A. Gregg, *J. Phys. Chem.*, **100**, 1996, 852.
57. V. Palermo, P. Samorì, *Angew. Chem. Int. Ed.*, **46**, 2007, 4428.
58. G. De Luca, W. Pisula, D. Credgington, E. Treossi, O. Fenwick, G. M. Lazzerini, R. Dabirian, E. Orgiu, A. Liscio, V. Palermo, K. Müllen, F. Cacialli, P. Samorì, *Adv. Funct. Mater.*, **21**, 2011, 1279–1295.
59. G. De Luca, A. Liscio, P. Maccagnani, F. Nolde, V. Palermo, K. Müllen, P. Samorì, *Adv. Funct. Mater.*, **17**, 2007, 3791.
60. H. Gilles, K. Fayçal, S. Peter, F. Denis, N. Claude, P. Xavier, G. Francis, *Adv. Mater.*, **8**, 1996, 242.
61. Z. Chen, Y. Zheng, H. Yan, A. Facchetti, *J. Am. Chem. Soc.*, **131**, 2008, 8.
62. J. H. Oh, H. W. Lee, S. Mannsfeld, R. M. Stoltenberg, E. Jung, Y. W. Jin, J. M. Kim, J. B. Yoo, Z. N. Bao, *Proc. Natl. Acad. Sci. USA*, **106**, 2009, 6065.
63. W. Su, Y. X. Zhang, C. T. Zhao, X. Y. Li, J. Z. Jiang, *Chem. Phys. Chem.*, **8**, 2007, 1857.
64. K. C. Dickey, J. E. Anthony, Y.-L. Loo, *Adv. Mater.*, **18**, 2006, 1721–1726.
65. I. Watanabe, K. Hong, M. F. Rubner, *J. Chem. Soc., Chem. Commun.*, 1989, 123–124.
66. I. Watanabe, K. Hong, M. F. Rubner, I. H. Loh, *Synth. Met.*, **28**, 1989, C473–C480.

67. A. Chyla, S. Kucharski, R. Janik, J. Sworakowski, M. Bienkowski, *Thin Solid Films*, **285**, 1996, 496–499.
68. M. Rikukawa, Y. Tabuchi, K. Ochiai, K. Sanui, N. Ogata, *Thin Solid Films*, **329**, 1998, 469–472.
69. J. Paloheimo, P. Kuivalainen, H. Stubb, E. Vuorimaa, P. Ylilahti, *Appl. Phys. Lett.*, **56**, 1990, 1157.
70. J. Paloheimo, H. Stubb, P. Ylilahti, P. Dyreklev, O. Inganas, *Thin Solid Films*, **210**, 1992, 283.
71. T. Bjornholm, D. R. Greve, N. Reitzel, T. Hassenkam, K. Kjaer, P. B. Howes, N. B. Larsen, J. Bogelund, M. Jayaraman, P. C. Ewbank, R. D. McCullough, *J. Am. Chem. Soc.*, **120**, 1998, 7643–7644.
72. N. Reitzel, D. R. Greve, K. Kjaer, P. B. Hows, M. Jayaraman, S. Savoy, R. D. McCullough, J. T. McDevitt, T. Bjornholm, *J. Am. Chem. Soc.*, **122**, 2000, 5788–5800.
73. H. Sirringhaus, P. J. Brown, R. H. Friend, M. M. Nielsen, K. Bechgaard, B. M. W. Langeveld-Voss, A. J. H. Spiering, R. A. J. Janssen, E. W. Meijer, P. Herwig, D. M. de Leeuw, *Nature*, **401**, 1999, 685–688.
74. M. J. Cook, J. McMurdo, D. A. Miles, R. H. Poynter, J. M. Simmons, S. D. Haslam, R. M. Richardson, K. Welford, *J. Mater. Chem.*, **4**, 1994, 1205.
75. O. Karthaus, H. Ringsdorf, V. V. Tsukruk, J. H. Wendorff, *Langmuir*, **8**, 1992, 2279.
76. J. Y. Josefowicz, N. C. Maliszewskyj, S. H. J. Idziak, P. A. Heiney, J. P. McCauley, A. B. Smith, *Science*, **260**, 1993, 323.
77. N. Reitzel, T. Hassenkam, K. Balashev, T. R. Jensen, P. B. Howes, K. Kjaer, A. Fechtenkötter, N. Tchebotareva, S. Ito, K. Müllen, T. Björnholm, *Chem. Eur. J.*, **7**, 2001, 4894.
78. J. E. Anthony, *Chem. Rev.*, **106**, 2006, 5028.
79. H. Ebata, T. Izawa, E. Miyazaki, K. Takimiya, M. Ikeda, H. Kuwabara, T. Yui, *J. Am. Chem. Soc.*, **129**, 2007, 15732.
80. P. Gao, D. Beckmann, H. N. Tsao, X. Feng, V. Enkelmann, M. Baumgarten, W. Pisula, K. Müllen, *Adv. Mater.*, **2**, 2009, 213.
81. H. N. Tsao, D. Cho, J. W. Andreasen, A. Rouhanipour, D. W. Breiby, W. Pisula, K. Müllen, *Adv. Mater.*, **21**, 2009, 209.
82. H. N. Tsao, K. Müllen, *Chem. Soc. Rev.*, **39**, 2010, 2372–2386.
83. H. N. Tsao, H. J. Räder, W. Pisula, A. Rouhanipour, K. Müllen, *Phys. Stat. Sol. (a)*, **205**, 2008, 421–429.
84. P. Miskiewicz, M. Mas-Torrent, J. Jung, S. Kotarba, I. Glowacki, E. Gomar-Nadal, D. B. Amabilino, J. Veciana, B. Krause, D. Carbone, C. Rovira, J. Ulanski, *Chem. Mater.*, **18**, 2006, 4724.
85. C. M. Duffy, J. W. Andreasen, D. W. Breiby, M. M. Nielsen, M. Ando, T. Minakata, H. Sirringhaus, *Chem. Mater.*, **20**, 2008, 7252.
86. W. Pisula, A. Menon, M. Stepputat, I. Lieberwirth, U. Kolb, A. Tracz, H. Sirringhaus, T. Pakula, K. Müllen, *Adv. Mater.*, **17**, 2005, 684.
87. A. Tracz, J. K. Jeszka, M. D. Watson, W. Pisula, K. Müllen T. Pakula, *J. Am. Chem. Soc.*, **125**, 2003, 1682.
88. A. Tracz, T. Makowski, S. Masirek, W. Pisula, Y. H. Geerts, *Nanotechnology*, **18**, 2007, 485303.
89. G. Xu, Z. Bao, J. T. Groves, *Langmuir*, **16**, 2000, 1834–1841.

3
Doping effects on charge transport in organic materials

K. LEO and M. HUMMERT,
Technische Universität Dresden, Germany

DOI: 10.1533/9780857098764.1.102

Abstract: In this chapter, we review recent basic work on molecular doping of organic semiconductors. We focus on doped layers prepared by co-evaporation in vacuum, which are utilised in highly efficient organic light-emitting diodes and organic solar cells. Various doping materials are compared, as well as their evaluation in devices. The fundamental processes during doping and the effort on layer properties are presented. Doping of organic semiconductors offers two striking advantages: The conductivity can be controlled and enhanced for orders of magnitude. Secondly, the Fermi energy levels are shifted towards the transport states, thus charge injection barriers are diminished.

Key words: organic semiconductors, doping, organic light-emitting diode (OLED), organic solar cells (OSC), conductivity.

3.1 Introduction

In this chapter, we review recent basic work on electrical doping of organic semiconductors, with the goal of improving the conductivity of the materials. We focus on doped layers achieved by co-evaporation in vacuum, where most of the recent work on doping has been done and which currently dominates commercial applications.

The article is organised as follows: In Section 3.1, the utilisation of organic semiconductors, in particular in emerging electronic consumer products such as organic light-emitting diodes (OLEDs) and organic solar cells (OSCs), is assessed. Differences from and similarities to inorganic semiconductors are discussed.

Section 3.2 introduces the co-evaporation technique and the fundamental processes during molecular doping. Different doping materials are presented, as well as their evaluation in doped layers. Doping of organic semiconductors offers two striking advantages: The conductivity can precisely be controlled and enhanced for orders of magnitude. Secondly, the Fermi energy levels are shifted towards the transport states, thus charge injection barriers are reduced.

The effect of increased conductivities and straightforward application in devices is presented in Section 3.3. Here, we discuss that doped layers in functional devices lead to superior properties. This involves charge transport as well as recombination contacts, which are necessary in tandem solar cells. Furthermore, transparent doped charge transport layers allow optimising optoelectronic devices with respect to thin film optics.

Finally, we comment on future trends and challenges in Section 3.4. Although doping of organic semiconductors is successfully applied in commercial products such as active matrix displays, further research at the border between materials chemistry and physics is still required.

3.1.1 Motivation

The most commonly used semiconducting material by some way, is silicon, offering high charge carrier mobility and device stability. These properties are responsible for use in the semiconductor industry, e.g. in consumer electronics.[1] Although silicon is one of the most prevalent elements in the Earth's crust, the production of highly pure elemental silicon is expensive.[2] Since silicon occurs exclusively in oxidised states the reduction requires large amounts of (electrical) energy, and energy-consuming purification steps have to follow. The total energy consumption during fabrication of solar cells, for instance, leads to a long energy payback time.[3] Environmental and social policies mean that, sustainable technologies increasingly become the focus. One approach to lower cost and energy consumption is to replace silicon by other semiconductors, such as organic materials. For instance, organic dyes usually have extremely high absorption coefficients, which allows for extremely thin photovoltaic devices.[4] Many fluorescent organic materials show a strong Stokes shift, leading to negligible absorption losses, which solves a key problem in inorganic LEDs. Furthermore, the huge diversity in organic chemistry allows straightforward synthesis and tailor-made organic materials. However, the most important reason for the replacement of brittle inorganic semiconductors is the flexibility and low weight of organic materials. Thus, the realisation of roll-to-roll production for large area and low cost modules is feasible.

Today, OLEDs and OSCs are commercially available. For these devices, thin films of organic semiconductors are required, which can be processed either by the deposition of small molecules from the vapour phase in vacuum or by wet deposition techniques.

Pure organic semiconducting materials have a very small charge carrier concentration. Thus their conductivity is low, which leads to unwanted ohmic resistance. The introduction of doped transport layers enables, the conductivity to be increased by orders of magnitude and the contact properties can be much improved.

In this chapter we review recent work on doping of organic semiconductors and focus on doped layers achieved by co-evaporation. This technique is used for the production of highly efficient OLED and OSC devices.

3.1.2 Comparison of organic and classical semiconductors

When comparing organic and classical inorganic semiconductors, one has to take into account that charge transport in organics cannot be described by the band model, but by a hopping mechanism.[5] Furthermore the correlation energies in organics (Frenkel exciton: 200–500 meV)[6] are much higher than for instance in GaAs (Wannier-Mott exciton: 4 meV)[7]. Due to the high exciton binding energy, the exciton is much smaller and usually localised on one organic molecule.

Organic semiconducting materials consist preferably of molecules, where atoms are fixed by discrete bonds within the molecule. Beside other elements, mostly sp^2 and/or sp hybridised carbon atoms are present. The remaining p orbitals form extended π-systems localised on the molecule, leading to two relevant frontier orbitals: the highest occupied molecular orbital (HOMO) and the lowest unoccupied molecular orbital (LUMO). Organic semiconductors have a high density of states of the HOMO/LUMO level and the coupling of the π-systems is weak. Thus, in most cases the semiconducting properties of organic materials are still remaining, even at high doping concentrations. In contrast, inorganic semiconductors show metallic behaviour at high doping levels. Different elements involved in the molecular structure determine properties like electron affinity ($E_A \approx -|E_{LUMO}|$), or in other words, whether the material is an electron acceptor or donor. However, electronic properties and packing performance of organic compounds are not entirely the sum of included elements, rather than the connectivity within the molecular structure. The molecular structure determines the material crystallinity that has a large influence on the layer morphology.

Molecular packing is a crucial factor for the conductivity of organic semiconducting layers. Due to a low intermolecular π–π interaction, charge carrier mobilities in organic layers are usually rather limited. On the other hand the mobility is strongly related to the thin film order and film morphology. Among other conditions (substrate effect, deposition rate, etc.), the substrate temperature during deposition is the key factor with respect to the phase crystallinity.[8] Single crystalline layers are often obtained at elevated temperatures, giving molecules the time to arrange in a well-ordered phase.[9] In contrast, amorphous layers are formed when the material is deposited on cooled substrates far below 0 °C. In the case of polycrystalline phases one has to consider the crystal grain size as well. This fact is known from organic field effect transistors (OFET) where the grain size is correlated with the mobility, which can be explained by a trapping mechanism and thermal release at the grains' boundaries.[10] Different pentacene phases obtained by the temperature method show a mobility decrease in the order of magnitudes from transition from single crystalline to polycrystalline films and finally amorphous layers.[11] However, since organic thin films are grown preferably at room temperature, organic semiconducting materials commonly appear in polycrystalline and amorphous layers. This fact gains significance with respect to the charge carrier mobility that is usually very small in such low ordered systems. Thus the doping strategy is the key for the realisation of efficient optoelectronic devices.

3.1.3 Doping of semiconductors

Generally, doping is an 'intended contamination' of semiconducting host materials with dopants. However, electrical doping is based on the same principle: mobile charge carriers are generated within the matrix, either electrons (e^-) or electron holes (h^+). The doping strategy leads to the fact that a given low charge carrier mobility (μ) is compensated by an increased charge carrier density (n), according to formula 3.1. The resulting conductivity (σ) is increased by order of magnitudes compared to the neat material.

$$\sigma(\mathrm{T}) = \mathrm{e} \cdot \mu(\mathrm{T}) \cdot n(\mathrm{T}) \qquad [3.1]$$

In the case of silicon (group 14 element), e.g. p-type doping is performed by the addition of a group 13 element atom (boron for instance), which replaces one silicon lattice atom. For charge generation, it captures an electron from the valence band and thus releases a mobile hole. For very high dopant concentrations, the dopant states interact and form a band-like structure, leading to metallic conductivity.

In organic molecules, the basic doping mechanism is the same. For p-type doping, one has to introduce a molecule which accepts an electron from the organic semiconductor and leaves a hole in the matrix. Owing to the weaker coupling of the electronic states in organic materials, the semiconducting properties of organic materials remain even at high doping concentrations, whereas inorganic semiconductors show metallic behaviour.

3.2 Basics of doping of organic semiconductors

Vacuum deposition offers a very reliable method for formation of neat and mixed layers. Mixed layers can either consist of dopant and matrix (prefix p- for p-doped, n- for n-doped) or they contain two intrinsic materials (prefix i-) that are un-doped, also called bulk. Intrinsic mixed layers and doped mixed layers are obtained by the same co-evaporation procedure, where two different materials are simultaneously evaporated from ceramic crucibles in (ultra) high vacuum chambers (Fig. 3.1). The evaporation rate of each material is controlled by the crucible's temperature and monitored by an oscillating crystal. Thus, the mixing ratio of two compounds can be adjusted as well. The structures of the compounds mentioned in this chapter can be found in the corresponding figures, namely molecular materials used for p-type doping (Fig. 3.2), n-type doping (Fig. 3.3), various matrix materials (Fig. 3.4), and hole transport materials (Fig. 3.5). Abbreviations, long

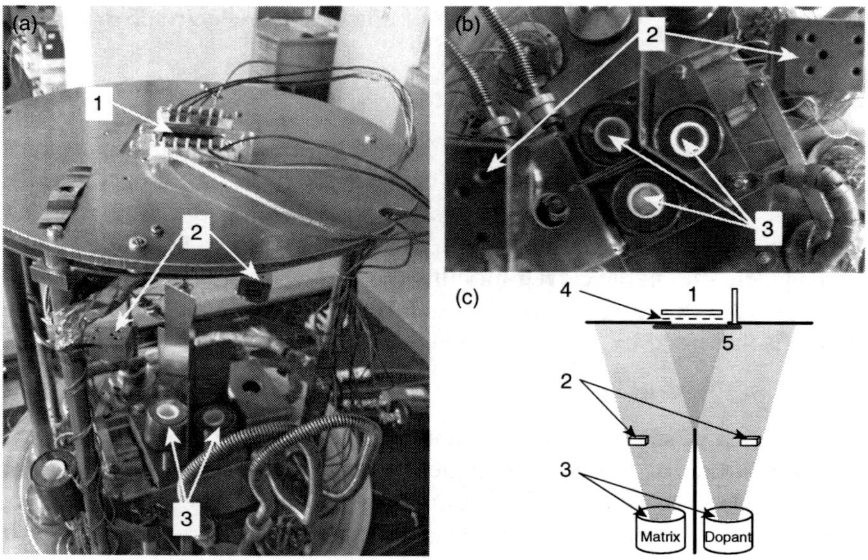

3.1 Photograph of the co-evaporation setup used in a simple one-chamber vacuum system (a, side view; b, view of the vaporiser from the substrates' position) and systematic co-evaporation scheme (c). 1 Substrate, 2 thickness monitors, 3 crucibles, 4 shadow mask, 5 shutter.

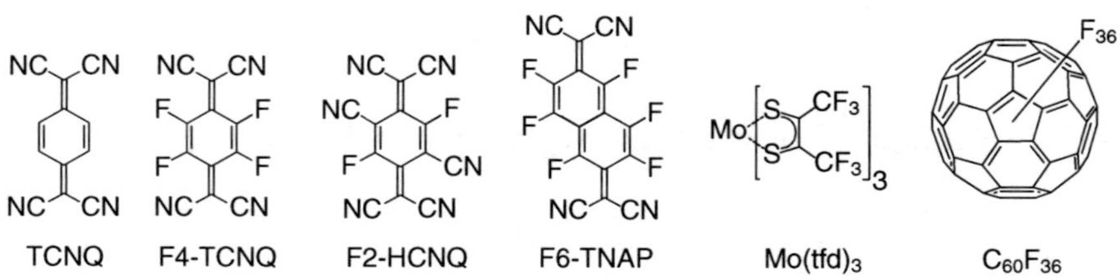

3.2 Structures of various p-type dopants. $C_{60}F_{36}$ consists of an isomer mixture; a general structure without any substitution pattern is depicted here. The molecular structure, obtained from x-ray diffraction, of a $C_{60}F_{36}$ isomers is depicted in Fig. 3.9.[44]

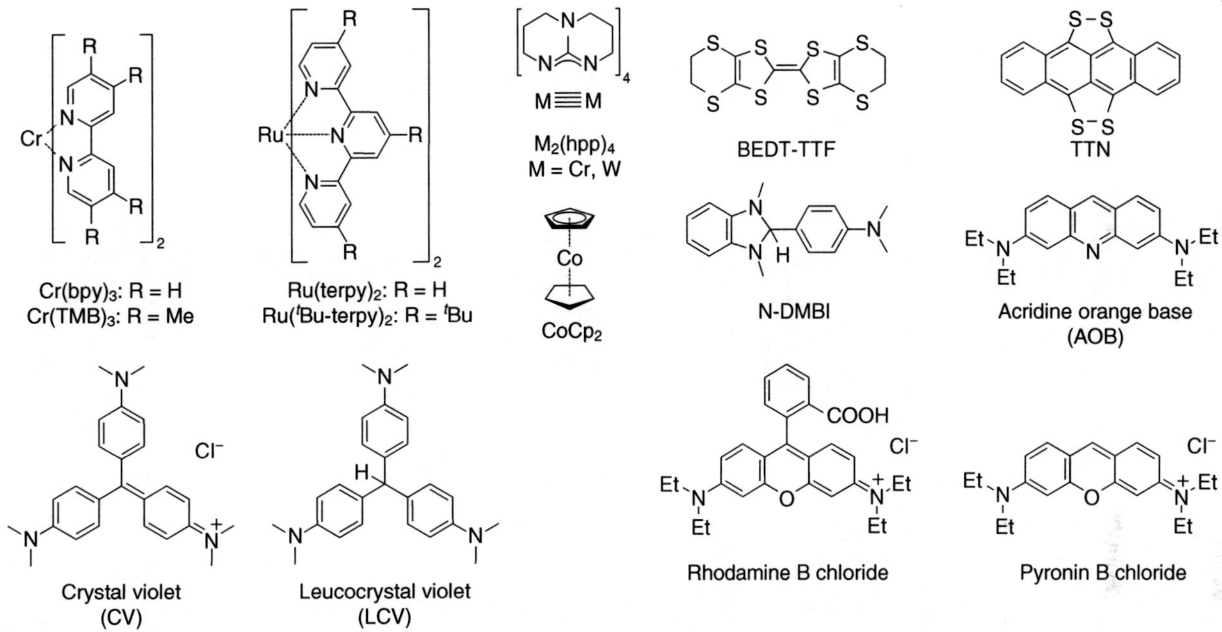

3.3 Structures of various molecular n-type dopants and precursors.

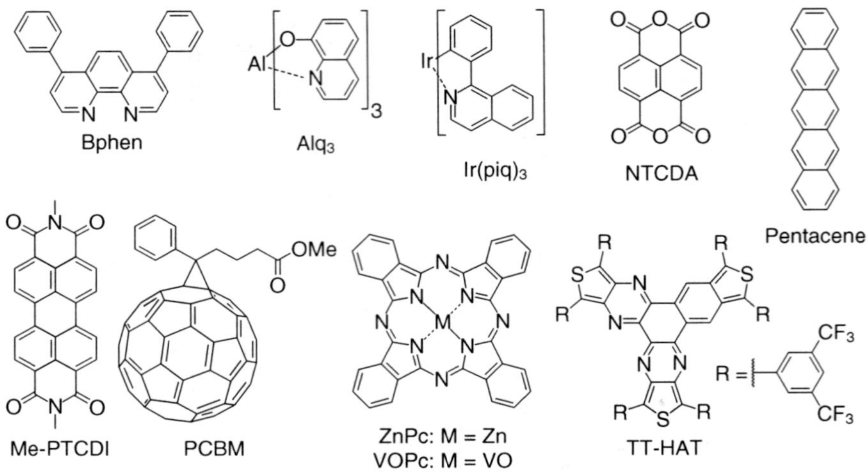

3.4 Structures of various matrix materials used as host.

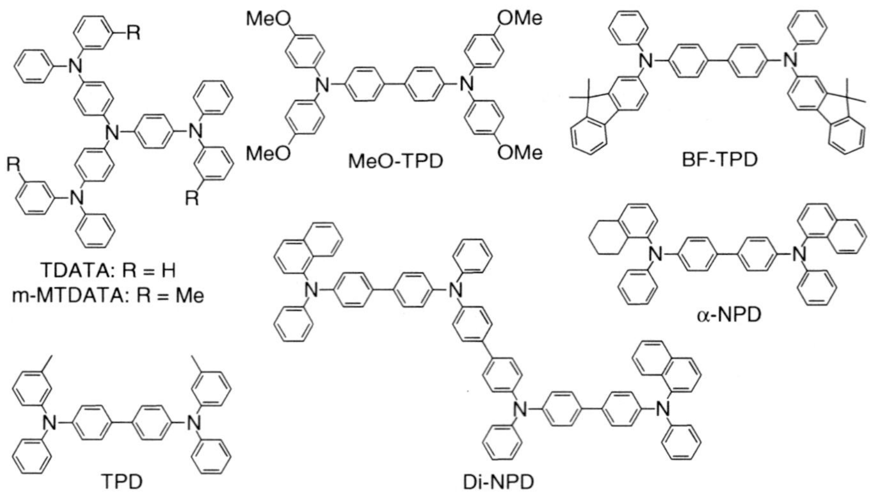

3.5 Structures of various hole transport materials, based on the triarylamine moiety.

names, and the corresponding CAS number of compounds mentioned in this review can be found in Section 3.7.

3.2.1 Doping mechanism

The doping mechanism becomes obvious by considering the electronic properties of the materials, namely the HOMO (E_{HOMO}) and LUMO

(E_{LUMO}) energy levels of host/matrix and dopant materials. A certain HOMO and LUMO energy level alignment necessary for doping is depicted in Fig. 3.6. The electron acceptor strength of a molecule working as p-dopant reflects a rather low LUMO energy level. On the other hand, an electron-rich molecule is characterised by a very high lying HOMO energy level, being an appropriate n-dopant. The corresponding matrix transport levels (electrons in LUMO, holes in HOMO) participate in reduction/oxidation processes and injected charge carriers can percolate inside the matrix film. Doping occurs if the energetics of the participating molecules is given in a certain manner: the LUMO energy level of the p-dopant must be aligned below or at least close to the HOMO energy level of the matrix molecule. Electron-transfer from the matrix molecule onto the p-dopant molecule (Fig. 3.6, left side (a)) results in a mobile hole. Analogously for n-type doping, the LUMO energy level of the matrix materials must be aligned below or close to the HOMO energy level of the n-dopant (Fig. 3.6, right side (b)), yielding mobile electrons within the host material.

For highly disordered amorphous phases, it is assumed that the charge transfer from the dopant to host material arises in donor–acceptor charge transfer complexes (CTC) analogously to well-ordered systems. CTC states of F4-TCNQ (Fig. 3.2) have already been revealed on single crystalline metal surfaces by spectroscopy,[12] microscopy and spectroscopic imaging at low temperatures.[13]

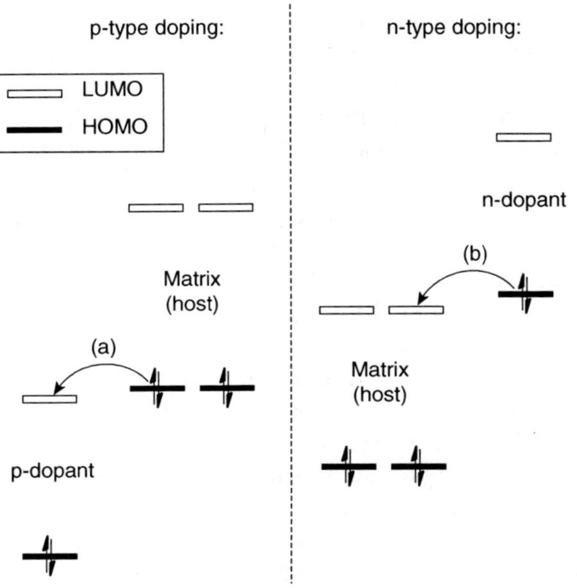

3.6 Mechanism of molecular doping. Efficient electron transfer requires suitable energy level alignment of p- and n-type dopants.

3.2.2 Doping materials and material criteria

Although organic materials can be easily p-doped by strongly oxidising inorganic reagents such as elemental iodine and bromine,[14] transition metal halogens and oxides (CuI,[15] $FeCl_3$, $SbCl_5$,[16] ReO_3,[17] MoO_x[18] or WO_3[19]) these techniques have only minor relevance in applications. These materials are either highly toxic, require high evaporation temperatures, or generate mobile ions that diffuse or drift at high electric field within the layer. Unfavourable diffusion effects get even more relevant in operating devices, where elevated temperatures are present. This leads to significantly lower device performance and lifetime.

It is amazing that the doping technology has been known for many years, but reports about structures and application of molecular p-dopants are still rare in the literature (Fig. 3.2). With respect to structure–property relations, the reported structures usually have one feature in common: they contain the most electron-deficient element in the periodic table, fluorine.[20] These materials are extremely strong electron accepting molecules and exhibit p-type doping in combination with corresponding host materials. The most important molecular organic p-dopants and whether they hold the balance between lifetime and performance will be discussed in Section 3.2.3.

The same is true for n-doping with strongly reductive metallic elements, which can easily diffuse in the fabricated thin films. Two alkali metal doping approaches are established, using the elemental metal or in form of its salts. Either the doping material is co-evaporated together with an organic material from a dispenser source forming a mixed layer or it is evaporated as electron-injecting layer additionally on the matrix material. In both cases diffusion processes are present, enhanced by the fact that deposition of the metal contact releases thermal energy. Thus the device performance is significantly reduced.[21]

A commonly applied doping method is the Alq_3/LiF/Al-system, where the different layers are deposited one after another. Here, it is worth noting that the deposited LiF is not primarily responsible for the doping effect. Hung and co-workers concluded from photoelectron spectroscopy (PES), in particular X-ray photoelectron spectroscopy (XPS), that LiF remains undissociated after deposition on Alq_3.[22] The doping effect occurs during subsequent deposition of the aluminium cathode in a secondary interfacial reaction within the Alq_3/LiF/Al system.[23]

A detailed study about evaporated surface layers of metallic lithium was published by Parthasarathy *et al.*[24] They found that a layer of Li below the cathode subsequently dope the electron transporting organic thin films by diffusion into the organic. Lithium is rather mobile in organic layers because of its low van der Waals radii of 181 pm.[25] Diffusion processes are reduced if doping is carried out with larger (alkali) metal atoms, such as caesium (van der Waals radius of 343 pm).[26] Indeed, caesium is widely used, for instance by co-evaporation of elemental caesium, caesium salts, or alloys. Cs doped devices are less sensitive to temperature, but migration cannot be eliminated entirely.

Bipolar devices like p-i-n and p-n junctions require stabile molecular dopants of a sufficient molecular size, to suppress the migration through the layer. Molecular dopants are suitable to overcome undesirable dopant diffusion. Another general aspect is the dopant volatility: although the dopant material must have sufficient vapour pressure for the co-evaporation process, a too high volatility can lead to problems during evaporation. Chamber contamination and pollution of subsequent layers can be the result. If dopant residuals are present in subsequently deposited layers, excitons can be quenched, leading to reduced device efficiency.[27] Volatile complex compounds of metals in low oxidation states, having an ionisation energy lower even than that of the caesium, are suitable candidates for this application.[28]

Beside molecular n-dopants a third class of compounds appeared. The initial substance is merely a precursor without having any doping effect. The precursor can be applied as usual by co-deposition, but needs a cofactor to be transformed into the active species. This happens *in situ* either by illumination or heating during co-evaporation. The precursors have E_{HOMO} values below the redox potential of oxygen (–4.3 eV). Since these are inert towards oxygen in air, handling of the precursor compounds is more convenient. For example, the triphenylmethane based leucocrystal violet is irreversibly transformed by electron release into its cationic form, confirmed by mass spectrometry and optical spectroscopy.[29] A possible precursor is leucocrystal violet (LCV), that requires irradiation with white light. Thus photo-induced charge transfer provides n-doping effect *in situ*. Alternately, crystal violet (CV) is converted by the evaporation process into its active form.[30] On one hand, these dopants do not tend to migrate through the layer, on the other hand they suffer from their limitation to relatively low-lying hosts. Electron transport materials ideally have a large HOMO–LUMO gap to avoid parasitic absorption. This fact leads to rather high LUMO energy levels which also depend on the application. Electron transport materials in OSC have transport levels of approximately –4 eV, whereas in OLED LUMO levels of –3 eV are required. Stronger donor molecules are needed to achieve higher conductivities and larger Fermi level shifts. Furthermore the doping efficiency is partially depending on additional parameters, such as the influence of the corresponding cofactors.

Discovering molecules with a proper HOMO level is still a challenging task. Owing to these limitations, only a couple of molecular structures are proven in literature to be useful as n-dopants in organic semiconducting devices. However, the advantages and drawbacks strongly influence the properties of doped layers. In addition to other factors, such as host material influences, the behaviour of the most important dopants will be discussed in the next subsection.

3.2.3 Properties of doped layers

Doping of organic semiconductors offers two striking advantages: by doping, the conductivity can be controlled and enhanced for orders of magnitude. Secondly, the Fermi energy level of a conducting electrode can be shifted towards the transport level of the corresponding layer, thus injection barriers are lowered. The first advantage of increased conductivities will be presented in the following.

Phthalocyanines supported by different central metal ions are appropriate models for studies of optoelectronic effects. This compound class is known as stable semiconductor and can be efficiently doped. In particular zinc phthalocyanine (ZnPc) is one of the best investigated materials in organic electronics. In one of our studies, ZnPc was doped with F4-TCNQ and investigated by different techniques.[31] The comprehensive conclusions of this model system are summarised here and are generally valid for doped organic semiconductors. A simple investigation is the lateral film conductivity, measured *in situ* in coplanar contact geometry while the (doped) layer is growing. The conductivities of doped layers cover a range of three orders of magnitude depending on the mixing ratio of F4-TCNQ, as shown in Fig. 3.7.[32] The highest conductivity is observed at 10 mol% doping ratio with 2×10^{-2} S/cm. Even the lowest doping concentration of 0.2% yields a conductivity of 3×10^{-5} S/cm that is still significantly higher than in an undoped layer (10^{-10} S/cm). Comparing crystalline and amorphous ZnPc phases, it has to be pointed out that the amorphous phase comprise lower intermolecular interaction and order.[33] This is indicated by the peak separation in the absorption spectrum of thin layers. The peak separation is smaller in amorphous phases than in crystalline layers (in Fig. 3.8 (a)).[34] This fact correlates with the mobility model of different phases as already discussed before. The comparison of doped amorphous and crystalline ZnPc films clearly reveals that crystalline layers generally provide higher conductivities. Despite the fact that charge transport in amorphous layers is hindered, other host material properties have to be considered with respect to application in devices: the strong parasitic absorbance makes ZnPc unsuitable for charge transport applications in optoelectronic devices.

Therefore we investigated several amorphous hole transport materials in combination with F4-TCNQ. It turned out that the doping concept works with a large number of hole transport materials as well. Some alternative amorphous charge transport materials are presented for comparison in Table 3.1, including a comparison with the TCNQ dopant. Conductivities

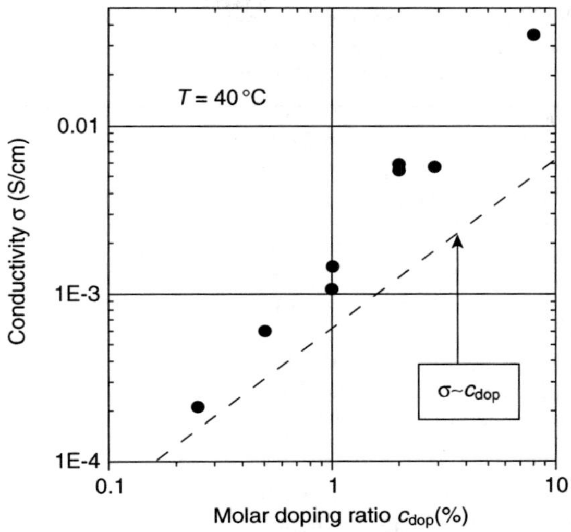

3.7 Conductivity of F4-TCNQ doped ZnPc as function of the doping concentration. The dashed line indicates linear behaviour of the conductivity versus the doping ratio. Reprinted from: Pfeiffer M, Leo K, Zhou X, Huang J S, Hofmann M, Werner A, Blochwitz-Nimoth J, 'Doped organic semiconductors: Physics and application in light emitting diodes', *Org Electron*, 2003, 4, 89–103. Copyright 2003, with permission from Elsevier.

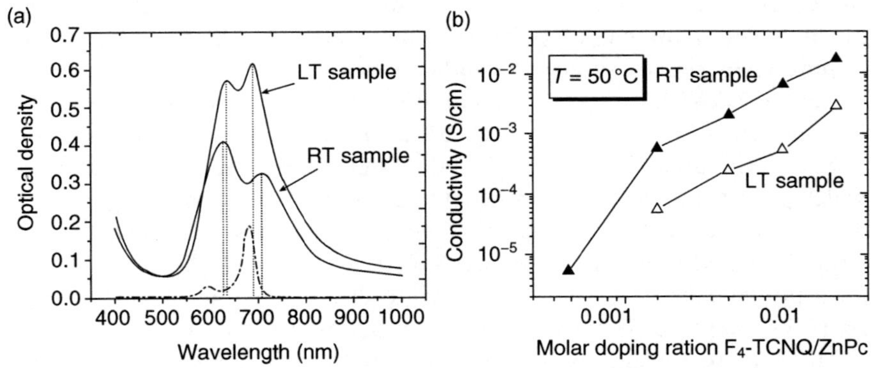

3.8 (a) Absorption spectra of a 50 nm thick ZnPc film prepared at room temperature (RT, 25 °C) and at low substrate temperature (LT, −150 °C). While the former preparation results in a crystalline film, an amorphous film is obtained for the latter. The lowest line (dash dot) shows the absorption spectrum of ZnPc in solution. (b) Dependence of the conductivity of the molar doping ratio for the RT and LT samples. Reprinted figure with permission from: Maennig B, Pfeiffer M, Nollau A, Zhou X, Leo K, Simon P, 'Controlled p-type doping of polycrystalline and amorphous organic layers: Self-consistent description of conductivity and field-effect mobility by a microscopic percolation model', *Phys Rev B*, 2001, 64, 195208.

Table 3.1 Comparison of the corresponding ionisation potential (IP), conductivities (σ), charge transfer degree (Z) in various combinations of hole transport materials and p-dopants. The doping ratio is always 2 mol%

Matrix p-dopant	ZnPc		m-MTDATA	TPD	MeO-TPD
	F4-TCNQ	TCNQ	F4-TCNQ		
IP	5.1	5.1	5.1	5.4	5.1
Z	1	0.2	1	0.64	0.74
σ (S/cm)	1×10^{-3}	1×10^{-6}	3×10^{-7}	1×10^{-7}	1×10^{-5}

Reprinted with permission from: Walzer K, Maennig B, Pfeiffer M, Leo K, 2007, 'Highly efficient organic devices based on electrically doped transport layers', *Chem Rev*, 107, 1233–1271. Copyright 2007, American Chemical Society.

ranging from 1×10^{-7} to 1×10^{-3} S/cm were obtained with the same doping ratio, but using diverse matrix materials. The differing conductivities are discussed in the following, considering energy levels and the charge transfer degree (Z).

Application in optoelectronic devices require wide gap materials with large energy difference between HOMO and LUMO ($E_{HOMO-LUMO}$), since light absorption in transport layers is undesirable in OLED and OSC. Given that these materials have to form amorphous layers, it is not surprising that the charge carrier mobility is much lower. Thus the conductivity is also much lower than that of polycrystalline materials. On the other hand, doping gives rise to sufficient conductivities in thin layers. For instance, doping of TDATA with F4-TCNQ in different molar concentrations results in conductivities between 4×10^{-7} S/cm and 6×10^{-6} S/cm (doping ratio 1–7%).[35] Conductivities of un-doped TDATA-films are below 10^{-9} S/cm, which proves that the doping concept works in amorphous materials as well. The change of TDATA absorption upon doping has also been demonstrated. Doped thin films on quartz substrates show additional very weak absorption peaks, that can be attributed to TDATA cations or charge transfer (CT) complexes, but overall the film remains transparent above 400 nm.

The doping efficiency is crucial, especially for transparent hole transport materials with low ionisation potential (IP = 5.0–5.7 eV; IP $\approx -|E_{HOMO}|$). The charge transfer degree can roughly be deduced from IR spectra, where the ratio of ionised to neutral dopant molecules can be observed on the base of C–N stretching oscillations. This absorption is quite sensitive to electronic changes of the molecule; hence both species have characteristic absorptions peaks at different stimulation energies. A completely disappearing absorption of the neutral molecule indicates full charge transfer with a relative doping yield of $Z = 1$, whereas small Z numbers indicate inhibited charge transfer, which can originate from less suitable orbital levels. Therefore two absorption peaks in the IR are present; their relative intensities are used for the estimation of Z. Table 3.1 gives an overview of the ionisation potential, the doping efficiency (Z) and conductivity (σ) for some material/p-dopant combinations. The role of the energy level of the participating frontier orbital becomes obvious in Table 3.1. Co-evaporation of ZnPc with 2% TCNQ ($Z = 0.2$, $\sigma = 1 \times 10^{-6}$ S/cm) results in much less efficient doping than using F4-TCNQ ($Z = 1$) in the same concentration.[36] Although both samples are polycrystalline, a three orders of magnitude lower conductivity (1×10^{-6} S/cm) is observed. The fluorine substitution in F4-TCNQ lowers the LUMO level for 0.5 eV and makes this dopant suitable for other transparent hole transport materials having higher ionisation potentials like TPD.[37] Owing to the higher ionisation potential of TPD (5.4 eV), only a moderate charge transfer ($Z = 0.64$) and lower conductivity (1×10^{-7} S/cm) is demonstrated than in the case of F4-TCNQ.[38] Slightly higher conductivity (3×10^{-7} S/cm) can be found for m-MTDATA.[39] The methoxy substituted hole transport material MeO-TPD shows structural analogy to TPD, but bears electron donating groups that cause a lower ionisation potential of 5.1 eV. In amorphous films of MeO-TPD doped with 2% F4-TCNQ high conductivities of 1×10^{-5} S/cm are found, which makes this material attractive for device applications.[40]

Although F4-TCNQ is verified to be a powerful p-type dopant, it has some drawbacks. Because of diffusion, the conductivity of F4-TCNQ doped layers is not sufficiently stable especially at elevated temperatures. Therefore F4-TCNQ doped layers exhibit reduced device lifetime.[41] The second drawback is related to the evaporation behaviour. The substitution of two fluorine atoms in F4-TCNQ by even more electron-withdrawing cyano groups (C≡N) results in the related structure F2-HCNQ.[42] The LUMO energy level is further lowered for 0.26 eV upon this modification. Comparing both dopants, the driving voltages are lower (4.9 V at 10 mA/cm^2) than the corresponding devices incorporating F4-TCNQ (5.0 V at 10 mA/cm^2) at given current densities. Overall the alternative dopant leads to higher power efficiency values. With respect to thermal stability of doped films, temperature-dependent conductivity measurements are performed. F2-HCNQ doped films keep on conducting up to 85 °C, whereas the conductivity of the F4-TCNQ doped system dramatically drops above 65 °C. Unfortunately no information on the device lifetime is given; this may not differ from F4-TCNQ due to only marginal changes of the molecular shape.

A further highly fluorinated dopant based on the 2,6-naphthoquinone motive appeared recently in the literature, named F6-TNAP.[43] Beside the synthesis, a detailed characterisation of doped thin films is reported. When applying doping with F6-TNAP in OLED devices, the current voltage characteristics (I/V) show reduced driving voltages. Within our search for other dopant materials, we investigated doping properties of the fullerene derivative $C_{60}F_{36}$.[44] It has been demonstrated that this material is well suited for p-type doping, showing OLED and OSC efficiencies comparable to devices with the commonly used F4-TCNQ. The advantages of lower volatility and therefore avoiding contamination of subsequently deposited layers are proven by UV photoelectron spectroscopy (UPS) and XPS (Fig. 3.9). Only if F4-TCNQ is present in the chamber, XPS measurements

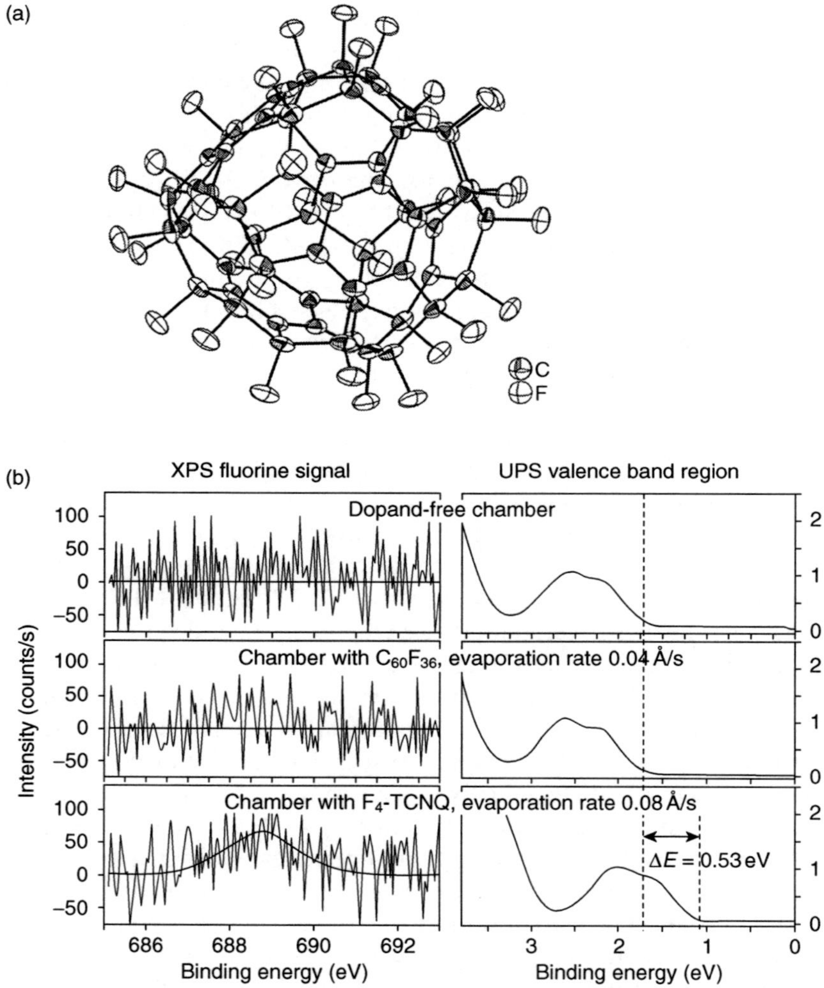

3.9 (a) Molecular structure of a $C_{60}F_{36}$ isomer determined from single-crystal X-ray diffraction data.[44] (b) UPS/XPS investigations of intrinsic MeO-TPD layers, made in a dopant-free chamber (top), or in a chamber equipped with either $C_{60}F_{36}$ (middle), or F4-TCNQ in unheated crucibles (bottom).

verify fluorine traces by a strong HOMO level shift (0.53 eV). Owing to the larger size of the fluorinated fullerene, dopant diffusion within the matrix is minimised.[45] This leads to higher thermal stability even at elevated temperatures, which is beneficial for device stability during operation. In combination with BF-DPB as the host material, doping with $C_{60}F_{36}$ is remarkably stable up to the glass transition temperature of the matrix material (160 °C).

The problem of dopant migration affects n-type doping as well. Metal-doped organic bulk layers as electron-injecting layers were investigated by the group of Kido.[21] Highly reactive metals (Li, Sr and Sm) were applied in the transport interlayer between the electrode and the emitter layer. They also point out problems of doping with Li: owing to dopant migration from the charge transport layer (Alq_3) into the emission layer, the luminescence is quenched.

First n-type doping with molecular dopant was achieved with NTCDA as host doped with BEDT-TTF.[46] The conductivity was increased by two orders of magnitude and a shift of the Fermi level of 0.2 eV towards the transport level was realised. In UPS studies doping of two kinds of semiconductors was investigated. While the doping effect of TTN on F_{16}ZnPc is revealed, mixed layers of Alq_3 and TTN showed no significant doping effect.[47] Considering the energy levels, these results indicate that the energy parameters match the TTN/F_{16}ZnPc couple, whereas the HOMO level of TTN is too low for efficient n-type doping of the Alq_3 matrix.

I–V measurements, UPS, XPS and inverse photoemission spectroscopy (IPES) were used to investigate the doping effect of bis(η^5-cyclopentadienyl) cobalt(II) ($CoCp_2$) in a TT-HAT matrix.[48] The matrix electron affinity ($E_A \cong -|E_{LUMO}|$) was determined to be 4.6 eV, which has a good fit with the HOMO energy level of $CoCp_2$ (–4.1 eV). In a two single-layer device structure matrix doping affected the current density to rise for three orders of magnitude. UPS data demonstrate a 0.56 eV shift of the Fermi level towards the matrix transport level. Electron-rich complexes are promising volatile n-type dopants, as demonstrated by Elliott and coworkers.[49] To deduce the HOMO energy level, cyclic voltammetry (CV) measurements were carried out on the complex compounds $Cr(bpy)_3$ (–3.14 eV) and $Cr(TMB)_3$ (–2.85 eV). Additionally, UPS and XPS were applied for the ruthenium compound $Ru(terpy)_2$. For the latter complex electrochemical data indicate a HOMO level of –3.10 eV, while the values from XPS and UPS are slightly higher, but in good agreement with each other (3.32 and 3.38 eV, respectively). Although $Ru(terpy)_2$ dopes narrow gap matrix materials having low LUMO energy levels, this material is inappropriate for doping of wide gap electron transport layers (–3 eV). Efficient n-type doping was also observed applying volatile complexes of the general formula $M_2(hpp)_4$ (M = Cr or W). In order to compare both dopants, conductivity and Seebeck (thermovoltage) measurements on doped C_{60} were performed.[50]

When applying LCV as doping agent in C_{60} matrices both activation procedures lead to conductivities around 1.3×10^{-2} S/cm and 8×10^{-3} S/cm, respectively.[27] In a further study, the cationic dye pyronin B chloride was utilised as dopant in a NTCDA matrix.[51] Analogously to earlier investigations of TTN in NTCDA conductivities of up to 1.9×10^{-4} S/cm were measured, two orders of magnitude higher than achieved by doping with TTN. The active doping species was found to be leuco pyronin B, generated during sublimation of pyronin B chloride.

Very recently the Bao-group presented N-DMBI as precursor for effective n-type doping.[52] Similar to the later compounds, hydrogen- and/or electron-transfer reaction mechanisms are assumed for effective doping of the well-known n-channel semiconductor PCBM in solution processed organic photovoltaic devices. In air stable n-type OFET the conductivity was significantly increased.

The potential of *in situ* generated dopants was evidenced by our group combining conductivity, field effect and Seebeck measurements with acridine orange base (AOB) as dopant.[53] The highest conductivity was observed after dopant activation by illumination during the film growth and thermal annealing of the sample. Doping in a molar ratio of 6%, the thin film conductivity of C_{60} was enhanced by more than six orders of magnitude compared to the undoped C_{60}. The irreversible electron transfer from the AOB dopant to the C_{60} matrix in codeposited films is demonstrated by near infrared (NIR) and Fourier transform infrared (FT-IR) spectroscopy. When exciting AOB by illumination, an AOB radical cation is formed and the electron is transferred to C_{60}.

3.2.4 Interfaces to doped semiconducting organic thin films

Substrate interfaces and surface effects are crucial for thin film conductivities. Ohmic contacts are generated if the work function of conductive electrode material aligns with the host's LUMO (electron transfer) or HOMO level (hole transfer), respectively. The work functions of most intrinsic organic semiconductors strongly depend on the substrate materials[54] as well as surface pre-treatment.[55] Thus the choice of substrates and materials is somewhat constrained. Here, the second advantage of doped layers is relevant for interface phenomena, for instance at an electrode surface. The UPS/XPS technique is a very useful tool to investigate such interface effects at metal/organic interfaces. An organic layer is deposited on the substrate in several steps and the layer is characterised by UPS/XPS after each growth step. The profile of the energetic structure inside a complete functional device stack can be determined in this way, as shown for an OLED.[56] The first study about metal/organic interface properties with doped layers was published in 2001, comparing intrinsic and doped ZnPc on ITO (indium tin oxide) and gold.[57] The most important observations for the differences of doped/undoped ZnPc on ITO are presented in the following. As shown in Fig. 3.10, the ionisation potential (IP) of the matrix material is not affected upon doping and remains unaltered at 5.5 eV. No accumulation of dopant molecules at the surface is observed since the interface dipole is of the same sign with similar value. Most important, the Fermi level (E_F) is shifted for 0.6 eV toward the transport states in the doped ZnPc layer (right) compared to the neat ZnPc layer (left). This is in accordance to the classical model known from inorganic semiconductors, where a thickness dependent energy alignment is found when a semiconductor is contacted to conducting substrates.[7]

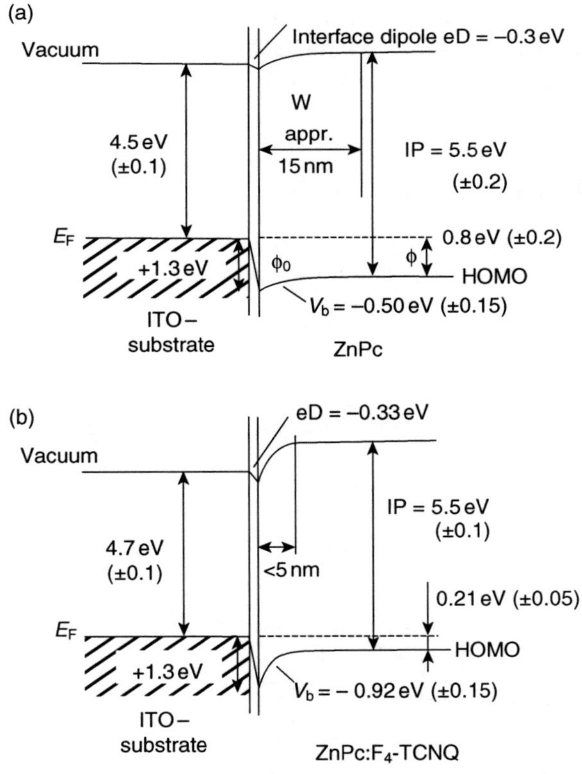

3.10 Energy level profile of the contact between ITO/i-ZnPc (a) and ITO/p-ZnPc (b) obtained from UPS and XPS measurements. Reprinted from: Blochwitz J, Fritz T, Pfeiffer M, Leo K, Alloway D M, Lee P A, Armstrong N R, 'Interface electronic structure of organic semiconductors with controlled doping levels', *Org Electron*, 2001, 2, 97–104. Copyright 2001, with permission from Elsevier.

A further impact of doping is a much stronger energy level bending in the space close to the interface. Additionally the space charge layer is significantly decreased from >15 nm for intrinsic ZnPc to <5 nm by doping of ZnPc. The space charge layer is now very thin so that charge carriers can easily tunnel through. The existence of a contact barrier is strongly reflected in the comparison of the device performance (Fig. 3.11). In a detailed study the energy level alignment using different substrates was investigated, while varying the dopant concentration. In this investigation, PES was also employed to gain thickness resolved information about interfaces to F4-TCNQ doped MeO-TPD layers.[40] The results of the energy alignment are summarised in Table 3.2. Independent of the substrate, even moderate doping ratios yield a charge carrier density high enough for energetic alignment. This means interface effects as well as the substrate work functions have less relevance, since the Fermi energy can be adjusted by the dopant concentration. Thus, the choice of substrates and materials is no longer constrained in order to get ohmic contacts. This fact has also relevance in the case of organic/organic junctions, which will be discussed in the next section.

3.3 Doped organic p-i-n devices

Our approach is to employ doped layers (p/n) for charge transport and undoped photoactive layers (i). The intrinsic active layers are sandwiched

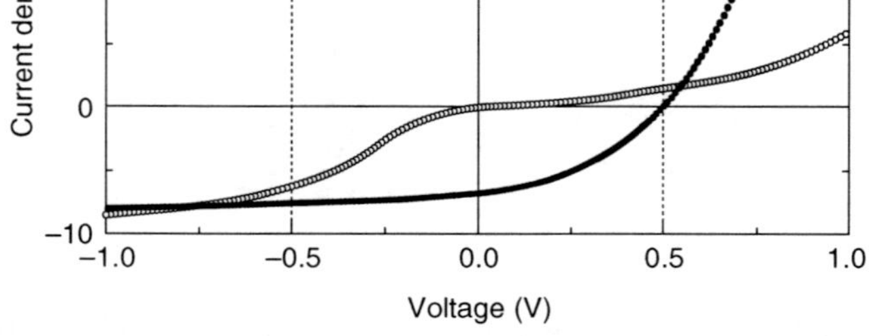

3.11 Comparison of *I*/*V* characteristics of ITO/p-ZnPc and ITO/i-ZnPc devices. Reprinted with permission from: Walzer K, Maennig B, Pfeiffer M, Leo K, 2007, 'Highly efficient organic devices based on electrically doped transport layers', *Chem Rev*, 107, 1233–1271. Copyright 2007, American Chemical Society.

Table 3.2 Experimental results from the measurement of the space charge layer thickness (*W*) and different energy level parameter depending on the doping concentration

Dopant conc. (mol%)	*W* (nm)	Φ_0 (eV)	Φ (eV)	eV_b (eV)
3	8 ± 1	1.5	0.84	0.66
6	5 ± 0.5	1.36	0.47	0.89
17	2.8 ± 0.3	1.63	0.38	1.25
27	0.6 ± 0.2	1.75	0.35	1.40

Reprinted with permission from: Olthof S, Tress W, Meerheim R, Lüssem B, Leo K, 'Photoelectron spectroscopy study of systematically varied doping concentrations in an organic semiconductor layer using a molecular p-dopant', *J Appl Phys*, 2009, 106, 103711. Copyright 2009, American Institute of Physics.

between wide gap p-doped hole and n-doped electron transport layers. This arrangement results in the p-i-n stack structure, according to Fig. 3.12. A general operation principle is shown there schematically as well. The use of transparent doped layers offers several advantages. The whole device can be made thicker and surface irregularities are levelled, reducing shortcuts (which are often a problem in organic devices). Additionally the wide gap layers can act as exciton blocking layers. The most striking issue is the optimisation with respect to thin film optics. Thickness variation of the transport layers leads to controlled placement of the photoactive layer in a position between the contacts, where constructive interference conditions are fulfilled. Merging all these aspects leads to significantly improved device performance. The optimised field distribution can reduce parasitic absorption and helps to harvest more photons in the light-absorbing layer of solar cells, or more photons are coupled out in case of OLED, respectively.

3.3.1 Organic p-n homojunctions

In the field of semiconducting devices, the p-n homojunction represents the archetypal device, which is also broadly used in inorganic silicon-based solar cells. In this context it is interesting to determine whether organic p-n junctions can also be described by the Shockley model. Such a device uses the same matrix material in both doped regimes and behaves as typical diodes. Though the device architecture seems to be quite simple, the realisation of a stable and reproducible ambipolar doped organic homojunction was challenging for a long time. In particular the lack of suitable dopants with proper energy levels in combination with the matrix material was a hindrance, accompanied by the problem of dopant diffusion. In view of the energy levels, the matrix E_{LUMO} must be aligned below the n-dopant's

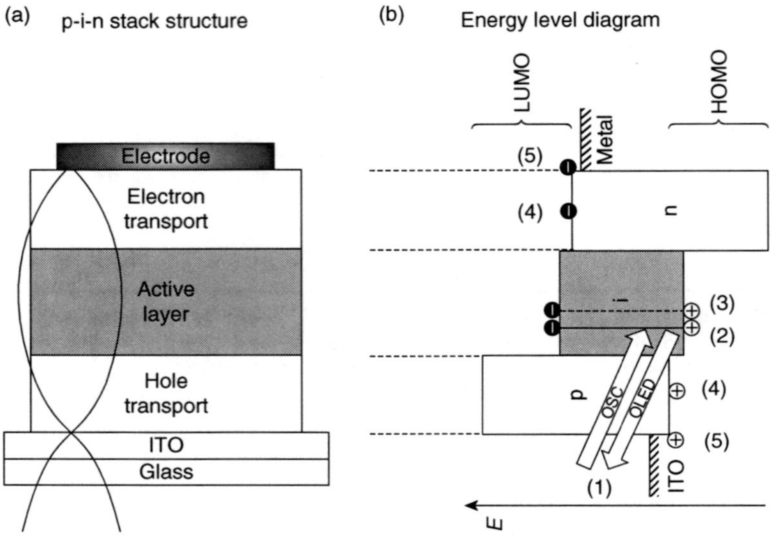

3.12 The figure shows the typical p-i-n structure (a), including an adumbrated standing wave. (b) The corresponding (simplified) energy level diagram is shown. (Note: the scheme gives an overview of the individual energy levels of the materials with respect to each other.) The operating principle of OSC and OLED including the corresponding single steps is shown therein. OSC: 1 light absorption; 2 exciton generation; 3 exciton separation and charge generation; 4 charge transport; 5 charge extraction. OLED (reverse order): 5 charge injection; 4 charge transport; 3 exciton formation; 2 radiating exciton recombination; 1 light emission.

HOMO level, while the matrix E_{HOMO} level must be close to the p-dopant's LUMO level, respectively.

An investigation of homo diodes holding these requirements was reported in 2005 by Harada and co-workers.[58] Since the matrix material must fit to the dopant energy levels, the organic matrix material should have a small energy gap ($\Delta E_{HOMO-LUMO}$). ZnPc was chosen which is additionally characterised by high mobility for both, holes and electrons. A p-i-n HOMO device is shown involving ZnPc doped with $Ru(terpy)_2$ for n-doping and F4-TCNQ for p-doping. A rather thick un-doped ZnPc layer was inserted between the p and n layers as a blocking junction to achieve good rectification of the diode. For a thin or no blocking layer, the narrow space-charge layers would cause tunnelling and lead to an ohmic junction. Apart from the temperature dependence of the diode parameters, the device characteristics of the p-i-n junctions are to some extent adequately described by the Shockley equation. Capacitance–voltage measurements of the organic HOMO junction imply a very high built-in voltage of at least 0.8 V, noticeably higher than the open circuit voltage (V_{OC}) in ZnPc/C60 p-i-n hetero junction structures ($V_{OC} = 0.5$ V).[59] Remarkably higher voltages ($V_{OC} = 1.1$ V) were observed in homo junctions containing pentacene as matrix material.[60] The energy gap ($\Delta E_{HOMO-LUMO}$) in pentacene is larger than in ZnPc and the LUMO level is slightly shifted upwards. Efficient doping becomes possible if the n-dopant's ionisation potential is further reduced by about 0.1–0.2 eV. This was achieved by attaching electron donating tert-butyl groups in the 4,4″-positions of the 2,2′:6′,2″-terpyridine ligand. The resulting Ru(^{t}Bu-terpy)$_2$ complex has been proven to be a stronger electron donor than Ru(terpy)$_2$.

Depending on the matrix material, light-emission can be observed as well. The first organic light-emitting p-n homojunction diode was realised using Ir(piq)$_3$ as matrix. The homo-OLED demonstrated emission around 660 nm at a low electroluminescence onset voltage of 1.8 V.[61] Unfortunately the overall performance was low compared to standard red heterojunction OLED. It is assumed that one of the reasons for the low current efficiency (0.2 cd/A at 100 cd/m^2) and quantum efficiency is the imbalanced mobilities of holes and electrons.

3.3.2 OLEDs with doped layers

The first organic thin film heterojunction OLED developed by Tang and Van Slyke consisted of a hole transporting and an electron transporting layer. The latter was also the light emitting zone of the device.[62] This simple stack was placed between ohmic electrodes, a high work function transparent anode and a reflective cathode with a low work function. This fundamental concept remained more or less unaltered to the present day. Meanwhile the device structure is somewhat more sophisticated with specialised layers for different intended purposes. This is the reason for rapidly grown efficiencies, especially compared with the first OLED. The efficiency, which is closely related to the driving voltage, is one of the most important OLED parameters. The voltage is basically corresponding to the energy of the emitted light plus a term that is related to the exciton binding energy. Furthermore injection barriers and charge transport losses have to be taken into account. Doping is one of the most promising ways to reduce the latter loss mechanisms.

Related to our first experiments with molecular dopants and their use in organic semiconductors, application in functional devices was the next step. In first experiments the simple structure used by Tang and Van Slyke was adapted by Blochwitz and co-workers to test the applicability of doping of evaporable small molecules.[63] An OLED including F4-TCNQ doped vanadyl phthalocyanine (VOPc) as hole transport layer showed a significant decrease of driving voltage, even at low doping concentrations ranging from

0.5% to 3% (Fig. 3.13). The materials back then quickly turned out to be unsuitable for application in OLED and material development started. For instance, the hole injection and transport materials used nowadays are amorphous materials, based on the triarylamine moiety. Since this compound class exhibits a high degree of freedom with respect to the rotation of each rigid phenyl ring (or other aromatic rings), predominantly amorphous glasses are formed by this substances, which is a benefit for formation of smooth amorphous layers. An overview about successfully used materials in hole transport layers is given in a work of Shirota.[64]

The next stage towards devices with higher efficiency is expanding the beneficial strategy by additional n-doped charge transport layers. Placing the undoped intrinsic emitter materials between p-doped hole and n-doped charge transport layers leads to the p-i-n structure. The concept of sandwiching the intrinsic layers between charge transport layers offers several advantages. However, some problems related with the structure are present as well. An imbalanced charge carrier injection can lead to accumulation of charge carriers and thus exciton quenching can occur. These problems can be solved by smart device architectures. Insertion of blocking layers between the charge transport and the emission layer is one approach to counteract these limitations. A balanced charge carrier concentration and lower driving voltage can be the result. Furthermore, formation of interface exciplexes and exciton loss can be avoided by using wide gap materials. The

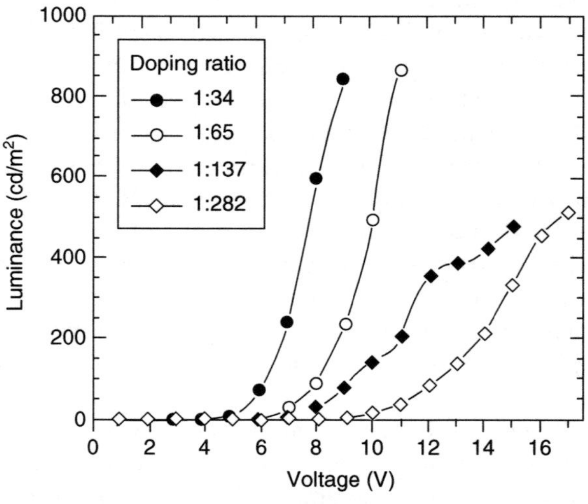

3.13 Luminance/voltage characteristics for several molecular doping ratios. With permission from: Blochwitz J, Pfeiffer M, Fritz T, Leo K, 'Low voltage organic light emitting diodes featuring doped phthalocyanine as hole transport material', *Appl Phys Lett*, 1998, 73, 729–731.

p-i-n structure is then extended to five different layers. An Alq_3 emission layer embedded between intrinsic Bphen and TPD for electron or hole blocking, respectively, was used with p-doped m-MTDATA and Li doped Bphen as charge transport layers.[65] The device showed excellent I/V characteristics with an exponential luminescence voltage dependence up to a brightness of 1000 cd/m^2. Brightness of 100 cd/m^2 was achieved at 2.55 V, only slightly above the corresponding photon energy. Going from the p-i-i to the p-i-n structure, one can clearly see in Fig. 3.14 that the same luminance is achieved at lower voltages, caused by improved electron transport in the p-i-n device.

Generally p-i-n devices can be improved by doping with more efficient and stable dopants. This concerns the morphologic stability of the doped layer and migration of dopant molecules, especially with rising temperature during operation.[43] An investigation of the Forrest group revealed the reduced lifetime of doped devices in comparison to undoped devices using a thermally stable emission layer. This indicates that p and n dopants can lead to accelerated device degradation. The extrapolated operational lifetime of Li doped p-i-n and undoped devices are estimated to 18000 and 60000 h, respectively.[66] Some demonstrations of more efficient dopants with commercial origin are published, but without disclosing the structure.[67] Meanwhile lifetimes in monochromic OLED are limited by the emission layer lifetime, indicating that doped layers are no longer the limiting device element. A red OLED having a lifetime of over 17000 h at a brightness of

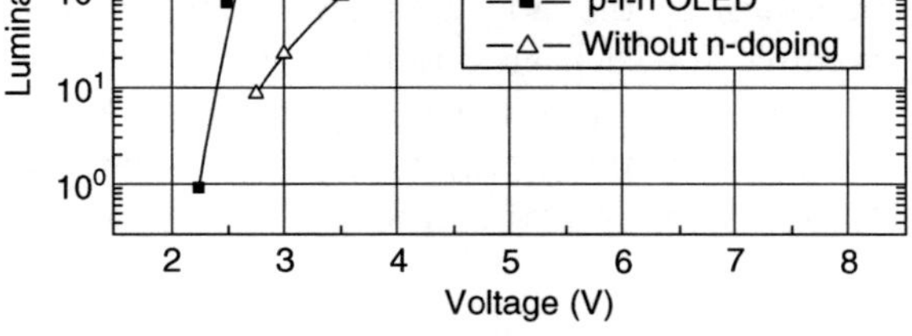

3.14 Luminance/voltage characteristics of an OLED with and without Li doping of the BPhen layer. Reprinted from: Pfeiffer M, Leo K, Zhou X, Huang J S, Hofmann M, Werner A, Blochwitz-Nimoth J, 'Doped organic semiconductors: Physics and application in light emitting diodes', *Org Electron*, 2003, 4, 89–103. Copyright 2003, with permission from Elsevier.

500 cd/m^2 and 80 °C is reported by Novaled. Even higher lifetime values up to 1 000 000 h can be expected at room temperature and 1000 cd/m^2 brightness. Applying the same conditions, blue and green emission OLED with lifetimes of 22 000 h and 200 000 h, respectively, are reported. It is expected that doped devices can have higher stability due to lower ohmic dissipation and more stable interfaces.

In white emitting OLED, doping of transport layers mainly enhances the device power efficiency, which is an important parameter for energy saving in the lighting sector. A breakthrough in overall device power efficiency of 90 lumens per watt at 1000 cd/m^2 brightness was reported in 2009.[68] At this value, OLED exhibit higher efficiency than fluorescent tubes (60–70 lm/W), the current benchmark for novel light sources.

The beneficial effect of doping on interface properties offers a wider choice of substrates, including cheaper and rougher ones. Meanwhile a large amount of publications deal with doped layers in OLED and only the most important work can be presented here. For more information, we refer to a review published in 2009.[69]

Top emitting (inverted) OLED, where the light is emitted away from the substrate, is rather interesting for large-scale and mass production, on metallic foils for instance. Due to the doping concept, thick charge transport layers can be realised in such devices, supported by two advantageous facts: deposition of the ITO top contact usually induces thermal stress onto the sub-layers and can damage the stack. If deposition takes place onto a thick transport layer, it acts as sacrificial layer and protects the sensitive emitter/blocker system from damage during the sputtering processes. Another approach is to use very thin transparent metal contacts. The choice of transparent cathode materials is expanded in context to the cathode's work function. On the other hand, the possibility of adjusting different layer thicknesses allows the formation of micro-cavities. Here, the emission layer is moved to a position between the contacts, where constructive interference conditions are fulfilled. The light outcoupling can be improved by optimisation of the hole and the electron transport layer thickness.[70] One example for such an optimisation is shown in Fig. 3.15. Here the simulated light emission is depending on the thickness of spacer layer and on the light-out-coupling capping layer. The variation of both thicknesses leads to formation of two cavities, where the device efficiency is maximised.

3.3.3 OSCs with doped layers

Unlike in classical silicon solar cells, in OSCs charge carriers are not directly generated by illumination. Instead an organic molecule in the light-absorbing layer gets excited und the created exciton migrates through the

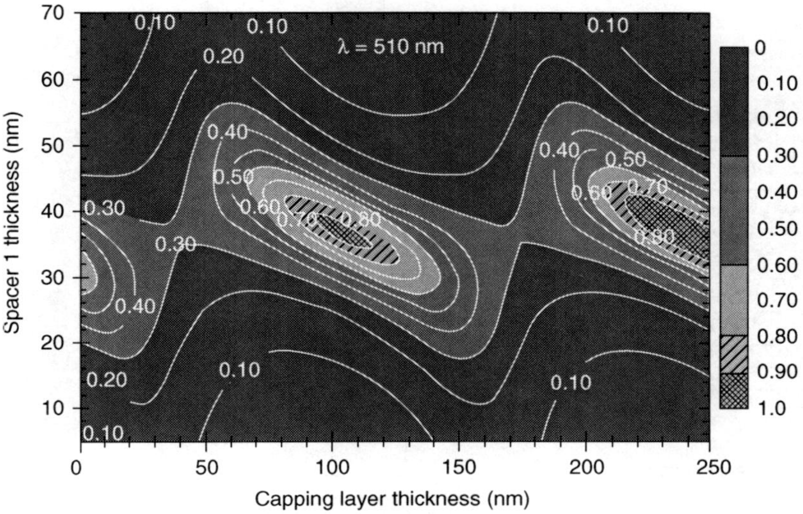

3.15 Normalised light transmission at the wavelength of 510 nm as a function of capping and spacer 1 layer thickness calculated based on the following OLED structure: Ag (80 nm)/spacer 1(*Y* nm)/emitter (15 nm)/spacer 2 (35 nm)/Ag (15 nm)/capping layer (*X* nm). The refractive index of all organic materials used is assumed to be 1.75. Reprinted with permission from: Huang Q, Walzer K, Pfeiffer M, Lyssenko V, He G, Leo K, 'Highly efficient top emitting organic light-emitting diodes with organic outcoupling enhancement layers', *Appl Phys Lett*, 2006, 88, 113515. Copyright 2006, American Institute of Physics.

absorber layer, as illustrated in Fig. 3.12.[71] To overcome the exciton binding energy (0.2–0.5 eV), field strengths of 10^6 V/cm would be necessary to enforce charge separation. Since such a large field is not available in OSC, an efficient hetero-junction is needed instead, where the charge separation takes place at the interfaces of two different organic materials. At these interfaces the exciton is separated and two charges (h^+/e^-) are generated. The electron is collected in the acceptor material, whereas the hole migrates into the donor material. The exciton diffusion length is limited in organic materials, thus the active light-absorbing layers are narrow.

Utilisation of undoped intrinsic materials in the middle of two transparent doped transport layers results in the p-i-n stack structure. Since photons are exclusively absorbed in the photoactive region, migration pathways to the hetero interface are short and parasitic absorption is avoided. The wide gap charge transport materials act as a membrane for the corresponding charge: excitons are reflected and can diffuse to the hetero junction interface. The major benefit of using transparent window layers addresses optical

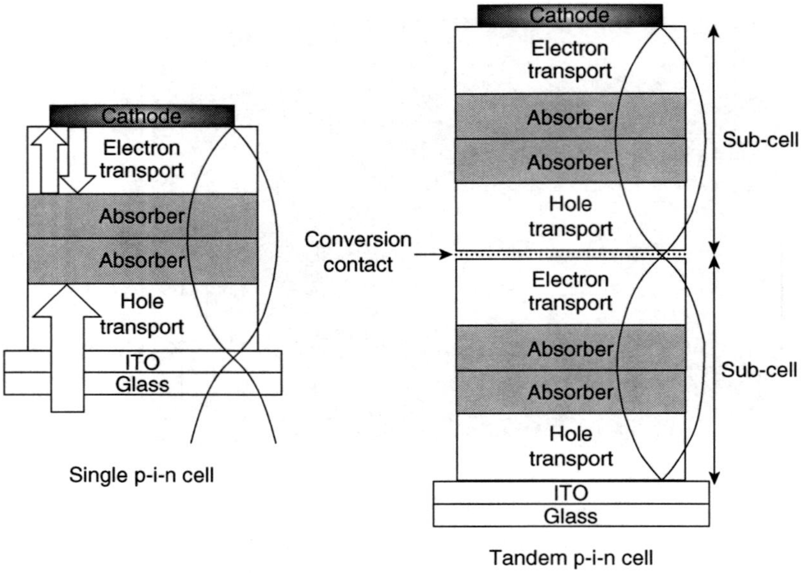

3.16 Ideal single cell and tandem stack structure with an indicated optical standing wave.

thin film phenomena (Fig. 3.16). The incoming light can partially pass through the photoactive region and is back reflected at the metallic cathode. Considering this interference state as standing wave within the solar cell, the photoactive layers can be placed in the maximum of the optical field distribution. The placement of the light-absorbing layers into the maximum of the standing wave is realised by thickness variation of the transparent charge transport layers. This concept works in single cells, as well as in multi-cells such as tandem cells.

Doped charge transport layers in single cells

A straightforward optimisation of such solar cell devices is motivated by simulation of the optical field distribution within the device, as demonstrated by Maennig and co-workers.[72] Using this approach several requirements are addressed to the charge transport materials, in particular the charge transport properties. Ohmic losses within thin transport layers are negligibly small, if the conductivities of the transport layers are higher than 10^{-5} S/cm. This is important for operation at low voltages and reflected in the corresponding current/voltage (I/V) relation of the cell. Since OSC show the typical diode behaviour, a low series resistance is indicated by high slopes with increasing voltages (second quadrant).

The formation of charge transport layers with sufficient and stable conductivities can be achieved by controlled doping, as demonstrated before. The proof that replacement of light-absorbing charge transport layers by wide gap charge transport layers leads to higher device efficiencies is discussed in detail here: A n-i-p stack with Me-PTCDI doped with rhodamine B for electron transport and F4-TCNQ doped ZnPc as hole transport material is used for this example. The photoactive bulk layer consists of a 30 nm thick ZnPc:C_{60} blend (1:1). The corresponding device involving the transparent charge transport materials contains MeO-TPD for hole conduction and C_{60} for electron conduction. A thin ZnPc interlayer between the hole transport layer and the reflecting gold top electrode ensures a good ohmic contact, but does not contribute to the photo current.[73]

Figure 3.17(a) depicts the *I*/*V* characteristics of both solar cells with the same active layer (30 nm ZnPc:C_{60}, 1:1). The cell including the low gap transport layers exhibits a fill factor of 45%, a short circuit current density of 3.7 mA/cm^2, and an open circuit voltage of 0.45 V. The power conversion efficiency of the low gap cell is 0.75%, whereas the wide-gap cell achieves 1.32% efficiency. Since the open circuit voltage (0.46 V) and fill factor (47%) are nearly the same, the strong efficiency increase originates from a nearly doubled short-circuit current of 6.1 mA/cm^2. The high current is mainly attributed to a significantly reduced parasitic absorption of the transparent charge transport layer, which leads to an increased charge generation in the light absorbing layer (Fig. 3.17(a)). The contribution of each absorber can be observed in the external quantum efficiency (EQE) of the wide-gap cell (Fig. 3.17(b)), where the EQE follows the absorption spectra of each absorber material. ZnPc contributes most to the photo current (up to 40% EQE) at its absorption peaks at 640 and 700 nm. The second absorption is assigned to the relatively low fullerene contribution (17%) which appears at the maximum 380 nm. Certainly, it is not ideal to use C_{60} in the ETL due to parasitic absorption, but the lower EQE can barely be explained by parasitic ETL absorption, since the optical field close to the cathode is weak. On the other hand, C_{60} exhibits generally low absorption in the visible range.

This fact is beneficial for the application of C_{60} in tandem cells, where this light absorbing acceptor material can be used in both photoactive sub-cell regions. Regarding the EQE, a second fact with respect to tandem devices gets obvious: there is a large gap in the EQE between C_{60} and ZnPc where the cell does not generate charges. These photons in the green region are lost, as long as this gap is not filled by another absorber. Besides thin film optics, the combination of two complementary absorber couples in one of the most important issues in the design of tandem devices. The valuable application of doped layers in such tandem devices is highlighted in the following subsection.

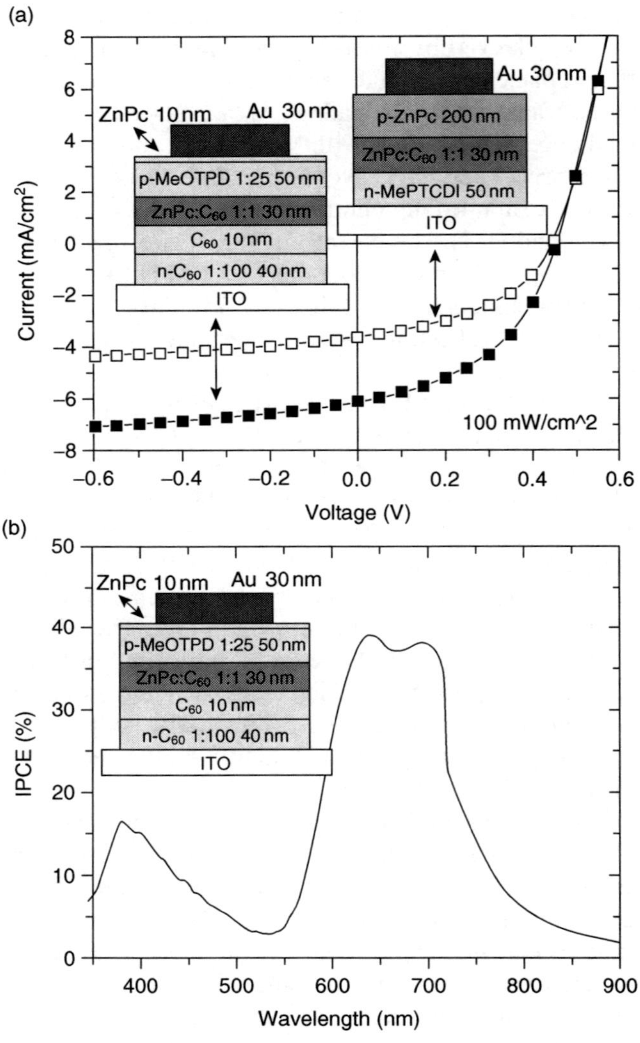

3.17 (a) I/V characteristics of both solar cells with the same active layer (30 nm ZnPc:C_{60}, 1:1). (b) EQE of the wide-gap cell. With kind permission from Springer Science+Business Media: Maennig B, Gebeyehu D, Simon P, Kozlowski F, Werner A, Li F, Grundmann S, Sonntag S, Koch M, Leo K, Pfeiffer M, Hoppe H, Meissner D, Sariciftci N S, Riedel I, Dyakonov V, Parisi J, Drechsel J, 'Organic p-i-n solar cells', *Appl Phys A Mat Sci & Proc*, 2004, 79, 1–14. Copyright 2004, Springer Science+Business Media.

Doped layers in stacked p-i-n cells

p-i-n OSC usually involve thin absorber layers covering only part of the sun spectrum. Thus the incident light is not completely absorbed in a single cell. To obtain entire photon conversion into electrons, improvements can be achieved by connecting two or more sub cells that are connected in series. If one assumes no losses, the open circuit voltage of multi-cells can in principle be the sum of each individual sub-cell. The total current in this arrangement is limited by the lowest sub-cell current.

In 1990 Hiramoto published the first organic tandem cell made of two combined sub-cells with an ultra-thin gold interlayer as recombination zone.[74] Here the electrons should be converted into holes without loss of energy (voltage), for instance by recombination of charge carriers or exciton quenching. They found an increase of photocurrent density when the Au layer was inserted and a dependence on the thickness of the recombination layer. Deposition of metal layers of 0.5 to 2 nm thickness does not lead to the formation of closed metal layers. Instead, metal clusters are formed that give rise to reduced reflectance compared to completely closed layers.[75]

The disadvantage of thin metal layers or metal clusters is the high absorbance and reflectance. To avoid losses and unbalanced absorption in the sub-cells, the recombination contact has to be highly transparent. Here highly doped layers can further improve the tandem device performance. The p-i-n architecture is also beneficial particularly for tandem devices. By comparing *I/V* characteristics of tandem cells using metal cluster based recombination contacts with those using highly doped contacts, it is found that the latter are superior (Table 3.3).[76] It turns out that the choice of matrix/dopants combinations influences the tandem device performance as well. In accordance to theory, the tandem cell with highly doped contacts exhibits exactly twice the V_{OC} of the single cell (Fig. 3.18). For the reference cells without recombination contact, gold or silver clusters

Table 3.3 Characteristic values from I-V measurements of prepared solar cells including different recombination contacts

Recombination contact	U_{OC} (V)	J_{SC} (mA/cm^2)	FF (%)
Single cell	0.85	2.39	48
Ag clusters	1.07	0.89	41
Au clusters	1.35	1.13	41
Highly doped layers	1.70	1.36	48

Reprinted with permission from: Timmreck R, Olthof S, Leo K, Riede M K, 'Highly doped layers as efficient electron–hole recombination contacts for tandem organic solar cells', *J Appl Phys*, 2010, 108, 033108. Copyright 2010, American Institute of Physics.

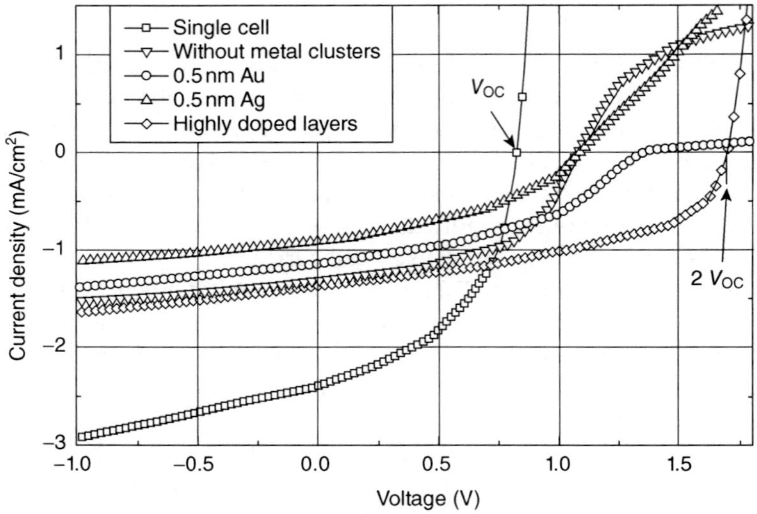

3.18 For comparison, current/voltage curves of not unoptimised tandem cells involving different conversion contacts are depicted. The tandem device with doped layers exhibits exactly double the V_{OC} of the single cell.

for recombination, significant losses in open circuit voltage are observed (37%, 20% and 37%, respectively). Using the approach of doped layers instead of metal clusters, losses by absorption or reflection at the recombination contact have been nullified.

Beside highly doped recombination contacts, doped charge transporting spacer layers play an important role for the device efficiency of tandem cells. The p-i-n concept incorporating transparent transport layers gives also the freedom to optimise a tandem cell with respect to the optical interference pattern. Using a metal-free lossless recombination contact between the sub cells, ohmic losses are negligible, but the current limitation is still unsolved. The impact of these spacer layers in p-i-n tandem cells on the device characteristics is reported as well and will be discussed in the following.[77] An ITO/p-i-n/p-i-i/Al structure was used, for adjusting balanced sub cell current consisting of two ZnPc and C_{60} sub cells (Fig. 3.19(a)). In this report current–voltage measurements confirmed optical simulations. The current of the second sub-cell near to the ITO electrode can be exactly tuned by changing the p-doped transparent layer thickness. Among the series of spacer thickness variations, almost constant device parameters (FF ≅ 59%, $V_{OC} \cong 1.06$ V) were observed. Figure 3.19(b) shows the systematically varied spacer layer thickness and its influence on the cell performance. If the sub cells are separated only by the recombination layers

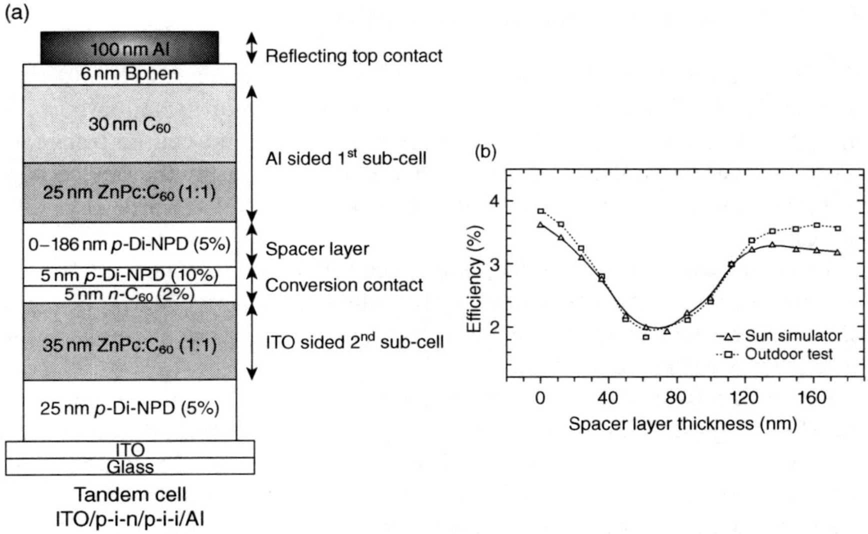

3.19 (a) Tandem device structure and (b) efficiency as function of the spacer thickness under illumination with a sun simulator (100 mW/ cm^2) and the real sun during outdoor measurements.

(d_{spacer} = 0 nm), both sub-cells are positioned in the first interference maximum and the currents of the individual sub-cells nearly match. Making the spacer layer thicker ($d_{spacer} > 0$ nm), the second sub-cell is moved out of the maximum and the currents drop, leading to deceased efficiency. The short-circuit current density of the ITO-sided sub-cell limits the performance and falls to a half of its initial value in the first interference minimum ($d_{spacer} \approx 75$ nm). Shifting the ITO-sided sub-cell into the second interference maximum ($d_{spacer} \approx 140$ nm) means J_{SC} gets back to nearly its original value.

Since the influence on the tandem device performance (energy losses, current matching) is well understood, recent investigations have been directed towards absorber material variations. To achieve even higher efficiencies, a smart combination of complementary absorber materials (C_{60}:green donor/C_{60}:red donor) covering the full visible spectrum is desirable. The straight forward application, namely merging the prior described concepts, allows efficiencies above 6%, currently reported in literature.[78] Presently achieved certified efficiencies, using polymers applied from solution (8.62%)[79] or small molecules deposited in vacuum (9.8%),[80] reasonably exceed this value, but without disclosing the stack architecture used and materials. However, it is worth noting that both values are achieved on an active surface area of >1.0 cm^2, since such device size is relevant for commercialisation of products.

3.4 Conclusion and future trends

We demonstrated that the use of doped layers in organic semiconducting devices like OLED and OSC leads to superior performance compared with undoped devices. Doped layers, in particular transparent charge transport layers, provide high conductivities, which is favourable for the device performance. Additionally these devices can be optimised towards thin film optics.

Today, doping is successfully applied in commercial OLED products such as cellular phones. It is expected to be broadly used in active matrix displays of consumer electronics. The same is true for lighting devices, where power efficiency is the key parameter.

Future trends in the field of doping direct towards the development of novel doping materials and alternative application methods. Although the synthesis and handling of air sensitive n-type dopants are controllable, current research is concentrating on more stable dopants. In the face of mass production of organic electronics, air stable dopants are desirable, due to a reduced expenditure during synthesis, purification and application. Modifications of dopant molecules with anchor groups lead to surfactants that are able to assemble on substrates in an ordered way. With respect to OLED and OFET surface assembled monolayers are expected to be of assistance.[81] These compounds can be applied by vapour deposition or from solution.

Merging both advantages, simple and cheap application of light-emitting layers from solution and the p-i-n concept using doped electron transport layers, new pathways towards mass production are opened.[82] It is expected that these hybrid OLED devices will lead to further improvements in power efficiency without additional manufacturing complexity. Although the vacuum technology is widely established in device manufacturing, molecular doping may also lead to improvement using alternative solution-based methods. Here, doping might be helpful for polymers and small molecule-based systems, as well. However, independently of the procedure, application of doping in a multitude of devices where the p-n junction plays a role is not realised so far.[83]

For advanced or postgraduate students who are interested in organic semiconductors, we recommend *Organic Molecular Solids* (Schwoerer, Wolf), providing a well-founded knowledge in this field.[84] Pope and Swenberg's *Electronic Processes of Organic Crystals* is the classic reference in the field.[6] Beside experimental and theoretical properties of aromatic hydrocarbon crystals the book is enriched with recent developments and latest practical applications.

The Organic Electronics Association (OE-A) is an information and communication platform for companies and research institutes active in the field of organic electronics in Germany and whole Europe.[85] Their goal is to tie science, technology and application together. OE-A assists collaboration of OE-A members by exchange of information and experiences and by cooperation in projects. They deliver information to interested parties, inform politics and the public, and thus increase visibility beyond the community. R&D activities are promoted while having an eye on future markets and technologies. The service ranges from market analysis to the point of establishing competitive production infrastructures for organic electronics. Additionally education and training course are organised.

Organic Electronics Saxony e.V. (OES) represent the large cluster in Saxony and the region of Dresden.[86] More than 15 companies and 10 research institutes are represented in Europe's largest cluster. The members cover the whole value chain for organic semiconductors, starting at universal fundamental research up to manufacturing technological products. An internationally accepted competence network has been formed, which focuses on vacuum based deposition technology. OES offers a communication platform where knowhow can be transferred and specific cooperations are supported.

The website orgworld.de is meant to give an overview about companies and research institutes active in the field of organic electronics.[87] International conferences in the field of organic electronics are listed. Worth mentioning are the frequently updated record charts in OPV and OLED efficiency. These charts reflect the evolution of these emerging technologies starting from the discovery of lab curios to commercial application.

3.5 Acknowledgements

We would like to thank the many coworkers (too many to mention by name) who have worked on the development of highly efficient organic devices in Dresden over the course of many years. We would like to thank Torben Menke, Debdutta Ray, Ronny Timmreck, Bjoern Luessem, Moritz Riede, Annette Petrich and Kentaro Harada for helpful discussion and support in preparing this review. Financial support was received from the German Federal Ministry of Education and Research (BMBF) through the InnoProfile program (03IP602).

3.6 References

1. Mizutori M, Yamada R, 'Semiconductors', *Ullmann's Encyclopedia of Industrial Chemistry*, 2000.
2. Zulehner W, Neuer B, Rau G, 'Silicon', *Ullmann's Encyclopedia of Industrial Chemistry*, 2000.
3. Alsema E A, 'Energy pay-back time and CO_2 emissions of PV systems', *Prog Photovolt Res Appl*, 2000, **8**, 17–25.

4. Tang C W, 'Two-layer organic photovoltaic cell', *Appl Phys Lett*, 1986, **48**, 183–185.
5. Coropceanu V, Cornil J, da Silva Filho D A, Olivier Y, Silbey R, Brédas J-L, 'Charge transport in organic semiconductors', *Chem Rev*, 2007, **107**, 926–952.
6. Pope M, Swenberg C E, *Electronic Processes in Organic Crystals and Polymers* (2th ed.), New York, Oxford University Press, 1999.
7. Seeger K, *Semiconductor Physics*, Berlin, Springer, 1997.
8. Goerlich P, Schneider H G, Hamann C, *Organische Festkoerper und duenne Schichten*, Berlin, Akademische Verlagsgesellschaft, 1978.
9. Borchardt-Ott W, *Kristallographie* (5th ed.), Berlin, Springer, 1997.
10. Horowitz G, Hajlaoui M E, 'Mobility in polycrystalline oligothiophene field-effect transistors dependent on grain size', *Adv Mat*, 2000, **12**, 1046–1050.
11. Dimitrakopoulos C D, Brown R, Pomp A, 'Molecular beam deposited thin films of pentacene for organic field effect transistor applications', *J Appl Phys*, 1996, **80**, 2501–2508.
12. Duhm S, Glowatzki H, Rabe J P, Koch N, Johnson R L, 'Spontaneous charge transfer at organic–organic homointerfaces to establish thermodynamic equilibrium', *Appl Phys Lett*, 2007, **90**, 122113.
13. Jaeckel F, Perera U, Iancu V, Braun K-F, Koch N, Rabe J, Hla S-W, 'Investigating molecular charge transfer complexes with a low temperature scanning tunneling microscope', *Phys Rev Lett*, 2008, **100**, 4–7.
14. Yamamoto Y, Yoshino K, Inuishi Y, 'Electrical properties of phthalocyanine-halogen complexes', *J Phys Soc Jpn*, 1979, **47**, 1887–1891.
15. Lee J-H, Leem D-S, Kim J-J, 'High performance top-emitting organic light-emitting diodes with copper iodide-doped hole injection layer', *Org Electron*, 2008, **9**, 805–808.
16. Ganzorig C, Fujihira M, 'Improved drive voltages of organic electroluminescent devices with an efficient p-type aromatic diamine hole-injection layer', *Appl Phys Lett*, 2000, **77**, 4211–4213.
17. Leem D-S, Park H-D, Kang J-W, Lee J-H, Kim J W, Kim J-J, 'Low driving voltage and high stability organic light-emitting diodes with rhenium oxide-doped hole transporting layer', *Appl Phys Lett*, 2007, **91**, 011113.
18. Matsushima T, Adachi C, 'Enhanced hole injection and transport in molybdenum-dioxide-doped organic hole-transporting layers', *J Appl Phys*, 2008, **103**, 034501.
19. Chang C-C, Hsieh M-T, Chen J-F, Hwang S-W, Chen C H, 'Highly power efficient organic light-emitting diodes with a p-doping layer', *Appl Phys Lett*, 2006, **89**, 253504.
20. Lide D R, *Handbook of Chemistry and Physics* (74th ed.), London, CRC Press, 1993.
21. Kido J, Matsumoto T, 'Bright organic electroluminescent devices having a metal-doped electron-injecting layer', *Appl Phys Lett*, 1998, **73**, 2866–2868.
22. Hung L S, Tang C W, Mason M G, 'Enhanced electron injection in organic electroluminescence devices using an Al/LiF electrode', *Appl Phys Lett*, 1997, **70**, 152–154.
23. Mason M G, Tang C W, Hung L S, Raychaudhuri P, Madathil J, Yan L, Le Q T, Gao Y, Lee S-T, Liao L S, Cheng L F, Salaneck W R, dos Santos D A, Brédas J L, 'Interfacial chemistry of Alq3 and LiF with reactive metals', *J Appl Phys*, 2001, **89**, 2756–2765.

24. Parthasarathy G, Shen C, Kahn A, Forrest S R, 'Lithium doping of semiconducting organic charge transport materials', *J Appl Phys*, 2001, **89**, 4986–4992.
25. Bondi A, 'van der Waals volumes and radii', *J Phys Chem*, 1964, **68**, 441–451.
26. Mantina M, Chamberlin A C, Valero R, Crame C J, Truhlar D G, 'Consistent van der Waals radii for the whole main group', *J Phys Chem A*, 2009, **113**, 5806–5812.
27. Arkhipov V I, Emelianova E V, Baessler H, 'Quenching of excitons in doped disordered organic semiconductors', *Phys Rev B*, 2004, **70**, 205205.
28. Cotton F A, Gruhn N E, Gu J, Huang P, Lichtenberger D L, Murillo C A, Van Dorn L O, Wilkinson C C, 'Closed-shell molecules that ionize more readily than cesium', *Science*, 2002, **298**, 1971–1974.
29. Li F, Werner A, Pfeiffer M, Leo K, Liu X, 'Leuco crystal violet as a dopant for n-doping of organic thin films of fullerene C_{60}', *J Phys Chem B*, 2004, **108**, 17076–17082.
30. Lewis E S, Perry J M, Grinstein R H, 'Mechanism of hydride transfer. III. Rates and isotope effects in the quinone oxidation of leuco triphenylmethane dyes', *J Am Chem Soc*, 1970, **92**, 899–905.
31. Pfeiffer M, Beyer A, Fritz T, Leo K, 'Controlled doping of phthalocyanine layers by cosublimation with acceptor molecules: A systematic Seebeck and conductivity study', *Appl Phys Lett*, 1998, **73**, 3202–3204.
32. Maennig B, Pfeiffer M, Nollau A, Zhou X, Leo K, Simon P, 'Controlled p-type doping of polycrystalline and amorphous organic layers: Self-consistent description of conductivity and field-effect mobility by a microscopic percolation model', *Phys Rev B*, 2001, **64**, 195208.
33. Hiller S, Schlettwein D, Armstrong N R, Woehrle D, 'Influence of surface reactions and ionization gradients on junction properties of $F_{16}PcZn$', *J Mater Chem*, 1998, **8**, 945–954.
34. Hoffmann M, Schmidt K, Fritz T, Hasche T, Agranovich V M, Leo K, 'The lowest energy Frenkel and charge-transfer excitons in quasi-one-dimensional structures: Application to MePTCDI and PTCDA crystals', *Chem Phys*, 2000, **258**, 73–96.
35. Zhou X, Blochwitz J, Pfeiffer M, Nollau A, Fritz T, Leo K, 'Enhanced hole injection into amorphous hole-transport layers of organic light-emitting diodes using controlled p-type doping', *Adv Funct Mat*, 2001, **11**, 310–314.
36. El-Khatib N, Boudjema B, Guillaud G, Maitrot M, Chermette H, 'Theoretical and experimental doping of molecular materials: P and N doping of zinc phthalocyanine', *J Less Com Met*, 1988, **143**, 101–112.
37. Fehse K, Olthof S, Walzer K, Leo K, Johnson R L, Glowatzki H, Broker B, Koch N, 'Energy level alignment of electrically doped hole transport layers with transparent and conductive indium tin oxide and polymer anodes', *J Appl Phys*, 2007, **102**, 073719.
38. Yoshida M, Fujii A, Ohmori Y, Yoshino K, 'Three-layered multicolor organic electroluminescent device', *Appl Phys Lett*, 1996, **69**, 734–736.
39. Adachi C, Kwong R, Forrest S R, 'Efficient electrophosphorescence using a doped ambipolar conductive molecular organic thin film', *Org Electron*, 2001, **2**, 37–43.
40. Olthof S, Tress W, Meerheim R, Lüssem B, Leo K, 'Photoelectron spectroscopy study of systematically varied doping concentrations in an organic semiconductor layer using a molecular p-dopant', *J Appl Phys*, 2009, **106**, 103711.

41. Wellmann P, Hofmann M, Zeika O, Werner A, Birnstock J, Meerheim R, He G, Walzer K, Pfeiffer M, Leo K, 'High-efficiency p-i-n organic light-emitting diodes with long lifetime', *J Soc Inform Displ*, 2005, **13**, 393–397.
42. Mi B X, Gao Z Q, Cheah K W, Chen C H, 'Organic light-emitting diodes using 3,6-difluoro-2,5,7,7,8,8-hexacyanoquinodimethane as p-type dopant', *Appl Phys Lett*, 2009, **94**, 073507.
43. Koech P K, Padmaperuma A B, Wang L, Swensen J S, Polikarpov E, Darsell J T, Rainbolt J E, Gaspar D J, 'Synthesis and application of 1,3,4,5,7,8-hexafluorotetracyanonaphthaquinodimethane (F6-TNAP): A conductivity dopant for organic light-emitting devices', *Chem Mater*, 2010, **22**, 3926–3932.
44. Meerheim R, Olthof S, Hermenau M, Scholz S, Petrich A, Tessler N, Solomeshch O, Luessem B, Riede M, Leo K, 'Investigation of $C_{60}F_{36}$ as low-volatility p-dopant in organic optoelectronic devices', *J Appl Phys*, 2011, **109**, 103102.
45. Avent A G, Clare B W, Hitchcock P B, Kepert D L, Taylor R, '$C_{60}F_{36}$: There is a third isomer and it has C1 symmetry', *Chem Comm*, 2002, 2370–2371.
46. Nollau A, Pfeiffer M, Fritz T, Leo K, 'Controlled n-type doping of a molecular organic semiconductor: Naphthalenetetracarboxylic dianhydride (NTCDA) doped with bis(ethylenedithio)-tetrathiafulvalene (BEDT-TTF)', *J Appl Phys*, 2000, **87**, 4340–4343.
47. Tanaka S, Kanai K, Kawabe E, Iwahashi T, Nishi T, Ouchi Y, Seki K, 'Doping effect of tetrathianaphthacene molecule in organic semiconductors on their interfacial electronic structures studied by UV photoemission spectroscopy', *Jpn J Appl Phys*, 2005, **44**, 3760–3763.
48. Chan C L, Fabrice A, Zhang Q, Barlow S, Marder S R, Kahn A, 'N-type doping of an electron-transport material by controlled gas-phase incorporation of cobaltocene', *Chem Phys Lett*, 2006, **431**, 67–71.
49. Bloom C J, Elliott C M, Schroeder P G, France C B, Parkinson B A, 'Low work function reduced metal complexes as cathodes in organic electroluminescent devices', *J Phys Chem B*, 2003, **107**, 2933–2938.
50. Menke T, Ray D, Meiss J, Leo K, Riede M, '*In-situ* conductivity and Seebeck measurements of highly efficient n-dopants in fullerene C60', *Appl Phys Lett*, 2012, **100**, 093304.
51. Werner A, Li F, Harada K, Pfeiffer M, Fritz T, Leo K, 'n-Type doping of organic thin films using cationic dyes', *Adv Funct Mater*, 2004, **14**, 255–260.
52. Wei P, Oh J H, Dong G, Bao Z, 'Use of a 1*H*-benzoimidazole derivative as an *n*-type dopant and to enable air-stable solution-processed *n*-channel organic thin-film transistors', *J Am Chem Soc*, 2010, **132**, 8852–8853.
53. Fenghong L, Pfeiffer M, Werner A, Harada K, Leo K, Hayashi N, Seki K, Xianjie L, Xuan-Dung, 'Acridine orange base as a dopant for n doping of C_{60} thin films', *J Appl Phys*, 2006, **100**, 2, 023716.
54. Hwang J, Wan A, Kahn A, 'Energetics of metal–organic interfaces: New experiments and assessment of the field', *Materials Science and Engineering: R: Reports*, 2009, **64**, 1–31.
55. Kim J S, Granstroem M, Friend R H, Johansson N, Salaneck W R, Daik R, Feast W J, Cacialli F, 'Indium–tin oxide treatments for single- and double-layer polymeric light-emitting diodes: The relation between the anode physical, chemical, and morphological properties and the device performance', *J Appl Phys*, 1998, **84**, 6859–6870.

56. Olthof S, Meerheim R, Schober M, Leo K, 'Energy level alignment at the interfaces in a multilayer organic light-emitting diode structure', *Phys Rev B*, 2009, **79**, 245308.
57. Blochwitz J, Fritz T, Pfeiffer M, Leo K, Alloway D M, Lee P A, Armstrong N R, 'Interface electronic structure of organic semiconductors with controlled doping levels', *Org Electron*, 2001, **2**, 97–104.
58. Harada K, Werner A, Pfeiffer M, Bloom C, Elliott C, Leo K, 'Organic homojunction diodes with a high built-in potential: Interpretation of the current–voltage characteristics by a generalized Einstein relation', *Phys Rev Lett*, 2005, **94**, 1–4.
59. Tress W, Leo K, Riede M, Influence of hole-transport layers and donor materials on open-circuit voltage and shape of I–V curves of organic solar cells', *Adv Funct Mat*, 2011, **21**, 2140–2149.
60. Harada K, Riede M, Leo K, Hild O, Elliott C, 'Pentacene homojunctions: Electron and hole transport properties and related photovoltaic responses', *Phys Rev B*, 2008, **77**, 1–9.
61. Harada K, Werner A G, Kuehl O, Pfeiffer M, Leo K, 'Organic p-i-n homojunctions: Fundamental properties and application in red phosphorescent OLED', *Book of Abstracts of the 8th European Conference on Molecular Electronics – ECME8*, Bologna, Oral contribution, 2005, 74.
62. Tang C W, Van Slyke S A, 'Organic electroluminescent diodes', *Appl Phys Lett*, 1987, **51**, 913–915.
63. Blochwitz J, Pfeiffer M, Fritz T, Leo K, 'Low voltage organic light emitting diodes featuring doped phthalocyanine as hole transport material', *Appl Phys Lett*, 1998, **73**, 729–731.
64. Shirota Y, 'Photo- and electroactive amorphous molecular materials – molecular design, syntheses, reactions, properties, and applications', *J Mater Chem*, 2005, **15**, 75–93.
65. Meerheim R, Luessem B, Leo K, 'Efficiency and stability of p-i-n type organic light emitting diodes for display and lighting applications', *Proceedings of the IEEE 97*, 2009, 1606–1626.
66. Huang J, Pfeiffer M, Werner A, Blochwitz J, Leo K, Liu S, 'Low-voltage organic electroluminescent devices using pin structures', *Appl Phys Lett*, 2002, **80**, 139–141.
67. D'Andrade B W, Forrest S R, Chwang A B, 'Operational stability of electrophosphorescent devices containing p and n doped transport layers', *Appl Phys Lett*, 2003, **83**, 3858–3860.
68. Novaled AG, 2011: http://novaled.de/oledcompetence/advantage_longlifetime.html/ [accessed 31.7.2011].
69. Reineke S, Lindner F, Schwartz G, Seidler N, Walzer K, Luessem B, Leo K, 'White organic light-emitting diodes with fluorescent tube efficiency', *Nature*, 2009, **459**, 234–238.
70. Huang Q, Walzer K, Pfeiffer M, Lyssenko V, He G, Leo K, 'Highly efficient top emitting organic light-emitting diodes with organic outcoupling enhancement layers', *Appl Phys Lett*, 2006, **88**, 113515.
71. Wuerfel P, *Physics of Solar Cells: From Basic Principles to Advanced Concepts* (2nd ed.), Weinheim, Wiley-VCH, 2009.
72. Maennig B, Gebeyehu D, Simon P, Kozlowski F, Werner A, Li F, Grundmann S, Sonntag S, Koch M, Leo K, Pfeiffer M, Hoppe H, Meissner D, Sariciftci N S,

Riedel I, Dyakonov V, Parisi J, Drechsel J, 'Organic p-i-n solar cells', *Appl Phys A Mat Sci & Proc*, 2004, **79**, 1–14.
73. Drechsel J, Maennig B, Gebeyehu D, Pfeiffer M, Leo K, Hoppe H, 'MIP-type organic solar cells incorporating phthalocynine/fullerene mixed layers and doped wide-gap transport layers', *Org Electron*, 2004, **5**, 175–186.
74. Hiramoto M, Suezaki M, Yokoyama M, 'Effect of thin gold interstitial-layer on the photovoltaic properties of tandem organic solar cell', *Chem Lett*, 1990, **19**, 327–331.
75. Westphalen M, Kreibig U, Rostalski J, Luth H, Meissner D, 'Metal cluster enhanced organic solar cells', *Sol Energy Mater Sol Cells*, 2000, **61**, 97–105.
76. Timmreck R, Olthof S, Leo K, Riede M K, 'Highly doped layers as efficient electron–hole recombination contacts for tandem organic solar cells', *J Appl Phys*, 2010, **108**, 033108.
77. Schueppel R, Timmreck R, Allinger N, Mueller T, Furno M, Uhrich C, Leo K, Riede M, 'Controlled current matching in small molecule organic tandem solar cells using doped spacer layers', *J Appl Phys*, 2010, **107**, 044503.
78. Riede M, Uhrich C, Widmer J, Timmreck R, Wynands D, Schwartz G, Gnehr W-M, Hildebrandt D, Weiss A, Hwang J, Sudharka S, Erk P, Pfeiffer M, Leo K, 'Efficient organic tandem solar cells based on small molecules', *Adv Funct Mater*, 2011, **21**, 3019–3028.
79. Dou L, You J, Yang J, Chen C-C, He Y, Murase S, Moriarty T, *et al.*, 'Tandem polymer solar cells featuring a spectrally matched low-bandgap polymer', *Nat Photon*, 2012, **6**, 180–185.
80. Press release: 'Heliatek achieves new world record for organic solar cells with certified 9.8% cell efficiency', http://www.heliatek.com/?p=1346&lang=en [Accessed 6.12.2011].
81. Lee J, Jung B-J, Lee J-I, Chu H Y, Do L-M, Shim H-K, 'Modification of an ITO anode with a hole-transporting SAM for improved OLED device characteristics', *J Mater Chem*, 2002, **12**, 3494–3498.
82. Press release: 'CDT, Sumitomo and Novaled to evaluate Novaled pin OLED structures' http://www.cdtltd.co.uk/press-release/novaled-structures, Cambridge Display Technology Ltd, 2008 [Accessed 31.7.2011].
83. Christiansen D, Alexander C K, 2006, *Standard Handbook of Electrical Engineering* (15th ed.), New York, McGraw-Hill.
84. Schwoerer M, Wolf H C, *Organic Molecular Solids*, Weinheim, Wiley-VCH, 2007.
85. Homepage: http://oe-a.com/ [Accessed 31.7.2011].
86. Homepage: http://www.oes-net.de/ [Accessed 31.7.2011].
87. Homepage: http://www.orgworld.de/ [Accessed 31.7.2011].

3.7 Appendix: compound abbreviations, full names and CAS numbers

α-NPD	*N*,*N*′-Diphenyl-*N*,*N*′-bis(α-naphthyl)-1,1′-biphenyl-4,4′-diamine	123847-85-8
Alq_3	Tris(8-hydroxyquinolinato)aluminium	2085-33-8
AOB	Acridine Orange Base	494-38-2

Bphen	Bathophenanthroline	1662-01-7
BF-DPB	*N*,*N*′-Bis(9,9-dimethyl-fluoren-2-yl)-*N*,*N*′-diphenyl-benzidine	361486-60-4
BEDT-TTF	Bis(ethylenedithio)tetrathiafulvalene	66946-48-3
$CoCp_2$	Cobaltocene, bis(cyclopentadienyl) cobalt(II)	1277-43-6
$Cr(bpy)_3$	Tris(2,2′-bipyridine)chromium	14751-89-4
$Cr_2(hpp)_4$	Tetrakis[μ-(1,3,4,6,7,8-hexahydro-2*H*-pyrimido[1,2-*a*]pyrimidinato] dichromium	200943-63-1
$Cr(TMB)_3$	Tris(4,4′,5,5′-tetramethyl-2,2′-bipyridine) chromium	661466-64-4
CV	Crystal violet	548-62-9
Di-NPD	$N^4,N^{4'}$-Bis[4′-(1-naphthalenylphenylamino)[1,1′-biphenyl]-4-yl]-$N^4,N^{4'}$-diphenyl-[1,1′-biphenyl]-4,4′-diamine	292827-46-4
F2-HCNQ	Difluorohexacyanoquinodimethane	29261-34-5
F4-TCNQ	Tetrafluorotetracyanoquinodimethane	29261-33-4
F6-TNAP	Hexafluorotetracyanonaphthoquinodimethane	912482-15-6
$Ir(piq)_3$	Tris(1-phenylisoquinoline)iridium	435293-93-9
LCV	Leuco Crystal Violet	603-48-5
m-MTDATA	Tris[4-[phenyl(3-methylphenyl)amino] phenyl]amine	124729-98-2
Me-NTCDI	*N*,*N*′-Dimethyl-1,4,5,8-naphthalenetetracarboxylic diimide	20958-66-1
MeO-TPD	*N*,*N*,*N*′,*N*′-Tetrakis(4-methoxyphenyl) benzidine	122738-21-0
$Mo(tfd)_3$	Molybdenum tris[1,2-bis(trifluoromethyl) ethane-1,2-dithiolene]	1494-07-1
N-DMBI	(4-(1,3-Dimethyl-2,3-dihydro-1*H*-benzoimidazol-2-yl)phenyl) dimethylamine	302818-73-1
NTCDA	Naphthalene tetracarboxylic anhydride	81-30-1
PCBM	[6,6]-Phenyl-C_{61}-butyricacidmethylester	160848-22-6
$Ru(^tBu\text{-}terpy)_2$	Bis[4,4′,4″-tris(1,1-dimethylethyl)-2,2′:6′,2″-terpyridine] ruthenium	1041858-64-3
$Ru(terpy)_2$	Bis(2,2′:6′,2″-terpyridine)ruthenium	86959-72-0
TCNQ	Tetracyanoquinodimethane	1518-16-7
TDATA	Tris(diphenylamino)triphenylamine	105389-36-4
TPD	*N*,*N*′-Diphenyl-*N*,*N*′-di(m-tolyl)benzidine	65181-78-4
TT-HAT	Tris(thieno)hexaazatriphenylene	872141-01-0

TTN	Tetrathianaphthacene	193-44-2
VOPc	Vanadyl(IV) phthalocyanine	13930-88-6
$W_2(hpp)_4$	Tetrakis[μ-(1,3,4,6,7,8-hexahydro-2*H*-pyrimido[1,2–*a*]pyrimidinato] ditungsten	463931-34-2
ZnPc	Zinc phthalocyanine	14320-04-8

4
Third-order nonlinear optical properties of π-conjugated polymers with thiophene units and molecular assembly of the polymers

H. KISHIDA, Nagoya University, Japan and
T. YAMAMOTO, Tokyo Institute of Technology, Japan

DOI: 10.1533/9780857098764.1.143

Abstract: Third-order optical nonlinearity in thiophene unit-based conjugated polymers is discussed. Optical nonlinearity is enhanced by improvement of the regioregularity of main chains and reduction of steric repulsion between main chains and side chains. The introduction of charge-transfer character to the electronic states also increases the figure of merit of optical nonlinearity. Moreover, it is explained that the manner of molecular assembly affects the electronic states. Piezochromism in poly(thiophenes) is demonstrated and discussed from the viewpoint of the packing of polymer chains.

Key words: third-order nonlinear optical properties, π-conjugated polymers, polythiophenes, charge transfer (CT), molecular assembly, π-stacking.

4.1 Introduction

In electronics, signal processing is achieved by the control of an electronic signal by another electric signal. The electrons conveying electronic signals are precisely controlled by external voltages or current signals. On the other hand, in optical systems, the control of the optical signal requires the precise control of the optical constants of the materials such as refractive indices and absorption coefficients. When the light is irradiated to the optical material which has large third-order optical nonlinearity, its optical constants can be controlled. Thus, the third-order optical nonlinearity has attractive physical properties for controlling the light signal. In fact, the exploration of the third-order optical nonlinearity has been done for various materials (Auston *et al.*, 1987; Nalwa, 1993).

In the study of the third-order optical nonlinearity, low-dimensional materials are good candidates because the oscillator strength concentrates on the discrete quantum states and the energy states near the van Hove singularity, which are characteristic of low-dimensional quantum systems (Haug and Koch, 2004). Among such low-dimensional materials, conjugated polymers are quite hopeful because of their ultrafast and large optical nonlinearity (Sauteret, 1976; Kersting *et al.*, 1993). In particular, poly(thiophenes) have been studied as promising third-order nonlinear optical materials (Torruellas *et al.*, 1990; Sakurai *et al.*, 1997). Of course, the general merits of conjugated polymers such as wide-range tunability of the gap energies and excellent processability hold as nonlinear optical materials.

In this chapter, we focus on the third-order optical nonlinearity of the thiophene-based conjugated polymers in which the chemical and electronic structures are finely tuned. Among various third-order nonlinear optical effects, we mainly discuss the third-order nonlinear susceptibility $\chi^{(3)}(-3\omega; \omega, \omega, \omega)$ evaluated by the third-harmonic generation (THG) method. THG is the triplicating process of photon energy. The pulsed laser with the photon energy $\hbar\omega$ is irradiated to the thin film samples. By comparing the intensity of the obtained third-harmonic (TH) light (photon energy is $3\hbar\omega$) with the intensity of TH from a reference sample, we can evaluate the magnitude of $\chi^{(3)}(-3\omega; \omega, \omega, \omega)$. In this way, we can discuss the coherent optical nonlinear processes and exclude the incoherent processes such as heat effects or saturation of absorption coming from the accumulation of the excited states. The nonlinear susceptibility obtained by THG gives us a good measure of the magnitude of the coherent optical nonlinearity.

4.2 Third-order nonlinear optical properties of π-conjugated polymers with thiophene units and related compounds

4.2.1 Poly(3-alkylthiophene-2,5-diyl)s and poly (3-alkynylthiophene-2,5-diyl)s

In this subsection, we discuss the optical nonlinearity dependent on the main chain structure in the thiophene-based conjugated polymers. In polythiophenes, the regioregularity of the main chains is controlled by the regularity of the side chains. The regioregularity can be varied by the head-to-tail coupling ratio (r) of alkylthiophene units (Fig. 4.1). The absorption spectra of spin-coated films of poly(3-hexylthiophenes-2,5-diyl)s (P3HT) with various regioregularity are shown in Fig. 4.2 by solid lines (Kishida *et al.*, 2005). With the increase of r, absorption peak energy shifts to the lower energy side. This reveals that the effective conjugation length increases through the π–π stacking. Moreover, in the highly regioregular sample (r = 0.985), a fine structure composed of phonon side bands appears. The energy positions of the zero-phonon, one-phonon and two-phonon bands are indicated by vertical lines. The improvement of the regioregularity

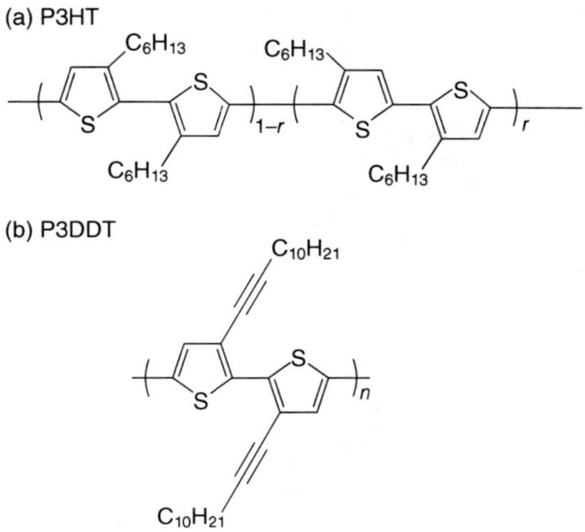

4.1 Chemical structures of (a) poly(3-hexylthiophene-2,5-diyl)s (P3HT) and (b) poly(3-dodecynylthiophene-2,5-diyl)s (P3DDT).

reduces the inhomogeneous broadening and each phonon side band is separated in the absorption spectra. The further improvement of the regioregularity is achieved in poly(3-alkynylthiophene-2,5-diyl)s, in which the steric hindrance between the main chain and the side chain is reduced by the introduction of alkynyl side chain (Fig. 4.1(b)). The absorption spectrum of spin-coated films of poly(3-dodecynylthiophene-2,5-diyl)s (P3DDT) is shown in Fig. 4.3 (Sato *et al.*, 2007). It shows clearer phonon structure up to the two-phonon side band.

The obtained $|\chi^{(3)}(-3\omega; \omega, \omega, \omega)|$ spectra (abbreviated as $|\chi^{(3)}|$ hereafter) of spin-coated films of P3HTs are shown in Fig. 4.2 by filled circles. The photon energies of the excitation laser (denoted as fundamental photon energy in Fig. 4.2) are in the region of 1/3 energy of the absorption peaks. The $|\chi^{(3)}|$ spectra show resonances at 1/3 energies of the absorption peaks, which are attributed to the three-photon resonance. The maximum values of $|\chi^{(3)}|$ increase with increase of r. Comparing the $r = 0.985$ sample with $r = 0.80$, it is noticed that the absorption coefficient α is not as enhanced while the $|\chi^{(3)}|$ values are enhanced by a factor of 3. This directly suggests that the improvement of regioregularity enhances the optical nonlinearity more significantly. Next, the spectral shapes are discussed. In P3HTs, the sample includes the distribution of the effective conjugation length, which affects the shapes of linear and nonlinear optical spectra. For the absorption spectra, the fine phonon structure is blurred for the $r = 0$ to 0.80 samples because the distribution of the conjugation length gives the dispersion of the absorption peak energies. On the other hand, for the $r = 0.985$ sample,

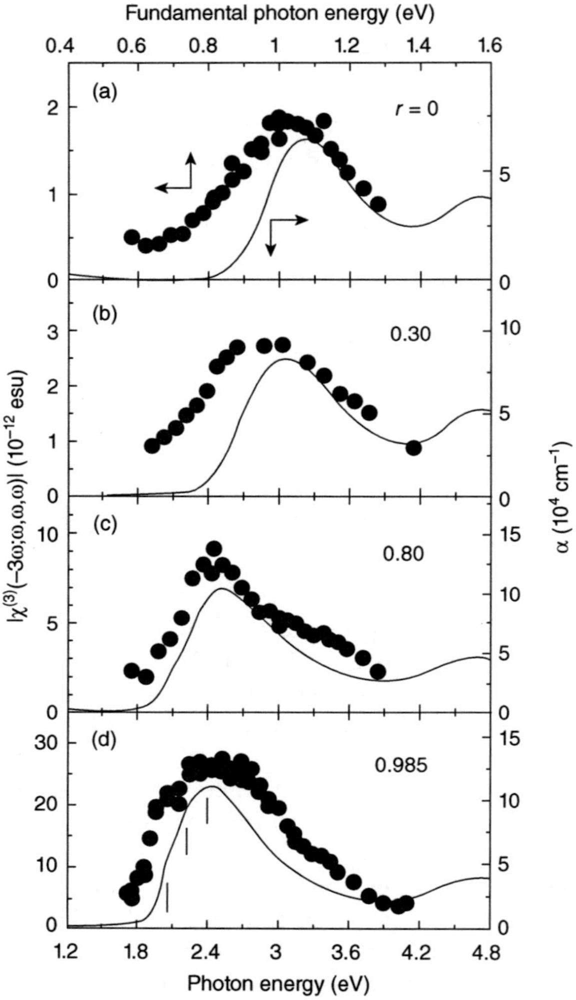

4.2 Absorption coefficient α (solid lines) and $|\chi^{(3)}|$ (dots) spectra of P3HT. Reprinted with permission from H. Kishida, K. Hirota, T. Wakabayashi, H. Okamoto, H. Kokubo and T. Yamamoto, *Appl. Phys. Lett.* 87, 121902 (2005). Copyright 2005, American Institute of Physics.

the effective conjugation length is sufficiently elongated and beyond the size of excited states, namely, excitons. Therefore, the distribution of the conjugation length does not give the dispersion of the absorption peak energies in the $r = 0.985$ sample. As a result, the phonon structure is discernible. For $|\chi^{(3)}|$ spectra, it should be noted that the peak energies of THG spectra are slightly lowered compared with the absorption spectra for the $r = 0$ and 0.30 samples, which include the wide distribution of the effective

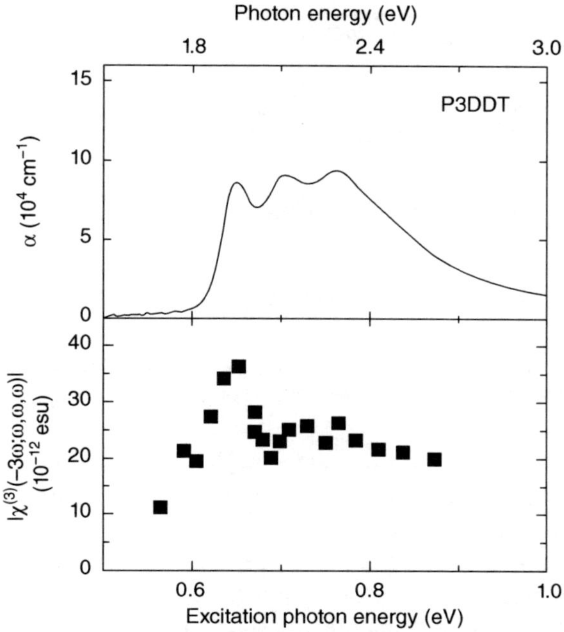

4.3 Absorption coefficient α (upper) and $|\chi^{(3)}|$ (lower) spectra of P3DDT. From Sato *et al.* (2007).

chain length as mentioned above. The chains with longer effective conjugation length, which have small optical gap energies, give larger optical nonlinearity. Therefore, $|\chi^{(3)}|$ are enhanced in the lower energy side of 1/3 energies of the absorption peaks and the $|\chi^{(3)}|$ peak positions are slightly shifted to the lower energy side for the $r = 0$ and 0.30 samples in Fig. 4.2.

The $|\chi^{(3)}|$ spectrum of P3DDT is shown in Fig. 4.3 by filled squares. The three-photon resonance to each phonon side band is clearly observed for P3DDT, while the clear phonon structure is not observed in the $|\chi^{(3)}|$ spectra of P3HTs. P3DDT shows the largest $|\chi^{(3)}|$ value at the zero-phonon band and smaller at one- and two-phonon side bands, which is in contrast with P3HTs.

The magnitude of the optical nonlinearity for various materials can be compared by the figure of merit. In this chapter, we have adopted a figure of merit, $|\chi^{(3)}|/\alpha$. Here, $|\chi^{(3)}|$ and α are the maximum values in each spectrum. $|\chi^{(3)}|$ and α are both proportional to the density of π electrons in conjugated polymers. The density of electrons is dependent on the volume of the side chains. In $|\chi^{(3)}|/\alpha$, the variation of the density of π-electrons is canceled. Therefore, $|\chi^{(3)}|/\alpha$ can be a good measure of the magnitude of third-order nonlinearity, which directly reflects the electronic states. The maximum

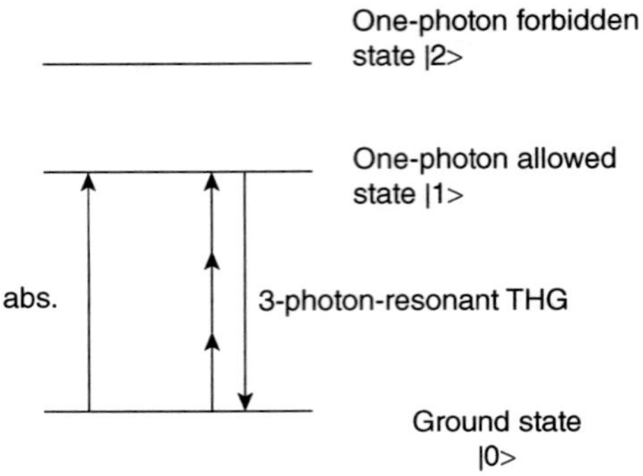

4.4 Energy structures of the excited states. Absorption and three-photon resonant THG processes are depicted.

values of $|\chi^{(3)}|$ and α are proportional to the relating transition dipole moments: $|\chi^{(3)}| \propto (\mu_{21} \cdot \mu_{10})^2$ and $\alpha \propto (\mu_{10})^2$, respectively. Here, μ_{mn} is the transition dipole moment between the states *m* and *n*. The level structure is shown in Fig. 4.4. The state 0 (|0>) is the ground state. The state 1 (|1>) is the first excited state, which is a one-photon allowed state and normally observed in absorption measurements. The state 2 (|2>) is the second excited state, which has the same symmetry with the ground state and is a one-photon forbidden state. In the THG process, three types of resonances such as one-photon, two-photon and three-photon resonances are possible. In Fig. 4.4, the three-photon resonant process to |1> is depicted, in which the first and second photons give virtual nonresonant transitions to the states |1> and |2>, respectively. The probability of such nonresonant transitions is also determined by the relevant transition dipole moments. Therefore, not only the state |1> but also the state |2> have significant roles in the THG process and the transition dipole moment μ_{21} governs the THG process. $|\chi^{(3)}|/\alpha$ is proportional to $(\mu_{21})^2$, indicating that the transition dipole moment between the excited states (|1> and |2>) determines the figure of merit. Thus, $|\chi^{(3)}|/\alpha$ reflects the magnitude of physical parameters of the electronic excited states governing the nonlinear optical processes.

The $|\chi^{(3)}|/\alpha$ values of P3HTs and P3DDT are plotted vs. the absorption peak energies in Fig. 4.5 (Yamamoto *et al.*, 2009a). With the increase of *r*, $|\chi^{(3)}|/\alpha$ increases accompanied by a decrease of the gap energy. Between *r* = 0.80 and 0.985, there is a sudden jump in $|\chi^{(3)}|/\alpha$ though the gap energies are similar. This fact strongly indicates that the improvement of the regioregularity enhances the transition dipole moments between the excited states, resulting in the enhancement of the optical nonlinearity. Moreover, P3DDT shows a larger $|\chi^{(3)}|/\alpha$ value with the decrease of the absorption peak energy. Thus, the improvement of the regioregularity shows two effects: the decrease of the optical gap energies and the increase of the transition dipole moments.

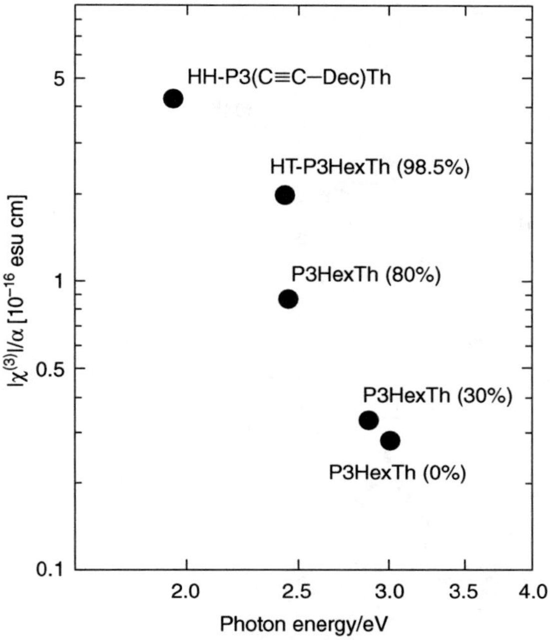

4.5 Values are plotted by the optical gap energies for P3HT (HT-P3HexTh) and P3DDT (HH-P3(C C-Dec)Th). Reprinted with permission from Yamamoto *et al.* (2009a).

The transition dipole moments between excited states are closely correlated with the spatial extent of the excited states. The elongation of the effective conjugation length due to the improvement of the regioregularity leads to the increase of the spatial extent of the excited states and the enhancement of transition dipole moments.

All the above-mentioned optical nonlinearity is for the non-doped states, in which the clear optical gap due to π–π^* transitions opens. Recently, the doped states of polythiophenes have been studied from the viewpoints of nonlinear optical properties (Kishida *et al.*, 2010). In the study, the spin-coated film of regioregular P3HT was dipped into the acetonitrile solution of $Cu(ClO_4)_2$ and the ClO_4^- film was obtained. In the doped states, two polaron states (P1 and P2) appear in between the optical gap and govern the electronic properties. The absorption spectrum for doped states is shown in Fig. 4.6 (Kishida *et al.*, 2010). The polaron absorption bands appear at 0.6 and 1.5 eV, while the absorption of π–π^* transition is suppressed. The $|\chi^{(3)}|$ spectrum by THG measurements is shown in Fig. 4.6 by filled circles. The clear resonant structure is observed at 0.75 eV in excitation photon energy. The double and triple energies of this excitation energy are 1.5 and 2.25 eV, respectively. Though 1.5 eV is equal to the P1–P2 transition energy, this

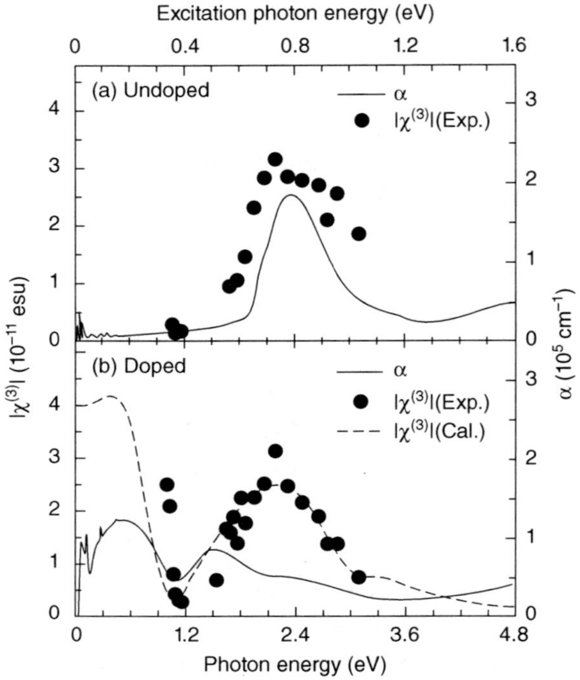

4.6 Absorption (solid line) and THG (dots) spectra of undoped and doped P3HT. The dotted line is the analysis by the discrete level model. Reprinted with permission from H. Kishida, K. Hibino, and A. Nakamura, *Appl. Phys. Lett.* 97, 103302 (2010). Copyright 2010, American Institute of Physics.

transition is not allowed for two-photon transition. 2.25 eV does not correctly correspond to the π–π^* transition energy of the doped P3HT. Therefore, we should ascribe the resonance at 0.75 eV to another resonant effect. Moreover, the THG spectrum for the doped P3HT shows another resonant structure at the energy lower than 0.4 eV in excitation energies. In order to clarify the origin of these resonances, the THG spectrum for the doped P3HT was analyzed by a simple discrete-level model. The analyzed result is shown in Fig. 4.6 by a dotted line and reproduces the experimental result at a satisfying level. It clarifies that the resonant structure at around 0.75 eV in excitation photon energy is due to the multiple resonance to the lower polaron state and π^* state. The resonant process and the level structure are shown in Fig. 4.7 (Kishida *et al.*, 2010). As shown in the figure, the one-photon energy (0.75 eV) is slightly larger than the energy difference between the π and P1 states. The three-photon energy (2.25 eV) is slightly smaller than the energy difference between π and π^* states. Thus, each of one-photon and three-photon energies does not satisfy the resonant

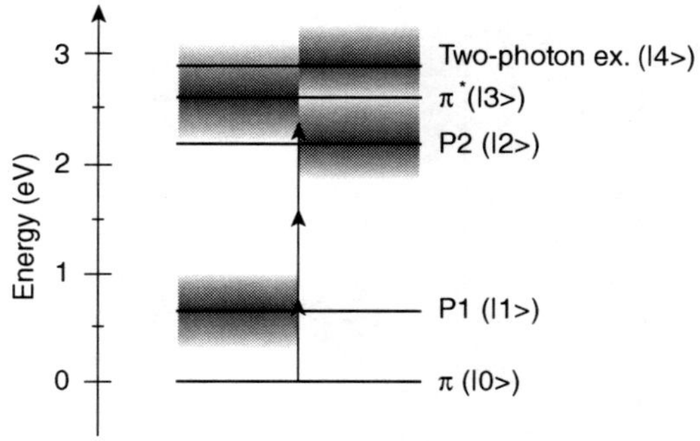

4.7 Multiple resonance in THG process. P1 and P2 are the polaron states. Reprinted with permission from H. Kishida, K. Hibino, and A. Nakamura, *Appl. Phys. Lett.* 97, 103302 (2010). Copyright 2010, American Institute of Physics. Numbering of the states in the figure does not correspond to that in the present article.

condition. However, one is lower than the resonant energy and the other is higher. As a result, quasi-resonance occurs and a clear resonant structure appears at 0.75 eV. The lower energy structure below 0.4 eV is assigned to the three-photon resonance to the transition between π and P1 states. Thus, in the doped conjugated polymers, not only π and π^* states but also polaron states give a dominant contribution to the third-order optical nonlinearity.

4.2.2 π-Conjugated polymers and compounds with charge-transfer (CT) electronic structure along the their main chains

The third-order optical nonlinearity in conjugated polymers results from their π–π^* transitions in most cases. In the simplest case, the photo-generated electron and hole are located at the same site and form excitons. This picture is experimentally and theoretically established and described as one-dimensional excitons (Hasegawa *et al.*, 1992; Abe *et al.*, 1992). However, the fact that an excited electron and a hole share the same site is not always favorable to the optical nonlinearity. Sharing the same site leads to a large oscillator strength of one-photon allowed state and a large absorption coefficient, but also to a decrease in the transition dipole moments between the one-photon allowed and forbidden states. Therefore, the inhibition of the share of the same site can possibly result in the enhancement of the figure of merit. One method to realize this strategy is the introduction of CT character to the electronic states. In this subsection, we discuss the third-order optical nonlinearity in CT conjugated polymers.

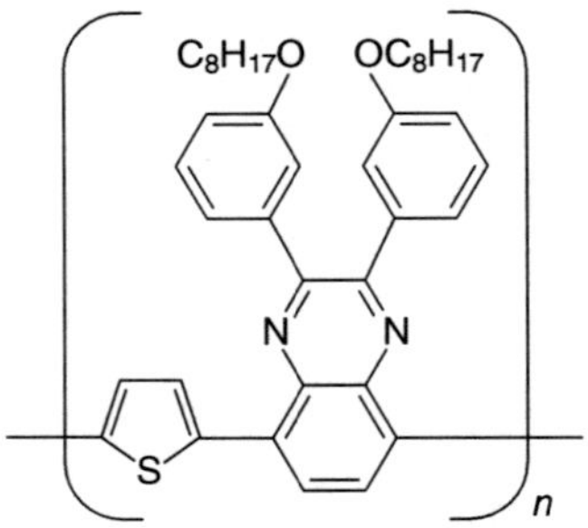

PThQx

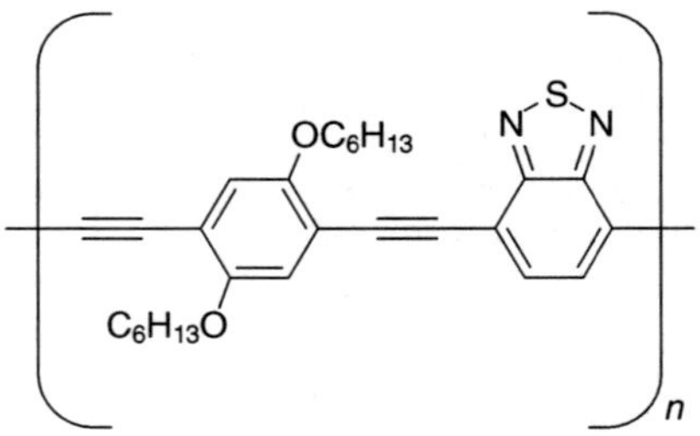

PAE

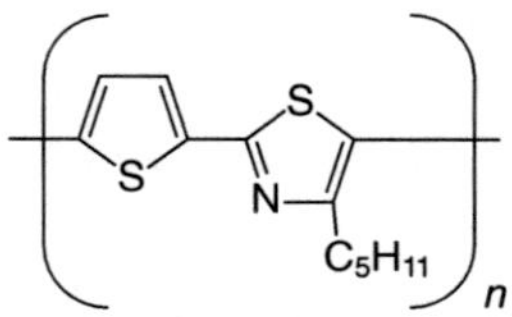

PThTz

4.8 **Chemical structures of charge-transfer conjugated polymers.**

Most conjugated polymers are likely to be a p-type semiconductor because the constituent unit of conjugated polymers has a relatively small ionization energy and is of electron-donating character. By the introduction of electron-accepting units to the conjugated chain, the alternative arrangement of electron-donating and electron-accepting units is achievable. In such polymers, the hole is likely to be on donor (D) molecules and electrons on acceptors (A). In fact, it was reported that the optical gap energy decreases in such polymers (Yamamoto *et al.*, 1996). This observation strongly indicates that the optical excitation has CT character. Hereafter, we call such D–A type polymers as CT polymers. After the first report, various CT polymers have been reported (Yamamoto *et al.*, 2005). The CT polymers studied for the nonlinear susceptibilities are shown in Fig. 4.8. In PThQx, thiophene (Th) units are the donor and quinoxaline (Qx) acceptor. This material can be regarded as acceptor-introduced polythiophenes. In PAE, dialkoxybenzene (BzO) is donor and benzothiaodiazole (BTz)

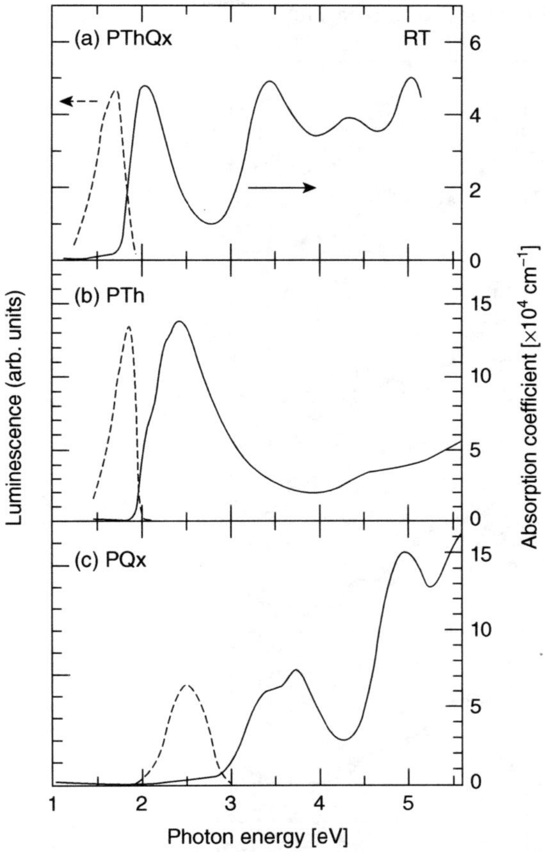

4.9 Absorption (solid lines) and luminescence (broken lines) spectra of PThQx, P3HT ($r = 0.985$) (labeled as PTh in the figure) and PQx. Reprinted with permission from Kishida *et al.* (2004b).

acceptor. In PThTz, Th is donor and thiazole (Tz) acceptor. Depending on the combination of donors and acceptors, the degree of the CT character is controlled. PThQx and PAE have strong CT character, while PThTz has weak CT character.

In Fig. 4.9 (Kishida *et al.*, 2004b), the absorption and luminescence spectra are shown for spin-coated films of PThQx, P3HT and poly(quinoxalines) (PQx). The latter two polymers are the homopolymers consisting of the single components of PThQx. The absorption peak energy of PThQx is lower than those of both P3HT and PQx. This clarifies that the electronic states of PThQx are not the intermediate states of P3HT and PQx but of CT character. Namely, the optical excitation is the transition from the highest occupied molecular orbital (HOMO) of Th to the lowest

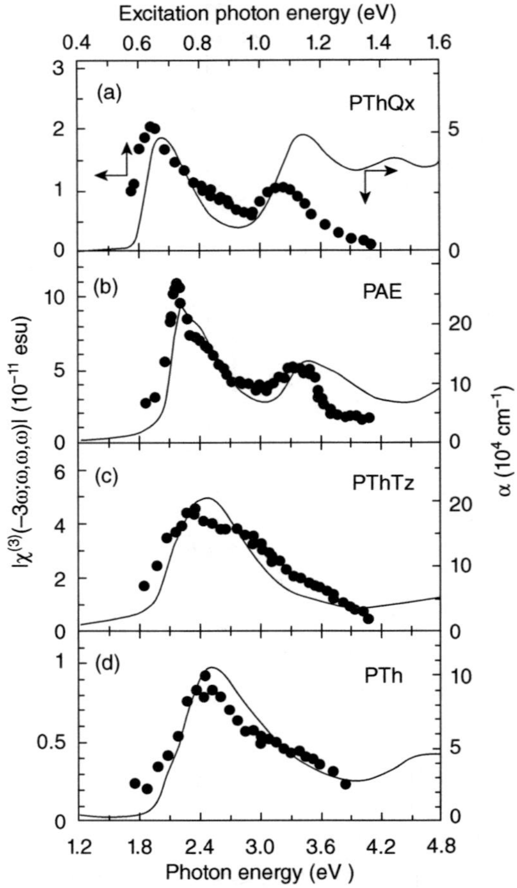

4.10 Absorption and THG spectra of CT polymers and P3HT ($r = 0.80$) (labeled as PTh in the figure). Reprinted with permission from H. Kishida, K. Hirota, H. Okamoto, H. Kokubo, and T. Yamamoto, *Appl. Phys. Lett.* 92, 033309 (2008). Copyright 2008, American Institute of Physics.

unoccupied molecular orbital (LUMO) of Qx. This is evidenced by the fact that a new luminescence peak is observed at the energy lower than those of P3HT and PQx.

In Fig. 4.10 (Kishida *et al.*, 2008), the absorption (solid line) and THG (filled circles) spectra are shown for spin-coated films of three CT polymers and P3HT. Each THG spectrum shows a strong resonance at one-third energy of the absorption peak. The resonances are assigned to a three-photon resonance (Yamamoto *et al.*, 2003a; Kishida *et al.*, 2004a). In order to discuss the effect resulting from the CT character in absorption and THG

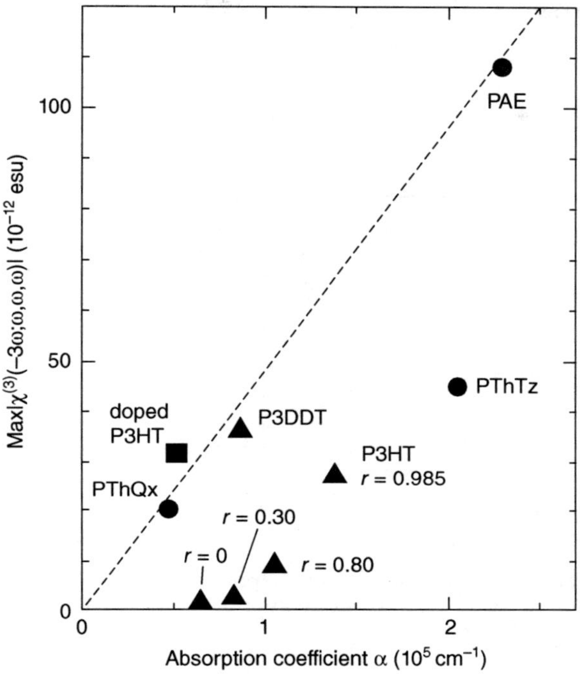

4.11 $|\chi^{(3)}|$ values are plotted vs. the absorption coefficient α . From Kishida (2009).

process, we plotted $|\chi^{(3)}|$ values vs. the absorption coefficients in Fig. 4.11 (Kishida, 2009). PThQx shows a smaller absorption coefficient than P3HTs. This is a result of the introduction of CT character. It inhibits the on-site excitation, namely, the share of the site by the excited electron and hole. The electron must be excited to the different position from the hole. This indicates the decrease of the transition dipole moments between the ground state and the one-photon allowed state, leading to a decrease in the oscillator strength contributing to the absorption coefficient. On the other hand, PThQx show comparable $|\chi^{(3)}|$ values with P3HT (the $r = 0.985$ sample). This indicates that μ_{21} increases though μ_{10} decreases. As a result that CT polymers get stronger CT character for the one-photon allowed state, it has similar spatial extent with the one-photon forbidden state, and the spatial overlap between the one-photon allowed and one-photon forbidden states increases. In Fig. 4.11, the slope of the line between the origin and the data point corresponds to the figure of merit. PAE has the similar figure of merit as PThQx. PThTz, which has weaker CT character, has a smaller figure of merit. Moreover, P3HTs show smaller figures of merit. This indicates that CT character enhances the optical nonlinearity evidently. One more thing

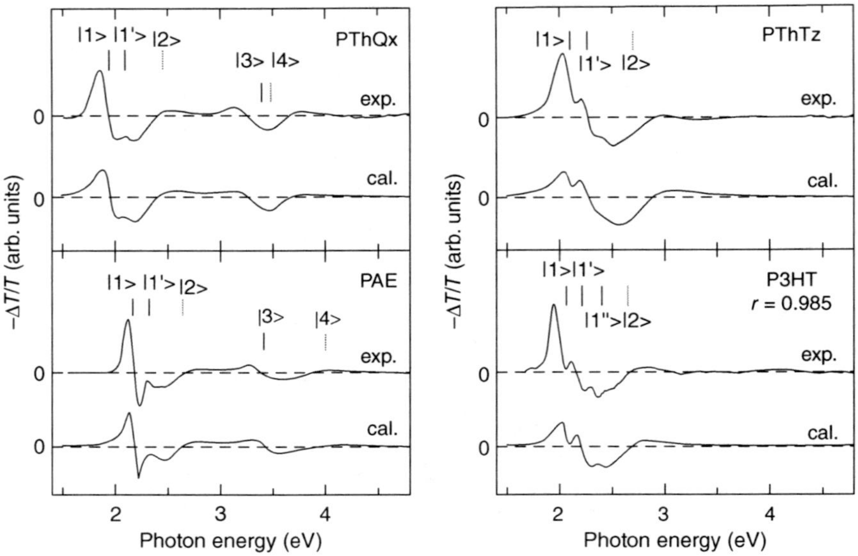

4.12 Electroabsorption spectra of CT polymers. Experimental spectra are compared with the ones calculated by a discrete level model. From Kishida *et al.* (2009).

to note is that P3DDT shows a large figure of merit, which is comparable with those of PThQx and PAE. This also indicates that, even if CT character is not introduced, the improvement of the main chain structure enhances the figure of merit.

As mentioned above, to obtain large $|\chi^{(3)}|/\alpha$ values, we can take two methods. The first one is to improve the regioregularity. The other one is the introduction of CT character. Both strategies aim to increase the transition dipole moment between the excited states. However, the detailed strategies are different. The former one is that the elongation of the effective conjugation length should lead to the expansion of the size of the excited states. In the latter strategy, the spatial extent of the one-photon allowed state is modified to be similar with one-photon forbidden states. In both strategies, the enhancement of the transition dipole moment between the excited states |1> and |2> is expected. To understand of the character of the state |2>, we should directly observe the state |2>. One method to observe it is electroabsorption measurements. In this method, the applied electric field breaks the inversion symmetry and makes the one-photon forbidden states optically allowed. As a result, the one-photon forbidden excited states are observed in absorption measurements under electric field and their energy position is obtained. The experimental results of the electroabsorption measurements in the CT polymers are shown in Fig. 4.12

(Kishida *et al.*, 2009). In the study, an alternating electric field of 1 kHz is applied to the spin-coated polymer film (typically a few hundred nm thick), which is sandwiched between a transparent ITO electrode and a semitransparent aluminum electrode. The typical electric field is from 0.5 to 2 MV/cm. The change of the transmittance (T), $-\Delta T/T$, is measured point by point in wavelength using the lock-in method and the difference spectra induced by the electric field are obtained. The positive signal is indicative of the increase of the absorption, while the negative signal the decrease of the absorption. All four spectra in Fig. 4.12 show positive and negative signals below 2.5 eV. These are mainly due to the Stark shift of the one-photon allowed states. Between 2.5 and 3.0 eV, positive signals appear. This signal suggests that the one-photon forbidden state is located there. The fitting results by a discrete level model are shown in the figure and reproduce the experimental results satisfactorily. The obtained energy positions are indicated by vertical bars in the figure. In the figure, |1> and |2> are the one-photon allowed state and the one-photon forbidden state, respectively. |1′> and |1″> are the one-phonon and two-phonon side bands of the state |1>, respectively. In PThQx and PAE, another structure is observed in higher energy region (3–4 eV), which are attributed to the transitions of constituent components.

In P3HT and CT polymers, the one-photon forbidden states |2> are located in the higher-energy side of the one-photon allowed state |1>. The energy separation between the states |1> and |2> is around 0.5–0.6 eV commonly in P3HT and CT polymers. If the CT character is strong, this energy separation should be reduced depending on the intensity of CT character. Considering that the present CT polymers have a relatively large energy separation, we can expect that further enhancement of CT character gives rise to the degeneracy of the one-photon allowed state and the one-photon forbidden state. Such degeneracy might lead to a further enhancement of the transition dipole moment between the excited states and improvement of the figure of merit.

4.3 Packing and molecular assembly of π-conjugated polymers

π-Conjugated polymers with long alkyl or alkoxy side chains often assume stacked structures in the solid state. When stiff π-conjugated polymers such as poly(*p*-phenylene) PPP, poly(thiophene-2,5-diyl) PTh (cf. Fig. 4.13), and poly(2,2′-bipyridine-5,5′-diyl) have no side chain, they usually assume a herringbone packing shown in Fig. 4.14 (Yamamoto, 2010). In the herringbone packing, there seems to be no effective electronic interaction between the π-conjugated polymer molecules because the π-electron cloud in the

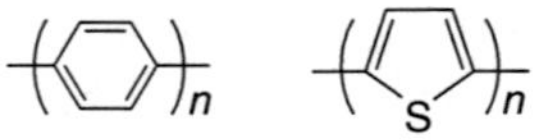

PPP PTh

4.13 Chemical structures of poly(*p*-phenylene) PPP and poly(thiophene-2,5-diyl) PTh.

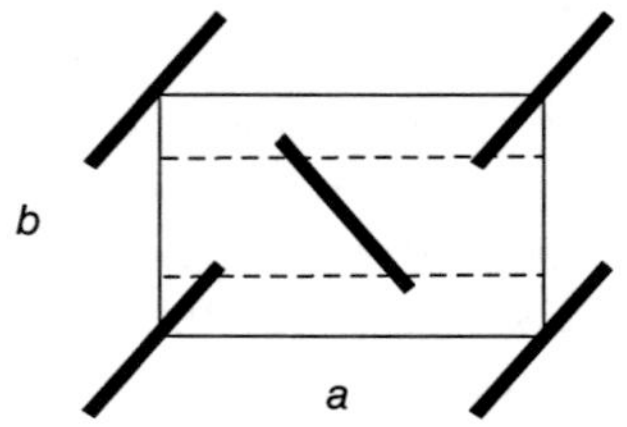

	a/nm	*b*/nm
PE	0.79	0.49
PPP	0.78	0.55
PTh	0.78	0.55

4.14 Herringbone packing of polyethylene (PE), poly(*p*-phenylene) (PPP) and poly(thiophene-2,5-diyl) (PTh). Seen from the direction of polymer main chain. From Yamamoto (2010).

polymer chain does not seem to overlap those in other polymer chains effectively.

4.3.1 HT-P3HT, P3DDT and rand-P3HT

When the π-conjugated polymer has long side chains such as long alkyl side chains and long alkoxy side chains, the polymer often assumes a π-stacked face-to-face packing structure to cause electronic interaction between the polymer molecules. The π-stacked face-to-face packing usually leads to a shift of the UV–vis peak of π-conjugated polymers to a longer wavelength, which is crucial for larger $|\chi^{(3)}|$. Typical examples of such polymers are head-to-tail regioregular poly(3-hexylthiophene-2,5-diyl), HT-P3HT (r = 1 or near 1 in Fig. 4.1(a)), and P3DDT shown in Fig. 4.1. They are thought to form π-stacked structures shown in Figs 4.15 and 4.16, respectively, in the solid state as well as in colloidal solutions. The packing structures have mainly been analyzed by powder X-ray diffraction (XRD) patterns of the polymers.

Such types of packing shown in Figs 4.15 and 4.16 require good ordering (or orientation) of the side chains, and the packing is possible only for polymers with controlled structures, which make the herringbone-type packing of the side chains possible as shown in Fig. 4.16(b). When the direction of the side chain along the polymer main chain is not controlled, as in regio random poly(3-hexylthiophene-2,5-diyl) (rand-P3HT), such clear packing shown in Fig. 4.15 is not possible. The comparison of UV–vis data of

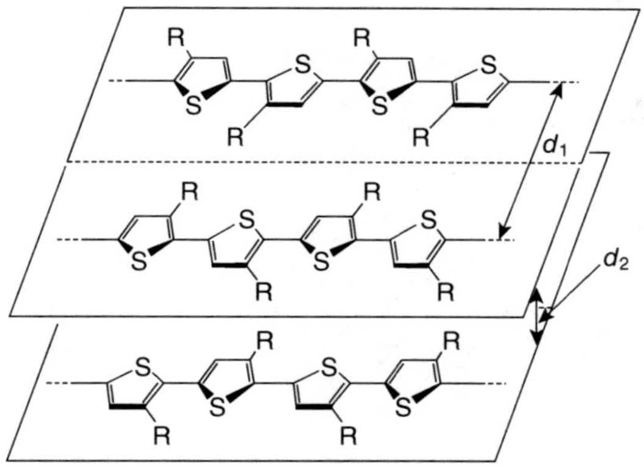

4.15 A packing model of HT-P3HT (R = hexyl). From Yamamoto (2010).

HT-P3HT and its corresponding regio random polymer, rand-P3HT, reveals the following:

- In good solvents (e.g., chloroform), HT-P3HT and rand-P3HT show analogous UV–vis spectra with UV–vis peaks at near positions.
- In poor solvents (e.g., a mixture of methanol and chloroform), HT-P3HT assembles molecularly to give a colloidal solution and the UV–vis peak of HT-P3HT shifts to a longer wavelength. On the other hand, the tendency for rand-P3HT to cause such a shift to a longer wavelength and to form the colloidal solution is weaker than that of HT-P3HT. Rand-P3HT with a HT/HH ratio ((head-to-tail content)/(head-to-head content) ratio) of 7/3, rand-P3HT(7/3), shows some shift of the UV–vis peak to a longer wavelength when methanol is added to the chloroform solution of the polymer, indicating that rand-P3HT(7/3) partly form colloidal particles in which several local parts (presumably HT-rich local parts) in a rand-P3HT(7/3) molecule form the π-stacked packing with other rand-P3HT(7/3) molecules. However, rand-P3HT with a HT/HH ratio of 1/2, rand-P3HT(1/2) does not show a significant shift of the UV–vis peak to a longer wavelength (Yamamoto *et al.*, 1998). The increase in the head-to-head content in poly(3-hexylthiophene-2,5-diyl) seems to suppress the face-to-face π-stacking molecular assembly of the polymer.
- In the solid state (cast film), HT-P3HT gives rise to a UV–vis spectrum similar to that of the colloidal solution described above. The UV–vis spectrum of rand-P3HT(1/2) in the solid state resembles that of rand-P3HT(1/2) in good solvent.

Similar results have been obtained with P3DDT, and UV–vis data of P3DDT in a good solvent (1,2-dichlorobenzene at 135 °C) and in the solid

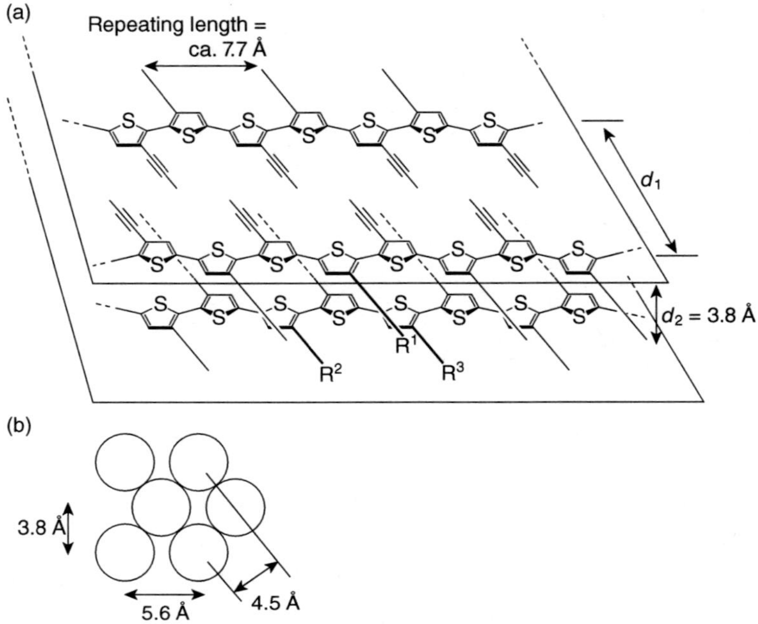

4.16 (a) Packing model of P3DDT (side chain = dodecynyl) in the solid state and (b) side view of the packed –C≡C– alkyl side chains, seen from the direction of the –C≡C– alkyl side chain. Only two layers in the π-stacked polymer solid are shown. The thiophene unit in the lower layer is believed to locate below the thiophene unit in the upper layer, with the opposite direction of S. In this situation, the alkyl groups in the side chains are presumed to be packed in a model shown in part (b). The repeating length was estimated from single-crystal X-ray data of thiophene oligomers, and aggregation of the alkyl groups in the side chain is assumed. The side chains are considered to be tilted from the polymer main chain. Modified from Yamamoto *et al.* (2009a) and Yamamoto (2010).

state are shown in Fig. 4.17. As shown in Fig. 4.17, the UV–vis peak of P3DDT in the good solvent (Fig. 4.17(a); 520 nm) shifts to a longer wavelength (Fig. 4.17(b); 584 nm with substructures). When the molecular weight of P3DDT is small (number average molecular weight M_n = 4400; cf. Fig. 4.17(b)), there do not seem to be enough π-interacting sites between the π-conjugated polymer molecules, and the self-assembly of the polymer does not seem to proceed enough to cause only a minor changes in the UV–vis spectrum in the solid state.

These results and the XRD data of HT-P3HT and P3DDT support a notion that in poor solvents and solid state, the π-conjugated polymers are self-organized and the π-electronic interaction between the face-to-face stacked polymer molecules leads to the shift of the UV–vis peak to a longer

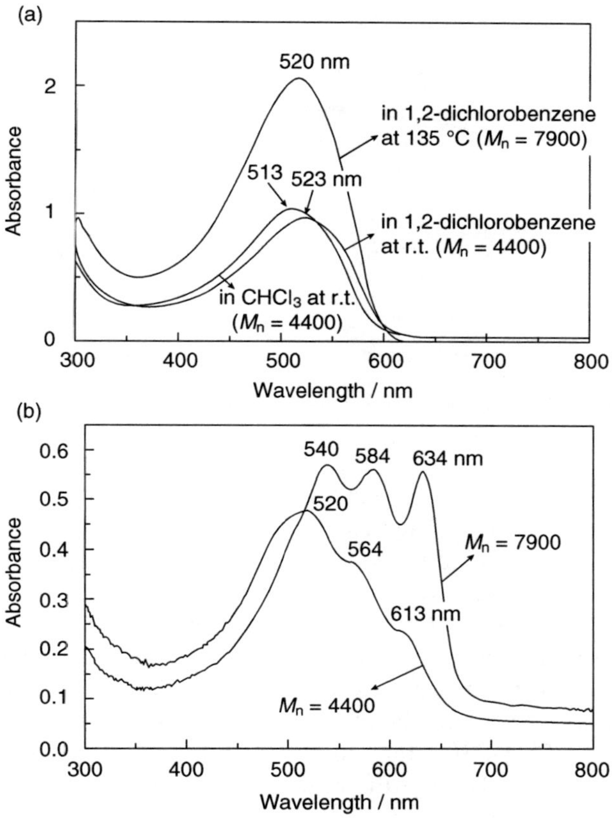

4.17 UV–vis spectra of original P3DDT with M_n of 7900 and a low-molecular-weight fraction of P3DDT with M_n of 4400: (a) in solutions; (b) in cast film. From Yamamoto *et al.* (2009a).

wavelength. The reason for the shift to a longer wavelength has not been elucidated enough. For low-molecular-weight compounds, electronic interaction between π-conjugated dye molecules has been explained in terms of H-aggregate and J-aggregate. F. C. Spano theoretically analyzed optical properties of HT-P3HT (Spano, 2005; Spano *et al.*, 2009).

The electronic interaction between the π-conjugated polymers may cause delocalization of π-electrons among the π-stacked polymer molecules and bring about new electronic states. For example, HOMO of single π-conjugated polymer with E(HOMO) may split into two orbits with E(HOMO) + ΔE and E(HOMO) – ΔE when it electronically interacts with HOMO of another π-conjugated polymer. LUMO of single π-conjugated polymer with E(LUMO) may also split into two orbits with E(LUMO) + ΔE and E(LUMO) – ΔE. This splitting resembles that caused by σ-bond

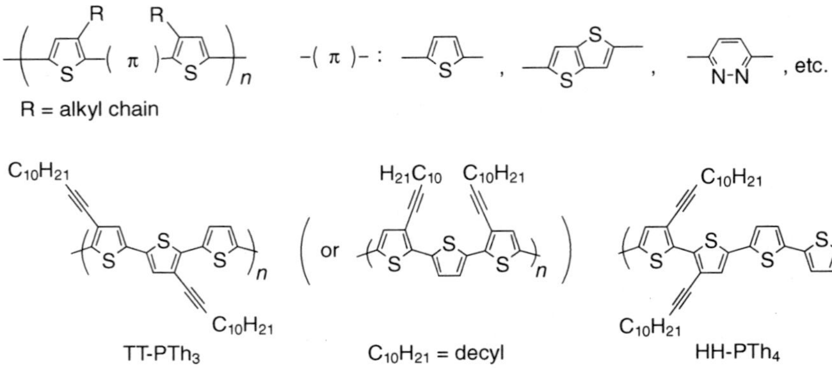

4.18 Examples of Th(R)-(π)-Th(R) type π-conjugated polymers. R = alkyl or alkynyl.

formation between two π-conjugated molecules, as splitting caused by formation of biphenyl from benzene. The selection rule for the electronic transition between the newly formed HOMOs and LUMOs is not clear for π-stacked polymer molecules. However, the electronic transition from the E(HOMO) + ΔE orbit to the E(LUMO) − ΔE orbit requires less energy than that required for the electronic transition occurring in single π-conjugated molecule, and this may be the reason for the bathochromic shift of the π–π^* transition by forming the face-to-face π-stacked molecular assembly. The larger $|\chi^{(3)}|$ of HT-P3HT than that of rand-P3HT (cf. Fig. 4.5) is thought to originate principally from the bathochromic shift caused by the π-stacked molecular assembly.

4.3.2 Th(R)–(π)–Th(R) type π-conjugated polymers

Recently various Th(R)–(π)–Th(R) type polymers (Th(R): thiophene-2,5-diyl unit having long side chain R (R = alkyl or alkynyl) at the β-position; –(π)–: connecting π-conjugated unit; examples of the polymers are shown in Fig. 4.18) are reported to self-assemble and form similar face-to-face π-stacked structures (e.g., Kokubo and Yamamoto, 2001; Kokubo *et al.*, 2006; McCulloch *et al.*, 2006; Ong *et al.*, 2005; Wu *et al.*, 2005; Chabinyc *et al.*, 2007; DeLongchamp *et al.*, 2007; Li *et al.*, 2008; Ohkita *et al.*, 2008). An example of a postulated packing structure of the polymer is shown in Fig. 4.19.

4.3.3 π-conjugated polymers consisting of –$C_6H_2(OR)_2$– *p*-phenylene units

When a *p*-phenylene unit in π-conjugated polymers has long alkoxy side chains at the 2,5-positions, the 2,5-dialkoxy-*p*-phenylene unit serves as an

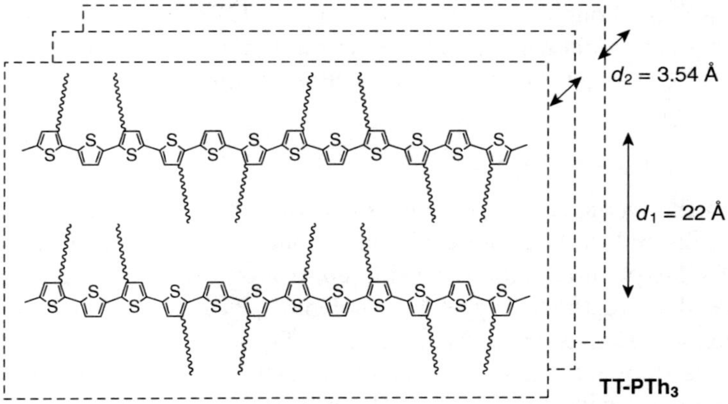

4.19 A postulated packing structure of TT-PTh3 shown in Fig. 4.18. From Yamamoto *et al.* (2009a) and Yamamoto (2010).

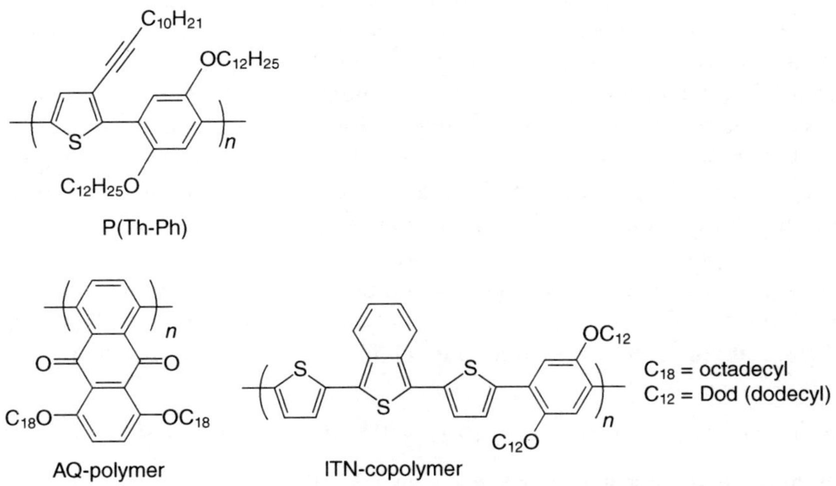

4.20 Examples of π-conjugated polymers with the $-C_6H_2(OR)_2$-p phenylene units in the main chain. An example of π-conjugated polymer (poly(anthraquinone): AQ-polymer) with a similar p-dialkoxy phenylene unit in the side part is also shown (left, bottom) (Yamamoto *et al.*, 2009a).

excellent building block which leads to self-assembly and π-stacking of the polymer (Yamamoto, 2009; Yamamoto *et al.*, 2010). Examples of such polymers are shown in Fig. 4.20, and a packing structure for the ITN-copolymer is proposed (Yamamoto *et al.*, 2010). PAE shown in Fig. 4.8 also has the 2,5-dialkoxy-*p*-phenylene unit and shows a strong tendency to self-assemble (Yamamoto *et al.*, 2003b); self-assembly of PAE shown in

Fig. 4.8 is thought to be assisted by a CT-electronic structure (the 2,5-dialkoxy-*p*-phenylene unit = electron-donor unit; the benzothiadiazole unit = electron-acceptor unit) as discussed below.

4.3.4 CT-polymers

When the π-conjugated polymer is constituted of donor (D) and acceptor (A) units and the polymer chain has a controlled structure, intermolecular D–A CT-type interaction is considered to assist the face-to-face π-stacking. Molecular structures of typical CT-polymers, PThQx and PThTz, are shown in Fig. 4.8. For example, thiazole unit (Tz) is an electron-accepting unit whereas thiophene unit (Th) is an electron-donating unit, and π-conjugated polymers constituted of Tz and Th units are thought to have an intramolecular CT electronic structure along the π-conjugated main chain. Generation of this type of CT electronic state is considered to bring about the UV–vis peak at a longer wavelength and larger $|\chi^{(3)}|$ as discussed above. In addition, intermolecular CT electronic interaction is considered to affect the face-to-face π-stacking molecular assembly of the CT polymers (Yamamoto *et al.*, 2003c). An example of postulated packing structure of such a CT polymer has been reported (Yamamoto *et al.*, 2006). Even when the side chain is somewhat more bulky with heterocyclic units such as –Th_n–R (Th = thiophene-2,5-diyl; R = alkyl), the alternating CT-type π-conjugated polymer can form the face-to-face π-stacked molecular assembly (Yamamoto *et al.*, 2011). This molecular assembly also leads to a shift of the UV–vis peak to a longer wavelength. CT-type π-conjugated oligomers and polymers tend to show larger two-photon excitation action cross-sections than π-conjugated oligomers and polymers without the CT electronic structure along the π-conjugated main chain (Yamamoto *et al.*, 2009b).

4.3.5 Poly(arylene ethynylene) PAE-type polymers

Poly(arylene)s and poly(heterocycle)s such as PPP shown in Fig. 4.13 often have steric repulsion (e.g., steric repulsion caused by *o*-hydrogen repulsion) between the adjacent aromatic (or heterocyclic) units along the polymer main chain and have a twisted main chain, at least as a single polymer molecule. In the case of PAE-type polymers containing the –C≡C– unit in the main chain (cf. Fig. 4.8), the polymer will not receive such a steric repulsion because of the presence of the –C≡C– space between the aromatic (or heterocyclic) units. Consequently PAE-type can easily assume a coplanar main chain, which is advantageous for the formation of the face-to-face π-stacked molecular assembly. In particular when the polymer is constituted of electron-donating aromatic (or heterocyclic) and electron-accepting (or heterocyclic) units, as in the case of PAE shown in Fig. 4.8,

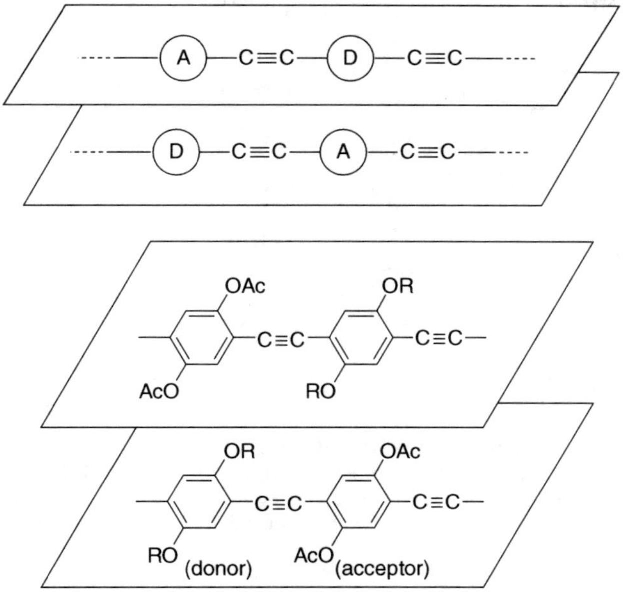

4.21 A face-to-face π-stacked packing model for molecular assembly of the PAE-type polymer with the CT-type electronic structure along the polymer (Yamamoto *et al.*, 1999).

the intermolecular CT interaction assists the face-to-face π-stacked molecular assembly, as shown in Fig. 4.21 (Yamamoto *et al.*, 1999, 2003b).

4.3.6 Piezochromism and alignment of the π-conjugated polymers on the surface of substrates

HT-P3HT and P3DDT are thought to assume the packing structures shown in Figs 4.15 and 4.16 in the solid state. When pressure is applied to the polymer solid, the π-stacking distance (d_2 in Figs 4.15 and 4.16) will be shortened. This shortening of d_2 has actually been revealed by XRD data of an analogue of P3HT, P3OT (alkyl side chain is octyl, instead of hexyl) (Samuelsen *et al.*, 1999). Application of high pressure (0.5 GPa) to a solid of HT-P3OT leads to a red shift of the UV–vis absorption peak by 0.04 eV (or 0.08 eV GPa^{-1}) (Kaniowski *et al.*, 1998), presumably because of an increase in the intermolecular electronic interaction between the π-conjugated polymers by shortening the π-stacking distance. The emission peak in the PL (photoluminescence) spectrum of HT-P3HT also shifts to a longer wavelength by about 0.15 eV (or 0.03 eV GPa^{-1}) when a high pressure (5 GPa) is applied (Muramatsu, 2001). Piezochromic data (shifts of

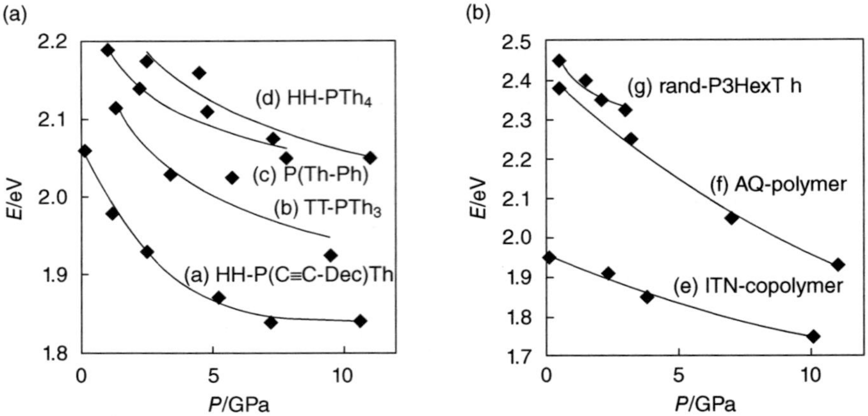

4.22 Piezochromic data (shifts of the UV–vis peak with the transition energy E) of π-conjugated polymers. For HH-PTh$_4$, P(Th-Ph), TT-PTh$_3$, AQ-polymer and ITN-copolymer, see Figs 4.18 and 4.20. HH-P(C≡C-Dec)Th = P3DDT. Rand-P3HexTh = rand-P3HT. From Yamamoto *et al.* (2009a).

UV–vis peaks caused by pressure) of π-conjugated polymers are shown in Fig. 4.22 (Yamamoto *et al.*, 2009a).

As described above, π-conjugated polymers often assume the face-to-face π-stacked structure in the solid state and colloidal solutions. Similar to cases of crystalline and amorphous silicons, morphology of π-conjugated polymers strongly affects on their electronic and optical properties. For example, HT-P3HT (film) gives rise to a larger carrier mobility (Sirringhause *et al.*, 1999) and a larger Faraday effect (Gangopadhyay *et al.*, 2011) than rand-P3HT (film), in addition to the above-described larger $|\chi^{(3)}|$ of HT-P3HT than that of rand-P3HT. Recently various π-conjugated polymer-based photovoltaic cells have been investigated, and it is reported that so-called bulk heterojunction solar cells show superior results; the cell uses molecularly self-assembled π-conjugated polymer layers (e.g., Chen *et al.*, 2008; Park *et al.*, 2009; Cowan *et al.*, 2010; Price *et al.*, 2011). π-Conjugated polymers are partly used in industry (e.g., capacitors) and are expected to support future industry. Revealing optical and electronic properties of π-conjugated polymers in relation to their morphology is important not only for basic interest but also for application of the polymer in industry.

4.4 Conclusions and future trends

We have demonstrated that the third-order optical nonlinearity in thiophene-based conjugated polymers can be enhanced by the control of chain structures and electronic structures. The chain structure is controlled by the head-to-tail coupling ratio in poly(3-hexylthiophene-2,5-diyl)s. The introduction of the alkynyl side chain reduces the steric repulsion between the main and side chains, which is demonstrated in poly(3-dodecynylthiophene-2,5-diyl)s. These structural improvements induce the enhancement of the transition dipole moments between the excited states and the increase of the figure of merit, $|\chi^{(3)}|/\alpha$. The improvement of the electronic structure is achieved by the introduction of the CT characters. The enhancement in the figure of merit is demonstrated in the three CT polymers. It results from the modification of the excitonic wavefunctions for one-photon-allowed and one-photon-forbidden states.

The structural improvements of π-conjugated polymers and the introduction of the CT character to π-conjugated polymers cause molecular assembly or face-to-face π-stacking of π-conjugated polymers. This π-stacking is thought to lead to an expansion of the electronic state in the single π-conjugated polymer to the whole of the molecular assembly. The expansion of the electronic state or delocalization of the wave function brings about not only a shift of the UV–vis absorption peak to a longer wavelength and enhancement of the third-order optical nonlinearity but also improvement of mobility of carrier in the π-conjugated polymer.

Thus, it is presented that the improvement of the chain structures and the introduction of CT character can be the strategy of the enhancement of third-order optical nonlinearity. These strategies might be applied to the other conjugated polymers. Moreover, the modification of the excitonic wavefunctions by the introduction of CT character can be widely applied to other systems such as organic molecular systems. We hope that the study shown in this chapter gives some clues and a model to the development of the third-order nonlinear optical materials.

4.5 Acknowledgments

This work is in collaboration with Professor H. Okamoto (University of Tokyo), Professor A. Nakamura (Nagoya University), and the members of their laboratories. T.Y. is grateful to his coworkers, especially to Professor T. Kanbara of University of Tsukuba and Professors T. Maruyama, T.-A. Koizumi, and H. Fukumoto of our laboratory, for helpful discussion and experimental supports.

4.6 References

Abe, S., Yu, J. and Su, W. P. (1992) *Phys. Rev. B.*, **45**, 8264–71.

Auston, D. H. *et al.* (1987) *Appl. Optics*, **26**, 211–34.

Chabinyc, M. L., Toney, M. F.; Kline, R. J., McCulloch and I., Heeney, M. (2007) *J. Am. Chem. Soc.*, **129**, 3226–3237.

Chen, C.-P., Chan, S.-H., Chao, T.-C., Ting, C. and Ko, B.-T. (2008) *J. Am. Chem. Soc.*, **120**, 12828–93.

Cowan, S. R., Roy, A. and Heeger, A. J. (2010) *Phys. Rev.* **82**, 245207/1–10.

DeLongchamp, D. M., Kline, R. J., Lin, E., Fischer, D. A., Richter, L. J., Lucas, L. A., Heeney, M., McCulloch, I. and Northrup, J. E. (2007) *Adv. Mater.*, **19**, 833–7.

Gangopadhyay, P., Koeckelberg, G. and Persoons, A. (2011) *Chem. Mater.*, **23**, 516–21.

Hasegawa, T., Iwasa, Y., Sunamura, H., Koda T., Tokura Y., Tachibana H., Matsumoto M. and Abe S. (1992) *Phys. Rev. Lett.*, **69**, 668–71.

Haug, H. and Koch, S. W. (2004) *Quantum theory of the optical and electronic properties of semiconductors*, pp. 184–92, Singapore: World Scientific.

Kaniowski, T., Nizioł, S., Sanetra, J., Trznadel, M. and Proń, A. (1998) *Synth. Met.*, **94**, 111–14.

Kersting, R., Lemmer, U., Mahrt, R. F., Leo, K., Kurz, H., Bässler, H. and Göbel, E. O. (1993) *Phys. Rev. Lett.*, **70**, 3820–3.

Kishida, H., Hirota, K., Wakabayashi, T., Lee, B. L., Kokubo, H., Yamamoto, T. and Okamoto, H. (2004a) *Phys. Rev. B.*, **70**, 115205-1–5.

Kishida, H., Okamoto, H., Lee, B. L., Kokubo, H. and Yamamoto, T. (2004b) *Trans. Mater. Res. Soc. Jpn.*, **29**, 981–4.

Kishida, H., Hirota, K., Wakabayashi, T., Okamoto, H., Kokubo, H. and Yamamoto, T. (2005) *Appl. Phys. Lett.*, **87**, 121902-1–3.

Kishida, H., Hirota, K., Okamoto, H., Kokubo, H. and Yamamoto, T. (2008) *Appl. Phys. Lett.*, **92**, 033309-1–3.

Kishida, H. (2009) *Control of super-hierarchical structures and innovative functions of next-generation conjugated polymers.*, eds. K. Akagi, Tokyo: CMC, pp. 257–62, in Japanese.

Kishida, H., Hirota, K., Okamoto, H., Kokubo, H. and Yamamoto, T. (2009) *Synth. Metals*, **159**, 868–70.

Kishida, H., Hibino, K. and Nakamura, A. (2010) *Appl. Phys. Lett.*, **97**, 103302-1- 3.

Kokubo, H. and Yamamoto, T. (2001) *Macromol. Chem. Phys.*, **202**, 1031–4.

Kokubo, H., Sato, T. and Yamamoto, T. (2006) *Macomolecules*, **39**, 3959–63.

Li, J., Qin, F., Li, C. M., Bao, Q., Chan-Park, M. B., Zhang, W., Qin, J. and Ong, B. S. (2008) *Chem. Mater.*, **20**, 2007–9.

McCulloch, I., Heeney, M., Bailey, C., Genevicius, K., MacDonald, I., Shkunov, M., Sparrowe, D., Tierney, S., Wagner, R., Zhang, W., Chabinyc, M. L. Kline, R. J., McGehee, M. D. and Toney, M. F. (2006) *Nature Mater.*, **5**, 328–33.

Muramatsu, Y., Yamamoto, T., Hasegawa, M., Yagi, T. and Koinuma, H. (2001) *Polymer*, **42**, 6673–5.

Nalwa, H. S. (1993) *Adv. Mater.*, **5**, 341–58.

Ohkita, H., Cock, S., Astuti, Y., Duffy, W., Tierney, S., Zhang, W., Heeney, M., McCulloch, I., Nelson, J., Bradley, D. D. C. and Durrant, J. R. (2008) *J. Am. Chem. Soc.*, **130**, 3030–42.

Ong, B. S., Wu, Y., Liu, P. and Gardner, S. (2005) *Adv. Mater.*, **17**, 1141–4.

Park, S. H., Roy, A., Beaupré, S., Cho, S., Coates, N., Moon, J. S., Moses, D., Leclerc, M., Lee, K. and Heeger, A. J. (2009) *Nature Photonics*, **3**, 297–303.

Price, S. C., Stuart, A. C., Yang, L., Zhou, H. and You, W. (2011) *J. Am. Chem. Soc.*, **133**, 4625–31.

Sakurai, K., Tachibana, H., Shiga, N., Terakura, C., Matsumoto, M. and Tokura, Y. (1997) *Phys. Rev. B.*, **56**, 9552–6.

Samuelsen, E. J., Mårdalen, J., Konestabo, O. R., Hanfland, M. and Lorenzen, M. (1999) *Synth. Met.*, **101**, 98–9.
Sato, T., Kishida, H., Nakamura, A., Fukuda, T. and Yamamoto, T. (2007) *Synth. Metals*, **157**, 318–22.
Sauteret, C., Hermann, J.-P., Frey, R., Pradère, F., Ducing, J., Baughman, R. H. and Chance, R. R. (1976) *Phys. Rev. Lett.*, **36**, 956–9.
Sirringhause, H., Brown, P. J., Friend, R. H., Nielsen, M. N., Beckgaard, K., Langeveld-Voss, B. M. M., Spiering, B. A. J. H., Janssen, R. A. J., Meijer, P., Herwig, P. and de Leeuv, P. D. M. (1999) *Nature*, **401**, 685–8.
Spano, F. C. (2005) *J. Chem. Phys.*, **122**, 234701-1–15.
Spano, F. C., Clark, J., Silva, C. and Friend, R. H. (2009) *J. Chem. Phys.*, **130**, 074904-1–16.
Torruellas, W. E., Neher, D., Zanoni, R., Stegeman, G. I., Kajzar, F. and Leclercc, M. (1990) *Chem. Phys. Lett.*, **175**, 11–6.
Wu, Y., Liu, P., Gardner, S. and Ong, B. S. (2005) *Chem. Mater.*, **17**, 221–3.
Yamamoto, T., Zhou, Z.-H., Kanbara, T., Shimura, M., Kizu, K., Maruyama, T., Nakamura, Y., Fukuda, T., Lee, B. L., Ooba, N., Tomaru, S., Kurihara, T., Kaino, T., Kubota, K. and Sasaki, S. (1996) *J. Am. Chem. Soc.*, **118**, 10389–99.
Yamamoto, T., Komarudin, D., Arai, M., Lee, B.-L., Suganuma, H., Asakawa, N., Inoue, Y., Kubota, K., Sasaki, S., Fukuda, T. and Matsuda, H. (1998) *J. Am. Chem. Soc.*, **120**, 2047–58.
Yamamoto, T., Kimura, T. and Shiraishi, K. (1999) *Macromolecules*, **32**, 8886–96.
Yamamoto, T., Lee, B. L., Kokubo, H., Kishida, H., Hirota, K., Wakabayashi, T. and Okamoto, H. (2003a) *Macromol. Rapid Commun.*, **24**, 440–3.
Yamamoto, T., Fang, Q. and Morikita, T. (2003b) *Macromolecules*, **36**, 4262–7.
Yamamoto, T., Arai., M., Kokubo, H. and Sasaki, S. (2003c) *Macromolecules*, **36**, 7986–93.
Yamamoto, T., Yamaguchi, I. and Yasuda, T. (2005) *Adv. Polym. Sci.*, **177**, 181–208.
Yamamoto, T., Otsuka, S.-I., Namekawa, K., Fukumoto, H., Yamaguchi, I., Fukuda, T., Asakawa, N., Yamanobe, T., Shiono, T. and Cai, Z. (2006) *Polymer*, **47**, 6038–41.
Yamamoto, T., Sato, T., Iijima, T., Abe, M., Fukumoto, H., Koizumi, T., Usui, M., Namamura, Y., Yagi, T., Tajima, H., Okada, T., Sasaki, S., Kishida, H., Nakamura, A., Fukuda, T., Emoto, A., Ushijima, H., Kurosaki, C. and Hirota, H. (2009a) *Bull. Chem. Soc. Jpn.*, **82**, 896–909.
Yamamoto, T., Kumagai, A., Saito, K. and Nagai, T. (2009b) *J. Nanosci. Nanotech.* **9**, 670–2.
Yamamoto, T. (2010) *Bull. Chem. Soc. Jpn.*, **83**, 431–55.
Yamamoto, T., Usui, M., Ootsuka, H., Iijima, T., Fukumoto, H., Sakai, Y., Aramaki, S., Yamamoto, H. M., Yagi, T., Yajima, H., Okada, Y., Fukuda, T., Emoto, A., Ushijima, H., Hasegawa, M. and Ohtsu, H. (2010) *Macromol. Chem. Phys.*, **211**, 2138–47.
Yamamoto, T., Otsuka, S.-I., Fukumoto, H., Sakai, Y., Aramaki, S., Fukuda, T., Emoto, A. and Ushijima H. (2011) *J. Polym. Sci. Part A; Polym. Chem.*, **49**, 1508–12.

5

Small molecule supramolecular assemblies for third-order nonlinear optics

I. BIAGGIO, Lehigh University, USA

DOI: 10.1533/9780857098764.1.170

Abstract: This chapter discusses organic molecules for third-order nonlinear optics – responsible for effects such as self- and cross-phase modulation, and all-optical switching – and the issues related to using such molecules to create a solid-state material. The chapter will review the figures of merit that help determine the potential of a small molecule, and discuss the problem of obtaining a bulk material with a large third-order nonlinearity that at the same time possesses a high optical quality compatible with photonic applications. As a concluding example, the chapter will end with a review of recent progress in optimizing the third-order optical nonlinearity of small molecules and in using those molecules to fabricate dense supramolecular assemblies for integrated nonlinear optics.

Key words: organic molecules, third-order nonlinear optics, self-assembly, small molecules, donor–acceptor substitution.

5.1 Introduction

The polarization induced in matter by the electric field of an optical wave has nonlinear components that depend on higher powers of the optical electric field. A second-order response leads to the linear electro-optic (Pockels) effect and to the generation of new optical frequencies through processes such as difference frequency or second harmonic generation. A third-order response allows the interaction between different optical waves and many other phenomena that can be seen in general as the combination of three photons with frequencies ω_1, ω_2, ω_3, to generate a fourth one, with frequency $\omega_4 = \omega_1 + \omega_2 \pm \omega_3$. This process is described by the frequency-dependent third-order nonlinear optical susceptibility $\chi^{(3)}(-\omega_4, \pm\omega_3, \omega_2, \omega_1)$ (Hellwarth, 1977), which is in general a complex function that gives the strength of the third-order response as a function of the wavelength of the interacting optical waves. Third-order nonlinear optical effects could enable a wide variety of applications that require light–light interaction. Examples include self-action phenomena such as self-focusing and nonlinear transmission (optical limiting), information transfer from one optical wave to another, and in general all optical switching fabrics that would route and process information without the need for electronic intermediaries.

The challenge to actually realize devices for these all optical switching applications is not only that of creating materials that possess a large third-order susceptibility, but also that such materials must at the same time have a very high optical quality and be compatible with integrated optics technology.

When working towards a larger nonlinear optical response, it is useful to consider the fact that both linear and nonlinear polarizabilities often grow hand in hand because of their fundamental microscopic origins. For materials that have a similar number density of polarizable units, linear and third-order susceptibilities are approximately related by

$$\frac{\chi^{(3)}(-\omega,-\omega,\omega,\omega)}{[\chi^{(1)}(-\omega,\omega)]^4}=\text{const} \qquad [5.1]$$

a quantity that will retain its order of magnitude among different materials of the same class. As an example, some chalcogenide glasses with a refractive index of ~2.7 reach a third-order susceptibility that is about three orders of magnitude larger than that of silica glass and share the same ratio of Eq. 5.1 with silica glass with a refractive index of 1.5 (Jin *et al.*, 2010). Thus, in these glasses high third-order nonlinearities are simply connected to high refractive indices.

A disproportionate increase in third-order susceptibility with respect to the linear susceptibility can be obtained in molecular materials, where the units responsible for the linear and nonlinear optical response are molecules with extended π-electron conjugation. From an ultra-simplified physics point of view, such molecules allow the wavefunction of the electron that interacts with the light to have a large volume, in some cases essentially as large as the molecule itself. The larger wavefunction raises the target area for interaction with a photon and consequently the linear and nonlinear optical polarizabilities. Because higher-order nonlinearities are proportional to higher-order powers of the volume of the wavefunction (see, e.g., Armstrong *et al.*, 1962, for the sum-over states expansion of nonlinear optical susceptibilities), it is possible for the ratio of Eq. 5.1 to become much larger in organic materials when compared with typical values for inorganic materials without delocalized electrons, such as oxide or chalcogenide glasses. This is one key basic advantage of organic molecular materials for applications in nonlinear optics.

In addition to this, organic molecules are attractive because their nonlinearities remain relatively large for off-resonant excitation. Their response is then electronic in origin and practically instantaneous. This is in contrast to other effects, like molecular reorientation, photorefraction based on photoinduced charge transport, or resonant carrier excitation in semiconductors, where strong optical nonlinearities can be obtained, but only at the expense of relatively long exposure times or relaxation times. Finally, organics are also of major interest because of their large flexibility for fine-tuning their chemical structure and properties towards the optimization of a given effect.

5.2 Fundamental principles of the third-order nonlinear optical response

The third-order nonlinear optical effects responsible for self-phase modulation, cross-phase modulation, and degenerate four-wave mixing are described by a frequency dependent third-order susceptibility $\chi^{(3)}(-\omega, -\omega, \omega, \omega)$ that gives the amplitude of the third-order polarization created at the same frequency by the interaction of up to three other optical waves. In the most general case, three optical fields with three different wavevectors and vectorial amplitudes $\boldsymbol{E}^a$, $\boldsymbol{E}^b$, $\boldsymbol{E}^c$ can create a nonlinear optical polarization at the same frequency that has an amplitude given by

$$P_i^{(3)} = \frac{3}{2}\varepsilon_0 \chi_{ijkl}^{(3)}(-\omega, -\omega, \omega, \omega) E_j^{a^*} E_k^b E_l^c \quad [5.2]$$

where ε_0 is the electric constant, and the amplitudes are defined so that the corresponding time-dependent field is given by $\boldsymbol{E}(t) = \mathrm{Re}[\boldsymbol{E}e^{-i\omega t}]$. The third-order susceptibility is also responsible for the change in the refractive index seen by an optical wave as a function of its intensity. When this change is expressed as $n(I) = n(0) + n_2 I$, with I the optical intensity, then the value of the nonlinear refractive index n_2 is

$$n_2 = \frac{3}{4\varepsilon_0 c n^2} \chi_{1111}^{(3)}(-\omega, -\omega, \omega, \omega) \quad [5.3]$$

where ε_0 is the electric constant, c the speed of light, and n is the linear refractive index of the nonlinear optical material.

Eq. 5.2 is in SI units, which will be used throughout in this chapter. This expression goes over to its equivalent in electrostatic units (esu) used by Maker and Terhune (1964), Hellwarth (1977), and a relevant part of the literature, with the substitution $\varepsilon_0\ \chi_{ijkl}^{(3)} \leftrightarrow 4c_{ijkl}$. Numerical values can then be converted using the rule

$$\chi_{ijkl}^{(3)}[\mathrm{m}^2\ \mathrm{V}^{-2}] = 4\frac{4\pi}{(10^{-4}c)^2} c_{ijkl}[\mathrm{esu}] \quad [5.4]$$

where c is the speed of light in vacuum in m/s and the first factor of four after the equal sign takes into account the factor of four that was originally included in the definition of their third-order susceptibilities c_{ijkl} by Maker and Terhune (1964). Not everyone in the 'esu world' uses this factor of 4, in which case it should be dropped from Eq. 5.4. But it is very important to be aware of it and consider its possible existence when comparing results from different authors.

The focus of this chapter is on *off-resonant* third-order effects, described by third-order susceptibilities that are real-valued. In this *zero frequency limit* the third-order susceptibilities do depend on the energy of the excited states, but their frequency dependence becomes negligible. In addition, the susceptibilities describing different nonlinear optical processes such as self phase modulation or third-harmonic generation – described by $\chi^{(3)}(-\omega, -\omega, \omega, \omega)$ and $\chi^{(3)}(-3\omega, \omega, \omega, \omega)$, respectively – become equal in the zero-frequency limit (see, e.g., Hellwarth, 1977). These facts mean values for off-resonant susceptibilities are the simplest way to compare different materials for their nonlinear optical performance.

What should the magnitude of the third-order susceptibility be in order to enable the use of third-order nonlinear optical effects in applications? One way to answer this question is to consider a light intensity of 1 GW/cm^2, which corresponds to a 30 W peak power focused on a circular surface with a 1 μm radius. Such a peak power is found in a 30 ps pulse at an energy of 1 nJ, and it can also be obtained in a train of 10 ps pulses at a repetition rate of 10 GHz with an average power of 3 W. This is the kind of power that one can reach at the output of a standard fiber amplifier. In a material with a real third-order susceptibility $\chi^{(3)} = 2 \times 10^{-19}$ m^2 V^{-2}, which is 1000 times larger than that of fused silica, a light intensity $I = 1$ GW/cm^2 induces a π phase shift in a probe wave after a propagation length

$$L = \frac{\varepsilon_0 c n^2 \lambda}{3\mathrm{Re}[\chi^{(3)}_{1111}]I} = 1.5\ \mathrm{mm} \qquad [5.5]$$

An off-resonant third-order susceptibility 1000 times larger than the one of silica glass can thus allow one wave to switch another one on and off by phase shifting the light traveling in one branch of a Mach Zehnder interferometer that is only a few millimeters long. It is reasonable to consider this third-order susceptibility value as the value that would enable practical nonlinear optical devices in integrated optics.

5.3 Macroscopic susceptibilities and microscopic polarizabilities

In the organic molecules of interest to this chapter, the π-electron wavefunction strongly couples to an optical electric field, dominating the nonlinear response in the visible and near-infrared. The way light–matter interaction happens can be tuned by modifying the π-electron wavefunction in various ways, e.g. by chemical substituents conjugatively attached elsewhere on a molecule. By changing the shape of the π-electron system and the substituents attached to it, the physical properties of the molecule, and in particular its linear and nonlinear optical properties, can be tuned with great flexibility.

Microscopically, the third-order susceptibility $\chi^{(3)}$ of a molecular material depends on the third-order polarizability (or second *hyper*polarizability) γ that describes the molecular polarization induced by the local optical electric field. The relationship between third-order susceptibility and polarizability is

$$\chi^{(3)}_{ijkl} = f_i f_j f_k f_l N \gamma_{ijkl} \qquad [5.6]$$

where the susceptibility and polarizability tensors are understood to be in the same reference frame, f_i are local field factors that assume the values $f_i = (n_i^2 + 2)/3$ in the Lorentz approximation (n_i is the refractive index for light polarized along the coordinate axis i), and N is the number density of molecules. Over the past few decades, lots of work has focused on understanding and optimizing the third-order polarizability γ. This has resulted in well-understood key principles that determine a larger nonlinear optical response. Typical examples are longer paths for π-electrons and the use of donor and acceptor groups (Brédas *et al.*,1994; Marder *et al.*, 1997). It has also been shown that in general molecular hyperpolarizabilities are connected to the difference in average length between single and double bonds in a conjugated path, i.e. bond length alternation (Marder *et al.*, 1993; Meyers *et al.*, 1994). General principles for optimizing molecular third-order polarizabilities have also been established when studying resonant two-photon absorption, which is connected to the imaginary part of the third-order polarizability, but a review of this work is outside the scope of this chapter.

Even though such structure–property relationship studies are very interesting, considering the problem of the nonlinear optical response of a molecule from a much more fundamental and simplified point of view can also be very valuable. One way of doing this that provides a good review of the efficiency and potential of the nonlinear optical response is to evaluate the fundamental quantum limit to the higher order polarizabilities of a molecule (Kuzyk, 2000a, 2000b, 2003a, 2003b, 2003c). For third-order nonlinear optics, Kuzyk showed that there is an upper limit to the molecular third-order polarizability that is basically given by the upper limit to the square of the dipole transition matrix element as determined through sum rules, which is $e^2\hbar^2 N_\pi/(2mE_{01})$, where e is the unit charge, $\hbar$ is Planck's constant divided by 2π, N_π is the number of delocalized electrons (estimated as twice the number of multiple bonds in the conjugated system), and E_{01} is the excitation energy to the lowest optically accessible state (Kuzyk, 2000a). For the zero-frequency limit of the third-order polarizability, its fundamental limit depends on the fourth power of the dipole transition matrix element – which is limited by the square of $e^2\hbar^2 N_\pi/(2mE_{01})$ – divided by the third power of the optical excitation energy (Armstrong *et al.*, 1962). These arguments are a summary explanation of the following expression (in SI units) for the fundamental limit, which applies to a centrosymmetric molecule (Kuzyk, 2000b, 2003a),

$$\gamma_{1111}^{c} = \gamma_c = \frac{e^4\hbar^4}{\varepsilon_0 m_e^2}\frac{N_\pi^2}{E_{01}^5} \qquad [5.7]$$

where ε_0 is the electric constant that appears in the SI definition of the photoinduced microscopic dipole. While the limit to the off-resonant third-order polarizability can theoretical reach $4\gamma_C$ for a noncentrosymmetric object that is an ideal two-level system (Kuzyk, 2000b), the quantity in Eq. 5.4 can serve as a useful guideline when evaluating the efficiency of a molecule (May *et al.*, 2005), and can be used to define an *intrinsic nonlinear polarizability* that is a scale-invariant measure of a molecule's non-linear optical properties (Zhou and Kuzyk, 2008). Here, we use Eq. 5.7 to define an intrinsic third-order polarizability $\gamma^I = \gamma_{rot}/\gamma_C$, where γ_{rot} is the experimental value for the rotational average of the third-order polarizability, defined in the next paragraph.

One simple case of molecular material is that in which the component molecules are randomly oriented with respect to each other, like for molecules in solution or diluted in a polymer matrix. Then, the effective third-order polarizability per molecule that determines the molecular contribution to the nonlinearity is obtained by an orientational average of the molecular third-order polarizability tensor γ_{ijkl}. The interaction between linearly polarized waves with the same polarization is then described by the rotational average $\gamma_{1111}^{rot}(-\omega, -\omega, \omega, \omega)$ (Andrews, 2004)

$$\gamma_{1111}^{rot} = \gamma_{rot} = \frac{1}{15}\sum_{j,k=1}^{3}(\gamma_{jjkk} + \gamma_{jkkj} + \gamma_{jkjk}) \qquad [5.8]$$

where the γ_{ijkl} are the individual components of the third-order polarizability tensor of a molecule in the molecule's reference frame. What is important to note here is that for similar values of the tensor elements of the third-order polarizability a one-dimensional molecule will have a significantly smaller γ^{rot} than a molecule that extends into two or three dimensions. Specifically, for a molecule with a one-dimensional conjugated system, $\gamma^{rot} \sim \gamma_{1111}/5$, while for a planar molecule with similar values of the tensor elements γ^{rot} would automatically become almost four times larger, and even better for non-planar molecules. In fact, broken conjugation in a non-planar molecule would not necessarily be a problem when considering orientational averaging because it would simply create partially independent, but potentially equally large, molecular tensor components that would efficiently contribute to γ^{rot}. In order to maximize the third-order susceptibility of amorphous assemblies of randomly oriented molecules it is better to have molecules where the conjugated system responsible for the nonlinearity extends in two or three dimensions, even if the individual tensor elements of the third-order polarizability are not expected to grow just because of that. Another reason to use non-planar molecules is that this characteristic tends to reduce intermolecular interactions, which would help avoid the occurrence of microcrystals in dense supramolecular assemblies. This will be discussed later on, in the next section.

The orientational average $\gamma_{1111}^{\mathrm{rot}}(-\omega, -\omega, \omega, \omega)$ determines the bulk third-order susceptibility contributed by the individual molecules to a molecular assembly in a manner equivalent to Eq. 5.6:

$$\chi_{1111}^{(3)} = f^4 N \gamma_{1111}^{\mathrm{rot}} \qquad [5.9]$$

with the local field factor f and the number density N. For a weak concentration of molecules in solution, the above expression can be re-written as

$$\chi_{1111}^{(3)} = f^4 C N_{\mathrm{A}} \frac{\rho_S}{M} \gamma_{1111}^{\mathrm{rot}} \qquad [5.10]$$

where C is the mass concentration of the molecule (as an example, C would be of the order of 0.01 when preparing solutions for nonlinear optical measurements), N_{A} is Avogadro's number, ρ_{S} is the density of the solvent, and M is the molar mass of the molecule. The total third-order susceptibility of the solution is then the sum of Eq. 5.10 and of the third-order susceptibility of the solvent.

Next, it is interesting to consider the case of a single-component dense supramolecular assembly of randomly oriented molecules. Such a structure is very similar to a glass made up of molecules. In such a case the number density of molecules N that appears in Eq. 5.6 is essentially determined by the space taken up by a single molecule in the assembly. In this context, obtaining a large third-order susceptibility for the bulk supramolecular assembly means that the third-order polarizability of a molecule must not only be large, but it must be large compared to the molecular volume, which determines how densely the molecules can be packed together. Since determining the average volume taken up by a molecule in a dense supramolecular assembly can be cumbersome, this point can be most easily expressed by defining a *specific third-order polarizability* $\tilde{\gamma} = \gamma_{1111}^{\mathrm{rot}} / m$, where $m = M/N_{\mathrm{A}}$ is the mass of the molecule. Eq. 5.6 can then be written as

$$\chi_{1111}^{(3)} = f^4 \rho \tilde{\gamma} \qquad [5.11]$$

where ρ is the density of the supramolecular assembly. This is useful because it allows deriving an estimate for the third-order susceptibility of a dense single-component assembly from the molecular third-order polarizability as it can be determined, e.g., for molecules in solution. Using $f^4 = 10$ (refractive index of ~1.8) and $\rho = 1000$ kg/m^3 (density of water) one finds the simple rule of thumb that to obtain a third-order susceptibility of the order of $\chi^{(3)} \sim 2 \times 10^{-19}$ m^2 V^{-2}, which is 1000 times larger than the $\chi^{(3)}$ of fused silica, one needs a specific third-order polarizability $\tilde{\gamma} \sim 2 \times 10^{-23}$ m^5 V^{-2} kg^{-1}.

When characterizing a nonlinear optical molecule it is useful to evaluate the magnitude of its third-order polarizability with respect to absolute quantities that give an idea of how efficient a given molecular design is. In this respect the *specific third-order polarizability* $\tilde{\gamma}$ discussed above is useful because it relates the size of the third-order polarizability to the size of the molecule. On the other hand, the discussion above also showed that the efficiency of a particular molecular design can additionally be judged from the proximity of a molecule's nonlinearity to the fundamental limit, i.e. by the dimensionless *intrinsic third-order polarizability* γ^I. For clarity, it is worthwhile to repeat here together the definition of these two figures of merit:

$$\tilde{\gamma} = \frac{\gamma_{1111}^{\text{rot}}}{m} \qquad [5.12]$$

$$\gamma^I = \frac{\gamma_{1111}^{\text{rot}}}{\gamma_C} \qquad [5.13]$$

where $m = M/N_A$ is the mass of the molecule and γ_C is the quantum limit given in Eq. 5.4. The specific and the intrinsic third-order polarizability offer complementary ways to assess how efficient the nonlinear optical response of a molecule is, and it is very informative to use them both when assessing the third-order response of a new molecule. Currently, good values for the intrinsic third-order polarizability in highly nonlinear molecules are of the order of 0.02, while good values for the specific third-order polarizability should be of the order of 10^{-23} m^5 V^{-2} kg^{-1}.

The interplay of intrinsic and specific third-order polarizabilities in different kinds of molecules is highlighted in Table 5.1. The first interesting observation derived from Table 5.1 is that the fundamental limit can indeed be almost reached in the smallest, simplest substances, like the helium atom and the hydrogen molecule, which have a nonlinearity within a factor of three of the fundamental limit (intrinsic third-order polarizability of the order of 0.3–0.4). Next, C_{60} represents a relatively large molecule with a sizable third-order nonlinearity, but its intrinsic and specific third-order polarizabilities remain relatively weak. Finally, the last two entries in the table are two recently developed small molecules with exceptionally high intrinsic and specific third-order polarizabilities. More details on the compounds in the last two rows of the table can be found in the literature (May

Table 5.1 Comparison of fundamental limit γ_C (from Eq. 5.9), experimental rotational average, and specific and intrinsic values of the third-order polarizability (see Eqs. 5.12–5.13) for some representative compounds. λ_{max} is the wavelength corresponding to the lowest energy optical excitation. N_π is an estimation for the number of electrons in the conjugated system. In the case of the 'DDMEBT' molecule (see text), which has a broken conjugation because of its non-planar structure, N_π is the sum of the number of electrons for the two conjugated systems contained in the molecule.

Molecule	m [10^{-23} kg]	N_π	λ_{max} [nm]	γ_C [10^{-48} m^5 V^{-2}]	γ_{rot} [10^{-48} m^5 V^{-2}]		$\tilde{\gamma}$ [10^{-23} m^5 V^{-2} kg^{-1}]	γ^I
He	0.66	2	62	1.3×10^{-4}	0.5×10^{-4}	(*a*)	7.5×10^{-4}	0.38
H_2	0.33	2	110	23×10^{-4}	8×10^{-4}	(*a*)	2.39×10^{-4}	0.35
C_{60}	119.7	60	465(*)	2807	4.6	(*b*)	0.38	1.6×10^{-3}
TDMEE	40.89	16	591	662	7	(*c*)	1.96	0.012
DDMEBT	69.16	26	527	985	6	(*d*)	0.87	0.006

(*a*) Zero-frequency limit from Shelton and Rice (1994), converted into SI units from atomic units, taking into account an additional factor of 6 stemming from a difference in the definition of third-order terms.
(*b*) Corresponds to γ_{1111} calculated using Eq. 5.6 and a refractive index of 2 from the bulk third-order susceptibility determined in the zero-frequency limit by Strohkendl *et al.* (1997).
(*c*) From May *et al.* (2005) but corrected for a calibration error that inflated the third-order polarizability values of May *et al.* (2005) by a factor 1.5.
(*d*) From Michinobu *et al.* (2005), but corrected for the same calibration error mentioned in footnote (c) above, and lowered by an additional 25% to reflect recent measurements (Beels *et al.*, 2012).
(*) The wavelength of maximum absorption for C_{60} was taken to correspond to the energy of the lowest lying electronic state that can only be accessed by a two-photon transition. This state is found at an energy of 2.67 eV (Strohkendl *et al.*, 1997), corresponding to a wavelength of 465 nm.

et al., 2005, 2007; Esembeson *et al.*, 2008; Koos *et al.*, 2009; Scimeca *et al.*, 2009; Beels *et al.*, 2012). In addition, DDMEBT will be discussed later on in this chapter. For our purposes in the present discussion, they serve as good examples of the kind of nonlinearities that can be obtained in relatively small organic molecules when they are optimized.

5.4 From molecules to bulk solid-state materials

While it is possible to develop and synthesize a large variety of molecules for nonlinear optics, a key challenge towards any application is the ability to create a solid-state material where the nonlinearity of a molecule can be efficiently expressed. One such example is the single crystal of poly(2,4-hexadiyne-1,6-diol di-*p*-toluenesulfonate) (PTS). The nonlinear refractive index of PTS is $2.2 \pm 0.3 \times 10^{-12}$ cm^2 W^{-1} at a wavelength of 1.6 μm (Lawrence *et al.*, 1994). Using Eq. 5.3 and a refractive index of 1.75 for PTS (Feldner *et al.*, 2001) one obtains a third-order susceptibility of ~2.4×10^{-18} m^2 V^{-1}, which is four orders of magnitude larger than fused silica. This is a very good value, but it should be noted that it is observed only for light polarized along the crystallographic axis corresponding to the orientation of the essentially one-dimensional polymer chains that make up the crystal. And while this third-order polarizability is relatively high, such a single-crystalline material is generally difficult to fabricate and to integrate with photonic circuitry.

One big advantage of third-order nonlinear optics is that no molecular ordering is necessary to translate the molecular nonlinearities into the bulk nonlinear susceptibility of a solid-state material. This facilitates the use of nonlinear optical molecules to create appropriate bulk materials: the molecules do not need to have any defined orientation when they are assembled together. One possibility to translate molecular nonlinearities into a bulk material is therefore to disperse nonlinear optical molecules in a polymer matrix. The resulting bulk susceptibility is then determined by Eq. 5.10, from the polarizability of the active molecules multiplied by their density. The general applicability of this technique combined with the many molecules that have been developed over the years has led to a large body of literature. A review of some of the results has been provided by Hales and Perry (2008).

The nonlinearity of a solid-state material obtained by dispersing nonlinear optical molecules in a polymer matrix is limited both by the maximum number density of molecules that can be reached and by the magnitude of their third-order polarizability. The practical maximum density that can be reached is limited by the necessity to avoid aggregation effects. Most of the work on third-order nonlinear optical molecules in the literature has until now focused on obtaining larger and larger third-order polarizabilities, which grow with molecular size. While this will increase the nonlinearity in a polymer matrix if the number density of the molecules can be maintained, it is important to ask if this is the optimum route towards obtaining the best bulk material. When molecules with stronger nonlinearity are embedded inside a polymer, their nonlinearity is always diluted, and the question then becomes whether the dilution effect of the nonlinearity can be compensated by a large third-order polarizability. If it cannot, then one has to look at other ways to combine molecules together that have the potential to deliver a larger number density.

An alternative to embedding into polymers is to create a dense supramolecular assembly where the nonlinear molecules are closely packed together in an essentially amorphous phase, with each molecule randomly oriented with respect to the others. In such a material there is no dilution effect and its maximum susceptibility would be directly determined by the nonlinearity of the molecule relative to its size (the *specific* third-order polarizability mentioned earlier). The ability to have a larger number density of molecules should allow such a dense single-component supramolecular assembly to display a third-order susceptibility that is competitive or larger than for the case of molecules embedded in a polymer matrix. Since bigger molecules are more difficult to handle, a natural idea is to use smaller molecules. Smaller molecules may not be able to compete with the 'best' absolute third-order polarizabilities, but larger third-order polarizabilities in larger molecules that are more difficult to handle and need to be diluted into polymers are also not useful.

The question for single-component assemblies is whether it is possible to obtain molecules with a *specific* third-order polarizability that is large enough to deliver the required third-order susceptibility in the bulk (see previous section). The focus is no longer on large third-order polarizabilities alone.

5.5 Small molecules with large third-order nonlinearities

Research on third-order nonlinear optics over the years has striven to make the third-order polarizability of individual organic molecules as large as possible. The result has often been relatively large molecules because the third-order polarizabilities increase rapidly with size. However, larger molecules are generally more difficult to combine into a dense solid state where their nonlinearity is not diluted and which is homogeneous with a good optical quality. This is a fundamental issue that has historically made it difficult to transfer high nonlinearities from molecules to solid state and to integrate organic third-order elements in integrated optics or other applications. Large molecules with large nonlinearities are only of academic interest if they cannot be combined in a bulk material that actually allows their use in practical applications.

On the other hand, molecules that do not necessarily have very large third-order polarizabilities in an absolute sense, but that have a large specific polarizability and can be more easily manipulated to obtain dense supramolecular assemblies, should be the focus of increased attention because of their potential to form bulk materials that reflect the nonlinearity of the constituent molecules while at the same time having a high optical quality.

Fortunately, investigations into the optimization of the third-order polarizability of smaller molecules with a compact conjugated system have revealed that a relatively high third-order polarizability can be maintained while reducing the size of the molecule if donor–acceptor substitution is used to control the relative energies of ground and excited states. While donor–acceptor substitution is well known for optimizing the second-order nonlinear optical response, it also plays a very simple and important role in increasing the third-order polarizability in small molecules.

In small molecules with donor-acceptor substitution, the wavefunction of the ground state is mostly centred on the donor-group(s), while the wavefunction of the first excited state is mostly centred on the acceptor-group(s) (Moonen *et al.*, 2003; Fernandez and Frenking, 2006). It is then the chemical structure of the substituents, in addition to the size of the molecule, which sets the difference between the energies of the highest occupied molecular orbital (HOMO) and of the lowest unoccupied molecular orbital (LUMO) (Fernandez and Frenking, 2006; May *et al.*, 2007). As the size of the conjugated system decreases, donor-acceptor substitution keeps the HOMO–LUMO gap smaller than what would be otherwise expected from a fundamental 'particle in a box' effect. As an example, for the polyynes studied by Slepkov *et al.* (2004), a linear conjugated system consisting of six consecutive triple bonds has a first optical transition from the ground state at ~4.08 eV (corresponding to a wavelength of 304 nm). Similarly, in the poly(triacetylene) wires investigated by Gubler *et al.* (1999), a linear conjugated system consisting of six double or triple bonds has a first optical transition at ~3.3 eV (wavelength of 377 nm). On the other hand, the TDMEE molecule (May *et al.*, 2005; see also Table 5.1) arguably has a smaller conjugated system than these compounds, but it has a first optical excitation energy of ~2.1 eV (591 nm wavelength). Finally, in another family of donor–acceptor substituted molecules the HOMO–LUMO gap has been shown to remain practically unchanged while the size of the conjugated system between donors and acceptor was varied (Bures *et al.*, 2007; May *et al.*, 2007).

Since the transition dipole matrix element between ground and excited state remains high in small donor–acceptor substituted molecules, the smaller HOMO–LUMO gap directly leads to a larger off-resonant third-order polarizability (Armstrong *et al.*, 1962). The importance of the excited state transition energies is also evident from the fact that the quantum-limit to the third-order polarizability is inversely proportional to the fifth power of the first optical excitation energy for a conjugated system of fixed size (see Section 5.3).

From the expression for the fundamental limit (Eq. 5.7), one sees that if excited state energies remain constant while the number of conjugated electrons increases, then the maximum third-order polarizability grows as the square of the number of delocalized electrons, but not faster (May *et al.*, 2007). Such an increase is less steep than what is generally observed in other molecular systems (Slepkov *et al.*, 2004), where the third-order polarizability vs. size of the conjugated system follows a much steeper power law that is caused by the simultaneous increase in transition dipole moment and decrease in HOMO–LUMO gap. The latter remained practically invariant for the molecules of May *et al.* (2007), as forced by the presence of the donors and acceptors, and kept the nonlinearity high as the size of the molecule shrank.

In summary, the effect of donor–acceptor substitution on the HOMO–LUMO gap leads to a much less steep decrease of the third-order polarizability as the size of a molecule shrinks (May *et al.*, 2007) when compared with unsubstituted molecules (Slepkov *et al.*, 2004; Gubler *et al.*, 1999). Clearly, the increase in third-order polarizability with donor–acceptor substitution can only be explained in such a straightforward way as long as the overlap between the HOMO and the LUMO can be maintained. As the size of the molecule increases past an optimum value, the centering of HOMO and LUMO on donor and acceptor groups should lead to smaller transition dipole matrix elements and the specific third-order polarizability should stop growing. This effect is different in physical origin from the saturation of the specific third-order polarizability with molecular size that is observed in unsubstituted compounds (Gubler *et al.*, 1999), which is more related to the transition from a molecular excitation to a localized excitation in what becomes increasingly like a polymer chain. A semi-quantitative summary and illustration of these principles is shown in Fig. 5.1. Donor–acceptor substituted molecules are able to maintain a large third-order nonlinearity at small sizes while unsubstituted molecules suffer a steep decrease of the third-order nonlinearity with molecular size. The dashed line that continues the trend of the donor-substituted systems towards larger sizes is a qualitative extrapolation that is as yet unsupported by direct experimental data but that is expected because of the decrease in the transition dipole moment for increasing space between donor and acceptor groups, towards which HOMO and LUMO gravitate.

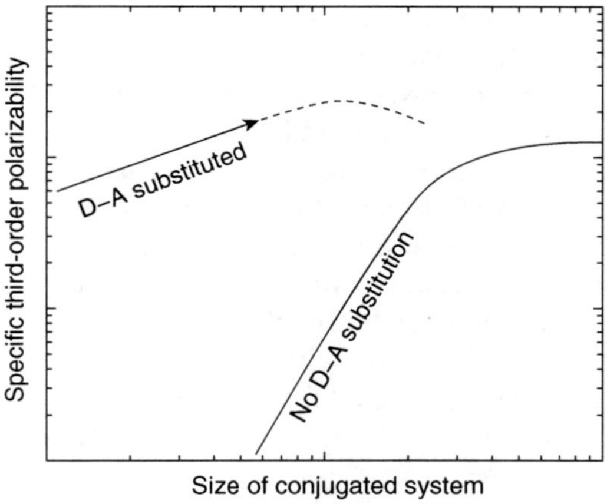

5.1 Qualitative representation of how the specific third-order polarizability (Eq. 5.12) of molecules is expected to vary with molecular size for the case of donor–acceptor substituted compact molecules and for other non-substituted conjugated systems such as those presented by Slepkov *et al.* (2004) and Gubler *et al.* (1999). Donor–acceptor substitution in smaller molecules keeps the variations of the HOMO–LUMO gap with size small, leading to a slower decrease of the third-order polarizability for shrinking size when compared to unsubstituted systems, where a combination of an increasing HOMO–LUMO gap and a smaller transition dipole moment leads to a steeper decrease for smaller molecules. The thick solid curves in the figure are obtained by interpolating and smoothing out the data published by May *et al.* (2005, 2007) for the donor–acceptor substituted small molecules, and by Slepkov *et al.* (2004) and Gubler *et al.* (1999) for the more extended systems without donor–acceptor substitution. The dashed curves represent a trend that is not yet supported by data, but that is expected because of HOMO and LUMO becoming increasingly separated for large distances between donor and acceptor groups.

Figure 5.1 shows that the bulk nonlinearity that can potentially be produced using small molecules (the specific third-order polarizability) can be as large as the best that can be expected for the largest molecules (assuming it was possible to combine them in a dense assembly with the required optical quality). In fact, small donor–acceptor substituted organic molecules were the first that clearly approached the fundamental quantum limit within less than two orders of magnitude, similar to what is observed in second-order nonlinear optics (Clays, 2001; Kuzyk, 2003c; May *et al.*, 2005, 2007). In addition, this relative closeness to the fundamental limit was accompanied by a large specific third-order polarizability and by the ability of many molecules, which can sublimate without decomposition, to create dense supramolecular assemblies by vapor deposition.

In summary, donor–acceptor substitution has been observed earlier to have an effect on third-order nonlinearities, but it plays a uniquely simple and important role in increasing the third-order response of small molecules; an effect that can be best visualized by the tuning of the HOMO–LUMO gap. This understanding of the effects of donor–acceptor substitution in small molecules directly connects to the fundamental sum-over-states expansion of the third-order polarizability in perturbation theory (Armstrong *et al.*, 1962), and to the fundamental limit developed by Kuzyk. Such a view delivers a particularly simple and direct guiding principle for understanding and optimizing the nonlinear optical response of small molecules.

5.6 Small molecule supramolecular assemblies with high optical quality and large third-order susceptibility

The key point of the previous section is that small molecules can deliver the same (or better) performance as large molecules from the point of view of creating a bulk nonlinear optical material, and as seen from the two figure of merit that describe the efficiency of the nonlinear optical response of a molecule, the intrinsic and the specific third-order susceptibility γ^I and $\tilde{\gamma}$ (see Eqs. 5.12 and 5.13). Two examples of such high performance small molecules have been presented in Table 5.1.

The next challenge once such molecules have been identified is to use them to create dense supramolecular assemblies. Molecules with a relatively small molecular mass that can sublimate without decomposition open up the possibility of creating supramolecular assemblies by physical vapor deposition or molecular beam deposition. These techniques have unique advantages. They operate in a clean environment (high vacuum), they do not need solvents, they can provide precise and *in situ* control of growth rate, substrate temperature, and film thickness, and are also fully compatible with vapor-phase microelectronic fabrication techniques that involve masking and dry-etching. The main issue to solve in order to obtain high-quality thin films in this way is the tendency of the molecules to crystallize. This is particularly a problem for planar molecules with donor–acceptor substitution where π–π stacking interactions and antiparallel molecular dipole alignment can be powerful forces towards forming crystals. In the case of molecular beam deposition on a substrate, such crystals would start growing independently at different locations on the substrate and lead to microcrystalline materials with too much light scattering to be used in applications. In fact, the TDMEE molecule (May *et al.*, 2005) has very

attractive properties from the nonlinear optical point of view (see Table 5.1) but forms very bad films (Esembeson *et al.*, 2008).

The breakthrough towards obtaining the first dense, homogeneous supramolecular assemblies for nonlinear optics that had a very high optical quality was a modification of the TDMEE molecule that consisted in making it non-planar (Esembeson *et al.*, 2008). This new molecule was the DDMEBT molecule also presented in Table 5.1, above. DDMEBT ([2-[4-(dimethylamino)phenyl]-3-([4-(dimethylamino)phenyl]ethynyl) buta-1,3-diene-1,1,4,4-tetracarbonitrile]) may surprise because it has a broken conjugation, but contains almost the complete TDMEE molecule as a conjugated sub-unit, which is the part most responsible for its nonlinearity. The rotational average of its third-order polarizability is only slightly smaller than that of TDMEE, but the non-planar structure of DDMEBT reduces intermolecular interactions in such a way that vapor depositing it on any substrate leads to a homogeneous bulk material with very high optical quality (Fig. 5.2).

The DDMEBT material has very attractive properties. It combines a high optical quality with a high third-order nonlinearity three orders of magnitude larger than fused silica, it has a relatively low refractive index of 1.8 at 1.5 μm, it has been found to be quite stable with a shelf life of years (Beels *et al.*, 2012), and it can be vapor deposited on any substrate. In order for it to be used in applications, it is necessary to control its interaction with light signals. While it would be possible to define waveguides out of the DDMEBT material alone, its properties render it quite attractive for integration with well-established integrated optics technologies. In particular, DDMEBT has been found to create thin homogeneous cover layers on top of nanostructured substrates and to be the ideal material to combine with silicon-on-oxide technology to create hybrid devices (Koos *et al.*, 2009; Scimeca *et al.*, 2009). Its use in so-called silicon-organic-hybrid (SOH) devices was the first demonstration of a practical device that combined organics and state-of-the-art silicon and optical telecom technology to create new functionalities that have never been possible before (Koos *et al.*, 2009; Leuthold *et al.*, 2009; Scimeca *et al.*, 2009; Vallaitis *et al.*, 2009). In particular, it was found that the DDMEBT material added only negligible losses to a slot-waveguide consisting of nanoscale silicon ridges ~200 nm tall and ~200 nm apart (Koos *et al.*, 2009), and that its nonlinearity created a waveguide with a record-high nonlinear parameter of 100 W^{-1} m^{-1}. This SOH device underwent state-of-the-art testing in a telecommunication setting, and it was successfully used for time-division demultiplexing of a 170 Gbit/s data stream using nonlinear optical frequency conversion in a 4 mm long SOH waveguide (Koos *et al.*, 2009). But since the nonlinear optical response of the organic material is electronic and off-resonant, much higher speeds are in principle possible (Vallaitis *et al.*, 2009).

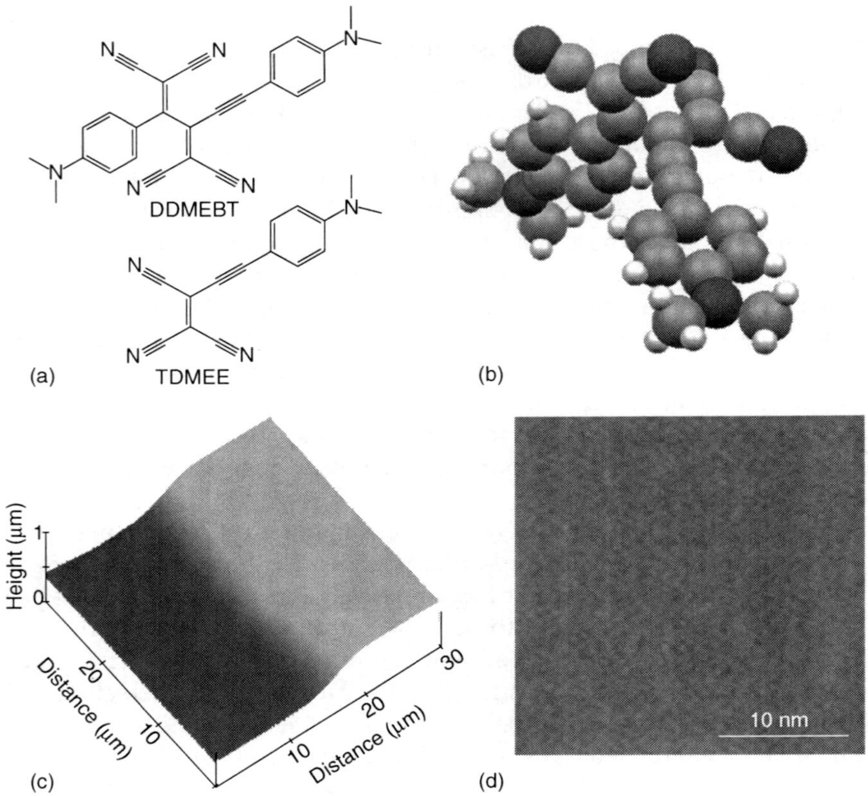

5.2 (a) The DDMEBT molecule compared to TDMEE, (b) a spatial model of the DDMEBT molecule, (c) atomic force microscopy (AFM) picture of the edge of a thin film obtained by molecular beam deposition of DDMEBT that shows the homogeneity and flatness of the film, and (d) SEM picture highlighting the homogeneity of the film on the nanoscale (adapted from Esembeson *et al.*, 2008).

5.7 Conclusion

In conclusion, dense small molecule supramolecular assemblies are the realization of a new paradigm for nonlinear optics, where the focus has been moved away from molecules that have higher and higher third-order polarizabilities and towards smaller molecules that have an optimized third-order polarizability closer to the quantum limit. Large molecules with large absolute third-order polarizabilities cannot necessarily be combined into a useful bulk material with correspondingly high nonlinearity, while the smaller molecules can be assembled into a dense, essentially amorphous material that optimally reflects the molecular properties, has a high optical quality, and can produce (by vapor deposition) wide area films (many square centimeters) that can be combined with any substrate or passive guided-wave technology. In this way it is possible to obtain a successful translation of the molecular properties to a stable bulk solid-state material with high optical quality, and to combine this material with existing passive integrated optic technology. This led to the first demonstration of ultrafast all-optical switching based on an organic supramolecular assembly and silicon-on-oxide technology, and to the realization of a new nonlinear optical functionality for the silicon photonics platform.

The outlook for the future is, as usual, not clear. Telecommunication technologies are evolving at a rapid rate and it is difficult to predict the role that organics-based systems might play in photonics platforms that in the second decade of this century are quickly moving towards coherent communication systems that rely heavily on electronic processing. Still, the variety of applications of optical systems is also increasing at a rapid place, and the availability of new nonlinear optical organic materials compatible with integrated optics technologies is yet another tool in the photonics toolbox, there to be picked up when needed.

5.8 References

S. S. Andrews (2004), 'Using rotational averaging to calculate the bulk response of isotropic and anisotropic samples from molecular parameters,' *J. Chem. Educ.*, **81**(6), 877.

J. A. Armstrong, N. Bloembergen, J. Ducuing, P. S. Pershan (1962), 'Interactions between light waves in a nonlinear dielectric,' *Phys. Rev.*, **127**, 1918–1939.

M. T. Beels, M. S. Fleischman, I. Biaggio, B. Breiten, M. Jordan, F. Diederich (2012), 'Compact TCBD based molecules and supramolecular assemblies for third-order nonlinear optics,' *Opt. Mater. Express*, **2**, 294.

J. L. Brédas, C. Adant, P. Tackx, A. Persoons, B. M. Pierce (1994), 'Third-order nonlinear optical response in organic materials: theoretical and experimental aspects,' *Chem. Rev.*, **94**, 243–278.

F. Bureŝ, W. Bernd Schweizer, J. C. May, C. Boudon, J.-P. Gisselbrecht, M. Gross, I. Biaggio, F. Diederich (2007), 'Property tuning in charge-transfer chromophores by systematic modulation of the spacer between donor and acceptor,' *Chemistry – A European Journal*, **13**(19), 5378–5387.

K. Clays (2001), 'Theoretical upper limits and experimental overestimates for molecular hyperpolarizabilities: a symbiosis,' *Opt. Lett.*, **26**, 1699–1701.

B. Esembeson, M. L. Scimeca, I. Biaggio, T. Michinobu, F. Diederich (2008), 'A high optical quality supramolecular assembly for third-order integrated nonlinear optics,' *Adv. Mater.*, **20**, 1–4.

A. Feldner, W. Reichstein, T. Vogtmann, M. Schwoerer, L. Friedrich, T. Pliŝka, M. Liu, G. I. Stegeman, S-H. Park (2001), 'Linear optical properties of polydiacetylene para-toluene sulfonate thin films,' *Optics Commun.*, **195**, 205–209.

I. Fernandez, G. Frenking (2006), 'π-Conjugation in donor-substituted cyanoethynylethenes: an EDA study,' *Chem. Commun.*, 5030.

U. Gubler, Ch. Bosshard, P. Günter, M. Y. Balakina, J. Cornil, J. L. Brédas, R. E. Martin, F. Diederich (1999), 'Scaling law for second-order hyperpolarizability in poly(triacetylene) molecular wires,' *Opt. Lett.*, **24**, 1599–1601.

J. M. Hales, J. W. Perry (2008), 'Organic and polymeric 3rd order nonlinear optical materials and device applications,' in: *Introduction to Organic Electronic and Optoelectronic Materials and Devices* (Eds: S.-S. Sun, L. Dalton), CRC, Orlando, pp. 521–579.

R.W. Hellwarth (1977), 'Third-order optical susceptibilities of liquids and solids,' *Prog. Quant. Electr.*, **5**, 1–68.

Z. Jin, I. Biaggio, J. Toulouse (2010), '*Ab initio* study of linear and nonlinear optical properties of mixed tellurite-chalcogenide glasses,' *J. Phys.: Condens. Matter*, **22**, 165903.

C. Koos, P. Vorreau, T. Vallaitis, P. Dumon, W. Bogaerts, R. Baets, B. Esembeson, I. Biaggio, T. Michinobu, F. Diederich, W. Freude, J. Leuthold (2009), 'All-optical high-speed signal processing with silicon-organic hybrid slot waveguides,' *Nature Photonics*, **3**, 216–219.

M. G. Kuzyk (2000a), 'Physical limits on electronic nonlinear molecular susceptibilities,' *Phys. Rev. Lett.*, **85**, 1218–1220.

M. G. Kuzyk (2000b), 'Fundamental limits on third-order molecular susceptibilities,' *Opt. Lett.*, **25**, 1183–1185.

M. G. Kuzyk (2003a), 'Fundamental limits on third-order molecular susceptibilities: erratum,' *Opt. Lett.*, **28**, 135.

M. G. Kuzyk (2003b), 'Erratum: Physical limits on electronic nonlinear molecular susceptibilities,' *Phys. Rev. Lett.*, **90**, 039902.

M. G. Kuzyk (2003c), 'Connecting at the speed of light,' *IEEE Circuits & Devices Magazine*, p. 8–17, September 2003.

B. L. Lawrence, M. Cha, J. U. Kang, W. Toruellas, G. Stegeman, G. Baker, J. Meth, S. Etemad (1994), 'Large purely refractive nonlinear index of single crystal P-toluene sulphonate (PTS) at 1600 nm,' *Electron. Lett.*, **30**, 447.

J. Leuthold, W. Freude, J.-M. Brosi, R. Baets, P. Dumon, I. Biaggio, M.L. Scimeca, F. Diederich, B. Frank, C. Koos (2009), 'Silicon organic hybrid technology: a Platform for practical nonlinear optics,' *Proceedings of the IEEE*, **97**(7) 1304–1316.

P. D. Maker and R. W. Terhune (1964), 'Study of optical effects due to an Induced polarization third-order in the electric field strength,' *Phys. Rev.*, **137**, A801–A818.

S. R. Marder, J. W. Perry, G. Bourhill, C. B. Gorman, B. G. Tiemann, K. Mansour (1993), 'Relation between bond-length alternation and second electronic hyperpolarizability of conjugated organic molecules,' *Science*, **261**, 186–189.

S. R. Marder, W. E. Torruellas, M. Blanchard-Desce, V. Ricci, G. I. Stegeman, S. Gilmour, J. L. Brédas, J. Li, G. U. Bublitz, S. G. Boxer (1997), 'Large molecular third-order optical nonlinearities in polarized carotenoids,' *Science*, **276**, 1233–1236.

J. C. May, J. H. Lim, I. Biaggio, N. N. P. Moonen, T. Michinobu, F. Diederich (2005), 'Highly efficient third-order optical nonlinearities in donor-substituted cyanoethynylethene molecules,' *Opt. Lett.*, **30**, 3057.

J. C. May, I. Biaggio, F. Bures, F. Diederich (2007), 'Extended conjugation and donor-acceptor substitution to improve the third-order optical nonlinearity of small molecules,' *Appl. Phys. Lett.*, **90**, 251106.

F. Meyers, S. R. Marder, B. M. Pierce, J. L. Brédas (1994), 'Electric field modulated nonlinear optical properties of donor–acceptor polyenes: Sum-over-states investigation of the relationship between molecular polarizabilities (α, β, γ) and bond length alternation,' *J. Am. Chem. Soc.*, **116**, 10703–10714.

T. Michinobu, J. C. May, J. H. Lim, C. Boudon, J.-P. Gisselbrecht, P. Seiler, M. Gross, I. Biaggio and F. Diederich, (2005) 'A new class of organic donor–acceptor molecules with large third-order optical nonlinearities,' *Chem. Commun.*, 737–739.

N. N. P. Moonen, R. Gist, C. Boudon, J.-P. Gisselbrecht, P. Seiler, T. Kawai, A. Kishioka, M. Gross, M. Irie, F. Diederich (2003), 'Donor-substituted cyanoethynylethenes: powerful chromophores for opto-electronic applications,' *Org. Biomol. Chem.*, **1**, 2032–2034.

M. Scimeca, I. Biaggio, B. Breiten, F. Diederich, T. Vallaitis, W. Freude, J. Leuthold (2009), 'Vapor deposition of organic molecules for ultrafast all-optical switching on silicon,' *Optics and Photonics News*, **20** (12), 39.

D. P. Shelton, J. E. Rice (1994), 'Measurements and calculations of the hyperpolarizabilities of atoms and small molecules in the gas phase,' *Chem. Rev.*, **94**(1), 3–29.

A. D. Slepkov, F. A. Hegmann, S. Eisler, E. Elliott, R. R. Tykwinski (2004), 'The surprising nonlinear optical properties of conjugated polyyne oligomers,' *J. Chem. Phys.*, **120**, 6807.

F. P. Strohkendl, T. J. Axenson, R. J. Larsen, L. R. Dalton (1997), 'Observation of the lowest lying electric-dipole-allowed two-photon resonance in C_{60},' *J. Phys. Chem. B*, **101**, 8802–8807.

T. Vallaitis, S. Bogatscher, L. Alloatti, P. Dumon, R. Baets, M. L. Scimeca, I. Biaggio, F. Diederich, C. Koos, W. Freude, J. Leuthold (2009), 'Optical properties of highly nonlinear silicon-organic hybrid (SOH) waveguide geometries,' *Opt. Expr.*, **17**(20), 17357.

J. Zhou, M. G. Kuzyk (2008), 'Intrinsic hyperpolarizabilities as a figure of merit for electro-optic molecules,' *J. Phys. Chem. C*, **112**, 7978–7982.

6

Molecular crystals and crystalline thin films for photonics

M. JAZBINSEK and P. GÜNTER,
ETH Zurich, Switzerland and Rainbow Photonics AG, Switzerland

DOI: 10.1533/9780857098764.1.190

Abstract: Organic materials with second-order nonlinear optical activity are interesting for many applications in optics and optoelectronics, such as light-frequency conversion, terahertz-wave generation, electric-field detection and electro-optic modulation. In a single crystalline form, such materials feature a high thermal and photochemical stability, which are important issues for the majority of device implementations using organic materials. This chapter presents state-of-the-art second-order nonlinear optical organic single crystalline materials, including molecular and crystal engineering approaches, processing in bulk and thin-film single crystalline forms, as well as most important present applications prospects.

Key words: organic nonlinear optics; single crystalline organic materials; single crystalline thin films; terahertz (THz) wave generation, electro-optics.

6.1 Introduction

Organic materials with second-order nonlinear optical (NLO) functionality are promising for various photonic applications, including light-frequency conversion, terahertz-wave generation, electric-field detection and electro-optic modulation. These materials are based on NLO molecules (chromophores) with a high molecular nonlinearity, which are most often dipolar. In order to achieve a macroscopic second-order NLO response, such molecules need to be arranged in a noncentrosymmetric way in a material, which can be done in a polymer matrix by electric-field poling, molecular self-assembly into amorphous acentric structures, as well as by self-assembly into single-crystalline acentric structures. This chapter is devoted to single-crystalline organic second-order NLO materials, which are attractive because of their high thermal and photochemical stability, owing to the highly stable chromophore packing, but involves challenging crystal growth in either bulk or thin-film form.

6.2 Second-order nonlinear optical (NLO) organic crystals

6.2.1 Microscopic and macroscopic NLO effects in organic materials

The basic design of organic nonlinear optical molecules for crystals is similar to that for poled polymers and is commonly based on π-conjugated polar chromophores with an asymmetric response to an external electric field (Zyss, 1994; Bosshard *et al.*, 1995; Bosshard and Günter, 1997; Bosshard *et al.*, 2000; Kuzyk, 2000; Clays and Coe, 2003; Dalton *et al.*, 2010). π-Conjugated structures are regions of delocalized electronic charge distribution resulting from the overlap of π orbitals. The electron distribution can be distorted by substituents (electron donor and acceptor groups) at both sides of the π-conjugated system as illustrated in Fig. 6.1(a). Upon applying an external electric field **E**, the total dipole moment **p** of such a molecule can be expanded in Taylor's series as

$$p_i = \mu_i + \varepsilon_0 \alpha_{ij} E_j + \varepsilon_0 \beta_{ijk} E_j E_k + \ldots, \quad [6.1]$$

where μ is the dipole moment, α the linear polarizability determining the linear electronic response to an applied field, and β the first-order

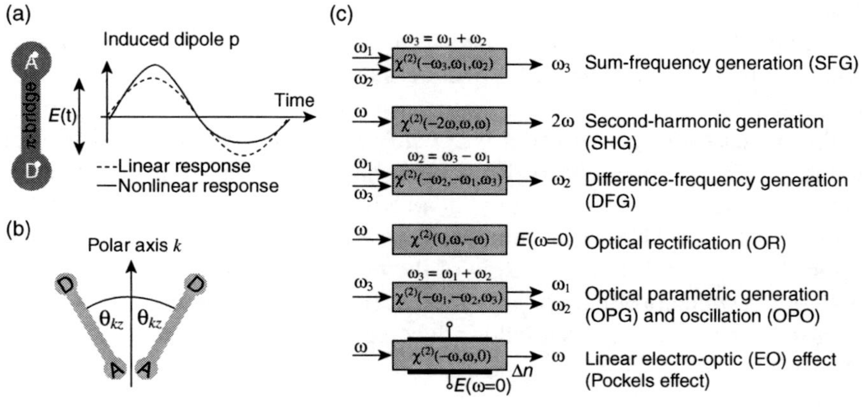

6.1 (a) Simple physical mechanism of the second-order nonlinearity of the donor–acceptor substituted π-conjugated molecules. If we excite such molecules with an optical field, we induce an asymmetric electronic response of the polarization. This is due to the nature of the substituents: the electron cloud (i.e. the electronic response) favors the acceptor A over the donor D. (b) Nonlinear optical molecules in the oriented-gas model, for optimizing the diagonal NLO susceptibility the optimal molecular-ordering angle $\theta_{kz} = 0°$. (c) Schematic representation of important nonlinear optical effects of the second-order, including the linear electro-optic effect.

hyperpolarizability determining the asymmetry of this response with respect to the polarity of the field; ε_0 is the electric constant (permittivity of vacuum).

For a macroscopic second-order nonlinear medium, the induced material polarization **P** can be expressed as a function of the applied field **E** as

$$P_i = \varepsilon_0 \chi_{ij}^{(1)} E_j + \varepsilon_0 \chi_{ijk}^{(2)} E_j E_k + \dots, \qquad [6.2]$$

where $\chi^{(1)}$ presents the linear susceptibility, which is related to the dielectric constant ε of the material as $\chi^{(1)} = \varepsilon - 1$, while $\chi^{(2)}$ presents the nonlinear susceptibility of the second order.

The linear susceptibility $\chi^{(1)}$ describes a linear polarization response of the material to an optical field $\mathbf{E}(\omega)$ oscillation at an angular optical frequency ω, which is most often measured by the refractive index n, where $\chi^{(1)}(\omega) = n^2(\omega) - 1$ and where the dispersion is reflected in its angular frequency ω dependence or dependence on the optical wavelength $\lambda = 2\pi c/\omega$; c is the speed of light in vacuum.

NLO effects of the second-order allow interaction of two optical electric fields $\mathbf{E}(\omega_1)$ and $\mathbf{E}(\omega_2)$ to produce polarization at a combination of their frequencies, which is the source of the field oscillating at a frequency ω_3. Based on the characteristics of the involved electric fields, several second-order NLO effects can be distinguished. A schematic representation of the most important ones is shown in Fig. 6.1(c). For example, sum-frequency generation describes the formation of the field at the sum frequency of the source fields $\omega_3 = \omega_1 + \omega_2$, with a special case of second-harmonic generation, where only one optical field at ω produces the field oscillation at its second-harmonic 2ω.

Special cases of second-order NLO effects involve static (or quasi-static) fields. Optical rectification (OR) is a special case of difference frequency generation $\omega_3 = \omega_1 - \omega_2$ with $\omega_1 = \omega_2$ producing a static field $\omega_3 = 0$ and is also used to for THz-wave generation as discussed later on. The electro-optic (EO) effect describes mixing of the optical field at ω and a static field with $\omega = 0$, which produces polarization at the same frequency ω. This effect is most often described as a field-induced change of the refractive index n at ω as

$$\Delta\left(\frac{1}{n^2}\right)_{ij} = r_{ijk} E_k \qquad [6.3]$$

where r_{ijk} is the electro-optic tensor and can be related to the second-order susceptibility as

$$r_{ijk}(\omega) = -\frac{2\chi_{ijk}^{(2)}(-\omega, \omega, 0)}{n_i^2(\omega) n_j^2(\omega)}. \qquad [6.4]$$

6.2.2 Molecular and crystal engineering approaches

In order to show a high macroscopic NLO response, the π-conjugated polar chromophores are designed to achieve a high hyperpolarizability response β, based on well-established molecular engineering principles, e.g. increasing the conjugation length and the strength of the donor and acceptor groups at the end of molecules (Zyss, 1994; Bosshard *et al.*, 1995; Bosshard and Günter, 1997; Bosshard *et al.*, 2000; Kuzyk, 2000; Clays and Coe, 2003; Dalton *et al.*, 2010). Optimization of the molecular nonlinear response by molecular engineering presents only one part in designing nonlinear optical or electro-optic materials. To show a macroscopic second-order nonlinearity, the arrangement of the molecules plays a very important role. The chromophores must be ordered in an acentric manner in a material to achieve a macroscopic second-order nonlinear optical response. This is most often realized by incorporating the chromophores into a polymer matrix and pole the composite under an influence of a strong DC electric field close to the glass transition temperature. Another possibility to obtain an efficient macroscopic second-order active organic material is to order the chromophores in an acentric structure by crystallization, offering the highest possible chromophore density and the best long-term orientational stability, which is the approach discussed in this chapter.

The relation between the macroscopic and the molecular coefficients is nontrivial because of interactions between neighboring molecules. However, most often the macroscopic second-order nonlinearities of organic materials can be well explained by the nonlinearities of the constituent molecules by the so-called oriented-gas model (Zyss and Oudar, 1982). In first approximation, the dipolar NLO chromophores, which are most common in state-of-the-art organic crystals, do only have one dominant tensor element $\beta = \beta_{zzz}$. In this case, the oriented-gas model gives relatively simple relations between the microscopic and the macroscopic NLO susceptibilities (Zyss and Oudar, 1982). For example, the diagonal element for the NLO susceptibility can be expressed as

$$\chi^{(2)}_{kkk} = N f_{\text{local}} \left\langle \cos^3 \theta_{kz} \right\rangle \beta_{zzz} \qquad [6.5]$$

showing the most important contributions to the NLO effect in organic materials: the chromophore number density N, the local-field correction factor f_{local}, the molecular nonlinearity β and the orientational factor, the so-called order parameter $\langle \cos^3\theta_{kz} \rangle$, which is averaged over all molecules in a macroscopic system. To maximize the diagonal NLO susceptibility $\chi^{(2)}_{kkk}$ along the polar axis k, the projection angles θ_{kz} should be close to zero, i.e. the charge transfer axes of the molecules should be close to parallel. Figure 6.1(b) schematically shows the molecules and projection angles θ_{kz} between the polar axis k and the molecular axis z. To maximize the off-diagonal susceptibility elements $\chi^{(2)}_{kll}$, the molecules should make an angle of about 54.7° with respect to the polar axis (Zyss and Oudar, 1982). The dispersion of the macroscopic NLO susceptibilities is usually also considered using Eq. (6.5), by assuming a simple two-level model (Oudar and Chemla, 1977) for the dispersion of hyperpolarizabilities β and local field factors f_{local} (Zyss, 1994; Bosshard *et al.*, 1995).

Achieving noncentrosymmetric arrangement of molecules in crystal in a way to achieve optimal NLO susceptibilities $\chi^{(2)}_{ijk}$ is a challenging topic. A major problem in achieving this is that most organic molecules will crystallize in a centrosymmetric space group, which is usually attributed to dipolar interaction forces that favor anti-parallel chromophore alignment. Crystal growth is the prototype of self-assembly in nature. The molecules pack into the crystalline lattice that corresponds to a minimum in the potential energy, which will in first approximation depend on the geometry of the molecules (close-packing principle). More precisely it will include electrostatic (Coulomb), van der Waals, and hydrogen-bond interactions. A correct determination of the structure requires a more fundamental quantum mechanical approach that is computationally very difficult. There have been several approaches identified to obtain noncentrosymmetric nonlinear optical organic crystals, including the use of molecular asymmetry, strong Coulomb interactions, non-rod-shaped π-conjugated cores and co-crystallization (Wong *et al.*, 1997; Bosshard *et al.*, 2000). Although the prediction of the crystal structure based on the π-conjugated donor–acceptor chromophores has not yet been made possible, these strategies have led to a number of organic crystals with a very high macroscopic NLO response.

Beside a high molecular nonlinearity β and a preferred acentric molecular packing leading to high macroscopic NLO susceptibilities $\chi^{(2)}$, it is very important that the material crystallization properties enable high-quality bulk or thin-film crystal growth, which is in particular challenging for organic crystalline materials with high macroscopic NLO susceptibilities. It is for all these reasons that out of the hundreds of organic NLO crystalline materials developed during the last 30 years, only a few of them have entered the application market. In the following section, we present some of the most successful examples of organic crystals and their properties.

6.2.3 Examples of organic NLO crystals

DAST (4-*N*,*N*-dimethylamino-4′-*N*′-methyl-stilbazolium tosylate) is the most well known and widely investigated organic electro-optic crystal. DAST was first reported in 1989 by Marder *et al.*, (1989) and Nakanishi *et al.* (1989) and is still recognized as a state of the art organic nonlinear optical crystal. High optical quality and large DAST crystals were grown

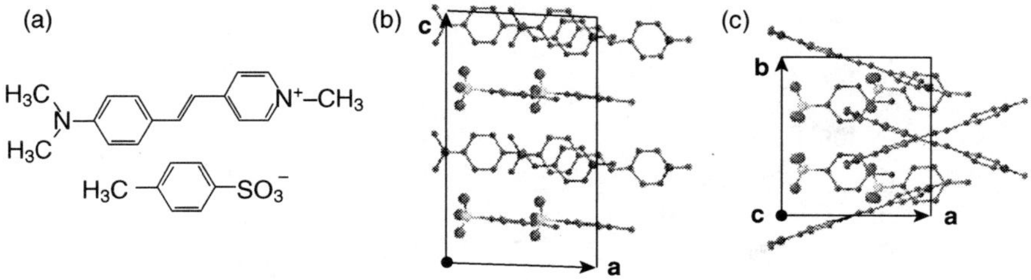

6.2 (a) Molecular units of the ionic DAST crystal. The positively charged, NLO active chromophore methyl-stilbazolium and the negatively charged tosylate. (b) and (c) X-ray structure of the ionic DAST crystal with the point group symmetry m showing molecules from one unit cell, projected along the crystallographic axes *b* and *c*; hydrogen atoms have been omitted for clarity.

from methanol solution by the slow cooling method (Pan *et al.*, 1996b; Ruiz *et al.*, 2008) which allowed their dielectric, linear, and nonlinear optical properties to be accurately determined. The reasons for the growing interest in obtaining high-quality DAST crystals are the high second-order NLO and the EO coefficients, being respectively ten times and twice as large as those of the inorganic standard $LiNbO_3$, in combination with a low dielectric constant, which allows for high-speed EO applications and broadband THz-wave generation.

DAST is an organic salt that consists of a positively charged stilbazolium cation and a negatively charged tosylate anion as shown in Fig. 6.2(a). The stilbazolium cation is one of the most efficient NLO active chromophores that pack in an acentric structure, whereas the counter ion tosylate is used to promote noncentrosymmetric crystallization (Marder *et al.*, 1989, 1994). The structure of DAST is shown in Fig. 6.2(b,c). The chromophores are packed with their main charge transfer axis oriented at about $\theta = 20°$ with respect to the polar axis *a*, resulting in a high-order parameter of $\cos^3\theta = 0.83$, which is close to the optimum for the EO applications.

As a result of the highly ordered packing of highly polarizable constituting molecules, DAST crystals are strongly anisotropic with a refractive index difference $n_1 - n_2 > 0.5$ in the visible and infrared wavelength range (Pan *et al.*, 1996a, 1996b). For applications in telecommunications, DAST crystals are well suited with a material absorption that is smaller than 1 cm^{-1} at 1.3 μm and 1.55 μm wavelengths. The dielectric constants of DAST in the low-frequency range, below acoustic and optical lattice vibrations, were determined as $\varepsilon_1^T = 5.2 \pm 0.4$, $\varepsilon_2^T = 4.1 \pm 0.4$, and $\varepsilon_3^T = 3.0 \pm 0.3$ (Pan *et al.*, 1996a), and are considerably lower than those of inorganic EO materials, which is very important for high-speed EO and phase-matched THz-wave generation and detection (Jazbinsek *et al.*, 2008b).

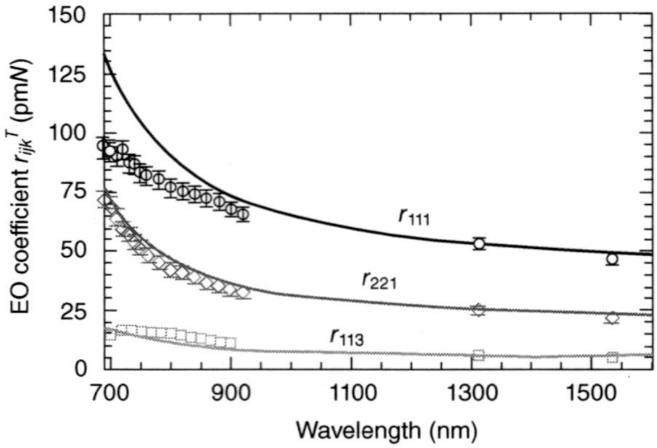

6.3 Dispersion of the largest EO tensor coefficients of DAST: r_{111}, r_{221}, and r_{113} (Pan *et al.*, 1996a).

The low-frequency (unclamped) EO coefficients r_{ijk}^T of DAST are shown in Fig. 6.3. The experimentally measured dispersion was modeled by the two-level model and is presented by solid curves in Fig. 6.3 (Pan *et al.*, 1996a); the deviation at shorter wavelengths stems from resonance effects when approaching the absorption edge. DAST exhibits large EO coefficients and refractive indices, resulting in a high EO figure of merit $n^3r = 455 \pm 80$ pm/V at wavelength $\lambda = 1535$ nm and therefore the reduced half-wave voltage $v_\pi = \lambda/(n^3r)$ compares favorably with inorganic single crystals and poled EO polymers.

Theoretical evaluations show that the upper limits of second-order optical nonlinearities of organic crystals have by far not been reached yet (Bosshard *et al.*, 2000); therefore, the research and development of novel NLO organic crystals is being continued. Several organic crystalline salts have been identified with similar or even superior NLO properties to DAST (Marder *et al.*, 1994; Coe *et al.*, 2002; Glavcheva *et al.*, 2005; Yang *et al.*, 2007a; Figi *et al.*, 2008b; Kim *et al.*, 2011; Kim *et al.*, 2012), also combined with better crystal growth possibilities (Yang *et al.*, 2007b; Ogawa *et al.*, 2008).

The chromophores for highly NLO crystals in general exhibit a limited temperature stability. In case of DAST (and similarly for other highly nonlinear organic salts), the chromophores start to decompose at about the melting temperature, which is for DAST at 256 °C. Therefore, the processing possibilities of stilbazolium salts are for most practical situations limited to solution-based techniques. On the other hand, melt growth is very attractive for several reasons, e.g. higher growth rates, higher purity, and very attractive waveguide processing possibilities, as will be discussed in the following section. The short π-conjugated chromophores with relatively low melting temperatures ($T_m < 150$ °C), but also relatively low first hyperpolarizabilities, such as COANP (2-cyclooctylamino-5-nitropyridine), were the only organic nonlinear optical crystals obtained by melt growth techniques until recently. Unfortunately, the EO figure of merit n^3r of these crystals may be one order of magnitude smaller as for the best stilbazolium salts. Therefore, to design organic EO materials with a broad spectrum of processing possibilities, the challenge is to simultaneously achieve a high thermal stability and nonlinearity in one compound. To evade the nonlinearity-thermal stability tradeoff of organic crystalline materials, different series of configurationally locked polyene chromophores have been recently developed (Kwon *et al.*, 2006, 2008). Several of these chromophores crystallize in a non-centrosymmetric structure with a high powder test efficiency in the same order of magnitude as that of DAST.

Configurationally locked polyene crystals DAT2 (2-(3-(2-(4-dimethylaminophenyl)vinyl)-5,5-dimethylcyclohex-2-enylidene)malononitrile) and OH1 (2-(3-(4-hydroxystyryl)-5,5-dimethylcyclohex-2-enylidene)malononitrile) have been already demonstrated to be particularly promising for integrated EO applications (Figi *et al.*, 2008a, 2010; Hunziker *et al.*, 2008a; Jazbinsek *et al.*, 2010) as well as THz-wave generation (Brunner *et al.*, 2008). The big advantage of DAT2 is the excellent range of possibilities for thin-film processing from all solution, vapor and melt, although its EO properties are not as high as for DAST due to a less optimal crystalline packing (Kwon *et al.*, 2006; Figi *et al.*, 2008a). On the other hand, OH1 crystals show the EO effect as high as DAST crystals (Hunziker *et al.*, 2008b). OH1 exhibits very good crystal processing possibilities from solution, for both high optical quality bulk crystals (Kwon *et al.*, 2010), as well as large-area single crystalline thin films on substrates (Kwon *et al.*, 2009).

Table 6.1 lists some of the presently best organic crystals, which have been already successfully used for applications such as frequency conversion, THz-wave generation, electro-optics and integrated optics.

6.2.4 Single crystal growth: solution, vapor, and melt-based techniques

Crystallization of organic materials is based on solution growth, melt growth or vapor growth. To produce either bulk, thin-film or wire crystals, different and in some cases rather complex growth techniques are required. The choice of an appropriate technique depends on different material properties, as well as a desired crystalline form.

For many of the highly NLO materials, such as DAST, DSTMS, DAPSH and OH1, best-quality bulk crystals are grown from solution. Most commonly, slow temperature lowering techniques or slow isothermal evaporation techniques are used, which can be combined with temperature gradients at the growth position. An example of the solution-growth process

Table 6.1 Some of the state-of-the-art molecular crystals: λ_c is the cut-off wavelength in the bulk, *d* is the nonlinear optical coefficient, *r* is the electro-optic coefficient, and T_m is the melting point

Material	Point group	λ_c (nm)	*d*, *r* (pm/V)	T_m (°C)
BNA(N-benzyl-2-methyl-4-nitroaniline) 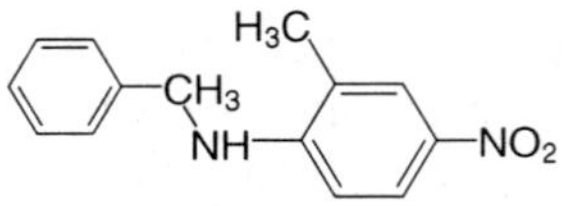 (Fujiwara *et al.*, 2007, 2006; Hashimoto *et al.*, 1997)	*mm*2	500	d_{33}(1064 nm) = 234 d_{32}(1064 nm) = 15.6	110
DAST(4′-dimethylamino-*N*-methyl-4-stilbazolium tosylate) 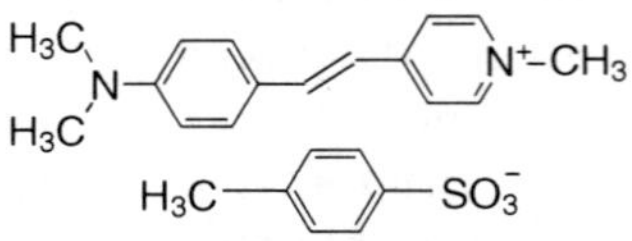 (Knöpfle *et al.*, 1995; Pan *et al.*, 1996a; Meier *et al.*, 1998)	*m*	700	d_{11}(1318 nm) = 1010 d_{11}(1542 nm) = 290 d_{11}(1907 nm) = 210 d_{12}(1907 nm) = 32 d_{26}(1907 nm) = 25 r_{11}(720 nm) = 92 r_{11}(1313 nm) = 53 r_{11}(1535 nm) = 47	256
DSTMS(4′-dimethylamino-*N*-methyl-4- stilbazolium 2,4,6-trimethylbenzenesulfonate) 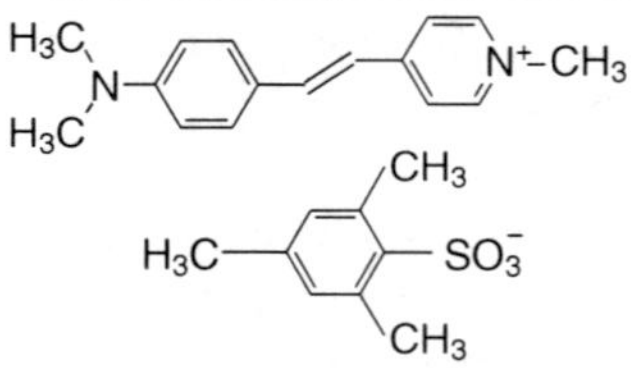 (Yang *et al.*, 2007b; Mutter *et al.*, 2007a)	*m*	690	d_{11}(1907 nm) = 210 d_{12}(1907 nm) = 31 d_{26}(1907 nm) = 35	258
DAPSH(*trans*-4-dimethylamino-*N*-phenyl-stilbazolium hexafluorophosphate) 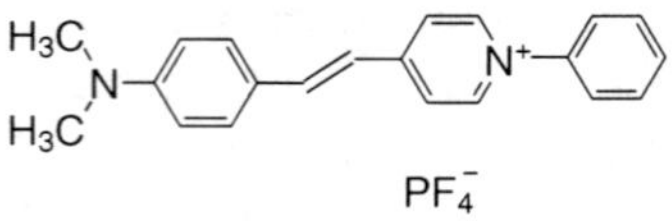 (Coe *et al.*, 2002, 2003; Figi *et al.*, 2008b)	*m*	700	d_{11}(1907 nm) = 290	–

Table 6.1 Continued

Material	Point group	λ_c (nm)	*d*, *r* (pm/V)	T_m (°C)
DAT2(2-(3-(2-(4-dimethylaminophenyl)vinyl)-5,5-dimethylcyclohex-2-enylidene) malononitrile) 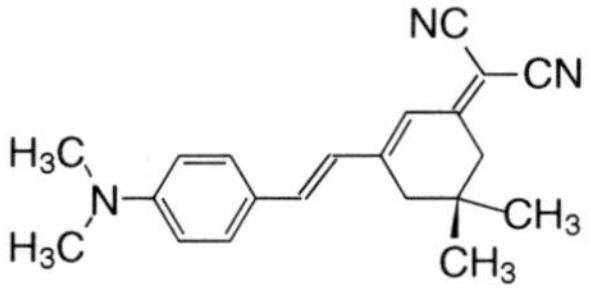 (Kwon *et al.*, 2006; Figi *et al.*, 2008a, 2010)	2	700	r_{12}(1550 nm) = 7.4 r_{22}(1550 nm) = 6.7	235
OH1(2-(3-(4-hydroxystyryl)-5,5-dimethylcyclohex-2-enylidene) malononitrile) 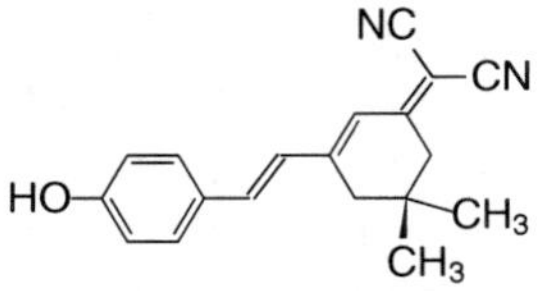 (Kwon *et al.*, 2008; Hunziker *et al.*, 2008b)	*mm*2	600	d_{33}(1907 nm) = 120 r_{33}(633 nm) = 109 r_{33}(1319 nm) = 52 r_{23}(1319 nm) = 30 r_{13}(1319 nm) = 6.8	212

optimization for obtaining high-optical quality bulk crystals of DAST can be found in Ruiz *et al.* (2008). Compared to DAST, growth of DSTMS crystals and OH1 crystals from solution may be faster and easier, which is due to the favorable thermodynamic properties of these materials (Yang *et al.*, 2007b; Kwon *et al.*, 2010).

High-quality single crystalline thin films of highly NLO materials are essential for the fabrication of integrated photonic devices. If one should start from the bulk crystals, then complicated, expensive, and time consuming cutting, polishing, and structuring procedures are required to fabricate waveguiding devices. Obviously thin films may be much more compatible with simpler and cheaper waveguiding structures for applications such as EO modulators. Various approaches have been investigated for the fabrication of single crystalline films, using either solution, melt or vapor growth techniques; an overview of different approaches for thin-film fabrication is reviewed in more detail in Jazbinsek *et al.* (2008a) and Jazbinsek and Gunter (2011).

One of the very attractive solutions for the integrated photonics devices presents a direct growth of the desired micro- and nano-structures at a

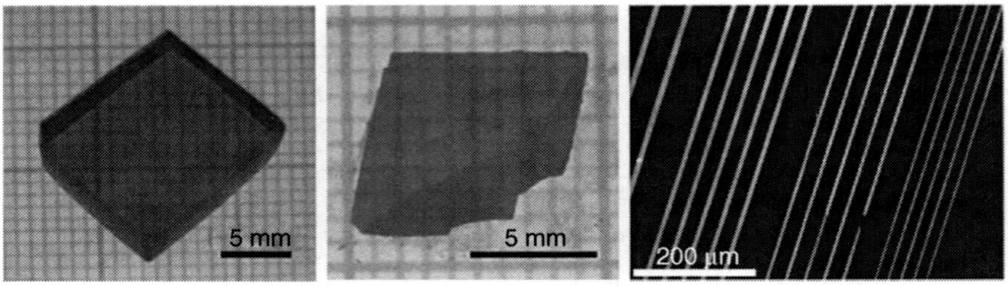

(a) (b) (c)

6.4 Examples of the grown crystals (a) bulk OH1 single crystal 4 mm thick (Kwon *et al.*, 2010), (b) thin film DSTMS single crystal 20 μm thick (Yang *et al.*, 2007b) and wire DAT2 single crystals 25 nm thick (Figi *et al.*, 2008a).

desired position. This can be done by first structuring standard inorganic templates, such as glass, silicon, electrodes and other materials, with void structures at positions where active organic crystalline materials are desired. This method was recently demonstrated by using melt-processable materials DAT2 (Figi *et al.*, 2008a), the small chromophore COANP (Figi *et al.*, 2009) and BNA (Figi *et al.*, 2011); these materials were chosen because of the favorable growth characteristics from melt as well as the tendency for thin-film formation. By this method, several-mm long single crystalline wires with a thickness up to several μm down to below 30 nm have been obtained (Figi *et al.*, 2008a). Figure 6.4 shows some examples of single-crystalline organic bulk, thin-film and wire crystals.

6.3 THz-wave generation and detection with organic crystals

The interest in generating THz waves stems from the unique interactions of these rays with matter, which can be exploited in various applications. For example, optical phonon resonances of crystalline materials and a part of vibrational and rotational excitations of molecules are in the THz range, which makes THz radiation very interesting for spectroscopy and material identification. Other applications include non-destructive material testing and imaging, various research material investigations such as carrier dynamics in semiconductors with sub-picosecond time resolution, medical diagnostics and pharmaceutical characterization (Dalton *et al.*, 2012).

6.3.1 THz-wave generation by difference-frequency mixing

Generating THz waves by optical difference-frequency generation (DFG) requires a pump source consisting of two frequencies ω_1 and ω_2 that are very close to each other, so that their difference frequency lies in the THz range: $\omega_{THz} = \omega_1 + \omega_2$. The phase matching condition should be also satisfied

$$\Delta\mathbf{k} = \mathbf{k}_{THz} - (\mathbf{k}_1 - \mathbf{k}_2). \qquad [6.6]$$

For collinear DFG and assuming that the optical frequencies are close together, so that in first approximation the dispersion in the optical range can be considered as $n_2 = n_1 + (\partial n/\partial\lambda)\Delta\lambda$, $\Delta\lambda = \lambda_2 - \lambda_1$, this leads to

$$\Delta k = \frac{\omega_{THz}}{c}(n_{THz} - n_g) \qquad [6.7]$$

and the following coherence length for THz generation

$$l_c = \frac{\lambda_{THz}}{2(n_{THz} - n_g)}, \qquad [6.8]$$

where $n_g = n - (\partial n/\partial\lambda)\lambda$ is the group index of the optical wave. Equation (6.8) is valid for a relatively small dispersion in the optical range, i.e. up to several THz if we use infrared pump light. For larger THz frequencies, it should be calculated as

$$l_c = \frac{1}{2}\left(\frac{n_{THz}}{\lambda_{THz}} - \frac{n_1}{\lambda_1} + \frac{n_2}{\lambda_2}\right). \qquad [6.9]$$

For the efficient generation of THz waves by difference-frequency generation, besides a high second-order nonlinear optical susceptibility, the most important parameter is the low refractive index mismatch $\Delta n = n_{THz} - n_g$ between the generated THz and the pump optical waves. This is where organic materials are of a big advantage compared to standard inorganic materials such as $LiNbO_3$. Because of the relatively low contribution of the lattice phonon vibrations to the dielectric constant, the dispersion of the refractive index between the optical and the THz frequency range is low and therefore the phase matching condition is almost naturally satisfied, while for inorganic materials such as $LiNbO_3$ special phase matching configurations are needed.

For DFG in the case of phase-matching and neglecting the pump-light absorption, the visible-to-THz conversion efficiency is given by (Sutherland, 2003)

$$\eta_{THz} = \frac{\omega_{THz}^2 d_{THz}^2 L^2 I_0}{2\varepsilon_0 c^3 n_0^2 n_{THz}} \exp(-\alpha_{THz} L/2) \frac{\sinh^2(\alpha_{THz} L/4)}{(\alpha_{THz} L/4)^2}, \qquad [6.10]$$

where

$$d_{THz} = \frac{1}{4} n_o^4 r \qquad [6.11]$$

Table 6.2 Organic and inorganic nonlinear optical materials that have been investigated for optical-to-THz frequency conversion and their most relevant parameters*. Where possible, the parameters close to the velocity-matched optical wavelengths and THz frequencies are given

	n_o	n_g	n_{THz}	r (pm/V)	d_{THz}[1] (pm/V)	FM_{THz}[2] (pm^2/V^2)	ν_{phonon} (THz)	α_{THz} (cm^{-1})	λ (nm)
DAST	2.13	2.3[3]	2.26	47	240	5600	22	20	1500
DSTMS	2.13	2.3	2.26	49	250	6100	22	15	1500
OH1	2.16	2.33	2.28[4]	52	280	7400	8	2[6]	1350
LAPC[5]	1.6	1.8	1.7	52	85	1700	>17	15	1500
GaAs	3.37	3.61	3.63	1.6	52	66	7.6	0.5	1560
ZnTe	2.83	2.18	3.16	4	64	160	5.3	1.3	840
InP	3.2	3.16	3.54[6]	1.45	38	40	10		
GaP	3.12		3.34	1	24	17	10.8	0.2	1000
ZnS	2.3		2.88	1.5	10	7	9.8		
CdTe	2.82		3.24	6.8	110	470		4.8	
$LiNbO_3$	2.2	2.18	4.96	28	160	1100		17	

[1] $d_{Thz} = \frac{1}{4} n_0^4 r$.

[2] $FM_{THz} = \frac{d_{THz}^2}{n_0^2 n_{THz}} = \frac{n_0^6 r^2}{16 n_{THz}}$.

[3] $\nu > 1.5$ THz.

[4] $\nu < 1.9$ THz.

[5] LAPC guest-host polymer (Zheng *et al.*, 2007).

[6] $\nu \approx 1$ THz.

*Refractive index n_o at the pump optical wavelength λ; group index n_g at λ; refractive index n_{THz} in the THz frequency range; the electro-optic coefficient r; the susceptibility d_{THz} for THz-wave generation; figure of merit FM_{THz} for THz generation by optical rectification; optical phonon frequency of the material ν_{phonon} in the THz range; the absorption α_{THz} in the THz frequency range.

is the nonlinear optical susceptibility for THz-wave generation, ω_{THz} the angular frequency of the generated THz wave, L the length of the THz-generation materials, I_0 the pump intensity, α_{THz} the absorption constant at the THz frequency, r the electro-optic coefficient, n_0 and n_{THz} the refractive indices at the pump optical and the generated THz frequencies, respectively. Besides phase-matching and minimal THz absorption, the main material figure of merit for THz generation (FM_{THz}) according to Eq. (6.10) is

$$FM_{THz} = \frac{d_{THz}^2}{n_0^2 n_{THz}} = \frac{n_0^6 r^2}{16 n_{THz}}. \quad [6.12]$$

Table 6.2 shows most of these parameters for a series of inorganic and organic crystals, as well as for an electro-optic polymer. As it can be seen in this table, the organic crystals OH1, DSTMS and OH1 show the largest figure of merit and can be also phase matched using pump lasers at telecommunication wavelengths 1.3–1.55 μm. OH1 in addition shows a very small absorption constant at THz frequencies, thus allowing large interaction lengths to be used (Brunner *et al.*, 2008). The optical damage threshold of these organic crystals mainly depends on the optical quality, both of the bulk crystal quality and the quality of surface polishing. Very slow cooling growth with high temperature stability of (±0.002 °C) has to be used for high damage threshold materials reaching $I_{damage} > 150$ GW/cm^2 for 150 fs pulses at 1550 nm (Rainbow Photonics, 2012).

6.3.2 THz-wave generation by OR

By using the process of OR, broadband THz radiation can be efficiently generated in noncentrosymmetric NLO crystals pumped by femtosecond pulses. An ultrashort laser pulse (10–200 fs) induces a quasi-static polarization in such materials through OR, which follows the amplitude of the pump pulse and thus acts as a source for the THz pulse. In other words, a short laser pulse has an intrinsically broad bandwidth, i.e. a laser beam with a pulse length of 10–200 fs has a bandwidth of roughly 40–2 THz, depending on the pulse shape. Different frequency components in such a pulse can mix with each other in a nonlinear crystal by difference-frequency generation, producing a broadband THz wave. In the simplest approximation (non-depleted pump approximation, negligible THz absorption,...) the intensity of the generated THz intensity is given by (Schneider *et al.*, 2006a)

$$I(\omega_{\mathrm{THz}}, L) = \frac{1}{2}\varepsilon_0 c n_{\mathrm{THz}} \left|E(\omega_{\mathrm{THz}}, L)\right|^2 \quad [6.13]$$

$$\approx \frac{d_{\mathrm{THz}}^2 I_0^2 \omega_{\mathrm{THz}}^2 L^2}{2\varepsilon_0 c^3 n_0^2 n_{\mathrm{THz}}} \quad [6.14]$$

and the conversion efficiency by

$$\eta = \frac{I(\omega_{\mathrm{THz}}, L)}{I_0} = \frac{d_{\mathrm{THz}}^2 I_0 \omega_{\mathrm{THz}}^2 L^2}{2\varepsilon_0 c^3 n_0^2 n_{\mathrm{THz}}} = \frac{1}{2\varepsilon_0 c^3}\mathrm{FM}_{\mathrm{THz}}\omega_{\mathrm{THz}}^2 I_0 L^2, \quad [6.15]$$

which the same as for DFG with the same figure of merit $\mathrm{FM}_{\mathrm{THz}}$ as the one listed in Table 6.2. For more general expressions considering the optical and THz-wave absorption, as well as velocity mismatching see e.g. Schneider *et al.* (2006a).

Figure 6.5 illustrates how velocity matching can be achieved in DAST. Due to the small absorption and the corresponding dispersion near 1.1 THz phase matching can be realized either for pump wavelengths between 830 and 1300 nm, yielding THz waves between 0.2 and 1 THz, or for pump waves with wavelengths between 1300 and 1700 nm for the generation of THz waves above 1.8 THz (Schneider *et al.*, 2006a, 2006b).

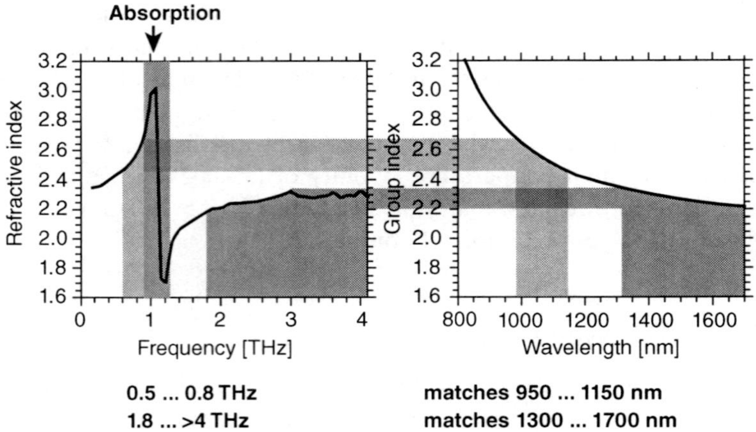

6.5 Refractive index dispersion in the optical and the group index dispersion in the THz frequency range for DAST, indicating the optimal ranges for velocity-matching.

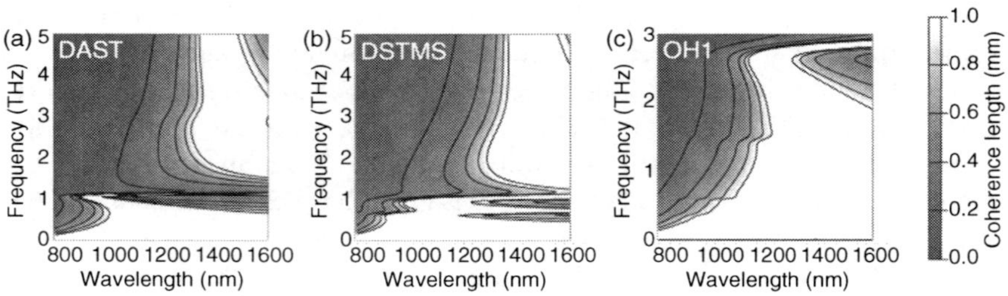

6.6 Coherence length for velocity matching $l_{c,\mathrm{VM}}$ for THz generation by using optical rectification in (a) DAST, (b) DSTMS, and (c) OH1 as a function of the pump optical wavelength and the generated THz frequency. The white area represents the range with best velocity matching leading to coherence lengths larger than 1 mm (Schneider *et al.*, 2006a).

Best conditions for velocity matching in DAST are also illustrated in Fig. 6.6(a), which shows the coherence length $l_{c,\mathrm{VM}}$ as a function of the pump optical wavelength and the generated THz frequency. The generation of THz waves with frequencies around ~1.1 THz is in DAST limited due to a transverse optical phonon (Walther *et al.*, 2000). Other organic materials have different velocity matching conditions and can be used for an available pump wavelength and the desired THz frequency range (Kwon *et al.*, 2007; Brunner *et al.*, 2009; Miyamoto *et al.*, 2009; Seo *et al.*, 2009; Kim *et al.*, 2011, 2012). Figures 6.6(b) and (c) show similar plots for some newer materials, DSTMS (Stillhart *et al.*, 2008) and OH1 (Brunner *et al.*, 2008). In DSTMS the optical phonon absorption near 1.1 THz is suppressed by a heavier counter anion and therefore this material is superior to DAST within this frequency range (Stillhart *et al.*, 2008). OH1 crystal is based on hydrogen bonds and has even higher figure of merit for THz-wave generation (see Table 6.2) compared to DAST and DSTMS and optimum velocity-matching between 1200 and 1460 nm for 0.3–2.5 THz, has no absorption at about 1 THz but at about 3 THz (Brunner *et al.*, 2008). THz frequencies from 0.5 THz to 10 THz have been generated in OH1 by difference-frequency generation (Uchida *et al.*, 2011).

6.4 Integrated electro-optic (EO) applications

The EO effect describes the change of the refractive index upon application of a static (or quasi-static) field and is defined by Eq. (6.3), which gives, assuming $\Delta n << n$, the following change of the refractive index Δn for light polarization parallel to the direction of the applied field E

$$\Delta n = -\frac{1}{2} n^3 r E \qquad [6.16]$$

where $r = r_{333}$ for the electric field and polarization along the x_3 material direction. This gives the following phase change $\Delta\phi$ for light traveling a distance L in the material upon application of the voltage V over the (electrode) distance D

$$\Delta\phi = \frac{2\pi}{\lambda}\Delta n L = -\frac{\pi n^3 r}{\lambda}\frac{L}{D}V. \qquad [6.17]$$

For the EO applications, an important parameter is the so-called half-wave voltage V_μ, which is the voltage required to change the phase of the optical field by π

$$V_\pi = \frac{\lambda}{n^3 r}\frac{D}{L}, \qquad [6.18]$$

and critically depends on the material EO figure of merit $FM_{EO} = n^3 r$, as well as the configuration of the EO material by the geometrical factor D/L. Waveguide configurations allow for small electrode distances D and for relatively large propagation distances L. This can decrease the half-wave voltages by more than three orders of magnitude compared to bulk materials, from kilovolts to less than 1 V for best EO materials. Therefore, integrated optics is the best solution for light modulation and switching needed in telecommunications.

For high data transmission rates or fast modulation and switching, the applied field or voltage V is modulated at high frequencies, up to GHz or even 100 GHz. Therefore, for larger propagation distances L, the applied field already changes during light propagation, which considerably reduces the final phase change $\Delta\phi$. The solution is in the so-called traveling-wave modulators, in which the electric field travels together with the optical field.

In organic materials the electric wave travels at about the same speed as the optical wave, due to the low dielectric constant in the low-frequency regime $\varepsilon \approx n^2$, which is not the case for most inorganic EO materials, where $\varepsilon >> n^2$. This kind of phase matching is important when building high-frequency EO modulators. The low dielectric constant of organic materials will also decrease the power requirement of the EO modulators. Another advantage of organic over inorganic materials is the almost constant EO coefficient over an extremely wide frequency range. This property is essential when building broadband EO modulators and field detectors.

The interest in organic crystals stems from the above advantages compared to inorganic materials, as well as the fact that the long-term orientational stability and photochemical stability, as well as the optical quality of molecular crystals may be significantly superior to those of polymers (Rezzonico *et al.*, 2008b). However, compared to polymers, processing of organic EO crystals in thin films and waveguides needed for integrated optics is much more challenging. In the following section we summarize several structuring techniques that have been used to build waveguides and optical modulators based on organic EO crystals.

6.4.1 Overview of the structuring techniques for organic crystals

As for general optical waveguides, organic crystals have to be structured with a (sub-)micron precision so that a suitable refractive index contrast for optical waveguiding is achieved. Although for optical waveguiding even very small refractive index changes of materials in the order of $\Delta n \sim 10^{-3}$ may be sufficient, for small waveguides needed for large-scale integration, as well as for reducing the half-wave voltage of EO modulators, larger index contrast is desired. For example, to fabricate microring resonators (Little *et al.*, 1997; Gheorma and Osgood, 2002; Rabiei *et al.*, 2002; Rezzonico *et al.*, 2008a) with a small radius below 10 μm, refractive index of the guiding medium should be at least by about $\Delta n \sim 0.5$ larger than the surrounding medium to avoid high losses. Organic EO crystals have a relatively high refractive index compared to poled polymers, which can reach the values of inorganic ferroelectric crystals such as $LiNbO_3$ ($n \sim 2.2$). This basically allows very efficient high-index contrast waveguiding with respect to substrate materials such as SiO_2. Organic EO crystals are also strongly anisotropic with birefringence as high as $\Delta n > 0.5$ at non-resonant wavelengths, which should be taken into account when designing the waveguides. The particular best orientation of the waveguides with respect to the propagation direction, optical and electric field orientation depends on the particular tensor properties of the material. Organic crystals are also suitable as active cladding materials to high-index silicon photonic passive waveguides (n_{Si} ~ 3.5), which can result in very compact EO modulators with high figures of merit, if organic crystals can be oriented in a suitable way (Jazbinsek *et al.*, 2010; Figi *et al.*, 2011).

The main challenges to build integrated EO modulators based on organic crystals are related to their processing possibilities: the organic crystal should be deposited on appropriate substrate materials and in a desired orientation to achieve planar light confinement, and then structured with an appropriate technique to achieve horizontal light confinement. Several approaches and techniques have been developed to fabricate optical waveguides in organic EO crystals. We can distinguish photolithography, photostructuring (including photobleaching and femtosecond laser ablation), ion implantation, electron-beam irradiation, and direct deposition into pre-structured inorganic templates. These techniques are schematically presented in Table 6.3 with some of their main features. A more detailed overview of organic single crystalline waveguides and modulators with examples can be found in Jazbinsek and Günter (2011) and Dalton *et al.* (2012). Presently most promising organic crystals, several of which

Table 6.3 Structuring methods investigated for organic single crystalline waveguides with second-order nonlinear optical activity. Δn is estimated assuming SiO_2 substrates (where applicable)

Technique	Max. index contrast Δn	Comments
Photolithography 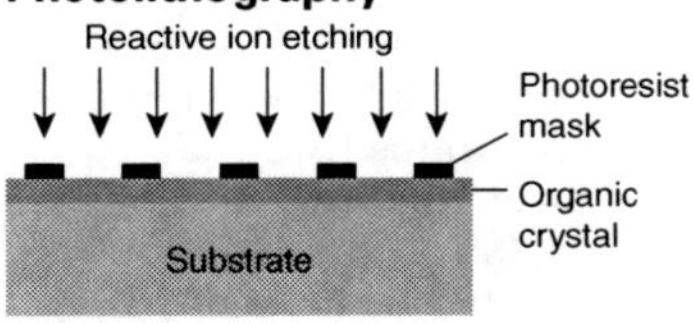	1.1 horizontal 0.6 vertical	• Thin films needed • Side-wall quality depends critically on the optimization of reactive ion etching (RIE)
(Tsuda *et al.*, 1992; Takayama *et al.*, 2001; Kaino *et al.*, 2002; Jazbinsek *et al.*, 2008b; Hunziker *et al.*, 2008a)		
Photobleaching 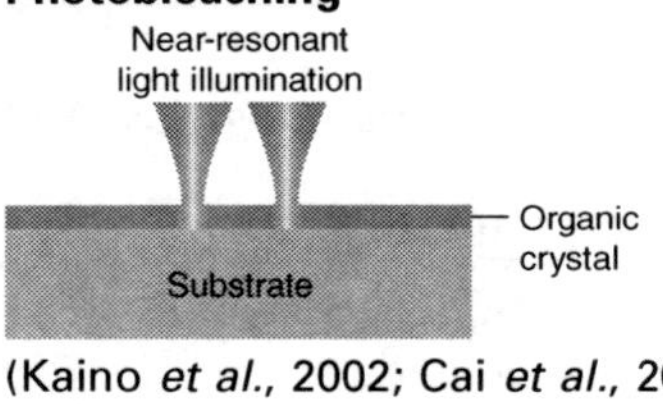	0.5 horizontal 0.6 vertical	• Thin films needed • Smooth side walls
(Kaino *et al.*, 2002; Cai *et al.*, 2003; Mutter *et al.*, 2003)		

Table 6.3 Continued

Technique	Max. index contrast Δn	Comments
Femtosecond laser ablation 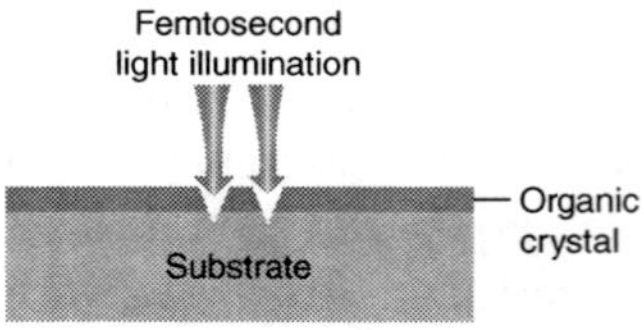 (Dittrich *et al.*, 2003)	1.1 horizontal 0.6 vertical	• Thin films needed • Side-wall quality depends critically on laser parameters
Ion implantation 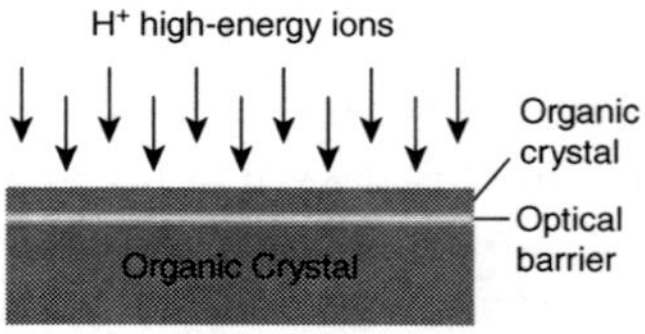 (Mutter *et al.*, 2007b, 2008)	0.1 vertical	• Provides vertical confinement if thin films are not available • Smooth refractive-index gradients
Electron-beam structuring 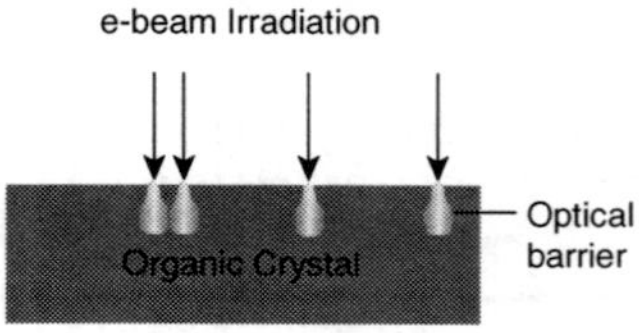 (Mutter *et al.*, 2007c)	0.1 horizontal 0.1 vertical	• Thin films not needed • Smooth side walls
Epitaxial growth 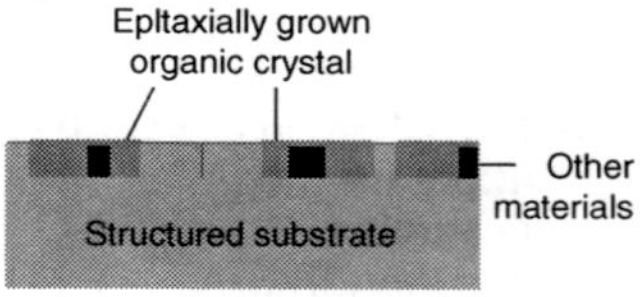 (Geis *et al.*, 2004; Figi *et al.*, 2008a, 2010, 2011; Jazbinsek *et al.*, 2010)	0.6 horizontal 0.6 vertical	• Very versatile • Structuring performed only in standard inorganic substrates (SiO_2, Si) • Easy electrode or other material integration • Limited by crystallization properties of organic material

have been already successfully used for building prototype integrated EO modulators are listed in Table 6.1.

Some of the techniques listed in Table 6.3 allow even for more complex waveguiding structures than simple straight waveguides. For example, by melt capillary growth, first single-cyrstalline organic EO

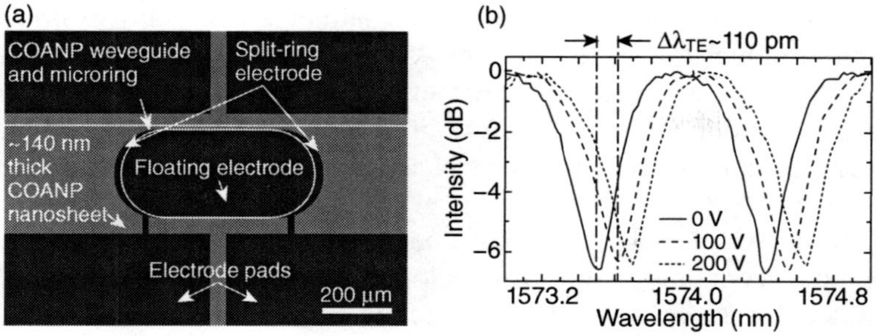

6.7 (a) Transmission microscope image between crossed polarizers of a COANP waveguide with a racetrack microring resonator grown by the melt capillary method in prefabricated channels. (b) Resonance curve of a TE mode at a wavelength around 1.574 μm (solid line); the dashed and dotted line are the corresponding electrooptically shifted curves by applying 100 V and 200 V voltage to the device electrodes (Figi *et al.*, 2009).

microring-resonator filters and modulators were demonstrated (Figi *et al.*, 2009). In this case organic material COANP (2-cyclooctylamino-5-nitropyridine) with very good melt-crystallization properties and a moderate EO coefficient $r_{33} = 15 \pm 2$ pm/V at 633 nm was employed. A top view transmission microscope image of a COANP crystal grown in a microring resonator channel waveguide is depicted in Fig. 6.7(a). Very high single-crystalline quality of these waveguides was confirmed by optical waveguiding characterization. Typical devices fabricated showed almost perfectly symmetric high extinction ratio resonance peaks of about 10 dB, ring losses $\alpha = 12 \pm 0.3$ dB/cm, and a finesse $F = 6.2 \pm 0.2$. The measured TE spectrum of the racetrack resonator shown in Fig. 6.7(a) showed a $\Delta\lambda = 110$ pm shift in response to an applied voltage of 100 V, corresponding to a frequency tunability of 0.11 GHz/V, which is comparable to what has been reported for ion-sliced $LiNbO_3$ microring resonators (Guarino *et al.*, 2007). A great improvement in performance is expected if materials with state-of-the-art EO figures of merit (n^3r of DAST or OH1 is more than one order of magnitude higher than for COANP) and higher index contrast (Δn with respect to borosilicate is at 1.55 μm about 0.15 for COANP and almost 0.7 for OH1 and DAST) can be used for melt growth.

6.5 Conclusions and future trends

We have presented the basics, present status and potential of organic nonlinear optical crystalline materials for applications, in particular for THz-wave generation and integrated EO modulators. These materials are composed of highly polar chromophores with a highly asymmetric, ultra-fast electronic response to external fields. They may exhibit high EO figures of merit of more than $n^3r = 450$ pm/V in the 1.55 μm telecommunication window, a low dielectric dispersion, as well as a high thermal stability of polar order, and are therefore promising for high-speed and highly efficient EO modulation, as well as phase-matched THz generation with a high conversion efficiency.

In the last decades, there have been several promising organic nonlinear optical crystalline materials developed for various applications. The basic molecular design is challenging mainly because the crystalline packing of highly nonlinear optical molecules is not yet possible to predict in order to achieve highly favorable non-centrosymmetric packing potentially useful for second-order nonlinear optical applications. Additionally, the crystal growth and processing of these materials is often very challenging. For the materials already identified in the past, the main progress achieved in the last few years has been in the growth of high optical quality bulk crystals for THz-wave generation and structuring of optical waveguides for EO applications. Several novel materials optimized for these applications have been also developed and characterized in the last few years. In particular, organic NLO crystals have established in the field of THz applications due to their figures of merit considerably superior to inorganic materials, which has enabled these materials to enter the commercial market, are used in many laboratories as the source of THz waves, and have been also implemented in commercial THz spectrometers and THz imaging devices.

Beside further development and implementation of organic NLO crystals in highly integrated photonic devices and new THz generation/detection schemes, the new material development is expected to continue to further extend the fundamental material knowledge on molecular and crystal engineering, as well as to identify optimized materials for novel applications and extend the possible application ranges, considering various pump laser sources, optical pump wavelengths, THz frequency ranges, and large-scale integrated photonics structures.

6.6 References

C. Bosshard and P. Günter. Electro-optic effects in organic molecules and polymers. In H. S. Nalwa and S. Miyata, editors, *Nonlinear Optics of Organic Molecules and Polymers*, page 391. CRC Press, Boca Raton, 1997.

C. Bosshard, M. Bösch, I. Liakatas, M. Jäger, and P. Günter. Second-order nonlinear optical organic materials: Recent developments. In P. Günter, editor, *Nonlinear Optical Effects and Materials*, volume 72, pages 163–300. Springer Series in Optical Science, Berlin Heidelberg NewYork, 2000.

Ch. Bosshard, K. Sutter, P. Prêtre, J. Hulliger, M. Flörsheimer, P. Kaatz, and P. Günter. *Organic Nonlinear Optical Materials*. Gordon and Breach Science Publishers, Amsterdam, 1995.

F. D. J. Brunner, O-P. Kwon, S. J. Kwon, M. Jazbinsek, A. Schneider, and P. Günter. A hydrogen-bonded organic nonlinear optical crystal for high-efficiency terahertz generation and detection. *Opt. Express*, **16** (21): 16496–16508, 2008.

F. D. J. Brunner, A. Schneider, and P. Günter. Velocity-matched terahertz generation by optical rectification in an organic nonlinear optical crystal using a Ti:sapphire laser. *Appl. Phys. Lett.*, **94** (6): 061119, February 2009.

B. Cai, K. Komatsu, and T. Kaino. Refractive index control and waveguide fabrication of DAST crystals by photobleaching technique. *Opt. Mater.*, **21** (1–3): 525–529, 2003.

K. Clays and B. J. Coe. Design strategies versus limiting theory for engineering large second-order nonlinear optical polarizabilities in charged organic molecules. *Chem. Mater.*, **15**: 642, 2003.

B. J. Coe, J. A. Harris, I. Asselberghs, K. Clays, G. Olbrechts, A. Persoons, J. T. Hupp, R. C. Johnson, S. J. Coles, M. B. Hursthouse, and K. Nakatani. Quadratic nonlinear optical properties of N-aryl stilbazolium dyes. *Adv. Funct. Mater.*, **12**: 110, 2002.

B. J. Coe, J. A. Harris, I. Asselberghs, K. Wostyn, K. Clays, A. Persoons, B. S. Brunschwig, S. J. Coles, T. Gelbrich, M. E. Light, M. B. Hursthouse, and K. Nakatani. Quadratic optical nonlinearities of *N*-methyl and *N*-aryl pyridinium salts. *Adv. Funct. Mater.*, **13** (5): 347–357, 2003.

L. Dalton, P. Günter, M. Jazbinsek, O-P. Kwon, and P. Sullivan. *Organic Nonlinear Optics: Molecules, Polymers and Crystals*. Cambridge University Press, Cambridge, 2012.

L. R. Dalton, P. A. Sullivan, and D. H. Bale. Electric field poled organic electro-optic materials: State of the art and future prospects. *Chem. Rev.*, **110** (1): 25–55, 2010.

P. Dittrich, R. Bartlome, G. Montemezzani, and P. Günter. Femtosecond laser ablation of DAST. *Appl. Surf. Sci.*, **220** (1–4): 88–95, 2003.

H. Figi, M. Jazbinsek, C. Hunziker, M. Koechlin, and P. Günter. Electro-optic single-crystalline organic waveguides and nanowires grown from the melt. *Opt. Express*, **16** (15): 11310–11327, 2008a.

H. Figi, L. Mutter, C. Hunziker, M. Jazbinsek, P. Günter, and B. J. Coe. Extremely large nonresonant second-order nonlinear optical response in crystals of the stilbazolium salt DAPSH. *J. Opt. Soc. Am. B*, **25** (11): 1786–1793, 2008b.

H. Figi, M. Jazbinsek, C. Hunziker, M. Koechlin, and P. Günter. Electro-optic tuning and modulation of single-crystalline organic microring resonators. *J. Opt. Soc. Am. B*, **26** (5): 1103–1110, 2009.

H. Figi, M. Jazbinsek, C. Hunziker, M. Koechlin, and P. Günter. Integrated electro-optic devices of melt-processable single-crystalline organic films. *Proc. SPIE*, **7599**: 75991N–1–6, 2010.

H. Figi, D. H. Bale, A. Szep, L. R. Dalton, and A. T. Chen. Electro-optic modulation in horizontally slotted silicon/organic crystal hybrid devices. *J. Opt. Soc. Am. B*, **28** (9): 2291–2300, 2011.

M. Fujiwara, K. Yanagi, M. Maruyama, M. Sugisaki, K. Kuroyanagi, H. Takahashi, S. Aoshima, Y. Tsuchiya, A. Gall, and H. Hashimoto. Second order nonlinear optical properties of the single crystal of N-benzyl 2-methyl-4-nitroaniline: Anomalous enhancement of the d(333) component and its possible origin. *Jpn. J. Appl. Phys. Part 1*, **45** (11): 8676–8685, 2006.

M. Fujiwara, M. Maruyama, M. Sugisaki, H. Takahashi, S. I. Aoshima, R. J. Cogdell, and H. Hashimoto. Determination of the d-tensor components of a single crystal of *N*-benzyl-2-methyl-4-nitroaniline. *Jpn. J. Appl. Phys. Part 1*, **46** (4A): 1528–1530, 2007.

W. Geis, R. Sinta, W. Mowers, S. J. Deneault, M. F. Marchant, K. E. Krohn, S. J. Spector, D. R. Calawa, and T. M. Lyszczarz. Fabrication of crystalline organic waveguides with an exceptionally large electro-optic coefficient. *Appl. Phys. Lett.*, **84** (19): 3729–3731, 2004.

I. L. Gheorma and R. M. Osgood. Fundamental limitations of optical resonator based high-speed eo modulators. *IEEE Photonics Technol. Lett.*, **14** (6): 795–797, 2002.

Z. Glavcheva, H. Umezawa, Y. Mineno, T. Odani, S. Okada, S. Ikeda, T. Taniuchi, and H. Nakanishi. Synthesis and properties of 1-methyl-4-2-[4-(dimethylamino) phenyl]ethenylpyridinium *p*-toluenesulfonate derivatives with isomorphous crystal structure. *Jpn. J. Appl. Phys. Part 1*, **44** (7A): 5231–5235, 2005.

A. Guarino, G. Poberaj, D. Rezzonico, R. Degl'Innocenti, and P. Günter. Electro-optically tunable microring resonators in lithium niobate. *Nat. Photonics*, **1** (7): 407–410, 2007.

H. Hashimoto, Y. Okada, H. Fujimura, M. Morioka, O. Sugihara, N. Okamoto, and R. Matsushima. Second-harmonic generation from single crystals of *N*-substituted 4-nitroanilines. *Jpn. J. Appl. Phys. Part 1*, **36** (11): 6754–6760, 1997.

C. Hunziker, S. J. Kwon, H. Figi, M. Jazbinsek, and P. Gunter. Fabrication and phase modulation in organic single-crystalline configurationally locked, phenolic polyene OH1 waveguides. *Opt. Express*, **16** (20): 15903–15914, 2008a.

C. Hunziker, S. J. Kwon, H. Figi, F. Juvalta, O-P. Kwon, M. Jazbinsek, and P. Gunter. Configurationally locked, phenolic polyene organic crystal 2-3-(4-hydroxystyryl)-5,5-dimethylcyclohex-2-enylidenemalononitrile: linear and nonlinear optical properties. *J. Opt. Soc. Am. B*, **25** (10): 1678–1683, 2008b.

M. Jazbinsek and P. Günter. Organic electro-optic crystal modulators. In A. Chen and E. Murphy, editors, *Broadband Optical Modulators: Science, Technology, and Applications*. CRC Press, 2011.

M. Jazbinsek, O-P. Kwon, C. Bosshard, and P. Günter. Organic nonlinear optical crystals and single crystalline thin films. In H. S. Nalwa, editor, *Handbook of Organic Electronics and Photonics*. American Scientific Publishers, 2008a.

M. Jazbinsek, L. Mutter, and P. Günter. Photonic applications with the organic nonlinear optical crystal DAST. *IEEE J. Sel. Top. Quantum Electron.*, **14** (5): 1298–1311, 2008b.

M. Jazbinsek, C. Hunziker, S. J. Kwon, H. Figi, O-P. Kwon, and P. Günter. Hybrid organic crystal/silicon-on-insulator integrated electro-optic modulators. *Proc. SPIE*, **7599**: 75990K–1–14, 2010.

T. Kaino, B. Cai, and K. Takayama. Fabrication of DAST channel optical waveguides. *Adv. Funct. Mater.*, **12** (9): 599–603, 2002.

P-J. Kim, J-H. Jeong, M. Jazbinsek, S-J. Kwon, H. Yun, J-T. Kim, Y. S. Lee, I-H. Baek, F. Rotermund, P. Gunter, and O-P. Kwon. Acentric nonlinear optical *N*-benzyl stilbazolium crystals with high environmental stability and enhanced molecular nonlinearity in solid state. *CrystEngComm*, **13** (2): 444–451, 2011.

P-J. Kim, J-H. Jeong, M. Jazbinsek, S-B. Choi, I-H. Baek, J-T. Kim, F. Rotermund, H. Yun, Y. S. Lee, P. Günter, and O-P. Kwon. Highly efficient organic THz generator pumped at near-infrared: Quinolinium single crystals. *Adv. Funct. Mater.*, **22**: 1750–1756, 2012.

G. Knöpfle, R. Schlesser, R. Ducret, and P. Günter. Optical and nonlinear optical properties of 4′-dimethylamino-*N*-methyl-stilbazolium tosylate (DAST) crystals. *Nonlinear Optics*, **9**: 143–149, 1995.

M. G. Kuzyk. Physical limits on electronic nonlinear molecular susceptibilities. *Phys. Rev. Lett.*, **85**: 1218, 2000.

O-P. Kwon, B. Ruiz, A. Choubey, L. Mutter, A. Schneider, M. Jazbinsek, V. Gramlich, and P. Günter. Organic nonlinear optical crystals based on configurationally locked polyene for melt growth. *Chem. Mater.*, **18** (17): 4049–4054, 2006.

O-P. Kwon, S-J. Kwon, M. Stillhart, M. Jazbinsek, A. Schneider, V. Gramlich, and P. Günter. New organic nonlinear optical verbenone-based triene crystal for terahertz applications. *Cryst. Growth Des.*, **7** (12): 2517–2521, 2007.

O-P. Kwon, S. J. Kwon, M. Jazbinsek, F. D. J. Brunner, J. I. Seo, C. Hunziker, A. Schneider, H. Yun, Y. S. Lee, and P. Günter. Organic phenolic configurationally locked polyene single crystals for electro-optic and terahertz wave applications. *Adv. Funct. Mater.*, **18** (20): 3242–3250, 2008.

S. J. Kwon, C. Hunziker, O-P. Kwon, M. Jazbinsek, and P. Günter. Large-area organic electro-optic single crystalline thin films grown by evaporation-induced local supersaturation with surface interactions. *Cryst. Growth Des.*, **9** (5): 2512–2516, 2009.

S. J. Kwon, M. Jazbinsek, O-P. Kwon, and P. Günter. Crystal growth and morphology control of OH1 organic electrooptic crystals. *Cryst. Growth Des.*, **10** (4): 1552–1558, 2010.

B. E. Little, S. T. Chu, H. A. Haus, J. Foresi, and J. P. Laine. Microring resonator channel dropping filters. *J. Lightwave Technol.*, **15** (6): 998–1005, 1997.

S. R. Marder, J. W. Perry, and W. P. Schaefer. Synthesis of organic salts with large 2nd-order optical nonlinearities. *Science*, **245** (4918): 626–628, 1989.

S. R. Marder, J. W. Perry, and C. P. Yakymyshyn. Organic salts with large 2nd-order optical nonlinearities. *Chem. Mater.*, **6** (8): 1137–1147, 1994.

U. Meier, M. Bosch, C. Bosshard, F. Pan, and P. Günter. Parametric interactions in the organic salt 4-*N*,*N*-dimethylamino-4′-*N*′-methyl-stilbazolium tosylate at telecommunication wavelengths. *J. Appl. Phys.*, **83**: 3486, 1998.

K. Miyamoto, S. Ohno, M. Fujiwara, H. Minamide, H. Hashimoto, and H. Ito. Optimized terahertz-wave generation using BNA-DFG. *Opt. Express*, **17** (17): 14832–14838, 2009.

L. Mutter, M. Jazbinsek, M. Zgonik, U. Meier, C. Bosshard, and P. Günter. Photobleaching and optical properties of organic crystal 4-*N*,*N*-dirnethylamino-4′-*N*′-methyl stilbazolium tosylate. *J. Appl. Phys.*, **94** (3): 1356–1361, 2003.

L. Mutter, F. D. J. Brunner, Z. Yang, M. Jazbinsek, and P. Günter. Linear and nonlinear optical properties of the organic crystal DSTMS. *J. Opt. Soc. Am. B*, **24** (9): 2556–2561, 2007a.

L. Mutter, A. Guarino, M. Jazbinsek, M. Zgonik, P. Günter, and M. Dobeli. Ion implanted optical waveguides in nonlinear optical organic crystal. *Opt. Express*, **15** (2): 629–638, 2007b.

L. Mutter, M. Koechlin, M. Jazbinsek, and P. Günter. Direct electron beam writing of channel waveguides in nonlinear optical organic crystals. *Opt. Express*, **15** (25): 16828–16838, 2007c.

L. Mutter, M. Jazbinsek, C. Herzog, and P. Günter. Electro-optic and nonlinear optical properties of ion implanted waveguides in organic crystals. *Opt. Express*, **16**: 731–739, 2008.

H. Nakanishi, H. Matsuda, S. Okada, and M. Kato. Organic and polymeric ion-complexes for nonlinear optics. In M. Doyama, S. Somiya, and R. P. H. Chang, editors, *Materials Research Society International Meeting on Advanced Materials*, pages 97–104. Materials Research Society, Pittsburgh, PA, 1989.

J. Ogawa, S. Okada, Z. Glavcheva, and H. Nakanishi. Preparation, properties and structures of 1-methyl-4-2-[4-(dimethylamino)phenyl]ethenylpyridinium crystals with various counter anions. *J. Cryst. Growth*, **310** (4): 836–842, 2008.

J. L. Oudar and D. S. Chemla. Hyperpolarizabilities of nitroanilines and their relations to excited-state dipole-moment. *J. Chem. Phys.*, **66**: 2664, 1977.

F. Pan, G. Knopfle, C. Bosshard, S. Follonier, R. Spreiter, M. S. Wong, and P. Günter. Electro-optic properties of the organic salt 4-*N*,*N*-dimethylamino-4′-*N*′-methyl-stilbazolium tosylate. *Appl. Phys. Lett.*, **69** (1): 13–15, 1996a.

F. Pan, M. S. Wong, C. Bosshard, and P. Günter. Crystal growth and characterization of the organic salt 4-*N*,*N*-dimethylamino-4′-*N*′-methyl-stilbazolium tosylate (DAST). *Adv. Mater.*, **8** (7): 592, 1996b.

P. Rabiei, W. H. Steier, C. Zhang, and L. R. Dalton. Polymer micro-ring filters and modulators. *J. Lightwave Technol.*, **20**: 1968–1975, 2002.

Rainbow Photonics Ltd. Unpublished, 2012.

D. Rezzonico, M. Jazbinsek, A. Guarino, O-P. Kwon, and P. Günter. Electro-optic Charon polymeric microring modulators. *Opt. Express*, **16** (2): 613–627, 2008a.

D. Rezzonico, S. J. Kwon, H. Figi, O-P. Kwon, M. Jazbinsek, and P. Günter. Photochemical stability of nonlinear optical chromophores in polymeric and crystalline materials. *J. Chem. Phys.*, **128**: 124713–1–6, 2008b.

B. Ruiz, M. Jazbinsek, and P. Günter. Crystal growth of DAST. *Cryst. Growth Des.*, **8** (11): 4173–4184, 2008.

A. Schneider, M. Neis, M. Stillhart, B. Ruiz, R. U. A. Khan, and P. Günter. Generation of terahertz pulses through optical rectification in organic DAST crystals: theory and experiment. *J. Opt. Soc. Am. B*, **23** (9): 1822–1835, 2006a.

A. Schneider, M. Stillhart, and P. Günter. High efficiency generation and detection of terahertz pulses using laser pulses at telecommunication wavelengths. *Opt. Express*, **14** (12): 5376–5384, 2006b.

J-Y. Seo, S-B. Choi, M. Jazbinsek, F. Rotermund, P. Günter, and O-P. Kwon. Large-size pyrrolidine-based polyene single crystals suitable for terahertz wave generation. *Cryst. Growth Des.*, **9** (12): 5003–5005, 2009.

M. Stillhart, A. Schneider, and P. Günter. Optical properties of 4-*N*,*N*-dimethylamino-4′-*N*′-methyl-stilbazolium 2,4,6-trimethylbenzenesulfonate crystals at terahertz frequencies. *J. Opt. Soc. Am. B*, **25** (11): 1914–1919, 2008.

R. L. Sutherland. *Handbook of Nonlinear Optics*. Dekker, New York, 2003.

K. Takayama, K. Komatsu, and T. Kaino. Serially grafted waveguide fabrication of organic crystal and transparent polymer. *Jpn. J. Appl. Phys. Part 1*, **40** (8): 5149–5150, 2001.

K. Tsuda, T. Kondo, F. Saito, T. Kudo, and R. Ito. New fabrication method of channel optical wave-guides of organic-crystal using an inorganic photoresist. *Jpn. J. Appl. Phys. Part 2*, **31** (2A): L134–L135, 1992.

H. Uchida, T. Sugiyama, K. Suizu, T. Osumi, and K. Kawase. Generation of widely tunable terahertz waves by difference-frequency generation using a configurationally locked polyene 2-[3-(4-hydroxystyryl)-5, 5-dimethylcyclohex-2-enylidene] malononitrile crystal. *Terahertz Sci. Technol.*, **4**: 132–136, 2011.

M. Walther, K. Jensby, S. R. Keiding, H. Takahashi, and H. Ito. Far-infrared properties of DAST. *Opt. Lett.*, **25** (12): 911–913, 2000.

M. S. Wong, F. Pan, V. Gramlich, C. Bosshard, and P. Günter. Self-assembly of an acentric co-crystal of a highly hyperpolarizable merocyanine dye with optimized alignment for nonlinear optics. *Adv. Mater.*, **9**: 554, 1997.

Z. Yang, M. Jazbinsek, B. Ruiz, S. Aravazhi, V. Gramlich, and P. Günter. Molecular engineering of stilbazolium derivatives for second-order nonlinear optics. *Chem. Mater.*, **19** (14): 3512–3518, 2007a.

Z. Yang, L. Mutter, M. Stillhart, B. Ruiz, S. Aravazhi, M. Jazbinsek, A. Schneider, V. Gramlich, and P. Günter. Large-size bulk and thin-film stilbazolium-salt single crystals for nonlinear optics and thz generation. *Adv. Funct. Mater.*, **17** (13): 2018–2023, 2007b.

X. M. Zheng, C. V. McLaughlin, P. Cunningham, and L. M. Hayden. Organic broadband terahertz sources and sensors, *Journal Of Nanoelectronics And Optoelectronics*, **2** (1), 58–76, 2007.

J. Zyss, editor. *Molecular Nonlinear Optics: Materials, Physics, Devices*. Academic Press, Boston, 1994.

J. Zyss and J. L. Oudar. Relations between microscopic and macroscopic lowest-order optical nonlinearities of molecular-crystals with one-dimensional or two-dimensional units. *Phys. Rev. A*, **26**: 2028, 1982.

Part II

(Opto)electronic and nonlinear optical properties of organic materials and their characterization

7
Charge generation and transport in organic materials

J. C. SANCHO-GARCIA,
Universidad de Alicante, Spain

DOI: 10.1533/9780857098764.2.219

Abstract: The generation of mobile charges and their transport across organic layers are commonly the most critical steps affecting the performance of organic-based electronic devices. At ambient temperatures, intermolecular hopping of self-localized charge carriers is expected to dominate the transport mechanism, whose properties can be accurately described by quantum-chemical calculations which, however, face a challenge when the nanostructure of the material has to be simultaneously addressed together with single-molecule aspects. Our recent work tries to understand the physico-chemical principles behind the performance of the theoretical methods commonly employed, as well as to pave the way towards full understanding of the transport mechanism by applying optimized theoretical methods. This would finally allow the performance of computationally guided molecular engineering of novel molecules, not yet synthesized, and anticipate the reasons for their expected performance in organic-based electronic devices.

Key words: charge carriers, transport modelling, crystalline organic semiconductors, charge transfer rates, structure-property guidelines.

7.1 Introduction

Chemistry, physics, and materials science have largely demonstrated an almost limitless capacity to drive wellbeing, economic growth and environmental sustainability during recent decades. Nonetheless, in a global age of continuous changes, some topics of the most interest represent real challenges for the world today, i.e. generating energy in a cost-effective, environmentally appropriate and truly sustainable manner, fabricating new sensors and devices at the cutting edge of miniaturization and detection processes, or manipulating matter at a submicroscopic size to create tailored materials with a vast envisioned range of applications, such as in electronics, biomaterials and energy production. Furthermore, some essential commodities, minerals, or fossil fuels might soon start to be in short supply. It is thus urgently needed to pave the way towards truly innovative new technologies for the creation of better and/or more effective products and processes able to achieve these challenges; that is, we hope substantial changes compared with incremental changes in the next generation of materials for these envisioned applications.

It is within this context where there have been tremendous developments lately in the use of organic materials, be they polymers, oligomers or small molecules, as active compounds in device applications, including organic field-effect transistors, light-emitting displays, memory devices, lasers, molecular machines, chemical sensors, and photovoltaic cells, among others. All these devices, largely independent of their functions, share a common structure consisting of a stack of thin films, such as anode/organic material/cathode. Since the pioneering discoveries of conducting and electroluminiscence properties in conjugated materials, the rational design of electronic devices based on organics is still a challenging issue that involves experimental and theoretical chemistry and physics with a (neccessarily) strong interdisciplinary component. Typically, the simulation of the mode of operation of an organic electronic device requires the detailed description of the steps involved (Bredas *et al.*, 2004) within charge- and energy-transfer processes, which can be hopefully estimated from the behaviour of few active molecules extracted from the bulk, and are usually: (i) charge injection from the electrodes, resulting in the appearance of charge carriers (electron and/or holes) and their associated dynamics; (ii) favourable exciton formation from these carriers; (iii) exciton migration to donor–acceptor (organic–organic) interfaces, or lower energy sites, and corresponding optical emission. The order of these steps is reversed in the case of organic solar cells (Brédas *et al.*, 2009) where dissociation of the exciton harvested leads to the formation of the charge carriers. Complementarily to this molecular picture, a multiscale approach connecting the different time and length scales relevant for the complex mechanism of operation of these devices may be also performed on top of the above treatment at the lowest size scale (Cheung and Troisi, 2008). The issue of theoretical accuracy, at both the quantum- and nanoscale, is thus of outmost importance to help in the interpretion of the large body of experimental results accummulated up to date.

Therefore, during the operation of any device made from organics, and largely independent of the final experimental architecture according to the expected utility, the final value of charge carriers' mobility determines the performance of the electronic device, and indeed the ultimate merit of the entire bottom-up approach used to build it. Note that modelling of the charge carrier dynamics turns out to be independent of how they are generated, whether through exciton dissociation or upon injection from the electrodes. Ideally, the performance of devices based on ordered organic crystals is believed to be mostly governed by the identity of the π-conjugated molecules used in the active layer, as well as by the substitution pattern used (if any) to improve the experimental working conditions; though, it is also well-known that charge carrier mobilities are ultimately influenced by many other factors (Coropceanu *et al.*, 2007) such as boundaries, the degree of disorder, impurities and crystal defects acting as traps, surface-induced effects from friendly substrates (Martinelli *et al.*, 2009), temperature, and device configuration to name just a few. It is thus clear that the mobilities, as well as other key properties, are known to depend strongly on the structural and morphological details of the samples. Therefore, the measured mobilities can thus significantly vary from one study to another so that an unambiguous interpretation of the data in the search of robust structure–property guidelines remains elusive. Despite these drawbacks, crystalline organic semiconductors made of small molecules, as oligoacenes and derivatives, have surpassed amorphous silicon in performance (Hasegawa and Takeya, 2009) and many technologically relevant applications are envisioned or are even commercially available now. Note that computational studies would prove very useful in paving the way towards new materials for organic-based applications, in view of the scarcity of directly comparable experimental results, provided they would be able to yield enough insight into the structure–property relationships governing its performance.

Evidently, the intrinsic accuracy of any proposed theoretical methodology needs a careful benchmarking, to avoid any bias on the final results due to an existing error bar, previously to any intended application to not-yet-synthesized materials or materials with unknown packing motifs. Therefore, at this stage, the strategy to estimate mobility values based on theory and simulations needs to provide a manifold of results: (i) have a predictive character; (ii) be largely independent of any experimental information; and (iii) be computationally cost-effective. Owing to the good compromise found between these factors, density functional theory (DFT) has now reached a leading role among theoreticians; however, while significant progress has been achieved in the understanding of variations in charge-carrier mobilities with the nature of the material, work still needs to be done concerning some technical issues influencing the final results.

7.2 Theoretical and computational framework

The charge injection/collection mechanism, after proper deposition of organic layers on the targeted electrodes, is a complex interfacial process (Heimel *et al.*, 2008) although the basic principle is well understood: there always exists a energy barrier, which needs to be minimized for efficient injection/collection, at the interface between the active layer and the electrode, arising from the expected non-alignment between the Fermi level of the metal and the frontier orbitals of the organic molecules (Ishii *et al.*, 1999). For instance, injection of holes (h^+) into an organic transport material relies on a close match between its ionization potential (I_p) and the work function (Φ_m) of the inorganic electrode material used as reservoir. This can be mathematically expressed as $\phi_{h^+} = I_p - \Phi_m$, and one ideally would search a barrierless process where carriers can freely inject into the organic solid. Indium tin oxide (ITO, In_2O_3-SnO_2) is normally used whose $\Phi_m \simeq 4.7$ eV. Injection on electrons, on the other hand, needs as close as possible a match between the work function and the electron affinity. The ease of charge injection from source electrodes, concomitantly with the intrinsic ability of the material to transport the generated carriers, often determines the p- and/or n-type behaviour of the material. It is also important to recall here that in real systems an interface dipole arises from (possibly) electron density redistributions and/or chemical reactions after deposition, with the corresponding complications for modelling, and with the further increase of the barrier with an extra energy term with respect to vacuum. Large efforts are also complementarily conducting to rule out the interfacial processes (Tse *et al.*, 2010).

Once charges are injected, their transport rate (k_{CT}) can be estimated (Nitzan, 2001) in first approximation by using expressions derived normally from Fermi's golden rule:

$$k_{CT} = \frac{2\pi}{\hbar}|V_{if}|^2 \rho(E_i - E_f) \quad [7.1]$$

where V_{if} is the electronic coupling between the two interacting initial (quasi-particle on donor) and final (quasi-particle on acceptor) states, and $\rho(E_i - E_f)$ is the corresponding density of states depending on the nuclear deformations associated with the charge transfer process. It is often appropriate to assume that thermal activation of intermolecular vibrations near room-temperature tends to strongly self-localize the charge carriers, after being generated in supramolecular assemblies, which leads to large molecular geometry relaxations upon charging and thus to localized states acting as traps (McMahon and Troisi, 2011). This situation can be ideally represented as a self-exchange electron/hole ($^{\cdot-}$ or $^{\cdot+}$, respectively) process between two neighbouring molecules (M and N) occupying well-defined sites on a crystalline lattice:

$$M^{\cdot+(-)} + N \xrightarrow{k_{CT}} M + N^{\cdot+(-)} \quad [7.2]$$

Actually, the renewed interest in theoretical models possibly relies on the fact that the key parameters affecting the value of k_{CT} start to be accessed from quantum-chemical calculations at the molecular scale. First, the intermolecular electronic coupling (Newton, 1991) is defined as:

$$V_{if} = \langle \Psi_i | \hat{H} | \Psi_f \rangle, \quad [7.3]$$

Ψ_j are the many-electron wavefunctions describing an excess charge localized on molecule M or N; i.e., the initial and final states. Since the wavefunction for an excess charge on unit **M** versus **N** only differs by the occupation of the involved frontier orbital, the above equation is easily rewritten in an one-electron picture as:

$$V_{if} = \left\langle \phi_M^{HOMO\,(LUMO)} \middle| \hat{H}' \middle| \phi_N^{HOMO\,(LUMO)} \right\rangle, \qquad [7.4]$$

where HOMO and LUMO mean the highest (lowest) occupied (unoccupied) molecular orbitals of an isolated molecule, for hole or electron transport respectively, and $\hat{H}'$ is now the corresponding effective Hamiltonian (Fock operator) of the two molecules that are considered. The expression can be further simplified although with some restrictions (Valeev *et al.*, 2006). Furthermore, the energy needed to swith from the initial $|\Psi_{M^{+(-)}}\Psi_N\rangle$ to the final $|\Psi_M\Psi_{N^{+(-)}}\rangle$ state, the molecular reorganization energy (Λ), is computed normally from the adiabatic potential energy surfaces of the reactants (Coropceanu *et al.*, 2003) as:

$$\Lambda = E_{M^{+(-)}//N} - E_{M^{+(-)}} + E_{N//M^{+(-)}} - E_N, \qquad [7.5]$$

where E_N or $E_{M^{+(-)}}$ indicates the total energy of the unionized or ionized molecule at its optimum geometry, respectively, and $E_{N//M^{+(-)}}$ or $E_{M^{+(-)}//N}$ the energy of the unionized or ionized molecule at the optimized geometry of the other state. This model thus assumes a full relaxation of the molecule supporting the charge before any jump to the closest neighbouring molecule occurs across the sample. Note that the expressions above further simplify in the case of identical reactants and products, as found normally in organic devices, and the thermodynamic free energy of reaction is no longer needed; however, sites' free-energy differences ($\Delta\varepsilon_{if}$) have to be incorporated into the treatment in case of less ordered or amorphous composition of the sample or due to electric-field induced polarization effects in the bulk (Olivier *et al.*, 2006). Note also that large electron- or hole-vibration (electronic) coupling helps to localize (delocalize) the state created upon arrival of the charge carrier. The interplay between the latter two magnitudes indeed determines the range of validity of generalized yet approximate (Picon *et al.*, 2007) expressions derived from Eq. (7.1), as it is for instance the case of the following (Marcus, 1993) and widely used hopping rate:

$$k_{CT} = \frac{2\pi}{\hbar}|V_{if}|^2 \frac{1}{\sqrt{4\pi\Lambda k_B T}} \exp - \frac{(\Lambda + \Delta\varepsilon_{if})^2}{4\Lambda k_B T}, \qquad [7.6]$$

since its main assumption is based on a relative weak electronic coupling between hopping sites compared with the (treated clasically as harmonic oscillators) vibronic coupling. The expression can be further refined by incorporating an effective quantum degree of freedom (the so-called Marcus–Levith–Jortner theory). Note that T denotes the temperature, and $\hbar$ and k_B are fundamental constants. The asociated charge carrier mobility

μ can be estimated from k_{CT} through random walk simulations (Deng and Goddard, 2004): once the charges are injected, or as the electron and hole separate from each other, note that the carrier–carrier interactions are usually neglected, their motion can be regarded as diffusive on a large length scale. Thus, to go a step further one inevitably needs to know the orientation and distance between neighbouring molecules in the crystal (Grozema and Siebbeles, 2008), either if this is experimentally known or if it needs to be computationally obtained.

One of the main advantages of the hopping model described above is the fact that transport parameters depend only on a pair of adjacent sites; note that this model can be further refined by incorporating the microscopic description of the degree of disorder experienced within the organic solid (McMahon and Troisi, 2010a). Essentially, there are two types of disorder; either energetic, due to differences in molecular environment, or positional, due to different relative positions of molecules. As an example of the former, irregular arrangements of molecules would lead to change in the polarization energies of charge carriers along the migration path (Martinelli *et al.*, 2010). Embedding models show also very promising to model this effect (Castet *et al.*, 2008) despite some encountered dependence on the technical details too (Bromley *et al.*, 2008). Furthermore, at room temperature the interacting molecules can instantaneously modify their mutual orientations with the corresponding fluctuation of the electronic coupling, thus having a different value at each of the snapshot analyzed after the corresponding dynamical simulation (Idé *et al.*, 2011), although a gaussian-like distribution of the values is often found. Admittedly, the localized or delocalized nature of charge carriers is still a partially unsolved issue. Note that models assuming highly localized carriers have been questioned recently (Madison and Hutchison, 2011) and thus more research insight is still needed (Konezny *et al.*, 2010). For instance, the relative stability of localized and delocalized charges in π-stacked triphenylenes at varying intermolecular orientations reveals a large dependence on the electronic coupling (Patwardhan *et al.*, 2011). Finally, if the basic assumption of charge localization does not longer remain, there exist other models with larger range of validity (Shuai *et al.*, 2011) which, however, exceeds the purpose of the present discussion (Troisi, 2010).

In short, the design or organic materials with high charge mobility has enormously benefited from the approach outlined here despite some encountered drawbacks (Wang *et al.*, 2010). Nonetheless, regardless of the adopted theoretical model and its associated difficulties (Stafström, 2010) we emphasize the following: (i) the magnitude of both inter- and intramolecular parameters needs to be ascertained before performing any approximate estimate of charge transfer rates; (ii) the carrier mobility increases with increased electronic coupling and decreased reorganization energy; and (iii) it seems always possible, independently of the operation mechanism, to establish a (at least) semiquantitative correlation between the measured mobility and the key molecular parameters.

We tackle now the computational evaluation of these two key quantities (Λ and V_{if}) by modern yet cost-effective DFT-based methods. Note that less (more) costly semiempirical (perturbative) approaches have been also applied in the past with, however, limited success in the case of Λ (Sancho-Garcia and Pérez-Jiménez, 2008a). Furthermore, the value of these quantities is expected to remain unaffected after application of an external electric field (Sancho-Garcia *et al.*, 2003) although electric field dependence of the charge carrier mobility is also found in common devices and can be thus modelled (Jakobsson, 2011) after solving Eq. (7.6). Among the existing challenges for any DFT-based method we remark the subtle, sometimes overlooked, interplay between some physical effects affecting the most the calculations (Reimers *et al.*, 2003) in π-conjugated systems, such as self-interaction or delocalization error, medium-range correlation signatures, dispersive-like weak interactions, static correlation effects, and size-extensivity of the results (i.e. the results should not degrade with system size), and the final mathematical form chosen for the corresponding electronic calculation. The importance of weak interactions (Grimme, 2011) is easy to understand in this contex: since they are known to drive the self-assembly or templating of any material, and organics are not an exception, they will determine the intermolecular orientation and thus the electronic coupling. Once the relative orientation of the molecules is known, the integral of Eq. (7.4) is not expected to significantly depend on the level of theory employed, either semiempirics or DFT-based, if this is of a reasonable enough quality (Huang and Kertesz, 2005). However, the large impact on Λ values of the so-called self-interaction error (Mori-Sánchez *et al.*, 2006), i.e. the spurious self-interaction of an electron with itself wrongly contained into the most common density functionals, is maybe not so admitted in the field despite charge transport being based on concerted migration of spin-unpaired particles. We will consequently explore next how to pragmatically incorporate these effects into accurate calculations of the key parameters entering into Eq. (7.6).

7.3 Single-molecule magnitudes

As simple as it seems, even calculation of molecular reorganization energies for simple π-conjugated systems is not exempted from difficulties. First, lattice relaxation around the charged molecules, expectedly smaller than the intramolecular relaxation upon charging, is usually neglected or taken as an adjustable parameter, although an explicit estimate is also possible (McMahon and Troisi, 2010b). Furthermore, we reiterate that standard DFT is plagued by a self-interaction error (SIE) due to the fact that the direct Coulomb term and the exchange-correlation potential involve the total density, including the particle on which the self-consistent field actually acts. Mathematically, for a one-electron system, two conditions are needed for non-SIE: $E_x[\rho] + J[\rho] = 0$ and $E_c[\rho] = 0$. That Coulomb self-interaction is nicely cancelled in a full exact-like exchange (EXX) treatment. However, the approximate treatment of exchange in common DFT treatments precludes this cancellation; additionally, not all the correlation functional are self-interaction free, so that a spurious self-interaction often remains in standard DFT calculations with independence of system, basis sets, computational code, or other technical details.

Generally speaking, functionals having minimal SIE should be always used for theoretical studies of conjugated systems (Johnson *et al.*, 2008) as those tackled here. As is commonly admitted in the field (Körzdörfer, 2011) the SIE is very often used interchangeably with the concept of delocalization error, which is maybe more intuitive: a function with large SIE would overstabilize delocalized densities while giving too high energies for localized densities. If one would like to completely isolate the role played by these unwanted physical effects, one should compute geometrical and electronic properties with self-interaction corrected (SIC) methods; the original SIC scheme, however, leads to an orbital-dependent mean-field which causes several formal and technical difficulties. This can be circumvented by treating SIC with optimized effective potentials or related methods although, however, with the corresponding increase in computational cost. Reaching this stage, a practical yet effective way to isolate if any particular result is governed by this all-pervasive error is to address the influence of EXX on the final properties. To illustrate this point, we rely next on the adiabatic connection method (Ernzerhof, 1998). We start by defining a generalized exchange-correlation functional connecting, through the non-negative dimensionless λ parameter (also named coupling-strength) a non-interacting particle system ($\lambda = 0$) and the real one ($\lambda = 1$) by means of the particle-particle operator $\hat{V}_{ee} = \sum_{i<j}^{N} \frac{\lambda}{r_{ij}}$; then the exchange-correlation energy can be expressed as:

$$E_{xc}[\rho] = \int_0^1 d\lambda W_\lambda[\rho], \quad \text{with} \quad W_\lambda[\rho] = \left\langle \Psi_\lambda \middle| \hat{V}_{ee} \middle| \Psi_\lambda \right\rangle - J[\rho] \qquad [7.7]$$

a particle–particle functional gathering all non-classical effects since $J[\rho]$ is the classical Coulomb functional. Ψ_λ is the unique wavefunction that yields to the ground state density of the interacting system. We can easily recognize two extreme situations:

$$W_0 = \left\langle \Psi_\lambda \middle| \hat{V}_{ee} \middle| \Psi_\lambda \right\rangle \Big|_{\lambda=0} - J[\rho] = E_x^{\mathrm{EXX}}, \qquad [7.8]$$

with E_x^{EXX} the exact exchange, which can be indeed written as a density function, $E_x^{\text{EXX}}[\rho]$, and

$$W_1 = \left\langle \Psi_\lambda \middle| \hat{V}_{ee} \middle| \Psi_\lambda \right\rangle\Big|_{\lambda=1} - J[\rho] = V_{ee} - J[\rho] \equiv E_{xc}^{\text{exact}}, \qquad [7.9]$$

with E_{xc}^{exact} the exact exchange-correlation term, which is known to be a density function too by the Hohenberg–Kohn theorem, $E_{xc}^{\text{exact}}[\rho]$. The task is now to deal with practical expressions, that is, trying to find a path interpolating these two extreme cases. The simplest form (Peach *et al.*, 2008) postulated (Becke, 1993a) was just a linear interpolation, $W_\lambda[\rho] = a[\rho] + b[\rho]\lambda$, which after being inserted in Eq. (7.6) gives $E_{xc}[\rho] = a[\rho] + \frac{1}{2}b[\rho]$ with $a[\rho] = E_x^{\text{EXX}}$ and $b[\rho] = E_{xc}^{\text{exact}}$, the latter being necessarily approximated as $E_{xc}^{\text{exact}}[\rho] \approx E_x[\rho] + E_c[\rho]$, and whose explicit form must be defined (see below). Since we are interested in a SIE-dependent property, that is Λ, we relax the linear λ dependence and deal with the following expression for a global hybrid function:

$$E_{xc}[\rho] = \alpha E_x^{\text{EXX}}[\rho] + 1(1-\alpha)E_x[\rho] + E_c[\rho], \qquad [7.10]$$

needing to find next the optimum value of α. Note that the existence of an accurate and well-behaved training set, composed of either experimental or very accurate theoretical values, is mandatory for the optimization process of α. We will use for this purpose the experimental Λ values for hole transport in (oligo)acenes (see Fig. 7.1) extracted from gas-phase ultraviolet photoelectron spectroscopy (Malagoli *et al.*, 2004), and thus without including any environmental contribution. Another strategy recently pursued to find an optimum value for α imposes the right curvature of the total energy of some (oligo)acenes, as a function of the electron number for different values of α (Sai *et al.*, 2011). Finally, a further degree of sophistication can be also incorporated through the use of a double-hybrid model, although however at higher computational cost (Sancho-Garcia and Pérez-Jiménez, 2009a).

We have selected, among others, the BLYP (Becke, 1988) and PBE (Perdew *et al.*, 1996) exchange-correlation models, and have thus

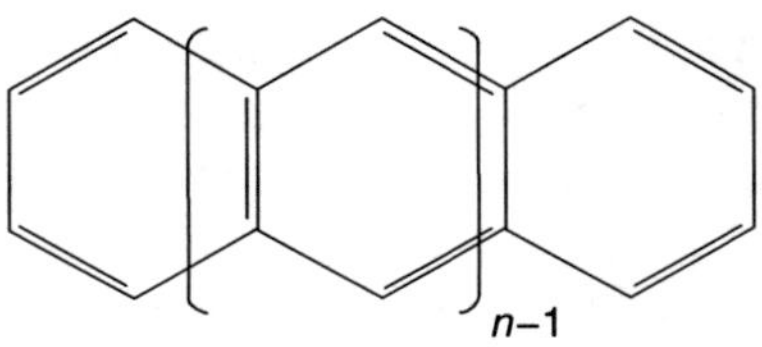

7.1 Chemical structure of (oligo)acenes: naphthalene (n = 1), anthracene (n = 2), tetracene (n = 3) and pentacene (n = 4). The hydrogen atoms and corresponding C–H bonds have been omitted for clarity.

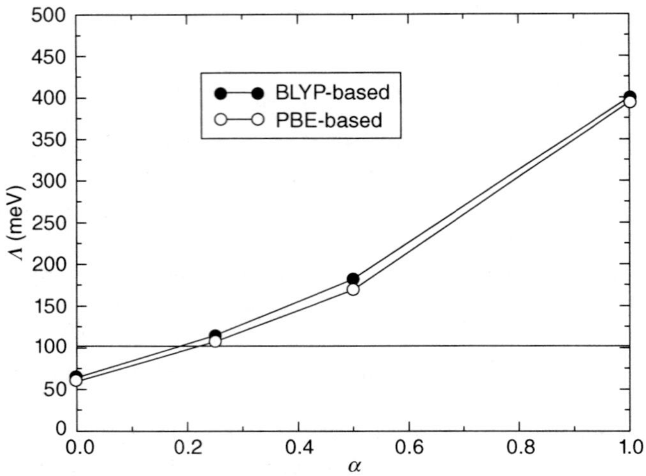

7.2 Evolution of reorganization energies of pentacene as a function of the α parameter (weight of EXX) for BLYP- and PBE-based hybrid models. The horizontal line indicates the experimental value.

systematically varied the value of α in Eq. (7.10) to find the evolution found in Fig. 7.2 for the pentacene case taken as example (Sancho-Garcia, 2007). First, the use of more sophisticated functionals led to similar conclusions and thus will not be discussed here. Second, the values of Λ hardly differ when sufficiently large basis sets are employed (actually by less than 1% between the 6-31G* and the cc-pVTZ). Third, and most importantly, the values for all (oligo)acenes shown in Fig. 7.1 are found to evolve smoothly with α, which allows us to optimize its value by minimizing the error function $\varepsilon = \sum_{ij} \Lambda_{\mathrm{exp},i} - P_{ij}(\alpha)^2$ ($\Lambda_{\mathrm{exp},i}$ are the experimentally determined reorganization energies for each molecule and $P_{ij}(\alpha)$ are the quadratically polynomial fitted values for each molecule i and basis sets j). Remarkably enough, α is found to vary only slightly within the assessed models ($\alpha \approx 0.25$) and be very close to the default value in the famous, and widely used within the field, B3LYP model (Becke, 1993b). Table 7.1 shows the values of Λ obtained with these optimized models; note that a mean absolute error of just 14 meV is obtained, which does not come at the expense of other related yet key properties for hole conduction as ionization potentials or fundamental gaps (see Table 7.2).

Viewing the promising results obtained with the newly reparameterized $\mathrm{DFT}_{\alpha\Lambda}$ models, we have also tackled (see Fig. 7.3) a diverse set of promising and related systems, which (might) form semiconducting crystals with (expectedly) high mobilities: (i) 5,6,11,12-tetraphenyltetracene, also known as rubrene (Sancho-Garcia and Pérez-Jiménez, 2008a); (ii)

Table 7.1 DFT-based theoretical estimates (cc-pCVDZ basis set) of intramolecular reorganization energies (meV)

Method	α	Naphthalene	Anthracene	Tetracene	Pentacene
$BLYP_{\alpha\Lambda}$	0.2533	194	148	122	103
$PBE_{\alpha\Lambda}$	0.2762	197	151	125	107
Exp.		182	174	138	102

Table 7.2 DFT-based theoretical estimates (cc-pCVDZ basis set) of energy difference between the frontier orbitals (gap, in eV) and adiabatic first ionization potential (I_p, in eV)

		Naphthalene		Anthracene		Tetracene		Pentacene	
Method	α	Gap	I_p	Gap	I_p	Gap	I_p	Gap	I_p
$BLYP_{\alpha\Lambda}$	0.2533	5.15	7.60	3.89	6.82	3.05	6.29	2.46	5.91
$PBE_{\alpha\Lambda}$	0.2762	5.35	7.85	4.05	7.06	3.19	6.53	2.58	6.15
Exp.		5.0–5.4	8.14	3.9–4.2	7.42	2.9–3.4	6.91	2.2–2.4	6.56

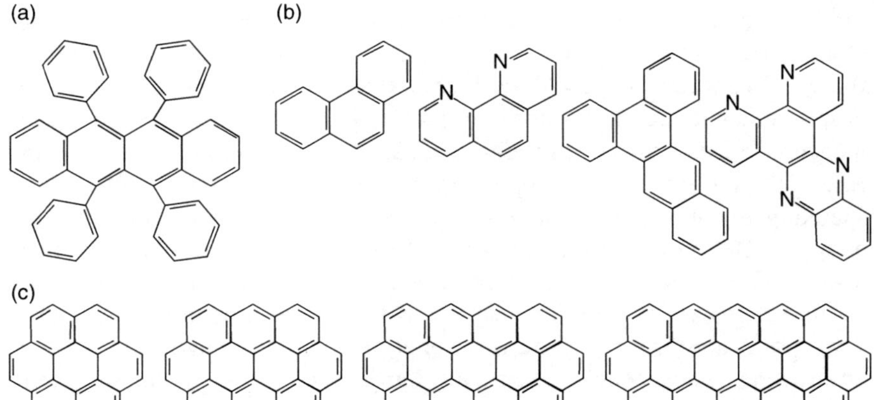

7.3 Chemical structure of (from top to bottom): rubrene, phenanthrene-type molecules, and circum(oligo)acenes (circumbenzene, circumnaphthalene, circumanthracene, and circumtetracene, from left to right).

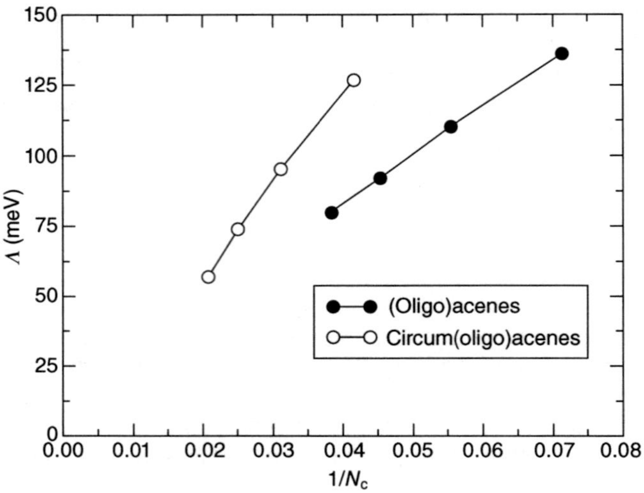

7.4 Evolution of reorganization energies of (oligo)acenes and circum(oligo)acenes as a function of the inverse number of carbon atoms.

phenanthrene-type homo- and heteroaromatic molecules (Sancho-Garcia, 2007); and (iii) circum(oligo)acenes of increasing size (Sancho-Garcia and Pérez-Jiménez, 2009b). Remarkably enough, even when field effect mobilites tend to correlate very well with the computed reorganization energies (Mas-Torrent *et al.*, 2004) for approximately similar electronic couplings, the Λ of rubrene (159 meV) is higher than for tetracene or pentacene, despite much higher mobilities in organic field-effect transistors are usually measured for the former (Sundar *et al.*, 2004): one can thus infer a leading role of the intermolecular coupling in this case. On the other hand, broadly speaking, extending the conjugated core seems to decrease reorganization energies. Therefore, by searching how to minimize the Λ values, we have investigated a set of prototype systems for graphene nanoribbons: circum(oligo)acenes containing circularly fused benzannelated rings. Figure 7.4 reports how the reorganization energies of both (oligo)acenes (from anthracene to hexacene) and circum(oligo)acenes (from circumbenzene to circumtetracene) linearly evolve as a function of the total number of carbon atoms, which is known to follow an inversely proportional relation due to strong electron–phonon coupling (Devos and Lannoo, 1998). A much more pronounced slope is observed for circum(oligo)acenes compared to (oligo) acenes; actually, we find one of the lowest Λ values reported to date for the largest circum(oligo)acenes considered here (circumanthracene and circumtetracene), which are expectedly due to a more rigid macrocyclic structure.

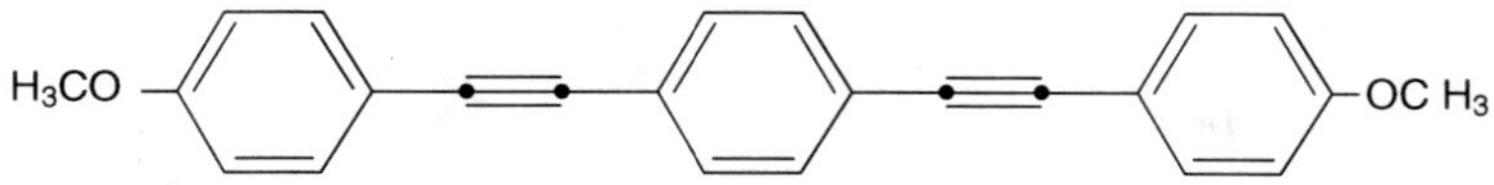

7.5 Chemical structures of 1,4-bis(phenylethynyl)benzene derivative. The hydrogen atoms on the central core have been omitted for clarity.

The prominent role played by SIE has been also recognized for a rigid rod-like molecule (see Fig. 7.5) intended to be used as a molecular wire in molecular electronics. The calculated intramolecular reorganization energy perfectly correlates again with the weight of the EXX introduced into any DFT functional, independently of the kernel and constraints used to develop its final mathematical expression. However, in this occasion, a much higher value of α was needed to match the experimental result, which unfortunately seems to even indicate a system-dependent value. Notwithstanding this feature, a simple attempt to correct the SIE using a N-electron average of the interelectronic repulsion energy as exchange-correlation functional, $E_{xc}[\rho] = -NJ\frac{\rho}{N}$, behaved very successfully (Sancho-Garcia and Pérez-Jiménez, 2008b) and signalled effectively this effect as the source of error.

7.4 Supramolecular organization of the samples

As said before, a deeper understanding of the relationship between the efficiency of charge transport and the chemical morphology of the samples is essential for the improvement of device performance (Mas-Torrent and Rovira, 2011). The molecular packing is thus of utmost importance for charge transport properties, and chemical substitution is known to be a very efficient way to modulate it. The supramolecular organization in thin films is intimately dominated by weak intermolecular interactions between adjacent molecules. This also holds true for the charge transport properties whose efficiency and anisotropy are dictated by intermolecular effects between close molecules. In the absence of experimental information about the crystalline structure or self-assembly of molecules, a state-of-the-art method for the automated study and design of new π-conjugated charge-transporting materials (Baumeier *et al.*, 2010) should be able to reliably predict the preferred crystal packing or self-assembly of the active organic molecules. However, this is a really challenging task for quantum-chemistry methods, and contrasts with most of the previous theoretical works where the electronic properties were inferred from available X-ray structures or from morphologies generated by force-field calculations. We would like to stress here the significant progress recently achieved in accurately predicting (Asmadi *et al.*, 2009) the crystal structure of a set of few small yet representative organic molecules in a DFT-based blind test, with nothing

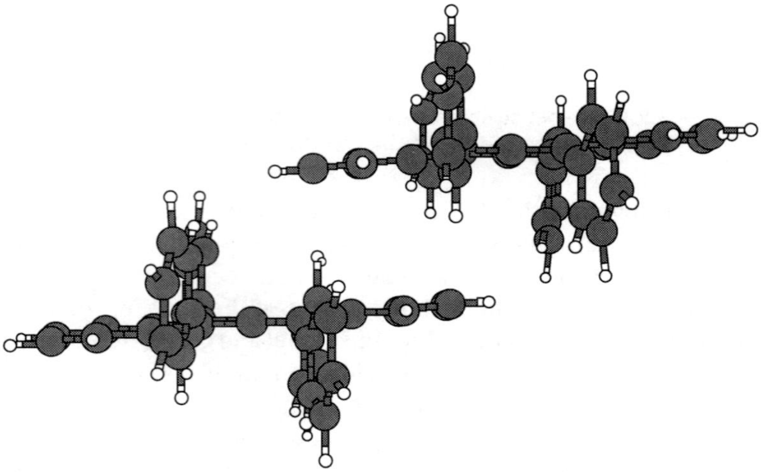

7.6 The structure of the rubrene dimer along the crystallographic *a* direction.

more than the molecular geometry as starting point, which can be considered as a major breakthrough in the field.

To illustrate current difficulties with state-of-the-art methods, Fig. 7.6 shows the microstructure of a rubrene dimer extracted from crystallographic data; this particular π-stacked arrangement (slipped-cofacial configuration) of rubrene molecules in the solid state happens to be optimal for charge transport (da Silva Filho *et al.*, 2005). The interaction energy of the dimer is calculated by the supermolecular approach using the expression $\Delta E = E_{\text{dimer}}(R) - 2E_{\text{monomer}}$, being $E_{\text{dimer}}(R)$ the total energy of the dimer at the fixed distance R between the centre of mass of the monomers. Unfortunately, all the $\text{DFT}_{\alpha\Lambda}$ methods tested gave very low or even repulsive energies ($\Delta E > 0$) and that, of course, precludes their use in algorithms for accurate searches along unknown potential energy surfaces.

To keep DFT-based methods as simple (cost-effective and without many technicalities) as possible, we decided to add dispersion interactions between weakly overlapping fragments (molecular pairs) through the modelling of atom-pair wise interactions between atoms A and B, separated by an internuclear distance R_{AB}, at essentially no extra computational cost. To do this, we rely on a separate calculation of this contribution (E_{disp}) to the electronic energy by resorting to a function (Grimme *et al.*, 2010) which explicitly depends on its well-known decay, as R_{AB}^{-n}:

$$E_{\text{disp}} = \sum_{A \neq B} \sum_{n=6,8\ldots} s_n \frac{C_n^{AB}}{R_{AB}^n} f_{d,n}(R_{AB}), \qquad [7.11]$$

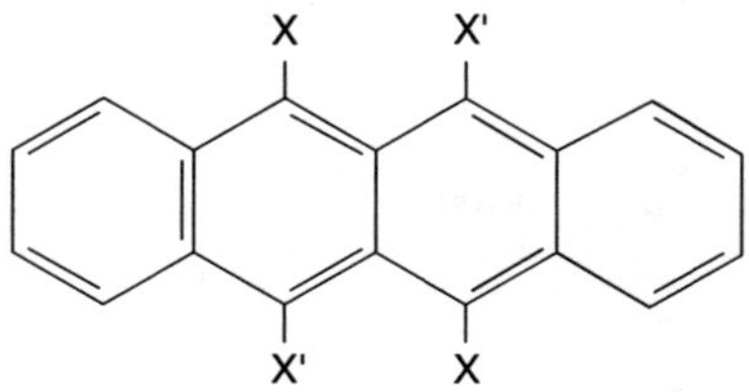

7.7 Chemical structures of the di- (X = substituents, X′ = H) and tetra- (X = X′ = substituents) functionalized tetracene derivatives. The hydrogen atoms and corresponding C–H bonds have been omitted for clarity.

where C_n^{AB} is the *n*th order dispersion coefficient for the atomic pair *AB*, and the s_n are the function-dependent scaling parameters used to efficiently couple the electronic and the *a posteriori* added dispersion energy. A damping function $f_{d,n}$ (R_{AB}) is also considered (Grimme *et al.*, 2011) to ensure adequate switching from short to long distances between atomic pairs. If we keep only the first term of the expansion, we might use $s_6 = 1.05$ and $s_6 = 0.70$ for $BLYP_{\alpha\Lambda}$–D and $PBE_{\alpha\Lambda}$–D, respectively, the –D means dispersion added, as determined previously (Grimme, 2006). Now, ΔE for the microstructure of Fig. 7.6 is predicted to be around −10–12 kcal/mol, in agreement with other calculations (Sancho-Garcia and Pérez-Jiménez, 2008a) taken into account the dispersion interactions more rigorously. Further extensions of the methodology, for instance a more refined form for the damping function or the influence of three-body interactions, are also envisioned and will hopefully be applied soon within the field.

Considering recent applications, it has been shown that controlled functionalization of tetracene by selective halogenation or cyanation alters the herringbone structure, typical of acene-type molecules, and favours an ultra-closely π-stacked arrangement; although the detailed layered structure might depend on the nature of the substrate used to grow the thin films. For instance, mono-(5-chlorotetracene) adopts a herringbone structure while 5,11-dichlorotetracene exhibits a face-to-face slipped π-stacked motif which translates into hole mobilites a few orders of magnitude higher than for 5-chlorotetracene (Moon *et al.*, 2004). The same trend is also found for 5,6,11,12-tetrachlorotetracene, which crystallizes in isolated stacks and thus gives a quasi one-dimensional (1D) crystal. Note that perhalogenation tune p-type towards n-type semiconductors while keeping the herringbone structure (Anthony, 2006). Hence, we have systematically investigated the evolution of charge-transport properties upon selected di- or tetra-functionalization of tetracene, see Fig. 7.7, employing the computational methodolody given by Eqs. (7.10)–(7.11). We will mostly concentrate on those (not yet synthesized) species achieving isolated stacks along the *z*-axis, giving rise to highly

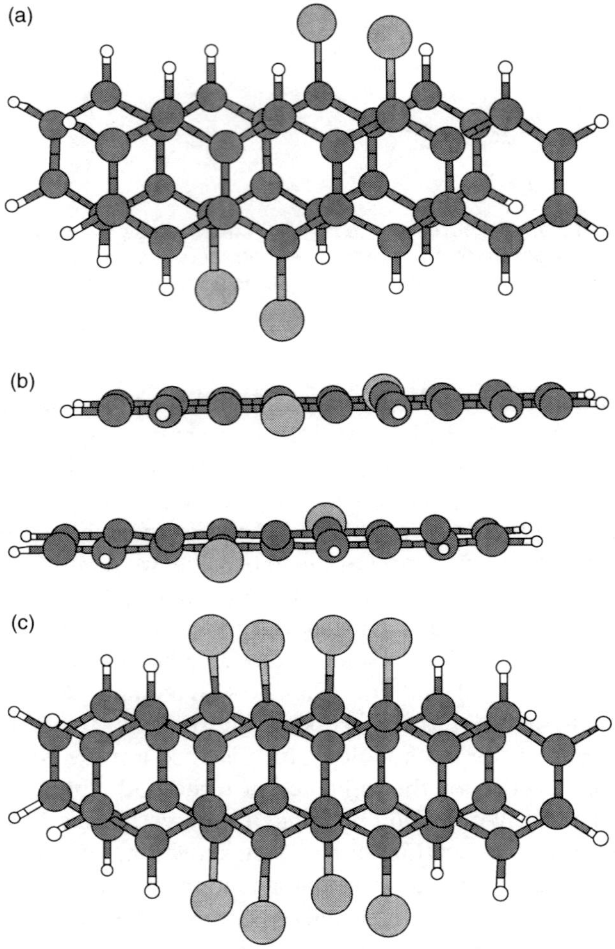

7.8 Optimized structure of a dimer of (from top to bottom) 5,11-dichlorotetracene, 5,6,11,12-tetrachlorotetracene, and 5,11-dicyanotetracene within a stack from perpendicular (left) and side (right) views.

directional transport, as shown in Fig. 7.8 (Fig. 7.9) for some selected dimer (trimer) systems. Table 7.3 summarizes the calculated intermolecular distance (*d*) between the layers as well as the shift of the upper molecule along the short and long molecular axes (*y*- and *x*-axis, respectively). As a performance measure for accuracy predictions, we mention the experimental results only available for the intermolecular distance in 5,6,11,12-tetrachlorotetracene (Chi *et al.*, 2008) and 5,11-dicyanotetracene (Li *et al.*, 2009): 3.58 and 3.40 Å, respectively, in close agreement with a calculated (Sancho-Garcia *et al.*, 2010) result of 3.44 and 3.18 Å, respectively. Finally,

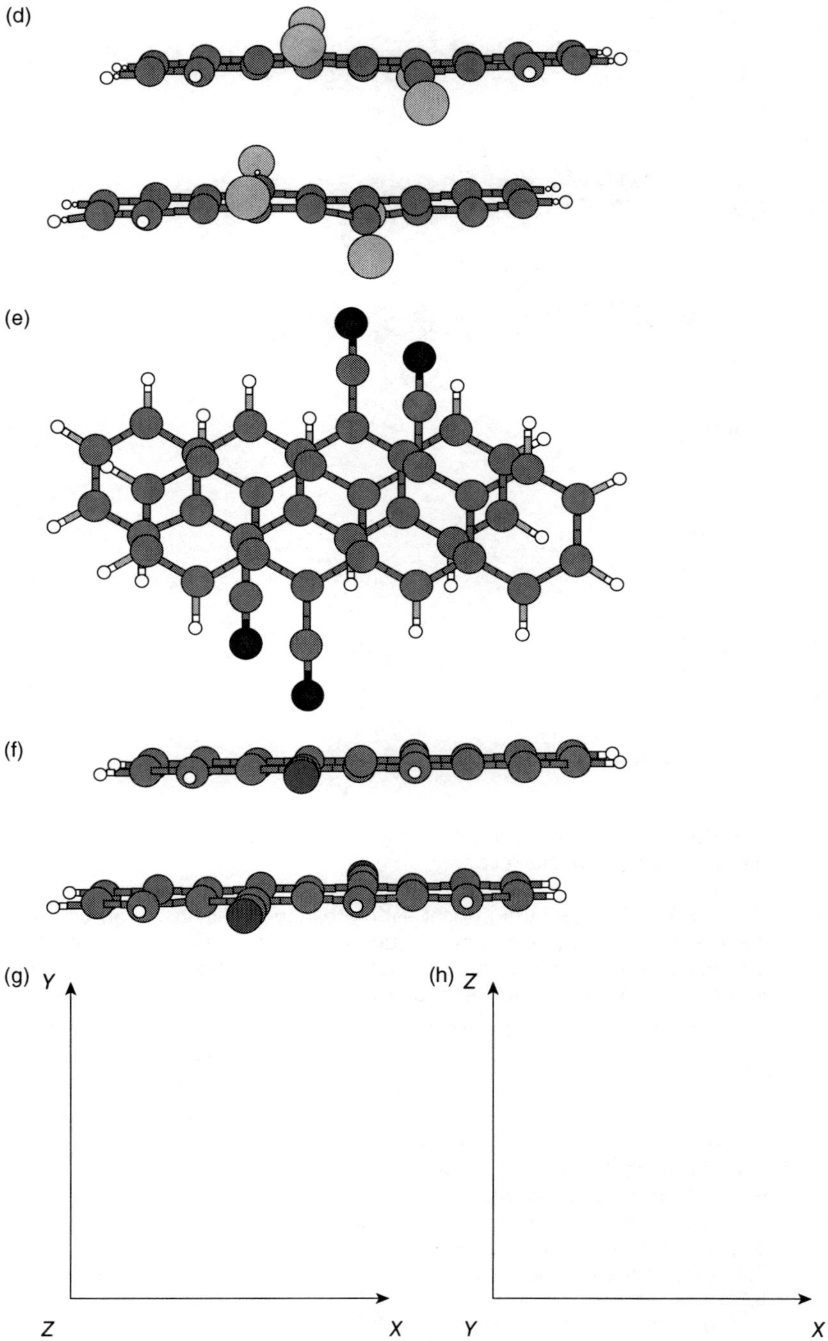

7.8 Continued.

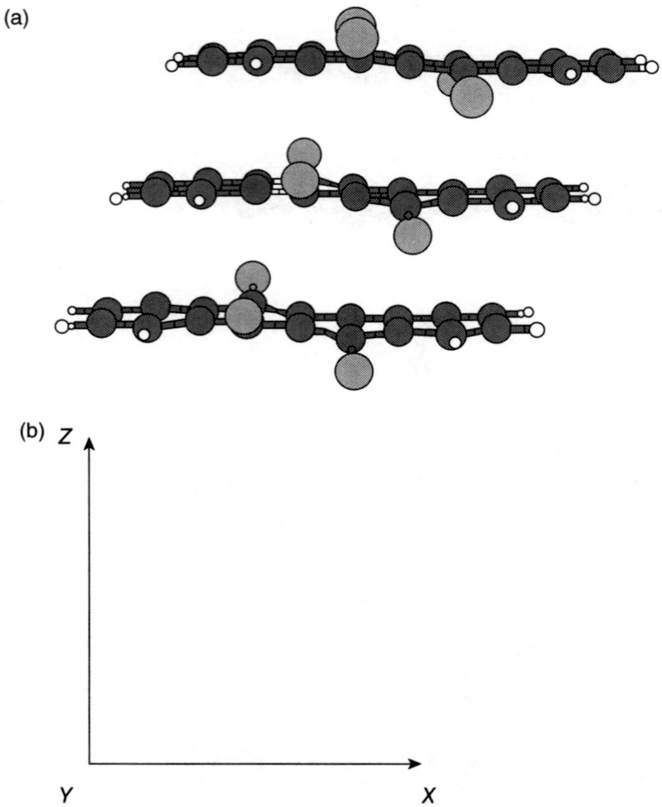

7.9 Side view of the optimized structure of a trimer of 5,6,11,12-tetrachlorotetracene.

Table 7.3 DFT-based theoretical estimates (cc-pVDZ basis set) of molecular packing parameters in functionalized tetracene derivatives: optimized intermolecular distance (*d*, in Å), and relative displacements along the short and long molecular axes (Å)

Derivative		Displacement	
	d	Short axis	Long axis
5,11-difluorotetracene	3.07	1.1	1.4
5,6,11,12-tetrafluorotetracene	3.03	1.0	1.4
5,11-dichlorotetracene	3.25	0.6	1.6
5,6,11,12-tetrachlorotetracene	3.44	0.3	1.2
5,11-dibromotetracene	3.42	0.1	1.6
5,6,11,12-tetrabromotetracene	3.55	0.3	1.3
5,11-dicyanotetracene	3.18	1.0	1.5
5,11-dicyano-6,12-difluorotetracene	3.20	0.8	1.4

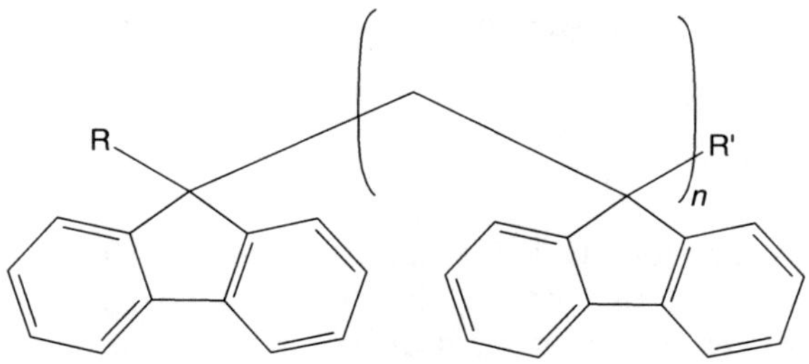

7.10 Chemical structures of (oligo)dibenzofulvenes. The hydrogen atoms and corresponding C–H bonds have been omitted for clarity.

the molecular electrostatic potential (MESP) has served as a tool to rationalize the noncovalent interactions ruling out the shape of a dimer (Sancho-Garcia and Pérez-Jiménez, 2010) and its variations upon substitution at central positions.

The aforesaid methodology can be also applied to other less ordered systems, as represented by a class of covalently bridged π-stacked fluorene systems called oligo(dibenzofulvenes), see Fig. 7.10, for studying and further engineering the involved intramolecular interactions (Sancho-Garcia, 2010). The orientation of the cofacial fluorene units evolves with the nature and size of the system, introducing undesired disorder effects for long oligomers approaching the polymer limit, which might largely drive the experimental hole drift mobilities measured. Our dispersion-corrected $DFT_{\alpha\Lambda}$–D studies for oligomers ranging from dimer to hexamer predict a final helical-like shape for the polymer, and have yielded intra-chain mobilities of charge carriers among several fluorene moieties belonging to the same chain higher than the experimental measures. The hopping between adjacent chains might be thus impeded by unfavourable inter-chain interactions, acting effectively as the limiting step, which would need further interdisciplinary investigation in order to ascertain the details of the samples.

7.5 Predicting relative and absolute values of mobilities

Admittedly, reaching this stage, we recognize that the dynamics of the intermolecular electronic coupling in crystalline oligoacenes has been shown (Troisi and Orlandi, 2006) to follow a Gaussian distribition, and that thus our approach is expected to provide an upper limit to the estimated mobility values. This is why we will discuss hereafter the calculated mobility values primarily on a relative basis despite the apparently good quantitative agreement reached with the experimental data. By assuming the same intermolecular distance within the dimers, a mobility ratio can be obtained from the following expression:

$$\frac{\mu_2}{\mu_1} \propto \frac{V_2^2}{V_1^2}\sqrt{\frac{\Lambda_1}{\Lambda_2}}\exp\frac{\Lambda_1-\Lambda_2}{4k_BT}. \qquad [7.12]$$

We estimate now the mobilities of circum(oligo)acenes (for an idealized packing motif composed of cofacial dimers separated by 3.5 Å), after noticing the low values of Λ found before, compared with the mobilities expected for tetracene or pentacene in the same condition, that is the ratio $\frac{\mu_{\text{circum(oligo)acene}}}{\mu_{\text{tetracene or pentacene}}}$. We infer 2–3 times higher mobilities, especially for circumtetracene, due to reduced reorganization energies together with larger electronic coupling; note, however, that the limit of applicability of Marcus theory (pertaining to the weak coupling regime) is believed to be approached (Troisi, 2011). Turning now the view to substituted (halogenated and/or cyanated) tetracenes, we remind readers that controlled functionalization at central positions with electron-withdrawing groups changed the crystalline pattern from herringbone to a quasi-1D packing. Although we observe an increase of both magnitudes, Λ and V_{if}, upon substitution, the overall impact on charge transfer rates is largely positive due to the square dependence on the electronic coupling, which is maximized for quasi-cofacially stacked molecules. Relation (7.12) leads now to a ratio of 2.9 between the calculated mobilities of 5,6,11,12-tetrachlorotetracene and 5,11-dichlorotetracene, in very good agreement with the ratio of 3.1 obtained from the experimental values (5 $cm^2 V^{-1} s^{-1}$ and 1.6 $cm^2 V^{-1} s^{-1}$, respectively).

For highly ordered systems, and if the relative position of the interacting molecules is known, the charge mobility can be estimated via the diffusion coefficient D by the Einstein–Smoluchowski equation via:

$$\mu = \frac{qD}{k_BT} \propto \frac{q}{k_BT}d^2\mathrm{k}_{\mathrm{CT}}, \qquad [7.13]$$

owing to the fact that site energies and electronic coupling keep the same values along the stack (Grozema and Siebbeles, 2008), as well as does the intermolecular distance d. Note that the above expression can be generalized for a set of ith specific hopping pathways (with d_i the distance between molecules and $\mathrm{k}_{\mathrm{CT}}^{(i)}$ its associated rate) although some difficulties associated with the use of the above expression have been recently highlighted (Stehr *et al.*, 2011). Feeding the results of our calculations into Eq. (7.13) we obtain a mobility of 6.2 $cm^2 V^{-1} s^{-1}$ for 5,6,11,12-tetrachlorotetracene and 1.9 $cm^2 V^{-1} s^{-1}$ for 5,11-dichlorotetracene, in close agreement with the experimental values (see above). Finally, Table 7.4 presents further applications to the entire set of compounds shown in Fig. 7.7; most of these values

Table 7.4 DFT-based theoretical estimates (cc-pVDZ basis set) of charge transport parameters in functionalized tetracene derivatives: Λ (meV), V (meV), k_{CT} ($\times 10^{13}$ s^{-1}), and μ (cm^2 V^{-1} s^{-1})

Derivative	Λ	V	k_{CT}	μ
5,11-difluorotetracene	167	75	4.6	1.7
5,6,11,12-tetrafluorotetracene	209	85	3.5	1.2
5,11-dichlorotetracene	153	70	4.7	1.9
5,6,11,12-tetrachlorotetracene	172	134	14	6.2
5,11-dibromotetracene	148	90	8.4	3.8
5,6,11,12-tetrabromotetracene	160	125	14	6.7
5,11-dicyanotetracene	97	52	5.7	2.2
5,11-dicyano-6,12-difluorotetracene	125	91	12	4.6

are predicted to be higher than experimental values reported for tetracene or related molecules (Yamashita, 2009). We believe that the computational protocol exposed here proves very attractive to: (i) rationalize experimental results; (ii) anticipate the performance of new materials for organic electronics; and (iii) provide useful structure–property guidelines. This protocol is believed to be of rather general applicability and to be exploitable with a wide variety of self-organizing as well as supramolecularly organized organic materials that are expected to play an increasing role in future applications. This also extends to materials with unknown packing motifs and/or polymorphs provided that the bottleneck of the computational cost is not achieved. Current attempts to develop codes specifically suited for carrier mobilities (Li *et al.*, 2011) or quantitative structure–property relationships (QSPR) models (Misra *et al.*, 2011) constitutes also very promising routes to be further explored.

7.6 From p-type to n-type semiconductors

The lack of n-type (electron conducting) high-performance semiconductors is a pitfall for current applications, and thus several strategies to promote ambipolar transport are being currently pursued (Cornil *et al.*, 2007). Most of the systems studied show poor behaviour after exposure to surrounding conditions even missing sometimes the desired field-effect performance (Meng *et al.*, 2011). Recent progress in n-type systems based on pentacenequinones (see Fig. 7.11) hoped to measure reasonable mobility values in thin-films transistors due again to an expectedly regular and compact packing while keeping a higher stability (Liang *et al.*, 2010). However, whereas 5,7,12,14-tetraaza-6,13-pentacenequinone (compound 1 in Fig. 7.11) reached mobilities higher than 0.1 cm^2 V^{-1} s^{-1}, N-substitution at the terminal rings (compound 2 in Fig. 7.11) led to amorphous films degrading

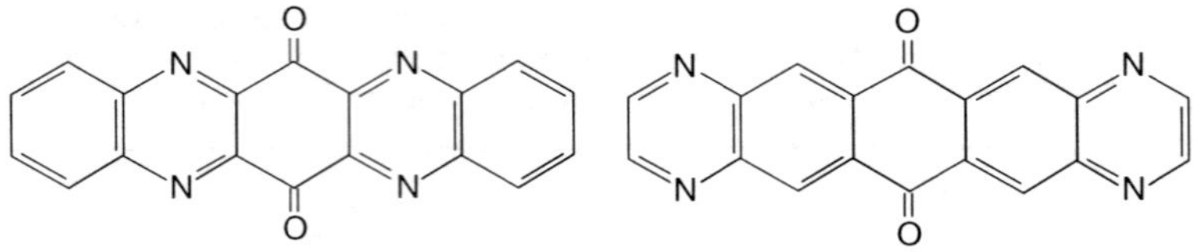

1 2

7.11 Chemical structures of pentacenequinones. The hydrogen atoms and corresponding C–H bonds have been omitted for clarity.

the expected mobility by few orders of magnitudes. Note that 6,13-pentacenequinone acts as an insulator, being indeed a major impurity in pentacene, and that π-deficient cores together with replacement of C with N atoms seems to be a successful strategy. Therefore, theoretical studies are further requested to successfully relate the molecular structure and packing of these molecules with their intrinsic semiconducting properties.

As expected, key information regarding the possible explotation of these compounds as n-type semiconductors is extracted from calculations employing the reparameterized $DFT_{\alpha\Lambda}$–D models. These systems keep: (i) $\Lambda_{h^+} \gg \Lambda_{e^-}$, the latter being 132 and 155 meV for compound 1 and 2, respectively, admittedly lower than for other n-type semiconductors; (ii) low adiabatic electron affinities of the order 1.8–2.0 ev, to be compared for instance with a (optimum) $\Phi_m \simeq 2.7$–2.9 eV for Ba or Ca, usually used as cathodes; (iii) a relationship between intermolecular electronic coupling and intramolecular reorganization energies ($V_{e^-} < \Lambda_{e^-}$) which is not expected to violate the range of validity of hopping theory at room temperature for the estimate of k_{CT}; (iv) a quasi-1D solid state packing expected from the stability of the cofacial dimeric structures sampled, with an intermolecular distance of around 3.15 Å for both compounds, being 3.32 Å the experimental (only available) value for compound 1; (v) a ratio between the mobilities of the order of $\mu_2/\mu_1 \approx 4$, which after feeding the results of the calculations into Eq. (7.13) becomes specifically 0.5 and 1.9 $cm^2\ V^{-1}\ s^{-1}$.

7.7 Conclusion

The rational design and construction of organic devices, in the worldwide effort to provide lightweight, clean, and renewable alternatives to current technologies, needs the concurrence of theoretical/computational guidelines: the selection of molecules and their self-assembly should be thoroughly motivated for maximizing efficiency and lifetime. Using theoretical tools to examine ways of improving the entire process requires prior study of charge propagation in model systems, after carefully benchmarking of the methods for good enough automated applications. Once this step is done, conditions need to be explosed for improving the expected performance of the materials engineered and to successfully establish a fair link between molecular identity and charge transport parameters from truly first-principles (with a minimized empirical input) calculations.

The microscopic mechanism widely used to understand charge transport in organic solids proposes that charge carriers, as they move across the active layer, can be localized by defects or disorder effects. Then, after the corresponding structural relaxation of the molecule and its accompanying environment, charges can hop between adjacent molecules thanks to thermally activated processes. The parameters dictating the efficiency of this mechanism can be estimated by accurate calculations at the molecular scale, leading thus to the corresponding charge hopping rate. The diffusion coefficient for charge carriers, and consequently the carrier mobility, needs additionally the average distance for hopping events, in case of highly ordered systems, or the distance for each individual path in case of inhomogeneities along the morphology. This feature implies a need to know the detailed structure of the aggregates bound by weak van der Waals forces, which can be also fortunately obtained from modern calculations performed with care.

To sum up, in the search of accurate estimates of the key molecular parameters governing charge transport properties, we have applied along the last years a customized method consisting of: (i) a specific reparameterization of a hybrid density functional to molecular reorganization energies; and (ii) the addition of a dispersion correction to the computational algorithm in order to reliably predict the most favourable packing of molecules, and thus the electronic coupling between neighbouring molecules. With the help of these new precisely calibrated methods, we have studied some promising functionalization of state-of-the-art molecules such as (oligo) acenes, before they are even synthesized, which could represent a gain of time for material screening, with encouraging results.

Having said that, we would like to provide next a brief outlook of some theoretical methods that could soon be complementarily applied to the problem at hand with (hopefully) expected success. Among the most promising strategies concerning calculations of key single-molecule parameters, we mention the existence of double-hybrid and long-range corrected density functions. These, again orbital-dependent functions, can be seen as an improvement of global hybrid functions defined by Eq. (7.10), either by allowing a mixing of correlation energies calculated by second-order perturbation theory and a well-matched density function complementarily to the mixing of a fixed quantity of exchange energies, or by overcoming this fixed quantity through a distance-dependent mixing of the latter terms. These two flexible strategies are known to reduce the SIE and are thus expected to provide accurate values of Λ too, although however at a higher computational and technical cost. Concerning the study of supramolecular entities, which opens the way towards inter-molecular electronic couplings and solid-state conducting paths, the existing corrections for dispersion energies could be possibly improved by using a more refined form for the damping function entering into Eq. (7.11), and thus able to accurately yield hydrogen-bonded as well as dispersion-bound dimers, or through the use of truly non-local van der Waals correlation density functions. The study of larger and larger molecular aggregates, providing a true connection to the atomistic scale, is also an expected accomplishment. However, we face a dilemma here since the formal computational cost of current (single- or double-) hybrid methods, N^4 and N^5 respectively where N is related to the size of the system, independently of the treatment of inter-molecular dispersion interactions, precludes this goal in the near future without further technical advances. As we can see, there is still 'plenty of room at the bottom' for further theoretical investigations; we thus foresee a fruitful and challenging field of research with chemistry, physics, and materials science taking part into its future achievements.

7.8 Acknowledgements

The work in recent years has been supported by the 'Ministerio de Educación y Ciencia' of Spain and the 'European Regional Development Fund' through projects CTQ2007-66461 and CTQ2011-27253. Discussions with Y. Olivier and J. Cornil (Mons, Belgium), R. F. Fink (Tübingen, Germany), J. Gierschner (Madrid, Spain) and A. J. Pérez-Jiménez (Alicante, Spain) are gratefully acknowledged.

7.9 References

Anthony J (2006), *Chem. Rev.*, **19**, 4535.

Asmadi A, Neumann M A, Kendrick J, Girard P, Perrin M-A, and Leusen F J J (2009), *J. Phys. Chem. C*, **113**, 16303.

Baumeier B, Kirkpatrick J, and Andrienko D (2010), *Phys. Chem. Chem. Phys.*, **12**, 11103.

Becke A D (1988), *Phys. Rev. A*, **38**, 3098.

Becke A D (1993a), *J. Chem. Phys.*, **98**, 1372.

Becke A D (1993b), *J. Chem. Phys.*, **98**, 5648.

Brédas J L, Beljonne D, Coropceanu V, and Cornil J (2004), *Chem. Rev.*, **104**, 4971.

Brédas J L, Norton J E, Cornil J, and Coropceanu V (2009), *Acc. Chem. Res.*, **42**, 1691.

Bromley S T, Illas F, and Mas-Torrent M (2008), *Phys. Chem. Chem. Phys.*, **10**, 121.

Castet F, Aurel P, Fritsch A, Ducasse L, Liotard D, Linares M, Cornil J, and Beljonne D (2008), *Phys. Rev. B*, **77**, 115210.

Cheung D and Troisi A (2008), *Phys. Chem. Chem. Phys.*, **10**, 5941.

Chi X, Li D, Zhang H, Garcia V, Garcia C, and Siegrist T (2008), *Organic Electronics*, **9**, 234.

Cornil J, Brédas J L, Zaumseil J, and Sirringhaus H (2007), *Adv. Mater.*, **19**, 1791.

Coropceanu V, André J M, Malagoli M, and Brédas J L (2003), *Theor. Chem. Acc.*, **110**, 59.

Coropceanu V, Cornil J, da Silva Filho D A, Olivier Y, Silbey R, and Brédas J L (2007), *Chem. Rev.*, **107**, 926.

da Silva Filho D A, Kim E-G, and Brédas, J L (2005), *Adv. Mater.*, **17**, 1072.

Deng W-Q and Goddard III W A (2004), *J. Phys. Chem. B*, **108**, 8614.

Devos A and Lannoo M (1998), *Phys. Rev. B*, **58**, 8236.

Ernzerhof M (1998), in Joubert D, *Density Functionals: Theory and Applications*, Heidelberg, Springer, 60.

Grimme S (2006), *J. Comput. Chem.*, **27**, 1787.

Grimme S (2011), *WIREs Comput. Mol. Sci.*, **2**, 211.

Grimme S, Antony J, Ehrlich S, and Krieg H (2010), *J. Chem. Phys.*, **132**, 154104.

Grimme S, Ehrlich S, and Goerigk L (2011), *J. Comput. Chem.*, **32**, 1456.

Grozema F C and Siebbeles L D A (2008), *Int. Rev. Phys. Chem.*, **27**, 87.

Hasegawa T and Takeya J (2009), *Sci. Technol. Adv. Mater.*, **10**, 024314.

Heimel G, Romaner L, Zojer E, and Brédas J L (2008), *Acc. Chem. Res.*, **41**, 721.

Huang J S and Kertesz M (2005), *J. Chem. Phys.*, **122**, 234707.

Idé J, Méreau R, Ducasse L, and Castet F (2011), *J. Phys. Chem. B*, **115**, 5593.

Ishii H, Sugiyama K, Ito E, and Seki K (1999), *Adv. Mater.*, **11**, 605.

Jakobsson M and Stafstöm S (2011), *J. Chem. Phys.*, **135**, 134902.

Johnson E R, Mori-Sánchez P, Cohen A J, and Yang W (2008), *J. Chem. Phys.*, **129**, 204112.

Konezny S J, Bussac M N, and Zuppiroli L (2010), *Phys. Rev. B*, **81**, 045313.

Körzdörfer T (2011), *J. Chem. Phys.*, **134**, 094111.

Li A, Wen S-H, Song J-L, and Deng W-Q (2009), *Organic Electronics*, **10**, 1054.

Li Z, Zhang X, and Lu G (2011), *Comput. Phys. Commun.*, **182**, 2632.

Liang Z, Tang Q, Liu J, Li J, Yan F, and Miao Q (2010), *Chem. Mater.*, **22**, 6438.

Madison T A and Hutchison G R (2011), *J. Phys. Chem. C*, **115**, 17558.

Malagoli M, Coropceanu V, da Silva Filho D A, and Brédas J L (2004), *J. Chem. Phys.*, **120**, 7490.

Marcus R A (1993), *Rev. Mod. Phys.*, **65**, 599.

Martinelli N, Savini M, Muccioli L, Olivier Y, Castet F, Zannoni C, Beljonne D, and Cornil J (2009), *Adv. Funct. Mater.*, **19**, 3254.

Martinelli N, Idé J, Sánchez-Carrera R, Coropceanu V, Brédas J L, Ducasse L, Castet F, Cornil J, and Beljonne D (2010), *J. Phys. Chem. C*, **114**, 20678.

Mas-Torrent M and Rovira C (2011), *Chem. Rev.*, **111**, 4833.

Mas-Torrent M, Hadley P, Bromley S T, Ribas X, Tarres J, Mas M, Molins M, Veciana J, and Rovira C (2004), *J. Am. Chem. Soc.*, **126**, 8546.

McMahon D P and Troisi A (2010a), *ChemPhysChem*, **11**, 2067.

McMahon D P and Troisi A (2010b), *J. Phys. Chem. Lett.*, **1**, 941.

McMahon D P and Troisi A (2011), *Phys. Chem. Chem. Phys.*, **13**, 10241.

Meng Q, Dong H, Hu W, and Zhu D (2011), *J. Mater. Chem.*, **21**, 11708.

Misra M, Andrienko D, Baumeier B, Faulon J-L, and von Lilienfeld O A (2011), *J. Chem. Theory Comput.*, **7**, 2549.

Moon H, Zeis R, Borkent E-J, Besnard C, Lovinger A J, Siegrist T, Kloc C, and Bao Z (2004), *J. Am. Chem. Soc.*, **126**, 15322.

Mori-Sánchez P, Cohen A J, and Yang W (2006), *J. Chem. Phys.*, **125**, 201102.
Newton M D (1991), *Chem. Rev.*, **91**, 767.
Nitzan A (2001), *Annu. Rev. Phys. Chem.*, **52**, 681.
Olivier Y, Lemaur V, Brédas J L, and Cornil J (2006), *J. Phys. Chem. A*, **110**, 6356.
Patwardhan S, Kocherzhenko A A, Grozema F C, and Siebbeles L D A (2011), *J. Phys. Chem. C*, **115**, 11768.
Peach M J G, Miller A M, Teale A M, and Tozer D J (2008), *J. Chem. Phys.*, **129**, 064105.
Perdew J P, Burke K, and Ernzerhof M (1996), *Phys. Rev. Lett.*, **77**, 3865.
Picon J-D, Bussac M N, and Zuppiroli L (2007), *Phys. Rev. B*, **75**, 235106.
Reimers J F, Cai Z L, Bilic′ A, and Hush N S (2003), *Ann. N.Y. Acad. Sci.*, **1006**, 235.
Sai N, Barbara P F, and Leung K (2011), *Phys. Rev. Lett.*, **106**, 226403.
Sancho-Garcia J C, Horowitz G, Brédas J L, and Cornil J (2003), *J. Chem. Phys.*, **119**, 12563.
Sancho-Garcia J C (2007), *Chem. Phys.*, **331**, 321.
Sancho-Garcia J C (2010), *Theor. Chem. Acc.*, **127**, 605.
Sancho-Garcia J C and Pérez-Jiménez A J (2008a), *J. Chem. Phys.*, **129**, 024103.
Sancho-Garcia J C and Pérez-Jiménez A J (2008b), *J. Phys. Chem. A*, **112**, 10325.
Sancho-Garcia J C and Pérez-Jiménez A J (2009a), *J. Chem. Phys.*, **131**, 134108.
Sancho-Garcia J C and Pérez-Jiménez A J (2009b), *Phys. Chem. Chem. Phys.*, **11**, 2741.
Sancho-Garcia J C and Pérez-Jiménez A J (2010), *Chem. Phys. Lett.*, **499**, 146.
Sancho-Garcia J C, Pérez-Jiménez A J, Olivier Y, and Cornil J (2010), *Phys. Chem. Chem. Phys.*, **12**, 9381.
Shuai Z, Wang L, and Li Q (2011), *Adv. Mater.*, **23**, 1145.
Stafström S (2010), *Chem. Soc. Rev.*, **39**, 2484.
Stehr V, Pfister J, Fink R F, Engels B, and Deibel C (2011), *Phys. Rev. B*, **83**, 155208.
Sundar V C, Zaumseil J, Podzorov V, Menard E, Willet R L, Someya T, Gershenson M E, and Rogers J A (2004), *Science*, **303**, 1644.
Troisi A (2010), *Adv. Polym. Sci.*, **223**, 259.
Troisi A (2011), *Organic Electronics*, **12**, 1988.
Troisi A and Orlandi G (2006), *J. Phys. Chem. A*, **110**, 4065.
Tse S C, Cheung C H, and So S K (2010), in So F, *Organic Electronics: Materials, Processing, Devices, and Applications*, CRC Press, 61.
Valeev E F, Coropceanu V, da Silva Filho D A, Salman S, and Brédas J L (2006), *J. Am. Chem. Soc.*, **128**, 9882.
Wang L, Nan G, Yang X, Peng Q, Li Q, and Shuai Z (2010), *Chem. Soc. Rev.*, **39**, 423.
Yamashita Y (2009), *Sci. Technol. Adv. Mater.*, **10**, 024313.

8

Optical, photoluminescent and electroluminescent properties of organic materials

J. GODLEWSKI and M. OBAROWSKA,
Gdansk University of Technology, Poland

DOI: 10.1533/9780857098764.2.245

Abstract: This chapter discusses optical, photoluminescence and electroluminescence properties of organic materials. First, the spectral features of individual molecules and molecular solid states are analysed. Next, the excitonic processes in organic materials are discussed. The chapter reviews experimental methods leading to the determination of basic excitonic parameters. Finally, the electroluminescence phenomena in organic materials are discussed.

Key words: organic materials, excitons, electroluminescence, organic light-emitting diode (OLED), recombination.

8.1 Introduction

Organic molecules are chemical compounds with complicated structures. Composed of many atoms, apart from electronic properties they also exhibit special physicochemical features. When organic molecules create molecular solid-state devices with crystal or amorphous structures, the properties of these devices follow from organic molecule interactions. Therefore in molecular solid-state structures, the energy levels of individual molecules form continuous bands of energy. Due to the weak interactions between the molecules, molecular solid-state structures exhibit the properties of the individual molecules to a greater degree than the properties characteristic for solid-state materials. A special feature of the molecular solid state is the fact that singlet and triplet states are excited due to light interaction being able to move across the material. These mobile quasi-particles are called excitons. Furthermore, excitons can be generated not only by light but also from the recombination process of charge carriers with opposite signs, electrons and holes injected into the system. This has important implications and enables the application of organic materials to light-emitting devices able to produce any colour. Owing to the excitation of the organic material by electromagnetic waves with energy, carriers are generated. Taking into account these features of organic materials, we can conclude that they have significant potential in many fields of science and technology. Therefore an understanding of their physicochemical properties is salient, and our current knowledge remains insufficient. There is thus a need for further research.

8.2 Electronic states of single molecule and molecular solid state

The term 'organic solids' is commonly used to describe materials containing carbon as one of many compounds. Carbon is a general compound, so there are a large number of materials which could legitimately be termed 'organic.' Organic solids are composed of molecules held together by weak van der Waals forces (Pope and Swenberg, 1999; Aradhya, *et al.*, 2012). The type of bonding determines their properties. Organic solids are generally soft with low melting points and poor electrical conductivity. Another fact also derives from the weak nature of bonding, namely that the properties of individual molecules are retained in a solid state. A study of optical properties of organic solids should therefore be prefigured by an explanation of individual molecules and their optical properties. Conjugated aromatic hydrocarbons, particularly the polyacenes, are generally used for this purpose. The term 'conjugated' in chemistry refers to those molecules in which atoms are connected through p-orbitals, with delocalized electrons in compounds with alternating single and multiple bonds, whereas the term 'aromatic' in organic chemistry refers to those organic compounds which have notable aromas. The polyacenes belong to a class of polycyclic aromatic hydrocarbon compounds, which are planar sets of linearly fused benzene rings, with the general formula C_{4n+2} H_{2n+4}. Focusing only on the properties of p-electrons is sufficient for the analysis of these molecules' electronic properties (Birks, 1973), which are in the highest energy occupied orbitals and so can be easily excited. For example, for a tetracene molecule, we find that p-electrons have the highest probability of being detected above and below the four aromatic rings, while the edge of the molecule (where the hydrogen atoms are located) is electron-deficient (Chi *et al.*, 2008). Electron distribution in tetracene molecules determines their crystal structure. It is worth noting that strong intermolecular coupling determines the high value of intrinsic mobility. This is an important parameter when selecting materials for electronic applications.

Owing to their complexity, organic molecules consist of many intermolecular bindings exhibiting vibronic, rotational and liberation moves. This set of possibilities causes a more complex molecular absorption spectrum than an atomic one, containing novel features. Molecular electronic states and the transitions between them are usually represented using a Jablonski diagram (Jablonski, 1933) shown in Fig. 8.1.

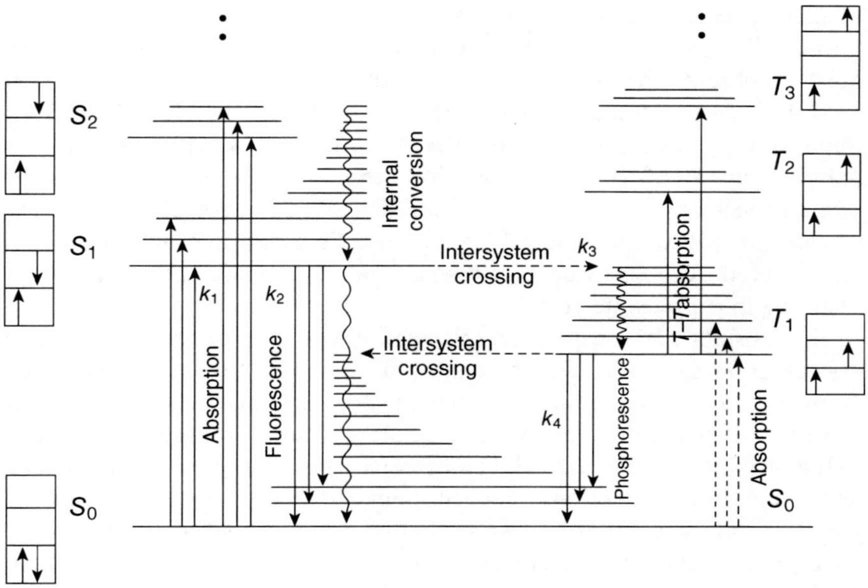

8.1 The schematic representation of a Jablonski diagram (modified based on Wolf and Haken, 1995). See text for a description of the symbols.

The singlet ground, first and second electronic states together with vibrational energy levels are depicted by S_0, S_1 and S_2, respectively. The first and second triplet states are depicted by T_1 and T_2. Following the absorption of light by the molecule, an electron ends up in a particular state of excitation, which is usually some higher vibrational level of either S_1 or S_2. This is indicated by an upward arrow. The rate constant is denoted as k_1. Generally, the rate constant k is defined as the reciprocal of the transition time. Next, relaxation of the excited state to its lowest vibrational level takes place.

This process is called internal conversion and generally occurs with a rate comparable to that of nuclear vibration. Internal conversion involves the energy dissipation from the molecule to its surroundings, and thus cannot occur for isolated molecules. When a molecule reaches the lowest energy vibrational state, the transition $S_1 \rightarrow S_0$ occurs, and fluorescence is emitted. The rate constant for this radiative process is denoted as k_2. Molecules in the S_1 state can also undergo a spin conversion to the first triplet state T_1. This radiationless transition is known as the intersystem crossing, and occurs with the rate constant represented by k_3. In molecules with large spin–orbit coupling, intersystem crossing is much more important than in molecules exhibiting only small spin–orbit coupling. Emission from T_1 is termed phosphorescence and occurs with the rate constant k_4. Here we must emphasize that for the electronically excited state of a molecular crystal, there are additional channels by which excited states decay. One important quenching process involves the collision of a triplet exciton with another triplet exciton, known as triplet exciton fusion (Rao *et al.*, 2010). The fusion reactions yield triplet and singlet states. Resulting from the singlet state relaxation, delayed fluorescence (Moore and Munro, 1965) can be observed. Another process is fission of a singlet state (Lee *et al.*, 2009), resulting in triplet state creation.

The existence of weak intermolecular interaction forces in molecular crystals and small molecular aggregates, causes this electronic configuration of molecules to remain practically unchanged in crystals. Spectral properties of small molecular aggregates and crystals are therefore directly traceable to the properties of the individual molecules. The arrangement of molecules and the mutual interactions between them causes the creation of common bands, responsible for collective excitation in molecular structure. Figure 8.2 represents the energetic levels of an isolated molecule and a molecular solid state.

The energy levels in a single molecule are denoted by *T*, triplet level, and *S*, singlet level. Molecular solid position of LUMO (lowest unoccupied molecular orbital), HOMO (highest occupied molecular orbital) and CT (charge transfer) exciton levels are shown. A_g denotes electron affinity of a molecule, A_C electron affinity of a crystal, E_g an energy gap, I_C ionization energy of a crystal, and I_g ionization energy of a molecule.

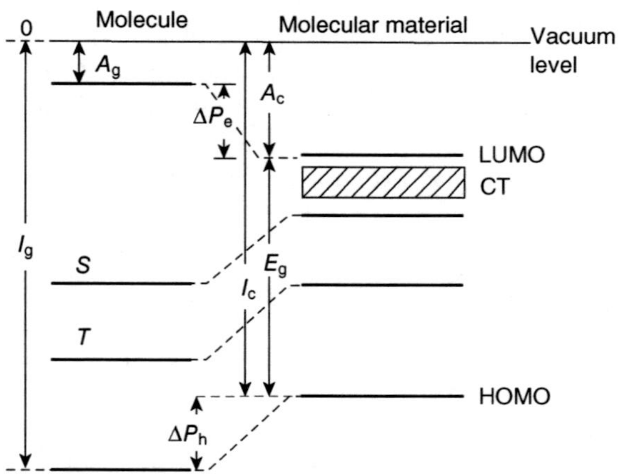

8.2 Illustration of energy levels for an isolated molecule and a molecular solid state.

The main energetic levels for a valence electron in an isolated molecule are: ionization energy (I_c), the energy required to remove an electron from a neutral molecule; and singlet (S) and triplet (T) excited levels and electron affinity (A_g), energy released upon the addition of an electron taken from infinity and placed on a neutral molecule. If the molecule becomes a negative ion due to the addition of a free electron, it is called an acceptor. A molecule is called a donor if the molecule gains the necessary energy to remove an electron from the highest filled valence level. In molecular crystals, molecules with excess electrons become crystal molecules, giving their energetic levels to the conduction band. On the other hand, those molecules with missing electrons become crystal molecules, and their energetic levels create the valence band. This situation is typical for amorphic organic materials and some molecular crystals at higher temperatures. Generally, instead of the term valence band, HOMO level is used, and instead of the term conduction band, LUMO is used. Energetic levels placed close to the conduction band may create CT excitons. The CT excitons are states that exist during the bimolecular recombination of charge carriers, and the optical excitation with energy below the band gap.

8.3 Absorption and emission spectroscopy

The energetic levels represented in the previous section can be observed experimentally due to absorption and emission spectra. The structure of absorption spectra depends on the phase in which molecules exist (Tanaka, 1964). The rotational spectra of molecules in solution cannot be resolved due to solvent interactions and thermal fluctuations. In the gas phase the spectra of individual rotational levels are resolvable, and their line widths are determined by other lifetime-limiting processes such as collisions with container walls (Pope and Swenberg, 1999). Many but not all molecular aggregate and crystal spectra are directly traceable to individual molecule properties. The crystal spectrum retains the spectral features of individual molecules, including their electronic-vibrational structure. However, certain new optical and electronic properties are found in molecular crystals, caused by collective molecular interaction. The intramolecular vibrational states in crystal spectra are apparent, arising from internal vibration of molecules. Vibrational levels for crystals move in the IR region of the electromagnetic wave, compared to those vibrational levels for molecules in solvent solution. This is due to molecular interaction in the crystalline phase, much stronger than the interaction between molecules and solvent. This is known as the solvation effect (Reichardt, 2003). Furthermore spectra depend on the direction of light polarization. The maxima of absorption spectra are also placed differently for isolated molecules and molecular solid states respectively (Tanaka, 1964). The energy tail for the fundamental absorption band,

and for the fluctuations of absorption bands, follows the Urbach rule. This rule states that the absorption coefficient is given by (Keil, 1966; Klafter and Jortner, 1977; Kurik, 1971):

$$\mu(\omega, T) = \mu_0 \exp[-\sigma_0(\hbar\omega_0 - \hbar\omega)/kT] \qquad [8.1]$$

where $\hbar\omega$ is the energy of incident radiation and μ_0, ω_0 and σ_0 are constants characteristic of the crystal. The above equation applies to high temperatures. For low temperatures the temperature dependence in the exponential disappears, and T should be replaced by T_0 – an experimentally determined parameter generally of the order of 100° K. This rule was first discovered by Urbach for silver halides, and has been found to apply to a large number of other crystals. The exponential dependence of the absorption coefficient has been found to be valid over a remarkably wide range of compounds. Excited molecules return to the ground state in relatively short time. The energy released in that process can appear as fluorescence, phosphorescence or heat. Figure 8.3 gives a schematic representation of absorption and emission spectra. The mirror symmetry relationship is evident. Such symmetry is a consequence of vibrational levels, identical for both absorption and emission processes. This phenomena is observed in organic material when there is no reabsorption of fluorescence (Kearwell and Wilkinson, 1969).

The new feature in the crystal spectra following from collective molecular interaction, is known as Davydov splitting (Davydov, 1971). Molecular exciton theory, developed by Alexander Davydov, predicts that in a crystal containing several molecules per unit cell, several branches of excited crystal states will correspond to each excited electronic molecular state. For example, in anthracene-type crystals containing two molecules per unit cell, two bands of excited states corresponding to each molecular term can be observed in spectra (Matsui, 1966). The width of each Davydov band depends on both types of intermolecular interaction. The effect of Davydov

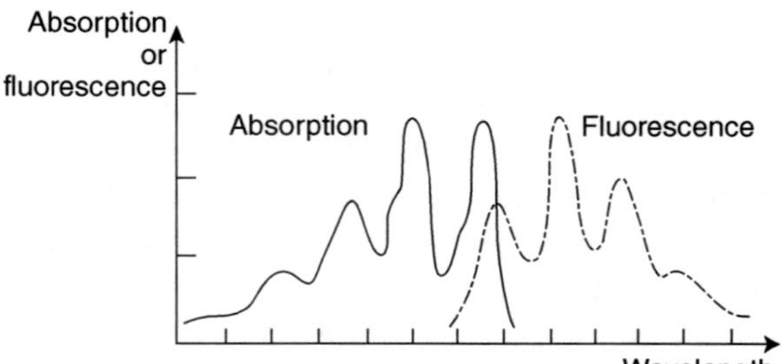

8.3 Mirror symmetry relationship between absorption (solid line) and emission spectra (dashed line).

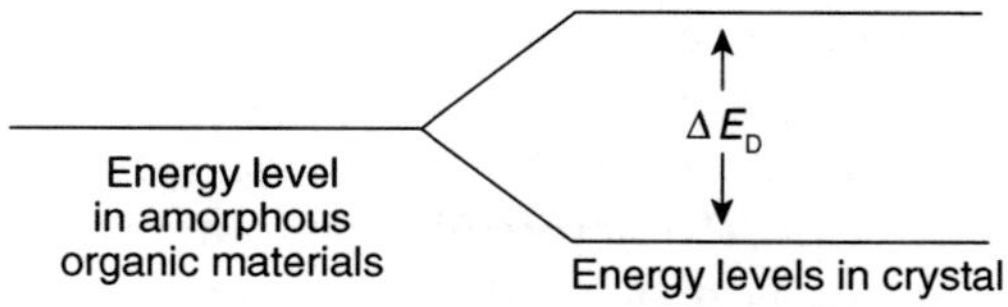

8.4 The Davydov splitting of crystal states. ΔE_D denotes energy gap between exciton levels in the crystal.

splitting provides direct evidence for the presence of collective excitation in the crystal. This can be described in terms of a delocalized exciton. A schematic representation of Davydov splitting is given in Fig. 8.4.

We must remember, however, that molecular crystals are not the only structure of organic matter and that layers of organic molecules are used for practical applications (Lee and Loo, 2010). There is no long distance order in organic layers – the structure of an organic layer is a mixture of amorphous and polycrystalline structure, and therefore Davydov splitting is not observed in absorption and emission spectra. We receive no information from optical spectra about the local structure of a layer. The absorption spectra of organic layers are a result of the lack of long distance order, and often consist of features connected with scattering of light.

8.4 Excitonic processes

In the previous section we looked at the electronic and spectral properties of individual molecules and molecular crystals. We now consider the behavior of excited states in molecular crystals. These states can be mobile and diffuse across the material. Created in molecular materials, these excited states are called excitons. Generally, an exciton is an electrically neutral particle, consisting of an electron and hole bound by the Coulomb force, and created due to photon absorption. The exciton was initially introduced by Frenkel (1931) and then generalized by Wannier (1937) and Peierls (1932). For molecular crystals, molecular exciton theory was developed by Davydov (1971), who extended the model of the small-radius exciton (Frenkel exciton) to create a consistent theory of light absorption, exciton formation and their coherent motion. His theory was experimentally verified and proven by many investigators (Agranovich, 2009). According to molecular exciton theory, excitons can be either mobile or trapped. In the former case, the exciton transport can be well described in the framework of macroscopic diffusion theory, regarding exciton migration as a form of random walk. The basic processes following from exciton motion will be analyzed later in this section.

After their creation in molecular materials due to light absorption, excitons can diffuse in all directions, taking part in different processes. The diffusive motion of excitons can be controlled in their coherent motion in band, or incoherent hopping process. It is usually assumed that where the scattering length is smaller than the lattice constant, the motion of excitons has a hopping character. The nature of exciton motion influences the diffusion coefficient, which is strongly dependent upon temperature. Regarding the nature of exciton motion, the model describing excitons transport in molecular materials is usually the same, namely the diffusive model of exciton migration. The equation describing the kinetics of exciton motion for one direction x and decay processes is given by (Mulder, 1968; Pope *et al.*, 2003; Michel-Beyerle and Haberkorn, 1978; Hofmann *et al.*, 2012):

$$\frac{\partial C(x,t)}{\partial t} = I_0 \kappa \exp(-\kappa x) - \frac{1}{\tau_0} C(x,t) + D \frac{\partial^2 C(x,t)}{\partial^2 x} - \gamma_{CC} C^2(x,t) - \gamma_{CD} C(x,t) D_0 + R(x,t) \qquad [8.2]$$

where x denotes the spatial dependence on the exciton density in direction of excitation, and $C(x, t)$ the concentration of excitons. The first term of the equation from the left side: $\frac{\partial C(x,t)}{\partial t}$ describes the evolution in time of exciton concentration, the second term: $I_0 \kappa \exp(-\kappa x)$ denotes the exciton generation rate, where I_0 is the incident photon flux at $x = 0$ and κ the linear absorption coefficient. The third term: $\frac{1}{\tau_0} C(x,t)$ is the monomolecular decay term of excitons, where τ_0 is exciton lifetime. The fourth term: $D \frac{\partial^2 C(x,t)}{\partial x^2}$ follows from the diffusive motion of excitons, where D is the diffusion coefficient. The fifth term $\gamma_{CC} C^2(x, t)$ describes collision between excitons and γ_{CC} is the bimolecular rate constant of exciton collisions. The results of exciton collisions can vary and will be discussed later. The sixth term in eq. [8.2]: $\gamma_{CD} C(x, t) D_0$ determines the transfer of exciton energy to doping or impurity centers, where γ_{CD} is the bimolecular rate constant of energy transfer and D_0 the concentration of doping or impurity centers. Finally, the last term of eq. [8.2]: $R(x, t)$ determines the change of exciton concentration due to the reabsorption of light coming from luminescence processes.

The solution of the above equation is quite complicated and can only be performed for particular cases. The purpose of our analysis is to find the exciton distribution in molecular material after absorption of light, and show methods for the determination of basic exciton parameters such as exciton lifetime (τ_0), diffusion length (ℓ) and diffusion coefficient (D).

For steady illumination and small reabsorption of light by the sample, monomolecular decay and diffusion terms must be taken into account and thus eq. [8.2] describing the excitons concentration is shown as follows (Mulder, 1968):

$$I_0\kappa\exp(-\kappa x)+\frac{1}{\tau_0}C(x)+D\frac{\mathrm{d}^2C(x)}{\mathrm{d}x^2}=0 \qquad [8.3]$$

The solution of the above equation must obey at least two boundary conditions. When the absorption coefficient value is larger than zero and the thickness of the illuminated material is infinitely long, it can be assumed that the exciton concentration tends to zero. This boundary condition can be represented as:

$$x\to\infty,\quad C(\infty)\to 0 \qquad [8.4]$$

From the other side, at short distances ($x\to 0$), the exciton concentration is strongly dependent on the exciton flux moving to the surface area, and the quenching rate. Taking into account the Fick law and assuming that the quenching rate equals s at the surface, the second boundary condition can be written:

$$D\left.\frac{\mathrm{d}C(x)}{\mathrm{d}x}\right|_{x\to 0}=sC(0) \qquad [8.5]$$

If the quenching rate of excitons flux $s = 0$, no annihilation of excitons takes place at the surface. On the other hand, when $s\to\infty$, every exciton that hits the front of the surface is annihilated.

The solution of the equation [8.3] under the boundary conditions mentioned above is given by (Mulder, 1968):

$$C(x)=\frac{I_0\kappa\ell^2}{D(1-\kappa^2\ell^2)}\left[\exp(-\kappa x)-\frac{\kappa\ell D+s\ell}{D+s\ell}\exp\left(-\frac{x}{\ell}\right)\right] \qquad [8.6]$$

where $\ell=\sqrt{D\tau_0}$ is the diffusion length of the exciton.

The graphical representation of equation [8.6] for singlet and triplet states is shown in Figs 8.5 and 8.6. From Figs 8.5 and 8.6 it can be seen that where no diffusion of excitons takes place in the sample ($\ell = 0$) the singlet and triplet concentration follows the absorption of light, in the sample (I). The formula describing exciton concentration in both cases can be represented:

$$C(x)\approx I_0\kappa\tau_0\exp(-\kappa x) \qquad [8.7]$$

When the absorption coefficient is large ($\kappa\to\infty$) the exciton concentration can be represented:

$$C(x)=\frac{I_0\ell}{D+s\ell}\exp\left(-\frac{x}{\ell}\right)=C(0)\exp\left(-\frac{x}{\ell}\right) \qquad [8.8]$$

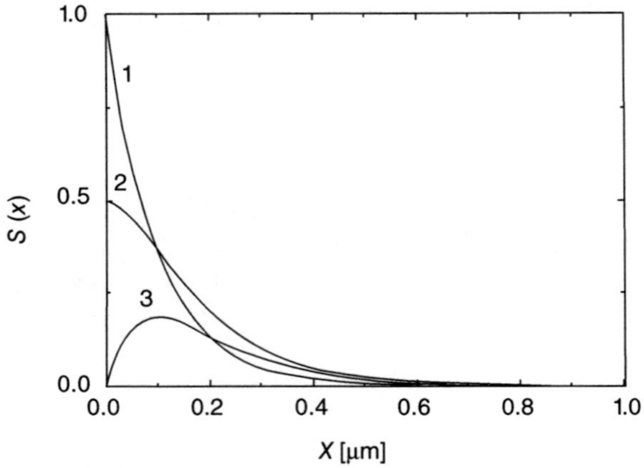

8.5 The relative singlet exciton concentration as a function of the thickness (x) calculated for absorption coefficient $\kappa = 10^5$ cm^{-1} and the singlet exciton lifetime $\tau_s = 10$ ns. The curve denoted as '1' determines the arrangement of singlet excitons in the case of the lack of diffusion ($\ell = 0$). Curves '2' and '3' were obtained for singlet exciton diffusion length $\ell = 100$ nm and the quenching rates $s = 0$ and $s = \infty$ respectively.

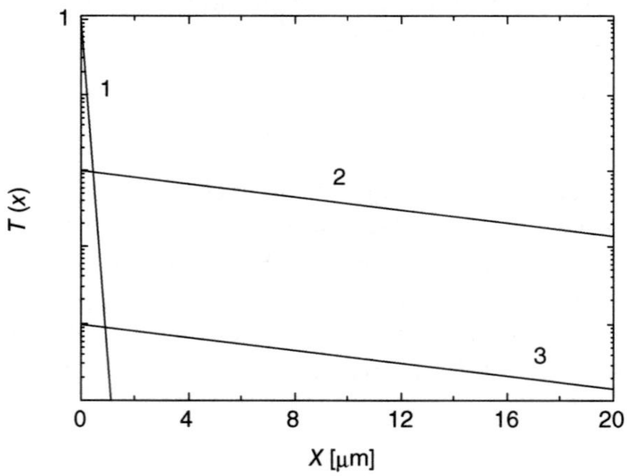

8.6 The relative triplet exciton concentration as a function of the thickness (x) calculated for absorption coefficient $\kappa = 10^5$ cm^{-1} and the singlet exciton lifetime $\tau_s = 10$ ms. The curve denoted as '1' determines the triplet excitons arrangement in the case of no diffusion. Curves '2' and '3' were obtained for triplet exciton diffusion length and the quenching rates $s = 0$ and $s = \infty$, respectively.

It follows from the above equation that the concentration of excitons decreases e-times at the distance $x = \ell$ for strong light absorption and long length of diffusion. This offers a good explanation of the physical sense of the exciton diffusion length. Finally, it is worth noting that for strong light absorption ($\kappa \to \infty$) and strong annihilation of excitons ($s \to \infty$) the exciton concentration is $C(x) = 0$.

When the diffusive motion takes place and excitons decay at the surface of the sample, the exciton concentration usually decreases close to the edge of a solid, and increases further from its edge. The rate of increase and decrease depends on the quenching rate (s) and the diffusion length (ℓ). When the quenching rate is high ($s \to \infty$), the exciton concentration is low at the surface. On the other hand, high values for diffusion length ($\ell \to \infty$) cause an increase of exciton concentration in the volume of the sample.

8.4.1 Methods of determination basic parameters of excitons

Useful formulas describing exciton concentration for particular cases follow from the general solution of eq. [8.3] represented by equation [8.6]. At the surface of a sample ($x = 0$), the exciton concentration can be represented by the formula (Mulder, 1968):

$$C(0) = \frac{I_0}{s\left(1+\frac{1}{\kappa\ell}\right)\left(1+\frac{D}{s\ell}\right)} \qquad [8.9]$$

This formula can be used to determine the value of the exciton flux annihilated at the surface. Furthermore, the exciton diffusion length can be determined from the measurement of sensibilized luminescence (Mulder, 1968) and electrode limited photocurrents (Godlewski, 2005). In sensibilized luminescence, the exciton annihilates due to energy transfer to the dye molecule located at the surface. Another possibility for exciton annihilation is exciton decay into a free electron and hole. This process leads to electrode limited photocurrent production. The decay of excitons at the surface is directly proportional to their concentration at the surface region, described by the formula [8.9]. The flux of excitons annihilated at the surface can be represented as (Mulder, 1968):

$$\Phi = s \cdot C(0) \qquad [8.10]$$

The decay of exciton flux Φ can be observed as electrode limited photocurrent and sensibillized luminescence. The reciprocal of eq. [8.10] after substituting as $C(0)$ the formula [8.9] is given by

$$\frac{1}{\Phi} = \frac{1}{b} + \frac{1}{b\kappa\ell} \qquad [8.11]$$

where

$$\frac{1}{b} = I_0\left(1 + \frac{D}{s\ell}\right) \qquad [8.12]$$

Thus, the exciton diffusion length can be determined from the graph showing the experimental relationship between the reciprocal of the quenching of exciton flux as a function of the reciprocal of the linear absorption coefficient, given by a straight line (Matsuse *et al.*, 2005; Kurrle and Pflauma 2008; Banerjee *et al.*, 2009; Signerski and Jarosz, 2011). The schematic representation of the method for determining exciton diffusion length based on experimental data of exciton quenching is shown in Fig. 8.7.

Lifetime is another important parameter characterizing excitons in molecular material. After excitation, the singlet exciton with lifetime value τ_0, decays through photon emission in a process known as fluorescence.

In some simple cases, when the generation of excitons in the sample is uniform and the exciton diffusion can be neglected, and the pulse of light is shorter than exciton lifetime, annihilation of excitons can be described by the equation (Pope and Swenberg, 1999):

$$\frac{\mathrm{d}C(t)}{\mathrm{d}t} = -\frac{1}{\tau_0}C(t) \qquad [8.13]$$

If we assume that after the excitation process ($t = 0$), the exciton density is $C(0)$ for every value of spatial distance x, the solution of the above equation can be represented as follows:

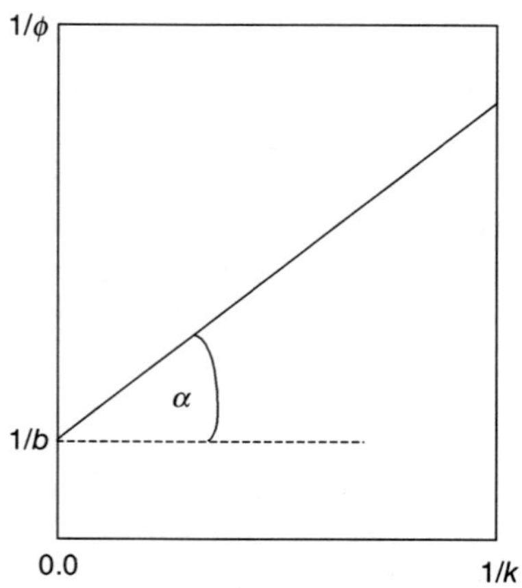

8.7 Plot of the reciprocal of the flux of excitons annihilated at the surface ($1/\phi$) versus the reciprocal of the linear absorption coefficient ($1/\kappa$).

$$C(t) = C(0)\exp\left(-\frac{t}{\tau_0}\right) \tag{8.14}$$

Experimentally observed luminescence is dependent on the probability of photon emission, which can be found with the formula (Wolf and Haken, 1995):

$$p_r = \frac{k_r}{k_t} = \frac{k_r}{k_r + k_{nr}} \tag{8.15}$$

where k_r is radiative transition rate, and k_{nr} is non-radiative transition rate.

If the excitons are created in the volume of the sample, the flux of fluorescence intensity (in quanta $cm^{-2}\ s^{-1}$) is a function of time and it can be calculated from the formula

$$\Phi(t) = p_r \int_0^V C(0)\exp\left(-\frac{t}{\tau_0}\right)dV = \Phi_0 \exp\left(-\frac{t}{\tau_0}\right) \tag{8.16}$$

where V is the total volume of the active sample and Φ_0 is the flux of fluorescence intensity at $t = 0$.

There are two major techniques for using fluorescence lifetime measurement in the nanosecond range: time-domain (Becker, 2005) and frequency-domain (Lakowicz and Gryczynski, 1991) data acquisition methods. Figure 8.8 presents the experimental setup for determination of the exciton lifetime.

In the time-domain, the sample is excited with a short pulse of light (pulsewidth < 1–2 ns) available from flash lamps, pulsed lasers, laser diodes, and light-emitting diodes (LEDs) with sufficient delay between pulses. A variety of fluorescence detection methods are available for lifetime measurements, with the most common currently time-correlated single photon counting (TCSPC) (O'Connor and Phillips, 1984; Becker, 2008). Photomultiplier tubes or avalanche photodiodes are used to record the time-dependent distribution of emitted photons after each pulse. In the frequency

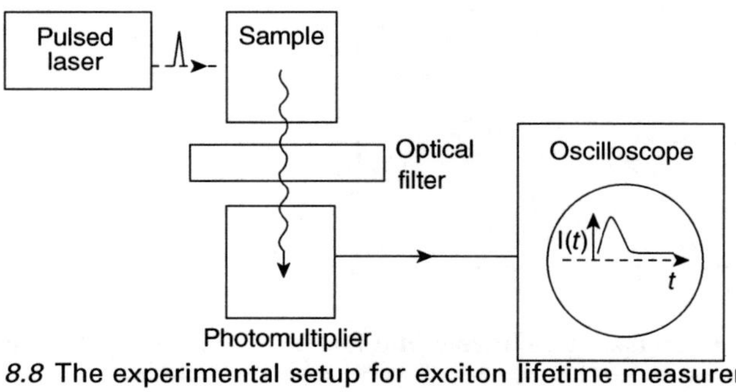

8.8 The experimental setup for exciton lifetime measurement.

domain technique (Lakowicz, 1991) incident light is sinusoidally modulated at high frequencies. In this configuration, emission occurs at the same frequency as the incident light but experiences a phase delay φ and change in the amplitude *M* relative to the excitation light (demodulation). Data are acquired with photomultipliers or charge-coupled devices equipped with a gain modulator.

In both methods, lifetime is calculated using curve fitting algorithms, with the least squares curve method being most common. If the fit is monoexponential, output provides a single fluorescent lifetime. The data analysis of multiexponential decay provides fluorescent lifetimes along with corresponding fractional contributions. The schematic representation of calculating the exciton lifetime is represented in Fig. 8.9.

8.4.2 Exciton interactions

Different annihilation processes exist for the crystal excited states. Singlet excitons usually decay due to photon emission, observed experimentally as luminescence. However, triplet excitons annihilate through phonon generation. These two processes of exciton annihilation are kinetically unimolecular. There are other channels for the decay of the excited states in addition to these, involving the interaction of the mobile exciton with free or trapped excitons, or with a foreign molecule or crystal defect. The analysis of these processes gives useful information about exciton lifetime and its diffusion length. In this part of the chapter we outline bimolecular exciton quenching processes.

An important exciton quenching process in organic solids from the kinetically bimolecular annihilation processes group involves the collision of excitons. Particularly interesting is the collision of a triplet exciton with another triplet exciton. A process often referred to as triplet exciton fusion

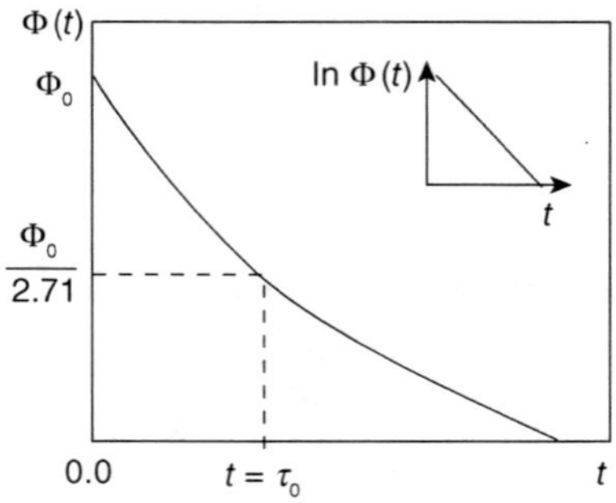

8.9 The relationship between the flux of fluorescence and time according to eq. [8.16] in linear and logarithmic scale.

reaction (Klein *et al.*, 1973; Arden *et al.*, 1976) leading to triplet and singlet states according to the formula (Pope *et al.*, 1970):

$$T_1 + T_1 \rightarrow \begin{cases} S_0 + T_n \\ S_0 + S_n \end{cases} \qquad [8.17]$$

where T_n and S_n represent a particular vibrational state in the nth triplet and singlet electronic manifolds, respectively. The excited triplet states T_n can decay without radiation or through phosphorescence or dissociation into two charge carriers. A similar situation can be observed for singlet excited state S_n, although here basic quenching is by delayed florescence. Although the spectral characteristic for delayed fluorescence is the same as for directly excited fluorescence, fluorescence lifetime changes dramatically, from nanoseconds to milliseconds.

Assuming that the triplet excitons are generated uniformly in the volume, and that the time of excitation is shorter than exciton lifetime, triplet exciton density obeys the equation (Pope and Swenberg, 1999):

$$\frac{\mathrm{d}T(t)}{\mathrm{d}t} = -\frac{1}{\tau_\mathrm{t}}T(t) - \gamma_{\mathrm{TT}}T^2(t) \qquad [8.18]$$

where τ_t is the triplet exciton lifetime and γ_{TT} is the total bimolecular annihilation rate constant. If, as the result of triplet excitons annihilation, the singlet excitons are generated with probability f, then the singlet exciton concentration can be represented (Pope and Swenberg, 1999):

$$\frac{\mathrm{d}S(t)}{\mathrm{d}t} = f\gamma_{\mathrm{TT}}T^2(t) - \frac{1}{\tau_\mathrm{s}}S(t) \qquad [8.19]$$

Due to the fact the triplet exciton lifetime is longer than singlet exciton lifetime, the decay of singlet states is controlled by the decay of triplet states. The following relationship based on eq. [8.19] can thus be stated:

$$f\gamma_{\mathrm{TT}}T^2(t) = \frac{1}{\tau_\mathrm{s}}S(t) \qquad [8.20]$$

If in eq. [8.18] the following relationship is paid:

$$\frac{1}{\tau_\mathrm{t}}T(t) \gg \gamma_{\mathrm{TT}}T^2(t) \qquad [8.21]$$

then the delayed fluorescence is described by the formula:

$$\Phi(t) = \Phi_0 \exp\left(-\frac{2t}{\tau_\mathrm{t}}\right) \qquad [8.22]$$

Where the relationship given by eq. [8.21] is not valid, the delayed fluorescence decreases according to hyperbolic relation. We can use eq. [8.22] to determine the triplet exciton lifetime. Taking into account experimental data for the delayed fluorescence, the triplet exciton lifetime can be obtained from the measurement of the slope of the plot of lnΦ versus time (t).

In addition to the exciton quenching process described above, there can also be singlet–singlet exciton fusion and singlet–triplet exciton reactions. In singlet–singlet exciton fusion, the end result is the production of a highly excited singlet state, which can convert internally to singlet state S_1 and decay back to the ground state S_0, giving off light, or ionize to the pair of electron and hole. In the case of singlet–triplet exciton reactions, the singlet exciton disappears, passing its energy to the triplet exciton, which can then be excited to an upper triplet state. The reverse process (the transfer of energy from triplet to singlet) is spin-forbidden.

If a molecular crystal contains imperfections, impurities or charge carriers, due to their interaction with excitons, exciton energy is transferred to them. This results in the value of exciton lifetime being reduced. Furthermore, if an impurity has luminescing properties, after the energy transfer from the exciton, luminescence from the impurity is possible. The kinetic of excitonic process based on eq. [8.2], neglecting the term describing collision between excitons and the reabsorption term, can be represented

$$\frac{\partial C(x,t)}{\partial t} = I_0 \kappa \exp(-\kappa x) - \frac{1}{\tau_0} C(x,t) - \gamma_{\mathrm{CD}} C(x,t) D_0 + D \frac{\partial^2 C(x,t)}{\partial^2 x} \qquad [8.23]$$

where γ_{CD} is the exciton-doping center interaction rate constant, and D_0 is the density of the reaction center inside the material. It can be seen that for steady state conditions, the solution of the above equation is formally the same as eq. [8.3] but the exciton lifetime is described by (Wolf and Haken, 1995; Wolf, 1967):

$$\frac{1}{\tau} = \frac{1}{\tau_0} + \gamma_{\mathrm{CD}} D_0 \qquad [8.24]$$

The exciton diffusion length of the host in the presence of impurities also changes and its value decreases. Similar diffusion length behavior can be observed in the process of exciton interaction with trapped charge carriers (Kalinowski and Godlewski, 1973, 1978). The energy transport process from host molecule to doping centers is used to alter color emitted in the luminescence or electroluminescence of organic materials. Exciton lifetime and diffusion length reduction can be observed experimentally. The energy transfer can be observed in the classical experiment with tetracene doped anthracene crystal (Braun *et al.*, 1982). When the density of tetracene molecules increases, the luminescence from mainly tetracene is observed. This observation supports the hypothesis that the energy transfer from anthrecene to tetracene molecules is complete.

8.5 Electroluminescence in organic materials

A. Bernanose and colleagues first observed electroluminescence in molecular materials over 50 years ago (Bernanose, 1953). The mechanism for electroluminescence was connected with molecule excitation due to charge carriers accelerated in a high electric field. A similar mechanism leading to electroluminescence in a system consisting of two electrodes attached to an anthracene crystal, was observed experimentally by M. Pope and co-workers (Pope, 1963). In this case, the electroluminescence was produced through injection of charge carriers from the point electrode made from silver paste, into the anthracene crystal. The creation of excited molecular states in the molecular solid by a high electric field is a poorly controlled process often leading to material destruction due to electrical breakdown. The results of the study on electroluminescence phenomena described above, and other publications, showed another way to obtain electroluminescence in organic materials. The first paper described the possibility of attaining electroluminescence using the recombination of charge carriers with opposite signs which were injected into the system, and was published by W. Helfrich and colleagues in 1965 (Helfrich, 1965). Nowadays this is the most popular way to obtain electroluminescence phenomena in organic materials (Scott, 2000; Forrest, 2003; Shinar, 2003; Kalinowski, 2004; Crawford, 2005; Kafafi, 2005; Li and Meng, 2006; Müllen and Scherf, 2006).

Figure 8.10 represents the classic 'sandwich' system to attain electroluminescence phenomena in organic materials This system consists of the organic material layer with two electrodes, one of which should be semitransparent.

High-yield electroluminescence is possible only when a few physical conditions are met. The most important is the efficient injection of charge carriers into the organic material. In practice this means that one electrode should inject holes, and the other should inject electrons. Electroluminescence in this case is only possible under the proper polarization of the

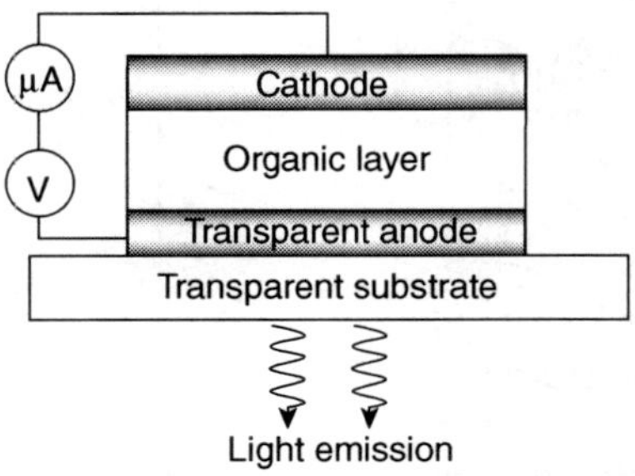

8.10 Schematic diagram of one layer organic light-emitting diode (OLED) device structure (Godlewski and Obarowska, 2007).

electrodes, thus the system is frequently called an electroluminescence diode, or an organic light-emitting diode (OLED). Figure 8.11 presents a schematic of energetic levels in a one-layer OLED.

Efficient charge carriers injection from metallic electrodes is possible when the work function for holes (Φ_p) and electrons (Φ_e) is relatively small. Charge carriers injected into the organic material may recombine with each other producing electroluminescence radiation. This type of recombination process, denoted in Fig. 8.11 as '1, 2 and 3,' is known as bimolecular recombination. The bimolecular recombination rate in a unit of volume and time can be represented in the formula described below (Kao and Hwang, 1981):

$$R_1(x) = \gamma_1 n_f(x) \cdot p_f(x) \qquad [8.25]$$

where γ_1 is the bimolecular recombination rate constant and $n_f(x)$ ($p_f(x)$) is free electrons (holes) concentration.

Recombination of free charge carriers represents only a fraction of all the recombination processes. This is because free charge carrier concentration is much lower than the trapped charge carrier concentration. Thus bimolecular recombination can largely be seen as the recombination of the free and trapped charge carriers. Where the free electrons recombine with trapped holes with concentration of $p_t(x)$ (process 2 in Fig. 8.11), the recombination rate in the volume and time unit can be represented as:

$$R_2(x) = \gamma_2 n_f(x) \cdot p_t(x) \qquad [8.26]$$

where γ_2 is the bimolecular recombination (between free electron and trapped hole) rate constant.

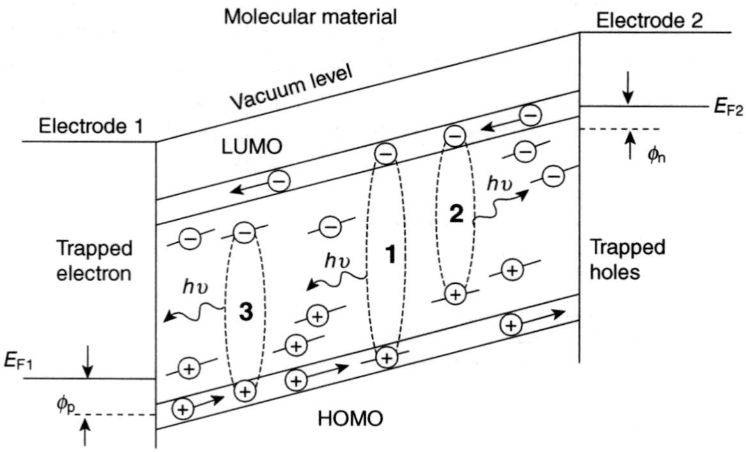

8.11 Schematic representation of energy levels for charge carriers in one-layer electroluminescence diode. The processes and symbols are described in the text.

When the recombination takes place between free holes and trapped electrons with a density of $n_t(x)$ (process 3 in Fig. 8.11), the recombination rate constant is given:

$$R_3(x) = \gamma_3 n_t(x) \cdot p_f(x) \quad [8.27]$$

The total recombination rate constant in molecular material results from free and trapped charge carriers of both signs decaying. The concentration of recombining states in a unit of time can be written:

$$R(x) = R_1(x) + R_2(x) + R_3(x) \quad [8.28]$$

The above formula represents the total rate by which excited states are created in the unit of volume, due to the recombination of charge carriers. The concentration of free and trapped charge carriers depends on injection efficiency and transport properties in molecular material. The transport and recombination processes for charge carriers of both signs in a one-layer OLED will be presented later.

8.5.1 Charge carrier injection and recombination

Figure 8.11 gives a schematic representation of the charge carrier current in a one-layer electroluminescence diode. Due to the charge carrier injection, inside the active layer of the diode the electron current $j_n(x)$ and the hole current $j_p(x)$ are created. A qualitative representation of their characteristics is given in Fig. 8.12.

A description of transport and charge carrier recombination process can be derived from charge carrier continuity equations. The current density for electrons and holes can be connected to the recombination process according to (Kao and Hwang, 1981; Godlewski *et al.*, 1994):

$$-\frac{1}{e}\frac{\mathrm{d}j_n(x)}{\mathrm{d}x} = R_1 + R_2 \quad [8.29]$$

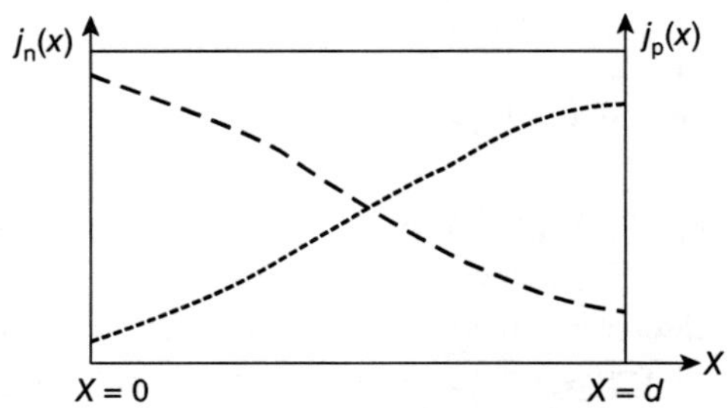

8.12 Schematic representation of electron (j_n) (dashed line) and hole (j_p) (dotted line) currents in a one-layer electroluminescence diode with thickness *d*.

and

$$\frac{1}{e}\frac{\mathrm{d}j_{\mathrm{p}}(x)}{\mathrm{d}x} = R_1 + R_3 \quad [8.30]$$

The total current density for charge carriers of both signs is the sum of the electron current and the hole current:

$$j = j_{\mathrm{n}}(x) + j_{\mathrm{p}}(x) \quad [8.31]$$

where

$$j_{\mathrm{n}}(x) = n_{\mathrm{f}}(x)e\mu_{\mathrm{n}}F(x) \quad [8.32]$$

and

$$j_{\mathrm{p}}(x) = p_{\mathrm{f}}(x)e\mu_{\mathrm{p}}F(x) \quad [8.33]$$

In the above equations μ_{n} and μ_{p} means the electron and hole motilities respectively, $F(x)$ is electric field.

Usually in organic materials under the typical transport conditions, the following relationship between free and trapped charge carriers applies: $n_{\mathrm{f}}(x) << n_{\mathrm{t}}(x)$ and $p_{\mathrm{f}}(x) << p_{\mathrm{t}}(x)$ thus the electric field can be described as follows (Kao and Hwang, 1981):

$$\frac{\mathrm{d}F(x)}{\mathrm{d}x} = \frac{e}{\varepsilon\varepsilon_0}[p_{\mathrm{t}}(x) - n_{\mathrm{t}}(x)] \quad [8.34]$$

To find the final equation describing the total current density as a function of applied voltage we must first establish the formula showing the relationship between free and trapped charge carriers. Generally, these expressions can be formulated as:

$$n_{\mathrm{t}}(x) = \int_0^{\infty} f_{\mathrm{n}}[E, n_{\mathrm{f}}(x), x, h_{\mathrm{n}}(E)]\mathrm{d}E \quad [8.35]$$

and

$$p_{\mathrm{t}}(x) = \int_0^{\infty} f_{\mathrm{p}}[E, p_{\mathrm{f}}(x), x, h_{\mathrm{p}}(E)]\mathrm{d}E \quad [8.36]$$

where $f_{\mathrm{n}}[E, n_{\mathrm{f}}(x), x, h_{\mathrm{n}}(E)]$ and $f_{\mathrm{p}}[E, p_{\mathrm{f}}(x), x, h_{\mathrm{p}}(E)]$ are the functions describing trap density; E denotes trap energy calculated regarding HOMO and LUMO levels; $n_{\mathrm{f}}(x)$ is free electron concentration as a function of a distance (x); $h_{\mathrm{p}}(x)$ is concentration of traps for holes; and $h_{\mathrm{n}}(x)$ is concentration of traps for electrons.

Further analysis of the current density as a function of external voltage must take the formula describing the relation between external electric field inside the sample with thickness d and applied voltage U into account. For planar sample geometry the equation can be represented as:

$$U = \int_0^d F(x)\mathrm{d}x \qquad [8.37]$$

The solution to this set of equations is difficult. The simple solution giving the relationship between current density and voltage under space charge limited current, without charge carriers trapped inside the system, has been reported (Parmenter and Ruppel, 1959):

$$j = \frac{9}{8}\varepsilon\varepsilon_0\mu_{\mathrm{eff}}\frac{U^2}{d^3} \qquad [8.38]$$

where μ_{eff} denotes effective mobility of charge carriers. The value of effective mobility is near the sum of electron and hole mobility, but depends on the recombination rate constant.

The above formula is almost identical to those obtained from Child's law for space charge limited currents of one sign (Kao and Hwang, 1981). In real molecular material, the formula is significantly complicated due to trapping sites. Finding the solution of the above set of equations allows us to describe the parameters which influence the recombination efficiency of charge carriers and hence electroluminescence intensity. Owing to the complexity of the problem, the analysis of equations describing the currents of both signs in insulators is considered for physical cases based on numerical calculations.

The bimolecular recombination rate constant is an important parameter, which decides the currents of both sign charge carriers, and electroluminescence phenomena. If this constant is relatively small, charge carriers can travel easily through the sample to the opposite electrode, avoiding recombination. In this case, only current is produced, and not electroluminescence. Later in this chapter, we will estimate the recombination rate constant.

In a low electric field the recombination rate constant can be estimated based on the Langevin theory. This states that two charge carriers with opposite signs will recombine upon reaching the distance that thermal energy kT (where k is the Boltzman constant) will be approaching.

The distance between charge carriers where recombination takes place is known as columbic radius (r_{C}) and the formula can be found from the equation:

$$\frac{e^2}{4\pi\varepsilon\varepsilon_0 r_{\mathrm{C}}} = kT \qquad [8.39]$$

where ε is dielectric permittivity of molecular material, and ε_0 is electric permittivity of vacuum.

Figure 8.13 gives the schematic representation of bimolecular recombination process according to Langevin model. The recombination process can be described as the average statistical current of holes (i) penetrating through a sphere with columbic radius:

$$S = 4\pi r_C^2 \quad [8.40]$$

Taking into account that the current density of holes flowing through the sphere S under the electric field F_C:

$$j_h = p_f(r_C)e\mu_p F(r_C) \quad [8.41]$$

where μ_p is hole mobility, $p_f(r_c)$ is hole concentration at the distance r_c from electron and $F(r_c)$ is electric field intensity at the distance r_C.

The current of holes flowing through the sphere limited by the columbic radius can be written as:

$$i = j_h S = p_f(r_C)\mu_p e^2 / \varepsilon\varepsilon_0 \quad [8.42]$$

The above expression has been found by taking the formula for electric field intensity from elementary electric charge e at a distance r_C:

$$F(r_C) = \frac{e}{4\pi\varepsilon\varepsilon_0 r_C^2} \quad [8.43]$$

The lifetime of electrons in recombination process is a function of the density of free holes. According to the formula [8.28]:

$$\tau_n = \frac{1}{\gamma_3 p_f(x)} \quad [8.44]$$

From the other side, if one electron is annihilated due to the recombination process at time τ_n, the hole current being the result of that process could be represented by:

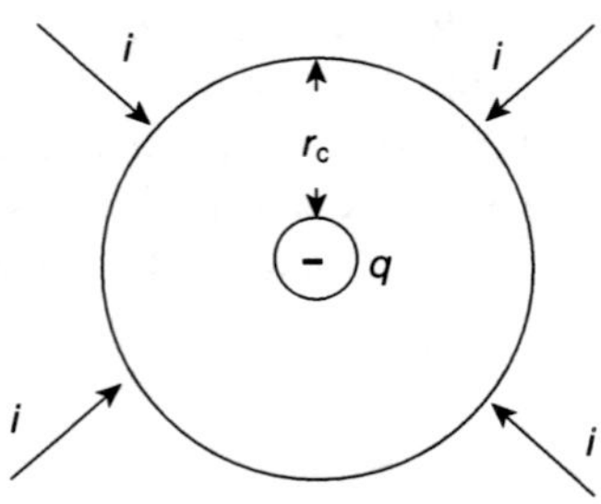

8.13 Schematic representation of a bimolecular recombination process. q denotes the charge carrier, r_C represents the columbic radius and i is the current flowing through the sphere with radius r_C.

$$i_h = \frac{e}{\tau_n} \quad [8.45]$$

Taking into account eq. [8.42], [8.44] and [8.45] we obtain the formula describing the bimolecular recombination (between trapped electrons and free holes) rate constant:

$$\gamma_3 = \frac{\mu_p e}{\varepsilon\varepsilon_0} \quad [8.46]$$

Similar analysis leading to the formula describing the bimolecular recombination (between free electrons and trapped holes) is given by:

$$\gamma_2 = \frac{\mu_e e}{\varepsilon\varepsilon_0} \quad [8.47]$$

where μ_e is electron mobility.

Based on the Langevin model, the bimolecular recombination rate constant between free electrons and free holes is given by (Pope and Swenberg, 1999):

$$\gamma_1 = \frac{\mu_{eff} e}{\varepsilon\varepsilon_0} \quad [8.48]$$

This is used to assume that the effective mobility of charge carriers is the sum of electron and hole mobility, and that the value of the bimolecular recombination rate in molecular materials is close to 10^{-12} m^3/s.

The Langevin model gives the simple description of recombination process for two charge carriers with opposite signs. More advanced models analyzing the recombination process, also take into account the diffusion of charge carriers in the external electric field (Hong and Nooland, 1978; Obarowska and Godlewski, 2000; Wojcik and Tachiya, 1998). The formula describing the bimolecular recombination rate based on them differs from those obtained from the Langevin model by about the factor 2/3 for a low electric field. For a high electric field they predict that the bimolecular recombination rate constant is a function of electric field intensity, and its value will decrease when the field increases. This behavior of the bimolecular recombination rate constant is usually the reason for the decrease in electroluminescence efficiency in a high electric field.

8.5.2 Excited states generation due to recombination of charge carriers

The bimolecular recombination process is crucial for electroluminescence processes. Excited states, created by charge carrier recombination, release energy due to radiation or radiationless processes.

The recombination of two charge carriers – electron and hole with a spin value of ½ (fermions) leads to singlet creation with probability of $p_S = 1/4$ and excited triplet states with probability of $p_T = 3/4$. According to the rules, only quants from singlet excited states are emitted. Taking this into account the maximum value for electroluminescence efficiency due to recombination process cannot exceed 25%. That value is possible only when every singlet state can emit one photon. Usually the efficiency of electroluminescence is much lower because the excited singlet states can release energy in other processes, such as intersystem crossing to a triplet state, decay into two triplet states, energy transfer to different molecules or charge carriers which stay in traps, or energy transfer to electrodes. Spin rules are a serious limitation for electroluminescence efficiency in organic materials. Therefore the achievement of efficient light emission from triplet states in electroluminescence diodes is a real challenge. One promising solution may be given in a study on electroluminescence utilizing emission from singlet and triplet states simultaneously, which gives an efficiency of up to 100% for some materials (Adachi *et al.*, 2001).

The achievement of high electroluminescence efficiency depends on many parameters. The basic condition is that the recombination process must take place inside the active layer and this condition is only fulfilled when the charge carrier concentration is high. In devices with a single layer, the maximum charge carrier concentration can be obtained in space charge limited conditions. In that case charge carriers' time of flight through the layer is longer than their lifetime. Therefore the probablility for biomolecular recombination is maximal. The creation of excited states due to recombination processes does not mean the same as the achievement of photon emission in electroluminescence. The material for the active layer in an OLED should be characterized by high quantum emission yield from singlet and triplet states if possible. In the singlet and triplet states, excitons can move through the material and transfer their energy to impurities, electrodes, charge carriers or lose energy by other processes. Material should be cleaned of impurities to avoid uncontrolled transfer. Energy transfer to impurities can be activated on purpose to achieve photon emission with an appropriate value of energy (an appropriate colour of light emission), a common technique in molecular materials.

To achieve efficient electroluminescence, devices with two or more organic material layers are used. The junction of organic materials allows the recombination process taking place in a region close to electrodes to be avoided, and thus limiting energy transfer to the electrodes. Furthermore, the high concentration of charge carriers is achieved on a potential barrier created in the junction (Tang and Vanslyke, 1987). The schematic representation of junction of two materials, ITO/NPB/Alq_3/Ca, is shown in Fig. 8.14.

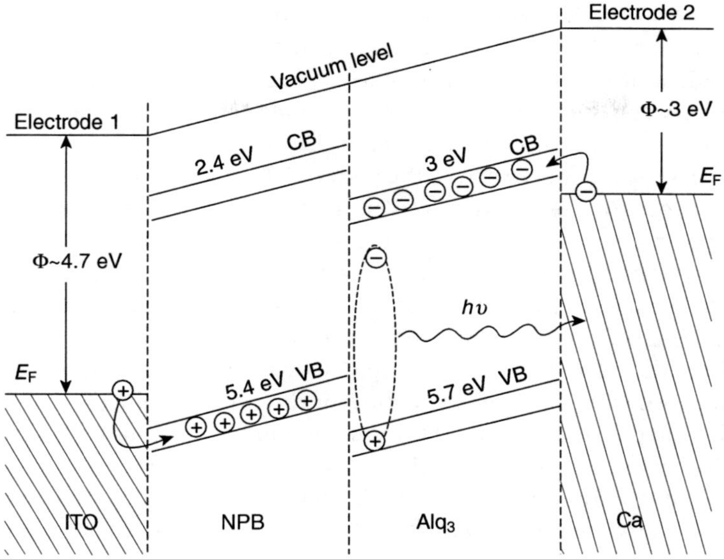

8.14 The model of a two-layer organic electroluminescence diode under forward bias. Recombination takes place in Alq_3 material near the junction of NPB with Alq_3 (Brutting *et al.*, 2001).

In this system, holes are injected from the ITO electrode into NPB layer and they next move to the junction of NPB with Alq_3. Electrons are also injected from Ca electrode and then transferred to the junction. Because the energy barrier height at the junction is higher for electrons, the recombination between electrons and holes takes place in Alq_3. Electroluminescence spectra thus reflect excited states in the material and is identical with photoluminescence spectra of Alq_3, which was shown by Brutting *et al.* (2001).

The electromagnetic wave emitted is from the visible region of the spectra. We can achieve the emission of light with any colour from organic electroluminescence diodes. This is useful in many devices as displays or sources of light (Pribat and Plais, 2001; Forrest, 2004; Rost *et al.*, 2004). OLEDs now achieve better parameters than LEDs made from inorganic materials (Kim, 2004).

8.6 Conclusion and future trends

The physiochemical features of organic materials are determined by their optical, photoluminescence and electroluminescence properties. A special attribute of the organic solid state is the low value of bonding energy between molecules. Domination of molecular properties over the solid state in organic solids is therefore observed. The creation of organic solids from individual molecules causes common energy levels to form. Weak interaction between molecules imposes a communal response upon molecular behavior in the molecular solid state. This collective response is known as exciton transfer energy across the material. Excitonic energy transfer causes many processes internally as well as at the surface. Owing to exciton decay, photon energy (usually in the visible region of the spectrum) can be emitted, although nonradioactive processes such as exciton interaction with impurities or dopants are also possible. It is worth noting that impurities usually limit energy transport, although the controlled energy transfer to impurities enables the colour of emitted light to change in organic light emitting devices, which is a particular feature of the organic solid state.

This chapter has described the connection between individual molecular excited states, and states in molecular solids. We have focused on a diffusive model of exciton migration in molecular solids, and thus the transport of excited energy in that system. It has been shown that excitons can be generated during optical absorption, as well as through bimolecular charge carrier recombination. Excitonic processes, referring to their interaction with themselves and dopings or impurities, have been demonstrated. Basic mathematical description was used to describe excitonic processes in the organic solid state. Experimental methods, allowing for the determination of basic excitonic parameters such as exciton lifetime and exciton diffusion length, have been presented. The process of bimolecular recombination was described, particularly using the Langevin model and the transport phenomenon in a one-layer electroluminescence diode.

To sum up, this chapter has presented a basic analysis of excitonic processes in organic materials. Furthermore, methods for determining exciton parameters have been shown. Optical, photoluminescent and electroluminescent properties contribute to an understanding of electronic processes taking place in molecular solids, and knowledge of them is essential for finding practical application for these types of materials.

8.7 References

Adachi C, Baldo M A, Thompson M E, Forrest S R (2001), 'Nearly 100% internal phosphorescence efficiency in a organic light-emitting devices', *J Appl Phys*, **90**, 5048.

Agranovich V (2009), *Excitations in organic solids*, Oxford University Press.

Aradhya S V, Frei M, Hybertsen M, Venkataraman L (2012), 'Van der Waals interactions at metal/organic interfaces at the single-molecule level', *Nature Materials* Advance Online Publication, Macmillan Publishers Limited.

Arden W, Kotani M, Peter L M (1976), 'Triplet exciton decay process in crystalline tetracene', *Phys Stat Sol (b)*, **75**, 621–631.

Banerjee S, Parhi A P, Iyer S S K, Kumar S (2009), 'Method of determining the exciton diffusion length using optical interference effect in Schottky diode', *Appl Phys Lett*, **94**, 223303.
Becker W (2005), *Advanced time-correlated single photon counting techniques*, Springer.
Becker W (2008), *The bh TCSPC Handbook*, 3rd ed. Becker & Hickl GmbH.
Bernanose A (1953), 'A new method of light emission by certain organic compounds', J *Chem Phys*, **50**, 64–68.
Birks J (1973), *Organic molecular photophysics*, John Wiley & Sons.
Braun A, Mayer U, Auweter H, Wolf H C, Schmid D (1982), 'Singlet-exciton energy transfer in tetracene-doped anthracene crystals as studied by time-resolved spectroscopy', *Zeitschrift Naturforschung Teil A*, **37**, 1013–1023.
Brutting W, Berleb S, Muckl A G (2001), 'Device physics of organic-light emitting diodes based on molecular materials', *Organic Electronics*, **2**, 1.
Chi X, Li D, Zhang H, Chen Y, Garcia V, Garcia C, Siegrist T (2008), '5,6,11,12-Tetrachlorotetracene, a tetracene derivative with p-stacking structure: The synthesis, crystal structure and transistor properties', *Organics Electronics*, **9**, 234–240.
Crawford G (2005), *Flexible flat panel display*, Wiley.
Davydov A S (1971), *Theory of molecular excitons*, Plenum Press.
Forrest S R (2003), 'The road to high efficiency organic light emitting devices', *Organic Electronics*, **4**, 45.
Forrest S R (2004), 'The path to ubiquitous and low-cost organic electronics appliances on plastic', *Nature*, **428**, 911.
Frenkel J (1931), 'On the transformation of light into heat in solids. II', *Phys Rev*, **37**, 1276–1294.
Godlewski J (2005), 'Current and photocurrent in organic materials determined by the interface phenomena', *Adv Coll Interface Sci*, **116**, 227.
Godlewski J, Obarowska M (2007), 'Organic light emitting devices', *Optoelectronics Rev*, **15**, 179.
Godlewski J, Signerski R, Kalinowski J, Stizza S, Berretoni M (1994), 'X-ray-excited photocurrent in anthracene crystals', *Molecular Crystals & Liquid Crystals*, **252**, 145.
Helfrich W (1965), 'Recombination radiation in anthracene crystals', *Phys Rev Lett*, **140**, 229.
Hofmann S, Rosenow T C, Gather M C, Lussem B, Leo K (2012), 'Singlet exciton diffusion length in organic light-emitting diodes', *Phys Rev B*, **85**, 245209–8.
Hong K M, Noolandi J (1978), 'Time-dependent escape rate from a potential well', *Surface Sci*, **75**, 561.
Jablonski A (1933), 'Efficiency of anti-stokes fluorescence in dyes', *Nature*, **131**, 839–840.
Kafafi Z H (2005), *Organic electroluminescence*, Taylor & Francis.
Kalinowski J (2004), *Organic light-emitting diodes*, Taylor & Francis.
Kalinowski J, Godlewski J (1973), 'Triplet-exciton charge carrier interaction in crystalline tetracene', *Phys Stat Sol (a)*, **20**, 403.
Kalinowski J, Godlewski J (1978), 'Spatial behavior of the charge carrier created at the illuminated interface molecular crystal/electrolyte', *Chem Phys*, **32**, 201.
Kao K C, Hwang W (1981), *Electrical transport in solids*, Pergamon Press.
Kearwell A, Wilkinson F (1969), *Transfer and storage of energy by molecules*, Wiley.
Keil T H (1966), 'Theory of the Urbach rule', *Phys Rev*, **144**, 582–587.

Kim W (2004), 'How to improve large sized TFT – LCDs for television', *Information Display*, **20**, 20.

Klafter J, Jortner J (1977), 'Urbach rule in the optical spectra of crystalline and amorphous organic solids', *Chem Phys*, **26**, 421–430.

Klein G, Voltz R, Schott M (1973), 'On singlet exciton fission in anthracene and tetracene at 77^0K', *Chem Phys Lett*, **19**, 391–394.

Kurrle D, Pflauma J (2008), 'Exciton diffusion length in the organic semiconductor diindenoperylene', *Appl Phys Lett*, **92**, 133306.

Kurik M V (1971), 'Urbach rule', *Phys Stat Sol (a)*, **8**, 9–45.

Lakowicz J R and Gryczynski I (1991), Frequency-domain fluorescence spectroscopy. In: *Topics in fluorescence spectroscopy*: Volume 1 *Techniques*, Plenum Press, 293–335.

Lee J, Jadhav P, Baldo M A (2009), 'High efficiency organic multilayer photodetectors based on singlet exciton fission', *Appl Phys Lett*, **95**, 033301.

Lee S S, Loo Y L (2010), 'Structural complexities in the active layers of organic electronics', *Ann Rev Chem and Biom Eng*, **1**, 59–78.

Li Z, Meng H (2006), *Organic light-emitting materials and devices*, Taylor & Francis.

Matsuse N, Ikame S, Suzuki Y, Naito H (2005), 'Charge-carrier transport and triplet exciton diffussion in a blue electro phosphorescent emitting layer', *J Appl Phys*, **97**, 123512.

Matsui A (1966), 'The polarized absorption edge and the Davydov splitting of anthracene', *J Phys Soc Jpn*, **21**, 2212–2222.

Michel-Beyerle M E, Haberkorn R (1978), 'Temperature-independent intersystem crossing in anthracene crystals', *Phys Stat Sol (b)*, **85**, 473–475.

Moore G F, Munro I H (1965), 'Delayed fluorescence from organic crystals', *Nature*, **208**, 772–773.

Mulder J B (1968), 'Diffusion and surface reaction of singlet excitons in antharcene', *Philips Res Rept Suppl*, **4**, 1.

Müllen K, Scherf U (2006), *Organic light emitting devices*, John Wiley & Sons.

Obarowska M, Godlewski J (2000), 'Electric field dependence of the bimolecular recombination rate of the charge carriers', *Synthetic Metals*, **109**, 219–222.

O'Connor D V, Phillips D (1984), *Time-correlated single photon counting*, Academic Press.

Parmenter R H, Ruppel W (1959), 'Two-carrier space-charge limited current in trap free insulator', *J Appl Phys*, **30**, 1548.

Peierls R (1932), 'On the theory of the absorption spectra of solids', *Ann Phys*, **13**, 905–952.

Pope M (1963), 'Electroluminescence in organic crystals', *J Chem Phys*, **38**, 2042.

Pope M, Swenberg Ch E (1999), *Electronic processes in organic crystals and polymers*, Oxford University Press.

Pope M, Brahms B J, Brenner H C (2003), Diffusion of excitons in systems with non-planar geometry: theory, *Chem Phys*, **288**, 105.

Pope M, Geacintov N E, Saperstein D, Vogel F (1970), 'Calculation of the diffusion length, diffusion coefficient and lifetime of triplet excitons in crystalline tetracene', *J Luminescence*, **12**, 224–230.

Pribat D, Plais F (2001), 'Matrix addressing for organic electroluminescent displays', *Thin Solid Films*, **383**, 25.

Rao A, Wilson M, Hodgkiss J, Albert-Seifried S, Bassler H, Friend R H (2010), 'Exciton fission and charge generation via triplet excitons in pentacene/C60 bilayers', *J Am Chem Soc*, **132**, 12698–12703.

Reichardt C (2003), *Solvents and solvent effects in organic chemistry*, Wiley VCH Verlag GmbH&Co.

Rost C, Karg S, Riess W, Loi M A, Murgia M, Muccini M (2004), 'Light-emitting ambipolar organic heterostructure field effect transistor', *Synthetic Metals*, **146**, 237.

Scott M (2000), 'Introduction to the physics of organic electroluminescence', *C. R. Acad. Sci.* Paris, t. 1, **4**, 381–402.

Shinar J (2003), *Organic light-emitting devices*, Springer.

Signerski R, Jarosz G (2011), 'Diffusion length of singlet excitons in copper phthalocyanine films', *Phot Lett Poland*, **3**, 107–109.

Tang C W, Vanslyke S A (1987), 'Organic electroluminescent diodes', *Appl Phys Lett*, **51**, 913.

Tanaka J (1964), 'The electronic spectra of pyrene, chrysene, azulene, coronene and tetracene crystals', *Bull Chem Soc Jpn*, **38**, 86–103.

Wannier G H (1937), 'The structure of electronic excitation levels in insulating crystals', *Phys Rev*, **52**, 191–197.

Wojcik M, Tachiya M (1998), 'Effect of an external electric field on diffusion-controlled bulk electron-ion recombination in high-mobility systems', *J Chem Phys*, **109**, 3999.

Wolf H C (1967), 'Energy transfer in organic molecular crystals: a survey of experiments', *Advances Atomic Molecular Phys*, **3**, 119–142.

Wolf H C, Haken H (1995), *Molecular physics and elements of quantum chemistry*, Springer-Verlag Telos.

9
Nonlinear optical properties of organic materials

J. PEREZ-MORENO, Skidmore College, USA and Washington State University, USA

DOI: 10.1533/9780857098764.2.274

Abstract: In this chapter, some of the basic concepts that are needed to understand the nonlinear optical properties of organic materials are introduced. The definitions, conventions and units that allow for a quantitative description of the nonlinear optical response at the molecular level are first discussed. The general approaches that are used to determine the macroscopic response of nonlinear optical materials in terms of the intrinsic response of the molecular units are illustrated by considering the second-order nonlinear response of organic crystals and poled-doped polymers. Finally, an overview of the quantum mechanical expressions for the molecular linear and nonlinear molecular (hyper) polarizabilities are introduced, after the qualitative response of a simpler damped driven harmonic oscillator is derived.

Key words: nonlinear optics (NLO), organic nonlinear optics, hyperpolarizabilities, poled-doped polymers, molecular photonics.

9.1 Introduction

Since the advent of conducting polymers in the late 1970s, significant efforts have been directed towards the generation of new organic-based materials that can outperform the response of materials based in inorganic semiconductors. At present, materials based in organics can be used for most optoelectronic applications, but in general, not as efficiently as inorganic semiconductor-based materials.

Polymer-based materials have been shown to be more efficient than materials based in semiconductors for the fabrication of electro-optic modulators (Dalton, 2001, 2002; Shi *et al.*, 2001). The electro-optic effect is a second-order nonlinear optical process that enables the modulation of the amplitude of an optical beam (with frequencies on the order of 10^{14} Hz) with an external alternating current (ac) electric field (in the frequencies range of 1 to 10^{11} Hz). Effectively, electro-optic modulators allow for the conversion of electronic output (such as the one generated by a computer) into a modulated optical beam that can then travel through an optical fiber and carry the encoded information to the desired location.

More efficient electro-optic modulators are generated when the nonlinear optical response occurs on the fastest time scale possible. This is achieved when the response is purely electronic (the time response of non-resonant electronic processes being in the order of 10^{-15} s). The fastest response is immediately realized in polymer-based materials, where the nonlinear optical effects are due to the changes on the (pi-) delocalized electronic cloud at the molecular levels. Also, in organic materials the nonlinear response of the bulk can be related to the microscopic molecular response by simple laws of addition. Therefore, in order to generate an optimal nonlinear optical material based in organics, one can focus on the optimization of the nonlinear molecular response first, and then use the optimized structures as the active building blocks to tailor the response of the bulk polymer-based material.

In addition to these advantages, organic-based materials are also easier and cheaper to produce, more versatile and more environmentally friendly than the materials based in inorganics. Furthermore, and since the organic response is directly related to the response of the molecules at the microscopic level, the biggest advantage of organic materials is the virtually infinite number of organic molecules that can be synthesized (Ertl, 2003; Lipinski and Hopkins, 2004), such as that a wealth of potential strategies for the tailoring and optimization of the material response are available.

Finally, the study and understanding of the linear and nonlinear optical response at the molecular level has lead to the development of non-invasive microscopic techniques (two-photon fluorescence, second- and third-harmonic imaging . . .) which are of great importance for medical and biological applications (Xu *et al.*, 1996; Squier and Müller, 2001; Helmchen and Denk, 2005; Salafsky, 2007).

9.2 Nonlinear optics (NLO) at the molecular level

9.2.1 Nonlinear optics (NLO)

Optics is the branch of science that studies the fundamental interactions between matter and light. The characterization and understanding of the optical properties of matter (that is, the properties that change by the interaction with light) are necessary in order to manipulate the optical properties of materials to our advantage.

Nonlinear optics (NLO) is the branch of optics that studies the interaction of matter and light in the regime where the response is not directly proportional to the strength of the electric field associated with the light source. Most of our daily life experience with optical phenomena occurs in the linear regime, and it is not until the invention of the laser – a source of high intensity monochromatic light (Maiman, 1960) – that the first generation of second-harmonic was reported (Franken *et al.*, 1962). In the experiment, the second-harmonic was generated after focusing a ruby laser (with a wavelength of 694 nm) into a sample made of quartz, and then, the output spectrum was recorded on a photographic paper with the aid of a spectrometer, which indicated that light at the second-harmonic frequency (wavelength of 347 nm) had been generated. Notoriously, the copy editor mistakenly took the dim spot at 347 nm on the photographic paper by a dust mark, and removed it from the final publication.

9.2.2 The electric-dipole approximation

The optical properties of a material are determined by the reaction of the material charges to the electromagnetic fields associated with the applied light beam(s). When an oscillating electric field is applied to the material, the charges accelerate in reaction to the fields. As the charges accelerate, radiation (electromagnetic waves) is produced. The resulting radiation field is made up from the contributions of all the accelerating charges at the microscopic level.

It can be shown that for most microscopic charge distributions, the first contribution to the radiation field is generated by the oscillating total dipole moment of the distribution (Griffiths, 1999). That is, if the charge distribution at the microscopic level gets an induced net oscillating dipole moment, the radiation field is then dominated by the dipolar contribution, provided that the following conditions apply:

- The typical size of the charge distribution is much smaller than the distance from the observer.
- The typical size of the charge distribution is much smaller than the wavelength associated with the incident beam(s), such as, at any given time, the amplitude of the associated electric field(s) is constant through the whole charge distribution.
- The radiating field is measured at distances that are much larger than the wavelengths associated with the fields (which allows the multipole expansion of the field to be truncated to the first non-zero contribution).

9.2.3 Mathematical description of the nonlinear response

Unless we indicate explicitly otherwise, we will be considering non-magnetic materials, within the electric-dipole approximation.

The linear regime

In the linear regime, the polarization $\vec{P}$ (defined as the electric dipole moment per unit of volume) is directly proportional to the amplitude of the applied electric field, $\vec{E}$:

$$\vec{P}(t) = \chi^{(1)} \cdot \vec{E}(t), \quad [9.1]$$

where the constant of proportionality, $\chi^{(1)}$, is defined as the linear susceptibility. Strictly speaking, the linear susceptibility is a two-rank tensor since it relates the components of two vectors:

$$P_i(t) = \chi_{ij}^{(1)} \cdot E_j(t), \quad [9.2]$$

where Einstein's notation convention is used, indicating an implicit sum over any repeated index.

The nonlinear regime

In the nonlinear regime, the induced polarization is not directly proportional to the amplitude of the applied electric field. Instead, the induced polarization of the medium is expressed as a series in the electric field:

$$P_i(t) = \chi_{ij}^{(1)} \cdot E_j(t) + \chi_{ijk}^{(2)} \cdot E_j(t) \cdot E_k(t) + \chi_{ijkl}^{(3)} \cdot E_j(t) \cdot E_k(t) \cdot E_l(t) + \ldots, \quad [9.3]$$

In practice, it is more convenient to express both the electric field and the induced polarization in the frequency domain. Assuming that the applied and radiated light is monochromatic, the electric field can be expressed as the sum over discrete frequencies, ω_k:

$$E_j(t) = \frac{1}{2} \sum_{\omega_k \geq 0}^{\omega_{max}} (E_j^{\omega_k} \cdot e^{-i\omega_k t} + c.c.), \quad [9.4]$$

where $E_j^{\omega_k}$ is the amplitude of jth component of the electric field oscillating at frequency ω_k, which can be complex and where (c.c.) means complex conjugated.

Expressing the induced polarization also as the sum of discrete frequencies:

$$P_j(t) = \frac{1}{2} \sum_{\omega'_k \geq 0}^{\omega'_{max}} (P_j^{\omega'_k} \cdot e^{-i\omega'_k t} + c.c.), \quad [9.5]$$

the relationship between the electric field and the induced polarization can be expressed in terms of the frequency amplitudes (Shi and Garito, 1998):

$$P_i^{\omega_\sigma} = K(\omega_1) \cdot \chi_{ij}^{(1)}(-\omega_\sigma; \omega_1) \cdot E_j^{\omega_1} + K(\omega_1, \omega_2) \cdot \chi_{ijk}^{(2)}(-\omega_\sigma; \omega_1, \omega_2) \cdot E_j^{\omega_1} \cdot E_k^{\omega_2}$$
$$+ K(\omega_1, \omega_2, \omega_3) \cdot \chi_{ijkl}^{(3)}(-\omega_\sigma; \omega_1, \omega_2, \omega_3) \cdot E_j^{\omega_1} \cdot E_k^{\omega_2} \cdot E_l^{\omega_3} + \ldots, \quad [9.6]$$

where, $\chi_{ij}^{(1)}(-\omega_\sigma; \omega_1)$ are the tensor components of the linear electric susceptibility, $\chi_{ijk}^{(2)}(-\omega_\sigma; \omega_1, \omega_2)$ are the tensor components of the second-order nonlinear susceptibility, $\chi_{ijkl}^{(3)}(-\omega_\sigma; \omega_1, \omega_2, \omega_3)$ are the tensor components of the third-order nonlinear susceptibility, and so on. The numerical factors,

$K(\omega_1), K(\omega_1, \omega_2), K(\omega_1, \omega_2, \omega_3)$ etc ... are determined by the ½ factors in the Fourier expansions for the fields (Eqs. 9.4 and 9.5), and are given by:

$$K(\omega_1, \omega_2, \omega_3, \ldots, \omega_n) = \frac{D}{2^{q-p}}, \quad [9.7]$$

where q is the number of non-zero input frequencies in the set; p is equal to 0 when the output frequency is 0, and 1 otherwise; and D is the number of different distinguishable orderings of the input frequencies.

The nth-order nonlinear susceptibility is a property of the material that tells us how 'easy' is to polarize the material through a nth-order nonlinear interaction with the electric fields. Conservation of energy implies that at any order, the input frequencies must add to match the output frequency:

$$\chi^{(n)}_{ijkl\ldots n}(-\omega_\sigma; \omega_1, \omega_2, \omega_3, \ldots, \omega_n) = 0 \quad \text{unless } \omega_1 + \omega_2 + \omega_3 + \ldots + \omega_n = \omega_\sigma. \quad [9.8]$$

Also, from the reality of the fields it follows that:

$$\chi^{(n)}_{ijkl\ldots n}(-\omega_\sigma; \omega_1, \omega_2, \omega_3, \ldots, \omega_n) = \chi^{(n)*}_{ijkl\ldots n}(\omega_\sigma; -\omega_1, -\omega_2, -\omega_3, \ldots, -\omega_n), \quad [9.9]$$

where (*) indicates the complex conjugated. Finally, and since the ordering in which one arranges the amplitudes of the electric fields in Eq. 9.6 should not affect the physical response, the nonlinear susceptibilities must always obey intrinsic permutation symmetry, that is the nonlinear susceptibility is unchanged if we simultaneously permute two indexes q and r; and the frequencies associated with those indexes:

$$\chi^{(n)}_{ij\ldots q\ldots r\ldots}(-\omega_\sigma; \omega_1, \ldots, \omega_q, \ldots, \omega_r, \ldots) = \chi^{(n)}_{ij\ldots r\ldots q\ldots}(-\omega_\sigma; \omega_1, \ldots, \omega_r, \ldots, \omega_q, \ldots). \quad [9.10]$$

9.2.4 The molecular nonlinear susceptibilities

In organic systems the optical nonlinearities of the medium are determined by the nonlinear properties of the molecular/monomer units which in the absence of net charge or intermolecular charge transfer interact weakly with each other. This regime is known as the weakly oriented gas model (Prasad and Williams, 1991).

Hence, in the same manner as the macroscopic nonlinear polarization can be expressed in terms of the components of the electric field (Eqs. 9.3 and 9.6), the induced dipole molecule of a molecule, $\vec{p}$, can be expanded as a series on the electric field. However, at the molecular level, the applied electric field is usually modified by the presence of dielectric medium surrounding the molecule. Different models can be used to relate the local field to the external applied field $\vec{E}(t)$ (Kuzyk, 1998). In general, if the external applied field is monochromatic (Eq. 9.4) the local field is also monochromatic:

$$f_j(t) = \frac{1}{2}\sum_{\omega_k \geq 0}^{\omega_{max}} (f_j^{\omega_k} \cdot e^{-i\omega_k t} + c.c.), \quad [9.11]$$

Such as, for every frequency, the frequency amplitude of the local field f_i^{ω} and the frequency amplitude of the applied electric field E_i^{ω} are related through the local field correction, $f(\omega)$:

$$f_i^{\omega} = f(\omega) \cdot E_i^{\omega}. \quad [9.12]$$

If the local field is monochromatic, the induced dipole moment on the molecule, $\vec{p}$, must also be monochromatic:

$$p_i(t) = \frac{1}{2}\sum_{\omega_\sigma \geq 0}^{\omega_{max}} (f_j^{\omega_\sigma} \cdot e^{-i\omega_\sigma t} + c.c.), \quad [9.13]$$

Such as we can define the molecular polarizabilities in the frequency domain through an expression similar to Eq. 9.6:

$$p_i^{\omega_k} = K(\omega_1) \cdot \alpha_{ij}^{(1)}(-\omega_k'; \omega_1) \cdot f_j^{\omega_1} + K(\omega_1, \omega_2) \cdot \beta_{ijk}^{(2)}(-\omega_k'; \omega_1, \omega_2) \cdot f_j^{\omega_1} \cdot f_k^{\omega_2} + K(\omega_1, \omega_2, \omega_3) \cdot \gamma_{ijkl}^{(3)}(-\omega_k'; \omega_1, \omega_2, \omega_3) \cdot f_j^{\omega_1} \cdot f_k^{\omega_2} \cdot f_l^{\omega_3} + \ldots, \quad [9.14]$$

where $\alpha_{ij}^{(1)}(-\omega_k'; \omega_1)$ is the molecular linear polarizability, $\beta_{ijk}^{(2)}(-\omega_k'; \omega_1, \omega_2)$ is the molecular second-order nonlinear susceptibility (also known as *the first hyperpolarizability*), $\gamma_{ijkl}^{(3)}(-\omega_k'; \omega_1, \omega_2, \omega_3)$ is the third-order nonlinear susceptibility (also known as *the second hyperpolarizability*), and so on. The first and second hyperpolarizabilities are a property of the molecule that tells us how 'easy' is to induce a dipole moment in the molecule through second- and third-order nonlinear interaction with the electric fields. The same generic relationships that apply to the macroscopic nonlinear susceptibilities (Eqs. 9.8, 9.9 and 9.10) are obeyed by the nonlinear hyperpolarizabilities.

9.2.5 Units

Notice that we have used cgs-gaussian units in our definition of the nonlinear macroscopic and microscopic susceptibilities (Eqs. 9.6 and 9.14), and will consistently use this set of units through the rest of the chapter. In the cgs-gaussian system, the polarization $\vec{P}$ and the electric field strength $\vec{E}$, have the same units (*statV/cm*), which simplifies the definition of the macroscopic nonlinear susceptibilities as well as the molecular polarizabilities. In the cgs-gaussian system, the second-order and third-order susceptibilities have units of *cm/statV* and $cm^2/statV^2$ respectively, while the linear susceptibility is dimensionless. Consequently, and since to convert from molecular polarizabilities to macroscopic susceptibilities it is necessary to multiply by the number density, N_d, which has units of cm^{-3}, the first hyperpolarizability has units of $cm^4/statV$, the second hyperpolarizability has units of $cm^5/statV^2$; and the molecular linear polarizability has units of cm^3.

In contrast, in SI units, the polarization has units of C/m^2 while the field strength has units of V/m, and there are two possible sets of SI units available, depending on which convention is used in the definition of the nonlinear susceptibilities. If we use the first convention (Convention I) the nonlinear susceptibilities are defined through:

$$P_i^{\omega_k'} = K(\omega_1)\cdot \varepsilon_0 \cdot \chi_{ij}^{(1)}(-\omega_k';\omega_1)\cdot E_j^{\omega_1} + K(\omega_1,\omega_2)\cdot \varepsilon_0 \cdot \chi_{ijk}^{(2)}(-\omega_k';\omega_1,\omega_2)\cdot E_j^{\omega_1}\cdot E_k^{\omega_2} + K(\omega_1,\omega_2,\omega_3)\cdot \varepsilon_0 \cdot \chi_{ijkl}^{(3)}(-\omega_k';\omega_1,\omega_2,\omega_3)\cdot E_j^{\omega_1}\cdot E_k^{\omega_2}\cdot E_l^{\omega_3} + \ldots, \quad [9.15]$$

where ε_0 is the vacuum permittivity with units of F/m. In Convention I, the second-order and third-order susceptibilities have units of m/V and m^2/V^2 respectively, and the linear susceptibility is dimensionless. Consequently, the first hyperpolarizability has units of m^4/V, the second hyperpolarizability has units of m^5/V^2; and the molecular linear polarizability has units of m^3.

Finally, if we use Convention II, the nonlinear susceptibilities are defined through:

$$P_i^{\omega_k'} = K(\omega_1)\cdot \varepsilon_0 \cdot \chi_{ij}^{(1)}(-\omega_k';\omega_1)\cdot E_j^{\omega_1} + K(\omega_1,\omega_2)\cdot \chi_{ijk}^{(2)}(-\omega_k';\omega_1,\omega_2)\cdot E_j^{\omega_1}\cdot E_k^{\omega_2} + K(\omega_1,\omega_2,\omega_3)\cdot \chi_{ijkl}^{(3)}(-\omega_k';\omega_1,\omega_2,\omega_3)\cdot E_j^{\omega_1}\cdot E_k^{\omega_2}\cdot E_l^{\omega_3} + \ldots, \quad [9.16]$$

such as the second-order and third-order susceptibilities have units of C^3/J^2 and $C^4 .m/J^3$ respectively, while the linear susceptibility has units of $C^2/m.\ J$. Then, the first hyperpolarizability has units of $C^3 .m^3/J$, the second hyperpolarizability has units of $C^4 .m^4/J^3$; and the molecular linear polarizability has units of $C.\ m$. A table with conversion units between the different SI conventions and cgs-gaussian units (also known as esu units) as well as a discussion on the different notations and conventions in NLO is can be found in Shi and Garito (1998).

Finally, most numeric calculations of the molecular polarizabilities use atomic units (a.u.). A reported *1 au* unit for the linear polarizability (α) converts to 0.014818 nm; a reported value of *1 au* for the first hyperpolarizability (β) converts to 8.641×10^{-33} $cm^4/statV$; and a reported value of *1 au* for the second hyperpolarizability (γ) converts to 5.0367×10^{-40} $cm^5/statV^2$.

9.2.6 Even-order and centrosymmetry

A system has a center of symmetry when the radiating charges are symmetrically arranged around the center of symmetry. Mathematically, this

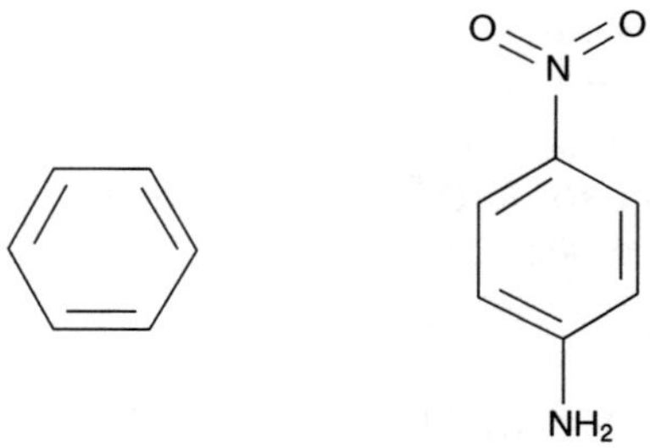

9.1 Schematic representation of the benzene ring (left), a perfect example of a conjugated system where the pi-delocalized electrons are shared equally by the six carbon atoms. However, since a pure benzene ring has a center of symmetry (at the center of the ring), the structure must be modified in order to be useful for second-order NLO applications. The centrosymmetry is traditionally broken by attaching an electron-donor species on one side and an acceptor in the other, such as in 4-nitroaniline (right), where the NH_2 group acts as a donor of electron, and NO_2 acts an electron acceptor.

means that there exists a system of Cartesian coordinates such as the distribution of charges is left invariant under the transformation $(x, y, x) \rightarrow (-x, -y, -z)$. A typical example of a molecule with a center of symmetry is the benzene ring (Fig. 9.1).

It can be shown that in order to generate even-order nonlinear response, a system must lack a center of symmetry (Boyd, 2003; Nalwa and Miyata, 1997). Hence, only molecules that lack a center of symmetry can be used to generate second-order nonlinear response. Furthermore, the molecules must also be arranged at the bulk level in such a manner that the bulk material also lacks a center of symmetry.

9.3 From microscopic (molecules) to macroscopic (materials)

As previously discussed, using the *oriented gas* model, organic materials can be considered to be made up of molecular units that interact weakly in the absence of net charge or intermolecular charge transfer. Within this model, the linear and nonlinear optical properties of the organic medium at the macroscopic level are determined principally by the linear and nonlinear optical properties of the molecular units. At the molecular level, the optical response is generated by the motion of the pi-delocalized electrons, typical of unsaturated organic compounds, which form orbitals that can extend over long distances. At the macroscopic level, the nonlinear response is related to the molecular response through addition rules. The oriented gas model has been successfully used for selective development of optimized second-order nonlinear optical crystals, when combined with experimental characterization of the first hyperpolarizability (Zyss and Oudar, 1982; Zyss *et al.*, 1993), or to determine experimentally the first hyperpolarizabilities of aromatic molecules (Oudar and Le Person, 1975), to cite just a few.

Here, for clarity we will focus on the second-order NLO response, but the same principles apply in the third- and higher-order regime (Kuzyk, 1998).

9.3.1 Second-order NLO organic crystals

The second-order nonlinear susceptibility, $\chi^{(2)}_{IJK}(-\omega_\sigma;\omega_1,\omega_2)$, it is related to the microscopic first hyperpolarizability through (Zyss and Oudar, 1982):

$$\chi^{(2)}_{IJK}(-\omega_\sigma;\omega_1,\omega_2) = N_{\mathrm{d}}\cdot f_I(\omega_\sigma)\cdot f_J(\omega_1)\cdot f_K(\omega_2)\cdot b_{IJK}(-\omega_\sigma;\omega_1,\omega_2), \qquad [9.17]$$

where N_{d} is the number of molecules per unit of volume in the crystal; $f_I(\omega)$ is the local field correction for the component I at frequency ω; and $b_{IJK}(-\omega_\sigma;\omega_1,\omega_2)$ is the orientational average over all possible configurations of the molecule in the unit cell:

$$b_{IJK}(-\omega_\sigma;\omega_1,\omega_2) = \frac{1}{N_g}\sum_{ijk}\left(\sum_{s=1}^{N_g}\cos\theta^{(s)}_{Ii}\cdot\cos\theta^{(s)}_{Jj}\cdot\cos\theta^{(s)}_{Kk}\right)\beta^{(2)}_{ijk}(-\omega_\sigma;\omega_1,\omega_2), \qquad [9.18]$$

with N_g being the number of equivalent sites, s, in the unit cell; such as $N_{\mathrm{d}} = N_g/V$, where V is the volume of the unit cell. In partially ordered materials, such as poled polymers of Langmuir–Blodget films where there is no unit cell, $N_g = 1$, such as N_{d} is directly the concentration of the active chromophore that is responsible for the nonlinear optical response.

In the case of crystals where the positions within the unit cell are fixed, the orientational average is simply the average of contributions from all the possible configurations of the molecule within the unit cell. Therefore, $b_{IJK}(-\omega_\sigma;\omega_1,\omega_2)$ is the tensor that transforms the contribution of a single molecule with first hyperpolarizability $\beta^{(2)}_{ijk}(-\omega_\sigma;\omega_1,\omega_2)$ into the macroscopic nonlinear susceptibility $\chi^{(2)}_{IJK}(-\omega_\sigma;\omega_1,\omega_2)$. Notice that we have differentiated between the molecular frame of reference $\{ijk\}$, and the crystal frame of reference $\{IJK\}$ (see Fig. 9.2). The angle $\theta^{(s)}_{Ii}$ is defined as the angle between the crystallographic axes I, and the molecular axis i. The scalar product between the corresponding unit vectors results in $\cos\theta^{(s)}_{Ii}$ and give the projection of the molecular coefficients $\beta^{(2)}_{ijk}(-\omega_\sigma;\omega_1,\omega_2)$ onto the crystal system of reference.

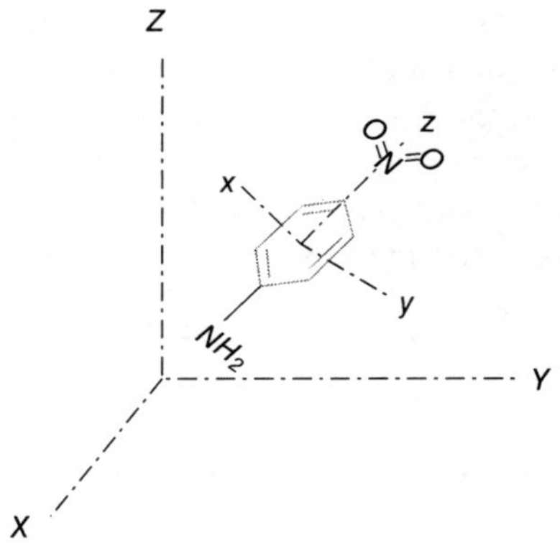

9.2 A schematic representation of the two different systems of coordinates that one must use to correlate the NLO response of the molecule with the macroscopic NLO response of the crystal or doped polymer. At the microscopic level, one chooses a molecular frame of reference that is adapted to the symmetry properties of the structure at the molecular level, {*ijk*}. For example, in the case of 4-nitroaniline, the molecular *z* axis is chosen along the direction of the dipole of the molecule. At the macroscopic level, one must use the crystal frame of reference {*IJK*}. The conversion factors are dependent on the angles between the unit vectors of each system of reference.

9.3.2 Poled-doped polymers

A different approach to the generation of a second-order nonlinear optical with active nonlinear chromophores is to produce a poled-doped polymer. A polymer host containing guest dipolar structures can have even-order response when the dipoles (the radiating structures within the polymer) are oriented to generate a noncentrosymmetric ordering at the macroscopic scale.

In the poled polymer the dipolar chromophores are guests inside of the rigid matrix provided by the polymer host. Initially, the dipoles are randomly oriented, but then the temperature of the host polymer film is raised near the glass transition temperature, T_g, while an static electric field is applied to orient the molecules. The resulting net orientation of the dipoles is achieved from a balance between the force that tends to align the dipoles in the direction of the field and the thermal fluctuations of the polymer. Finally, the polymer is cooled slowly while the aligning field is still applied, such as the net orientation is preserved in the final rigid state of the poled doped polymer.

The poling process induces a polar axis in the polymer film, such as the dipoles are distributed cylindrically around the Z-axis (defined by the direction of the applied aligning electric field). There exists an infinite-fold rotation symmetry around the Z-axis, with infinite mirror planes ($C_{\infty v}$ or ∞mn symmetry). In principle, for each molecule, the angle between the dipole moment of the molecule and the polar axis, Z might be slightly different, and as a consequence, the macroscopic response is proportional to the average angle between the molecular axis (along the direction of the dipole moment of the molecule) and the macroscopic polar axis (Z):

$$\chi^{(2)}_{IJK}(-\omega_\sigma;\omega_1,\omega_2) = N_\mathrm{d}\cdot f_I(\omega_\sigma)\cdot f_J(\omega_1)\cdot f_K(\omega_2)\cdot\langle b_{IJK}(-\omega_\sigma;\omega_1,\omega_2)\rangle, \qquad [9.19]$$

with

$$\langle b_{IJK}(-\omega_\sigma;\omega_1,\omega_2)\rangle = \sum_{ijk}\langle\cos\theta_{Ii}\cdot\cos\theta_{Jj}\cdot\cos\theta_{Kk}\rangle\beta^{(2)}_{ijk}(-\omega_\sigma;\omega_1,\omega_2), \qquad [9.20]$$

where the brackets $\langle\ldots\rangle$ denote the average over the angular distribution.

With polar symmetry, only the projections along the Z-axis do not average to zero, and there are only two independent non-zero components of the second-order susceptibility tensor, $\chi^{(2)}_{ZXX}$ and $\chi^{(2)}_{ZZZ}$. Since the X- and Y- axes are equivalent, it is convenient to choose the geometry in such a way that the X-axis is coincident with the component of the incident light beam into the plane perpendicular to the Z-axis (the polar axis), such as the X-axis corresponds with the p-component of the beam. With this choice of geometry, the two independent non-zero components of $\langle b_{IJK}(-\omega_\sigma;\omega_1,\omega_2)\rangle$ are given by:

$$\langle b_{ZZZ}\rangle = \langle\cos^3\theta_{Zz}\rangle\cdot\beta_z, \qquad [9.21]$$

and

$$\langle b_{ZXX}\rangle = \langle\cos\theta_{Zz}\cdot\sin^2\theta_{Zz}\cdot\sin^2\delta\rangle\cdot\beta_z, \qquad [9.22]$$

with

$$\beta_z = (\beta_{zxx}+\beta_{zyy}+\beta_{zzz}) \approx \beta_{zzz}. \qquad [9.23]$$

The molecular coordinates have been chosen such as the z-axis coincides with the direction of the dipole moment of the structure, and δ is the angle between the X-axis and the projection of the dipole moment into the XY-plane (see Fig. 9.3). For simplicity, the frequency dependence of the b_{IJK} tensor and of the first hyperpolarizability tensor components has been omitted.

The angular averages are calculated by using a Boltzmann distribution function:

$$F(\theta) = \mathrm{e}^{-\frac{U(\theta)}{k_\mathrm{B}T}}, \qquad [9.24]$$

where k_B is the constant of Boltzmann, T is the temperature (in K) and $U(\theta)$ is the potential energy of the molecule in the medium, such as the

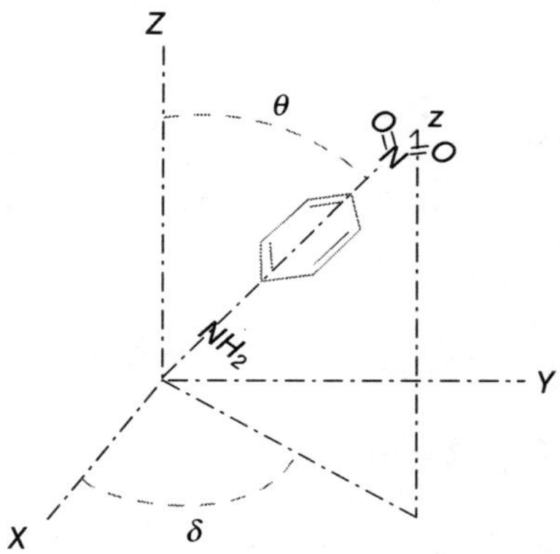

9.3 A schematic representation of the choice of Cartesian coordinates {*XYZ*} with respect to the molecular principal axis. The molecular coordinates have been chosen such as the dipole moment of the structure coincides with the *z*-axis. The Cartesian coordinates have been chosen such as *Z* is the polar axis, and *X* coincides with the p-polarization component of the beam. With this geometry, the two independent components of the macroscopic second-order NLO susceptibility are functions of the averages of two angles θ and δ.

probability of finding the molecule oriented within a small angular segment $\mathrm{d}\theta$ is given by:

$$\mathrm{prob}(\theta) = F(\theta)\sin\theta\mathrm{d}\theta. \qquad [9.25]$$

For an isotropic medium the potential energy function of a dipolar molecule is given by:

$$U(\theta) = -f(0)\cdot\mu_0\cdot E_{\mathrm{dc}}\cdot\cos\theta, \qquad [9.26]$$

where E_{dc} is the magnitude of the applied static field, $f(0)$ is the local field correction for the static field, and μ_0 is the magnitude of the molecule's dipole moment. Henceforth, we can combine Eqs. 9.24, 9.25 and 9.26 to explicitly calculate the orientational average:

$$\langle\cos\theta\rangle = \frac{\int_0^{\pi}\cos\theta\cdot e^{p}\cdot\sin\theta\mathrm{d}\theta}{\int_0^{\pi}e^{p}\cdot\sin\theta\mathrm{d}\theta} = L_1(p) = \coth(p) - \frac{1}{p}, \qquad [9.27]$$

where $L_1(p)$ has been defined as the first-order Langevin function, and p has been defined through:

$$p = \frac{f(0)\cdot\mu_0\cdot E_{\mathrm{dc}}\cdot\cos\theta}{k_B T}. \qquad [9.28]$$

Similarly:

$$\langle \cos^3\theta \rangle = \frac{\int_0^{\pi} \cos^3\theta \cdot e^p \cdot \sin\theta \mathrm{d}\theta}{\int_0^{\pi} e^p \cdot \sin\theta \mathrm{d}\theta} = L_3(p) = \left(1+\frac{6}{p^2}\right) L_1(p) - \frac{2}{p}, \quad [9.29]$$

and

$$\langle \cos\theta \sin^2\theta \cos^2\delta \rangle = \frac{1}{2}[L_1(p) - L_3(p)], \quad [9.30]$$

where L_3 is defined as the third-order Langevin function.

Hence, in terms of the Langevin functions, the only two independent non-zero components of the second-order susceptibility tensor are given by:

$$\chi^{(2)}_{ZZZ}(-\omega_\sigma; \omega_1, \omega_2) = N_d \cdot f_Z(\omega_\sigma) \cdot f_Z(\omega_1) \cdot f_Z(\omega_2) \cdot L_3(p) \cdot \beta_z(-\omega_\sigma; \omega_1, \omega_2), \quad [9.31]$$

and

$$\chi^{(2)}_{ZXX}(-\omega_\sigma; \omega_1, \omega_2) = N_d \cdot f_Z(\omega_\sigma) \cdot f_X(\omega_1) \cdot f_X(\omega_2) \cdot \frac{1}{2}[L_1(p) - L_3(p)] \cdot \beta_z(-\omega_\sigma; \omega_1, \omega_2). \quad [9.32]$$

Figure 9.4 shows the behavior of the two relevant Langevin functions, L_3 and ($L_1/2 - L_3/2$). In the low poling field limit (when $p = 1$), Eqs. 9.31 and 9.32 reduce to:

$$\chi^{(2)}_{ZZZ}(-\omega_\sigma; \omega_1, \omega_2) = N_\mathrm{d} \cdot f_Z(\omega_\sigma) \cdot f_Z(\omega_1) \cdot f_Z(\omega_2) \cdot \frac{p}{5} \cdot \beta_z(-\omega_\sigma; \omega_1, \omega_2), \quad [9.33]$$

and

$$\chi^{(2)}_{ZXX}(-\omega_\sigma; \omega_1, \omega_2) = N_\mathrm{d} \cdot f_Z(\omega_\sigma) \cdot f_X(\omega_1) \cdot f_X(\omega_2) \cdot \frac{p}{15} \cdot \beta_z(-\omega_\sigma; \omega_1, \omega_2). \quad [9.34]$$

Notice how in the low-poling field limit the ratio between the two independent components is approximately 3, a fact that can be used to check experimentally whether or not the approximation holds. When the approximation holds, Eqs. 9.31 and 9.32 can be used to estimate the value of β_z (as defined in Eq. 9.23) if the components of the second-order nonlinear susceptibilities are known; or to estimate the macroscopic response when the values of β_z are given.

In the high-poling field limit (when $p > 30$), on the other hand, the aligning force from the external direct current (dc) field dominates over

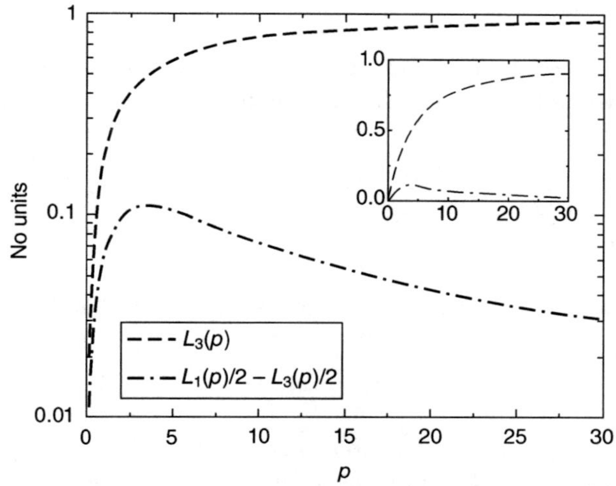

9.4 The behavior of the two relevant Langevin functions for a poled-doped polymer, $L_3(p)$ and $L_1(p)/2 - L_3(p)/2$. For better comparison, the vertical scale in the main figure is logarithmic, while a linear scale is used in the inset. In the low poling field limit, $L_3(p) = p/5$ and $L_1(p)/2 - L_3(p)/2 = p/15$. In the high field limit, $L_3(p)$ saturates and $L_1(p)/2 - L_3(p)/2$ goes to zero.

the thermal fluctuations, and consequently, $\chi^{(2)}_{ZXX}$ goes to zero while $\chi^{(2)}_{ZZZ}$ saturates.

9.4 Quantum mechanical expressions for the molecular (hyper)polarizabilities

9.4.1 The harmonic oscillator analogy

The driven damped harmonic oscillator

Qualitative insight on the behavior of the molecular linear and nonlinear polarizabilities can be gained using the harmonic and anharmonic oscillator analogies. To derive a relationship for the induced polarization as a function of the strength of the external electric field, we first assume that the electrons are bound to the positive sites through a harmonic restoring force:

$$F_k = -k \cdot x_{\mathrm{d}}, \qquad [9.35]$$

where x_{d} is the displacement from the equilibrium position, and k is a constant that characterizes the strength of the restoring force. As the electric field is applied, the electron also experiences a force given by:

$$F_E = -e \cdot E_0, \tag{9.36}$$

where E_0 is the magnitude of the electric field and $-e$ is the charge of an electron. A new equilibrium is achieved when the two forces are equal to each other, which yields following expression for the induced dipole moment:

$$p = e \cdot x_d = \frac{e^2 E_0}{k}. \tag{9.37}$$

Therefore, the linear polarizability of the harmonic oscillator is given by:

$$\alpha = \frac{e^2}{k}. \tag{9.38}$$

As we would expect, in this simple model, when the binding force is increased, the electron is less influenced by the external field, resulting in lower polarizabilities. However, this model is highly unrealistic, as it predicts that once the electron is removed from equilibrium, it will oscillate around the equilibrium point indefinitely with frequency $\omega_0 = \sqrt{\frac{k}{m}}$, where m is the mass of the electron. A more realistic model must include a damping force F_d, that is proportional to the speed $v_d = \frac{dx_d}{dt}$, and acts against the motion:

$$F_d = -\frac{2\Gamma}{m}\frac{dx_d}{dt}, \tag{9.39}$$

where Γ is a constant that characterizes the damping force. Also, instead of a static electric field, we assume that the field is harmonic oscillating at frequency ω:

$$E(t) = \frac{1}{2} E_0^{\omega} \cdot e^{-i\omega t} + c.c. \tag{9.40}$$

This model is known as the driven damped harmonic oscillator. Substituting Eqs. 9.35, 9.39 and 9.40 into Newton's second law we obtain:

$$m\left[\frac{d^2 x_d}{dt^2} + 2\Gamma\frac{dx_d}{dt} + \omega_0^2 x_d + e\left(\frac{1}{2} E_0^{\omega} e^{-i\omega t} + c.c.\right)\right] = 0, \tag{9.41}$$

which yields the following expression for the time-dependent induced polarization:

$$p(t) = -e \cdot x_d(t) = \frac{e^2}{m(\omega_0^2 - 2i\Gamma - \omega^2)}\left(\frac{1}{2} E_0^{\omega} e^{-i\omega t} + c.c.\right). \tag{9.42}$$

Clearly, the induced polarization is also oscillating at frequency ω:

$$p(t) = \frac{1}{2} p_0^{\omega} \cdot e^{-i\omega t} + \text{c.c.} \qquad [9.43]$$

Comparison of Eqs. 9.40 and 9.43 in combination with Eqs. 9.11, 9.13 and 9.14 yields the expression for the linear polarizability in the frequency domain:

$$\alpha(-\omega; \omega) = \frac{e^2}{m(\omega_0^2 - 2i\Gamma\omega - \omega^2)} = \frac{e^2}{m} D(\omega). \qquad [9.44]$$

The dispersion (i.e. the dependence on the frequency) of the linear polarizability is determined by the linear dispersion function $D(\omega)$, which is plotted in Fig. 9.5 and does remarkably match the behavior of real systems, and the dispersion function that is predicted by exact quantum perturbation theory (also plotted in Fig. 9.6), provided that we associate the oscillator's natural frequency ω_0 with the energy difference between excited and ground state, $E_{eg} = \hbar\omega_0$, where $\hbar$ is the reduced Plank constant. Notice that in the linear case, the polarizability is resonant when $\omega_0 = \omega$ (Fig. 9.5).

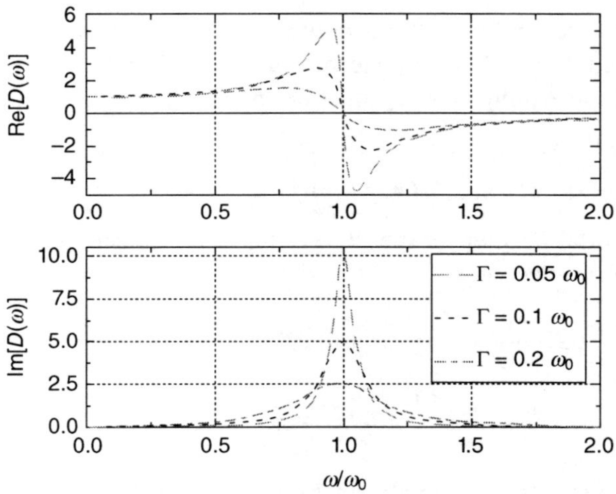

9.5 The behavior of the real and imaginary part of the dispersion function that is predicted from the driven damped harmonic oscillator model, $D(\omega)$, as a function of different values of the damping constant, Γ. Clearly the expressions are resonant in the vicinity of the natural frequency of the linear oscillator, when $\omega_0 = \omega$.

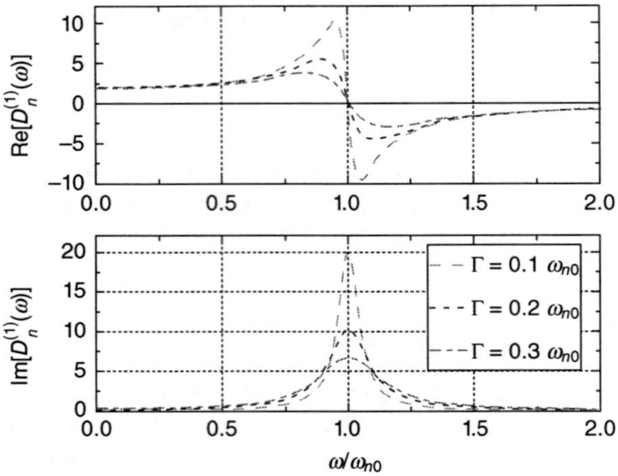

9.6 The behavior of the real and imaginary part of the dispersion function that is predicted from time-dependent quantum perturbation theory, $D_n^{(1)}(\omega)$ as a function of different values of the linewidths, $\Gamma \equiv \Gamma_n$. Clearly the expressions are resonant when the photon energy, $\hbar\omega$, is in the vicinity of the energy difference between ground and excited state, $E_{n0} = \hbar\omega_{n0}$, when $\omega_{n0} = \omega$.

The driven damped anharmonic oscillator

In order to generate nonlinear terms in the induced polarization, the restoring force must contain anharmonic terms, such as higher-order terms in the displacement from equilibrium, x_d must be included:

$$\begin{aligned} F_k &= -k^{(1)} \cdot x_\mathrm{d} - k^{(2)} \cdot (x_\mathrm{d})^2 - k^{(3)} \cdot (x_\mathrm{d})^3 \\ &= -m\omega_0^2 \cdot x_\mathrm{d} - m\xi^{(2)} \cdot (x_\mathrm{d})^2 - m\xi^{(3)} \cdot (x_\mathrm{d})^3 + \ldots. \end{aligned} \quad [9.45]$$

Using this model (known as the driven damped anharmonic oscillator), Newton's equation becomes:

$$\begin{aligned} m\Bigg[&\frac{\mathrm{d}^2 x_\mathrm{d}}{\mathrm{d}t^2} + 2\Gamma \frac{\mathrm{d}x_\mathrm{d}}{\mathrm{d}t} + m\omega_0^2 \cdot x_\mathrm{d} + m\xi^{(2)} \cdot (x_\mathrm{d})^2 + \\ &m\xi^{(3)} \cdot (x_\mathrm{d})^3 + \ldots + e\left(\frac{1}{2} E_0^\omega \mathrm{e}^{-i\omega t} + \mathrm{c.c.}\right)\Bigg] = 0. \end{aligned} \quad [9.46]$$

Equation 9.43 is solved using classical perturbation theory, and yields an expression for the equilibrium displacement x_d, that is dominated by a contribution which is directly proportional to the driving electric field, but contains also contributions that depend on higher-order terms. The linear contribution yields the same expression for the linear polarizability in the frequency domain as from the harmonic case (Eq. 9.42). From the first nonlinear contribution, an expression for the first hyperpolarizability can be deduced:

$$\beta(-2\omega;\omega,\omega)=\xi^{(2)}(-e)\left(\frac{e}{m}\right)^2 D(\omega)D(\omega)D(2\omega). \quad [9.47]$$

We notice how the dispersion of the first hyperpolarizability, given by the term $D(\omega)D(\omega)D(2\omega)$, has now a new kind of resonance, when $\omega_0 = 2\omega$. This is achieved when the energy of two photons (each one with energy $\hbar\omega$) matches the transition energy of the molecule, E_{eg}, and can only be achieved through the nonlinear interaction. Similarly, the expression for the second hyperpolarizability has a resonance when $\omega_0 = 3\omega$:

$$\gamma(-3\omega;\omega,\omega,\omega)=\xi^{(3)}(e)\left(\frac{e}{m}\right)^3 D(\omega)D(\omega)D(\omega)D(3\omega). \quad [9.48]$$

9.4.2 Quantum mechanical expressions

Exact expressions for the molecular linear polarizability and the nonlinear hyperpolarizabilities can be obtained using time-dependent quantum perturbation theory. Here we will introduce the sum-over-states (SOS) expressions derived by Orr and Ward (1971). These expressions apply to the effects produced by the changes of the electronic configuration of the molecule induced by the perturbing field and do not include vibronic coupling (i.e. the interaction between the nuclear and the electronic motions). Although these expressions apply strictly within the Born–Oppenheimer approximation they can be easily generalized.

The expressions are derived under the dipole approximation, assuming that the state wave function of the molecule interacting with light has a slowly varying component and a rapidly varying component. The molecule is initially in the ground state, and upon perturbation by the oscillating electric field, the state wave function changes into a superposition of ground and excited states. By assumption, the state functions of the unperturbed molecule are known:

$$\hat{H}(0)|n\rangle=\left(E_n - i\hbar\frac{\Gamma_n}{2}\right)|n\rangle, \quad [9.49]$$

where $\hat{H}(0)$ is the Hamiltonian corresponding to the unperturbed molecule; E_n is the energy of the nth eigenstate, $|n\rangle$; and Γ_n is the linewidth of the nth eigenstate, which is inversely proportional to the lifetime of the eigenstate. We notice that the linewidths need to be introduced in order to account from the effects of the environment. If the molecule is completely isolated, once it reaches an eigenstate, it should stay in that eigenstate indefinitely. In reality, a molecule is interacting continuously with its environment, such as when it reaches an excited state, the interaction with the environment will provide for a pathway to decay back to the ground state, resulting in a finite lifetime for the excited state.

When the perturbing electric field (associated with the beam) interacts with the molecule, a new time-dependent term, $V_{int}(t)$, must be added to the Hamiltonian, which within the dipole approximation is given by:

$$V_{int}(t) = -\vec{\mu} \cdot \vec{E}(t), \qquad [9.50]$$

where $\vec{E}(t)$ is the perturbing (local) electric field, and the dipole moment operator is given by $\vec{\mu} = -e\vec{r}$, where $\vec{r}$ is the electronic position operator, which for a molecule with N electrons contributing to the polarization is defined as $\vec{r} \equiv \vec{r}^1 + \vec{r}^2 + \ldots + \vec{r}^N$, where $\vec{r}^i$ is the position operator of the ith electron.

Assuming that the perturbing electric field can be expressed as the sum over discrete frequencies (Eq. 9.4), the expressions for the molecular polarizabilities can be expressed in the frequency domain. The tensor components of the molecular linear polarizability are given by:

$$\begin{aligned}\alpha_{ij}(-\omega;\omega) &= e^2 \sum_{n\neq 0} \langle \mu_i \rangle_{0n} \langle \mu_j \rangle_{n0} \left[\frac{1}{\left(E_{n0} - i\hbar\frac{\Gamma_n}{2} - \hbar\omega\right)} + \frac{1}{\left(E_{n0} + i\hbar\frac{\Gamma_n}{2} + \hbar\omega\right)} \right] \\ &= e^2 \sum_{n\neq 0} \langle \mu_i \rangle_{0n} \langle \mu_j \rangle_{n0} D_n^{(1)}(\omega)\end{aligned} \qquad [9.51]$$

where the sum is performed over all the excited states of the unperturbed molecule, and the transition moments are defined as:

$$\langle \mu_i \rangle_{nm} = -e\langle r_i \rangle_{nm} = -e\langle n|r_i|m\rangle, \qquad [9.52]$$

with r_i being the ith component of the electronic position operator. When $n = m$, $\langle \mu_i \rangle_{nn}$ measures directly the dipole moment of the excited state $|n\rangle$; while when $n \neq m$, $\langle \mu_i \rangle_{nm}$ is associated with the probability of making a transition between the states with polarization in the ith direction. Notice how, the dispersion behavior of each contributing excited state, $D_n^{(1)}(\omega)$, resembles qualitatively the behavior that is derived from the damped driven harmonic oscillator analogy (Eq. 9.41). However, the classical model misses the dependence on the transition dipole moments predicted by the quantum mechanical description.

Similarly, the expression for the first hyperpolarizability is given by:

$$\beta_{ijk}(-\omega_1 - \omega_2; \omega_1, \omega_2) = -e^3 \sum_{n,m\neq 0} \langle \mu_i \rangle_{0n} \langle \bar{\mu}_j \rangle_{nm} \langle \mu_k \rangle_{m0} D_{nm}^{(2)}(\omega_1, \omega_2), \qquad [9.53]$$

where the dispersion factors are given by:

$$D_{nm}^{(2)}(\omega_1,\omega_2)=\frac{1}{\left(E_{m0}-i\hbar\frac{\Gamma_m}{2}-\hbar\omega_1-\hbar\omega_2\right)\left(E_{n0}-i\hbar\frac{\Gamma_n}{2}-\hbar\omega_1\right)}$$
$$+\frac{1}{\left(E_{m0}+i\hbar\frac{\Gamma_m}{2}+\hbar\omega_2\right)\left(E_{n0}-i\hbar\frac{\Gamma_n}{2}-\hbar\omega_1-\hbar\omega_2\right)}$$
$$+\frac{1}{\left(E_{m0}+i\hbar\frac{\Gamma_m}{2}+\hbar\omega_2\right)\left(E_{n0}-i\hbar\frac{\Gamma_n}{2}-\hbar\omega_1\right)}$$
$$+\text{(Permutations of }\omega_1\leftrightarrow\omega_2\text{ for the previous terms)},\qquad [9.54]$$

and the bar operator is defined as:

$$\langle\bar{\mu}_i\rangle_{nm}=\begin{cases}\langle\mu_i\rangle_{nn}-\langle\mu_i\rangle_{00} & \text{if } n=m,\\ \langle\mu_i\rangle_{nm} & \text{if } n\neq m.\end{cases}\qquad [9.55]$$

Finally, the expression for the second hyperpolarizability is given by:

$$\gamma_{ijkl}(-\omega_1-\omega_2-\omega_3;\omega_1,\omega_2,\omega_3)$$
$$=e^4\sum_{n,m,s\neq 0}\langle\mu_i\rangle_{0n}\langle\bar{\mu}_j\rangle_{nm}\langle\bar{\mu}_k\rangle_{ms}\langle\mu_l\rangle_{s0}\,D_{nms}^{(3)}(\omega_1,\omega_2,\omega_3)$$
$$-e^4\sum_{n,m\neq 0}\langle\mu_i\rangle_{0n}\langle\mu_j\rangle_{n0}\langle\mu_k\rangle_{0m}\langle\mu_l\rangle_{m0}\,D_{nm}^{(3)}(\omega_1,\omega_2,\omega_3),\qquad [9.56]$$

where the dispersion terms are given by:

$$D_{nms}^{(3)}(-\omega_1-\omega_2-\omega_3;\omega_1,\omega_2,\omega_3)$$
$$=\frac{1}{\left(E_{n0}-i\hbar\frac{\Gamma_n}{2}-\hbar\omega_1-\hbar\omega_2-\hbar\omega_3\right)\left(E_{m0}-i\hbar\frac{\Gamma_m}{2}-\hbar\omega_1-\hbar\omega_2\right)\left(E_{s0}-i\hbar\frac{\Gamma_s}{2}-\hbar\omega_1\right)}$$
$$+\frac{1}{\left(E_{n0}+i\hbar\frac{\Gamma_n}{2}+\hbar\omega_3\right)\left(E_{m0}-i\hbar\frac{\Gamma_m}{2}-\hbar\omega_1-\hbar\omega_2\right)\left(E_{s0}-i\hbar\frac{\Gamma_s}{2}-\hbar\omega_1\right)}$$
$$+\frac{1}{\left(E_{n0}+i\hbar\frac{\Gamma_n}{2}+\hbar\omega_3\right)\left(E_{m0}+i\hbar\frac{\Gamma_m}{2}+\hbar\omega_1+\hbar\omega_2\right)\left(E_{s0}-i\hbar\frac{\Gamma_s}{2}-\hbar\omega_3\right)}$$
$$+\frac{1}{\left(E_{n0}+i\hbar\frac{\Gamma_n}{2}+\hbar\omega_1\right)\left(E_{m0}+i\hbar\frac{\Gamma_m}{2}+\hbar\omega_1+\hbar\omega_2\right)\left(E_{s0}+i\hbar\frac{\Gamma_s}{2}+\hbar(\omega_1+\omega_2+\omega_3)\right)}$$
$$+\text{Permutations of }\omega_1\leftrightarrow\omega_2\leftrightarrow\omega_3,\qquad [9.57]$$

and:

$$D_{nm}^{(3)}(-\omega_1 - \omega_2 - \omega_3; \omega_1, \omega_2, \omega_3)$$

$$= \frac{1}{\left(E_{m0} - i\hbar\frac{\Gamma_m}{2} - \hbar(\omega_1 + \omega_2 + \omega_3)\right)\left(E_{m0} - i\hbar\frac{\Gamma_m}{2} - \hbar\omega_3\right)\left(E_{n0} - i\hbar\frac{\Gamma_s}{2} - \hbar\omega_1\right)}$$

$$+ \frac{1}{\left(E_{m0} - i\hbar\frac{\Gamma_n}{2} - \hbar\omega_3\right)\left(E_{n0} + i\hbar\frac{\Gamma_m}{2} + \hbar\omega_2\right)\left(E_{n0} - i\hbar\frac{\Gamma_s}{2} - \hbar\omega_1\right)}$$

$$+ \frac{1}{\left(E_{m0} + i\hbar\frac{\Gamma_n}{2} + \hbar(\omega_1 + \omega_2 + \omega_3)\right)\left(E_{m0} + i\hbar\frac{\Gamma_m}{2} + \hbar\omega_3\right)\left(E_{n0} + i\hbar\frac{\Gamma_s}{2} + \hbar\omega_1\right)}$$

$$+ \frac{1}{\left(E_{m0} + i\hbar\frac{\Gamma_n}{2} + \hbar\omega_3\right)\left(E_{n0} - i\hbar\frac{\Gamma_m}{2} - \hbar\omega_2\right)\left(E_{n0} + i\hbar\frac{\Gamma_s}{2} + \hbar\omega_1\right)}$$

$$+ \text{Permutations of } \omega_1 \leftrightarrow \omega_2 \leftrightarrow \omega_3. \qquad [9.58]$$

Notice that, in order to satisfy Eq. 9.10, the permutations of $\omega_1 \leftrightarrow \omega_2 \leftrightarrow \omega_3$ must be simultaneously performed with permutations of corresponding *jkl* indexes of the transition dipole moments on Eq. 9.56 (that is, if for example we permute $\omega_1 \leftrightarrow \omega_2$, we must also permute $\mu_j \leftrightarrow \mu_k$; if we permute $\omega_1 \leftrightarrow \omega_3$, we must also permute $\mu_j \leftrightarrow \mu_l$, and so on). Similarly, the permutations of $\omega_1 \leftrightarrow \omega_2$ in Eq. 9.54 must be performed by simultaneously permuting the *jk* indexes of the transition dipole moments on Eq. 9.53.

The importance of Eqs. 9.51 to 9.56 in the field of molecular nonlinear optics cannot be over-emphasized. When dealing with macroscopic quantities, it is possible to use classical models like the anharmonic oscillator to get an intuition of the optical response of the bulk. However, when dealing with molecules, we are forced to use a full quantum mechanical picture, and most of our understanding of the molecular response is built upon the use of sum-over-states expressions such as the ones given by Eqs. 9.51 to 9.56. For example, by truncating the expression for the first hyperpolarizability (Eq. 9.53) to the contribution of only two levels (ground and first excited), useful structure-relationship properties have been derived (Marder *et al.*, 1991, 1993). More recently, Perez-Moreno *et al.* (2011) have introduced a self-consistent method for the modeling of complex planar octupolar molecules. The authors were able to reduce the number of required physical quantities needed for the full modeling of the molecular nonlinear optical response by applying the sum rules formalism and symmetry principles to the quantum mechanical sum-over-states expressions (Eqs. 9.51 to 9.56).

9.5 Conclusion and future trends

In this chapter we have introduced the fundamental definitions and concepts needed to describe and understand the light–matter interaction at the molecular level; and the basic tools that allow connection of the nonlinear optical response at the microscopic level with the properties of the bulk in polymer-based materials.

Most of our understanding of the nonlinear optical response at the molecular level has been derived from the study and modeling of molecules where only one tensor component dominates the response. Similarly, the behavior of poled-doped polymers is well understood when the guest molecules are dipolar with only one dominant tensor component, as described in Section 9.3.2. Dipolar compounds are normally described using simplified two-level models to avoid the complexity of the full sum-over-states expressions.

However, at present, molecules have become more complex and new geometries are being explored. This has brought new challenges into the field, adding complexity to the analysis of the polymer–guest interaction; and demanding a better modeling of the quantum mechanical response that goes beyond the traditional two-level model. Understanding the nonlinear optical response of molecules of *any* type of geometry will lead to the development of better optoelectronic materials, and to the development of new applications in other fields such as microscopy and biomedical imaging.

9.6 References

Boyd R (2003), *Nonlinear Optics*, San Diego, Elsevier Science.

Dalton L (2001), 'Realization of sub 1 volt polymeric EO modulators through systematic definition of material structure–function relationships', *Synth. Met.*, **124**, 3–7.

Dalton L (2002), 'Nonlinear optical materials: from chromophore design to commercial applications', *Polymers for Photonics Applications I*, 158, Springer-Verlag, Berlin.

Ertl P (2003), 'Cheminformatics analysis of organic substitutents: identification of the most common substituents, calculation of substituent properties, and automatic identification of drug-like biososteric groups', *J. Chem. Inf. Comput. Sci.*, **43**, 474–380.

Franken P, Hill A, Peters A and Weinreich G (1962), 'Generation of optical harmonics', *Phys. Rev. Lett.*, **7**, 118–119.

Griffiths D (1999), *Introduction to electrodynamics*, New Jersey, Prentice Hall.

Helmchen F and Denk W (2005), 'Deep tissue two-photon microscopy', *Nature Methods*, **2**, 932–940.

Kuzyk M (1998), 'Relationship between the molecular and bulk response', in Kuzyk M and Dirk C, *Characterization techniques and tabulations for organic nonlinear optical materials*, New York, Marcel Dekker, 111–220.

Lipinski C and Hopkins A (2004), 'Navigating chemical space for biology and medicine', *Nature*, **432**, 855–861.

Maiman T (1960), 'Stimulated optical radiation in ruby', *Nature*, **187**, 493–494.

Marder S, Beratan D and Cheng L (1991), 'Approaches for optimizing the first hyperpolarizability of conjugated organic molecules', *Science*, **252**, 103.

Marder S, Gorman C, Tiemann B and Cheng L (1993), 'Stronger acceptors can diminish nonlinear optical response in simple donor-acceptor polyenes', *J. Am. Chem. Soc.*, **115**, 3005–3007.

Nalwa H S and Miyata S (1997), *Nonlinear optics of organic molecules and polymers*, Boca Raton, CRC Press.

Orr B and Ward J (1971), 'Perturbation theory of the non-linear optical polarization of an isolated system', *Mol. Phys.*, **20**, 513–526.

Oudar J L and Le Person H (1975), 'Second-order polarizabilities of some aromatic molecules', *Opt. Commun.*, **15**, 258–262.

Perez-Moreno J, Hung S-T, Kuyzk M G, Zhou Z, Ramini S K and Clays K (2011), 'Experimental verification of a self-consistent theory of the first-, second-, and third-order (non)linear optical response', *Phys. Rev. A*, **84**, 033837.

Prasad P and Williams D (1991), *Introduction to nonlinear optical effects in molecules and polymers*, New York, Wiley.

Salafsky J (2007), 'Second-harmonic generation for studying structural motion of biological molecules in real time and space', *Phys. Chem. Chem. Phys.*, **42**, 5704–4711.

Shi R and Garito A (1998), 'Introduction: conventions and standards for Nonlinear Optical processes', in Kuzyk M and Dirk C, *Characterization techniques and tabulations for organic nonlinear optical materials*, New York, Marcel Dekker, 1–36.

Shi Y, Zhang C, Zhan H, Bechtel J, Dalton L, Robinson B and Steier W (2001), 'Low (sub-1-volt) halfwave voltage polymer electro-optic modulators achieved by controlling chromophore shape', *Science*, **288**, 119–122.

Squier J and Müller M (2001), 'High resolution nonlinear microscopy: A review of sources and methods for achieving optimal imaging', *Rev. Sci. Instr.*, **72**, 2855.

Xu C, Zipfel W, Shear J, Williams R and Webb W (1996), 'Multiphoton fluorescence excitation: new spectral windows for biological nonlinear microscopy', *Proc. Nat. Acad. Sci.*, **93**, 10763.

Zyss J and Oudar J (1982), 'Relationships between microscopic and macroscopic lowest-order optical nonlinearities of molecular crystals with one- or two-dimensional units', *Phys. Rev. A*, **26**, 2028–2048.

Zyss J, Pecaut J, Levy J P and Masse R (1993), 'Synthesis and crystal structure of guanidinium l-monohydrogentartrate: encapsulation of an optically nonlinear octupolar cation', *Acta Crystallogr. Sect. B*, **49**, 334–342.

10
Ultrafast intrachain exciton dynamics in π-conjugated polymers

Z. V. VARDENY, University of Utah, USA and C.-X. SHENG, Nanjing University of Science and Technology, China

DOI: 10.1533/9780857098764.2.297

Abstract: We have used the femtosecond transient photomodulation spectroscopy in a broad spectral range for studying the ultrafast dynamics of the primary photoexcitations in various polymers. An emerging universal picture is presented based on the C_{2h} point group symmetry of the polymer chain, where the electron–electron interaction is dominant. In this picture the $1B_u$ exciton is the primary photoexcitation. However, the $1B_u$ exciton may decay within a few hundreds fs to the more stable 'dark' exciton ($2A_g$) in 'dark' polymers where the photoluminescence (PL) efficiency is extremely low. In contrast, in polymers with high PL emission efficiency, the $1B_u$ exciton is the lowest excited state, and consequently may show stimulated emission that can be utilized for laser action.

Key words: intrachain exciton, π-conjugated polymers, ultrafast dynamics.

10.1 Introduction

10.1.1 Optical properties of photoexcitations in π-conjugated polymers

Π-conjugated polymers (PCPs) owe their name to the delocalization of π-electrons along the polymer backbone. PCPs have found a number of applications in the fields of organic light-emitting diodes (OLEDs) [1–3], thin-film transistors [4, 5], organic solar cells [6, 7], and organic spin-valve devices [8, 9]. Therefore unraveling the properties of photoexcitations in PCPs is of great interest for achieving fundamental understanding of their materials' physics, which is essential for designing efficient organic optoelectronic devices. A particular research focus is the primary intrachain photoexcitations and their optical properties that take place upon photon absorption; after all the primary photoexcitations are the foundation core of the optoelectronic applications of these materials [10, 11].

Although many kinds of PCPs have been introduced in last three decades, PCP falls in one of two separate groups [12]; namely polymers having degenerate ground state (DGS) represented by trans-polyacetylene [*t-(CH)$_x$*] (Table 10.1(a)), and polymers having non-degenerate ground state (NDGS) such as polyfluorene (*PFO*) (Table 10.1(d)). An alternative way to categorize π-conjugated polymers is related with their photoluminescence (PL) quantum efficiency (PLQE) [13, 14]; where polymers with high PLQE have been used as active materials for OLED applications. Originally it was thought that DGS polymers should be non-luminescent, as a result of soliton photogeneration [12], but later on it was discovered that some DGS polymers such as di-substituted polyacetylene (*DPA*) (Table 10.1(c)) have in fact high PLQE (>50% in both film and solution) that may also lead to laser action [15]. By contrast, NDGS polymers have been assumed to have higher PLQE than DGS polymers, unless circumvented by defect-related non-radiative recombination; however, a NDGS polymer, namely regio-random poly (thienylene-vinylene) (Table 10.1(b)) with intrinsic low PLQE was also recently reported [16]. Therefore in spite of intense studies of the linear and nonlinear optical properties of PCPs, the basic model for the proper description of the intrachain photoexcitations in this family is still somewhat controversial.

Table 10.1 Backbone structure of four π-conjugated polymers

	Degenerated ground state	Non-degenerated ground state
	trans-polyacetylene	regiorandom poly (thienylene-vinylene)
Low PLQE	(a) 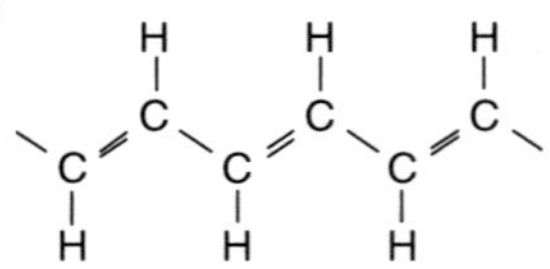	(b) 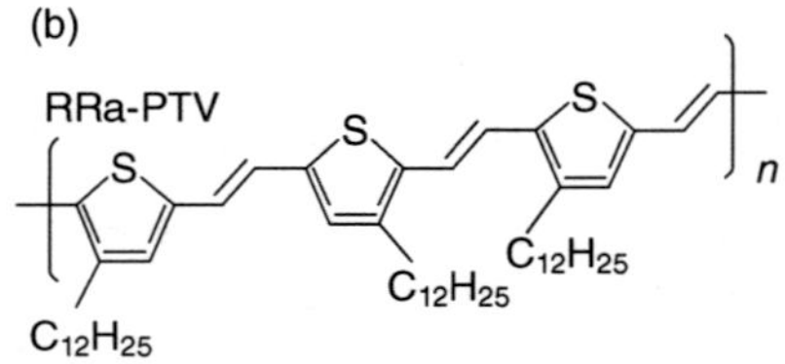
	disubstituted polyacetylene	polyfluorene
High PLQE	(c) 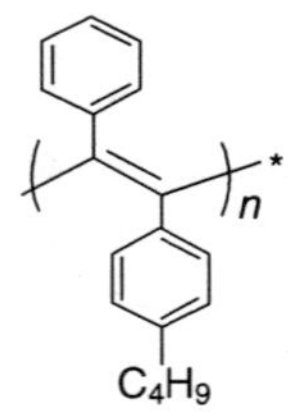	(d) 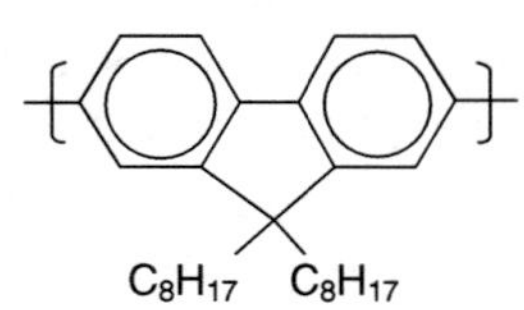

PLQE : photoluminescence quantum efficiency.

Described by the Su, Schrieffer and Heeger (SSH) model [17], the soliton excitation in *t-(CH)*$_x$ is an amphoteric defect that can accommodate zero, one, or two electrons [12, 18, 19]. In *t-(CH)*$_x$ solitons have a reversed spin-charge relationship; i.e. the neutral soliton (S^0) has spin ½; whereas charged solitons ($S^{\pm}$) have spin zero [20]. The SSH model, which includes the electron–phonon (e-h) interaction but neglects electron–electron (e-e) interaction, has been used to show that a photoexcited electron–hole pair is unstable toward the formation of charged soliton–antisoliton ($S^{\pm}$) pair [20]. For the NDGS π-conjugated polymers, adding an extrinsic gap component to the electron–phonon Hamiltonian (SSH model) results in polarons and bipolarons as the proper descriptions of their primary charge excitations [12].

During the last three decades the study of photoexcitations in *t-(CH)*$_x$ has revealed several unexpected phenomena that were not predicted by the SSH model [21, 22]. Most important, an overall neutral state, as well as charged excitation has been observed; the neutral state has been correlated with S^0 (or $2A_g$) transitions [23]. This finding, together with the absence of optical transitions at the midgap level in undoped *t-(CH)*$_x$ have indicated that electron–electron interaction in *t-(CH)*$_x$ cannot be ignored. Under these circumstances, the nature of the photoexcitations in this polymer may be very different from that predicted by the SSH Hamiltonian.

Moreover, singlet and triplet excitons have been shown to play a crucial role in the photophysics of most PCPs [24–31]. In NDGS polymers such as *PFO* and poly(*p*-phenylene-vinylene) derivatives, the binding energy of singlet intrachain excitons has been experimentally shown to be larger than 0.5 eV [32–34], which is much higher than room temperature as well as the most coupled vibration of ~0.2 eV [35–37]. The existence of excitons with relatively large binding energy can be justified in theoretical studies only when the electron–hole interaction and electron correlation effects are included in the Hamiltonian; in other words, the SSH model is not suitable to describe the intrachain photophysics of PCPs.

To describe the intramolecular excitations (and vibrations) in π-conjugated polymers, the symmetry of the single polymer chain usually dictates the wavefunction nomenclatures. Most π-conjugated polymers belong to the C_{2h} point group symmetry, which has a center of symmetry [12]. The properties of odd and even wavefunctions are then determined by the properties of the C_{2h} group irreducible representations of which character table is given in Table 10.2. Notations such as $1B_u$, mA_g, etc. are reserved for the intrachain excitons, where the index before the nomenclature denotes the state order within the exciton manifold. In this notation singlet excitons are distinguished from triplet excitons by the number on the upper left; for example 1^1B_u stands for the lowest energy singlet state, whereas the corresponding 1^3B_u denotes the lowest triplet state. Because of the weak

Table 10.2 Character table of the C_{2h} symmetry group

C_{2h}	E	C_2	i	σ_x
A_g	1	1	1	1
B_g	1	−1	1	−1
A_u	1	1	−1	−1
B_u	1	−1	−1	1

spin–orbit interaction in the four polymers discussed in this chapter, we will focus our attention on photogenerated singlet excitons, which are the primary photoexcitations in PCPs. Therefore all the notations such as $1B_u$, mA_g etc. refers to singlet states. The dipole moment component is the strongest in the direction of the chain; its irreducible representation is B_u. Therefore optical transitions involving B_u and A_g pair of states are symmetry allowed. On the contrary, dipole transitions involving two A_g states are symmetry forbidden. However a two-photon process, such as two-photon absorption (TPA), which is an optical nonlinear process, allows optical transition between two A_g states. In particular transitions from $1A_g$ to other A_g states in the singlet exciton manifold are allowed, and thus TPA spectroscopy has been used to map the excited A_g states.

In the exciton notation in PCPs, the ground state is $1A_g$, the excited states are even symmetry excitons, mA_g, or odd symmetry excitons, nB_u [13, 38]. The importance of the e-e interaction in PCPs such as polyacetylene can be concluded from the excited states order of $2A_g$ (the lowest two photon allowed state) and $1B_u$ (the lowest one photon allowed state) [22, 39]. If the magnitude of the C–C bond alternation in the polymer chain is relatively small, then $E(2A_g) < E(1B_u)$ [39, 40]. However when the 'effective' C–C bond alternation magnitude is relatively large, the order is reversed, namely $E(2A_g) > E(1B_u)$ that results in strong PL emission. An excellent example is the polymer poly(di-acetylene), or *DPA*, where the polymer chain has a $t\text{-}(CH)_x$ backbone structure, but also contains bulky side-groups. It has been shown [15, 41] that in spite of being a polymer with DGS structure that supports soliton excitations, DPA is highly luminescent. This happens since, unlike $t\text{-}(CH)_x$ the large bond alternation in *DPA* pushes the $2A_g$ exciton to be above the $1B_u$ exciton; this allows strong PL emission. Also in PPV-type polymers, as another example, the benzene ring in the backbone structure gives rise to a large 'effective' C–C bond alternation for the extended π electrons [34], and therefore leads to high PL efficiency.

In this chapter we review the studies of ultrafast photoexcitations in π-conjugated polymers shown in Table 10.1. In general we have studied photoexcitations in such polymers in a broad time interval from femtoseconds to milliseconds and spectral range from 0.1 to 3.5 eV [26, 32, 41].

However, in this chapter we review only our transient studies in the femtosecond to picosecond time domains (t < 200 ps). The main experimental technique described here is transient photomodulation (PM), which gives information complementary to that obtained by transient PL, which is limited to radiative processes, or transient photoconductivity (PC), which is sensitive to high mobility photocarriers. The PM method, in contrast, is sensitive to non-equilibrium excitations in *all* states.

The scheme of the PM experiments is the following: the polymer sample is photoexcited with above-gap light pulses, and the changes in the optical absorption of the sample are probed in a broad spectral range from the IR to visible using several light pulse sources. The PM spectrum is essentially a difference spectrum, i.e., the difference in the optical absorption ($\Delta\alpha$) spectrum of the polymer when it contains a non-equilibrium photoexcitation concentration and that in the equilibrium ground state. Therefore the optical transitions of the various photoexcitations are clearly revealed in the experiment.

If the electron–electron interaction dominates the photophysics of PCPs, a bound e-h pair or an exciton is immediately generated upon excitation. By definition the exciton is a neutral, spinless excitation of the polymer. In Fig. 10.1 we summarize the optical transitions of singlet excitons in polymers shown in Table 10.1. From our measurements a universal picture emerges: the two important exciton levels ($1B_u$ and mA_g) are shown to play an important role in all four polymers. The mA_g level is known to have strong dipole moment coupling to $1B_u$ as deduced form the various optical nonlinear spectra of π-conjugated polymers analyzed in terms of the 'four essential states' model [13, 32]. We therefore expect two strong optical transitions to form following the $1B_u$ photogeneration, PA_1 from $1B_u$ to mA_g and PA_2 from $1B_u$ to another A_g state, dubbed kA_g. For polymers with low PL efficiency, $2A_g$ lies below the $1B_u$, and fast internal conversion $1B_u$ *to* $2A_g$ is expected that results in an excited state transition from $2A_g$ to higher B_u state (PA_3).

10.1.2 Experimental set up for measuring ultrafast transient response of photoexcitations

We have used the femtosecond (fs) two-color pump-probe technique for measuring the transient photoexcitation response in the fs to ns time domain. Two fs Ti : sapphire laser systems have been utilized; a low power (energy/pulse ~0.1 nJ) high repetition rate (~80 MHz) [42] laser system for the mid-IR spectral range; and a high power (energy/pulse ~10 μJ) low repetition rate (~1 kHz) laser system for the near-IR/visible spectral range [43]. For both laser systems the excitation photon energy, $\hbar\omega$ was set at 3.1 eV.

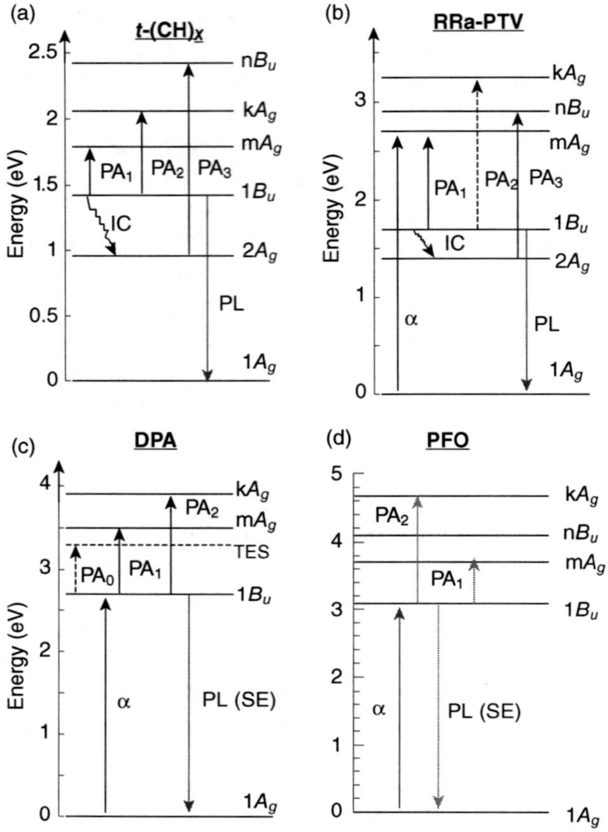

10.1 The most important intrachain states and allowed transitions for polymers shown in Table 10.1, respectively. The PA bands are excited state absorptions; α is linear absorption; PL is photoluminescence which may also be replaced by SE (stimulated emission) in the pump-probe experiment. (c) Adapted from Sheng, C.-X *et al., Phys. Rev. B*, **81**, 205103, 2010; (d) Adapted from Tong M. *et al., Phys. Rev. B*, **75**, 125207, 2007.

For the transient mid-IR measurements, both signal and idler output of an optical parametric oscillator (Tsunami, Opal, Spectra Physics) were respectively used as probe beams that generate $\hbar\omega$ (probe) from 0.55 to 1.05 eV. Moreover, a difference frequency set up with a NLO crystal was used to extend the spectral range from ~0.13 to ~0.43 eV. For the transient near-IR and visible measurements, white light super-continuum was generated having $\hbar\omega$ (probe) spectrum ranging from 1.2 to 2.7 eV. The transient PM signal, $\Delta T/T(t)$ is the fractional change, ΔT in transmission, T, which is negative for PA, and positive for photobleaching (PB) and stimulated emission (SE). The transient PM spectra from the two laser systems were normalized to each other in the near-IR/visible spectral range, for which $\hbar\omega$ (probe) from the low-power laser system was doubled. The pump and probe beams were carefully adjusted to get complete spatial overlap on the film, which was kept under dynamic vacuum. In addition, the pump/probe beam-walk with the translation stage was carefully monitored and the transient response was adjusted by the beam walk measured response.

10.2 Ultrafast dynamics in π-conjugated polymers

As explained above PCP photophysics is determined by a series of excited states with alternating odd (B_u) and even (A_g) parity that correspond to one-photon and two-photon allowed transitions, respectively [13]. Optical excitation into these states is followed by sub-picosecond nonradiative relaxation to the lowest excited state [44]. This hot energy relaxation process is due to either vibrational cooling within vibronic sidebands of the same electronic state, or phonon-assisted transitions between two different electronic states that is termed internal conversion in molecular spectroscopy [45]. Usually internal conversion is the fastest relaxation channel that provides efficient nonradiative transfer from a higher excited state into the lowest excited state of the same spin multiplicity [46, 47].

Before focusing on the ultrafast intrachain exciton dynamics of various polymers, we note that there are two other ultrafast relaxation processes which may interfere and successfully compete with the internal conversion process. These processes include singlet exciton fission and exciton dissociation [42]. The former process creates two triplet excitons with opposite spins from one singlet exciton in time faster than the usual intersystem crossing time from the singlet to the triplet manifold [48, 49] whereas the second process directly generates charge carriers [50, 51]. These two processes are not included in our present discussion. Another important effect, the one-dimensional vs. three-dimensional (or intrachain vs. interchain) interplay will also not be discussed in this chapter. In the most primitive model, intrachain photoexcitation results in an exciton with relatively large binding energy, whereas interchain photoexcitation, for some PCPs with strong interchain interaction such as MEH-PPV, may produce separate charges (polaron pair) or excimers on coupled neighboring chains [52, 53].

10.2.1 Exciton dynamics in trans-polyacetylene: a degenerate ground state polymer with weak photoluminescence

Trans-polyacetylene (t-$(CH)_x$) is the simplest example of PCPs with the intrinsic low PL efficiency of $<10^{-4}$ [54]. According to the SSH model [17], upon photon absorption charged soliton–antisoliton ($S^{\pm}$) pair is created in the middle of energy gap [20] within ~100 fs. Therefore we expect to detect a photoinduced absorption band from either the highest occupied molecular orbitals (HOMO) to S^+ or S^- to the lowest unoccupied molecular orbital (LUMO) [20].

However, the nature of the photoexcitations in $t\text{-}(CH)_x$ may be very different from that predicted by the SSH Hamiltonian, because the electron–electron interaction cannot be ignored. In this case, neutral singlet excitons may be the primary photoexcitations; these photoexcitations are characterized by two transient photoinduced absorption bands, similar to many other polymers that belong to the PCP family (see Fig. 10.1(a)).

Figure 10.2(a) shows transient PM spectra of a $t\text{-}(CH)_x$ film pumped at the absorption edge (1.55 eV), at 0 and 1 ps, respectively. The figure reveals two prominent PA bands in the mid-infrared, namely PA_1 at 0.37 eV and PA_2 at 0.64 eV, as well as another PA band, PA_3 at 1.4 eV. The transient dynamics of the three PA bands in the PM spectra are shown in Fig. 10.2(b). PA_1 and PA_2 are instantaneously generated and have the same fast decay component of about 110 fs. However PA_3 band is photogenerated with a time delay of ~100 fs and that follows a much slower decay [55]. The similarity between the transient decays of PA_1 and PA_2 bands considered together with the build-up dynamics of PA_3 suggest that PA_3 has a different character than that of PA_1 and PA_2. We also note that the transient PM spectrum of the $t\text{-}(CH)_x$ film down to 0.13 eV does not show any other important band, such as due to IR-active vibrations (IRAV) that are related to charge photogeneration[56, 57]. We thus conclude that upon excitation at the band edge only neutral photoexcitations are generated in $t\text{-}(CH)_x$.

Our results shown in Fig. 10.2(a) can be fully explained by the exciton picture, similar to in other polymers in the PCP family. We claim that singlet excitons are instantaneously photogenerated in $t\text{-}(CH)_x$ upon photon absorption with energy close to the optical gap. In this picture, PA_1 is the transition from $1B_u$ to mA_g, wheras PA_2 is the transition from $1B_u$ to another higher two photon state, the so-called kA_g [58]. Although the photogenerated excitons in $t\text{-}(CH)_x$ have two characteristic PA bands similar to many other polymers in the family, they decay in record time (~100 fs) to a 'dark' exciton ($2A_g$ state) that does not emit light (is not radiative). Both transient PM spectra and decay dynamics clearly show that the generation of $2A_g$ exciton comes at the expense of $1B_u$ exciton. Since the $2A_g$ state in non-luminescent polymers lies below the $1B_u$ and is non-emissive, then according to Kasha's rule there need be an internal conversion from the $1B_u$ to the $2A_g$. Thus we interpret PA_3 as due to the transition from $2A_g$ to higher B_u state, namely nB_u with slower decay [55].The absence of IRAVs is in agreement with the assignment of the three PA bands as due to transitions involving intrachain excitons.

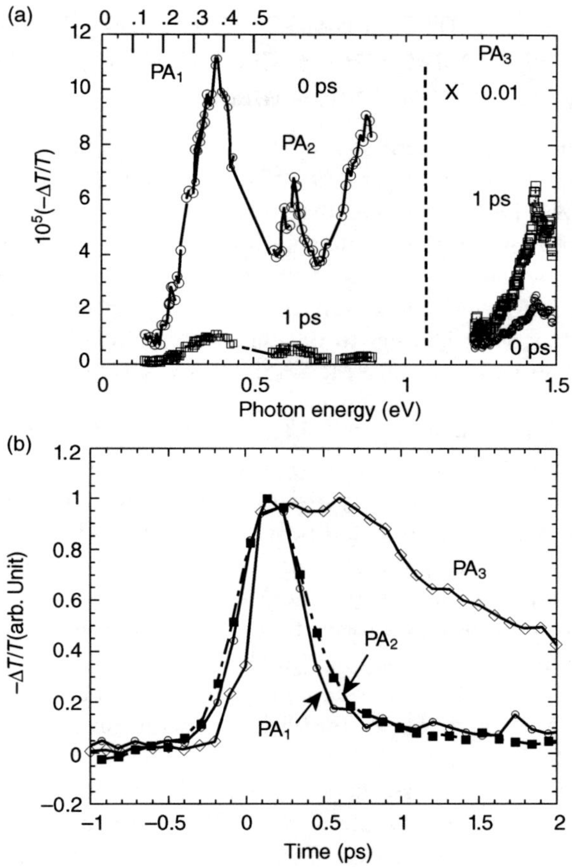

10.2 Transient PM spectra and decay dynamics of *t-(CH)ₓ* film pumped at 1.55 eV at room temperature. (a) The $t = 0$ (circles) and $t = 1$ ps (squares) transient PM spectra in the spectral range of 0.15 to 1.5 eV. Various bands are assigned; PA_1 and PA_2 are due to $1B_u$ excitons; PA_3 is due to $2A_g$ excitons. (b) The transient decay dynamics of the main bands assigned in (a).

From the PA_1 transition, the exciton binding energy in $t\text{-}(CH)_x$ may be estimated as ~0.35 eV, which is smaller than the binding energy of excitons in many other polymers in the family of PCPs [34], but much higher than 0.026 eV of room temperature. The relatively large binding energy in $t\text{-}(CH)_x$ is in agreement with the absence of IRAVs indicating that free charge carriers are not generated in polyacetylene; also it refutes the picture advanced by the SSH model of charge soliton generation immediately following photon absorption.

10.2.2 Exciton dynamics in region-random poly(thienylene-vinylene): a non-degerate ground state polymer with weak photoluminescence

A new class of non-luminescent π-conjugated polymers, namely region-random poly(thienyle vinylene), [or *RRa-PTV*] (see Table 10.1(b)) has been recently synthesized [16]. We found that this polymer has very small PL efficiency of $<2 \times 10^{-4}$ in both films and solutions, and thus the PL weakness cannot be caused by the film quality [59], but rather is *intrinsic* in nature. Properties of the long-lived photoexcitations and doping induced absorption have shown that *PTV* belongs to the class of NDGS polymers [60]. In order to explain the weak PL in *PTV* within Kasha's rule, the photogenerated $1B_u$ exciton should undergo an internal conversion into a 'dark' exciton at lower energy level [14]. The 'dark' exciton in PCPs is an even-parity state (A_g) [13]. We thus conclude that the dark exciton in *PTV* is the $2A_g$ state, which has been extensively discussed in the literature in relation with the photophysics of π-conjugated polymers [13, 14, 61].

For verifying the proposed photoexcitation scenario in *PTV* we measured the ultrafast transient response. Figure 10.3(a) and (b) show the transient PA spectra in *RRa-PTV* film at various times, *t* following the pulse excitation; similar response was obtained in *RRa-PTV* in solution. The PA spectrum is dominated by two PA bands; PA_1 in the mid-infrared spectral range with peak at ~0.95 eV, and PA_3 in the visible/near-infrared range with a broad peak at ~1.6 eV. Figure 10.3(c) and (d) show the PA transient decay. The two PA bands do not share the same dynamics; whereas PA_1 decays almost completely within ~0.5 ps, the decay of PA_3 is longer (~2.5 ps) into a plateau that indicates the formation of a relatively stable photoexcitation. We thus conclude that the two PA bands do not belong to the same photoexcitation. A closer inspection at the PA response at '$t = 0$' (insets to Fig. 10.3(c) and (d)), reveals that PA_1 is instantaneously generated, whereas PA_3 is formed with a delay of ~200 fs. We therefore conclude that PA_3 is not related to the primary photoexcitation in *PTV*; but is formed on the expense of PA_1, during its decay process.

PA_1 spectrum is generic to many π-conjugated polymers, which was previously identified [53] as due to optical transitions from the photogenerated $1B_u$ exciton into an even-parity state (mA_g) that is part of the essential states in these polymers [13]. Therefore we infer that PA_1 here is also due to the photogenerated $1B_u$ exciton. Its ultrafast decay kinetics indicates that there is another state lower in energy ($<E(1B_u)$), into which the photogenerated $1B_u$ decays. As a check of this proposed scenario we estimate the PL efficiency from the fast PA_1 decay and compare it to the measured PL efficiency, η measured with an integrated sphere. For this estimate we used the relation: $\eta = \tau/\tau_{rad}$, where τ is the exciton lifetime and τ_{rad} (~1 ns [62]) is the

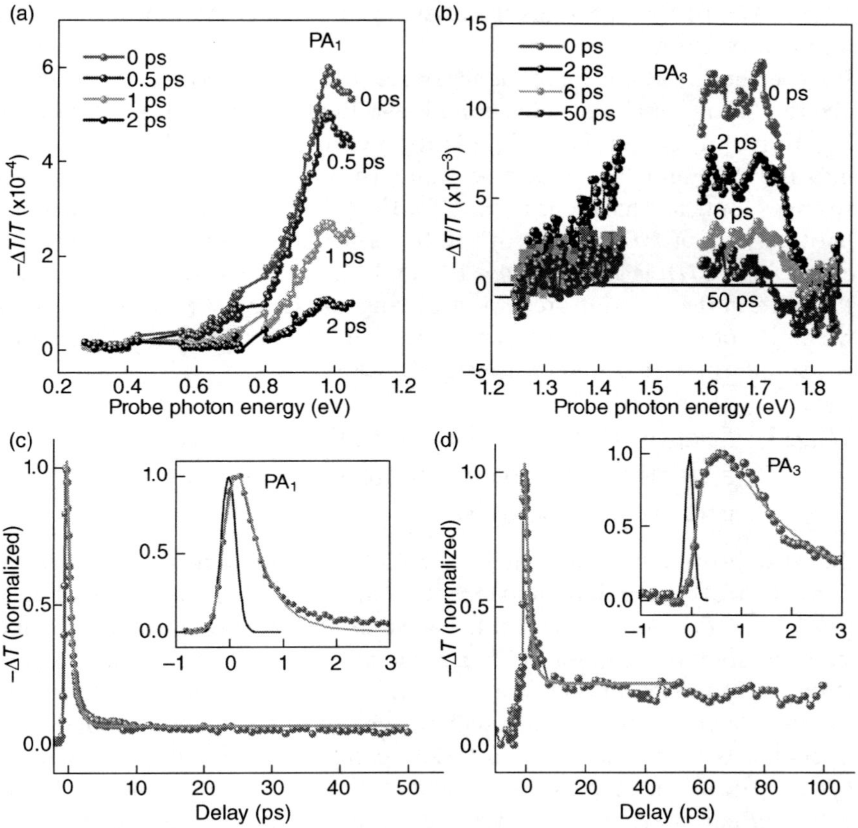

10.3 Transient PA spectra of *RRa-PTV* film measured at various delay time, *t* in (a) mid-infrared and (b) visible/near-infrared spectral range; PA_1 and PA_3 bands are assigned. (c) and (d) The respective decay dynamics of PA_1 at 0.95 eV and PA_3 at 1.6 eV. The insets to (c) and (d) focus on the PA formation and decay evolution around '$t = 0$', where the data (solid circle) is fitted with an exponential formation and decay processes (line), taking into account the pump/probe cross-correlation trace (line).

$1B_u$ radiative lifetime. Using PA_1 lifetime of ~0.5 ps we thus estimate $\eta(RRa\text{-}PTV) \approx 5 \times 10^{-4}$, which is in good agreement with the measured η-value (2×10^{-4}). In the proposed decay scenario PA_3 is a transition from the $2A_g$ state. However its optical transition should be into an odd-parity exciton, namely the nB_u; which is also part of the four essential states in PCPs [13]. The stabilization of PA_3 at later time ($t > 5$ ps) into a plateau, indicates that the $2A_g$ exciton becomes trapped. The PA_2 from $1B_u$ to kA_g, which is normally observed in other PCPs, is not well separated from PA_3 because of spectral overlap; however, the difference in dynamics at 1.4 and

1.7 eV (Fig. 10.3(b)) presents the existence of another PA transition, which we identify as PA_2.

Nevertheless, as the consequence of PA_1 and PA_3 bands, the excited state diagram of the essential states in *PTV* was then constructed, as depicted in Fig. 10.1(b). It shows the initial photogeneration of the $1B_u$ and its decay into the $2A_g$, and the respective states (mA_g and nB_u) into which their strongest optical transitions occur (Fig. 10.3(a)). The similarity between the photophysics of *PTV* to that of $t\text{-}(CH)_x$ is striking. As in *PTV* the PL efficiency in $t\text{-}(CH)_x$ is very low, $\eta < 10^{-4}$ [54]. Also, $t\text{-}(CH)_x$ shows an ultrafast PA band in the near-infrared spectral range at ~1.35 eV (PA_3 in Fig. 10.2), which is not instantaneously generated, and is stabilized at later time, that is also interpreted as due to transitions from the $2A_g$.

10.2.3 Exciton dynamics in disubstituted polyacetylene: a degenerate ground state polymer with strong photoluminescence

The di-substituted polyacetylene (*DPA*) (backbone structure is shown in Table 10.1(c)) is a unique polymer [63–65]. Similar to trans-polyacetylene [$t\text{-}(CH)_x$], *DPA* has been shown to have a degenerate ground state [15] that supports soliton excitations [12]. But in contrast to $t\text{-}(CH)_x$, *DPA* has very strong PL emission in both solution and film forms, which make it suitable for optoelectronic applications such as light-emitting diodes and solid-state lasers [66, 67]. The strong PL in *DPA* compared with the very weak PL in $t\text{-}(CH)_x$ shows that the lowest excited state in the former polymer is $1B_u$ with *odd* parity symmetry; otherwise the transition from the lowest lying exciton state to the ground state would be forbidden, and this would have weakened the PL. We note, however, that the reverse excited state order in *DPA* compared to $t\text{-}(CH)_x$ is not in conflict with its degenerate ground state property.

Several theories have been advanced for explaining the reverse excited state order in *DPA* compared to that in $t\text{-}(CH)_x$ [68–72]. One explanation is that the strong electron–electron interaction in this polymer causes delocalization of the excited state among the backbone polyene-like monomers on the chain; but also includes the phenyl side groups (see Table 10.1(c) inset). This 'transverse delocalization', in turn pushes up $E(2A_g)$ above $E(1B_u)$ resulting in strong PL emission [72]. Therefore, in addition to the four essential excited states known to determine the NLO spectra of π-conjugated polymers, namely $1A_g$, $1B_u$, mA_g, and nB_u [38], in *DPA* there is another important excited state complex, dubbed 'transverse excited state' or *TES* [73].

Figure 10.4(a) shows the transient PM spectrum of *DPA* film at $t = 0$ ps. PM spectra contains three PA bands: PA_0 at ~0.6 eV, PA_1 at ~0.9 eV, and a

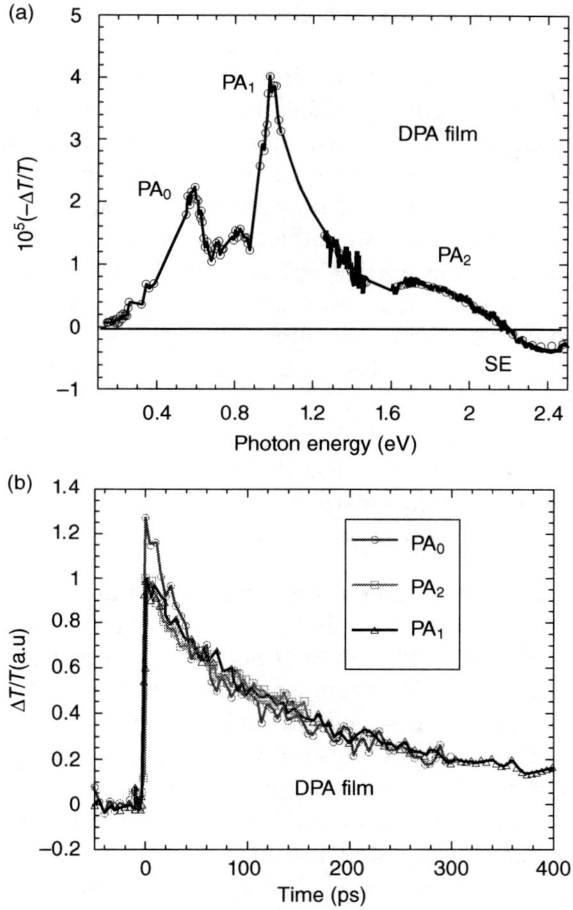

10.4 (a) The transient PM spectrum of *DPA* film at $t = 0$ ps. The PA bands PA_0, PA_1, PA_2 and SE are assigned. (b) The transient decay dynamics of the main PA bands assigned in (a). (Adapted from Sheng C-X *et al.*, *Phys. Rev. B*, **81**, 205103, 2010.)

broad PA_2 band in the near-IR spectral range; the PM spectra also contain a strong SE band peaked at ~2.6 eV. All PA bands show approximately the same decay dynamics (Fig. 10.4(b)), and therefore originate from the same photoexcitation species. Since the SE is part of this species' characteristic, we therefore identify it as intrachain singlet exciton, or $1B_u$. The PA_1 and PA_2 bands are consequently interpreted as optical transitions from the relaxed $1B_u \rightarrow \mathrm{m}A_g$ and $1B_u \rightarrow \mathrm{k}A_g$, respectively; similar to excitons in other PCP [45]. However PA_0 does not occur in the PM spectrum of excitons in other PCPs. We therefore propose that PA_0 is due to the optical transition from the relaxed $1B_u$ to the *TES*, that is also absent in other PCPs. The *TES* is probably reminiscent of the $2A_g$ state in unsubstituted *t-(CH)*$_x$, which lies in between the $1B_u$ and mA_g as shown in energy diagram of Fig. 10.1(c). Additional optical spectroscopies studies have shown that *TES* dipole moment has a strong component perpendicular to the chain direction, which contains almost degenerate states with odd and even parity symmetry [73]. Nevertheless, besides the unique properties of *TES* state, the primary photoexcitations is $1B_u$ excitons, as in other luminescent PCPs [73].

We also note that the transient PM spectrum of the *DPA* film down to 0.13 eV does not show any other important band, such as due to IRAVs that are related to charge photogeneration [15, 56]. This is in agreement with the assignment of the three PA bands as due to transitions involving intrachain excitons.

10.2.4 Exciton dynamics in polyfluorene: a non-degenerate ground state polymer with strong photoluminescence

Polyfluorene is an attractive material for display applications due to efficient blue emission [74], and relatively large hole mobility with trap free transport [75]. Poly(9,9-dioctylfluorene) [*PFO*] (Table 10.1(d)) exhibits a complex morphological behavior where a pristine sample with glassy structure (dubbed α phase) can be changed into a film with more superior order (dubbed β phase). The structural versatility of *PFO* can be exploited in manipulating the sample electronic and optical properties [76, 77].

Nevertheless, the disordered α phase *PFO* with the bulky side group should provide a clean case for studying intrachain photoexcitations, and characteristic excited states with only a small contribution due to interchain interaction. However, several early studies of ultrafast photoexcitation dynamics in *PFO* led to confusing results. In one study of oriented PFO [78], three types of photoexcitations were invoked including hot carriers, excitons, and charge polarons. In two other studies [79, 80], both excitons and bound polaron pairs were shown to simultaneously coexist. More recent studies focused on three beams' excitation [81–83], where the role of even parity states was emphasized. In these studies it was realized that when excited deeper into the singlet manifold, then charge species are primarily photogenerated; however, most of them geminately recombine within few ps. Under these circumstances it is difficult to decide whether PFO excited states are 'band-like' or 'excitonic' in nature.

Figure 10.5(a) shows the full PM spectrum of a PFO film at $t = 0$ ps. The PM spectrum contains two PA bands (PA_1 and PA_2) at 0.55 and 1.65 eV, respectively, and a SE band that peaks at 2.5 eV. The three bands decay dynamics are shown in Fig. 10.5(b); the decay dynamics of all three bands

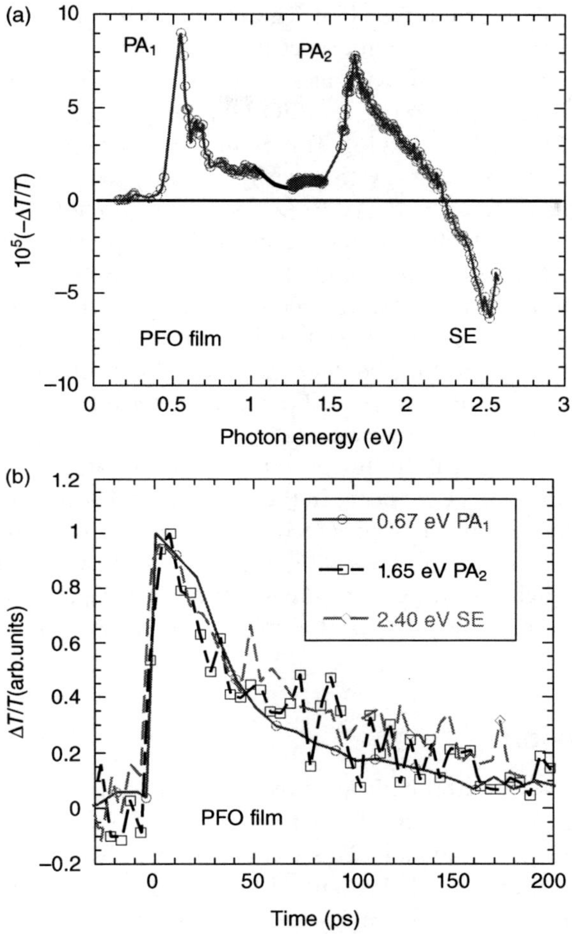

10.5 (a) The transient PM spectra of *PFO* at $t = 0$. The PA bands PA_1, PA_2 and SE are assigned. (b) The transient decay dynamics of the main bands assigned in (a). (Adapted from Tong M. *et al.*, *Phys. Rev. B*, **75**, 125207, 2007.)

are equal, with a lifetime of ~100 ps. This shows that the PM bands belong to the same photogenerated species, in contrast to earlier conclusions [78]. Since SE is related to excitons, we thus attribute this species to photogenerated singlet excitons, namely $1B_u$. We therefore conclude that the two PA bands are optical transitions from $1B_u$ to the two most strongly coupled A_g excitons in the singlet manifold, namely mA_g and kA_g, respectively. Same with the other three polymers in this chapter, the transient PM spectrum of the *PFO* film down to 0.13 eV does not show any IRAV bands that are related to charge photogeneration [56]. Therefore, no hot excitons [78] or polaron-pairs [79] are needed to interpret this spectrum.

Our results are in agreement with lack of long-lived polaron photogeneration in as spun (i.e. glassy, α phase) PFO film measured by the continuous wave (CW) PM technique [76]. This is in agreement with the large exciton intrachain binding energy that we found [34] for *PFO*. Also our results clearly indicate that the bulky *PFO* side groups do not allow strong interchain interaction, which is largely responsible for polaron photogeneration in PCPs [84]. Another possibility for the lack of polaron photogeneration here is that the laser excitation photon energy (~3.1 eV) is close to the absorption onset of *PFO* (~3.0 eV) [83]. From the excitation dependence of charge photogeneration in *PFO* using cw PM spectroscopy it was found [76] that the quantum efficiency of polaron photogeneration dramatically increases at photon energies close to $E(\mathrm{m}A_g)$ ~3.7 eV; up to this photon energy there is very little steady state photogenerated polarons. We thus expect dramatic changes to occur in the transient PM spectrum of *PFO* at *higher* excitation photon energies [83], and/or when the glassy phase changes into a more ordered phase [80], and when a strong electric field such as in organic light-emitting diodes made of *PFO* is capable of exciton dissociation even at relatively low excitation photon energy close to $\mathrm{E}(1B_u)$ [85].

10.3 Conclusion

In this chapter we reviewed the ultrafast dynamics of the primary photoexcitations in four typical polymers that represent both degenerate and non-degenerate ground state polymers. This include trans-polyacetylene (DGS, low intrinsic PL efficiency), polyfluorene (NDGS, high PL efficiency), *DPA* (DGS, high PL efficiency), and regio-random poly (thienylene-vinylene) (NDGS, low intrinsic PL efficiency). In all of these polymers, we show that the $1B_u$ exciton is the primary photoexcitation when the laser excitation photon energy is close to the absorption onset of polymer. However, the obtained transient PA dynamics show that the primary exciton ($1B_u$) decays within few hundreds fs to the more stable 'dark' exciton ($2A_g$) when the PL emission of the polymer is intrinsic low; this include *t-(CH)*$_x$ and *RRa-PTV*. Therefore the photophysics of these two polymers is governed by the dark excitons ($2A_g$). On the contrary, in polymers with high PL efficiency such as the DGS polymer *DPA* and the NDGS polymers *PFO*, the $1B_u$ exciton is the lowest excited state, which decays back to ground state with strong photoluminescence emission, and show stimulated emission that can be utilized for laser action.

This chapter shows that broadband ultrafast photomodulation spectroscopy is a powerful tool in analyzing the photophysics of π-conjugated polymers. Our results present the universality in the photophysics of π-conjugated polymers, which is based on C_{2h} point group symmetry of the single polymer chain. Pursuing broader probing spectra, adjustable pumping photon energy as well as better spatial and temporal resolution, is always the experimental desire. At the same time, the effects of external fields, such as electric field on the photophysics of PCPs can deepen our understanding of their photophysics [85, 86].

Other directions for future work include:

- *Interchain effects on PCP photophysics.* The work in this chapter focuses only on intrachain photoexcitations in PCPs. However, various interchain photoexcitations appear to be very important in thin films of many PCPs. Ordered PCP films exhibit photophysics remarkably different from dilute solutions or disordered films. The branching of intrachain singlet excitons and polarons occurs instantaneously in many systems [53, 56]. This topic has been controversial for more than two decades [52]. Very recently, our joint theory-experiment work provided new insight to this subject. We reported pressure-dependent transient picosecond and CW PM studies of disordered and ordered films of 2-methoxy-5-(2-ethylhexyloxy) poly(*para*-phenylenevinylene). Photoinduced absorption (PA) bands of intrachain excitons and polarons in the disordered film exhibit weak pressure dependence. In contrast, the ordered film exhibits two additional transient PA bands in the mid-infrared that blueshift dramatically with pressure. Based on high-order configuration interaction calculations, we ascribed the PA bands in the ordered film to excimers [52]. Systematical photophysics study of polymers as well as monomers under high hydrostatic pressure could bring more information in this topic.
- *Charge photogeneration dynamics in polymer/fullerene blend films.* Organic solar cells that utilize π-conjugated polymers have attracted widespread interests in both academic and commercial communities in recent years. One of the most studied bulk heterojunction (BHJ) organic photovoltaic materials, the blend of region-regular poly (3-hexylthiophene) (RR-P3HT) and [6,6]-phenyl-C61-butyric acid methyl ester (PCBM), is often considered as a 'model system' [87–88] for OPV applications. Numerous steady-state and time-resolved spectroscopic studies have been conducted in pristine P3HT as well as P3HT/PCBM blend in order to understand the properties and evolution of neutral and charged photoexcitations, but many aspects of the photophysics remain unclear [89–91]. One of the fundamental reasons is that additional complications arise from the strong interchain interactions in the P3HT self-organized π-stacked two-dimensional lamellae in the film [92]. Consequently it has been shown in P3HT films that both neutral and charged photoexcitations are at least partially delocalized in the lamellae [92, 93]. Although

the delocalized polaron in RR-P3HT/PCBM blend has been proven to play a critical role in the optoelectronic devices [94], the mechanism for the photogeneration of delocalized polarons is still debated [89, 94, 95]. Furthermore, evidences for the involvement of charge transfer complex (CTC) as an intermediate state in the charge photogeneration process have been reported [87, 96, 97]. More thorough investigation of the CT excitation and their interaction with free polaron excitations may improve the power conversion efficiency of organic solar cell, and drive the development of novel photoactive materials.

10.4 Acknowledgments

We would like to thank our collaborators at the University of Utah Physics Department over the years 2000–2011, without whom this work would have never been completed. These are T. Basel, J. Holt, E. Olejnik, B. Pandit, R.C. Polson, S. Singh, and M. Tong. We also acknowledge useful collaboration with X. Jiang from the University of South Florida, S. Mazumdar from the University of Arizona, and E. Ehrenfreund from the Technion in Israel.

This work was supported in part over the years by the Department of Energy grant No. FG-04-ER46109. C.-X.Sheng thanks the support of NSF of China No. 61006014, NHTRDP of China, No. 2011AA050520, NUST Research Funding No.2011ZDJH23, and NUST 'Zijin star' project.

10.5 References

1. Burroughes J.H. *et al.*, *Nature*, **347**, 539, 1990.
2. Forrest S.R., *Nature*, **428**, 911, 2004.
3. Malliaras G. and Friend R.H., *Phys. Today*, **58**, 53, 2005.
4. Garnier F. *et al.*, *Science*, **265**, 1684, 1994.
5. Sekitani T., Zschieschang U., Klauk H. and Someya T., *Nature Mater.*, **9**, 1015, 2010.
6. Granström M. *et al.*, *Nature*, **395**, 257, 1998.
7. Dennler G. *et al.*, *Adv. Mater.*, **21**, 1223, 2009.
8. Xiong Z.H. *et al.*, *Nature*, **427**, 821, 2004.
9. Nguyen T.D., Hukic-Markosian G., Wang F., Wojcik L., Li X., Ehrenfreund E. and Vardeny Z.V., *Nature Mater.*, **9**, 345, 2010.
10. Sariciftci S. (ed.), *Primary Photoexcitations in Conjugated Polymers: Molecular Exciton versus Semiconductor Band Model*, World Scientific, Singapore, 1997.
11. Rothberg L., Photophysics of Conjugated Polymers, in *Semiconducting Polymers: Chemistry, Physics and Engineering*, Vol. I, edited by G. Hadziioannou and G.G. Malliaras, John Wiley, Weinheim, 2006. pp. 179–204.
12. Heeger A.J., Kivelson S., Schrieffer J.R. and Su W.P., *Rev. Modern Phys.*, **60**, 781, 1988.
13. Dixit S.N., Guo D. and Mazumdar S., *Phys. Rev. B*, **43**, 6781, 1991.

14. Soos Z.G., Elemad S., Galvao D.S. and Ramasesha S., *Chem. Phys. Lett.*, **194**, 341, 1992.
15. Gontia E. *et al.*, *Phys. Rev. Lett.*, **82**, 4058, 1999.
16. Zhang C. *et al.*, *Polymer Chem.*, **1**, 663, 2010.
17. Su W.P., Schrieffer J.R. and Heeger A.J., *Phys. Rev. Lett.*, **42**, 1698, 1979.
18. Vardeny Z.V. and Tauc J., *Phys. Rev. Lett.*, **54**, 1844, 1985.
19. Vardeny Z.V. and Tauc J., *Philos. Mag. B*, **52**, 313, 1985.
20. Su W.P. and Schrieffer J.R., *Proc. Natl. Acad. Sci. USA*, **77**, 5626, 1980.
21. Tavan P. and Schulten K., *Phys. Rev. B*, **36**, 4337, 1987.
22. Kohler B.E., Spangler C. and Westerfield C., *J. Chem. Phys.*, **89**, 5422, 1988.
23. Orenstein J., in *Handbook of Conducting Polymers*, Skotheim T.A., Elsenbaumer R.L. and Reynolds J., editors, 2nd Ed., Marcel Dekker Inc., New York, 1998, p.1297.
24. Robins L., Orenstein J. and Superfine R., *Phys Rev. Lett.*, **56**, 1850, 1986.
25. Swanson L.S., Shinar J. and Yoshino K., *Phys. Rev. Lett.*, **65**, 1140, 1990.
26. Wei X. *et al.*, *Phys. Rev. Lett.*, **68**, 666, 1992.
27. Kersting R. *et al.*, *Phys. Rev. Lett.*, **70**, 3820, 1993.
28. Leng J.M. *et al.*, *Phys. Rev. Lett.*, **72**, 156, 1994.
29. Yan M. *et al.*, *Phys. Rev. Lett.*, **72**, 1104, 1994.
30. Frolov S.V. *et al.*, *Phys. Rev. Lett.*, **78**, 4285, 1997.
31. Frolov S.V. *et al.*, *Phys. Rev. B*, **65**, 205209, 2002.
32. Tong M. *et al.*, *Phys. Rev. B*, **75**, 125207, 2007.
33. Chandross M., Mazumdar S., Jeglinski S., Wei X., Vardeny Z.V., Kwock E.W. and Miller T.M., *Phys. Rev. B*, **50**, 14702, 1994.
34. Zhao H., Mazumdar S., Sheng C.-X., Tong M. and Vardeny Z.V., *Phys. Rev. B*, **73**, 075403, 2006.
35. Vardeny Z.V. *et al.*, *Phys. Rev. Lett.*, **51**, 2326, 1983.
36. Horovitz B., Gutfreund H. and Weger M., *Phys. Rev. B*, **17**, 2796, 1978.
37. Ehrenfreund E. *et al.*, *Phys. Rev. B*, **36**, 1535, 1987.
38. Guo D. *et al.*, *Phys. Rev. B.*, **48**, 1433, 1993.
39. Soos Z.G., Ramasesha S. and Galvao D.S., *Phys. Rev. Lett.*, **71**, 1609, 1993.
40. Soos Z.G. and Ramasesha S., *Phys. Rev. B*, **29**, 5410, 1984.
41. Gontia I.I. *et al.*, *Phys. Rev. B*, **66**, 075215, 2002.
42. Sheng C.X. *et al.*, *Phys. Rev. B*, **71**, 125427, 2005.
43. Singh S., Drori T. and Vardeny Z.V., *Phys. Rev. B*, **77**, 195304, 2008.
44. Kersting R. *et al.*, *Phys. Rev. Lett.*, **73**, 1440, 1994.
45. Birks J.B., *Photophysics of Aromatic Molecules*, Wiley-Interscience, London, 1970.
46. Harison N.T. *et al.*, *Phys. Rev. Lett.*, **77**, 188, 1996.
47. Pope M. and Swenberg C.E., *Electronic Processes in Organic Crystals and Polymers*, 2nd Ed., Oxford University Press, Oxford, 1999.
48. Wilson M.W.B., Rao A., Clark J., Kumar R.S.S., Brida D., Cerullo G. and Friend R.H., *J. Am. Chem. Soc.*, **133**, 11830, 2011.
49. Zimmerman P.M., Zhang Z. and Musgrave C.B., *Nature Chem.*, **2**, 648, 2010.
50. Miranda P., Moses D. and Heeger A.J., *Phys. Rev. B*, **70**, 085212, 2004.
51. Gadermaier C., Cerullo G., Sansone G., Leising G., Scherf U. and Lanzani G., *Phys. Rev. Lett.*, **89**, 117402, 2002.
52. Aryanpour K., Sheng C.-X., Olejnik E., Pandit B., Psiachos D., Mazumdar S. and Vardeny Z.V., *Phys. Rev. B*, **83**, 155124, 2011.

53. Sheng C.-X., Tong M., Singh S. and Vardeny Z.V., *Phys. Rev. B*, **75**, 085206, 2007.
54. Carter P.W. and Porter J.D., *Phys. Rev. B*, **43**, 14438, 1991.
55. Vardeny Z., Strait J., Pfost D., Tauc J. and Abeles B., *Phys. Rev. Lett.*, **48**, 1132, 1982.
56. Miranda P.B., Moses D. and Heeger A.J., *Phys. Rev. B*, **64**, 081201, 2001.
57. Vardeny, Z.V., Orenstein, J. and Baker, G.L., *Phys. Rev. Lett.*, **50**, 2032, 1983.
58. Frolov, S.V. *et al.*, *Phys. Rev. Lett.*, **78**, 729, 1997.
59. Brassett A.J. *et al.*, *Phys. Rev. B*, **41**, 10586, 1990.
60. Lafalce I., Jiang X.M., Zhang C. *et al.*, *unpublished*.
61. Apperloo J.J. *et al.*, *J. Phys. Chem. A*, **106**, 21, 2002.
62. Harrison N.T. *et al.*, *Phys. Rev. Lett.*, **77**, 1881, 1996.
63. Shirakawa H., Masuda T. and Takeda K., in *The Chemistry of Triple-Bonded Functional Groups*, edited by S. Patai, Wiley, New York, 1994.
64. Masuda T., Hamano T., Tsuchihara K. and Higashimura T., *Macromolecules*, **23**, 1374, 1990.
65. Lüer L., Manzoni C., Cerullo G., Lanzani G. and Vardeny Z.V., *Chem. Phys. Lett.*, **444**, 61, 2007.
66. Tada K., Sawada H., Kyokane J. and Yoshino K., *Jpn. J. Appl. Phys.*, Part 2, **34**, L1087, 1995.
67. Frolov S.V., Fujii A., Chinn D., Hirohata M., Masahiro R.H., Masuda T., Yoshino K. and Vardeny Z.V., *Adv. Mater.*, **10**, 871, 1998.
68. Shukla A. and Mazumdar S., *Phys. Rev. Lett.*, **83**, 3944, 1999.
69. Shukla A., *Chem. Phys.*, **300**, 177, 2004.
70. Shukla A., *Phys. Rev. B*, **69**, 165218, 2004.
71. Shukla A. and Sony P., *Synth. Met.*, **135**, 368, 2005.
72. Ghosh H., Shukla A. and Mazumdar S., *Phys. Rev B*, **62**, 12763, 2000.
73. Sheng C.-X., Tong M. and Vardeny Z.V., *Phys. Rev. B*, **81**, 205103, 2010.
74. Grice A., Bradley D.D.C., Bernius M.T., Inbasekaran M., Wu W.W. and Woo E.P., *Appl. Phys. Lett.*, **82**, 629, 1998.
75. Redecker M., Bradley D.D.C., Inbasekaran M. and Woo E.P., *Adv. Mater.*, **9**, 798, 1997.
76. Cadby A.J. *et al.*, *Phys. Rev. B*, **62**, 15604, 2000.
77. Redecker M., Bradley D.D.C., Inbasekaran M. and Woo E.P., *Appl. Phys. Lett.*, **74**, 1400, 1999.
78. Xu S., Klimov V.I., Kraabel B., Wang H. and McBranch D.W., *Phys. Rev. B*, **64**, 193201, 2001.
79. Stevens M.A., Silva C., Russel D.M. and Friend R.H., *Phys. Rev. B*, **63**, 165213, 2001.
80. Korovyanko O.J. and Vardeny Z.V., *Chem. Phys. Lett.*, **356**, 361, 2002.
81. Gambetta A., Virgili T. and Lanzani G., *Appl. Phys. Lett.*, **86**, 253509, 2005.
82. Zhang X., Xia Y. and Friend R.H., *Phys. Rev. B*, **73**, 245201, 2006.
83. Lüer L., Manzoni C., Egelhaaf H.-J. *et al.*, *Phys. Rev. B*, **73**, 035216, 2006.
84. Hendry E., Koeberg M., Schins J.M. *et al.*, *Phys. Rev. B*, **71**, 125201, 2005.
85. Virgili T., Cerullo G., Lüer L. *et al.*, *Phys. Rev. Lett.*, **90**, 247402, 2003.
86. Cabanillas-Gonzalez J., Virgili T., Gambetta A. *et al.*, *Phys. Rev. Lett.*, **96**, 106601, 2006.
87. Clarke T.M. and Durrant J.R., *Chem. Rev.*, **110**, 6736, 2010.

88. Li G., Shrotriya V., Huang J. *et al.*, *Nat. Mater.*, **4**, 864, 2005.
89. Banerji N., Cowan S., Vauthey E. and Heeger A.J., *J. Phys. Chem. C*, **115**, 9726, 2011.
90. Marsh R.A., Hodgkiss J.M., Albert-Seifried S. and Friend R.H., *Nano Lett.*, **10**, 923, 2010.
91. Guo J., Ohkita H., Benten H. and Ito S., *J. Am. Chem. Soc.*, **132**, 6154, 2010.
92. Österbacka R., An C.P., Jiang X.M. and Vardeny Z.V., *Science*, **287**, 839, 2000.
93. Jiang X. *et al.*, *Adv. Funct. Mater.*, **12**, 587, 2002.
94. Hwang W., Moses D. and Heeger A.J., *J. Phys. Chem. C*, **112**, 4350–4354, 2008.
95. Howard I.A., Mauer R., Meister M. and Laquai F., *J. Am. Chem. Soc.*, **132**, 14866, 2010.
96. Drori T., Holt J. and Vardeny Z.V., *Phys. Rev. B*, **82**, 075207, 2010.
97. Hallermann M., Kreigel I., Da Como E. *et al.*, *Adv. Funct. Mater.*, **19**, 3662, 2009.

11
Ultrafast charge carrier dynamics in organic (opto)electronic materials

H. DIESINGER, E. A. CHAN, J. YIN and C. SOCI,
Nanyang Technological University, Singapore

DOI: 10.1533/9780857098764.2.318

Abstract: The chapter is a summary of ultrafast time-resolved pump-probe spectroscopy techniques for the study of charge carriers in organic semiconductors. Four methods featuring sub-nanosecond time resolution are distinguished according to their probe wavelength: infrared active vibrational (IRAV) mode spectroscopy, time-resolved THz spectroscopy (TRTS), time-resolved microwave conductivity (TRMC), and transient photocurrent (TPC) spectroscopy. Examples of the application of these techniques to the study of charged photoexcitations in molecular crystals, pristine polymers and polymer donor–acceptor blends are presented. Presently accepted knowledge gathered with these methods is summarized, and the aptitude of each method for discriminating between free and localized charge carriers is discussed in view of its implication in the design and optimization of materials for organic photovoltaic devices.

Key words: infrared active vibrational (IRAV) modes, transient photocurrent (TPC) spectroscopy, time-domain terahertz spectroscopy (TRTS), time-resolved microwave conductance (TRMC).

11.1 Introduction

The nature of primary photoexcitations in organic semiconductors, particularly in conjugated polymers, has been the subject of an active debate in the scientific community for more than three decades (Sariciftci, 1997). The mechanisms of charge carrier photogeneration and thermalization promptly after photoexcitation are indicative of the strength of electron–electron interactions relative to the bandwidth (as manifested by the exciton binding energy), while carrier transport, trapping, and recombination processes are intimately related to the degree of order, purity, and morphology of the materials. Besides their fundamental interest, these processes determine the overall efficiency of organic electronic devices, such as organic photovoltaic cells; as such, there has been a lot of effort to develop charge-specific ultrafast techniques to probe their dynamics. Non-resonant probing spectroscopic methods allow the characterization of charged excitations (e.g. polarons) in organic semiconductors (for a discussion on ultrafast probes of excitonic species see Chapter 10). In general, they rely on either all-optical or electro-optical techniques where charge carriers are photogenerated with an ultrashort (~100 fs) pump pulse and their dynamics is detected with a delayed probe. We limit this summary to a choice of techniques that: (i) are highly selective to charge carrier density and dynamics (i.e., we do not consider photoinduced absorption measurements probing the visible to near-infrared spectral region where polaronic signatures overlap with excitonic features); (ii) have temporal resolution between nanoseconds and femtoseconds, that is the timescale most relevant to study photogeneration and prompt recombination processes which are indicative of intrinsic electronic and transport properties; (iii) can discriminate between charge carriers with different degrees of localization. The techniques presented here operate at distinct probe wavelengths: infrared active vibrational (IRAV) modes are probed in the medium infrared optical region, time-resolved terahertz spectroscopy, or TRTS (also referred to as optical-pump terahertz-probe spectroscopy) detects carrier signatures in the far-infrared, time-resolved microwave conductivity (TRMC) is a contactless conductance probing method that uses free space radiation at a specific microwave frequency, whereas transient photocurrent (TPC) spectroscopy consists in detecting photocurrent transients in a waveguide configuration, covering a spectral range from direct current (DC) to upper microwave frequencies.

As the field of organic semiconductors evolved through the initial fundamental studies of molecular crystals to the more application-oriented investigation of conjugated polymers, donor-acceptor and hybrid organic–inorganic systems, these techniques have been widely employed toward the experimental characterization of the charge generation and transport properties of new materials and material combinations. In the early days, polydiacetylene has been widely considered as a model system for highly ordered organic semiconductors, where the availability of high-quality single crystal allowed the observation of intrinsic material properties with limited influence of trapping. The temperature and field dependence of the peak transient photocurrent measured at about 100 ps after photoexcitation have shown the need for modifying the assumption that carrier generation in low-mobility semiconductors is limited by geminate recombination (Moses *et al.*, 1987, 1989). As experimental techniques with higher time resolution and materials with less disorder became available, the charge carrier generation and transport behavior observed in the polydiacetylene model system was found to be of general validity for the whole class of low-mobility semiconductors. A classical example of highly ordered conjugated polymer system is stretch-oriented poly(*para*-phenylene vinylene), or PPV. In this system photogeneration and transport are greatly enhanced compared with amorphous PPV films, and carrier dynamics can be studied on an extended timescale. Polarons and excitons in PPV were shown to be independent primary excitations by IRAV spectroscopy and fast TPC measurements (Soci *et al.*, 2005). Further studies of this system by TPC under high fluence showed that the initial photogeneration dynamics is dominated by bimolecular recombination, whereas the photocurrent tail that persists for a long time after photoexcitation is explained by a combination of bimolecular recombination, by recombination with trapped charges, and by trapping itself (Soci *et al.*, 2005). Geminate recombination is unnecessary to interpret the carrier dynamics.

The availability of ultrafast, all-optical probing techniques such as femtosecond optical-pump THz-probe allowed the experimental proof of direct polaron generation in a variety of organic semiconductors. The work by the groups of Hegmann and Ostroverkhova confirmed that, within the picosecond timescale, the initial photoconductance peak in ordered molecular crystals pentacene, tetracene and rubrene is field-independent and increases at low temperature, indicative of bandlike transport.

Recent studies of charged photoexcitations were mainly focused on low-bandgap polymers with a particular high yield of direct polaron generation, such as regioregular poly(3-hexylthiophene), rr-P3HT. P3HT was studied at timescales up to hundreds of nanoseconds by time-resolved microwave conductivity, the response of which is mostly determined by the Q-factor of the microwave resonance cavity containing the sample to typically 40 nanoseconds (De Haas and Warman, 1982). Although this method is too slow to resolve initial free-carrier photogeneration, it still allows monitoring the temperature dependence of the peak intensity. Temperature-dependent pulse-radiolysis TRMC of P3HT films with different molecular weight (MW) revealed positive temperature dependence of the conductivity that was more pronounced on medium MW films than on high MW ones. Since it was known that the conductivity of the material increases upon heating to the melting point and the increase of phoconductivity was observed below the melting point, the observed temperature dependence was assigned to a 'pre-melting' effect. Optical-pump THz probe spectroscopy featuring picosecond time resolution was also used to compare the complex carrier mobilities of a PPV derivative, poly(2-methoxy-5(2′-ethylhexyloxy)*p*-phenylene vinylene) (MEH-PPV), and rr-P3HT films (Hendry *et al.*, 2006). The study revealed that in P3HT, a real component of the photoconductivity, attributed to free charges, was still present after 150 ps, while in PPV the purely imaginary photoconductivity could be attributed to exciton polarizability.

Low-bandgap polymers blended with fullerene derivatives play a key role in solar energy conversion in polymer solar cells. In so-called bulk heterojunction systems, the photoexcited polymer absorber serves as electron donor, and charge transfer toward the fullerene acceptor takes place at ultrafast ($t \sim 50$ fs) time scales. This process leads to extremely efficient charge separation, as first demonstrated by Sariciftci *et al.* (1992) using photoinduced absorption (PIA) and light-induced electron spin resonance (LESR). Photoluminescence is quenched by adding the fullerene acceptors, whereas a turnover in the PIA spectrum indicates a transition from stimulated emission to photoinduced absorption. More direct evidence was given by an ESR spectrum that showed the occurrence of ionic radicals. In 2001 Brabec *et al.* studied the dynamics of charge separated state generation by time-resolved PIA, observing both the excitonic stimulated emission band and the polaronic absorption band on a femtosecond timescale. The decay dynamics of the excitonic charge transfer state has recently been studied by a 'pump-push' technique (Bakulin *et al.*, 2012). The charge separated state increases photocurrent yield and slows down recombination, so that the slow carrier recombination time makes it accessible to all three methods (Soci *et al.*, 2005): TRTS, TRMC, and TPC. Pristine polymers and their blends were compared by transient photocurrents, showing an increase of the photocurrent decay timescale from 100 ps to some nanoseconds due to decreased recombination after the charge separation (Murthy *et al.*, 2012). Drude-like transport models (Cunningham and Hayden, 2008) and transport models involving barriers and wells (Nemec *et al.*, 2009) have been derived from optical-pump THz-probe experiments. Energy levels of traps in these systems were obtained from TRMC measurements (Savenije *et al.*, 2011). Such models reveal the existence of a distribution of carriers that are mobile on different length scales, which has tremendous importance for photovoltaic and photodetecting devices.

The organization of this chapter will loosely follow the material-driven research path highlighted above for the study of charged photoexcitations in organic semiconductors: Sections 11.2 to 11.5 are dedicated to the spectroscopic methods IRAV, TPC, TRTS, and TRMC, respectively; in each of these sections the principles and development of the spectroscopic method is first introduced, followed by its application to the study of organic crystal model systems, pristine polymers, and polymer donor–acceptor blends where applicable. The characteristic time resolution and the timescale of typical observation intervals of these methods are also summarized at the end of each Section. Section 10.6 puts emphasis on the extent of carrier localization and the aptitude of each of the methods presented to detect charge carriers that are free to contribute to galvanic currents or localized to a certain degree. To conclude, Section 10.7 summarizes generally accepted knowledge on photoexcitation, charge carrier conduction, trapping, and decay mechanisms in organic semiconductors.

11.2 Infrared-active vibrational (IRAV) modes

The pristine π-conjugated polymer chain is neutral and has a set of Raman-active vibrations that are strongly coupled to the electronic states. When introduced onto the backbone of a conjugated polymer either by chemical doping or photoexcitation, charge carriers (solitons, polarons and bipolarons) break the local symmetry and thereby transform the even parity Raman-active vibrational modes into odd-parity IRAV modes. These IRAV modes are distinguished from conventional IR-active modes in organic polymer by having a one-to-one correspondence with the strongest Raman-active modes of the polymer, as observed in resonant Raman scattering (Raman modes are referred to as 'amplitude modes'). The observation of IRAV modes indicates (Heeger *et al.*, 2010): (i) the existence of local structural distortion; (ii) the self-localization of charges; (iii) the strong coupling between self-localized charges and the local chain distortion. IRAV modes describe charge oscillations along the polymer chain (hence are also denoted as 'phase modes') and show unusually large oscillator strength. Their intensity is proportional to the charge carrier density induced (either by chemical, photo, or interfacial doping) on the polymer chain; therefore IRAV modes provide a unique probe for charge carrier density and dynamics in conjugated polymers. The first successful theoretical model for IRAV 'phase modes' was developed by Horovitz and colleagues (Horovitz, 1982). This model included effects of disorder, which inhibits electronic motion along the polymer chain, assuming a restoring force due to a 'pinning' potential. Introduction of the pinning potential renormalizes the bare frequencies of the skeleton chain and allows key properties of the IRAV modes to be derived, including their frequency (and the higher-frequency of IRAV modes induced by chemical doping compared to photo-doping), the one-to-one correspondence with even parity Raman modes, and their oscillator strength and relative intensities.

11.2.1 Experimental observations of IRAV modes in conjugated polymers

The first experimental observation of IRAV modes was made in doped polyacetylene, trans-(CH)x and trans-(CD)x , as shown in Fig. 11.1. Fincher and coworkers (Fincher *et al.*, 1979) found that upon doping two new absorption modes (1370 and 900 cm^{-1}) appeared in the IR region. The spectra were correctly attributed to molecular vibrations made IR-active by doping, and were considered an evidence for charged solitons.

In a later experiment, for doped polyacetylene the lowest frequency IRAV mode was found at approximately 900 cm^{-1} whereas upon photoexcitation the lowest frequency mode was at approximately 450 cm^{-1}

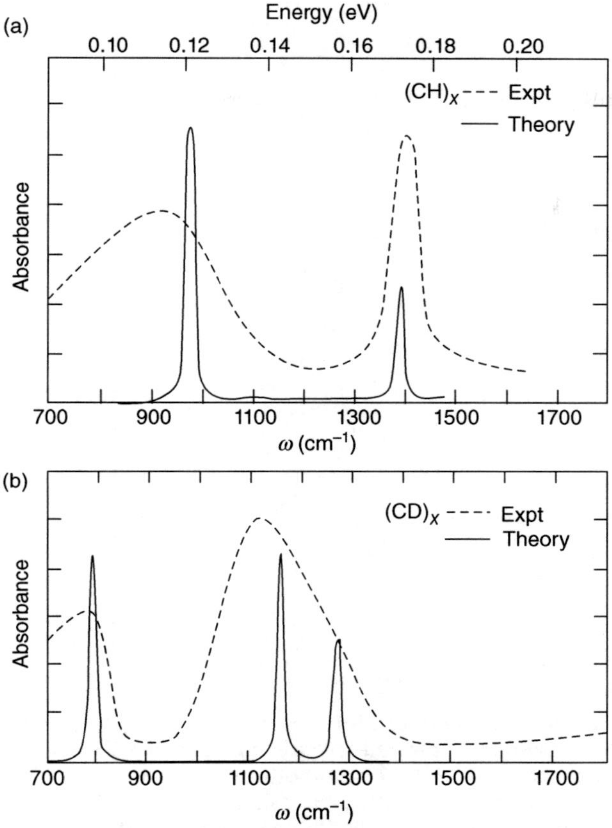

11.1 Doping-induced IRAV modes for *trans*-polyacetylene (a) *trans*-$(CH)_x$; (b) *trans*-$(CD)_x$. Reprinted with permission from Etemad, S., Pron, A., Heeger, A. J., Macdiarmid, A. G., Mele, E. J. & Rice, M. J. 1981. Infrared-active vibrational modes of charged solitons in $(CH)_x$ and $(CD)_x$. *Physical Review B*, **23**, 5137–5141. Copyright (1981) by the American Physical Society (DOI: 10.1103/PhysRevB.23.5137).

(Schaffer *et al.*, 1987). For doping-induced charged solitons, pinning arises from a combination of the coulomb attraction of the counter-ion and structural disorder. For photoinduced solitons the weaker pinning arises only from structural disorder due to the absence of translational symmetry.

Four localized IRAV modes (1088, 1161, 1200, 1354 cm^{-1}) associated with bipolaron distortions of the poly(3-hexylthiophene) (P3HT) chains and made infrared-active through coupling to the uniform translation of the bipolaron (see Fig. 11.2) were observed in both the doping-induced and PIA spectra, as for instance in polythiophene (PT) and P3HT (Kim *et al.*, 1988). An additional mode (1396 cm^{-1}) was identified from an IR-active

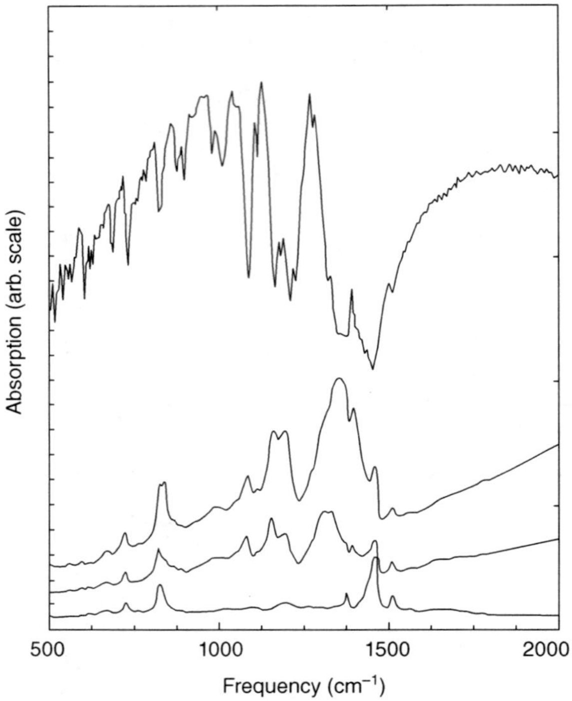

11.2 Detailed photoinduced infrared spectral changes of P3HT (top), and doping-induced infrared spectral changes of P3HT (bottom, undoped; second and third from bottom, 1 and 3 mol/% PF6-, respectively). Reprinted with permission from Kim, Y. H., Spiegel, D., Hotta, S. & Heeger, A. J. 1988. Photoexcitation and doping studies of poly(3-hexylthienylene). *Physical Review B*, **38**, 5490–5495. Copyright (1988) by the American Physical Society (DOI: 10.1103/PhysRevB.38.5490).

localized mode associated with the nonuniform translation of the bipolaron. Comparison of the energies of the photoinduced and doping-induced electronic transitions yields an estimate of the change in Coulomb energy of the bipolaron on photoexcitation, $U_B \approx 0.25$ eV; this relatively small value of U_B is consistent with bipolaron formation in P3HT.

11.2.2 Carrier density and dynamics investigated by ultrafast photoinduced IRAV modes absorption in conjugated polymers

In doped conjugated polymers, the strength of the IRAV modes is proportional to the doping level. Upon photoexcitation of semiconducting polymers (Fincher *et al.*, 1979; Voss *et al.*, 1991), there is a one-to-one correspondence between the photoinduced IRAV modes (steady-state and transient) and the doping-induced IRAV modes of the same polymers (Fincher *et al.*, 1979; Voss *et al.*, 1991; Soos *et al.*, 1994; Ehrenfreund and Vardeny, 1997). Thus, the strength of the photoinduced IRAV modes is directly proportional to the density of photogenerated charge carriers on the polymer chain.

The first utilization of photoinduced IRAV modes for measuring photocarrier density was reported by Mizrahi and coworkers (Mizrahi *et al.*, 1999) for MEH-PPV/C60. With a temporal resolution on the order of 100 ps, they demonstrated PIA with the characteristic signatures of the IRAV modes, identical to those observed in steady-state experiments. Moses *et al.* (2000) utilized an approach for quantitatively estimating the density of photoinduced charge carriers at subpicosecond times by transient excited-state absorption probed in the spectral region spanning IRAV modes in the prototypical luminescent polymers, PPV and MEH-PPV (see Fig. 11.3). In MEH-PPV, the lifetime of the photoinduced IRAV modes and the exciton lifetime are different; the IRAV lifetime is more than an order of magnitude shorter than the lifetime of the neutral excitons as inferred from the photoluminescence decay time (Sariciftci *et al.*, 1992; Sariciftci and Heeger, 1994). When mixed with acceptors such as C60, conjugated polymers (PPV, MEH-PPV, etc.) undergo ultrafast photoinduced electron transfer with an

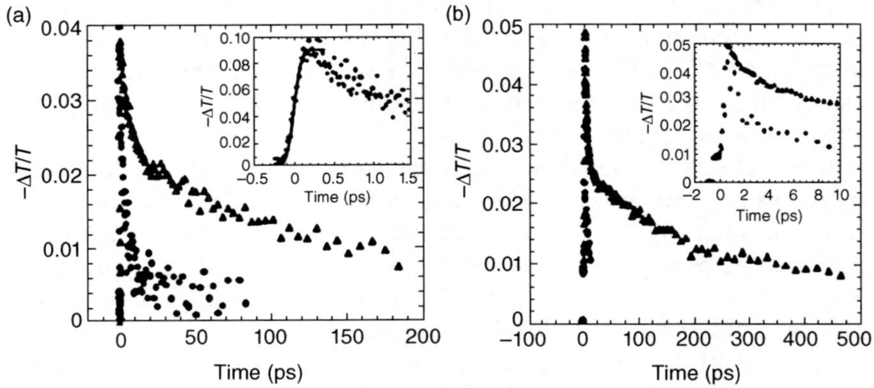

11.3 (a) Comparison of the PIA waveform in MEH-PPV and MEH-PPV/C60 (50%, C60, by weight), when pumped at the same light intensity at 800 nm and probed at 7 μm. (b) Comparison of the PIA waveform in MEH-PPV and MEH-PPV/C60 (10%, C60, by weight), when pumped at the same light intensity at 400 nm and probed at 9 μm. Reprinted with permission from Moses, D., Dogariu, A. & Heeger, A. J. 2000. Ultrafast photoinduced charge generation in conjugated polymers. *Chemical Physics Letters*, **316**, 356–360. Copyright (2000) by Elsevier (DOI: 10.1016/s0009-2614(99)01316-0).

associated increase in the PIA (both steady-state and transient) (Sariciftci *et al.*, 1992; Sariciftci and Heeger, 1994) and the photoconductivity signals (Lee *et al.*, 1993; Kraabel *et al.*, 1996), with an associated decrease (quenching) of the photoluminescence. As expected for photoinduced electron transfer, the strength of the photoinduced IRAV modes is proportional to the concentration of C60 in polymer/C60 composites. These observations provide direct experimental evidence that neutral excitons do not generate IRAV mode absorption. If excitons did generate IRAV mode absorption, addition of C60 would reduce the strength of the photoinduced IRAV modes, since the exciton density is quenched by efficient photoinduced electron transfer. In contrast, a significant increase in the IRAV mode PIA signal is measured upon addition of C60. So it is concluded that the strength of the photoinduced IRAV mode absorption provides a direct, all-optical, ultrafast probe to the charge carrier density at the short timescales ($t < 100$ fs) typical of carrier thermalization in disordered semiconductors.

Miranda *et al.* (2001) detected charged photoexcitations (polarons) in conjugated luminescent polymers by ultrafast photoinduced IRAV absorption. Figure 11.4 shows the spectra of the IRAV modes obtained with steady-state excitation for the MEH-PPV/C60 blend, and with the ultrafast setup for both MEH-PPV and the MEH-PPV/C60 blend. The steady-state and ultrafast spectra are in reasonable agreement, with all the spectral features present in both. They are also in good agreement with the data obtained at 100 ps time resolution. The ultrafast photoinduced IRAV modes demonstrate that polarons are produced in less than 100 fs, consistent with the early predictions of Su and Schrieffer. From the ratio of IRAV signals in pristine MEH-PPV and the MEH-PPV/C60 blend, the quantum efficiency for charge pair generation in pristine MEH-PPV when pumped at 400 nm was estimated as 10%.

Soci *et al.* (2005) studied the charge carrier density in highly ordered PPV by means of photoinduced absorption probed at IRAV modes. Figure 11.5 shows the intensity dependence of the IRAV signal, as obtained with pump polarized perpendicular and probe polarized parallel to the polymer chain axis. The ultrafast onset of the IRAV signals confirms that charge-carrier photogeneration happens on a time scale faster than the experimental resolution (300 fs). However, by taking into account the laser photon energy, intensity, laser beam size, reflectivity, and absorption coefficient of the sample, and assuming a carrier quantum yield of 10% (according to Miranda *et al.*), an absolute value for the initial carrier density consistent with direct fast transient photocurrent measurements was deduced.

Holt *et al.* (2009) used a variety of optical probes to study the long-lived photoexcitations in blends of MEH-PPV with a strong electron acceptor molecule, 2,4,7-trinitro-9-fluorenone (TNF). They found a strong PIA signal when the MEH-PPV/TNF blend is pumped using excitation below the

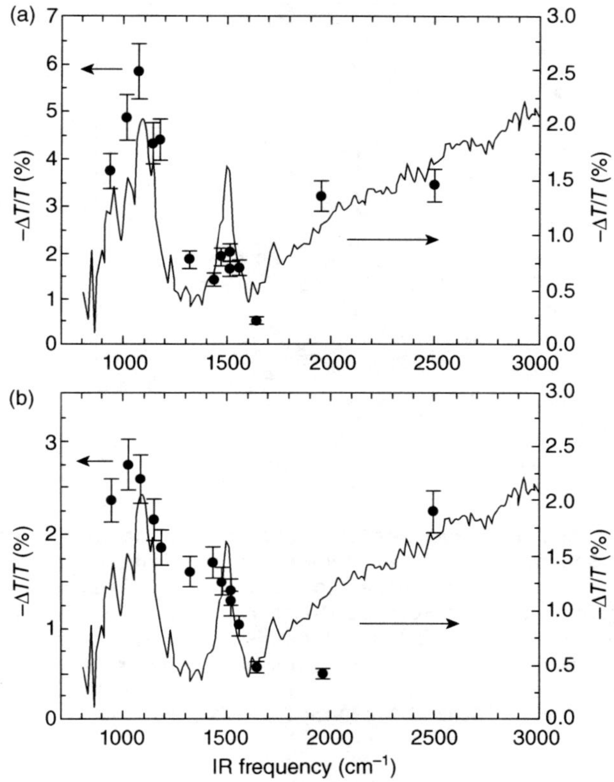

11.4 Photoinduced IRAV spectrum for (a) MEH-PPV/C60 blend and (b) pristine MEH-PPV. Reprinted with permission from Miranda, P. B., Moses, D. & Heeger, A. J. 2001. Ultrafast photogeneration of charged polarons in conjugated polymers. *Physical Review B*, **64**, 081201(R). Copyright (2001) by the American Physical Society (DOI: 10.1103/PhysRevB.64.081201).

pristine MEH-PPV bandgap. Fourier transform infrared (FTIR) spectroscopy PIA reveals the existence of various IRAV modes below 0.2 eV. The IRAV transitions indicate that P1 (0.3 eV) and P2 (1.2 eV) PIA bands are associated with charge species and confirm that charges indeed develop along the polymer backbone upon photoexcitation.

11.3 Transient photocurrent (TPC) spectroscopy

11.3.1 Auston switch

In 1975, D.H. Auston proposed a device for 'Picosecond optoelectronic switching and gating in silicon' (Auston, 1975). The principle was to create,

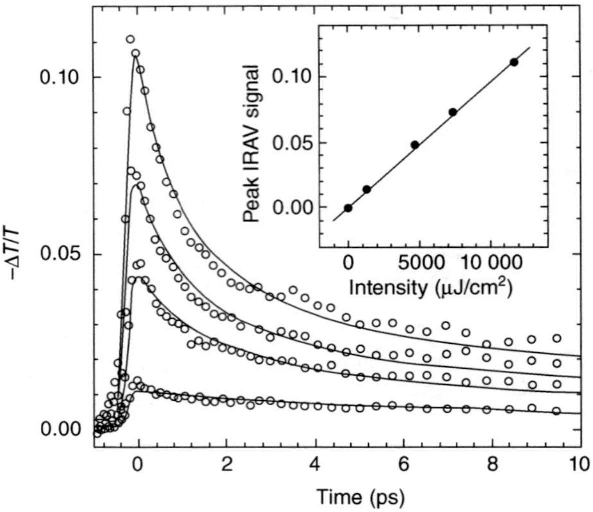

11.5 Intensity dependence of the transient IRAV signal in oriented PPV, measured with perpendicular pump polarization and parallel probe polarization with respect to the polymer chain axis. Reprinted with permission from Soci, C., Moses, D., Xu, Q. H. & Heeger, A. J. 2005. Charge-carrier relaxation dynamics in highly ordered poly(*p*-phenylene vinylene): Effects of carrier bimolecular recombination and trapping. *Physical Review B*, **72**. Copyright (2005) by the American Physical Society (DOI: 10.1103/PhysRevB.72.245204).

by photocarrier generation, a short circuit between two ends of a gap in a microstrip waveguide, as shown in Fig. 11.6(a). A short wavelength 'on' pulse absorbed near the surface of the silicon substrate is used for photoexcitation. A consecutive 'off' pulse with higher wavelength and hence higher penetration depth creates a short circuit between the ground plane and the waveguide. With this configuration, pulses with a length controlled by the delay between on and off pulses can be generated. If only the 'on' pulse was used, the length of the generated electrical pulse would be determined by the carrier lifetime of the silicon substrate. However, the delay between 'on' and 'off' pulses can be controlled on a timescale orders of magnitude smaller.

The same device can be used for pulse generation, by applying a DC bias on the input, or for optical sampling of an incoming arbitrary waveform. Evolutions of the device consisted in replacing the silicon substrate with other materials having shorter-lived carrier lifetimes, to eliminate the need for the off pulse. Instead of shining the off pulse, cross-correlation of the generated electrical pulse can be obtained by a sampling probe pulse (Fig. 11.6(b)). Radiation damaged or implanted materials (Smith *et al.*, 1981; Ketchen *et al.*, 1986), amorphous materials (Auston *et al.*, 1980) and advanced

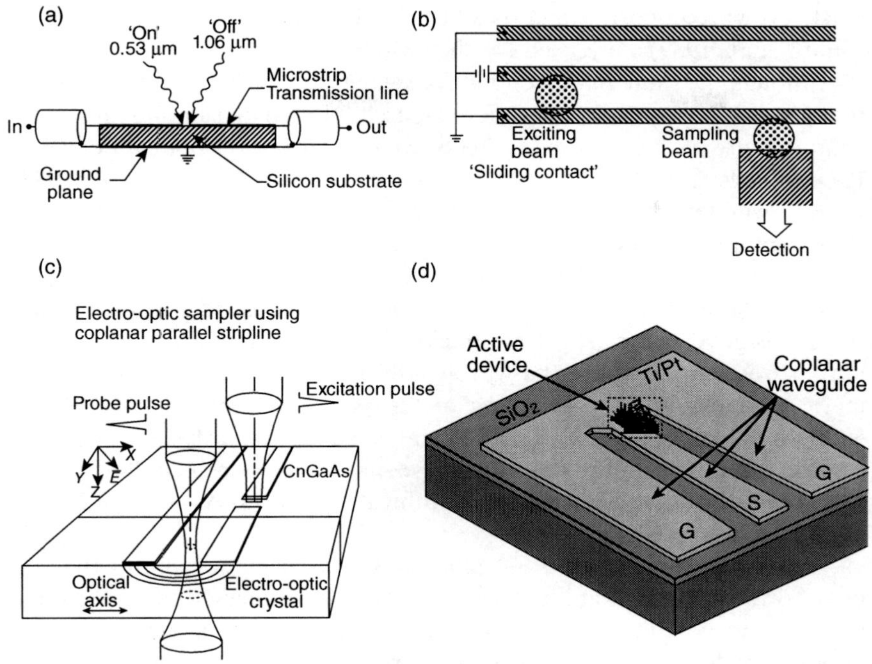

11.6 (a) An optoelectronic gate switches on by absorption of 0.53 μm pulse and switches off by a delayed 1.06 μm pulse. (b) Coplanar transmission line consisting of three parallel 5 μm wide Al lines separated by 10 μm. Also shown is the 25 μm wide sampling probe separated by a 10 μm gap from the transmission line. The exciting and sampling laser beams are indicated by the circles. The DC bias was 20 V. (c) Electro-optic sampler arrangement with coplanar strip geometry. (d) Illustration of the InP-nanowire high-speed photoconductor. Reprinted with permission from: (a) Auston, D. H. 1975. Picosecond optoelectronic switching and gating in silicon. *Applied Physics Letters*, **26**, 101–103, Copyright (1975), American Institute of Physics (DOI: 10.1063/1.88079.); (b) Ketchen, M. B., Grischkowsky, D., Chen, T. C., Chi, C. C., Duling, I. N., Halas, N. J., Halbout, J. M., Kash, J. A. & Li, G. P. 1986. Generation of subpicosecond electrical pulses on coplanar transmission-lines. *Applied Physics Letters*, **48**, 751–753, Copyright (1986), American Institute of Physics (DOI: 10.1063/1.96709); (c) Mourou, G. A. & Meyer, K. E. 1984. Subpicosecond electro-optic sampling using coplanar strip transmission-lines. *Applied Physics Letters*, **45**, 492–494, Copyright (1984), American Institute of Physics (DOI: 10.1063/1.95312); (d) Logeeswaran, V. J., Sarkar, A., Islam, M. S., Kobayashi, N. P., Straznicky, J., Li, X., Wu, W., Mathai, S., Tan, M. R. T., Wang, S.-Y. & Williams, R. S. 2008. A 14-ps full width at half maximum high-speed photoconductor fabricated with intersecting inp nanowires on an amorphous surface. *Applied Physics A: Materials Science & Processing*, **91**, 1–5. Copyright (2008), with permission from Springer (DOI: 10.1007/s00339-007-4394-x).

MBE grown composite materials such as low-temperature grown GaAs (Smith *et al.*, 1989) were used for the switch.

Although the same photoconductive switch is suitable for both photocurrent generation and sampling, later research focused on sampling the propagating pulse by electro-optic effects, which enabled all-optical detection. Pockels cells were first built on $LiNbO_3$ crystals (Valdmanis *et al.*, 1982); $LiTaO_3$ was later used as optoelectric material (Valdmanis *et al.*, 1983). Furthermore, the performance is increased by switching to thinner substrates (Holzman *et al.*, 2000) or to coplanar transmission line design (Fig. 11.6(c)) (Mourou and Meyer, 1984) since in microstrip design, the line spacing is governed by substrate thickness and dispersion becomes important at frequencies where the wavelength drops below the line spacing. A major progress was the sampling of the pulse by exploiting the Franz–Keldysh effect (Lampin *et al.*, 2001) which is a particularly sensitive optoelectric effect and offers the advantage that the device including emitter and detector can be fabricated monolithically on SI GaAs substrate carrying a LT-GaAs epilayer.

Auston switches are used for generating few-oscillation pulses that can be used for transmission line spectroscopy or transformed into free-space radiation in combination with antennas. An application with particular importance for organic electronics is an Auston switch implemented by depositing a transmission line gap over an organic semiconductor with unknown optoelectric properties. This method is often referred to as TPC spectroscopy and will be described in detail in the next section. The photoconductive switch is mostly implemented as microstrip design since organic semiconductors are often incompatible with the use of standard photolithographic methods because of the need of chemical resists. Examples of coplanar design do, however, exist (Logeeswaran *et al.*, 2008) and yield performance gains of an order of magnitude. Their schematic is shown in Fig. 11.6(d).

In recent work, we applied electromagnetic modeling to Auston switch structures (Diesinger *et al.*, 2011, 2012) with two main objectives:

1. Assessing the potential of a switch structure featuring a transition from microstrip gap to coplanar accesses: with this approach, a gap can be defined on organic semiconductors by compatible deposition techniques and the performance of coplanar design could still be approached by optimization of the transition.
2. Decomposing the Auston switch into a hybrid model, consisting of intrinsic material response and structural response. The signal deterioration due to the structural response can then be eliminated by deconvolution, yielding the intrinsic material response with the fastest achievable time resolution.

The photoconductive switch was modeled as a 4-port S-matrix (see Fig. 11.7). The S-parameters were calculated by the electromagnetic modeling tool High Frequency Structure Simulator (HFSS) and used to describe the structural response of the switch as transimpedance function between the current generated in the unknown material and the output signal propagating along the transmission line. Previous approaches based on a description of the switch by S-parameters (Green and Sobolewski, 2000; Tripon-Canseliet *et al.*, 2006) were fundamentally different since they aimed at computing the transfer between input and output ports rather than representing the active area by a current generator and describing the effect of the structure as a transimpedance.

11.3.2 Acquisition of photocurrent transients

The TPC characterization method consists of incorporating an unknown semiconductor into an Auston switch that is DC biased on one side, and in observing, on a fast oscilloscope or boxcar integrator, a photocurrent transient generated in response to a negligibly short laser pulse incident on the switch. A summary of application of the method on molecular crystal, polymers, polymer blend is given in following paragraphs.

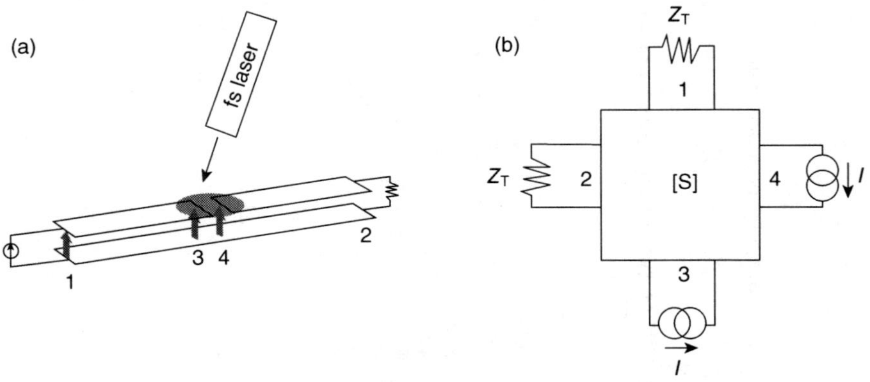

11.7 Approach of modeling a photoconducive gap in a microstrip as 4-port S-matrix: (a) schematic of the microstrip gap with active material (oval spot) with port assignment; (b) equivalent circuit; outer ports 1 and 2 are considered to be terminated with the transmission line impedance Z_T, while inner ports 3 and 4, located at the transmission line edges, are open. Adapted from Diesinger, H., Panahandeh-Fard, M., Baillargeat, D. & Soci, C. 2011. Electromagnetic modeling and optimization of photoconductive switches for terahertz generation and photocurrent transient spectroscopy. *Proceedings of the 2011 IEEE MWP*, 373–376 (DOI: 10.1109/MWP.2011.6088749).

Molecular crystals

Models for the carrier photoexcitation mechanism in molecular crystals have been established initially on the bases of measurements on oligoacenes and later applied to conjugated polymers as well. These models (e.g. the Onsager model) emphasize the localized nature of photoexcitations and describe carrier generation as a secondary process involving exciton dissociation. Moses *et al.* (2006) showed that for crystalline tetracene, the predictions of the Onsager model are at variance with experimental data. The peak height of photocurrent transients increased for lower temperatures, in contrast to the DC photocurrent. By normalizing PC transients to a constant peak height, it was shown that the T dependence of the tail was opposite to the peak T dependence, revealing two transport mechanisms corresponding to an initial peak due to carriers occupying extended states and a slower decay corresponding to thermally activated hopping. This is shown in Figs 11.8 and 11.9. A study on functionalized pentacene, comparing optical pump THz probe and PC transient, has been presented by Hegmann *et al.* (2004). The conclusions are similar to the case of tetracene, where the height of the initial peak that rises within 0.5 ps (limited by the resolution of the THz probe) increases with decreasing temperature.

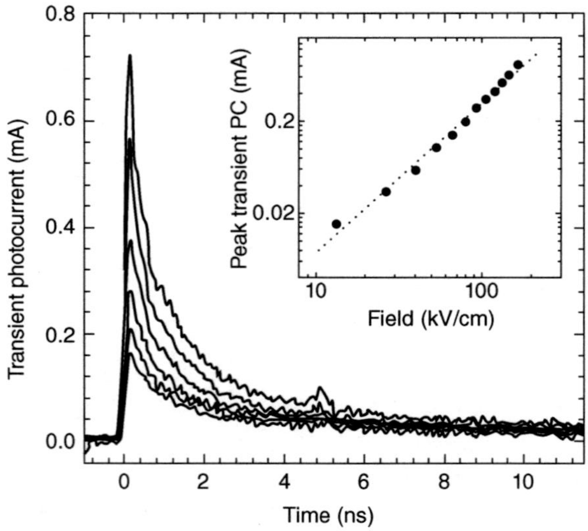

11.8 Photocurrent transients measured at different temperatures, *T* increasing from top to bottom; the inset shows the field dependence of the peak height. Reprinted with permission from Moses, D., Soci, C., Chi, X. L. & Ramirez, A. P. 2006. Mechanism of carrier photogeneration and carrier transport in molecular crystal tetracene. *Physical Review Letters*, **97**, 067401. Copyright (2006) by the American Physical Society (DOI: 10.1103/PhysRevLett.97.067401).

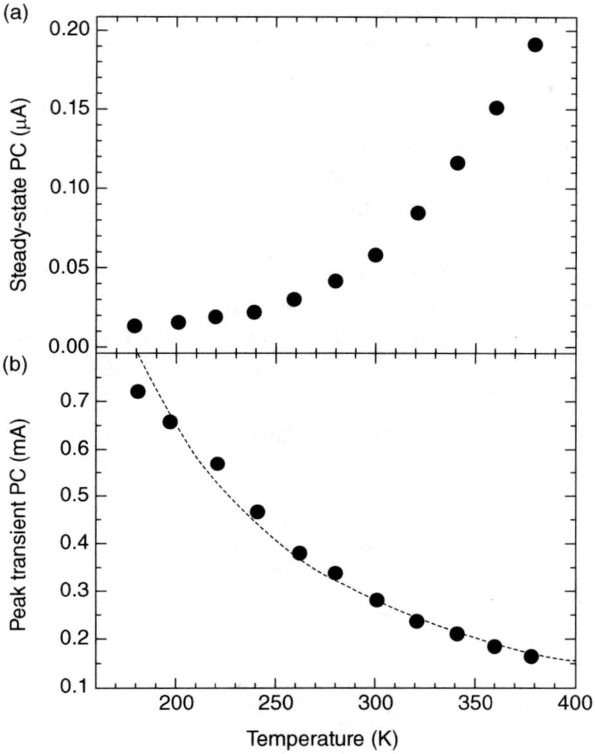

11.9 Temperature dependence of steady state PC vs temperature dependence of the transient peak height. Reprinted with permission from Moses, D., Soci, C., Chi, X. L. & Ramirez, A. P. 2006. Mechanism of carrier photogeneration and carrier transport in molecular crystal tetracene. *Physical Review Letters*, **97**, 067401. Copyright (2006) by the American Physical Society (DOI: 10.1103/PhysRevLett.97.067401).

Polymers

Moses, Sinclair, and Heeger first used transient photocurrent spectroscopy to study the photogeneration and transport properties of polydiacetylene-(bis *p*-toluene sulfonate) (PDA-TS), showing that the carrier generation is in disagreement with the Onsager model (Moses *et al.*, 1987). Temperature and field-dependent transients showed a temperature-independent initial peak due to hot carriers and a thermally activated tail attributed to trap dominated transport. The temperature and field independence of initial generation quantum yield shows that the Onsager model is also not applicable in this organic polymer crystals. The time resolution of the measurement was 3 ns and the transient, including the tail, around 30 ns.

A study by Moses *et al.* (2006) on PPV and its derivatives using a combination of IRAV PIA spectroscopy and photocurrent transient spectroscopy shows that polarons and excitons are both independent primary excitations. An apparent increase of photoconductivity at photon energies above 3–4 eV is due to external currents induced by electron photoemission, an experimental artifact. Using a quenching gas to suppress photoemitted electrons, the photoconductivity becomes nearly independent on photon energy.

Soci *et al.* (2005) studied charge carrier dynamics in highly ordered PPV by IRAV PIA and photoconductance transient on a 10 ps and 20 ns timescale respectively. At short times after the photoexcitation (10 ps, as measured by IRAV), the carrier dynamics follows a decay largely dominated by bimolecular recombination, as shown in Fig. 11.10. At longer times (20 ns, as obtained by TPC measurements), the carrier dynamics is the result of a system of differential equations that include bimolecular recombination

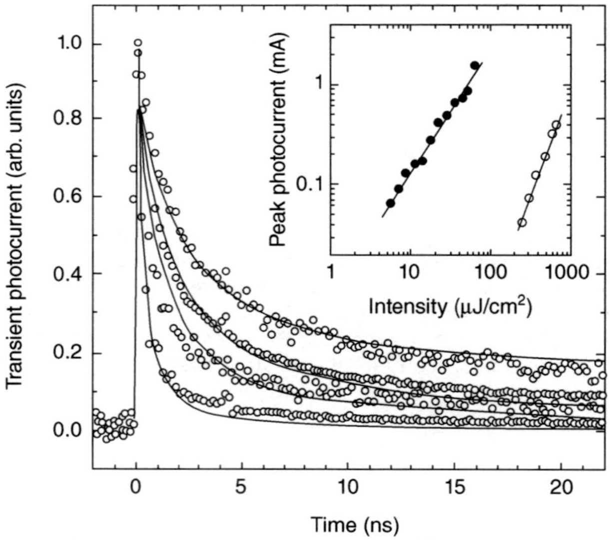

11.10 Photocurrent transient of highly ordered PPV sample for different intensities increasing from top to bottom, normalized at $t = 0$; the solid lines are the solutions of the carrier dynamic differential equation; the inset shows the peak height as function of excitation intensity for one- and two-photon excitation (solid and open circles respectively). Reprinted with permission from Soci, C., Moses, D., Xu, Q. H. & Heeger, A. J. 2005. Charge-carrier relaxation dynamics in highly ordered poly(*p*-phenylene vinylene): Effects of carrier bimolecular recombination and trapping. *Physical Review B*, **72**, 245204. Copyright (2005) by the American Physical Society (DOI: 10.1103/PhysRevB.72.245204).

(with the same coefficient as determined for first 10 ps), trapping, detrapping, and recombination of trapped carriers with free carriers of opposite sign. It is noteworthy that monomolecular recombination is not needed to describe the carrier dynamics. It is suggested that in other conjugated polymers where the free carrier density is described by an initial fast exponential decay followed by a slower stretched exponential, the mechanism responsible of the fast exponential is trapping rather than monomolecular recombination. Due to the highly ordered material used in this study, the initial decay is dominated by bimolecular recombination and the trapping becomes important at timescales observed by TPC. The time resolution of IRAV and PC transient were 300 fs and 100 ps, respectively, and the observation intervals of the above examples were 10 ps for IRAV and 20 ns for PC transient.

Polymer blends

Soci *et al.* (2007) compared photoresponsivities and photocurrent transients of pristine polymers P3HT and poly[2,6-(4,4-bis-(2-ethylhexyl)-4H-cyclopenta[2,1-b;3,4-b′]dithiophene)-alt-4,7-(2,1,3-benzothiadiazole)] (PCPDTBT) with corresponding bulk heterojunction blends of these polymers with a fullerene derivative (phenyl-C61-butyric acid methyl ester, PCBM). They found a 100-fold increase in DC photoresponsivities and the appearance of a photocurrent tail that decays within few nanoseconds, whereas the prisitine polymers show a TPC peak of less than 100 ps wide and limited by instrumental resolution. The identical effect of the blend on PCPDTBT as on P3HT, in combination with the lower bandgap of PCPDTBT, makes this polymer promising for high energy conversion efficiency solar cells.

Timescales

The time resolution of the technique is currently about 100 ps. The limitation is essentially due to the structural response of an Auston switch implemented as gap in a microstrip line, as discussed by Diesinger *et al.* (2012). By optimizing the switch design and deconvolving its response from the experimental transient, a resolution of 10 ps and less can be achieved. The time interval of the observation is limited only by the memory and sampling interval of the storage oscilloscope typically used.

11.4 Time-resolved terahertz spectroscopy (TRTS)

Time-resolved terahertz (THz) spectroscopy of organic semiconductors typically employs an optical pump pulse and a THz probe which is non-resonant with the bandgap to monitor the charge carriers generated upon photoexcitation. As in conventional pump-probe techniques, the relative change in transmission of the THz pulse induced by the pump beam, $\Delta T/T_0$, is monitored as a function of the delay time of the THz probe with respect to the pump pulse. Next, the change in transmission coefficient is related to the refractive index or, via Fresnel-like equations, to the conductivity change of the sample. Often, it is found that the response of the sample is independent of frequency in the spectral range of the THz pulse, hence the differential transmission is directly related to the DC conductivity. In the simplest case (Hegmann *et al.*, 2002), a linear relation between the relative differential transmission and the DC conductivity is assumed (Fig. 11.11). We will now mention applications of this technique to the study of charged photoexcitations in organic semiconductors; for a general review on Terahertz spectroscopy applied to the study of carrier dynamics in semiconductors we refer the reader to Ulbricht *et al.* (2011).

11.4.1 Molecular crystals

Optical-pump THz-probe spectroscopy was performed on a number of molecular crystals, including pentacene, rubrene, tetracene and pentacene bulk crystals (Thorsmolle *et al.*, 2004; Ostroverkhova *et al.*, 2005a, 2006) and their films (Ostroverkhova *et al.*, 2005b).

Hegmann *et al.* (2002) suggest that in functionalized pentacene, mobile carriers are a primary photoexcitation because of the subpicosecond rise-time of the photoconductivity. Furthermore, it was found that the measured photoconductance is independent on the amplitude of the THz field peak in contradiction with field induced exciton dissociation; the generation of mobile carriers in less than 0.5 ps, i.e. below the system resolution, is in agreement with previous studies by Miranda *et al.* (2001) and Moses *et al.* (2000). At times $t > 4$ ps, the photoconductivity decay was found to follow a $t^{-\beta}$ law, indicative of dispersive transport. The exponent beta was found to be nearly temperature independent between 10 and 300 K. This is inconsistent with multiple trapping models. Temperature dependence does exist for the amplitude of the initial peak that increases with decreasing temperature, indicative of bandlike transport. The time resolution of the photoconductivity measurement is limited by the THz probe pulse to some picoseconds, and the observation interval here is about 40 ps due to the fast decay time. The observation interval could be made longer by using longer delay stages (e.g. a mechanical delay of 30 cm would correspond to a nanosecond optical delay). Thorsmolle *et al.* (2004) found similar results in pentacene and provide an estimate for the carrier mobility based on the number of absorbed photons and the observed conductivity peak.

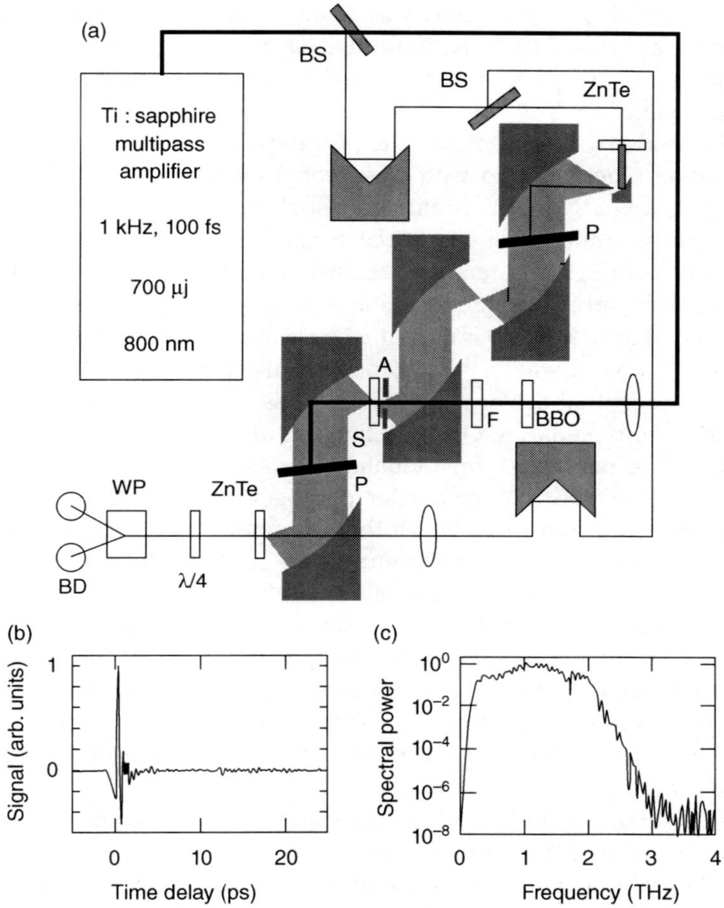

11.11 (a) Schematic of the THz apparatus. BS: beam splitter; P: polyethylene visible beam block; F: blue filter; S: sample; A: 1 mm aperture; l/4: quarter-wave plate; WP: Wollaston prism; and BD: balanced detector. Gold coated parabolic mirrors are used to direct the THz pulse through the system. (b) Single-scan THz trace; full width at half maximum 250 fs and (c) spectrum obtained under dry N_2. The oscillatory nature of the spectrum is an artifact resulting from multiple reflections of the THz pulse. Reprinted with permission from Lui, K. P. H. & Hegmann, F. A. 2001. Ultrafast carrier relaxation in radiation-damaged silicon on sapphire studied by optical-pump-terahertz-probe experiments. *Applied Physics Letters*, **78**, 3478–3480, Copyright (2001), American Institute of Physics (DOI: 10.1063/1.1375841).

Carrier mobility in pentacene was again observed to increase from room temperature down to 30 K, indicative of bandlike transport. It was also found that the initially free carriers became trapped within few picoseconds.

The study by Ostroverkhova *et al.* (2005a) on crystals of a pentacene derivative functionalized with triisopropylsilylethynyl (TIPS) side groups and pentacene thin films confirmed initial bandlike transport by the temperature dependence of the initial peak. It is emphasized that bandlike mobility increases with temperature, where temperature activated mobility had not been previously reported in polyacene thin films. In thin films, the photoconductivity decay is described by a single exponential. In contrast to the single crystal where the photoconductivity follows a $t^{-\beta}$ behavior, the single exponential decay is attributed to fast carrier capture by deep traps at the grain boundaries. Ostroverkhova *et al.* (2006) compared the carrier dynamics of pentacene, functionalized pentacene, tetracene, and rubrene single crystals. Fast charge carrier photogeneration and bandlike charge transport were confirmed for all these samples. The decay dynamics were studied as function of temperature down to 20 K. A transition to faster decay dynamics in pentacene single crystals was observed below 70 K. At low temperature, the power law exponent of functionalized pentacene was found to be temperature independent, while in pentacene the exponent beta went from 0.6 to 1.2 for temperatures from 100 K down to 5 K. These findings suggest that in pentacene carriers are trapped by shallow traps at very low temperature, whereas such traps are absent in functionalized pentacene.

Ostroverkhova *et al.* (2005b) then studied pentacene and functionalized pentacene thin films with respect to their fabrication methods. One of the films of functionalized pentacene evaporated at 25 °C, exhibiting the fastest decay dynamics described by a single-exponential $e^{t/\tau}$ with $\tau = 0.6$ to 1.4 ps. In contrast, in films evaporated at 85 °C and in solution grown functionalized pentacene, the decay could be fitted by the power-law function $t^{-\beta}$ with $\beta = 0.53$ over at least 100 ps, as shown in Fig. 11.12. The different decay dynamics are attributed to the structure: functionalized pentacene evaporated at 25 °C shows small interconnected domains in atomic force microscpy (AFM) images. The fast single exponential decay is attributed to fast trapping on grain boundaries, whereas the larger crystallites obtained by evaporation at 85 °C and solution grown films lead to longer carrier lifetimes and power law decays similar to single crystal over more than 100 ps. Pristine pentacene samples showed a decay that could be fitted by a biexponential function. The deep-level trapping on grain boundaries, defects, and/or chemical impurities prevented the observation of the carrier transport on time scales above 20–30 ps after photoexcitation.

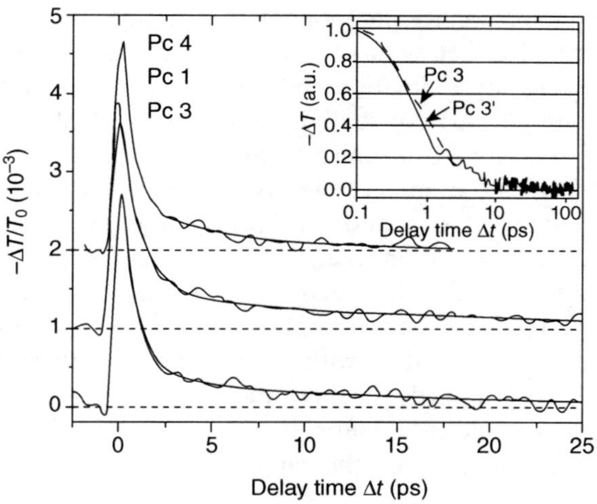

11.12 Differential transmission $-\Delta T/T_0$ as a function of pump-probe delay time Δt obtained in Pc 1 (evaporated on KCl at 35 °C), Pc3 (evaporated on mica at 35 °C), and Pc4 (evaporated on glass at 25 °C) thin films under optical excitation at 580 nm in air at room temperature. Fits with a biexponential function are also shown. The inset shows the ΔT transients normalized to their values at Δt=0 obtained under the same conditions in the Pc 3 and Pc 3′ films. Reprinted with permission from Ostroverkhova, O., Shcherbyna, S., Cooke, D. G., Egerton, R. F., Hegmann, F. A., Tykwinski, R. R., Parkin, S. R. & Anthony, J. E. 2005. Optical and transient photoconductive properties of pentacene and functionalized pentacene thin films: Dependence on film morphology. *Journal of Applied Physics*, **98**, 033701. Copyright (2005), American Institute of Physics (DOI: 10.1063/1.1949711).

11.4.2 Polymers

In contrast to the above cases where the response of the investigated film to the photoexcitation is a purely real conductivity, examples exist in the literature where a phase shift of the transmitted THz pulse and hence complex conductivity has been observed. For instance, the transient photoconductivity of GaAs (bulk and nanowires) (Beard *et al.*, 2000; Parkinson *et al.*, 2007) has been interpreted in terms of Drude-like charge carrier dynamics.

The method has been applied by Hendry *et al.* (2006) to the polymers MEH-PPV and rr-P3HT. It was found that in rr-P3HT, the complex conductivity has real and imaginary components that both increase with frequency between 0.3 and 1.3 THz, and decay similarly on the observation timescale of some nanoseconds. MEH-PPV has a frequency-dependent real conductivity that decays within some picoseconds and a frequency-dependent imaginary component that decays over some hundreds of picoseconds. The behavior of rr-P3HT is interpreted by a model based on tight binding approximation and attributed to free charges. The purely imaginary (long-term) conductivity component of MEH-PPV has been attributed to exciton polarizability. Here the time resolution was 3 ps, similar to the previous case of molecular crystals; due to the relatively slow decay of the photoinduced free and localized charges in these partially ordered polymers, the observation interval in this example was 10 ns.

Unuma *et al.* (2010) used THz spectroscopy to measure the complex conductivity of doped polythiophenes (P3HT and PEDOT) without optical pumping by interpreting their data within the Drude–Smith model. They concluded that carriers are partly localized and that the degree of localization depends on carrier density. Here, time resolution and observation interval are not applicable because the carriers in the sample are induced by static doping.

11.4.3 Polymer donor–acceptor blends

Cunningham and Hayden (2008) performed optical pump Thz probe measurements on P3HT and a P3HT:PCBM blends and modeled the carrier dynamics by the Drude–Smith model. They found that the blend has a higher photocarrier generation yield but a lower mobility than the pristine P3HT. Observation interval is up to 100 ps and time resolution is claimed to be subpicosecond.

Nemec *et al.* (2009) studied a low band-gap polyphenylene copolymer (LBPP-1) blended with PCBM for which the complex conductivity is measured by optical-pump THz-probe spectroscopy. Here the analysis was not based on the assumption that the decay of the photoconductance is small during the THz probing interval; instead, the data were analyzed in 2D frequency space, in which one frequency represents the spectrum of the decay and the other the spectrum of the probe pulse. The complex conductivity was compared to the response predicted by Monte-Carlo simulations rather than from the Drude–Smith model which, according to the authors, is a purely phenomenological description. Time resolution was again on the order of 2.3 ps, and the pump probe delay was swept from 2.5 to 50 ps.

11.5 Time-resolved microwave conductivity (TRMC)

TRMC is a contactless conductivity probing technique that consists in measuring the transmission of a microwave signal across the sample as a function of delay between a pump pulse (visible laser in flash-photolysis TRMC or gamma radiation in the MeV range in pulse-radiolysis TRMC) and the microwave probe pulse. A polymer-coated quartz plate is placed in a microwave resonant cavity at a position of maximum electric field strength, as shown in Fig. 11.13. The photoinduced change in the conductance of the sample upon flash photolysis, ΔG, is monitored as a change in the microwave power, $\Delta P/P$, reflected by the cavity at resonance using microwave circuitry and detection equipment. As microwave source, a Gunn oscillator operating at about 9 GHz and with an output power of about 100 mW is typically used. Microwaves reflected by the cell are detected by a Schottky barrier diode detector. The change in reflected microwave power is in the simplest case and within certain limits proportional to the change in conductance of the polymer film. The response time of such cavities is typically around 40 ns, due to the Q-factor of the resonance. The excitation can hence be performed by a ns laser rather than a fs laser. In contrast to optical pump terahertz probe with ps resolution, the microwave signal is CW at 9 GHz, so observation of microwave reflectance is performed electrically by a detection circuitry and an oscilloscope rather than by optical delay stages, and the interval of observation is typically over hundreds of nanoseconds, out of the typical range of delay stages. Application of this somewhat slower technique is to the observation of charges generated in polymers; the lifetime of the singlet exciton is shorter than the excitation pulse (hence steady state of singlet excitons establishes during the excitation), but the free charges, whether dissociated excitons or directly generated, do have lifetimes in the range of this long timescale.

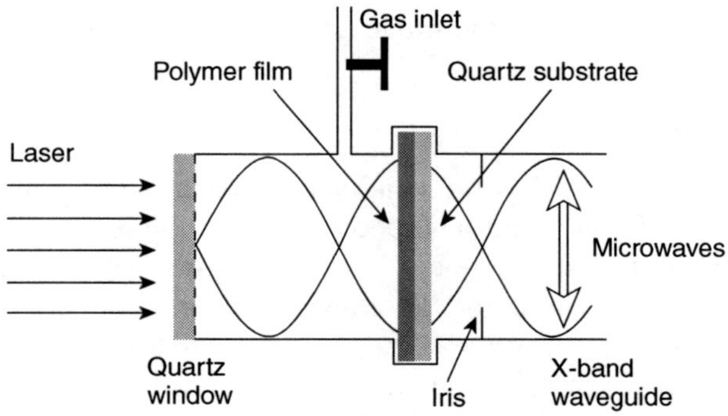

11.13 Schematic representation of the microwave resonant cavity containing a thin-layer sample (not to scale). The sinusoidally varying dashed lines represent the standing-wave pattern of the microwave electric field. Reprinted with permission from Dicker, G., De Haas, M. P., Siebbeles, L. D. A. & Warman, J. M. 2004. Electrodeless time-resolved microwave conductivity study of charge-carrier photogeneration in regioregular poly(3-hexylthiophene) thin films. *Physical Review B*, **70**, 045203. Copyright (2004) by the American Physical Society (DOI:10.1103/PhysRevB.70.045203).

Historically, TRMC was developed for detection of transient ionic conductivity of pulse irradiated dielectric liquids (Infelta *et al.*, 1977) and derived from pulse radiolysis with optical detection. Later, the evolution from pulse radiolysis to flash photolysis (De Haas and Warman, 1982) consisted of extending the detection from charge separation induced by high energetic 'ionizing' pulses to photon absorption-induced charge separation. Even before this, the sensitivity and time resolution was increased to enable studying the geminate decay of molecular ions in liquids (Bao *et al.*, 1996; Sirringhaus *et al.*, 1998; Brown *et al.*, 2001). In the case of polymer films, the photoinduced conductance of the sample proportionally reduces the microwave power reflected from the microwave chamber. The conductance is then plotted as function of time, which typically yields a multiexponential decay. The temporal decay function is not always analyzed.

11.5.1 Molecular crystals

Saeki *et al.* (2006) measured the photoinduced transient conductivity of pentacene by TRMC. The real and imaginary components of the conductivity were separated. As excitation source, an ArF excimer laser, 193 nm wavelength, with 35–30 ns FWHM was used. The conductivity values were compared to molecular orbital calculations. The decay constant was found to be much faster than expected triplet lifetime. An explanation is that triplets might be quenched by interaction with oxygen and that a high incident photon density may promote triplet–triplet annihilation. The observation interval was limited here to 2 μs due to the fast decay. It can be made as long as desired by choosing the range on the oscilloscope.

11.5.2 Polymers

Dicker *et al.* (2004) found that the conductance change of a regioregular P3HT film is proportional to the absorbed photons at the onset of the curve and proportional to the square root of it above a critical value (Fig. 11.14). This behavior was attributed to the transition from monomolecular to bimolecular recombination, either of the excitons (exciton annihilation) or charge carrier bimolecular recombination. Furthermore, it is related to the lamellar character of the regioregular P3HT: below the threshold of one excited state per lamellae, the recombination is monomolecular. This is in agreement with other works that attribute the reduced bimolecular recombination rate to the two-dimensional character of the conduction (Sirringhaus *et al.*, 1999; Osterbacka *et al.*, 2000; Brown *et al.*, 2001; Juška *et al.*, 2009; Osterbacka *et al.*, 2010). The decay of the photoinduced conductivity was shown over some hundred nanoseconds. The decay waveform, however, was not analyzed in detail.

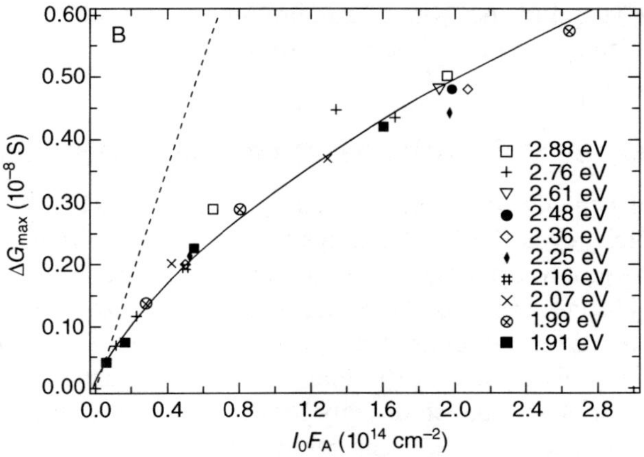

11.14 The dependence of the maximum photoconductance, ΔG_{max}, for a 110 nm film on the product of the incident intensity, I_0, and the fraction of photons absorbed in the sample, *FA*, for photon energies below 3 eV. The data for I_0F_A values are expanded. The full line was calculated and the dashed straight line corresponds to the limiting, low-intensity linear dependence. Reprinted with permission from Dicker, G., De Haas, M. P., Siebbeles, L. D. A. & Warman, J. M. 2004. Electrodeless time-resolved microwave conductivity study of charge-carrier photogeneration in regioregular poly(3-hexylthiophene) thin films. *Physical Review B*, **70**, 045203. Copyright (2004) by the American Physical Society (DOI: 10.1103/PhysRevB.70.045203).

11.5.3 Polymer donor–acceptor blends

The method has also been applied to blends of P3HT:PCBM and pDA2T:PCBM, showing that, upon thermal annealing, the photocarrier generation yield is reduced but the carrier lifetime increases, which is overall beneficial for the photoconversion efficiency of solar cells (Grzegorczyk *et al.*, 2008). Decay times for the pDA2T blend are on the order of microseconds.

Savenije *et al.* (2011) found that, in blends of P3HT:PCBM, charges do not undergo post-nanosecond relaxation into deep traps. By temperature-dependent TRMC, they determined activation energies of 10 meV and 80 meV attributed to hole and electron transport.

Rance *et al.* (2011) studied blends of the polymer poly(2,5-bis(3-tetradecylthiophen-2-yl)thieno[3,2-b]thiophene) (pBTTT) with $PC_{71}BM$ and bis-$PC_{61}BM$, for which the high-frequency carrier mobility is determined from the peak conductance of the TRMC. It is found that the pBTTT:$PC_{71}BM$ blend shows enhanced electron mobility as a result of its intercalated structure with pure PCBM domains, while the pBTTT:bis-$PC_{61}BM$ blend shows hole-dominated conductance with carrier mobility independent of bi-PCBM loading, due to the non-intercalated structure.

11.6 Experimental evidence of charge localization

In the field of organic photovoltaics, a major issue which affects power conversion efficiency is the localization of charges at different scales. Initially, excitons and free charges in band-like states are photogenerated simultaneously. Some excitons dissociate, while extended band-like states thermalize, take part in dispersive transport, or become trapped in more or less shallow traps, giving rise to thermally activated transport. While PIA spectroscopy cannot distinguish directly between polarons and excitons (if not coupled to magnetic spin measurements), photocurrent measurements have the opposite limitation of detecting exclusively charge carriers that are free on the entire device scale. In the following, it shall be discussed to which extent the three methods of TRTS, TRMC and TPC described previously for the characterization of prompt (sub-nanosecond) transport properties are appropriate for detecting partially located charges.

The response of a material containing free and localized charges can be described by a simple expression. The conductivity due to the free carriers is often described by the Drude model containing a force to overcome the inertial moment of the carriers and a viscous drag force:

$$\sigma_f(\omega) = \frac{N_f e^2 \tau_f}{m(1 - i\omega\tau_f)} \qquad [11.1]$$

where N_f is the carrier density, e the elementary charge, τ_f the momentum relaxation time, and m the carrier mass. If $\omega << 1/\tau_f$, the response of the material to an external electric field is described by a purely real conductivity:

$$\sigma_f = \frac{N_f e^2 \tau_f}{m} \qquad [11.2]$$

The localized charges contribute a permittivity that is described by several, or in the most primitive case a single, oscillator strength:

$$\epsilon_l(\omega) = \frac{N_l e^2}{m(\omega_0^2 - \omega^2)} \qquad [11.3]$$

where ω_0 is the resonance frequency. For simplicity the expression contains no dissipative term.

When the response of the photoexcited material is measured in terms of a 'complex conductivity', the contribution from localized charges is reflected by an imaginary term $i\omega\varepsilon_l$, and the entire conductivity writes:

$$\sigma = \sigma' + i\sigma'' = \frac{N_f e^2 \tau_f}{m} + i\omega \frac{N_l e^2}{m(\omega_0^2 - \omega)} = \sigma_f + i\omega\varepsilon_l \qquad [11.4]$$

This corresponds to the description of a lossy dielectric medium and predicts a frequency-independent real part and a frequency-dependent imaginary part of the complex conductivity. Stronger carrier localization translates into a higher ω_0, and the higher the probing frequency ω needs to be to make the imaginary conductance discernible. In the literature, the Drude–Smith model, an extension of the Drude model with the objective of taking into account partly localized carriers, is more frequently seen than the above oscillator model (Smith, 2001). It has been used in THz probe spectroscopy (Unuma *et al.*, 2010) and optical-pump THz-probe spectroscopy (Cunningham and Hayden, 2008). The Drude–Smith model also predicts a rise of the real part of the conductance with increasing frequency. An alternative approach is given by the interpretation of experimental data in terms of the Kubo formalism (Prins *et al.*, 2006).

We illustrate the measurement of photoinduced complex conductivity with examples from the literature: Hendry *et al.* (2006) have studied the complex photoconductivity of rr-P3HT films and MEH-PPV films by TRTS and observed that the THz pulse transmitted through the photoexcited material is subject to a phase shift (Fig. 11.15(a)). Spectra for the real and imaginary parts of the conductivities are derived by Fourier analysis and repeated for different pump-probe delay times. Both real and imaginary components increase for increasing frequency throughout the spectrum in the range from 0.3 to 1.3 THz (Fig. 11.15(c) and (e)). For P3HT real and imaginary components at the probing frequency of 0.6 THz are of the same order of magnitude throughout the observation time interval (Fig. 11.15(b–d)), while for MEH-PPV the real part of the conductivity decays within picoseconds and afterwards the conductivity is almost purely imaginary (Fig. 11.15(e, f)). This difference is explained by the presence of long-lived mobile carriers in P3HT, whereas in MEH-PPV carriers decay within few picoseconds and the purely imaginary part is ascribed to exciton polarizability.

In the case of TRMC, the complex conductivity was measured in pentacene by Saeki *et al.* (2006). The complex conductivity of the film obtained after determining the evolution of the Q-factor of the resonance chamber (by fitting a resonance curve over a Gunn diode frequency interval that changes around 9 GHz, with a relative change of 10^{-5}) is shown in Fig. 11.16. In this case the imaginary component of the conductivity is small compared to the real part; Moreover, these data do not allow verifying the frequency dependence of the imaginary part of the conductivity (as predicted by Eq. 11.5) over a wide range due to the very narrow frequency range imposed by the fitting procedure. In contrast, the center frequency of a THz burst is

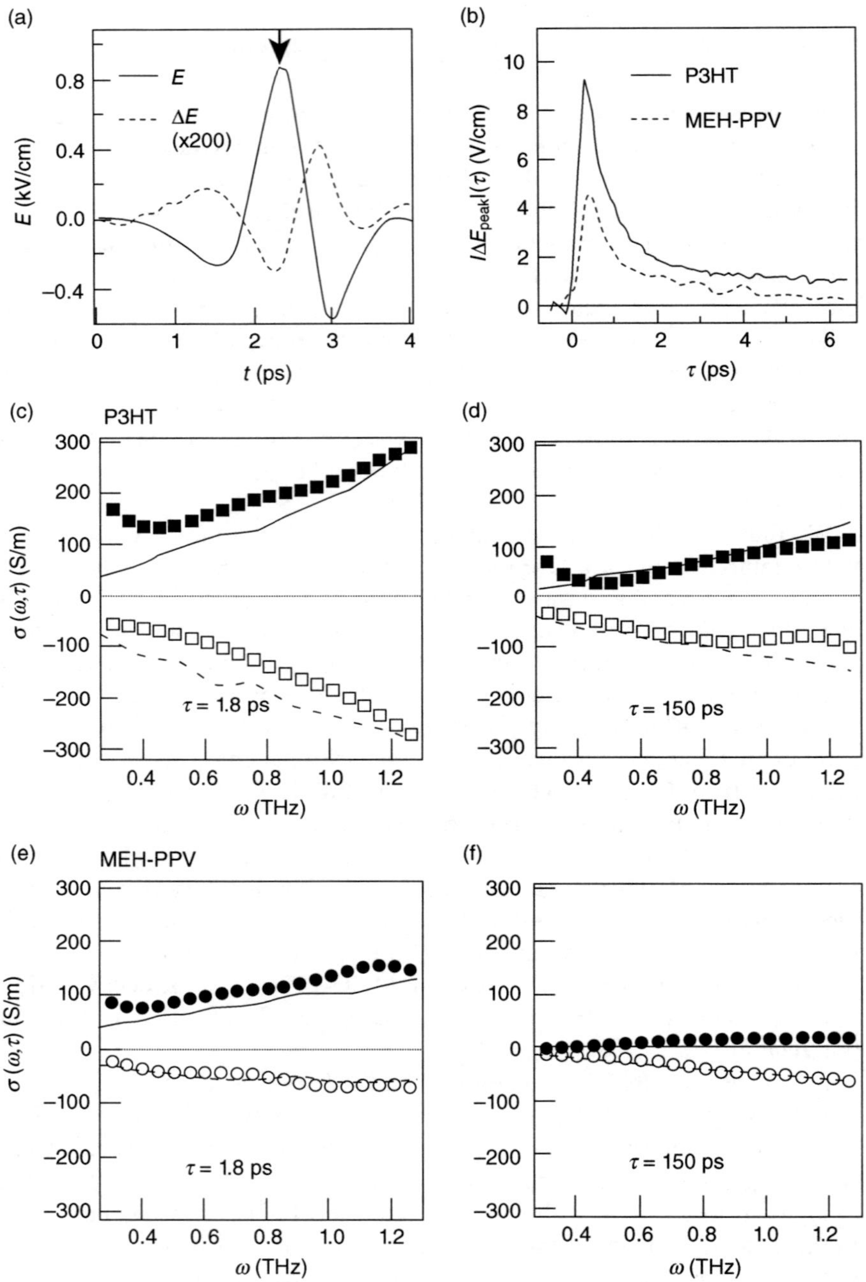
(a)
E
ΔE
(x200)
E (kV/cm)
t (ps)
(b)
P3HT
MEH-PPV
IΔE_peak I(τ) (V/cm)
τ (ps)
(c)
P3HT
σ (ω,τ) (S/m)
τ = 1.8 ps
ω (THz)
(d)
τ = 150 ps
ω (THz)
(e)
MEH-PPV
σ (ω,τ) (S/m)
τ = 1.8 ps
ω (THz)
(f)
τ = 150 ps
ω (THz)

11.15 (a) The electric field $E(t)$ of the THz pulse transmitted through the rr-P3HT sample (solid line), with the modulation $\Delta E(t)$ measured 1 ps after photoexcitation (dotted line) and (b) decay of the modulation measured at the peak field strength. Real and imaginary conductivities for two different times: (c) and (d) rr-P3HT and (e) and (f) MEH-PPV (filled and open for real and imaginary parts respectively). The lines through the data points are to guide the eye. Adapted from Hendry, E., Koeberg, M., Schins, J. M., Siebbeles, L. D. A. & Bonn, M. 2006. Free carrier photogeneration in polythiophene versus poly(phenylene vinylene) studied with thz spectroscopy. *Chemical Physics Letters*, **432**, 441–445. Copyright (2006) with permission from Elsevier (DOI: 10.1016/j.cplett.2006.10.105).

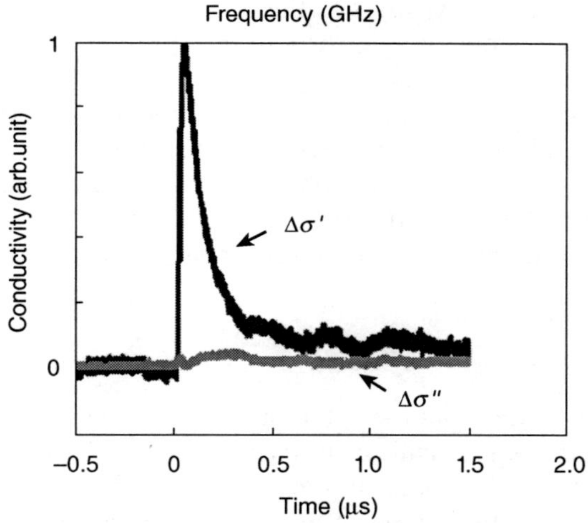

11.16 Normalized complex conductivity of a pentacene sample determined by microwave photoconductivity from analyzing transient Q-curve around 9 GHz center frequency. Reprinted with permission from Saeki, A., Seki, S. & Tagawa, S. 2006. Electrodeless measurement of charge carrier mobility in pentacene by microwave and optical spectroscopy techniques. *Journal of Applied Physics*, **100**, 023703, Copyright (2006), American Institute of Physics (DOI: 10.1063/1.2214638).

about 100 times higher than the 9 GHz Gunn oscillator frequency, hence THz spectroscopy using a few oscillations burst with THz bandwidth allows determining the complex conductivity on a larger frequency range.

In contrast, TPC spectroscopy probes the carrier mobility under DC bias since the electric field applied to the sample is essentially constant. The current density reads:

$$j = E\sigma + \varepsilon\frac{dE}{dt} + E\frac{d\varepsilon}{dt} \quad [11.5]$$

The temporal evolution of the carrier response currently accessible by the technique ranges from DC up to some 10 s of GHz. We emphasize that the technique does not probe any AC response.

The Drude Lorentz model suggests that stronger localization is detectable with higher probing frequency and that for each contributing oscillator, a certain range of frequency is needed to match it. The validation of more realistic carrier localization models involving potential barriers or wells can also benefit from experimental data over large ranges of frequency spaces, probing frequency and carrier evolution as function of pump-probe delay. A challenge is the disentanglement of both frequency spaces, since phenomena related to carrier evolution (generation, collection, recombination) may occur at same time or frequency scales as the ones due to localization. An approach to resolve this ambiguity consists in interpreting TRTS data in two-dimensional frequency space (Nemec *et al.*, 2009).

Therefore we emphasize the importance of combining different subnanosecond spectroscopic methods with specific sensitivity to charge carriers in order to span a wide range of probe frequencies and pump-probe delays, so that the cumulated spectral response allows matching the most adequate carrier dynamics model with the least assumptions.

11.7 Conclusion

To date, the nature of primary photoexcitation in organic crystals and polymers, and in particular the mechanisms of charge carrier photogeneration and transport are better understood, although some of their aspects remain controversial. As observed by several of the spectroscopic techniques presented here, the initial carrier photogeneration is independent of temperature, field, and wavelength, which is in contradiction with the predictions of the Onsager model (Moses *et al.*, 2001; Hegmann *et al.*, 2002). In bulk heterojunctions, photoconductivity observed by the relatively slow TRMC method does increase with temperature, however with low activation energies of tens of millielectronvolts (Savenije *et al.*, 2011; Murthy *et al.*, 2012). This has been attributed to thermal activation of detrapping from shallow traps.

In conjugated polymers, it is now accepted that a percentage of free carriers is generated directly (Moses *et al.*, 2002; Bakulin *et al.*, 2012) and that some polymer donor–acceptor blends reach high and even unitary internal quantum yield (Park *et al.*, 2009). For the remaining carriers that generate from secondary processes, electric-field assisted dissociation of excitons into polarons in about 10 ps has been reported

(Graupner *et al.*, 1998), and also excess excitation energy can possibly be involved in the hot-exciton dissociation process (Murthy *et al.*, 2012 and references therein).

More attention has been paid in recent years to establish relationships between structure/morphology and electronic/transport properties of organic semiconductors in view of organic photovoltaic applications. For instance, a recent time-resolved PIA study has shown that copolymers with different donor and acceptor moieties can acquire large yields of polaron pair formation. Enlarging the distance between center of masses of donor and acceptor can increase the recombination time of polaron pairs (Tautz *et al.*, 2012). The morphology of donor–acceptor blends, e.g. the degree of intercalation and the size of the crystalline domains, is also of paramount importance to achieve balanced electron and hole charge transport and carrier extraction (Rance *et al.*, 2011). It seems to be accepted that the transport in the lamellae of regioregular P3HT is somehow two-dimensional with a reduced 'Langevin' recombination rate as compared to the three-dimensional case. One approach consists of assuming monomolecular recombination and showing that bimolecular processes appear above a critical carrier density accounting for the sub-linear carrier density as function of excitation, and hence sub-linear evolution of observed TRMC (Dicker *et al.*, 2004) at a critical density that corresponds to more than one excitation per lamella. The other approach consists in deriving a bimolecular recombination coefficient for the two-dimensional case as suggested by Juška *et al.* (2009). This accounts for an overall 2.5 exponent power law of the recombination as function of carrier density. The overall behavior described by the two approaches is equivalent. The effect of dimensionality on charge transport properties is in agreement with the observation that thermal annealing and chemical additives that are known to increase donor–acceptor domain sizes induce higher charge carrier photogeneration and photovoltaic power conversion efficiencies (Ma *et al.*, 2005; Peet *et al.*, 2006).

Bimolecular recombination seems to become an important process when a solar cell is operated near open circuit (OC) voltage (Cowan *et al.*, 2012), although this point is still under debate (Deibel and Wagenpfahl, 2010; Street, 2010; Street and Schoendorf, 2010). The increase of recombination and hence the drop of photocurrent while approaching the OC voltage could then be responsible for the low fill factors of organic photovoltaic devices.

11.8 Acknowledgments

We gratefully acknowledge financial support from the CNRS-NTU-Thales UMI 3288 (CINTRA) and the NTU NAP startup Grant No. M4080511 as well as the Singapore Ministry of Education Academic Research Fund Tier 3 Grant No. MOE2011-T3-1-005.

11.9 References

Auston, D. H. 1975. Picosecond Optoelectronic Switching and Gating in Silicon. *Applied Physics Letters*, **26**, 101–103, DOI: 10.1063/1.88079.

Auston, D. H., Lavallard, P., Sol, N. & Kaplan, D. 1980. An Amorphous Silicon Photodetector for Picosecond Pulses. *Applied Physics Letters*, **36**, 66–68, DOI: 10.1063/1.91276.

Bakulin, A. A., Rao, A., Pavelyev, V. G., Van Loosdrecht, P. H. M., Pshenichnikov, M. S., Niedzialek, D., Cornil, J., Beljonne, D. & Friend, R. H. 2012. The Role of Driving Energy and Delocalized States for Charge Separation in Organic Semiconductors. *Science*, **335**, 1340–1344, DOI: 10.1126/science.1217745.

Bao, Z., Dodabalapur, A. & Lovinger, A. J. 1996. Soluble and Processable Regioregular Poly(3-Hexylthiophene) for Thin Film Field-Effect Transistor Applications with High Mobility. *Applied Physics Letters*, **69**, 4108, DOI: 10.1063/1.117834.

Beard, M. C., Turner, G. M. & Schmuttenmaer, C. A. 2000. Transient Photoconductivity in Gaas as Measured by Time-Resolved Terahertz Spectroscopy. *Physical Review B*, **62**, 15764–15777, DOI: 10.1103/PhysRevB.62.15764.

Brabec, C. J., Zerza, G., Cerullo, G., De Silvestri, S., Luzzati, S., Hummelen, J. C. & Sariciftci, S. 2001. Tracing Photoinduced Electron Transfer Process in Conjugated Polymer/Fullerene Bulk Heterojunctions in Real Time. *Chemical Physics Letters*, **340**, 232–236, DOI: 10.1016/s0009-2614(01)00431-6.

Brown, P. J., Sirringhaus, H., Harrison, M., Shkunov, M. & Friend, R. H. 2001. Optical Spectroscopy of Field-Induced Charge in Self-Organized High Mobility Poly(3-Hexylthiophene). *Physical Review B*, **63**, 125204, DOI: 10.1103/PhysRevB.63.125204.

Cowan, S. R., Banerji, N., Leong, W. L. & Heeger, A. J. 2012. Charge Formation, Recombination, and Sweep-out Dynamics in Organic Solar Cells. *Advanced Functional Materials*, **22**, 1116–1128, DOI: 10.1002/adfm.201101632.

Cunningham, P. D. & Hayden, L. M. 2008. Carrier Dynamics Resulting from above and Below Gap Excitation of P3ht and P3ht/Pcbm Investigated by Optical-Pump Terahertz-Probe Spectroscopy. *Journal of Physical Chemistry C*, **112**, 7928–7935, DOI: 10.1021/jp711827g.

De Haas, M. P. & Warman, J. M. 1982. Photon-Induced Molecular Charge Separation Studied by Nanosecond Time-Resolved Microwave Conductivity. *Chemical Physics*, **73**, 35–53, DOI: 10.1016/0301-0104(82)85148-3.

Deibel, C. & Wagenpfahl, A. 2010. Comment on 'Interface State Recombination in Organic Solar Cells'. *Physical Review B*, **82**, 207301, DOI: 10.1103/PhysRevB.82.207301.

Dicker, G., De Haas, M. P., Siebbeles, L. D. A. & Warman, J. M. 2004. Electrodeless Time-Resolved Microwave Conductivity Study of Charge-Carrier Photogeneration in Regioregular Poly(3-Hexylthiophene) Thin Films. *Physical Review B*, **70**, 045203, DOI: 10.1103/PhysRevB.70.045203.

Diesinger, H., Panahandeh-Fard, M., Baillargeat, D. & Soci, C. 2011. Electromagnetic Modeling and Optimization of Photoconductive Switches for Terahertz Generation and Photocurrent Transient Spectroscopy. *Proceedings of the 2011 IEEE MWP*, 373–376, DOI: 10.1109/MWP.2011.6088749.

Diesinger, H., Panahandeh-Fard, M., Wang, Z., Baillargeat, D. & Soci, C. 2012. Enhancing Photocurrent Transient Spectroscopy by Electromagnetic Modeling. *Review of Scientific Instruments*, **83**, 053103–6, DOI: 10.1063/1.4710996.

Ehrenfreund, E. & Vardeny, Z. V. 1997. Phonon Spectroscopy in Pi-Conjugated Polymers; the Role of the Excited Electronic States. *Optical Probes of Conjugated Polymers*, **3145**, 324–332, DOI: 10.1117/12.295532.

Fincher, C. R., JR., Ozaki, M., Heeger, A. J. & MacDiarmid, A. G. 1979. Donor and Acceptor States in Lightly Doped Polyacetylene, (Ch)_{X}. *Physical Review B*, **19**, 4140–4148, DOI: 10.1103/PhysRevB.19.4140.

Graupner, W., Cerullo, G., Lanzani, G., Nisoli, M., List, E. J. W., Leising, G. & De Silvestri, S. 1998. Direct Observation of Ultrafast Field-Induced Charge Generation in Ladder-Type Poly(Para-Phenylene). *Physical Review Letters*, **81**, 3259–3262, DOI: 10.1103/PhysRevLett.81.3259.

Green, K. & Sobolewski, R. 2000. Extending Scattering-Parameter Approach to Characterization of Linear Time-Varying Microwave Devices. *Microwave Theory and Techniques, IEEE Transactions on*, **48**, 1725–1731, DOI: 10.1109/22.873902.

Grzegorczyk, W. J., Savenije, T. J., Heeney, M., Tierney, S., McCulloch, I., Bavel, S. V. & Siebbeles, L. D. A. 2008. Relationship between Film Morphology, Optical, and Conductive Properties of Poly(Thienothiophene): [6,6]-Phenyl C-61-Butyric Acid Methyl Ester Bulk Heterojunctions. *The Journal of Physical Chemistry C*, **112**, 15973–15979, DOI: 10.1021/jp8044548.

Heeger, A. J., Sariciftci, N. S. & Namdas, E. B. 2010. *Semiconducting and Metallic Polymers*, New York, Oxford University Press.

Hegmann, F. A., Tykwinski, R. R., Lui, K. P. H., Bullock, J. E. & Anthony, J. E. 2002. Picosecond Transient Photoconductivity in Functionalized Pentacene Molecular Crystals Probed by Terahertz Pulse Spectroscopy. *Physical Review Letters*, **89**, 227403, DOI: 10.1103/PhysRevLett.89.227403.

Hegmann, F. A., Ostroverkhova, O., Gao, J. B., Barker, L., Tykwinski, R. R., Bullock, J. E., Anthony, J. E. 2004. Picosecond Transient Photoconductivity in Organic Molecular Crystals. *Ultrafast Phenomena in Semiconductors and Nonstructure Materials VIII*, **5352**, 196–207, DOI: 10.1117/12.526915.

Hendry, E., Koeberg, M., Schins, J. M., Siebbeles, L. D. A. & Bonn, M. 2006. Free Carrier Photogeneration in Polythiophene Versus Poly(Phenylene Vinylene) Studied with Thz Spectroscopy. *Chemical Physics Letters*, **432**, 441–445, DOI: 10.1016/j.cplett.2006.10.105.

Holt, J., Singh, S., Drori, T., Zhang, Y. & Vardeny, Z. V. 2009. Optical Probes of Pi-Conjugated Polymer Blends with Strong Acceptor Molecules. *Physical Review B*, **79**, 195210, DOI: 10.1103/PhysRevB.80.049902.

Holzman, J. F., Vermeulen, F. E., Arnold, B. W. & Elezzabi, A. Y. 2000. Photoconductive Gating of Picosecond Electrical Pulses on Thinned-Silicon Substrates. *Electronics Letters*, **36**, 1225–1226, DOI: 10.1049/el:20000895.

Horovitz, B. 1982. Infrared Activity of Peierls Systems and Application to Polyacetylene. *Solid State Communications*, **41**, 729–734, DOI: 10.1016/0038-1098(82)91126-7.

Infelta, P. P., De Haas, M. P. & Warman, J. M. 1977. The Study of the Transient Conductivity of Pulse Irradiated Dielectric Liquids on a Nanosecond Timescale Using Microwaves. *Radiation Physics and Chemistry*, **10**, 353–365, DOI: 10.1016/0146-5724(77)90044-9.

Juška, G., Genevičius, K., Nekrašas, N., Sliaužys, G. & Österbacka, R. 2009. Two Dimensional Langevin Recombination in Regioregular Poly(3-Hexylthiophene). *Applied Physics Letters*, **95**, 013303, DOI: 10.1063/1.3141513.

Ketchen, M. B., Grischkowsky, D., Chen, T. C., Chi, C. C., Duling, I. N., Halas, N. J., Halbout, J. M., Kash, J. A. & Li, G. P. 1986. Generation of Subpicosecond Electrical Pulses on Coplanar Transmission-Lines. *Applied Physics Letters*, **48**, 751–753, DOI: 10.1063/1.96709.

Kim, Y. H., Spiegel, D., Hotta, S. & Heeger, A. J. 1988. Photoexcitation and Doping Studies of Poly(3-Hexylthienylene). *Physical Review B*, **38**, 5490–5495, DOI: 10.1103/PhysRevB.38.5490.

Kraabel, B., Hummelen, J. C., Vacar, D., Moses, D., Sariciftci, N. S., Heeger, A. J. & Wudl, F. 1996. Subpicosecond Photoinduced Electron Transfer from Conjugated Polymers to Functionalized Fullerenes. *Journal of Chemical Physics*, **104**, 4267–4273, DOI: 10.1063/1.471154.

Lampin, J. F., Desplanque, L. & Mollot, F. 2001. Detection of Picosecond Electrical Pulses Using the Intrinsic Franz-Keldysh Effect. *Applied Physics Letters*, **78**, 4103–4105, DOI: 10.1063/1.1381030.

Lee, C. H., Yu, G., Moses, D., Pakbaz, K., Zhang, C., Sariciftci, N. S., Heeger, A. J. & Wudl, F. 1993. Sensitization of the Photoconductivity of Conducting Polymers by C-60 – Photoinduced Electron-Transfer. *Physical Review B*, **48**, 15425–15433, DOI: 10.1103/PhysRevB.48.15425.

Logeeswaran, V. J., Sarkar, A., Islam, M. S., Kobayashi, N. P., Straznicky, J., Li, X., Wu, W., Mathai, S., Tan, M. R. T., Wang, S.-Y. & Williams, R. S. 2008. A 14-Ps Full Width at Half Maximum High-Speed Photoconductor Fabricated with Intersecting Inp Nanowires on an Amorphous Surface. *Applied Physics A: Materials Science & Processing*, **91**, 1–5, DOI: 10.1007/s00339-007-4394-x.

Ma, W., Yang, C., Gong, X., Lee, K. & Heeger, A. J. 2005. Thermally Stable, Efficient Polymer Solar Cells with Nanoscale Control of the Interpenetrating Network Morphology. *Advanced Functional Materials*, **15**, 1617–1622, DOI: 10.1002/adfm.200500211.

Miranda, P. B., Moses, D. & Heeger, A. J. 2001. Ultrafast Photogeneration of Charged Polarons in Conjugated Polymers. *Physical Review B*, **64**, 081201(R), DOI: 10.1103/PhysRevB.64.081201.

Mizrahi, U., Shtrichman, I., Gershoni, D., Ehrenfreund, E. & Vardeny, Z. V. 1999. Picoseconds Time Resolved Photoinduced Absorption by Infrared Active Vibrations as a Probe for Charge Photogeneration in Meh-Ppv/C-60 Composites. *Synthetic Metals*, **102**, 1182–1185, DOI: 10.1016/s0379-6779(98)00316-6.

Moses, D., Sinclair, M. & Heeger, A. J. 1987. Carrier Photogeneration and Mobility in Polydiacetylene – Fast Transient Photoconductivity. *Physical Review Letters*, **58**, 2710–2713, DOI: 10.1103/PhysRevLett.58.2710.

Moses, D., Sinclair, M., Phillips, S. & Heeger, A. J. 1989. Transient Photoconductivity in Polyacetylene and Polydiacetylene. *Synthetic Metals*, **28**, D675–D681, DOI: 10.1016/0379-6779(89)90760-1.

Moses, D., Dogariu, A. & Heeger, A. J. 2000. Ultrafast Detection of Charged Photocarriers in Conjugated Polymers. *Physical Review B*, **61**, 9373–9379, DOI: 10.1103/PhysRevB.61.9373.

Moses, D., Soci, C., Miranda, P. & Heeger, A. J. 2001. The Role of Electron Photoemission in the 'Photoconductivity' of Semiconducting Polymers. *Chemical Physics Letters*, **350**, 531–536, DOI: 10.1016/S0009-2614(01)01317-3.

Moses, D., Miranda, P. B., Soci, C. & Heeger, A. J. 2002. Mechanism of Carrier Photoexcitation in Semiconducting Polymers: The Role of Electron Photoemission in 'Photoconductivity' Measurements. In: Jabbour, G. E. & Sariciftci, N. S. (eds.)

Electronic, Optical and Optoelectronic Polymers and Oligomers. Warrendale: Materials Research Society, DOI: 10.1117/12.456939.

Moses, D., Soci, C., Chi, X. L. & Ramirez, A. P. 2006. Mechanism of Carrier Photogeneration and Carrier Transport in Molecular Crystal Tetracene. *Physical Review Letters*, **97**, 067401-067404, DOI: 10.1103/PhysRevLett.97.067401.

Mourou, G. A. & Meyer, K. E. 1984. Subpicosecond Electro-Optic Sampling Using Coplanar Strip Transmission-Lines. *Applied Physics Letters*, **45**, 492–494, DOI: 10.1063/1.95312.

Murthy, D. H. K., Gao, M., Vermeulen, M. J. W., Siebbeles, L. D. A. & Savenije, T. J. 2012. Mechanism of Mobile Charge Carrier Generation in Blends of Conjugated Polymers and Fullerenes: Significance of Charge Delocalization and Excess Free Energy. *The Journal of Physical Chemistry C*, **116**, 9214–9220, DOI: 10.1021/jp3007014.

Nemec, H., Nienhuys, H. K., Perzon, E., Zhang, F. L., Inganas, O., Kuzel, P. & Sundstrom, V. 2009. Ultrafast Conductivity in a Low-Band-Gap Polyphenylene and Fullerene Blend Studied by Terahertz Spectroscopy. *Physical Review B*, **79**, 245326, DOI: 10.1103/Physrevb.79.245326.

Osterbacka, R., An, C. P., Jiang, X. M. & Vardeny, Z. V. 2000. Two-Dimensional Electronic Excitations in Self-Assembled Conjugated Polymer Nanocrystals. *Science*, **287**, 839–842, DOI: 10.1126/science.287.5454.839.

Osterbacka, R., Pivrikas, A., Juska, G., Poskus, A., Aarnio, H., Sliaužys, G., Genevičius, K., Arlauskas, K. & Sariciftci, N. S. 2010. Effect of 2-D Delocalization on Charge Transport and Recombination in Bulk-Heterojunction Solar Cells. *IEEE Journal of Selected Topics in Quantum Electronics*, **16**, 1738–1745, DOI: 10.1109/JSTQE.2010.2048746.

Ostroverkhova, O., Cooke, D. G., Shcherbyna, S., Egerton, R. F., Hegmann, F. A., Tykwinski, R. R. & Anthony, J. E. 2005a. Bandlike Transport in Pentacene and Functionalized Pentacene Thin Films Revealed by Subpicosecond Transient Photoconductivity Measurements. *Physical Review B*, **71**, 035204, DOI: 10.1103/Physrevb.71.035204.

Ostroverkhova, O., Shcherbyna, S., Cooke, D. G., Egerton, R. F., Hegmann, F. A., Tykwinski, R. R., Parkin, S. R. & Anthony, J. E. 2005b. Optical and Transient Photoconductive Properties of Pentacene and Functionalized Pentacene Thin Films: Dependence on Film Morphology. *Journal of Applied Physics*, **98**, 033701, DOI: 10.1063/1.1949711.

Ostroverkhova, O., Cooke, D. G., Hegmann, F. A., Anthony, J. E., Podzorov, V., Gershenson, M. E., Jurchescu, O. D. & Palstra, T. T. M. 2006. Ultrafast Carrier Dynamics in Pentacene, Functionalized Pentacene, Tetracene, and Rubrene Single Crystals. *Applied Physics Letters*, **88**, 162101, DOI: 10.1063/1.2193801.

Park, S. H., Roy, A., Beaupre, S., Cho, S., Coates, N., Moon, J. S., Moses, D., Leclerc, M., Lee, K. & Heeger, A. J. 2009. Bulk Heterojunction Solar Cells with Internal Quantum Efficiency Approaching 100%. *Nature Photonics*, **3**, 297–302, DOI: 10.1038/nphoton.2009.69.

Parkinson, P., Lloyd-Hughes, J., Gao, Q., Tan, H. H., Jagadish, C., Johnston, M. B. & Herz, L. M. 2007. Transient Terahertz Conductivity of GaAs Nanowires. *Nano Letters*, **7**, 2162–2165, DOI: 10.1021/nl071162x.

Peet, J., Soci, C., Coffin, R. C., Nguyen, T. Q., Mikhailovsky, A., Moses, D. & Bazan, G. C. 2006. Method for Increasing the Photoconductive Response in Conjugated

Polymer/Fullerene Composites. *Applied Physics Letters*, **89**, 252105–3, DOI: 10.1063/1.2408661.

Prins, P., Grozema, F. C., Schins, J. M., Savenije, T. J., Patil, S., Scherf, U. & Siebbeles, L. D. A. 2006. Effect of Intermolecular Disorder on the Intrachain Charge Transport in Ladder-Type Poly(p-Phenylenes). *Physical Review B*, **73**, 045204, DOI: 10.1103/PhysRevB.73.045204.

Rance, W. L., Ferguson, A. J., McCarthy-Ward, T., Heeney, M., Ginley, D. S., Olson, D. C., Rumbles, G. & Kopidakis, N. 2011. Photoinduced Carrier Generation and Decay Dynamics in Intercalated and Non-Intercalated Polymer:Fullerene Bulk Heterojunctions. *ACS Nano*, **5**, 5635–5646, DOI: 10.1021/nn201251v.

Saeki, A., Seki, S. & Tagawa, S. 2006. Electrodeless Measurement of Charge Carrier Mobility in Pentacene by Microwave and Optical Spectroscopy Techniques. *Journal of Applied Physics*, **100**, 023703, DOI: 10.1063/1.2214638.

Sariciftci, N. S. 1997. *Primary Photoexcitations in Conjugated Polymers: Molecular Exciton Versus Semiconductor Band Model*, Singapore, World Scientific Publishing Co.

Sariciftci, N. S. & Heeger, A. J. 1994. Reversible, Metastable, Ultrafast Photoinduced Electron-Transfer in Conjugated Polymer and Buckminsterfullerene Composites and Heterojunctions. *Molecular Crystals and Liquid Crystals Science and Technology Section A – Molecular Crystals and Liquid Crystals*, **256**, 317–326, DOI: 10.1080/10587259408039262.

Sariciftci, N. S., Smilowitz, L., Heeger, A. J. & Wudl, F. 1992. Photoinduced Electron-Transfer from a Conducting Polymer to Buckminsterfullerene. *Science*, **258**, 1474–1476, DOI: 10.1126/science.258.5087.1474.

Savenije, T. J., Murthy, D. H. K., Gunz, M., Gorenflot, J., Siebbeles, L. D. A., Dyakonov, V. & Deibel, C. 2011. Absence of Postnanosecond Charge Carrier Relaxation in Poly(3-Hexylthiophene)/Fullerene Blends. *The Journal of Physical Chemistry Letters*, **2**, 1368–1371, DOI: 10.1021/jz200569h.

Schaffer, H. E., Friend, R. H. & Heeger, A. J. 1987. Localized Phonons Associated with Solitons in Polyacetylene – Coupling to the Nonuniform Mode. *Physical Review B*, **36**, 7537–7541, DOI: 10.1103/PhysRevB.36.7537.

Sirringhaus, H., Tessler, N. & Friend, R. H. 1998. Integrated Optoelectronic Devices Based on Conjugated Polymers. *Science*, **280**, 1741–1744, DOI: 10.1126/science.280.5370.1741.

Sirringhaus, H., Brown, P. J., Friend, R. H., Nielsen, M. M., Bechgaard, K., Langeveld-Voss, B. M. W., Spiering, A. J. H., Janssen, R. A. J., Meijer, E. W., Herwig, P. & de Leeuw, D. M. 1999. Two-Dimensional Charge Transport in Self-Organized, High-Mobility Conjugated Polymers. *Nature*, **401**, 685–688, DOI: 10.1038/44359.

Smith, F. W., Q. Le, H., Diadiuk, V., Hollis, M. A., Calawa, A. R., Gupta, S., Frankel, M., Dykaar, D. R., Mourou, G. A. & Hsiang, T. Y. 1989. Picosecond Gaas-Based Photoconductive Optoelectronic Detectors. *Applied Physics Letters*, **54**, 890–892, DOI: 10.1063/1.100800.

Smith, N. V. 2001. Classical Generalization of the Drude Formula for the Optical Conductivity. *Physical Review B*, **64**, 155106, DOI: 10.1103/PhysRevB.64.155106.

Smith, P. R., Auston, D. H., Johnson, A. M. & Augustyniak, W. M. 1981. Picosecond Photoconductivity in Radiation-Damaged Silicon-on-Sapphire Films. *Applied Physics Letters*, **38**, 47–50, DOI: 10.1063/1.92128.

Soci, C., Moses, D., Xu, Q. H. & Heeger, A. J. 2005. Charge-Carrier Relaxation Dynamics in Highly Ordered Poly(P-Phenylene Vinylene): Effects of Carrier

Bimolecular Recombination and Trapping. *Physical Review B*, **72**, 5204, DOI: 10.1103/PhysRevB.72.245204.

Soci, C., Hwang, I. W., Moses, D., Zhu, Z., Waller, D., Gaudiana, R., Brabec, C. J. & Heeger, A. J. 2007. Photoconductivity of a Low-Bandgap Conjugated Polymer. *Advanced Functional Materials*, **17**, 632–636, DOI: 10.1002/adfm.200600199.

Soos, Z. G., Hayden, G. W., Girlando, A. & Painelli, A. 1994. Pariser-Parr-Pople Force Field for π-Electrons – Raman and Infrared Shifts of Transpolyacetylene. *Journal of Chemical Physics*, **100**, 7144–7152.

Street, R. 2010. Reply to 'Comment on "Interface State Recombination in Organic Solar Cells" '. *Physical Review B*, **82**, DOI: 10.1103/PhysRevB.82.207302.

Street, R. A. & Schoendorf, M. 2010. Interface State Recombination in Organic Solar Cells. *Physical Review B*, **81**, DOI: 10.1103/PhysRevB.81.205307.

Tautz, R., Da Como, E., Limmer, T., Feldmann, J., Egelhaaf, H.-J., Von Hauff, E., Lemaur, V., Beljonne, D., Yilmaz, S., Dumsch, I., Allard, S. & Scherf, U. 2012. Structural Correlations in the Generation of Polaron Pairs in Low-Bandgap Polymers for Photovoltaics. Nature Communications, **3**, 970, DOI: 10.1038/ncomms1967.

Thorsmolle, V. K., Averitt, R. D., Chi, X., Hilton, D. J., Smith, D. L., Ramirez, A. P. & Taylor, A. J. 2004. Ultrafast Conductivity Dynamics in Pentacene Probed Using Terahertz Spectroscopy. *Applied Physics Letters*, **84**, 891–893, DOI: 10.1063/1.1644046.

Tripon-Canseliet, C., Faci, S., Blary, K., Deshours, F., Alquie, G., Formont, S. & Chazelas, J. 2006. Optically-Controlled Microwave Phase Shifting and Sampling by Efficient Photoconductive Switching on Lt-Gaas Substrate Integrated Technology. *Proceedings of the Society of Photooptical Instrumentation Engineers (SPIE)*, 6343, Photonics North 2006, 63432K, DOI: 10.117/12,707688.

Ulbricht, R., Hendry, E., Shan, J., Heinz, T. F. & Bonn, M. 2011. Carrier Dynamics in Semiconductors Studied with Time-Resolved Terahertz Spectroscopy. *Reviews of Modern Physics*, **83**, 543–586, DOI: 10.1103/RevModPhys.83.543.

Unuma, T., Fujii, K., Kishida, H. & Nakamura, A. 2010. Terahertz Complex Conductivities of Carriers with Partial Localization in Doped Polythiophenes. *Applied Physics Letters*, **97**, 03308, 1–3, DOI: 10.1063/1.3466916.

Valdmanis, J., Mourou, G. & Gabel, C. 1983. Subpicosecond Electrical Sampling. *IEEE Journal of Quantum Electronics*, **19**, 664–667, DOI: 10.1109/jqe.1983.1071915.

Valdmanis, J. A., Mourou, G. & Gabel, C. W. 1982. Picosecond Electro-Optic Sampling System. *Applied Physics Letters*, **41**, 211–212, DOI: 10.1063/1.93485.

Voss, K. F., Foster, C. M., Smilowitz, L., Mihailovic, D., Askari, S., Srdanov, G., Ni, Z., Shi, S., Heeger, A. J. & Wudl, F. 1991. Substitution Effects on Bipolarons in Alkoxy Derivatives of Poly(1,4-Phenylene-Vinylene). *Physical Review B*, **43**, 5109–5118, DOI: 10.1103/PhysRevB.43.5109.

12

Short-pulse induced photocurrent and photoluminescence in organic materials

I. BIAGGIO, Lehigh University, USA

DOI: 10.1533/9780857098764.2.356

Abstract: This chapter discusses short pulse photoexcitation in organic materials, mostly organic molecular crystals, and the resulting time dynamics of excited states as can be observed by detecting photoluminescence and photoconductivity. The chapter will review the electronic detection of a transient photoconductivity after photoexcitation and the relevant experimental limits. It will then discuss the typical photoluminescence dynamics expected in molecular crystals when initially photoexcited singlet excitons can undergo fission into triplet excitons, and the related onset of photoconductivity because of exciton dissociation. Rubrene single crystals will be used to illustrate how exciton dynamics is reflected in different ways in the time-evolution of the photocurrent and the photoluminescence.

Key words: excitons, singlet, triplet, photoconductivity, photoluminescence, exciton fission, exciton fusion.

12.1 Introduction

A well-defined initial condition of electronic excited states can be set up by illumination with a laser pulse much shorter than the excited state lifetime. In organic materials, the initial photoexcitation is often a localized transition that results in a molecular exciton. The relaxation processes that follow a single-pulse illumination can then be studied by detecting a number of physical observables. Of the many possible physical observables, this chapter will discuss two. One is the luminescence emitted upon radiative recombination of the photoexcited states. The other one is the electrical current that arises when excitons dissociate into free carriers.

Pulsed exposure allows us to study the dynamics of both photoluminescence and photocurrent over multiple time scales, and to selectively identify the different physical processes that may be active at different times. The simultaneous investigation of both photoluminescence and photocurrent is attractive because they are complementary to each other: one arises when the exciton recombines radiatively, the other one arises when the exciton follows another route, which ultimately leads to dissociation into free carriers. While not specific about everything that can happen on the way to the generation of free carriers, the onset of an electrical current is one of the clearest and most direct measures of the generation of free carriers by exciton dissociation.

The following section gives a general introduction to photocurrent detection, with a special emphasis on how a current and its time-dependence can be observed both for fast and slow charge carrier dynamics. Two important regimes of carrier dynamics will be discussed: the most common case of a charge carrier's dynamics being so slow that the corresponding current can be directly converted into a detected voltage proportional to it, and the case when large capacitances lead to a response time of the circuit that is longer than the dynamics of charge carrier generation or relaxation in the sample.

The next section will then review the typical photoluminescence dynamics that is observed in molecular materials when the photoexcited singlet excitons can spontaneously transform into triplet excitons and vice versa. In particular, the chapter will discuss the interesting case of singlet–triplet interaction in molecular crystals and the peculiar photoluminescence dynamics that arises from it.

Measurements recently performed in the rubrene molecular crystals will be used along the way as an example of both luminescence and photocurrent dynamics after pulsed illumination, and of their relationship. Rubrene is peculiar because after pulsed photoexcitation it shows both a strong delayed photoluminescence with a decay on multiple timescales, as well as a strong delayed photocurrent that builds up on the microsecond time scale. I will discuss both pulsed photocurrent and pulsed photoluminescence observations in rubrene and show how they both arise from the details of exciton dynamics.

None of the effects discussed in this chapter are in principle exclusive to rubrene. Similar observations have been made in other molecular crystals such as tetracene, anthracene, or pentacene. However, rubrene is a material where several different aspects of singlet and triplet exciton dynamics are nicely observable at the same time and where enough data is available to provide for adequate examples that highlight the relationships between the data collected using the two observables discussed here: photoluminescence and photocurrent.

12.2 Photocurrent response after short pulse excitation

This section is dedicated to the simplest basic principles of photocurrent detection after illumination with a pulse so short that any charge transport during the duration of the light pulse can be neglected. In other words, the role of the laser pulse is to set up the initial condition for the experiment, as an example the initial number of charge carriers between the contacts. It is worth repeating that the focus of this discussion is on the evolution of the transient photoconductivity after exposure to a single light pulse. Such a *single-shot* experiment is realized using laser pulse trains with a repetition rate much longer than the relaxation time of the sample back to its state just prior to the illumination.

For a transient experiment where one follows the conductivity of a sample after illumination with a single pulse, the photoexcited carriers move in an applied field E, and every individual carrier contributes an amount $e\mu E$ to the current, where e is the unit charge and μ is the carrier mobility. This contribution does not depend on the position of the carrier between two extended contacts. Hence, a general expression for the total transient current measured after pulsed photoexcitation that is valid for any given space- and time-dependent density $n(t,\mathbf{r})$ of charge carriers between two parallel contacts of surface area S is

$$I(t)=\int_S\left[\frac{U_0}{d}e\mu\frac{1}{d}\int_0^d n(x,y,z,t)\mathrm{d}z\right]\mathrm{d}x\mathrm{d}y=\frac{N(t)e\mu}{d^2}U_0=G(t)U_0, \quad [12.1]$$

where the first integral is over the surface area S of the contacts, d is the distance between the contacts, and U_0 is the applied voltage that gives an homogeneous electric field $E = U_0/d$. $N(t)=\int n(\mathbf{r},t)\mathrm{d}V$ is the total number of carriers between the contacts. As an example, this number can vary because of charge carrier recombination or trapping, or also because of carriers reaching a contact and exiting the sample through it. The electric field can be assumed homogeneous because we are interested in a one-shot experiment where a small density of carriers moves between the contacts, but the number of photoexcited carriers is not enough to achieve any screening of the electric field during the lifetime of the carriers. The time-dependent conductance of the sample is

$$G(t)=\frac{N(t)e\mu}{d^2}. \quad [12.2]$$

Its inverse is the electrical resistance of the sample, $R_s(t) = 1/G(t)$. Note that the conductance $G(t)$ depends on the *total* number of carriers that are moving between the contacts at any given time (it is the conductivity integrated over the contact area and divided by the distance between the contacts). The transient electric current produced after illumination with a single pulse does not depend on the spatial distribution of carriers between the contacts. Most importantly, for such one-shot experiments observing a photocurrent does *not* require that the photoinduced charge carriers actually reach a contact, and it is *not* necessary to homogeneously illuminate the volume between the contacts. A transient current is seen as long as charges are drifting between the contacts also when the current density itself may vanish in certain regions. A prime example of this situation are the transient currents observed in time-of-flight experiments (Haynes and Schockley, 1951; Spear, 1957; Kepler, 1960; Dolezalek, 1976) where a carrier packet created by a single pulse close to one contact drifts towards the other contact, which gives a current through the sample up to the moment when the carriers disappear from the sample, typically when they reach the other contact.

It is essential to point out this simple fact of transient experiments because it has not been uncommon to incorrectly expect that photoconductivity measurements detect selectively photocarriers that can reach electrodes even when discussing the results of transient absorption compared to transient photocurrents (Tao *et al.*, 2011). While such an expectation can intuitively arise from experience with bulk photoconductors under continuous wave illumination, it is important to underline that it does not correspond to reality in the case of transient single-shot photocurrents induced by short light pulses.

In the following, the main features of short-pulse transient photocurrent experiments are discussed by considering the simple case of photocarriers that have a constant probability per unit time of being trapped, recombine, or otherwise leave their mobile state. This gives a well-defined lifetime with the mobile carrier density decaying exponentially, which is the simplest possible situation in a photoconductor. For small electric fields that do not allow the photoexcited carriers to reach a contact, the number of photocarriers found between the contacts will decay as $N(\mathrm{t}) = N(0)\exp(-t/\tau)$ where τ is the carrier lifetime. Using Eq. 12.1, one then finds that the time-integral of the total transient current is simply given by the number of initially photoexcited carriers multiplied by the ratio between drift length $L_\mathrm{d} = \tau\,\mu\,E = \tau\,\mu\,U_0/d$ and the distance between the contacts:

$$\int_0^\infty I(t)\mathrm{d}t = eN(0)\frac{L_\mathrm{d}}{d} \qquad [12.3]$$

This will grow linearly with the applied electric field as long as the drift length remains smaller than the distance between the contacts and will then saturate at higher electric fields when the regime typical of time-of-flight experiments is reached. Since for a short pulse experiment the total number of initially photoexcited carriers is simply given by the total number of absorbed photons multiplied with the quantum yield for photocarrier generation, determination of the time-integral of the photocurrent is a useful way to relate drift length and photocarrier generation quantum yield.

The photocurrent discussed above is the ideal current that appears in a sample where photocarriers have been created by a laser pulse. Next, it is necessary to discuss how a *photocurrent* experiment can determine *photoconductivity* dynamics. A simple way to do this is to measure the time-dependent voltage observed in a circuit like that depicted in Fig. 12.1. Here, a strong direct current (DC) voltage is applied to a high-resistivity sample in series with a probe resistor over which a voltage is measured (e.g., through an oscilloscope). Both sample and detection system can be modelled by a resistor in parallel with a capacitor. For the detection system, the capacitance typically comes from a coaxial cable and the resistor can be the built-in resistor of an oscilloscope. For the sample, the capacitance comes from the applied contacts, and can be particularly large if the sample is a thin-film sandwiched between large area contacts.

Assuming that the applied voltage U_0 is constant, one can equate the currents flowing through the sample and the detection system to obtain a differential equation describing the voltage $U(t)$ that is detected over the probe resistor:

$$\frac{U_o - U(t)}{R_s(t)} - C_S \frac{\mathrm{d}}{\mathrm{d}t} U(t) = \frac{U(t)}{R} + C \frac{\mathrm{d}}{\mathrm{d}t} U(t). \qquad [12.4]$$

When the induced photoconductivity is so small that the voltage drop over the measurement resistor remains much smaller than the applied voltage [$R \ll R_s(t)$], the term $U(t)/R_s(t)$ on the left-hand side is negligible compared with the other terms, and the equation simplifies to

$$\frac{U(t)}{\tau_{RC}} + \frac{\mathrm{d}}{\mathrm{d}t} U(t) = \frac{U_0}{C_{TOT}} \frac{1}{R_s(t)} = \frac{I(t)}{C_{TOT}}, \qquad [12.5]$$

where $C_{TOT} = C_s + C$, $\tau_{RC} = R\, C_{TOT}$, and $I(t) = U_0/R_s(t) = G(t)U_0$ is the 'underlying' time-dependent current – given by Eq. 12.1 – that one would measure in an ideal set-up not limited by parasitic capacitances. The initial condition $U(0) = 0$ delivers the solution

$$U(t) = \frac{1}{C_{TOT}} \int_0^t I(s) \mathrm{e}^{(s-t)/\tau_{RC}} \mathrm{d}s, \qquad [12.6]$$

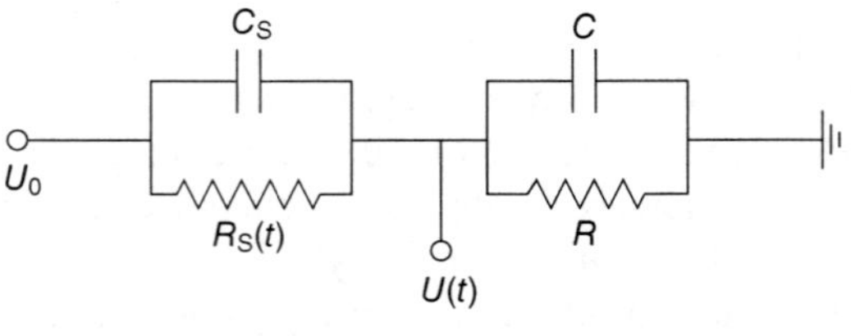

12.1 Electric equivalent of a typical simple measuring system for transient photoconductivity. The left-hand part of the diagram is the sample to investigate, which has a capacitance C_S and a time-dependent resistance $R_S(t)$. The right-hand part of the diagram may be the combination of a coaxial cable and an oscilloscope.

where $I(s)$ is given by Eq. 12.1. Depending on the time dynamics of the photoinduced conductance $G(t)$ (see Eq. 12.2), this gives two characteristically different response functions for the relationship between the detected $U(t)$ and the underlying current $I(t)$. When changes in $G(t)$ occur on timescales much longer than τ_{RC}, the detected voltage is simply proportional to the photocurrent and $U(t) = R\ I(t)$. But when changes in $G(t)$ occur on timescales much shorter than τ_{RC}, then the detected voltage is proportional to the integral of the photocurrent, that is to the total electric charge that has accumulated on the contacts up to the measurement time t. The integral of the photocurrent after the build-up of $U(t)$ is complete, is then simply the final voltage reached multiplied by the total capacitance in the circuit. This provides a very simple method to directly obtain the time dynamics of the photocurrent experimentally even when the relevant time dynamics is confined to times shorter than τ_{RC}. This was recognized early on (Kolomiets and Lebedev, 1966) and has been applied to various material systems (Antoniadis and Schiff, 1992; Tapponnier *et al.*, 2003).

For the case mentioned above where the photoinduced charge carriers have a well-defined lifetime and $N(t) = N(0)\exp(-t/\tau)$, the time-dependent conductance is zero before illumination, jumps to a finite value $G_0 = N(0)\ e\ \mu/d^2$ right after pulsed illumination, and then decays exponentially. Eq. 12.6 then becomes

$$U(t) = G_0 U_0 \frac{1}{C_{\mathrm{TOT}}} \frac{\mathrm{e}^{-t/\tau} - \mathrm{e}^{-t/\tau_{\mathrm{RC}}}}{1/\tau_{\mathrm{RC}} - 1/\tau}. \qquad [12.7]$$

When the lifetime of the photocarriers is much shorter than τ_{RC}, the detected signal at short times is an exponential build-up with a time constant given by the carrier lifetime,

$$U(t) = G_0 U_0 \frac{\tau}{C_{\mathrm{TOT}}}(1 - \mathrm{e}^{-t/\tau}). \qquad [12.8]$$

When the lifetime of the photocarriers is much longer than τ_{RC}, the initial time-dependence of the detected signal is an exponential build-up with a time constant τ_{RC},

$$U(t) = G_0 U_0 R(1 - \mathrm{e}^{-t/\tau_{\mathrm{RC}}}), \qquad [12.9]$$

followed by a time-dependence at longer times that is simply

$$U(t) = G_0 U_0 R \mathrm{e}^{-t/\tau} = RI(t) \qquad [12.10]$$

and tracks the time-dependent conductance $G(t)$ (see Eq. 12.2). Note that an instantaneous release of a number of free carriers is seen as an exponential buildup both for the case when the carrier lifetime is much shorter than τ_{RC} (build-up time τ) and for the case when the lifetime is much longer than τ_{RC} (build-up time τ_{RC}). The difference is that build-up time and final amplitude of the first case are by a factor τ/τ_{RC} smaller than for the second case. Making this distinction is important for the common situation where, while tracking the time-dependence of a current at times longer than τ_{RC}, one observes what appears to be an initial step-wise increase of the detected voltage. Proper analysis based on the simple relationships given above can help distinguish between the two cases that can give rise to such a step, and decide if the initial step is, e.g., caused by an instantaneous release of carriers that live only a short time and do not contribute to the current seen later on, or else if it is caused by an instantaneous release of long-lived carriers. As an example, the ratio between the photocurrent caused by N_2 carriers at times longer than τ_{RC} and the step-amplitude due to an initial release of N_1 carries with a lifetime $\tau << \tau_{RC}$ would be proportional to the ratio between the corresponding conductivities ($e\mu_1N_1$ and $e\mu_2N_2$ for charge carriers with different mobilities μ_1 and μ_2), multiplied by τ_{RC}/τ. This means that the relative size of any initial step in a photoconductivity signal that is caused by short-lived high-mobility carriers would need to depend on τ_{RC}, which offers an easy way to distinguish such a case from the case where there is simply an initial fast release of long-lived photocarriers. The latter would produce an initial build-up with an amplitude that remains in the same proportion to later photoconductivity signals no matter what the value of τ_{RC} is (Najafov *et al.*, 2010). An example of such a photocurrent signal after impulsive excitation is shown in Fig. 12.2, which plots the photocurrent dynamics measured in a rubrene crystal after impulsive excitation with an intense 20 ps pulse (Najafov *et al.*, 2010). This photocurrent trace shows a step-wise build-up that occurs within a time τ_{RC} followed by a slower build-up dynamics. The measurement can be repeated using different values of τ_{RC} and the shape of the curve remains the same, which proves that the initial step is caused by an initial release of charge carriers that is probably related to dissociation of a fraction of the singlet exciton population that has been initially created by the laser pulse. Such a dissociation happens within the singlet lifetime but is too fast to be detected by the current measurement shown in Fig. 12.2.

Another important message from this simple overview is that it is possible to determine the dynamics of a fast photocurrent – by simply measuring the time dynamics of the integrated photocurrent – even when the sample has a geometry that makes the corresponding capacitance too large. This has been demonstrated early on in time-of-flight experiments (Kolomiets and Lebedev, 1967), and it is a quite advantageous technique when working with thin-film samples (Tapponnier *et al.*, 2003; Najafov *et al.*, 2006b). By analyzing the detected voltage in the $\tau_{RC} \rightarrow \infty$ limit of Eq. 12.6 it is possible to study the time dynamics of photocarriers at times much shorter than τ_{RC} using relatively simple contacting geometries. This obviates

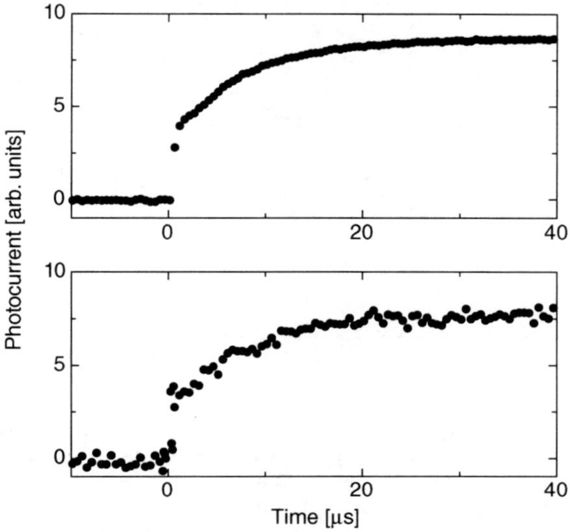

12.2 Example of photocurrent dynamics as seen for times longer than τ_{RC} in a rubrene sample under relatively high exposure conditions that produce a current that has two components, a 'prompt' one that arises in a time of τ_{RC} or less, and a later one that grows with a slower time constant. Two measurements performed with two measurement resistors (1 kΩ for the top plot, 50 Ω for the bottom plot) show the same dynamics, proving that the initial step is caused by a quick release of long-lived charge carriers (data taken in the context of the measurements presented by Najafov *et al.*, 2010).

the necessity of designing contacting situations with low capacitance similar to those used in photoconductive switches (Auston, 1975), which can be quite complicated for organic materials (Diesinger *et al.*, 2011).

An example of how dynamics of photoexcited charge carriers can be measured by determining this integrated pulsed photoconductivity (IPP, see Tapponnier *et al.*, 2003) in experimental set-ups when carrier relaxation occurs on timescales shorter than τ_{RC} is shown in Fig. 12.3. The figure shows the time dependence of the voltage measured, after pulsed illumination, over a resistor in series with a high-capacitance organic light-emitting diode, at times much shorter than τ_{RC}. For one sample the photoexcited carriers relax quickly with a simple exponential decay, followed by a slower decrease of the remaining carrier density. This data is shown by the open symbols in Fig. 12.3. The corresponding sample conductivity is given by the dashed curve, with an initial component decaying exponentially and a more than 10 times smaller delayed component decaying as a stretched exponential (Tapponnier *et al.*, 2003). For another sample, on the other hand,

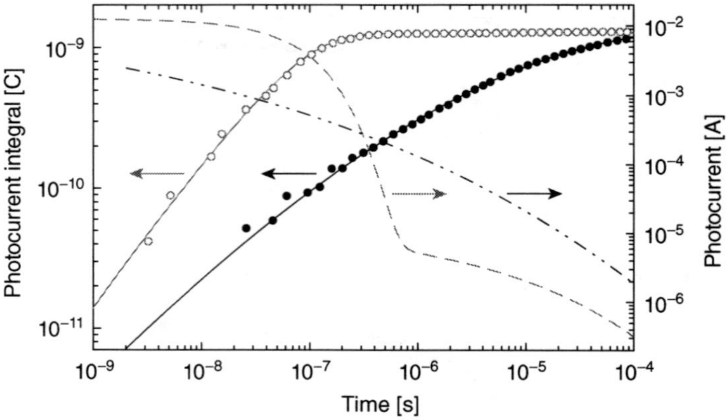

12.3 Example of a measurement of the integrated current ($U(t)C_{TOT}$ from Eq. 12.6) for charge carrier dynamics occurring at times much shorter than τ_{RC} in a thin film sample with high capacitance. The open data points correspond to a sample where the carrier relaxation is basically exponential, while the full data points belong to a stretched exponential decay of the carrier density. The solid lines are fits to the experimental data, while the dashed and dash-dotted lines represent the underlying time-dependent photoconductivity for the open and full data points, respectively. The dashed line is the sum of a simple exponential decay dominant at short times below 1 μs and of a stretched exponential decay of much weaker amplitude that slightly influences the data at times longer than 1 μs (adapted from Tapponnier *et al.*, 2003).

the photoexcited carriers relaxed much more slowly with a dynamics dominated from the beginning by a stretched exponential (full symbols in Fig. 12.3). Even though the sample capacitance in this experiment was too big to directly observe photocurrent dynamics on the sub-microsecond time scale, the IPP experiment clearly allows the photoconductivity dynamics to be easily extracted down to about ~10 ns after photoexcitation.

An example of the information that can be extracted from an IPP measurement is the linear increase in the integrated current that is observed in Fig. 12.3 (open symbols) starting around 4 ns after photoexcitation. This linear increase in the IPP signal means that all photocarriers contributing to the current appear at times shorter than 4 ns, meaning that any excitons dissociate in less than 4 ns. If there was a slower build-up of the photocarrier number, then the IPP signal would grow *superlinearly* with time. In general, a slope larger than +1 in the log–log plot of Fig. 12.3 corresponds to a photoconductivity that increases with time, a slope of +1 corresponds to a constant photoconductivity, and a slope of less than +1 corresponds to a photoconductivity that is decreasing with time.

12.3 Exciton dynamics and photoluminescence in organic molecular crystals

The simplest result of the optical excitation of small organic molecules to an excited state is a photoluminescence signal that decays exponentially, corresponding to the lifetime of the photoexcited state. This behavior is typical of many organic dyes. However, a richer time dynamics can be observed when the singlet exciton does not immediately decay, but instead transforms into longer-lived states. This section is dedicated to the case where the photoexcited singlet exciton can efficiently transform into triplet excitons.

An efficient way for singlet excitons to generate triplet excitons arises when the photoexcited state of a given molecule, which is in a singlet spin state, has an energy that is at least double the energy of the corresponding triplet spin state. When the photoexcited singlet state occurs on a molecule that is loosely bound to a twin molecule, as is the case when a larger molecule consists of two identical sub-units, or in molecular crystals, the photoexcited singlet state can undergo fission into two triplet states whose combined spin is still zero (Müller *et al.*, 2007). This effect will compete with radiative recombination as a way for the photoexcited singlet to relax. In addition, the two triplet states on the two neighboring molecules can combine again to produce again a singlet state that has the same characteristics as the initially photoexcited one and can then radiatively recombine. Because of this effect the time-dependence of the photoluminescence will acquire a delayed component that can have a lifetime much longer than the initial 'prompt' photoluminescence caused by the radiative decay of the singlet state. This has been seen in molecules that consist of two sub-units similar to tetracene (Müller *et al.*, 2007) where a normal photoluminescence decay with an exponential decay time of a few nanoseconds was followed by a much longer decay with an exponential decay time of hundreds of nanoseconds or longer, determined by the triplet lifetime.

A similar situation can occur in van der Waals bonded organic molecular crystals, with the added feature that the triplet states in these crystals are mobile (Irkhin and Biaggio, 2011), meaning that the probability that they can interact with each other again will depend on their concentration. Triplet–triplet interaction can then lead to *fusion* of two triplet excitons to form a singlet exciton again. Because of this bimolecular effect, the delayed photoluminescence in such a case will not be exponential anymore, but will become a power-law at higher densities when triplet–triplet interactions becomes important (Pope and Swenberg, 1999). This is a well-known effect peculiar to organic molecular crystals that has been one of the first dynamic effects observed early on in 1963 using the recently invented Q-switched laser (Hall *et al.*, 1963; Kepler *et al.*, 1963).

The following simple set of rate equations describes the dynamics of singlet excitons (with number density S) and triplet excitons (with density T) in the presence of singlet to triplet and triplet to singlet conversion occurring via fission and fusion (Pope and Swenberg, 1999),

$$\frac{\mathrm{d}T}{\mathrm{d}t} = -\frac{T}{\tau_T} + 2f_S\frac{S}{\tau_S} - \gamma T^2, \qquad [12.11]$$

$$\frac{\mathrm{d}S}{\mathrm{d}t} = -\frac{S}{\tau_S} + \frac{1}{2}f_T\gamma T^2, \qquad [12.12]$$

where f_S is the probability that singlet decay results in fission into two triplet excitons, f_T is the probability that triplet–triplet collision results in the creation of a singlet through a fusion process, τ_T is the triplet exciton lifetime, τ_S is the singlet exciton lifetime, and γ is a bimolecular recombination coefficient. Here I have neglected any other singlet to triplet transformation process that is not fission, which would be represented by another term proportional to S/τ_S in Eq. 12.11. And finally, it should be noted that the above equations do not depend on space, meaning that all concentrations are assumed to be homogeneous. This assumption is valid as long as exciton diffusion lengths are smaller than any significant modulation in exciton densities, and could potentially fail if the excitation is localized to a region smaller than exciton diffusion length. Such an effect has been observed in rubrene and used to determine the diffusion length of triplet excitons (Irkhin and Biaggio, 2011), but the following discussion will assume that both photoexcitation and detection occur over regions much larger than any exciton diffusion lengths.

Before discussing some limiting cases for which analytical solutions can be found, it is useful to consider an example of the full time-dependence of the exciton densities as it can be obtained from a full numerical solution of the above system of equations. This is done in Fig. 12.4 for an initial condition of $T(0) = 0$ and a given singlet exciton density $S(0)$, as it would be typically created by an ultrashort laser pulse. The parameters chosen for this calculation are those that allow highlighting different typical behaviors on different timescales: I chose more than four orders of magnitudes between singlet lifetime and triplet lifetime, so that the effect of fusion becomes clearly visible at longer time scales and higher exciton densities. The evolution of the exciton densities are shown over a wide range of initial conditions that give rise to values of the dimensionless parameter $T\gamma\tau_T$ between 1 and 10^6. But to understand the results in Fig. 12.4, it is best to give the numerical values of the parameters individually, as is done in the figure caption. These numerical values must be understood in self-consistent units of density and time as they can be directly inserted in Eqs. 12.11 and 12.12. They can be partially rescaled by keeping the products $S(0)\ \gamma\ \tau_S$ and

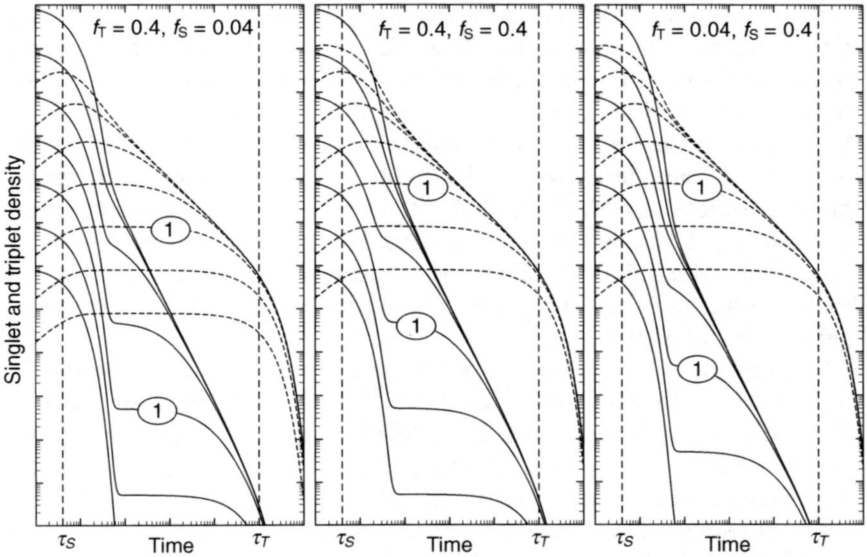

12.4 Singlet exciton densities (solid curves) and triplet exciton densities (dashed curves) as predicted by Eqs. 12.11 and 12.12. Initial conditions for this calculation are $T = 0$ and S from 0.01 to $S = 10^4$ with a factor of 10 between the curves (the curves for $S = 1$ are marked with the label '1'), with exciton lifetimes $\tau_T = 100$ and $\tau_S = 4 \times 10^{-3}$ indicated by the dashed vertical lines, and $\gamma = 1$ (please see text for a discussion of the units and relative scaling of these parameters). The graph in the center belongs to the case where $f_T = f_S = 0.4$. The graphs on the left and right show the prediction for 10 times smaller fission and fusion efficiency, respectively. Note that in these graphs, with equal-length decades on horizontal and vertical axes, a power law $\sim t^{-1}$ is a straight line with a slope of −45°.

$T(0)\,\gamma\,\tau_T$ constant: if time constants are multiplied by a factor, densities must be multiplied by the same factor, and the bimolecular recombination coefficient must be divided by that factor squared.

Figure 12.4 shows that the initial photoexcited singlet density S generally decays to an intermediary value that depends on the existence of both fission and fusion. At lower excitation densities, this intermediary value is well defined and approximately equal to $2 f_S^2 f_T \gamma \tau_S S(0)^2$, obtained from the steady-state solution of Eq. 12.12 using a triplet density equal to $2\,f_S\,S(0)$, which is the total number of triplets generated in the absence of exciton interaction. The initial decay is exponential with time constant τ_S and it is accompanied by an exponential build-up in triplet density. After having reached a maximum, the triplet exciton density decays. At high densities this decay first follows a power law before ultimately becoming exponential, while at lower densities there is no triplet interaction and the decay is exponential all the way. In both cases the exponential decay constant is the triplet lifetime. Figure 12.4 also shows that – in contrast to what happens for the singlet exciton dynamics – the bundle of curves describing the time dynamics of the triplet exciton density is always the same, independent on the value of fission and fusion probabilities. The reason for this is that for this choice of parameters the singlet density rapidly becomes much smaller than the triplet density. It follows that the latter does not depend on fusion probability (which describes the probability of forming a singlet instead of other forms of annihilation upon triplet–triplet collision), and a change of fission probability only changes the initial triplet density in the same way a higher value of $S(0)$ would.

Interestingly, both the singlet and triplet exciton densities observed at longer times no longer increase for higher $S(0)$. They tend instead to limiting values that are independent from the initial density. For increasing initial densities the importance of the power law decay increases, while the number of excitons found at later times remains invariant. Responsible for this peculiar result are the quadratic terms in Eqs. 12.11 and 12.12. Higher values of S(0) lead to higher bimolecular recombination and the exciton densities self-regulate to arrive at a fundamental value that is only dependent upon fission and fusion efficiencies, bimolecular recombination coefficient, and triplet exciton lifetime. As sown in Fig. 12.4, triplet densities can initially be higher or lower than this limiting value but they will ultimately converge towards it. A full expression for this limit, as well as a general analytical solution valid for times past the singlet exciton lifetime, are given below.

A relatively simple analytical solution to Eqs. 12.11 and 12.12 can be found under the reasonable assumption that the triplet lifetime is much longer than the singlet lifetime, $\tau_T >> \tau_S$. This solution is valid at times much larger than the singlet exciton lifetime τ_S, when the rate of change of the population of singlet excitons is determined by the dynamics of the triplet population. In this regime the left-hand side of Eq. 12.12 is much smaller than any of the terms on the right-hand side, and the rate at which singlet excitons are created by fusion must be similar to the rate at which triplet excitons are created by fission. It follows that $S = \tau_S f_T \gamma T^2/2$. Inserting this into Eq. 12.11 delivers

$$\frac{dT}{dt} = -(1 - f_S f_T)\gamma T^2 - \frac{T}{\tau_T}, \qquad [12.13]$$

which has the analytical solution

$$T(t) = T(0)[(1+r)e^{t/\tau_T} - r]^{-1}, \qquad [12.14]$$

where $r = T(0)\,(1 - f_S f_T)\,\gamma\,\tau_T$. Equation 12.14 is a very good approximation, as shown by how closely it follows the numerical solution in Fig. 12.4 for times larger than τ_S. The parameter r determines the dynamics of the triplet population, which goes from an exponential decay for $r << 1$ or $t > \tau_T$ to a power law $T(t) = T(0)/(1 + r\ t/\tau_T)$ for $r >> 1$ and $t < \tau_T$. The power law becomes essentially proportional to t^{-2} for t/τ_T between $1/r$ and 1. Unless both fission and fusion processes are very efficient at the same time, it is generally safe to use $r = T(0)\ \gamma\ \tau_T$. However, one should note that the case where both fission and fusion are extremely efficient ($f_S f_T$ –> 1) is equivalent to the case where the density of excitons is small and r – 0. Extremely high fission and fusion probabilities would effectively remove the bimolecular recombination term for Eqs. 12.11 and 12.13 because a triplet interaction event would lead to the formation of a singlet that would immediately undergo again fission to create two triplets, thus rendering the exciton collision effectively elastic.

The photoluminescence emitted after photoexcitation with a short laser pulse is caused by radiative recombination of the singlet excitons, and its power is therefore proportional to the total number of singlet excitons present in the sample at any given time. For times larger than the singlet lifetime, this number is, as shown above, proportional to $T(t)^2$. Thus, the power of the delayed photoluminescence is proportional to $T(t)^2$ as given by Eq. 12.14, integrated over the volume. This photoluminescence power is shown in Fig. 12.5 for varying initial numbers of triplet excitons. The figure shows two plots that apply to the two cases where the exciton density changes because of increased photoexcitation over the same volume, or where the exciton density changes because the photoexcitation volume is changed while keeping the total amount of photoexcited excitons constant. It is interesting to compare these two cases because they reflect two typical experimental situations: varying the total energy of the photoexcitation pulse, or varying the diameter of the illuminated area without changing the energy of the photoexcitation pulse. These two cases correspond to exciton density variation at constant volume and at constant exciton numbers. One sees that the initial photoluminescence power grows with the square of the initial exciton density in the constant volume case (and saturates towards later times), while it grows linearly with the initial exciton density for the constant number case. On the other hand, when keeping the same total number of initial excitons, one finds that the delayed PL intensity increases for decreasing exciton densities. This is to be expected because higher densities imply faster triplet exciton fusion, so that a larger PL is emitted early after excitation at the expense of what remains at later times.

The rubrene molecular crystal shows a particularly clear delayed photoluminescence after impulsive excitation, and provides a good example of the full evolution of the photoluminescence from the moment of photoexcitation to the final decay caused by the triplet lifetime. The initial photoluminescence decay in a rubrene crystal is shown in Fig. 12.6. The initial

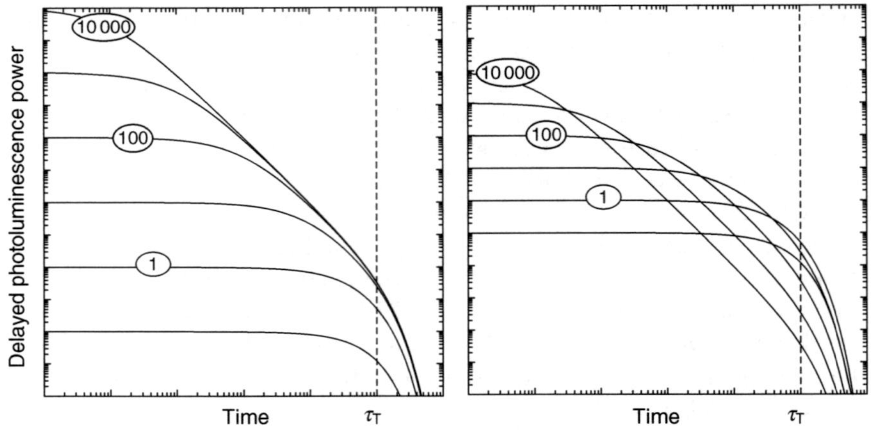

12.5 Dynamics of the detected photoluminescence power arising from triplet exciton fusion after impulsive excitation in a molecular crystal. The two graphs show the evolution of the photoluminescence dynamics in the two cases where the initial triplet density is varied by varying the number of photons in the excitation pulse (left) or by varying the density of photoexcitation through, e.g. different focusing conditions of the laser beam (right). Each curve is labelled with the value of the dimensionless parameter *r*. Note that in these plots the length of one decade on the time-axis is twice the length of one decade on the vertical axes, so that a power law of $\sim t^{-2}$ has a slope of −45°.

exponential decay of the photoluminescence is ~4 times faster in the rubrene crystal than for molecules in solution, hinting at an additional decay mechanism in the crystal. In addition, the curve for the crystal in Fig. 12.6 clearly shows that the initial exponential decay with a decay time of 4 ns leaves behind a delayed photoluminescence signal. From this data one immediately concludes that in this experiment the lifetime of singlet excitons in the rubrene crystal was of the order of 4 ns, and that during that time the singlet excitons underwent fission into triplet excitons. To confirm the presence of triplet excitons it is necessary to follow the decay of the delayed photoluminescence and confirm that it can be described by the exciton dynamics model based on fission and fusion that was presented above.

The evolution of the delayed photoluminescence in rubrene is shown in Fig. 12.7 over 7 orders of magnitude and four time decades (Ryasnyanskiy and Biaggio, 2011). The photoluminescence power and its time-evolution can be very well fitted by the square of Eq. 12.14, as expected for a photoluminescence dominated by triplet dynamics and bimolecular recombination (compare Fig. 12.7 with the theoretical predictions in Fig. 12.5). In addition to the typical power-law decay of the photoluminescence connected with triplet-triplet interaction, Fig. 12.7 also shows the deviation

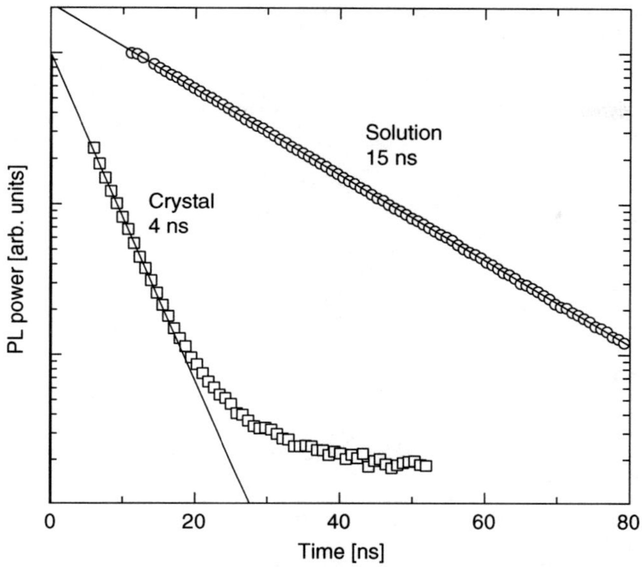

12.6 Time dynamics of the photoluminescence after impulsive excitation for a solution of rubrene molecules and for a rubrene crystal (Mamedov and Becker, 2008). Round and square symbols are experimental results for solution and crystal, respectively. The solid lines are exponential fits, labeled with the corresponding exponential time constants.

from this power law that occurs at times longer than the triplet lifetime. Inspection of the data in Fig. 12.7 delivers a triplet lifetime of 100 ± 20 μs (Ryasnyanskiy and Biaggio, 2011).

12.4 Exciton dynamics and delayed photocurrent

Measurements of photocurrent dynamics in rubrene single crystals (Najafov *et al.*, 2006a, 2008, 2010) have shown that in addition to a small initial release of charge carriers possibly due to singlet exciton dissociation, seen as an initial step in photocurrent dynamics curves such as those shown in Fig. 12.2, there is also a strong delayed photocurrent. At low excitation densities this delayed photocurrent has a build-up time of 100 μs (Najafov *et al.*, 2008, 2010). The fact that in the rubrene crystal the photocurrent build-up time is the same as the triplet lifetime (Ryasnyanskiy and Biaggio, 2011) is a strong hint that the delayed photocurrent may be caused by triplet exciton dissociation. While the exact mechanisms for this effect have not been described yet in rubrene, it is interesting to consider what the exciton dynamics presented above predicts for a delayed photocurrent caused by

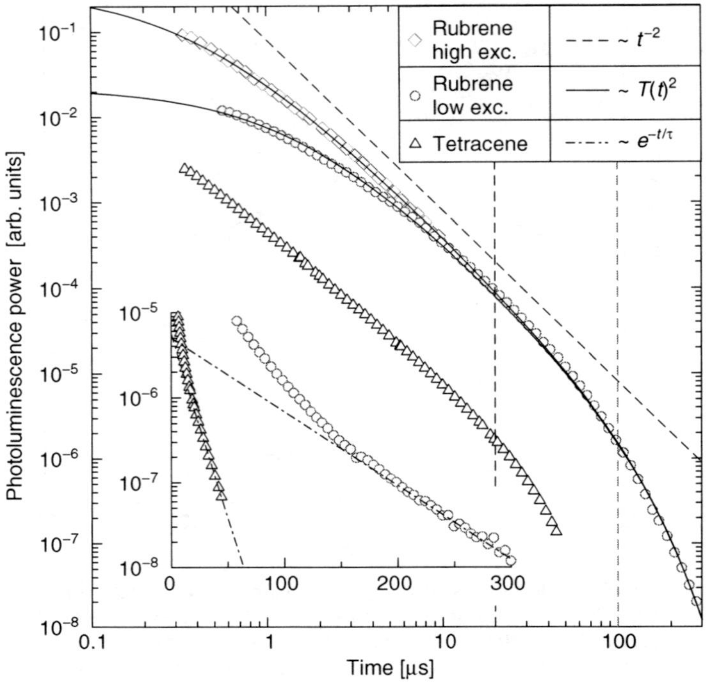

12.7 Photoluminescence dynamics after impulsive excitation in rubrene (circles) and in tetracene (triangles). The solid lines are fits to Eq. 12.14. A dashed line with a slope of –2 is shown for reference. The vertical dashed lines correspond to the triplet lifetimes as determined from this data, and they show the times at which the transition from the power law to an exponential decay occurs for the two data sets. The inset is a semilogarithmic plot of the same data, showing the final exponential decay caused by the triplet exciton lifetime. The dash-dotted lines in the inset are perfect exponential decays with time constants of 10 μs for tetracene and 50 μs for rubrene. These times correspond to half the triplet lifetime. Note that this graph, like Fig. 12.5, has a vertical axis with decades exactly half as long as those on the horizontal axis, and a power law of t^{-2} is a straight line with a slope of –45° (adapted from Ryasnyanskiy and Biaggio, 2011).

triplet excitons dissociation into free carriers. Because of their large binding energy, one cannot assume that triplet excitons in molecular crystals can spontaneously dissociate thermally. However, one can envision a process in which a triplet exciton interacts with a defect state which traps the electron and releases the hole in the valence band by a process similar to the Auger effect (Hangleiter, 1988; Chen and Meng, 2001; Gregg, 2003; Arkhipov and Bässler, 2004). This would use the energy released by trapping the electron in the defect to excite the hole into the valence band.

For the case where exciton dissociation results in charge carriers that have a lifetime much longer than the triplet lifetime, the amount of photoconductivity observed as a function of time is simply proportional to the total density of carriers released by exciton dissociation. This is the time-integral of the triplet density of Eq. 12.14,

$$\sigma(t) \propto \int_0^t T(u)\mathrm{d}u = \frac{1}{\gamma}\ln[(1+r)\mathrm{e}^{t/\tau_T} - r] - \frac{t}{\gamma\tau_T}, \qquad [12.15]$$

with a final value $\sigma(\infty) \sim \gamma^{1}\ln(1 + r)$. This function is plotted in Fig. 12.8 for different values of the parameter r. Again, similar to the discussion of the delayed photoluminescence in the previous section, two cases are shown: variation of r by changing the total number of excitons, and variation of r by changing the photoexcitation volume while keeping the total number of initially created excitons constant. In both cases one can see how the build-up speed of the photocurrent starts increasing in coincidence with the appearance of a power law in the dynamics of the triplet excitons. As more triplet excitons fall prey to triplet–triplet annihilation, the build-up speed of the photocurrent increases, and the final amplitude of the photocurrent starts to saturate. Note how increasing the initial triplet density while

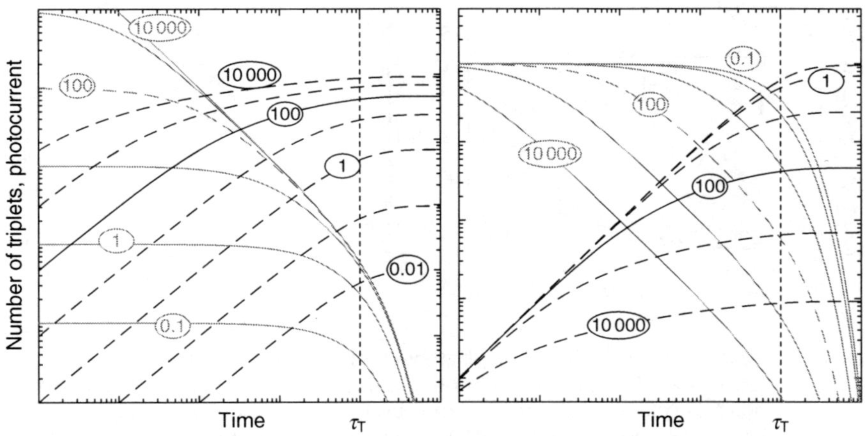

12.8 Evolution of the total number of triplet excitons (solid curves) and of the photocurrent caused by triplet exciton dissociation (dashed curves) for different excitation densities. The vertical dashed line again indicates the triplet exciton lifetime. The two graphs show the evolution of the photocurrent dynamics in the two cases where the initial triplet density is varied by varying the number of photons in the excitation pulse (left) or by varying the density of photoexcitation through, e.g., different focusing conditions of the laser beam (right). In the latter case the initial total number of triplets is constant. Each curve is labeled with the value of the dimensionless parameter r.

keeping the total number constant (Fig. 12.8, right) *decreases* the number of triplets and the photocurrent found at later times, while it is the other way around when the triplet density is varied by changing the total number of initial triplets in a constant volume (Fig. 12.8, left). For the right-hand plot in Fig. 12.8, the initial total number of triplet excitons is held constant, while for the left-hand plot the effects of triplet–triplet annihilation at high densities are overcompensated by the increase in the total number of triplets.

A build-up rate for the photocurrent can be defined by the inverse of the time t_0 at which a fraction q of the final photocurrent amplitude is reached. This can be readily calculated from Eq. 12.15 and is

$$t_0 = -\tau_T \ln[1 + r^{-1}(1 - (1 + r)^q)], \qquad [12.16]$$

The build-up rate $1/t_0$ is plotted in Fig. 12.9 (left) for various initial triplet densities, together with the corresponding final amplitude of the photocurrent. The delayed photoconductivity amplitude depends linearly on the triplet density for low densities, and makes a transition to a sublinear

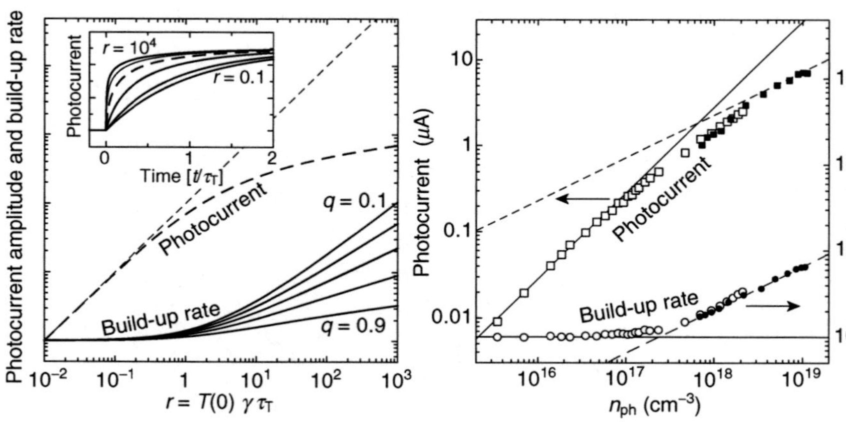

12.9 Left: Dependence of photocurrent dynamics on the excitation density as defined by the dimensionless parameter r. The inset shows a set of normalized photocurrent build-up curves as predicted by Eq. 12.15, plotted from $r = 0.1$ (slowest build-up) to $r = 10\,000$ (faster build-up) in increments of a factor of 10. The dashed curve in the inset corresponds to $r = 100$. The main graph shows the excitation density dependence (at constant excitation volume) of the final photocurrent amplitude (dashed curve) and of the build-up rate as given by the inverse of Eq. 12.16 for $q = 0.1$ to $q = 0.9$ in steps of 0.2. The slope of the middle build-up rate curve on the log–log plot is 0.5. Right: Experimental data for photocurrent amplitude and photocurrent build-up rate after pulsed illumination in a rubrene crystal. Different data point symbols represent different experiments (adapted from Najafov *et al.*, 2008).

logarithmic growth at higher densities where $r > 1$. The build-up rate, on the other-hand, starts increasing at higher triplet densities, becoming proportional to the square root of the triplet density when choosing $q = 0.5$. While rubrene's delayed photocurrent and its saturation behavior were initially explained by a different model based on quadratic recombination of charge carriers (Najafov *et al.*, 2008), one now sees that that interpretation must be replaced with the model presented here, based on triplet exciton dynamics. In fact, Fig. 12.9 shows the good agreement that exists between the predictions from Eqs. 12.15 and 12.16 and experimental results.

12.5 Conclusion

This chapter presented simple models for two dynamic quantities that can be measured after impulsive excitation in molecular crystals, the photoluminescence and the photocurrent. It reviewed specific limiting cases in the detection of impulsively created photoconductivity, and discussed the influence of singlet exciton fission into triplet excitons and triplet exciton fusion into singlet excitons on both photoconductivity and photocurrent. Under pulsed conditions, the photoluminescence and photocurrent dynamics resulting from singlet–triplet interaction have well-defined characteristic dependences on the initial photoexcitation density making the effect easily recognizable. While the delayed photoluminescence is characterized by an initial power-law decay followed by a final exponential decay, the delayed photocurrent is characterized by an excitation strength dependent build-up that becomes an exponential build-up with a time constant given by the triplet lifetime at low excitation densities. Thus, in the presence of triplet exciton dissociation there is a one-to-one relationship between the final exponential decay of the photoluminescence dynamics and the exponential build-up of a delayed photocurrent observed at low excitation densities.

12.6 References

H. Antoniadis and E. Schiff (1992), 'Transient photocharge measurements and electron emission from deep levels in undoped *a*-Si:H,' *Phys. Rev. B*, **46**, 9482.

V. I. Arkhipov and H. Bässler (2004), 'Exciton dissociation and charge photogeneration in pristine and doped conjugated polymers,' *Phys. Status Solidi A*, **201**, 1152.

D. H. Auston (1975), 'Picosecond optoelectronic switching and gating,' *Appl. Phys. Lett.*, **26**, 101.

C.-H. Chen and H.-F. Meng (2001), 'Defect auger exciton dissociation and impact ionization in conjugated polymers,' *Phys. Rev. B*, **64**, 125202.

H. Diesinger, M. Panahandeh-Fard, D. Baillargeat, and C. Soci (2011), 'Electromagnetic modeling and optimization of photoconductive switches for terahertz generation and photocurrent transient spectroscopy,' *Proceedings of the IEEE*, 373–376. doi 10.1109/MWP.2011.6088749

F.~K. Dolezalek (1976), in: *Photoconductivity and related phenomena*, J. Mort and D. M. Pai, Eds., Elsevier/North Holland.
B. A. Gregg (2003), 'Excitonic solar cells,' *J. Phys. Chem. B*, **107**, 4688.
J. L. Hall, D. A. Jennings, and R. M. McClintock (1963), 'Study of anthracene fuorescece excited by the ruby giant-pulse laser,' *Phys. Rev. Lett.*, **11**, 364.
A. Hangleiter (1988), 'Nonradiative recombination via deep impurity levels in semiconductors: The excitonic auger mechanism,' *Phys. Rev. B*, **37**, 2594.
J.-R. Haynes and W. Shockley (1951), 'The mobility and life of injected holes and electrons in germanium,' *Phys. Rev.*, **81**, 835.
P. Irkhin and I. Biaggio (2011), 'Direct imaging of anisotropic exciton diffusion and triplet diffusion length in rubrene single crystals,' *Phys. Rev. Lett.*, **107**, 017402.
R. G. Kepler (1960), 'Charge carrier production and mobility in anthracene crystals,' *Phys. Rev.*, **119**, 1226.
R. G. Kepler, J. C. Caris, P. Avakian, and E. Abramson (1963), 'Triplet excitons and Delayed Fluorescence in Anthracene Crystals', *Phys. Rev. Lett.*, **10**, 400.
B. T. Kolomiets and E. A. Lebedev (1967), 'Drift mobility of carriers in glassy arsenic selenide,' *Soviet Physics – Semiconductors*, **1**, 244.
F. Mamedov and H-C. Beker (2008), Uppsala University, private communication.
A. M. Müller, Y. S. Avlasevich, W. W. Schoeller, K. Mullen, and C. J. Bardeen (2007), 'Exciton fission and fusion in bis(tetracene) molecules with different covalent linker structures,' *J. Am. Chem. Soc.*, **129**, 14240.
H. Najafov, I. Biaggio, V. Podzorov, M. F. Calhoun, and M. E. Gershenson (2006a), 'Primary photoexcitations and the origin of the photocurrent in rubrene single crystals,' *Phys. Rev. Lett.*, **96**, 056604.
H. Najafov, I. Biaggio, T. Chuang, and M. K. Hatalis (2006b), 'Exciton dissociation by a static electric field followed by nanoscale charge transport in PPV polymer films,' *Phys. Rev. B*, **73**, 125202.
H. Najafov, B. Lyu, I. Biaggio, and V. Podzorov (2008), 'Investigating the origin of the high photoconductivity of rubrene single crystals,' *Phys. Rev. B*, **77**, 125202.
H. Najafov, B. Lyu, I. Biaggio, and V. Podzorov (2010), 'Two mechanisms of exciton dissociation in rubrene single crystals,' *Appl. Phys. Lett.*, **96**, 183302.
M. Pope and C. Swenberg (1999), *Electronic Processes in Organic Crystals and Polymers*, Oxford University Press.
A. Ryasnyanskiy and I. Biaggio (2011), 'Triplet exciton dynamics in rubrene single crystals,' *Phys. Rev. B.*, **84**, 193203.
W. E. Spear (1957), 'Transit time measurements of charge carriers in amorphous selenium films,' *Proc. Phys. Soc. London, Sect. B*, **70**, 669.
S. Tao, H. Matsuzaki, H. Uemura, H. Yada, T. Uemura, J. Takeya, T. Hasegawa, and H. Okamoto (2011), 'Optical pump-probe spectroscopy of photocarriers in rubrene single crystals,' *Phys. Rev. B.*, **83**, 075204.
A. Tapponnier, I. Biaggio, P. Günter (2003), 'Integrated pulsed photoconductivity of organic light emitting diodes,' *Appl. Phys. Lett.*, **83**, 5473–5475.

13

Conductivity measurements of organic materials using field-effect transistors (FETs) and space-charge-limited current (SCLC) technique

O. D. JURCHESCU, Wake Forest University, USA

DOI: 10.1533/9780857098764.2.377

Abstract: This chapter describes in detail two major techniques used to estimate the charge carrier mobility in organic semiconductors: field-effect transistors (FET) and space-charge-limited current (SCLC) measurements. While the calculations using FET structures provide details about material properties at the interface with gate dielectrics, the bulk-mobility can be extracted from SCLC data. The values of mobility obtained from the two methods are often very different due to anisotropy arising from molecular packing along various crystallographic directions, or as a result of different structure or chemistry of the surface versus the bulk. For this reason, these techniques address different aspects of the organic solid, thus are complementary.

Key words: organic field-effect transistors (FETs), space-charge-limited current (SCLC), mobility, organic semiconductors.

13.1 Introduction

Organic electronics research has thrived and expanded over recent years due to its promise to encourage development of products which can complement silicon-based applications with novel functionalities that cannot be addressed by silicon technologies. Plastic electronics may perhaps even replace silicon when this is economically feasible, in applications that operate under modest performance requirements (e.g. tens of kHz). Organic semiconductors promise to impact the industry by reduced complexity processing, compatibility with arbitrary substrates and great chemical tunability. The seemingly infinite variety of organic compounds presents the prospect of tailoring materials to have any desired properties. Organic-based applications can thus potentially address a broad market, which includes flexible displays, electronic papers, sensors, disposable or wearable electronics, medical applications, and more[1–7]. There is a tremendous excitement in both academia and industry towards the study of organic field-effect transistors (OFETs), organic photovoltaics (OPVs), organic light-emitting diodes (OLEDs), and sensors. A key parameter in device operation is the *field-effect mobility* (μ) of the charge carriers within the semiconductor layer. In OFETs, for example, μ is the figure of merit for device quality, with a high mobility being desired to increase the operational speed. Mobility is a complex property, as it is dependent on both the molecular structure of the organic semiconductor and its microstructure, but also the quality of the contacts, interfaces, and other factors of intrinsic or extrinsic nature[8–17]. The highest reported mobilities for organic semiconductors have increased five orders of magnitude in the last 25 years as a result of improved knowledge on the factors that govern the transport in organic materials and devices[3]. In spite of this spectacular growth, the mobility is still modest (1–10 cm^2/Vs)[3] and the current state-of-the art performance is still limited to low operation frequencies, 1 MHz being the highest reported[18]. (The operation frequency is directly proportional with the transconductance, which in turn is determined by the mobility value, as we will show in Section 13.2.2.[3]) Flexible active-matrix displays or sensor arrays, however, require frequencies higher than 10 MHz[7].

This chapter will focus on two major techniques frequently used to estimate the charge carrier mobility in organic semiconductors. Mobility calculations using FET structures are able to provide details about material properties at the interface with gate dielectrics. The bulk-mobility can be extracted from space-charge-limited current (SCLC) measurements. The values of mobility obtained from these two methods are often very different due to anisotropy arising from molecular packing along various crystallographic directions, or as a result of different structure or chemistry of the surface versus the bulk[13,19–22]. For this reason, these techniques address different aspects of the organic solid and thus are complementary.

13.2 Field-effect transistor (FET) measurements

13.2.1 Device structure

The OFET is a three-terminal device (Fig. 13.1), with the three contacts being referred to as gate, drain, and source, and an organic semiconductor as the active layer. The channel forms at the organic semiconductor/gate insulator interface and the electric field applied at the gate electrode determines the density of the charge carriers accumulated at the interface. The current between source and drain (I_D) is modulated by the gate-source voltage (V_{GS}) and the drain-source voltage (V_{DS}). This offers a great control over the electrical conductivity of the channel by tuning the density of charge carriers over a wide range. OFETs are referred to as 'single crystal-OFETs' (SC-OFETs) when the active semiconducting layer consists of an organic single crystal. If the semiconductor is in a thin-film form, the devices are referred to as 'organic thin-film transistors' (OTFTs). Generally SC-

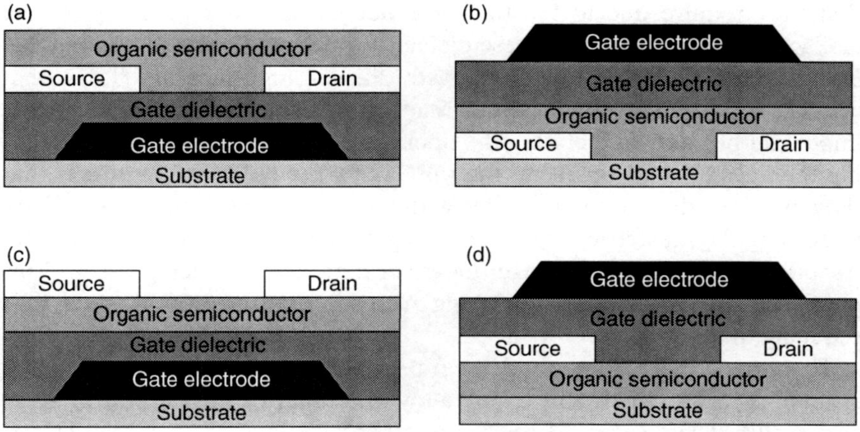

13.1 Cross-section of FET structures. Each layer is represented in a different shade. (a) Bottom-gate, bottom contacts FET. (b) Top-gate, bottom contacts FET. (c) Bottom-gate, top-contacts FET. (d) Top-gate, top-contacts FET.

OFETs yield superior performance given the improved order present in the channel, and mobilities as high as 10–40 cm^2/V s have been reported in high-purity organic single crystals[19,23,24]. In spite of their modest performance, OTFTs present increased technological appeal due to their reduced complexity processing. The use of OTFTs is particularly attractive in conjunction with solution-based manufacturing techniques. Fast deposition by using roll-to-roll processing, inkjet printing or spray deposition[25–27] may allow their production in large volumes and at low cost per unit area, and thus the introduction of the concept of 'electronics everywhere'[28,29]. The performance of OTFTs has improved significantly since the first reports, which date from the 1980s[30–32], and field-effect mobilities rivaling hydrogenated amorphous silicon (α-Si:H) – near 1 cm^2/Vs – were reported in vacuum sublimed[33,34] and solution deposited OTFTs[35–42].

Several device architectures can be manufactured depending on the sequence in which the layers are deposited, as presented in Fig. 13.1a–d. Bottom contact, bottom gate structures (Fig. 13.1a) are generally preferred for quick screening of newly synthesized organic semiconductor compounds, or testing new deposition methods due to their ease of fabrication and the minimum processing required on the easily degradable organic layers. Here, substrates consisting of gate electrode/gate dielectric/source and drain contacts are pre-fabricated and the organic semiconductor is deposited during the last step of the fabrication process. Gold or silver are commonly used as contact materials, and SiO_2, AlO_3, or other inorganic oxides or nitrides as gate dielectrics. Their drawback is the high operating voltages (>20 V) that they require due to the thickness needed in order to minimize the leakage currents. Replacing these dielectrics with molecular self-assembled monolayers (SAMs) allows demonstration of high-mobility TFTs that operate with supply voltages lower than 2 V[34,43]. The choice of dielectrics is important not only in dictating the operating voltages, but also in modulating the field effect mobility of the OFETs by the polarization effects that they may introduce at the interface with the organic semiconductors[11,44]. By increasing the dielectric constant of the gate dielectric material, the dielectric polarizability increases, resulting in formation of Fröhlich polarons. This polaronic coupling induces scattering events in the transistor channel, thus reducing mobility[11,44].

Theoretical calculations and several experimental reports predict that structures in Fig. 13.1b and c may allow for superior device performance as a result of less severe contact effects[10,45–47]. Indeed, measurements performed on OFET structures consistent with Fig. 13.1b, with Cytop gate dielectric, demonstrated low hysteresis and mobilities as high as 2.4 cm^2/Vs[48–50]. Fabrication of FETs with top contacts presents manufacturing challenges related to the degradation of the organic semiconductors upon their exposure to high-energy metal atoms when methods such as e-beam or thermal evaporation are used. Nevertheless, several reports show good electrical properties in devices of Fig. 13.1c[10,51]. A recently developed soft-contact deposition method called 'flip-chip lamination' may allow for further progress in development of top-contact OFETs[52,53]. Mobilities as high as 8 cm^2/Vs were reported in transistors of Fig. 13.1d fabricated on rubrene single crystals with parylene top-gate insulator[54]. Top-gate TFTs (Fig. 13.1b and d) also offer encapsulation of organic layer by the gate dielectric and gate electrode, therefore improving environmental stability.

13.2.2 Device operation

The operation of an OFET is significantly different from that of a silicon metal oxide semiconductor field-effect transistor (MOSFET), which operates in *inversion* mode. In Si MOSFETs the regions adjacent to the source and drain contacts are heavily doped, and consequently a flow of minority carriers is induced to form the I_D current[55]. OFETs work on intrinsic semiconductors and operate in *accumulation* mode. Here the transistor channel forms in the organic semiconductor at its interface with the gate insulator as follows: holes are accumulated when a negative V_{GS} is applied (p-type conduction), and electrons when a positive V_{GS} is applied (n-type conduction). The type of conduction originates from the intrinsic properties of the semiconductor and from the choice of the contact material, as will be described in Section 13.2.3. Organic semiconductors are predominantly

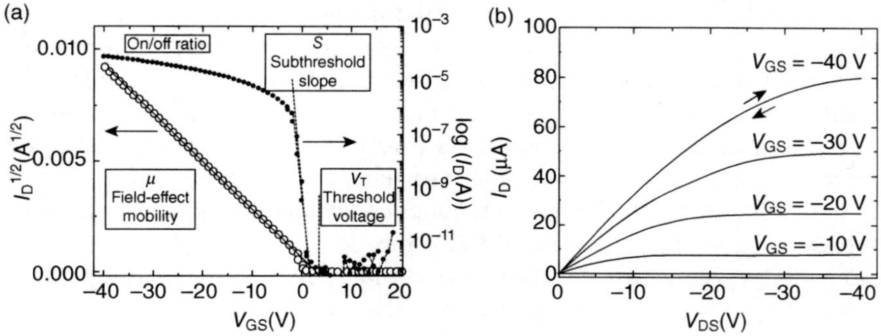

13.2 **Electrical characteristics of organic FETs: (a) evolution of the drain current I_D with the gate voltage in the saturation regime. (b) Evolution of the drain current I_D with the drain voltage V_{DS} for various gate voltages, V_{GS}.**

p-type[3,26,42,56–59], and only a few n-type conductors have been discovered[41,60–64]. Ambipolar FETs operate in both voltage regimes, and were demonstrated on donor-acceptor blends[65,66] or alternating layers[67,68], and in a limited number of small-band gap organic semiconductors capable of transporting both electrons and holes[69–71].

The device is active (a drain current is formed) when the gate voltage value exceeds the value of the threshold voltage (V_T). Nevertheless, even at lower gate voltages, a small current is present as a result of diffusion of charge carriers due to their thermal energy. The 'sub-threshold regime' is quantitatively characterized by the subthreshold swing, *S*. *S* can be determined by taking the inverse of the slope of the *log* (I_D) versus V_{GS} curve – subthreshold slope (Fig. 13.2a, right axis, see dashed lines for the slope) in the subthreshold regime:

$$S = \frac{\partial V_{GS}}{\partial(\log I_D)} \qquad [13.1]$$

The subthreshold swing quantifies the speed with which the device turns on (the increase in I_D with V_{GS}), with a low value corresponding to better device performance. Values of 0.88 V/dec were estimated in pentacene TFTs on Cytop dielectric[72] and 0.1 V/dec with a SAM dielectric[34]. In single crystal FETs typical values are 0.3–0.5 V/dec[54,56,72,73]. As the subthreshold swing is determined by the quality of the insulator/semiconductor interface, this parameter allows for estimation of the interface trap density (N_i) from OFET data by using eq. 13.2[55]:

$$S = \frac{kT}{e}\ln 10\left(1 + \frac{e}{C_i}N_i\right) \qquad [13.2]$$

Here k is Boltzmann's constant, e is the elementary charge and C_i is the gate dielectric capacitance per unit area. For SiO_2, for example, the number of traps at the organic/dielectric interface can be mitigated by chemically treating this surface with SAMs of organic trichlorosilanes prior to the deposition of the semiconductor. As a result, a higher mobility and less hysteresis in the I–V characteristics are generally obtained[58,61,74,75].

In the operation of a FET, two regimes can be distinguished. In the *linear regime* ($V_{DS} \ll V_{GS} - V_T$, Fig. 13.2b), the device acts as a gate voltage-controlled variable resistor: the source-drain current (I_D) increases with the applied V_{DS} voltage, following the expression in eq. 13.3:

$$I_D = \frac{W}{L} C_i \mu \left[(V_{GS} - V_T) V_{DS} - \frac{V_{DS}^2}{2} \right] \qquad [13.3]$$

Here W and L are the channel length and width respectively. At higher V_{DS} ($V_{DS} \gg V_{GS} - V_T$, Fig. 13.2b), a depletion area forms at the drain contact. Beyond this point, called 'pinch-off', the transistor operates in the *saturation regime* (Fig. 13.2b), and I_D varies quadratically with the gate voltage (eq. 13.4):

$$I_D = \frac{W}{L} \frac{C_i}{2} \mu (V_{GS} - V_T)^2 \qquad [13.4]$$

It can be observed from eq. 13.3 and 13.4 that in OFETs the drain current is controlled by the applied gate-source voltage. This is quantitatively described by the transconductance g_m:

$$g_m = \left. \frac{\partial I_D}{\partial V_{GS}} \right| V_{DS} \qquad [13.5]$$

The transconductance is determined from the transfer characteristics of a device, at constant V_{DS} (Fig. 13.2a). An increase in the drain current I_D with increased negative gate-source voltage V_{GS} is a signature for a hole-transporting organic layer. For an electron-transporting material the drain current increases when a positive gate-source voltage is applied, and for the case of ambipolar transport the current is increased for both positive and negative voltages. Figure 13.2a represents typical transfer transistor characteristics in the saturation regime of an OFET fabricated on an organic single crystal at the surface of octyltrichlorosilane (OTS) treated SiO_2 gate dielectric, with bottom Au contacts (Fig. 13.1a)[58]. Following eq. 13.3 and 13.4, the field-effect mobility μ can be calculated from the slope of the I_D vs V_{GS} in the linear regime, and $(I_D)^{1/2}$ vs V_{GS} (curve in open circle in Fig. 13.2a) in the saturation regime respectively, using the following expressions:

$$\mu_{lin} = \frac{L}{W} \frac{1}{C_i V_{DS}} \frac{\partial I_D}{\partial V_{GS}} \quad \text{linear regime} \qquad [13.6]$$

$$\mu_{sat} = \frac{L}{W}\frac{2}{C_i}\left(\frac{\partial\sqrt{I_D}}{\partial V_{GS}}\right)^2 \quad \text{saturation regime} \qquad [13.7]$$

The intercept of this slope with the V_{GS} axis allows for determination of V_T, with a threshold voltage near zero being desired to minimize the power consumption. Increased negative or positive V_T indicates the presence of trapping or doping effects, respectively, in the transistor channel. A V_T shift may also be created during device operation as a result of interaction with the environment or bias stress effects. These effects may be of irreversible or (partially) reversible nature[14,76].

13.2.3 Contact effects

Generally, calculations using the expressions for the linear regime (eq. 13.6) yield lower mobility values than those extracted from eq. 13.7, corresponding to the saturation regime. This originates from the critical effect that the contact resistance has on the device properties, pointing to the fact that the mobility extracted from the current–voltage characteristics is a device property rather than an intrinsic property of the organic semiconductor material. The electrode material is generally chosen so that the energy level of the metal work-function aligns reasonably well with the highest occupied molecular orbital (HOMO) of the organic semiconductor for p-type conduction and the lowest unoccupied molecular orbital (LUMO) for n-type. The contact material is in direct contact with the organic semiconductor, to form a Schottky contact. It is not always straightforward, however, to choose the proper electrode material. A monotonic relationship between the contact work-function and resulting device performance has not been established, due to the very complex phenomena that occur at the metal/organic interface. For example, in addition to the Schottky barriers and interface dipoles which form at these interfaces, the position of the HOMO and LUMO levels changes slightly with molecular orientation, which in turn is highly dependent on the processing details (surface energy, solvent, deposition method, etc.)[8,77–79]. For this reason, it is difficult to relate the metal work-function with the organic semiconductor HOMO/LUMO levels determined by cyclic voltammetry (CV) or ultraviolet photoelectron spectroscopy (UPS), and to subsequently provide working schemes for contact choices.

The parasitic contact effects limit the device performance and become more severe with decreases in channel length and for materials with higher intrinsic mobility, where the contact resistance becomes comparable with the channel resistance. They can yield non-ideal *I-V* characteristics, reduced mobility, increased threshold voltage, and sometimes a lack of current saturation in the output characteristics (Fig. 13.2b) and current saturation in the transfer characteristics (Fig. 13.2a). Reducing contact effects is recognized as one of the most stringent problems in present organic semiconductor research as, for example, OTFTs are operated in the linear regime in applications such as active matrix displays. Intense effort dedicated to understanding and minimizing these effects includes scanning potential imaging studies, investigations focused on gated transmission line model and gated-four-terminal test structures, and capacitance–voltage measurements[54,80–88]. Exceptionally low contact resistances, of the order of 10–100 Ω cm were achieved in electrolyte-gated OTFTs and single crystal transistors[89,90]. It is worth noting that there is an increasing effort biased towards replacing traditional metal electrodes with alternative materials such as reduced graphene oxide[91] or carbon nanotube arrays[92].

OFETs represent, on one hand, powerful tools to investigate electrical properties of organic semiconductors, with great control of the charge carrier density. On the other hand, the OTFT technology may offer a viable approach to low-cost flexible electronics industry. For this reason, it is critical that improvements in device performance are implemented without compromising the cost-efficiency and ease of manufacturability.

13.3 Space-charge-limited current (SCLC) measurements

13.3.1 Phenomenological description of SCLC

SCLC theory was developed in 1940 by Mott and Gurney for single carrier injection in a trap-free insulator[93], modified by Rose and Lampert for an insulator with localized trapping states in the gap[94,95], and first used on organic crystals in 1962 by Helfrich and Mark[96]. Since then, this method was adopted for measurements in a wide variety of organic crystalline and amorphous semiconductors[13,54,97,98].

In SCLC measurements, an organic material of thickness L is sandwiched between two contacts (Fig. 13.3a), and the evolution of the current with the applied voltage is recorded. Three regimes can be distinguished, as can be observed in Fig. 13.3b, where the current density, J, is plotted versus applied voltage, V. At small electric fields (low V), the transport is ohmic, and the current increases linearly with the voltage: $J \alpha V$. It is consequently named the *ohmic regime*. At higher fields the current becomes space-charge limited, which is reflected in a quadratic dependence on the applied voltage, $J \alpha V^2$, as will be demonstrated later. In the first part of the *SCLC regime*, a fraction of the injected carriers is trapped and the steady-state SCL current is reduced by a factor θ, which, in a simplified picture, represents the ratio between the number (concentration) of free (n_f) and total (n_{tot}) charge carriers present in the solid. Nevertheless, θ may present more complicated

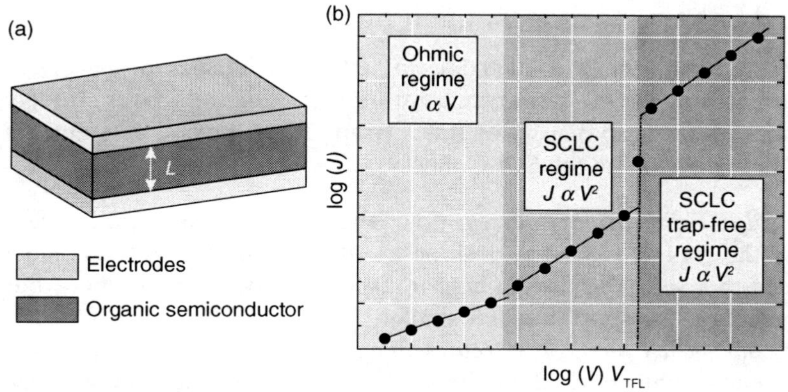

13.3 (a) Schematic representation of the capacitor-type configuration used for development of SCLC theory. (b) Current density (*J*) versus applied voltage (*V*). The ohmic, SCLC and SCLC-trap-free regimes are indicated.

dependencies on the applied voltage, device geometry, spatial and energetic distribution of traps, and more. As the voltage increases, the Fermi level moves towards the semiconductor valence band for the case of p-type conduction, or the conduction band for the case of n-type conduction and traps are gradually filled, yielding an increase in the density of free charge carriers. At the trap-filling limit voltage (V_{TFL}) all traps are filled ($\theta = 1$) and the current rises abruptly. After this point, the system is in the *SCLC trap-free regime*. The transition point is determined by the density of trapping states and the current can sometimes increase by orders of magnitude for solids with a high density of traps. The value of the V_{TFL} contains important information about the concentration of traps, N_t, as the two quantities are related via a very simple expression:

$$N_t \cong \frac{\varepsilon_0 \varepsilon_r}{e} \frac{V_{TFL}}{L^2} \qquad [13.8]$$

where ε_0 is the permittivity of free space, and ε_r the dielectric constant of the semiconductor. Note that N_t represents the volume concentration of traps, which is an intrinsic property of the organic semiconductor, while N_i from eq. (13.2) quantifies the surface trap densities at the organic semiconductor/dielectric interface.

The mobility estimated from the trap-free SCLC regime represents the intrinsic mobility of the material, as it will be discussed later. Using this model, trap densities as low as 1.74×10^{11} cm^{-3} and a mobility of 35 cm^2/Vs were estimated in ultra-pure pentacene single crystals[13].

13.3.2 Mathematical description of SCLC theory

The current in an organic solid is determined by the intrinsic properties of the material as well as the carrier concentration gradients, which are strongly dependent on the trap density of states (trap DOS). Several assumptions are made in developing the SCLC theory:

- *Unipolar, 1-dimensional current flow*: The charge carriers are injected from the contact placed at position $x = 0$ and collected at the contact placed at $x = L$. The case of hole-transport will be discussed here, but the electron-transport model is similar.
- *The contacts are ohmic: they represent an infinite source of charge carriers*. Ohmic contacts yield an ohmic regime at low applied voltage (Fig. 13.3b). Deviations from a slope of 1 in the double logarithmic plot of J versus V at low voltages are a clear indication that the contacts are not ohmic. In polycrystalline and amorphous silicon TFTs, for example, ohmic contacts are obtained by selective contact doping. Achieving ohmic injection in organic TFTs is not trivial, although attempts of contact doping were reported[50,99]. Here Schottky contacts are generally formed. This limits the use of SCLC model to a very few systems, where ohmic contacts are achieved[13,100,101].
- *Only the drift component of the current is considered*, and the diffusion current is neglected:

  $$J(x) = e\mu n_{\mathrm{f}}(x)E(x) \qquad [13.9]$$

  where $J(x)$ *and* $E(x)$ represent the local current density and electric field respectively, at distance x from the injecting electrode, $n_{\mathrm{f}}(x)$ is the local density of free charges and μ the bulk mobility.
- *The mobility of free charge carriers is independent of the magnitude of the applied electric field.* The dependence of mobility of the applied field originating from effects such as Poole–Frenkel is generally very small[102], and thus here is ignored for simplicity.
- *All traps are homogeneously distributed in space and correspond to one discrete energy level.* Note that in a wide variety of systems, an exponential or Gaussian distribution of trapping states has been reported[103,104].
- *The density of free charge carriers (n_{f}) follows Boltzmann statistics (eq. 13.10), and the density of the trapped (localized) carriers (n_{t}) follows the Fermi–Dirac statistics (eq. 13.11).* The free and trapped charge carriers are in equilibrium under SCLC flow, in the dark, and the photo-effects are not considered.

$$n_{\mathrm{f}} = N_{\mathrm{V}} \mathrm{e}^{-\frac{E_{\mathrm{F}}(x)}{kT}} \qquad [13.10]$$

$$n_t = \frac{h(E)}{1+e^{\frac{E_i - E_F(x)}{kT}}} \quad [13.11]$$

Here N_V represents the effective density of states in the valence band, E_F is the Fermi level, E_i is the trap energy, $h(E)$ describes the energetic distribution of localized states, and k is Boltzmann's constant.

The expression for the SCLC current–voltage characteristics is obtained by combining the continuity equation (eq. 13.9) with Poisson's equation (eq. 13.12):

$$\frac{dE(x)}{dx} = \frac{e}{\varepsilon_0 \varepsilon_r} n_{tot}(x) = \frac{e}{\varepsilon_0 \varepsilon_r}[n_f(x) + n_t(x)] \quad [13.12]$$

where $E(x)$ is the local electric field, and n_{tot}, n_f, *and* n_t are the total, free and trapped space charges at position x. The expression for the current as a function of position becomes:

$$J(x) = \mu\theta\varepsilon_0\varepsilon_r E \frac{dE}{dx} \quad [13.13]$$

By integrating eq. 13.13 across the organic material, and using the boundary conditions described earlier, an expression for the density of current versus electric field is obtained (eq. 13.14). This equation represents the mathematical formulation of Mott–Gurney law:

$$J = \frac{9}{8}\varepsilon_0\varepsilon_r\mu\theta\frac{V^2}{L^3} \quad [13.14]$$

A current–voltage response $J \alpha V^2$ in the SCLC regime, corresponding to eq. 13.14, was reported in pentacene single crystals and anthradithiopene films[13,105]. In the case of pentacene, for example, a unique discrete trap-state originated from the presence of pentacenequinone impurity molecules. In general, the localized states (traps) in the gap are generated by crystal defects, impurities, or thermal molecular motion[11,13,15,17,56,101,106]. Extended defects, such as edge dislocations or screw dislocations may also modify the available energy levels in their vicinity, often leading to the presence of vacant orbitals in the band gap. Regardless of their origin, the traps are characterized by their energies, with the shallow traps being positioned closer to the band-edge and the deep traps extending into the gap.

If discrete trapping energies separated by an energy barrier greater than kT are present, or if the trap energies follow more complicated distributions, the current–voltage characteristics will deviate from a $J \alpha V^2$ to a $J \alpha V^n$ dependence, with $n > 2$[94,107]. The faster increase in current as a function of voltage for the case of traps distributed in energy compared to single trap case described in SCLC originates from the differences in the shift in

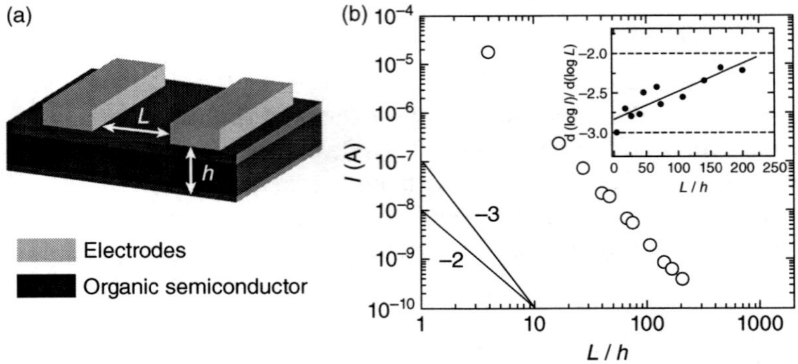

13.4 (a) Schematic representation of coplanar electrode configuration, with *L* being the contact separation and *h* the crystal thickness. (b) Current versus the ratio of the electrode distance *L* with respect to the crystal thickness *h* at constant bias. The lines are guides for the eyes and correspond to 1D model (slope –3) and 2D model (slope –2) respectively. The inset shows the derivative d(log*I*)/d(log*L*) versus *L/h*. Reproduced with permission from Ref. 109.

Fermi level, which is proportional to the space charge that is in turn, proportional to the applied voltage. For this reason, the shape of the *I-V* curve can offer important details about density of trap states in the band-gap of an organic semiconductor. The trap density of states in the band gap of small-molecule organic semiconductors was estimated by Kalb *et al.* from temperature-dependent space-charge-limited current measurements[100,101].

13.3.3 SCLC – the case of planar contacts

While the Mott–Gurney theory presented in eq. 13.14 describes the current–voltage characteristics for the 1D case of a sandwich-type contact geometry (Fig. 13.3a), Geurst analyzed the 2D case of SCLCs in thin-films with coplanar contacts and negligible thickness (Fig. 13.4a)[108]. In the latter model, the current *I* depends quadratically on the applied voltage *V*, which is similar to Mott–Gurney theory, but inversely proportional to L^2, as compared with L^3 present in the 1D model:

$$I = \frac{2}{\pi}\varepsilon_0\varepsilon_r\mu\frac{V^2}{L^2} \qquad [13.15]$$

The Mott–Gurney and Geurst models describe the two limiting geometries to evaluate the mobility from SCLC measurements and do not provide a solution for the case of a thin film with a well-defined thickness, *h*, which is in the order of the electrode separation length, *L*. It was further demonstrated experimentally that for planar contacts with short electrode separation, the space charge is confined to the *in plane* longitudinal direction of the applied field and the system responds to Mott–Gurney model. When L becomes much larger than the crystal thickness ($L \gg h$), the transverse component of the electric field dominates and Geurst model is more appropriate[109]. The transition from a 1D to a 2D space-charge-limited conduction takes place gradually, as L/h increases (Fig. 13.4b).

The charge transport in organic semiconductors at high voltages is limited by the emergence of space charges. SCLC measurements may serve as a sensitive tool for simultaneously estimate the intrinsic bulk mobility μ and the bulk density of traps N_t in an organic semiconductors. The challenges come from the difficulty in establishing ohmic contacts and the high voltages generally required in order to access the trap-free regime necessary for mobility determination.

13.4 Future trends

The measurement of conductivity in organic semiconductors can offer insights into the relevant physical processes that govern the electrical conduction in this fascinating class of materials. This knowledge may accelerate the needed improvements in performance, a critical step before the migration of organic-based devices from a laboratory effort to a viable technology. Strategies for improving the mobility include design and synthesis of new materials[38,60,63,110–114], thin-film microstructure control by careful processing[8,9,115], but also device scaling to reduce the parasitic effects at contact overlap[3]. Equally important is achieving this improved performance while maintaining or optimizing the cost, and even allowing other advantages, such as large-area electronics[27,116].

13.5 References

1. Forrest, S. R., The path to ubiquitous and low-cost organic electronic appliances on plastic. *Nature* **2004**, 428, (6986), 911–918.
2. Sekitani, T.; Nakajima, H.; Maeda, H.; Fukushima, T.; Aida, T.; Hata, K.; Someya, T., Stretchable active-matrix organic light-emitting diode display using printable elastic conductors. *Nature Materials* **2009**, 8, (6), 494–499.
3. Klauk, H., Organic thin-film transistors. *Chemical Society Reviews* **2010**, 39, (7), 2643–2666.
4. Mannsfeld, S. C. B.; Tee, B. C. K.; Stoltenberg, R. M.; Chen, C. V. H. H.; Barman, S.; Muir, B. V. O.; Sokolov, A. N.; Reese, C.; Bao, Z., Highly sensitive flexible pressure sensors with microstructured rubber dielectric layers. *Nature Materials* **2010**, 9, (10), 859–864.
5. Sekitani, T.; Yokota, T.; Zschieschang, U.; Klauk, H.; Bauer, S.; Takeuchi, K.; Takamiya, M.; Sakurai, T.; Someya, T., Organic nonvolatile memory transistors for flexible sensor arrays. *Science* **2009**, 326, (5959), 1516–1519.

6. Brabec, C. J., Organic photovoltaics: technology and market. *Solar Energy Materials and Solar Cells* **2004**, 83, (2–3), 273–292.
7. Gelinck, G. H.; Huitema, H. E. A.; Van Veenendaal, E.; Cantatore, E.; Schrijnemakers, L.; Van der Putten, J.; Geuns, T. C. T.; Beenhakkers, M.; Giesbers, J. B.; Huisman, B. H.; Meijer, E. J.; Benito, E. M.; Touwslager, F. J.; Marsman, A. W.; Van Rens, B. J. E.; De Leeuw, D. M., Flexible active-matrix displays and shift registers based on solution-processed organic transistors. *Nature Materials* **2004**, 3, (2), 106–110.
8. Gundlach, D. J.; Royer, J. E.; Park, S. K.; Subramanian, S.; Jurchescu, O. D.; Hamadani, B. H.; Moad, A. J.; Kline, R. J.; Teague, L. C.; Kirillov, O.; Richter, C. A.; Kushmerick, J. G.; Richter, L. J.; Parkin, S. R.; Jackson, T. N.; Anthony, J. E., Contact-induced crystallinity for high-performance soluble acene-based transistors and circuits. *Nature Materials* **2008**, 7, (3), 216–221.
9. Salleo, A.; Kline, R. J.; DeLongchamp, D. M.; Chabinyc, M. L., Microstructural characterization and charge transport in thin films of conjugated polymers. *Advanced Materials* **2010**, 22, (34), 3812–3838.
10. Gundlach, D. J.; Zhou, L.; Nichols, J. A.; Jackson, T. N.; Necliudov, P. V.; Shur, M. S., An experimental study of contact effects in organic thin film transistors. *Journal of Applied Physics* **2006**, 100, (2).
11. Hulea, I. N.; Fratini, S.; Xie, H.; Mulder, C. L.; Iossad, N. N.; Rastelli, G.; Ciuchi, S.; Morpurgo, A. F., Tunable Frohlich polarons in organic single-crystal transistors. *Nature Materials* **2006**, 5, (12), 982–986.
12. Jurchescu, O. D.; Baas, J.; Palstra, T. T. M., Electronic transport properties of pentacene single crystals upon exposure to air. *Applied Physics Letters* **2005**, 87, (5), 052102.
13. Jurchescu, O. D.; Baas, J.; Palstra, T. T. M., Effect of impurities on the mobility of single crystal pentacene. *Applied Physics Letters* **2004**, 84, (16), 3061–3063.
14. Goldmann, C.; Gundlach, D. J.; Batlogg, B., Evidence of water-related discrete trap state formation in pentacene single-crystal field-effect transistors. *Applied Physics Letters* **2006**, 88, (6).
15. Krellner, C.; Haas, S.; Goldmann, C.; Pernstich, K. P.; Gundlach, D. J.; Batlogg, B., Density of bulk trap states in organic semiconductor crystals: Discrete levels induced by oxygen in rubrene. *Physical Review B* **2007**, 75, (24).
16. Pernstich, K. P.; Oberhoff, D.; Goldmann, C.; Batlogg, B., Modeling the water related trap state created in pentacene transistors. *Applied Physics Letters* **2006**, 89, (21).
17. Jurchescu, O. D.; Popinciuc, M.; van Wees, B. J.; Palstra, T. T. M., Interface-controlled, high-mobility organic transistors. *Advanced Materials* **2007**, 19, (5), 688–692.
18. Myny, K.; Steudel, S.; Smout, S.; Vicca, P.; Furthner, F.; van der Putten, B.; Tripathi, A. K.; Gelinck, G. H.; Genoe, J.; Dehaene, W.; Heremans, P., Organic RFID transponder chip with data rate compatible with electronic product coding. *Organic Electronics* **2010**, 11, (7), 1176–1179.
19. Sundar, V. C.; Zaumseil, J.; Podzorov, V.; Menard, E.; Willett, R. L.; Someya, T.; Gershenson, M. E.; Rogers, J. A., Elastomeric transistor stamps: reversible probing of charge transport in organic crystals. *Science* **2004**, 303, (5664), 1644–1646.

20. Lee, J. Y.; Roth, S.; Park, Y. W., Anisotropic field effect mobility in single crystal pentacene. *Applied Physics Letters* **2006**, 88, (25).
21. Reese, C.; Bao, Z., High-resolution measurement of the anisotropy of charge transport in single crystals. *Advanced Materials* **2007**, 19, (24), 4535–4538.
22. Ostroverkhova, O.; Cooke, D. G.; Hegmann, F. A.; Tykwinski, R. R.; Parkin, S. R.; Anthony, J. E., Anisotropy of transient photoconductivity in functionalized pentacene single crystals. *Applied Physics Letters* **2006**, 89, (19).
23. Podzorov, V.; Menard, E.; Borissov, A.; Kiryukhin, V.; Rogers, J. A.; Gershenson, M. E., Intrinsic charge transport on the surface of organic semiconductors. *Physical Review Letters* **2004**, 93, (8).
24. Takeya, J.; Kato, J.; Hara, K.; Yamagishi, M.; Hirahara, R.; Yamada, K.; Nakazawa, Y.; Ikehata, S.; Tsukagoshi, K.; Aoyagi, Y.; Takenobu, T.; Iwasa, Y., In-crystal and surface charge transport of electric-field-induced carriers in organic single-crystal semiconductors. *Physical Review Letters* **2007**, 98, (19).
25. Sekitani, T.; Noguchi, Y.; Zschieschang, U.; Klauk, H.; Someya, T., Organic transistors manufactured using inkjet technology with subfemtoliter accuracy. *Proceedings of the National Academy of Sciences of the United States of America* **2008**, 105, (13), 4976–4980.
26. Cho, J. H.; Lee, J.; Xia, Y.; Kim, B.; He, Y. Y.; Renn, M. J.; Lodge, T. P.; Frisbie, C. D., Printable ion-gel gate dielectrics for low-voltage polymer thin-film transistors on plastic. *Nature Materials* **2008**, 7, (11), 900–906.
27. Azarova, N. A. O., J. W.; McLellan, C. A.; Grimminger, M. A.; Chapman, E. K.; Anthony J. E.; Jurchescu, O. D., Fabrication of organic thin-film transistors by spray-deposition for low-cost, large-area electronics. *Organic Electronics* **2010**, 11, (12), 1960–1965.
28. Nathan, A.; Chalamala, B. R., Flexible electronics technology, part II: Materials and devices. *Proceedings of the IEEE* **2005**, 93, (8), 1391–1393.
29. Nathan, A.; Chalamala, B. R., Special issue on flexible electronics technology, part 1: Systems and applications. *Proceedings of the IEEE* **2005**, 93, (7), 1235–1238.
30. Tsumura, A.; Koezuka, H.; Ando, T., Macromolecular electronic device – field-effect transistor with a polythiophene thin-film. *Applied Physics Letters* **1986**, 49, (18), 1210–1212.
31. Ebisawa, F.; Kurokawa, T.; Nara, S., Electrical-properties of polyacetylene polysiloxane interface. *Journal of Applied Physics* **1983**, 54, (6), 3255–3259.
32. Kudo, K.; Yamashina, M.; Moriizumi, T., Field-effect measurement of organic-dye films. *Japanese Journal of Applied Physics Part 1 – Regular Papers Short Notes & Review Papers* **1984**, 23, (1), 130–130.
33. Lin, Y. Y.; Gundlach, D. J.; Nelson, S. F.; Jackson, T. N., Stacked pentacene layer organic thin-film transistors with improved characteristics. *Ieee Electron Device Letters* **1997**, 18, (12), 606–608.
34. Halik, M.; Klauk, H.; Zschieschang, U.; Schmid, G.; Dehm, C.; Schutz, M.; Maisch, S.; Effenberger, F.; Brunnbauer, M.; Stellacci, F., Low-voltage organic transistors with an amorphous molecular gate dielectric. *Nature* **2004**, 431, (7011), 963–966.
35. Park, S. K.; Anthony, J. E.; Jackson, T. N., Solution-processed TIPS-pentacene organic thin-film-transistor circuits. *IEEE Electron Device Letters* **2007**, 28, 877–879.

36. Park, S. K.; Jackson, T. N.; Anthony, J. E.; Mourey, D. A., High mobility solution processed 6,13-bis(triisopropyl-silylethynyl) pentacene organic thin film transistors. *Applied Physics Letters* **2007**, 91, (6).
37. Park, S. K.; Mourey, D. A.; Subramanian, S.; Anthony, J. E.; Jackson, T. N., High-mobility spin-cast organic thin film transistors. *Applied Physics Letters* **2008**, 93, (4).
38. Subramanian, S.; Park, S. K.; Parkin, S. R.; Podzorov, V.; Jackson, T. N.; Anthony, J. E., Chromophore fluorination enhances crystallization and stability of soluble anthradithiophene semiconductors. *Journal of the American Chemical Society* **2008**, 130, (9), 2706–2707.
39. Becerril, H. A.; Roberts, M. E.; Liu, Z. H.; Locklin, J.; Bao, Z. N., High-performance organic thin-film transistors through solution-sheared deposition of small-molecule organic semiconductors. *Advanced Materials* **2008**, 20, (13), 2588–2594.
40. Liu, S. H.; Wang, W. C. M.; Briseno, A. L.; Mannsfeld, S. C. E.; Bao, Z. N., Controlled deposition of crystalline organic semiconductors for field-effect-transistor applications. *Advanced Materials* **2009**, 21, (12), 1217–1232.
41. Yan, H.; Chen, Z. H.; Zheng, Y.; Newman, C.; Quinn, J. R.; Dotz, F.; Kastler, M.; Facchetti, A., A high-mobility electron-transporting polymer for printed transistors. *Nature* **2009**, 457, (7230), 679–686.
42. Mas-Torrent, M.; Durkut, M.; Hadley, P.; Ribas, X.; Rovira, C., High mobility of dithiophene-tetrathiafulvalene single-crystal organic field effect transistors. *Journal of the American Chemical Society* **2004**, 126, (4), 984–985.
43. Facchetti, A.; Yoon, M. H.; Marks, T. J., Gate dielectrics for organic field-effect transistors: New opportunities for organic electronics. *Advanced Materials* **2005**, 17, (14), 1705–1725.
44. Coropceanu, V.; Bredas, J. L., Organic transistors – A polarized response. *Nature Materials* **2006**, 5, (12), 929–930.
45. Necliudov, P. V.; Shur, M. S.; Gundlach, D. J.; Jackson, T. N., Contact resistance extraction in pentacene thin film transistors. *Solid-State Electronics* **2003**, 47, (2), 259–262.
46. Necliudov, P. V.; Shur, M. S.; Gundlach, D. J.; Jackson, T. N., Modeling of organic thin film transistors of different designs. *Journal of Applied Physics* **2000**, 88, (11), 6594–6597.
47. Street, R. A.; Salleo, A., Contact effects in polymer transistors. *Applied Physics Letters* **2002**, 81, (15), 2887–2889.
48. Hamilton, R.; Smith, J.; Ogier, S.; Heeney, M.; Anthony, J. E.; McCulloch, I.; Veres, J.; Bradley, D. D. C.; Anthopoulos, T. D., High-performance polymer-small molecule blend organic transistors. *Advanced Materials* **2009**, 21, (10–11), 1166–1171.
49. Cheng, X. Y.; Caironi, M.; Noh, Y. Y.; Wang, J. P.; Newman, C.; Yan, H.; Facchetti, A.; Sirringhaus, H., Air stable cross-linked cytop ultrathin gate dielectric for high yield low-voltage top-gate organic field-effect transistors. *Chemistry of Materials* **2010**, 22, (4), 1559–1566.
50. Whiting, G. L.; Arias, A. C., Chemically modified ink-jet printed silver electrodes for organic field-effect transistors. *Applied Physics Letters* **2009**, 95, (25).
51. Walser, M. P.; Kalb, W. L.; Mathis, T.; Batlogg, B., Low-voltage organic transistors and inverters with ultrathin fluoropolymer gate dielectric. *Applied Physics Letters* **2009**, 95, (23).

52. Coll, M.; Goetz, K. P.; Conrad, B. R.; Hacker, C. A.; Gundlach, D. J.; Richter, C. A.; Jurchescu, O. D., Flip chip lamination to electrically contact organic single crystals on flexible substrates. *Applied Physics Letters* **2011**, 98, (16).
53. Coll, M.; Miller, L. H.; Richter, L. J.; Hines, D. R.; Jurchescu, O. D.; Gergel-Hackett, N.; Richter, C. A.; Hacker, C. A., Formation of silicon-based molecular electronic structures using flip-chip lamination. *Journal of the American Chemical Society* **2009**, 131, (34), 12451–12457.
54. Podzorov, V.; Sysoev, S. E.; Loginova, E.; Pudalov, V. M.; Gershenson, M. E., Single-crystal organic field effect transistors with the hole mobility similar to 8 cm(2)/V s. *Applied Physics Letters* **2003**, 83, (17), 3504–3506.
55. S. M. Sze, Physics of Semiconductor Devices, (Wiley, New York, 1981), 2nd ed.
56. Gershenson, M. E.; Podzorov, V.; Morpurgo, A. F., Colloquium: Electronic transport in single-crystal organic transistors. *Reviews of Modern Physics* **2006**, 78, (3), 973–989.
57. McCulloch, I.; Heeney, M.; Chabinyc, M. L.; DeLongchamp, D.; Kline, R. J.; Coelle, M.; Duffy, W.; Fischer, D.; Gundlach, D.; Hamadani, B.; Hamilton, R.; Richter, L.; Salleo, A.; Shkunov, M.; Sporrowe, D.; Tierney, S.; Zhong, W., Semiconducting thienothiophene copolymers: design, synthesis, morphology, and performance in thin-film organic transistors. *Advanced Materials* **2009**, 21, (10–11), 1091–1109.
58. Jurchescu, O. D.; Subramanian, S.; Kline, R. J.; Hudson, S. D.; Anthony, J. E.; Jackson, T. N.; Gundlach, D. J., Organic single-crystal field-effect transistors of a soluble anthradithiophene. *Chemistry of Materials* **2008**, 20, (21), 6733–6737.
59. Loo, Y. L.; McCulloch, I., Progress and challenges in commercialization of organic electronics. *MRS Bulletin* **2008**, 33, (7), 653–662.
60. Anthony, J. E.; Facchetti, A.; Heeney, M.; Marder, S. R.; Zhan, X. W., n-Type organic semiconductors in organic electronics. *Advanced Materials* **2010**, 22, (34), 3876–3892.
61. Chua, L. L.; Zaumseil, J.; Chang, J. F.; Ou, E. C. W.; Ho, P. K. H.; Sirringhaus, H.; Friend, R. H., General observation of n-type field-effect behaviour in organic semiconductors. *Nature* **2005**, 434, (7030), 194–199.
62. Lim, Y. F.; Shu, Y.; Parkin, S. R.; Anthony, J. E.; Malliaras, G. G., Soluble n-type pentacene derivatives as novel acceptors for organic solar cells. *Journal of Materials Chemistry* **2009**, 19, (19), 3049–3056.
63. Tang, M. L.; Oh, J. H.; Reichardt, A. D.; Bao, Z. N., Chlorination: a general route toward electron transport in organic semiconductors. *Journal of the American Chemical Society* **2009**, 131, (10), 3733–3740.
64. Newman, C. R.; Frisbie, C. D.; da Silva, D. A.; Bredas, J. L.; Ewbank, P. C.; Mann, K. R., Introduction to organic thin film transistors and design of n-channel organic semiconductors. *Chemistry of Materials* **2004**, 16, (23), 4436–4451.
65. Shkunov, M.; Simms, R.; Heeney, M.; Tierney, S.; McCulloch, I., Ambipolar field-effect transistors based on solution-processable blends of thieno[2,3-b] thiophene terthiophene polymer and methanofullerenes. *Advanced Materials* **2005**, 17, (21), 2608–2612.
66. Meijer, E. J.; De Leeuw, D. M.; Setayesh, S.; Van Veenendaal, E.; Huisman, B. H.; Blom, P. W. M.; Hummelen, J. C.; Scherf, U.; Klapwijk, T. M., Solution-processed ambipolar organic field-effect transistors and inverters. *Nature Materials* **2003**, 2, (10), 678–682.

67. Lin, Y. Y.; Dodabalapur, A.; Sarpeshkar, R.; Bao, Z.; Li, W.; Baldwin, K.; Raju, V. R.; Katz, H. E., Organic complementary ring oscillators. *Applied Physics Letters* **1999**, 74, (18), 2714–2716.
68. Kuwahara, E.; Kubozono, Y.; Hosokawa, T.; Nagano, T.; Masunari, K.; Fujiwara, A., Fabrication of ambipolar field-effect transistor device with heterostructure of C-60 and pentacene. *Applied Physics Letters* **2004**, 85, (20), 4765–4767.
69. Chikamatsu, M.; Mikami, T.; Chisaka, J.; Yoshida, Y.; Azumi, R.; Yase, K.; Shimizu, A.; Kubo, T.; Morita, Y.; Nakasuji, K., Ambipolar organic field-effect transistors based on a low band gap semiconductor with balanced hole and electron mobilities. *Applied Physics Letters* **2007**, 91, (4).
70. Anthopoulos, T. D.; Setayesh, S.; Smits, E.; Colle, M.; Cantatore, E.; de Boer, B.; Blom, P. W. M.; de Leeuw, D. M., Air-stable complementary-like circuits based on organic ambipolar transistors. *Advanced Materials* **2006**, 18, (14), 1900–1904.
71. Isik, D.; Shu, Y.; Tarabella, G.; Coppede, N.; Iannotta, S.; Lutterotti, L.; Cicoira, F.; Anthony, J. E.; Santato, C., Ambipolar organic thin film transistors based on a soluble pentacene derivative. *Applied Physics Letters* **2011**, 99, (2).
72. Kalb, W. L.; Mathis, T.; Haas, S.; Stassen, A. F.; Batlogg, B., Organic small molecule field-effect transistors with Cytop (TM) gate dielectric: Eliminating gate bias stress effects. *Applied Physics Letters* **2007**, 90, (9).
73. de Boer, R. W. I.; Gershenson, M. E.; Morpurgo, A. F.; Podzorov, V., Organic single-crystal field-effect transistors. *Physica Status Solidi A – Applied Research* **2004**, 201, (6), 1302–1331.
74. Salleo, A.; Chabinyc, M. L.; Yang, M. S.; Street, R. A., Polymer thin-film transistors with chemically modified dielectric interfaces. *Applied Physics Letters* **2002**, 81, (23), 4383–4385.
75. Takeya, J.; Goldmann, C.; Haas, S.; Pernstich, K. P.; Ketterer, B.; Batlogg, B., Field-induced charge transport at the surface of pentacene single crystals: a method to study charge dynamics of two-dimensional electron systems in organic crystals. *Journal of Applied Physics* **2003**, 94, (9), 5800–5804.
76. Street, R. A.; Salleo, A.; Chabinyc, M. L., Bipolaron mechanism for bias-stress effects in polymer transistors. *Physical Review B* **2003**, 68, (8).
77. Duhm, S.; Heimel, G.; Salzmann, I.; Glowatzki, H.; Johnson, R. L.; Vollmer, A.; Rabe, J. P.; Koch, N., Orientation-dependent ionization energies and interface dipoles in ordered molecular assemblies. *Nature Materials* **2008**, 7, (4), 326–332.
78. Kim, C. S.; Lee, S.; Gomez, E. D.; Anthony, J. E.; Loo, Y. L., Solvent-dependent electrical characteristics and stability of organic thin-film transistors with drop cast bis(triisopropylsilylethynyl) pentacene. *Applied Physics Letters* **2008**, 93, (10), 103302.
79. Meredig, B.; Salleo, A.; Gee, R., Ordering of poly(3-hexylthiophene) nanocrystallites on the basis of substrate surface energy. *ACS Nano* **2009**, 3, (10), 2881–2886.
80. Nichols, J. A.; Gundlach, D. J.; Jackson, T. N., Potential imaging of pentacene organic thin-film transistors. *Applied Physics Letters* **2003**, 83, (12), 2366–2368.
81. Burgi, L.; Sirringhaus, H.; Friend, R. H., Noncontact potentiometry of polymer field-effect transistors. *Applied Physics Letters* **2002**, 80, (16), 2913–2915.

82. Mathijssen, S. G. J.; Coelle, M.; Mank, A. J. G.; Kemerink, M.; Bobbert, P. A.; de Leeuw, D. M., Scanning Kelvin probe microscopy on organic field-effect transistors during gate bias stress. *Applied Physics Letters* **2007**, 90, (19).
83. Teague, L. C.; Hamadani, B. H.; Jurchescu, O. D.; Subramanian, S.; Anthony, J. E.; Jackson, T. N.; Richter, C. A.; Gundlach, D. J.; Kushmerick, J. G., Surface Potential Imaging of Solution Processable Acene-Based Thin Film Transistors. *Advanced Materials* **2008**, 20, (23), 4513–4516.
84. Luan, S. W.; Neudeck, G. W., An experimental-study of the source drain parasitic resistance effects in amorphous-silicon thin-film transistors. *Journal of Applied Physics* **1992**, 72, (2), 766–772.
85. Hamadani, B. H.; Richter, C. A.; Gundlach, D. J.; Kline, R. J.; McCulloch, I.; Heeney, M., Influence of source-drain electric field on mobility and charge transport in organic field-effect transistors. *Journal of Applied Physics* **2007**, 102, (4).
86. Pesavento, P. V.; Chesterfield, R. J.; Newman, C. R.; Frisbie, C. D., Gated four-probe measurements on pentacene thin-film transistors: contact resistance as a function of gate voltage and temperature. *Journal of Applied Physics* **2004**, 96, (12), 7312–7324.
87. Chesterfield, R. J.; McKeen, J. C.; Newman, C. R.; Frisbie, C. D.; Ewbank, P. C.; Mann, K. R.; Miller, L. L., Variable temperature film and contact resistance measurements on operating n-channel organic thin film transistors. *Journal of Applied Physics* **2004**, 95, (11), 6396–6405.
88. Hamadani, B. H.; Richter, C. A.; Suehle, J. S.; Gundlach, D. J., Insights into the characterization of polymer-based organic thin-film transistors using capacitance–voltage analysis. *Applied Physics Letters* **2008**, 92, (20).
89. Braga, D.; Ha, M.; Xie, W.; Frisbie, C. D., Ultralow contact resistance in electrolyte-gated organic thin film transistors. *Applied Physics Letters* **2010**, 97, (19).
90. Hulea, I. N.; Russo, S.; Molinari, A.; Morpurgo, A. F., Reproducible low contact resistance in rubrene single-crystal field-effect transistors with nickel electrodes. *Applied Physics Letters* **2006**, 88, (11).
91. Wobkenberg, P. H. W. P. H.; Eda, G.; Leem, D. S.; de Mello, J. C.; Bradley, D. D. C.; Chhowalla, M.; Anthopoulos, T. D., Reduced graphene oxide electrodes for large area organic electronics. *Advanced Materials* **2011**, 23, (13), 1558–1562.
92. Cicoira, F.; Aguirre, C. M.; Martel, R., Making contacts to n-type organic transistors using carbon nanotube arrays. *ACS Nano* **2011**, 5, (1), 283–290.
93. Mott, N. F.; Gurney, R.W., *Electronic Processes in Ionic Crystals*. Oxford University Press, Oxford: 1940.
94. Rose, A., Space-charge-limited current in solids. *Physical Review* **1955**, 97, 1538–1544.
95. Rose, A.; Lampert, M. A., Photoconductor performance, Space-charge currents, and the steady-state fermi level, *Physical Review* **1959**, 113, 1227.
96. Mark, P.; Helfrich, V., Space-charge limited currents in organic crystals, *J. Appl. Phys*, **1962**, 33, 205.
97. Blom, P. W. M.; deJong, M. J. M.; Vleggaar, J. J. M., Electron and hole transport in poly(p-phenylene vinylene) devices. *Applied Physics Letters* **1996**, 68, (23), 3308–3310.

98. Campbell, A. J.; Bradley, D. D. C.; Lidzey, D. G., Space-charge limited conduction with traps in poly(phenylene vinylene) light emitting diodes. *Journal of Applied Physics* **1997**, 82, (12), 6326–6342.
99. Schroeder, R.; Majewski, L. A.; Grell, M., Improving organic transistor performance with Schottky contacts. *Applied Physics Letters* **2004**, 84, (6), 1004–1006.
100. Kalb, W. L.; Batlogg, B., Calculating the trap density of states in organic field-effect transistors from experiment: A comparison of different methods. *Physical Review B* **2010**, 81, (3).
101. Kalb, W. L.; Haas, S.; Krellner, C.; Mathis, T.; Batlogg, B., Trap density of states in small-molecule organic semiconductors: A quantitative comparison of thin-film transistors with single crystals. *Physical Review B* **2010**, 81, (15).
102. Blom, P. W. M.; deJong, M. J. M.; vanMunster, M. G., Electric-field and temperature dependence of the hole mobility in poly(*p*-phenylene vinylene). *Physical Review B* **1997**, 55, (2), R656–R659.
103. Pasveer, W. F.; Cottaar, J.; Tanase, C.; Coehoorn, R.; Bobbert, P. A.; Blom, P. W. M.; de Leeuw, D. M.; Michels, M. A. J., Unified description of charge-carrier mobilities in disordered semiconducting polymers. *Physical Review Letters* **2005**, 94.
104. Tanase, C.; Meijer, E. J.; Blom, P. W. M.; de Leeuw, D. M., Unification of the hole transport in polymeric field-effect transistors and light-emitting diodes. *Physical Review Letters* **2003**, 91, (21).
105. Platt, A. D.; Day, J.; Subramanian, S.; Anthony, J. E.; Ostroverkhova, O., Optical, fluorescent, and (photo)conductive properties of high-performance functionalized pentacene and anthradithiophene derivatives. *Journal of Physical Chemistry C* **2009**, 113, (31), 14006–14014.
106. Troisi, A.; Orlandi, G., Charge-transport regime of crystalline organic semiconductors: Diffusion limited by thermal off-diagonal electronic disorder. *Physical Review Letters* **2006**, 96, (8).
107. Bonham, J. S.; Jarvis, D. H., New approach to space-charge-limited conduction theory. *Australian Journal of Chemistry* **1977**, 30, (4), 705–720.
108. Geurst, J. A., Theory of space-charge limited currents in thin semiconductor layers, *Phys. Stat. Sol.* **1966**, 15, 107.
109. Jurchescu, O. D.; Palstra, T. T. M., Crossover from one- to two-dimensional space-charge-limited conduction in pentacene single crystals. *Applied Physics Letters* **2006**, 88, (12).
110. Anthony, J. E., The larger acenes: versatile organic semiconductors. *Angewandte Chemie-International Edition* **2008**, 47, (3), 452–483.
111. Swartz, C. R.; Parkin, S. R.; Bullock, J. E.; Anthony, J. E.; Mayer, A. C.; Malliaras, G. G., Synthesis and characterization of electron-deficient pentacenes. *Organic Letters* **2005**, 7, (15), 3163–3166.
112. Moon, H.; Zeis, R.; Borkent, E. J.; Besnard, C.; Lovinger, A. J.; Siegrist, T.; Kloc, C.; Bao, Z. N., Synthesis, crystal structure, and transistor performance of tetracene derivatives. *Journal of the American Chemical Society* **2004**, 126, (47), 15322–15323.
113. Oh, J. H.; Lee, H. W.; Mannsfeld, S.; Stoltenberg, R. M.; Jung, E.; Jin, Y. W.; Kim, J. M.; Yoo, J. B.; Bao, Z. N., Solution-processed, high-performance n-channel organic microwire transistors. *Proceedings of the National Academy of Sciences of the United States of America* **2009**, 106, (15), 6065–6070.

114. Okamoto, T.; Jiang, Y.; Qu, F.; Mayer, A. C.; Parmer, J. E.; McGehee, M. D.; Bao, Z. N., Synthesis and characterization of pentacene- and anthradithiophene-fluorene conjugated copolymers synthesized by Suzuki reactions. *Macromolecules* **2008**, 41, (19), 6977–6980.
115. Lee, S. S.; Loo, Y. L., Structural complexities in the active layers of organic electronics. *Annual Review of Chemical and Biomolecular Engineering* **2010**, 1, 59–78.
116. Arias, A. C.; MacKenzie, J. D.; McCulloch, I.; Rivnay, J.; Salleo, A., Materials and applications for large area electronics: solution-based approaches. *Chemical Reviews* **2010**, 110, (1), 3–24.

14

Charge transport features in disordered organic materials measured by time-of-flight (TOF), xerographic discharge (XTOF) and charge extraction by linearly increasing voltage (CELIV) techniques

A. PIVRIKAS, University of Queensland, Australia
and Johannes Kepler University of Linz, Austria

DOI: 10.1533/9780857098764.2.398

Abstract: In conjunction with various photophysical processes the charge transport largely determines the performance of electronic devices. In the past a vast number of techniques have been used to study transport in various types of semiconductors. Although some techniques are relatively simple to use, the interpretation of results is not so straightforward. Therefore, it is important to understand the limitations and applicability of these techniques when measuring the charge carrier mobility and recombination. In this work, the most commonly used charge transport techniques are presented. The advantages and disadvantages of time-of-flight (TOF), xerographic discharge (XTOF) and charge extraction by linearly increasing voltage (CELIV) techniques are discussed in detail. Various possible and technique-specific experimental conditions are used to demonstrate the quirks of each technique. It is shown how experimental data obtained using these techniques represent the charge carrier transport and how the results compare. Dispersive charge transport models in the framework of Gaussian density-of-states are used to explain the observed features of charge transport in disordered organic semiconductors.

Key words: charge transport, charge carrier mobility, time-of-flight (TOF), charge extraction by linearly increasing voltage (CELIV), organic semiconductors.

14.1 Introduction

In general, organic electronics has created immense academic interest due to its unlimited and flexible molecular engineering possibilities, allowing new organic materials with tailored physical properties to be synthesized [1, 2]. The constituent molecules of organic semiconductor materials are kept together mainly by weak van der Waals interactions, where electronic transfer integrals are expected to be small, electron–phonon coupling and carrier scattering is strongly pronounced, where π-electron delocalization, inter-chain interaction, band gap, carrier density, extent of disorder, film morphology, and device processing conditions determine the electrical and optical properties [3]. Understanding of the basic transport mechanisms in this special class of electronic materials is not only of fundamental academic interest, but also carries great technical relevance in that more professional knowledge may help to increase transport efficiency with the consequences of increased device speed, reduced power and heat losses in such promising fields as organic light-emitting diode (OLED) displays, organic field-effect transistor (OFET) used in integrated circuits, or photovoltaic cells [4]. Electron and hole mobilities are the fundamental parameters determining the charge transport phenomena in semiconductors. Typically, hopping-type transport within localized density of states (DOS) is observed in disordered organic semiconductors [5]. In hopping transport, mobility should be thermally activated, whereas in band transport in inorganic crystals, in contrast, carrier mobility follows the opposite temperature dependence, e.g. decreases with rising temperature [5, 6].

The most popular techniques used in the past to experimentally measure the charge carrier mobility in various semiconductors are:

- Hall effect (magneto-resistance, cyclotron resonance) [6]
- field-effect-transistor (FET) measurements [6]
- steady-state and transient space-charge-limited-current (SCLC) methods [7]
- time-of-flight (TOF) [8, 9]
- xerographic discharge time-of-flight (XTOF) [10–12]
- charge extraction by linearly increasing voltage (CELIV) technique [13, 14]
- charge extraction with applied voltage techniques [15, 16]
- transient electroluminescence [17, 18]
- impedance spectroscopy [19, 20]
- pulse-radiolysis time-resolved microwave conductivity [21]
- Auston–Switch photoconductivity [22]
- conductivity/concentration method [23, 24]
- surface acousto-electric traveling wave (SAW) [25, 26]
- Photoinduced transient stark spectroscopy [27]

Each of the above methods has its advantages and disadvantages. The Hall effect is a classic technique used to study the charge transport and film conductivity of inorganic semiconductors; however, it is not suitable for low conductivity (as well as low-mobility) disordered organic materials due to experimentally difficult to register low Hall voltages. TOF and XTOF techniques have been widely used to study various semiconductors, while special attention has been paid to organic materials suitable for xerography. However, this technique has a number of limitations and requirements for studied films (as discussed in the following section) which often render this technique inapplicable to low-mobility semiconductors with strongly pronounced dispersive (time-dependent) transport. CELIV, as a relatively novel technique has shown great promise in being able to reliably characterize charge transport and recombination processes in a very broad range of semiconductors; however, lately it has been shown that the interpretation of CELIV transient signals is not straightforward, requiring adequate correction factors and experimental conditions to be taken into account. The FET technique is very widely used to measure the carrier mobility at the interface between the semiconductor and dielectric, where the interface properties often determine the charge transport, in contrast to bulk transport properties of the films [28]. Since it is a steady-state measurement technique, the amount of trapped charge must be taken into account in order to correctly estimate the carrier mobility. Similar problems arise with SCLC (Mott's law) and other steady-state measurement techniques, while transient methods usually have the great advantage of being able to probe the dynamics of carrier trapping if it is present. Many other techniques have been reported in the literature that are able to determine the charge carrier mobility, but they fall outside the scope of this chapter.

This review is on different experimental methods, namely TOF, XTOF and CELIV, that have been widely used in the past to directly measure charge carrier mobilities in organic semiconductors. The advantages and disadvantages of discussed techniques are specified. As an outcome of each presented technique, typical experimental results of charge transport in disordered organic materials are highlighted.

14.2 Measurement techniques

14.2.1 TOF technique

Since the work of Spear, Leblanc, and Kepler, TOF measurements have been routinely used to determine the mobilities of photogenerated charge carriers in crystalline and disordered, organic and inorganic semiconductors [29–31]. In the TOF method, the sample is irradiated by a short light pulse, and a current transient signal arising from the photogenerated charge carriers moving in the applied electric field is monitored as a function of time. The current transients in disordered organic materials typically show various levels of dispersion as compared to non-dispersive signals in pure crystalline semiconductors [32]. The dispersion implies that the photocurrent is non-constant and it decays prior to transit time (as would be expected for a sheet of charges propagating with constant velocity), and the spread of arrival times is much broader than expected from the Einstein relation between mobility and diffusion. It is generally agreed that dispersive transport arises from wide variations in site-to-site hopping rates, which is caused by spatial (off-diagonal) and/or energetic (diagonal) disorder [5]. If charge transport is strongly dispersive it cannot be measured by TOF because of the absence (or difficulty in resolving) of the transit time kink in the photocurrent decay.

A schematic TOF circuitry, timing diagram, and transport of photogenerated charge carriers inside the film are shown in Fig. 14.1. The transport of photogenerated charge carriers in the applied electric field creates a time-dependent photocurrent in the external circuit which is measured as a voltage drop over the load resistance (R_{load}) of the oscilloscope.

The necessary conditions for TOF techniques to be applicable may be summarized as follows [33, 34]:

- Light absorption depth $\alpha d >> 1$, charge carriers shall be generated at the surface of the film. Typical TOF measurements are taken at low light intensities where the thin sheet of photogenerated carriers ($\alpha d >> 1$) is transported through the film with thickness d under an applied external voltage U. When the packet of charge carriers reaches the opposite contact the photocurrent drops forming a characteristic kink at time equal to transit time t_{tr} which is observed in the oscilloscope, as seen in Fig. 14.1. The charge carrier mobility is then estimated $\mu = d^2/t_{tr}U$.
- Duration of excitation $t_{laser} < t_{tr}$, the duration of photogenerating pulse must be shorter that carrier transit time.
- Dielectric relaxation time $\tau_\sigma > t_{tr}$, and $t_{del} < t_{tr}$. The laser pulse is applied after some delay time (t_{del}) to photogenerate the charge carriers in the film. The use of the voltage pulse instead of constant applied voltage is important when studied films have slight intrinsic conductivity, because it is important to ensure that t_{del} is much shorter than dielectric relaxation time ($\tau_\sigma = \varepsilon\varepsilon_0/\sigma$), otherwise the applied external electric field will be screened by thermally generated charge carriers in the film and the transport of photogenerated carriers will be hindered.
- The carrier lifetime $\tau > t_{tr}$, otherwise moving carriers will disappear during transport.

Two boundary cases for light intensity can be distinguished in TOF experiment: low intensity (small amount of charge carriers compared to charge carriers stored on the contacts CU, $Q_0 <$ CU), when an externally applied electric field inside the film is disturbed only fractionally and high light intensity, when the applied external electric field is completely screened by charge carriers inside the film. Typically, the small charge mode is used, when $Q_0 <<$ CU. In this case the photo-injected charge Q_0 does not perturb the applied electric field ensuring the small signal condition, where C is the sample capacitance. However, high light intensity TOF is also possible, when the large amount of photogenerated carriers disturbs the external

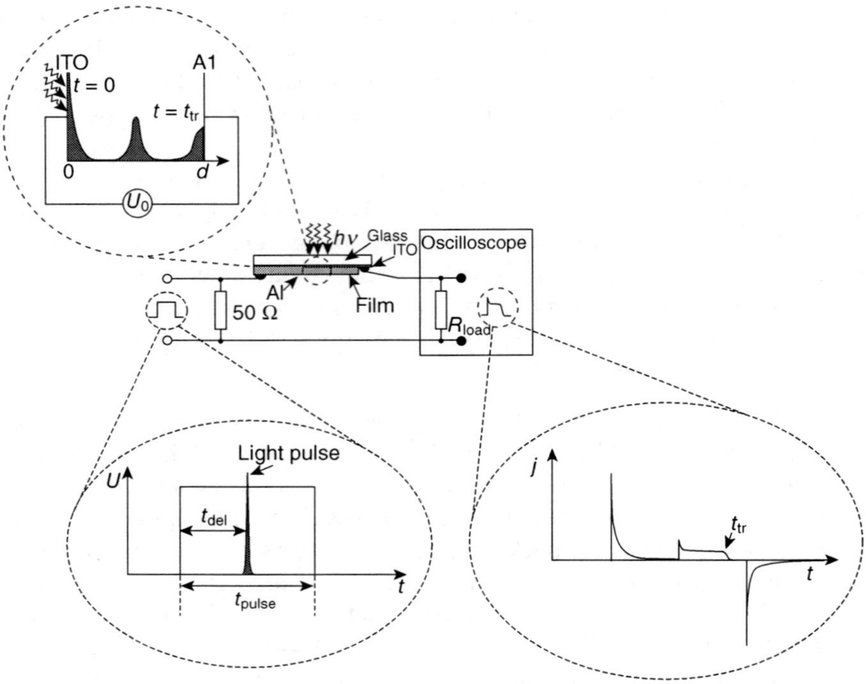

14.1 Experimental TOF setup with pulse timing. Reprinted with permission from [49].

applied electric field, the charge carrier transit time becomes shorter ($t_{cusp} = 0.8\ t_{tr}$) than the small charge transit time. Additional information about the charge carrier recombination can be accessed using high light intensity is used in TOF, because a high-intensity laser photogenerates a large amount of charge carriers inside the film ($Q_0 >> CU$), creating a large reservoir of charge carriers inside the film which further recombine during extraction. At these high concentrations the recombination dynamics is typically determined by bimolecular carrier recombination [35].

In Fig. 14.2(I) ideal schematic TOF transients are shown for different light intensities without accounting for trapping, diffusion and recombination during transport. Typical small charge current (SCC) at low light intensity, space charge perturbed current (SCPC) at moderate intensity, when $Q_0 = CU$ and space charge limited current (SCLC) at high intensity shows three distinguishing cases of the current transients. SCPC transients are typically recorded in low mobility materials, where Langevin bimolecular recombination dominates. The extracted charge will saturate as a function of light intensity to a value stored on the electrodes simply due to fast bimolecular recombination which empties the excess photogenerated

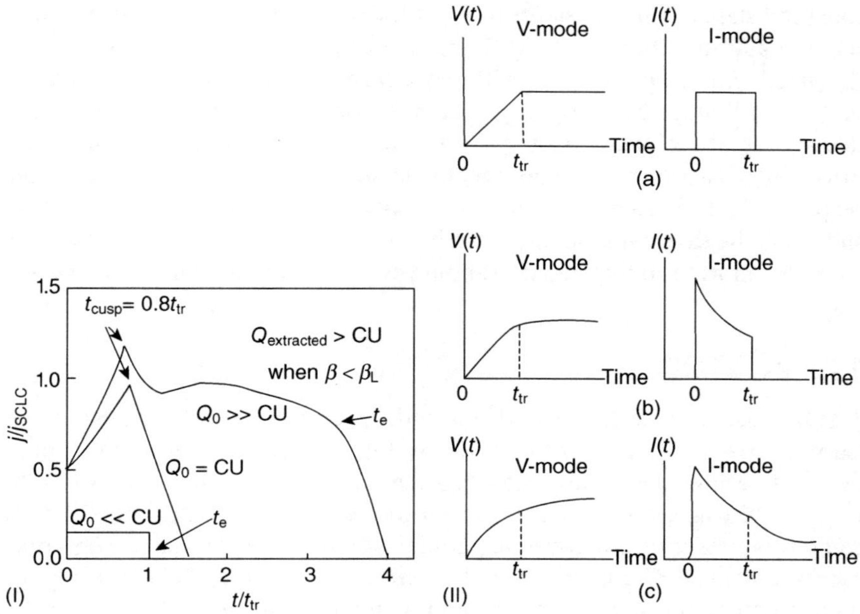

14.2 (I) ideal TOF transients without carrier recombination for different laser light intensities, demonstrating typical small charge mode (low light intensity, $Q_0 < \text{CU}$), space charge perturbed mode (moderate light intensity, $Q_0 \approx \text{CU}$), and space charge limited mode (high light intensity, $Q_0 > \text{CU}$) current transients. The reservoir extraction time t_e (note that it is only present in non-Langevin carrier bimolecular recombination case) is much larger than carrier transit time t_{tr} or t_{cusp}. The extracted charge will also exceed CU value. (II) Small charge mode TOF transient current (*I*-mode) and voltage (*V*-mode) waveforms showing effect of trapping and detrapping: (a) no trapping, (b) trapping no detrapping, (c) trapping and detrapping. Reprinted with permission from [12].

charge carriers in the reservoir during extraction, since only approximately CU can be extracted per transit time (space charge limited current condition). SCLC transients with a second extraction kink (t_e) are only possible when non-Langevin charge carrier bimolecular recombination is present in studied material. In this case, the excess photogenerated charge will recombine as fast and it will be extracted at time (t_e), which is much longer than the transit time (t_{tr}) and should not be confused with it, as seen from Fig. 14.2(I).

Depending on the capacitance of the studied devices and total load resistance, TOF can be done in either typical differential mode (RC << t_{tr}) or in integral mode, but the transient signals in the latter mode do not show the kink at transit time, therefore result interpretation is not as direct, as shown for small signal TOF transients in Fig. 14.2(II). Charge carrier trapping and recombination influence the TOF transient photocurrent signal as well. In the presence of trapping some of the carriers are removed from the transmitting charge sheet by trapping. If carrier transport involves trapping and detrapping of the injected charge, the waveforms in this case will have two prior and post-transit time components, called the fast and slow components respectively. The fast component is associated with the drifting of the charge sheet and the slow component, which has a longer time constant, corresponds to one or more trapping and detrapping events occurring during the transit.

14.2.2 XTOF

XTOF was successfully carried out and developed at the IBM Research Laboratories [10] as well as at Xerox [36, 37]. It is a very useful non-destructive method of investigating charge carrier photogeneration efficiency and transport properties of photoreceptors used in printers [38]. It can be conveniently employed in parallel with the conventional xerographic measurements for photoreceptor characterization [12]. While a TOF experiment requires the sample preparation with blocking electrodes, there is no such requirement in XTOF measurements, which can sometimes be a great advantage. Also, this technique is useful when the high threshold for electric breakdown of the films is required in order to measure the charge transport.

Figure 14.3(II) shows a schematic XTOF experimental setup to demonstrate the essence of the technique. The floating surface of the sample resting on an earthed electrode is charged in the dark by a corona charging device. How well the sample is charged depends upon several variables, the most important of which are:

- the magnitude of the corona-produced current;
- the amount of this current reaching the sample surface; and
- the uniformity with which the charge is deposited on the sample surface.

A well-designed corona device should both charge the sample to any pre-selected voltage and evenly distribute the charge over the floating surface, and should do so consistently for a long period of time. Schematic sketches of three corona devices are shown in Fig. 14.3(a).

Once the sample under test is charged, its exact potential at various stages of the experiment may be required. These are as follows:

- The initial voltage of the sample for calculation of drift mobility.
- The sample voltage after optical excitation. When this is subtracted from the initial voltage, it provides an estimate of total charge injected into the sample.

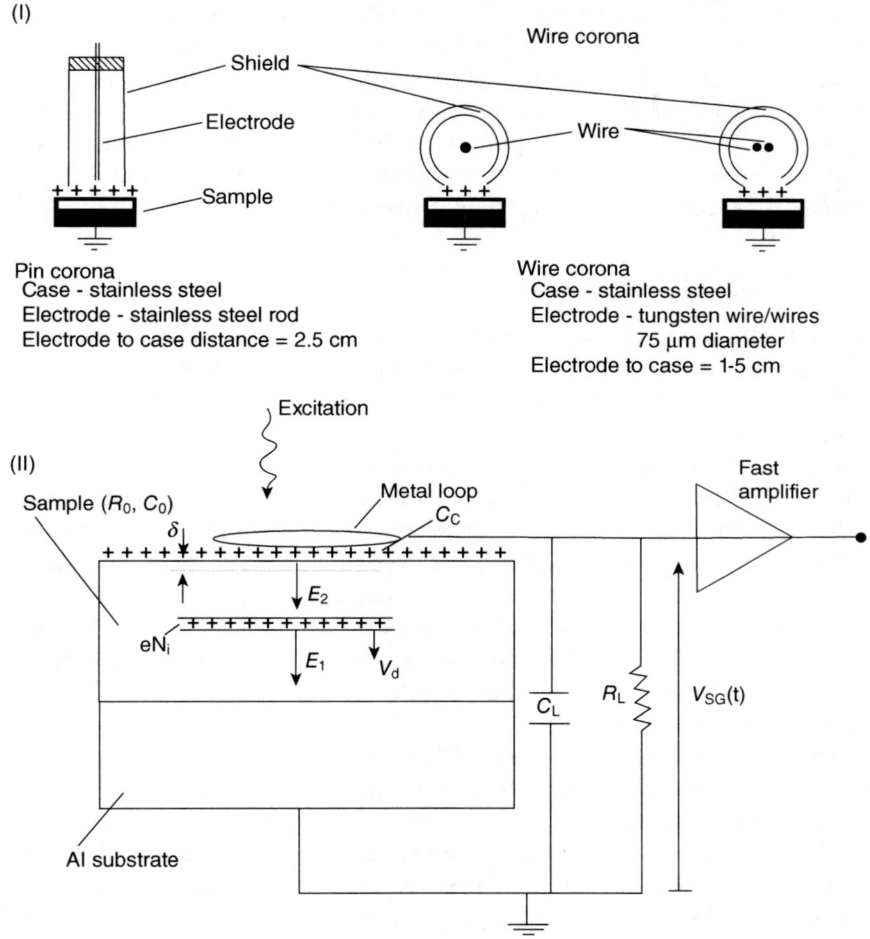

14.3 (I) Schematic sketches of different corona devices. (II) Xerographic time of flight experiment for the determination of drift mobility in high resistivity semiconductors. Reprinted with permission from [12].

For the measurement, a short pulse of strongly absorbed light is used to generate a thin sheet of electron–hole pairs near the surface of the sample. As in TOF, the wavelength of the light is chosen so that the absorption depth is smaller than the thickness of the sample. The charge injected into the sample induces a surface voltage change which is detected by a circular metallic loop positioned above the floating surface of the sample. The current in the external circuit can be further amplified before being displayed on the oscilloscope.

For the special case shown in Fig. 14.3(I), the electrons neutralize some of the positive charges on the surface, whereas holes move towards the conductive substrate under the influence of the applied field. The typical TOF transient current that is monitored via a rear electrode contains information on flight times of the photo-injected charge carriers in the sample and, hence, on mobility. The theory involved in the interpretation of the XTOF transient signals is the same as that for TOF.

14.2.3 CELIV technique

The CELIV technique was suggested by G. Juška [13, 14]. Previously, it has been used to study the charge carrier drift mobility in microcrystalline solar cells, because TOF is not applicable due to high sample conductivity. Results obtained by using the CELIV technique were found to be in a good agreement with TOF and other techniques [39, 40]. In the past it was shown that the CELIV technique was suitable to measure the mobility and density relaxation in π-conjugated polymers and other organic semiconductors [41–45]. Also, the CELIV technique is less susceptible to dispersive transport which often makes it the only applicable technique for charge transport studies in strongly disordered organic materials [46–53]. Another advantage of CELIV is that films with any conductivity (or equilibrium charge carrier concentration) can be measured as opposed to TOF as discussed in the above section.

The experimental setup in the CELIV technique is the same as in TOF. CELIV measurements are performed in the low RC regime, e.g. $t_{tr} >> RC$. A significant advantage of the CELIV technique is that it makes it possible to directly and simultaneously measure the charge carrier mobility and concentration as a function of time, electric field and temperature [54]. Lately it has been shown that a proper correction factor must be used in order to correctly estimate the photogenerated charge carrier mobility [14]. This is especially pronounced at high carrier concentrations (so that the total photogenerated charge Q_0 is higher than CU) and it strongly depends on the depth of carrier photogeneration arising in the films with various thickness or light absorption coefficient.

A schematic timing diagram of the applied voltage pulse and current response is shown in Fig. 14.4. A linearly increasing voltage pulse with the slope $A = V_{max}/t_{pulse}$ is applied to the sample to extract the equilibrium charge carriers. The charge carriers inside the film can also be photogenerated by a light pulse (sometimes called photo-CELIV technique). Then, the photogenerated charge carriers are extracted by the linearly increasing voltage pulse after some delay time t_{del}. At the beginning of the CELIV current transient a differentiating RC circuit current density step j_0 is seen (Fig. 14.4). The following current increase Δj is related to the conductivity

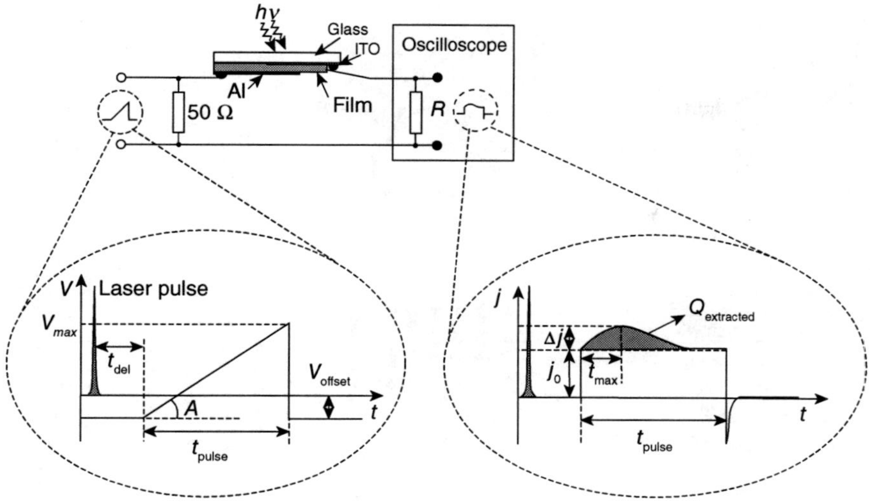

14.4 CELIV experimental setup with pulse timing.

current caused by the transported charge carriers in the film. The current continues to increase as the voltage increases until the charge carriers are extracted from the film and the current drops down to the capacitive step. In practice, if the duration of the applied triangle voltage pulse t_{pulse} is not long enough, then there might be some carriers left in the film and the current will end at a higher level than the capacitive step.

14.3 Experimental results of charge carrier mobility determination

14.3.1 Electric field and temperature-dependent carrier mobility measured by TOF

Figure 14.5(I) shows experimentally measured TOF photocurrent transients at various applied voltages and various temperatures (at 293K, 180K, and 120 K) in the purified regioregular P3HT. The transit time is seen to increase by approximately 3 orders of magnitude upon cooling to 120 K, as well as shifting with applied voltage, confirming that the kink represents the arrival of the carriers to the opposite electrode and therefore it can be used for determination of charge carrier drift mobility. As can be seen from the plateau of current transients and rather short post-transit time decay, the photocurrent transients are non-dispersive at room temperature, but gradually become more dispersive at lower temperatures. Dispersive transport occurs if the mean hopping rate of the charge carriers decreases as a

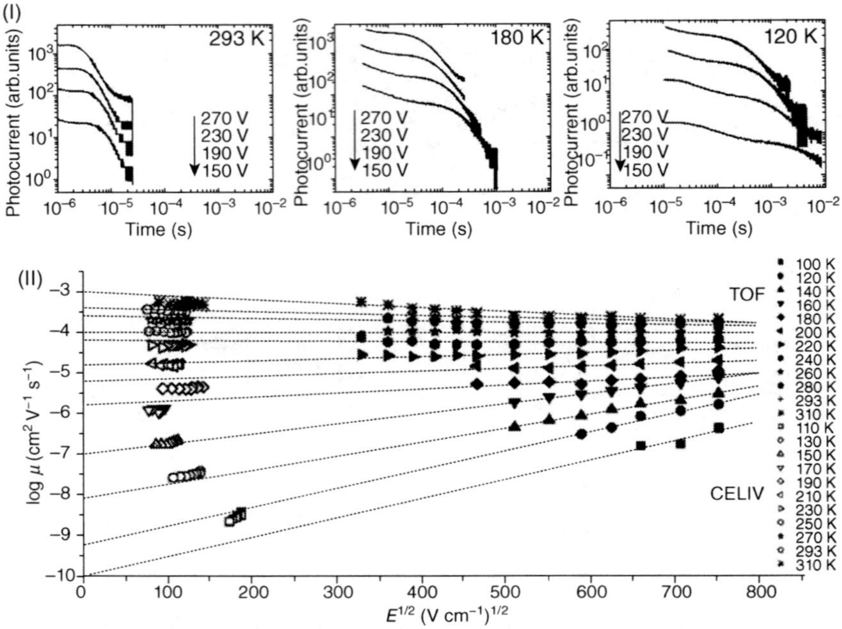

14.5 (I) TOF current transients measured in regioregular poly(3-hexylthiophene) at various electric fields and temperatures. (II) Temperature and electric-field dependence of the mobility determined by both TOF and CELIV experimental techniques. The dotted lines represent linear fits of the TOF data. Inset shows the monomer of studied poly(3-hexylthiophene-2,5-diyl). Reprinted with permission from [39].

function of time [32, 55]. Dispersive transport has been observed in conjugated polymers probed by impedance spectroscopy [56] and a transient electroluminescence technique [57]. Since the energy relaxation of the charge carriers is faster at higher temperatures, the photocurrent transient settles to a plateau, indicating that non-dispersive transport has been reached prior to the arrival of the charge carriers at the exit electrode. The time to reach quasi-equilibrium, however, increases faster than the transit time with decreasing temperature; therefore, dispersive transport starts dominating at lower temperatures as indicated by dispersive photocurrent transients.

The shape of the transients was analyzed using Scher–Montroll formalism [58] to describe the charge transport in disordered media. The prediction of that formalism is that the current slopes prior and after the transit time sums to 2, while allowing the current transients to be recorded at various electric fields and, film thicknesses to be superimposed when they are normalized to t_{tr}. It was found that that the Scher–Montroll formalism is not applicable to describe charge transport in studied films.

The temperature and electric-field dependence of the hole mobility obtained by both CELIV and TOF experimental techniques are plotted in Fig. 14.5(II). The overall agreement between the experiments shows that the two different experiment techniques are mutually consistent. The mobility follows Poole–Frenkel-type dependence at all temperatures, whereas the slope of the electric-field dependence of mobility decreases with increasing temperature, and becomes temperature independent at around 250–270 K and even negative at room temperature. Disorder formalism was used to explain the temperature and electric field dependence of mobility [5]. Owing to the presence of positional disorder, a carrier can avoid an energetically unfavorable site by executing a detour around that site which leads to a negative electric field dependent mobility at low fields. Typically, a hopping concept involving Miller–Abrahams-type jump rates is used; however, the phenomenological transport characteristics can similarly be rationalized in terms of Marcus-type jump rates [59, 60]. The theoretical work by Fishchuk *et al.* employing the effective medium approach shows that the shape of $\mu(E)$ is similar for both types of jump rates [61]. Considering the principal differences by the two applied methods, this finding suggests that charge carrier motion does not depend on the mode of charge generation (surface or volume) in the regioregular P3HT samples.

14.3.2 Time and carrier concentration-dependent mobility measured by CELIV

It has been shown in the past that the charge carrier mobility in disordered organic semiconductors depends on the charge carrier concentration [62]. In contrast to inorganic crystals where carrier mobility decreases with increasing concentration due to carrier-carrier scattering [63], the several orders of magnitude higher charge carrier mobility observed in FETs as compared to LEDs can originate from the increased charge carrier mobility at higher charge carrier concentrations [64]. Arkhipov *et al.* argued that the mobility may also decrease in disordered materials with increasing charge carrier concentration due to the effect of additional Coulomb traps introduced by charge carrier doping [65, 66]. It was shown that typically observed enhanced SCLC in poly(*p*-phenylene vinylene) diodes at higher applied voltages is due to the enhancement of the charge carrier mobility at higher charge carrier densities rather than the effect of the positive electric field dependence of mobility, although the later contribution cannot be neglected [67].

Since the CELIV technique is able to independently determine whether carrier mobility is concentration or time-dependent, it was used in the past to determine the charge transport in regiorandom P3HT [44, 68]. In Fig. 14.6(I) the photo-CELIV response as a function of (a) the intensity of the

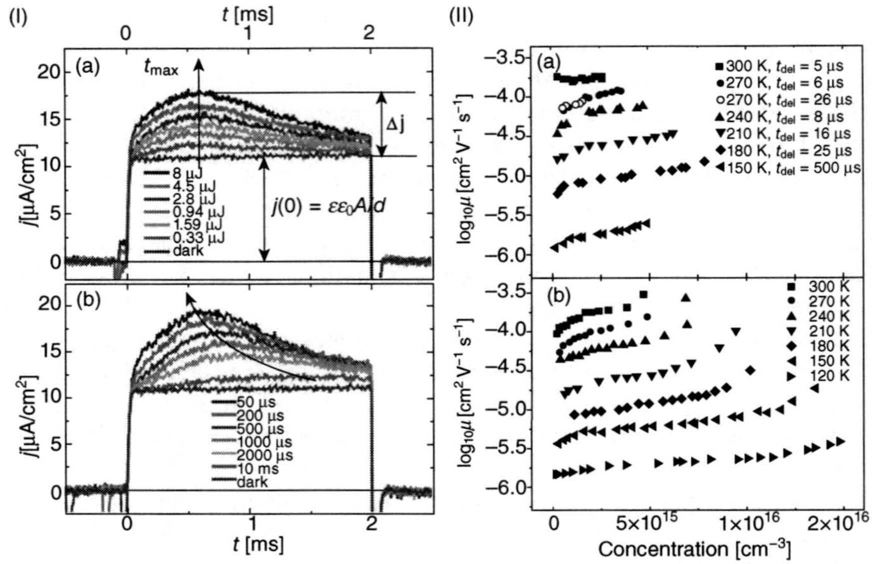

14.6 (I) Photo-CELIV curves for a 355 nm thick regiorandom P3HT film where the extraction current transient (D_j) depending on the (a) photoexcitation intensity and (b) the delay time (t_{del}) after the photoexcitation is shown. The voltage rise speed was in both cases 2 V/2 ms. The thin arrows shows the trend in the extraction maximum t_{max}; note the intensity-independent t_{max} in (a). Reprinted with permission from [44]. (II) Carrier mobility versus the concentration of the extracted charge carriers calculated from (a) light intensity-dependent measurements at fixed delay times; (b) delay time-dependent measurements at fixed light intensity. The delay time used during the light intensity-dependent studies is displayed in the key. Inset shows the monomer of studied poly(3-hexylthiophene-2,5-diyl). Reprinted with permission from [40].

laser light pulse and (b) delay time between the laser pulse and applied voltage studied in regiorandom P3HT films. The shift in extraction maximum t_{max} is clearly seen as a function of laser delay time t_{del}, but not as a function of light intensity, directly indicating that the time-dependent (not the concentration) carrier relaxation within DOS governs the carrier transport.

Figure 14.6(II) shows the mobility of photogenerated carriers versus the concentration of extracted charge carriers estimated from (a) light intensity-dependent measurements at fixed delay times (b) from delay time-dependent measurements at fixed light intensity. The weak dependence of mobility seems to agree well with results obtained in the past in the low carrier concentration range [62, 65, 66, 69].

The time-dependent mobility cannot be attributed to the concentration dependence of the mobility alone. The model of disordered hopping transport within localized DOS can well explain the time-dependent mobility under non-equilibrium conditions in disordered semiconductors in which the charge transport sites are both energetically and spatially disordered [5]. Theoretical studies, employing the Monte Carlo simulation technique [5], an effective medium approach [70], have been used to describe charge transport in the presence of both energetic and spatial disorder. Miller–Abrahams carrier hopping rates between localized sites [71] describe the thermally activated carrier hopping process. The activated jump rate is determined by a jump attempt frequency multiplied by the Boltzmann factor $e^{-\Delta E/kT}$. The rate for downward jumps with localized density of states (this scenario is described for electrons while for holes it is opposite) is more probable than upward jump, since the energy can always be dissipated rapidly via the large number of high-frequency phonons typical for organic conjugated materials. Due to the asymmetric hopping rates, a dynamic equilibrium may be attained when the number of even lower lying sites is so low that, on average, all the jumps will be thermally activated upward jumps. Whether the mean velocity of the packet of migrating charge carriers generated under non-equilibrium conditions reaches a quasi-static value depends on the shape of the distribution of the localized states. The relaxation process occurs simultaneously with charge extraction; photogenerated charge carriers drift across the sample under the externally applied electric field. Since the time to reach quasi-equilibrium has a stronger temperature dependence that the transit time with decreasing temperature, eventually charge transport without quasi-equilibrium is measured leading to a frequently observed non-dispersive to dispersive transition in transport.

14.3.3 Charge transport in organic semiconductors measured by XTOF

Numerous organic materials have been studied using XTOF in the past [38]. As a typical example, the hole transport measured in binary solid solutions of triphenylamine and bisphenol-A-polycarbonate is presented in Fig. 14.7(I) [72]. A photo-emitting electrode of amorphous selenium was used for charge injection. Under space-charge-limited conditions, hole transport was found to be linearly dependent on the electric field, as seen in Fig. 14.7(I). These results were found to be in good agreement with previous TOF studies.

Figure 14.7(II) shows the temperature dependence of the carrier mobility for different electric fields. It was found that below –45 °C, the potential discharge does not show a well-defined transit time, whereas above 50 °C the measurements were not possible due to dark discharge. In experimental temperature interval, the drift mobility was described by an exponential

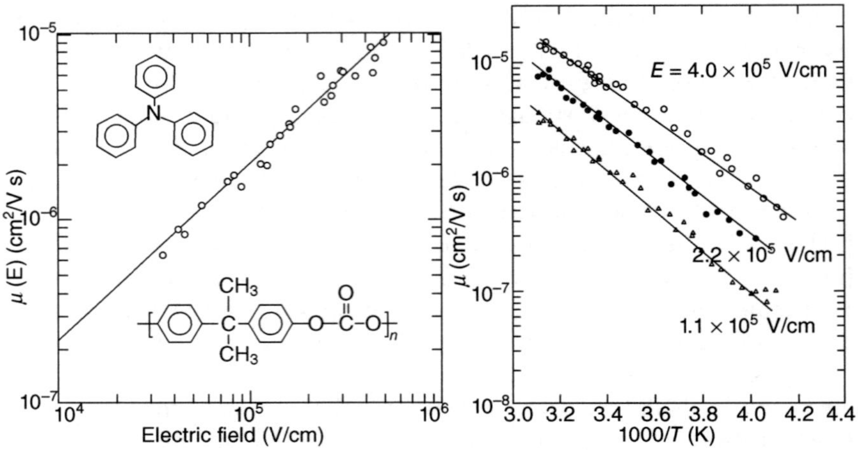

14.7 (I) Electric field dependence of the hole mobility in the films with 40% triphenylamine concentration. Top chemical structure is for triphenylamine and bottom structure for bisphenol-A-polycarbonate. (II) Temperature dependence of the hole mobility fordifferent electric fields in the films with 40% triphenylamine concentration. Reprinted with permission from [72].

dependence upon reciprocal temperature (Arrhenius plot) with activation energy of approximately 0.30 eV. The temperature dependence was found to be not greatly affected by the electric field possibly due to small range of applied electric field, since Gill's law (which demonstrates the dependence of the activation energy on electric field) is expected to be observed in systems where disordered transport prevails [73].

14.4 Charge transport models in disordered organic semiconductors

Bässler's formalism is often used to explain the time-dependent features of carrier transport in disordered solids [5]. A particle started at arbitrary energy within a Gaussian DOS of an undiluted system of hopping sites executes a random walk relaxing into lower energy tail states. It is a specific feature of a Gaussian as compared with an exponential DOS that the mean energy saturates over a long time, indicating attainment of dynamic equilibrium. Contrary to the case of steady state carrier hopping at the Fermi level in amorphous inorganic semiconductors, the thermodynamic equilibrium is not established in case of photogenerated carrier transport because they are moving in an unoccupied DOS [74].

Figure 14.8(I) portrays the temporal evolution of a packet of carriers at low concentration (carrier interaction with each other is neglected) relaxing within a Gaussian DOS after photogeneration. At longer times the shape of the packet becomes slightly asymmetric, with the high-energy side of the occupational DOS (ODOS) equilibrating earlier. At longer times the shape of the ODOS becomes Gaussian and acquires a width σ, indicating that dynamic equilibrium has been established.

Many investigators have used the Poole–Frenkel model to interpret high-field transport phenomena in various materials [75, 76]. Usually, a Poole–Frenkel (PF)-type dependence of the mobility at high electric field is observed ($\ln \mu \sim SE^{1/2}$, where S is the slope) [39, 77]. The application of electric field results in a lowering of the potential barrier in the direction of the electric field [78], shown schematically in Fig. 14.8(II) [79]. This facilitates charge carrier motion from a trapping center. This holds true for a large number of systems along with other features such as sign-reversal of the coefficient S at a particular temperature, deviation from Poole–Frenkel law at low electric fields [5]. There are many other modifications of the original Poole–Frenkel model based on various assumptions, such as those by Hill [76], Hartke [80], Antula [81], and Adamec and Calderwood [82].

The charge carrier concentration-dependent transport is analyzed using the concept of concentration-dependent carrier population within Gaussian

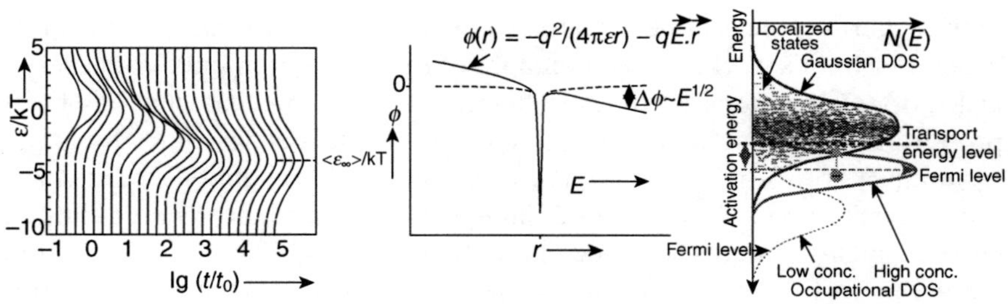

14.8 (I) Temporal evolution of the distribution of carrier energies in a Gaussian DOS. All profiles are broken at the same carrier density illustrating the different relaxation patterns for mobile and immobile carriers. E denotes the theoretical mean energy in the long-time limit. Reprinted with permission from [5]. (II) Potential barrier in the Poole–Frenkel model. The application of an electric field E lowers the potential barrier in the direction of the field by an amount proportional to $E^{1/2}$. Reprinted with permission from [78]. (III) Hopping transport within localized Gaussian density-of-states (DOS) distribution at various electron concentrations in disordered system. Transport level (equivalence of the mobility edge) marks the energy above which carrier mobility strongly increases. Carrier concentration dependent occupational DOS defines the carrier distribution within localized states and determines the activation energy – the energy which is required for carrier excitation above transport level. Dashed arrows show the position of Fermi level at high and low carrier concentrations.

DOS. Effective Medium Approach theory is able to well describe the dynamics and outcomes of carrier concentration distribution within ODOS [83]. Fig. 14.8(III) shows the energetic distribution of electrons within Gaussian DOS at various carrier concentrations. At low carrier concentrations, the ODOS is poorly populated with the Fermi level being below the ODOS center. The majority of the activated jumps start from upper portion of the ODOS rather than from its center. When the states are filled at higher concentrations, the Fermi level is gradually lifted and it gets located within the ODOS, while the ODOS distribution progresses toward the DOS center. Therefore, activated jumps, on average, require less activation energy and are more probable at higher concentrations, resulting in higher charge carrier mobilities at higher concentrations. Meyer-Neldel rule or compensation law, usually observed in organic semiconductors, can be explained using the above reasoning [54, 73, 84–86]. This law demonstrates that the activation energy of carrier mobility is dependent on carrier concentration (or electric field), while it usually predicts the existence of an isokinetic temperature point, at which charge carrier mobility is concentration-independent.

14.5 Conclusion

Device performance depends largely on the nature of charge transport processes. In this work, a brief summary of the most commonly used charge transport techniques and diverse charge transport mechanisms has been presented. Different modes of time-of-flight and xerographic discharge techniques are shown and discussed for charge carrier mobility and trapping measurements. Advantages of high light intensity space-charge-limited current measurements for recombination are highlighted. The charge extraction by linearly increasing voltage technique allows the charge carrier mobility and concentration to be studied simultaneously. Time and concentration-dependent transport measurements are possible using this technique, although careful attention must be dedicated to experimental conditions, since the current transient signal depends on these parameters as well.

Mobility and its dependence on electric field, carrier concentration and temperature are typically observed to be different in disordered organic materials compared to ordered structures. Present theoretical models employing hopping-type transport within Gaussian distribution of density-of-states are able to explain the observed experimental properties to a large extent. Due to either the technique limitations or the material properties, experimental data for non-dispersive transport are generally available only over a narrow electric field, carrier concentration and temperature ranges, which complicates the interpretation. A broader experimentally accessible range of electric field and carrier concentration-dependent mobility is required for further model development.

14.6 References

1. C.J. Brabec, N.S. Sariciftci, J.C. Hummelen, *et al.* Plastic solar cells. *Advanced Functional Materials*, **11**(1):15–26, 2001.
2. S.R. Forrest. The path to ubiquitous and low-cost organic electronic appliances on plastic. *Gravitational, electric, and magnetic forces: an anthology of current thought*, page 120, 2005.
3. M. Pope and C.E. Swenberg. *Electronic processes in organic crystals and polymers*. Oxford University Press, 1999.
4. A. Pivrikas, H. Neugebauer, and N.S. Sariciftci. Charge carrier lifetime and recombination in bulk heterojunction solar cells. *IEEE Journal of Selected Topics in Quantum Electronics*, **16**(6):1746–1758, 2010.
5. H. Bassler. Charge transport in disordered organic photoconductors: a Monte Carlo simulation study. *Physica Status Solidi (B)*, **175**(1):15–56, 1993.
6. S.M. Sze and K.K. Ng. *Physics of semiconductor devices*. Wiley-Blackwell, 2007.
7. A. Rose. Space-charge-limited currents in solids. *Physical Review*, **97**(6):1538, 1955.
8. W.E. Spear. Drift mobility techniques for the study of electrical transport properties in insulating solids. *Journal of Non-Crystalline Solids*, **1**(3):197–214, 1969.
9. J. Schottmiller, M. Tabak, G. Lucovsky, and A. Ward. The effects of valency on transport properties in vitreous binary alloys of selenium. *Journal of Non-Crystalline Solids*, **4**:80–96, 1970.
10. I.P. Batra, K.K. Kanazawa, and H. Seki. Discharge characteristics of photoconducting insulators. *Journal of Applied Physics*, **41**(8):3416–3422, 1970.
11. J. Mort and G. Pfister. *Electronic properties of polymers*. John Wiley & Sons, 1982.
12. S.M. Vaezi-Nejad. Xerographic time of flight experiment for the determination of drift mobility in high resistivity semiconductors. *International Journal of Electronics*, **62**(3):361–384, 1987.
13. G. Juška, K. Arlauskas, M. Viliunas, and J. Kocka. Extraction current transients: New method of study of charge transport in microcrystalline silicon. *Physical Review Letters*, **84**(21):4946–4949, 2000.
14. G. Juška, N. Nekrašas, V. Valentinavicius, P. Meredith, and A. Pivrikas. Extraction of photogenerated charge carriers by linearly increasing voltage in the case of langevin recombination. *Physical Review B*, **84**(15):155202, 2011.
15. J.B. Arthur, W. Bardsley, M. Brown, and A.F. Gibson. Carrier extraction in germanium. *Proceedings of the Physical Society. Section B*, **68**:43, 1955.
16. R. Bray. Minority carrier extraction in germanium. *Physical Review*, **100**(4):1047, 1955.
17. C. Hosokawa, H. Tokailin, H. Higashi, and T. Kusumoto. Transient behavior of organic thin film electroluminescence. *Applied Physics Letters*, **60**(10):1220–1222, 1992.
18. D.J. Pinner, R.H. Friend, and N. Tessler. Transient electroluminescence of polymer light emitting diodes using electrical pulses. *Journal of Applied Physics*, **86**:5116, 1999.

19. E. Barsoukov and J.R. Macdonald. *Impedance spectroscopy: theory, experiment, and applications*. LibreDigital, 2005.
20. G. Garcia-Belmonte, A. Munar, E.M. Barea, J. Bisquert, I. Ugarte, and R. Pacios. Charge carrier mobility and lifetime of organic bulk heterojunctions analyzed by impedance spectroscopy. *Organic Electronics*, **9**(5):847–851, 2008.
21. A.M. van de Craats, J.M. Warman, M.P. de Haas, D. Adam, J. Simmerer, D. Haarer, and P. Schuhmacher. The mobility of charge carriers in all four phases of the columnar discotic material hexakis (hexylthio) triphenylene: Combined tof and pr-trmc results. *Advanced Materials*, **8**(10):823–826, 1996.
22. D. Moses and A.J. Heeger. Fast transient photoconductivity in polydiacetylene: carrier photogeneration, carrier mobility and carrier recombination. *Journal of Physics: Condensed Matter*, **1**:7395, 1989.
23. P.H. Nguyen, G. Paasch, W. Brütting, and W. Riess. Analysis of the dc conductivity of the quasi-one-dimensional charge-density-wave conductor (fluoranthene). *Physical Review B*, **49**(8):5172, 1994.
24. W. Brütting, P.H. Nguyen, W. Rieβ, and G. Paasch. dc-conduction mechanism and peierls gap in organic and inorganic charge-density-wave conductors. *Physical Review B*, **51**(15):9533, 1995.
25. N. Karl. Charge carrier mobility in organic crystals. In *Organic electronic materials*, R. Farchioni and G. Grosso (eds), Springer, 2001.
26. N. Karl, K.H. Kraft, and J. Marktanner. Charge carrier mobilities in dark-conductive organic thin films determined by the surface acoustoelectric travelling wave (saw) technique. *Synthetic metals*, **109**(1–3):181–188, 2000.
27. J. Cabanillas-Gonzalez, T. Virgili, A. Gambetta, G. Lanzani, T.D. Anthopoulos, and D.M. De Leeuw. Photoinduced transient stark spectroscopy in organic semiconductors: A method for charge mobility determination in the picosecond regime. *Physical Review Letters*, **96**(10):106601, 2006.
28. S. Grecu, M. Roggenbuck, A. Opitz, and W. Brütting. Differences of interface and bulk transport properties in polymer field-effect devices. *Organic Electronics*, **7**(5):276–286, 2006.
29. W.E. Spear. Transit time measurements of charge carriers in amorphous selenium films. *Proceedings of the Physical Society. Section B*, **70**:669, 1957.
30. O.H. LeBlanc Jr. Electron drift mobility in liquid n-hexane. *The Journal of Chemical Physics*, **30**:1443, 1959.
31. R.G. Kepler. Charge carrier production and mobility in anthracene crystals. *Physical Review*, **119**(4):1226, 1960.
32. T. Tiedje and A. Rose. A physical interpretation of dispersive transport in disordered semiconductors. *Solid State Communications*, **37**(1):49–52, 1981.
33. R. Österbacka, G. Juška, K. Arlauskas, A.J. Pal, K.M. Källman, and H. Stubb. Electric field redistribution and electroluminescence response time in polymeric light-emitting diodes. *Journal of Applied Physics*, **84**:3359, 1998.
34. G. Juška, K. Genevicius, K. Arlauskas, R. Öšterbacka, and H. Stubb. Charge transport at low electric fields in π-conjugated polymers. *Physical Review B*, **65**(23):233208, 2002.
35. A. Pivrikas, G. Juška, A.J. Mozer, M. Scharber, K. Arlauskas, N.S. Sariciftci, H. Stubb, and R. Österbacka. Bimolecular recombination coefficient as a sensitive testing parameter for low-mobility solar-cell materials. *Physical Review Letters*, **94**(17):176806, 2005.

36. S.B. Berger and R.C. Enck. Xerographic time-of-flight: A powerful technique for space charge measurement in selenium photoreceptors. In *Proceedings of the International Symposium on Industrial Uses of Selenium and Tellurium, October 21–23, 1980, the Sheraton Centre Hotel, Toronto, Ontario, Canada*, page 166. Selenium-Tellurium Development Association, 1980.
37. S.B. Berger, R.C. Enck, M.E. Scharfe, and B.E. Springett. The application of selenium and its alloys to xerography. In *The Physics of Selenium and Tellurium, Proc. Conf., Konigstein, W. Germany, May 1979*, pages 256–266, 1979.
38. P.M. Borsenberger and D.S. Weiss. *Organic photoreceptors for xerography*, volume 59. CRC, 1998.
39. A.J. Mozer, N.S. Sariciftci, A. Pivrikas, R. Österbacka, G. Juška, L. Brassat, and H. Bässler. Charge carrier mobility in regioregular poly(3-hexylthiophene) probed by transient conductivity techniques: A comparative study. *Physical Review B*, **71**(3):035214, 2005.
40. A.J. Mozer, G. Dennler, N.S. Sariciftci, M. Westerling, A. Pivrikas, R. Österbacka, and G. Juška. Time-dependent mobility and recombination of the photoinduced charge carriers in conjugated polymer/fullerene bulk heterojunction solar cells. *Physical Review B*, **72**(3):035217, 2005.
41. A. Pivrikas, R. Österbacka, G. Juska, K. Arlauskas, and H. Stubb. Time-dependent Langevin-type bimolecular charge carrier recombination in regiorandom poly (3-hexylthiophene). *Synthetic Metals*, **155**(2):242–245, 2005.
42. A. Pivrikas, P. Stadler, H. Neugebauer, and N.S. Sariciftci. Substituting the post-production treatment for bulk-heterojunction solar cells using chemical additives. *Organic Electronics*, **9**(5):775–782, 2008.
43. G. Sliauzys, G. Juska, K. Arlauskas, A. Pivrikas, R. Österbacka, M. Scharber, A. Mozer, and N.S. Sariciftci. Recombination of photogenerated and injected charge carriers in [pi]-conjugated polymer/fullerene blends. *Thin Solid Films*, **511**:224–227, 2006.
44. R. Österbacka, A. Pivrikas, G. Juka, K. Geneviius, K. Arlauskas, and H. Stubb. Mobility and density relaxation of photogenerated charge carriers in organic materials 1. *Current Applied Physics*, **4**(5):534–538, 2004.
45. G. Dennler, A.J. Mozer, G. Juska, A. Pivrikas, R. Österbacka, A. Fuchsbauer, and N.S. Sariciftci. Charge carrier mobility and lifetime versus composition of conjugated polymer/fullerene bulk-heterojunction solar cells. *Organic Electronics*, **7**(4):229–234, 2006.
46. D.A.M. Egbe, S. Türk, S. Rathgeber, F. Kühnlenz, R. Jadhav, A. Wild, E. Birckner, G. Adam, A. Pivrikas, V. Cimrova, G. Knör, N.S. Sariciftci, and H. Hoppe. Anthracene based conjugated polymers: Correlation between π-π-stacking ability, photophysical properties, charge carrier mobility, and photovoltaic performance. *Macromolecules*, **43**(3):1261–1269, 2010.
47. D.A.M. Egbe, E. Tekin, E. Birckner, A. Pivrikas, N.S. Sariciftci, and U.S. Schubert. Effect of styryl side groups on the photophysical properties and hole mobility of PPE-PPV systems. *Macromolecules*, **40**(22):7786–7794, 2007.
48. D.A.M. Egbe, G. Adam, A. Pivrikas, A.M. Ramil, E. Birckner, V. Cimrova, H. Hoppe, and N.S. Sariciftci. Improvement in carrier mobility and photovoltaic performance through random distribution of segments of linear and branched side chains. *J. Mater. Chem.*, **20**(43):9726–9734, 2010.
49. A. Pivrikas, N.S. Sariciftci, G. Juška, and R. Österbacka. A review of charge transport and recombination in polymer/fullerene organic solar

cells. *Progress in Photovoltaics: Research and Applications*, **15**(8):677–696, 2007.

50. A. Pivrikas, G. Juška, M. Scharber, N.S. Sariciftci, and R. Österbacka. Charge transport and recombination in bulk-heterojunction solar cells. In *Organic Photonics and Electronics*. Optical Society of America, 2006.
51. A. Pivrikas, G. Juska, K. Arlauskas, M. Scharber, A. Mozer, N.S. Sariciftci, H. Stubb, and R. Österbacka. Charge carrier transport and recombination in bulk-heterojunction solar-cells. In *Proceedings of SPIE*, volume 5938, page 59380N, 2005.
52. G. Adam, A. Pivrikas, A.M. Ramil, S. Tadesse, T. Yohannes, N.S. Sariciftci, and D.A.M. Egbe. Mobility and photovoltaic performance studies on polymer blends: effects of side chains volume fraction. *J. Mater. Chem.*, Royal Society of Chemistry, **21**(8): 2594–2600, 2011.
53. S. Günes, A. Wild, E. Cevik, A. Pivrikas, U.S. Schubert, and D.A.M. Egbe. Effect of shifting of aromatic rings on charge carrier mobility and photovoltaic response of anthracene and thiophene-containing MEH-PPE-PPVS. *Solar Energy Materials and Solar Cells*, **94**(3):484–491, 2010.
54. A. Pivrikas, M. Ullah, T.B. Singh, C. Simbrunner, G. Matt, H. Sitter, and N.S. Sariciftci. Meyer-neldel rule for charge carrier transport in fullerene devices: A comparative study. *Organic Electronics*, **12**(1):161–168, 2011.
55. P.M. Borsenberger, L.T. Pautmeier, and H. Bässler. Nondispersive-to-dispersive charge-transport transition in disordered molecular solids. *Physical Review B*, **46**(19):12145, 1992.
56. H.C.F. Martens, H.B. Brom, and P.W.M. Blom. Frequency-dependent electrical response of holes in poly(*p*-phenylene vinylene). *Physical Review B*, **60**(12):8489–8492, 1999.
57. P.W.M. Blom and M. Vissenberg. Dispersive hole transport in poly(*p*-phenylene vinylene). *Physical Review Letters*, **80**(17):3819–3822, 1998.
58. H. Scher and E.W. Montroll. Anomalous transit-time dispersion in amorphous solids. *Physical Review B*, **12**(6):2455, 1975.
59. R.A. Marcus. Electron transfer reactions in chemistry. theory and experiment. *Reviews of Modern Physics*, **65**(3):599, 1993.
60. N.S. Hush. Distance dependence of electron transfer rates. *Coordination Chemistry Reviews*, **64**:135–157, 1985.
61. I.I. Fishchuk, A. Kadashchuk, H. Bässler, and M. Abkowitz. Low-field charge-carrier hopping transport in energetically and positionally disordered organic materials. *Physical Review B*, **70**(24):245212, 2004.
62. C. Tanase, P.W.M. Blom, and D.M. De Leeuw. Origin of the enhanced space-charge-limited current in poly(*p*-phenylene vinylene). *Physical Review B*, **70**(19):193202, 2004.
63. L.S. Pan, D.R. Kania, P. Pianetta, and O.L. Landen. Carrier density dependent photoconductivity in diamond. *Applied Physics Letters*, **57**(6):623–625, 1990.
64. C. Tanase, E.J. Meijer, P.W.M. Blom, and D.M. De Leeuw. Unification of the hole transport in polymeric field-effect transistors and light-emitting diodes. *Physical Review Letters*, **91**(21):216601, 2003.
65. V.I. Arkhipov, P. Heremans, E.V. Emelianova, G.J. Adriaenssens, and H. Bassler. Charge carrier mobility in doped semiconducting polymers. *Applied Physics Letters*, **82**(19):3245–3247, 2003.

66. V.I. Arkhipov, P. Heremans, E.V. Emelianova, and H. Baessler. Effect of doping on the density-of-states distribution and carrier hopping in disordered organic semiconductors. *Physical Review B*, **71**(4):045214, 2005.
67. P.W.M. Blom, M.J.M. De Jong, and J.J.M. Vleggaar. Electron and hole transport in poly(*p*-phenylene vinylene) devices. *Applied Physics Letters*, **68**(23):3308–3310, 1996.
68. A. Pivrikas, G. Juška, R. Österbacka, M. Westerling, M. Viliunas, K. Arlauskas, and H. Stubb. Langevin recombination and space-charge-perturbed current transients in regiorandom poly (3-hexylthiophene). *Physical Review B*, **71**(12):125205, 2005.
69. Y. Roichman and N. Tessler. Generalized Einstein relation for disordered semiconductors – implications for device performance. *Applied Physics Letters*, **80**:1948, 2002.
70. I.I. Fishchuk, D. Hertel, H. Bässler, and A.K. Kadashchuk. Effective-medium theory of hopping charge-carrier transport in weakly disordered organic solids. *Physical Review B*, **65**(12):125201, 2002.
71. A. Miller and E. Abrahams. Impurity conduction at low concentrations. *Physical Review*, **120**(3):745, 1960.
72. P.M. Borsenberger, W. Mey, and A. Chowdry. Hole transport in binary solid solutions of triphenylamine and bisphenol-a-polycarbonate. *Journal of Applied Physics*, **49**(1):273–279, 1978.
73. A. Pivrikas, M. Ullah, H. Sitter, and N.S. Sariciftci. Electric field dependent activation energy of electron transport in fullerene diodes and field effect transistors: Gill's law. *Applied Physics Letters*, **98**:092114, 2011.
74. L. Pautmeier, R. Richer, and H. Bässler. Hopping in a gaussian distribution of energy states: transition from dispersive to non-dispersive transport. *Philosophical Magazine Letters*, **59**(6):325–331, 1989.
75. J. Frenkel. On pre-breakdown phenomena in insulators and electronic semiconductors. *Physical Review*, **54**(8):647, 1938.
76. R. Hill. Poole-frenkel conduction in amorphous solids. *Philosophical Magazine*, **23**:59–86, 1971.
77. C. Vijila, A. Pivrikas, H. Chun, C. Zhikuan, R. Österbacka, and C.S. Jin. A study of charge transport in a novel electroluminescent poly(phenylene vinylene-co-fluorenylene vinylene) based [pi]-conjugated polymer. *Organic Electronics*, **8**(1):8–13, 2007.
78. K.C. Kao. *Dielectric phenomena in solids*. Academic Press, 2004.
79. M. Jaiswal and R. Menon. Polymer electronic materials a review of charge transport. *Polymer International*, **55**(12):1371–1384, 2006.
80. J.L. Hartke. The three-dimensional Poole–Frenkel effect. *Journal of Applied Physics*, **39**(10):4871–4873, 1968.
81. J. Antula. Hot-electron concept for Poole-Frenkel conduction in amorphous dielectric solids. *Journal of Applied Physics*, **43**(11): 4663–4668, 1972.
82. V. Adamec and J.H. Calderwood. Electrical conduction in dielectrics at high fields. *Journal of Physics D: Applied Physics*, **8**:551, 1975.
83. I.I. Fishchuk, A.K. Kadashchuk, J. Genoe, M. Ullah, H. Sitter, T.B. Singh, N.S. Sariciftci, and H. Bässler. Temperature dependence of the charge carrier mobility in disordered organic semiconductors at large carrier concentrations. *Physical Review B*, **81**(4):045202, 2010.

84. M. Ullah, A. Pivrikas, I.I. Fishchuk, A. Kadashchuk, P. Stadler, C. Simbrunner, N.S. Sariciftci, and H. Sitter. Electric field and grain size dependence of meyer-neldel energy in C60 films. *Synthetic Metals*, Elsevier, **161**(17): 1987–1990, 2011.
85. M. Ullah, II Fishchuk, A. Kadashchuk, P. Stadler, A. Pivrikas, C. Simbrunner, V.N. Poroshin, N.S. Sariciftci, and H. Sitter. Dependence of Meyer–Neldel energy on energetic disorder in organic field effect transistors. *Applied Physics Letters*, **96**:213306, 2010.
86. M. Ullah, A. Pivrikas, I.I. Fishchuk, A. Kadashchuk, P. Stadler, C. Simbrunner, N.S. Sariciftci, and H. Sitter. Effect of source-drain electric field on the Meyer–Neldel energy in organic field effect transistors. *Applied Physics Letters*, **98**:223301, 2011.

15
Surface enhanced Raman scattering (SERS) characterization of metal–organic interactions

K. WILLETS, The University of Texas at Austin, USA
and K. MAYER, Tufts University, USA

DOI: 10.1533/9780857098764.2.421

Abstract: Surface enhanced Raman scattering (SERS) is emerging as an important analytical technique for the characterization of metal–organic interactions and interfaces. This chapter describes the history and theory of SERS and details recent developments in its use towards gaining a fundamental understanding of metal/molecule interactions. We also summarize several examples of practical applications of SERS, and review progress towards the incorporation of SERS into optoelectronic devices.

Key words: surface enhanced Raman scattering, nanoparticles, optoelectronics, metal–organic interactions, microscopy, plasmonics.

15.1 Introduction

15.1.1 Surface enhanced Raman scattering (SERS) history

Conventional Raman spectroscopy (named for its discoverer in 1928, Chandrasekhara Venkata Raman) has long been a popular technique for measuring the vibrational energy levels of molecules. The technique is based on the inelastic scattering that occurs when an incident photon interacts with a molecule, resulting in one of two cases: (1) the photon imparts some energy to the molecule, resulting in a higher-energy vibrational state of the molecule and a lower-energy (red-shifted) scattered photon (this is called a Stokes shift, Fig. 15.1); (2) the photon picks up energy from the molecule, resulting in a lower-energy vibrational state of the molecule and a higher-energy (blue-shifted) scattered photon. (this is called an anti-Stokes shift, Fig. 15.1). The Stokes or anti-Stokes shifts from the excitation wavelength are typically in the range of 10–60 nm, or just a few tenths of an electron volt for visible light. Spectroscopists usually find it more convenient to express these shifts in terms of wavenumbers, defined relative to the incident laser wavelength, rather than in nm. The useful Raman spectrum of most molecules lies in the range between 500 and 2000 wavenumbers (cm^{-1}). Typically, Raman scattering is measured using laser illumination because of

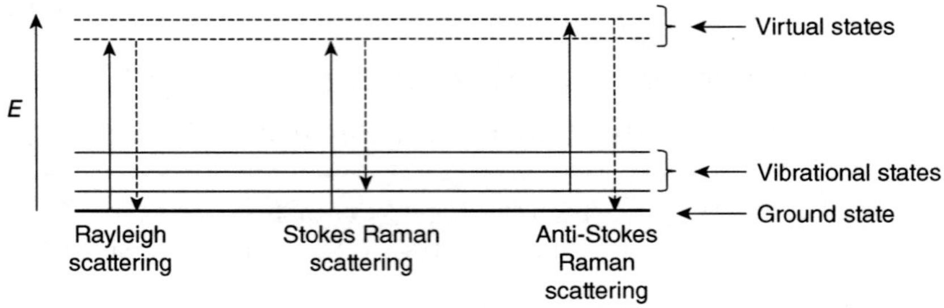

15.1 Schematic of Rayleigh (elastic) scattering and Raman (inelastic) scattering. Raman scattering can be Stokes shifted (to the blue) or anti-Stokes shifted (to the red).

the narrow linewidth required to obtain the requisite resolution for Raman scattering.

After Raman's 1928 description of this scattering phenomenon, conventional Raman spectroscopy developed alongside the technology of spectroscopy throughout the mid-20th century. With the advent of modern grating spectrometers in the 1950s, it was possible to obtain detailed Raman spectra of organic molecules in the 'fingerprinting' region of 500–2000 wavenumbers for the first time. The other key development in Raman spectroscopy was the invention of the laser in 1960, allowing much improved signal intensity, precision, and resolution in measurements of scattering. Around the same time, Raman spectroscopy was combined with confocal microscopy in the technique termed Raman microspectroscopy or hyperspectroscopy. This allows the useful correlation of high-resolution spatial images with spectral information. In the 1960s and 1970s Raman spectroscopy was applied to many problems in chemistry and solid-state physics, including assigning the vibrational modes of countless organic molecules, measuring the temperature and phonon modes of solid crystals, and sensing gases in real time. In many of these cases, Raman spectroscopy was used as a complementary method to IR spectroscopy – the two techniques probe the same spectral region but detect different subsets of modes. The main limitation of Raman spectroscopy is the fact that Raman scattering produces a relatively low signal, necessitating high analyte concentrations. Also, the Raman scattering signal is sometimes overwhelmed by fluorescence of the analyte.

The next major development in the field was the discovery in the mid-1970s of surface enhanced Raman scattering or SERS (Albrecht and Creighton, 1977; Jeanmaire and Van Duyne, 1977). In SERS, molecules adsorbed to a nano-structured metal surface show enhanced Raman scattering, by as much as 11 orders of magnitude. While the effect was originally reported on electrochemically roughened noble metal films,

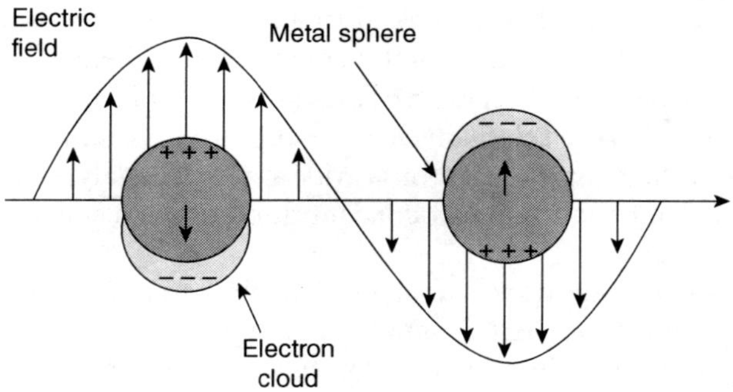

15.2 Metal nanoparticles support a localized surface plasmon resonance (LSPR) in which the delocalized electrons oscillate in response to light. (Adapted from Willets & Van Duyne, *Ann. Rev. Phys. Chem.*, 2007.)

SERS substrates now range from metal island films to controlled nanoparticle arrays. The SERS phenomenon on all of these varied substrates is due in large part to the excitation of a localized surface plasmon in the metal, wherein the free electrons oscillate collectively in response to the electric field of incident light (see Fig. 15.2) (Willets and Van Duyne, 2007). The localized surface plasmon resonance (LSPR) results in strongly enhanced electromagnetic fields at the metal surface; these enhanced electromagnetic fields are greatest at sharp nanoscale features and in small gaps between adjacent nanostructures (Halas *et al.*, 2005; Hao *et al.*, 2007; Kelly *et al.*, 2003). These regions of greatest enhancement are known as 'hot spots.' (The exact mechanisms by which plasmonic E-field enhancements lead to increased Raman scattering cross-sections, as well as chemical contributions to the increased scattering, will be discussed in Section 15.2.1.) After the initial discovery of SERS in the 1970s, interest plateaued until the early 2000s, when SERS saw a resurgence in research interest due to the dramatic improvements in the synthesis and fabrication of metallic nanomaterials. SERS has now been demonstrated for a wide variety of analytes and nanostructured substrates, and has shown detection sensitivities down to the single molecule level.

15.1.2 Role of SERS in the field of optoelectronics

In order for the development of new optical and optoelectronic devices incorporating organic materials to proceed, scientists need a detailed fundamental understanding of processes happening at the metal–organic interface. For example, do organic molecules adsorbed to a metal surface have different conformations from those in bulk solids or free in solution? In the case of biomolecules adsorbed or linked to a surface, do they retain their functions? To what extent do electron transfer and other chemical processes between molecule and metal occur? SERS is an ideal tool for studying such processes because it is a relatively non-perturbative, label-free, far-field optical technique that can be used on a wide range of analytes, and especially because it can be used on low-concentration or even single-molecule samples.

In addition to its importance as a fundamental tool for the study of metal–organic interactions, SERS signals from organic materials have been incorporated into several interesting prototype devices and applied techniques, which will be discussed in Sections 15.3 and 15.4.

15.2 Surface enhanced Raman scattering (SERS) background

15.2.1 SERS theory

In experimental practice, the SERS enhancement factor (EF) is described as the intensity ratio between SERS and normal Raman scattering for a given analyte normalized by the number of molecules probed:

$$\mathrm{EF}_{\mathrm{SERS}}(\omega_s) = \frac{[I_{\mathrm{SERS}}(\omega_s)/N_{\mathrm{surf}}]}{[I_{\mathrm{NRS}}(\omega_s)/N_{\mathrm{vol}}]} \qquad [15.1]$$

Here, I_{SERS} and I_{NRS} are the scattering intensities of SERS and normal Raman scattering, respectively; N_{surf} is the number of molecules adsorbed onto the SERS substrate in the area being probed; and N_{vol} is the number of molecules in the excitation volume of the laser used in normal Raman scattering. Note that to determine this type of quantitative enhancement factor, conventional Raman measurements must always be taken in parallel with SERS. The measurement of N_{surf} remains an experimental challenge, due to its dependence on the available surface area of the metal substrate, the adsorption probability of the analyte, and the geometry of the adsorbed analyte. As such, reported SERS EF values can vary wildly based on estimates of N_{surf}, although EF values typically range from 10^6 to 10^9.

There are two main sources of the enhancement seen in SERS: electromagnetic (plasmonic) enhancement and chemical (charge transfer) enhancement. It is thought that the electromagnetic enhancement gives an electric field enhancement on the order of E^4, and that any additional enhancement, which varies greatly among different combinations of substrate and analyte, depends on the particular chemical interaction between the adsorbed molecule and the metal substrate.

Electromagnetic enhancement in SERS, as mentioned above, arises from excitation of surface plasmons in the metal. The presence of plasmon-enhanced electric fields in the region of a metal particle will enhance both the incident and Raman scattered fields; therefore let us write the enhancement factors for each of these (EF_{inc} and EF_{sca}) as follows, keeping in mind that intensity is defined as the magnitude of the electric field squared:

$$\mathrm{EF}_{\mathrm{inc}} = \frac{\left|E_{\mathrm{SPR}}(\omega)^2\right|}{E_0^2}, \quad \mathrm{EF}_{\mathrm{sca}} = \frac{\left|E_{\mathrm{SPR}}(\omega \pm \omega_{\mathrm{s}})^2\right|}{E_0^2} \qquad [15.2]$$

Here ω is the frequency of the excitation laser, ω_s is the frequency of the Stokes or anti-Stokes (Raman) light, E_{SPR} is the E-field in the neighborhood of the plasmonic particle, and E_0 is the E-field of the incident laser. To calculate the total electromagnetic enhancement factor (EF_{EM}), we multiply the expressions above:

$$\mathrm{EF}_{\mathrm{EM}}(\omega_{\mathrm{s}}) = \frac{\left|E_{\mathrm{SPR}}(\omega)^2\right|\left|E_{\mathrm{SPR}}(\omega - \omega_{\mathrm{s}})^2\right|}{E_0^4} \qquad [15.3]$$

It is because both the incident and scattered fields are enhanced (and therefore both the absorption and scattering from nearby molecules) that the electromagnetic effect enhances Raman scattering so powerfully. Also, the maximum enhancement has been shown to occur when the substrate is chosen such that the plasmon resonance matches the laser wavelength (Willets and Van Duyne, 2007). Because ω and ω_s are typically separated by only 500–2000 cm^{-1}, the enhancement terms for the excitation and Raman scattered light (e.g. $E(\omega)^2$ and $E(\omega - \omega_s)^2$ respectively) are often approximated to be the same, leading to the E^4 dependence often associated with SERS.

When additional scattering enhancement beyond what is predicted by EF_{EM} is measured, it is attributed to chemical enhancement. Two main factors contribute to this chemical enhancement: charge transfer (CT) and molecular resonances (Jensen and Morton, 2009). In the first case, light at the laser wavelength excites not only the surface plasmon of the metal substrate, but also a metal/molecule CT interaction. In the other case, termed surface enhanced resonance Raman scattering (SERRS), a separate chemical enhancement arises from resonance of the laser with an electronic transition in the analyte. There has been some debate in the literature as to whether these two factors can account for the entirety of the chemical enhancement observed in SERS, or whether there is a third, non-resonant source of chemical enhancement (Jensen *et al.*, 2008; Lombardi and Birke, 2008). Therefore it is very important to use clear and consistent terminology when describing the various contributions to SERS, and to report EFs in the standardized form of Equation 15.1.

Lombardi and Birke (2008) have described a 'unified approach' to SERS that takes into account the EM, CT, and SERRS contributions. The EM enhancement in this approach is determined by Equation 15.3 above, with electric fields calculated from appropriate dielectric functions. The latter two factors are calculated through a Herzberg–Teller expansion of the polarizability. The resulting frequency-dependent expression for the total SERS enhancement (where the EF is proportional to $|R_{IFK}(\omega)|^2$) is:

$$R_{IFK}(\omega) = \frac{\mu_{KI}\mu_{FK}h_{IF} < i|Q_k|f >}{\left[(\varepsilon_1(\omega) + 2\varepsilon_0)^2 + \varepsilon_2^2\right]\left[(\omega_{FK}^2 - \omega^2) + \gamma_{FK}^2\right]\left[(\omega_{IK}^2 - \omega^2) + \gamma_{IK}^2\right]} \quad [15.4]$$

Here, the three factors in the denominator express the resonance conditions for EM, CT, and molecular enhancements, in that order. ε_1 and ε_2 are the real and imaginary parts respectively of the dielectric function of the metal, and ε_0 is the dielectric constant of the surrounding medium. ω_{FK} and ω_{IK} are the frequencies of the charge transfer resonance and the molecular resonance, respectively. γ_{FK} and γ_{IK} are the damping terms associated with these resonances. The dipole transition moments μ_{KI} and μ_{FK} describe the allowed molecular transition from the molecule's ground state I and excited state K, and the charge-transfer transition from the metal's Fermi level F to K. h_{IF} is the Herzberg–Teller coupling term between I and F (see Fig. 15.3). $|i>$ and $|f>$ are vibrational states of the system and Q_k are its normal modes.

15.2.2 SERS examples

SERS vs. normal Raman

SERS has been carried out on a wide range of substrates including electrochemically roughened silver electrodes, island films, metal films deposited over nanospheres (FONs) (Dick *et al.*, 2002), and synthetically and lithographically prepared nanoparticles, but the most successful substrate for SERS measurements to date is randomly assembled silver colloid clusters. It is thought that the tight junctions between nanoparticles in a cluster produce extremely high electric fields at hot spots, leading to significant Raman scattering enhancements. For the example SERS spectrum in Fig. 15.4, silver colloids were aggregated with NaCl and incubated with berberine (Lombardi and Birke, 2008). By comparing the resulting SERS spectrum with the conventional solution-phase Raman spectrum of berberine, also shown in the figure, we can identify several interesting features. First, the SERS spectrum clearly shows greater signal-to-noise than the conventional Raman spectrum, even though the intensities between the two spectra cannot be directly compared, as the spectra are not normalized to the number of molecules sampled (see Equation 15.1). In addition, the relative intensity of different modes varies widely among the spectral peaks from normal Raman to SERS. For example, the peak at 729 cm^{-1} becomes dominant in the SERS spectrum, while showing only mid-level intensity relative

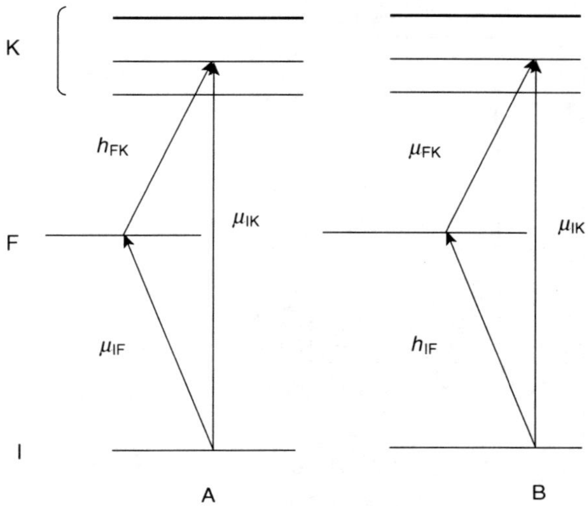

15.3 Energy level diagram describing all possible transitions in the metal–molecule SERS system. I and K are the ground state and excited states of the molecule, respectively. F is the Fermi level of the metal. Columns A and B show the possibilities of a molecule → metal and metal → molecule charge transfer, respectively. The μ terms are electronic transition moments and the h terms are vibronic coupling terms, connecting the energy levels as shown. (Reprinted with permission from John R. Lombardi *et al.*, A unified approach to surface enhanced Raman spectroscopy, *Journal of Physical Chemistry C*. Copyright 2008 American Chemical Society.)

to the other peaks in the normal Raman data. On the other hand, the peak at 1203 cm^{-1} is barely observed above the noise in the SERS data, despite a strong signal in the normal Raman spectrum. These differences are because different Raman modes originate from different bonds within the analyte molecule, and those closer to the metal surface are more strongly enhanced (via both electromagnetic and charge transfer processes) than those further away. Also, for some modes, the frequencies are shifted slightly between the normal Raman and SERS spectra. This is explained by slight differences in molecule conformation and bond lengths when the molecule is adsorbed on a surface versus free in solution.

SERRS (resonance Raman)

As discussed in Section 15.2.1, one of the contributions to the SERS signal is from electronic resonances of the analyte at the laser wavelength; this is referred to as surface enhanced resonance Raman scattering or SERRS. In resonance Raman scattering, rather than being excited to an instantaneous

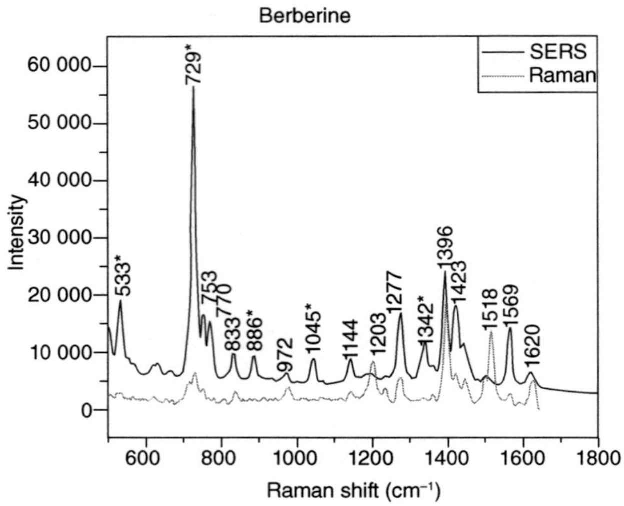

15.4 Comparison of surface-enhanced and normal Raman spectra for berberine on silver colloid clusters. Note the differences in relative peak heights and peak positions. (Reprinted with permission from John R. Lombardi *et al.*, A unified approach to surface enhanced Raman spectroscopy, *Journal of Physical Chemistry C*. Copyright 2008 American Chemical Society.)

virtual state by the laser (as shown in Fig. 15.1), the molecule is excited to an electronic excited state. As the molecule relaxes back to the ground state, vibrational energy levels in the neighborhood of the excited electronic state are probed, and the emitted photons have corresponding energy shifts. Thus, the enhancement due to resonance Raman is actually not related to the presence of a metal surface at all, even though it is taken into account when calculating the total EF in SERS. The main advantage of resonance Raman is generally that it greatly enhances a small subset of vibrational modes of the molecule, allowing specific structures to be targeted. The combination of resonance Raman and SERS has proven to be a powerful technique. For example, SERRS has been used to study photobleaching of R6G (Maher *et al.*, 2002) and to probe the vibrational modes of carbon nanotubes (CNTs) (Kneipp *et al.*, 2001).

Single molecule SERS

The enhancement offered by SERS can be significant enough to enable detection of Raman scattering from even a single molecule. To date, the only substrate on which single-molecule SERS detection has been reliably shown is silver colloid clusters (Michaels *et al.*, 2000; Nie and Emery, 1997). When Raman spectra are taken from a molecule (or few molecules)

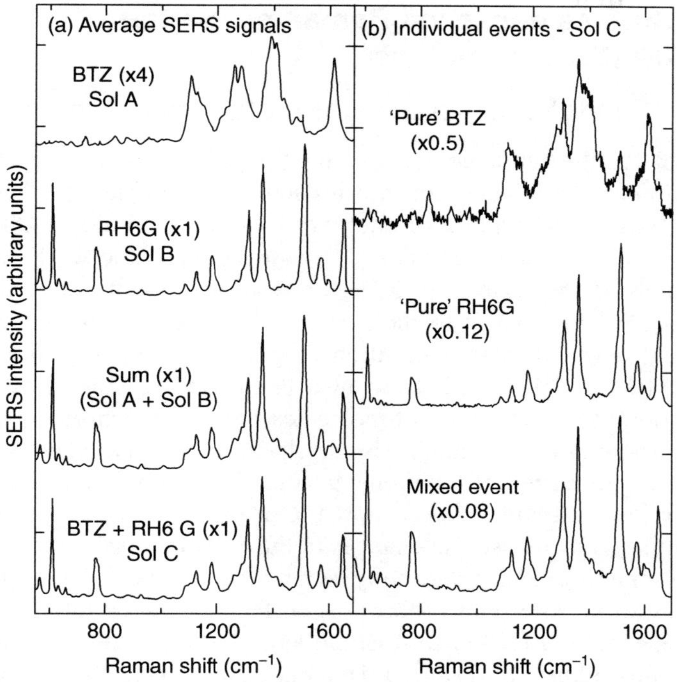

15.5 Bianalyte SERS. (a) Averaged SERS spectra for R6G alone, BTZ alone, and mixtures of the two analytes. (b) Single SERS spectra from a sample with both analytes present, showing the characteristics of BTZ only, R6G only, and both. (Reprinted with permission from E. C. Le Ru *et al.*, Proof of single-molecule sensitivity in surface enhanced Raman scattering (SERS) by means of a two-analyte technique, *The Journal of Physical Chemistry B*. Copyright 2006 American Chemical Society.)

adsorbed to a single cluster, fluctuations are seen in both the total intensity of the scattering and the relative intensities among the various peaks. This 'blinking' behavior is very indicative of single or few-molecule activity. Single molecule SERS can be further proven by employing a bi-analyte technique, in which a silver colloid cluster substrate is incubated with a mixture of two analytes at sufficiently low concentration that only one analyte should adsorb to a single cluster (Le Ru *et al.*, 2006; Van Duyne *et al.*, 2007, 2011). In the single molecule regime, spectral signatures are observed from either one or the other analyte, but rarely both; through statistical analysis, single molecule behavior can be confirmed. See Fig. 15.5 for an example of such data for benzotriazole (BTZ) and rhodamine dyes (Le Ru *et al.*, 2006).

15.3 Surface enhanced Raman scattering (SERS) applications

15.3.1 Fundamental studies of metal–organic interfaces

As a tool for the fundamental investigation of metal–organic interfaces, SERS can provide a wealth of information. For example, SERS has been used to study the adsorption geometry of 1,4-benzenedithiol (BDT) on the surface of gold and silver colloids (Joo *et al.*, 2001). While BDT forms two Ag–S bonds on the surface of the silver, resulting in a planar geometry of the molecule with respect to the surface, the molecule can access both a planar and perpendicular orientation on gold surfaces, depending on the concentration of BDT. SERS has also been used to study how surface charge affects the orientation of molecules such as cytochrome-*c* at a gold surface (Yu and Golden, 2007). Using self-assembled monolayers (SAMs) terminated with different functional groups, the authors tuned the surface charge of the gold and followed how the orientation of the cytochrome-*c* was modulated by measuring changes in the SERS intensity ratio between totally symmetric and non-totally symmetric modes of the molecule. SERS has also been used to follow the surface-induced photoreduction of 4-nitrobenzenethiol (NBT) to 4-aminobenzenethiol (ABT) on nitric acid etched copper films (Shin *et al.*, 2007). While adsorbed NBT has stable SERS peaks on gold and silver films, a signature NBT SERS peak at 1330 cm^{-1} disappears over time and the final spectrum matches ABT, indicating surface-induced photoconversion of NBT to ABT on the copper surface.

In addition to probing adsorbate geometry and photoconversion at interfaces, SERS is also useful for following analyte dynamics at the metal–molecule interface. For example, by combining SERS with super-resolution imaging, the diffusion behavior of a single dye molecule within the tightly confined volumes of a SERS hot spot can be followed in real time (Willets and Stranahan, 2010). (See Fig. 15.6.) These studies revealed analyte mobility on the surface of the metal nanoparticles and showed that the SERS intensity depends strongly on the spatial position of the mobile analyte on the nanoparticle surface.

15.3.2 Detection of explosives

One of the most promising applications of SERS is in the detection of trace amounts of explosives such as trinitrotoluene (TNT). For example, P. C. Ray *et al.* reported detection of TNT with picomolar sensitivity via SERS of cysteine-modified gold nanoparticle clusters (Ray *et al.*, 2009). In this method, the presence of TNT causes particle aggregation due to complexation of TNT with cysteine; the aggregation then induces a great SERS enhancement (up to 9 orders of magnitude). (See Fig. 15.7.) The fact that the sensor is aggregation-based causes an easily seen color change with

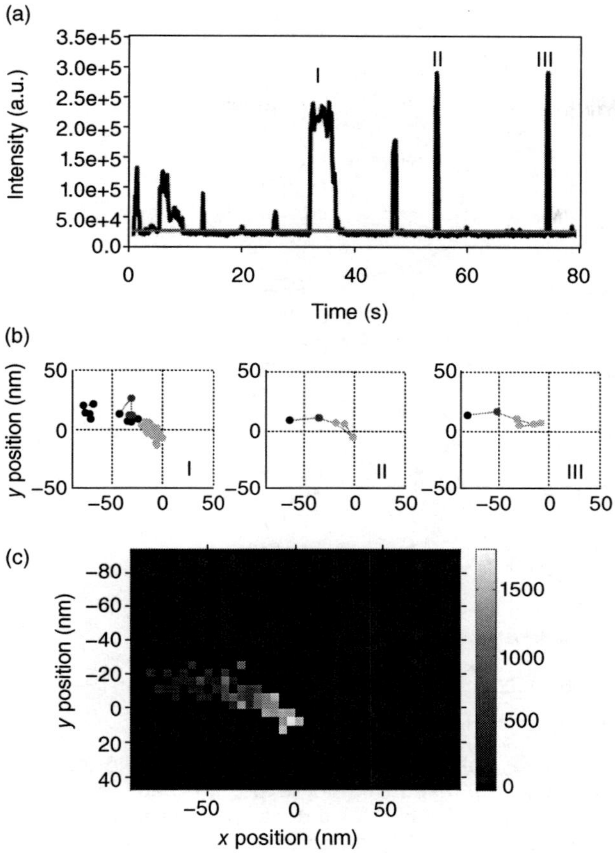

15.6 Super-resolution imaging applied to SERS. (a) Time trace of SERS intensity at a single hot spot. (b) Trajectories of the point from which scattering originates for each of the three on-events in (a). (c) Histogram of the intensity of SERS scattering at each location. (Reprinted with permission from Sarah M. Stranahan *et al.*, Super-resolution optical imaging of single-molecule SERS hot spots, *Nano Letters*. Copyright 2010 American Chemical Society.)

TNT detection. The sensor is specific; i.e., the high SERS response is seen only for TNT and not for other explosives or heavy metals. Later, Tsukruk *et al.* (2009) demonstrated an explosives sensor based upon porous alumina membranes coated with CTAB-capped gold nanoparticle clusters. This sensor was not aggregation-based, so it lacked the colorimetric aspect of the sensor discussed above, but it was sensitive to as little as 15–30 molecules of TNT or DNT (a related explosive). This great improvement in sensitivity was attributed to the increased gold surface area made available by using the porous membranes with densely deposited nanoparticles, and to the waveguiding effect of the alumina.

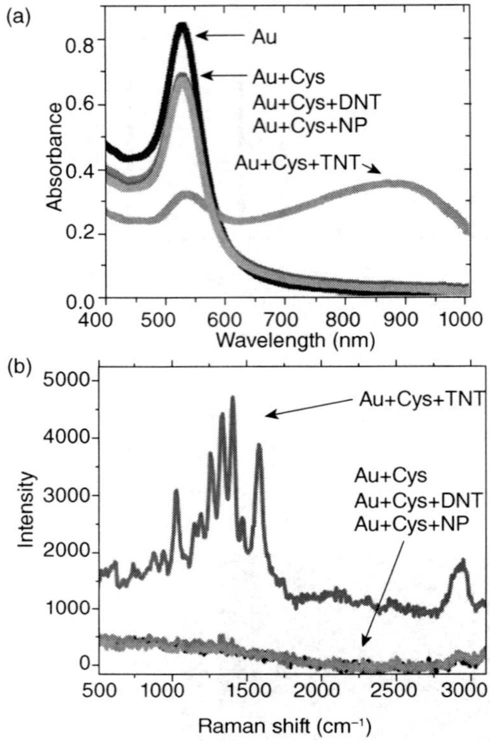

15.7 Explosives detection using SERS. (a) LSPR/colorimetric and (b) Raman spectral response upon exposing cysteine-modified gold nanoparticles to TNT. The TNT induces particle aggregation, producing both an LSPR shift and enhanced Raman scattering. (Reprinted with permission from Samuel S. R. Dasary *et al.*, Gold nanoparticle based label-free SERS probe for ultrasensitive and selective detection of trinitrotoluene, *Journal of the American Chemical Society*. Copyright 2009 American Chemical Society.)

15.3.3 Detection of small bioanalytes

Another important application of SERS related to metal–organic interactions is the detection of small bioanalytes. For example, Liz-Marzan and coworkers have demonstrated SERS detection of drug metabolites for application to drug testing (Correa-Duarte *et al.*, 2009). Their sensor was based upon carbon nanotubes (CNT) coated with silver nanoparticles, and then modified with an antibody specific to the target, benzoylecgonine (BCG), the main metabolite of cocaine. The authors detected BCG down to the nM range through a combination of BCG's unique Raman modes and shifts in the Raman modes of the antibody that occur when BCG binds.

As another example, Van Duyne *et al.* have carried out several studies of glucose sensing both *in vitro* and *in vivo* by SERS. The authors designed sensor chips based on gold FONs coated with a self-assembled monolayer of mercaptooctayltri(ethylene glycol) (EG3). The glucose can reversibly partition within the EG3 monolayer and the SERS from the glucose can be measured over time. The authors created a calibration curve for the SERS spectrum over a range of glucose concentrations and could reliably detect glucose from 5 to 44 mM (10–800 mg/dL), in the meaningful range for diabetes monitoring (Stuart *et al.*, 2005). Later, these sensors were implanted under the skin in rats and the signal monitored through spatially offset Raman spectroscopy (SORS), in which the scattering is measured at a distance offset from the laser incidence. In SORS, the signal originates at a greater depth, allowing scattering to be collected from under the skin. In this case, the glucose concentrations measured were in the correct range but did not quantitatively agree with established methods; however, this should be possible with improved sensor design (Van Duyne *et al.*, 2010).

15.3.4 SERS of DNA and proteins

Many SERS applications have been developed for DNA detection and identification. Perhaps the best-known of these is the DNA sensor of Mirkin and coworkers (Cao *et al.*, 2002). This sensor is based on single-stranded DNA sequences A1 and A2 that are complementary to the two ends of target sequence B. A1 is immobilized on a substrate, while A2, modified with a SERS probe molecule, decorates the surface of gold nanoparticles in solution. If B is introduced into the solution, one end of the target hybridizes to the tethered A1 sequence, while the other end hybridizes to the SERS-tagged A2 sequence, effectively tethering the SERS probe and the gold nanoparticles to the surface. Silver nanoparticles are then grown on the surface of the tethered gold nanoparticles to induce SERS from the attached probe. This method has been shown to detect specific DNA sequences at fM concentrations and can be easily multiplexed to sense multiple DNA sequences, simply by introducing different SERS probes.

SERS has also proven effective for protein detection, for example in the work of Porter *et al.*, who have reported an immunosensor for prostate specific antigen (PSA), sensitive to the pg/mL level (Grubisha *et al.*, 2003). In this scheme, anti-PSA is immobilized on a gold surface and the sample is flowed over the surface to capture PSA. Then, gold nanoparticles labeled with both a SERS probe dye and a second antibody specific to the target are added, forming SERS-active junctions between the gold nanoparticles and substrate. SERS spectra were then measured via a scanned fiber bundle and the concentration of PSA was determined from the intensity of the SERS peaks of the probe dye. SERS measurements on proteins have also helped to elucidate aspects of the SERS mechanism itself. For example, Hofkens and coworkers have measured single molecule SERS of green fluorescent protein (GFP) on silver colloid clusters (Habuchi *et al.*, 2003). They were able to observe single-molecule blinking as well as transitions between the protonated and deprotonated forms of individual molecules. This experiment also revealed important information about the size of SERS hot spots. Because GFP is ~4 × 8 nm in size, whereas previous single-molecule SERS studies were carried out with small dyes (<1 nm in size), the size of hot spots was shown to be larger than previously believed. Etchegoin *et al.* (2003) have used SERS to monitor photoinduced oxygen release from hemoglobin, identified by transient increases in certain relative peak heights in the SERS spectrum. The authors attribute this to a chemical enhancement often seen in organic molecules in the presence of oxygen – using hemoglobin allows for a better understanding of the charge transfer mechanism at play because it produces a single oxygen molecule at a known location in the structure.

15.3.5 SERS applied to thin films

SERS has also been used to characterize thin film materials used in optoelectronic devices. For example, Muraki and Yoshikawa (2010), among others, have used SERS to characterize tris (8-hydroxyquinoline) aluminum (Alq3) films with gold and silver films deposited on top. They were able to measure the stability of the films, the effects of several different spacer layers between Alq3 and gold, and the presence of the $Alq3^-$ anion that appeared when silver films were used. Hesketh and coworkers have used SERS to characterize films of HKUST-1, a metal–organic framework composed of copper ions linked by benzenetricarboxylate. SERS spectra were measured on both extended gold films and gold-coated cantilevers to be used in gas sensing, and compared with normal Raman (Allendorf *et al.*, 2008). Finally, Paez and coworkers have used SERS to investigate the formation of thin films of perylene tetra-carboxylic dianhydride (PTCDA) coated with silver, indium, and magnesium (Salvan *et al.*, 2004). By examining the peaks associated with the outside of the film, the authors found that PTCDA forms sharp boundaries with Ag and Mg, but that In actually penetrates the film. They also observed that certain modes indicating charge transfer were enhanced in the cases of Ag and In, but in the case of Mg, they found evidence that the metal had formed bonds to the film.

15.4 Active and passive control of surface enhanced Raman scattering (SERS) signals

While much of the SERS literature is focused on sensing applications, as described above, recent work has addressed the passive and active control of light emission via SERS. As the understanding of SERS phenomena has increased, researchers have recognized the potential for SERS signals to be modulated, opening the door for controlling SERS emission properties, which is an important first step towards the realization of SERS-based optical devices.

15.4.1 Controlling the polarization of SERS emission

An important feature of SERS is that the Raman scattering from a molecule is not radiated directly by the analyte, but is instead coupled to and re-radiated by the plasmon modes of the nanoparticle. The result of this plasmon re-radiation is that the polarization of the emitted light is controlled by the orientation of the nanoparticle, rather than the orientation of the molecule. For example, in SERS-active nanoparticle dimers, studies have shown that the polarization of the emitted light will be aligned along the longitudinal (long) axis of the dimer (Xu *et al.*, 2009). This polarization alignment of the SERS emission is observed even when the SERS excitation and emission are resonant with the transverse (short-axis) plasmon mode of the dimer (Shegai *et al.*, 2011). Thus, by fabricating arrays of SERS-active nanoparticle dimers with controlled orientation, it is possible to control the output polarization of the emitted light on the nanometer length scale.

This effect can be further exploited through nanoparticle trimers (Shegai *et al.*, 2008; Xu *et al.*, 2009). In this case, a third nanoparticle is introduced to the original dimer pair. By changing the position of the third nanoparticle relative to the original dimer, the polarization of the emission can be rotated away from the alignment of the dimer long axis. The extent to which the output polarization is rotated by the third nanoparticle is extremely sensitive to the emission wavelength, due to coupling of the plasmon modes of the third nanoparticle with the original dimer. However, unlike nanoparticle dimers, in which the output polarization is independent of the number of adsorbed analytes, the use of nanoparticle trimers requires analyte to be preferentially adsorbed in a single junction. In the case when multiple analytes are adsorbed to the nanoparticle trimer, the emission appears isotropic, due to labeling of all junction regions within the aggregate (Shegai *et al.*, 2011). Thus, single-molecule SERS is favored for controlling emission polarization using nanoparticle trimers, because an individual molecule can only occupy one junction region within the nanostructure.

15.4.2 Active control of SERS intensity through nanoparticle structure modulation

The sensitivity of the SERS signal intensity to the structure of the underlying metal substrate is well established and is the subject of active investigation. In particular, nanoparticle aggregates are targeted as SERS substrates because of the strong relationship between the size of the gap between adjacent nanoparticles and the intensity of the SERS response. Recent work has exploited elastometric substrates to actively control the gap spacing, and thus the SERS intensity, between adjacent nanoparticles (Alexander *et al.*, 2010). By fabricating nanorod dimers on pre-strained silicone rubber films, Lopez and coworkers were able to modulate the SERS signal from adsorbed analytes, simply by stretching and relaxing the underlying elastomer. The authors noted that the SERS signal intensity disappeared at nanoparticle gaps greater than 20 nm and reached a maximum when the gap size was close to 15 nm. Thus, the authors demonstrated a simple 'on-off' optical switch based on controlling the strain applied to an elastomeric substrate.

Plasmon-enhanced optical force gradients in highly focused laser beams offer another strategy for controlling the distance between adjacent nanoparticles and thus, the resulting SERS signal. Käll and coworkers have used optical tweezers to capture silver nanoparticles in a microfluidic cell and form SERS-active aggregates at the laser focus (Käll *et al.*, 2009). By premixing the nanoparticles and an analyte of interest at a T-junction within the microfluidic cell, the aggregates show a strong spectral signature associated with the analyte. Upon shuttering the trapping laser, the nanoparticles are released and the SERS signal drops. The experiment can be repeated multiple times, varying the analyte identity at the T-junction, and forming and releasing nanoparticle aggregates with the trapping laser.

15.4.3 Electrochemical control of SERS intensity

Another approach for actively controlling the intensity of SERS signals is to use electrochemical modulation of SERS analytes (Brolo *et al.*, 2009; Cortes *et al.*, 2010; Flood *et al.*, 2011; Haran *et al.*, 2009). Two major strategies have been described, the first relying on changes in the electronic resonance of adsorbed analyte during reduction and oxidation, and the second based on structural changes that lead to new vibrational modes as the potential is scanned.

Figure 15.8 shows an example of a simple 'on-off' SERS switch, based on the reduction and oxidation of Nile Blue (Cortes *et al.*, 2010). In the oxidized form, Nile Blue has an electronic resonance in the red region of the spectrum, leading to strong SERRS. However, as the potential is scanned negative, the reduced form of the dye is produced, and the electronic resonance – as well as the SERRS signal – disappears. The SERRS signal can be modulated between high and low intensity by simply sweeping the potential from negative to positive as shown in the figure. Interestingly, this experiment also provides a means of identifying single molecule SERS because the intensity goes from high to low in a single, digital step in the single molecule case (reminiscent of single-step photobleaching in

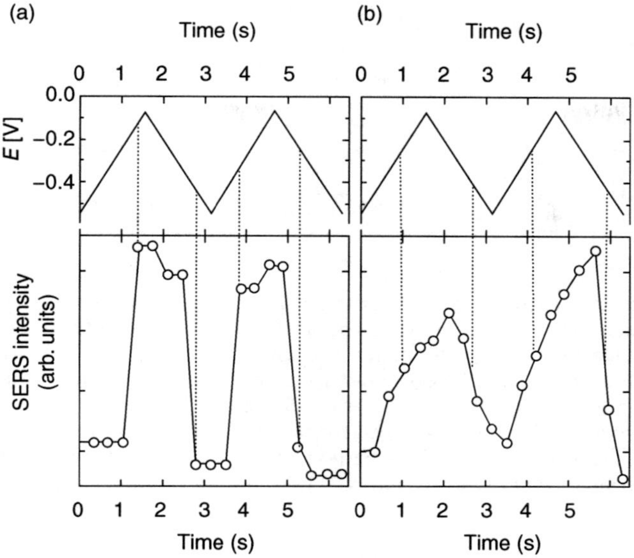

15.8 Electrochemically switchable SERS. (a) As the sample potential is ramped, Nile Blue is oxidized and reduced. Because the oxidized form has a significant added SERS enhancement due to molecular resonance but the reduced form does not, the SERS intensity undergoes switching. (Reprinted with permission from Emiliano Cortés *et al.*, Electrochemical modulation for signal discrimination in surface enhanced Raman scattering (SERS), *Analytical Chemistry*. Copyright 2010 American Chemical Society.)

single-molecule fluorescence, Fig. 15.8a). In the case of multiple Nile Blue molecules on the nanoparticle surface, the SERRS intensity changes gradually as the potential is swept, due to the heterogeneity in the oxidation/reduction potentials of the different dyes on the surface (Fig. 15.8b).

A second example of electrochemically modulated SERS is shown in Fig. 15.9 (Flood *et al.*, 2011). In this case, the original analyte can undergo two oxidation steps (Fig. 15.9a), each of which generates a new species with a distinct SERS spectrum (Fig. 15.9b). The authors demonstrated that the three electrochemically distinct vibrational modes at 529, 508 and 1640 cm^{-1} could represent molecular logic gates and encode AND, XOR, and NOR gates, respectively. This work represents the first example of a molecular logic device based upon SERS and also demonstrates the potential of SERS for future optical and optoelectronic device applications.

15.5 Conclusion

Owing to its capacity to probe the detailed interactions between organic materials and metal surfaces in a sensitive and label-free manner, SERS

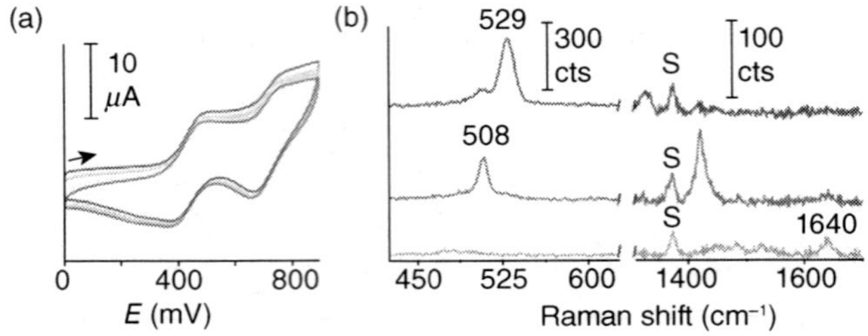

15.9 Electrochemically modulated SERS measured via distinct spectral peaks appearing in the oxidized and reduced forms of the analyte. (a) Overlay of 40 cyclic voltammograms, showing two oxidation/reduction steps. (b) SERS spectra at each oxidation state. Applied voltages: 0 V (bottom), 0.6 V (middle) and 0.9 V (top). (Reprinted with permission from Edward H. Witlicki *et al.*, Molecular logic gates using surface-enhanced Raman-scattered light, *Journal of the American Chemical Society*. Copyright 2011 American Chemical Society.)

can be very useful for the characterization of materials to be used in optoelectronic devices. In the coming years, we expect to see the development of more sophisticated SERS substrates (Banholzer *et al.*, 2008) and an increasingly detailed understanding of the effects of plasmonic and molecular resonance, charge transfer (Zhou *et al.*, 2006, 2007), and molecular orientation (Joo *et al.*, 2001; Yu and Golden, 2007) on SERS signals. SERS has already been incorporated in some proof-of-concept devices, including switches actively controlled through polarization (Shegai *et al.*, 2008, 2011, Xu *et al.*, 2009), force modulation (Alexander *et al.*, 2010; Käll *et al.*, 2009), or electrochemical reaction (Brolo *et al.*, 2009; Cortes *et al.*, 2010; Flood *et al.*, 2011; Haran *et al.*, 2009).

15.6 References

Albrecht, M. G. & Creighton, J. A. (1977) Anomalously intense raman-spectra of pyridine at a silver electrode. *Journal of the American Chemical Society,* **99**, 5215–5217. DOI: 10.1021/ja00457a071

Alexander, K. D., Skinner, K., Zhang, S. P., Wei, H. & Lopez, R. (2010) Tunable SERS in gold nanorod dimers through strain control on an elastomeric substrate. *Nano Letters,* **10**, 4488–4493. DOI: 10.1021/nl102.3172

Allendorf, M. D., Houk, R. J. T., Andruszkiewicz, L., Talin, A. A., Pikarsky, J., Choudhury, A., Gall, K. A. & Hesketh, P. J. (2008) Stress-induced chemical detection using flexible metal–organic frameworks. *Journal of the American Chemical Society,* **130**, 14404–14405. DOI: 10.1021/ja805235k

Banholzer, M. J., Millstone, J. E., Qin, L. D. & Mirkin, C. A. (2008) Rationally designed nanostructures for surface-enhanced Raman spectroscopy. *Chemical Society Reviews,* **37**, 885–897. DOI: 10.1039/B710915f

Brolo, A. G., Dos Santos, D. P., Andrade, G. F. S. & Temperini, M. L. A. (2009) Electrochemical control of the time-dependent intensity fluctuations in surface-enhanced Raman scattering (SERS). *Journal of Physical Chemistry C,* **113**, 17737–17744. DOI: 10.1021/jp907389v

Cao, Y. W. C., Jin, R. C. & Mirkin, C. A. (2002) Nanoparticles with Raman spectroscopic fingerprints for DNA and RNA detection. *Science,* **297**, 1536–1540. DOI: 10.1126/science.297.5586.1536

Correa-Duarte, M. A., Sanles-Sobrido, M., Rodriguez-Lorenzo, L., Lorenzo-Abalde, S., Gonzalez-Fernandez, A., Alvarez-Puebla, R. A. & Liz-Marzan, L. M. (2009) Label-free SERS detection of relevant bioanalytes on silver-coated carbon nanotubes: The case of cocaine. *Nanoscale,* **1**, 153–158. DOI: 10.1039/b9nr00059c

Cortes, E., Etchegoin, P. G., Le Ru, E. C., Fainstein, A., Vela, M. E. & Salvarezza, R. C. (2010) Electrochemical modulation for signal discrimination in surface enhanced Raman scattering (SERS). *Analytical Chemistry,* **82**, 6919–6925. DOI: 10.1021/ac101152t

Dick, L. A., Mcfarland, A. D., Haynes, C. L. & Van Duyne, R. P. (2002) Metal film over nanosphere (MFON) electrodes for surface-enhanced Raman spectroscopy (SERS): Improvements in surface nanostructure stability and suppression of irreversible loss. *Journal of Physical Chemistry B,* **106**, 853–860. DOI: 10.1021/jp0136381

Etchegoin, P., Liem, H., Maher, R. C., Cohen, L. F., Brown, R. J. C., Milton, M. J. T. & Gallop, J. C. (2003) Observation of dynamic oxygen release in hemoglobin using surface enhanced Raman scattering. *Chemical Physics Letters,* **367**, 223–229. DOI: 10.1016/S0009-2614(02)01705-0

Flood, A. H., Witlicki, E. H., Johnsen, C., Hansen, S. W., Silverstein, D. W., Bottomley, V. J., Jeppesen, J. O., Wong, E. W. & Jensen, L. (2011) Molecular logic gates using surface-enhanced Raman-scattered light. *Journal of the American Chemical Society,* **133**, 7288–7291. DOI: 10.1021/ja200992x

Grubisha, D. S., Lipert, R. J., Park, H. Y., Driskell, J. & Porter, M. D. (2003) Femtomolar detection of prostate-specific antigen: An immunoassay based on surface-enhanced Raman scattering and immunogold labels. *Analytical Chemistry,* **75**, 5936–5943. DOI: 10.1021/ac034356f

Habuchi, S., Cotlet, M., Gronheid, R., Dirix, G., Michiels, J., Vanderleyden, J., De Schryver, F. C. & Hofkens, J. (2003) Single-molecule surface enhanced resonance Raman spectroscopy of the enhanced green fluorescent protein. *Journal of the American Chemical Society,* **125**, 8446–8447. DOI: 10.1021/ja03533f1

Halas, N. J., Talley, C. E., Jackson, J. B., Oubre, C., Grady, N. K., Hollars, C. W., Lane, S. M., Huser, T. R. & Nordlander, P. (2005) Surface-enhanced Raman scattering from individual AU nanoparticles and nanoparticle dimer substrates. *Nano Letters,* **5**, 1569–1574. DOI: 10.1021/nl050928v

Hao, F., Nehl, C. L., Hafner, J. H. & Nordlander, P. (2007) Plasmon resonances of a gold nanostar. *Nano Letters,* **7**, 729–732. DOI: 10.1021/nl062969c

Haran, G., Shegai, T., Vaskevich, A. & Rubinstein, I. (2009) Raman spectroelectrochemistry of molecules within individual electromagnetic hot spots. *Journal of the American Chemical Society,* **131**, 14390–14398. DOI: 10.1021ja904480r

Jeanmaire, D. L. & Van Duyne, R. P. (1977) Surface Raman spectroelectrochemistry. 1. Heterocyclic, aromatic, and aliphatic-amines adsorbed on anodized silver electrode. *Journal of Electroanalytical Chemistry,* **84**, 1–20. DOI: 10.1016/S0022-0728(77)80224-6

Jensen, L. & Morton, S. M. (2009) Understanding the molecule-surface chemical coupling in SERS. *Journal of the American Chemical Society,* **131**, 4090–4098. DOI: 10.1021/ja809143c

Jensen, L., Aikens, C. M. & Schatz, G. C. (2008) Electronic structure methods for studying surface-enhanced Raman scattering. *Chemical Society Reviews,* **37**, 1061–1073. DOI: 10.1039/b706023h

Joo, S. W., Han, S. W. & Kim, K. (2001) Adsorption of 1,4-benzenedithiol on gold and silver surfaces: Surface-enhanced Raman scattering study. *Journal of Colloid and Interface Science,* **240**, 391–399. DOI: 10.1006/jcis2001.7692

Käll, M., Tong, L. M., Righini, M., Gonzalez, M. U. & Quidant, R. (2009) Optical aggregation of metal nanoparticles in a microfluidic channel for surface-enhanced Raman scattering analysis. *Lab on a Chip,* **9**, 193–195. DOI: 10.1039/b813204f

Kelly, K., Coronado, E., Zhao, L. & Schatz, G. (2003) The optical properties of metal nanoparticles: The influence of size, shape, and dielectric environment. *Journal of Physical Chemistry B,* **107**, 668–677. 10.1021/jp026731y

Kneipp, K., Jorio, A., Kneipp, H., Brown, S. D. M., Shafer, K., Motz, J., Saito, R., Dresselhaus, G. & Dresselhaus, M. S. (2001) Polarization effects in surface-enhanced resonant Raman scattering of single-wall carbon nanotubes on colloidal silver clusters. *Physical Review B,* **63**. DOI: 10.1103/PhysRevB.63.081401

Le Ru, E. C., Meyer, M. & Etchegoin, P. G. (2006) Proof of single-mokeule sensitivity in surface enhanced Raman scattering (SERS) by means of a two-analyte technique. *Journal of Physical Chemistry B,* **110**, 1944–1948. DOI: 10.1021/jp054732v

Lombardi, J. R. & Birke, R. L. (2008) A unified approach to surface-enhanced Raman spectroscopy. *Journal of Physical Chemistry C,* **112**, 5605–5617. DOI: 10.1021/jp800167v

Maher, R. C., Cohen, L. F. & Etchegoin, P. (2002) Single molecule photo-bleaching observed by surface enhanced resonant Raman scattering (SERRS). *Chemical Physics Letters,* **352**, 378–384. DOI: 10.1016/S0009-2614(01)01474-9

Michaels, A. M., Jiang, J. & Brus, L. (2000) Ag nanocrystal junctions as the site for surface-enhanced Raman scattering of single Rhodamine 6G molecules. *Journal of Physical Chemistry B,* **104**, 11965–11971. DOI: 10.1021/jp0025476

Muraki, N. & Yoshikawa, M. (2010) Characterization of the interface between metal and tris (8-hydroxyquinoline) aluminum using surface-enhanced Raman scattering with glass cap encapsulation. *Chemical Physics Letters,* **496**, 91–94. DOI: 10.1016/j.cplett.2010.07.011

Nie, S. M. & Emery, S. R. (1997) Probing single molecules and single nanoparticles by surface-enhanced Raman scattering. *Science,* **275**, 1102–1106. DOI: 10.1126/science.275.5303.1102

Ray, P. C., Dasary, S. S. R., Singh, A. K., Senapati, D. & Yu, H. T. (2009) Gold nanoparticle based label-free SERS probe for ultrasensitive and selective detection of trinitrotoluene. *Journal of the American Chemical Society,* **131**, 13806–13812. DOI: 10.1021/ja905134d

Salvan, G., Zahn, D. R. T. & Paez, B. (2004) Surface enhanced Raman scattering in organic thin films covered with silver, indium and magnesium. *Journal of Luminescence,* **110**, 296–302. DOI: 10.1016/j.lumin.2004.08.024

Shegai, T., Li, Z. P., Dadosh, T., Zhang, Z. Y., Xu, H. X. & Haran, G. (2008) Managing light polarization via plasmon-molecule interactions within an asymmetric metal nanoparticle trimer. *Proceedings of the National Academy of Sciences of the United States of America,* **105**, 16448–16453. DOI: 10.1073/pnas.0808365105

Shegai, T., Brian, B., Miljkovic, V. D. & Käll, M. (2011) Angular distribution of surface-enhanced Raman scattering from individual AU nanoparticle aggregates. *ACS Nano,* **5**, 2036–2041. DOI: 10.1021/nn1031406

Shin, K. S., Lee, H. S., Joo, S. W. & Kim, K. (2007) Surface-induced photoreduction of 4-nitrobenzenethiol on Cu revealed by surface-enhanced Raman scattering spectroscopy. *Journal of Physical Chemistry C,* **111**, 15223–15227. DOI: 10.1021/jp073053c

Stuart, D. A., Yonzon, C. R., Zhang, X. Y., Lyandres, O., Shah, N. C., Glucksberg, M. R., Walsh, J. T. & Van Duyne, R. P. (2005) Glucose sensing using near-infrared surface-enhanced Raman spectroscopy: Gold surfaces, 10-day stability, and improved accuracy. *Analytical Chemistry,* **77**, 4013–4019. DOI: 10.1021/ac0501238

Tsukruk, V. V., Ko, H. & Chang, S. (2009) Porous substrates for label-free molecular level detection of nonresonant organic molecules. *Acs Nano,* **3**, 181–188. DOI: 10.1021/nn800569f

Van Duyne, R. P., Dieringer, J. A., Lettan, R. B. & Scheidt, K. A. (2007) A frequency domain existence proof of single-molecule surface-enhanced Raman spectroscopy. *Journal of the American Chemical Society,* **129**, 16249–16256. DOI: 10.1021/ja077243c

Van Duyne, R. P., Yuen, J. M., Shah, N. C., Walsh, J. T. & Glucksberg, M. R. (2010) Transcutaneous glucose sensing by surface-enhanced spatially offset Raman spectroscopy in a rat model. *Analytical Chemistry,* **82**, 8382–8385. DOI: 10.1021/ac101951j

Van Duyne, R. P., Kleinman, S. K., Ringe, E., Valley, N., Wustholz, K. L., Phillips, E., Scheidt, K. A. & Schatz, G. C. (2011) Single-molecule surface-enhanced Raman spectroscopy of crystal violet isotopologues: Theory and experiment. *Journal of the American Chemical Society,* **133**, 4115–4122. DOI: 10.1021/ja110964d

Willets, K. A. & Stranahan, S. M. (2010) Super-resolution optical imaging of single-molecule SERS hot spots. *Nano Letters,* **10**, 3777–3784. DOI: 10.1021/nl102559d

Willets, K. A. & Van Duyne, R. P. (2007) Localized surface plasmon resonance spectroscopy and sensing. *Annual Review of Physical Chemistry,* **58**, 267–297. DOI: 10.1146/annurev.physchem.58.032806.104607

Xu, H. X., Li, Z. P., Shegai, T. & Haran, G. (2009) Multiple-particle nanoantennas for enormous enhancement and polarization control of light emission. *Acs Nano,* **3**, 637–642. DOI: 10.1021/nn800906c

Yu, Q. M. & Golden, G. (2007) Probing the protein orientation on charged self-assembled monolayers on gold nanohole arrays by SERS. *Langmuir,* **23**, 8659–8662. DOI: 10.1021/la7007073

Zhou, Q., Li, X. W., Fan, Q., Zhang, X. X. & Zheng, J. W. (2006) Charge transfer between metal nanoparticles interconnected with a functionalized molecule probed by surface-enhanced Raman spectroscopy. *Angewandte Chemie-International Edition,* **45**, 3970–3973. DOI: 10.1021/Jp067045s

Zhou, Q., Zhao, G., Chao, Y. W., Li, Y., Wu, Y. & Zheng, J. W. (2007) Charge-transfer induced surface-enhanced Raman scattering in silver nanoparticle assemblies. *Journal of Physical Chemistry C,* **111**, 1951–1954.

16

Second harmonic generation (SHG) as a characterization technique and phenomological probe for organic materials

K. D. SINGER and Y. WU, Case Western Reserve University, USA

DOI: 10.1533/9780857098764.2.442

Abstract: Optical second harmonic generation (SHG) is a powerful technique to study nonlinear optical properties of materials. SHG is notable for its requirement for the absence of a material inversion center. In an electric field induced SHG (EFISHG) experiment, symmetry is broken by an electric field allowing SHG. Hyper-Rayleigh scattering (HRS) is another technique that probes incoherent SHG in isotropic solution. Both EFISHG and HRS are used to study molecular nonlinear optical mechanisms and characterize new materials.

Key words: electric yield induced second harmonic generation (EFISHG), hyper-Rayleigh scattering (HRS), Kleinman symmetry.

16.1 Introduction

The invention of high-intensity pulsed lasers around 1960 paved the way for the observation of optical harmonics. Franken and co-workers first observed second harmonic generation (SHG) in crystalline quartz using a Q-switched ruby laser (Franken *et al.*, 1961). While SHG was of interest in its own right for extending the tuning range of lasers, it also became a favored method to characterize the second order nonlinear optical (NLO) response in new materials for use in the linear electro-optic effect as well as for SHG. In particular, optical SHG has become an important tool in the study of organic NLO materials (Singer *et al.*, 1998). Because a very large number of possible molecular structures for organic second order nonlinear optics have been investigated, SHG is routinely used to measure the NLO response both at the molecular level and in condensed media. More recently, the important symmetry constraints on second order NLO processes has opened the way for a number of applications including biomedical and other imaging and to probe bulk structure and dynamics of molecules.

A paramount consideration in SHG, and any even order NLO process, is that the material must not exhibit an inversion center. This requirement of noncentrosymmetry applies at both the molecular and condensed phase levels. For organic nonlinear optics, dipolar molecules arranged in a polar condensed phase have received the most attention as they do meet this requirement. At the molecular level, a two-level quantum model indicates that the dipolar NLO response is largely dependent on the inner product of the molecules permanent dipole and transition dipole. Compounds with donor and acceptor substituted conjugated π-electron systems are the staple of investigations into new nonlinear optical materials (Zyss, 1994). A number of approaches for generating polar macroscopic structures from polar molecules have been investigated. Nonlinear molecules are grown into a crystal usually by solution or melt techniques with the resulting crystal formed by dipolar (van der Waals) or hydrogen bonding interactions. Owing to the high molecular density and the highly polar structure, in some cases, a very high optical nonlinearity can result. These high nonlinearities surpassed previously studied inorganic crystals, and sparked interest in the potential application of organic materials. Other approaches to forming polar materials include monomolecular growth or deposition, and poled polymer films. Poled polymers have received the most attentions of these approaches (Singer *et al.*, 1986; Dalton *et al.*, 2010). In this case, NLO molecules are dissolved or covalently attached in amorphous polymers. In a state of high molecular mobility (usually the polymer is heated above its glass transition temperature), a strong DC electric field is applied to break the center of symmetry and impart polar order. With the field applied, the polymer is put into an immobile state by cooling or chemical reaction, thus locking in the polar order.

Other important symmetry aspects have driven research in SHG. Most notably, the lack of inversion symmetry at dielectric discontinuities has suggested the importance of second harmonic as probe of surfaces and other interfaces (Heinz *et al.*, 1982). Recently, more exotic forms of noncentrosymmetric materials have been investigated including chiral and octupolar media (Zyss, 1993; Verbiest *et al.*, 1994; Ostroverkhov *et al.*, 2000b; Sioncke *et al.*, 2003; Haupert and Simpson, 2009).

16.2 Second harmonic generation (SHG) in bulk media

There are several experimental techniques studied for measuring the NLO response of molecules and materials. Intense fundamental optical beams and sensitive detection of SHG signal are always the key to these measurements.

The second harmonic coefficient d_{ij} ($i = 1$–3, $j = 1$–6) can be determined through absolute and relative measurements. Absolute measurements include second harmonic phase matching and parametric fluorescence (Kurtz, 1975). While such measurements were needed in the early days, the reference materials studied by these methods have been used almost exclusively since to carry out measurements relative to accepted standard materials.

Two relative techniques, slab rotation Maker fringes (Maker *et al.*, 1962) and the wedge translation technique, are typically favored. In relative techniques, high-powered laser pulses are incident on the sample while the intensity of the transmitted frequency-doubled light is monitored with a sensitive detector (e.g., photo multiplier tobe (PMT). Band-pass filters are carefully placed in the beam path in order to block the fundamental and pass the SHG to the detector.

The wedge technique has been fully detailed previously by Chemla and Kupecek (1971). In this technique, measured samples are shaped as a wedge, where the relative phase between the free and bound harmonic waves is varied by thickness variation upon translation. During the experimental run, the fundamental light normally incidents on the sample and the SHG signal against the sample translation distance is recorded. The wedge technique is useful in the infrared region where the long coherence lengths prevent the observation of the fringes in the Maker fringe technique and it is also very convenient to analyze (Kurtz, 1975). Another advantage of the wedge method is that multiple reflections of the fundamental and second harmonic waves are eliminated.

Jerphagnon and Kurtz (JK) (Jerphagnon and Kurtz, 1970) presented in detail a method, Maker fringe technique, for determination of the SHG coefficients, with the assumption of isotropic media with no absorption or dispersion. In this technique, samples are slabs or thin uniform films. The SHG signal versus the angle of incidence is recorded. Over the years, this technique has been applied to inorganic and organic materials. However, many of the second order NLO organic materials under investigation today such as liquid crystals, Langmuir–Blodgett films, poled polymers, and organic crystals absorb at the wavelength of the typical SHG experiment, exhibit a fair degree of anisotropy and have dispersion in the index of refraction. Further development or correction to this technique has been done later (Kuzyk *et al.*, 1989).

A complete description of Maker fringe experiments has been given (Herman and Hayden, 1995), and recently updated to include multiple reflections in multilayer materials (Park and Herman, 2012). Nonlinear measurement taken at far from resonance of the nonlinear material is always preferred in order to minimize the absorption. Otherwise, one should have an accurate measurement of the thickness and absorption coefficient in order to get meaningful results in the case of absorption of the SHG. Absorption at the fundamental frequency is even more complicated. Concomitant with the lack of inversion symmetry, most of the NLO materials exhibit birefringence. Ignoring birefringence of the materials can result in significant error in the nonlinear coefficient with even modest dispersion (Herman and Hayden, 1995).

In a poled system, the molecules are aligned in a polar manner along the external field having $C_{\infty v}$ symmetry. This symmetry group yields a third-rank $\chi^{(2)}$ tensor with a limited number of terms. In addition, second harmonic implies permutation symmetry in the last two Cartesian indices. As a result, only the d_{31} and d_{33} terms in the reduced tensor notation are independent (Boyd, 2003). When considering poled polymers, JK's model, the ratio of d_{31}/d_{33} decreases as the dispersion increases. This can lead to errors in the determination of the induced molecular alignment distribution. In the standard rigid-gas model (Williams, 1987) describing the electric field induced orientation of dipoles predicts $d_{31}/d_{33} = 1/3$ for low poling fields and ratios smaller than 1/3 for high poling fields. In general, this ratio provides information on the orientational distribution and ratios of $d_{31}/d_{33} < 1/3$ are generally attributed to more order (Singer *et al.*, 1987). In these cases, correct analysis of the SHG experiments taking absorption and birefringence into account is necessary; otherwise the orientational distribution derived may be subject to systematic errors, and the degree of order overestimated. Even with the correct consideration of dispersion, large errors occur as the thickness approaches an integer multiple of twice the coherence length. Thus, a complete angular scan in Maker fringes experiment and subsequent fit of the data to the theory is the only way to ensure accuracy.

We focus now on molecular hyperpolarizability measurements involving SHG relating to macroscopic material measurements. Two important techniques for measuring molecular nonlinearities: electric field induced second harmonic generation (EFISHG) and hyper-Rayleigh scattering (HRS) will be described. The nature of the susceptibilities, the method for deducing the molecular susceptibility tensor components, measurement conventions, the role of local fields, and experimental techniques are described in the limit of weakly interacting molecules. We will also discuss the application of the NLO techniques in bulk material structure and molecule dynamics studies.

16.3 Electric field induced second harmonic generation (EFISHG)

EFISHG is a well-established technique to determine the first-order NLO response of dipolar molecules. The schematic of a typical EFISHG experiment is shown in Fig. 16.1. A strong external electric field is applied to a liquid solution in order to re-orient the dissolved molecules. As a result, the center of inversion of a gas or liquid is broken allowing bulk second harmonic generation. Original EFISHG experiments were carried out on

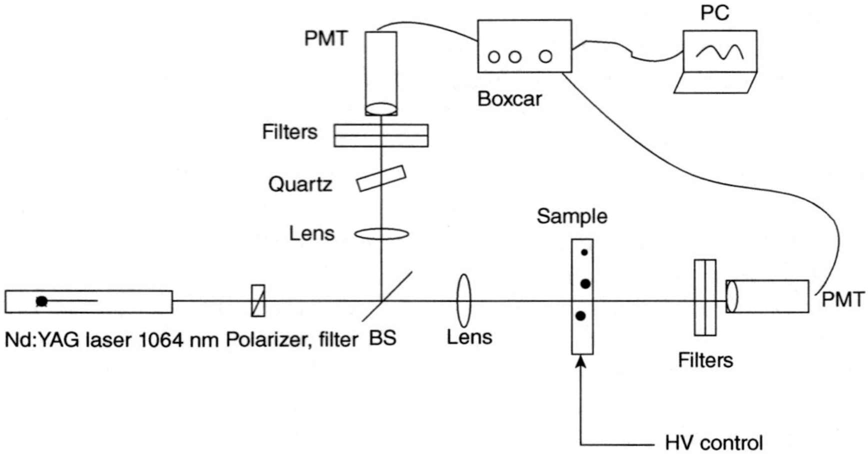

16.1 Schematic layout of a typical EFISHG experiment. The SHG from a quartz crystal is usually used as reference. Alternating or static electric field is introduced by applying high voltage on the sample while performing SHG measurement. The SHG from quartz crystal is used as a reference to monitor the fluctuation of the intensity of the laser.

molecules in the gas phase (Finn and Ward, 1974; Ward and Bigio, 1975). Though the contributions of local fields and interactions are minimized in the gas phase, many compounds of interest cannot develop the necessary vapor pressure for such measurements. The technique was then adapted for use in measurements in liquid phase (solution of molecules) (Levine and Bethea, 1975; Oudar, 1977; Singer and Garito, 1981). Such measurements can give reliable results, but care must be taken to minimize the effects of local fields and interactions. The solvent is usually carefully selected to minimize both absorption and second harmonic response. The relevant third order macroscopic polarization density at frequency 2ω is given by,

$$P_I^{2\omega} = \Gamma_{IJKL}^{2\omega} E_J^{\omega} E_K^{\omega} E_L^{0} \quad [16.1]$$

where $\Gamma_{IJKL}^{2\omega}$ is the EFISH susceptibility.

The microscopic contributions contributing to Γ include the two lowest order hyperpolarizabilities. Isotropic averaging of the molecular quantities is performed because only dilute liquid solution is involved. An infinite dilution technique was developed to address the effects of solvent–solute interactions (Singer and Garito, 1981). Later developments of this technique included dispersion measurements (Teng and Garito, 1983), determination of the real and imaginary parts of the response (Kajzar *et al.*, 1987), and simultaneous measurement of the third harmonic coefficient (Cheng *et al.*, 1991). Recently, this has been used for probing the structure and symmetry of bulk material or molecules (Xiao *et al.*, 2006; Lim *et al.*, 2009; Satou *et al.*, 2011)

16.3.1 Characterizing the molecular hyperpolarizability

The macroscopic susceptibility, $\Gamma^{2\omega}_{IJKL}$ can be related to the molecular quantities of interest. Since EFISHG measurements involve dilute liquid solutions, isotropic averaging of the molecular quantities is performed. Thus, the macroscopic polarization density is given by:

$$P_I(t) = N\langle p_i(t)\rangle_I \qquad [16.2]$$

where $\langle p_i(t)\rangle_I$ is the ensemble average of all the dipolar molecules, indices I and i denote dimensions of the laboratory frame and molecular frame. Assuming a Boltzmann distribution, we can write:

$$\langle p_i(t)\rangle_I = \frac{1}{\int \mathrm{d}\Omega}\int\left(1+\frac{\mu_l a_{lL} E_L^0}{kT}\right)[\beta_{ijk}(t;\tau_1,\tau_2)a_{il}a_{jJ}a_{kK}E_J(t-\tau_1)E_K(t-\tau_2)$$
$$+\gamma_{ijlk}(t;\tau_1,\tau_2,\tau_3)E_J(t-\tau_1)E_K(t-\tau_2)E_L(t-\tau_3)]\mathrm{d}\Omega \qquad [16.3]$$

where Ω denotes the solid Euler angles, $\boldsymbol{a}$ is the rotation matrix that connects the molecular dipole frame to the laboratory frame, μ is the dipole moment, β, γ are second- and third-order molecular optical susceptibilities (two lowest-order hyperpolarizabilities), and E^0, E^ω are the static field and optical field. The orientation average then becomes the average of products of the direction cosines. These have been tabulated by Cyvin, Rauch and Decius (1965). Having the expressions for the macroscopic polarization, we can now define a macroscopic susceptibility, Γ_{IJKL}:

$$\Gamma_{IJKL}E_L = N(\chi_1\delta_{IJ}E_k + \chi_2\delta_{IK}E_J + \chi_3\delta_{JK}E_I) \qquad [16.4]$$

where,

$$\chi_1\frac{1}{30} = (4\gamma_{uuvv} - \gamma_{uvuv} - \gamma_{uvvu}) + \frac{1}{30kT}(4\beta_{uuv}\mu_v - \beta_{uvu}\mu_v - \beta_{uvv}\mu_u) \qquad [16.5]$$

$$\chi_2 = \frac{1}{30}(4\gamma_{uvuv} - \gamma_{uvvu} - \gamma_{uuvv}) + \frac{1}{30kT}(4\beta_{uvu}\mu_v - \beta_{uvv}\mu_u - \beta_{uuv}\mu_v) \qquad [16.6]$$

$$\chi_3\frac{1}{30} = (4\gamma_{uvvu} - \gamma_{uuvv} - \gamma_{uvuv}) + \frac{1}{30kT}(4\beta_{uvv}\mu_u - \beta_{uuv}\mu_v - \beta_{uvu}\mu_v) \qquad [16.7]$$

In order to select the component of the polarization density oscillating at the optical second harmonic, a Fourier transform is applied. We also assume the optical waves are quasi-monochromatic. Experiments generally measure only the Γ_{1111} component of the EFISH susceptibility, which can be written as:

$$\Gamma_{1111} = \frac{Nf^0 f^{2\omega} f^{\omega} f^{\omega}}{15}\left[\frac{3}{2}(2\gamma_{uuvv} + \gamma_{uvvu}) + \frac{1}{2}\frac{1}{kT}(2\beta_{uuv}\mu_v + \beta_{uvv}\mu_u)\right] \quad [16.8]$$

where u and v are summed over the molecular coordinates. The numerical factors inside the parentheses can be different if a different convention is used (Singer *et al.*, 1998). The additional factors, f coefficients, are local field factors and will be discussed below. To simplify the expression, we define the scalar part of γ_{uvwx} as:

$$\gamma = \frac{1}{15}\sum_{u,\,v=x,\,y,\,z}(2\gamma_{uuvv} + \gamma_{uvvu}) \quad [16.9]$$

and the inner product of the vector part of β_{uvw} and μ_u as:

$$\beta\mu = \frac{1}{3}\sum_{u,\,v=x,\,y,\,z}(2\beta_{uuv}\mu_v + \beta_{uvv}\mu_u) \quad [16.10]$$

With these new definitions:

$$\Gamma_{1111} = Nf^0 f^{2\omega} f^{\omega} f^{\omega}\left(\frac{3}{2}\gamma + \frac{1}{10kT}\beta\mu\right) \quad [16.11]$$

It is usually assumed that the contribution from γ is negligible compared with the other term. This, however, is not routinely tested, but should be reasonable to assume given the trend toward molecules with increasingly larger values of $\beta\mu$. In any case, the magnitude of this term can be assessed using third harmonic generation or other third-order processes.

16.3.2 Local field effects

The electric force between the external field and the polar molecules is affected by the ambient dielectric, which is also polarized by the external field, creating local field effects. In order to extract the nonlinear coefficients correctly from EFISHG measurement, local field effects should be considered (Singer and Garito, 1981). The effects of local fields can be accurately determined experimentally by carrying out concentration-dependent measurements on dilute solutions of the molecule of interest. A related and equivalent technique was introduced by Burland *et al.* (1991). To measure β accurately, we will require concentration dependent measurements of not only the NLO susceptibility but also the static dielectric constant and linear polarizability, and develop a concentration extrapolation procedure. As local fields increase with dielectric constant, taking these into account is most important in high dielectric solvents. Even more important, molecular associations in liquid solution between solute molecules can occur at higher concentration. The concentration extrapolation technique described below will reveal this through nonlinear dependence and minimize by infinite dilution extrapolation. Consider the EFISHG susceptibility corresponding to a two component solution:

$$\Gamma_{1111} = N_0 f^0 f^{2\omega} f^{\omega} f^{\omega} \gamma_0' + N_1 f^0 f^{2\omega} f^{\omega} f^{\omega} \gamma_1' \qquad [16.12]$$

where $\gamma' = \frac{3}{2}\gamma + \frac{1}{10kT}\beta\mu$.

It has been shown that good results are obtained when one chooses Debye type local field factors at optical frequencies and Onsager type in the static limit (Singer and Garito, 1981). With Onsager model, we have result of local field factors

$$f_i^0 = \frac{\varepsilon(n_i^2+2)}{2\varepsilon+n_i^2} \qquad [16.13]$$

$$f_i^* = \frac{\varepsilon(2\varepsilon+1)(n_i^2+2)^2}{3(2\varepsilon+n_i^2)^2} \qquad [16.14]$$

where, n_i refers to a 'index' of refraction for the species. Normally, this will apply to the solute. One need not use the crystal index of refraction; this number can be determined by the extrapolation procedure explained below. For the Debye-type local fields,

$$f_i^\infty = \frac{n^2+2}{3} \qquad [16.15]$$

where this applies to optical frequencies and n is the refractive index.

By measuring the concentration dependence of the EFISHG susceptibility, the static dielectric constant, and the index of refraction, one obtains linear dependence at low concentration. From the slopes and intercepts of the curves, μ and γ' can be determined. The quantity γ can be measured independently using four-wave mixing, third harmonic generation, or some other third-order process, but as pointed out above, the term is often neglected. With this, β can be determined using the reported techniques (Singer and Garito, 1981).

16.3.3 Limitations

The EFISHG technique has been widely used for characterizing uncharged, dipolar chromophores. As an external field interacts only with molecules having permanent dipole moment, it is limited to such molecules. Because of the presence of a high-voltage direct current (DC) field, this technique is not applicable for measuring conducting or ionic materials.

In the EFISHG experiment, the external field aligns the nonlinear optical molecules to remove the inversion symmetry of the solution being studied. The measured quantity is the scalar product of the dipole moment and the vector portion of the susceptibility tensor, which are assumed to be parallel. This limitation has made the technique mainly useful for determining a single hyperpolarizability component, β_{zzz}, appropriate for linear dipolar chromophores having C_{2v} symmetry, where the z-direction is along the molecular dipole. The HRS technique has much more general applicability.

16.4 Hyper-Rayleigh scattering (HRS)

HRS, incoherent second harmonic scattering from an isotropic material, is an alternative to EFISHG as a method of measuring the second-order molecular susceptibility of NLO molecules. As mentioned earlier, SHG is not normally allowed in liquid with inversion symmetry; however, local fluctuations can result in small amounts of incoherent second harmonic scattering whose polarization and angle dependence contain a significant amount of information on the β tensor. This nonlinear optical phenomenon was first reported in 1965, soon after the development of high peak power lasers. In HRS experiment, the second harmonic yield is much smaller than other techniques mentioned above. Thus, ultra-high powered laser pulses are tightly focused on a liquid solution in order to create a very strong optical field. The practical application of this technique only came after development of the theoretical work. Most recently, the HRS technique has been successfully used in characterizing the nonlinearity of the dipolar and multipolar molecules (Clays *et al.*, 1991; Zyss and Ledoux, 1994; Hubbard *et al.*, 1996; Ostroverkhov *et al.*, 2000b). In the basic HRS experiment, the scattered SHG signal is collected at 90° with respect to the propagation direction of the fundamental light (Fig. 16.2). In this scheme, only few terms of the β tensor are measured (Clays *et al.*, 1991). In more sophisticated HRS experiments, where the SHG is collected at arbitrary direction, the incoherent scattered SHG signal can be decoded and the six rotational invariants can be calculated.(Hubbard *et al.*, 1996; Ostroverkhov *et al.*, 2000b; Wu *et al.*, 2008)

HRS is applicable for measuring both dipolar and multipolar molecules in uniform phase. Prior to 1990, EFISHG was the standard method of measuring molecular nonlinear coefficients, with HRS gaining use thereafter (Clays *et al.*, 1991). HRS has important distinctions from EFISHG, most of which arise from the lack of an externally applied electric field in HRS. This allows measurements with molecules that lack a permanent dipole moment but have only higher order moments such as octupolar (Zyss, 1993). It also allows experiments with ionic or conductive species which are

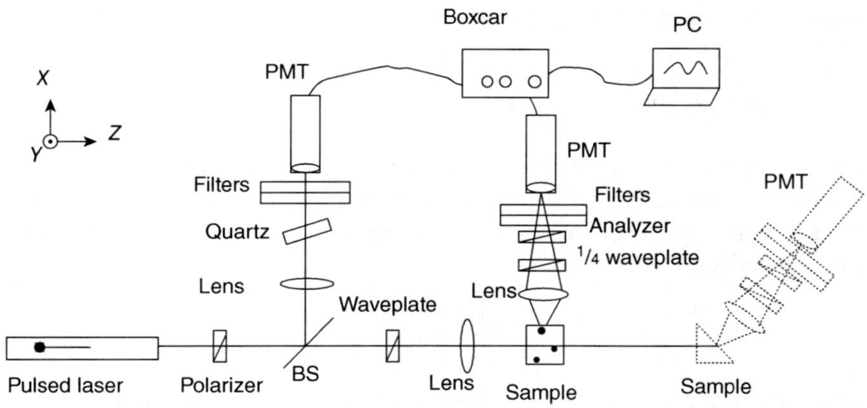

16.2 Typical experiment layout of the hyper-Rayleigh scattering at 90 degrees and 45 degrees (KD-HRS, dashed configuration for sample signal collection). Elliptically polarized fundamental light is tightly focused on sample solution in quartz cuvette in order to generate very intense field. The polarization of the fundamental laser light is controlled by a polarizer and a waveplate. The harmonic scattering is collected with lens and checked with ¼ waveplate and polarizer. The SHG from quartz crystal is used as a reference to monitor the fluctuation of the intensity of the laser.

not suitable for an EFISHG experiment. HRS can measure several of the tensor components of the susceptibility while EFISHG can measure only the portion parallel to the dipole moment. The HRS technique can also measure nonlinear response close to resonance, where the Kleinman symmetry is no long valid. The two techniques combined can determine the angle between the dipole moment and the vector portion of the nonlinear susceptibility.

In isotropic media, the average value of second order response, B_{IJK}, vanishes because the scattering units are randomly oriented, and fields emitted from each individual source (the NLO molecules in solution) interfere destructively. However, fluctuations in orientation can generate a measurable amount of incoherently scattered second harmonic light. Assuming that values of B_{IJK} are correlated only over distances much smaller than the wavelength, the SHG intensity is proportional to the average value of the square of B_{IJK} as follows

$$I_{IL}^{2\omega} \propto \int < B_{IJK,1} B_{LMN,2}^{*} > \mathrm{d}^3 x_1 \int E_J E_K E_M^{*} E_N^{*} \mathrm{d}^3 x_2 \qquad [16.16]$$

where the brackets indicate averaging over all orientations. The subscripts 1 and 2 on B_{IJK} refer to different positions within the scattering media. The assumption that B_{IJK} is uncorrelated over long distances allows those subscripts to be dropped. If the scatterers are randomly oriented molecules, the total intensity, $I^{2\omega}$, is the sum of the contributions from individual molecules and can be written as

$$I^{2\omega} = GN \sum_{IJKLMN} < B_{IJK} B^{*}_{LMN} > I^{\omega} I^{\omega} \qquad [16.17]$$

where N is the number density of scattering units, β the molecular second harmonic coefficient, I^{ω} the incident intensity, and G is a factor that incorporates corrections for local fields and experimental geometry. According to Zyss and Ledoux (1994),

$$G = \frac{2^{11}\pi^5}{cR^2\lambda^4}\left(\frac{n_{2\omega}+2}{3}\right)\left(\frac{n_{\omega}+2}{3}\right)^2 \qquad [16.18]$$

where the n's are the refractive indices of the solvent at the indicated frequencies, λ the wavelength, and R the distance from the scattering center to the detector. Internal referencing to the solvent or to a solution of known nonlinearity determines G. Eq. (16.18) can be extended to include scattering from two species in the limit of low absorption:

$$I^{2\omega} = G\left(N_{\text{solvent}} < \beta^2_{\text{solvent}} > + N_{\text{dye}} < \beta^2_{\text{dye}} >\right)\left(I^{\omega}\right)^2 \qquad [16.19]$$

which predicts a linear relationship between $I^{2\omega}$ and N_{dye} in the limit of a dilute solution ($N_{\text{dye}} << N_{\text{solvent}}$). It is important to note that the values of $\langle\beta^2\rangle$ in Eq. (16.19) are defined with respect to laboratory-fixed axes. The transformation to molecule-fixed axes, i, j, and k as well as the method for determining β_{ijk} will be reviewed later.

In an HRS experiment, it is easier to determine the linear relationship described in Eq. (16.19) by using alternating half-wave plates and linear polarizers to set the polarization and continuously vary the incident intensity of I^{ω}. It is worth pointing out that at very high fundamental intensities, additional NLO processes such as self-focusing lead to subquadratic behavior in the intensity of the scattered light (Mao *et al.*, 2007). Right angle scattering is the favored geometry, where the SHG signal is collected perpendicular to the direction of the incident fundamental light (Fig. 16.2). This configuration measures the Kleinman symmetric contributions to the β tensor (Hubbard *et al.*, 1996; Ostroverkhov *et al.*, 2000b; Wu *et al.*, 2008). With incident light polarized perpendicular to the plane of incidence, an analyzer with the same orientation will measure

$$I_Z^{2\omega} = GN\langle\beta^2_{ZZZ}\rangle\left(I^{\omega}\right)^2 \qquad [16.20]$$

where Z indicates the laboratory axis perpendicular to the plane of the incidence. In the case of an analyzer parallel to the direction of travel of incident beam, the signal will be

$$I_X^{2\omega} = GN\langle\beta^2_{ZXX}\rangle\left(I^{\omega}\right)^2 \qquad [16.21]$$

Although local fluctuation model for SHG in uniform solution has successfully explained many experiments or observations, some discrepancy was found in some experiments. Recently, a non-local model was discussed by Shelton (2010) in order to solve the discrepancy.

16.4.1 Dipolar materials

According to group theory, the third-rank β tensor can be decomposed into rotational invariants. In an HRS experiment, the incident fields are indistinguishable, so β is symmetric with respect to the interchange of two indices. In this case, the tensor decomposition has six rotational invariant components: two vectors, one rank-2 pseudotensor, one rank-3 tensor, and a complex term describing the interference between the vectors.

$$\beta \sim 1 \otimes (1 \otimes 1)_{\text{sym}} \sim 1 \otimes (0 \oplus 2) \sim 1 \oplus 1 \oplus 2 \oplus 3 \qquad [16.22]$$

According to the result of the dipole oscillator model, if all of the frequencies are much smaller than any resonant frequency, the dispersion of the susceptibility can be neglected, and permutation symmetry among all three of the tensor Cartesian indices holds. This is known as Kleinman symmetry. In this case, only the vector and rank-3 tensor in β tensor decomposition are non-zero, leaving the Kleinman-symmetric terms. That is,

$$\beta_{\text{Kleinman}} \sim (1 \otimes 1 \otimes 1)_{sym} \sim 1 \oplus 3 \qquad [16.23]$$

The vector component is a dipolar contribution measurable with EFISHG technique while the tensor part has multipolar symmetry measurable only with HRS. Many HRS measurements have been carried out for molecules that can be approximated as being one-dimensional dipolar molecules, possessing $C_{\infty v}$ symmetry. If the π-electrons in a molecule are conjugated in one dimension, Kleinman symmetry is automatically satisfied since all the β terms vanishes except β_{zzz} obviously symmetric in all three indices. Therefore, the measurement directly gives the value of $\langle \beta_{ZZZ}^2 \rangle$. Conversely, multidimensional molecules will generally be Kleinman nonsymmetric given that most measurements include visible second harmonic in visible-absorbing chromophores.

Recall that in EFISHG, the second-order NLO polarizability is linearly proportional to the local field factors and permanent dipole moment of the molecules (Eq. (16.11)). Comparison of the HRS data and EFISHG data has been shown to verify the theory (Clays *et al.*, 1991). In the EFISHG measurement of a dipolar solution, attention should be paid to the increase of effective molecular dipole moment caused by the dipole moments themselves (Clays *et al.*, 1991). However, the HRS signal is directly proportional to β^2, the knowledge of the dipole moment is not required at all, but the sign of β can only be determined with EFISHG.

16.4.2 Multipolar materials

Dipolar molecules usually tend to form an anti-parallel molecular arrangement when the density of the solution is high and there is no external aligning field. This arrangement, where dipoles tend to cancel, inhibits the second order nonlinear process and thus limits its application to electro-optical configurations. However, researchers have expanded their investigations on NLO molecules whose nonlinearity origins from octupole or other multi-pole contributions. These materials are called multipolar materials, but possessing structures that are noncentrosymmetric.

There has been considerable interest in HRS experiment in two-dimensional octupolar molecules, which can provide a different route to non-centrosymmetric second-order nonlinear optical materials (Verbiest *et al.*, 1994; Zyss and Ledoux, 1994). Schematic structures are shown in Fig. 16.3. These effects arise from the third order Kleinman symmetric term (3) in Eq. (16.22). As discussed above, the HRS measures the rotational invariants in tensor decomposition.

According to Eq. (16.23), only the 3rd rank tensor is allowed in β if all dipolar-like quantities vanish. A molecule or material which has such properties belongs to an octupolar group, that is, a point group consistent with octupolar symmetry. Some examples are shown in Fig. 16.3. In this case, there is no dipole moment, but the structure is noncentrosymmetric, nonetheless. From analysis of the symmetry, certain point groups, such as C_{2v}, D_{3h}, T_d, orthorhombic 222 group are octupolar groups. When a non-centrosymmetric molecule belongs to a multipolar group of order strictly higher than 3, the β tensor must strictly vanish. For example, a 5-fold symmetry molecule cannot have nonzero β although it is noncentrosymmetric. The octupolar planar symmetric (i.e., disk-like molecule, e.g. crystal violet) tensors under rotation can be generalized to lower order tensors. However, octupolar molecules can also have out-of plane structures, such as chiral

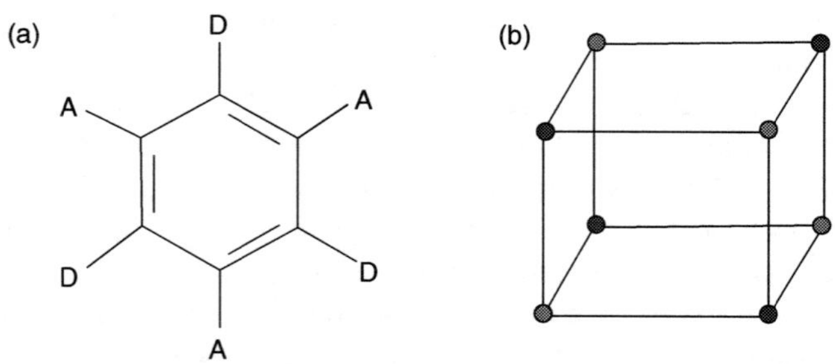

16.3 Schematic structure for two- and three-dimensional octupolar nonlinear optical response.

geometries. The average value of the rank-6 tensor $\beta \otimes \beta$ is measured in the HRS experiment.

We note that one must carefully consider the applicability of Kleinman symmetry in order to determine the type of HRS to perform. Much work has focused on approximately one-dimensional dipolar molecules. In this case, donor and acceptor substitution on a bridging conjugated π-electron system yields such a one-dimensional dipolar molecule. In this case, Kleinman symmetry applies regardless of the relationship of the experimental wavelengths to the electronic states. In the more general case, of multidimensional molecules the nominal far-from-resonance requirement for Kleinman symmetry to hold must be assessed. The cases of two- and three-dimensional conjugated electron systems of octupolar or other symmetry can yield Kleinman-disallowed (KD) contributions to the scattered signal. Given the fact that many of the nonlinear optical chromophores of interest absorb in the visible, it may not be possible to perform HRS measurements in Kleinman symmetric regime with visible second harmonic light. The conditions for far-from-resonance Kleinman symmetry for HRS are much more difficult to attain than is widely appreciated. This being the case, 90° scattering measurements may, in fact, have KD contributions in the measured signal. The degree to which this is the case will determine systematic errors in determination of the β tensor for multi-dimensional molecules.

This is a particular issue with the reports of octupolar molecules in the literature. As noted above, the octupolar tensor invariant (3 ss), is a Kleinman symmetric contribution, and it is likely that HRS measurements on chromophores with obvious octupolar symmetry contain negligible Kleinman non-symmetric components. In this case, 90° scattering depolarization ratio measurements are appropriate. Indeed, the fact that many of these measurements obtain depolarization ratios expected for octupolar molecules reflects this (Zyss and Ledoux, 1994). However, HRS measurements in multidimensional molecules without octupolar symmetry may contain non-Kleinman symmetric contributions. This can result in systematic errors in reported values.

Axially ordered chiral media can exhibit second-order nonlinear optical properties arising from the electric-dipole coupling (Hubbard *et al.*, 1998; Ostroverkhov *et al.*, 2001a, 2001c). Such media require breaking of Kleinman symmetry at both the molecular and supramolecular levels. However, while chirality is required at the supramolecular level, it is not at the molecular level. A chiral arrangement of non-chiral molecules is sufficient. The mixed symmetry first rank tensor (1) and the second rank (2) pseudotensor terms in Eq. (16.22) are the KD components. The second rank tensor, of particular interest, requires a molecule of two or three dimensions as one-dimensional molecules are necessarily Kleinman symmetric. Such multidimensional molecules aligned in a chiral manner with a symmetry axis exhibit second-order nonlinear optical effects. Two-dimensional molecules with C_{2v} symmetry, as well as chiral disk-like chromophores could contribute to large nonlinear optical response in chiral axial materials (Ostroverkhov *et al.*, 2000a, 2000b, 2001b; Sanguinet *et al.*, 2005, 2006).

When measuring the Kleinman allowed components (the vector and rank-3 tensor components), the HRS scattering output is perpendicular to the incident beam. In order to study the KD and cross terms, a scattering experiment, where the signal collection is at 45° to the incident beam (Fig. 16.2), has been developed (Hubbard *et al.*, 1996; Ostroverkhov *et al.*, 2000b; Wu *et al.*, 2008). The SHG signal can be collected at any angle, but 45° is a convenient value.

In this case, careful selection of the input and output polarizations enables one to measure the complete collection of the relevant nonlinear optical figures of merit applicable in the most general case. As was shown, one measures six figures of merit, one for each of the six rotationally invariant tensor components of the molecular hyperpolarizability, namely (Ostroverkhov *et al.*, 2000b):

$$I^{2\omega}(\omega_i, \omega_o) = GN\left(I^{\omega}\right)^2 \sum_L \beta_L^2 \Delta_L(\omega_i, \omega_o) \qquad [16.24]$$

where β_L^2 are the tensor invariant figures of merit (where L = 1 ss, 1 mm, 1 sm+, 1 sm−, 2 mm, 3 ss). These components are 1 ss, 1 mm, 1 sm+, the vector contributions that are symmetric (s) or of mixed symmetry (m), the second-rank pseudotensor of mixed symmetry (2 mm) and the third rank symmetric tensor (octupolar). The Δ_L are the corresponding invariant tensors. We note that the 1 ss and 3 ss are the Kleinman symmetric parts normally measured in 90° HRS, where the two components can be assessed from the depolarization ratio. In this case, Eq (16.24) then becomes Eqs (16.20) and (16.21) regarding different polarization.

In this general case, one measures elliptically polarized incident light and scattered second harmonic light. In principle, either the incident or outgoing relative phase can be varied during the experiment. An analysis of the information on the tensor invariant figures of merit reveals that continuously varying the elliptical input polarization for a fixed HRS elliptical polarization yield the most information. The six invariants can be obtained by varying the input polarization (or relative phase) continuously for two different settings of the output polarization. The two sets of polarization-dependent data can be simultaneously fit to yield the six invariants. The optimum output ellipticity and orientation of the outgoing fields have been derived (Hubbard *et al.* 1998; Ostroverkhov *et al.* 2000b). The illustration of the polarization of the fundamental and SHG is shown in Fig. 16.4.

The polarization of the incident light can also be controlled either by rotating a waveplate or, better yet, electrically through a liquid crystal

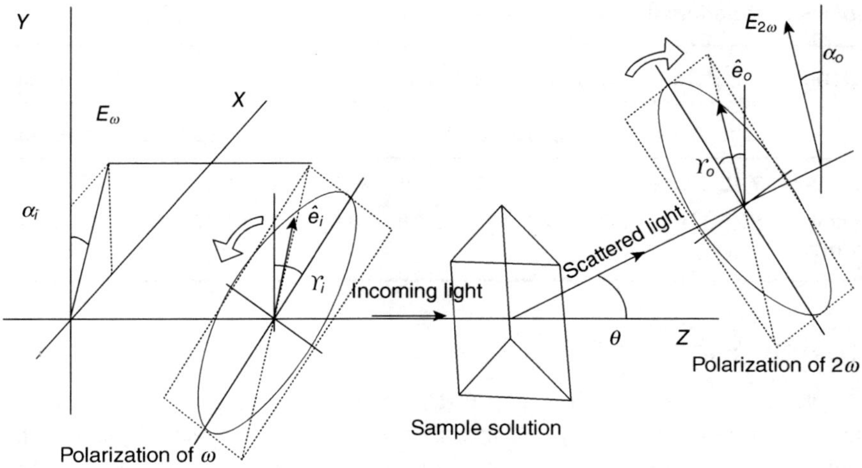

16.4 Illustration of light polarization in KD-HRS experiment. The fundamental light is elliptically polarized by employing polarizer, waveplates or retarder. The scattered SHG is collected at 45 degrees. Combination of polarizer and ¼ waveplate is used to analyze the polarization of the signal.

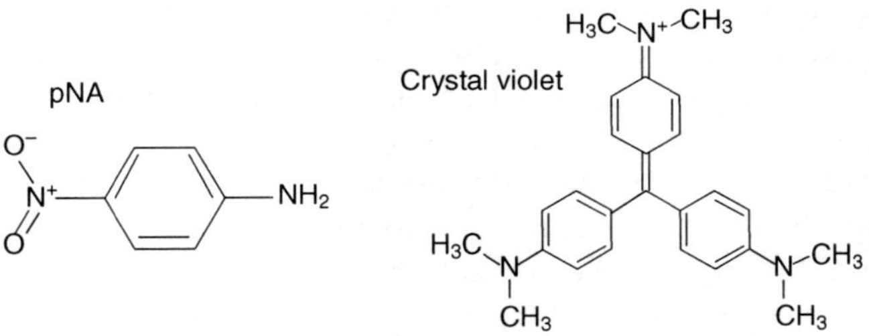

16.5 Molecular structure of a quasi-one-dimensional chromophore (pNA) with approximate Kleinman symmetry and crystal violet, a two-dimensional chromophores not expected to be Kleinman symmetric.

retarder in order to remove the possible mechanical errors. This method has improved the measurement reproducibility and sensitivity (Wu *et al.*, 2008).

As an example, consider the chromophores in Fig. 16.5. The chromophores' p-NA is quasi-one-dimension and would be expected to exhibit small KD components (1 mm and 2 mm in Table 16.1), while crystal violet (CV) could exhibit substantial response in all components since it is two-dimensional. These properties are indeed reflected in the results of HRS shown in Table 16.1 (Ostroverkhov *et al.* 2000a, 2001a).

Table 16.1 Rotationally invariant figures of merit for two chromophores in Fig. 16.5

Material	$\Vert\beta_{1ss}\Vert$	$\Vert\beta_{1mm}\Vert$	$\Vert\beta_{2mm}\Vert$	$\Vert\beta_{3ss}\Vert$	Depolarization ratio	
	(esu × 10^{-30})				90° exp	45° exp
pNA (ref)	7.12 ± 0.16	0.0 ± 2.1	3.0 ± 2.6	4.95 ± 0.23	0.20 ± 0.01	024 ± 0.01
Crystal violet	83.5 ± 2.1	72.2 ± 2.1	84.1 ± 6.5	76.0 ± 4.2	0.63 ± 0.01	0.65 ± 0.06

16.4.3 Experimental issues

As we pointed out earlier, the HRS process is extremely inefficient. It largely depends on the peak intensity of the light pulse. Typically, the number of scattered photons is a factor of about 10^{-14} lower than the number of incident photons. Thus, it requires very high incident power of the fundamental beam. For this reason, high peak power lasers such as Q-switched Nd:YAG, modelocked Ti:sapphire lasers, and chirped parametric amplifier lasers, were employed as the fundamental light. Tunable optical sources are very desirable to conduct experiments such that the second harmonic light is not in the linear absorption band of the chromophore or to probe resonant behavior of NLO chromophores. Similar to the EFISHG technique, the HRS experiment is also a relative experiment which requires a known reference. This can be accomplished using an external or internal (solvent) reference (Hubbard *et al.*, 1998).

Since a very intense optical field is used, other nonlinear process such as multi-photon fluorescence occurs in many dyes across the visible spectrum and has an $(I^{\omega})^2$ dependence which competes with HRS. In some cases, fluorescence can be separated from HRS either temporally or spectrally (Olbrechts *et al.*, 1999; Wu *et al.*, 2008). However, when the fluorescence is largest near the two-photon absorption maximum of the molecule, the hyper-Rayleigh signal can be completely extinguished due to dephasing of the molecule's virtual excited state. A tunable band-pass filter, such as a grating monochromator, in front of the detector allows spectral isolation of the narrow hyper-Rayleigh line from broad-band fluorescence when dephasing is not too severe. High-intensity tunable sources such as optical parametric oscillator (OPO) and optical parametric amplifier (OPA) can be used to minimize fluorescence by conducting experiments in far-off resonant regimes.

Recently, a single photon detection technique was successfully used in an HRS experiment. For molecules that have nano-second fluorescence, time-correlated single photon HRS can easily differentiate the SHG and two-photon fluorescence (Wu *et al.*, 2008). In the KD HRS scheme, it has also shown that method of using a liquid crystal retarder to control the polarization of fundamental wave improves data reproducibility compared to conventional rotating quarter wave method (Wu *et al.*, 2008). It should be noted that it is possible that the fluorescence decay is very fast, so that multiphoton fluorescence and SHG occur on similar time scales, in which case temporal separation methods are not sufficient (Olbrechts *et al.*, 1999; Wu *et al.*, 2008). In this case, a combination of techniques is required (Wu *et al.*, 2008).

16.4.4 New applications

Since HRS is an instantaneous process, it can be useful for monitoring ultrafast processes. It is possible to attain temporal resolution at the same order of the pulse duration of the incident light using a pump-probe technique or similar optical methods. Moreover, because the HRS polarization factor is very sensitive to the local symmetry, it is a very powerful optical method for determining the structure and dynamics of the molecules or particles.

Self-assembly/aggregation of the molecules in equilibrium solution is important for understanding many biological phenomena such as the formation of protein. The size and symmetry of the formed supramolecular structure (e.g. the cluster of melamine, cyanuric acid, or sym-triphenyl cyanuric acid) can be detected effectively by measuring the HRS signal (Ray and Das, 1997). For example, the HRS technique was used to determine the critical micelle concentration (CMC) of surfactants that contain an NLO chromophore in solution (Ghosh *et al.*, 2003).

The NLO responses of metallic nanoparticles have been extensively investigated over the past years (Dadap *et al.*, 1999; 2004). Free electron in the conduction-band, or surface plasmas, are generated on the nanoparticles, where the free election model of linear and nonlinear index of refraction applies. Related linear optics can be explained by Mie scattering theory. Recently, a report showed that aggregation of metallic particles exhibits a strong NLO effect due to the quadrature contribution to the first hyperpolarizability. Zhang *et al.* (2003) have investigated by HRS the aggregation of protein-modified gold nanoparticles. Russier-Antoine *et al.* (2008a, 2008b) used a different protein aggregation method and further confirmed that HRS could become a potential powerful technique in determining small aggregation of proteins in aqueous samples. Aggregations of gold and Au-Ag alloy nano-particles were both studied. The enhancement of SHG is attributed to both the size and surface inhomogeneity of the nanoparticles. The non-uniformity breaks down the local inversion symmetry.

A new kind of fluid, ionic liquid, which is usually composed of asymmetric organic cations (usually, imidazolium, pyridinium, pyrrolidinium, or alkylammoniumion derivatives) with bulky inorganic anions, has attracted interest. Understanding the local structure and interaction is necessary for knowing the potential application of such liquids. HRS spectroscopy on such a system was recently reported (Rodriguez *et al.*, 2010). The HRS measurement results have provided significant physical insights into the local structure (orientation, cluster size, etc.) of the ionic liquids (Revillod *et al.*, 2009; Rodriguez *et al.*, 2010).

Inter-molecule charge transfer can possibly enhance the nonlinear response of the system because the change of the electronic structure of each individual molecule. Study of charge transfer between aromatic π donors (methyl substitute benzenes) and archetypical organic π acceptors was reported (Pandey *et al.*, 2011).

The HRS technique was also reported to be an effective method for determining the fluid partition coefficient (Ray and Das, 1996). A solute has different concentration in two immiscible liquids if the two liquids are combined. The ratio between the two concentrations tends to be constant. HRS can be used to determine this ratio by measuring the SHG signal in different liquids because SHG signal has a well-defined relationship with the concentration. HRS technique was also used to measure the ferroelectric molecular order in liquids and their relaxation time (Shelton, 2005, 2008), monitoring real-time nucleation and growth of nonparticles (Segets *et al.*, 2009), investigating the conformation of polymers (Vandendriessche *et al.*, 2009), etc.

16.5 Second harmonic generation (SHG) probing structure and dynamics

16.5.1 Molecular orientation at surfaces and interfaces

Because of the dielectric discontinuity, the inversion center is necessarily broken at surfaces and interfaces. As a consequence, the interfacial region between two uniform (centro-symmetric) media is expected to have an SHG signal. It has been theoretically shown by Heinz *et al.* (1982), that the SHG signal generated from the surface provides information on the molecular orientation at the surface (Heinz *et al.*, 1983). Since then, surface SHG has become an important technique for studying surfaces, interfaces, and surface-deposited materials (Shen, 2000; Simpson, 2001; Madden *et al.*, 2011).

One application has been for the study of liquid crystals deposited on surfaces. The orientation of molecules depends largely on the treatment of the substrate and is the extremely critical in liquid crystal devices. A variety of techniques (pretransitional optical birefringence, contact angle measurement, anchoring energy measurement, etc.) have been employed to study the liquid crystal alignment in different situations (surface treatments, external field, etc.). Optical SHG has been demonstrated to be a powerful method for probing the orientation distribution of the molecules adsorbed at an interface (Chen *et al.*, 1989; Marowsky *et al.*, 1990; Feller *et al.*, 1991). This technique was successfully applied to the study of liquid crystal monolayers at various interfaces. At the interface of the substrate and the liquid crystal, it is obvious that the symmetry is broken. Thus, SHG is allowed and determined by the macroscopic nonlinear susceptibilities of both the substrate and the molecules. The coherent SHG signal relates not only to the effective second-order susceptibility but also to the incident angle (Feller *et al.*, 1991). Nonlinear measurements performed on quartz substrate covered with nematic liquid crystal 5CB have been presented and the tilt angle of the molecules was determined from the low anisotropy and large birefringence of the observed SHG signal (Reznikov *et al.*, 2000). NLO ellipsometry and applications to microscopy provide a powerful tool for studying interfaces (Madden *et al.*, 2011).

16.5.2 Molecular orientation in the bulk

In the development of new second-order NLO materials, a great deal of work has involved the study and optimization of molecular orientation within the bulk NLO material. Crystal engineering optimization includes the crystal is grown with the constituent molecules oriented favorably for second-order NLO processes (Sutter *et al.*, 1988; Kondo *et al.*, 1989). On the other hand, SHG can also be a probe technique for studying the orientation and symmetry of the molecules in bulk materials. Studies of third-order NLO susceptibilities have been carried out on liquid crystals, and the relationship between molecular and bulk susceptibilities has been enumerated for some susceptibility components. Theoretical calculation of the second-order nonlinear susceptibility in unpoled liquid crystals has been conducted (Zhong-Can and Yu-Zhang, 1985). Symmetry breaking associated with molecular reorientations can dramatically increase in non-crystalline materials, such as liquid crystals and polymer glasses. The relationship between the second-order NLO susceptibility of the oriented bulk material and molecular constituents was derived for orientationally ordered materials belonging to the point group ∞mm (Singer *et al.*, 1987). The nonlinear susceptibility was shown to depend on microscopic order parameters for the isotropic system. All nonzero tensor components of χ_{ijk} can be related to components of the molecular hyperpolarizability and the microscopic order parameters corresponding to the statistical average of the first three even-order Legendre polynomials (Singer *et al.*, 1987). These oriented gas models have been subsequently extending to other types and orders of nonlinearity (Kuzyk, 1998) and to include the effects of anisotropy. The optical measurement technique used to determine orientation is the polarization dependence of the Maker fringe technique. This technique was applied to measure the molecular distribution of poled chromophores under stress (Kuzyk *et al.*, 1989).

16.5.3 Orientational dynamics

Poled polymer systems normally exhibit orientational relaxation which decreases the NLO susceptibility over time. The ultimate applicability of polymers for second-order nonlinear optics may depend on the ability to produce polymers whose orientation has sufficient stability for extended use in photonic and electronic systems.

In the poled polymer film, although the film is already in solid state when the external field is removed, the molecular orientation relaxes because it is in a metastable glass state. In order to understand the relaxation mechanisms, SHG has been used as a probe (Hampsch *et al.*, 1990). In fact, the relaxing SHG from such system can be used in studying the dynamics of the molecules (Singer and King, 1991; Sugihara *et al.*, 1996; Dureiko *et al.*, 1998; Herman and Cline, 1998). Early studies on the isothermal decay of the second-order nonlinear optical susceptibility of a guest–host azo–dye (disperse red 1) poly methy methacrylate (PMMA) film show that a common underlying process involving a distribution of local relaxations governs the film behavior during temperature changing (Singer and King, 1991).

The relaxation of orientation of chromophores in a polymer is closely related to the general relaxation mechanisms in the polymers, which have been the subject of much study using both mechanical and dielectric spectroscopy. Nonlinear optics provides another approach to these spectroscopies, which directly probes the orientational dynamics of the NLO chromophores by second harmonic generation. These measurements, in the frequency domain are called chielectric spectroscopy (Sugihara *et al.*, 1996; Dureiko *et al.*, 1998; Herman and Cline, 1998). Temperature-dependent chielectric spectra provide information on the relevant time and energy scales for relaxation processes and were found to be consistent with previously derived models describing relaxation in thermoplastic polymers. The time dependence of the isothermal decay is modeled by use of chromophore-reorientation models. The statistical physics is applied in deriving the temperature dependence of the chromophore-reorientation parameters (Dureiko *et al.*, 1998). In practice, the Dissado–Hill (D-H) model was chosen for the chromophore reorientation and the Adam–Gibbs model for the relaxation. The temperature dependence of the parameters that describe the chromophore reorientation is obtained through a fit to the isothermal time dependent data. Chielectric spectroscopy near the glass transition temperature of the host polymer has also been shown and agreement between theory and experiment was found (Dureiko *et al.*, 1998).

Photorefractive (PR) polymers are potentially interesting for a number of image processing and display technologies (Christenson *et al.*, 2010). A general requirement for a PR material is that the material must exhibit a linear electro-optic effect, that is, that the material lacks inversion symmetry (Moerner *et al.*, 1997). The figure of merit of PR material is the diffraction efficiency. The centrosymmetry of the polymer is broken through applying an electric field above the glass transition temperature of the polymer resulting electro-optic response. NLO chromophores will rotate inside the polymer liquid. The Debye rotational diffusion model was adapted to describe the reorientation of chromophores, which can be treated as a rigid dipolar molecule in an isotropic medium. Both the EFISHG and FWM (four-wave mixing) experimental data show that onset (i.e., external field turned on) in the diffraction efficiency has contributions from photoconductive (induces charge distribution, i.e. electric field) and molecular reorientation (breaks symmetry) (Ostroverkhova *et al.*, 2002). The EFISHG signal is quadratically proportional to the electric field. Thus, the average signal is sensitive to space charge distribution. Polymers with different glass transition temperature were studied with EFISHG technique (Ostroverkhova *et al.*, 2002).

16.5.4 Chirality (second harmonic ellipsometry)

In the mid-1990s, very large circular and linear chiral dichroic ratios in optical SHG measurements of chiral surface films were reported (Petralli-Mallow *et al.*, 1993; Kauranen *et al.*, 1994). A new nonlinear spectroscopy method, second harmonic ellipsometry, was developed based on this effect. Similar to the linear circular dichroism experiment, left- and right-circularly polarized light, in this case, intense fundamental laser pulses, is incident on the sample through a prism coupler. The SHG is collected at the direction of the specular reflection of the fundamental. Rich information of the structure of the chiral material and chiral molecules can be inferred with proper analysis of the SHG signal, such as including null ellipsometry, rotating quarter wave plate ellipsometry, and rotating halfwave plate ellipsometry (Plocinik *et al.*, 2005). Orientational effects were shown to be the dominant factor for generating large SHG chiral dichroic ratios (40% in magnitude) in several distinct uniaxial surface systems (Burke *et al.*, 2003). Compared with the traditional linear optical methods (including absorbance circular dichroism and optical rotary dispersion), the new NLO technique is several orders of magnitude more sensitive to chirality in oriented systems (Burke *et al.*, 2003). Second harmonic ellipsometry has been applied in probing surface anisotropy and biological imaging (Haupert and Simpson, 2009; Madden *et al.*, 2011).

16.5.5 Electron/charge dynamics

We know that second harmonic generation of the bulk material is affected by an electric field through EFISH. Such an electric field can be generated either outside or inside the material through charge redistribution. Therefore, the SHG can be used as a tool to study the internal electric field. Solid-state EFISHG is thus useful in measuring charge carriers in semiconductors (Miragliotta and Wickenden, 1996; Manaka *et al.*, 2007; Taguchi *et al.*, 2011). For example, in one study of organic solar cells, SHG in specular reflection was used as a probe of photogenerated carriers. Analysis from the data suggests that the SHG is actually induced from the internal electric field (Taguchi *et al.*, 2011). Therefore, this method provides a direct way to investigate the electric field, i.e. carrier distribution, inside the organic solar cells. The time-resolved SHG technique was also used to measure the relaxation rate of the hot-electron transfer from lead selenide (PbSe) nanocrystals to a titanium dioxide (TiO_2) electron acceptor (Tisdale *et al.*, 2010). A pump light generates charge carriers and a probe beam (fundamental) is used to study the $\chi^{(2)}$ change by looking at the SHG. The same experimental technique was also used to study multiexciton generation in pentacene (Chan *et al.*, 2011).

16.6 Conclusion

Second harmonic generation, the first nonlinear optical effect discovered after the invention of Q-switched lasers, remains an important technique for tunable lasers and as a tool for understanding materials. Second harmonic generation techniques are still widely used in the development of new second order nonlinear optical materials and will continue to be as long as new nonlinear optical materials are being sought. The requirement for an absence of inversion center is at the core of the power of second harmonic techniques for studying materials. Because of this, we expect it to continue to be a powerful technique for biological membrane and material interface studies, especially imaging and nonlinear ellipsometry. In addition, EFISHG has been shown to be a powerful method for probing electric fields surrounding charges in electronic materials. Thus, we expect that EFISHG both in the imaging and dynamics modalities will become increasingly important tools for studying semiconductors and devices in the future.

16.7 Acknowledgments

The authors wish to acknowledge funding from the National Science Foundation Center for Layered Polymer Systems under grant number DMR-0423914.

16.8 References

Boyd, R. W. (2003) *Nonlinear optics* (2nd Edition), New York, Academic Press.

Burke, B. J., Moad, A. J., Polizzi, M. A. & Simpson, G. J. (2003) Experimental confirmation of the importance of orientation in the anomalous chiral sensitivity of second harmonic generation. *J. Am. Chem. Soc.*, **125**, 9111.

Burland, D. M., Walsh, C. A., Kajzar, E. & Sentein, C. (1991) Comparison of hyperpolarizabilities obtained with different experimental methods and theoretical techniques. *J. Opt. Soc. Am. B*, **8**, 2269.

Chan, W.-L., Ligges, M., Jailaubekov, A., Kaake, L., Miaja-Avila, L. & Zhu, X. Y. (2011) Observing the multiexciton state in singlet fission and ensuing ultrafast multielectron transfer. *Science*, **334**, 1541–1545.

Chemla, D. & Kupecek, P. (1971) Analyse des experiences de generation de second harmonique. *Rev. Phys. Appl. (Paris)*, **6**, 31–50.

Chen, W., Feller, M. B. & Shen, Y. R. (1989) Investigation of anisotropic molecular orientational distributions of liquid-crystal monolayers by optical second-harmonic generation. *Physical Rev. Lett.*, **63**, 2665.

Cheng, L. T., Tam, W., Stevenson, S. H., Meredith, G. R., Rikken, G. & Marder, S. R. (1991) Experimental investigations of organic molecular nonlinear optical polarizabilities. 1. Methods and results on benzene and stilbene derivatives. *J. Phys. Chem.*, **95**, 10631.

Christenson, C. W., Blanche, P.-A., Tay, S., Voorakaranam, R., Gu, T., Lin, W., Wang, P., Yamamoto, M., Thomas, J., Norwood, R. A. & Peyghambarian, N. (2010) Materials for an updatable holographic 3D display. *J. Display Technol.*, **6**, 510.

Clays, K., Persoons, A. & Eacute (1991) Hyper-Rayleigh scattering in solution. *Phys. Rev. Lett.*, **66**, 2980.

Cyvin, S. J., Rauch, J. E. & Decius, J. C. (1965) Theory of hyper-raman effects (nonlinear inelastic light scattering): selection rules and depolarization ratios for the second-order polarizability. *J. Chem. Phys.*, **43**, 4083.

Dadap, J. I., Shan, J., Eisenthal, K. B. & Heinz, T. F. (1999) Second-harmonic rayleigh scattering from a sphere of centrosymmetric material. *Phys. Rev. Lett.*, **83**, 4045.

Dadap, J. I., Shan, J. & Heinz, T. F. (2004) Theory of optical second-harmonic generation from a sphere of centrosymmetric material: small-particle limit. *J. Opt. Soc. Am. B*, **21**, 1328.

Dalton, L. R., Sullivan, P. A. & Bale, D. H. (2010) Electric field poled organic electro-optic materials: state of the art and future prospects. *Chem. Rev.*, **110**, 25.

Dureiko, R. D., Schuele, D. E. & Singer, K. D. (1998) Modeling relaxation processes in poled electro-optic polymer films. *J. Opt. Soc. Am. B*, **15**, 338.

Feller, M. B., Chen, W. & Shen, Y. R. (1991) Investigation of surface-induced alignment of liquid-crystal molecules by optical second-harmonic generation. *Phys. Rev. A*, **43**, 6778.

Finn, R. S. & Ward, J. F. (1974) Measurements of hyperpolarizabilities for some halogenated methanes. *J. Chem. Phys.*, **60**, 454.

Franken, P. A., Hill, A. E., Peters, C. W. & Weinreich, G. (1961) Generation of optical harmonics. *Phys. Rev. Lett.*, **7**, 118.

Ghosh, S., Krishnan, A., Das, P. K. & Ramakrishnan, S. (2003) Determination of critical micelle concentration by hyper-rayleigh scattering. *J. Am. Chem. Soc.*, **125**, 1602.

Hampsch, H. L., Yang, J., Wong, G. K. & Torkelson, J. M. (1990) Dopant orientation dynamics in doped second-order nonlinear optical amorphous polymers. 1. Effects of temperature above and below T_g in corona-poled films. *Macromolecules*, **23**, 3640.

Haupert, L. M. & Simpson, G. J. (2009) Chirality in nonlinear optics. *Ann. Rev. Phys. Chem.*, **60**, 345–365.

Heinz, T. F., Chen, C. K., Ricard, D. & Shen, Y. R. (1982) Spectroscopy of molecular monolayers by resonant second-harmonic generation. *Phys. Rev. Lett.*, **48**, 478.

Heinz, T. F., Tom, H. W. K. & Shen, Y. R. (1983) Determination of molecular orientation of monolayer adsorbates by optical second-harmonic generation. *Phys. Rev. A*, **28**, 1883.

Herman, W. N. & Hayden, L. M. (1995) Maker fringes revisited: second-harmonic generation from birefringent or absorbing materials. *J. Opt. Soc. Am. B*, **12**, 416.

Herman, W. N. & Cline, J. A. (1998) Chielectric relaxation: chromophore dynamics in an azo-dye-doped polymer. *J. Opt. Soc. Am. B*, **15**, 351.

Hubbard, S. F., Petschek, R. G. & Singer, K. D. (1996) Spectral content and dispersion of hyper-Rayleigh scattering. *Opt. Lett.*, **21**, 1774.

Hubbard, S. F., Petschek, R. G., Singer, K. D., Dsidocky, N., Hudson, C., Chien, L. C., Henderson, C. C. & Cahill, P. A. (1998) Measurements of Kleinman-disallowed hyperpolarizability in conjugated chiral molecules. *J. Opt. Soc. Am. B*, **15**, 289.

Jerphagnon, J. & Kurtz, S. K. (1970) Maker fringes: a detailed comparison of theory and experiment for isotropic and uniaxial crystals. *J. Appl. Phys.*, **41**, 1667.

Kajzar, F., Ledoux, I. & Zyss, J. (1987) Electric-field-induced optical second-harmonic generation in polydiacetylene solutions. *Physical Rev. A*, **36**, 2210.

Kauranen, M., Verbiest, T., Maki, J. J. & Persoons, A. (1994) Second-harmonic generation from chiral surfaces. *J. Chem. Phys.*, **101**, 8193.

Kondo, T., Morita, R., Ogasawara, N., Umegaki, S. & Ito, R. (1989) A nonlinear optical organic crystal for waveguiding SHG devices: (-)2-(α-methylbenzylamino)-5-nitropyridine (MBANP). *Jap. J. Appl. Phys.*, **28**, 1622.

Kurtz, S. (1975) *Quantum electronics*, Academic Press.

Kuzyk, M. G. (1998) *Relationship between the molecular and bulk response*, Marcel Dekker.

Kuzyk, M. G., Singer, K. D., Zahn, H. E. & King, L. A. (1989) Second-order nonlinear-optical tensor properties of poled films under stress. *J. Opt. Soc. Am. B*, **6**, 742.

Levine, B. F. & Bethea, C. G. (1975) Second and third order hyperpolarizabilities of organic molecules. *J. Chem. Phys.*, **63**, 2666.

Lim, E., Shibata, Y., Manaka, T. & Iwamoto, M. (2009) Probing the electric field in organic double layer-system by optical second harmonic generation. *Thin Solid Films*, **518**, 893.

Madden, J. T., Hall, V. J. & Simpson, G. J. (2011) Mining the polarization-dependence of nonlinear optical measurements. *Analyst*, **136**, 652.

Maker, P. D., Terhune, R. W., Nisenoff, M. & Savage, C. M. (1962) Effects of dispersion and focusing on the production of optical harmonics. *Phys. Rev. Lett.*, **8**, 21.

Manaka, T., Kohn, H., Ohshima, Y., Lim, E. & Iwamoto, M. (2007) Direct observation of trapped carriers in polydiacetylene films by optical second harmonic generation. *Appl. Phys. Lett.*, **90**, 171119.

Mao, G., Wu, Y. & Singer, K. D. (2007) Third harmonic generation in self-focused filaments in liquids. *Opt. Express*, **15**, 4857.

Marowsky, G., Lüpke, G., Steinhoff, R., Chi, L. F. & Möbius, D. (1990) Anisotropic second-order nonlinearities of organic monolayers. *Phys. Rev. B*, **41**, 4480.

Miragliotta, J. & Wickenden, D. K. (1996) Nonlinear electroreflectance from gallium nitride using optical second-harmonic generation. *Phys. Rev. B*, **53**, 1388.

Moerner, W. E., Grunnet-Jepsen, A. & Thompson, C. L. (1997) Photorefractive polymers. *Ann. Rev. Mater. Sci.*, **27**, 585–623.

Olbrechts, G., Wostyn, K., Clays, K. & Persoons, A. (1999) High-frequency demodulation of multiphoton fluorescence in long-wavelength hyper-Rayleigh scattering. *Opt. Lett.*, **24**, 403.

Ostroverkhov, V., Ostroverkhova, O., Petschek, R. G., Singer, K. D., Sukhomlinova, L., Twieg, R. J., Wang, S. X. & Chien, L. C. (2000a) Optimization of the molecular hyperpolarizability for second harmonic generation in chiral media. *Chem. Phys.*, **257**, 263.

Ostroverkhov, V., Petschek, R. G., Singer, K. D., Sukhomlinova, L., Twieg, R. J., Wang, S. X. & Chien, L. C. (2000b) Measurements of the hyperpolarizability tensor by means of hyper-Rayleigh scattering. *J. Opt. Soc. Am. B*, **17**, 1531.

Ostroverkhov, V., Ostroverkhova, O., Petschek, R. G., Singer, K. D., Sukhomlinova, L. & Twieg, R. J. (2001a) Prospects for chiral nonlinear optical media. *IEEE J. Selected Topics in Quantum Electron.*, **7**, 781.

Ostroverkhov, V., Petschek, R. G., Singer, K. D. & Twieg, R. J. (2001b) Λ-like chromophores for chiral non-linear optical materials. *Chem. Phys. Lett.*, **340**, 109.

Ostroverkhov, V., Singer, K. D. & Petschek, R. G. (2001c) Second-harmonic generation in nonpolar chiral materials: relationship between molecular and macroscopic properties. *J. Opt. Soc. Am. B-Opt. Phys.*, **18**, 1858–1865.

Ostroverkhova, O., Stickrath, A. & Singer, K. D. (2002) Electric field-induced second harmonic generation studies of chromophore orientational dynamics in photorefractive polymers. *J. Appl. Phys.*, **91**, 9481.

Oudar, J. L. (1977) Optical nonlinearities of conjugated molecules. Stilbene derivatives and highly polar aromatic compounds. *J. Chem. Phys.*, **67**, 446.

Pandey, R., Ghosh, S., Mukhopadhyay, S., Ramasesha, S. & Das, P. K. (2011) Geometry and quadratic nonlinearity of charge transfer complexes in solution using depolarized hyper-Rayleigh scattering. *J. Chem. Phys.*, **134**, 044533.

Park, D. H. & Herman, W. N. (2012) Closed-form Maker fringe formulas for poled polymer thin films in multilayer structures. *Opt. Express*, **20**, 173.

Petralli-Mallow, T., Wong, T. M., Byers, J. D., Yee, H. I. & Hicks, J. M. (1993) Circular dichroism spectroscopy at interfaces: a surface second harmonic generation study. *J. Phys. Chem.*, **97**, 1383.

Plocinik, R. M., Everly, R. M., Moad, A. J. & Simpson, G. J. (2005) Modular ellipsometric approach for mining structural information from nonlinear optical polarization analysis. *Phys. Rev. B*, **72**, 125409.

Ray, P. C. & Das, P. K. (1996) Measurement of partition coefficients of substituted benzoic acids between two immiscible solvents by hyper-rayleigh scattering. *J. Phys. Chem.*, **100**, 15631.

Ray, P. C. & Das, P. K. (1997) Probing molecular self-assembly by hyper-Rayleigh scattering in solution. *Chem. Phys. Lett.*, **281**, 243.

Revillod, G., Nishi, N. & Kakiuchi, T. (2009) Orientation correlation of sulfosuccinate-based room-temperature ionic liquids studied by polarization-resolved hyper-rayleigh scattering. *J. Phys. Chem. B*, **113**, 15322.

Reznikov, Y., Ostroverkhova, O., Singer, K. D., Kim, J.-H., Kumar, S., Lavrentovich, O., Wang, B. & West, J. L. (2000) Photoalignment of liquid crystals by liquid crystals. *Phys. Rev. Lett.*, **84**, 1930.

Rodriguez, V., Grondin, J., Adamietz, F. & Danten, Y. (2010) Local structure in ionic liquids investigated by hyper-rayleigh scattering. *J. Phys. Chem. B*, **114**, 15057.

Russier-Antoine, I., Bachelier, G., Sabonière, V., Duboisset, J., Benichou, E., Jonin, C., Bertorelle, F. & Brevet, P.-F. (2008a) Surface heterogeneity in Au-Ag nanoparticles probed by hyper-Rayleigh scattering. *Phys. Rev. B*, **78**, 035436.

Russier-Antoine, I., Huang, J., Benichou, E., Bachelier, G., Jonin, C. & Brevet, P.-F. (2008b) Hyper Rayleigh scattering of protein-mediated gold nanoparticles aggregates. *Chem. Phys. Lett.*, **450**, 345.

Sanguinet, L., Twieg, R. J., Wiggers, G., Mao, G., Singer, K. D. & Petschek, R. G. (2005) Synthesis and spectral characterization of bisnaphthylmethyl and trinaphthylmethyl cations. *Tetrahedron Lett.*, **46**, 5121.

Sanguinet, L., Williams, J. C., Yang, Z., Twieg, R. J., Mao, G., Singer, K. D., Wiggers, G. & Petschek, R. G. (2006) Synthesis and characterization of new truxenones for nonlinear optical applications. *Chem. Mater.*, **18**, 4259.

Satou, H., Ohshima, Y., Kohn, H., Manaka, T. & Iwamoto, M. (2011) Direct observation of space charge field in tetracene field-effect transistor using time-resolved microscopic optical second harmonic generation. *J. Appl. Phys.*, **109**, 054506.

Segets, D., Martinez Tomalino, L., Gradl, J. & Peukert, W. (2009) Real-time monitoring of the nucleation and growth of ZnO nanoparticles using an optical hyper-rayleigh scattering method. *J. Phys. Chem. C*, **113**, 11995.

Shelton, D. P. (2005) Slow polarization relaxation in water observed by hyper-Rayleigh scattering. *Phys. Rev. B*, **72**, 020201.

Shelton, D. P. (2008) Polar domain fluctuations in doped liquid nitrobenzene. *J. Chem. Phys.*, **129**, 134501.

Shelton, D. P. (2010) Nonlocal hyper-Rayleigh scattering from liquid nitrobenzene. *J. Chem. Phys.*, **132**, 154506.

Shen, Y. R. (2000) Surface nonlinear optics: a historical perspective. *IEEE J. Selected Topics Quantum Electron.*, **6**, 1375.

Simpson, G. J. (2001) New tools for surface second-harmonic generation. *Appl. Spectrosc.*, **55**, 16A.

Singer, K. D. & Garito, A. F. (1981) Measurements of molecular second order optical susceptibilities using dc induced second harmonic generation. *J. Chem. Phys.*, **75**, 3572.

Singer, K. D. & King, L. A. (1991) Relaxation phenomena in polymer nonlinear optical materials. *J. Appl. Phys.*, **70**, 3251.

Singer, K. D., Sohn, J. E. & Lalama, S. J. (1986) Second harmonic generation in poled polymer films. *Appl. Phys. Lett.*, **49**, 248.

Singer, K. D., Kuzyk, M. G. & Sohn, J. E. (1987) Second-order nonlinear-optical processes in orientationally ordered materials: relationship between molecular and macroscopic properties. *J. Opt. Soc. Am. B*, **4**, 968.

Singer, K. D., Hubbard, S. F., Schober, A., Hayden, L. M. & Johnson, K. (1998) *Second harmonic susceptibility tables and measurement techniques*, Marcel Dekker.

Sioncke, S., Verbiest, T. & Persoons, A. (2003) Second-order nonlinear optical properties of chiral materials. *Mater. Sci. Eng.: R: Reports*, **42**, 115.

Sugihara, T., Haga, H. & Yamamoto, S. (1996) Electric field response of second order optical nonlinearity in dye doped poled polymer. *Appl. Phys. Lett.*, **68**, 144.

Sutter, K., Bosshard, C., Wang, W. S., Surmely, G. & Gunter, P. (1988) Linear and nonlinear optical properties of 2-(N-prolinol)-5-nitropyridine. *Appl. Phys. Lett.*, **53**, 1779.

Taguchi, D., Shino, T., Chen, X., Zhang, L., Li, J., Weis, M., Manaka, T. & Iwamoto, M. (2011) Analyzing carrier lifetime of double-layer organic solar cells by using optical electric-field-induced second-harmonic generation measurement. *Appl. Phys. Lett.*, **98**, 133507.

Teng, C. C. & Garito, A. F. (1983) Dispersion of the nonlinear second-order optical susceptibility of organic systems. *Phys. Rev. B*, **28**, 6766.

Tisdale, W. A., Williams, K. J., Timp, B. A., Norris, D. J., Aydil, E. S. & Zhu, X. Y. (2010) Hot-electron transfer from semiconductor nanocrystals. *Science*, **328**, 1543–1547.

Vandendriessche, A., Asselberghs, I., Clays, K., Smet, M., Dehaen, W., Verbiest, T. & Koeckelberghs, G. (2009) Investigation of the conformation of hyperbranched poly(arylene oxindole)s using hyper-Rayleigh scattering. *J. Polym. Sci. Part A: Polym. Chem.*, **47**, 3740.

Verbiest, T., Clays, K., Samyn, C., Wolff, J., Reinhoudt, D. & Persoons, A. (1994) Investigations of the hyperpolarizability in organic molecules from dipolar to octopolar systems. *J. Am. Chem. Soc.*, **116**, 9320.

Ward, J. F. & Bigio, I. J. (1975) Molecular second- and third-order polarizabilities from measurements of second-harmonic generation in gases. *Phys. Rev. A*, **11**, 60.

Williams, D. J. (1987) *Nonlinear optical properties of guest-host polymer structures*, Academic Press.

Wu, Y., Mao, G., Li, H., Petschek, R. G. & Singer, K. D. (2008) Control of multiphoton excited emission and phase retardation in Kleinman-disallowed hyper-Rayleigh scattering measurements. *J. Opt. Soc. Am. B*, **25**, 495.

Xiao, D., Ramsay, E., Reid, D. T., Offenbeck, B. & Weber, N. (2006) Optical probing of a silicon integrated circuit using electric-field-induced second-harmonic generation. *Appl. Phys. Lett.*, **88**, 114107.

Zhang, C. X., Zhang, Y., Wang, X., Tang, Z. M. & Lu, Z. H. (2003) Hyper-Rayleigh scattering of protein-modified gold nanoparticles. *Anal. Biochem.*, **320**, 136.

Zhong-Can, O.-Y. & Yu-Zhang, X. (1985) Theory of second-harmonic generation in nematic liquid crystals. *Phys. Rev. A*, **32**, 1189.

Zyss, J. (1993) Molecular engineering implications of rotational invariance in quadratic nonlinear optics: from dipolar to octupolar molecules and materials. *J. Chem. Phys.*, **98**, 6583.

Zyss, J. (1994) *Molecular nonlinear optics: materials, physics, and devices*, Academic Press.

Zyss, J. & Ledoux, I. (1994) Nonlinear optics in multipolar media: theory and experiments. *Chem. Rev.*, **94**, 77.

Part III

Applications of (opto)electronic and nonlinear optical organic materials in devices

17
Organic solar cells (OSCs)

M. HÖSEL, D. ANGMO and F. C. KREBS,
Technical University of Denmark, Denmark

DOI: 10.1533/9780857098764.3.473

Abstract: The organic solar cell (OSC) is the only solar cell technology that fully addresses the challenges of manufacturing on a scale corresponding to an energy production capacity of 1 GW_{peak} per day based solely on abundant elements. This ability arises through the use of printing techniques and photoactive layers comprising carbon-based conjugated polymer materials. The technology is briefly reviewed in this chapter by describing the common device structures and materials. The second half of the chapter gives an overview on suitable roll-to-roll (R2R) processing techniques for the fabrication of OSCs. Finally some demonstration projects and future trends are illustrated.

Key words: organic solar cell (OSC), roll-to-roll (R2R) processing, inks, materials, polymer solar cells, devices.

17.1 Introduction

Organic solar cells (OSCs) are one of the emerging photovoltaic (PV) technologies and are classified as third-generation solar cells with organic polymer material as the light absorbing layer. This chapter gives a brief review of OSCs with a short classification of the application. In addition, the working principles and device structures are described. The main organic materials in the active layer of the photovoltaic devices are summarized as well. Solubility of the organic material is the key advantage in the production of OSCs. Solution processing by roll-to-roll (R2R) techniques under ambient conditions makes it possible to fabricate OSCs at low cost and in large quantities. The first part of this chapter gives an overview about OSCs, how they work and which materials are used. The second part is focused on the production of OSCs and shows demonstration projects and future trends as well.

17.2 Organic solar cells (OSCs)

The Earth receives more energy from the sun in one hour than humankind uses in one year. On a cloudless day, 1000 W of solar radiation per square meter reaches the surface of the Earth according to the ASTM G173-03 standard reference for the AM1.5G spectrum (ASTM Standard G173 2008). The utilization of this reliable and renewable energy source is one opportunity to overcome the present and future human energy demand. PV devices can convert solar radiation directly into electrical energy and the most common known systems are mono- and polycrystalline PN-junction silicon solar cells with efficiencies up to 25% on laboratory scale (Green *et al.* 2011). Despite these first-generation PV systems, so-called thin film solar cells or second-generation PV emerged with amorphous silicon, cadmium telluride (CdTe) or copper indium-gallium diselenide (CIGS) as major material systems, with best efficiencies of 19.6% for CIGS (Green *et al.* 2011). The newest third-generation solar cells are characterized by their main material component based on organic matter. The various technologies are dye-sensitized solar cells (DSSC or Grätzel cell), small molecule organic solar cells prepared by evaporation under vacuum and polymer solar cells. The latter is often called organic solar cell as well, but you have to keep in mind that polymeric material is part of the light-absorbing layer and can be prepared by solution processing. The focus of this chapter is on polymeric OSCs, which are characterized by their simple, fast, low-cost and high-volume processability.

Today's polymer solar cells with record efficiencies of 8.3% (Green *et al.* 2011) are based on a bulk heterojunction of polymer and carbon fullerene derivatives. The history of OSCs goes back to 1986 with first reports on small molecule heterojunction PV (Tang 1986), followed in 1991 with the first dye/dye bulk heterojunction PV by Hiramoto and later in 1993, the first polymer-C_{60} heterojunction PV was made (Sariciftci *et al.* 1993). A brief history on the development organic PV was published by Spanggaard & Krebs (2004) and numerous reviews, progress reports and books have been published that cover the broad spectrum and development of OSCs (Brabec *et al.* 2001; Coakley & McGehee 2004; Hoppe & Sariciftci 2004; Dennler & Sariciftci 2005; Brabec *et al.* 2008; Krebs 2008, 2010; Thompson & Frechet 2008; Deibel & Dyakonov 2010; Helgesen *et al.* 2010; Po *et al.* 2010; Nelson 2011; Servaites *et al.* 2011; Zhang *et al.* 2011).

The advantage of OSCs based on polymers is the possibility of solution processing using coating and printing techniques (Krebs 2009b; Søndergaard *et al.* 2012) such as slot-die coating (Krebs *et al.* 2010c), screen (Krebs *et al.* 2009a), gravure (Kopola *et al.* 2011) and inkjet printing (Eom *et al.* 2010) on flexible substrate. The fabrication with vacuum-free R2R processes on flexible substrates under ambient conditions is a major subject of this chapter which will lower the cost of production dramatically. Modules costs of less than 1 €/W_p (watt peak) are expected (Nielsen *et al.* 2010). The main disadvantages of solution processed OSCs are still present and can be seen in low power conversion efficiency (PCE) and limited operational stability compared to silicon solar cells. As already written, the efficiency of lab-scale devices reached 8.3% but modules are still only at 3.5% (Green *et al.* 2011). Theoretical limits for single junction cells are in the range of 10% and 15% for tandem (multi-junction) solar cells (Koster *et al.* 2006; Scharber *et al.* 2006; Dennler *et al.* 2008). The published lifetimes of OSCs are in excess of 1000 h including reports of more than one year of outdoor stability (Krebs & Spanggaard 2005; Hauch *et al.* 2008; Zimmermann *et al.* 2009; Voroshazi *et al.* 2011). Achieving combination of large-scale processing, efficiency, and stability for the same material is the main challenge for OSCs to finally leave the lab and niche markets for industrialization.

17.3 Working principle and device structures

The field of solution-processed OSCs covers various types of polymer : acceptor material systems and device structures where the acceptor part can be fullerenes, polymers, semiconductor nanoparticles, or metal oxides. The latter are often named hybrid solar cells. In this chapter we only focus on the widely studied polymer : fullerene material system based on the bulk heterojunction concept. In short, the soluble donor and acceptor material is mixed in an organic solvent and deposited on the substrate with conductive layers. After evaporation of the solvent and post-treatment steps, an interpenetrated network is formed by microphase separation. The large interface area between donor and acceptor is important for charge separation after the exciton generation due to illumination. The interconnected domains with continuous paths to the electrodes allow efficient charge transport to anode and cathode. Material combination and generation of a stable network configuration with several treatment processes is critical for increasing the efficiency of OSCs.

The working principle of an OSC (Blom *et al.* 2007; Deibel & Dyakonov 2010) can be described in four fundamental steps and is illustrated in Fig. 17.1 for bulk heterojunction devices:

1. *Exciton generation*: Upon illumination of the active material, a photon is absorbed and promotes the electron to the lowest unoccupied molecular orbital (LUMO). The same time, a positive charge carrier, or so-called hole, stays in the highest occupied molecular orbital (HOMO). Both charge carriers are attracted to each other and bound by Coulomb forces forming an exciton, also referred to as electron–hole pair.
2. *Exciton diffusion*: The generated exciton diffuses inside the donor phase to the interface of the donor and acceptor material. Exciton decay or charge carrier recombination can appear if they are generated too far from the interface. The BHJ concept of two intermixed materials decreases the diffusion length compared to a stacked bilayer structure

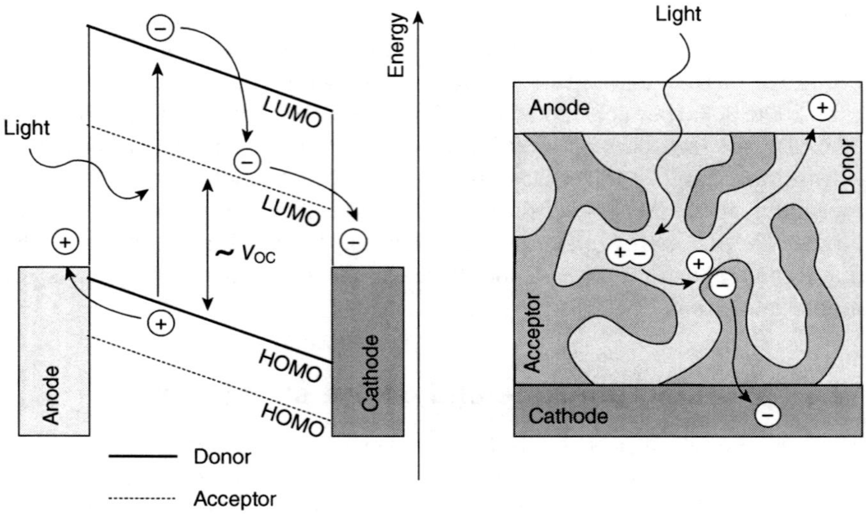

17.1 Left: Simplified band diagram of photocurrent generation due to photon absorption, exciton generation, exciton diffusion, exciton dissociation and charge carrier transport to the electrodes in a BHJ solar cell. Right: corresponding BHJ structure with charge carrier transportation paths.

and reduces the decay rate of the exciton. Ideally, the dimensions of the two phases should be in the range or smaller than the diffusion length of 4–20 nm (Halls *et al.* 1996; Haugeneder *et al.* 1999; Pettersson *et al.* 1999; Piris *et al.* 2009).

3. *Exciton dissociation*: The exciton dissociates into a free electron and hole at the interface of donor and acceptor material.
4. *Charge carrier transport*: The free charge carriers are separated by an internal electric field caused by electrodes with different work functions. They are transported through the donor and acceptor material, where the electrons are collected at the cathode and the holes at the anode. The photocurrent is generated by short circuiting or applying a load to an external circuit.

Each step is important for efficient power generation. Hereby, the difference between HOMO and LUMO, or optical band gap is a crucial parameter. Organic materials with a large absorption range (low band gap) can be synthesized and do directly influence the first step of the working principle. The microphase separation can be largely affected by processing parameters.

Fig. 17.2 shows the current density–voltage characteristics (J-V curve) for a typical solar cell with the key parameters, which are the open-circuit

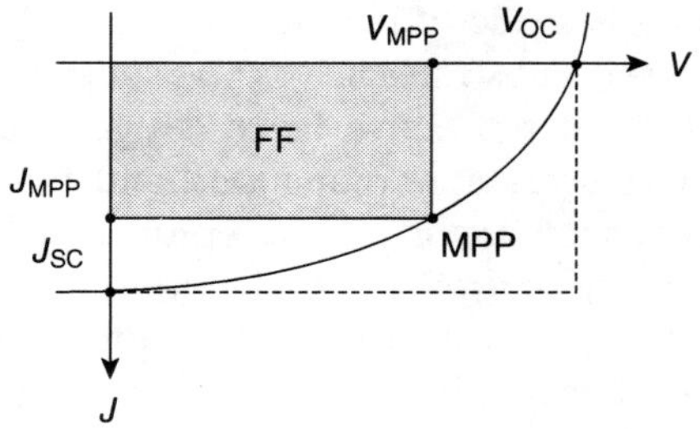

17.2 Characteristic *J-V* curve of an OSC with its open circuit voltage V_{oc} in V, short circuit current density J_{sc} in mA/cm^2 and fill factor FF in %.

voltage V_{oc}, short-circuit current density J_{sc}, the fill factor FF and the maximum power point MPP.

The power conversion efficiency (PCE, η) is the most important parameter and shows the ratio of the maximum electrical power (= $I_{MPP} \cdot V_{MPP}$) produced by the cell and the power of the incident light P_{in} on a given active area A:

$$\eta = \frac{I_{MPP} \cdot V_{MPP}}{P_{in} \cdot A} = FF \frac{I_{SC} \cdot V_{OC}}{P_{in} \cdot A} = \mathrm{FF} \frac{J_{SC} \cdot V_{OC}}{P_{in}}$$

where FF is defined as:

$$\mathrm{FF} = \frac{I_{MPP} \cdot V_{MPP}}{I_{SC} \cdot V_{OC}} = \frac{J_{MPP} \cdot V_{MPP}}{J_{SC} \cdot V_{OC}}$$

The FF is a value for the practical produced power compared to the theoretically possible value and should be in the range of 60–65% for a good performing organic solar cell. It is highly influenced by parasitic resistances R_s and R_{sh}. The series resistance R_s includes all the resistances at the interfaces in the layers, the conductivity of the semiconductors and the electrodes. It needs to be low for a good performance of the device. The shunt resistance R_{sh} needs to be high and includes all the leakage of current through shunts as a result of defects in the layers. Shunting should be avoided for a highly efficient cell.

The quantum efficiency (QE) is another parameter for the energy conversion performance in solar cells. The external quantum efficiency (EQE) defines the ratio of collected charge carriers at the electrodes and the number of incident photons on the solar cell at a given wavelength. The ratio of the collected charges carriers to all absorbed photons at a given wavelength is defined as internal quantum efficiency (IQE).

The design of an OSC comprises a layer stack of two electrodes with a sandwiched photoactive layer in between. At least one of the electrodes must be transparent and is typically made of indium tin oxide (ITO) sputtered or evaporated on a transparent substrate such as glass or polyethylene terephthalate (PET). The history of the active layer design began with single layer, then double layer (bilayer heterojunction), and today intermixed layers (bulk heterojunction of donor and acceptor, Fig. 17.1) are widely used. The latter one largely improves the charge separation and transport by its nanoscale morphology in the range of the diffusion length of the excitons. Recombination losses of the charge carriers are successfully decreased. The design of structures with pillars in the distance and size of the diffusion length is the next step in optimizing the efficiency of OSCs (Kim *et al.* 2011). An additional buffer layer between the active layer and the electrodes serve as charge selective transport layer, either blocking holes or electrons.

Two reference device geometries, namely normal and inverted geometry, are actually used to build OSCs in lab-scale size and full solution processing. Fig. 17.3 illustrates the layer stacks with the most common materials, whereas their specific function is summarized in Table 17.1. The normal geometry has been used as the traditional device structure for a long time and is known for relative efficiencies and its relatively simple production (Li *et al.* 2005; Ma *et al.* 2005). One drawback is the use of vacuum steps to evaporate the cathode electrode on top of the active layer and its limited operational lifetime. Flipping the layer stack and introducing a charge transport layer allows full solution processing of the anode. This inverted structure avoids vacuum steps, except for ITO, and has been proved a better choice for large-scale processing (Krebs *et al.* 2009b). Therefore this chapter focuses more on the inverted geometry and its fabrication.

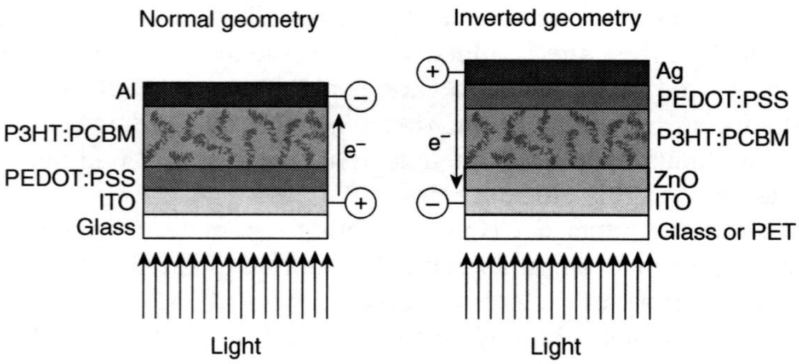

17.3 OSC devices with a bulk heterojunction active layer in normal and inverted geometry. In the normal geometry the transparent ITO electrode acts as anode, whereas it is a cathode in the inverted structure.

Table 17.1 The function of common materials in normal and inverted device geometries. The materials in **bold** are widely used standard combinations for each geometry

Normal geometry					
Substrate	Anode	Hole transport layer	Active layer	Electron transport layer	Cathode
Glass	**ITO**	**PEDOT:PSS**	**P3HT:PCBM**	TiO_x	LiF / **Al**
PET		MoO_3		ZnO	LiF / Au
		V_2O_5			
Inverted geometry					
Substrate	Cathode	Electron transport layer	Active layer	Hole transport layer	Anode
Glass	**ITO**	**ZnO**	**P3HT:PCBM**	**PEDOT:PSS**	**Ag**
PET	Ag-solid	TiO_x			Ag-grid
PEN	Al/Cr	Cs_2CO_3			

PET = polyethylene terephthalate; PEN = polyethylene naphthalate; ITO = indium tin oxide;
PEDOT:PSS = poly(3,4-ethylenedioxythiophene):poly(styrenesulfonate); MoO_3 = molybdenum trioxide; V_2O_5 = vanadium pentoxide;
P3HT = poly(3-hexylthiophene); PCBM = [6,6]-phenyl C61 butyric acid methyl ester (fullerene derivate);
TiO_x = titanium oxide; ZnO = zinc oxide; LiF = lithium fluoride; Cs_2CO_3 = cesium carbonate;
Al = aluminum; Au = gold; Ag = silver; Cr = chromium.

17.4 Materials

The active photoconversion layer and the hole transport layer PEDOT:PSS are in principle the only organic layers in an OSC. The active layer can be polymer-based, small molecule-based or a hybrid organic–inorganic structure. All other layers, except the substrate, are metals or metal oxides. Here we briefly describe the several layer materials and focus on the organic polymer-based photoactive layer at the end.

17.4.1 Substrate and front electrode

Glass or polymeric materials such as polyethylene terephthalate (PET) or polyethylene naphthalate (PEN) are the basic substrates to built on the subsequent layer structure. PET or PEN is thin and flexible and makes it the first choice in large-scale R2R processing. Indium tin oxide (ITO) is widely used as a transparent electrode on glass or PET because of its excellent properties as a hole conductor. The drawback is the price and scarcity of indium. In addition, it uses vacuum-based processes for deposition, which shows up in an huge embodied energy of more than 80% in the final device (Espinosa *et al.* 2011). Avoiding indium and finding alternative transparent conducting electrodes is highly demanding. One promising approach without using vacuum steps is the use of printed silver grids in combination with highly conductive poly(3,4-ethylenedioxythiophene) : poly(styrenesulfonate) (PEDOT:PSS) (Galagan, *et al.* 2011a). The chemical structure of PEDOT:PSS is shown in Fig. 17.4.

17.4.2 Intermediate layers (ILs)

The intermediate layer (IL) between the active layer and the electrodes acts as a charge selective conductor, either blocking electrons or holes and conducting the opposite charge and vice versa. It might also improve the

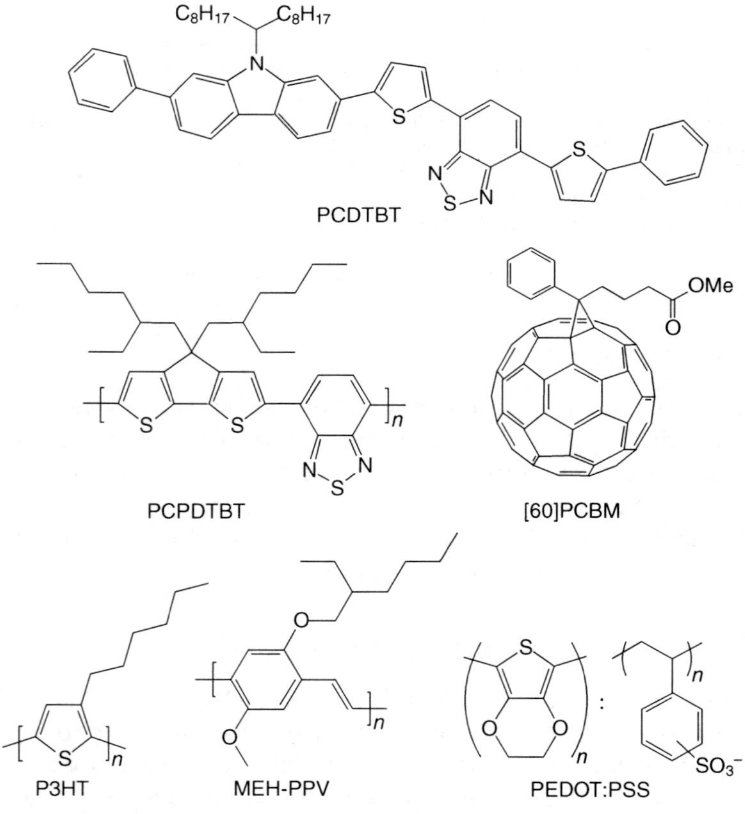

17.4 Chemical structure of organic material used in OSC.

alignment of the energy levels and compensate for the roughness of the underlying surface to remove some of the shunts. A myriad of different materials have been studied, but it is beyond the scope of this chapter to mention them all (Po *et al.* 2011). In normal geometries, PEDOT:PSS is the most used electron-blocking material and is dissolved or dispersed in aqueous solution. It is spin-coated on ITO, forming a thin layer with thickness of 60–100 nm and is dried at approximately 150 °C for 5–10 min.

Most of the R2R produced OSCs are based on inverted structures, where electron-conducting materials are necessary as first IL. Typical materials are TiO_x or ZnO. They can easily be coated from solution-based nanoparticles or precursors of the metal oxides. An environmental side effect is obtained by using aqueous ZnO solution and additionally it alleviates the inflection point in the J-V curve caused by photodoping (Søndergaard *et al.* 2010). The layer thickness of the aqueous ZnO is in the range of 20 nm and needs heat treatment for 5–40 min at 140 °C to become insoluble and obtain electron-conducting characteristics.

The second IL between active layer and back electrode acts as charge selecting layer similar to the first IL and has to have the opposite blocking characteristics than the first one. In OSCs with normal geometry a second IL is not mandatory but materials like TiO_x and ZnO improve the efficiency and act as environmental barrier, inducing stability (Gilot *et al.* 2007; Hayakawa *et al.* 2007; Lee *et al.* 2007). At the same time, the thin oxide layer can improve the optical absorption by shifting the field distribution inside the cell. Hole conducting PEDOT:PSS has been typically used in OSCs with inverted geometry, but MoO_3 and V_2O_5 have also been reported (Krebs *et al.* 2009b; Huang *et al.* 2010). Applying water-based PEDOT:PSS on top of a hydrophobic PCBM:PCBM film, which causes dewetting and inhomogeneous layers is almost impossible. Therefore, a special screen-printing formulation of PEDOT:PSS diluted in isopropanol is used for R2R coating (Krebs *et al.* 2009b). For P3HT:PCBM-based OSCs, the interface between the active layer and PEDOT:PSS was found to be the weakest (Dupont *et al.* 2012). Delamination and thermomechanical stresses may result in poor device performance. Annealing time and temperature increase the adhesion in this interface.

17.4.3 Back electrode and encapsulation

A metallic back electrode completes the OSC structure and acts as either anode or cathode depending of the geometry. The most commonly used electrode material is aluminum, silver and gold, but calcium has been reported too. Aluminum is often applied with a thin layer of lithium fluoride (2–10 nm), which improves contact to the active layer. Furthermore it protects the active layer from damage during the evaporation (Brabec *et al.*

2002). In normal devices the electrode is thermally evaporated and therefore not well suited for R2R processes. Silver electrodes can be easily applied by screen printing on inverted devices. The silver paste is commercially available and easy to process, but the influence of different solvents in the ink on the active layer has to be considered (Krebs *et al.* 2011).

Encapsulation is necessary to prevent exposure to humidity and oxygen. A barrier material with a sufficient oxygen transmission rate (OTR) of at least 10^{-3} cm^3 m^{-2} day^{-1} atm^{-1} and a water-vapor transmission rate (WVTR) of at least 10^{-4} g m^{-2} day^{-1} are required (Dennler *et al.* 2005; Lungenschmied *et al.* 2007). Devices prepared on rigid substrate are typically encapsulated with glass or metal using epoxy. Flexible barrier foils are used for large-scale flexible encapsulation and can be applied by lamination with pressure-sensitive adhesives (PSA) (Krebs *et al.* 2010c; Tanenbaum *et al.* 2012). Alternating layers of inorganic oxides such as SiO_x and polymers in the barrier foil are used to achieve high OTR and WVTR.

17.4.4 Active layer

Light absorption and charge carrier generation happens in the photoactive layer and therefore huge research efforts are being made to develop high-performance donor and acceptor materials. The main challenges are good stability, material abundance, cost efficiency and large-scale processability, although not everything is fulfilled by one material at the moment. The following section outlines some of the more successful materials for OSCs extracted from countless reports. Table 17.2 summarizes current state-of-the-art OSC lab-scale devices and shows the solar cell parameters for different materials and device structures.

Table 17.2 Solar cell performance parameters of state-of-the-art bulk heterojunction OSC. Active layer is highlighted in bold

Structure	V_{OC} (V)	J_{SC} (mA/cm^2)	FF (%)	PCE (%)	Reference
ITO/PEDOT:PSS/**P3HT:[60] PCBM**/Al	0.63	9.5	68	5	Ma *et al.* (2005)
ITO/PEDOT:PSS/**PCDTBT:[70] PCBM**/BCP/Al	0.91	11.8	66	7.1	Chu *et al.* (2011)
ITO/PEDOT:PSS/**PCDTBT:[70] PCBM**/TiO_x/Al	0.88	10.6	66	6.1	Park *et al.* (2009)
ITO/PEDOT:PSS/ **P3HT:bisPCBM**/Sm/Al	0.724	9.14	68	4.5	Lenes *et al.* (2008)
ITO/PEDOT:PSS/**PTB7:[70] PCBM**/Ca/Al	0.74	14.5	68.97	7.4	Liang *et al.* (2010)

The bulk heterojunction, interpenetrating network by blending donor and acceptor material was introduced in 1995 and showed great improvements in charge separation and efficiencies (Yu *et al.* 1995). They used 2-methoxy-5-(2-ethylhexyloxy)-polyphenylenevinylene (MEH-PPV) as electron donor and the soluble fullerene derivate [6,6]-phenyl C61 butyric acid methyl ester (PCBM) as electron acceptor in the intermixed active layer (Fig. 17.4).

A bulk heterojunction active layer with material combinations of poly(3-hexylthiophene) (P3HT) (Fig. 17.4) and [60]PCBM or [70]PCBM is state-of-the-art and well studied. Efficiencies are in the range of 5% (Li *et al.* 2005; Ma *et al.* 2005). The mismatch of the absorption spectrum of P3HT with the solar emission spectrum limits further improvement in efficiency. P3HT has a band gap of around 1.9 eV and absorbs only wavelength below 650 nm. The photon flux reaching the surface of the Earth has a maximum of approximately 1.8 eV (700 nm) and therefore P3HT can harvest only 22.4% of available photons (Bundgaard & Krebs 2007).

A way to overcome this physical barrier is the synthesis of polymer material with low band gaps collecting as many photons as possible. The offset of the HOMO and LUMO levels between donor and acceptor becomes important as well, whereas the open circuit voltage of the device is defined by the difference between the energy level of the HOMO in the donor and the energy level of the LUMO in the acceptor. The lowest band gap of the two materials defines the maximum current. In case of PCBM as acceptor, the optimum band gap has to be in the range of 1.2–1.7 eV. Absorption of more photos leads to potentially higher efficiencies.

The preparation of low band gap (LBG) polymers follows the donor–acceptor approach, in which the polymer backbone has electron-rich and electron-poor domains. One of the most promising and efficient LBG polymers is poly[2,6- (4,4-bis-(2-ethylhexyl)-4*H*-cyclopenta[2,1-b;3,4-b′]-dithiophene)-*alt*-4,7-(2,1,3-benzothiadiazole)] (PCPDTBT) (Fig. 17.4), which is based on a benzothiadiazole unit (acceptor) and a 4,4-bis (2-ethylhexyl)-4*H*-cyclopenta[2,1-b;3,4-b′]dithiophene unit (donor). That band gap is around 1.46 eV. Reported power conversion efficiencies are up to 4.5% in combination with [60]PCBM and 6.5% with [70]PCBM (Boland *et al.* 2010). High efficiencies of up to 7.1% were achieved with the LBG polymer poly[*N*-9″-hepta-decanyl-2,7-carbazole-*alt*-5,5-(4′,7′-di-2-thienyl-2′,1′,3′-benzothiadiazole)] (PCDTBT) (Fig. 17.4) and [70]PCBM dissolved in dichlorobenzene and 13% dimethyl sulfoxide (Chu *et al.* 2011). The cell had very good characteristics with V_{OC} of 0.91 V, J_{SC} of 11.8 mA/cm^2 and a FF of 66%. Further LBG polymers are reviewed in detail in Bundgaard & Krebs (2007) and Kumar & Chand (2011).

The active layer is processed out of a blend solution of donor and acceptor. In case of P3HT:PCBM the optimal ratio is around 1:1 (Li *et al.* 2005) with concentration of 20–40 mg/ml. The range of solvents is large, but chlorobenzene or dichlorobenzene is typically used (Dang *et al.* 2011). A certain dry layer thickness is achieved after deposition of the ink and evaporation of the solvent. The theoretical maximum J_{SC} with a 5 μm thick active layer is calculated with 15.2 mA/cm^2 for an IQE value of 100%. For more realistic thicknesses of 400 nm and an IQE value of 80% the J_{SC} decreases to 10.2 mA/cm^2 (Dennler *et al.* 2009).

Improving the efficiency of the device is done by thermal annealing the active layer. This can drastically change the structure and morphology of the material (Padinger *et al.* 2003; Benanti & Venkataraman 2006). P3HT can crystallize when the temperature is above the glass transition temperature. The concentration of PCBM has also an influence on the morphology upon annealing (Chirvase *et al.* 2004). It can improve the phase separation due to PCBM cluster growth. Other possibilities to improve the morphology of the active layer are solvent vapor treatment (Zhao *et al.* 2007) and additives in the active layer ink.

Polymers are made soluble by attaching solubilizing side chains such as alkyl groups onto the conjugated polymer backbone. They do not contribute to the light harvesting and make the material soft, which is related to the instability of OSCs (Krebs & Spanggaard 2005). After solution processing and drying, the side chains are in principle useless and can be removed. This interesting application can be achieved with thermocleavable materials. Thermocleavable ester groups are attached to the polymer backbone with a branched alkyl chain as solubilizing group. After heating to 200 °C the solubilizing groups are eliminated leaving the polymer insoluble. Heterojunction devices with poly-3-(2-methyl-hexan-2-yl)-oxy-carbonylbithiophene (P3MHOCT) and C_{60} showed an improved stability after thermal treatment (Krebs & Spanggaard 2005). The P3MHOCT is converted to the more rigid and insoluble poly-3-carboxydithiophene (P3CT) at 200 °C and with a further heating at 300 °C it converts to native polythiophene (PT) (Fig. 17.5). The cleaving is visible due to a color change from red over orange to purple-blue. OSC devices prepared with

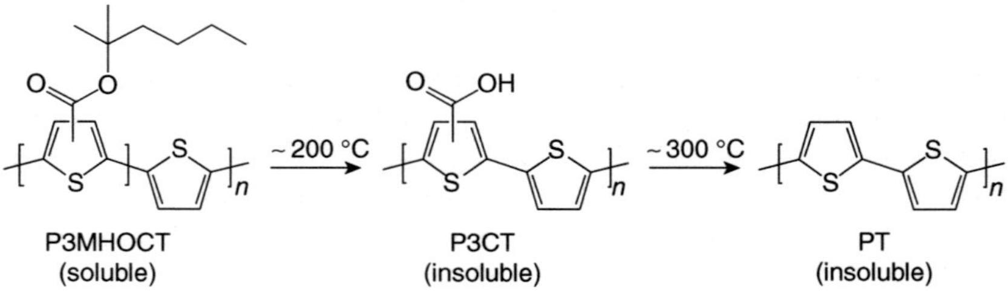

17.5 Preparation of OT via a themolytic reaction.

P3MHOCT:PCBM showed a large improvement in efficiency from P3MHOCT over P3CT to PT yielding at 1.5% for PT:[70]PCBM (Gevorgyan & Krebs 2008). Thermocleaving can be seen as a breakthrough in processing of polymers and adopting this technique to other polymers offers several advantages such as only having the active components are in the final layer. Detailed information about thermocleavable polymers can be found in Helgesen *et al.* (2010); Petersen *et al.* (2008) and Jørgensen *et al.* (2009).

Polymer material development and new active layer concepts such as polymer–polymer solar cells, inorganic–organic hybrid solar cells and nanostructured inorganics filled with polymer are extremely fast moving fields of research. Tandem OSCs improve the cell efficiency by stacking different band gaps materials on top of each other. The challenge is not only to find good material combinations to harvest as much photos as possible. From the processing point of view, the solvents and solubility of each subsequent material must match to prevent negative interaction like dissolving. Thermocleaving and protecting intermediate layer can improve the development of these multijunction solar cells. Covering all of the further technologies in this chapter is not possible and we refer the reader to extensive review reports (Deibel & Dyakonov 2010; Sista *et al.* 2011; Weickert *et al.* 2011).

For the sake of completeness it shall be mentioned that small-molecule organic solar cells (Peumans *et al.* 2003; Rand *et al.* 2007) are another group of organic solar cells and typically fabricated by vacuum processing in the preferred p-i-n structure (Maennig *et al.* 2004; Riede *et al.* 2008, 2011), employing either exciton blocking layers, or p-doped and n-doped electron transport layers. The intrinsic active layer is either a stacked planar heterojunction or a bulk heterojunction of donor and acceptor materials. The acceptor is typically the fullerene C_{60}. Donor materials used are polyacenes (Yoo *et al.* 2004), oligothiophenes (Schulze *et al.* 2006), and metal phthalocyanines as benchmark materials (Rand *et al.* 2007). Copper phthalocyanine (CuPc) and zinc phthalocyanine (ZnPc) are the most studied materials. Red-absorbing fluorinated ZnPc (F4-ZnPc) can lead to 0.1–0.15 V higher open circuit voltage than standard ZnPc. Combined with green-absorbing dicyanovinyl-capped sexithiophene (DCV6T) in a tandem-cell structure it can lead to an absorption over the whole spectrum with efficiencies up to 5.6% (Riede *et al.* 2011). Efficiency of more than 8% in a tandem structure has already been achieved (Green *et al.* 2011). A drawback of small-molecule OSCs is the energy-intensive vacuum processing. Recent developments demonstrate solution-based small-molecule solar cells (Zhang *et al.* 2012). Efficiency of 6.7% was achieved with 5,5′-*bis*{(4-(7-hexylthiophen-2-yl) thiophen-2-yl)-[1,2,5]thiadiazolo[3,4-c]pyridine}-3,3′-di-2-ethylhexylsilylene-2,2′-bithiophene, $DTS(PTTh_2)_2$ as donor and [70]PCBM as acceptor dissolved in chlorobenzene. The normal structured bulk heterojunction device still comprises an evaporated Al cathode.

17.5 Roll-to-roll (R2R) processing of organic solar cells (OSCs)

Solubility of the photoactive polymers is the biggest advantage in competing against other solar cell technologies. The low-cost and high-throughput production capabilities can compensate for the shorter lifetime and lower efficiencies compared with traditional silicon solar cells. This section briefly describes the structure of serial interconnected solution processed OSCs and then introduces common coating and printing techniques used in fabricating OSCs in large scale. Table 17.3 lists performance parameters of P3HT:PCBM solar cells, in which the active layer is fabricated with various process technologies, either full R2R or R2R-compatible.

Standard single solar cells are usually prepared in the lab by spin coating and printing or evaporating of the electrodes. The active area is small in the range of square millimeters to square centimeters and the material lost during spin coating is high. Using spin-coated cells in electrical applications

Table 17.3 Solar cell parameters of OSCs with P3HT:PCBM layers fabricated by coating and printing processes

Process (active layer)	Active area (cm^2)	V_{OC} (V)	I_{SC} (mA)	FF (%)	PCE (%)	Reference
Slot-die (a)	4.8	3.62	6.86	44	2.33	Krebs *et al.* (2009b)
Slot-die (b)	96	7.56	60	37.9	1.79	Krebs *et al.* (2010c)
Slot-die (c)	13.2	6.47	>6	64	2.2	Zimmermann *et al.* (2011)
Gravure (d)	9.65	3.02	13.91	44	1.92	Kopola *et al.* (2011)
Gravure (e)	0.045	0.56	0.216	45	1.21	Voigt *et al.* (2011)
Screen printing (f)	1.44	0.59	21.07	29.78	2.59	B. Zhang *et al.* (2009)
Inkjet (g)	0.09	0.628	0.9612	55.27	3.71	Eom *et al.* (2010)
Inkjet (h)	0.16	0.573	1.49	45	2.4	Lange *et al.* (2010)

a) Inverted structure, full R2R, module of 8 serial connected cells.
b) Inverted structure, full R2R, module of 16 serial connected cells, average value over 600 modules, max. PCE 2%.
c) Flatbed process, 11 serial connected cells, ITO-free, Cr/Al/Cr on flexible plastic substrate, evaporated Au grid.
d) Flatbed process, normal geometry, 5 serial connected cells.
e) Flatbed process, inverted structure, single cell, evaporated Au electrode.
f) Flatbed process, normal structure, evaporated Al electrode.
g) Normal structure, evaporated LiF/Al electrode.
h) Normal structure, evaporated Al electrode.

is not suitable because of its limited size, productivity, and an open circuit voltage below 1 V. Increasing the size of the cell is influenced due to the conductivity of the electrodes and therefore the collection of photocurrent is limited. To increase the output of OSCs and use them for powering practical applications, the cells have to be interconnected and produced in a larger scale. R2R coating and printing techniques allow deposition and patterning of the necessary layers and utilize the materials to a maximum. The inverted layer structure allows the printing of silver electrodes without using vacuum, unlike the evaporation of aluminum or silver in normal structured OSCs. Almost all materials of the complete solar cell are soluble and available as printable paste or ink. Only the normally used transparent conductive electrode ITO is evaporated or sputtered on the flexible substrate. ITO is costly and life cycle assessment (LCA) analyses revealed that the embodied energy in OSC modules is more than 80% (Espinosa *et al.* 2011). A lot of research is going on to avoid ITO and replace it with alternative materials, such as silver or grids and highly conductive polymers (Krebs 2009a; Galagan, *et al.* 2011a). The use of flexible substrates is essential for R2R processing and transparent foils like polyethylene terephthalate (PET) and encapsulation with flexible barrier material makes the final device lightweight and thin.

17.5.1 Serial interconnected device structure

A typical R2R produced OSC module consists of numerous serial interconnected single cells to sum up the open circuit voltages of each cell, as illustrated in Fig. 17.6. For example, if an application requires around 8 V to charge a battery, 16 single cells with an estimated open circuit voltage of 0.5 V have to be connected. On the other hand, the extracted current is equally to the photogenerated current of a single cell. The cell with the smallest short-circuit current weakens the total module performance. The serial connection of solar cells is only suitable for cells with equal short-circuit currents. The coating and printing allow homogeneous layer quality continuously in web- and cross-direction of the production process. The similarity in the single cells is given under fixed process conditions and changes during the fabrication should be avoided – at least in an industrialized process. A R2R process with continuously changing parameters is useful for research purposes and will be described later.

The width of a single cell is limited to the conductivity and ohmic loss of the electrode material and upscaling minimizes the photogenerated current. The stripe width of R2R produced OSCs is in the range of 5–18 mm for ITO as electrode material (Krebs *et al.* 2010c), and calculations of the size dependence are published in Hoppe *et al.* (2012). On the other hand, the small cell width and the slightly shifted subsequent layers decrease the

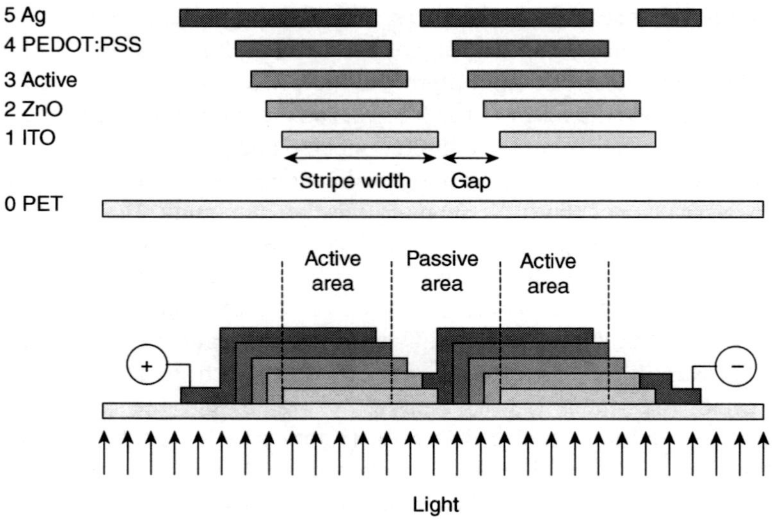

17.6 A serial connected OSC is illustrated for two single cells. The inverted structure consists of the substrate PET (0), ITO (1) ZnO (2), Active layer (3), PEDOT:PSS (4) and finally the Ag layer (5) for serial connection. Only a certain proportion of the module is generating photocurrent (active area).

overall relative active area of the module and a lot of the deposited material does not generate photocurrent. By using highly conductive electrode material, a monolithic design architecture with a high active to total area ratio is feasible and a lot of research is going on (Galagan *et al.* 2011a; Manceau *et al.* 2011). An advantage of such a device structure is that no patterning is required and the manufacturing is simplified with less registration effort.

17.5.2 Printing and coating processes

The availability of polymers and other essential materials for solution processing makes it optimal to convert the lab-scale spin coating process to large-scale industrial like R2R coating and printing processes. Patterning and additive deposition of material where it is needed is a requirement for cost-effective production of OSC devices. The adaptation of known thin film formation technologies from other industrial sectors such as photographic film fabrication or graphical printing is a feasible route to upscale to fabrication of OSCs.

The OSC is based on various layers as shown in prior sections, which may require different processing technologies, intermediate treatments, or different parameters like drying time and web speed. To avoid slow and costly vacuum steps, an inverted layer structure with printable electrodes is preferred. Integrated R2R processing on a single factory line is desired, but the varying requirements for each step induce discrete processing steps for the multilayer OSC, as shown in Fig. 17.7. The advantage of single step processing is the specific optimization and quality control. A sudden failure during the run would not have impact on subsequent steps and eventually ruin the whole production.

Commercially available coating and printing machines can be optimized and upgraded to the needs of OSC manufacturing. In principle every printing method is compatible to R2R processing and several technologies can be combined parallel or sequentially in a plant to fulfill the fabrication requirements for all layers, inks and process parameters. A machine used for R2R processing of OSCs based in the labs of DTU Energy Conversion is shown in Fig. 17.8 and comprises flexographic printing, slot-die coating, rotary screen printing and two ovens each of 2 m length.

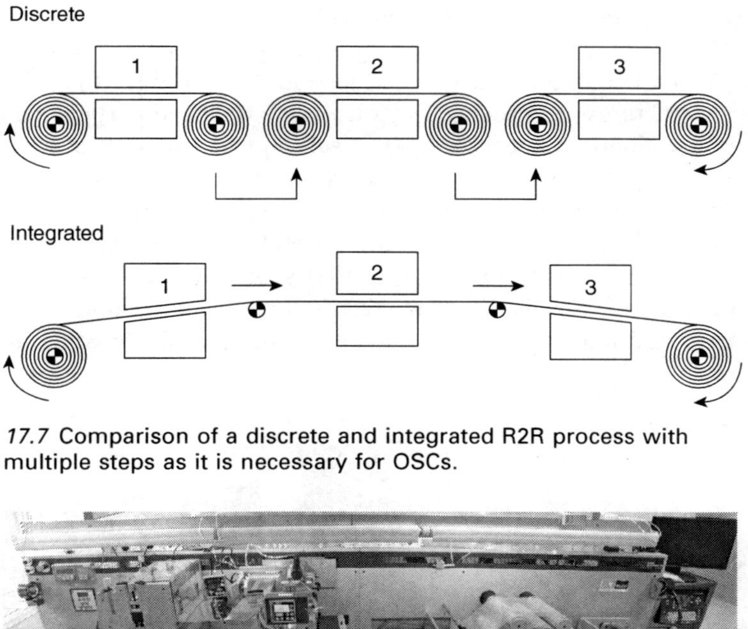

17.7 Comparison of a discrete and integrated R2R process with multiple steps as it is necessary for OSCs.

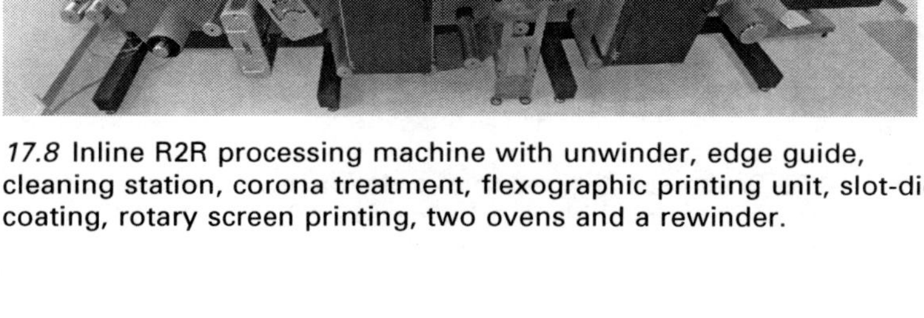

17.8 Inline R2R processing machine with unwinder, edge guide, cleaning station, corona treatment, flexographic printing unit, slot-die coating, rotary screen printing, two ovens and a rewinder.

The most suitable and commonly used coating and printing processes with their application in OSCs are now described. The main requirement for the deposition of the thin film functional layer is homogeneity without pinholes over large area. For detailed views on the technologies we refer to specialized books such as Gutoff & Cohen (2006), Krebs (2008), Tracton (2007), Kipphan (2001) and Krebs (2009b).

17.5.3 Slot-die coating

Slot-die coating belongs to the premetered coating technologies and provides smooth layers with homogeneous thicknesses in web direction and cross-direction. The operation of the coating and the system itself is quite simple as shown in Fig. 17.9. The wet layer thickness is controlled by the flow rate of the supplied ink and the coating speed. The coating head is precision engineered and can be quite complex to provide the same pressure distribution along the coating width.

Being a one-dimensional coating technique, it can enable patterning in the form of continuous stripes of functional material for the OSCs. Multilayer structures are produced by displacing the head perpendicular to the web direction to align the stripes of subsequent materials in each discrete process step. The stripes are formed by inserting stripe masks into the head, building a stripe-wide meniscus at the outlet of the ink. The thickness of the mask depends on the ink viscosity and is in the range of 10–200 μm. Low-viscous polymer solutions for OSCs with viscosities below 20 mPa s require masks with thicknesses of 20–50 μm. The masks may also be varied in case of different wetting behaviors of the ink. The coating head lip is

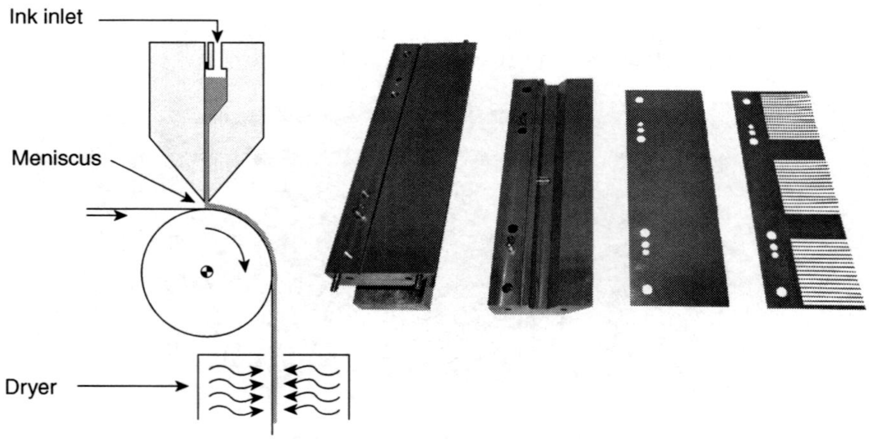

17.9 Schematic illustration of a slot-die coating process and images of the disassembled head including the stripe mask (right).

never in contact with the substrate. Pre-coated layers or the substrate itself cannot be influenced of contact-pressure like in other printing technologies. The ink is pumped precisely into the head with either piston, gear or syringe pumps. Higher viscous inks above 100 mPa s may require pressure vessels with equipment for flow control. The whole coating system with the ink reservoir is a closed systems, making it ideal for critical solvents. The ink is not exposed to the environment except for the moment it is applied to the moving substrate. Web speed and ink flow define the wet layer thickness. The final dry layer thickness can be estimated with the following formula:

$$d = \frac{f}{Sw} \cdot \frac{c}{\rho}$$

where d is the thickness in cm, f is the ink flow rate in cm^3/min, S is the web speed in cm/min, w the coated width in cm, c is the solid concentration in the ink in g/cm^3 and ρ is the density of the dried ink material in g/cm^3.

Typical coating speeds for the fabrication of OSCs with slot-die coating are in the range of 0.5–2 m/min, depending on the viscosity and wetting of the ink. The drying time has an impact on web speed as well, because the maximum drying temperature is limited to the substrate properties and the length of the oven, which itself is limited by the machine size and processiblity.

Extensive efforts were made in the processing of OSCs with slot-die coating. The deposition of the hole blocking layer (ZnO), photoactive layer (P3HT:PCBM, Fig. 17.10) or the hole injection layer (PEDOT:PSS) in discrete steps is reported in Krebs, *et al.* (2009b, 2010c), Zimmermann *et al.*

17.10 Slot-die coating of P3HT:PCBM solution for the active layer of OSCs. 32 stripes of 5 mm width are coated parallel. The gap between each stripe is 1 mm.

(2011) and Galagan, *et al.* (2011b). For instance, the process known as ProcessOne developed at Risø DTU (now DTU Energy Conversion) uses a pre-patterned ITO-PET substrate and all three subsequent layers, ZnO, P3HT:PCBM, and PEDOT:PSS, are slot-die coated. ZnO solution was coated on top of ITO at a speed of 2 m/min with a wet layer thickness of 3.125 μm and a dry layer thickness of 23 nm. The active layer was coated at a speed of 1.4 m/min with a wet layer thickness of 4.85 μm and a dry layer thickness of 127 nm. The PEDOT:PSS layer with a viscosity of 270 mPa s was coated at 0.3 m/min, resulted in a wet layer thickness of 75 μm and a dry layer thickness of 20 μm. After each individual coating the layers were dried using hot air convection ovens at 140 °C installed in the R2R machine. The modules had 16 individual stripes and the final interconnecting silver electrode layer was screen-printed. The best efficiency of 2% was achieved on a module size of 96 cm^2, with interconnected stripes of 13 mm width (Krebs *et al.* 2010c). A photograph of slot-die coated OSC modules with screen-printed silver electrodes is shown in Fig. 17.11.

Slot-die can also be used for the coating of silver nanoparticle ink to form electrode layers as a substitute for ITO (Krebs 2009a). In principle, the majority of functional solutions or dispersions, such as nanoparticle inks, can be coated through slot-die coating technology. The ink properties and coating parameter may have to be optimized against each other to achieve a satisfying layer quality.

R2R slot-die coating can also be deployed as a research tool to study ink and material compositions in the entire processing parameter space (Alstrup *et al.* 2010). Gradients of donor, acceptor, and solvents can be applied by differential pumped slot-die coating using two pumps and a miniaturized slot-die head, as shown in Fig. 17.12. This method allows the material screening and characterization of OSCs with very small amounts of material in a

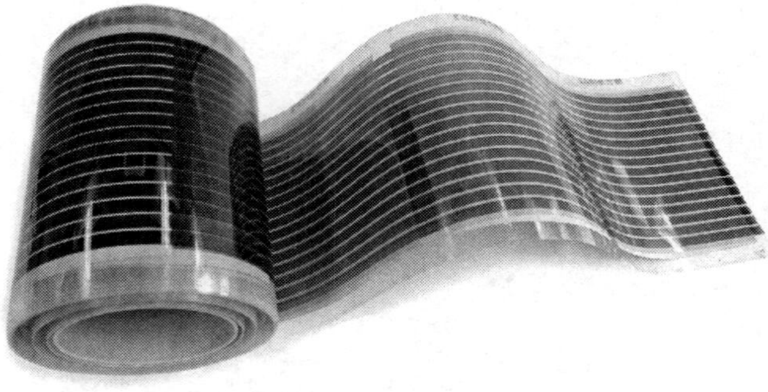

17.11 A roll of flexible OSCs modules fabricated by R2R slot-die coating with screen-printed silver electrodes. The area of each module is 360 cm^2 having 16 serial connected stripes. The OSCs were produced at DTU Energy Conversion, Roskilde, Denmark.

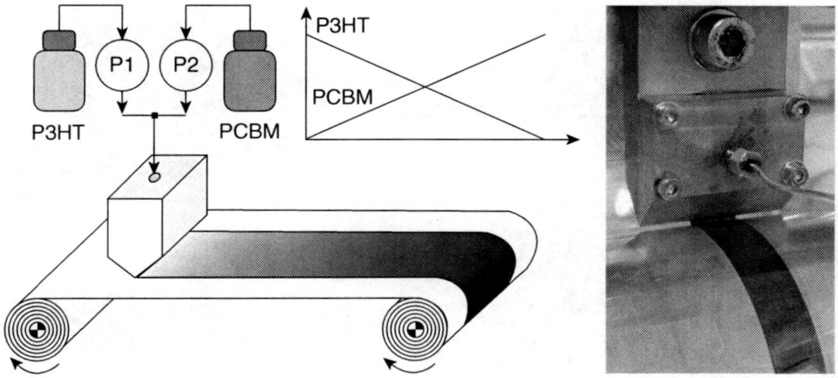

17.12 Principle of differential pumped slot-die coating for material screening of ink compositions. The small coating head is shown on the right.

R2R process. Finding optimum material ratios, such as P3HT to PCBM, and best working layer thicknesses is much faster than individual spin coating. Processing parameters from spin coating studies cannot directly be transferred to the R2R fabrication process, whereas the differential pumped slot-die coating exactly simulates the final R2R fabrication process.

17.5.4 Screen printing

Screen printing is a low-cost production technique which is suitable for lab-scale research activities but also as an industrial process with high throughput in the form of rotary screen printing or continuous flatbed screen printing. The process itself is relatively simple, and the equipment comprises a screen and a squeegee. The screen is a woven fabric made out of polyester, nylon or stainless steel covered with an emulsion layer. It is mounted to a metal frame under tension. The printing pattern in the screen is developed by photochemical processes and kept open (without emulsion). During the printing process (Fig. 17.13), the screen is placed above the substrate at a certain snap-off distance of a few millimeters. The ink is placed on the screen and is spread by moving the squeegee. The squeegee moves over the screen with a sufficient pressure downward to the substrate, forcing the ink through the open areas of the screen. The ink is left behind on the substrate as the screen snaps back. In rotary screen printing the screen has a cylindrical form and rotates around a fixed squeegee.

The resulting wet layer thickness is large compared to other printing techniques and the paste-like ink needs to be high viscous with thixotropic behavior. The large exposed area of the ink to the environment requires a low volatility of the solvents to prevent clogging of the screen. The main

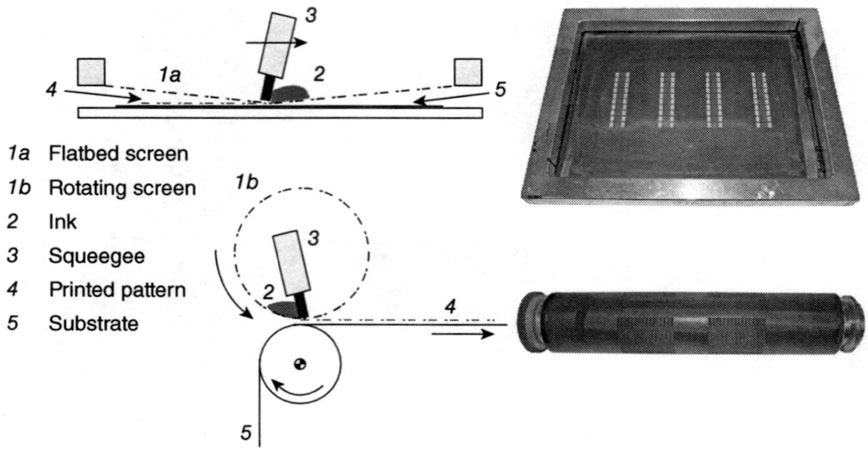

17.13 Schematic illustration of flatbed and rotary screen printing process. Images of corresponding screen are shown on the right.

parameters of a screen are the mesh number (wire per unit length), mesh opening and the wire diameter. The theoretical paste volume V_{screen} defines the wet layer thickness of the printed film in first approximation. It is characterized by the volume between the threads of the mask and the thickness of the emulsion. The volume of ink per open screen (cm^3/m^2) that is transferred to the substrate is influenced by process parameters such as squeegee force, squeegee angle, snap-off distance, and ink rheology, summarized by the pick-out ratio, k_p. The final dry layer thickness d can be calculated with the relationship:

$$d = V_{\text{screen}} \cdot k_p \cdot \frac{c}{\rho}$$

where c is the solid concentration in the ink in g/cm^3 and ρ is the density of the dried ink material in g/cm^3.

Screen printing is widely used in the industry for graphical applications, but also for printing of conductors for flexible electronics. Furthermore it is used to print electrodes in the traditional silicon solar cell industry. In the field of OSCs, screen printing of the active layer polymer MEH-PPV with a suitable rheology was successfully demonstrated (Krebs *et al.* 2007; Krebs 2008). There are some publications utilizing screen printing for P3HT mixtures, which show the possibility of overcoming the challenges of ink rheology and large wet thickness (Jørgensen *et al.* 2009; Krebs *et al.* 2009a; Zhang *et al.* 2009).

Nowadays, screen printing is used for the deposition of conductors such as PEDOT:PSS and in majority for silver ink in full solution processes

(Krebs *et al.* 2010c; Galagan, *et al.* 2011a). In stripe-based OSC modules, the printed silver electrodes are used for serial interconnection and current collection, as seen in Fig. 17.14. They may be printed in grid-like structure, be full layer or serve only for the serial connection in the form of a thin stripe. Screen printing is two-dimensional and so the printing image can have any shape. Thin lines are also possible, which can be used to pattern grid structures with a honeycomb-like design, as used for monolithic OSCs (Manceau *et al.* 2011). The silver inks can be solvent or water-based having a high solid load of silver flakes or nanoparticles. They are typically dried using hot air convection ovens. Another type is ultra-violet (UV) curable ink using UV light to cure the printed layer in seconds. The influence of different types of silver ink on the performance of OSCs is reported in Krebs *et al.* (2011). Additionally, screen printing is used to pattern the ITO layer on PET by printing etch resist and further etching, stripping and washing of the ITO layer (Krebs *et al.* 2009b).

17.5.5 Gravure printing

Traditionally, gravure printing is used to print graphical products at very fast speeds of up to 15 m/s. The two-dimensional process offers possibilities for high-resolution patterning and homogeneous layers by using low-viscosity ink. The technique comprises basically a chromium coated gravure cylinder, a doctor or chambered blade and a soft impression cylinder, as shown in Fig. 17.15. The patterned gravure cylinder with engraved cells of a certain pick-up volume (cm^3/m^2), depth, width, density, and screen angle

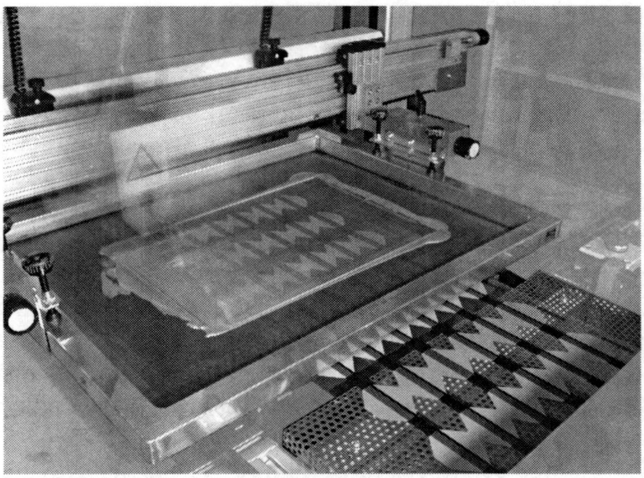

17.14 Flatbed screen printing of silver electrodes.

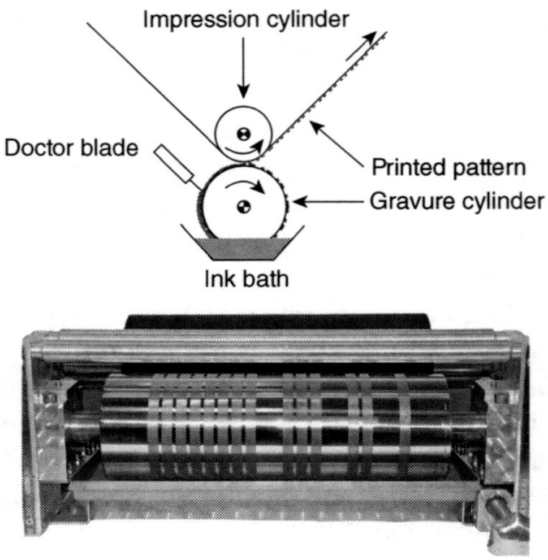

17.15 Schematic illustration of gravure printing process (top). Photograph of the chromium coated gravure cylinder with engraved stripe pattern (bottom). The impression cylinder in the background has a soft rubber surface (black). The doctor blade is not visible.

can be adjusted to optimize the printing result. New pattern designs require a new engraved gravure cylinder and this is a time-consuming and costly process. The cells of the gravure cylinder are continuously filled in an ink bath or with a closed chambered doctor blade system, as illustrated in Fig. 17.15. Excess ink is doctored off just before the nip of the gravure and impression cylinder. The nip pressure forces the transfer of the ink to the substrate. The cells are not emptied entirely which results in a certain ink transfer rate. The wet layer thickness is defined by the pick-up volume and ink transfer rate from the cells to the substrate. Optimization of the printing ink is quite important with respect to the viscosity, solid load, and surface tension. The printing quality is highly dependent on the printing speed, ink rheology, and pressure of the impression cylinder. As a rule of thumb, higher speeds are better.

Gravure printing was demonstrated as a suitable process for the deposition of P3HT:PCBM and PEDOT:PSS in OSCs with a normal structured geometry (Kopola *et al.* 2010, 2011). The printing was performed in a sheet-fed gravure printability and the cathode electrode was thermally evaporated. The printing speed was up to 18 m/min with highly concentrated inks for the photoactive P3HT:PCBM layer (up to 150 mg/ml in o-DCB). A power conversion efficiency of 2.8% was obtained for a single cell and 1.92% for module with five connected stripes.

Inverted structured solar cells having three subsequent gravure printed layers were successfully manufactured too (Voigt *et al.* 2011). Solution processed TiO_x, P3HT:PCBM, and PEDOT:PSS were sheet-to-sheet gravure printed on patterned ITO-PET substrate with a speed of 40 m/min. The efficiency was measured with 0.58% on a device area of 4.5 mm^2 and gold evaporated contacts. Organic solar cells where only the P3HT:PCBM active layer was successfully processed using R2R gravure printing are described in Hübler *et al.* (2011). In this case, the substrate was paper with a ZnO/Zn/ZnO layer as electrode and the active layer was gravure printed with a speed of 12 m/min. Later on, PEDOT:PSS was R2R flexo printed and resulted in an efficiency of 1.31% (area: 9 mm^2) at an illumination level of 600 W/m^2.

In all cases, and especially the PEDOT:PSS ink was heavily optimized for gravure printing using surfactants, wetting agents and solvent mixtures. The fabrication was primarily sheet-to-sheet to show the compatibility of the R2R gravure process and electrodes were evaporated. The conversion to a full R2R process where all layers are gravure printed is still questionable because of the high speed and the oven length required for drying and annealing. Further research will show if the gravure printing process can compete against slot-die coating in the manufacturing of OSCs.

17.5.6 Flexographic printing

Flexography is a R2R printing technology with a cylindrical relief plate carrying the printing pattern. The image stands up from the printing plate, typically made from rubber or photopolymer. Printing speeds can be above 100 m/min. The flexo system comprises a fountain roller that fills the ceramic anilox roller with ink, which has engraved cells all over the surface. The anilox roller has a certain cell volume and meters the ink in a uniform thickness and equally to the printing plate cylinder. Excess ink is scraped off with a doctor blade. Afterwards, ink is transferred to the substrate running between the plate cylinder and impression cylinder, as illustrated in Fig. 17.16. A chambered doctor blade inking system can be used to avoid exposure of the ink to the atmosphere.

So far, flexographic printing has not been used for the deposition of active layer ink in OSCs. It was shown as an application method for patterned PEDOT:PSS on top of P3HT:PCBM on a paper-based OSCs (Hübler *et al.* 2011). The R2R printing of modified PEDOT:PSS was performed a with speed of 30 m/min using an anilox cylinder with a volume of 25 cm^3/m^2. The cell had an efficiency of 1.31%, as mentioned before in the gravure printing section. In Krebs *et al.* (2010b), flexo technology was used for pre-wetting the surface of P3HT:PCBM with n-octanol prior the coating of PEDOT:PSS. The whole flexographic process is highly applicable to organic photovoltaics with the requirement of low volatility of the inks. Chambered doctor blade systems for the application of the ink are beneficial. Nevertheless, no further

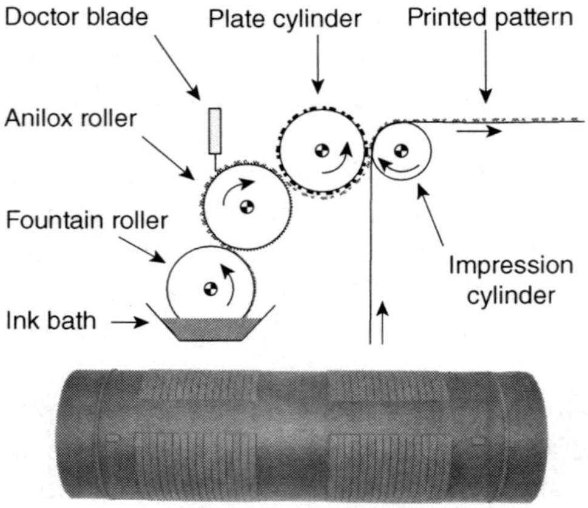

17.16 Schematic illustration of the flexographic printing process (top) and a photograph of a flexo plate cylinder, here seen as sleeve design (bottom).

publications of flexography with respect to OSCs were found. It might be used for electrode production, either grids or full layer, because it is capable of producing fine line structures and very thin layers. Silver grid structures with a line width of 20–50 μm and distance of 0.8–2 mm were successfully R2R printed at a speed of 5 m/min on PET (Deganello *et al.* 2010).

17.5.7 Inkjet

Inkjet printing is well known from home office applications and recently found its way into high-throughput R2R industrial processes for graphical applications. The most used technology nowadays is the drop-on-demand inkjet head, where the ink droplets are generated according to demand. The drop generation can be induced thermally by heating up the ink and generating a tiny bubble pushing out a droplet. More often used, especially with respect to the choice of solvents, is the piezoelectric printhead. Here, the reverse piezoelectric effect is used to deform a piezo actuator in each single ink chamber of an ink nozzle to generate pressure, forcing out a drop, as illustrated in Fig. 17.17. The interaction of the inkjet head and ink is crucial to build perfect drops. In all cases, the ink requires a special voltage waveform driving the piezos. The inkjet printing process can reach high resolution over 1200 dpi and generates typical drop volumes of 1–80 pL, depending on print head type and manufacturer. The firing rate of a printhead can easily reach tens of kilohertz. Inkjet ink must not dry at the nozzle and therefore it must have a low evaporation rate. Other parameters for ink formulations are viscosity (typically below 30 mPa s) and surface tension, which has to be in the range of 30 mN/m. The ink can either be

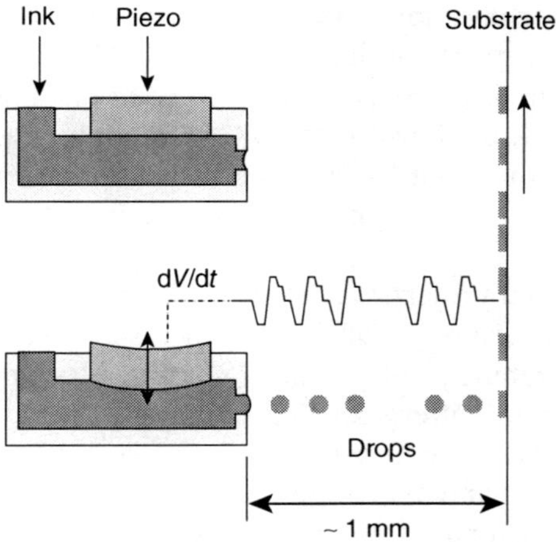

17.17 Schematic illustration of the drop on demand inkjet printing process using a time-dependent voltage waveform. The piezo is deformed by the reverse piezoelectric effect, which generates pressure forcing out the ink.

aqueous, solvent-based, or UV curable and the inks are a complex mixture of many solvents and surfactants. Ink parameters can vary heavily and are dependent on the application and inkjet head manufacturer. Further information about technology and the chemistry of inkjet ink can be found in Derby (2010), Pond (2000) and Magdassi (2010).

Inkjet technology has found its way into OSC fabrication, seen in an increasing amount of scientific reports (Hoth *et al.* 2007, 2008; Kim *et al.* 2008; Lange *et al.* 2010). Efficiencies of up to 3.71% were achieved with inkjet printed P3HT:PCBM and PEDOT:PSS layer (Eom *et al.* 2010). Additives dramatically affect the morphology and the solar cell performance of the device. So far, no R2R inkjet fabrication of OSCs has been reported, but it is just a question of time, especially for the fabrication of complex patterns.

17.6 Demonstration projects and conclusion

The development of organic solar cells is relatively young compared with traditional silicon solar cell technologies. The learning curve for manufacturing costs with targets of 1 €/W_p and below is very steep (Nielsen *et al.* 2010). The OSC material development and process upscaling is rapidly moving forward, which is obvious in the amount of publications, research collaborations, funded R&D project and a growing number of companies trying to commercialize OSC technology. Some of the companies working in the field of OSC fabrication are Solarmer (http://www.solarmer.com),

Plextronics (http://www.plextronics.com), Mekoprint (http://mekoprint.dk/electronics/solar), Eight19 (http://www.eight19.com) and Heliatek (http://www.heliatek.com), where the last one is strongly focused on small molecule organic solar cells. Flexible but low-efficiency OSC modules are already commercially available and the companies trying to leave niche markets. Applications like building integrated photovolatics (BIPV) are seen as promising possibilities.

A lot of R&D demonstration projects have been made to make the OSC and their benefits public. For example, fully R2R processed OSC modules were manufactured and combined with printed circuits, blocking diode, light-emiting diode (LED) and battery for the 'Lighting Africa' initiative in Zambia (Krebs *et al.* 2010a). A solar cell module of size 22.5 × 30.5 cm can be easily folded together and the integrated lithium ion battery is charged by the sun during daylight. The LED could be used to light the workspace at night. Furthermore, organic solar cells were prepared in large numbers using R2R methods and integrated into the Organic Electronic Association (OE-A) (http://www.oe-a.org) demonstrator, which is shown in Fig. 17.18. The credit card sized lamp contains a full R2R processed OSC module built by DTU Energy Conversion. The whole thin package includes R2R printed circuitry, a lithium-ion battery, a LED and diodes and was assembled in partnership with Mekoprint. More than 10 000 units were manufactured and handed out at the Large Area Organic and Printed Electronic Conference (LOPE-C) 2011 in Frankfurt, Germany. The demonstration showed the possibility of producing OSCs and the package with

17.18 Credit card sized lamp with build in organic solar cell, lithium ion battery, diodes and LED. Danish research institute Risø DTU (now DTU Energy Conversion) and Mekoprint built the demonstrator in more than 10 000 units.

a high technical yield of 89% at reasonable prices. The demonstration was a huge success.

Despite demonstration projects, a lot of research has to be done to leave niche markets and make OSCs more valuable. Some of the main research areas that have already been addressed are higher efficiencies, improvements in operational stability and lifetime, multijunction and hybrid structures, low band gap polymers, thermocleavable polymers, controlled morphology, aqueous processing, indium-free electrodes, process upscaling, minimizing of embodied energy, and the reduction of energy payback time. Especially the synthesis of countless new organic materials leads to steady progress in efficiency and solar cell performance. A strategy to narrow the band gap of the polymer is the use of polyisothionaphthene (PITN) that favors a quinoid structure. Alternating copolymers of electron-accepting unit and electron-donating units is another way to promote the quinoid character. The energy levels can be fine-tuned by fluorine substitution on the polymer backbone as demonstrated on low band gap copolymers benzo[1,2-*b*:3-4,*b′*]dithiophene and thieno[3,4-*b*]thiophene (PTBs). A deeper insight and further strategies in the development of new polymers for efficient OSCs are reviewed in Son *et al.* (2012). All record cell efficiencies around 8% that are published are manufactured under special conditions and on very small cell sizes. Achieving high efficiencies on large areas under full R2R processed and ambient conditions is a very challenging task. But hopefully in the near future, the strong research efforts will make this happen, so that OSC can compete against other solar cell technologies.

17.7 Acknowledgments

This work was supported by the European Commission's Seventh Framework Programme (FP7/2007–2013) in the project LARGECELLS (Grant No. 261936) and by the Framework 7 ICT 2009 collaborative project HIFLEX (Grant No. 248678).

17.8 References

Alstrup, J. *et al.*, 2010. Ultra fast and parsimonious materials screening for polymer solar cells using differentially pumped slot-die coating. *ACS Applied Materials & Interfaces*, **10**(2), pp.2819–2827.

ASTM Standard G173 ed., 2008. *Standard Tables for Reference Solar Spectral Irradiances: Direct Normal and Hemispherical on 37° Tilted Surface*, ASTM International. Available at: http://www.astm.org/Standards/G173.htm [Accessed October 14, 2011].

Benanti, T. & Venkataraman, D., 2006. Organic solar cells: an overview focusing on active layer morphology. *Photosynthesis Research*, **87**(1), pp.73–81.

Blom, P.W.M. *et al.*, 2007. Device physics of polymer:fullerene bulk heterojunction solar cells. *Advanced Materials*, **19**(12), pp.1551–1566.
Boland, P., Lee, K. & Namkoong, G., 2010. Device optimization in PCPDTBT:PCBM plastic solar cells. *Solar Energy Materials and Solar Cells*, **94**(5), pp.915–920.
Brabec, C.J. *et al.*, 2002. Effect of LiF/metal electrodes on the performance of plastic solar cells. *Applied Physics Letters*, **80**(7), p.1288.
Brabec, C.J., Sariciftci, N.S. & Hummelen, J., 2001. Plastic solar cells. *Advanced Functional Materials*, **11**(1), pp.15–26.
Brabec, C.J., Dyakonov, V. & Scherf, U., 2008. *Organic Photovoltaics*, Wiley-VCH.
Bundgaard, E. & Krebs, F.C., 2007. Low band gap polymers for organic photovoltaics. *Solar Energy Materials and Solar Cells*, **91**(11), pp.954–985.
Chirvase, D., Parisi, J. & Hummelen, J., 2004. Influence of nanomorphology on the photovoltaic action of polymer–fullerene composites. *Nanotechnology*, **15**, pp.1317–1323.
Chu, T.-Y. *et al.*, 2011. Morphology control in polycarbazole based bulk heterojunction solar cells and its impact on device performance. *Applied Physics Letters*, **98**(25), p.253301.
Coakley, K. & McGehee, M.D., 2004. Conjugated polymer photovoltaic cells. *Chemistry of Materials*, **16**(23), pp.4533–4542.
Dang, M.T. *et al.*, 2011. Polymeric solar cells based on P3HT:PCBM role of the casting solvent. *Solar Energy Materials and Solar Cells*, **95**(12), pp.3408–3418.
Deganello, D. *et al.*, 2010. Patterning of micro-scale conductive networks using reel-to-reel flexographic printing. *Thin Solid Films*, **518**(21), pp.6113–6116.
Deibel, C. & Dyakonov, V., 2010. Polymer–fullerene bulk heterojunction solar cells. *Reports on Progress in Physics*, **73**, p.096401.
Dennler, G. & Sariciftci, N.S., 2005. Flexible conjugated polymer-based plastic solar cells: from basics to applications. *Proceedings of the IEEE*, **93**(8), pp.1429–1439.
Dennler, G., Lungenschmied, C. & Neugebauer, H., 2005. Flexible, conjugated polymer-fullerene-based bulk-heterojunction solar cells: Basics, encapsulation and integration. *J. Mater. Res*, **20**(12), pp.3224–3233.
Dennler, G. *et al.*, 2008. Design rules for donors in bulk-heterojunction tandem solar cells? towards 15% energy-conversion efficiency. *Advanced Materials*, **20**(3), pp.579–583.
Dennler, G., Scharber, M.C. & Brabec, C.J., 2009. Polymer-fullerene bulk-heterojunction solar cells. *Advanced Materials*, **21**(13), pp.1323–1338.
Derby, B., 2010. Inkjet printing of functional and structural materials: fluid property requirements, feature stability, and resolution. *Annual Review of Materials Research*, **40**, pp.395–414.
Dupont, S.R. *et al.*, 2012. Interlayer adhesion in roll-to-roll processed flexible inverted polymer solar cells. *Solar Energy Materials and Solar Cells*, **97**, pp.171–175.
Eom, S.H. *et al.*, 2010. High efficiency polymer solar cells via sequential inkjet-printing of PEDOT:PSS and P3HT:PCBM inks with additives. *Organic Electronics*, **11**(9), pp.1516–1522.
Espinosa, N. *et al.*, 2011. A life cycle analysis of polymer solar cell modules prepared using roll-to-roll methods under ambient conditions. *Solar Energy Materials and Solar Cells*, **95**(5), pp.1293–1302.

Galagan, Y., J M Rubingh, J.-E., *et al.*, 2011a. ITO-free flexible organic solar cells with printed current collecting grids. *Solar Energy Materials and Solar Cells*, **95**(5), pp.1339–1343.

Galagan, Y., de Vries, I.G., *et al.*, 2011b. Technology development for roll-to-roll production of organic photovoltaics. *Chemical Engineering and Processing: Process Intensification*, **50**(5–6), pp.454–461.

Gevorgyan, S.A. & Krebs, F.C., 2008. Bulk heterojunctions based on native polythiophene. *Chemistry of Materials*, **20**(13), pp.4386–4390.

Gilot, J. *et al.*, 2007. The use of ZnO as optical spacer in polymer solar cells: theoretical and experimental study. *Applied Physics Letters*, **91**(11), pp.113520–4.

Green, M.A. *et al.*, 2011. Solar cell efficiency tables (Version 38). *Progress in Photovoltaics: Research and Applications*, **19**(5), pp.565–572.

Gutoff, E.B. & Cohen, E.D., 2006. *Coating and drying defects*, 2nd ed, Hoboken, NJ: John Wiley & Sons.

Halls, J. *et al.*, 1996. Exciton diffusion and dissociation in a poly (p-phenylenevinylene)/C60 heterojunction photovoltaic cell. *Applied Physics Letters*, **68**, p.3120.

Hauch, J. *et al.*, 2008. Flexible organic P3HT: PCBM bulk-heterojunction modules with more than 1 year outdoor lifetime. *Solar Energy Materials and Solar Cells*, **92**(7), pp.727–731.

Haugeneder, A. *et al.*, 1999. Exciton diffusion and dissociation in conjugated polymer/fullerene blends and heterostructures. *Physical Review B*, **59**(23), p.15346.

Hayakawa, A. *et al.*, 2007. High performance polythiophene/fullerene bulk-heterojunction solar cell with a TiO_x hole blocking layer. *Applied Physics Letters*, **90**(16), pp.163517–4.

Helgesen, M., Søndergaard, R. & Krebs, F.C., 2010. Advanced materials and processes for polymer solar cell devices. *Journal of Materials Chemistry*, **20**(1), pp.36–60.

Hoppe, H. & Sariciftci, N.S., 2004. Organic solar cells: An overview. *Journal of Materials Researd*, **19**(7), pp.1924–1945.

Hoppe, H., Seeland, M. & Muhsin, B., 2012. Optimal geometric design of monolithic thin-film solar modules: architecture of polymer solar cells. *Solar Energy Materials and Solar Cells*, **97**, pp.119–126.

Hoth, C.N. *et al.*, 2007. High photovoltaic performance of inkjet printed polymer:fullerene blends. *Advanced Materials*, **19**(22), pp.3973–3978.

Hoth, C.N. *et al.*, 2008. Printing highly efficient organic solar cells. *Nano Letters*, **8**(9), pp.2806–2813.

Huang, J.-S., Chou, C.-Y. & Lin, C.-F., 2010. Efficient and air-stable polymer photovoltaic devices with WO_3-V_2O_5 mixed oxides as anodic modification. *Electron Device Letters, IEEE*, **31**(4), pp.332–334.

Hübler, A.C. *et al.*, 2011. Printed paper photovoltaic cells. *Advanced Energy Materials*, **1**(6), pp.1018–1022.

Jørgensen, M., Hagemann, O. & Alstrup, J., 2009. Thermo-cleavable solvents for printing conjugated polymers: application in polymer solar cells. *Solar Energy Materials and Solar Cells*, **93**(4), pp.413–421.

Kim, J. *et al.*, 2011. Optimum design of ordered bulk heterojunction organic photovoltaics. *Solar Energy Materials and Solar Cells*, **95**(11), pp.3021–3024.

Kim, J.-M. *et al.*, 2008. Polymer based organic solar cells using ink-jet printed active layers. *Applied Physics Letters*, **92**, p.033306.

Kipphan, H., 2001. *Handbook of print media*, Berlin, Heidelberg, New York: Springer Verlag.

Kopola, P. *et al.*, 2010. High efficient plastic solar cells fabricated with a high-throughput gravure printing method. *Solar Energy Materials and Solar Cells*, **94**(10), pp.1673–1680.

Kopola, P. *et al.*, 2011. Gravure printed flexible organic photovoltaic modules. *Solar Energy Materials and Solar Cells*, **95**(5), pp.1344–1347.

Koster, L., Mihailetchi, V. & Blom, P.W.M., 2006. Ultimate efficiency of polymer/fullerene bulk heterojunction solar cells. *Applied Physics Letters*, **88**, p.093511.

Krebs, F.C., 2008. *Polymer photovoltaics: a practical approach*, SPIE Publications.

Krebs, F.C., 2009a. All solution roll-to-roll processed polymer solar cells free from indium-tin-oxide and vacuum coating steps. *Organic Electronics*, **10**(5), pp.761–768.

Krebs, F.C., 2009b. Fabrication and processing of polymer solar cells: A review of printing and coating techniques. *Solar Energy Materials and Solar Cells*, **93**(4), pp.394–412.

Krebs, F.C., 2010. *Polymeric solar cells:materials, design, manufacture*, Lancaster, PA.: DEStech Publications.

Krebs, F.C. & Spanggaard, H., 2005. Significant improvement of polymer solar cell stability. *Chemistry of Materials*, **17**(21), pp.5235–5237.

Krebs, F.C. *et al.*, 2007. Large area plastic solar cell modules. *Materials Science and Engineering: B*, **138**(2), pp.106–111.

Krebs, F.C., Jørgensen, M., Norrman, K., Hagemann, O., *et al.*, 2009a. A complete process for production of flexible large area polymer solar cells entirely using screen printing – first public demonstration. *Solar Energy Materials and Solar Cells*, **93**(4), pp.422–441.

Krebs, F.C., Gevorgyan, S.A. & Alstrup, J., 2009b. A roll-to-roll process to flexible polymer solar cells: model studies, manufacture and operational stability studies. *Journal of Materials Chemistry*, **19**(30), pp.5442–5451.

Krebs, F.C., Nielsen, T.D., Fyenbo, J., Wadstrøm, M., *et al.*, 2010a. Manufacture, integration and demonstration of polymer solar cells in a lamp for the 'Lighting Africa' initiative. *Energy & Environmental Science*, **3**(5), p.512.

Krebs, F.C., Fyenbo, J. & Jørgensen, M., 2010b. Product integration of compact roll-to-roll processed polymer solar cell modules: methods and manufacture using flexographic printing, slot-die coating and rotary screen printing. *Journal of Materials Chemistry*, **20**(41), pp.8994–9001.

Krebs, F.C., Tromholt, T. & Jørgensen, M., 2010c. Upscaling of polymer solar cell fabrication using full roll-to-roll processing. *Nanoscale*, **2**(6), pp.873–886.

Krebs, F.C., Søndergaard, R. & Jørgensen, M., 2011. Printed metal back electrodes for R2R fabricated polymer solar cells studied using the LBIC technique. *Solar Energy Materials and Solar Cells*, **95**(5), pp.1348–1353.

Kumar, P. & Chand, S., 2011. Recent progress and future aspects of organic solar cells. *Progress in Photovoltaics: Research and Applications*, **20**, pp.377–415.

Lange, A. *et al.*, 2010. A new approach to the solvent system for inkjet-printed P3HT:PCBM solar cells and its use in devices with printed passive and active layers. *Solar Energy Materials and Solar Cells*, **94**(10), pp.1816–1821.

Lee, K. *et al.*, 2007. Air-stable polymer electronic devices. *Advanced Materials*, **19**(18), pp.2445–2449.

Lenes, M. *et al.*, 2008. Fullerene bisadducts for enhanced open-circuit voltages and efficiencies in polymer solar cells. *Advanced Materials*, **20**(11), pp.2116–2119.

Li, G. *et al.*, 2005. High-efficiency solution processable polymer photovoltaic cells by self-organization of polymer blends. *Nature Materials*, **4**, pp.864–868.

Liang, Y. *et al.*, 2010. For the bright future – bulk heterojunction polymer solar cells with power conversion efficiency of 7.4%. *Advanced Materials*, **22**(20), pp.E135–E138.

Lungenschmied, C., Dennler, G. & Neugebauer, H., 2007. Flexible, long-lived, large-area, organic solar cells. *Solar Energy Materials and Solar Cells*, **91**(5), pp.379–384.

Ma, W. *et al.*, 2005. Thermally stable, efficient polymer solar cells with nanoscale control of the interpenetrating network morphology. *Advanced Functional Materials*, **15**(10), pp.1617–1622.

Maennig, B. *et al.*, 2004. Organic p-i-n solar cells. *Applied Physics A: Materials Science & Processing*, **79**(1), pp.1–14.

Magdassi, S., 2010. *The chemistry of inkjet inks*, Singapore: World Scientific Publishing.

Manceau, M. *et al.*, 2011. ITO-free flexible polymer solar cells: From small model devices to roll-to-roll processed large modules. *Organic Electronics*, **12**(4), pp.566–574.

Nelson, J., 2011. Polymer:fullerene bulk heterojunction solar cells. *Materials Today*, **14**(10), pp.462–470.

Nielsen, T. *et al.*, 2010. Business, market and intellectual property analysis of polymer solar cells. *Solar Energy Materials and Solar Cells*, **94**(10), pp.1553–1571.

Padinger, F., Rittberger, R. & Sariciftci, N.S., 2003. Effects of postproduction treatment on plastic solar cells. *Advanced Functional Materials*, **13**(1), pp.85–88.

Park, S.H. *et al.*, 2009. Bulk heterojunction solar cells with internal quantum efficiency approaching 100%. *Nature Photonics*, **3**(5), pp.297–302.

Petersen, M., Gevorgyan, S.A. & Krebs, F.C., 2008. Thermocleavable low band gap polymers and solar cells therefrom with remarkable stability toward oxygen. *Macromolecules*, **41**(23), pp.8986–8994.

Pettersson, L.A.A., Roman, L.S. & Inganäs, O., 1999. Modeling photocurrent action spectra of photovoltaic devices based on organic thin films. *Journal of Applied Physics*, **86**, p.487.

Peumans, P., Yakimov, A. & Forrest, S.R., 2003. Small molecular weight organic thin-film photodetectors and solar cells. *Journal of Applied Physics*, **93**(7), pp.3693–3723.

Piris, J. *et al.*, 2009. Photogeneration and ultrafast dynamics of excitons and charges in P3HT/PCBM blends. *Journal of Physical Chemistry C*, **113**(32), pp.14500–14506.

Po, R., Maggini, M. & Camaioni, N., 2010. Polymer solar cells: recent approaches and achievements. *Journal of Physical Chemistry C*, **114**(2), pp.695–706.

Po, R. *et al.*, 2011. The role of buffer layers in polymer solar cells. *Energy & Environmental Science*, **4**(2), pp.285–310.

Pond, S.F., 2000. *Inkjet technology and product development strategies*, Carlsbad: Torrey Pines Research.

Rand, B.P. *et al.*, 2007. Solar cells utilizing small molecular weight organic semiconductors. *Progress in Photovoltaics: Research and Applications*, **15**(8), pp.659–676.

Riede, M. *et al.*, 2008. Small-molecule solar cells – status and perspectives. *Nanotechnology*, **19**(42), p.424001.

Riede, M. *et al.*, 2011. Efficient organic tandem solar cells based on small molecules. *Advanced Functional Materials*, **21**(16), pp.3019–3028.

Sariciftci, N.S. *et al.*, 1993. Observation of a photoinduced electron transfer from a conducting polymer (MEHPPV) onto C60. *Synthetic Metals*, **56**(2–3), pp.3125–3130.

Scharber, M.C. *et al.*, 2006. Design rules for donors in bulk-heterojunction solar cells – towards 10% energy-conversion efficiency. *Advanced Materials*, **18**(6), pp.789–794.

Schulze, K. *et al.*, 2006. Efficient vacuum-deposited organic solar cells based on a new low-bandgap oligothiophene and fullerene C60. *Advanced Materials*, **18**(21), pp.2872–2875.

Servaites, J.D., Ratner, M.A. & Marks, T.J., 2011. Organic solar cells: A new look at traditional models. *Energy & Environmental Science*, **4**, pp.4410–4422.

Sista, S. *et al.*, 2011. Tandem polymer photovoltaic cells – current status, challenges and future outlook. *Energy & Environmental Science*, **4**(5), p.1606.

Son, H.-J. *et al.*, 2012. Overcoming efficiency challenges in organic solar cells: rational development of conjugated polymers. *Energy & Environmental Science*, **5**, pp.8158–8170.

Søndergaard, R. *et al.*, 2010. Fabrication of polymer solar cells using aqueous processing for all layers including the metal back electrode. *Advanced Energy Materials*, **1**(1), pp.68–71.

Søndergaard, R. *et al.*, 2012. Roll-to-roll fabrication of polymer solar cells. *Materials Today*, **15**(1–2), pp.36–49.

Spanggaard, H. & Krebs, F.C., 2004. A brief history of the development of organic and polymeric photovoltaics. *Solar Energy Materials and Solar Cells*, **83**(2–3), pp.125–146.

Tanenbaum, D.M. *et al.*, 2012. Edge sealing for low cost stability enhancement of roll-to-roll processed flexible polymer solar cell modules. *Solar Energy Materials and Solar Cells*, **97**, pp.157–163.

Tang, C., 1986. Two-layer organic photovoltaic cell. *Applied Physics Letters*, **48**, p.183.

Thompson, B. & Frechet, J.M.J., 2008. Polymer – fullerene composite solar cells. *Angewandte Chemie International Edition*, **47**(1), pp.58–77.

Tracton, A.A., 2007. *Coatings technology*, Boca Raton, FL: CRC Press.

Voigt, M.M. *et al.*, 2011. Gravure printing for three subsequent solar cell layers of inverted structures on flexible substrates. *Solar Energy Materials and Solar Cells*, **95**, pp.731–734.

Voroshazi, E. *et al.*, 2011. Long-term operational lifetime and degradation analysis of P3HT:PCBM photovoltaic cells. *Solar Energy Materials and Solar Cells*, **95**(5), pp.1303–1307.

Weickert, J. *et al.*, 2011. Nanostructured organic and hybrid solar cells. *Advanced Materials*, **23**(16), pp.1810–1828.

Yoo, S., Domercq, B. & Kippelen, B., 2004. Efficient thin-film organic solar cells based on pentacene/C60 heterojunctions. *Applied Physics Letters*, **85**(22), p.5427.

Yu, G. *et al.*, 1995. Polymer photovoltaic cells: enhanced efficiencies via a network of internal donor-acceptor heterojunctions. *Science*, **270**(5243), pp.1789–1791.

Zhang, B., Chae, H. & Cho, S., 2009. Screen-printed polymer:fullerene bulk-heterojunction solar cells. *Japanese Journal of Applied Physics*, **48**(2), pp.020208–1–020208–3.

Zhang, F. *et al.*, 2011. Recent development of the inverted configuration organic solar cells. *Solar Energy Materials and Solar Cells*, **95**(7), pp.1785–1799.

Zhang, L. *et al.*, 2012. Triisopropylsilylethynyl-functionalized dibenzo[def,mno] chrysene: a solution-processed small molecule for bulk heterojunction solar cells. *Journal of Materials Chemistry*, **22**(10), p.4266.

Zhao, Y. *et al.*, 2007. Solvent-vapor treatment induced performance enhancement of poly (3-hexylthiophene): methanofullerene bulk-heterojunction photovoltaic cells. *Applied Physics Letters*, **90**, p.043504.

Zimmermann, B., Würfel, U. & Niggemann, M., 2009. Longterm stability of efficient inverted P3HT: PCBM solar cells. *Solar Energy Materials and Solar Cells*, **93**(4), pp.491–496.

Zimmermann, B. *et al.*, 2011. ITO-free flexible inverted organic solar cell modules with high fill factor prepared by slot die coating. *Solar Energy Materials and Solar Cells*, **95**(7), pp.1587–1589.

18
Organic light-emitting diodes (OLEDs)

T. SCHWAB, B. LÜSSEM, M. FURNO, M. C. GATHER and K. LEO, TU Dresden, Germany

DOI: 10.1533/9780857098764.3.508

Abstract: The basics of highly efficient, organic light-emitting diodes (OLEDs) based on small molecules are discussed. All factors influencing OLED efficiency are explained and the benefits of pin-type OLEDs are discussed. It is shown how the pin concept can be used to realize highly efficient monochrome and white OLEDs. An extensive list summarizing the current state-of-the-art in emitter material systems for these devices is provided. The analysis of light modes in OLEDs is discussed in detail and experimental methods to enhance the outcoupling efficiency are introduced. Finally, a short overview on the different microscopic mechanisms of OLED degradation and techniques to study these mechanisms is given.

Key words: organic light-emitting diodes (OLEDs), pin-type, highly efficient, outcoupling, degradation.

18.1 Introduction

Ever since the invention of organic light emitting diodes (OLEDs) by Tang and VanSlyke in 1987[1], the field has developed rapidly. The efficiency of OLEDs has increased steadily, making OLEDs a promising alternative to many of today's display and lighting technologies.

Tang and VanSlyke were not the first to observe electroluminescence of organic materials. In an early study in the 1950s, Bernarose studied light emission from a single crystal of an organic compound[2,3]. These early devices suffered from a high driving voltage and consequently a low power efficiency[4,5]. The use of thin amorphous films led to thinner active layers and lower driving voltages[6,7], but it was not until Tang and VanSlyke used an organic hetero-structure that the power efficiency of OLEDs reached reasonable levels of above 1 lm/W.

In recent years, two key developments drove OLED research. The discovery of highly efficient phosphorescent emitters led to a quadruplication of the achievable quantum efficiency, and internal quantum efficiencies of nearly 100%[8,9] were reached. Furthermore, the use of electrically doped layers reduced the driving voltage and thus increased the power efficiency[10,11]. Nowadays, monochrome OLEDs with more than 50% external quantum efficiency and white OLEDs reaching the luminous efficacy of fluorescent tubes have been reported[12,13].

The success of OLEDs has triggered significant interest from industry. Currently, two main applications are envisaged: OLED displays and OLED lighting. Today, several companies such as Samsung, Sony, and LG are selling OLED displays for handheld devices or even small TV sets. The strengths of OLED displays compared with the established LCD technology are high-contrast ratios, wide viewing angles, extremely thin displays, and lower power consumption. However, despite their advantages, OLED manufacturing turned out to be challenging, and the production technology is still much in its infancy and limited to rather small substrates.

For OLED lighting, OLEDs can profit even more from their high power efficiency. Currently, commercially produced OLEDs are more efficient than a normal incandescent lamp by a factor of 2–3, and the efficiency is still increasing. For example, OSRAM (ORBEOS lighting panels) and Philips (Philips lumiblade) are selling white OLEDs as demonstrator kits. Besides their high power efficiency, OLEDs are truly flat area devices generating soft high-quality light without further light distribution, can be made transparent, and can be produced on flexible substrates, which enables many novel and exciting design opportunities.

In this chapter, the advances of pin-type OLEDs are described. Starting with a short introduction into the OLED technology (Section 18.2) and the pin concept (Section 18.3), highly efficient monochrome OLEDs are described (Section 18.4). Emphasis is put on the optimization of the optical microcavity of the OLED and the improvement of the outcoupling efficiency, which has the highest potential to improve OLED efficiency. A detailed list of current state-of-the-art emitter material systems (guest/host systems) and of the performance reached by devices based on these is also provided. In Section 18.5, the design principles used for monochrome OLEDs will be applied to white OLEDs and one approach – the triplet harvesting OLED – is discussed in detail. Again, currently used material systems are listed, together with device characteristics and references. The chapter closes with a short discussion of OLED degradation (Section 18.6).

18.2 Basics of organic light-emitting diodes (OLEDs)

In its simplest form, an OLED consists of a single layer of organic material, contacted by a reflecting (metal) contact – the cathode – and a transparent anode, which often consists of indium doped tin oxide (ITO). Upon application of a positive voltage, electrons and holes are injected at the cathode and anode, respectively. The charge carriers drift towards the center of the device, recombine and form excitons, i.e. bound electron–hole pairs. Usually, excitons in organic materials are of the Frenkel-type, i.e. due to the strong coulomb interaction both electron and hole are located on the same molecule. The exciton can relax towards the ground-state of the molecule and can emit a photon during this process.

To quantify the efficiency of all these processes involved in light generation, several definitions have been proposed in the past. The most commonly used ones are the external quantum efficiency (EQE or η_q) and the power efficiency or luminous efficacy (η_P). The EQE is defined as the ratio of the number of emitted photons to the number of injected charge carriers. It can be expressed as a product of the so-called *electrical efficiency* γ, *the exciton spin fa*ctor χ, the *efficiency of radiative decay* η_r, and the *outcoupling efficiency* ξ[12,14]:

$$\eta_q = \gamma\chi\int_{\lambda}\eta_r(\lambda)\xi(\lambda)s_{PL}(\lambda)\mathrm{d}\lambda \qquad [18.1]$$

where $s_{PL}(\lambda)$ denotes the normalized luminescence spectrum of the emitter $\left(\int_{\lambda} s_{PL}(\lambda)\mathrm{d}\lambda = 1\right)$.

The *electrical efficiency* γ is defined as the ratio of injected charge carriers to the number of formed excitons, i.e. γ includes all electrical losses. A significant parameter determining the electrical efficiency is the charge carrier balance, which is the ratio of injected electrons to injected holes. If there is an imbalance between electron and hole injection, not all charge carriers will recombine with a charge carrier of opposite polarity and the energy is lost. Thus, for highly efficient devices, the injection of holes and electrons has to be balanced.

The *exciton spin factor* χ is defined as the ratio of radiative excitons to the total number of excitons. Due to spin statistics, three out of four excitons formed in the device are triplets and only one exciton is a singlet. In fluorescent emitters, only singlets can radiatively relax to the ground state and thus χ equals 0.25. However, if phosphorescent molecules are used, all excitons can decay radiatively, and an exciton spin factor of unity can be reached[8].

η_r is the *efficiency of radiative decay* or photoluminescence (PL) efficiency. It depends on the radiative (Γ_{rad}) and non-radiative (Γ_{nrad}) decay rates of the emitter as follows:

$$\eta_r = \frac{\Gamma_{rad}}{\Gamma_{rad} + \Gamma_{nrad}}. \qquad [18.2]$$

As the flat OLED structure forms an optical micro-cavity, the radiative decay rate Γ_{rad} depends strongly on the optical environment and, according to Fermi's golden rule, on the optical mode density at the position of the emitter[15].

$$\Gamma_{\text{rad},ij} \propto |M_{ij}|\rho(\nu_{ij}) \qquad [18.3]$$

with $\Gamma_{\text{rad},ij}$: transition rate from state i to j, M_{ij} the matrix element connecting state i and j, and $\rho(\nu_{ij})$ the photonic mode density at the transition frequency ν_{ij}.

The photonic mode density can be influenced by the cavity length and the position of the emitter. If the cavity is in resonance at the peak wavelength of the emission and if the emitter is placed into an antinode of the optical field, the photonic density is large, resulting in an increase of radiative rate (cf. Eq. 18.3) and radiative efficiency (cf. Eq. 18.2)[14]. In contrast, if the cavity is detuned or if the emitter is placed in a node of the optical field, the emission is inhibited.

This effect, known as Purcell effect[16], can be described by the Purcell factor F, which is the ratio of the power radiated at the wavelength λ by the dipole in the cavity to the power radiated by the same emitter in infinite space. It can be expressed as[17]

$$F(\lambda) = \int_0^\infty K(\lambda, u)\,\mathrm{d}u^2 = 2\int_0^\infty uK(\lambda, u)\,\mathrm{d}u \qquad [18.4]$$

where $uK(\lambda, u)$ is the spectral power density, i.e. the power the emitter radiates in the cavity at a given wavelength. u is the in-plane wave vector, i.e. the photon wave vector k is split in an in-plane component u and an out-of-plane component v. If normalized to the full wave vector $k = \frac{2\pi}{\lambda}n(\lambda)$, with n index of the material where photon generation occurs, the following relation holds for u and v:

$$1 = u^2 + v^2 \qquad [18.5]$$

Once the Purcell factor has been calculated, the radiative rate of the emitter inside the cavity Γ_{rad} can be determined by

$$\Gamma_{\text{rad}} = F(\lambda)\Gamma_{\text{rad,inf}} \qquad [18.6]$$

i.e. the radiative rate of the emitter in infinite space (outside the cavity) $\Gamma_{\text{rad,inf}}$ is multiplied by the Purcell factor. If the Purcell factor F is high – corresponding to a high photonic mode density (cf. Eq. 18.3) – the radiative rate increases. Thus, the radiative efficiency η_r of the emitter depends strongly on the optical microcavity and the photonic mode density at the position of the emitter (cf. Eq. 18.2).

Finally, ξ is the *outcoupling efficiency*, i.e. the ratio photons emitted to outer space to the total number of photons generated within the OLED cavity. In the general OLED stack, there are two interfaces with significantly differing refractive indices: the interface between the organic layers including the anode made of indium tin oxide ITO (n~1.7–1.8) and the glass substrate (n~1.5) and the interface between glass and air. A large share of photons is trapped in the cavity due to total reflection at these interfaces. The critical angle for total reflection at these interfaces can be expressed in terms of the in-plane wave vector $u_{crit}(\lambda)$. Rewriting Snell's law yields

$$u_{\text{crit}}(\lambda) = \frac{n_0(\lambda)}{n_e(\lambda)} \qquad [18.7]$$

i.e. only the part of the power spectrum with an in-plane wave-vector smaller than $u_{crit}(\lambda)$ can be transmitted from the emitter material (refractive index $n_e(\lambda)$) into the layer with refractive index $n_0(\lambda)$. Light reflected at the interface between the OLED stack and the glass substrate feeds so-called organic modes, and light reflected at the interface glass substrate/air is lost in substrate modes.

Therefore, only a fraction of the power emitted by the molecule escapes the OLED cavity. If $U(\lambda)$ denotes the outcoupled power (normalized to the power emitted by the molecule in infinite space), the outcoupling efficiency can be written as

$$\xi(\lambda) = \frac{U(\lambda)}{F(\lambda)}. \qquad [18.8]$$

Simple calculations show that only approximately 20% of all light is extracted from the cavity, if no special measures are taken[18].

The external quantum efficiency is only a photon generation probability, i.e. the ratio between the flux of injected charge carriers and the flux of generated photons, and does not account for the power conversion efficiency of the device. To obtain the power efficiency or luminous efficacy, the external quantum efficiency has to be multiplied by the photon energy $h\nu$, where h is the Planck constant and ν the photon frequency, and divided by the energy of an injected electron eV_a, where V_a is the applied bias:

$$\eta_p = K \int_\lambda V(\lambda)\gamma\chi\eta_r(\lambda)\xi(\lambda)\frac{h\nu}{eV_a}\,d\lambda. \qquad [18.9]$$

As the luminous efficacy is a photometric quantity, the efficiency is weighted by the sensitivity of the human eye $V(\lambda)$. K is a constant and amounts to 683 lm/W.

According to Eq. 18.9, OLEDs should be driven at the lowest possible operating voltage to optimize not only the quantum efficiency, but also the power efficiency of OLEDs. One concept that ensures both low driving voltage and high quantum efficiency is the pin concept consisting of p- and n-doped charge carrier transport layers and intrinsic emission layers.

18.3 Pin organic light-emitting diodes (OLEDs)

To reach higher efficiency, an OLED usually needs more than one layer as described in the preceding section. The reason is that one layer cannot usually serve all functions involved in the light generation process, i.e. charge injection and transport, charge recombination, and optimized out-coupling. In the pin concept discussed here, the OLED consists of at least five organic layers (cf. Fig. 18.1), i.e. a p- and n-doped hole and an electron transport layer (HTL (p) and ETL (n)), an emission layer (EML (i)) and two blocking layers (hole blocking layer, HBL, and electron blocking layer, EBL). The charge transport layers ensure a good injection and transport of charge carriers towards the emission layer, where the excitons are formed and decay. To optimize the charge carrier balance, the EML is enclosed by two charge blocking layers, which confine the charges inside the EML and thus ensure that all charges recombine with a charge of opposite polarity.

p- and n-type doping of the HTL and the ETL leads to optimized charge carrier injection and transport[19,20]. Doping increases the number of free charges in the organic layers so that the conductivity rises by several orders of magnitude[10]. This minimizes the voltage drop across the transport layers and allows the OLEDs to be driven close to the thermodynamic limit[11]. Furthermore, the use of doped transport layers improves injection at the contacts significantly[21]. A space charge region is formed at the interface, which leads to strong level bending and thin injection barriers, which can be easily be tunneled through by the charge carriers[21]. Thus, a quasi-ohmic contact is formed and charge carriers can be injected into the device without additional voltage loss.

Doped organic transport layers have a further advantage since they can be used to optimize the OLED optical cavity. As the doped transport layers have a much higher conductivity compared to the intrinsic emission and blocking layers, a very small voltage drops across the doped layers and the thickness of the doped layers can be adjusted to optimize the OLED cavity

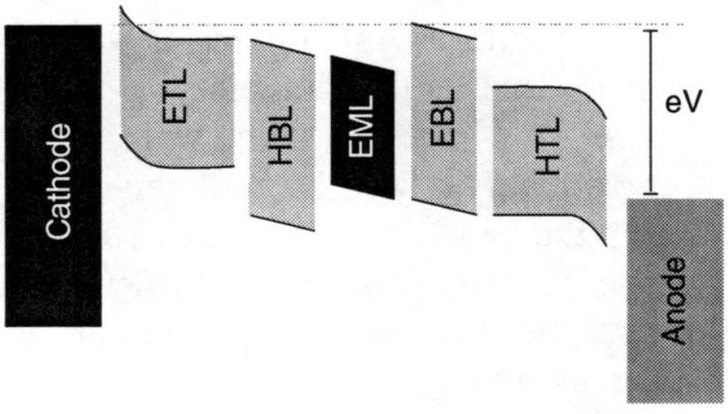

18.1 General setup of a pin OLED consisting of electron and p- and n-doped charge transport layer (ETL and HTL), electron and hole blocking layer (EBL and HBL) and the emission layer (EML).

and to shift the emitter position into the antinode (i.e. maximum) of the optical field with no influence on the electrical characteristics. Hence, the radiative rate of the emitter increases (cf. Eq. 18.6) and, as will be shown in the next section on monochrome OLEDs, the outcoupling efficiency increases as well.

18.4 Highly efficient monochrome organic light-emitting diodes (OLEDs)

The strength of organic electronics is based on the near endless number of different organic compounds. In particular, organic emitters span the whole visible spectrum and monochrome OLEDs in all colors are available.

For monochrome devices, external quantum efficiencies beyond 20% are well established[22–25]. He *et al.* reached an EQE of close to 20% by using two different host materials – one electron and one hole conductive host[25]. Tanaka *et al.*[22] even reached 29% external quantum efficiency for a green OLED at a brightness of 100 cd/m^2 or 26% at 1000 cd/m^2. Tanaka *et al.* also reached 21% EQE for a phosphorescent blue device[26] using matrix materials with a high triplet level. Chopra *et al.* reported on an increase in EQE of blue OLEDs to 23% if the excitons are well confined in the EML by blocking layers with a high triplet energy[24]. Table 18.1 gives an overview of different monochromatic emitter systems currently used and lists the EQE, the luminous efficiency and the color of the generated light for each material system. Most emitter systems consist of a phosphorescent emitted molecule embedded in a matrix with suitable triplet energy. In blue-emitting devices, FIrpic is probably the most widely used emitter. Green emission is provided by $Ir(ppy)_3$ and $Ir(ppy)_2(acac)$. In the red, $Ir(MDQ)_2(acac)$ is widely used. Phosphorescent emitter materials are thus dominated by iridium-based complexes with platinum and osmium-based materials playing a relatively minor role. However, in particular for generation of blue light, fluorescent emitter molecules are still frequently used (due to their often more favorable longevity and more saturated blue emission). We have thus listed a few particularly interesting fluorescent systems as well, albeit they generally reach lower efficiencies due to their inability to harvest triplet excitons.

Considering that only approximately one-fifth of all generated photons are extracted from the device[18], the EQEs listed in Table 18.1 show that a nearly perfect charge to photon conversion or internal efficiency has been reached in monochrome devices, but that still a large amount of photons remain trapped in the OLED microcavity.

Thus, the largest improvement in OLED efficiency can be reached by increasing the outcoupling efficiency ξ. To increase the outcoupling efficiency, a detailed understanding of the optical properties of the OLED

Table 18.1 Monochrome OLEDs

	Emitting layer			PE [lm/W]		EQE [%]		CIE	Reference	
	Dopant	c [wt.%]	Host	(max)	(@ cd/m^2)	(max)	(@ cd/m^2)	(x;y)	Year	Key
blue										
phosph.	FCNIrpic	3.0	mcPPO1	29.8	15.0 (@1000)	25.1	24.0 (@1000)	(0.14;0.18)	2011	27
phosph.	FCNIr	15.0	PPO 21	19.0	9.0 (@ 100)	19.2	12.1 (@1000)	(0.14;0.16)	2010	28
phosph.	FCNIr	15.0	PPO 2	16.6	/	18.4	11.8 (@1000)	(0.14;0.15)	2009	29
phosph.	FIr6	8.0	BCPO	33.1	/	19.8	/	(0.14;0.26)	2010	30
phosph.	FIr6 / FIr6	4.0 / 25.0	mCP / UGH2	25.0	20.0 (@1000)	17.0	/	(0.16;0.28)	2008	31
phosph.	FIrpic	11.0	26DCzPPy	46.1	34.5 (@1000)	24.3	22.6 (@1000)	/	2008	32
phosph.	FIrpic	20.0	TCTA	56.0	42.0 (@1000)	23.0	21.0 (@1000)	/	2008	33
phosph.	FIrpic	8.0	BCPO	40.6	/	23.5	19.4 (@1000)	(0.13;0.30)	2010	30
phosph.	FIrpic	8.0	TCTA: TmPyPb	55.4	40.3 (@1000)	20.4	18.3 (@1000)	/	2012	34
phosph.	FIrpic	15.0	BTPS	/	31.4 (@1000)	/	17.7 (@1000)	/	2012	35
phosph.	FIrpic	5.0	TAPC: PO15	59.0	49.0 (@1000)	/	/	/	2010	36
fluores.	DPAVBi	6.0	MADN	/	/	8.2	7.7 (@1000)	/	2012	37
fluores.	BD-6MDPA	5.0	MADN	6.1	/	8.2	/	(0.13;0.15)	2010	38
fluores.	/		TAT	1.9	/	7.2	/	(0.156;0.088)	2011	39,40
fluores.	FLBD	6.0	FLBH	/	/	5.4	/	(0.15;0.09)	2009	41
green										
phosph.	Ir(ppy)3 / Ir(ppy)3	8.0 / 8.0	TPBI / TCTA	/	101 (@100)	/	21.5 (@100)	(0.37;0.61)	2008	42
phosph.	Ir(ppy)3	8.0	CBP	133	107 (@1000)	29.0	26.0 (@1000)	/	2007	22
phosph.	Ir(ppy)3	10.0	BTPS	/	82.0 (@1000)	/	26.0 (@1000)	/	2012	35
phosph.	Ir(ppy)3	8.0	BCPO	87.5	/	21.6	19.1 (@1000)	(0.28;0.65)	2010	30
phosph.	Ir(ppy)3	9.0	o-CzOXD	59.3	/	20.2	17.9 (@10000)	/	2008	43
phosph.	Ir(ppy)3	9.0	O-TPA-m-OXD	66.5	/	19.6	/	(0.36;0.60)	2010	44
phosph.	Ir(ppy)3	7.0	CBP	/	/	16.8	/	/	2012	45
phosph.	Ir(ppy)2(acac)	8.0	CBP	/	115 (@1000)	/	40.0 (@1000)	/	2011	46

Table 18.1 Continued

	Emitting layer			PE [lm/W]		EQE [%]		CIE	Reference	
	Dopant	c [wt.%]	Host	(max)	(@ cd/m²)	(max)	(@ cd/m²)	(x;y)	Year	Key
phosph.	Ir(ppy)2(acac)	8.0	CBP	97.0	79.0 (@1000)	29.2	29.2 (@1000)	/	2011	47
phosph.	Ir(ppy)2(acac)	8.0	CBP	/	65.0 (@1000)	/	24.5 (@1000)	/	2011	48
phosph.	Ir(ppy)2(acac)	9.0	m-TPA-o-OXD	105	/	23.7	22.4 (@1000)	(0.31;0.64)	2010	49
phosph.	Ir(ppy)2(acac)	9.0	p -BISiTPA	72.4	47.3 (@1000)	22.7	22.4 (@1000)	(0.35;0.62)	2011	50,51
phosph.	Pt-BppyA	8.0	CBP / TPBi	79.3	/	20.9	17.0 (@1000)	/	2011	52
fluores.	C545T	0.8	TPBA	33.2	/	9.9	/	(0.24;0.62)	2006	53
fluores.	C545T	1.0	MADN	/	/	/	7.5 (@1000)	/	2012	37
yellow										
phosph.	Ir(fbi)2(acac)	7.0	m-TPA-o-OXD	61.7	/	20.6	18.6 (@1000)	(0.50;0.50)	2010	49
phosph.	Ir(fbi)2(acac)	7.0	P-BISiTPA	51.9	29.3 (@1000)	20.5	18.9 (@1000)	(0.52;0.48)	2011	50,51
phosph.	Ir(pbi)2(acac)	10.0	CphBzIm	62.0	35.0 (@1000)	19.2	18.5 (@1000)	(0.42;0.56)	2010	54
red										
phosph.	Ir(MDQ)2(acac)	10.0	NPB	/	/	29.0	/	/	2010	55
phosph.	Ir(MDQ)2(acac)	10.0	NPB	/	81.0 (@100)	/	25.5 (@100)	(0.66;0.34)	2008	42
phosph.	Ir(mphmq)2(tmd)	3.0	Bebq2	32.0	25.4 (@1000)	24.6	/	(0.648;0.35)	2011	56
phosph.	Os(bpftz)2(PPhMe2)2	10.0	TICNBI	22.1	/	22.0	20.8 (@1000)	(0.65;0.35)	2012	57
phosph.	Os(fptz)2(PPh2Me)2	7.0	POAPF	34.5	26.1 (@1000)	19.9	18.6 (@1000)	(0.64;0.36)	2009	58
phosph.	Ir(piq)2(acac)	8.0	BBTC	13.0	/	19.3	/	/	2012	59
phosph.	Ir(piq)3	7.0	BCPO	20.4	/	17.0	/	(0.67;0.33)	2010	30
phosph.	Ir(pq)2(acac)	10.0	SPPO2	/	/	14.1	/	/	2009	60
phosph.	Ir(pq)2(acac)	8.0	SPPO21	/	/	20.0	12.1 (@1000)	/	2010	61
fluores.	Aβ-ADN	40.0	Rubrene	/	/	11.3	/	/	2009	62

Abbrevations used in the table: phosph. = phosphorescent; fluores. = fluorescent; c = concentration; EQE = external quantum efficiency; PE = power efficacy.

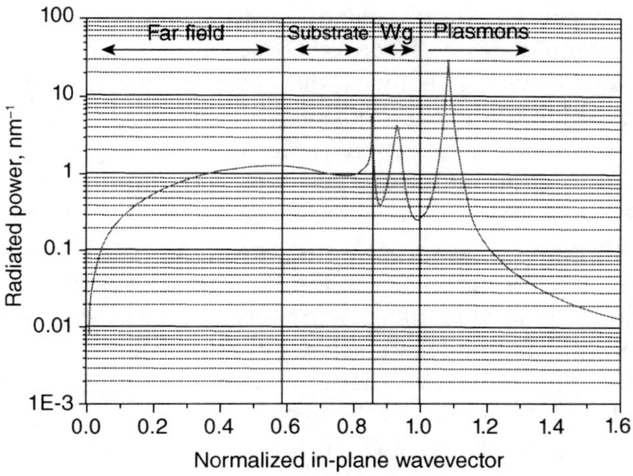

18.2 Spectral power density as a function of the normalized in-plane wavevector for a red pin OLED.

cavity is essential. Several reports on modeling of the different loss-channels in OLEDs have been published[63–65], which quantify the different loss mechanisms in OLEDs.

The results of such studies as published by Meerheim *et al.* are discussed in the following[12]. The simulation results for a monochrome red OLED, i.e. the different modes excited in the microcavity, are shown in Fig. 18.2. The pin-OLED consists of an ITO anode, 60 nm of NHT-5 doped with 4 wt% NDP-2 (Novaled AG) as HTL, 10 nm of 2,2′,7,7′-tetrakis-(*N*,*N*-diphenylamino)-9,9′-spirobifluorene (Spiro-TAD) as EBL, 20 nm of *N*,*N*′-di(naphthalen-2-yl)-*N*,*N*′-diphenyl-benzidine (NPB) doped with the red phosphorescent emitter iridium(III)bis(2-methyldibenzo-[f,h]quinoxaline) (acetylacetonate) ($Ir(MDQ)_2(acac)$) (10wt%) as EML, 10 nm of bis-(2-methyl-8-quinolinolato)-4-(phenyl-phenolato)aluminum (III) (BAlq) as HBL, and finally BPhen doped with Cs as ETL. To study interference effects and the effect of the cavity on the efficiency of the devices, the thickness of the ETL has been varied between 30 and 260 nm.

The simulations are based on an optical model treating the emitting molecule as forced damped harmonic oscillator[14]. In Fig. 18.2 the normalized radiated power, i.e. $uK(\lambda, u)$ (cf. Eq. 18.4) of a dipole in the cavity at a wavelength of 610 nm for an ETL thickness of 70 nm is plotted vs. the normalized in-plane wave vector.

The normalized radiated power peaks at several in-plane wave vectors, which correspond to the various resonant modes of the cavity[15]. For normalized in-plane wave vectors u in the EML larger than 1, the out-of-plane wave vector v is imaginary $\left(v = \sqrt{1-u^2}\right)$, which means that the waves are evanescent. In particular, there is one dominating mode at $u \sim 1.1$, which is the surface plasmon polariton localized at the cathode surface.

For in-plane wave vectors below 1, the out-of-plane wave vector is real and the corresponding waves are plane waves. However, as discussed in Section 18.2, due to total reflection at the interface organic layers/glass substrate (at $u_{crit} \sim 0.88$) and at the interface glass substrate/air (at $u_{crit} \sim 0.6$), not the whole in-plane wave vector range from 0 to 1 couples to external modes. Only emission at wave vectors below the critical angle for total reflection at the interface glass/air (cf. Eq. 18.7) can be eventually extracted from the device. In particular, there is one waveguided mode in the organic layers ($u \sim 0.9$), which limits the efficiency of the device.

In conclusion, Fig. 18.2 shows that only a small fraction of the emitted power is coupled into the far-field. In particular, coupling to the plasmonic mode is strong and thus much power is lost at the cathode. However, coupling to these evanescent modes is a near-field effect, i.e. an effect occurring only at relatively short distances from the emitter. If the distance between the emitter and the cathode is increased, coupling to the plasmonic mode is strongly reduced.

This effect is shown in Fig. 18.3. The optical simulation is repeated for varying ETL thickness, i.e. the distance between the emitter and the cathode is varied. In Fig. 18.3, the strength of the different loss modes (plasmons, waveguided and substrate, losses due to a radiative efficiency η_r below 1 (non-radiative losses, Eq. 18.2), and electrical losses) is summarized[12].

Figure 18.3a shows that the number of outcoupled photons oscillates with ETL thickness, which is due to the different resonances of the OLED cavity. At the peaks of the plot of the fraction of outcoupled photons vs. the ETL thickness, the cavity is in resonance and the emitter is positioned at an antinode of the optical field of the cavity, ensuring efficient light outcoupling.

Furthermore, coupling to the plasmonic mode is strongly reduced if the thickness of the ETL is increased. However, due to the thicker cavity, coupling to waveguided modes increases and, overall, the outcoupling efficiency is only increased slightly.

To couple out waveguided modes, OLEDs can be processed on index-matched glass, i.e. glass with a refractive index matching the refractive index of the organic layers ($n \sim 1.7–1.8$, cf. Eq. 18.7). Index matching ensures that no interface where total reflection can occur is present in the device and, as consequence, virtually all internally generated photons can reach the outer surface of the substrate. Furthermore, to avoid reflections at the interface to air, a large outcoupling lens can be applied, so that all light can exit the substrate. The corresponding distribution of loss channels is shown in Fig. 18.3b. As the substrate is index-matched, all waveguide modes are suppressed and the outcoupling efficiency increases significantly. In

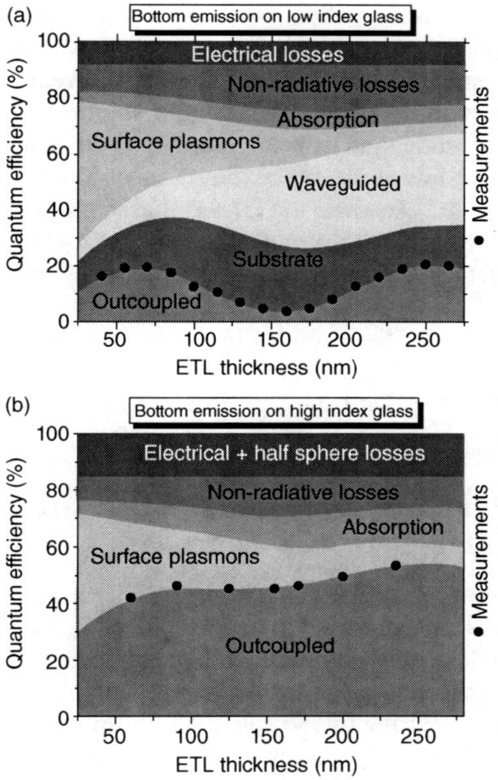

18.3 Different loss channels for the red OLED on low index glass (a) and on high index glass (b) measured with an outcoupling lens[12]. Reprinted with permission from Meerheim *et al.*, *Applied Physics Letters* **97**, 253305 (2010). Copyright 2010, American Institute of Physics.

particular, the thickness of the ETL can be increased without increasing losses due to waveguided modes. Thus, the outcoupling efficiency becomes larger for thicker ETLs, where coupling to plasmonic modes is weaker.

The comparison of these numerical results with measured EQE data is shown as black dots in Fig. 18.3. For both OLEDs on low-index (Fig. 18.3a) and high-index glass (Fig. 18.3b), the correspondence between experiment and calculations is excellent. For OLEDs on low index glass, the highest EQE is reached at 250 nm ETL thickness (second order cavity) and reaches 21%. For the high-index glass the EQE is significantly increased to 54%, i.e. more than half of the injected electrons generate a photon which is coupled out from the cavity.

In summary, highly efficient monochrome OLEDs can be designed, if the optical structure of the OLED is optimized. However, although devices with an EQE beyond 50% are reached, these results are lab results as the application of a large outcoupling lens to extract all substrate modes is not practicable for commercial applications. Therefore, great efforts is being made to find other outcoupling enhancement solutions that avoid waveguided and substrate modes and will increase OLED efficiency further. Concepts discussed in the literature are textured substrates[66], microlens foils attached to the substrate[67], layers in the OLED that scatter waveguided modes[68], OLEDs on photonic crystals, or the use of horizontally aligned emitters[69,70]. For more details, the reader is referred to a review article of Saxena *et al.*[71].

Finally, the same concepts as discussed for monochrome OLEDs can be applied for the generation of white light. However, to generate white light with a good color quality, emission from red, green and blue emitters has to be well balanced, which is described in the next section.

18.5 Highly efficient white organic light-emitting diodes (OLEDs)

The first white OLEDs were published by Kido *et al.*[72,73]. In these OLEDs, fluorescent materials were used, limiting the overall efficiency of these devices. The use of phosphorescent emitters significantly increased the efficiency[74–76]. In particular, the highest efficiency can be obtained by using triplet emitters for all colors. Using this so-called all-phosphorescent setup, Sun and Forrest were able to reach an efficiency of 68 lm/W by using a low-index grid to enhance the outcoupling efficiency[77]. However, at a lighting relevant brightness (500 cd/m^2) the power efficiency dropped to 38 lm/W. Already in 2006, Konica-Minolta had announced an efficiency of 64 lm/W at a brightness of 1000 cd/m^2. In 2009, Reineke *et al.* used high-index glass substrates and a high-index outcoupling structure to increase the efficiency of white OLEDs. An efficiency of 90 lm/W at 1000 cd/m^2 (70 lm/W at 5000 cd/m^2) was reached[13]. Recently, Sasabe *et al.* reported on a white OLED using a blue iridium carbene complex, which reaches without outcoupling enhancement methods approximately 60 lm/W at low brightness and 43.4 lm/W at 1000 cd/m^2.[78]

However, although stable phosphorescent red and green emitters are disclosed in the literature, there is a lack of stable phosphorescent blue emitters. In particular, the lifetime of the phosphorescent blue emitter FIrpic (iridium(III)-bis(4,6,-di-uorophenyl-pyridinato-N,C2)-picolinate) is rather low (1–2 hours). Therefore, there is a high need to find alternative approaches, which avoid the use of unstable phosphorescent emitters.

One of these approaches are so-called hybrid white stacks combining fluorescent blue with phosphorescent red and green emission[79]. Using an ambipolar interlayer between the fluorescent blue and the phosphorescent green emitter to avoid quenching of triplets, Schwartz *et al.* were able to reach high efficiencies (14.4 lm/W at 1000 cd/m^2 without outcoupling enhancement)[80]. However, these approaches are always limited in efficiency since three out of four excitons formed on the blue emitter decay non-radiatively and are therefore lost.

To avoid these losses of excitons on the fluorescent blue emitter, the triplet harvesting concept has been proposed[81–85]. The triplet harvesting concept is based on the idea that an internal quantum efficiency of 100% is still possible in hybrid white OLEDs, if the triplets formed on the blue emitter are transferred to the phosphorescent red or green emitter (i.e. the triplets are 'harvested'). To transfer the triplets from the blue emitter to a phosphorescent emitter, the triplet energy of the blue emitter has to be larger than the triplet energy of the phosphorescent emitter. A candidate for triplet harvesting is 4P-NPD (*N*,*N*′-di-1-naphthalenyl-*N*,*N*′-diphenyl-[1,1′:4′,1″:4″,1‴-quaterphenyl]-4,4‴-diamine). It has a triplet level of 2.30 eV[81] and thus allows harvesting on yellow or red emitters.

A triplet harvesting OLED as proposed by Schwartz *et al.*[82] and Rosenow *et al.*[84] is shown in Fig. 18.4. The OLED is a normal pin OLED. The emission

(a)

Material	Thickness
ITO	90 nm
MeO-TPD:NDP2	20 nm
Spiro-TAD	10 nm
4P-NPD Ir(MDQ)$_2$acac	10 nm: 8 wt%
4P-NPD	0...30 nm
BPhen	10 nm
BPhen:Cs	30 nm
Al	100 nm

(b)

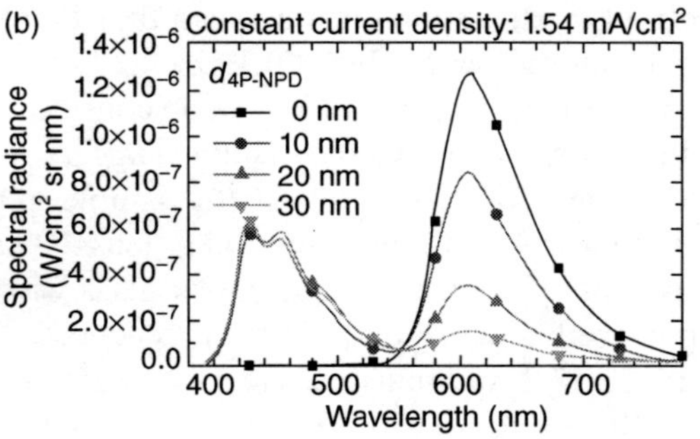

18.4 (a) Red/blue triplet harvesting OLED (MeO-TPD: *N,N,N′,N′*-tetrakis(4-methoxyphenyl)-benzidine). The excitons are formed at the interface between the undoped 4P-NPD layer and the hole blocker (BPhen). Starting from this interface, triplets diffuse towards the 4P-NPD layer doped with the red emitter Ir(MDQ)$_2$acac, where they are harvested[84]. (b) Spectrum of the triplet harvesting OLED for varying thickness of the intrinsic 4P-NPD layer. At a constant current density, the emission of the red emitter decreases, whereas the emission of the blue emitter remains constant. Reprinted with permission from Rosenow *et al.*, *Journal of Applied Physics* **108**, 113113 (2010). Copyright 2010, American Institute of Physics.

layer consists of a neat layer of the blue emitter 4P-NPD with varying thickness (0–30 nm) and a layer of 4P-NPD doped with the red emitter $Ir(MDQ)_2acac$. 4P-NPD has a higher hole than electron conductivity, so that excitons are mainly formed at the interface between the undoped 4P-NPD and the hole blocking layer BPhen. Starting from this interface, the triplets diffuse towards the 4P-NPD layer doped with the red emitter, where the triplets can be harvested and can recombine radiatively. By plotting the spectrum in dependence of the undoped 4P-NPD layer thickness (Fig. 18.4b), triplet harvesting can be proven. For increasing undoped 4P-NPD layer thickness, the red emission decreases as less triplets reach the doped layer, but the emission in the blue spectral region remains constant.

In the device shown in Fig. 18.4, the green part of the visible spectrum is missing and the emission is not white. As the molecule 4P-NPD does not allow harvesting onto green phosphorescent emitters directly, it is challenging to include a green emitter in the stack. However, it is possible to stack two OLEDs, one triplet harvesting blue/green and one green/yellow OLED as shown in Fig. 18.5.

The stack is shown in Fig. 18.5a. The two individual OLED units are connected by a p-n junction consisting of p-doped HTL (MeO-TPD:NDP2), a thin layer of Al and n-doped ETL (BPhen:Cs). Operated in the reverse direction, electrons and holes are generated in the ETL and HTL, respectively, by Zener-Tunneling from the HOMO of the HTL to the LUMO of the ETL[86]. The complete stack can thus be seen as a simple series connection of the two individual OLEDs.

The spectrum shown in Fig. 18.5b consists of the emission of both OLED units and spans the whole visible range. A value of 90 lm/W is reached at a luminance of 1000 cd/m^2, if the sample is prepared on high-index glass and measured with an outcoupling lens. If a scalable outcoupling solution consisting of a sheet of high index glass with small pyramid-structures on top is used, a figure of 47 lm/W is reached[84].

Table 18.2 summarizes the key performance data for a selection of recently reported white-emitting material systems. Apart from the blue component, the emission is almost exclusively generated by phosphorescent emitters in state-of-the-art white OLEDs. The blue component is then either contributed through a hybrid, triplet harvesting arrangement or by adding a phosphorescent blue emitter. It remains to be seen which of the two approaches can achieve better device performance under practical conditions. Currently, triplet harvesting OLEDs hold the record in terms of EQE, whereas the highest luminous efficacies in devices without additional optical outcoupling structures are observed in fully phosphorescent devices. These contain blue and orange emitters or a tri-layer emitters system comprising of different matrices doped with red, green, and blue phosophorescent emitters.

(a)

Material	Thickness
ITO	90 nm
MeO-TPD:NDP2	40 nm: 2 wt%
Spiro-TAD	10 nm
4P-NPD: $Ir(MDQ)_2acac$	5 nm: 5 wt%
4P-NPD:	5 nm
BPhen	10 nm
BPhen:Cs	90 nm
Al	0.5 nm
MeO-TPD:NDP2	65 nm: 2 wt%
Spiro-TAD	10 nm
TCTA: $Ir(ppy)_3$: $Ir(dhfpy)_2acac$	10 nm: 8 wt% 1 wt%
TPBi	10 nm
BPhen:Cs	75 nm
Al	100 nm

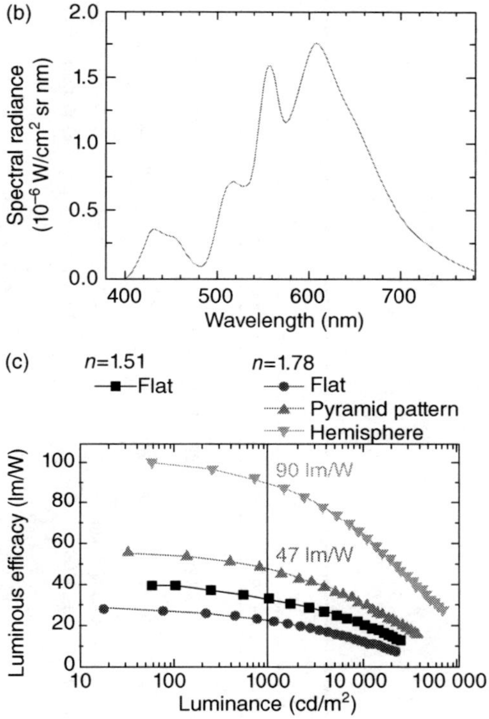

18.5 Stacked white triplet-harvesting OLEDs (TCTA: 4,4′,4″ tris(*N*-carbazolyl)-triphenylamine, TPBi: 2,2′2″-(1,3,5-benzenetriyl)-tris[1-phenyl-1H-benzimidazole], Ir(dhfpy)2acac: bis(2-(9,9-dihexylfluorenyl)-1-pyridine) (acetylacetonate). Two OLEDs – a yellow one and a blue/red triplet harvesting OLED – are stacked (a), so that a white spectrum is reached (b). On high-index glass, 90 lm/W are reached with a large outcoupling lense[84]. Reprinted with permission from Rosenow *et al.*, *Journal of Applied Physics* **108**, 113113 (2010). Copyright 2010, American Institute of Physics.

In conclusion, it is possible to generate white light very efficiently using OLEDs. The use of unstable phosphorescent blue emitters can be avoided by replacing them with fluorescent emitters. However, it is still challenging harvesting on a phosphorescent green emitter, so that complex stacked OLEDs have to be used to reach a white spectrum. Therefore, further development has to be focused on either the development of stable phosphorescent blue emitters or on the development of fluorescent blue emitters with a low singlet/triplet splitting that allows harvesting on a green emitter.

However, even for the triplet harvesting OLEDs discussed in this section the lifetimes are too short for commercial applications. Although white OLEDs with 100 000 h of lifetime have already been reported[98], these high

Table 18.2 White OLEDs

Emitting Layer			PE [lm/W]			EQE [%]		CIE	Reference	
Dopants	*c* [wt.%]	Hosts	(max)	(@ cd/m²)	with outcoupl.*	(max)	(@ cd/m²)	(x;y)	Year	Key
Hybrid										
– / Ir(MDQ)2(acac) / Ir(ppy)3:Ir(dhfpy)2(acac)	– / 5.0 / 8.0:1.0	4p-NPD/ 4p-NPD / TCTA	/	33.0 (@1000)	90.0 (@1000) (a),(d)	/	26.0 (@1000)	(0.505;0.422)	2010	84
– / Ir(MDQ)2(acac) / Ir(ppy)3:Ir(dhfpy)2(acac)	– / 5.0 / 8.0:1.0	4p-NPD/ 4p-NPD / TCTA	/	/	30.1 (@1000) (b)	/	23.7 (@1000)	(0.542;0.416)	2012	87
Ir(MDQ)2(acac) / – / Ir(ppy)3	5.0 / – / 4.0	NPB / 4p-NPD / TPBi	/	22.0 (@1000)	/	14.9	10.4 (@1000)	(0.49;0.41)	2009	81
BczVBi / Ir(pq)2(acac) / Ir(ppy)3 / BczVBi	5.0 / 5.0 / 4.0 / 5.0	CBP / CBP / CBP / CBP	37.6	23.8 (@500)	/	18.7	/	(0.40;0.41)	2006	83
BczVBi / Ir(pq)2(acac) / Ir(ppy)3 / BczVBi	5.0 / 5.0 / 6.0 / 5.0	CBP / CBP / CBP / CBP	/	/	21.6 (b)	/	/	(0.396;0.418)	2006	77
Full phosph.										
FIrpic:Ir(fbi)2(acac)	6.5:0.75	mCP	43.3	/	/	19.3	16.1 (@100)	(0.33;0.39)	2009	88
FIrpic:Ir(FIpy-CF3)3	19.0:1.0	H2	/	23.3 (@1000)	/	/	16.4 (@1000)	(0.383;0.432)	2012	89
FIr6:Ir(ppy)3:Ir(pq)2(acac)	20.0:0.5:2.0	UGH2	26.0	11.0 (@1000)	/	12.0	/	(0.43;0.45)	2004	74
FIrpic:Ir(fbi)2(acac)	8.0:0.67	p -BISiTPA	42.7	25.2 (@1000)	/	19.1	17.4 (@1000)	(0.38;0.44)	2011	50,51

FIr6:Ir(ppy)3:Ir(pq)2(acac) / FIr6:Ir(ppy)3:Ir(pq)2(acac)	25.0:2.5:2.3 / 5.0:1.3:1.0	UGH2 / mCP	40.0	26.0 (@1000)	/	18.0	/	(0.35;0.40)	2009	90
Os(fptz)2(PPh2Me)2 / FIrpic	7.0 / 7.0	POAPF / POAPF	43.9	29.5 (@1000)	/	18.4	14.9 (@1000)	(0.38;0.34)	2009	58
FIrpic / Ir(pq)2(acac)	15.0 / 3.0	4CzPBP / 4CzPBP	/	46.0 (@100)	62.8 (@100) (c)	/	20.6 (@100)	(0.31;0.40)	2006	91
FCNIrpic / Ir(ppy)3:Ir(pq)2(acac)	5.0 / 3.0:0.7	DFCz:DFPO / BPSBF	43.7	22.8 (@1000)	/	19.5	17.9 (@1000)	(0.39;0.46)	2008	92
FIrpic:Ir(pq)2(dpm) / FIrpic:Ir(pq)2(dpm)	7.0:3.0 / 20.0:3.0	TCTA / DczPPy	/	44.0 (@1000)	/	/	25.0 (@1000)	(0.335;0.396)	2008	93
Ir(pq)2(acac) / Ir(ppy)3 / FIr6	4.0 / 2.0 / 20.0	TCTA / mCP / UGH2	38.6	/	64.0 (max.) (b)	15.3	/	/	2007	94
Ir(pq)2(acac) / Ir(ppy)3 / FIr6	4.0 / 2.0 / 20.0	TCTA / mCP / UGH2	37.0	22.0 (@500)	63.0 (max.) (b)	15.3	13.6 (@500)	(0.37;0.41)	2008	95
Ir(pq)2(dpm) / Ir(ppy)3 / Ir(dbfmi)	2.0 / 6.0 / 10.0	CBP / CBP / PO9	59.9	43.3 (@1000)	/	21.6	21.5 (@1000)	(0.43;0.43)	2010	78
Ir(MDQ)2(acac) / FIrpic / Ir(ppy)3	10.0 / 20.0 / 8.0	TCTA / TPBi / TPBi	/	30.0 (@1000)	124 (@1000) (a),(d)	/	13.1 (@1000)	(0.45;0.47)	2009	13
Ir(MDQ)2(acac) / FIr6 / Ir(ppy)3:Ir(dhfpy)2(acac)	10.0 / 20.0 / 20.0 / 8.0:1.0	TCTA / TCTA / SPPO1 / SPPO1	/	17.4 (@1000)	/	/	10.0 (@ 1000)	(0.444;0.409)	2012	96
Proprietary commercial OLED stack	N/A	N/A	/	/	87 (@1000) (b),(d)	/	/	(0.463;0.436)	2012	97

Abbrevations used in the table: phosph. = phosphorescent; c: concentration; EQE = external quantum efficiency; PE = power efficacy.
* Outcoupling techniques used: (a) half sphere; (b) micro lens array; (c) diffusive optical film; (d) high index substrate.

lifetimes have only been reached for hybrid white approaches without triplet harvesting. Therefore, some mechanisms that lead to the low lifetimes of OLEDs are discussed in the following section and it will be shown, how long-living OLEDs can be realized.

18.6 Degradation of organic light-emitting diodes (OLEDs)

Although much progress has been made and the lifetime of OLEDs has significantly increased over the years, little is known about the precise degradation mechanism. There are many possible mechanisms of OLED degradation, related to intrinsic (such as degradation of emitter molecules) and extrinsic (such as contact delamination due to water vapor intake). Of particular interest are the intrinsic decay processes, for instance those related to chemical reactions occurring in the devices.

Techniques which provide microscopic information about these reactions are especially useful since they allow optimization of devices based on a detailed understanding of the degradation processes.

One technique, which has been shown to be highly valuable in determining degradation pathways and reaction products in OLEDs, is matrix-assisted laser desorption/ionization time-of-flight mass spectrometry (MALDI-TOF-MS)[99,100]. Using this technique it is possible to deduce a common pathway for OLED degradation based on the dissociation of the Ir-emitter and the formation of complexes of the individual fragments with the hole blocking layer[23,99]. For example, the proposed reaction pathway of the degradation of the phosphorescent blue emitter FIrpic is shown in Fig. 18.6. Two dissociation reactions are postulated – a reversible (pathway 1) and an irreversible one (pathway 2). The fragments can interact with the blocking materials, i.e. either with TPBi or with the charged complex BPhen+Cs, and build the final degradation product[100].

In red OLEDs, the strength of the complexation was shown to correlate with the lifetime[99], presenting the possibility of finding optimized materials without the need to process full OLED structures for testing. Furthermore, it was shown that the lifetime of OLEDs can be increased significantly, if blocking materials are used that do not form complexes with the Ir-emitter. For example, the use of an alternative hole blocking layer, BAlq (bis-(2-methyl-8-quinolinolato)-4-(phenyl-phenolato) aluminum-(III)), instead of BPhen was shown to yield extremely long-living red OLEDs[23]. Extrapolated lifetimes of more than 10^6 at an initial brightness of 100 cd/m^2 were reached.

In conclusion, OLEDs can be made long-living. However, the lifetime of highly efficient white OLEDs is still limited by the blue emitter. Further development has to be done to overcome these limitations.

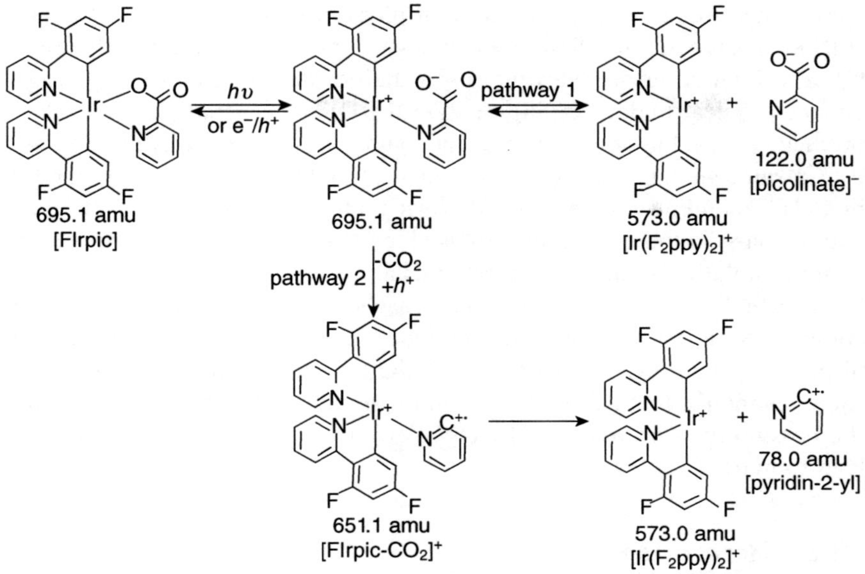

18.6 Proposed mechanism of the dissociation of the phosphorescent blue emitter FIrpic[100]. Reprinted from Moraes *et al.*, *Organic Electronics* **12**, 341–347 (2011), with permission from Elsevier.

18.7 Future trends

The development of OLEDs has been impressive in all respects and both efficiency and lifetime have significantly improved over time. Currently, several companies are preparing for the mass-production of OLEDs for lighting or display applications or are already producing OLED devices. However, there are still challenges to be solved before OLEDs can reach their full potential.

One challenge is the improvement of outcoupling efficiency. Still, the majority of light generated inside the device is trapped in the OLED cavity. Although the use of high-index glass combined with an outcoupling lens improves the outcoupling efficiency significantly, this approach is less suitable for commercial applications. Scalable and cost-effective solutions have to be found, which avoid total reflection or scatter light inside the cavity to surpass the efficiency limit.

A second limitation of OLED research is the lack of a stable and efficient blue phosphorescent emitter. Although the triplet harvesting approach can avoid the use of a phosphorescent emitter, the available triplet harvesting molecules are also limited in their lifetime. Further materials development is thus necessary and novel emitters have to be designed.

Finally, a predictive simulation of all physical processes of the OLED is still lacking. Up to now, all progress in device optimization was purely driven by experimental efforts, slowing down the development and hampering a deeper physical understanding of the OLED. The reasons for the lack of predictive modeling are manifold. The main reason is arguably the hopping transport occurring in the devices, which can only be described by mobility models depending on several unknown materials parameters. Combined with the large number of materials used in OLED research, the uncertainty in any simulation is large and forecloses a predictive simulation.

Nevertheless, OLED research has reached a level where OLEDs can compete with conventional technologies in both fields of application, display and lighting. In particular for lighting applications, OLEDs have some distinctive features such as a slim form factor, flexible substrates and the possibility of producing OLEDs by roll-to-roll manufacturing that will lead to market success.

18.8 References

1. Tang, C. W. & VanSlyke, S. A. Organic electroluminescent diodes. *Applied Physics Letters* **51**, 913 (1987).
2. Bernanose, A., Comte, M. & Vouaux, P. A new method of emission of light by certain organic compounds. *Journal de Chimie Physique et de Physico-Chimie Biologique* **50**, 64 (1953).
3. Bernanose, A. Electroluminescence of organic compounds. *British Journal Of Applied Physics* S54–S56 (1955).
4. Helfrich, W. & Schneider, W. G. Recombination radiation in anthracene crystals. *Physical Review Letters* **14**, 229–231 (1965).
5. Williams, D. & Schadt, M. A simple organic electroluminescent diode. *Proceedings of the Institute of Electrical And Electronics Engineers* **58**, 476 (1970).
6. Kampas, F. J. & Gouterman, M. Porphyrin films – electroluminescence of octaethylporphin. *Chemical Physics Letters* **48**, 233–236 (1977).
7. Vincett, P. S., Barlow, W. A., Hann, R. A. & Roberts, G. G. Electrical-conduction and low-voltage blue electro-luminescence in vacuum-deposited organic films. *Thin Solid Films* **94**, 171–183 (1982).
8. Adachi, C., Baldo, M. A., Thompson, M. E. & Forrest, S. R. Nearly 100% internal phosphorescence efficiency in an organic light emitting device. *Journal of Applied Physics* **90**, 5048 (2001).
9. Baldo, M. A., Lamansky, S., Burrows, P. E., Thompson, M. E. & Forrest, S. R. Very high-efficiency green organic light-emitting devices based on electro-phosphorescence. *Applied Physics Letters* **75**, 4–6 (1999).
10. Pfeiffer, M., Beyer, A., Fritz, T. & Leo, K. Controlled doping of phthalocyanine layers by cosublimation with acceptor molecules: a systematic Seebeck and conductivity study. *Applied Physics Letters* **73**, 3202 (1998).
11. Meerheim, R., Walzer, K., Pfeiffer, M., Leo, K. & He, G. Highly efficient organic light emitting diodes (OLED) for displays and lighting. *Proceedings of SPIE* **89**, 22 (2006).

12. Meerheim, R., Furno, M., Hofmann, S., Lüssem, B. & Leo, K. Quantification of energy loss mechanisms in organic light-emitting diodes. *Applied Physics Letters* **97**, 253305 (2010).
13. Reineke, S. *et al.* White organic light-emitting diodes with fluorescent tube efficiency. *Nature* **459**, 234 (2009).
14. Furno, M., Meerheim, R., Hofmann, S., Lüssem, B. & Leo, K. Efficiency and rate of spontaneous emission in organic electroluminescent devices. *Physical Review B* **85**, 115205 (2012).
15. Barnes, W. L. Fluorescence near interfaces: the role of photonic mode. *Journal of Modern Optics* **45**, 661–699 (1998).
16. Purcell, E. Spontaneous emission probabilities at radio frequencies. *Physical Review* **69**, 681 (1946).
17. Neyts, K. A. Simulation of light emission from thin-film microcavities. *Journal of the Optical Society of America A* **15**, 962 (1998).
18. Greenham, N. C., Friend, R. H. & Bradley, D. D. C. Angular dependence of the emission from a conjugated polymer light-emitting diode: implications for efficiency calculations. *Advanced Materials* **6**, 491 (1994).
19. Walzer, K., Maennig, B., Pfeiffer, M. & Leo, K. Highly efficient organic devices based on electrically doped transport layers. *Chemical Reviews* **107**, 1233 (2007).
20. Meerheim, R., Lüssem, B. & Leo, K. Efficiency and stability of p-i-n type organic light emitting diodes for display and lighting applications. *Proceedings of the IEEE* **97**, 1606–1626 (2009).
21. Blochwitz, J. *et al.* Interface electronic structure of organic semiconductors with controlled doping levels. *Organic Electronics* **2**, 97 (2001).
22. Tanaka, D. *et al.* Ultra high efficiency green organic light-emitting devices. *Japanese Journal Of Applied Physics Part 2 – Letters & Express Letters* **46**, L10–L12 (2007).
23. Meerheim, R. *et al.* Influence of charge balance and exciton distribution on efficiency and lifetime of phosphorescent organic light-emitting devices. *Journal of Applied Physics* **104**, 014510 (2008).
24. Chopra, N. *et al.* High efficiency blue phosphorescent organic light-emitting device. *Applied Physics Letters* **93**, 143307 (2008).
25. He, G. *et al.* High-efficiency and low-voltage p-i-n electrophosphorescent organic light-emitting diodes with double-emission layers. *Applied Physics Letters* **85**, 3911 (2004).
26. Tanaka, D., Agata, Y., Takeda, T., Watanabe, S. & Kido, J. High luminous efficiency blue organic light-emitting devices using high triplet excited energy materials. *Japanese Journal of Applied Physics* **46**, L117–L119 (2007).
27. Jeon, S. O., Jang, S. E., Son, H. S. & Lee, J. Y. External quantum efficiency above 20% in deep blue phosphorescent organic light-emitting diodes. *Advanced Materials* **23**, 1436–1441 (2011).
28. Jeon, S. O., Yook, K. S., Joo, C. W. & Lee, J. Y. High-efficiency deep-blue-phosphorescent organic light-emitting diodes using a phosphine oxide and a phosphine sulfide high-triplet-energy host material with bipolar charge-transport properties. *Advanced Materials* **22**, 1872–1876 (2010).
29. Jeon, S. O., Yook, K. S., Joo, C. W. & Lee, J. Y. Phenylcarbazole-based phosphine oxide host materials for high efficiency in deep blue phosphorescent organic light-emitting diodes. *Advanced Functional Materials* **19**, 3644–3649 (2009).

30. Chou, H.-H. & Cheng, C.-H. A highly efficient universal bipolar host for blue, green, and red phosphorescent OLEDs. *Advanced Materials* **22**, 2468–2471 (2010).
31. Eom, S.-H. *et al.* Low voltage and very high efficiency deep-blue phosphorescent organic light-emitting devices. *Applied Physics Letters* **93**, 133309 (2008).
32. Su, S.-J., Sasabe, H., Takeda, T. & Kido, J. Pyridine-containing bipolar host materials for highly efficient blue phosphorescent OLEDs. *Chemistry of Materials* **20**, 1691–1693 (2008).
33. Sasabe, H. *et al.* Wide-energy-gap electron-transport materials containing 3,5-dipyridylphenyl moieties for an ultra high efficiency blue organic light-emitting device. *Chemistry of Materials* **20**, 5951–5953 (2008).
34. Chen, Y., Chen, J., Zhao, Y. & Ma, D. High efficiency blue phosphorescent organic light-emitting diode based on blend of hole- and electron-transporting materials as a co-host. *Applied Physics Letters* **100**, 213301 (2012).
35. Sasabe, H., Seino, Y., Kimura, M. & Kido, J. A m-terphenyl-modified sulfone derivative as a host material for high efficiency blue and green phosphorescent OLEDs. *Chemistry of Materials* **24**, 1404–1406 (2012).
36. Chopra, N. *et al.* High efficiency and low roll-off blue phosphorescent organic light-emitting devices using mixed host architecture. *Applied Physics Letters* **97**, 033304 (2010).
37. Pu, Y.-J. *et al.* Optimizing the charge balance of fluorescent organic light-emitting devices to achieve high external quantum efficiency beyond the conventional upper limit. *Advanced Materials* **24**, 1765–1770 (2012).
38. Jeon, Y.-M., Lee, J.-Y., Kim, J.-W., Lee, C.-W. & Gong, M.-S. Deep-blue OLEDs using novel efficient spiro-type dopant materials. *Organic Electronics* **11**, 1844–1852 (2010).
39. Kim, S.-K., Yang, B., Ma, Y., Lee, J.-H. & Park, J.-W. Exceedingly efficient deep-blue electroluminescence from new anthracenes obtained using rational molecular design. *Journal of Materials Chemistry* **18**, 3376–3384 (2008).
40. Yokoyama, D. *et al.* Dual efficiency enhancement by delayed fluorescence and dipole orientation in high-efficiency fluorescent organic light-emitting diodes. *Applied Physics Letters* **99**, 123303 (2011).
41. Matsumoto, N., Miyazaki, T., Nishiyama, M. & Adachi, C. Efficient deep-blue organic light-emitting diodes based on 9,9-bis(4-biphenylyl)fluorene derivatives. *The Journal of Physical Chemistry C* **113**, 6261–6266 (2009).
42. Meerheim, R., Nitsche, R. & Leo, K. High-efficiency monochrome organic light emitting diodes employing enhanced microcavities. *Applied Physics Letters* **93**, 043310 (2008).
43. Tao, Y. *et al.* A simple carbazole/oxadiazole hybrid molecule: an excellent bipolar host for green and red phosphorescent OLEDs. *Angewandte Chemie* (International ed. in English) **47**, 8104–8107 (2008).
44. Tao, Y. *et al.* Molecular design of host materials based on triphenylamine/oxadiazole hybrids for excellent deep-red phosphorescent organic light-emitting diodes. *Journal of Materials Chemistry* **20**, 1759–1765 (2010).
45. Cao, X. A. & Zhang, Y. Q. Performance enhancement of organic light-emitting diodes by chlorine plasma treatment of indium tin oxide. *Applied Physics Letters* **100**, 183304 (2012).
46. Wang, Z. B. *et al.* Unlocking the full potential of organic light-emitting diodes on flexible plastic. *Nature Photonics* **5**, 753–757 (2011).

47. Helander, M. G. *et al.* Chlorinated indium tin oxide electrodes with high work function for organic device compatibility. *Science* **332**, 944–947 (2011).
48. Wang, Z. B. *et al.* Highly simplified phosphorescent organic light emitting diode with > 20% external quantum efficiency at >10,000 cd/m^2. *Applied Physics Letters* **98**, 073310 (2011).
49. Tao, Y. *et al.* Multifunctional triphenylamine/oxadiazole hybrid as host and exciton-blocking material: high efficiency green phosphorescent OLEDs using easily available and common materials. *Advanced Functional Materials* **20**, 2923–2929 (2010).
50. Gong, S. *et al.* Bipolar tetraarylsilanes as universal hosts for blue, green, orange, and white electrophosphorescence with high efficiency and low efficiency roll-off. *Advanced Functional Materials* **21**, 1168–1178 (2011).
51. Gong, S. *et al.* De novo design of silicon-bridged molecule towards a bipolar host: all-phosphor white organic light-emitting devices exhibiting high efficiency and low efficiency roll-off. *Advanced Materials* **22**, 5370–5373 (2010).
52. Wang, Z. B. *et al.* Pt(II) complex based phosphorescent organic light emitting diodes with external quantum efficiencies above 20%. *Applied Physics Letters* **98**, 213301 (2011).
53. Okumoto, K., Kanno, H., Hamaa, Y., Takahashi, H. & Shibata, K. Green fluorescent organic light-emitting device with external quantum efficiency of nearly 10%. *Applied Physics Letters* **89**, 063504 (2006).
54. Hung, W.-Y. *et al.* A new benzimidazole/carbazole hybrid bipolar material for highly efficient deep-blue electrofluorescence, yellow–green electrophosphorescence, and two-color-based white OLEDs. *Journal of Materials Chemistry* **20**, 10113–10119 (2010).
55. Hofmann, S. *et al.* Top-emitting organic light-emitting diodes: influence of cavity design. *Applied Physics Letters* **97**, 253308 (2010).
56. Kim, D. H. *et al.* Highly efficient red phosphorescent dopants in organic light-emitting devices. *Advanced Materials* **23**, 2721–2726 (2011).
57. Ting, H.-C. *et al.* Indolo[3,2-b]carbazole/benzimidazole hybrid bipolar host materials for highly efficient red, yellow, and green phosphorescent organic light emitting diodes. *Journal of Materials Chemistry* **22**, 8399–8407 (2012).
58. Chien, C.-H., Hsu, F.-M., Shu, C.-F. & Chi, Y. Efficient red electrophosphorescence from a fluorene-based bipolar host material. *Organic Electronics* **10**, 871–876 (2009).
59. Kwak, J. *et al.* New carbazole-based host material for low-voltage and highly efficient red phosphorescent organic light-emitting diodes. *Journal of Materials Chemistry* **22**, 6351–6355 (2012).
60. Jeon, S. O. *et al.* High efficiency red phosphorescent organic light-emitting diodes using a spirobenzofluorene type phosphine oxide as a host material. *Organic Electronics* **10**, 998–1000 (2009).
61. Jeon, S. O., Yook, K. S., Joo, C. W. & Lee, J. Y. Theoretical maximum quantum efficiency in red phosphorescent organic light-emitting diodes at a low doping concentration using a spirobenzofluorene type triplet host material. *Organic Electronics* **11**, 881–886 (2010).
62. Kondakov, D. Y., Pawlik, T. D., Hatwar, T. K. & Spindler, J. P. Triplet annihilation exceeding spin statistical limit in highly efficient fluorescent organic light-emitting diodes. *Journal of Applied Physics* **106**, 124510 (2009).

63. Mladenovski, S. *et al.* Integrated optical model for organic light-emitting devices. *Journal of Applied Physics* **109**, 083114 (2011).
64. Kim, S.-Y. & Kim, J.-J. Outcoupling efficiency of organic light emitting diodes and the effect of ITO thickness. *Organic Electronics* **11**, 1010–1015 (2010).
65. Mladenovski, S., Neyts, K., Pavicic, D., Werner, A. & Rothe, C. Exceptionally efficient organic light emitting devices using high refractive index substrates. *Optics Express* **17**, 7562–70 (2009).
66. Cheng, Y.-H., Wu, J.-L., Cheng, C.-H., Syao, K.-C. & Lee, M.-C. M. Enhanced light outcoupling in a thin film by texturing meshed surfaces. *Applied Physics Letters* **90**, 091102 (2007).
67. Möller, S. & Forrest, S. R. Improved light out-coupling in organic light emitting diodes employing ordered microlens arrays. *Journal of Applied Physics* **91**, 3324 (2002).
68. Yamasaki, T., Sumioka, K. & Tsutsui, T. Organic light-emitting device with an ordered monolayer of silica microspheres as a scattering medium. *Applied Physics Letters* **76**, 1243 (2000).
69. Flämmich, M. *et al.* Oriented phosphorescent emitters boost OLED efficiency. *Organic Electronics* **12**, 1663–1668 (2011).
70. Frischeisen, J., Yokoyama, D., Endo, A., Adachi, C. & Brütting, W. Increased light outcoupling efficiency in dye-doped small molecule organic light-emitting diodes with horizontally oriented emitters. *Organic Electronics* **12**, 809–817 (2011).
71. Saxena, K., Jain, V. K. & Mehta, D. S. A review on the light extraction techniques in organic electroluminescent devices. *Optical Materials* **32**, 221–233 (2009).
72. Kido, J., Hongawa, K., Okuyama, K. & Nagai, K. White light-emitting organic electroluminescent devices using the poly(*N*-vinylcarbazole) emitter layer doped with three fluorescent dyes. *Applied Physics Letters* **64**, 815 (1994).
73. Kido, J., Kimura, M. & Nagai, K. Multilayer white light-emitting organic electroluminescent device. *Science* **267**, 1332 (1995).
74. D'Andrade, B. W., Holmes, R. J. & Forrest, S. R. Efficient organic electrophosphorescent white-light-emitting device with a triple doped emissive layer. *Advanced Materials* **16**, 624 (2004).
75. D'Andrade, B. W., Thompson, M. E. & Forrest, S. R. Controlling exciton diffusion in multilayer white phosphorescent organic light emitting devices. *Advanced Materials* **14**, 147 (2002).
76. Tokito, S., Iijima, T., Tsuzuki, T. & Sato, F. High-efficiency white phosphorescent organic light-emitting devices with greenish-blue and red-emitting layers. *Applied Physics Letters* **83**, 2459 (2003).
77. Sun, Y. & Forrest, S. R. Organic light emitting devices with enhanced outcoupling via microlenses fabricated by imprint lithography. *Journal of Applied Physics* **100**, 073106 (2006).
78. Sasabe, H. *et al.* High-efficiency blue and white organic light-emitting devices incorporating a blue iridium carbene complex. *Advanced Materials* **22**, 5003–7 (2010).
79. Schwartz, G., Fehse, K., Pfeiffer, M., Walzer, K. & Leo, K. Highly efficient white organic light emitting diodes comprising an interlayer to separate fluorescent and phosphorescent regions. *Applied Physics Letters* **89**, 083509 (2006).

80. Schwartz, G., Walzer, K., Pfeiffer, M. & Leo, K. High-efficiency white organic-light-emitting diodes combining fluorescent and phosphorescent emitter systems. *Proceedings of SPIE* **6192**, 61920Q (2006).
81. Schwartz, G., Reineke, S., Rosenow, T. C., Walzer, K. & Leo, K. Triplet harvesting in hybrid white organic light-emitting diodes. *Advanced Functional Materials* **19**, 1319–1333 (2009).
82. Schwartz, G., Pfeiffer, M., Reineke, S., Walzer, K. & Leo, K. Harvesting triplet excitons from fluorescent blue emitters in white organic light-emitting diodes. *Advanced Materials* **19**, 3672 (2007).
83. Sun, Y. R. *et al.* Management of singlet and triplet excitons for efficient white organic light-emitting devices. *Nature* **440**, 908 (2006).
84. Rosenow, T. C. *et al.* Highly efficient white organic light-emitting diodes based on fluorescent blue emitters. *Journal of Applied Physics* **108**, 113113 (2010).
85. Kondakova, M. E. *et al.* Highly efficient fluorescent-phosphorescent triplet-harvesting hybrid organic light-emitting diodes. *Journal of Applied Physics* **107**, 014515 (2010).
86. Kleemann, H. *et al.* Organic zener diodes: tunneling across the gap in organic semiconductor materials. *Nano Letters* **10**, 4929–4934 (2010).
87. Thomschke, M., Reineke, S., Lüssem, B. & Leo, K. Highly efficient white top-emitting organic light-emitting diodes comprising laminated microlens films. *Nano Letters* **12**, 424–428 (2012).
88. Wang, Q. *et al.* Harvesting excitons via two parallel channels for efficient white organic leds with nearly 100% internal quantum efficiency: fabrication and emission-mechanism analysis. *Advanced Functional Materials* **19**, 84–95 (2009).
89. Zhang, B. *et al.* High-efficiency single emissive layer white organic light-emitting diodes based on solution-processed dendritic host and new orange-emitting iridium complex. *Advanced Materials* **24**, 1873–1877 (2012).
90. Eom, S.-H. *et al.* White phosphorescent organic light-emitting devices with dual triple-doped emissive layers. *Applied Physics Letters* **94**, 153303 (2009).
91. Ide, N., Komoda, T. & Kido, J. Organic light-emitting diode (OLED) and its application to lighting devices. *Proceedings of SPIE* **6333**, 63330M–1–10 (2006).
92. Seo, C. W. & Lee, J. Y. Highly efficient white phosphorescent organic light emitting diodes using a mixed host structure in deep blue emitting layer. *Thin Solid Films* **520**, 5075–5079 (2012).
93. Su, S.-J., Gonmori, E., Sasabe, H. & Kido, J. Highly efficient organic blue-and white-light-emitting devices having a carrier- and exciton-confining structure for reduced efficiency roll-off. *Advanced Materials* **20**, 4189–4194 (2008).
94. Sun, Y. & Forrest, S. R. High-efficiency white organic light emitting devices with three separate phosphorescent emission layers. *Applied Physics Letters* **91**, 263503 (2007).
95. Sun, Y. & Forrest, S. R. Multiple exciton generation regions in phosphorescent white organic light emitting devices. *Organic Electronics* **9**, 994–1001 (2008).
96. Weichsel, C., Reineke, S., Furno, M., Lüssem, B. & Leo, K. Organic light-emitting diodes for lighting: High color quality by controlling energy transfer processes in host–guest-systems. *Journal of Applied Physics* **111**, 033102 (2012).
97. Komoda, T., Yamae, K., Kittichungchit, V., Tsuji, H. & Ide, N. Extremely high performance white OLEDs for lighting. *SID Symposium Digest of Technical Papers* **43**, 610–613 (2012).

98. Birnstock, J., He, G., Murano, S., Werner, A. & Zeika, O. White stacked OLED with 35 lm/W and 100,000 hours lifetime at 1000 cd/m^2 for display and lighting applications. *SID Symposium Digest of Technical Papers* **39**, 822–825 (2008).
99. Scholz, S., Meerheim, R., Lüssem, B. & Leo, K. Laser desorption/ionization time-of-flight mass spectrometry: a predictive tool for the lifetime of organic light emitting devices. *Applied Physics Letters* **94**, 043314 (2009).
100. Moraes, I. R. D., Scholz, S., Lüssem, B. & Leo, K. Analysis of chemical degradation mechanism within sky blue phosphorescent organic light emitting diodes by laser-desorption/ionization time-of-flight mass spectrometry. *Organic Electronics* **12**, 341–347 (2011).

19
Organic spintronics

Z. V. VARDENY and T. D. NGUYEN,
University of Utah, USA and E. EHRENFREUND,
Technion–Israel Institute of Technology, Israel

DOI: 10.1533/9780857098764.3.535

Abstract: Recent advances in organic spin response have unraveled long polaron spin coherence time, substantive magnetic field effect (MFE) in organic light emitting diodes (OLED) and films, and spin-polarized carrier injection from ferromagnetic electrodes in organic spin valves (OSV). Here we compare spin response in films, OLED and OSV devices based on π-conjugated polymers made of protonated, and deuterated (D-) hydrogen having weaker HFI strength. We show that the HFI plays a crucial role in determining the spin response in all devices. OLEDs (and films) based on the D-polymers show substantial narrower MFE response, and the larger spin diffusion length leads to higher magnetoresistance. We also show that the dominant MFE (B) response in devices and films is the polaron pair model.

Key words: organic spintronics, organic semiconductors, organic light-emitting diodes (OLED,), organic spin-valves (OSVs), spin-polarized carrier injection, giant magneto resistance (GMR), optically detected magnetic resonance (ODMR), hyperfine interaction (HFI), magnetic field effect, magneto-electroluminescence (MEL), organic magneto-resistance.

19.1 Introduction

Over the past two decades the electron spin has transformed from an exotic subject in classroom lectures to a degree of freedom that materials scientists and engineers exploit in new electronic devices. This interest has been motivated from the prospect of using the spin degree of freedom in addition to the charge, as an information-carrying physical quantity in electronic devices; thus changing the device functionality in an entirely new paradigm, which has been dubbed spintronics (Wolf *et al.*, 2001; Zutic *et al.*, 2004). This interest culminated by awarding the 2007 Nobel Prize in Physics to Drs. Fert and Grünberg for the discovery and application of giant magneto resistance (GMR). More recently the spintronics field has focused on hybrids of ferromagnetic (FM) electrodes and semiconductors, in particular spin injection and transport in the classic semiconductor gallium arsenide

(Awschalom and Flatte, 2007). However, spin injection into semiconductors has been a challenge because of the impedance matching problem between the FM and semiconductor (Yunus *et al.*, 2008). In any case this research has yielded much original physical insight, but no successful applications yet. For a few years, interest has also risen in similar phenomena in organics.

The organic spintronics field was launched via articles by Dediu *et al.* (2002) that suggested spin injection in organic semiconductors and Xiong *et al.* (2004) that demonstrated spin-polarized currents in thin films of small organic molecule, namely Alq_3. Organic semiconductors can absorb and emit light while transporting charge, and this leads to optoelectronic devices such as photovoltaic cells, organic light-emitting diodes (OLEDs), and organic field-effect transistors (OFETs). It is thus expected that adding control of the electron spin to the multifunctional characteristics of these versatile materials should yield novel magnetic devices in the near future. However, there are serious challenges to be faced in understanding the properties of spin-polarized current in organic semiconductors and obtaining high-quality devices. Most importantly the approach in studying spin injection and transport in organic thin films is fundamentally different from that used for inorganic counterparts.

Organic semiconductors are composed of light elements that have weak spin–orbit interaction; consequently they should possess long spin relaxation times (Ruden and Smith, 2004; Pramanik *et al.*, 2007; McCamey *et al.*, 2008). Moreover, hyperfine interaction (HFI) has been thought to play an important role in organic magneto-transport (Bobbert *et al.*, 2009). For example, if the HFI constant, $\underline{a}$ determines the spin lattice relaxation time T_{SL} of the injected carriers, and consequently also their spin diffusion length in a device setting, then the device performance may be enhanced simply by manipulating the nuclear spins of the organic spacer atoms. Moreover the HFI may also play an important role in other organic magneto-electronic devices such as two-terminal devices (Pulizzi, 2009). In addition to spin polarized carrier injection into organic semiconductors, another fascinating effect has been observed when applying a small external magnetic field to OLEDs that was dubbed 'magnetic field effect', MFE (Hayashi, 2004), in which both the current and electroluminescence increase by as much as 40% at room temperature in a relatively small magnetic field of ~100 gauss. In fact, MFE in organics was discovered about four decades ago (Ern and Merrifield, 1968); however, renewed interest in it has recently risen since it was realized that the MFE in OLEDs has tremendous potential application. In spite of the recent research effort in organic MFE, the underlying mechanism and basic experimental findings are still hotly debated. Finally, HFI can influence other spin response processes such as optically detected magnetic resonance (ODMR) in organic semiconductor films (McCamey *et al.*, 2008). The ODMR is actually an MFE induced by microwave (MW) absorption that mixes the spin sublevels of polaron pair species. Therefore, understanding of ODMR in organic films and electronic devices may shed light on the organic MFE phenomenon. Moreover, ODMR spectroscopy is unique in that it readily measures the spin relaxation time in organic semiconductors, and thus information may be obtained about the electronic spin interaction strength. We have therefore used this technique to obtain the HFI strength in the same materials that are used for spin injection and MFE studies.

In this chapter we review some of the research highlights achieved at the University of Utah in the field of organic spintronics (Nguyen *et al.*, 2010a, 2010b, 2011a, 2011b; Wang and Vardeny, 2010); Specifically, we investigate the role of the HFI in various organic magneto-electronic devices and films by replacing all strongly coupled hydrogen atoms (^{1}H, nuclear spin I = ½) in the organic π-conjugated polymer poly(dioctyloxy) phenyl vinylene (DOO-PPV) spacer (dubbed here H-polymer), with deuterium atoms (^{2}H, I = 1) (hereafter D-polymer; see Fig. 19.1) having much smaller a_{HF}, namely $a_{HF}(D) = a_{HF}(H)/6.5$ (Carrington and McLachlan, 1967). We studied the influence of this hydrogen isotope exchange on the magnetic response of *three spin-dependent processes*. These are:

- magneto-electroluminescence (MEL) and magneto-conductance (MC) response in OLEDs, which are presented in Section 19.2. In this section we also summarize the newly discovered ultra small MFE, such as in MEL and MC at $B < 1$ mT.
- GMR in organic spin-valves (OSVs), where the spin transport in the device determines its performance (Xiong *et al.*, 2004)· is presented in Section 19.3.
- ODMR in thin films, where spin-dependent recombination of photogenerated spin ½ polaron pairs is enhanced under magnetic resonance conditions (Yang *et al.*, 2007), which is discussed in Section 19.4.

19.2 Magneto-conductance (MC) and magneto-electroluminescence (MEL) in organic light-emitting diodes (OLEDs)

The MFE involving photoconductivity (PC) in poly-phenylene OLEDs was first conducted by Frankewich *et al.* in 1996. They detected a significant change of a few percent in PC when a small magnetic field of ~100 mT was applied to the device. Later on Kalinowski *et al.* showed that the electroluminescence (EL) can be modulated in OLEDs made of small molecule such as tris(8-hydroxyquinline aluminum) (Alq_3; structure given in Fig. 19.1) by a small applied magnetic field (Kalinowski *et al.*, 2003, 2004). In 2004

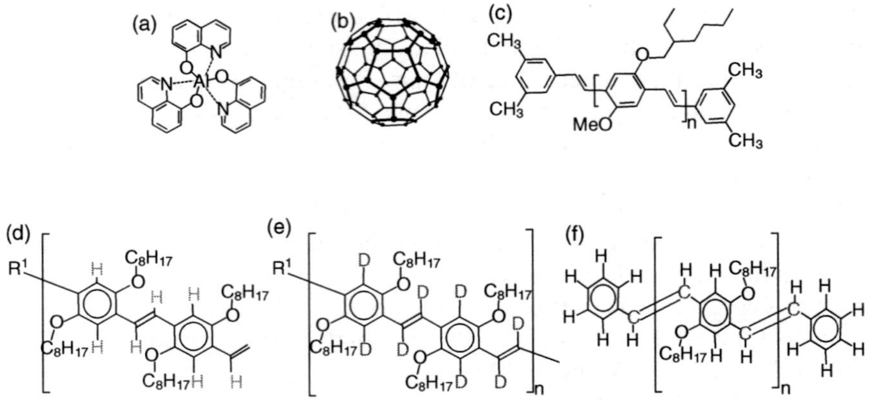

19.1 Chemical structures of several organic semiconductors that we used for the organic spintronics research at the University of Utah: small molecules (a) tris(8-hydroxyquinolinato)aluminium (Alq_3) (b) fullerene C_{60} and π-conjugated polymers (c) 2-methoxy-5-(2′-ethylhexyloxy) (MEHPPV) (d) poly(dioctyloxy)phenylenevinylene that contains hydrogen isotope (H-polymer) (e) deuterated DOO-PPV where some hydrogen atoms in the DOO-PPV skeleton are replaced by deuterium atoms (D-polymer) and (f) C13-rich DOO-PPV (C13-polymer) where some ^{12}C in the polymer chain backbone were replaced by ^{13}C.

Wohlgenannt and coworkers demonstrated a very large magneto-resistance (MR) up to 10% at a characteristic field of 10 mT in OLEDs made of the polymer poly(9,9-dioctylfluorenyl-2,7-diyl) (PFO) (Francis *et al.*, 2004). The effect was dubbed organic magneto-resistance (OMAR). Later, they also found that OMAR with similar magnitude occurs in small molecules such as Alq_3 (Mermer *et al.*, 2005). Since OMAR is among the largest obtained magneto-resistive effects in any bulk materials, it has attracted a number of research groups to this novel field.

Figure 19.2 shows a typical OLED structure and the largest reported MEL and MC (which is essentially an inverse of OMAR) magnitudes of an OLED (Nguyen *et al.*, 2008). The MEL (MC) response may reach ~ 60% (30%) at an applied magnetic field of ~100 mT. The effect has been found to be essentially independent of the angle between film plane and applied magnetic field (Mermer *et al.*, 2005).

Various models have been put forward to explaining the MFE in OLED devices based on π-conjugated organic solids (Prigodin *et al.*, 2006; Bobbert *et al.*, 2007; Desai *et al.*, 2007b; Hu and Wu, 2007; Wang *et al.*, 2008). These models are based on the role of the HFI between the spin ½ of the injected charge carriers and the proton nuclear spins closest to the backbone structure of the active layer. The general understanding is that the spin mixing in bound pairs of charge polarons, dubbed here polaron-pair (PP) becomes

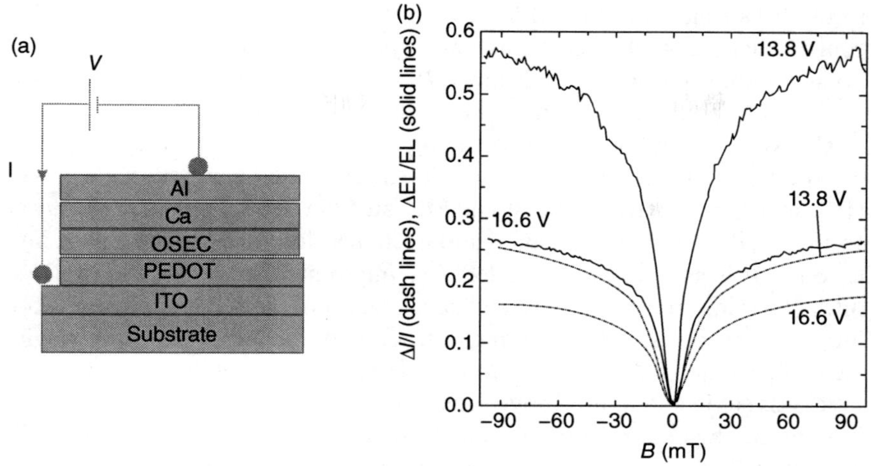

19.2 (a) A typical OLED device structure. (b) Room temperature magneto-conductance (ΔI/I) and magneto-electroluminescence (ΔEL/EL) in an OLED device made of ITO(30 nm)/PEDOT(~100 nm)/Alq_3(~100 nm)/Ca(~30 nm)/Al(30 nm) at two different bias voltages (from Nguyen *et al.*, 2008 with permission).

less effective as the magnetic field increases, thereby causing a MFE of the measured physical quantity. The PP species consist of negative (P^-) and positive (P^+) polarons; whereas same-charge pairs constitute pairs of P^- or P^+, which may be termed as 'loosely bound' negative or positive bipolarons, respectively.

Prior to light emission from singlet and triplet exciton recombination in OLEDs based on π-conjugated polymers, some of the injected free carriers form PP species consisting of P^- and P^+ separated by a distance of few nm, each having spin ½. The free carriers and PP excitations are in dynamic equilibrium in the device active layer, which is determined by the processes of formation/dissociation, and recombination of PP via intrachain excitons. A PP excitation can be either in spin singlet (PP_S) or in spin triplet state (PP_T), depending on the mutual polarons' spin configuration. The steady state PP density depends on the PP_S and PP_T 'effective rate constant', k, which is the sum of the formation, dissociation and recombination rate constants, as well as the triplet–singlet (T-S) mixing via the intersystem crossing (ISC) interaction. If the effective rates k_S for PP_S and k_T for PP_T are not identical to each other, then any disturbance of the T-S mixing rate, such as by the application of an external magnetic field, B, would perturb the dynamical steady state equilibrium that results in MEL and MC. In the absence of an external magnetic field the T-S mixing is caused by both the exchange interaction between the unpaired spins of the polarons belonging to the PP species, and the HFI between the polaron spin and several adjacent H nuclei. For PP excitation in π-conjugated polymer chains, where the polarons are separated by a distance $R > \sim 1–2$ nm, the HFI with protons is expected to be dominant, and actually determines the ISC rate.

The MFE in organic devices is analogous to the MFE observed in chemical and biochemical reactions, where it was explained using the radical pair (RP) mechanism (Brocklehurst and McLauchlan, 1996; Timmel *et al.*, 1998). In this model the HFI, Zeeman, and exchange interactions are taken into account. It is assumed that the RPs are immobile, and hence the radicals' diffusion is ignored; but the overall decay rate, k of the RP is explicitly taken into account. The steady-state singlet fraction of the RP population (singlet yield, Φ_S) is then calculated from the coherent time evolution of RP wavefunctions subjected to the above interaction. It is clear that when $\hbar k$ is much larger than the energy of the HFI and exchange interactions, then the MFE is negligibly small; since the pairs disappear before any spin exchange between the spin sublevels can occur. In contrast, for a relatively small decay rate $\hbar k$, the MFE becomes substantially larger, and for a negligible exchange interaction the singlet yield behaves as $\Phi_S \sim B^2/(B_0^2 + B^2)$ ('upside-down' Lorentzian function of B), with $B_0 \sim a_{HF}/g\mu_B$, where a_{HF} is the HFI constant. The half width at half maximum (HWHM) is then $B_{1/2} = B_0$. In organic devices, the PP species play the role of RPs, and the calculated MC (and MEL) response may be then expressed in terms of the singlet Φ_S and triplet Φ_T PP yields in an external magnetic field, B.

In this section, we show our recent advance in understanding the MFE in organic semiconductor two-terminal devices using both experiment and theoretical methods. Specifically the role of the HFI is readily demonstrated.

19.2.1 Experimental methods and results

The MEL and MC measurements in OLEDs have been generally conducted on devices with typical area of 5 mm^2, where the organic spacers are deposited on a hole transport layer: poly(3,4-ethylenedioxythiophene) [PEDOT]-poly(styrene sulphonate) [PSS]. For *bipolar* devices we capped the bilayer structure with a transparent anode: indium tin oxide [ITO], and a cathode: calcium (protected by aluminum film). For MEH-PPV we also fabricated *unipolar* devices: the hole-unipolar device was in the form of ITO/PEDOT-PSS/MEH-PPV/Au; whereas the electron-unipolar device was Al/LiF(~2 nm)/MEH-PPV/Ca/Al; these devices did not show EL. The organic diodes were transferred to a cryostat that was placed in between the two poles of an electromagnet producing magnetic fields up to ~300 mT. The devices were driven at constant bias, V, using a Keithley 236 apparatus; and the current, I, was measured while sweeping B.

The MC and MEL responses are defined, respectively, via:

$$\mathrm{MC}(B) = \Delta I(B)/I(0) = \frac{I(B) - I(B=0)}{I(B=0)}$$

$$\mathrm{MEL} = \Delta\mathrm{EL}(B)/\mathrm{EL}(0) = \frac{\mathrm{EL}(B) - \mathrm{EL}(B=0)}{\mathrm{EL}(B=0)}. \qquad [19.1]$$

where ΔI and ΔEL are the field-induced changes in the current and EL intensity, respectively.

Figure 19.3a shows the MEL response of two OLED devices based on H- and D-polymers having the same thickness d_f, measured at the same bias V; very similar MC response was also measured simultaneously with MEL (Fig. 19.3b). The MEL and MC responses are narrower in the D-polymer device; in fact the field, $B_{1/2}$ for the MEL is about double in the H-polymer device than in the D-polymer device. We also found that $B_{1/2}$ increases with V (Fig. 19.3a inset) (Nguyen *et al.*, 2010b); actually $B_{1/2}$ increases ~ linearly with the device electric field, $E = (V-V_{bi})/d_f$, where V_{bi} is the built-in potential in the device that is related to the onset bias voltage where EL and MEL are observed (Bloom *et al.*, 2007, 2009). Importantly, in all cases we found that $B_{1/2}(\mathrm{H}) > B_{1/2}(\mathrm{D})$ for devices having the same value of the electric field, E (Fig. 19.3a inset).

MEL in OLEDs may be considered as an example of a much broader research field that deals with MFE in physics (Groff *et al.*, 1974), chemistry

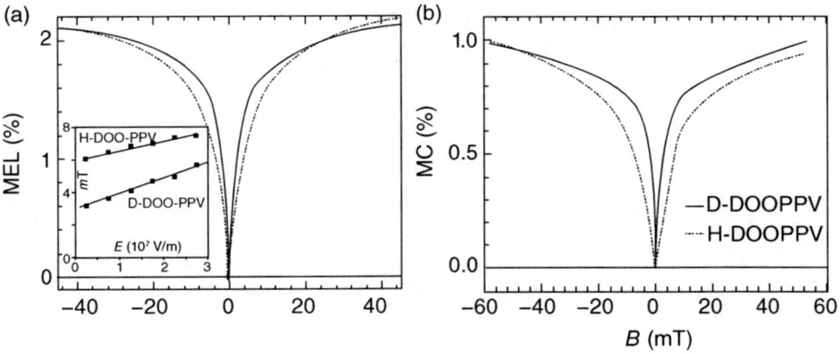

19.3 Isotope dependence of magneto-electroluminescence (MEL) (a) and maneto-conductivity (MC) (b) responses in OLEDs based on DOO-PPV polymers. Room temperature MEL (MC) response of D- and H-polymers (solid and dash lines, respectively) measured at bias voltage $V = 2.5$ volt. Inset to (a) the field, $B_{1/2}$ at half the MEL maximum for the two polymers, plotted vs the applied bias voltage, V, that is given in terms of the internal electric field in the polymer layer, $E = [V-V_{bi}]/d_f$, where V_{bi} is the built-in potential in the device and d_f is the active layer thickness; the lines are linear fits to 'guide the eye' (from Nguyen *et al.*, 2010b, Nguyen *et al.*, 2010a with permission).

and biology (Timmel *et al.*, 1998; Hayashi, 2004). For MFE that results from pairs of radical ions it was empirically realized (Weller *et al.*, 1983) that $B_{1/2}$ scales with the HFI constant a_{HF}. In fact a semi-empirical law was advanced (Weller *et al.*, 1983), which for opposite charge radicals with same a_{HF} reads:

$$B_{1/2} \approx 2a_{HF}[I(I+1)]^{1/2}/g\mu_B, \qquad [19.2]$$

where I is the nuclear spin quantum number. In fact the nuclear spin in the analysis that led to Eq. (19.2) was treated classically (Weller *et al.*, 1983); and therefore it should be considered to be a crude approximation. Nevertheless, using Eq. (19.2) we get $B_{1/2}(H) \approx a_{HF}(H)\sqrt{3}/\, g\mu_B$ for the H-polymer with I = ½, and $B_{1/2}(D) \approx 2a_{HF}(D)\sqrt{2}/\, g\mu_B$ for the D-polymer with $I = 1$. Taking $a_{HF}(H)$ and $a_{HF}(D)$ from the fits to the ODMR lines obtained in Section 19.4, we find that $B_{1/2}(H) \approx 6$ mT for the H-polymer, which is reduced to $B_{1/2}(D) \approx 1.5$ mT for the D-polymer; compared with the experimental $B_{1/2}$ values of ~6 mT and ~3 mT, respectively obtained at small E (Fig. 19.4a, inset). Nonetheless, from the obtained reduction in $B_{1/2}$, we conclude that the MEL response in OLED of π-conjugated polymers *is mainly due to the HFI* (Sheng *et al.*, 2006; Desai *et al.*, 2007a).

Surprisingly, Figs 19.4a and b show that the MEL and MC have yet another component at low B (dubbed 'ultra-small-field MEL/MC', namely USMEL and USMC), which has an opposite sign to that of the positive MEL (MC) at higher fields. A similar low-field component was also observed in some biochemical reactions (Brocklehurst and McLauchlan, 1996) and anthracene crystals (Belaid *et al.*, 2002) with probably the same underlying mechanism as in OLED devices. The USMEL (USMC) component is also due to the HFI, since its width is isotope dependent; it is indeed seen that

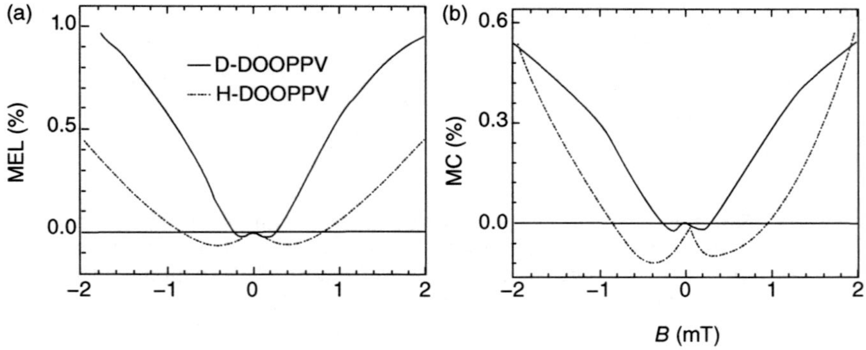

19.4 Room temperature MEL (MC) response of D- and H-polymers (solid and dash lines, respectively) measured at bias voltage $V = 2.5$ volt, plotted for $|B| < 3$ mT (from Nguyen *et al.*, 2010b, Nguyen *et al.*, 2010a with permission).

the dip in the USMEL response occurs at $B_{min} \sim 0.7$ mT in H-polymer, whereas it is at $B_{min} = 0.2$ mT in the D-polymer. The positive, high field MEL component was interpreted as due to less effective ISC between PP spin sublevels at $B > 0$ (Sheng *et al.*, 2006; Bergeson *et al.*, 2008); therefore the negative USMEL component should be due to ISC *rate increase* between PP spin sublevels at low B. This might happen, for example if there is a level-crossing (LC) between PP_S and PP_T spin sublevels (Hayashi, 2004). But since the two PP_T spin sublevels with $m_s = \pm 1$ split linearly with B (Bergeson *et al.*, 2008), then an isotope dependent LC in the PP spin sublevels at very low field cannot be easily accounted for with the four basic spin wavefunctions of PP_S and PP_T that are traditionally considered in the simple MEL models (Sheng *et al.*, 2006; Bergeson *et al.*, 2008). We were therefore led to conclude that *additional spin wavefunctions* are needed to explain the USMEL response.

The USMFE response is not limited to bipolar devices. In Figs 19.5a and b we show MC(B) responses of hole-only and electron-only MEH-PPV diodes; similar responses were measured for DOO-PPV devices (Nguyen *et al.*, 2011a). The high-field MC in unipolar devices is *negative* (Fig. 19.5a) (Wang *et al.*, 2008), and thus the USMFE response here appears as

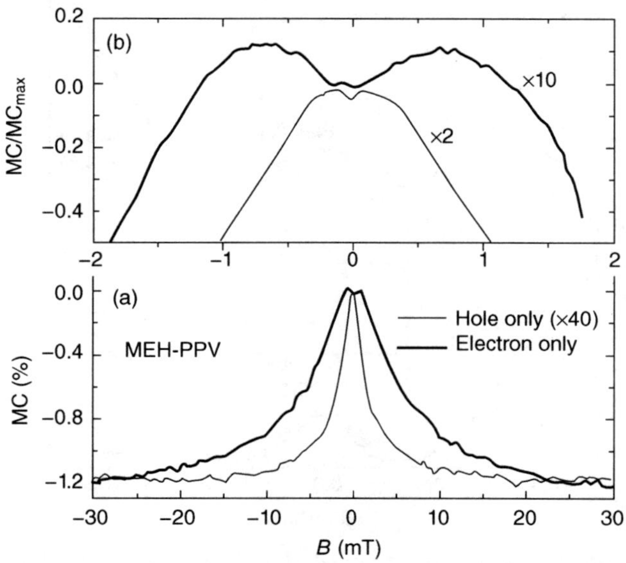

19.5 Normalized MC(B)/USMC(B) response for (a) $|B| < 30$ mT, and (b) $|B| < 2$ mT in hole- and electron-only *unipolar* diodes based on MEH-PPV, measured at room temperature and $V = 3$ volt and 20 volt, respectively. The USMC(B) responses are somewhat shifted in (b) for clarity (from Nguyen *et al.*, 2010a with permission).

'negative-to-positive' sign reversal with a *maximum* at $B_m \sim 0.8$ mT for the electron-only device, and $B_m \sim 0.1$ mT for the hole-only device (Fig. 19.5b). Importantly, the HWHM ΔB is smaller in the hole-only device compared with that in the electron-only device; this is consistent with *smaller* a_{HF} for hole-polaron than for electron-polaron in MEH-PPV, in agreement with recent measurements using transient spin response (McCamey *et al.*, 2010). We therefore conclude that B_m increases with ΔB in unipolar devices similar to bipolar devices (Nguyen *et al.*, 2010a).

The USMFE response depends on both bias voltage and temperature; an example is shown in Fig. 19.6 for D-polymer device. At 10 K we found that $|MC_{min}|$ decreases by a factor of 2 as the bias increases from 3.4 to 4.4 V, whereas B_m does not change much. At $V = 3.4$ V we found that $|MC_{min}|$ increases as the temperature increases from 10 to 300 K, whereas B_m is not affected by the temperature. Importantly, the dependence of MC_{min} of V and T is found to follow the same dependencies as the saturation value, MC_{max}; so that the ratio, MC_{min}/MC_{max} is *independent* of V and T (Fig. 19.6 insets). This indicates that the USMFE component is *correlated* with the

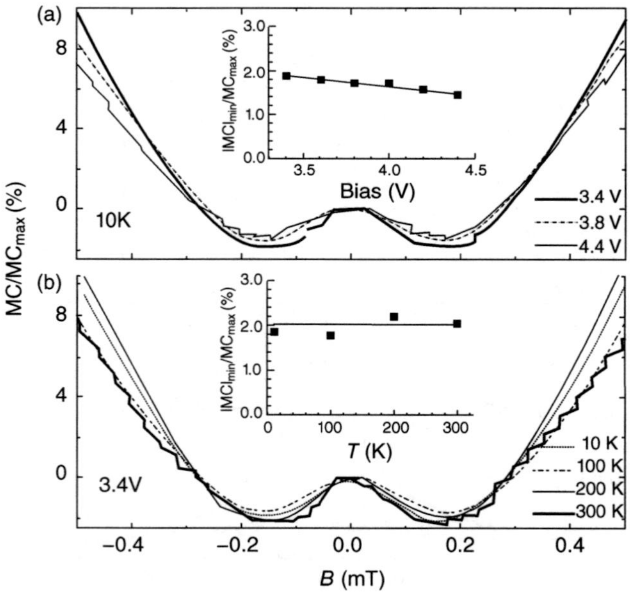

19.6 Normalized MC(B) response of a bipolar diode based on D-polymer for $|B| < 0.5$ mT at (a) various bias voltages at $T = 10$ K, and (b) various temperatures at $V = 3.4$ volt; MC_{max} is defined in Fig. 19.1. The insets in (a) and (b), respectively, summarize MC_{min}/MC_{max} at various voltages at 10 K, and various temperatures at 3.4 volt (from Nguyen *et al.*, 2010a with permission).

normal MC response, and thus is also determined by the HFI. We thus conclude that any viable model describing the normal MC(B) response *needs also explain the USMFE response component.*

19.2.2 MFE models; HFI, Zeeman and exchange interactions

In the absence of an external magnetic field, the triplet–singlet ISC is caused by both the exchange interaction between the unpaired spins of each polaron belonging to the PP excitation, and the HFI between each polaron spin and the adjacent nuclei. For PP excitation in π-conjugated polymer chains, where the polarons are separated by a distance, $R > \sim 1–2$ nm, the HFI with protons is expected to be dominant. In deuterated conjugated polymers, the strongly coupled protons, ^1H (nuclear spin $I = ½$), are replaced by deuterium, ^2H ($I = 1$) having considerably smaller HFI constant, a_{HF}(D); thus the role of the HFI on MC and MEL responses may be readily tested.

The PP model for MEL and MC response that we developed is based on radical pair models introduced previously to describe the effect of magnetic fields on chemical and biochemical reactions (Timmel *et al.*, 1998). We take into account the HFI, Zeeman and exchange interactions. We assume that the PP excitations are immobile, hence PP diffusion is ignored, but we take into account the overall rate of PP decay, e.g. through exciton recombination and/or dissociation into free polarons that contribute to the device current. The steady-state singlet fraction of the PP population ('singlet yield', Φ_S) is then calculated from the coherent time evolution of PP wavefunctions subject to the above interactions. The calculated MC (MEL) response is then expressed as a weighted average of the singlet (Φ_S) and triplet (Φ_T) PP yields in an external magnetic field, B. In the following we use the term SP rather than PP in order to include also the case of like-charge spin polaron pair.

In the traditional view of organic MEL and MC (dubbed MFE) in OLED, as B increases, the intermixing between the SP singlet/triplet spin configurations decreases due to the increased Zeeman contribution, thereby affecting their respective populations; this leads to a monotonous, $MFE_M(B)$ response (Bergeson *et al.*, 2008; Wang *et al.*, 2008). However, if the exchange interaction constant, J is finite, then a new $MFE_{LC}(B)$ component emerges at $B \approx B_{LC} = J$, where a singlet–triplet LC occurs, giving rise to *excess* spin intermixing between the singlet and triplet SP manifolds. The $MFE_{LC}(B)$ component has therefore an *opposite* sign with respect to the regular $MFE_M(B)$ response, which results in a strong MFE(B) modulation response at $B = B_{LC}$ (Hayashi, 2004). We show below that by explicitly taking into account the HFI between each of the SP constituents and N (≥ 1) strongly coupled neighboring nuclei, we can explain the USMFE component response as due to a LC response at $B = 0$.

Our model is based on the time evolution of the SP spin sublevels in a magnetic field. For bipolar devices the SP species is the polaron-pair, whereas for unipolar devices the SP species is a π-dimer (i.e. bi-radical, or bipolaron (Bobbert *et al.*, 2007; Wang *et al.*, 2008)). The spin Hamiltonian, H, includes exchange interaction (EX), HFI and Zeeman terms: $H = H_Z + H_{HF} + H_{ex}$; where $H_{HF} = \sum_{i=1}^{2}\sum_{j=1}^{Ni}[S_i \cdot \tilde{A}_{ij} \cdot I_j]$ is the HFI term, $\tilde{A}$ is the hyperfine tensor describing the HFI between polaron (i) with spin S_i (= ½) and N_i neighboring nuclei, each with spin I_j, having isotropic a_{HF} constant; $H_Z = g_1\mu_B BS_{1z} + g_2\mu_B BS_{2z}$ is the electronic Zeeman interaction component; g_i is the g-factor of each of the polarons in the SP species (we chose here $g_1 = g_2$); μ_B is the Bohr magneton; $H_{ex} = JS_1 \cdot S_2$ is the isotropic exchange interaction; and B is along the z-axis. All parameters in the Hamiltonian H are given in units of magnetic field (mT). An example of the SP spin sublevels using the Hamiltonian H for $N_1 = N_2 = 1$, and I = ½ (namely, overall 16 wavefunctions) is shown in Fig. 19.7a. Note the multiple level-crossings that occur at $B = 0$. Other level-crossings appear at larger B, but those are

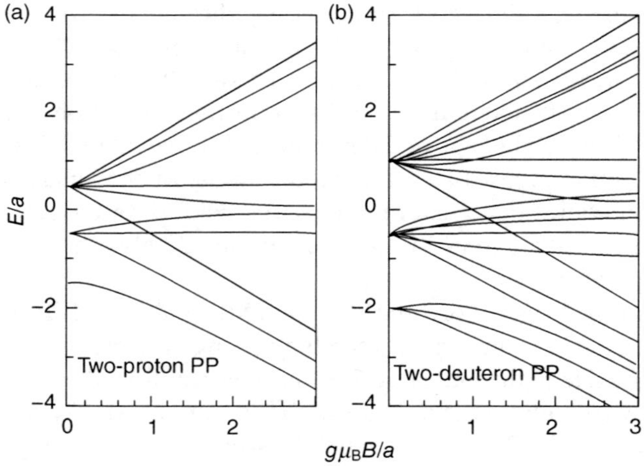

19.7 (a) Energy levels (E) of the 16 spin sublevels of a polaron-pair where each of the two polarons couples to a single proton in the H-polymer (nuclear spin, I = ½), based on the spin Hamiltonian that includes HF (a), exchange (J_{ex}) and Zeeman interactions, as a function of the applied magnetic field, B for the case $J_{ex} \ll a$. Both E and B are given in units of a. (b) Same as in (a) but for the 36 spin sublevels of a polaron-pair coupled to two ^{2}H nuclei in the D-polymer (I = 1) (from Nguyen *et al.*, 2010b with permission).

between mostly triplet sublevels that hardly change the S-T intermixing rate and related $(SP)_S$ and $(SP)_T$ populations. The same SP spin sublevels using D for $N_1 = N_2 = 1$ and $I = 1$ are shown in Fig. 19.7b.

The steady-state $(SP)_S$ and $(SP)_T$ populations are determined by the spin-dependent generation rate and 'effective recombination rate' that includes dissociation (which contributes to the device current density) and recombination (which contributes to the device EL intensity) rates. The SP spin sublevel populations are also influenced by the S-T ISC. Any change of the S-T intermixing rate, such as produced by B may perturb the overall relative steady state spin sublevel populations; and via the SP dissociation mechanism it may consequently contribute to MFE(B) response. To obtain sizable MFE, the SP recombination rate need be smaller than the S-T intermixing rate by the HFI. The USMFE response in this model results from the competition between the coherent S-T inter-conversion of nearly degenerate levels at small B ($B \ll a_{HF}$) and the SP spin coherence decay rate, k (Timmel *et al.*, 1998), as explained below.

The relevant time evolution of the S-T intermixing that determines the steady state SP_S population is obtained in our model via the time-dependent density matrix, ρ(t). Solving the spin Hamiltonian, H for the energies E_n and wavefunctions Ψ_n, we express the time evolution of the singlet population ρ_S(t) as (Timmel *et al.*, 1998; Hayashi, 2004):

$$\rho_S(t) = \mathrm{Tr}[\rho(t)P^S] = \frac{4}{M}\sum_{m,n=1}^{M} |P^S_{mn}|^2 \cos\omega_{mn}t, \qquad [19.3]$$

where $P^S{}_{mn}$ are the matrix elements of the $(SP)_S$ projection operator, $\omega_{mn} = (E_n - E_m)/\hbar$, and M is the number of spin configurations included in the SP species (for $I = ½$ $M = 2^{N+2}$). In the absence of a spin decay mechanism, Eq. (19.3) yields for the $(SP)_S$ steady-state population (apart from the rapidly oscillating terms): $<\rho_S(t = \infty)\geq 4\Sigma_m|P^S{}_{mm}|^2/M + 4\Sigma_{m\neq n}|P^S{}_{mn}|^2/M$, where the summations are restricted to *degenerate levels*, for which $\omega_{mn}(B) = 0$. Here, the first term contributes to $MFE_M(B)$ response, whereas the second term contributes to $MFE_{LC}(B)$ response that modulates $< \rho_S\ (t = \infty) >$ *primarily* at $B = 0$, where the S-T degeneracy is relatively high (see Fig. 19.6). The combination of the monotonous $MFE_M(B)$ and $MFE_{LC}(B)$ components at B~0 explains, in principle, the USMFE response in organic devices.

When allowing for SP spin decay, $\rho_S(t)$ in Eq. (19.3) should then be revised to reflect the disappearance of SP with time. Furthermore, for MFE to occur the decay rates of singlet and triplet configuration must be different from each other. Thus, in a decaying system the population in each of the M levels would decay at a different rate, γ_n ($n = 1,\ldots,M$). Under these conditions, Eq. (19.3) for the singlet fraction is given by (Timmel *et al.*, 1998)

$$\rho_S(t) = \mathrm{Tr}[\rho(t)P^S] = \frac{4}{M}\sum_{m,n=1}^{M} |P^S_{mn}|^2 \cos(\omega_{mn}t)\mathrm{e}^{-\gamma_{nm}t} \qquad [19.4]$$

where $\gamma_{nm} = \gamma_n + \gamma_m$. Eq. (19.4) expresses the fact the singlet (or triplet) time evolution contains both a coherent character (through the $\cos(\omega_{mn}t)$ factor) and an exponential decay factor. The measured MFE (that is MC and MEL) may be calculated using Eq. (19.4). For instance, if the dissociation yields are k_{SD} and k_{TD} for the singlet and triplet configurations, respectively, then the time-dependent dissociated fraction of either the singlet or triplet is $k_{aD}\rho_a(t)$ ($\alpha = S,T$) and thus the dissociation yield is (Ehrenfreund and Vardeny, 2012),

$$\Phi_{\alpha D} = \int_0^{\infty} k_{\alpha D}\rho_{\alpha}(t)\,dt = \frac{4}{M}\sum_{n,m} P^{\alpha}_{n,m}\sigma_{m,n}(0)\frac{k_{\alpha D}\gamma_{nm}}{\gamma^2_{nm}+\omega^2_{nm}}. \qquad [19.5]$$

The total dissociation yield is $\Phi_D = \Phi_{SD} + \Phi_{TD}$ and the *MC(B)* response is then given by

$$MC(B) = \frac{\Phi_D(B) - \Phi_D(0)}{\Phi_D(0)}. \qquad [19.6]$$

For a slow decay such that $k \ll a_{HF}/\hbar$, the abrupt $MFE_{LC}(B)$ obtained at $B = 0$ in the absence of the spin decay is spread over a field range of the order of $\hbar k/g\mu_B$, after which $\Phi_S(B)$ increases again due to the more dominant $MFE_M(B)$ component at large B.

For the MEL response, the final expression depends on the radiative recombination path of the SE and the detailed relaxation route from *PP* to the SE. For instance, in polymers where the SE–TE gap is relatively large (say, >10% of the SE energy), there is a substantial SE–TE intersystem crossing through the spin orbit coupling. As a result, PP^T (PP^S) may transform not only to TE (SE) but also to SE (TE). Let us denote the effective SE (TE) generation rates, from the PP^{α} ($\alpha = S,T$) configuration, as $k_{\alpha,SE}$ ($k_{\alpha,TE}$). Then, similar to MC, we can define the 'SE generation yield', $\Phi_{SE} = \Phi_{S,SE} + \Phi_{T,SE}$ where $\Phi_{\alpha,SE}$ is given by Eq. (19.5) in which $k_{\alpha D}$ is replaced by $k_{\alpha,SE}$. Since the EL is proportional to the SE density, the MEL response is still given by Eq. (19.6), in which Φ_D is replaced by Φ_{SE}.

Figure 19.8 shows the singlet yield and resulting MEL(B) response of the H-polymer OLED. Importantly, the calculated MEL response captures the experimental USMEL response comprising of a negative component having minimum at $B_{min} \sim 0.5$ mT, that changes sign to positive MEL with an approximate $B^2/(B_0^2 + B^2)$ shape with $B_0 \approx 4.5$ mT. The high field shape, namely $B^2/(B_0^2 + B^2)$, is the generic feature in this model. For small values of the exchange interaction B_0 is determined mainly by the HFI constant a_{HF}; also the USMEL response is a strong function of the decay constant k, as shown in Fig. 19.10. The negative component with B_{min} appears only for relatively long decay times, e.g., $\hbar k/a_{HF} \leq 0.1$. For $J_{ex}/a_{HF} > 1$, the characteristic SMEL response is no longer distinguishable.

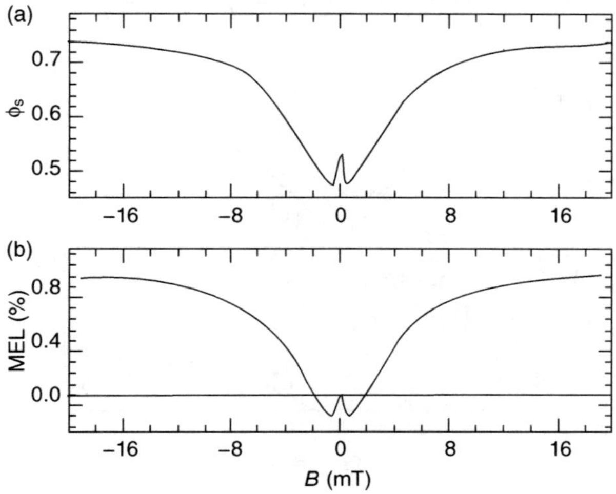

19.8 Calculated magnetic field response of the singlet yield (a) and magneto-conductance (b) for a two-proton PP, where $g_1 = g_2 = g \sim 2$, $a_1 = a_2 = a$, with $a/g\mu_B = 3.5$ mT, $J = 0$, $\delta_{TS} = 0.96$ and $\hbar k/a = 2 \times 10^{-3}$. The resulting MEL response HWHM is ~4.5 mT, and $B_{min} \sim 0.5$ mT (from Nguyen *et al.*, 2010b with permission).

We note that in Fig. 19.8 the MEL HWHM (=4.5 mT) is not exactly equal to $a_{HF}/g\mu_B$, presumably because of the contribution of the USMEL component at low B. We also note that high field resolution is needed for observing the USMEL component, and this might be the main reason why this component has not been observed so far in organic magneto-transport.

The calculated MEL response for various decay rate constant, k (given here in units of a_{HF}) is shown in Fig. 19.9 in which B_{min} is strongly dependent on k. Moreover the model calculation obtained using $\hbar k/a_{HF} \sim 0.002$, also nicely reproduces the USMEL effect (Fig. 19.10), where the calculated B_{min} occurs at ~0.7 and 0.3 mT in the H- and D-polymer, respectively; in excellent agreement with the experiment (Fig. 19.4a).

Figure 19.11 shows the calculated MC(B) response together with its energy sublevels for an axially symmetric anisotropic HFI with $N_1 = N_2 = 1$ ($I = ½$; $M = 16$), where a_{HF}(electron)/$g\mu_B$ = 3a_{HF}(hole)/$g\mu_B$ = 3 mT [parameters extracted from the unipolar MEH-PPV MC(B) response], $J = 0$, δ_{TS} = 0.96, and an exponential SP decay $\hbar k/a_{HF} = 0.001$. The calculated MC(B) response captures the experimental USMC response comprising a negative component having minimum at $B_m \sim a_{HF}/6g\mu_B = 0.5$ mT, with an approximate positive $B^2/(B_0^2 + B^2)$ shape at large B, and $B_0 \approx 4.5$ mT. The excellent agreement between theory and experiment, including both B_m, and the USMC intricate response and relative amplitude validates the used model.

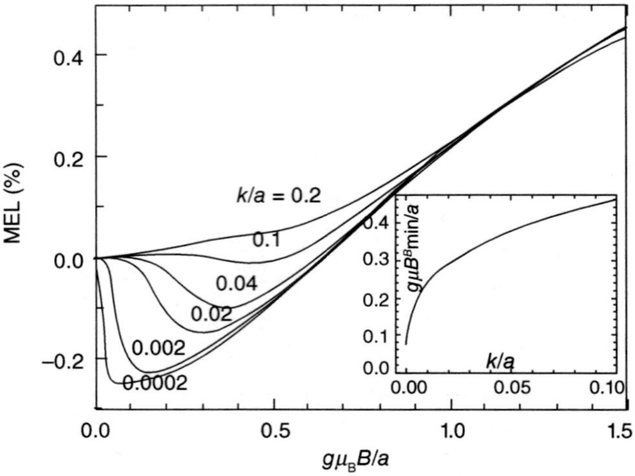

19.9 Calculated MEL response for the two-polaron two proton model, for various decay rate constant, *k* (given here in units of *a*). The inset shows the calculated dependence of B_{min} on *k*; it approximately follows the functional dependence, $B_{min}/a \sim (\hbar k/a)^{0.28}$ (from Nguyen *et al.*, 2010b with permission).

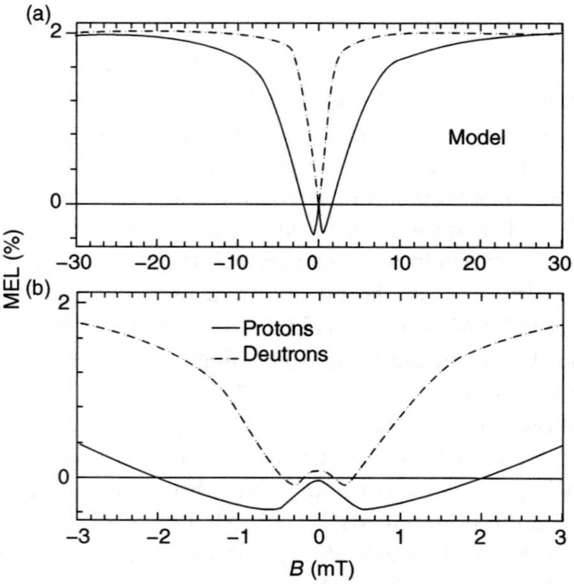

19.10 Simulations of the MEL response in the two polymers, that reproduces the response data in Figs. 19.3(a) and 19.4(a) based on the described model (Timmel *et al.*, 1998) using the calculated spin sublevels given in Fig. 19.6 (from Nguyen *et al.*, 2010b with permission).

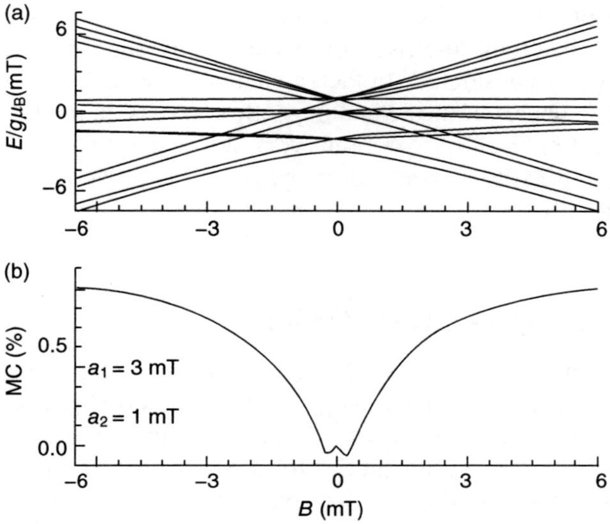

19.11 (a) Example of calculated spin energy levels vs B using the Hamiltonian H in the text for a spin pair with isotropic HFI; $a_1 = 3a_2 = 3$ mT, and $J = 0$. Note the multiple level-crossing at $B = 0$. (b) Calculated MC(B) response for a SP with axially symmetric HFI averaged over all magnetic field directions. The isotropic HFI is the same as in (a), whereas the anisotropic HFI component is $a_{zz} = 0.15a_i$ for the respective spin ½ specie (from Nguyen *et al.*, 2010a with permission).

19.2.3 Conclusions

In our research investigations described in this section we elucidated the role of HFI in the MFE response of conductivity and electroluminescence in various OLEDs based on DOO-PPV isotopes which have different hyperfine coupling constants. We showed that both MEL and MC are strongly dependent on the HFI constant of the materials; the stronger is the HFI constant, the broader is the MFE response. Our findings may be useful for material scientists for designing the novel materials for MFE applications. Furthermore, we found a novel USMFE response at $B \ll a_{HF}/g\mu_B$ in many bipolar and unipolar organic devices, which demonstrate that MEL(B) and MC(B) response are much richer than anticipated before. The USMFE component scales with the normal MC(B) response, and is thus also due to the HFI of the SP pair with B. Our simple model to simulate the MFE explicitly includes in the SP Hamiltonian the most strongly interacting nuclear spins, and is capable of reproducing the entire MFE(B) response, including the USMFE component. Our findings show that via the USMFE component relatively small B is capable of substantially alter both electrical and optoelectronic response in organic diodes, as well as chemical, and biological reactions discussed elsewhere (Hayashi, 2004); and thus should be seriously considered. In fact, a chemical USMFE has been proposed to be at the heart of the 'avian magnetic compass' in migratory birds (Maeda *et al.*, 2008). In this respect our work shows that the USMFE component appears in MFE response of many more organic compounds that has been thought before.

19.3 Organic spin-valves (OSVs)

The electronic spin sense was basically ignored in charge-based electronics, until a few decades ago. The technology of spintronics (or spin-based electronics), where the electron spin is used as the information carrier in addition to the charge, offers opportunities for a new generation of electronic devices combining standard microelectronics with spin-dependent effects that arise from the interaction between the carrier spin and externally applied magnetic fields. Adding the spin degree of freedom to conventional semiconductor charge-based electronics increases substantially the functionality and performance of electronic products. The advantages of these new devices are increased data processing speed, decreased electric power consumption, and increased integration densities compared with conventional semiconductor electronic devices – which are nearly at their physical limits nowadays. The discovery of the GMR effect in 1988 is considered to be the beginning of the new generation of spin-based electronics (Baibich *et al.*, 1988); this discovery led to the Nobel Prize in Physics (2008). Since then, the role of the electron spin in solid-state devices and possible technology that specifically exploits spin rather than charge properties have been studied extensively (Prinz, 1998). A good example of rapid transition from discovery to commercialization for spintronics is the application of GMR and tunneling magnetoresistance (TMR) (Miyazaki and Tezuka, 1995; Moodera *et al.*, 1995) in magnetic information storage. Since the first laboratory demonstration of GMR in 1988, the first GMR device as a magnetic field sensor was commercialized in 1994; and 'read-heads' for magnetic hard disk drives were announced in 1997 by IBM. Major challenges in the field of spintronics are the optimization of electron spin lifetimes, the detection of spin coherence in nano-scale structures, transport of spin polarized carriers across relevant length scales and hetero-interfaces, and the manipulation of both electron and nuclear spins on sufficiently fast timescales (Zutic *et al.*, 2004). The success of these ventures depends on a deeper understanding of fundamental spin interactions in solid-state materials as well as the roles of dimensionality, defects and semiconductor band structure in modifying the spin properties. With proper understanding and control of the spin degrees of freedom in semiconductors and heterostructures the potential for realization of high-performance spintronic devices is excellent. Research in this field so far has led to the understanding that the future of spintronics relies mainly on successful spin injection into multilayer devices and optimization of spin lifetimes in these structures. Hence, for obtaining multi-functional spintronic devices operating at room temperature, different materials suitable for efficient spin injection and spin transport have to be studied thoroughly.

In recent years spintronics has benefited from the class of emerging materials, mainly semiconductors. The III-V and II-VI systems and also magnetic atom doped III-V and II-VI systems (dilute magnetic semiconductors) have been studied extensively as either promising spin transport materials or as spin injecting electrodes (Tang *et al.*, 2002; Zutic *et al.*, 2004). Little attention has been paid so far to the use of organic semiconductors such as small molecules or π-conjugated polymers (PCP) as spin transporting materials. The conducting properties of the PCPs was discovered in the late seventies, and later on the semiconducting properties of the organic semiconductors (OSEC) and PCPs have given birth to a completely new field of electronics, namely 'organic, or plastic electronics'. OSEC and PCPs are mainly composed of light atoms such as carbon and hydrogen, which leads to large spin correlation length due to weak spin orbit coupling. This makes the small molecules and PCPs more promising materials for spin transport than their inorganic counterparts (Ruden and Smith, 2004; Rocha *et al.*, 2005). The ability to manipulate the electron spin in organic molecules offers an alternative route to spintronics.

Key requirements for success in engineering spintronics devices using spin injection via ferromagnetic (FM) electrodes in a two-terminal device such as diode (or junction) include the following processes (Zutic *et al.*, 2004; Yunus *et al.*, 2008): (i) efficient injection of spin-polarized (SP) charge carriers through one device terminal (i.e. FM electrode) into the semiconductor interlayer; (ii) efficient transport and sufficiently long spin relaxation time within the semiconductor spacer; (iii) effective control and manipulation of the SP carriers in the structure; and (iv) effective detection of the SP carriers at a second device terminal (i.e. another FM electrode). Usually FM metals have been used as injectors of SP charge carriers into semiconductors, and they can also serve to detect a SP current. However, more recently magnetic semiconductors have been sought to serve as the spin injector because of the conductivity mismatch between the metallic FM and the semiconductor interlayer (see below). The conductivity mismatch was thought to be less severe using OSEC as the medium in which spin-aligned carriers are injected from the FM electrodes, since carriers are injected into the OSEC mainly by tunneling, and the tunnel barrier may be magnetic field-dependent that defuses the conductivity mismatch (Yunus *et al.*, 2008).

Spin relaxation lifetimes in conventional, inorganic semiconductors are primarily limited by the spin–orbit interaction (Wolf *et al.*, 2001; Zutic *et al.*, 2004). However, OSEC are composed of light elements such as carbon and hydrogen that have weak spin–orbit interaction for the relevant electronic states that participate in the electrical conductance process; and consequently are thought to possess long spin relaxation times (Dediu *et al.*, 2002; Sanvito and Rocha, 2006). Therefore, OSEC appear to offer significant potential applications for spintronic devices (Sanvito and Rocha, 2006; Naber *et al.*, 2007). Figure 19.12a schematically shows the canonical example of a spintronic device, the spin valve (SV) (Jonker, 2003; Xiong *et al.*, 2004). Two FM electrodes (in this example $La_{2/3}Sr_{1/3}MnO_3$ (LSMO) and Co, respectively (Xiong *et al.*, 2004)), used as spin injector and spin detector, respectively are separated by a nonmagnetic spacer (which in organic SV is an OSEC layer). By engineering the two FM electrodes so that they have different coercive fields (Hc), their relative magnetization directions may switch from parallel (P) to anti-parallel (AP) alignment (and vice versa) upon sweeping the external magnetic field, H (see Fig. 19.13). The FM electrode capability for injecting SP carriers depends on its interfacial spin-polarization value, P, which is defined in terms of the density, n, of carriers close to the FM metal Fermi level with spin up, $n\uparrow$ and spin down, $n\downarrow$; and is given by the relation: $P = [n\uparrow - n\downarrow]/[n\uparrow + n\downarrow]$. The spacer decouples the FM electrodes, while allowing spin transport from one contact to the other. The device electrical resistance depends on the relative orientation of the magnetization in the two FM injecting electrodes; this has been dubbed magnetoresistance (MR). The electrical resistance is usually higher for the AP magnetization orientation, an effect referred to as GMR (Wolf *et al.*, 2001), which is due to spin injection and transport through the spacer interlayer. The spacer usually consists of a non-magnetic metal, semiconductor, or a thin insulating layer (in the case of a magnetic tunnel junction). The MR

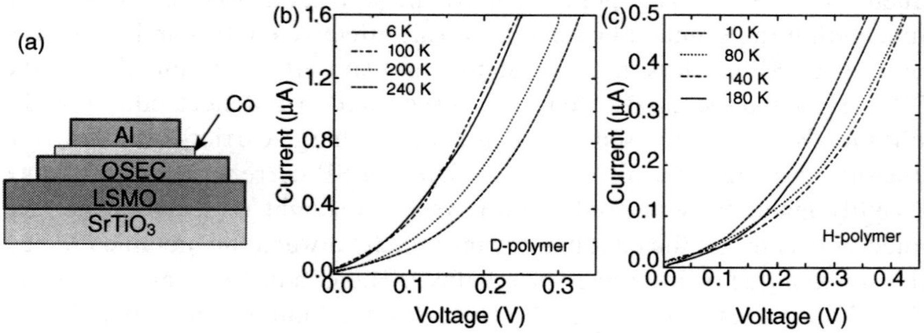

19.12 Typical organic spin valve structure (a) and I-V characteristic response of (b) D-polymer, and (c) H-polymer OSV devices at various temperatures (from Nguyen *et al.*, 2010b with permission).

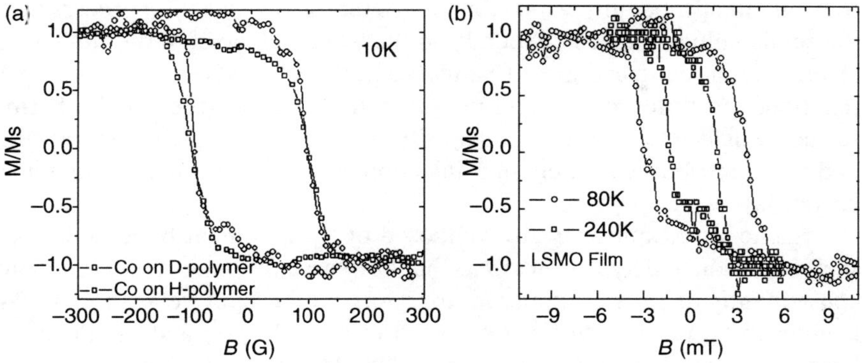

19.13 Magneto-optic Kerr effect of (a) colbalt on H- and D-polymer and (b) LSMO (from Nguyen *et al.*, 2010b with permission).

effect in the latter case is referred to as tunnel MR, and does not necessarily show spin injection into the spacer interlayer as in the case of GMR response, but rather the SP carriers are injected through the nonmagnetic overlayer.

Semiconductor spintronics is very promising because it allows for electrical control of the spin dynamics; and due to the relatively long spin relaxation time, multiple operations on the spins can be performed when they are out of equilibrium (i.e. transport via SP carriers occurs) (Awschalom and Flatte, 2007). This type of spin valve (for example based on GaAs as a spacer; Lou *et al.*, 2007) may have other interesting optical properties, such as circular polarized emission that may be controlled by an external magnetic field (Sanvito, 2007). Significant spin injection from FM metals into nonmagnetic semiconductors is challenging, though, because at thermal equilibrium condition the carrier density with spin-up and spin-down are equal, and no spin polarization exists in the semiconductor layer. Therefore, in order to achieve SP carriers the semiconductor needs be driven far out of equilibrium and into a situation characterized by different quasi-Fermi levels for spin-up and spin-down charge carriers.

Early calculations of spin injection from a FM metal into a semiconductor showed (Smith and Silver, 2001; Albrecht and Smith, 2002; Sanvito, 2007) that the large difference in conductivity of the two materials inhibits the creation of such non-equilibrium situation, and this makes efficient spin injection from metallic FM into semiconductors difficult; this has been known in the literature as the 'conductivity mismatch' hurdle. However, a tunnel barrier contact between the FM metal and the semiconductor may effectively achieve significant spin injection (Wang *et al.*, 2005). The tunnel barrier contact can be formed, for example by adding a thin insulating layer between the FM metal and the semiconductor (Dediu *et al.*, 2008). Tunneling through a potential barrier from a FM contact is spin selective because the barrier transmission probability, which dominates the carrier injection process into the semiconductor spacer, depends on the wave functions of the tunneling electron in the contact regions (Yunus *et al.*, 2008). In FM materials the wave functions are different for spin-up and spin-down electrons at the Fermi surface, which are referred to as majority and minority carriers, respectively; and this contributes to their spin injection capability through a tunneling barrier layer.

Organic semiconductors are composed of light element building blocks such as carbon and hydrogen atoms that have a weak spin–orbit interaction; consequently they are supposed to possess long spin relaxation times leading to be ideal materials for spin transport (Ruden and Smith, 2004; Pramanik *et al.*, 2007; McCamey *et al.*, 2008). Indeed, GMR has been measured in OSV devices based on small organic molecule and polymer spacers, both as thick films and thin tunnel junctions (Xiong *et al.*, 2004; Wang *et al.*, 2005; Majumdar, 2006; Hueso *et al.*, 2007; Pramanik *et al.*, 2007; Santos *et al.*, 2007; Tombros *et al.*, 2007; Dediu *et al.*, 2008; Vinzelberg *et al.*, 2008; Drew, 2009). The role of HFI has been theoretically studied on organic magneto-transport (Bobbert *et al.*, 2009). It is noteworthy that if the HFI determines the spin lattice relaxation time, T_{SL} of the injected carriers, and consequently also their spin diffusion length in OSVs, then the device performance may be enhanced simply by manipulating the nuclear spins of the organic spacer atoms. Moreover the HFI also plays an important role in other organic magneto-electronic devices such as two-terminal devices (see Section 19.2), and other spin response processes such as optically detected magnetic resonance in organic semiconductor films (see Section 19.4).

In this section we summarize our studies of the HFI role in polymer OSV devices and films by replacing various atoms in the organic interlayer composed of a π-conjugated polymer DOO-PPV by different isotopes. First, we replaced all strongly coupled hydrogen atoms (^{1}H, nuclear spin I = ½; dubbed here H-polymer), with deuterium atoms (^{2}H, I = 1) [hereafter, D-polymer; see Fig. 19.1] having much smaller HFI constant, a_{HF}, namely $a_{HF}(D) = a_{HF}(H)/6.5$ (Carrington and McLachlan, 1967). We also replaced some of the ^{12}C atoms ($I = 0$) in the polymer chains by ^{13}C ($I = ½$; hereafter C13-rich polymer, see Fig. 19.2). We studied the influence of the isotope exchange on the GMR in OSVs, where the spin transport in the device determines its performance (Xiong *et al.*, 2004). We also report GMR in OSV devices fabricated with C_{60} spacer that has very small HFI, and thus shows exceptional magnetoresistance response properties.

19.3.1 Experimental methods

The OSVs were fabricated using the DOO-PPV polymers and C_{60} molecule as spacers in between two FM electrodes (Xiong *et al.*, 2004); these were

$La_{0.67}Sr_{0.33}MnO_3$ (LSMO) [bottom electrode, FM_1], and cobalt (Co) [top electrode, FM_2]. The LSMO films with thickness of ~200 nm and area of 5 × 5 mm^2, were grown epitaxially on <100> oriented $SrTiO_3$ substrates at 735 °C using direct current (DC) magnetron sputtering technique, with Ar and O_2 flux in the ratio of 1 : 1. The films were subsequently annealed at 800 °C for ~10 hours before slow cooling to room temperature. The LSMO films were subsequently patterned using standard photolithography and chemical etching techniques. Unlike cobalt, the LSMO films are already stable against oxidation; consequently the LSMO films have been cleaned and re-used multiple times as substrates without serious degradation. Following the LSMO substrate cleaning using chloroform, we deposited the DOO-PPV polymer layer by spin casting from a 6 mg/ml 1,2-dichlorobenzene solution. Subsequently the hybrid organic/inorganic junction was introduced into an evaporation chamber with a base pressure of 5×10^{-7} torr, where we deposited a thin (10 – 15 nm) Co film capped by an aluminum (Al) film using a shadow mask. The obtained active device area was typically about 0.2 × 0.4 mm^2.

We fabricated several OSV devices having various DOO-PPV layer thickness, d_f controlled by the spin angular speed, and calibrated against measurements using thickness profilometry methods (KLA Tencor). Films with $d_f > 80$ nm appeared segregated, and were subsequently discarded, whereas films having $20 < d_f < 80$ nm were smooth. D- and H-polymer-based OSV with various d_f were measured and compared at several bias voltage, V and temperature, T. For the d_f dependence of the MR, we used the same LSMO substrate and device structure in which d_f was physically reduced using a doctor blade and subsequently measured using the profilometer. The *I-V* characteristics of the OSV devices were non-linear with weak temperature dependence (see Fig. 19.12). Typical device resistance was in the range of 100–500 kΩ, and depended on d_f. Most devices showed a 'metallic' behavior, where R increases with T; however, some devices showed mixed behavior (Fig. 19.12). The OSV MR was measured in a closed-cycle refrigerator with T in the range from 10 to 300 K, using the 'four probe' method, while varying an external in-plane magnetic field. The magnetization properties of the FM electrodes were measured by the magneto-optic Kerr effect (MOKE) (Fig. 19.13); from these measurements we determined typical low temperature coercive fields for the electrodes alone as B_{c1}~4 mT and B_{c2}~10 mT, for the LSMO and Co films, respectively.

19.3.2 Results and discussion

Giant MR in OSVs; isotope dependence

For the spin injection and transport investigations comparing the three polymer isotopes we used OSV devices, where the polymer film was sandwiched between two FM electrodes having different coercive fields, B_c. In our design these were LSMO and Co thin films, with low temperature $B_{c1} \approx 4$ mT and $B_{c2} \approx 15$ mT, respectively (Xiong *et al.*, 2004; Nguyen *et al.*, 2010b), which depend on the LSMO and Co film thicknesses; and nominal spin injection polarization degree $P_1 \approx 95\%$ and P_2 that depends on the environment (Santos *et al.*, 2007; Dediu *et al.*, 2008; Vinzelberg *et al.*, 2008; Barraud *et al.*, 2010). Since $B_{c1} \neq B_{c2}$, then it is possible to switch the relative magnetization directions of the FM electrodes between P and AP alignments, upon sweeping the external magnetic field, B (Fig. 19.14); where the device resistance, R is dependent on the relative magnetization orientations. When $R(\text{AP}) > R(\text{P})$ the maximum MR value, $[\Delta R/R]_{max}$ (or MR_{SV}) is given by the ratio: $[R(\text{AP})–R(\text{P})]/R(\text{P})$. According to a modified Jullière formula (Jullière, 1975), MR_{SV} is related to the polarization degree of the FM electrodes P_1 and P_2 by (Xiong *et al.*, 2004):

$$[\Delta R/R]_{max} = 2P_1P_2D/(1 - P_1P_2D). \quad [19.7]$$

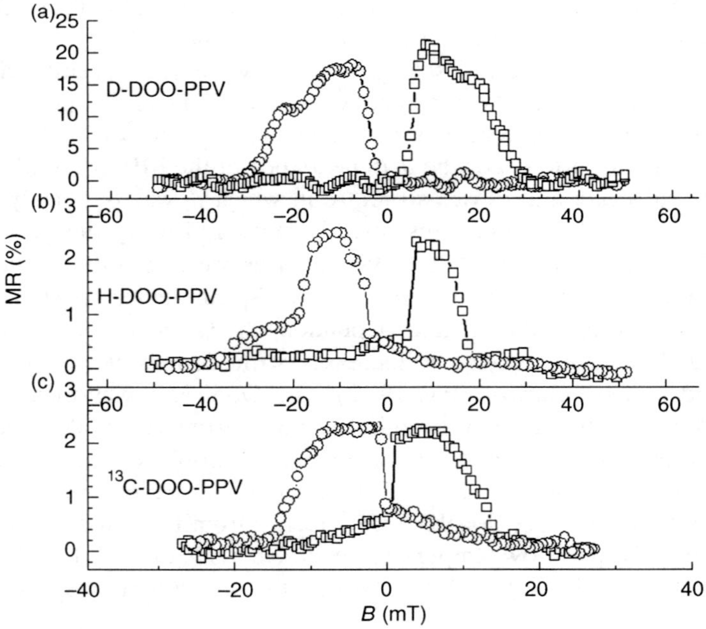

19.14 Magnetoresistance loop of an OSV device with the configuration of LSMO (200 nm)/DOO-PPV(~30 nm)/Co(15 nm) spin-valve devices measured at T = 10 K and V = 20 mV, based on (a) D-, (b) H-, and (c) C13-rich polymers. The circle (square) curve denotes MR measurements made upon increasing (decreasing) B (from Nguyen *et al.*, 2011b with permission).

In Eq. (19.7) $D = \exp[-(d_f)/\lambda_s]$, where λ_s is the carrier spin diffusion length in the organic interlayer (Xiong *et al.*, 2004). We measured the MR hysteresis loop to obtain the MR_{SV} value in OSVs based on the three polymer isotopes at various bias voltage, V, and temperature, T, using the same LSMO substrate, but different devices. In most cases we fitted the MR response using the equation (Bobbert *et al.*, 2009):

$$MR(B) = \frac{1}{2} MR_{max}[1 - m_1(B)m_2(B)]\exp[-d_f / l_s(B)], \qquad [19.8]$$

where MR_{max} is the MR when neglecting spin relaxation, and $l_s(B)$ is a field-dependent spin diffusion length parameter given by the relation: $l_s(B) = l_s(0)[1 + (B/B_0)^2]^{3/8}$, where B_0 is a characteristic field related to $a_{HF}/g\mu_B$. The functions $m_{1,2}(B)$ in Eq. (19.8) stand for the normalized magnetizations of the FM electrodes, which are used here as free parameters (for simplicity we used for $m(B)$ the 'error function' centered at the respective 'turn on' and 'turn off' fields (Bobbert *et al.*, 2009)).

Figure 19.14 shows representative MR hysteresis loops for three similar OSVs (d_f ~ 25 nm) based on D-, H- and C13-rich polymers measured at $T = 10$ K and $V = 10$ mV. The following points are worth emphasizing:

- First, when using freshly prepared LSMO substrates we obtained *positive* MR, where $R(AP) > R(P)$, in agreement with results on another polymer OSV (Majumdar, 2006). Since $P_1 > 0$, then the positive MR may be readily explained using Eq. (19.7) only if $P_2 > 0$ for the Co electrode, in agreement with recent measurements using organic magnetic tunnel junctions (Santos *et al.*, 2007).
- Importantly, all devices based on the D-polymer have *much larger* MR_{SV} values than those based on the H- and C13-polymers. This holds true for similar devices at all V, T and d_f. The improved magnetic properties of OSVs based on the D-polymer may be explained using Eq. (19.7) by a larger λ_s. Indeed, the major difference between the injected spin ½ carriers in D-polymer and the other two polymer isotopes, namely H- and C13-rich polymers is their spin relaxation time T_{SL}; which was shown to be much longer in the D-polymer (Nguyen *et al.*, 2010b, 2011b). In order to examine this assumption we studied the MR response of OSVs from D- and H-polymers at various d_f, but otherwise same LSMO substrate, which were measured at the same temperature and applied voltage (Fig. 19.15) (Nguyen *et al.*, 2010b). From the obtained exponential $MR_{max}(d_f)$ dependence we got $\lambda_s(D)$ = 49 nm, whereas $\lambda_s(H)$ = 16 nm; in agreement with the increase in T_{SL} of the D-polymer measured using ODMR. The larger obtained $\lambda_s(D)$ is also reflected in the fitting parameters to the MR(B) response when using Eq. (19.7); we indeed found (Nguyen *et al.*, 2010b) that $l_s(0)$ is ~3 times larger in the

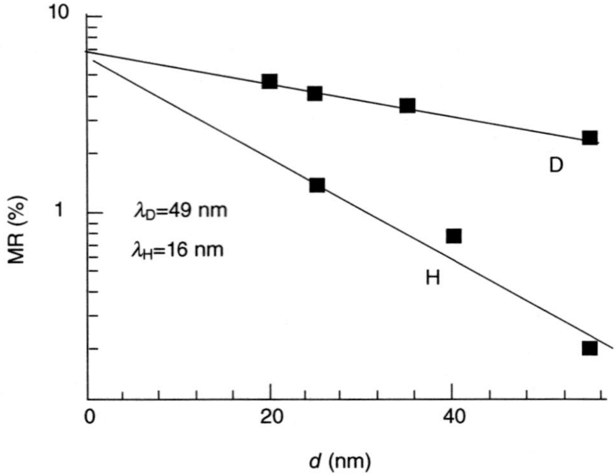

19.15 The maximum MR value (MR_{SV}) of D- and H-polymer OSV having various interlayer thicknesses, d_f measured at $T = 10$ K and $V = 80$ mV. The lines are fits to the data points, where $MR_{SV}(d_f) = 6.7\%$ $\exp(-d_f/\lambda_s)$, with spin diffusion lengths, $\lambda_s(D) = 49$ nm and $\lambda_s(H) =$ 16 nm, respectively (from Nguyen *et al.*, 2010b with permission).

D-polymer compared to the H-polymer, whereas B_0 (which is related to a_{HF}, Nguyen *et al.*, 2010b) is ~2.5 times smaller. Based on our results we therefore conclude that the improved spin transport in the organic layer is the main advantage of the D- polymers to form more efficient OSVs.

- The MR(V) dependence of the three polymer OSVs (Fig. 19.16) shows a power law dependence, $MR_{SV}(V) \propto V^{-p}$, where the exponent $p < 1$. It was suggested (Zhang and White, 1998) that there are two different tunneling processes from the FM electrode; a direct tunneling that conserves spin, and a two-step tunneling (involving hopping) that does not conserve spin. Since the latter process has a less steep voltage dependence, then it dominates at large bias voltage. This may explain the MR(V) decrease at large V, because the device MR is mainly determined by the two-step, non-spin conserving tunneling process. Inelastic tunneling via 'magnetic impurities' accompanied by phonon emission gives rise to a power law dependence (Sheng *et al.*, 2004), $MR_{SV}(V) \propto V^{-p}$. The exponent p was predicted to be isotope dependent (Sheng *et al.*, 2004), namely $p \sim M^{3/2}$, where M is the effective isotope mass which determines the phonon spectrum that participate in the hopping process. In Fig. 19.16 we plot $MR_{SV}(V)$ for the three isotopes in double logarithmic scale; it is indeed seen that $MR_{SV}(V)$ obeys a power law decay with V, having $p < 1$ that is *isotope dependent*. We obtained a larger p for the

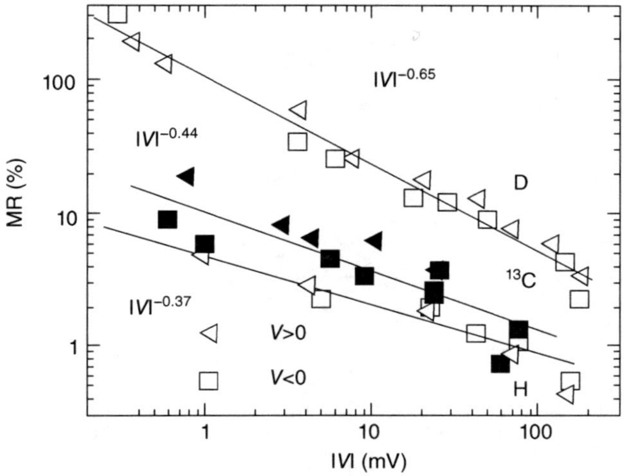

19.16 The maximum MR value (MR_{SV}) of three isotope OSV devices having $d_f = 35 \pm 5$ nm, as a function of the applied bias voltage, V, measured at $T = 10$ K. Note that data are plotted on a log-log scale showing a power law behaviour, V^{-p} with isotope-dependent exponent, p (from Nguyen *et al.*, 2011b with permission).

deuterated device, as predicted by the model (Sheng *et al.*, 2004); we found the ratio $p(D)/p(H) \sim 1.7$. This ratio is larger than $[M(CD)/M(CH)]^{3/2} \sim 1.1$ expected from the model. However, for the C13-rich polymer we found $p(C13)/p(C12) \sim 1.18$, which is in very good agreement with the predicted ratio $[M(C13H)/M(C12H)]^{3/2} \sim 1.12$.

- We also note that MR_{SV} strongly decreases at high T for all OSV devices, irrespective of the MR_{SV} value at low T (Fig. 19.17); thus it is mainly caused by the spin injection properties of the LSMO electrode into the organic layer (Wang *et al.*, 2007; Dediu *et al.*, 2009), rather than by the organic interlayer. We therefore conclude that different spin injectors need be found in order to achieve a significant advance in organic spintronics. In this regard the use of *deuterated* organic semiconductors as the device interlayer, both as evaporated small molecules or spin-cast polymers, should substantially improve the OSV device performance.

Giant MR of C_{60}-based OSVs

Since the GMR response of D-polymer OSVs shows an improved response that is mainly caused by the small HFI of this polymer isotope, it is tempting to fabricate OSV devices made of C_{60} film interlayer. The C_{60} molecule is composed of 60 carbon atoms that include two isotopes; 99% ^{12}C with nuclear spin $I = 0$, and 1% ^{13}C with $I = ½$ (Abragam, 1961) (see Fig. 19.1 for

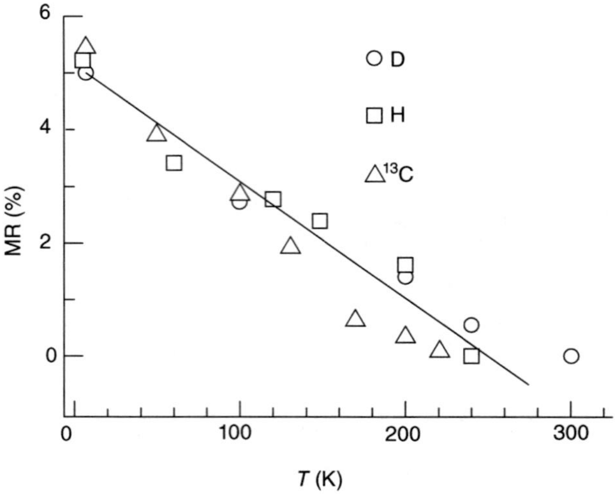

19.17 Normalized MR_{SV} of three isotope OSVs similar to those shown in Fig. 19.16, but as a function of temperature, T measured at $V = 80$ mV (from Nguyen *et al.*, 2011b with permission).

the chemical structure). From these values we can estimate the average effective HFI per C_{60} molecule to be about $a(^{13}C)/60 < 0.1$ mT. It should therefore be beneficial to fabricate C_{60}-based OSV devices (Wang, 2009).

Figure 19.18 shows the GMR response of a LSMO/C_{60}/Co/Al OSV device measured at 120 K and 60 mV bias, where the C_{60} interlayer was thermally evaporated with a measured thickness, $d_f = 40$ nm. At this relatively large thickness we do not expect tunneling MR to dominate the OSV MR response. It is seen that the MR(B) response is very sharp, and maintains a large value even at these relatively high temperature and bias voltage. The GMR is still reversed [namely R(antiparallel) < R(parallel)] showing that one of the Ps in Eq. (19.7) is negative (Xiong *et al.*, 2004; Barraud *et al.*, 2010). Also the sharp response shows that the ratio $d_f/l_s(B)$ when using Eq. (19.8) to fit the MR(B) response is rather small (Bobbert *et al.*, 2009). It is also noteworthy that the turn-off coercive field due to the Co electrode on the C_{60} spacer is *much larger* than the Co coercive fields obtained in the polymer-based OSV devices (Fig. 19.14); this shows that the Co coercive field is extremely sensitive to the environment (Schultz *et al.*, 2010). The large and sharp GMR response is very encouraging emphasizing the important role that C_{60} might play in the field of organic spintronics.

Conclusions and future trends

In this section we reviewed some of the latest achievements in OSV research at the University of Utah and Technion, related to the important role of the

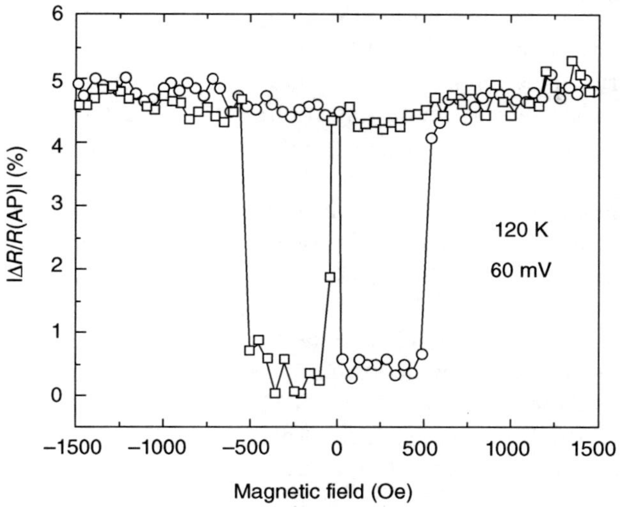

19.18 MR(V) loop of a LSMO(200 nm)/C_{60}(40 nm)/Co(5 nm) spin-valve device measured at T = 120 K and V = 60 mV. The circle (square) curve denotes MR measurements made upon increasing (decreasing) B (from Nguyen *et al.*, 2011b with permission).

HFI in determining the spin response. By replacing the building block atoms in the polymer chains such as carbon and hydrogen, with different isotopes, it is then possible *to tune the HFI* and thus study in detail its influence in the material spin response. We showed elsewhere (Nguyen *et al.*, 2010b, 2011b) that the HFI has a crucial role in determining the spin ½ ODMR linewidth and saturation properties, from which we inferred that the HFI governs the spin ½ relaxation rate in semiconducting polymers. This led us to fabricate OSV devices based on polymers having different isotopes, with the anticipation that tuning the HFI would also lead to better control over the spin diffusion length in the organic interlayer (Nguyen *et al.*, 2010b). We found that this is indeed the case for OSV based on the three DOO-PPV isotopes, where the D-polymer showed the best GMR performance, and longer obtained diffusion length. Surprisingly, we also found that the GMR voltage dependence in OSV is *isotope dependent*, where lighter isotopes show steeper voltage dependence. We proposed a model to explain this phenomenon based on the tunneling properties of spin-polarized carriers at the organic/FM interface. However our model is by no means conclusive; much more experimental and theoretical work needs to be done before a more definitive understanding of the GMR(V) response in OSV devices is achieved.

The organic spintronics field is in its infancy; much more work has to be accomplished before the field matures. At the present time controversies regarding the exact operation of OSV still exist; especially the MR signs in these devices (Barraud *et al.*; 2010, Schultz *et al.*, 2010). Recently spin-polarized carrier injection into an organic semiconductor has been directly observed using low-energy muon spin rotation (Drew, 2009); so the doubts raised at the beginning of the organic spintronics field (Xu *et al.*, 2007; Jiang *et al.*, 2008) could be defused. In general, spin injection achieved in OSV is of the same magnitude as in spin-valves fabricated using inorganic semiconductor spacers. In particular both organic and inorganic semiconductor interlayer suffer from the same conductivity mismatch problem (Schmidt, 2005), and show only little MR at room temperature. A noticeable difference, however, is the ability to change the LUMO and HOMO levels of the organic interlayer in OSV by depositing a thin insulator layer, such as LiF or SiO_x prior to the deposition of the FM electrode (Dediu *et al.*, 2008; Schultz *et al.*, 2010; Yunus *et al.*, 2010). This may lead to a variety of magneto-transport effect that cannot be achieved with inorganic spin-valves.

19.4 Optically detected magnetic resonance (ODMR) in poly (dioctyloxy) phenyl vinylene (DOO-PPV) isotopes

Resonant absorption resulting from the application of a MW radiation with a frequency matching the energy difference of the Zeeman split levels of a spin system under magnetic field is the basis for various related magnetic resonance techniques. In conventional electron spin resonance (ESR) the resonant absorption of the MW is directly detected measuring thus properties of the ground electronic state in thermal equilibrium (Carrington and McLachlan, 1967). ODMR techniques are extensions of the more common ESR techniques (Cavenet, 1981). ODMR measures changes in the optical absorption (photo-induced absorption detected magnetic resonance: PA-DMR) and/or emission (photoluminescence detected magnetic resonance: PL-DMR) that occur as a result of electron spin transitions. Since the ground state of conjugated polymers is almost always spin singlet, only spin-bearing electronic excited states are studied. When applied to π-conjugated semiconductor films, ODMR techniques can be used for assigning the correct spin quantum number to the optical absorption bands of long-lived photoexcitations and/or for studying spin-dependent reactions that may occur between these photoexcitations (Vardeny and Wei, 1997). One major advantage of ODMR over ESR is a much higher sensitivity in the detection of resonant processes, because the energy range of the detected photons in ODMR is much higher (order of eV) than the microwave energies that monitor ESR (tens of μeV). Much better detectors with higher detectivity exist for photons in the eV- range than in the μeV range. A significant increase in the sensitivity of ODMR when compared to ESR also results from differences in the spin polarization; in ODMR experiments in π-conjugated polymers the typically obtained spin polarizations exceeds the thermal equilibrium polarization by at least one order of magnitude (Vardeny and Wei, 1997).

19.4.1 Experimental methods

As an illustrative example, we show in Fig. 19.19 the basic experimental setup for PA-DMR in organic semiconductor thin films. The sample is mounted in a MW cavity equipped with windows for optical transmission, and between the poles of a superconducting DC magnet. The cavity is a liquid helium cooled cryostat. As in photo-modulation, the sample is illuminated by both pump and probe beams. The pump beam produces photoexcitations in the sample film. After being modulated, the moderately strong MW at f_{MW} = 3 GHz are introduced into the cavity through a waveguide. Then the changes ΔT in transmission (T) of the probe associate with the MW-induced changes in the photo-excitation recombination kinetics in the DC magnetic field are detected with lock-in amplifier using a phase-sensitive technique. For the PL-DMA measurement, the photoluminescence changes induced by MW transitions from the film are detected without using the probe beam. This technique is complementary to the PA-DMR.

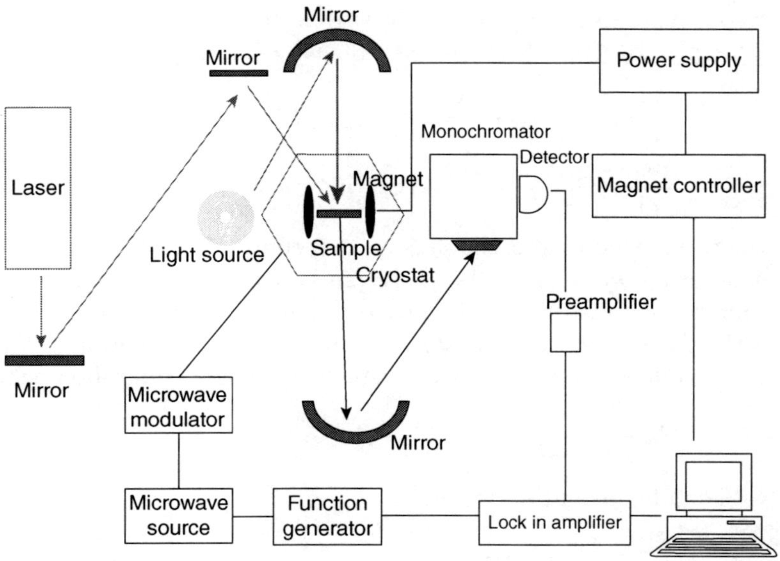

19.19 ODMR experimental set-up.

Discussion

ODMR in organics is in fact an MFE that occurs under resonance conditions with MW radiation that induces spin sublevel mixing among the PP spin sublevels (Vardeny and Wei, 1997). It is thus not surprising that models similar to those used to explain the MFE without MW radiation have been advanced to explain spin ½ ODMR in the organics.

In the following analysis we assume that the ODMR response dynamics exclusively occurs when all spin levels actively participate in determining the measured MFE physical quantity (Yang *et al.*, 2007). When PP_S decay rate (γ_S) is effectively faster than PP_T decay rate (γ_T), then at steady state when MW is off $n_{S,off} < n_{T,off}$, where $n_{S,off}$ ($n_{T,off}$) is the PP_S (PP_T) density when the MW is off (Vardeny and Wei, 1997). Under resonant MW radiation (MW with the frequency f_{MW} is 'on') a net transfer from PP_T to PP_S takes place (via the $m_T = 0$ spin sublevel in the PP_T manifold), bringing the system to a new quasi-equilibrium state, where the total PP density, $n_{PP} = n_S + n_T$, is reduced; i.e. $n_{PP,on} < n_{PP,off}$. Under square wave MW modulation at frequency f, both $\Delta n_S(f) = n_{S,on} - n_{S,off}$ and $n_T(f) = n_{T,on} - n_{T,off}$ responses decrease with increasing f (for example in the form of Lorentzians in f, if their time decays are exponentials (McCamey *et al.*, 2008).

In order to quantify the $\Delta n(f)$ response we make use of the fact that the two triplet spin sub-levels ($ms = \pm 1$) should have a common dynamics. We also take the limit of strong singlet to $ms = 0$ triplet mixing (Yang *et al.*, 2007), thus reducing the coupled set of four rate equations to a coupled set of two rate equations for the PP in the triplets and singlet/triplet states, respectively. These equations are written for the experimental conditions $T \gg hf_{MW}/k_B \approx 0.14$ K as follows (Yang *et al.*, 2007):

$$\frac{dn_S}{dt} = G_S - \gamma_S n_S - \frac{\gamma_{SL}}{2}(n_S - n_T) - P(n_S - n_T)$$
$$\frac{dn_T}{dt} = G_T - \gamma_T n_T - \frac{\gamma_{SL}}{2}(n_T - n_S) - P(n_T - n_S) \quad [19.9]$$

where n_S denotes the coupled $m_S = 0$ sublevels of PP_T and PP_S and n_T denotes the $m_S = \pm 1$ in PP_T, γ_{SL} is spin-lattice relaxation rate, P is the MW spin flip rate (assumed to be proportional to the MW power, P_{MW}), and G is the generation rate. The steady-state solution of equation 19.9 (i.e., $dn_S/dt = dn_T/dt = 0$) show a typical magnetic resonance saturation behavior:

$$\Delta_n \sim \frac{P}{\Gamma_{eff} + P} \quad [19.10]$$

with an effective rate constant

$$\Gamma_{eff} = \frac{\gamma_{SL}}{2} + \frac{1}{\gamma_S^{-1} + \gamma_T^{-1}}$$

where Δ_n is the MW induced change in the polaron density.

Thus, Γ_{eff} can be directly obtained by measuring the ODMR as a function of microwave power, P_{MW}.

19.4.2 ODMRs of DOO-PPV isotopes: role of the hyperfine interaction

For the ODMR investigations we used a *S*-band resonance system at ~3 GHz and 10 K, where the MW power, P_{MW} was modulated at ~200 Hz (Yang *et al.*, 2007). The PL intensity increases due to enhancement of photogenerated PP recombination in the polymer film upon MW absorption at resonance; therefore the ODMR signal appears at B ~100 mT that corresponds to spin ½ species with $g \approx 2$.

Figures 19.20a and b show the spin ½ ODMR of polaron-pairs in H- and D-polymer films, respectively. Importantly, the resonance in D-polymer is *narrower* than that in H-polymer; at small P_{MW} the resonance full width at half maximum (FWHM) is $\delta B_D \approx 0.7$ mT in the D-polymer compared to $\delta B_H \approx 1.2$ mT in the H-polymer (Fig. 19.20). Since the HFI constant $a_{HF}(D) = a_{HF}(H)/6.5$ (Carrington and McLachlan, 1967) we expect to obtain a much narrower δB_D; however δB is also determined by the wavefunction extent of the polaron on the polymer chain, which determines its spin density spread on the number of hydrogen nuclei, N of the polymer chain (Weinberger *et al.*, 1980). Thus a_{HF} and N *together* determine an 'intrinsic' HFI linewidth, δB_{HF}.

The spin ½ density of the polaron excitation in π-conjugated polymer chains is usually spread over several repeat units, which may consist of about $N = 10$ CH building units (Fesser *et al.*, 1983). Consequently the polaron spin ½ ODMR (and ESR) line shape at resonance conditions is susceptible to interactions with N protons in the immediate vicinity of the backbone intrachain carbon atoms. Having nuclear spin ½, each coupled proton splits the two electron-polaron spin levels ($m_s = \pm½$) into two electron-polaron/proton levels via the HFI; and these lines are further split by the other N–1 coupled protons (Carrington and McLachlan, 1967). Therefore, the energy levels of the ensemble system containing one electron-polaron and N protons, depends on the HFI constant a_{HF} and field B, which is given by the spin Hamiltonian:

$$H = H_{HFI} + H_Z. \qquad [19.11]$$

In Eq. (19.11) $H_Z = g\mu_B B S_z$ is the electronic Zeeman component (we ignore the nuclear Zeeman interaction), and H_{HFI} $(= \Sigma_n a_{HFn}\, S \cdot I_n)$ is the isotropic HFI term. Here a_{HFn}, I_n ($I_n = ½$) are the HFI constant and nuclear spin operators of the nth nucleus ($n = 1,\ldots,N$), and S ($S = ½$) is the electron-polaron spin operator. Solving the Schrodinger equation with the Hamiltonian in the form of Eq. (19.11) in the high field limit ($\mu_B B \gg a_{HF}$) we obtain that each of the $m_s = \pm½$ electron-polaron energy level splits into 2^N levels

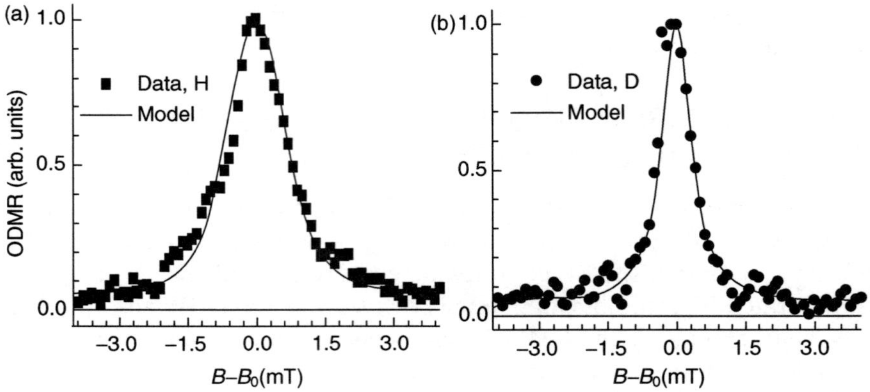

19.20 Isotope dependence of spin ½ optically detected magnetic resonance (ODMR) response in DOO-PPV polymers. (a) and (b) are the normalized spin ½ ODMR response in the H- and D-polymers; the lines though the data points are based on a model fit taking into account the HFI splitting and the polaron spin density spread on the polymer chain, as well as inhomogeneous and natural line broadening contributions to the resonance width (from Nguyen *et al.*, 2010b with permission).

(for $I = ½$). As a result there are 2^N $m_s = ½$ to $m_s = -½$ allowed optical transitions in the GHz range (hereafter, 'satellite lines') that leave the nuclear spins unchanged. In general, the satellite lines do not have equal spacing since a_{HFn} in Eq. (19.11) are not equal to each other in the most general case. The satellite lines form a distribution in B of which width, Δ_{HF}, can be conveniently calculated from the square root of the second moment (Carrington and McLachlan, 1967):

$$\Delta_{HF} = 2[\Sigma_n (B_n - B_0)^2 / 2^N]^{1/2}, \quad [19.12]$$

where B_0 is the resonance field in the absence of HFI, and B_n are the resonance fields for the satellite lines. For a uniform spin distribution we take $a_{HFn} = a_{HF}/N$; and consequently there are N+1 non-degenerate electron/proton energy levels; which, when using Eq. (19.12) give $\Delta_{HF} \sim a_{HF}/\sqrt{N}$. Under this condition we find for the polaron excitation in DOO-PPV polymer chains with $N = 10$, $\Delta_{HF} \sim 0.3a_{HF}$.

In addition the inhomogeneous broadening causes an 'extrinsic' linewidth, δB_{inh}. When $\delta B_{inh} < \delta B_{HF}$ the resonance linewidth, δB can be approximately written as:

$$\delta B \approx \delta B_{HF} + \delta B_{inh} + \delta B_0, \quad [19.13]$$

where δB_0 ($<\delta B_{HF}$) is the homogeneous linewidth (~$1/T_2$, the dephasing rate). The spin density distribution among N CH units decreases the effective HFI, a_{HFeff} of each nucleus; a uniform polaron spin distribution leads to $a_{HFeff} = a_{HF}/N$ (Weinberger *et al.*, 1980). For the H-polymer HFI with one spin ½ nucleus splits the resonant line into two separate lines, with $\Delta B = a_{HF}/g\mu_B$ (Abragam, 1961). With electronic spin distribution on N nuclei, the two lines are further split into $(N+1)$ satellite lines with unequal intensity that are separated from each other by $\Delta B_{eff} = a_{HFeff}/g\mu_B$; and can be readily calculated. The solid line in Fig. 19.20a for the H-polymer is a model calculation for the spin ½ resonance assuming $N = 10$, where $a(\mathrm{H}) = 3.5$ mT, and $\delta B_{inh} + \delta B_0 = 0.65$ mT (Eq. (19.13)); it is in excellent agreement with the data. For the D-polymer with spin 1 nucleus, each resonant line is split into three separate lines; which are further split into $(2N + 1)$ satellite lines for N CH units. The solid line in Fig. 19.20b for the D-polymer was similarly calculated using $N = 10$ and $\delta B_{inh} + \delta B_0 = 0.65$ mT; but with $a(\mathrm{D})/g\mu_B = a(\mathrm{H})/6.5g\mu_B = 0.54$ mT. The excellent agreement with the data using the known $a(\mathrm{D})/a(\mathrm{H})$ ratio indicates that the model calculation captures the main effect of the hydrogen isotope exchange on the $g \approx 2$ ODMR line in the polymer.

In Fig. 19.21 we show the dependence of the $g \approx 2$ ODMR line intensity, δPL, on P_{MW} for both H- and D-polymers. The ODMR intensity shows a saturation behavior (Yang *et al.*, 2007), namely $\delta\mathrm{PL} \sim P_{MW}/(P_{MW} + P_S)$, where $P_S = \alpha\Gamma_{eff}$, α is an experimentally determined constant, and Γ_{eff} is given by the relation (Yang *et al.*, 2008):

$$\Gamma_{eff} = \gamma + \frac{1}{2}\gamma_{SL}, \qquad [19.14]$$

where γ is an average PP recombination rate, and $\gamma_{SL} = T_{SL}^{-1}$. Since the PP recombination process at $B \sim 0.1$ T is not expected to depend on the isotope exchange, then variation in P_S upon deuteration reflects changes in γ_{SL} (via Eq. (19.14)). Figure 19.21 shows that P_S is smaller for the D-polymer; in fact from the data fit we obtain the ratio: $P_S(\mathrm{H})/P_S(\mathrm{D}) = 1.4$; and using Eq. (19.14) we conclude that γ_{SL} is *smaller* in the D-polymer due to the weaker HFI. In the following we more precisely determine the ratio, η of the γ_{SL} rates for the polarons in the two isotope polymers, using a complementary ODMR method.

In Fig. 19.22 we show that the ODMR linewidth, δB nonlinearly increases with P_{MW} in both polymers, and the increase is steeper for the H-polymer. The increase in δB may be related to the magnetic field strength, B_1, of the MW electromagnetic radiation, which influences the homogeneous linewidth δB_0 through the relation (Abragam, 1961):

$$\delta B_0(P_{MW}) = \delta B_0(0)[1 + (\beta/\gamma_{SL})P_{MW}]^{1/2}, \qquad [19.15]$$

where $\beta = (T_2\delta B_0(0)\alpha)^{-1}$ and $P_{MW} = \alpha(B_1)^2$. Since $\delta B_0(0)$ and β are independent of the hydrogen isotope, then from the excellent fit to the data in

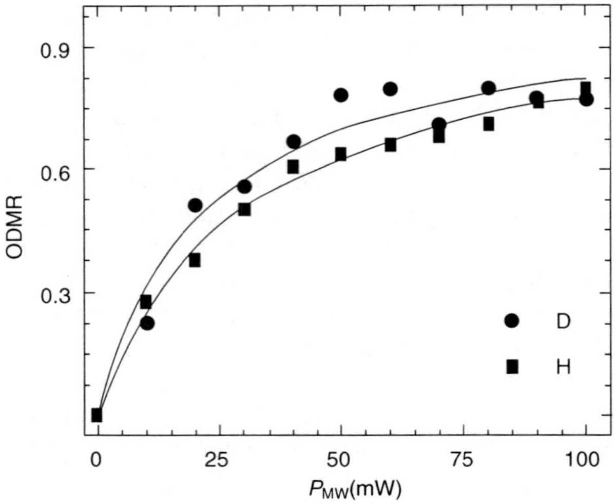

19.21 The microwave power (P_{MW}) dependence of the integrated ODMR signal intensity for the two polymers; the lines through the data points are a model fit using a saturation equation (see text) (from Nguyen *et al.*, 2010b with permission).

Fig. 19.22 using Eq. (19.15) we determine the ratio η to be $\eta \equiv \gamma_{SL}(H)/\gamma_{SL}(D) \approx 4$. This indicates that the spin relaxation time of polarons in the D-polymer is *substantially longer* than in H-polymer; and this should increase the spin diffusion length of spin-polarized carriers in OSV devices based on this polymer (see Section 19.3).

19.5 Conclusion

In this chapter we reviewed the latest achievements in organic spintronics research at the University of Utah and the Technion, related to the important role of the HFI in determining the spin response. By replacing the skeleton atoms in the polymer chains such as carbon and hydrogen, with different isotopes, it was then possible to tune the HFI and thus study in detail its influence on the material spin response. We showed compelling evidence that the HFI has a crucial role in determining the magnetic response of three spin-dependent processes in organic semiconductors, namely MFE in OLEDs, GMR in OSV devices, and electron resonance response in thin films. In particular, we showed that (i) both MEL(B) and MC(B) responses in OLEDs are strongly dependent on the HFI constant of the materials; the stronger is the HFI constant, the broader is the magnetic field response. Furthermore, we found a novel ultra-small response at $B \ll a_{HF}/g\mu_B$ in many bipolar and unipolar organic devices. This MFE response component scales

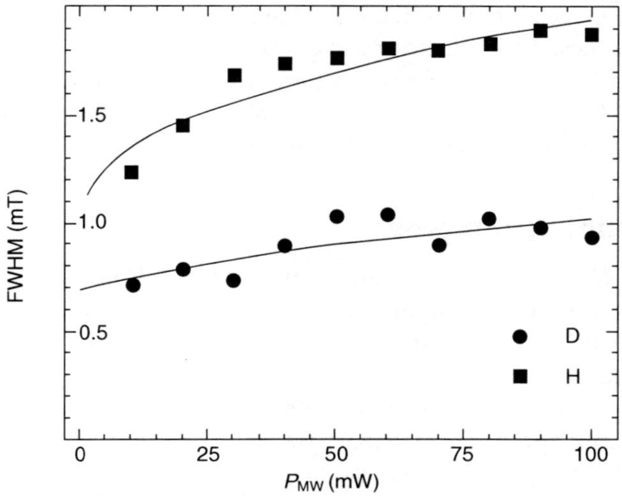

19.22 The microwave power dependence of the ODMR linewidth for the two polymers; the lines through the data points are model fits based on Eq. (4.7) (from Nguyen *et al.*, 2010b with permission).

with the normal MC(B) response, and is thus also due to the HFI of the spin-pairs with B. We introduced a simple model to simulate the MFE response that explicitly includes in the spin-pair Hamiltonian the most strongly interacting nuclear spins, and is capable of reproducing the entire MFE(B) response, including the ultra-small component; (ii) the HFI has an important role in determining the polaron spin ½ resonance linewidth and saturation properties; and (iii) the HFI determines the spin diffusion length of the injected spin polarized electrons; the weaker is the HFI, the longer is the spin diffusion length, and the larger is the GMR in the device.

Our findings may be useful to material scientists for designing novel materials for organic spintronics applications. For example, the ideal organic semiconductors for OSV application should have miniature HFI for supporting electron-polarized spin transport. Among the existing materials, the fullerenes, for example C_{60} might be an excellent candidate for spin-polarized transport since it naturally contains 99% ^{12}C with nuclear spin $I = 0$, and only 1% ^{13}C with $I = ½$ and thus some small HFI. At the present time the application of MFE in OLEDs as a magnetic sensor is still pending. One of the reasons is that substantial MFE response only occurs at a few mT, which is still too large for MFE-based magnetic sensor application. Our findings pave the way to reduce the necessary field for the MFE simply by exchanging isotopes to lower the HFI in the active material. However, if the HFI is too small, as in the case of C_{60}, then the MFE response is very weak since the HFI is also the origin of the MFE response itself. An alternative solution to improve the MFE-based sensor is to use the USMFE. In fact, a chemical USMFE has been proposed to be at the heart of the 'avian magnetic compass' in migratory birds. In this respect our work shows that the USMFE component appears in MFE response of many more organic compounds that has been thought before.

19.6 Acknowledgments

We thank L. Wojcik for the synthesis of the DOO-PPV polymers with different isotopes. We also thank Prof. Li for the LSMO films that he supplied for the OSV studies. In addition we that Drs. F. Wang and G. Hukic for the ODMR work, and the graduate students in Vardeny's research group for many useful discussions. The work reviewed in this work was supported by the DOE grant DE-FG02-04ER46109 over the years, and NSF (Grant No. DMR-1104495 and CEMRI program at the UoU); and the Israel Science Foundation (ISF 472/11; E.E.), and the Israel-USA BSF (Grant # 2010135).

19.7 References

Abragam, A. 1961. *The Principles of Nuclear Magnetism*. Nature Publishing Group.

Albrecht, J. D. & Smith, D. L. 2002. Electron spin injection at a Schottky contact. *Phys. Rev. B,* **66**, 113303–113306.

Awschalom, D. D. & Flatte, M. E. 2007. Challenges for semiconductor spintronics. *Nature Phys.,* **3**, 153–159.

Baibich, M., Broto, J. M., Fert, A., Nguyenvandan, F., Petroff, F., Etienne, P., Creuzert, G., Friederich, A. & Chazelas, J. 1988. Giant magnetoresistance of (001)Fe/(001) Cr magnetic superlattices. *Phys. Rev. Lett.,* **61**, 2472–2475.

Barraud, C., Seneor, P., Mattana, R., Fusil, S., Bouzehouane, K., Deranlot, C., Graziosi, P., Hueso, L., Bergenti, I., Dediu, V., Petroff, F. & Fert, A. 2010. Unravelling the role of the interface for spin injection into organic semiconductors. *Nat. Phys.,* **6**, 615–620.

Belaid, R., Barhoumi, T., Hachani, L., Hassine, L. & Bouchriha, H. 2002. Magnetic field effect on recombination light in anthracene crystal. *Synth. Metals,* **131**, 23–30.

Bergeson, J. D., Prigodin, V. N., Lincoln, D. M. & Epstein, A. J. 2008. Inversion of magnetoresistance in organic semiconductors. *Phys. Rev. Lett.,* **100**, 067201.

Bloom, F. L., Wagemans, W., Kemerink, M. & Koopmans, B. 2007. Separating positive and negative magnetoresistance in organic semiconductor devices. *Phys. Rev. Lett.,* **99**, 257201.

Bloom, F. L., Kemerink, M., Wagemans, W. & Koopmans, B. 2009. Sign inversion of magnetoresistance in space-charge limited organic devices. *Phys. Rev. Lett.,* **103**, 066601.

Bobbert, P. A., Nguyen, T. D., Oost, F. W. A. V., Koopmans, B. & Wohlgenannt, M. 2007. Bipolaron mechanism for organic magnetoresistance. *Phys. Rev. Lett.,* **99**, 216801.

Bobbert, P. A., Wagemans, W., Oost, F. W. A. V., Koopmans, B. & Wohlgenannt, M. 2009. Theory for spin diffusion in disordered organic semiconductors. *Phys. Rev. Lett.,* **102**, 156604.

Brocklehurst, B. & Mclauchlan, K. A. 1996. Free radical mechanism for the effects of environmental electromagnetic fields on biological systems. *Int. J. Radiat. Biol.,* **69**, 3–24.

Carrington, A. & Mclachlan, A. D. 1967. *Introduction to Magnetic Resonance*. Nature Publishing Group.

Cavenet, B. C. 1981. Optically detected magnetic resonance (O.D.M.R.) investigations of recombination processes in semiconductors. *Advances in Physics,* **30**, 475–538.

Dediu, V., Margia, M., Matacotta, F. C., Taliani, C. & Barbanera, S. 2002. Room temperature spin polarised injection in organic semiconductor. *Solid State Commun.,* **122**, 181–184.

Dediu, V., Hueso, L. E., Bergenti, I., Riminucci, A., Borgatti, F., Graziosi, P., Newby, C., Casoli, F., Jong, M. P. D., Taliani, C. & Zhan, Y. 2008. Room temperature spintronic effects in Alq_3-based hybrid devices. *Phys. Rev. B,* **78**, 115203–115208.

Dediu, V. A., Hueso, L. E., Bergenti, I. & Taliani, C. 2009. Spin routes in organic semiconductors. *Nature Mater.,* **8**, 707–716.

Desai, P., Shakya, P., Kreouzis, T. & Gillin, W. P. 2007a. Magnetoresistance in organic light-emitting diode structures under illumination. *Phys. Rev. B,* **76**, 235202.

Desai, P., Shakya, P., Kreouzis, T., Gillin, W. P., Morley, N. A. & Gibbs, M. R. J. 2007b. Magnetoresistance and efficiency measurements of Alq_3-based OLEDs. *Physical Review B,* **75**, 094423.

Drew, A. J. 2009. Direct measurement of the electronic spin diffusion length in a fully functional organic spin valve by low-energy muon spin rotation. *Nature Mater.,* **8**, 109–114.

Ehrenfreund, E. & Vardeny, Z. V. 2012. Effects of magnetic field on conductance and electroluminescence in organic devices. *Israel Journal of Chemistry,* **52**, 552–562.

Ern, V. & Merrifield, R. E. 1968. Magnetic field effect on triplet exciton quenching in organic crystals. *Phys. Rev. Lett.,* **21**, 609–611.

Fesser, K., Bishop, A. R. & Campbell, D. K. 1983. Optical absorption from polarons in a model of polyacetylene. *Phys. Rev. B,* **27**, 4805–4825.

Francis, T. L., Mermer, O., Veeraraghavan, G. & Wohlgenannt, M. 2004. Large magneto-resistance at room-temperature in small-molecular weight organic semiconductor sandwich devices. *New J. Phys.,* **6**, 185–192.

Frankevich, E. L., Zakhidov, A., Yoshino, K., Maruyama, Y. & Yakushi, K. 1996. Photoconductivity of poly(2,5-diheptyloxy-p-phenylene vinylene) in the air atmosphere: magnetic-field effect and mechanism of generation and recombination of charge carriers. *Phys. Rev. B,* **53**, 4498–4508.

Groff, R. P., Suna, A., Avakian, P. & Merrifield, R. E. 1974. Magnetic hyperfine modulation of dye-sensitized delayed fluorescence in organic crystals. *Phys. Rev. B,* **9**, 2655–2660.

Hayashi, H. 2004. *Introduction to Dynamic Spin Chemistry; Magnetic Field Effects on Chemical and Biochemical Reactions*. Nature Publishing Group.

Hu, B. & Wu, Y. 2007. Tuning magnetoresistance between positive and negative values in organic semiconductors. *Nature Mater.,* **6**, 985–991.

Hueso, L. E., Bergenti, I., Riminucci, A., Zhan, Y. & Dediu, Y. 2007. Multipurpose magnetic organic hybrid devices. *Adv. Mater.,* **19**, 2639–2642.

Jiang, J. S., Pearson, J. E. & Bader, S. D. 2008. Absence of spin transport in the organic semiconductor Alq_3. *Physical Review B,* **77**, 035303.

Jonker, B. T. 2003. Progress toward electrical injection of spin-polarized electrons into semiconductors. *Proc. IEEE,* **91**, 727–740.

Jullière, M. 1975. Tunneling between ferromagnetic films. *Phys. Lett. A,* **54**, 225–226.

Kalinowski, J., Szmytkowski, J. & Stampor, W. 2003. Magnetic hyperfine modulation of charge photogeneration in solid films of Alq_3. *Chemical Physics Letters,* **378**, 380–387.

Kalinowski, J., Cocchi, M., Virgili, D., Fattori, V. & Marco, P. D. 2004. Magnetic field effects on organic electrophosphorescence. *Phys. Rev. B,* **70**, 205303–205309.

Lou, X., Adelmann, C., Crooker, S. A., Garlid, E. S., Zhang, J., Reddy, S. M., Flexner, S. D., Palmstrøm, C. J. & Crowell, P. A. 2007. Electrical detection of spin transport in lateral ferromagnet–semiconductor devices. *Nature Phys.,* **3**, 197–202.

Maeda, K., Henbest, K. B., Cintolesi, F., Kuprov, I., Rodgers, C. T., Liddell, P. A., Gust, D., Timmel, C. R. & Hore, P. J. 2008. Chemical compass model of avian magnetoreception. *Nature,* **453**, 387–390.

Majumdar, S. 2006. Application of regioregular polythiophene in spintronic devices; effect of interface. *Appl. Phys. Lett.,* **89**, 122114.

McCamey, D. R., Seipel, H. A., Paik, S. Y., Walter, M. J., Borys, N. J., Lupton, J. M. & Boehme, C. 2008. Spin Rabi flopping in the photocurrent of a polymer light-emitting diode. *Nature Mater.,* **7**, 723–727.

McCamey, D. R., Van Schooten, K. J., Baker, W. J., Lee, S. Y., Paik, S. Y., Lupton, J. M. & Boehme, C. 2010. Hyperfine-field-mediated spin beating in electrostatically bound charge carrier pairs. *Phys. Rev. Lett.,* **104**, 017601.

Mermer, Ö., Veeraraghavan, G., Francis, T. L., Sheng, Y., Nguyen, D. T., Wohlgenannt, M., Köhler, A., Al-Suti, M. K. & Khan, M. S. 2005. Large magnetoresistance in nonmagnetic π-conjugated semiconductor thin film devices. *Phys. Rev. B*, **72**, 205202.

Miyazaki, T. & Tezuka, N. 1995. Giant magnetic tunneling effect in Fe/Al2O3/Fe junction. *J. Magn. Magn. Mater.,* **139**, L231–L234.

Moodera, J. S., Kinder, L. R., Wong, T. M. & Meservey, R. 1995. Large magnetoresistance at room temperature in ferromagnetic thin film tunnel junctions. *Phys. Rev. Lett.,* **74**, 3273–3276.

Naber, W. J. M., Faez, S. & Van der Wiel, W. G. 2007. Organic spintronics. *J. Phys. D,* **40**, R205–R228.

Nguyen, T. D., Sheng, Y., Rybicki, J. & Wohlgenannt, M. 2008. Magnetic field-effects in bipolar, almost hole-only and almost electron-only tris-(8-hydroxyquinoline) aluminum devices. *Phys. Rev. B,* **77**, 235209.

Nguyen, T. D., Gautam, B. R., Ehrenfreund, E. & Vardeny, Z. V. 2010a. Magnetoconductance response in unipolar and bipolar organic diodes at ultrasmall fields. *Phys. Rev. Lett.,* **105**, 166804.

Nguyen, T. D., Hukic-Markosian, G., Wang, F., Wojcik, L., Li, X.-G., Ehrenfreund, E. & Vardeny, Z. V. 2010b. Isotope effect in spin response of [pi]-conjugated polymer films and devices. *Nat. Mater,* **9**, 345–352.

Nguyen, T. D., Gautam, B. R., Ehrenfreund, E. & Vardeny, Z. V. 2011a. Magnetoconductance of π-conjugated polymer based unipolar and bipolar diodes. *Synthetic Metals,* **161**, 604–607.

Nguyen, T. D., Hukic-Markosian, G., Wang, F., Wojcik, L., Li, X.-G., Ehrenfreund, E. & Vardeny, Z. V. 2011b. The hyperfine interaction role in the spin response of π-conjugated polymer films and spin valve devices. *Synthetic Metals,* **161**, 598–603.

Pramanik, S., Stefanita, C. G., Patibandla, S., Bandyopadhyay, S., Garre, K., Harth, N. & Cahay, M. 2007. Observation of extremely long spin relaxation times in an organic nanowire spin valve. *Nature Nanotech.,* **2**, 216–219.

Prigodin, V. N., Bergeson, J. D., Lincoln, D. M. & Epstein, A. J. 2006. Anomalous room temperature magnetoresistance in organic semiconductors. *Synthetic Metals,* **156**, 757–761.

Prinz, G. A. 1998. Magnetoelectronics. *Science,* **282**, 1660–1663.

Pulizzi, F. 2009. Why going organic is good. *Nature Mater.,* **8**, 691.

Rocha, A. R., Garcia-Suarez, V. M., Baily, S. W., Lambert, C. J., Ferrer, J. & Sanvito, S. 2005. Towards molecular spintronics. *Nat Mater*, **4**, 335–339.

Ruden, P. & Smith, D. L. 2004. Theory of spin injection into conjugated organic semiconductors. *J. Appl. Phys.,* **95**, 4898–4904.

Santos, T. S., Lee, J. S., Migdal, P., Lekshmi, I. C., Satpati, B. & Moodera, J. S. 2007. Room temperature tunnelling magnetoresistance and spin-polarized tunnelling through an organic semiconductor barrier. *Phys. Rev. Lett.,* **98**, 016601.

Sanvito, S. 2007. Organic electronics: Spintronics goes plastic. *Nat Mater,* **6**, 803–804.

Sanvito, S. & Rocha, A. R. 2006. Molecular-spintronics: The art of driving spin through molecules. *J. of Comput. & Theoretical Nanoscience,* **3**, 624.

Schmidt, G. 2005. Concepts for spin injection into semiconductors, *J. Phys. D,* **38**, R107–R122.

Schultz, L., Nuccio, L., Willis, M., Desai, P., Shakia, P., Kreouzis, T., Malik, V. K., Bernhard, C., Pratt, F. L., Morley, N. A., Suter, A., Nieuwenhuys, G. J., Prokscha, T., Morenzoni, E., Gillin, W. P. & Drew, A. J. 2010. Engineering spin propagation across a hybrid organic/inorganic interface using a polar layer. *Nat. Mater.,* **10**, 39.

Sheng, L., Xing, D. Y. & Sheng, D. N. 2004. Theory of the zero-bias anomaly in magnetic tunnel junction. *Phys. Rev. B,* **70**, 094416.

Sheng, Y., Nguyen, T. D., Veeraraghavan, G., Mermer, Ö., Wohlgenannt, M., Qiu, S. & Scherf, U. 2006. Hyperfine interaction and magnetoresistance in organic semiconductors. *Phys. Rev. B,* **74**, 045213.

Smith, D. L. & Silver, R. N. 2001. Electrical spin injection into semiconductors. *Phys. Rev. B,* **64**, 045323–045336.

Tang, H. X. Monzon, F. G., Jedemu, F. J., Filip, A. T., van Wees, B. J. & Roukes, M. L. 2002. Spin injection and transport in micro- and nanoscale devices. *In:* Awschalom, D., Loss, D. & Samarth, N. (eds.). *Semiconductor Spintronics and Quantum Computation*. New York: Springer, 31–92.

Timmel, C. R., Till, U., Brocklehurst, B., McLachlan, K. A. & Hore, P. J. 1998. Effects of weak magnetic fields on free radical recombination reactions. *Mol. Phys.,* **95**, 71–89.

Tombros, N., Jozsa, C., Popinciuc, M. & Jonkman, H. T. 2007. Electronic spin transport and spin precession in single graphene layers at room temperature. *Nature,* **448**, 571–575.

Vardeny, Z. V. & Wei, X. 1997. *Handbook of Conducting Polymers II.* New York: Marcel Dekker, 639–666.

Vinzelberg, H., Schumann, J., Elefant, D., Gangineni, R. B., Thomas, J. & Büchner, B. 2008. Low temperature tunnelling magnetoresistance on $La_2/3Sr1/3MnO_3$/Co junctions with organic spacer layers. *J. Appl. Phys.,* **103**, 093720.

Wang, F. 2009. PhD thesis, University of Utah.

Wang, F. & Vardeny, Z. V. 2010. Recent advances in organic spin-valve devices. *Synthetic Metals,* **160**, 210–215.

Wang, F. J., Xiong, Z. H., Wu, D., Shi, J. & Vardeny, Z. V. 2005. Organic spintronics: The case of Fe/Alq3/Co spin-valve devices. *Synth. Metals,* **155**, 172–175.

Wang, F. J., Yang, C. G., Vardeny, Z. V. & Li, X. 2007. Spin response in organic spin valves based on $La_2/3Sr1/3MnO_3$ electrodes. *Phys. Rev. B,* **75**, 245324.

Wang, F., Bassler, H. & Vardeny, Z. V. 2008. Studies of magnetoresistance in polymer/fullerene blends. *Phys. Rev. Lett.,* **101**, 236805.

Weinberger, B. R., Ehrenfreund, E., Pron, A., Heeger, A. J. & MacDiarmid, A. G. 1980. Electron spin resonance studies of magnetic soliton defects in polyacetylene. *J. Chem. Phys.,* **72**, 4749–4755.

Weller, A., Nolting, F. & Staerk, H. 1983. A quantitative interpretation of the magnetic field effect on hyperfine-coupling-induced triplet formation from radical ion pairs. *Chem. Phys. Lett.,* **96**, 24–27.

Wolf, S. A., Awschalom, D. D., Buhrman, R. A., Daughton, J. M., Von Molnár, S., Roukes, M. L., Chtchelkanova, A. Y. & Treger, D. M. 2001. Spintronics: a spin-based electronics vision for the future. *Science,* **294**, 1488–1495.

Xiong, Z. H., Wu, D., Vardeny, Z. V. & Shi, J. 2004. Giant magnetoresistance in organic spin-valves. *Nature,* **427**, 821–824.

Xu, W., Szulczewski, G. J., Leclair, P., Navarette, I., Schad, R., Miao, M., Guo, H. & Gupta, A. 2007. Tunneling magnetoresistance observed in $La_{0.67}Sr_{0.33}MnO_3$/organic molecule/Co junctions *Appl. Phys. Lett.,* **90**, 072506.

Yang, C. G., Ehrenfreund, E. & Vardeny, Z. V. 2007. Polaron spin-lattice relaxation time obtained from ODMR dynamics in [pi]-conjugated polymers. *Phys. Rev. Lett.,* **99**, 157401.

Yang, C. G., Ehrenfreund, E., Wang, F. & Vardeny, Z. V. 2008. Spin dependent kinetics of polaron pairs in organic light emitting diodes studied by ELDMR dynamics. *Phys. Rev. B,* **78**, 205312.

Yunus, M., Ruden, P. P. & Smith, D. L. 2008. Ambipolar electrical spin injection and spin transport in organic semiconductors. *J. Appl. Phys.,* **103**, 103714–103721.

Yunus, M., Ruden, P. P. & Smith, D. L. 2010. Macroscopic modeling of spin injection and spin transport in organic semiconductors. *Synthetic Metals,* **160**, 204–209.

Zhang, J. & White, R. M. 1998. Voltage dependence of magnetoresistance in spin dependent tunnelling junctions. *J. Appl. Phys.,* **83**, 6512–6514.

Zutic I., Fabian J. & Das-Sarma, A. S. 2004. Spintronics: Fundamentals and applications. *Rev. Mod. Phys.,* **76**, 323–410.

20
Organic semiconductors (OSCs) for electronic chemical sensors

T. J. DAWIDCZYK, H. KONG, and H. E. KATZ,
Johns Hopkins University, USA

DOI: 10.1533/9780857098764.3.577

Abstract: This chapter discusses an assortment of organic semiconductors used as sensors. Both p- and n-channel semiconductors are used to create architectures such as field-effect transistors. Active layers of small molecules, polymers, carbon nanotubes and graphene are included. Depending on the analyte and organic semiconductor used in the active layer, different levels of responses are achieved, allowing for better selectivity. Receptors have been incorporated into the sensors to further enhance the response and selectivity to specific analytes.

Key words: organic electronics, sensors, organic transistors, carbon nanotubes, graphene.

20.1 Introduction to organic semiconductors (OSCs)

20.1.1 Covalent molecular structures

Organic semiconductors (OSCs) are organic materials that share two features. First, their molecular orbitals are at energy levels accessible for redox processes and/or photoexcitation. The second is that the subunits hosting these orbitals are in close enough proximity that electrical charge can be transported among them. Like most organic materials, OSCs are insulators under ordinary ambient conditions, but have the property that their conductivity can increase markedly on the application of voltage, electromagnetic radiation, or heat. The change in conductivity possible for an OSC is the product of the charge density induced by the energetic input multiplied by the mobilities of the charges under electric fields.

The term 'organic semiconductor' includes both polymers and lower molecular weight molecular solids. The lower molecular weight molecules dominated the early history of OSC devices, especially organic field-effect transistors (OFETs). They were mostly extensively conjugated heterocyclic and fused ring oligomers. Eventually, self-organizing side chains were appended. Later, it was found that if the side chains were effective enough at organizing crystal lattices, the conjugated portions could be as small as three rings long. Also, the earliest molecular solid semiconductors were hole

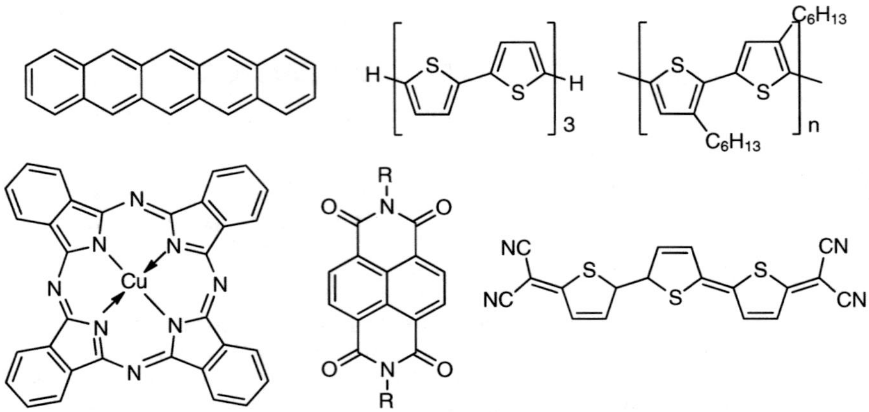

20.1 Typical molecular structures leading to OSC materials. Top, from left: pentacene, α-sexithiophene (α-6T), poly(3-hexylthiophene) or P3HT. Bottom, from left, copper phthalocyanine, naphthalenetetracarboxylic diimide (NTCDI), terthiophene tetracyanoquinodimethane. The first four are hole carriers, and the last two electron carriers. There are thousands of possible variations in these structures, including substitution of some carbons for heteroatoms and appending vast libraries of side chains, that enable tuning of charge transport, processing conditions, and chemical interactions.

carriers (p-channel materials in OFETs), though great progress has been made recently in designing electron transporting (n-channel) OSCs. Polymeric semiconductors have also been developed to a high level in the last five years with the addition of numerous heterocyclic subunits to complement the thiophene rings used in the earliest examples. These new subunits conferred better stabilization of charge carriers and molecular organization that approaches that of the molecular solids, increasing charge carrier mobility. Prototypical OSC molecular structures are shown in Fig. 20.1.

20.1.2 Carbon nanotubes

Single-wall carbon nanotubes (SWNTs) constitute a special example of a conjugated, and arguably organic, electronic material. They are hollow cylindrical tubes, which are based on monolayer graphite ('graphene') sheets, rolled up into cylinders with diameters ranging between 1 and 100 nm, and lengths ranging from tens of nm to cm. SWNTs have been widely investigated for a variety of potential applications such as high mobility transistors, integral logic circuits (inverters and ring oscillators), energy conversion and storage devices, electrodes for signal transmission, and detectors for chemical and biological sensing. In addition, due to their high polarizability and smooth surface, SWNTs will strongly bond to each other through van der Waals forces (~500 eV per 1 μm of SWNT length) to form bundles, whereas some synthetic methods will also form concentric cylinders of carbon that are called multi-walled carbon nanotubes (MWNTs) (Khlobystov and Britz, 2006).

The electronic properties of a SWNT depend on the direction (chiral vector) in which the graphene was rolled up to form the nanotube. Figure 20.2 shows the SWNT structures which can be formed by roll-up of a graphene sheet along a chiral vector (Hersam, 2008; Sloan *et al.*, 2002). The chiral vector C_h is specified by a pair of integers (n, m) and a pair of unit vectors of graphene lattice (a_1, a_2); thus, $C_h = na_1 + ma_2$. The diameter (d) and the chiral angle (θ) of an SWNT can be calculated as follows:

$$d = \left(\frac{\sqrt{3}}{\pi}\right) a_{cc}\left(\sqrt{m^2 + mn + n^2}\right);$$

where a_{cc} is the nearest-neighbor carbon atom distance (1.421 Å),

$$\theta = \tan^{-1}\left\{\frac{\sqrt{3}m}{2n+m}\right\}.$$

The chiral vector determines the characteristics of the SWNT as a semiconductor or metal (including semi- or quasi-metal). When $n - m \neq 3q$ (q is an integer) (if $m = 0$, the nanotubes can be called 'zigzag nanotubes'), the electronic density of states in the SWNT exhibits a significant band gap near the Fermi level and the nanotube is thus semiconducting; when $n - m = 3q$ (if $q = 0$, the nanotubes can be called 'armchair nanotubes'), the conductance and valence bands in the SWNT overlap and the nanotube is thus metallic (or semimetallic when $n \neq m$). Statistically, there are twice as many ways to roll a graphene sheet into a semiconducting SWNT (67%) as there are ways to roll the same sheet into a metallic SWNT (33%) (Lu *et al.*, 2011). For semiconducting SWNTs, the band gap varies inversely with diameter. Furthermore, for both metallic and semiconducting SWNTs, the dominant optical transitions vary with diameter and chiral vector.

20.2 Sensitive organic semiconductor (OSC) devices

20.2.1 Organic chemiresistors and OFETs

Because of the ability to modulate the conductivities of OSCs and the reactivity of both charge carriers and domain boundaries with extrinsic chemical species, OSCs can be highly responsive to environmental chemical agents ('analytes'). When OSCs are positioned between a pair of electrodes and their native resistivities are modulated by chemical exposure, the device

(a)

(b)

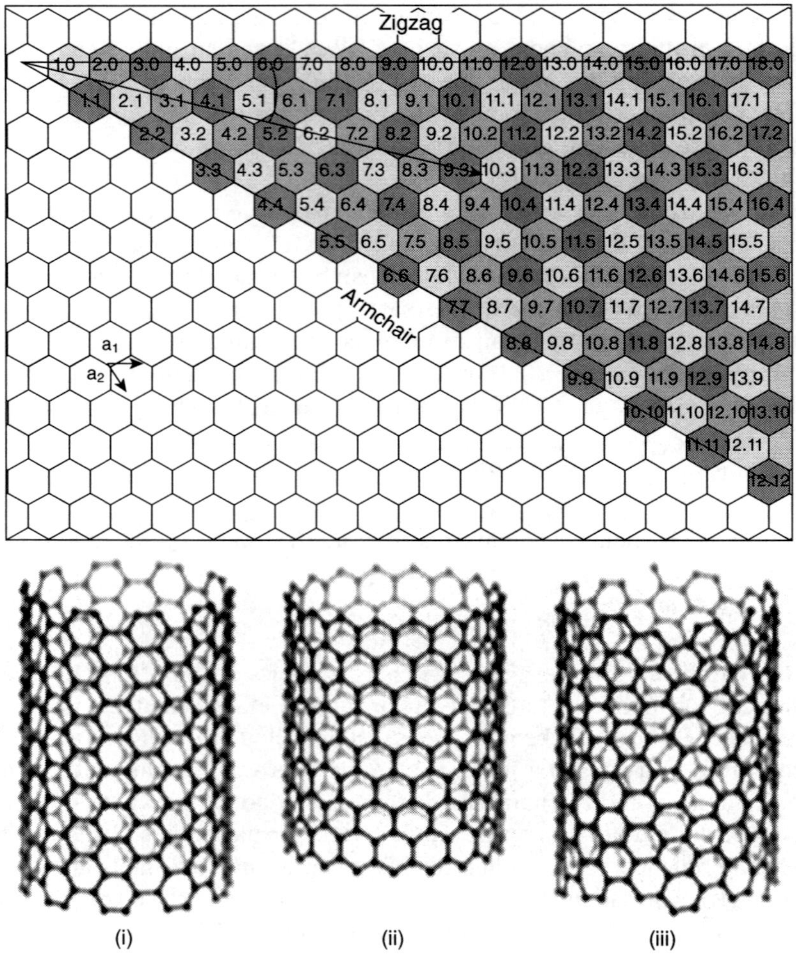

20.2 A SWNT is formed by roll-up of a graphene sheet along a chiral vector. (a) Chiral vector, $C_h = na_1 + ma_2$, where (n, m) are integers known as the chiral indices and a_1 and a_2 are the unit vectors of the graphene lattice. Nanotubes with $n = m$ (known as 'armchair nanotubes') and those with $n - m = 3q$, where $q = 0, 1, 2, 3, \ldots$, are metallic at room temperature. Carbon nanotubes with $n - m = 3q + 1$ and $n - m = 3q + 2$ are semiconductors with a band gap that varies inversely with diameter. Nanotubes with $m = 0$ are known as 'zigzag nanotubes' and can be either metallic or semiconducting. θ is the chiral angle, and (b) (i) armchair, (ii) zigzag, and (iii) chiral nanotubes.

is termed a chemiresistor. When a third, 'gate' electrode is used to preset, or scan, a range of conductivities that are further perturbed by an analyte, the device becomes and OFET sensor. A schematic of an OFET sensor, integrated with a gaseous analyte delivery system, is shown in Fig. 20.3. Table 20.1 shows some systems and the response to various analytes.

Unpackaged OFETs will generally have their output current change upon exposure to polar vapors, including relative humidity from water vapor (Li *et al.*, 2005). Pentacene OFETs were shown to even have their electrical contacts susceptible to humidity. Shorter channel lengths resulted in higher sensitivity, while the on/off ratio (conductance with versus without the gate voltage) could vary by an order of magnitude with varying humidity levels. Another pentacene-based OFET showed a current decrease of 80% when the relative humidity was varied from 0% to 30% (Zhu *et al.*, 2002). The device was deactivated at a relative humidity of 75%.

N-channel materials are susceptible to oxygen as well as humidity, because of the ability of both to quench, or trap, electrons. For example, in perylenetetracarboxylic diimide (PTCDI), oxygen was shown to decrease the electron mobility and density (Graaf and Schlettwein, 2006).

Copper phthalocyanine (CuPc) devices showed a greater response for more polar vapors. Furthermore, the response was also enhanced using thinner OSC films (Yang *et al.*, 2007). An array of CuPc and other phthalocyanines were exposed to various analytes (Bohrer *et al.*, 2009). By changing the copper core to cobalt, nickel, zinc and H_2 the array was seen to have a sensitivity that correlated well to the Lewis base binding enthalpy of the analyte, for the metal phthalocyanines, and to the analytes' hydrogen bonding enthalpy, for the phthalocyanine with no metal core.

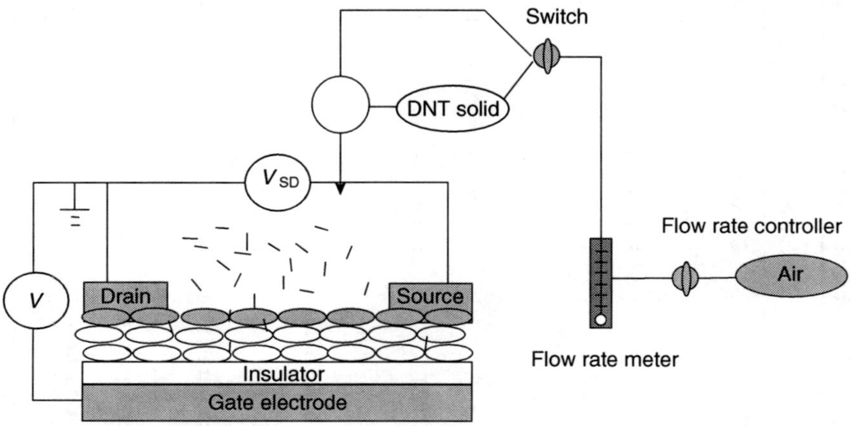

20.3 A schematic of a gas sensing setup, where the analyte (denoted by arrows) is dinitrotoluene (DNT) (Huang *et al.*, 2010).

Table 20.1 Conductance and current responses of selected carbon-based and organic semiconductors to various analytes

Active layer	Receptor	Analyte	Concentration	Response	Reported
DDFTTF	Calixarene	Ethyl acetate	200 ppm	Change of 0.53 uA/s	Sokolov *et al.*, *Adv. Mater.* (2010)
Bis CF3 NTCDI	None	DNT	~200 ppm	8% decrease in current	Huang *et al.*, *J. Mater. Chem.* (2010)
PTO	PTA or PTG	(S)-(−)-β-citronellol	30.6 ppm	2.5 nA decrease	Torsi *et al.*, *Nat. Mater.* (2008)
Bulk SWNTs	None	NH_3 and NO_2	200 ppm (NH_3) and 1% (NO_2)	100% increase (NH_3) and 30% decrease (NO_2) in conductance	Kong *et al.*, *Science* (2000)
SWNTs	Polyethyleneimine	NO_2	less than 1 ppb	~50% increase in conductance	Qi *et al.*, *Nano Lett.* (2003)
SWNT-FET	Polydiacetylene	TNT	1 fM	10% (4 nA) increase in current	Kim *et al.*, *ACS Nano* (2011)
SWNTs	None	TNT	0.5 ~ 5000 ppb	100% (7 μA) increase in current	Park *et al.*, *Biosens Bioelectron* (2010)
TNT with SWNTs	None	DMMP and TNT	2 ppb	4 ~ 15% increase in current	Roberts *et al.*, *ACS Nano* (2009)
SWNT-FET	Single-stranded DNA (ss-DNA)	Methanol, TMA, and PA	150~20 000 ppm	+8 ~ −30 current changes	Staii *et al.*, *Nano Lett.* (2005)
Reduced graphene oxide	None	HCN, DNT, CEES, and DMMP	0.5 ppb (DNT)	0.2% decrease in conductance (DNT)	Robinson *et al.*, *Nano Lett.* (2008)
Reduced graphene oxide	None	NO_2, NH_3, and 2,4-DNT	5 ~ 52 ppm	~ 25% resistance change	Fowler *et al.*, *ACS Nano* (2009)
SWNTs network	None	DMMP	$P = 0.0004P_0$ to $0.003P_0$	12% decrease in conductance	Snow and Perkins, *Nano Lett.* (2005)

Using multiple OSCs and 16 analytes of differing polarities, an electric fingerprint was developed for an array of OFETs, which would give distinct responses for the various analytes (Crone *et al.*, 2001). These devices were then shown to operate in circuits and the sensitivity was calculated for 1 ppm (Crone *et al.*, 2002). Another application of the electronic fingerprint approach has been used to detect volatile organic compounds for food quality analysis (Liao *et al.*, 2005). Pentacene, poly(3-hexylthiophene) (P3HT) and poly(3-octylthiophene) devices were exposed to acetic acid, octanoic acid, ethanol, propanol and other vapors giving unique responses. The carbonyl group was shown to have a greater response, while longer side chains had less of an effect on the current because of their lower diffusivity in the OSC film. Other studies have shown that the sensing array method could give lower detection limits compared to single measurements (Chang *et al.*, 2006).

Contact effects were investigated in an α,ω-dihexylsexithiophene (DHα6T) film exposed to butanol vapor (Torsi *et al.*, 2009). The on state showed that the sensitivity enhancement was from the channel transport in the transistor and the contact resistance and leakage current had lower roles. While at lower gate voltages, which exhibited lower sensitivities, the contact resistance was shown to play a larger role. The OSC will interact more with longer alkyl side chains (Torsi *et al.*, 2003). The longer the side chain, the higher the mass absorbed onto the OSC.

A decrease in film thickness, increases sensor response to the analyte, from the faster diffusion to the conduction channel (Huang *et al.*, 2010; Jung *et al.*, 2011). The authors shows the response of a 20 nm film of a NTCDI derivative to DNT. The device was fabricated on bare silicon with only the native oxide (~2 nm) as the gate dielectric.

Adsorption of polar molecules introduces traps into the OSC film, which alters the source drain current. While these molecules may adhere to the grain boundaries or other defect sites in the film, there is no specific binding to the film. Increasing the binding strength will increase the number of analytes on the film, resulting in a stronger signal. One of the attractive features of OSCs is the ability to covalently bond a receptor group to an existing semiconducting material. This allows the receptor to get closer to the charge conduction channel. Placing an ether group on a polythiophene side chain was found to increase the response to ethanol compared to a polythiophene which just had an alkyl side chain (Torsi *et al.*, 2003).

To better sense dimethyl methylphosphonate (DMMP), a simulant for the nerve agent sarin, a hydroxylated bithiophene oligomer was used as a receptor (Huang *et al.*, 2007). The OSC used was 5,5′-bis(4-hexylphenyl)-2,2′-bithiophene (6PTTP6) in a 1 : 1 ratio with the receptor that seemed to give a single phase film, when examined with scanning electron microscopy and X-ray diffraction. The current decrease occurred faster upon exposure to DMMP and was greater in magnitude for the device with the incorporated receptor. The decrease in current can be attributed to stronger bonding of the DMMP to the receptor and a change in the local electric field from binding to the receptor. By incorporating a receptor with the 6PTTP6 and decreasing the film thickness, the detection limit was reduced from 150 ppm to 5 ppm (Huang *et al.*, 2008). Fig. 20.4 shows the response of the sensor with the reduced thickness to DMMP.

Electron transporting materials have also been used with receptors to detect DMMP (See *et al.*, 2007). An OSC film of NTCDI had receptors of a hydroxphenylated NTCDI deposited on top of the film (See *et al.*, 2007). This receptor formed islands due to the difference in surface energy with the NTCDI. Even though the receptor was not fully incorporated into the film it still showed a more selective response to DMMP. An array of hole and electron conducting OSC were tested with DNT, a by-product of trinitrotoluene (TNT) production (Huang *et al.*, 2010). Hole conducting 6PTTP6 films were made with and without the presence of hydrogen bonding receptor groups. The 6PTTP6 device without the receptor showed a decrease in current upon exposure to DNT while the device with the receptor saw an increase in current with DNT exposure. Other OSC films showed varying responses in either mobility, or mobility and threshold voltage combined, while some, like α-6T did not register a response to DNT. This further illustrates the pattern response of various OFETs to different analytes.

To further show how receptor groups can influence the response of a sensor, two different side chain receptors were incorporated onto an OSC core and placed on top of an existing OSC film (Torsi *et al.*, 2008). These receptors made the sensor respond differently to two enantiomers of a chiral alcohol citronellol, seen in Fig. 20.5. This shows the selectivity that is achievable by carefully selecting the receptor molecule. In another study,

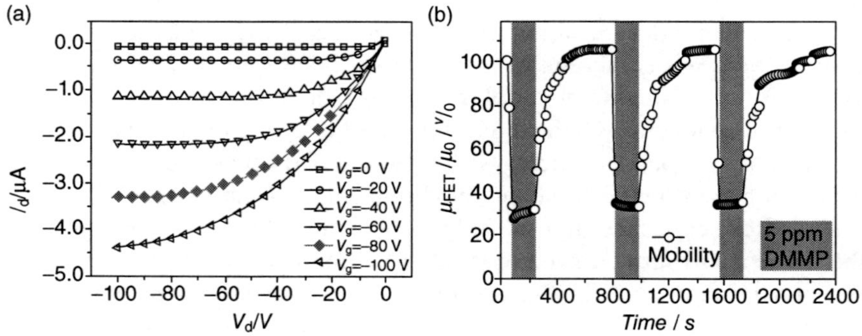

20.4 The output characteristics of the ultra-thin sensor (a) and the change in mobility upon exposure to DMMP (b) (Huang *et al.*, 2008).

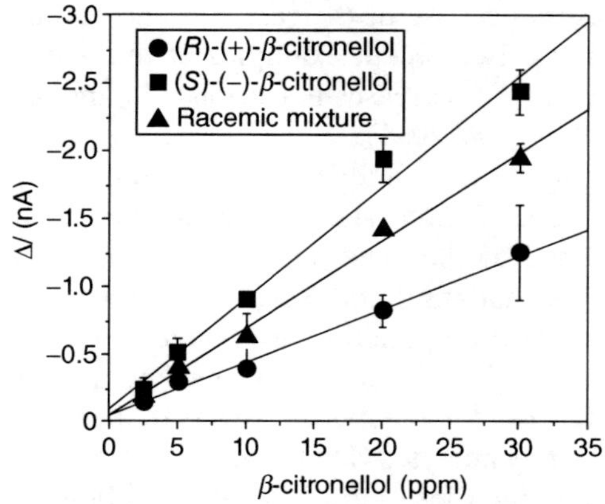

20.5 The sensor response (V_g of −100 V and V_d of −50 V) to different chiralities of citronellol (Torsi *et al.*, 2008).

two monolayers (10 nm) films of 5,50-bis-(7-dodecyl-9H-fluoren-2-yl)-2,20-bithiophene (DDFTTF) were exposed to various volatile organic compounds (Sokolov *et al.*, 2010). The receptors used were calix[8]arene (C[8]A) and c-methyl calyx[4]-resorcinarene (CM[4]RA), which were previously shown to work as size selective cavities in other host–guest sensors. C[8]A showed a ethyl acetate detection limit five times lower than then pure DDFTTF film.

20.2.2 Mechanism of the OFET response

Since the charge transport in longer channel devices is dominated by the charge carriers at the boundaries between grains, the sensing mechanism is dominated by the formation of dipole-induced traps at the grain boundaries when analyte is adsorbed. This analyte-induced trap decreases the source–drain current for most polar analytes, because there is always a region around the dipole where charge carriers are at lower energy, and there is an activation barrier for them to move away from that region. Short channel devices have fewer grain boundaries, so the analyte molecules that diffuse to the electrode/OSC interface alter the source–drain current; that interface is the site of the dominating resistance in such devices. When interfacial resistances do not determine the current–voltage relationships, then dipole adsorption or diffusion into the bulk OSC results in current-lowering traps. On the other hand, analytes with redox activity can act as quenchers or dopants for the main semiconducting domains, lowering or raising current levels, respectively. An electron withdrawing analyte can act as a dopant for a p-type OSC and a trap for an n-type OSC. The doping mechanism is particularly useful for sensors because of the rareness and distinction of a turn-on signaling for analytes. Two recent examples of this are the use of n-channel OSCs to obtain current increases in response to ammonia vapor (Huang *et al.*, 2011; Tremblay *et al.*, 2011).

The extent to which an analyte can alter the charge transport varies on the analyte's location in the OSC film. The charge carriers furthest away from the gate can be screened by the lower OSC layer from up to 90% of the applied gate voltage, which means that the charge carriers are operating at a voltage much lower than the applied gate voltage (Horowitz, 2004). Electrostatic modeling has suggested that a cluster of polar analytes can have a potential of a few tenths of a volt, which means that carriers further from the gate can be trapped by analytes (Huang *et al.*, 2007). Analytes may be more prevalent at grain boundaries of the OSC than the bulk, depending on the ability of the analyte to intercalate into the bulk OSC. The main benefit of a transistor sensor over a chemiresistor is the signal amplification from the gate electrode. This allows multiple parameters to be measured, instead of a pure current measurement as in the case of a chemiresistor. A transistor allows for current, threshold voltage (V_T), and mobility (μ) to be measured (Horowitz *et al.*, 2004; Huang *et al.*, 2010).

20.3 Sensitive carbon nanotube and graphene devices

The application of carbon nanotubes to chemical sensing often depends on the isolation of those with semiconducting behavior, because it is these kinds of nanotubes whose current will be more responsive to environmental chemical changes. For example, Kong and Franklin reported electrical resistance changes upon exposure to gaseous molecules such as NO_2 or NH_3. These responses were both faster and of greater magnitude than those of previously used solid state analogues. In this case, the SWNT is a hole-doped semiconductor, as can be gleaned from the current versus gate voltage (I-V_g) curve shown in Fig. 20.6 (middle plot), where a positive gate voltage led to current decreases. The I-V_g curve recorded after the S-SWNT sample was exposed to NH_3 exhibits a shift of 24 V, which is caused by shifts the valence band of the nanotube away from the Fermi level, resulting in hole depletion and reduced conductance (Fig. 20.6, left plot). In contrast, the I-V_g curve was shifted by 14 V after NO_2 exposure, because of the nanotube Fermi level shifting closer to the valence band (Fig. 20.6, right plot). When bulk materials made from multiple SWNTs were used, effects were averaged over metallic and semiconducting tubes (Kong *et al.*, 2000).

The coating of SWNTs for selective penetration by gaseous analytes is a route to improved selectivity of the response. For example, Dai showed that polyethyleneimine coating affords n-type nanotube devices capable of detecting NO_2 at less than 1 ppb concentrations while being insensitive to

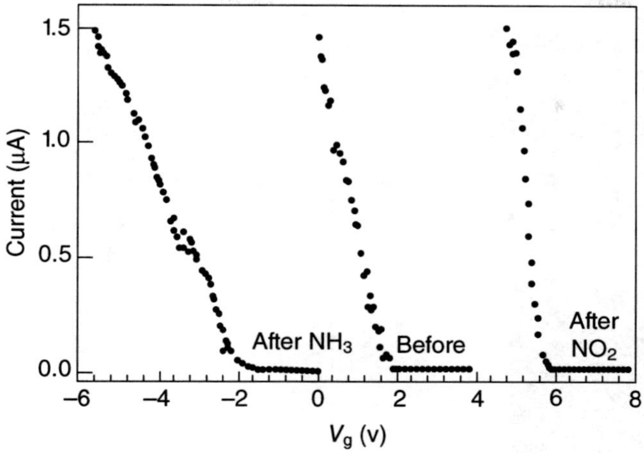

20.6 **Chemical gating effects to the semiconducting SWNT. Current versus gate voltage curves before NO_2 (middle plot), after NO_2 (right plot), and after NH_3 (left plot) exposures. The measurements with NH_3 and NO_2 were carried out successively after sample recovery.**

NH_3 (Fig. 20.7) (Pengfei *et al.*, 2003). On the other hand, coating Nafion (a polymeric perfluorinated sulfonic acid ionomer) on nanotubes prevents NO_2 from even arriving, and allows for selective sensing of NH_3.

A second example of coating was reported by Lee and coworkers, who reported selective and sensitive TNT sensors using biomimetic polydiacetylene (PDA)-coated nanotube-FETs (Fig. 20.8) (Kim *et al.*, 2011). Selective binding events between the TNT molecules and phage display derived TNT receptors were effectively transduced by conductance modulation, with the effect being realized through the PDA coatings. The resulting sensors exhibited an unprecedented 1 fM sensitivity toward TNT in real time, with excellent selectivity over various similar aromatic compounds (Fig. 20.9).

A tremendous chemical amplification of the effect of a small analyte molecule reaching an SWNT was discovered by Mulchandani and coworkers. In their system, a SWNTs network was first modified with trinitrophenol (TNP), an analog of TNT, and then ligated with the anti-TNP single chain antibody (Park *et al.*, 2010). This antibody, with its many dipoles and ionizable functional groups, greatly perturbed the electronic environment of the SWNT. Upon exposure to TNT or closely related molecules, the bound antibodies were displaced, reversing its influence on the SWNT conductance and yielding a marked electrical signal. The sensor detected between 0.5 ppb and 5000 ppb TNT with good selectivity to other nitroaromatic explosives and demonstrated good accuracy for monitoring TNT in aqueous solution. The Bao group also noted the robustness of SWNTs to water

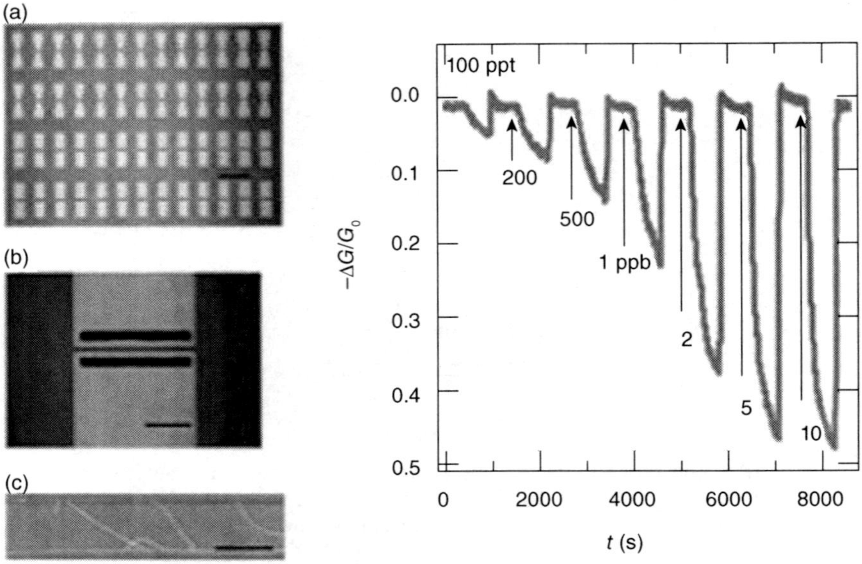

20.7 (a) Optical image of an array of multiple-SWNT devices. (b) Optical image of one device. The black regions (100 μm long) contain catalyst patterned on top of opposing Mo source and drain electrodes. (c) Scanning electron microscopy (SEM) image of several nanotubes bridging two opposing Mo electrodes in a device and change in conductance normalized by initial conductance (G_0) at $V_g = 0$ as a function of time for a PEI functionalized n-type MT device exposed to various concentrations of NO_2 gas. The device was exposed to each concentration of NO_2 for 10 min, after which recovery was made by UV light (254 nm) desorption of NO_2. The concentration of NO_2 was varied by diluting 100 ppm of NO_2 (in Ar) with air by using four mass-flow controllers. The diluted gas was then flown into a homemade chamber that houses the sensor chip.

(Roberts *et al.*, 2009). SWNT TFTs responded to trace concentrations, down to 2 ppb, of DMMP and TNT in aqueous solutions, when mounted on a polymer dielectric. Along with reliable cycling underwater, the TFT sensors fabricated with aligned, sorted nanotube networks (enriched with semiconductor SWNTs) showed a higher sensitivity to analytes than those fabricated with random, unsorted networks with predominantly metallic charge transport (Fig. 20.10).

The Johnson group has reported two kinds of biomacromolecule modifications, DNA and olfactory receptor protein, to SWCNTs to produce altered electronic properties (Johnson and Staii, 2005). Their aim was to create diversity in responses of SWNTs modified with variations on the DNA and protein motifs. In the first case, by choosing the base sequence

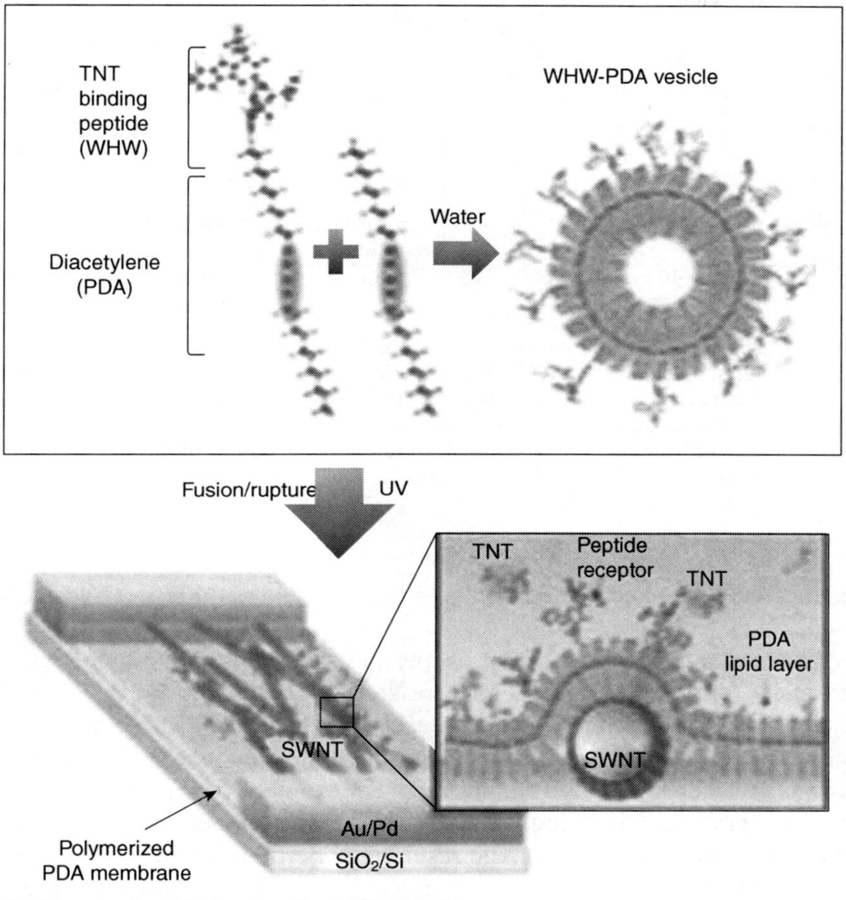

20.8 Schematic diagram depicting TNT sensors based on WHW-PDA (TNT-binding peptide conjugated with polydiacetylene polymer) functionalized SWNT-FETs. After formation of WHW-PDA/PDA vesicles in H_2O, the vesicles were applied to a SWNT-FET, which resulted in rupture, fusion, and UV polymerization of the WHW-PDA/PDA membranes on the SWNT-FET. The molecules and SWNT are not drawn to scale.

of the single stranded DNA appended to the SWNTs, different responses to gases were obtained (Fig. 20.11). In the second case, nanotubes were coated with olfactory receptor proteins, known to bind specific vapors as part of the mammalian nose. While a nose has hundreds of these receptor proteins, the initial set of experiments utilized only three. Response patterns for organic analytes with different polarities and molecular shapes were

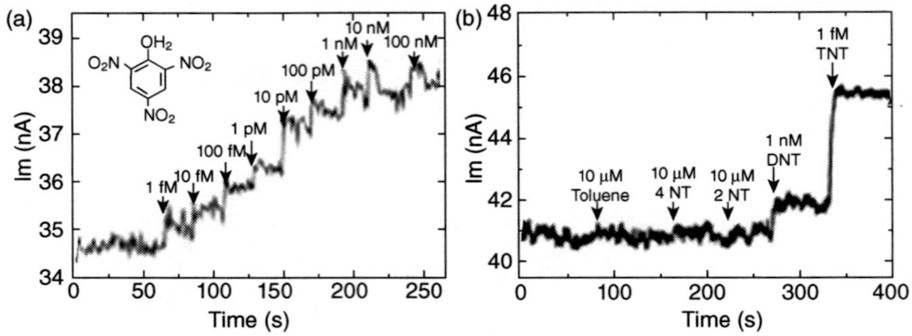

20.9 **Specific and sensitive detection of TNT molecules with TNT sensors based on WHW-PDA-functionalized SWNTFETs. (a) Real-time conductance measurements obtained from the WHW-PDA-functionalized SWNT-FETs after the introduction of TNT at various concentrations. Arrows indicate the injection points of target molecules. (b) Selective response to TNT in mixed solution of toluene, 4NT, 2NT, and DNT.**

distinct when these coated nanotubes were analyzed as OFETs (Goldsmith *et al.*, 2011).

While not a major focus of this chapter, a brief mention of graphene, the 'unrolled' form of SWCNTs, is appropriate, as early results on its chemical responsiveness are being reported. Because it is generally highly conductive and subject to facile doping, only the most pristine graphene specimens will show chemically modulated conductivity. For example, Robinson and Snow obtained graphene with limited extrinsic dopants by reducing graphene oxide (Robinson *et al.*, 2008). The resulting material responded in just ten seconds to simulants of the three main classes of chemical-warfare agents and one explosive at parts-per-billion concentrations (Fig. 20.12).

Kaner and coworkers used spin-coated films of reduced graphene oxide on interdigitated electrode arrays. Hydrazine was both the solvent delivery vehicle and reducing agent (Fowler *et al.*, 2009). The sensor response is consistent with a charge transfer mechanism between the analyte and graphene with a limited role of the electrical contacts.

Finally, the Snow group utilized simultaneous and synergistic conductance and capacitance measurements on a single-walled carbon nanotube (SWNT) network for detection of dilute chemical vapors (Fig. 20.13) (Snow and Perkins, 2005). Adsorbates from chemical vapors produce a rapid response in both the capacitance and the conductance of the SWNT network. These responses are caused by a combination of two distinct physiochemical properties of the adsorbates: charge transfer (which is associated with the semiconductor behavior) and polarizability, which is a governing property of dielectrics. The ratio of the conductance response to the

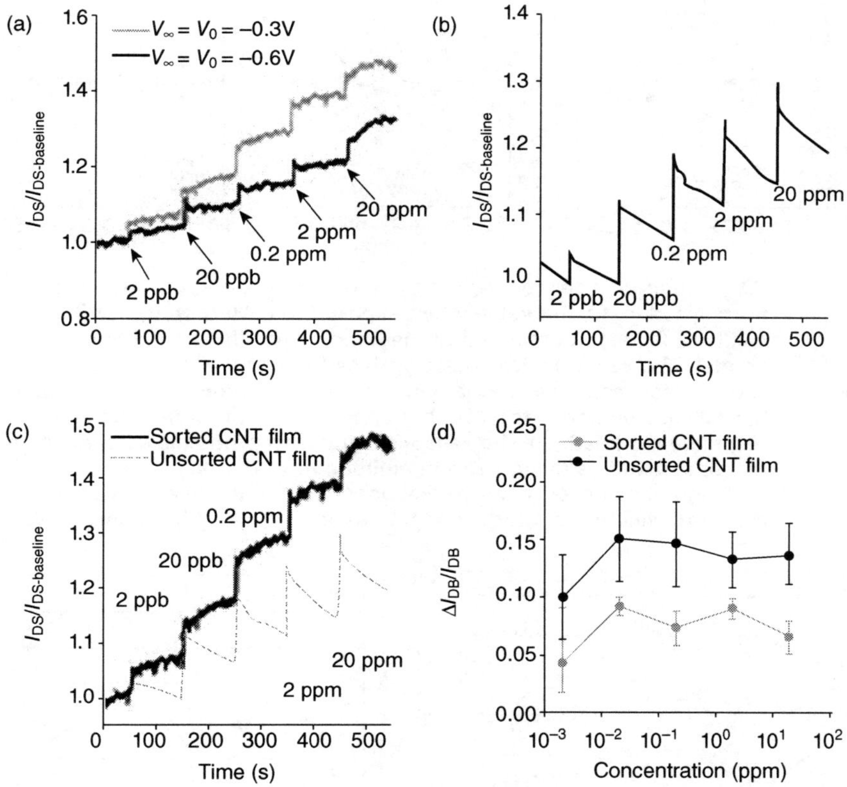

20.10 Sensor response to TNT for sorted (sc) and unsorted (met) SWNT networks. (a) $I_{DS}/I_{DS\text{-}0}$ of a sorted SWNT network upon exposure to TNT solutions ranging from 2 ppb to 200 ppm for $V_{DS} = V_{GS} = -0.3$ V (gray) and $V_{DS} = V_{GS} = -0.6$ V (black). (b) $I_{DS}/I_{DS\text{-}0}$ of an unsorted SWNT network upon exposure to TNT solutions ranging from 2 ppb to 200 ppm for $V_{DS} = V_{GS} = -0.3$ V. (c) Composite plot of A and B showing the $I_{DS}/I_{DS\text{-}0}$ for TFTs with sorted (black) and unsorted (gray) SWNT networks. (d) The average relative response for TNT solutions with sorted (black) and unsorted (gray) networks with standard deviations shown by error bars.

capacitance response is a concentration-independent parameter associated with a chemical vapor that can assist in its identification (Fig. 20.14).

20.4 Conclusion

It is clear from the above examples that conjugated organic (and all-carbon) materials have strong and in many cases controllable responses to chemical analytes. Great progress has been made in deducing the mechanisms of the

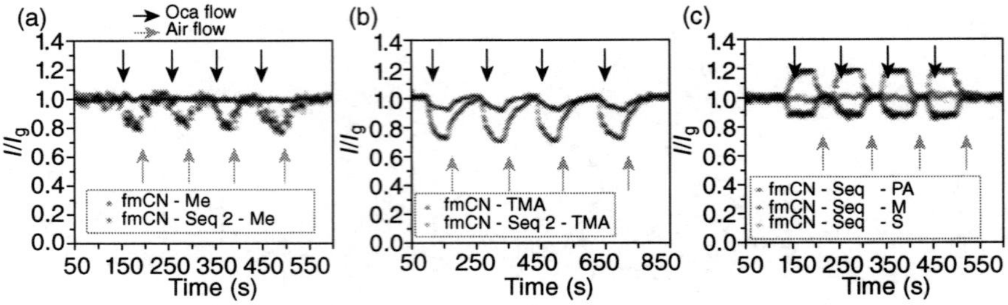

20.11 Change in sensor current upon odor exposure. Currents are normalized to 10, the value when exposed to air (no odor). (a) Bare swCN-FET does not respond to methanol vapor (horizontal line points). The same device coated with ss-DNA sequence 2 (Seq 2) shows clear responses to methanol (wavy line points). (b) A second bare device responds to TMA, but after application of Seq 2, the response is tripled. (c) The sensor response to propionic acid (positive responses) differs in sign and magnitude from the response to methanol (shallower negative respones). Horizontal line data are the current baseline (no odor). V_B = 100 mV and V_G = 0 V for all data sets.

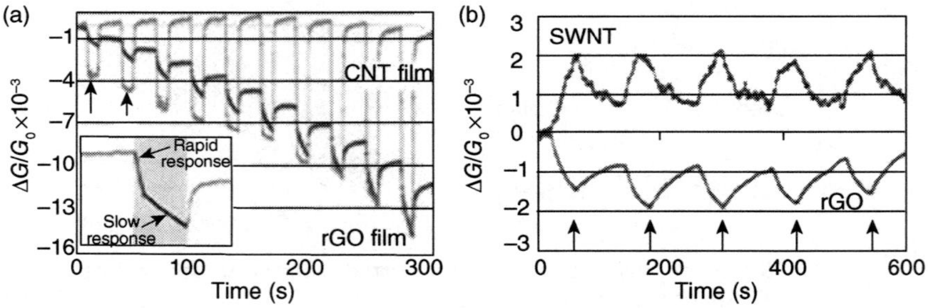

20.12 (a) Real-time conductance response to 5 s acetone pulses of increasing concentration for a SWNT network sensor (tall, lighter color) and a rGO network sensor (shallow, darker color).
(b) Conductance response ($\Delta G/G_o$) of a SWNT and rGO device to 30 s pulses of 0.5 ppb DNT (dinitrotoluene). Arrows mark the end of each 30 s pulse.

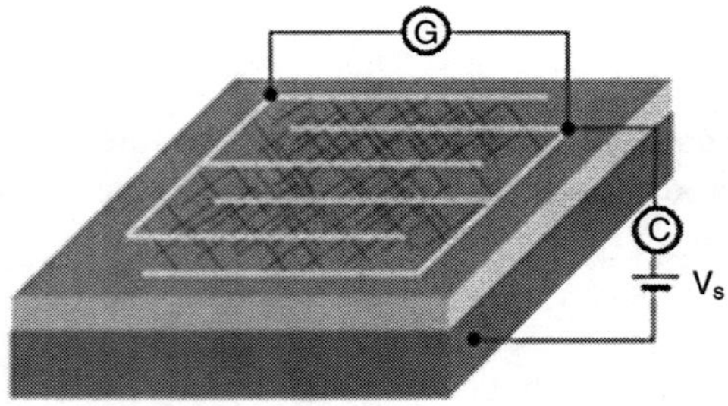

20.13 Schematic of the device, which was designed for simultaneous measurement of the SWNT network capacitance and conductance.

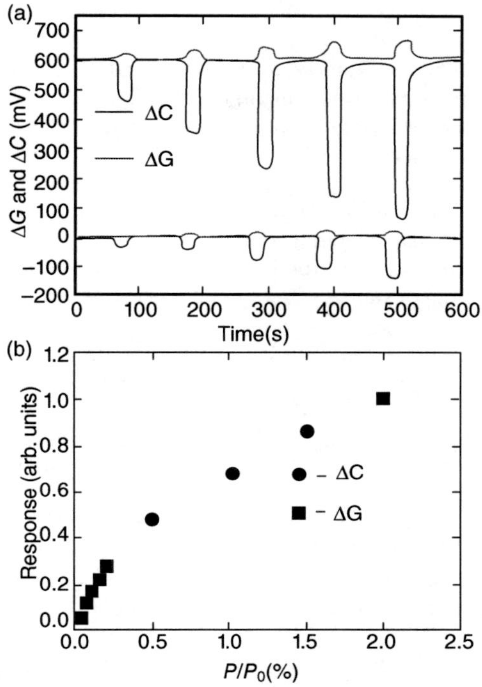

20.14 (a) Normalized capacitance (large, negative responses) and conductance (small, positive responses) response to 20 s doses of DMMP at concentrations ranging from $P = 0.0002P_0$ to $0.02P_0$. (b) Magnitude of the capacitance and conductance response as a function of P/P_0.

responses, associating the responses selectively with analytes of interest, and fabricating devices that best capture the transduced electronic signals. Major challenges for this technology include diminishing the signals that arise from interferents, statistical treatments of responses to quantify the certainty of 'positive' signals for analytes, and integrating the sensitive resistors and OFETs into circuits that digitize, amplify and convert the signals into forms that can be processed logically in concert with other sensory inputs, and transmitted effectively to other sensors in networks and to interested users.

20.5 Acknowledgments

We are grateful to NSF Electronic, Communication, and Cyber Systems (grant number 0730926) and to the Johns Hopkins University Applied Physics Laboratory for support of this work. We also warmly thank authors whose figures are reproduced here with permission.

20.6 References

Bohrer, F. I., Colesniuc, C. N., Park, J., Ruidiaz, M. E., Schuller, I. K., Kummel, A. C. & Trogler, W. C. 2009. Comparative gas sensing in cobalt, nickel, copper, zinc, and metal-free phthalocyanine chemiresistors. *Journal of the American Chemical Society*, **131**, 478–485.

Chang, J. B., Liu, V., Subramanian, V., Sivula, K., Luscombe, C., Murphy, A., Liu, J. S. & Frechet, J. M. J. 2006. Printable polythiophene gas sensor array for low-cost electronic noses. *Journal of Applied Physics*, **100**, 014506.

Crone, B., Dodabalapur, A., Gelperin, A., Torsi, L., Katz, H. E., Lovinger, A. J. & Bao, Z. 2001. Electronic sensing of vapors with organic transistors. *Applied Physics Letters*, **78**, 2229–2231.

Crone, B. K., Dodabalapur, A., Sarpeshkar, R., Gelperin, A., Katz, H. E. & Bao, Z. 2002. Organic oscillator and adaptive amplifier circuits for chemical vapor sensing. *Journal of Applied Physics*, **91**, 10140–10146.

Fowler, J. D., Allen, M. J., Tung, V. C., Yang, Y., Kaner, R. B. & Weiller, B. H. 2009. Practical chemical sensors from chemically derived graphene. *ACS Nano*, **3**, 301–306.

Goldsmith, B. R., Mitala, J. J., Josue, J., Castro, A., Lerner, M. B., Bayburt, T. H., Khamis, S. M., Jones, R. A., Brand, J. G., Sligar, S. G., Luetje, C. W., Gelperin, A., Rhodes, P. A., Discher, B. M. & Johnson, A. T. 2011. Biomimetic chemical sensors using nanoelectronic readout of olfactory receptor proteins. *ACS Nano*, **5**, 5408–5416.

Graaf, H. & Schlettwein, D. 2006. Influence of gas molecules on the charge carrier mobility in thin films of semiconducting perylene tetracarboxylic imides. *Journal of Applied Physics*, **100**, 126104.

Hersam, M. C. 2008. Progress towards monodisperse single-walled carbon nanotubes. *Nature Nanotechnology*, **3**, 387–394.

Horowitz, G. 2004. Organic thin film transistors: from theory to real devices. *Journal of Materials Research*, **19**, 1946–1962.

Horowitz, G., Lang, P., Mottaghi, M. & Aubin, H. 2004. Extracting parameters from the current–voltage characteristics of field-effect transistors. *Advanced Functional Materials*, **14**, 1069–1074.

Huang, J., Miragliotta, J., Becknell, A. & Katz, H. E. 2007. Hydroxy-terminated organic semiconductor-based field-effect transistors for phosphonate vapor detection. *Journal of the American Chemical Society*, **129**, 9366–9376.

Huang, J., Sun, J. & Katz, H. E. 2008. Monolayer-dimensional 5,5′-Bis(4-hexylphenyl)-2,2′-bithiophene transistors and chemically responsive heterostructures. *Advanced Materials*, **20**, 2567–2572.

Huang, J., Dawidczyk, T. J., Jung, B. J., Sun, J., Mason, A. F. & Katz, H. E. 2010. Response diversity and dual response mechanism of organic field-effect transistors with dinitrotoluene vapor. *Journal of Materials Chemistry*, **20**, 2644–2650.

Huang, Y., Fu, L., Zou, W., Zhang, F. & Wei, Z. 2011. Ammonia sensory properties based on single-crystalline micro/nanostructures of perylenediimide derivatives: core-substituted effect. *Journal of Physical Chemistry C*, **115**, 10399–10404.

Johnson, A. T. & Staii, C. 2005. DNA-decorated carbon nanotubes for chemical sensing. *Nano Letters*, **5**, 1774–1778.

Jung, B. J., Hardigree, J. F. M., Dhar, B. M., Dawidczyk, T. J., Sun, J., See, K. C. & Katz, H. E. 2011. Naphthalenetetracarboxylic diimide layer-based transistors with nanometer oxide and side chain dielectrics operating below one volt. *ACS Nano*, **5**, 2723–2734.

Khlobystov, A. N. & Britz, D. A. 2006. Noncovalent interactions of molecules with single walled carbon nanotubes. *Chemical Society Reviews*, **35**, 637–659.

Kim, T. H., Lee, B. Y., Jaworski, J., Yokoyama, K., Chung, W. J., Wang, E., Hong, S., Majumdar, A. & Lee, S. W. 2011. Selective and sensitive TNT sensors using biomimetic polydiacetylene-coated CNT-FETs. *ACS Nano*, **5**, 2824–2830.

Kong, J., Franklin, N. R., Zhou, C., Chapline, M. G., Peng, S., Cho, K. & Dai, H. 2000. Nanotube molecular wires as chemical sensors. *Science*, **287**, 622–625.

Li, D. W., Borkent, E. J., Nortrup, R., Moon, H., Katz, H. & Bao, Z. N. 2005. Humidity effect on electrical performance of organic thin-film transistors. *Applied Physics Letters*, **86**, 042105.

Liao, F., Chen, C. & Subramanian, V. 2005. Organic TFTs as gas sensors for electronic nose application, *Sensors and Actuators B-Chemical*, **107**, 849–855.

Lu, F., Meziani, M. J., Cao, L. & Sun, Y. P. 2011. Separated metallic and semiconducting single-walled carbon nanotubes: opportunities in transparent electrodes and beyond. *Langmuir*, **27**, 4339–4350.

Park, M., Cella, L. N., Chen, W., Myung, N. V. & Mulchandani, A. 2010. Carbon nanotubes-based chemiresistive immunosensor for small molecules: detection of nitroaromatic explosives. *Biosensors and Bioelectronics*, **26**, 1297–1301.

Pengfei, Q. F., Vermesh, O., Grecu, M., Javey, A., Wang, O., Dai, H. J., Peng, S. & Cho, K. J. 2003. Toward large arrays of multiplex functionalized carbon nanotube sensors for highly sensitive and selective molecular detection. *Nano Letters*, **3**, 347–351.

Qi, P., Vermesh, O., Grecu, M., Javey, A., Wang, Q., Dai, H., Peng, S. & Cho, K. J. 2003. Toward large arrays of multiplex functionalized carbon nanotube sensors for highly sensitive and selective molecular detection, *Nano Letters* **3**, 347–351.

Roberts, M. E., Lemieux, M. C. & Bao, Z. 2009. Sorted and aligned single-walled carbon nanotube networks for transistor-based aqueous chemical sensors. *ACS Nano*, **3**, 3287–3293.

Robinson, J. T., Perkins, F. K., Snow, E. S., Wei, Z. Q. & Sheehan, P. E. 2008. Reduced Graphene Oxide Molecular Sensors. *Nano Letters*, **8**, 3137–3140.

See, K. C., Becknell, A., Miragliotta, J. & Katz, H. E. 2007. Enhanced response of n-channel naphthalenetetracarboxylic diimide transistors to dimethyl methylphosphonate using phenolic receptors. *Advanced Materials*, **19**, 3322–3327.

Sloan, J., Kirkland, A. I., Hutchison, J. L. & Green, M. L. 2002. Structural characterization of atomically regulated nanocrystals formed within single-walled carbon nanotubes using electron microscopy. *Accounts of Chemical Research*, **35**, 1054–1062.

Snow, E. S. & Perkins, F. K. 2005. Capacitance and conductance of single-walled carbon nanotubes in the presence of chemical vapors. *Nano Letters*, **5**, 2414–2417.

Sokolov, A. N., Roberts, M. E., Johnson, O. B., Cao, Y. D. & Bao, Z. A. 2010. Induced sensitivity and selectivity in thin-film transistor sensors via calixarene layers. *Advanced Materials*, **22**, 2349–2353.

Staii C., Johnson, A.T., Jr, Chen, M. & Gelperin, A. 2005. DNA-decorated carbon nanotubes for chemical sensing, *Nano Letters* **5**, 1774–1778.

Torsi, L., Tanese, M. C., Cioffi, N., Gallazzi, M. C., Sabbatini, L., Zambonin, P. G., Raos, G., Meille, S. V. & Giangregorio, M. M. 2003. Side-chain role in chemically sensing conducting polymer field-effect transistors. *Journal of Physical Chemistry B*, **107**, 7589–7594.

Torsi, L., Farinola, G. M., Marinelli, F., Tanese, M. C., Omar, O. H., Valli, L., Babudri, F., Palmisano, F., Zambonin, P. G. & Naso, F. 2008. A sensitivity-enhanced field-effect chiral sensor. *Nature Materials*, **7**, 412–417.

Torsi, L., Marinelli, F., Angione, M. D., Dell'aquila, A., Cioffi, N., De Giglio, E. & Sabbatini, L. 2009. Contact effects in organic thin-film transistor sensors. *Organic Electronics*, **10**, 233–239.

Tremblay, N. J., Jung, B. J., Breysse, P. & Katz, H. E. 2011. Digital inverter amine sensing via synergistic responses by n and p organic semiconductors. *Advanced Functional Materials*, **21**, 4314–4319.

Yang, R. D., Gredig, T., Colesniuc, C. N., Park, J., Schuller, I. K., Trogler, W. C. & Kummel, A. C. 2007. Ultrathin organic transistors for chemical sensing. *Applied Physics Letters*, **90**, 263506.

Zhu, Z. T., Mason, J. T., Dieckmann, R. & Malliaras, G. G. 2002. Humidity sensors based on pentacene thin-film transistors. *Applied Physics Letters*, **81**, 4643–4645.

21
Organic bioelectronics

G. TARABELLA, N. COPPEDÈ and
S. IANNOTTA, Institute of Materials
for Electronics and Magnetism, Italy and F. CICOIRA,
P. KUMAR and C. SANTATO,
École Polytechnique de Montréal, Canada

DOI: 10.1533/9780857098764.3.597

Abstract: In this chapter we focus on a novel boosting development of organic electronics that exploits the mixed electronic and ionic conduction that certain organic materials can sustain, to produce a wide range of devices with potential applications in bioelectronics. Examples of such devices are organic electrochemical transistors and miniaturized bio-laboratories making use of conducting polymers, whose physico-chemical properties are controlled electronically. Organic electrochemical transistors can be operated in aqueous environment as efficient ion-to-electron converters, thus providing an interface between the worlds of biology and electronics. For this reason, they present exciting opportunities for applications in biosensing. Biocompatible organic conducting polymers able to operate in aqueous environment can be interfaced with biological systems. By exploiting conducting polymer properties such as electrochemical switching and combined electronic and ionic transport, it is possible to use the polymers as platforms where cells can adhere and proliferate.

Key words: organic electronics, electrochemical transistors, bioelectronics, conductive polymers.

21.1 Introduction to organic bioelectronics

Organic bioelectronics refers to the application of organic electronics devices at the interface with biology (Mabeck and Malliaras, 2006; Berggren and Richter-Dahlfors, 2007; Martin, 2007; Nikolou and Malliaras, 2008; Owens and Malliaras, 2010; Richter-Dahlfors *et al.*, 2011; Lin and Yan, 2012), including applications in medicine and nanomedicine (Richter-Dahlfors *et al.*, 2011), neuroscience (Asplund *et al.*, 2010), and recently *in-vivo* implantation (Khodagholy *et al.*, 2011a). Organic bioelectronic devices exploit the property of conductive polymers to conduct both electrons and ions, which enables them to be interfaced to biological systems. Moreover, the soft nature of organic materials is exploitable to achieve the mechanical compatibility with tissues, as well as the conformability to non-planar flexible substrates (e.g. skin), required for biomedical devices and implants. The possibility of modifying the conductivity of organic polymers upon biological/electrochemical effects permits a real-time sensor with high sensitivity and selectivity. Conductive organic polymers are a special class of polymers, whose distinct chemical property is the *conjugation*: the chemical bonds along the polymer backbone consists of alternating single and double carbon–carbon bonds. (Shirakawa *et al.*, 1977; Malliaras and Friend, 2005). The conjugated molecular structure leads to one unpaired electron per carbon atom (the π electron). In π bonding the carbon orbitals are in the sp^2 configuration and the p_z orbitals of successive carbon atoms along the polymer backbone overlap. The overlapping leads to the electron delocalization that permits the mobility of charge carriers along the polymer chain. The most widely used conducting polymers are polyaniline, polypyrrole, polycarbazole and polythiophene, and their derivatives, with applications covering a broad spectrum of chemical and biological sensing. The electrical conductivity of conducting polymers can be controlled over several orders of magnitudes, and can be achieved by a reversible doping or dedoping of the polymers. The doping or dedoping could be of different types, the most well established being the chemical, electrochemical, photochemical and interfacial doping (Fig. 21.1).

Organic bioelectronics aims at exploiting the key advantages of organic semiconductors, such as ease of processing, optoelectronic properties that can be tuned by chemical synthesis, and compatibility with mechanically flexible substrates. Nowadays organic electronics devices are almost ubiquitous, being present in many different areas, ranging from plastic electronics, organic thin films transistors (Horowitz, 1998; Facchetti *et al.*, 2005), organic light-emitting diodes (Friend *et al.*, 1999), e-textile technology (De Rossi, 2007) and organic solar cells (JY Kim *et al.*, 2007).

Devices that are particularly interesting for applications in sensing and bioelectronics are organic thin film transistors (OTFTs). In OTFTs, the switching and amplifying properties of transistors can be exploited to detect low signals generated by analytes: a small applied gate voltage (about 1 V) can results in a significant variation in the current of the transistor channel (the source–drain current). OTFTs have been already exploited for many different applications, especially in bioelectronics and drug delivery (Berggren and Richter-Dahlfors, 2007), diagnostics (Owens and Malliaras, 2010), or environment (Horowitz, 1998; Sirringhaus *et al.*, 1998; Johnson *et al.*, 2007).

OTFTs can be divided in two main classes: organic field effect transistors (OFETs) and organic electrochemical transistors (OECTs). In our discussion we focus on OECTs, a new promising class of OTFTs. OECTs exploit the electrochemical doping/dedoping process in conductive polymers like poly(3,4-ethylenedioxythiopene)-doped with poly(styrene sulphonate), PEDOT:PSS, the most used and commercially available conductive polymer.

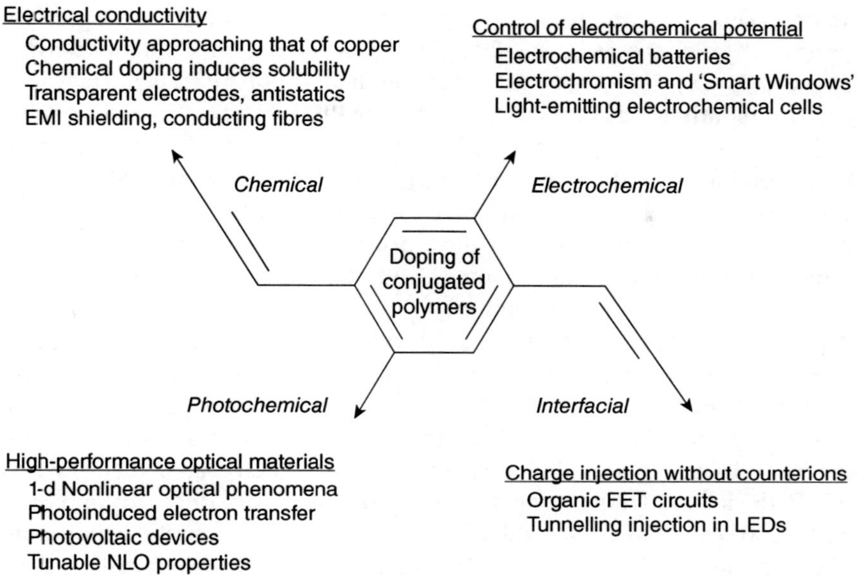

21.1 Schematic diagram for various types of doping of conjugated polymers and corresponding applications of these doped conjugated polymers (NLO, nonlinear optical; FET, field-effect transistor; LED, light-emitting diode) From Heeger, A. J. (2001). Semiconducting and Metallic Polymers: The Fourth Generation of Polymeric Materials (Nobel Lecture). *Angewandte Chemie International Edition,* **40,** 2591–2611.

In OECTs, the polymer is in contact with the source and drain electrodes. The application of a source–drain voltage (V_{ds}) generates the drain–source current (I_{ds}) upon application of an electrical bias to the gate electrode (V_{gs}), which is immersed into the electrolyte. V_{gs} controls the doping/dedoping of the conducting polymer, resulting in a modulation of I_{ds}.

The first section of this chapter explains in detail the working principle of OECTs and reviews the most advanced results on the device physics of OECTs. In the second section, we discuss OECTs used as enzymatic biosensors, in particular as glucose sensors. The application of OECTs in glucose sensing gives also the opportunity the gain insight into the mechanism of the doping or dedoping processes in conducting polymers. The third section deals with bio-applications of OECTs such as cell-based biosensor.

21.2 Organic electrochemical transistors (OECTs)

Organic electrochemical transistors (OECTs) can work in aqueous environments, thus providing an opportunity to interface electronics with biological systems. The liquid interface is realized through confining the electrolyte reservoir by a micro-well, usually made up of the polymer polydimethylsiloxane (PDMS), in contact with the conductive polymer (Fig. 21.2). The most frequently used conductive polymer is PEDOT:PSS. The overlapping of the liquid with the polymer defines the OECT channel, which is where doping/dedoping of the polymer takes place (DeFranco *et al.*, 2006; Cicoira *et al.*, 2010). In recent years, extensive exploration of OECT device physics has improved the knowledge of the working mechanism of these devices. In particular, insight has been gained about the role of gate electrode geometry (Cicoira *et al.*, 2010) and material (Tarabella *et al.*, 2010) on the OECT response, as well as how different kinds of electrolytes, such as ionic liquids (Yang *et al.*, 2010) or micellar electrolytes (Tarabella *et al.*, 2012), can affect the OECT performance. Theoretical works have also supported these efforts, with modelling analysis able to address specific issues for the device structure (Bernards and Malliaras, 2007; Yaghmazadeh *et al.*, 2011). The first OECT device modelling has been analysed in detail by Bernards and Malliaras (2007) and has been used also in other papers to describe experimental observations (Cicoira *et al.*, 2010; Lin and Yan, 2012). We recall here the equations governing the OECT response:

$$I_{\mathrm{ds}} = \frac{q\mu p_0 t W}{L V_{\mathrm{p}}}\left(V_{\mathrm{p}} - V_{\mathrm{g}}^{\mathrm{eff}} + \frac{V_{\mathrm{ds}}}{2}\right)V_{\mathrm{ds}} \qquad [21.1]$$

$$V_{\mathrm{p}} = \frac{q p_0 t}{c_1} \qquad [21.2]$$

$$V_{\mathrm{g}}^{\mathrm{eff}} = V_{\mathrm{G}} + V_{\mathrm{offset}} \qquad [21.3]$$

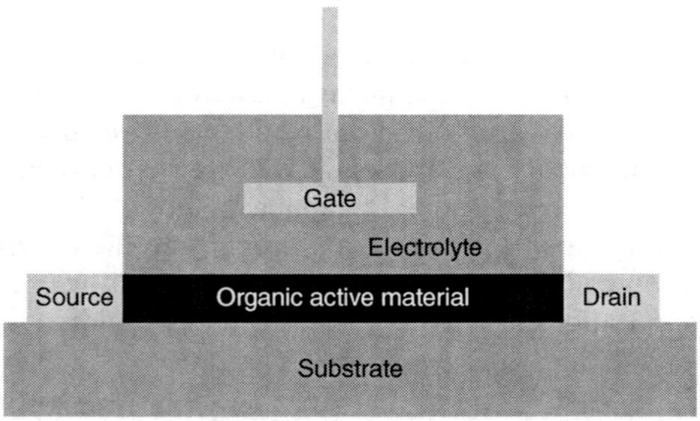

21.2 Schematic diagram of an OECT device. The gate is immersed in the electrolyte. The overlapping of the electrolyte with the organic active material (black) defines the transistor channel.

where q is electronic charge, μ is the charge carrier mobility, p_0 is the initial hole density in the organic semiconductor before application of V_{gs}, t is the thickness of the organic semiconductor film, W and L are the width and length of the OECT channel, respectively, V_p is the pinch-off voltage, V_g^{eff} is the effective gate voltage applied (i.e. the gate voltage required to produce the same electrolyte potential in the absence of faradaic effects), and V_{offset} is an offset voltage related to the potential drop at the interfaces gate electrode/electrolyte and electrolyte/channel. It is worth noting that c_i is the effective capacitance per unit area of the transistor that is not only related to the capacitance of the interface between the electrolyte and the organic semiconductor channel (C_d) but also to that of the interface between the electrolyte and the gate electrode (C_g).

A successive work (Yaghmazadeh *et al.*, 2011) improved Bernard's model, identifying optimal OECT parameters for specifc OECT applications. OECTs have been fruitfully exploited with two operational modes *ion-to-electron converters* (Nilsson *et al.*, 2002; Bernards *et al.*, 2006) and *electrochemical sensors* (ECSs) (Zhu *et al.*, 2004; Yang *et al.*, 2010; Cicoira *et al.*, 2010) using PEDOT:PSS as the conducting polymer. In the former case, upon application of a positive V_{gs}, cations from the electrolyte move toward the polymer channel. Part of these ions enter the conducting polymer and dedope it. The consequence is a decrease of I_{ds}. The transistor can work as an *electrochemical sensor* (ECS) when a redox reaction occurs at the gate electrode, as a result of a charge transfer between the analyte in the electrolyte and the gate electrode. Numerical modelling provides an opportunity to shed light on the optimum design layout of an OECT, in order to maximize the transistor response for both the operational modes. It has been found that for an OECT working as *ion-to-electron converter* a gate electrode much larger than the channel gives a higher modulation of I_{ds}, and hence an optimized value of the *transconductance*, expressed as the ratio $\partial I_d / \partial V_g$.

Recently, a compact 'inductance-like' (RLC) model has been proposed for PEDOT:PSS OECTs, to interpret the persistent oscillating currents observed during the experiments (Tu and Forchheimer, 2012). This model predicts that analogue circuit functions can be realized with 'inductor-like' electrochemical devices based on the energy conversion in the overall electrochemical process. By utilizing the above model, the authors concluded that PEDOT:PSS OECTs are promising for analogue circuit functions and even artificial biological systems. Likewise, data modelling showed that, when working as an electrochemical sensor, a gate with a smaller area with respect to the channel yields higher OECT modulations of I_{ds}, and hence a better sensitivity (Cicoira *et al.*, 2010). A conductive polymer with an improved hole mobility and capacitance per unit area could further increase the ECS-OECT response. The explanation of these design rules relies on the distribution of the potential dropping at the gate electrode/electrolyte and electrolyte/channel interfaces. For an ion-to-electron converter with a large-area gate electrode, the potential drop at the gate/electrolyte interface is almost negligible and the potential of the electrolyte (V_{sol}) is nearly at the same potential as the gate voltage applied. Since the potential of the channel is zero (a correct approximation when $V_{ds} << V_{gs}$), there is a large potential drop at the electrolyte/channel interface that in turn leads to a strong modulation of I_{ds}. For an OECT working as an electrochemical sensor and making use of small-area gate electrode, the potential drop at the gate/electrolyte interface is close to V_{gs}, leading to a maximum OECT sensitivity (defined as the difference between the signals in absence of analyte and in presence of analyte).

Recently, the first OECT working in enhancement-mode making use of PEDOT:PSS as the channel material and Al as the gate electrode was reported (Knoll and Thamer, 2011).

The effect of the OECT device geometry has been explored (Cicoira *et al.*, 2010) for devices with different channel area (A_{ch})/gate electrode area (A_g) ratios (Fig. 21.3) and fabricated by an unconventional patterning approach. To confine the electrolyte on the polymer channel, the surface of the substrate was treated with a patterned hydrophobic self-assembled monolayer of perfluorooctyltrichlorosilane, (FOTS), before the liquid was gently dropped on the channel (Yang *et al.*, 2010). The devices were tested using as the analyte hydrogen peroxide (H_2O_2), the species often detected during glucose sensing (Zhu *et al.*, 2004; Bernards *et al.*, 2008). The response of PEDOT:PSS OECTs has been investigated varying the concentration of H_2O_2 between 10^{-6} and 1 M, for V_{gs} in the range between 0–1 V. OECTs with small-area gate showed lower background signal and a higher sensitivity. The minimum and maximum detectable analyte concentration were independent of A_{ch}/A_g. Successively, the role of the gate material on the OECT response was studied (Tarabella *et al.*, 2010). OECTs making use of Ag and Pt electrodes were operated in halide electrolytes, such as sodium chloride (NaCl). Pt and Ag gate electrodes give very different responses: using Ag (Ag OECTs) yields to a larger current modulation compared with OECTs using Pt (Pt OECTs). The results were explained considering the different regimes of operation describing Ag and Pt OECTs: Faradaic for Ag OECTs and non-Faradaic for Pt OECTs. Under the Faradaic regime, the OECT response is governed by a Nernst-like equation (Bernards and Malliaras, 2007) and a steady state gate current is recordable at the gate electrode. The non-Faradaic regime refers to a capacitive mode of operation, where the OECT response depends only on the geometry of the device and the gate current (I_{gs}) is several orders of magnitude lower than I_{ds}. Since the OECT response depends directly on the potential drop occurring at the gate electrode/electrolyte and electrolyte/channel

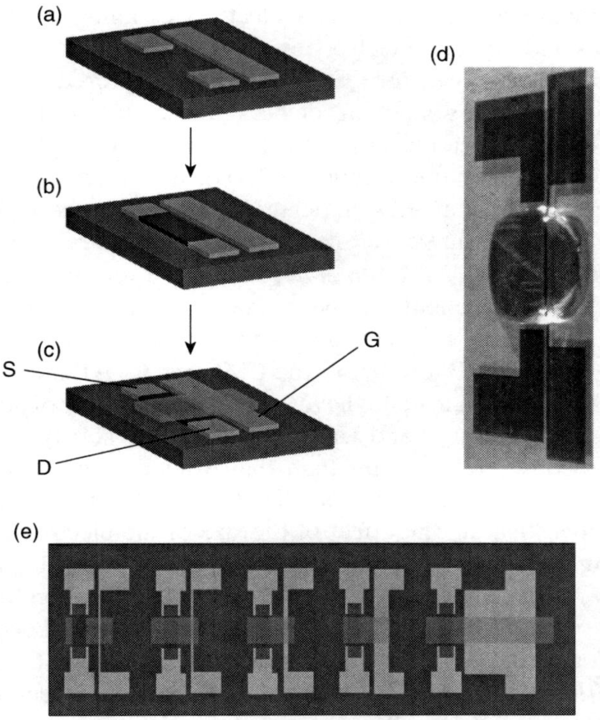

21.3 (a–c) Schematics of the patterning process, (d) image of a device with $\gamma\,(A_{ch}/A_g) = 40$, and (e) layout of OECTs with $\gamma = 40$, 10, 5, 1 and 0.2 (from left to right) on the same substrate. The patterning process involves the definition of Pt source (S), drain (D), and gate (G) electrodes (a), a PEDOT:PSS channel (b), and a hydrophobic SAM which confines the electrolyte over the channel and gate electrode. Reprinted with permission from Cicoira *et al.* (2010) Copyright @ 2010 Wiley-VCH.

interfaces, the electrolyte solution potential (V_{sol}) determines the OECT response. The different way the materials affect V_{sol} explains the origin of the different behaviours observed: under the non-Faradaic regime $V_{sol} < V_{gs}$, due to the formation of an electrostatic double layer (EDL) formed at the gate–electrolyte interface, whereas under the Faradaic regime $V_{sol} \approx V_{gs}$, and no potential drop occurs at the interface.

The electrolyte also has a strong influence on the response of OECTs. Electrolytes can be liquid but also solid or gels. The degree of dissociation of the ions of the electrolyte defines the strength of the electrolyte: strong electrolytes are almost completely dissociated whereas weak electrolytes are only partially dissociated. Organic electronics make use of different kind of electrolytes, such as conventional electrolytes, ionic liquids, ion gels, polyelectrolytes and polymer electrolytes.

A new class of electrolytes for OECTs has been explored very recently (Tarabella *et al.*, 2012) by studying devices where the electrolyte is an aqueous solution of a cationic surfactant hexadecyl trimethyl ammonium bromide (or cetyl ammonium bromide, CTAB). When dissolved in water, CTAB forms positively charged spherical micelles above a critical micellar concentration (CMC, about 9×10^{-4} M, at room temperature) (Knock *et al.*, 2003; Cano-Sarabia *et al.*, 2010; Shi *et al.*, 2011). The use of surfactants as electrolytes offers unprecedented opportunities to study the working mechanism of OECTs because they give access to two distinct types of electrolytes, i.e. dissociated CTAB ions below the CMC and CTAB micelles above the CMC. CTAB spherical micelles have an average diameter of about 4 nm (Valstar *et al.*, 1999; Riisager and Hanson, 2002; Chakraborty and Sarkar, 2004), which is considerably larger than that of electrolyte ions typically used in OECTs (Lin *et al.*, 2010a; Tarabella *et al.*, 2010). Micelles are therefore ideal for investigating the effect of the size of the electrolyte *ions* on channel doping/dedoping. In addition, micelles are of primary interest for drug delivery systems, which might be a new area of application for organic bioelectronics (Berggren and Richter-Dahlfors, 2007; Leger, 2008; Richter-Dahlfors *et al.*, 2011). It has been found that the OECT response (defined as $\Delta I/I_0 = |I_{ds,off} - I_{ds,0}|/I_{ds,0}$, where $I_{ds,off}$ is the *off* current at $V_{gs} \neq 0$ and $I_{ds,0}$ the *on* current at $V_{gs} = 0$) increased above the critical micellar concentration of CTAB, revealing that positively charged CTAB micelles dedope PEDOT:PSS more efficiently than dissociated ions (Fig. 21.4). Results also indicated that, by monitoring the OECT response as a function of CTAB concentration, OECTs can provide a simple method to detect the formation of micelles in aqueous solutions, or in other terms, the transition phase of surfactants. The OECT response achieved with a commonly used electrolyte such as NaCl within the same range of concentrations showed weaker concentration dependence, confirming the role of CTAB micelles in OECT operation. The electrochromic properties of PEDOT:PSS shed light on the role that micelles had as dedoping elements of PEDOT:PSS. An *in situ* UV-Vis spectroscopy experiment was performed during OECT operation. When working with a CTAB concentration beyond the CMC, the optical absorption of PEDOT:PSS between 400 and 700 nm increased upon application of a positive V_{gs}, corresponding to the electrochromic switching between the oxidized (light blue) and reduced (dark blue) state of PEDOT:PSS. This result demonstrated that CTAB micelles enter the PEDOT:PSS layer and dedope it, resulting in a OECT response.

Recently, the fabrication by inkjet-printing of OECTs based on PEDOT:PSS has been reported (Basiricò *et al.*, 2012). OECTs used NaCl aqueous solutions as the electrolyte and showed output transistor

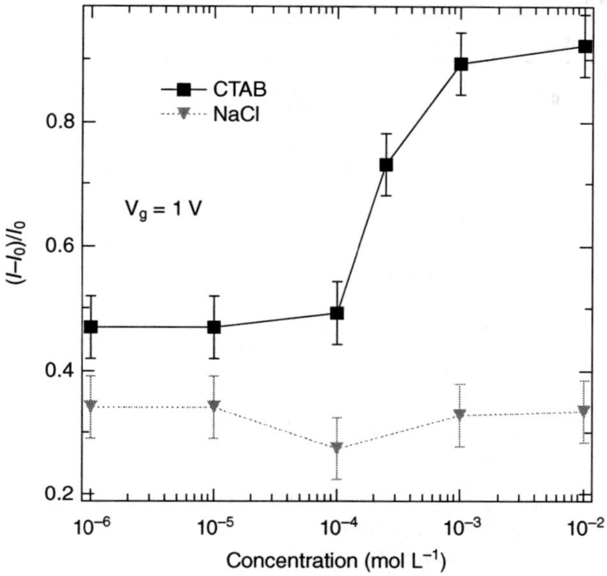

21.4 The OECT modulation increased above the critical micellar concentration of CTAB, shows that positively charged CTAB micelles dedope PEDOT:PSS more efficiently than dissociated NaCl ions.

characteristics typical of operation in depletion regime. This work paves the way for low-cost, print-on-demand fabrication of circuits for applications such as biosensors and disposable electronics.

21.3 Enzymatic sensing with organic electrochemical transistors (OECTs)

OECTs have been successfully applied as enzyme-based biosensors for the detection of glucose and lactate (Table 21.1). A glucose sensor based on OECTs was developed for the first time in 2004 (Zhu *et al.*, 2004). The sensitivity of the OECT glucose sensor reached the μM level (Macaya *et al.*, 2007) and the mechanism of operation has been successively investigated, in order to clarify the glucose sensing mechanism (Bernards *et al.*, 2008). The performance of the OECT glucose sensor was dramatically improved by a modification of the gate electrode with multi-walled carbon nanotubes (MWCNTs) modified with chitosane (CHIT) and Pt-nanoparticles (Tang *et al.*, 2011). The mechanism is based on the improved electrocatalytic properties of the gate electrode and the larger surface area available for enzyme immobilization. The detection limits of the OECTs

Table 21.1 Key characteristics of OECT devices used in enzymatic sensing. Acronyms are explained in the text

Conducting polymer	Analyte	Detection limit	Reference
PEDOT:PSS	Glucose and lactate	1 mM	Zhu *et al.* (2004) Macaya *et al.* (2007)
PEDOT:PSS	Glucose	5 nM	Tang *et al.* (2011)
PEDOT:PSS	Glucose, enzyme and mediator stored in ionic liquid	10^{-7} M	Yang *et al.* (2010)
Vapour phase-polymerized PEDOT	Glucose	below 10 μM	Y. Kim *et al.* (2010)
PEDOT:PSS	Lactate (enzyme dissolved in an ion gel)	100 mM	Khodagholy *et al.* (2012)
PEDOT:PSS	Microfluidic channel. Glucose and lactate are detectable simultaneously	10^{-5} M	Yang *et al.* (2009)
PEDOT:PSS	Prostate specific antigen antichymotrypsin (PSA–ACT) complex	1 pg/mL	DJ Kim *et al.* (2010)

reached 0.5 μM with MWCNT-CHIT/GOx/Pt and 5 nM with CHIT/GOx/Pt-NPs/Pt.

A new strategy for glucose sensing has been investigated (Yang *et al.*, 2010) by using room temperature ionic liquids, integrated in planar OECT entirely made of PEDOT:PSS. The room temperature ionic liquid (in this case the hydrophilic triisobutyl(methyl)-phosphonium tosylate) including the enzyme glucose oxidase (GOx) and the redox mediator bis (n5-cyclopentandienyl)iron (ferrocene, Fc) was placed on the *virtual* FOTS well. The OECT detected glucose concentrations in the range from 10^{-7} to 10^{-2} M. Successively, another glucose sensor OECT with high sensitivity, below 10 μM, was demonstrated (Y. Kim *et al.*, 2010). The OECT made use of vapour phase polymerized PEDOT and Pt gate electrode. An acid sensitive fluorescent layer was coated on the vapour polymerized PEDOT. The optical sensitivity of the *double-layered* OECT sensor correlated linearly with the electrochemical sensitivity. The fluorescence intensity change of the double-layered OECT sensor was linearly dependent on the pH, providing a dual sensitivity (electrochemical and optical) capable to widening the methodology of simple glucose sensing. A simple architecture for enzymatic sensing, fabricated using a one-layer patterning process, has been previously proposed (Malliaras *et al.*, 2009). This kind of OECT employed a ferrocene mediator to shuttle electrons between the enzyme glucose oxidase and a PEDOT:PSS gate electrode. The device was fabricated using a one-layer patterning process and offered glucose detection down to the μM range, consistent with levels present in human saliva.

Recently, a flexible lactate sensor based on an OECT has been reported (Khodagholy *et al.*, 2012). The work was motivated by the need to measure lactate concentration in physiological fluids, such as sweat, in real time. Indeed, flexibility leads to wearable biosensors that can even be applied conformably to skin. The electrolyte was a mixture of a room temperature ionic liquid, an ion gel and a solution of the lactate enzyme (LO_x) in PBS. The room temperature ionic liquid was useful to solubilize biomolecules, like enzymes and proteins. The ion gel enabled the development of a solid-state electrolyte that finds application in a wet environment, such as sweat analysis. The ion gel consisted of two monomeric units: *N*-isopropylacrylamide (NIPAAm) and *N*,*N* 0-methylene-bis(acrylamide) (MBAAm), hydrated. 1-Ethyl-3-methylimidazolium ethyl-sulphate, (C2mIm)(EtSO4), was chosen as the room temperature ionic liquid, miscible with water. With ferrocene as redox mediator, the OECT was able to detect lactate concentrations in the range 10–100 mM, compatible with lactate concentration in blood (Fig. 21.5).

It has been shown that glucose and lactate are detectable simultaneously inside a microfluidic multi-analyte OECT sensor based on PEDOT:PSS (Yang *et al.*, 2009). The sensor layout consisted of one input reservoir and four separated measurement reservoirs, each containing a planar OECT (Fig. 21.6). The microfluidic channels, where the aqueous analytes flow without application of external pressures, are defined by a patterning technique consisting of the sequential deposition of FOTS layer, photolithography, oxygen plasma etching, and lift off. The *microfluidic OECT* was completed by patterning a PEDOT:PSS channel with a parylene lift-off technique (DeFranco *et al.*, 2006). The sensors were exposed to different solutions (PBS, with glucose or lactate) and showed good sensing properties as well as good selectivity. The application of microfluidics to OECTs has been further investigated (Khodagholy *et al.*, 2011b) to develop a high-density arrays of 64 OECTs with channel length of 6 μm and width of 15 μm. An example of application of the array-layout will be given later in the section concerning the *in vivo* application of OECTs.

A first application of OECTs as immunosensors has been reported to detect the prostate specific antigen antichymotrypsin (PSA–ACT) complex,

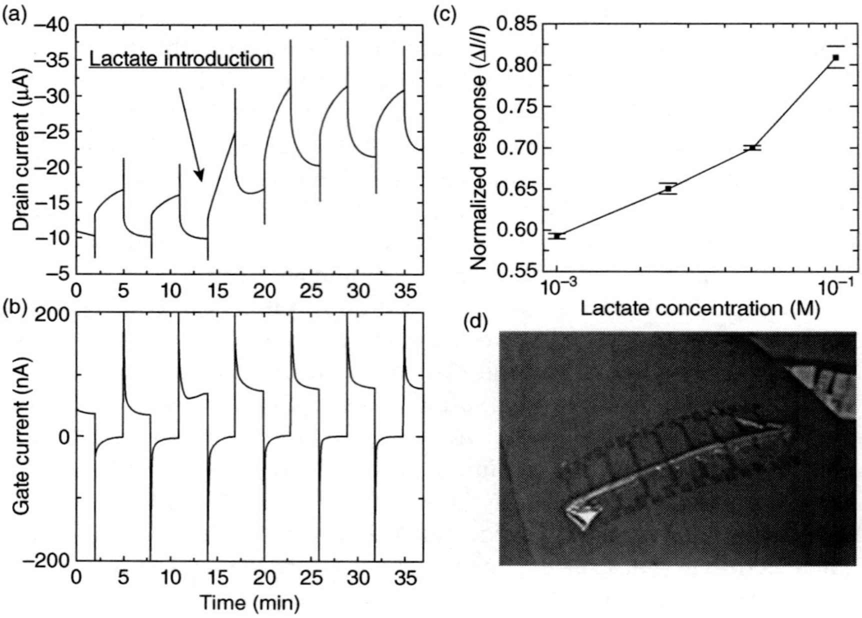

21.5 (a) Drain current vs. time with addition of 25 mM lactate indicated by an arrow, (b) corresponding gate current vs. time, (c) normalized response of the OECT vs. lactate concentration (d) conformal OECT with gel shown on a forearm. Reprinted with permission from Khodagholy *et al.*, 2012 Copyright @ Royal Society of Chemistry.

with a detection limit down to 1 pg/mL (DJ Kim *et al.*, 2010) PSA is currently the best serum marker for the preoperative diagnosis and screening of prostate cancer. Gold nanoparticles (AuNPs) were conjugated with PSA polyclonal antibody (pAb) (AuNPs–PSA pAb). The improved sensitivity of the device, with respect to other methods, is attributable to the role of AuNPs in the signal amplification. The sensor performances were particularly improved in the lower concentration range where the detection is clinically important for the screening of prostate cancer. This result shows that the OECT immunosensors can be used as transducer platforms acceptable to the point-of-care diagnostic systems and demonstrates adaptability of organic electronics to clinical applications. The device is a low-cost and flexible alternative with respect to current optical detection systems.

21.4 Cell-based organic electrochemical transistors (OECTs)

In this section, we report a first a series of experiments performed with OECTs based on PEDOT doped with Tosylate (PEDOT:Tosylate).

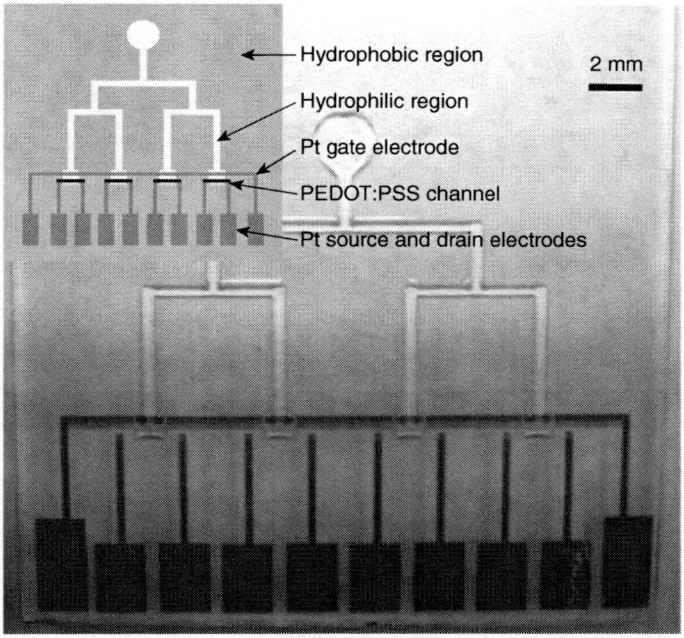

21.6 Photograph of the finished chip with a drop of water spread across the microfluidic system. The inset explains the various features. Reprinted with permission from Yang *et al.* (2009) Copyright @ Royal Society of Chemistry.

PEDOT:Tosylate helps the biocompatibility and stability under biological conditions of OECTs. The experiments were carried out to control and analyse the adhesion and the density of different kinds of cells on the PEDOT:Tosylate surface. Adhesion is an essential event for cells, since the seeding density is regulated by the adhesive properties of the substrate. In particular stem cells (Saltò *et al.*, 2008), epithelia (Bolin *et al.*, 2009a) and cancer cells (Wan *et al.*, 2009) were investigated. The control of the adhesion and density of stem cells (c17.2 cell line) on PEDOT:Tosylate was achieved through the modulation of the PEDOT:Tosylate oxidation state (Saltò *et al.*, 2008). To this purpose PEDOT:Tosylate films were patterned into two equally sized electrodes. A bias applied between these electrodes induced an electrochemical reaction; the negatively (positively) biased electrode was reduced (oxidized) to the neutral (fully oxidized) state. Water contact angles, measured along the reduced and oxidized film surfaces, were 30° and 58°, respectively. The origin of the electronic control of the wettability of PEDOT:Tosylate surfaces was attributed to a change in the binding characteristics between Tosylate and PEDOT at the surface (Fig. 21.7). Experiments performed with c17.2 neural stem cells revealed that oxidized,

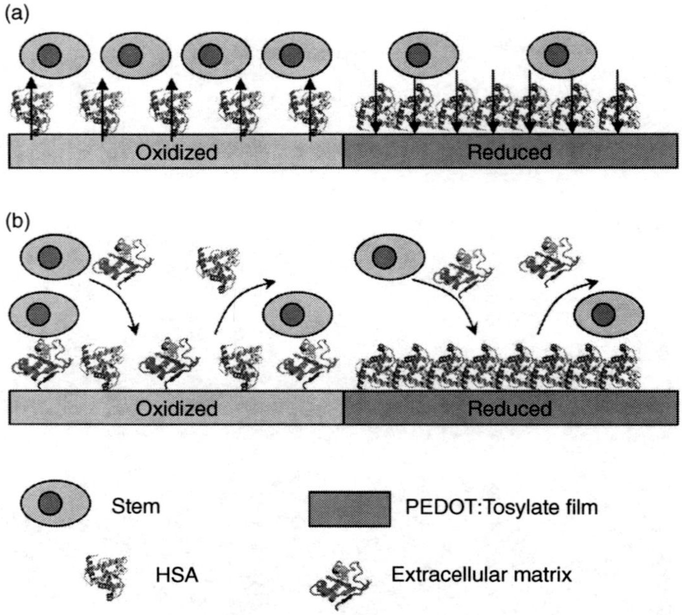

21.7 Schematic diagram of proposed mechanisms for the difference in adhesion and density of stem cells achieved between the reduced and oxidized PEDOT:Tosylate electrode surfaces: (a) on the oxidized PEDOT surface, a less dense layer of heat stable antigen (HSA) proteins is formed; however, the proteins are oriented in a favourable direction that promotes good stem cell adhesion; (b) stem cells that are approaching a potential host surface launch proteins that form an extracellular matrix. A dense and tightly bound HSA protein layer prevents the formation of an optimal extracellular matrix along the surface. Reprinted with permission from Saltò *et al.*, (2008) Copyright (2008) American Chemical Society.

hydrophilic PEDOT:Tosylate surfaces have an increased cell adhesion compared to reduced surfaces.

Electronic control of the cell density on PEDOT:Tosylate was also demonstrated for epithelial cells (Madin-Darby canine kidney), using an OECT employing a PEDOT:Tosylate channel and an aqueous NaCl electrolyte (Bolin *et al.*, 2009a). V_{ds} and V_{gs} were used to control the degree of PEDOT:Tosylate reduction along the transistor channel. The electrochromism of PEDOT:Tosylate permitted the visual monitoring of the change of electrochemical state (reduced or oxidized) along the OECT channel. Fluorescence microscopy revealed that the cell density decreased while moving towards the more oxidized region of PEDOT:Tosylate. The control of the cell density on a PEDOT:Tosylate surface along which an electrical bias gradient was applied has been reported for normal and cancerous cell lines (Wan *et al.*, 2009; C.M. Nelson, 2007). Conducting polymers, such as polypyrrole or PEDOT:PSS/Tosylate, can be reversibly switched between their oxidized and reduced state. In these studies, cell adhesion was mediated by the adhesion protein fibronectin. By applying an electrical bias at the opposite sides of an indium tin oxide (ITO) substrate (Fig. 21.8) where a PEDOT:Tosylate film had been previously deposited a redox gradient was established within the PEDOT:Tosylate stripe. After establishing the electrical bias gradient, the adhesions of normal and cancerous cell lines was tested. Images acquired with fluorescence microscopy revealed that the density of both normal and cancerous cells were higher towards the oxidized side of the PEDOT:Tosylate stripe, despite the higher concentration of adsorbed fibronectin on the reduced side.

The same device layout was exploited to electrically control cell migration, controlling the redox state of the PEDOT:Tosylate film (Gumus *et al.*, 2010). The control of the cell migration is important for tissue engineering applications (Carter, 1967; Brandley and Schnaar, 1989; Berggren and Richter-Dahlfors, 2007). PEDOT:Tosylate was electrically biased to affect the migration of bovine aortic endothelial cells. To this purpose, a linear potential gradient was induced at opposite sides of an ITO substrate: the polymer film was reduced near the negatively biased contact and oxidized at the positively biased one. It is worth noting that the presence of the ITO substrate ensures the uniformity of the bias gradient along the polymer film,

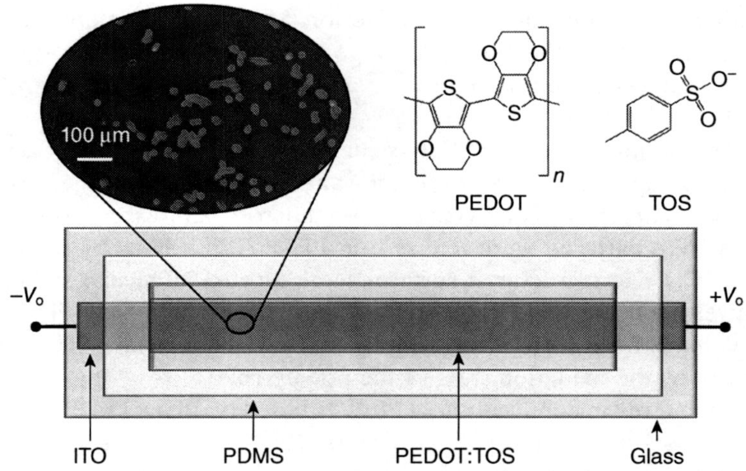

21.8 Electrochromism in the PEDOT: tosylate transistor channel corresponding to different degrees of reduction due to variations in applied electrical bias. Reprinted with permission from Bolin *et al.*, 2009a Copyright 2009 @ Wiley-VCH.

leading to a continuum microenvironment for cell growth. After establishing the redox gradient in the polymer film, cells were seeded and cell migration was observed. Cell adhesion, affected by the adsorption of proteins like fibronectin that act as adhesion mediators, affects the movement of the cells on the surface. An investigation of the density of fibronectin adsorbed on the PEDOT:Tosylate film revealed the density is higher at the reduced side of the ITO such that cells located on this side make short distances at slower speed with respect to cells located on the opposite side during the migration.

Fibronectin has a central role also in cellular growth and differentiation. Fibronectin alterations could be responsible of diseases like cystic fibrosis or cancer (Hynes, 2002; Chandler *et al.*, 2011) such that achieving the precise control of the fibronectin conformation is of extreme importance (Wan *et al.*, 2012). Forster resonance energy transfer imaging (Smith *et al.*, 2007) has been used to directly observe the fibronectin surface conformation onto the PEDOT:Tosylate polymer film.

PEDOT:Tosylate has been employed to cover uniformly poly(ethyleneterephthalate) nanofibres by vapour phase polymerization. The electrospun organic nanofibre structures were then used to study cells growth in a complex three-dimensional environment (Bolin *et al.*, 2009b). Contact angle measurements were used to investigate the wettability of the surfaces. Uncoated nanofibre surfaces were found to be strongly hydrophobic (contact angle 147°), whereas coated nanofibre surfaces were found to be more hydrophilic (contact angle 62°). Experiments showed that SH-SY5Y cells, cultured on the three-dimensional surfaces, formed neuritis, thus indicating neural cell viability and adhesion. Moreover, upon application of an electrical bias in the 1.5–3.0 V range, the coated nanofibres were electrochemically switched to induce Ca^{2+} signalling in SH-SY5Y neuroblastoma cells.

The origin of the wettability switching in electrochemically active polymers for applications in cell biology and medicine was further investigated (Wang *et al.*, 2008). Polypyrrole surface microstructures, including pillar and mesh-like patterns, were realized on PEDOT:PSS films by photolithography. The microstructured surfaces were electrochemically switched from dewetting to wetting (water contact angle from 129° to 44°). The switching originates from surface topography and surface tension effects, also influenced by the oxidation state of the polypyrrole.

An interesting application of OECTs based on PEDOT:PSS as cell-based biosensors has been reported recently (Fig. 21.9). (Lin *et al.*, 2010b) Cell activity has been monitored *in vitro* for two cells lines: human esophageal squamous epithelial cancer cells (indicated as KYSE30) and fibroblast cells (HFF1). OECTs showed excellent biocompatibility in the culture medium for both substrates: the cell density on PEDOT:PSS was comparable than

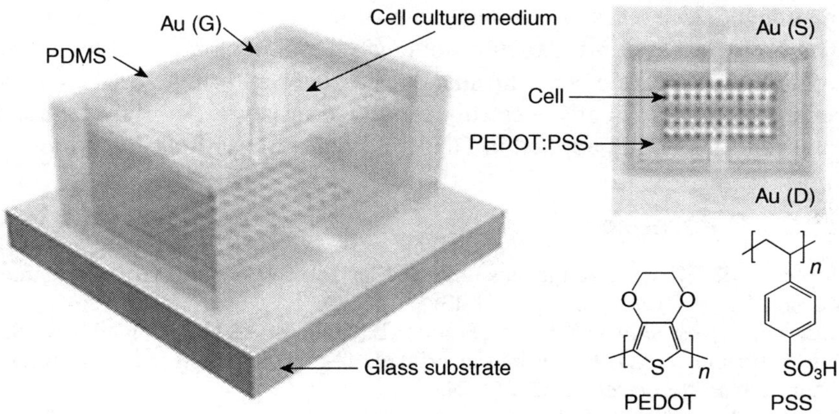

21.9 Schematic diagram of an electrochemical cell with an OECT based on PEDOT:PSS fabricated on the bottom. Cell culture medium was used as an electrolyte during the electrical measurements. Reprinted with permission from Lin *et al.*, 2010b Copyright @ 2010 Wiley-VCH.

on silica. The detachment of the cells under trypsin (an enzyme inducing the detachment) treatment and the morphological changes of the cells during OECT response under addiction of retinoic acid were investigated. Retinoic acid is an anti-cancer drug that can induce morphological changes of cancer cell lines (and it can also induce the rupture of cells) more slowly than trypsin, another anti-cancer drug. It was found that OECTs are sensitive to the change of the surface charge and the morphology of the cells. The transfer characteristics of the OECTs, acquired before and after trypsin treatment, revealed a V_{gs} shift of –150 mV for cancer cells and –70 mV for fibroblast cells. On the one hand, upon addiction of retinoic acid in absence of cells on the polymer OECT channel, there was no OECT response. On the other hand, after 7 hours from the retinoic acid addiction it was possible to observe that most of the cells detached from the OECT and V_{gs} had a shift.

21.5 Conclusions and future trends

Currently a great deal of attention is being paid to emerging technologies that exploit the ability of organics to conduct ions in addition to electron and holes. Mixed ionic/electronic transport opens exciting scenarios for organic electronics, which has been so far considered mostly as a flexible and low-cost alternative to well-established inorganic semiconductor technologies. In particular, because of the importance of ion fluxes in biology, mixed ionic/electronic transport is the underpinning of the new field of organic bioelectronics, which deals with the coupling of organic electronics with biological systems. Examples of devices that exploit mixed ionic/electronic transport are electrochromic devices, light-emitting electrochemical cells, and more recently electrolyte-gated transistors. A bright future is foreseen for OECTs applied in DNA and neurotransmitter detection.

21.6 References

Asplund, M., Nyberg, T. & Inganas, O. 2010. Electroactive polymers for neural interfaces. *Polymer Chemistry*, **1**, 1374–1391.

Basiricò, L., Cosseddu, P., Scidà, A., Fraboni, B., Malliaras, G. G. & Bonfiglio, A. 2012. Electrical characteristics of ink-jet printed, all-polymer electrochemical transistors. *Organic Electronics*, **13**, 244–248.

Berggren, M. & Richter-Dahlfors, A. 2007. Organic bioelectronics. *Advanced Materials*, **19**, 3201–3213.

Bernards, D. A. & Malliaras, G. G. 2007. Steady-state and transient behavior of organic electrochemical transistors. *Advanced Functional Materials*, **17**, 3538–3544.

Bernards, D. A., Malliaras, G. G., Toombes, G. E. S. & Gruner, S. M. 2006. Gating of an organic transistor through a bilayer lipid membrane with ion channels. *Applied Physics Letters*, **89**, 053505.

Bernards, D. A., Macaya, D. J., Nikolou, M., Defranco, J. A., Takamatsu, S. & Malliaras, G. G. 2008. Enzymatic sensing with organic electrochemical transistors. *J. Materials Chemistry*, **18**, 116–120.

Bolin, M. H., Svennersten, K., Nilsson, D., Sawatdee, A., Jager, E. W. H., Richter-Dahlfors, A. & Berggren, M. 2009a. Active control of epithelial cell-density gradients grown along the channel of an organic electrochemical transistor. *Advanced Materials*, **21**, 4379–4382.

Bolin, M. H., Svennersten, K., Wang, X., Chronakis, I. S., Richter-Dahlfors, A., Jager, E. W. H. & Berggren, M. 2009b. Nano-fiber scaffold electrodes based on PEDOT for cell stimulation. *Sensors and Actuators B: Chemical*, **142**, 451–456.

Brandley, B. K. & Schnaar, R. L. 1989. Tumor cell haptotaxis on covalently immobilized linear and exponential gradients of a cell adhesion peptide. *Developmental Biology*, **135**, 74–86.

Cano-Sarabia, M., Angelova, A., Ventosa, N., Lesieur, S. & Veciana, J. 2010. Cholesterol induced CTAB micelle-to-vesicle phase transitions. *Journal of Colloid Interface Science*, **350**, 10–15.

Carter, S. B. 1967. Effects of cytochalasins on mammalian cells. *Nature*, **213**, 261–264.

Chakraborty, H. & Sarkar, M. 2004. Optical spectroscopic and TEM studies of catanionic micelles of CTAB/SDS and their interaction with a NSAID. *Langmuir*, **20**, 3551–3558.

Chandler, E. M., Saunders, M. P., Yoon, C. J., Gourdon, D. & Fischbach, C. 2011. Adipose progenitor cells increase fibronectin matrix strain and unfolding in breast tumors. *Physical Biology*, **8**, 015008.

Cicoira, F., Sessolo, M., Yaghmazadeh, O., Defranco, J. A., Yang, S. Y. & Malliaras, G. G. 2010. Influence of device geometry on sensor characteristics of planar organic electrochemical transistors. *Advanced Materials*, **22**, 1012–1016.

De Rossi, D. 2007. Electronic textiles: A logical step. *Nature Materials*, **6**, 328–329.

DeFranco, J. A., Schmidt, B. S., Lipson, M. & Malliaras, G. G. 2006. Photolithographic patterning of organic electronic materials. *Organic Electronics*, **7**, 22–28.

Facchetti, A., Yoon, M. H. & Marks, T. J. 2005. Gate dielectrics for organic field-effect transistors: New opportunities for organic electronics. *Advanced Materials*, **17**, 1705–1725.

Friend, R. H., Gymer, R. W., Holmes, A. B., Burroughes, J. H., Marks, R. N., Taliani, C., Bradley, D. D. C., Santos, D. a. D., Bredas, J. L., Logdlund, M. & Salaneck, W. R. 1999. Electroluminescence in conjugated polymers. *Nature*, **397**, 121–128.

Gumus, A., Califano, J. P., Wan, A. M. D., Huynh, J., Reinhart-King, C. A. & Malliaras, G. G. 2010. Control of cell migration using a conducting polymer device. *Soft Matter*, **6**, 5138–5142.

Horowitz, G. 1998. Organic field-effect transistors. *Advanced Materials*, **10**, 365–377.

Hynes, R. O. 2002. Integrins: Bidirectional, allosteric signaling machines. *Cell*, **110**, 673–687.

Johnson, K. S., Needoba, J. A., Riser, S. C. & Showers, W. J. 2007. Chemical sensor networks for the aquatic environment. *ChemInform*, **107,** 623–640.

Khodagholy, D., Doublet, T., Gurfinkel, M., Quilichini, P., Ismailova, E., Leleux, P., Herve, T., Sanaur, S., Bernard, C. & Malliaras, G. G. 2011a. Highly conformable conducting polymer electrodes for *in vivo* recordings. *Advanced Materials*, **23**, H268–H272.

Khodagholy, D., Gurfinkel, M., Stavrinidou, E., Leleux, P., Herve, T., Sanaur, S. & Malliaras, G. G. 2011b. High speed and high density organic electrochemical transistor arrays. *Applied Physics Letters*, **99**, 163304.

Khodagholy, D., Curto, V. F., Fraser, K. J., Gurfinkel, M., Byrne, R., Diamond, D., Malliaras, G. G., Benito-Lopez, F. & Owens, R. M. 2012. Organic electrochemical transistor incorporating an ionogel as a solid state electrolyte for lactate sensing. *Journal of Materials Chemistry*, **22**, 4440–4443.

Kim, D.-J., Lee, N.-E., Park, J.-S., Park, I.-J., Kim, J.-G. & Cho, H. J. 2010. Organic electrochemical transistor based immunosensor for prostate specific antigen (PSA) detection using gold nanoparticles for signal amplification. *Biosensors and Bioelectronics*, **25**, 2477–2482.

Kim, J. Y., Lee, K., Coates, N. E., Moses, D., Nguyen, T.-Q., Dante, M. & Heeger, A. J. 2007. Efficient tandem polymer solar cells fabricated by all-solution processing. *Science*, **317**, 222–225.

Kim, Y., Do, J., Kim, J., Yang, S.Y., Malliaras, G. G., Ober, C. K. & Kim, E. 2010. A glucose sensor based on an organic electrochemical transistor structure using a vapor polymerized poly(3,4-ethylenedioxythiophene) layer. *Japanese Journal of Applied Physics*, **49**.

Knock, M. M., Bell, G. R., Hill, E. K., Turner, H. J. & Bain, C. D. 2003. Sum-frequency spectroscopy of surfactant monolayers at the oil–water interface. *Journal of Physical Chemistry B*, **107**, 10801–10814.

Knoll, M. & Thamer, M. 2011. An enhancement-mode electrochemical organic field-effect transistor. *Electrochemistry Communications*, **13**, 597–599.

Leger, J. M. 2008. Organic electronics: The ions have it. *Advanced Materials*, **20**, 837–841.

Lin, P. & Yan, F. 2012. Organic thin-film Transistors for chemical and biological sensing. *Advanced Materials*, **24**, 34–51.

Lin, P., Yan, F. & Chan, H. L. W. 2010a. Ion-sensitive properties of organic electrochemical transistors. *ACS Applied Material Interfaces*, **2**, 1637–1641.

Lin, P., Yan, F., Yu, J., Chan, H. L. W. & Yang, M. 2010b. The application of organic electrochemical transistors in cell-based biosensors. *Advanced Materials*, **22**, 3655–3660.

Mabeck, J. T. & Malliaras, G. G. 2006. Chemical and biological sensors based on organic thin-film transistors. *Analytical Bioanalytical Chemistry*, **384**, 343–353.

Macaya, D. J., Nikolou, M., Takamatsu, S., Mabeck, J. T., Owens, R. M. & Malliaras, G. G. 2007. Simple glucose sensors with micromolar sensitivity based on organic electrochemical transistors. *Sensors and Actuators B–Chemical*, **123**, 374–378.

Malliaras, G. & Friend, R. 2005. An organic electronics primer. *Physics Today*, **58**, 53–58.

Malliaras, G. G., Shim, N. Y., Bernards, D. A., Macaya, D. J., Defranco, J. A., Nikolou, M. & Owens, R. M. 2009. All-plastic electrochemical transistor for glucose sensing using a ferrocene mediator. *Sensors*, **9**, 9896–9902.

Martin, D. C. 2007. Organic electronics: Polymers manipulate cells. *Nature Materials*, **6**, 626–627.

Nelson, C. M., Liu, W. F. & Chen, C. S. 2007. *Methods in Molecular Biology*, Berlin, Springer, http://www.springerprotocols.com/Abstract/doi/10.1007/978-1-59745-353-0_1.

Nikolou, M. & Malliaras, G. G. 2008. Applications of poly(3,4-ethylenedioxythiophene) doped with poly(styrene sulfonic acid) transistors in chemical and biological sensors. *Chemical Record*, **8**, 13–22.

Nilsson, D., Kugler, T., Svensson, P.-O. & Berggren, M. 2002. An all-organic sensor–transistor based on a novel electrochemical transducer concept printed electrochemical sensors on paper. *Sensors and Actuators B: Chemical*, **86**, 193–197.

Owens, R. M. & Malliaras, G. G. 2010. Organic electronics at the interface with biology, *MRS Bulletin*, **35**, 449.

Richter-Dahlfors, A., Svennersten, K., Larsson, K. C. & Berggren, M. 2011. Organic bioelectronics in nanomedicine. *Biochimica et Biophysica Acta, General Subjects*, **1810**, 276–285.

Riisager, A. & Hanson, B. E. 2002. CTAB micelles and the hydroformylation of octene with rhodium/TPPTS catalysts: Evidence for the interaction of TPPTS with micelle surfaces. *Journal of Molecular Catalysis A: Chemical*, **189**, 195–202.

Saltò, C., Saindon, E., Bolin, M., Kanciurzewska, A., Fahlman, M., Jager, E. W. H., Tengvall, P., Arenas, E. & Berggren, M. 2008. Control of neural stem cell adhesion and density by an electronic polymer surface switch. *Langmuir*, **24**, 14133–14138.

Shi, Y., Luo, H. Q. & Li, N. B. 2011. Determination of the critical premicelle concentration, first critical micelle concentration and second critical micelle concentration of surfactants by resonance Rayleigh scattering method without any probe. *Spectrochimica. Acta, Part A*, **78**, 1403–1407.

Shirakawa, H., Louis, E. J., Macdiarmid, A. G., Chiang, C. K. & Heeger, A. J. 1977. Synthesis of electrically conducting organic polymers: Halogen derivatives of polyacetylene, (CH). *Journal of the Chemical Society, Chemical Communications*, 578–580.

Sirringhaus, H., Tessler, N. & Friend, R. H. 1998. Integrated optoelectronic devices based on conjugated polymers. *Science*, **280**, 1741–1744.

Smith, M. L., Gourdon, D., Little, W. C., Kubow, K. E., Eguiluz, R. A., Luna-Morris, S. & Vogel, V. 2007. Force-induced unfolding of fibronectin in the extracellular matrix of living cells. *PLoS Biology*, **5**, e268.

Tang, H., Yan, F., Lin, P., Xu, J. B. & Chan, H. L. W. 2011. Highly sensitive glucose biosensors based on organic electrochemical transistors using platinum gate electrodes modified with enzyme and nanomaterials. *Advanced Functional Materials*, **21**, 2264–2272.

Tarabella, G., Santato, C., Yang, S. Y., Iannotta, S., Malliaras, G. G. & Cicoira, F. 2010. Effect of the gate electrode on the response of organic electrochemical transistors. *Applied Physics Letters*, **97**, 123304.

Tarabella, G., Nanda, G., Villani, M., Coppedè, N., Mosca, R., Malliaras, G. G., Santato, C., Iannotta, S. & Cicoira, F. 2012. Organic electrochemical transistors monitoring micelle formation. *Chemical Science*, **3** (12), 3432–3435.

Tu, D. & Forchheimer, R. 2012. Self-oscillation in electrochemical transistors: An RLC modeling approach. *Solid-State Electronics*, **69**, 7–10.

Valstar, A., Almgren, M., Brown, W. & Vasilescu, M. 1999. The interaction of bovine serum albumin with surfactants studied by light scattering. *Langmuir*, **16**, 922–927.

Wan, A. M. D., Brooks, D. J., Gumus, A., Fischbach, C. & Malliaras, G. G. 2009. Electrical control of cell density gradients on a conducting polymer surface. *Chemical Communications*, 5278–5280.

Wan, A. M. D., Schur, R. M., Ober, C. K., Fischbach, C., Gourdon, D. & Malliaras, G. G. 2012. Electrical control of protein conformation. *Advanced Materials*, **24**, 2501–2505.

Wang, X., Berggren, M. & Inganaããs, O. 2008. Dynamic control of surface energy and topography of microstructured conducting polymer films. *Langmuir*, **24**, 5942–5948.

Yaghmazadeh, O., Cicoira, F., Bernards, D., Yang, S. Y., Bonnassieux, Y. & Malliaras, G. G. 2011. Optimization of organic electrochemical transistors for sensor applications. *Journal of Polymer Science Part B: Polymer Physics*, **49**, 34–39.

Yang, S. Y., Defranco, J. A., Sylvester, Y. A., Gobert, T. J., Macaya, D. J., Owens, R. M. & Malliaras, G. G. 2009. Integration of a surface-directed microfluidic system with an organic electrochemical transistor array for multi-analyte biosensors. *Lab Chip*, **9**, 704–708.

Yang, S. Y., Cicoira, F., Byrne, R., Benito-Lopez, F., Diamond, D., Owens, R. M. & Malliaras, G. G. 2010. Electrochemical transistors with ionic liquids for enzymatic sensing. *Chemical Communications*, **46**, 7972–7974.

Zhu, Z. T., Mabeck, J. T., Zhu, C., Cady, N. C., Batt, C. A. & Malliaras, G. G. 2004. A simple poly(3,4-ethylenedioxythiophene)/poly(styrene sulfonic acid) transistor for glucose sensing at neutral pH. *Chemical Communications*, 1556–1557.

22
Organic electronic memory devices

M. C. PETTY, Durham University, UK

DOI: 10.1533/9780857098764.3.618

Abstract: This chapter provides a review of the different concepts that have been proposed for the realisation of electronic memory structures based on electro-active organic materials. One option is to develop the organic equivalents of existing (e.g. silicon-based) random access memory devices. Resistive memories also offer an attractive technology by which devices can easily be scaled. However, the operational principles of these deceptively simple memory elements have yet to be firmly established. A major challenge is the development of molecular-based memories, i.e. exploiting individual molecules or small groups of molecules. An even more radical approach is to look beyond the von Neumann architecture, which forms basis of most computers, and to emulate the memory processes that are found in nature.

Key words: organic electronics, plastic electronics, organic memory devices.

22.1 Introduction

The past decade has seen an upsurge in academic and commercial interest in the field of organic electronics (Petty, 2007; Sun and Dalton, 2008; Cuevas and Scheer, 2010). Devices such as organic light-emitting displays, and chemical and physical sensors are already in the marketplace, while others, including photovoltaic cells and smart cards, are developing fast. Much of the work concerns the replacement of silicon and other inorganic semiconductors, which are currently used in electronic devices, with organic compounds. The motivation is the reduced cost and large-area capability of many organic thin film technologies. Tremendous progress has been achieved, but much more of the work demonstrated in the laboratory needs to be translated into reliable and reproducible products, which can be manufactured cheaply.

The current progress on organic transistors is particularly impressive (Bao and Locklin, 2007; Braga and Horowitz, 2009; Sirringhaus, 2009). Field-effect mobility values in the range 0.1 to 1 $cm^2\ V^{-1}\ s^{-1}$ are routinely demonstrated and field-effect transistors (FETs) can conveniently be fabricated using cost-effective techniques such as inkjet printing. Low voltage operation (<10 V) is also possible with an appropriate choice of the gate insulator (Noh and Sirringhaus, 2009; Yun *et al.*, 2009). Although most organic FETs are based on p-type materials, n-type devices are becoming available, opening up the possibility of an all-organic technology equivalent to silicon complementary metal-oxide-semiconductor (CMOS) circuitry (Graz and Lacour, 2010; Baeg *et al.*, 2011; Roelofs *et al.*, 2011). These signal processing systems will not compete directly with single crystal silicon technology in terms of operation speed but are likely to form key components in display drivers in portable computers and pagers, and as components in transaction cards and radio frequency identification (RFID) tags. Integration into textiles ('smart' clothing) is also envisaged.

For organic electronics to realise its full potential, it is essential that a further basic circuit element is developed – a memory cell. Although memories represent by far the largest part of conventional (silicon-based) electronic systems, work on organic memories is lagging significantly behind the development of organic FETs. Many different concepts have been proposed, but there is currently no consensus on the way forward (Claverie *et al.*, 2005; Ling *et al.*, 2008; ITRS, 2011; Heremans *et al.*, 2011; Pershin and Di Ventra, 2011). Although many memories are entirely electronic in nature (i.e. these are both written to and read from by electrical means), some devices, compact discs (CDs), digital video discs (DVDs) and hard discs, require additional transducers (drivers) to convert optical, magnetic or other physical quantities to electronic signals. Proposed memories also exploit other physical effects in materials such as photochromism, chemical hole burning or molecular chirality. This review will focus on electronic memories because of their potential for high writing and access speeds at low cost.

22.2 Memory types

The essential requirement for a memory is that the individual memory elements, or bit-cells, possess at least two stable sates (or metastable states, but stable for a period appropriate for the data storage) and that these can be switched by an external stimulus (the writing process). It is also important that the states can be distinguished by applying a further external signal (the reading process). Information storage devices are commonly grouped into random access devices and sequential access devices. In the former, the storage cells are organised in an array of memory elements (an early example is the magnetic core memory). This architecture provides short access times, which are independent of the location of a particular memory cell. In contrast, the access time for a cell in a sequential device will depend on its physical location relative to the position of the read and write units. Such mass storage devices (a common example is the hard disc drive found in computer systems) are usually used as secondary memory for information archiving and retrieval.

Memory devices are inherently volatile or non-volatile. For volatile devices, the information is lost if the power supply is removed. Figure 22.1 shows how these two broad electronic memory classifications are further subdivided. In silicon technology, examples of volatile memory are the static random access memory (SRAM) and the dynamic random access memory (DRAM) (Prince, 2000). Currently, DRAM is used as the main memory in most computers. This technology exploits the charge stored on the plates of a capacitor; two different charge storage levels represent the binary information. A DRAM cell comprises one capacitor and a single access transistor (to select a particular memory element during the read and write operations). This memory is relatively efficient in its use of area on a silicon chip. However, due to leakage of charge from the capacitor, the data stored in DRAM cells must periodically be refreshed. The minimum required cell capacitance is about 50 fF (50×10^{-15} F). This is governed by so-called soft errors, i.e. the effect of the Earth's bombardment by cosmic rays, which can lead to charged particles affecting the memory cell (Streetman and Banerjee, 2000). Maintaining a capacitance of 50 fF as the cell dimensions are reduced from one chip generation to the next presents an enormous challenge. There is a limit to how much the gate insulator thickness can be reduced. Alternative approaches focus on increasing the device area (e.g. using trenches and multi-capacitor stacks) and exploiting high permittivity insulators (e.g. ferroelectrics).

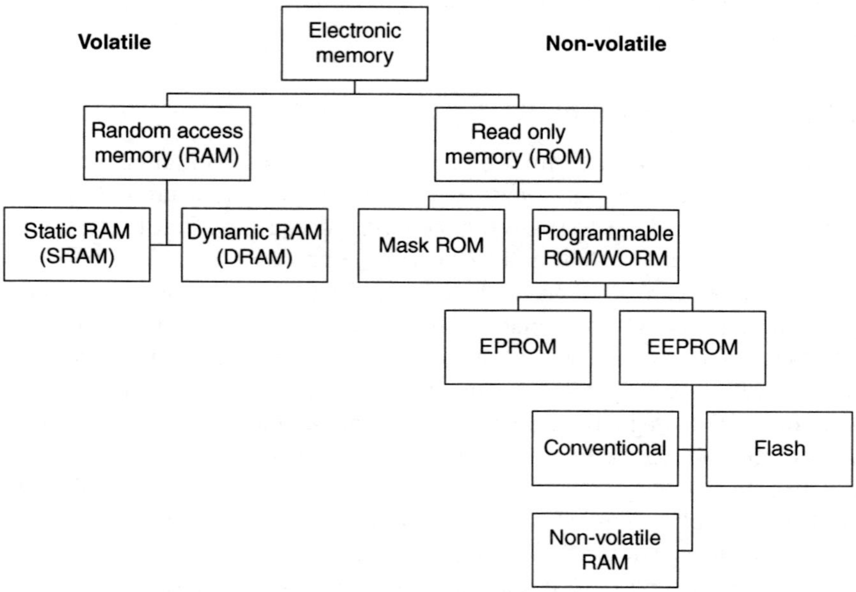

22.1 Classification of electronic memory devices.

SRAM is a memory technology that uses flip-flop-based logic gates. This memory is also volatile as the information stored (as a result of the logic gates being locked in a particular configuration) will be lost if the power is removed. A single SRAM cell consists of two cross-coupled inverters and two access transistors, resulting in six transistors for each cell. Such memory takes up additional space on the Si wafer and is more expensive than DRAM. However, SRAM has faster access times than DRAM and is used as cache (temporary) memory to store frequently used instructions and data for quicker access by the system processor.

Non-volatile read only memories (ROMs) can be further divided into those which are only programmed once and devices that are re-programmable. In the case of the first group, the programming of silicon-based memories can be achieved during chip manufacturing by appropriate layout of the final metallisation mask (Mask ROM). This may also be accomplished by the user in an initial programming step in which metal bridges can be fused or left intact to represent the binary information (programmable ROMs or PROMs). Using organic materials, write-once/read-many-times (WORM) memory can be realised by burning polymer 'fuses' (Möller *et al.*, 2003). A thin conductive polymer layer, or fuse, of poly(3,4-ethylenedioxythiphene) (PEDOT) is sandwiched between two electrodes. The as-deposited device shows high conductance due to the polymer, but when a burning-voltage pulse is applied (the write process) the PEDOT fuses 'blow' and cause the device to be in an open-circuit condition. The write process can be as short as microseconds, depending on the thickness of the PEDOT layer and the amplitude of the voltage pulse. A nanomechanical sequential data storage concept based on an atomic force microscope (AFM) – called the 'Millipede' – that has potentially ultrahigh density has been developed by IBM (Vettiger *et al.*, 2000; Gotsmann *et al.*, 2010). Indentations, 4–8 nm, are made in a thin (100 nm) polymer layer. Demonstrations of endurance and retention performed at a storage density of 1 Tbit in^{-2} show 10^8 write cycles using the same tip, 10^3 erase cycles and 3×10^5 read cycles of the media, and extrapolated to 10 years of retention at 85 °C.

Electronically programmable ROMs or EPROMs can exploit charge storage on a floating gate of a FET. In the well-established configuration, shown in Fig 22.2(a), the gate electrode of a silicon metal oxide semiconductor FET (MOSFET) is replaced by a floating (i.e. electrically isolated) gate, usually made from polycrystalline silicon (polysilicon). The device is programmed by exciting electrons over the potential barrier associated with the silicon dioxide layer. As a result, the gate of the MOSFET becomes negatively charged and is no longer floating.

Once sufficient charge has been built up, the transistor is turned off and there is no current between the drain and the source contacts. In order to

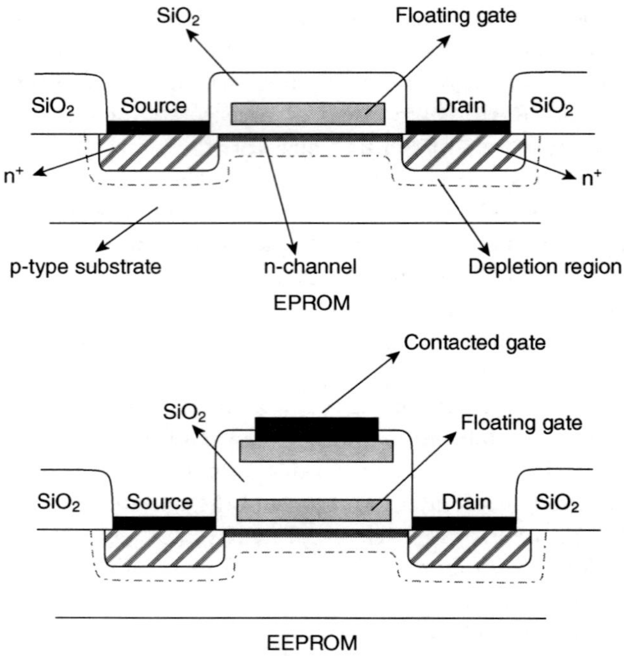

22.2 Schematic diagrams of electronic programmable memories. Top: electronic programmable memory (EPROM). Bottom: electronic programmable and erasable memory (EEPROM or E^2PROM).

remove the charge, the electrons located on the polysilicon gate need to acquire large energies to pass back over the polysilicon-SiO_2 barrier. This can be achieved by illuminating the memory device with UV light. Thus, the EPROM is electrically programmable and optically erasable.

If a MOSFET has both a conventional (contacted gate) and a floating gate, it has the potential for both electrical programming and erasing. The structure of such an EEPROM (or E^2PROM) is also shown in Fig 22.2(b). Figure 22.3 shows the effect of the charge on the floating gate on the transfer characteristics of the transistor. The voltage V_t represents the threshold voltage, or turn-on voltage, of the transistor. The drain-to-source current, I_{ds}, versus gate voltage, V_g, characteristic is effectively moved along the voltage axis as the charge accumulates on the floating gate. The voltage shift, ΔV, is related to the stored charge, ΔQ, by

$$\Delta V = \frac{d\Delta Q}{\varepsilon} = \frac{\Delta Q}{C_i} \qquad [22.1]$$

where d is the oxide thickness beneath the floating gate of the EEPROM, ε is its permittivity and C_i is the gate insulator capacitance per unit area.

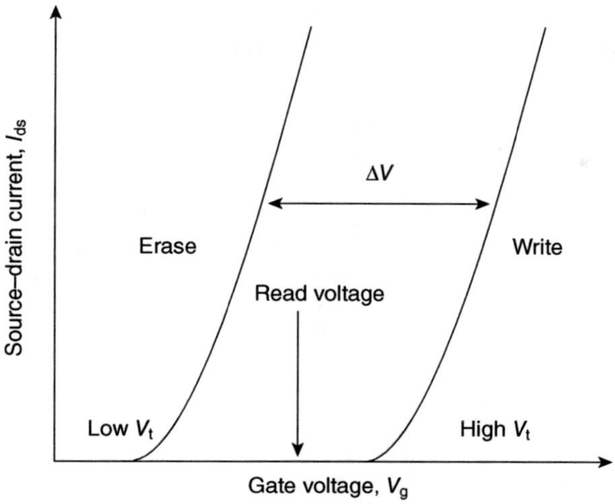

22.3 Effect of charge storage on the floating gate of an EEPROM. The charge results in a shift in the drain-to-source current versus gate voltage characteristics by an amount ΔV. The voltage V_t is the threshold voltage.

In a floating gate memory device, the charge is injected to the floating gate either by hot-carrier injection or by Fowler–Nordheim (FN) tunnelling. Both processes are depicted in the potential energy versus distance diagrams shown in Fig 22.4. The energy corresponding to the potential barrier that the electrons must surmount is shown as ΔE. In the hot-carrier process, the high longitudinal field in the pinch-off region of the transistor channel accelerates electrons towards the drain. If the electrons acquire sufficient energy from the field, they become 'hot' and can surmount the 3.1 eV energy barrier ($=\Delta E$) that exists between the conduction band of the silicon and the SiO_2. The Fowler–Nordheim mechanism is a quantum mechanical tunnelling process. The application of a strong electric field (in the range 8–10 MV cm^{-1}) across the oxide layer results in a significant tunnelling current (current transport *through* rather than *over* the potential energy barrier) without destroying the dielectric properties.

Flash memory is a simple, cost-effective EEPROM architecture, in which the devices are re-programmed in blocks of information rather than by individual memory cells (the term 'flash' refers to this erase process). In addition, flash memory offers fast read access times (although not as fast as volatile DRAM memory) and better shock resistance than hard disk memories. Flash memory is categorised by two basic approaches, dependent on how the individual cells are interconnected. In NOR gate flash, each cell has one end connected directly to ground, and the other end connected to a bitline. The memory acts in a similar fashion to a NOR logic gate. In

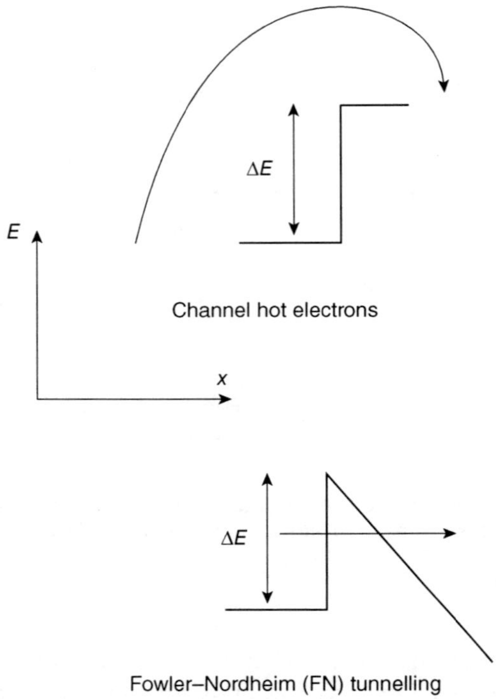

22.4 Charging/discharging mechanisms in floating gate memories. Top: excitation of hot carriers over a potential energy barrier ΔE. Bottom: Fowler–Nordheim tunnelling through the barrier.

contrast, the transistors in NAND flash are connected in a manner that resembles a NAND gate, with several transistors connected in series. NOR flash is characterised by fast initial access for high read performance, while NAND flash has slow initial access and high write performance. This latter flash memory configuration also permits a denser layout and greater storage capacity per chip and is the main form of universal serial bus (USB) flash devices that are used for high capacity data storage. One advantage, however, of the NOR approach is that the technology provides full address and data buses, allowing random access to any memory location. Hot-carrier injection is usually used to write data to NOR cells whereas Fowler–Nordheim tunnelling is used to write and erase NAND cells, and to erase NOR cells.

Multilevel flash technology allows more than one bit to be stored per memory cell. For example, rather than a '0' and a '1' being associated with the bit states, values may be interpreted as four distinct combinations, 00, 01, 10 and 11 (in this particular case, providing two bits of information).

However, in these devices, it is necessary be be able to distinguish between multiple charge levels on the floating gate.

Key factors for any electronic memory technology are: the operating voltage; the time for changing the memory state; the read time; the retention time of the state; the amount of energy to change the state; and the number of memory cycles. At present (2013), Moore's Law continues to drive transistor-based memory scaling but the complexity will increase. Transistor technology based around a 22 nm process (the 22 nm node) was introduced by semiconductor companies in 2011. The potential for scaling the technology for multiple generations beyond this is therefore important. Current memory technologies, such as DRAM, SRAM and NAND flash, are approaching very difficult issues related to their continued scaling. Flash scaling is easier due to its relatively simple structure. However, scaling will eventually become difficult due to the high electric fields required for the programming and erase operations, and the stringent requirements for long-term charge storage. These demands are imposing fundamental scaling limitations on the memory cell operating voltages and on the physical thickness of the tunnelling dielectric. Overcoming such restrictions will require innovations in cell structure and device materials. A future scenario can be envisaged where logic is implemented using MOSFET technology (or by using polymer FETs, in the case of plastic electronics) and the memory is added to the system using a different technology.

Research over the past decade has led to the discovery of several new memory technologies. The International Technology Roadmap for Semiconductors has identified six promising candidates (ITRS, 2011): ferroelectric memory (including ferroelectric FET and ferroelectric polarisation ReRAM), nanoelectromechanical memory, nanoionic or redox memory, Mott memory, macromolecular memory (polymer or organic resistive memory) and molecular memory. Some of the technologies are interrelated. Many of these devices can be classified as non-volatile RAM rather than re-programmable ROM. The distinction is blurred, but is generally based on write-times, which are in the nanosecond to microsecond range for the non-volatile RAMs and of the order of seconds for the re-programmable ROM. Non-volatile RAM represents the ideal memory. The following sections focus of some of these ideas, in as far as these have been developed using organic materials.

22.3 Resistive memory

Resistive memory devices rely on an appropriate physical mechanism to switch the resistance of a material between high resistance and low resistance states (Cho *et al.*, 2001a; Lee and Chen, 2012; Strukov and Kohlstedt, 2012). The absolute value of the ON/OFF ratio need not be that high – it has been suggested that a factor of 10 or more is sufficient (Scott and Bozano, 2007) – so long as the distributions of the ON/OFF values over a large number of memory cells is narrow. The intense activity in such technologies is motivated by their very simple structure and ability for scaling to small dimensions. There is also the possibility that the devices can be stacked in multiple levels in 3-D.

Organic resistive memory structures are generally formed by interposing thin layers of organic molecules between two electrodes, as shown in Fig 22.5. The cross-point (or crossed-bar) architecture permits the closest packing of bit-cells, with each occupying an area of $4F^2$, where F is the minimum feature size (the line-width and spacing of the electrodes). A variation uses the tip of a scanning probe microscope instead of a (thermally evaporated) metallic top electrode, leading to ultrahigh data storage density.

A very wide range of organic materials that exhibit resistive switching has been reported. At least six different types of switching characteristics have been identified, according to the nature of the current, I, versus voltage, V, behaviour (Scott and Bozano, 2007). These are illustrated in Fig 22.6. Figure 22.6(a) shows relatively symmetric hysteresis; there is no sharp threshold voltage, or stable states, which might be used as the basis for a memory device. In contrast, the I-V characteristics in Fig. 22.6(b) reveal a sharp increase in current at a particular positive applied voltage; the device can then be turned off by reversing the polarity of the bias. Similar behaviour is observed in the characteristics of Fig. 22.6(c); however, the device turns off as the positive voltage is reduced and before zero bias is reached. The I-V curves in Fig. 22.6(d) show a similar sharp turn on at a positive

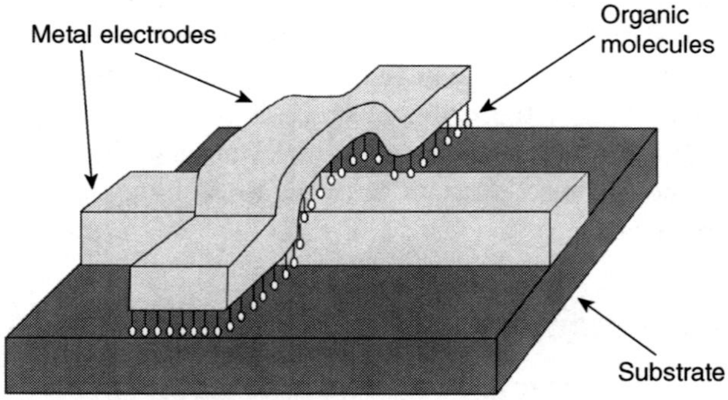

22.5 Crossed-point architecture in which organic molecules are sandwiched between metallic electrodes.

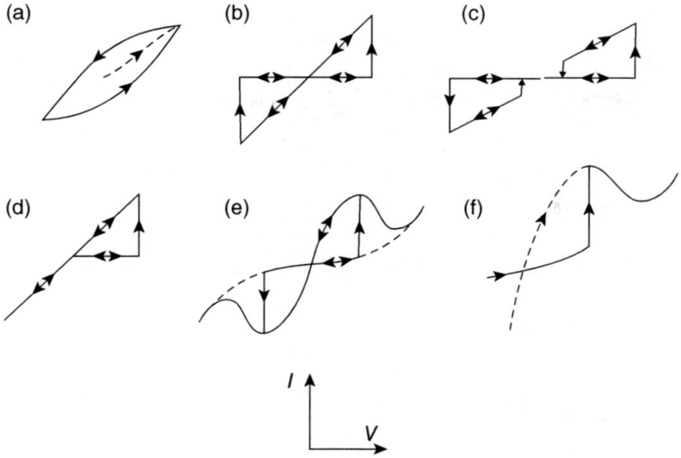

22.6 Current, *I*, versus voltage, *V*, behaviour reported in resistive switching/memory devices. Scott and Bozano (2007).

voltage, but the device does not turn off again. This type of characteristic is typical of a WORM memory. A negative differential resistance region (NDR) is noted in the characteristics shown in Figs. 22.6(e) and (f). Here, as the voltage is increased, the current decreases.

Over the past 40–50 years, there have been many reports of switching and memory effects in thin films. The film thicknesses are generally less than 1 μm and the phenomena are observed in different types of material (inorganic compounds, such as silicon dioxide and metal oxides, and organic compounds, such as polymers and charge-transfer complexes). Furthermore, the thin films have been formed using a variety of techniques (e.g. spin-coating, thermal evaporation). The only experimental parameter that is common to all the structures studied is the presence of metallic electrodes (e.g. Al) below and on top of the thin film. Reviews of the switching types of various organic devices sandwiched between metallic electrodes have been given by Scott and Bozano (2007) and Cho *et al.* (2011a). It is important to note that researchers undertake their experiments in different ways, e.g. different voltage ranges and/or different measurement protocols. Therefore, it is often difficult to compare directly the data reported by individual groups.

Figure 22.7(a) shows the chemical structure of an organic material – 7-{4-[5-4-*tert*-butylphenyl-1,3,4-oxadiazol-2-yl]phenyl}-9,9-dihexyl-*N*,*N*-diphenyl-fluoren-2-amine – that has been used in an organic resistive memory (Dimitrakis *et al.*, 2008). The device was formed by sandwiching a thin film (approximately 50 nm in thickness and formed by spin-coating) of this compound between aluminium electrodes. The current versus voltage

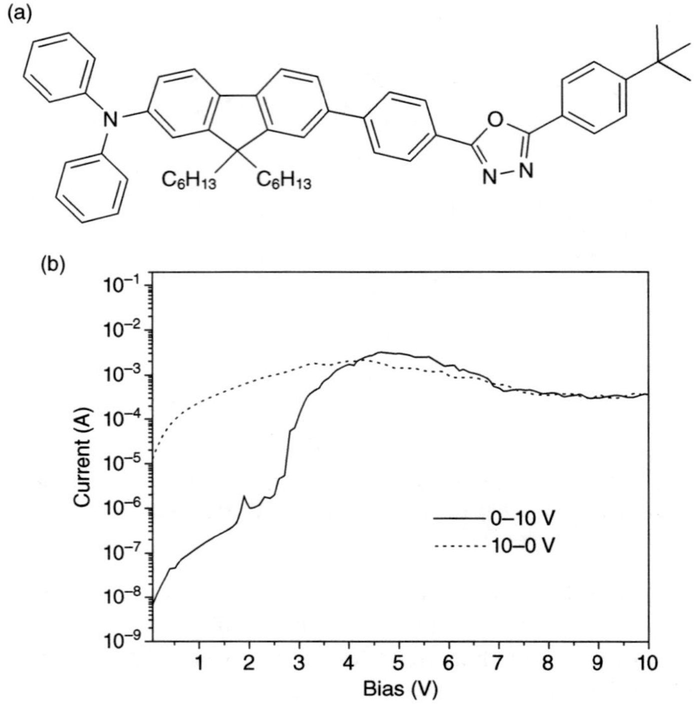

22.7 (a) Polyfluorene derivative used in resistive memory device. (b) Current versus voltage characteristics of resistive memory device based on polyfluorene molecule shown in (a). After Dimitrakis *et al.* (2008). Reprinted with permission from Dimitrakis, Normand, Tsoukalas, Pearson, Ahn, Mabrook, Zeze, Petty, Kamtekar, Wang, Bryce and Green, *Journal of Applied Physics*, Vol. 104, Article Number 044510 (2008). Copyright (2008), American Institute of Physics.

behaviour, shown in Fig 22.7(b), is reasonably symmetric to the polarity of the applied voltage and exhibits both NDR and memory effects, i.e. the features shown by the *I-V* characteristics of Fig. 22.6(e). The data are very similar to those reported by many others for different organic thin films (e.g. Bozano *et al.*, 2004; Tseng *et al.*, 2005; Chen and Ma, 2006; Lauters *et al.*, 2006; Simon *et al.*, 2006). For the device shown in Fig. 22.7(b), the ON state is obtained by applying a voltage close to the current maximum (just before the NDR region ~4.5 V) and reducing it to zero. In contrast, switching from the high conductivity ON state to the OFF state is accomplished by selecting a voltage corresponding to the current minimum in the negative differential resistance region (~8 V) and reducing this rapidly to zero (erase). The state of the device (ON or OFF) is determined by measuring the current at a voltage of 1 V (read).

There is currently no agreement on how resistive thin film memories, such as those described above, operate. The explanations fall into two distinct categories: (i) the injection and storage of charge in the thin film; and (ii) metallic filament formation. Certain devices exploit the coupling of electronic and ionic dopant transport (Ng *et al.*, 2011a). A device 'formation' step is often required before switching effects are observed, but details of the process are sometimes sketchy. There is some confusion in the literature between the electrical behaviour of the unformed (virgin) device and that of the ON and OFF states. Three different states for resistive memory devices are often reported, each exhibiting different current versus voltage behaviour: unformed; formed ON state; and formed OFF state (Cölle *et al.*, 2006; Verbakel *et al*, 2007; Dimitrakis *et al.*, 2008).

Very similar electrical data to those shown in Fig. 22.7(b) were reported by Simmons and Verderber (1967). These authors used thin (20–300 nm) films of silicon oxide sandwiched between aluminium and gold electrodes. The forming process (described in detail by the authors) involved the application of a positive voltage to the gold electrode. This was thought to result in Au ions being injected into the insulating film. However, subsequent experiments by Dearnaley *et al.* (1970a) contradicted this idea. A theory was developed in terms of the growth and thermal rupture of many conducting filaments through the insulating layer. Other mechanisms were proposed by the same authors to explain the forming process (Dearnaley *et al.*, 1970b). Direct observation of silver filaments in Ag/polymer/highly-doped Si devices by electron microscopy and energy dispersive X-ray spectroscopy has been reported by Cho *et al.* (2011b). Furthermore, conductivity images of the high and low conductive states were observed using AFM. In the work of Cölle *et al.* (2006) and Verbakel *et al.*, (2007), it is argued that the device formation results from a 'soft' breakdown in the underlying aluminium oxide layer. The role of the organic film in the subsequent switching is only as a limiting series resistance. Rozenberg *et al.* (2004) have suggested a phenomenological model based on the assumption that the semiconducting part of the memory device possesses a non-percolating domain structure; in particular, regions adjacent to the electrodes together with a central domain. The electrical behaviour is then characterised by domain-domain and domain-electrode tunnelling.

Many of the recipes for switching are based on the incorporation of metallic nanoparticles within an organic thin film (Bozano *et al.*, 2004; Yang *et al.*, 2006). The presence of the nanoparticles is thought to influence the transfer of charge (e.g. by a trapping/hopping process). It is now clear that the intentional addition of such nanoparticles is not necessary to observe memory effects, although these may improve the switching performance (Mahapatro *et al.*, 2004; Chen and Ma, 2006; Pearson *et al.*, 2007; Scott and Bozano, 2007; Dimitrakis *et al.*, 2008). Bistable devices that do not include nanoparticles are preferable as their presence will restrict the device scaling, i.e. when the cell dimensions become comparable to those of the nanoparticles.

Figure 22.8 shows how the two-terminal resistive elements can be incorporated into a two-dimensional array (random access memory). The terms bitline and wordline refer to the column and row organisation, respectively, of the memory. There is one memory cell at the intersection of the wordline and the bitline. The most common way to read the resistive elements is to apply a voltage to the selected wordline and to couple a sense amplifier to each bitline, as shown in Fig 22.9(a). The voltage increase at each sense amplifier depends on the resistance of the element and the (parasitic) capacitance at the input. Thus, the time taken for the voltage signal to rise above a threshold level is directly proportional to the resistance. Unfortunately, the cross-point architecture will have many parallel paths connecting each wordline to each bitline; only one of these passes directly through the addressed node. These additional currents can easily overwhelm the desired signal (Scott and Bozano, 2007). However, if the nodes are rectifying, for example, by the addition of a diode to each resistive memory element, the excess current can be minimised. Figure 22.9(b) illustrates this effect: one of the additional diodes (top left in the figure) will be reverse biased, thereby reducing significantly the current flowing though the unwanted

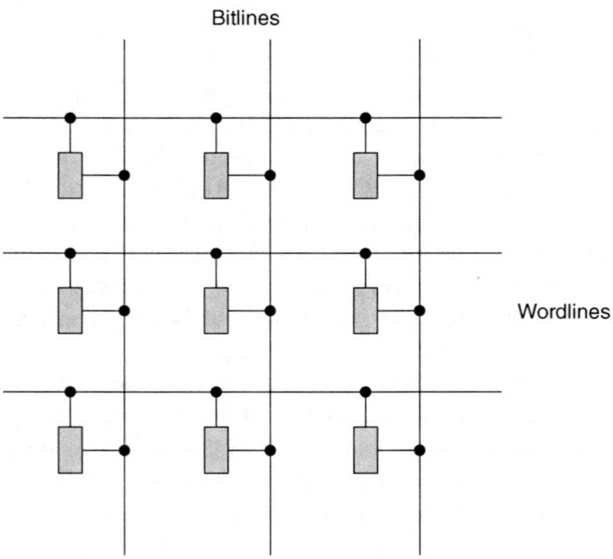

22.8 Incorporation of two-terminal resistive memory elements in an array.

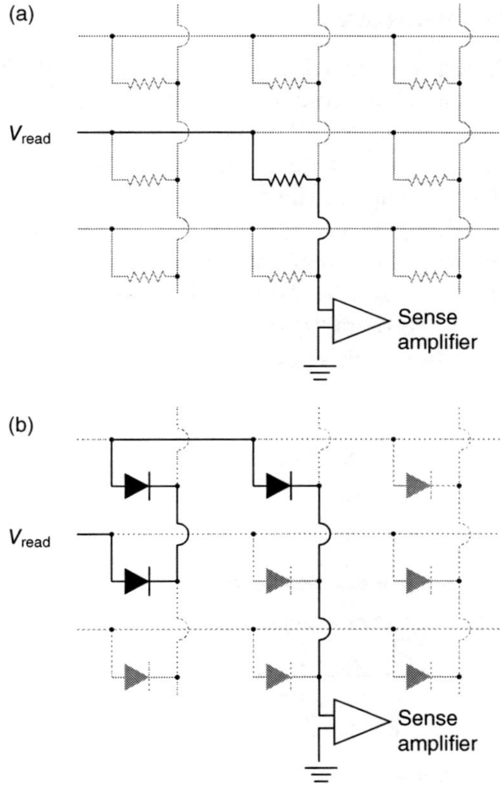

22.9 Reading state of resistive memory element in an array: (a) basic array and (b) resistive memory elements with rectification properties. Scott and Bozano (2007).

path. It is envisaged that high rectification ratios (approaching 10^8 for a 10 Megabyte memory) will be needed for practical devices (Scott and Bozano, 2007).

A three-dimensional 8 × 8 crossed-bar array of polymer resistive devices based on a composite of polyimide and 6-phenyl-C_{61} butyric acid methyl ester has been reported by Song *et al.* (2010). The particular architecture that was fabricated possessed three active layers, providing a total of 64 × 3 = 192 memory cells. The memory assembly exhibited a high device yield: specifically, 56 cells out of 64 were working in the first and second active layers while 48 cells were operative as memory devices in the third layer. The overall yield – 160/192 or 83.3% – suggests that such a three-dimensional array may provide a viable technology for the high-density integration of organic memory cells.

22.4 Organic flash memory

As noted in Section 22.2, charge storage is the basis of the operation of flash memory. A flash device is similar in structure to a MOSFET, except that it has two gate electrodes, one above the other. The top electrode forms the control gate, below which a floating gate is capacitatively coupled to the control gate and the underlying silicon. The memory cell operation involves putting charge on the floating gate or removing it, corresponding to two logic levels. Nanoflash devices utilise single or multiple nanoparticles as the charge storage elements. These are usually embedded in the gate oxide of an FET and located in close proximity to the transistor channel. Figure 22.10(a) depicts a simple version of this device – a metal/insulator/semicon-

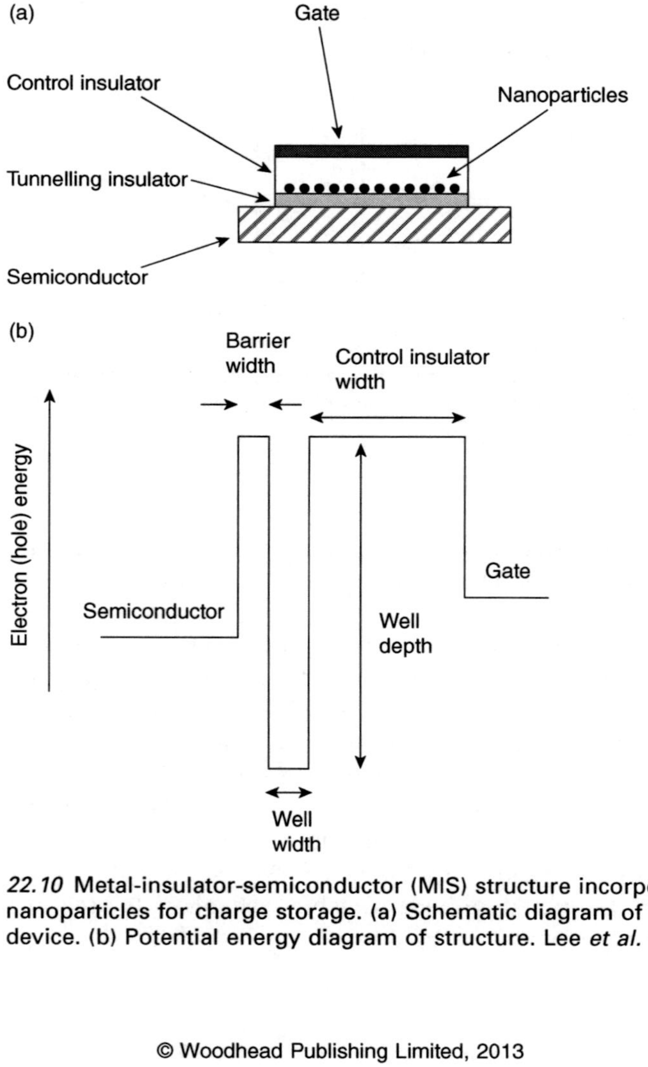

22.10 Metal-insulator-semiconductor (MIS) structure incorporating nanoparticles for charge storage. (a) Schematic diagram of MIS device. (b) Potential energy diagram of structure. Lee *et al.* (2005).

ductor or MIS structure; the potential energy variation (for either electrons or holes) across the structure is given in Fig 22.10(b) (Lee *et al.*, 2005). A very thin (<3 nm) barrier between the substrate and the nanoparticles allows charge to move between the substrate and the nanoparticles by the process of quantum mechanical tunnelling. This charge is stored on the nanoparticles, resulting in a shift in the capacitance versus voltage characteristics of the MIS device (Paul *et al.*, 2003). The charge retention time is key to the successful development of any flash memory. This is dependent on the leakage current that can flow through the gate insulator. These currents must be less than 10^{-22} A if a device is to lose less than 50% of its charge over a 10-year period.

Figure 22.11 provides details of a hybrid silicon/organic MISFET device incorporating a self-assembled layer of gold nanoparticles (Kolliopoulou *et al.*, 2004): Figure 22.11(a) shows a schematic diagram of the self-assembled layer while Fig 22.11(b) reveals a photograph of the final device. In the self-assembly process, an oxidised silicon surface is first functionalised with a long chain amine derivative. This surface is exposed to a solution of carboxylic acid-derivatised gold nanoparticles (Au-NPs), which will assemble onto the positively charged amine surface at an appropriate solution pH. Langmuir–Blodgett layers of cadmium arachidate (Petty, 1996) are then used to provide the gate insulator. Testing the memory characteristics of the devices consists of applying positive or negative voltage pulses successively to the gate of a previously unstressed device, and keeping the source and drain electrodes grounded (i.e. at 0 V). The voltage pulse height is progressively increased while the pulse duration is kept constant. The injected (rejected) charges into (out of) the gold nanoparticles causes a shift of the transistor threshold voltage, V_t (Fig. 22.3) to higher or lower voltages.

The high V_t state is usually called the 'write' state while the low V_t is the 'read' state (Fig. 22.3). The final device exhibits a clear memory 'window' in Fig 22.12, which is a plot of the threshold voltage shift ΔV_t as a function of the write/erase (programming) voltage, $V_{W/E}$. Since no device hysteresis was observed for a gate–voltage cycle for samples without the Au nanoparticles, it was concluded that these were responsible for the charge storage effects. The application of positive pulses to the gate lowers the threshold voltage of an unstressed device, indicating that electrons are extracted from the nanoparticles to the gate. Negative pulses increase V_t due to electron injection from the gate metal onto the gold nanoparticles. This mechanism reveals that the transistor channel does not contribute to the charge exchange and hence cannot affect the memory properties of the hybrid device. Although the SiO_2 layer used in the work was relatively thin (5 nm), the distance between the surface of the underlying silicon and the gold particles is effectively increased to over 10 nm because of the presence of

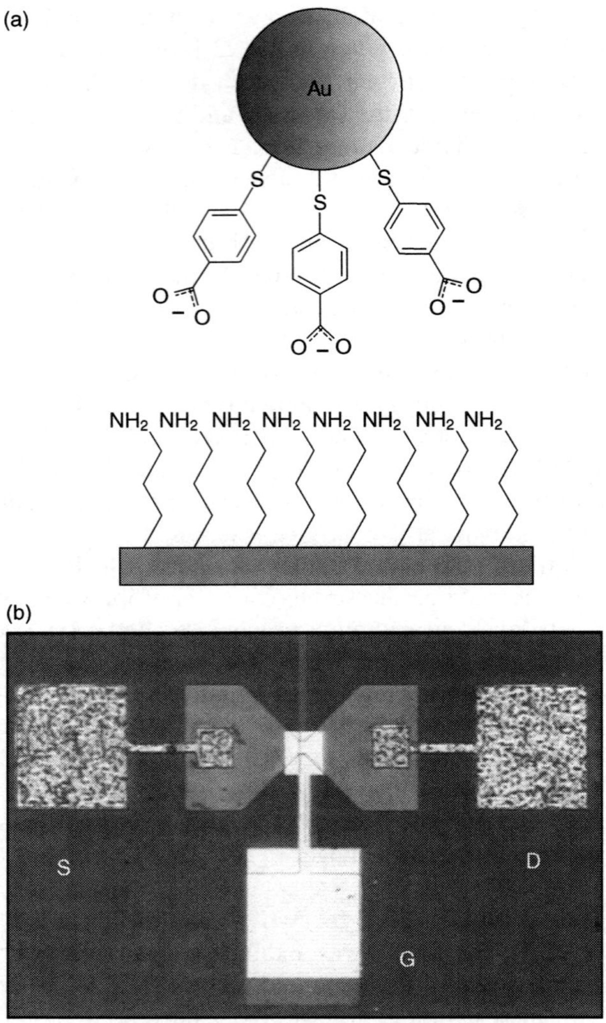

22.11 (a) Self-assembly of gold nanoparticles onto functionalised silicon oxide surface. (b) Optical micrograph of transistor device incorporating the nanoparticles, where S = source, D = drain and G = gate. Reproduced from *Microelectronic Engineering*, **73–74** (2004) 725–729, Kolliopoulou *et al.* (2004), with permission from Elsevier.

the amine functionalising layer and the organic capping layer associated with the Au nanoparticles. This prevents easy charge transfer via tunnelling from the semiconductor (channel) to the Au.

The above work has been extended to other insulators, in particular poly(methyl methacrylate) (PMMA), which can be deposited conveniently

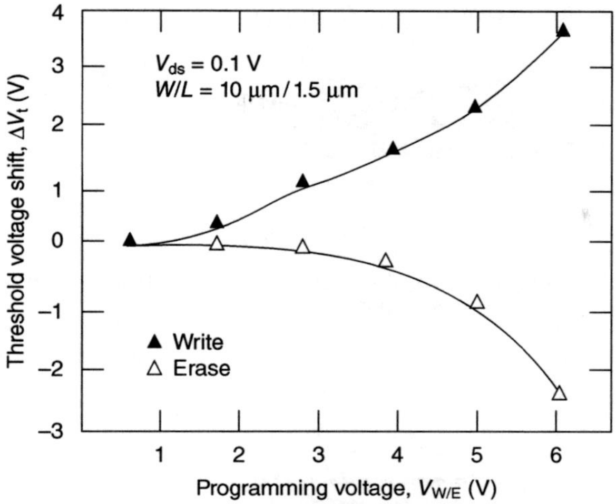

22.12 Memory 'window' for hybrid organic-Si flash memory device. Reproduced from *Microelectronic Engineering*, **73–74** (2004) 725–729, Kolliopoulou *et al.* (2004), with permission from Elsevier.

by spin-coating (Mabrook *et al.*, 2008). Different charge storage elements have also been studied, including metallic nanoparticles, C_{60} and single wall carbon nanotubes (Mabrook *et al.*, 2009a). MIS devices based on PMMA/Au-NPs/pentacene were shown to exhibit behaviour indicating that charging and discharging of the nanoparticle layer takes place via the organic semiconductor surface, rather than through the gate insulator (observed in other MIS-based organic memories) (Mabrook *et al.*, 2009b).

An 'all-organic' memory FET based on pentacene/Au-NP/PMMA has also been developed (Mabrook *et al.*, 2009c). The device architecture is depicted in Fig 22.13. The dependence of the drain-to-source current I_{ds} on the drain-source voltage V_{ds} (transistor output characteristics) is shown for the memory device and also for control device (i.e. with no Au-NPs) in Fig 22.14. In each measurement, both forward and reverse voltage scans are shown. Whereas both the Au-NP and control devices exhibit typical transistor behaviour, the addition of the nanoparticles produces a distinct hysteresis in the output characteristics. This is attributed to the charging and discharging of the Au-NPs with the applied voltage. Figure 22.15 shows the effect of programming pulses on the value of I_{ds} when a voltage was applied to the gate electrode. The write state of the memory was achieved by the application of a pulse of –30 V for 1 s while, for the erase state, a voltage pulse of 30 V was used for the same period of time. With a gate voltage of 10 V, it was possible to distinguish whether the device was in the write or erase state from the value of the drain-to-source current, evident from Fig. 22.15.

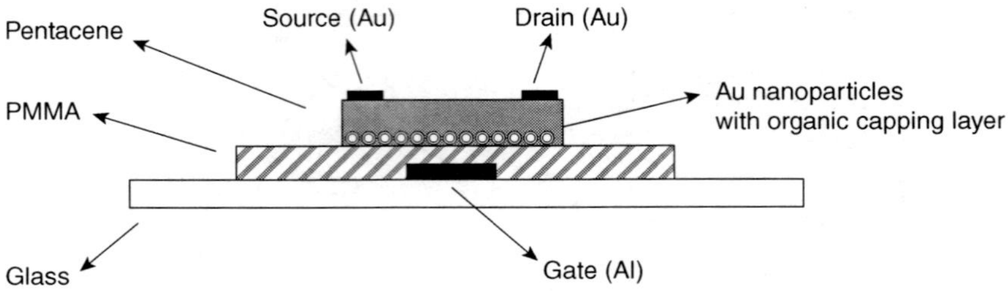

22.13 Schematic diagram of pentacene organic flash memory transistor. Mabrook *et al.* (2009b).

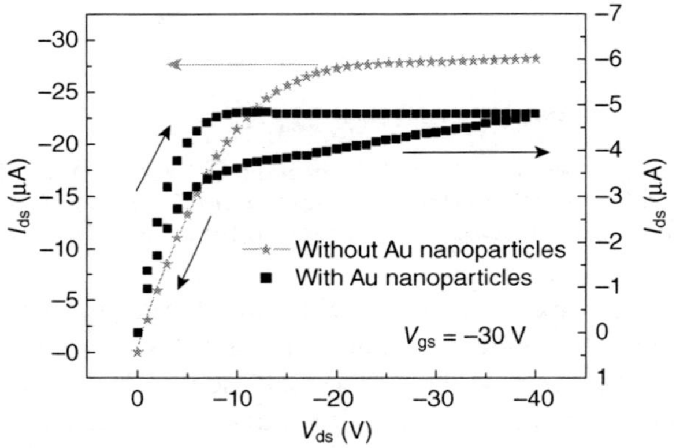

22.14 Transistor output characteristics for pentacene memory transistor. Data are shown for devices with and without gold nanoparticles. After Mabrook *et al.* (2009c). Reprinted with permission from Mabrook, Yun, Pearson, Zeze and Petty, *Applied Physics Letters,* Vol. 94, Article number 173302 (2009). Copyright (2009), American Institute of Physics.

Pentacene memory transistor arrays (26 × 26 memory elements) on plastic substrates and operating at relatively low voltages and have been achieved using thermally evaporated aluminium as the floating gate and a 2 nm thick alkyl-phosphonic acid-based self-assembled monolayer as the dielectric (Sekitani *et al.*, 2009). The small thickness of the dielectric allows very small programming and erase voltages (≤6 V) to produce a large, non-volatile, reversible threshold–voltage shift. Furthermore, the transistors endured more than 1000 program and erase cycles. Other notable achievements in this category of organic flash memory include the development of solution processing in order to build up memory devices on flexible supports (Kim and Lee, 2010; Leong *et al.*, 2011a, b). The challenge now is to exploit manufacturing routes such as inkjet printing, screen printing and spray coating to achieve low-cost memories compatible with flexible

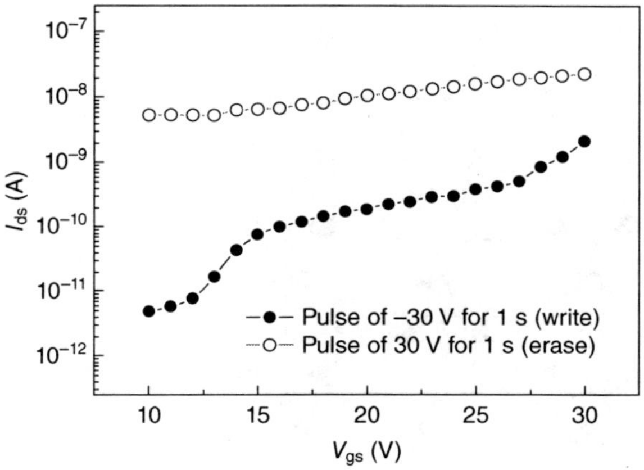

22.15 Effect of applying write and erase pulses to pentacene organic memory transistor. After Mabrook *et al.* (2009c). Reprinted with permission from Mabrook, Yun, Pearson, Zeze and Petty, *Applied Physics Letters*, Vol. 94, Article number 173302 (2009). Copyright (2009), American Institute of Physics.

substrates, which can be integrated into devices such as display drivers, RFID tags and electronic paper (e-paper) (Leong *et al.*, 2011a).

22.5 Ferroelectric random access memory (RAM)

Longer retention times in capacitative DRAM-type memory cells can be achieved if the dielectric layer has ferroelectric properties. Reviews of ferroelectric memories have been given by Ling *et al.* (2008) and Heremans *et al.* (2011). A ferroelectric material is one that has an in-built electric polarisation; furthermore, this can be reversed by an applied electric field. One important organic material that exhibits such behaviour is poly(vinylidene difluoride) (PVDF). This polymer, the chemical structure of which is shown in Fig 22.16(a), consists of the mer unit CH_2CF_2 and has at least four different crystalline phases. Two of these, the α-phase and the β-phase, are shown in Fig 22.16(b) and (c), respectively. The crystalline structures are both based on the orthorhombic unit cell; the figures give the projection of the carbon atoms (small open circles) and fluorine atoms (large filled circles) on the *ab* planes of the unit cells. The α-phase is the most common structure and the other forms can be obtained from this parent phase by applications of mechanical stress, heat and electric field. The polymer chain in the α-phase results in the dipole moments associated with the carbon–fluorine bonds arranged in opposite directions, so that there is no net polarisation within the crystal.

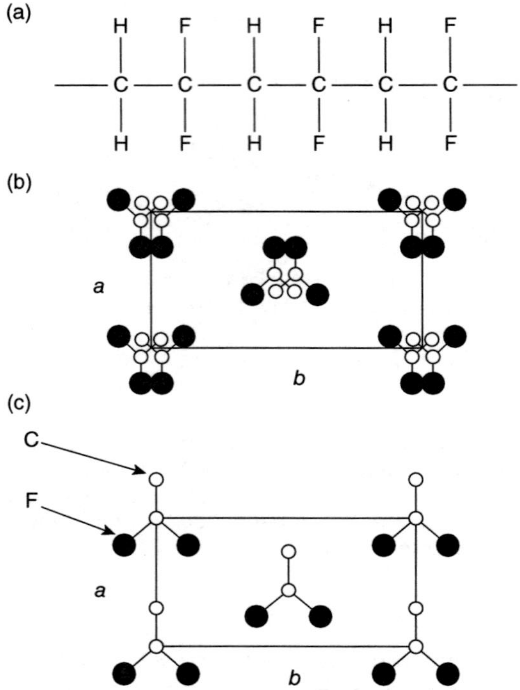

22.16 Structural forms of poly(vinylidene difluoride), PVDF. (a) Section of $(CH_2CF_2)_n$ chain; (b) α-phase; (c) β-phase. The projections of the C atoms (open circles) and F atoms (large full circles) on to the *ab* planes are shown. The H atoms have been omitted for clarity.

When the α-phase of PVDF is mechanically deformed by stretching or rolling at temperatures below 100 °C, the β-phase of PVDF is formed. The unit cell of this structure, Fig. 22.16(c), has a net dipole moment (normal to the chain direction). However, because of the random orientation of the crystallites, there is no net polarisation in the material. The application of a strong electric field, a process called poling, is needed to confer the PVDF with an overall dipole moment. Furthermore, the material can be taken though a hysteresis loop, shown in Figure 22.17 in the form of a plot of the polarisation as a function of the applied electric field. The applied field required to reduce the polarisation to zero, E_c, is called the coercive field while the polarisation retained by the material with no applied field, P_r, is the remanent or residual polarisation.

Co-polymers of PVDF with trifluoroethylene (TrFE) and tetrafluoroethylene (TFE) have also been shown to exhibit strong ferroelectric effects. These can possess a large spontaneous polarisation of about 0.1 C m^{-2}, excellent polarisation stability, high resistivity (low leakage) of up to

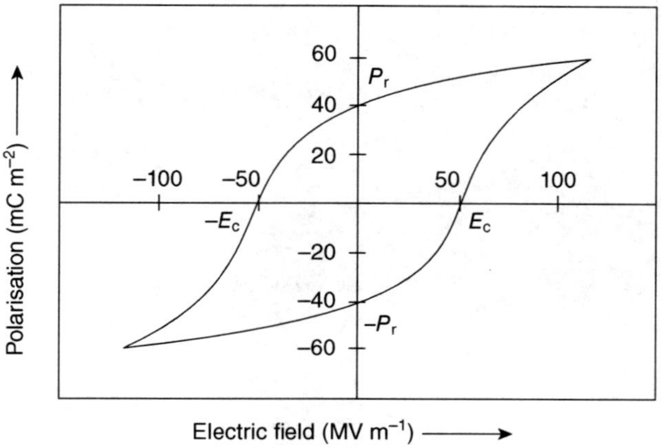

22.17 Typical polarisation versus electric field hysteresis loop for poly(vinylidene difluoride), PVDF. P_r = remanent polarisation; E_c = coercive force.

10^{12} Ω cm and switching times as short as 0.1 μs (Ling *et al.*, 2008). An attractive morphological feature of the co-monomers is that they force the polymer into an all-*trans* configuration that has a polar crystalline phase. This eliminates the need for mechanical stretching to yield a polar phase. Poly(VDF-TrFE) crystallises to a much greater extent than PVDF (up to 90% crystalline), resulting in a higher P_r, a lower E_r and a much sharper hysteresis loop.

Writing a '1' or a '0' to the ferroelectric memory element requires the application of a high voltage across the ferroelectric capacitor. Depending on the polarity of the voltage, this polarises the capacitor in one particular state. To read the cell, a further voltage is applied. If the polarity of this voltage matches the polarisation direction in the ferroelectric material, nothing happens. However, in the case that the applied voltage polarity opposes the dipole direction, the ferroelectric will be re-polarised and charge will flow to the sensing circuit. The reading process – destructive readout – results in the loss of data, which must be rewritten. Complex addressing circuitry is therefore needed.

A significant problem in the reliability of ferroelectric memories is that of fatigue. This is the tendency of the polarisation to decrease as a result of repeated switching. This phenomenon can have a number of physical origins including the pinning of ferroelectric domain walls as a result of the migration and trapping of free electron carriers, the inhibition of opposite domain wall nucleation by charge injection and the formation of a passive surface layer that reduces the electric field in the ferroelectric film. The relatively large number of programming cycles required for writing data to and reading data from ferroelectric capacitative memories (see above)

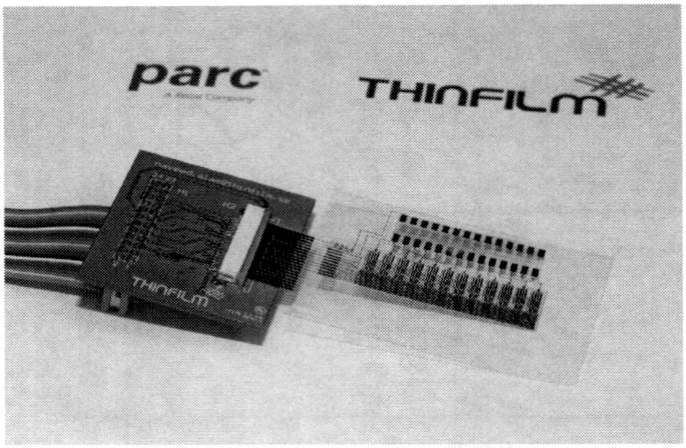

22.18 Photograph of Thin Film/PARC plastic memory and organic interface circuitry. Kindly provided by Dr Janos Verez, PARC Inc.

exacerbates the fatigue problem. A further issue of exploiting ferroelectric memories relates to scaling; a reduction in the capacitor size will reduce the average current during a switching event.

A number of the problems outlined above may be circumvented by combining a ferroelectric polymer with a transistor – to form a ferroelectric FET or FeFET (Ng *et al.*, 2009, 2011b; Heremans *et al.*, 2011). The different polarisation states of the ferroelectric gate insulator lead to different values of the drain-to-source current of the transistor. Reading the memory state is therefore a non-destructive process, so the device lifetime is determined only by the number of times that the memory is written.

Thin Film Electronics ASA together with PARC, a Xerox company, recently announced they have produced a working prototype of a printed non-volatile ferroelectric memory device addressed with complementary organic circuits, the organic equivalent of CMOS circuitry (Thin Film, 2012). A photograph of the device is shown in Fig 22.18. The memory has a retention time greater than 10 years, an endurance of more than 10^9 cycles with a read/write time of microseconds. The device operates over the temperature range –20 °C to 60 °C. The inkjet-printed decoder array has approximately 200 transistors, with a smallest printed feature size of 35 microns. The building blocks include NAND, OR and transmission gates. The circuit uses complementary printed thin film transistor logic (with p- and n-type devices, both with charge carrier mobilities of approximately 0.15 $cm^2 V^{-1} s^{-1}$).

22.6 Molecular memories

There are many proposals to develop memories that exploit changes in individual molecules or groups of molecules. Bistable rotaxane molecules have been used in a number of these studies; an example is shown in Fig 22.19 (Chen *et al.*, 2003; Flood *et al.*, 2004). This molecule is amphiphilic and the ring component can move between the polar (right-hand side of the molecule) and nonpolar (central sulphur-rich group) regions of the main part of the molecule. The molecules can be assembled on an electrode using the Langmuir–Blodgett approach and a top electrode then deposited to form a cross-bar structure. If the molecules are deposited onto non-metallic substrates (e.g. Si or carbon nanotubes), the absence of switching using 'control' compounds suggests that the rotaxane molecule itself is responsible for the bistability (Diehl *et al.*, 2003). A 160 000 bit molecular electronic memory circuit, fabricated at a density of 10^{11} bits cm^{-2} has been fabricated from such bistable rotaxanes (Green *et al.*, 2007). Although the circuit possesses a large number of defects, these could be readily identified through electronic testing, and isolated using software coding. The working bits were then configured to form a fully functional random access memory.

Memory devices can also exploit the movement of atoms. For example, it has been proposed to fabricate a high density memory using the electrostatic attraction between suspended, crossed carbon nanotubes (Rueckes *et al.*, 2000). A further device, based on silver sulphide (Ag_2S), works by controlling the formation and annihilation of an atomic bridge at the crossing point between two electrodes (Ruitenbeek, 2005; Terabe *et al.*, 2005). Figure 22.20 shows a schematic diagram of one such memory cell. A 1 nm thick silver layer deposited on top of the Ag_2S layer disappears into the sulphide layer when a current flows from the platinum electrode to the silver electrode, Fig 22.20(a). This results in loss of contact between the two electrodes and initialises the device. A bridge of silver atoms is locally formed by applying a voltage of the opposite sign, re-establishing contact between the silver sulphide and the platinum, Fig 22.20(b). The conductance through the device can be as small as one quantum unit of conductance, suggesting that the silver bridge can touch the platinum electrode with just one atom. Although this structure is not strictly a molecular electronic

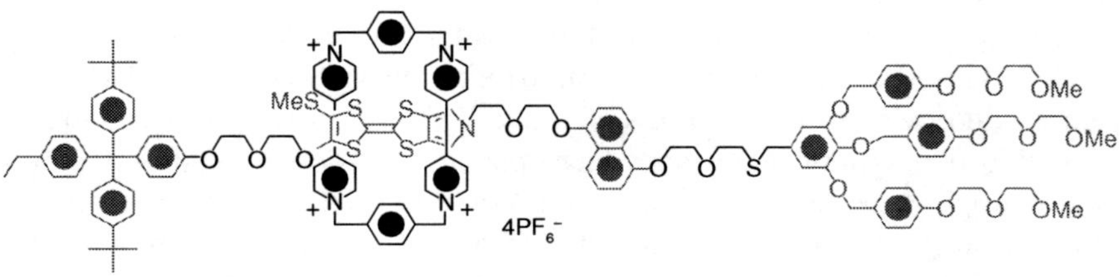

22.19 Bistable rotaxane structure. The 'ring' component of the molecule can move between the hydrophobic sulphur-rich region shown centrally and the hydrophilic 'stopper' region on the far right of the molecule (Chen *et al.*, 2003).

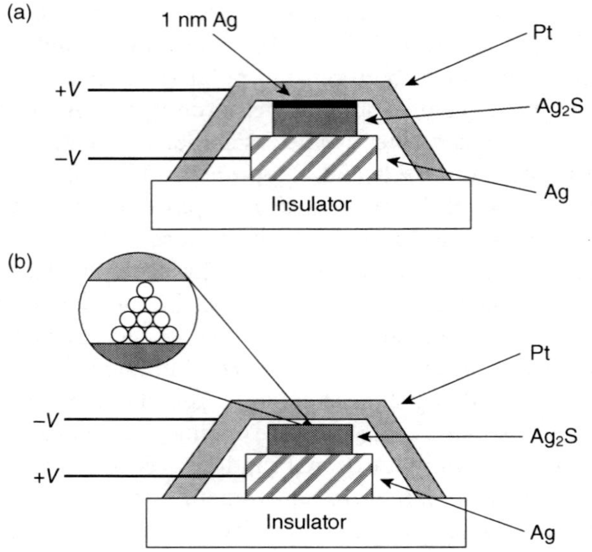

22.20 A rewritable memory bit based on the properties of a silver sulphide mixed ionic conductor. (a) A 1 nm thick silver layer deposited on top of the Ag_2S layer is incorporated into the sulphide layer when the Pt electrode is positively biased. (b) A bridge of Ag atoms is formed locally when the Pt electrode is negatively biased. (Ruitenbeek, 2005). Reprinted by permission from Macmillan Publishers Ltd: Ruitenbeek J V (2005), *Nature*, **433**, 21–22.

device as it is based on inorganic compounds, the principle of operation may have wider applicability.

Spintronics (an acronym for SPIN TRansport electrONICS) exploits the spin of an electron to process and store digital information (Rocha *et al.*, 2005; Wolf *et al.*, 2006, 2010); the subject is also called magnetoelectronics. The simplest method of generating a spin polarised current is to inject the current through a ferromagnetic material. Much of the interest in the area started with the company IBM exploiting magnetically-induced resistance or magnetoresistance in the early 1990s. Magnetic random access memory (MRAM) is the first mainstream spintronic non-volatile RAM. The most successful device to date is the spin valve. This uses a layered structure of thin films of magnetic materials, which changes electrical resistance depending on applied magnetic field direction. When the magnetisation vectors of the ferromagnetic layers are aligned, then a relatively high current will flow, whereas if the magnetisation vectors are antiparallel then the resistance of the system is much higher. The magnitude of the change ((antiparallel resistance – parallel resistance) / parallel resistance) is called the giant magnetoresistance (GMR) ratio. Devices have been demonstrated with

GMR ratios as high as 200%, with typical values greater than 10%. This is a vast improvement over the anisotropic magnetoresistance effect in single layer materials which is usually less than 3%.

Figure 22.21 shows a schematic diagram for a non-volatile memory device based on a magnetic tunnel junction. This has been developed by IBM and other companies and exploits a spin-dependent tunnelling phenomenon. The devices have two ferromagnetic layers separated by a thin insulating barrier. The first layer polarises the spins of current-carrying electrons, which cross the barrier to the second layer by quantum mechanical tunnelling when both layers are magnetically aligned. When the magnetism of the second layer is reversed, the tunnelling is reduced. The magnitude of the current in the memory cell can be used to indicate a '0' or a '1'.

A MRAM chip product prototype was announced in 2003 and the first ever MRAM commercial products were launched in 2006 (Wolf *et al.*, 2010). Such devices possessed read and write speeds of tens of nanoseconds. Early

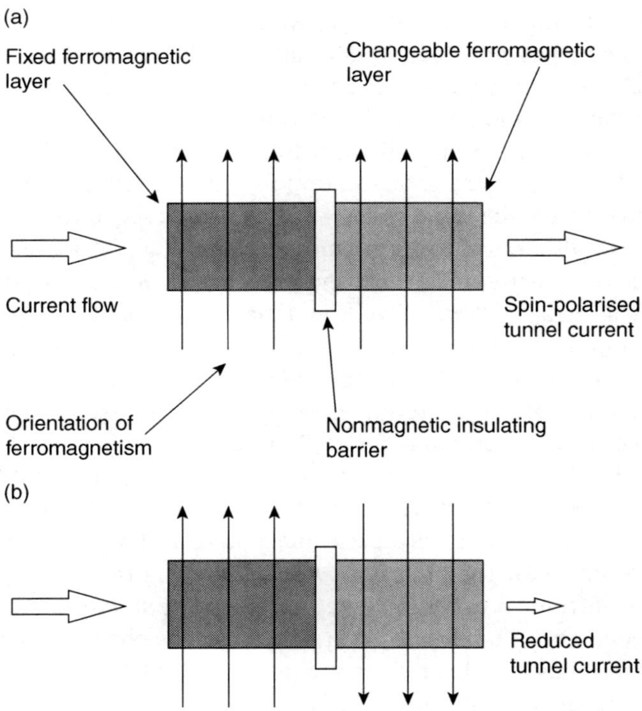

22.21 Schematic diagram of a magnetic tunnel junction. (a) The fixed ferromagnetic layer polarises the spins of current-carrying electrons, which cross the barrier to the second layer by quantum mechanical tunnelling when the magnetism of both layers is aligned. (b) When the direction of magnetism in the second layer is reversed, the tunnelling is reduced.

devices were based on 180 nm technology. A number of developments suggest that scaling below 60 nm should be possible.

The most notable is the discovery of the spin-momentum-transfer effect (Wolf *et al.*, 2010). This exploits the net angular momentum that is carried by a spin-polarised current and the transfer of this momentum to the magnetisation of the second layer. It offers the potential of orders of magnitude lower switching currents and therefore a much lower energy for writing. The spin-momentum-transfer effect becomes important when the minimum dimension of the memory cell is less than 100 nm and becomes more efficient as the cell size is reduced (the opposite to what occurs with the use of conventional magnetic field switching). The related memory technology is called spin transfer torque random access memory, or STT-RAM.

MRAM has similar performance to SRAM, similar density of DRAM but much lower power consumption than DRAM, and is significantly faster and suffers no degradation over time in comparison to flash memory. It is this combination of features that some suggest make it the 'universal memory,' able to replace SRAM, DRAM, EEPROM and flash. However, to date, MRAM has not been widely adopted in the market. This may simply be related to the reluctance of companies to invest enormous amount of money in new technologies at the present time.

There are also many proposals for a transistor that works by switching electrons between spin states. In one option, a gate controls whether most of the electrons have the same spin state. It is also possible to exploit magnetic domain wall motion to form computational logic elements. For this case, information is stored through the presence or absence of a domain wall in a linear array of domain walls in a magnetic thin film loop confined to a channel in a silicon chip. Such a memory might eventually provide a solid-state replacement for the magnetic hard drive used in computers.

Although most of the materials developed for spintronic devices are inorganic, a case has been made for exploiting organic compounds (Rocha *et al.*, 2005). This is based on the weak spin-orbit and hyperfine interactions in organic molecules, which leads to the possibility of preserving spin-coherence over times and distances much longer than in conventional metals or semiconductors (Dediu *et al.*, 2009). Although reports on all-organic spintronic devices have yet to appear, spin-valves using a tris-(8-hydroxyquiniline) aluminium (Alq_3) layer sandwiched between ferromagnetic electrodes have been shown to possess a GMR value of about 5% at low temperatures (Wang *et al.*, 2005).

22.7 Future trends

The rationale for developing organic memory devices is that if plastic electronics is to become a fully integrated and significant technology, then some memory capability is essential. Although memories represent the largest part of silicon-based electronic systems, work on organic memories is lagging behind the development of FETs. Clearly, options for organic memory are to develop the organic equivalents of silicon DRAM, SRAM and flash, i.e. based around organic transistors. Some success in this direction has already been achieved, with the demonstration of arrays of pentacene-based memory architectures (Sekitani *et al.*, 2009). As organic electronics is not likely to compete directly, in terms of operational speed and device density, with inorganic-based technologies, the specifications for the memory elements can, to some extent, be relaxed. However, the devices must be robust and operate for a useful number of read and write cycles ($>10^{12}$ cycles, with a lifetime of >10 years). The polymer ferroelectric memory developed by the Thin Film Company (Thin Film, 2012) is very attractive. Such devices should be inexpensive and ideal for use in standalone consumer applications, including personalised toys and online-enabled games. These can also be integrated with logic elements, sensors, batteries, and displays for mass market applications such as all-printed RFID tags. There are other simple concepts for memory devices, such as those based on resistive switching. The Hewlett-Packard Co. has recently announced its intention to market a two-terminal resistive memory device – probably based on a metal oxide – in 2013 (EE Times, 2011). However, research on organic compounds has quite a long way to go – in terms of understanding the physics behind the device operation and developing arrays of reproducible memory cells that can be addressed in cross-bar architectures – before the technology can find its way into the marketplace. Magnetic random access memories based around spintronics seem poised to provide a 'universal' memory, although the technology has yet to be demonstrated adequately with organic compounds. Table 22.1 compares the important memory parameters (Heremans *et al.*, 2011) that have been achieved to date with organic memory *arrays*, based on flash technology, resistive switching and ferroelectric polymers.

A major challenge is the realisation of molecular-based memories, i.e. exploiting individual molecules or small groups of molecules. Current memory technologies, such as DRAM, SRAM, and NAND Flash, are approaching very difficult issues related their continued scaling to and beyond the 16 nm generation. The availability of memory elements on the molecular scale would revolutionise not only organic electronics but solve a potential 'bottleneck' Moore's Law scaling issue for inorganic-based technologies.

But, maybe even more radical approaches to the organic memory problem are needed? Most of organic electronics technology is focused on emulating the operation of existing silicon and gallium arsenide electronic devices using organic materials, such as polymers. These organic FETs and memory

Table 22.1 Memory parameters for thin-film non-volatile organic memory arrays

Parameter	Ideal (Heremans *et al.*, 2011)	Flash organic transistor (Sekitani *et al.*, 2009)	Resistive polymer switch (Song *et al.*, 2010)	Ferroelectric polymer FET (Ng *et al.*, 2011b)	Ferroelectric polymer (Thin Film, 2012)
Retention	>10 years at maximum operational temperature	>12 h	>14 h >1 year (extrapolated)	>8 h	>10 years
Read time	μs to ms	100 ms		130 μs	μs
Write time	μs to ms	1 s	100 ms	5 ms	μs
Write voltage	<24 V	6 V	5 V (erase 10 V; read 1 V)	30 V (read, 5 V)	
Read/write/ erase endurance	Application dependent: 10^3 to $>10^6$ write/erase cycles	$>10^4$ write/erase cycles			$>10^9$ cycles
Temperature range	Storage: –40 °C to +85 °C Operational: –20 °C to +50 °C				Storage: –20 °C to 85 °C Operational: –20 °C to 60 °C
Memory size	Application dependent: RFID, 8 to 128 bit; e-tickets, toys, badges, wireless sensor nodes, 1 to 64 kbit	26 × 26 array	8 × 8 array in 3 stacked layers = 192 cells	7 × 8 array	

structures can then be utilised in a conventional computer, which exploits the stored-program concept designed by John Von Neumann. The von Neumann architecture is the basis for most modern computers. This uses a single storage structure to hold both the set of instructions on how to perform the computation and the data required or generated by the computation. Such machines are also known as stored-program computers; a schematic diagram is given in Fig 22.22. The separation of storage from the processing unit is implicit in this model. In it, programs and data are stored in a slow-to-access storage medium (such as a hard disk) and work on these is undertaken in a fast-access, volatile storage medium (RAM).

Nature's computer is the brain. The building blocks are neurons, or nerve cells, rather than transistors, and brains utilise parallel processing instead of the serial approach in von Neumann systems. This means that the brain can send a signal to hundreds of thousands of other neurons in less than 20 milliseconds, even though it takes a million times longer to send a signal than a computer switch. The brain works mainly by non-linear computation using the rate of pulse production by a neuron as the information signal being sent to another cell. In the brain, the processes of signal processing and memory are intertwined. There are about 10^{11} neurons in the human brain, and each is connected to 10^3–10^4 others. This gives a crude 'bit-count' of 10^{11} to 10^{15}. An equivalent artificial 'brain' might therefore be built from $10^5 \times 8$ Gbit chips, with a power dissipation of many MW!

Neural logic may be either analogue or digital. In the latter form, the neuron is designed to respond to the sum of its N inputs (which may be inhibitory or excitory). Figure 22.23 shows a schematic diagram of such a

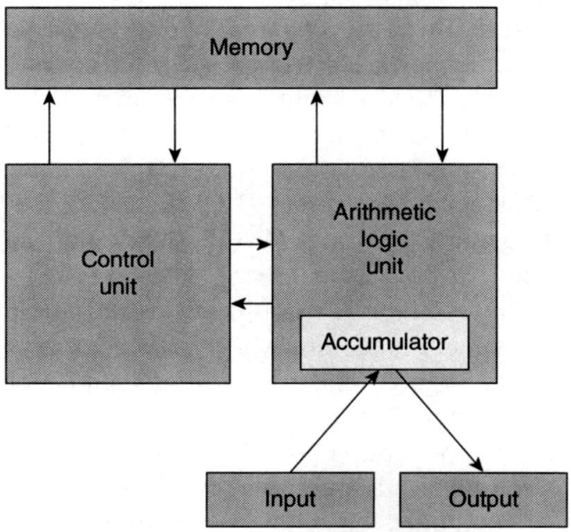

22.22 Von Neumann computer architecture.

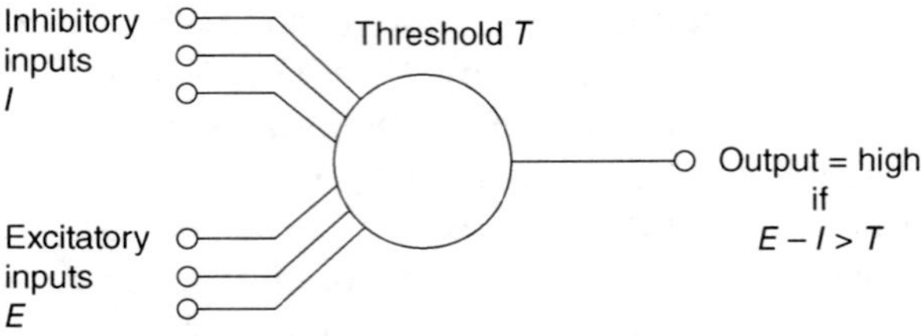

22.23 Representation of neural logic.

gate. Provided that the sum exceeds a given threshold T, the neuron will output logic 1, otherwise it outputs logic 0. By combining neurons into totally connected networks – neural networks – it is possible to construct adaptive learning systems, control systems and pattern recognition systems. Artificial neural networks may be exploited in devices such as the electronic nose, emulating the human olfactory system. Some elementary properties of neurons can be mimicked by simple chemical systems, although no computational system exploiting these has been built. A different (ultimately, more successful?) approach to developing a molecular electronics computer is, then, not to assemble piecemeal signal processing and memory elements, but to look beyond von Neumann architectures and, thereby, to produce systems similar to the processing and memory devices found in nature.

22.8 Sources of further information

The International Technology Roadmap for Semiconductors (ITRS, 2011) is an invaluable source for discussions on the progress of electronic devices and predictions for the future. There are a number of international meetings that currently focus on organic/plastic electronics and often contain sessions devoted to organic electronic devices, including memories. These include the Materials Research Society Spring and Summer Meetings (MRS, 2012), the International Thin Film Transistor Conference (ICT, 2012), the International Conference on Molecular Electronics (ECME, 2013), the International Conference on Organic Electronics (ICOE, 2011) and the Printed Electronics USA Meetings (Printed Electronics USA, 2011).

The Plastic Electronics Foundation is a non-profit organisation with the main objective to bring science and industry together to develop and promote the technology of organic electronics jointly and globally. An annual conference focuses on new developments in the science and technology of organic electronics components (Plastic Electronics, 2012). A related magazine – *+Plastic Electronics* – is published by Pira International, to analyse the strategies of companies launching and commercialising products based on these technologies (Plus Plastic Electronics, 2012).

22.9 Acknowledgement

Many thanks are due to Dr Janos Verez, PARC Inc., for discussions on the Thin Film/PARC memory and for the provision of the photograph shown in Fig. 22.18.

22.10 References

Baeg K-J, Kim J, Khim D, Caironi M, Kim D-Y, You I-K, Quinn J R, Facchetti A and Noh Y-Y (2011), 'Charge injection engineering of ambipolar field-effect transistors for high-performance organic complementary circuits', *ACS Appl. Mater. Interf.*, **3**, 3205–3214.

Bao Z and Locklin J (2007), *Organic Field Effect Transistors*, Boca Raton, CRC Press.

Bozano L D, Kean B W, Deline V R, Salem J R and Scott J C (2004), 'Mechanism for bistability in organic memory elements', *Appl. Phys. Lett.*, **84**, 607–609.

Braga D and Horowitz G (2009), 'High-performance organic field-effect transistors', *Adv. Mater.*, **21**, 1473–1486.

Chen J and Ma D (2006), 'Performance improvement by charge trapping of doping fluorescent dyes in organic memory devices', *J. Appl. Phys.*, **100**, 034512.

Chen Y, Ohlberg D A A, Li X, Stewart D R, Williams R S, Jeppesen J O, Nielsen K A, Stoddart J F, Olynick D L and Anderson E (2003), 'Nanoscale molecular-switch devices fabricated by imprint lithography', *Appl. Phys. Lett.*, **82**, 1610–1612.

Cho B, Song S, Yongsung J, Kim J-W and Lee T (2011a), 'Organic resistive memory devices: performance enhancement, integration, and advanced architectures', *Adv. Funct. Mater.*, **21**, 2806–2829.

Cho B, Yun J-M, Song S, Ji Y, Kim D-Y and Lee T (2011b), 'Direct observation of Ag filamentary paths in organic resistive memory devices', *Adv. Funct. Mater.*, **21**, 3976–3981.

Claverie A, Tsoukalas D, King T-J and Slaughter J M (Editors) (2005), *Materials and Processes for Nonvolatile Memories*, MRS Symposium Proceedings Volume 830, Warrendale, Materials Research Society.

Cölle M, Büchel M and De Leeuw D M (2006), 'Switching and filamentary conduction in non-volatile organic memories', *Organic Electronics*, **7**, 305–312.

Cuevas J C and Scheer E (2010), *Molecular Electronics: An Introduction to Theory and Experiment*, Singapore, World Scientific.

Dearnaley G, Morgan D V and Stoneham A M (1970a), 'A model for filament growth and switching in amorphous oxide films', *J. Non-Cryst. Solids*, **4**, 593–612.

Dearnaley G, Stoneham A M and Morgan D V (1970b), 'Electrical phenomena in amorphous oxide films', *Rep. Prog. Phys.*, **33**, 1129–1191.

Dediu V A, Hueso L E, Bergenti I and Taliani C (2009), 'Spin routes in organic semiconductors', *Nature Mater.*, **8**, 707–716.

Diehl M R, Steuerman D W, Tseng H-R, Vignon S A, Star A, Celestre P C, Stoddart J F and Heath J R (2003), 'Single-walled carbon nanotube-based molecular switch tunnel junctions', *ChemPhysChem*, **4**, 1335–1339.

Dimitrakis P, Normand P, Tsoukalas D, Pearson C, Ahn J H, Mabrook M F, Zeze D A, Petty M C, Kamtekar K T, Wang C, Bryce M R and Green M (2008),

'Electrical behaviour of memory devices based on fluorene-containing organic thin films', *J. Appl. Phys.*, **104**, 044510.

ECME (2013), European Conference on Molecular Electronics. Home page of 2013 meeting in Imperial College, London. Available from: http://www.mmsconferencing.com/ecme13/ [accessed 6 December 2012].

EE Times (2011), 'HP, Hynix plan to launch memristor memory in 2013'. Available from: http://www.eetimes.com/electronics-news/4229171/HP-Hynix-to-launch-memristor-memory-2013 [accessed 6 December 2012].

Flood A H, Stoddart J F, Steuerman D W and Heath J R (2004), 'Whence molecular electronics?', *Science*, **306**: 2055–2056.

Gotsmann B, Knoll A W, Pratt R, Frommer J, Hedrick J L and Duerig U (2010), 'Designing polymers to enable nanoscale thermomechanical data storage', *Adv. Funct. Mater.*, **20**, 1276–1284.

Graz I M and Lacour S P (2010), 'Complementary organic thin film transistor circuits fabricated directly on silicone substrates', *Organic Electronics*, **11**, 1815–1820.

Green J E, Choi J W, Boukai A, Bunimovich Y, Johnston-Halperin E, Delonno E, Luo Y, Sheriff B A, Xu K, Shin Y S, Tseng H-R, Stoddart J F and Heath J R (2007), 'A 160-kilobit molecular electronic memory patterned at 10^{11} bits per square centimetre', *Nature*, **445**, 414–417.

Heremans P, Gelinck G H, Müller R, Baeg K-J, Kim D-Y and Noh Y-Y (2011), 'Polymer and organic non-volatile memory devices', *Chem. Mater.*, **23**, 341–358.

ICOE (2011), International Conference on Organic Electronics. Homepage for 2011 conference in Rome, Italy. Available from: http://icoe2011.artov.imm.cnr.it/ [accessed 06 December 2012].

ICT (2012), International Conference on Thin Film Transistors. Homepage for 2012 conference in Lisbon, Portugal. Available from: http://www.fct.unl.pt/itc2012 [accessed 06 December 2012].

ITRS (2011), International Technology Roadmap for Semiconductors. (2011 Edition: Emerging Research Devices). Available from: http://www.itrs.net/reports.html [accessed 6 December 2012].

Kim S-J and Lee J-S (2010), 'Flexible organic transistor memory devices', *Nano Lett.*, **10**, 2884–2890.

Kolliopoulou S, Dimitrakis P, Normand P, Zhang H L, Cant N, Evans S D, Paul S, Pearson C, Molloy A, Petty M C and Tsoukalas D (2004), 'Integration of organic insulator and self-assembled gold nanoparticles on Si MOSFET for novel non-volatile memory cells', *Microelectronic Engineering*, **73–74**, 725–729.

Lauters M, McCarthy B, Sarid D and Jabbour G E (2006), 'Multilevel conductance switching in polymer films', *Appl. Phys. Lett.*, **89**, 013507.

Lee C, Meteer J, Narayan V and Kan E C (2005), 'Self-assembly of metal nanocrystals on ultrathin oxide for non-volatile memory applications', *J. Electronic Mater.*, **34**, 1–11.

Lee T and Chen Y (2012), 'Organic resistive non-volatile memory materials', *MRS Bull.*, **37**, 144–149.

Leong W L, Mathews N, Tan B, Vaidyanathan S, Dötz F and Mhaisalkar S (2011a), 'Towards printable organic thin film transistor based flash memory devices', *J. Mater. Chem.*, **21**, 5203–5214.

Leong W L, Mathews N, Tan B, Vaidyanathan S, Dötz F and Mhaisalkar S (2011b), 'Solution processed non-volatile top-gate polymer field-effect transistors', *J. Mater. Chem.*, **21**, 8971–8974.

Ling Q-D, Liaw D-J, Zhu, C, Chan D S-H, Kang E-T and Neoh K-G (2008), 'Polymer electronic memories: materials, devices and mechanisms', *Prog. Polym. Sci.*, **33**, 917–978.

Mabrook M F, Pearson C, Kolb D, Zeze D A and Petty M C (2008), 'Memory effects in hybrid silicon-metallic nanoparticle-organic thin film structures', *Organic Electronics*, **9**, 816–820.

Mabrook M F, Jombert A S, Machin S E, Pearson C, Kolb D, Coleman K S, Zeze D A and Petty M C (2009a), 'Memory effects in MIS structures based on silicon and polymethylmethacrylate with nanoparticle charge-storage elements', *Mater. Sci. Eng. B*, **159–160**, 14–17.

Mabrook M F, Yun Y, Pearson C, Zeze D A and Petty M C (2009b), 'Charge storage in pentacene/polymethylmethacrylate memory devices', *IEEE Elec. Dev. Lett.*, **30**, 632–634.

Mabrook M F, Yun, Y, Pearson, C, Zeze, D A and Petty M C (2009c), 'A pentacene-based organic thin film memory transistor', *Appl. Phys. Lett.*, **94**, 173302.

Mahapatro A K, Agrawal R and Ghosh S (2004), 'Electric-field-induced conductance in 8-hydroxyquinoline aluminium (Alq_3)', *J. Appl. Phys.*, **96**, 3583–3585.

Möller S, Perlov C, Jackson W, Taussig C and Forrest S R (2003), 'A polymer/semiconductor write-once read-many-times memory', *Nature*, **426**, 166–169.

MRS (2012), Materials Research Society. Available from: http://www.mrs.org/home/ [accessed 06 December 2012].

Ng T N, Russo B and Arias A C (2009), 'Degradation mechanisms of organic ferroelectric field-effect transistors used as a non-volatile memory', *J. Appl. Phys.*, **106**, 094504.

Ng T N, Russo B and Arias A C (2011a), 'Solution-processed memristive junctions used in a threshold indicator', *IEEE Trans. Electron. Dev.*, **58**, 3435–3443.

Ng T N, Russo B, Krusor B, Kist R and Arias A C (2011b), 'Organic inkjet-patterned memory array based on ferroelectric field-effect transistors', *Organic Electronics*, **12**, 2012–2018.

Noh Y-Y and Sirringhaus H (2009), 'Ultra-thin polymer gate dielectrics for top-gate field-effect transistors', *Organic Electronics*, **10**, 174–180.

Paul S, Pearson C, Molloy A, Cousins M A, Green M, Kolliopoulou S, Dimitrakis P, Normand P, Tsoukalas D and Petty M C (2003), 'Langmuir–Blodgett film deposition of metallic nanoparticles and their application to electronic memory structures', *Nano Lett*, **3**, 533–536.

Pearson C, Ahn, J H, Mabrook M F, Zeze D A, Petty M C, Kamtekar K T, Wang C, Bryce M R, Dimitrakis P and Tsoukalas D (2007), 'Electronic memory device based on a single-layer fluorene-containing organic thin film', *Appl. Phys. Lett.*, **91**, 123506.

Pershin Y V and Di Ventra M (2011), 'Memory effects in complex materials and nanoscale systems', *Adv. Phys.*, **60**, 145–227.

Petty M C (1996), *Langmuir–Blodgett films*, Cambridge, Cambridge University Press.

Petty M C (2007), *Molecular Electronics: From principles to practice*, Chichester, Wiley.

Plastic Electronics (2012), Plastic Electronics Conference. Homepage of 2012 conference in Dresden, Germany. Available from: http://www.plastic-electronics.org/ [accessed 6 December 2012].

Plus Plastic Electronics (2012), Plus Plastic Electronics magazine. Available from: http://www.plusplasticelectronics.com/ [accessed 6 December 2012].

Prince B (2000), *High Performance Memories*, Chichester, Wiley.

Printed Electronics USA (2011), Printed Electronics USA Conference. Homepage of 2011 conference in Santa Clara, USA. Available from: http://www.idtechex.com/printed-electronics-usa-11/pe.asp [accessed 6 December 2012].

Rocha A R, García-Suárez V M, Bailey S W, Lambert C J, Ferrer J and Sanvito S (2005), 'Towards molecular spintronics', *Nature Mater.*, **4**, 335–339.

Roelofs W S C, Mathijssen S G J, Bijleveld J C, Raiteri D, Geuns T C T, Kemerink M, Cantatore E, Janssen R A J and de Leeuw D M (2011), 'Fast ambipolar integrated circuits with poly(diketopyrrolopyrrole-terthiophene)', *Appl. Phys. Lett.*, **98**, 203301.

Rozenberg M J, Inoue I H and Sánchez M J (2004), 'Nonvolatile memory with multilevel switching: a basic model', *Phys. Rev. Lett.*, **92**, 178302.

Rueckes T, Kim K, Joselevich E, Tseng G Y, Cheung C L and Lieber C M (2000), 'Carbon nanotube-based nonvolatile random access memory for molecular computing', *Science*, **289**, 94–97.

Ruitenbeek J V (2005), 'Silver nanoswitch', *Nature*, **433**, 21–22.

Scott J C and Bozano L D (2007), 'Nonvolatile memory elements based on organic materials', *Adv. Mater.*, **19**, 1452–1463.

Sekitani T, Yokota T, Zschieschang U, Klauk H, Bauer S, Takeuchi K, Takamiya M, Sakurai T and Someya T (2009), 'Organic nonvolatile memory transistors for flexible sensor arrays', *Science*, **326**, 1516–1519.

Simmons J G and Verderber R R (1967), 'New conduction and reversible memory phenomena in thin insulating films', *Proc. R. Soc. London, Ser. A.*, **301**, 77–102.

Simon D T, Griffo M S, DiPietro R A, Swanson S A and Carter S A (2006), 'Admittance spectroscopy of polymer-nanoparticle non-volatile memory devices', *Appl. Phys. Lett.*, **89**, 133510.

Sirringhaus H (2009), 'Reliability of organic field-effect transistors', *Adv. Mater.*, **21**, 3859–3873.

Song S, Cho B, Kim T-W, Ji Y, Jo M, Wang G, Choe M, Kahng Y-H, Hwang H and Lee T (2010), 'Three-dimensional integration of organic resistive memory devices', *Adv. Mater.*, **22**, 5048–5052.

Streetman B G and Banerjee S (2000), *Solid State Electronic Devices*, New Jersey, Prentice Hall.

Strukov D B and Kohlstedt H (2012), 'Resistive switching phenomena in thin films: materials, devices and applications', *MRS Bulletin*, **37**, 108–114.

Sun S-J and Dalton L R (Editors) (2008), *Introduction to Organic and Optoelectronic Materials and Devices*, Boca Raton, CRC Press.

Terabe K, Hasegawa T, Nakayama T and Aono M (2005), 'Quantised conductance atomic switch', *Nature*, **433**, 47–50.

Thin Film (2012), Home page of Thin Film Electronics ASA. Available from http://www.thinfilm.no/ [accessed 6 December 2012].

Tseng R J, Huang J, Ouyamg J, Kaner R B and Yang Y (2005), 'Polyaniline nanofiber/gold nanoparticles non-volatile memory', *Nano Lett.*, **5**, 1077–1080.

Verbakel F, Meskers S C J, Janseen R A J, Gomes H L, Cölle M, Büchel M and De Leeuw D M (2007), 'Reproducible resistive switching in non-volatile organic memories', *Appl. Phys. Lett.*, **91**, 192103.

Vettiger P, Despont M, Dreschsler U, Dürig U, Häberle W, Lutwyche M I, Rothuizen H E, Stutz R, Widmer R and Binnig G K (2000), 'The 'Millipede' – more than one thousand tips for future AFM data storage', *IBM J. Res. Dev.*, **44**, 323–340.

Wang F J, Xiong Z H, Wu D, Shi J and Vardeny Z V (2005), 'The case for Fe/Alq_3/Co spin-valve devices', *Synthetic Metals*, **155**, 172–175.

Wolf S A, Chtchelkanova A Y and Treger DM (2006), 'Spintronics – a retrospective and perspective', *IBM Journal of Research and Development*, **50**, 101–110.

Wolf S A, Lu J W, Stan M R, Chen E and Treger D M (2010), 'The promise of nanomagnetics and spintronics for future logic and universal memory', *Proc. IEEE*, **98**, 2155–2168.

Yang Y, Ouyang J, Ma L, Tseng R J-H and Chu C-W (2006), 'Electrical switching and bistability in organic/polymeric thin films and memory devices', *Adv. Funct. Mater.*, **16**, 1001–1014.

Yun Y, Pearson, C, Cadd D H, Thompson R L, Petty M C (2009), 'A cross-linked poly(methyl methacrylate) gate dielectric by ion-beam irradiation for organic thin film transistors', *Organic Electronics*, **10**, 1596–1600.

23
Unconventional molecular scale logic devices

M. OSZAJCA, Jagiellonian University, Poland, A. PODBORSKA, AGH University of Science and Technology, Poland and K. SZACIŁOWSKI, AGH University of Science and Technology, Poland and Jagiellonian University, Poland

DOI: 10.1533/9780857098764.3.654

Abstract: Semiconducting nanoparticles offer a versatile platform for various logic devices. Wide band gap semiconductors modified with molecular species are materials with unique optical and electronic properties. The most intriguing property of such systems is photoelectrochemical photocurrent switching (PEPS) effect. The polarity of photocurrent generated within these materials depends on many variables (light, electrode polarization, redox processes). Materials showing the PEPS effect can be used for construction of simple logic gates and other devices.

Key words: nanoparticles, semiconductors, photo electrochemical photocurrent switching (PEPS) effect, logic devices.

23.1 Introduction

In modern societies based on information technologies the permanent need for more efficient information processing systems is becoming the key issue for continuous development. Therefore leading companies compete with each other by manufacturing constantly smaller electronic components and logic gates on silicon wafers. Eventually this competition may lead to utilization of logic gates based on single molecules.

The story of molecular scale logic devices started a long time ago with the work of Arieh Aviram and Mark Ratner [1, 2] where they suggested application of molecular species for information processing. This idea was embodied by A.P. de Silva, who was the first to report a working molecular logic gate [3]. Now there are thousands of different logic gates and other, more complex, logic devices implemented in chemical systems and processing data encoded in molecules [4–9].

These molecular devices, although very small, are quite impractical. Application of numerous chemical/optical stimuli to molecular processors may rapidly cause crosstalk errors with an increasing number of parallel data streams [10]. On the other hand, application of numerous electrical contacts for a single (even large) molecule is almost impossible – the most sophisticated devices reach nowadays separation of ca. 100 nm between individually actuated electrodes [11].

While it is hard to access devices based on numerous single molecules (just one isolated molecule can be easily addressed by means of scanning probe microscopy), self-assembly of molecules on active surfaces is a tempting compromise. Although this approach does not provide addressing of single molecules, it facilitates signal transduction from the molecular scale to electrical signals that can be recorded using macroscopic devices. Current development of chemistry and physics of semiconducting nanoparticles renders them good candidates for such computing systems. While computational functions performed by such devices are simple and the devices themselves are relatively slow, this is not a serious obstacle against their applications. Recent advances in information processing utilized massively parallel architectures of simple computational modes. The best example are the HTM (hierarchical temporal memory) networks [12, 13], where advanced information processing occurs within large hierarchical network of simple computing nodes. This approach mimics the structure of the cerebral cortex, where information is processed in layered circuits (usually six main layers) which undergo characteristic bottom-up, top-down, and horizontal interactions. Such computational approach does not require high numerical efficiency of a single node, but rather a large number of simple, interconnected nodes.

This idea somehow resembles the concept of 'smart dust', first developed by Kristof Pister in 1997 [14]. The idea is based on a large number of small (~1 mm^3) ubiquitous and autonomous computing devices, equipped with sensor devices, processing unit, memory and communication facility. These devices form, in turn, dynamic mobile networks. This idea was further developed and various sensing and labeling nanostructures became available [15, 16], and finally achieved the complexity of 'lab-on-a-particle' via integration of nanoparticles with functional biomolecules [17–21]. Apart from sensing applications, similar concepts were applied in construction of simple logic devices based on micellar and liposomic structures with embedded molecular sensors, switches and logic gates [22–24].

23.2 Properties of nanoparticles and their applications in molecular scale logic devices

Semiconducting nanoparticles offer a versatile platform for various logic devices. They can be easily prepared in high quantities (and using environmentally friendly protocols) [25] and exhibit tunable electronic [26, 27] and optical properties [28, 29]. Furthermore, owing to very large specific surface area (in the case of small nanoparticles most of the atoms are on the surface) and coordinative unsaturation of surface atoms, semiconductor nanoparticles strongly adsorb molecules on their surfaces [30]. Thus obtained objects combine collective transport properties of solids with structural versatility of molecules [31, 32]. Furthermore, mutual interplay between these two components leads to emergent optical and electrical properties [33]. These specific interactions generate novel properties, which can dominate the immanent properties of components, i.e. nanoparticles and molecules.

Organic and organometallic molecules assembled at semiconducting surfaces modify the surface properties when covalently linked to substrates. These layers can fine-tune the work functions of inorganic material, thereby minimizing the energy barriers for injection or extraction of charge carriers into or out of semiconductor. Furthermore, due to the emergent properties, the interface between inorganic semiconductor and deposited molecules carries the entire device functionality; the interface then essentially becomes the device and the coupling of the molecular energy levels with those of the semiconducting support defines the overall charge-transport characteristics of such hybrid devices [34]. These systems can also be understood as interfaces between the world of molecules and macroscopic systems. Information processing occurs at the molecule–semiconductor interface and this process is governed by the emergent features of semiconductor–electric dipole interactions and interfacial electronic coupling, while the semiconducting support mechanically stabilizes the fragile molecular entities and acts as a charge collector/actuator generating electric output, which in turn can be detected by a user interface. In contrast to classical electronic and optoelectronic devices these novel hybrid systems can respond not only to optical and electronic signals, but also to subtle chemical changes. It opens new possibilities of computing at the nanoscale, including neuromimetic systems. These materials, upon appropriate chemical modification, can constitute an interface between electric devices and neurons and directly monitor chemical synaptic communication [35, 36]. These systems are also quite different from molecular logic gates. While information processing takes place at the molecular level, these systems require Avogadro number of molecules and reading the state of a particular molecule within solution is virtually impossible. The nanoparticle-based systems may utilize a much smaller number of molecules (e.g. complete coverage of 10 nm nanoparticle with small molecules requires ~1500 small organic molecules, but only ~100 bulkier ones). As photocurrent switching phenomenon does not require any interaction between nanoparticles and occurs at single nanoparticle level, in principle it should be possible to confine a simple logic device within a single nanoparticle.

Another advantage of chemically engineered semiconducting nanoparticles is the possibility of advanced control over their morphology by application of appropriate surfactants and/or capping agents [37–39]. These additives modulate the rate of crystal growth of different crystallographic planes and thus affect the shape of the crystallite. Preferential adsorption can be used to induce a systematic shape evolution if relative growth rates along two or more crystallographic axes can be finely tuned. The highest degree of morphology control can be achieved via sequential modulation of growth conditions, especially during the electrodeposition processes [40, 41]. This flexibility allows the synthesis of virtually all possible geometries within single semiconducting (nano)particle. On application of diffusion-controlled growth even higher complexity is easily achievable [42].

However, the main effects responsible for the unique electronic and optical properties of nanocrystals are quantum size effect and quantum confinement. The electric structure of bulk materials can be described by energy bands. These bands can be understood as a continuum of discrete energy levels with negligible energy difference. When the size of crystals decreases to the nanoscale, the electronic structure is no longer described by continuous bands, but by discrete electronic levels characteristic for molecular species. When the radius of the particle decreases, the energy difference between the lowest electron and hole quantum states (highest occupied molecular orbital – lowest unoccupied molecular orbital, HOMO-LUMO, gap) increases. As a result, the optoelectronic properties are very strongly size- and shape-dependent [43, 44]. The very important parameter which describes the relation of the size of semiconducting particle with the degree of quantum confinement is the Bohr radius (r_B) of the exciton defined as (23.1):

$$r_{\mathrm{B}} = \frac{\varepsilon\hbar^2}{\mu e^2}, \qquad [23.1]$$

where ε is the dielectric constant and μ is the reduced mass of the exciton (23.2):

$$\mu = \frac{m_{\mathrm{e}} m_{\mathrm{h}}}{m_{\mathrm{e}} + m_{\mathrm{h}}}, \qquad [23.2]$$

where m_{e} and m_{h} are the effective masses of electron and hole [45]. Depending on the Bohr radius value we can distinguish three different situations which determine the confinement regime (23.3–23.5):

$$r \gg r_{\mathrm{B}} \qquad [23.3]$$

$$r_{\mathrm{e}} > r > r_{\mathrm{h}} \qquad [23.4]$$

$$r \ll r_{\mathrm{B}} \qquad [23.5]$$

r is the radius of the particle while r_e and r_h are the electron and hole Bohr radii given by (23.6 and 23.7):

$$r_e = \frac{\varepsilon \hbar^2}{m_e e^2} \quad [23.6]$$

$$r_h = \frac{\varepsilon \hbar^2}{m_h e^2} \quad [23.7]$$

In the first case the Bohr radius is much smaller than the radius of the particle (23.3) and the system is characterized by weak confinement regime. In this case the binding energy of an exciton is larger than the quantization energies of electrons and holes.

In the opposite situation to that described above, when the particle radius is smaller than the Bohr radius (23.6) the system is in strong confinement regime.

The case of an intermediate confinement regime is a special situation. In such a system an electron moves much more quickly than a hole and as a result the hole is confined in the average potential of a rapidly moving electron in the centre of the nanocrystal. This situation arises if the electron mass is much smaller than the hole mass, and the dot size is small compared with the electron Bohr radius but large compared with that of the hole. The electron is then quantized by the confinement potential and the hole moves adiabatically in the attractive electrostatic potential caused by the electron orbital. In semiconductor composites with intermediate confinement regime optical spectra are dominated by transitions between electron and hole quantized energy levels.

The quantum confinement may vary along different directions depending on the nanocrystal shape [43, 44]. For example the dimensions of a quantum dot are comparable or smaller than the Bohr radius and characterized a small sphere, confining in all three dimensions. The quantum wire confines in two dimensions whereas a quantum well confines in one dimension [43].

Synthesis of the various composite semiconductor nanoclusters is another step to obtain more advanced and useful materials. Linking two or more nanocrystals we get new materials with unique properties [27, 43, 44, 46, 47]. Composite semiconductor nanoclusters can be classified into two categories, namely, capped (core–shell) and coupled (dumbbell) heterostructures. In the former, the confinement effects within the components may be different, which gives rise to five different types of heterostructure nanoparticles (Fig. 23.1).

In semiconductor–semiconductor core–shell heterostructures we can distinguish five different band alignment types, depending on the value of the band gap and relative band edge positions. This in turn provides different spatial confinement modes. Type-I heterostructures (Fig. 23.1a) are

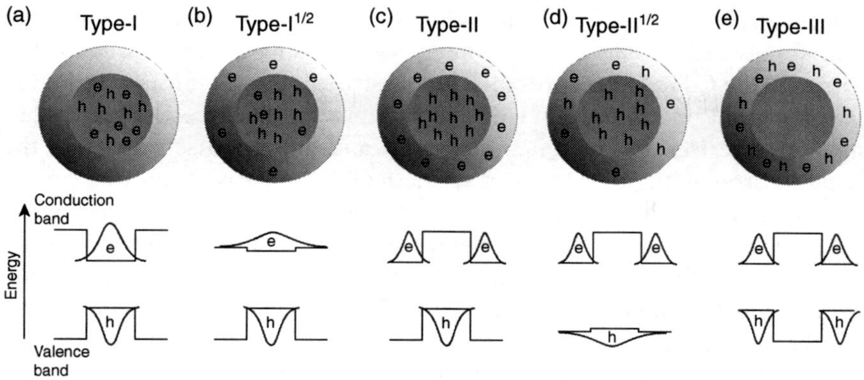

23.1 Different types of the core–shell semiconductor heterostructures with schematic representation of charge carrier confinement and the radial probability functions for the lowest energy electron (e) and hole (h) wave functions.

combining one narrow band gap and one wide band gap semiconductor [27, 43]. If a narrow gap semiconductor is covered by a shell of semiconductor with wide band gap, both electron and hole wave functions will be strongly confined to the core (Fig. 23.1a), which provides the lowest energy states for both electrons and holes. Moreover, the excitons in the core are protected from interaction with the surface and the environment. This type of heterostructure can be fabricated in the CdSe/ZnS, InAs/ZnS and InP/ZnS systems. Confinement of the exciton to the inner sphere results in high quantum efficiency and stability of photo- and electroluminescence.

In some cases, for example in the CdSe/CdS core–shell nanostructures, the band gap of the shell material is not wide enough to confine both electron and hole wave functions. In this case the electron can easily move between CdSe and CdS phases, whereas the hole is confined to the CdSe core due to its large effective mass and substantial offset of the valence band energies (Fig. 23.1b). This system is called type-$I^{1/2}$ (also known as *quasi* type-II) [27, 43].

In type-II heterostructures the energies of the conduction and valence bands of the shell are either both higher or both lower than those of the core. As a result, one of the carriers is mostly confined in the core of the heterostructures, while the other one remains in the shell (Fig. 23.1c). Their effective band gaps are heavily affected by the band offsets of the cores and shells. In the lowest excited state, the electron and hole wave functions are spatially separated, which results in the electron wavefunction mainly residing in one semiconductor, whereas the hole wavefunction is localized more in the second semiconductor. The e–h recombination transition energy is smaller than the band gap of either of the constituent material components.

Therefore, radiative recombination of the e–h pair produces a red-shifted emission, which would not be available with a single semiconductor. For example the CdTe/CdSe core–shell heterostructures emit infrared radiation which cannot be achieved in CdTe or CdSe materials. Type-II nanocrystals are characterized by low quantum yields and enhanced sensitivity to the local environment. Low oscillator strength of the radiative transition caused by reduced overlap of the electron and hole wavefunctions following the relaxation of the carriers to the band edge leads to longer lifetimes of the excited state. This characteristic of type-II heterostructures suggests that they could be highly attractive for applications in photovoltaics, where charge separation in space domain is desirable.

In the previous examples the narrow-band gap semiconductor was covered with a layer of semiconductor with a wider band gap. The opposite situation is, however, also possible and leads to interesting phenomena. Here one can distinguish two additional types of confined heterostructures. The most typical case can be observed when the valence band edges are of comparable energies, but the conduction band edge of the core semiconductor is higher than the shell one. Such system is called type-$II^{1/2}$ (or 'inverted' type-II) and it was observed in ZnSe/CdSe nanocrystals [48]. This heterostructure can only support partial spatial separation between electrons and holes. As a result electrons are confined in the shell region while holes are delocalized over the whole volume of the heterostructure (Fig. 23.1d). Such situation leads to reduced (but nonzero) overlap between the electron and hole wavefunctions. ZnSe/CdSe heterostructures is characterized by high emission of quantum yields, even 80–90% [48].

It is possible also to obtain such a system in which both electron and hole are localized in the shell of the heterostructure (Fig. 23.1e). This example can be called type-III heterostructure. In principle such heterostructures should behave just like empty shells. An additional effect may, however, result from plasmonic effects, as the plasmon oscillation of an empty shell and a core–shell system will be very different.

Furthermore, chemical properties of nanoparticle surfaces (or their surface molecules) can be used to modulate their ability to self-assemble into complex structures, including circuit-like assemblies [49–51]. Anchoring appropriate molecules may facilitate self-assembly via the switching-on of various molecular interactions, including electrostatic, van der Waals, magnetic, and hydrophobic interactions. Before developing a self-assembly system one should consider the general rules of designing self-assembling components. Firstly these components should be agitated either by inherent thermal noise at small scales, or by external means at larger scales, usually to overcome strong surface forces. Secondly, the assembling particle needs to recognize only specific interactions and interact in proper directions to result in a desired structure. Additionally, a self-assembling system exhibits the most structural diversity when the energy is more or less equally partitioned between the interactions present in this system. Therefore one should look for self-assembled structures that are not dominated by one specific interaction as it leads to greater control over their shape and morphology.

23.3 Photoelectrochemical photocurrent switching (PEPS) effect

The most intriguing emergent property of molecule-semiconductor systems is photoelectrochemical photocurrent switching (PEPS) effect. This effect, initially regarded as a scientific curiosity, has become a field of intense research [52]. In addition to the classic p-n, p-i-n or Schottky junctions, which generate photocurrent on illumination and changes in bias potential or incident photon [53], in the PEPS effect also the polarity of photocurrent can be easily changed from anodic to cathodic and vice versa. Since first being described in 2006, numerous materials have been shown to exhibit this kind of behavior, but the most typical ones are wide band gap semiconductors (TiO_2, CdS) with surfaces modified with appropriate molecular species. Surface binding, redox and optical properties of these surface species as well as the degree of electronic coupling between surface and bulk of the hybrid materials are key factors influencing the performance of PEPS-based devices.

The effect of photocurrent switching depends on many variables which should be considered in order to fully understand the nature of this phenomenon. They involve the redox properties of a semiconductor surface and its bulk, the availability of electron donors and acceptors in the electrolyte solution, and certainly the applied potential and the energy of incident light. The direction of photogenerated net currents is a derivative of competition between various redox processes – anodic photocurrents are observed when oxidation reactions prevail at the working electrode while cathodic photocurrents require good efficiencies of reduction processes.

The key point of all the photocurrent switching devices is the PEPS effect. Due to specific electronic structure, resulting from surface modification, these materials are able to generate both anodic and cathodic photocurrent, while formally they are n-type semiconductors. The changes observed on chemisorption of cyanoferrate complexes onto titanium dioxide resemble those related to the surface transfer doping [54]. Cyanoferrate moieties interact with empty surface states and serve as electron donors. Strong titanium–nitrogen bonding facilitates formation of two-dimensional surface complexes with the HOMO–LUMO gap smaller than the band gap of the neat semiconductor (Fig. 23.2). The strength of this interaction can be quantified by the shift in the redox potential of iron species on binding to the

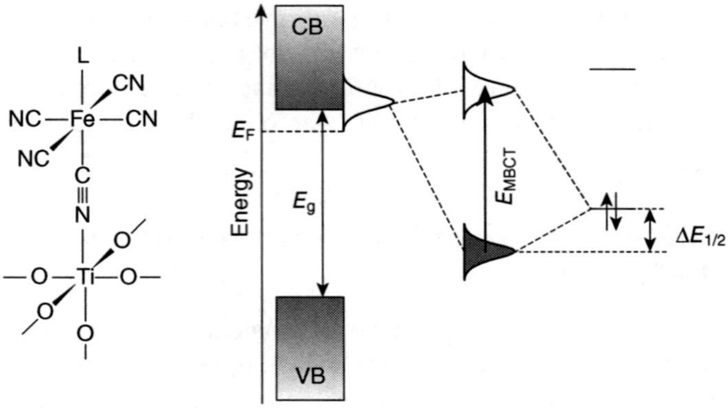

23.2 Tentative structure of the surface complex and its electronic structure. Cyanoferrate anions are bound to the TiO_2 surface via axial cyano ligand (left). Interaction of the HOMO orbital of the complex with empty d orbitals of titanium constituting the surface states results in formation of new energy levels of bonding (localized mostly on iron complex) and antibonding (localized at titanium centres) character. $\Delta E_{½}$ is the change of the redox potential of the iron complex due to surface binding.

surface [55, 56]. Furthermore, the LUMO level of the surface complex formally belongs to the conduction band. Therefore excitation of the surface complex within the MBCT (*metal-to-band charge transfer*) transition leads to photocurrent generation. The photocurrent switching phenomena originate from the ability of iron complexes to change reversibly their oxidation state. Iron(II) species act as efficient electron donors, while iron(III) does not. Furthermore, at sufficiently low photoelectrode potential the cyanoferrate layer prevents electron injection into the conducting support and only cathodic photocurrents can be observed. This effect is even more pronounced in the case of ferrocene-modified titania [57]. The photocurrent switching mechanism is shown in Fig. 23.3.

Similar behavior can be seen for the organic chromophores (anthraquinone derivatives [59], folic acid [60], carminic acid [61]), and organometallic species (e.g. ferrocene [57]) adsorbed at the TiO_2 surface. Depending on the photoelectrode potential they can donate an electron to the conduction band or to the sacrificial electron acceptor in the electrolyte, thus changing the photocurrent direction.

Analogous behavior was also observed in the case semiconducting materials modified within the bulk, where the acceptor levels are located just below the edge of the conduction band [62]. Sulfur doped cadmium sulfide is a good example of such a system. The dopant levels associated with sulfur interstitials are located within the band gap (Fig. 23.4).

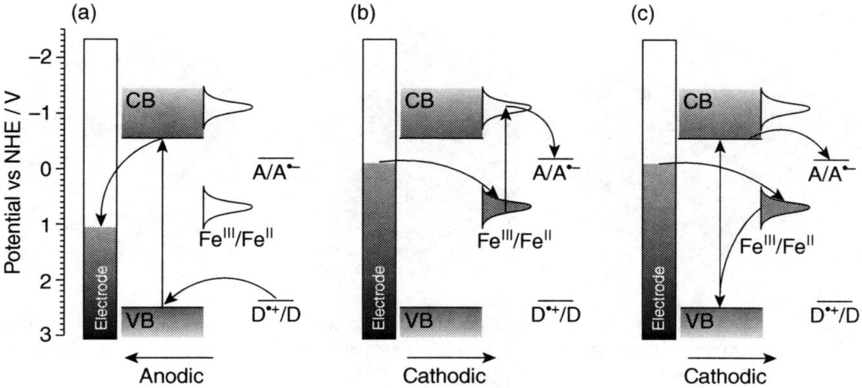

23.3 The mechanism of photocurrent generation at electrodes made of TiO_2 modified with iron cyanide complexes upon UV (a, c) and visible (b) light generation. High electrode potentials (a) ensure presence of Fe^{III} surface species, while Fe^{II} prevails at low potentials (b, c). NHE stands for normal hydrogen electrode. Adapted from [58].

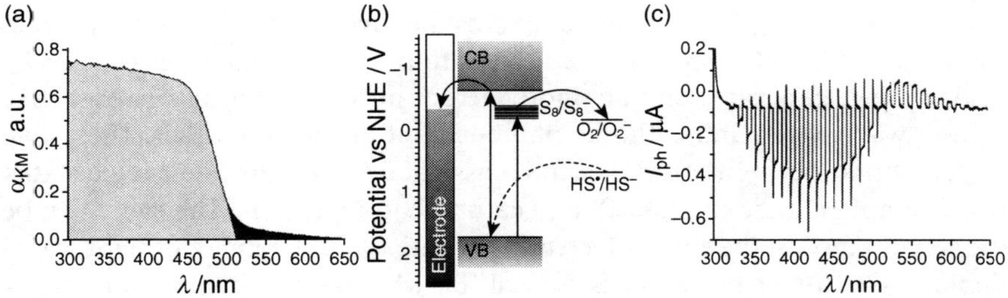

23.4 Diffuse reflectance spectrum of sulfur-doped CdS nanopowder, the sub-band gap absorption is marked in dark gray (a), schematic of the photocurrent generation mechanism (b) and photocurrent action spectrum recorded at switching potential (c). Adapted from [62].

At appropriate potential (cf. Fig. 23.4b) excitation within fundamental transition results in anodic photocurrent. At the same time excitation of the sub-band gap transition, which involves the interstitial sulfur atoms, results in cathodic photocurrents. Detailed potential-dependent photocurrent profiles are shown in Fig. 23.5. This peculiar photocurrent generation phenomenon can be used for construction of various logic gates.

23.4 Logic devices based on photoelectrochemical photocurrent switching (PEPS) effect

Any chemical system which can exist in at least two stable states can be regarded as a molecular switch. In the case of systems which show the PEPS

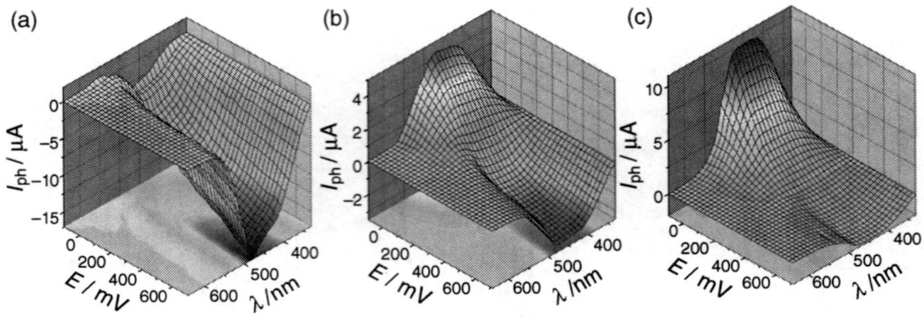

23.5 Potential-dependent photocurrent action spectra of stoichiometric cadmium sulfide (a) and of sulfur doped CdS at lower (b) and higher (c) dopant loading. Adapted from [62].

effect the wavelength of incident light and photoelectrode potential can be used as switching stimuli [4, 63, 64]. For example the photoelectrode made of folic acid modified TiO_2 could be used to construct simple logic gates [60, 65]. In order to analyze the photoelectrochemical behavior in terms of Boolean logic the photocurrent map must be divided into zones of different photoactivity; in the current case it gives four 2×2 subdiagrams (Fig. 23.6). This system generates anodic photocurrent at high potentials irrespective of the wavelength and cathodic photocurrent at low potentials. These two types of photocurrent outputs can be assigned to two different logic states, for example anodic to logic '0' and cathodic to logical '1'. The same can be done with inputs: the photoelectrode potential and the incident light wavelength. The lower potential is logical '0', whereas the higher potential – logical '1'. Analogously the lower and higher wavelength regions were assigned to logic '0' and '1'. The four subdiagrams show four different logic devices: PASS 1, YES, INH and IMP. Depending on the range of potentials and wavelength applied, this system can work in four different modes. It is therefore possible to get different functions in the same system.

More complex behavior, leading also to more advanced Boolean functionality, was observed in the case of titanium dioxide modified with $[Fe(CN)_6]^{4-}$ complexes (or other cyanoferrates) [66]. Competition between cathodic and anodic photocurrent in this system can be applied to mimic the XOR logic gate (Fig. 23.7). It was the first example of a optoelectronic XOR gate operated by optical inputs and electrical outputs [66]. Illumination with violet (400 nm) and/or blue (460 nm) light sources caused the cathodic, anodic, or null photoresponses of the system, depending on the sequence of pulses and photoelectrode polarization. At the positive polarization of the electrode, when the surface complex was oxidized, only the anodic photocurrent was observed, but applying a negative potential (reduced the complex) results in a cathodic photocurrent. The interesting

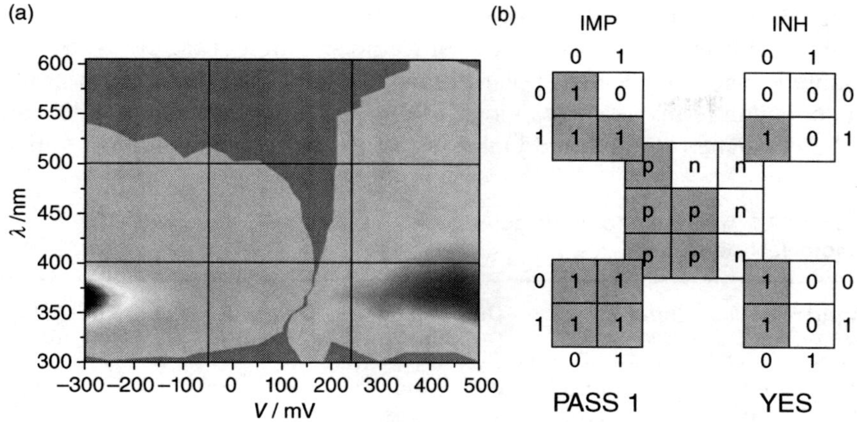

23.6 Maps of photocurrent generation for folic acid modified TiO_2 photoelectrode (a) and diagram of logic gates (b).

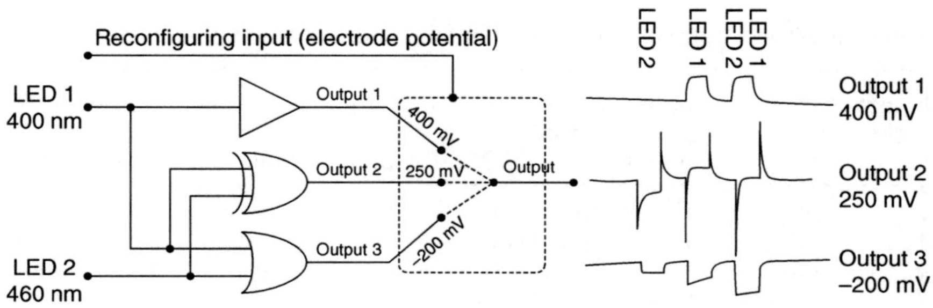

23.7 Programmable logic system based on $[Fe(CN)_6]^{4-}$ modified TiO_2 photoelectrode.

effect was observed at potential of 250 mV, under conditions of partial reduction of the surface complex. Pulsed irradiation with violet light results in an anodic photocurrent while irradiation with blue light generates the cathodic photocurrent. Moreover when the system is illuminated with both light sources no photocurrent was observed. To analyze the system in term of Boolean logic it was necessary to assign logic values to input and output signal. Logical '0' and '1' were assigned to OFF and ON states of the lights. Two different wavelengths correspond to two different inputs of the logic gates. In analogous way the logical '0' was assigned for the state when the photocurrent was not generated and logic '1' when the photocurrent (anodic or cathodic) was observed (Table 23.1).

In the case of sulfur-doped CdS it was possible to mimic two different complex systems: 1 : 2 demultiplexer [67] and binary half-adder [68]. The implementation of logic devices on the PEPS-based platform is simple

(Fig. 23.8). Sulfur-doped CdS generates anodic photocurrents on the excitation of fundamental transition at positive polarization of the electrode, but cathodic ones at negative polarization (cf Fig. 23.5). This can be easily described in terms of Boolean logic (Table 23.2). Boolean values of '0' and '1' are assigned to OFF and ON states of the light source and to positive

Table 23.1 True table for logic gates based on $[Fe(CN)6]^{4-}$ modified TiO2 photoelectrode

Input 1 (460 nm)		**Input 2** (400 nm)		**Output 1** (–200 mV)		**Output 2** (250 mV)		**Output 3** (400 mV)	
0	(off)	**0**	(off)	**0**	(no)	**0**	(no)	**0**	(no)
1	(on)	**0**	(off)	**1**	(yes)	**1**	(yes)	**0**	(no)
0	(off)	**1**	(on)	**1**	(yes)	**1**	(yes)	**1**	(yes)
1	(on)	**1**	(on)	**1**	(yes)	**0**	(no)	**1**	(yes)

Table 23.2 The truth table of the 1 : 2 optoelectronic demultiplexer

Input1 (light)		**Input 2** (photoelectrode potential)		**Output 1** (anodic photocurrent)		**Output 2** (cathodic photocurrent)	
0	(off)	**0**	(positive)	**0**	(no)	**0**	(no)
1	(on)	**0**	(positive)	**1**	(yes)	**0**	(no)
0	(off)	**1**	(negative)	**0**	(no)	**0**	(no)
1	(on)	**1**	(negative)	**0**	(no)	**1**	(yes)

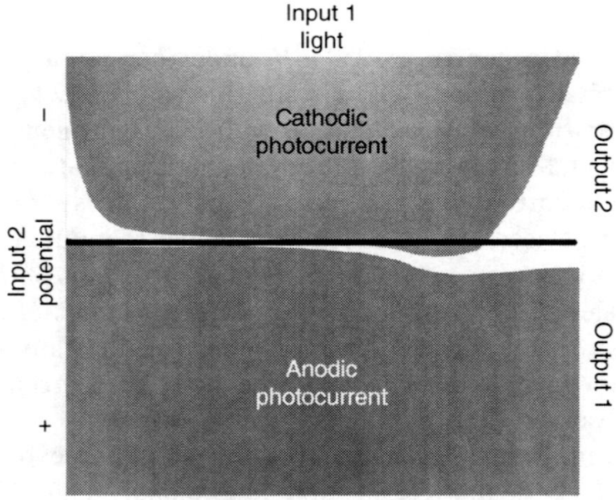

23.8 The phase diagram of the S-doped CdS photoactivity.

and negative polarization of the photoelectrode, respectively. The same concerns the output: anodic and cathodic photocurrents were assigned to Boolean signals of '1' in two different output channels, while Boolean '0' was assigned to null photocurrent. Upon these assumptions the CdS photoelectrode behaved as a 1 : 2 demultiplexer: input signal (light pulse) was directed to one of the two outputs in the form of electric pulse. The systems presented here is one of the smallest demultiplexers and is definitely the smallest semiconductor-based device. While experiments were performed on much larger devices (containing billions of nanoparticles), the photocurrent switching does not require any interactions between individual particles. Therefore the operation of the demultiplexer can be reduced to a single CdS nanoparticle, in this particular case of a diameter of 7 nm [67].

Even more complex devices have been constructed when aqueous electrolyte was replaced with ionic liquid electrolyte [68]. The basic unit was assembled using two ITO electrodes (one covered with thin film of CdS) sandwiching a layer of semisolid electrolyte ((polyoxyethene[20] hexadecyl ether, methyltridodecylammonium chloride and tetrabutylammonium iodide). Such a device behaves as a Schottky photodiode. The combination of two photodiodes gives AND (Fig. 23.9 a) or XOR (Fig. 23.9 b) logic gates with optical input (456 nm light from high power light-emitting, diode, LED) and electrical (photocurrent) output. Photocurrent profiles recorded during pulsed illumination are shown in the middle section of Fig. 23.9. Connection of these two logic gates in parallel and sharing the input light sources resulted in optoelectronic binary half adder. The system was able to take two bits in the form of light pulses and yield the arithmetic sum of these data in the form of photocurrent pulses on two outputs of the device [68]. It is important to note that this system works without external polarization and power supply. Application of external bias potential may result in more complex switching patterns and may lead to ternary logic. This

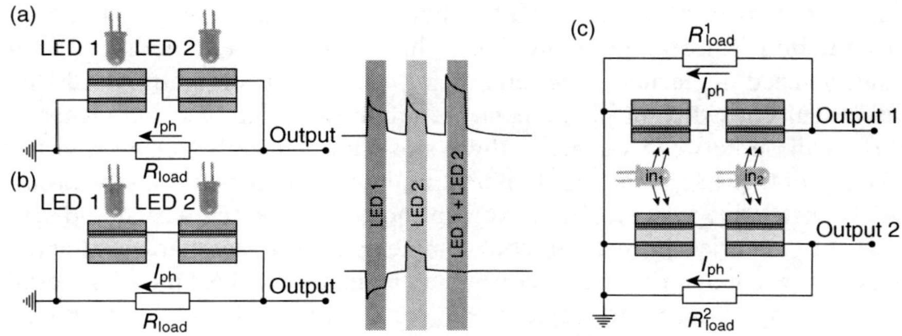

23.9 Connection diagrams of AND (a) and XOR (b) logic gates along with their photocurrent responses. These gates can be wired to built binary half-adder (c).

system provides a universal platform for the construction of optoelectronic logic devices based on unconventional semiconducting materials. Appropriate selection of semiconductors (many different materials can be used in this type of devices) and appropriate connection of simple two-electrode photoelectrochemical cells may lead to an infinite diversity of logic systems. Table 23.3 presents logic devices reported to date, which are based on the photocurrent switching phenomenon. In some cases different Boolean functions are obtained by appropriate choice of output thresholds, reconfiguration by potential pulses, or by specific design of the device, e.g. layer sequence and surface modifiers.

23.5 Conclusions and future trends

Wide band gap semiconductors (oxides and chalcogenides) modified with molecular species are unique materials. Owing to mutual interaction between the lattice and the molecular species they show peculiar emergent properties, among which the PEPS effect is perhaps the most intriguing. While the conductivity type is not affected by surface modification (i.e. the Fermi level potential and doping density remain almost intact), their photoelectrochemical properties are completely changed. Polarity of photocurrent generated on such electrodes is a function of conducting substrate potential, incident light wavelength and electrolyte composition. Therefore these materials are naturally suited for application in chemical sensors and switching devices, including Boolean logic gates. Information-processing characteristics of devices based on hybrid semiconductors are usually not the result of complex structural engineering (as in the case of silicon electronic devices) but rather molecular-scale tailoring and fine-tuning of electronic properties of interfaces. Their performance is defined by the nature of the chemical species adsorbed at the surface (or immobilized within the nanocrystals) and is a result of a subtle interplay between the components of the system. Furthermore, their function can be modified by simple chemical methods, for example, oxidation, reduction, or covalent modification of immobilized molecules. The underlying physics is straightforward: the photocurrent switching phenomena are simply the result of thermodynamic competition between various pathways of photoinduced electron transfer. This simple principle offers an almost unlimited number of optoelectronically switchable systems. The next step should involve the design and synthesis of materials presenting more complex photocurrent maps, containing several switching processes occurring in one material. This kind of material would give an access to ternary logic (to date, there is only one systems reported which may qualify for ternary switching [92]). This is not a trivial task, as surface complexes with several stable redox states do not provide the desired functionality. Another important issue involves integrating

Table 23.3 Molecular logic devices, based on photocurrent switching phenomenon, reported to date

System	Switching stimuli	Logic device	References
2 MIM (metal–insulator–metal) structures	λ^a, E^b	AND, OR	[69, 70]
Multilayer structures based on dye-modified *tert*-pentacrylamide deposited on Au surface	λ	AND, XOR	[71, 72]
Helical oligopeptide chains with ruthenium bipyridine and carbazole chromophores	λ	XOR	[73, 74]
Self assembled monolayers of pentaaryl[60]fullerene	R^c	XOR	[75]
Palladium(II) phthalocyanine and $[Ru(bpy)_3]^{2+}$ derivatives on Au electrodes	λ, R	XOR	[76, 77]
TiO_2 modified with hexacyanoferrates	λ, E, R	XOR/OR/AND/YES/DEMUX[d]	[55–58, 63, 66, 78]
TiO_2 modified with folic or carminic acid	λ, E, R	OR/AND/DEMUX	[60, 61]
TiO_2 modified with polyvinylcarbazole and carotenoid	λ, R	XOR/AND/DEMUX	[79]
CdSe/ZnS nanoparticles and *i*-motif motor DNA on Au electrode	λ, R, pH	AND	[80]
CdS nanoparticles assembled with double-stranded DNA	λ, DNA intercalators	AND	[81]
CdS nanoparticles and Cu^{2+} in polyacrylic acid matrix	λ, E, R	AND	[82]
Se, CdTe, CdSe, CdS	λ, E	DEMUX	[62, 83–88]
PbS nanoparticles	λ, R	DEMUX	[89]
CdS nanoparticles embedded in poly-(diallydimethyl-ammonium chloride)	λ, E, R	DEMUX	[90]
Films of poly(*N*-dodecylacrylamide) modified with $Ru(bpy)_3^{2+}$ and ferrocene	λ, R	OR/XOR/AND	[91]
CdS/ionic liquid/iodide	λ	AND/XOR/HA	[68]
TiO_2 modified with ruthenium complexes	λ, R	AND/OR/DEMUX/ ternary logic devices	[92]

[a] Incident light wavelength.
[b] Photoelectrode potential.
[c] Sacrificial reagents.
[d] 1 : 2 Demultiplexer.

memory functions with Boolean switching. This should result in a new quality of molecular computing devices. The real problem, however, is not to find the system of more complex switching characteristics, but to perform the switching at an isolated nanoparticle and subsequently wire several nanoparticles within a more complex circuit. In such a circuit all nanoparticles should be independently stimulated with light and/or electrical stimuli. This will lead to devices with unprecedented integration scale, complexity and performance.

23.6 Acknowledgments

Financial support from AGH University of Science and Technology within the contract No. 11.11.180.509/11, Ministry of Science and Higher Education (grants Nos. UMO-2011/03/B/ST5/01495, 1609/B/H03/2009/36 and 0117/B/H03/2010/38), National Centre for Research and Development (grants No. NCBiR/ENIAC-2009-1/1/2010) and European Nanoelectronics Initiative Advisory Council ENIAC (grant No. 120122-MERCURE) is gratefully acknowledged. A.P. thanks the Foundation for Polish Science for the research grant within the VENTURES initiative. M.O. thanks Foundation for Polish Science for the MPD Programme fellowship co-financed by the EU European Regional Development Fund.

23.7 References

1. Aviram A. Molecules for memory, logic and amplification. *J Am Chem Soc*. 1988;**110**:5687–5692.
2. Aviram A, Ratner MA. Molecular rectifiers. *Chem Phys Lett*. 1974;**29**:277–283.
3. de Silva AP, Gunaratne HQN, McCoy CP. A molecular photonic AND gate based on fluorescent signalling. *Nature*. 1993;**364**:42–44.
4. Szaciłowski K. Digital information processing in molecular systems. *Chem Rev*. 2008;**108**:3481–3548.
5. Pischel U. Chemical approaches to molecular logic elements for addition and subtraction. *Angew Chem Int Ed*. 2007;**46**:4026–4040.
6. Pischel U. Digital operations with molecules – advances, challenges, and perspectives. *Aust J Chem*. 2010;**63**:148–164.
7. Pischel U, Hellerc B. Molecular logic devices (half-subtractor, comparator, complementary output circuit) by controlling photoinduced charge transfer processes. *New J Chem*. 2008;**32**:369–556.
8. Amelia M, Zou L, Credi A. Signal processing with multicomponent systems based on metal complexes. *Coord Chem Rev*. 2010;**254**:2267–2280.
9. Balzani V, Credi A, Venturi M. Processing energy and signals by molecular and supramolecular systems. *Chem Eur J*. 2008;**14**:26–39.
10. Privman V, Strack G, Solenov D, Pita M, Katz E. Optimization of enzymatic biochemical logic for noise reduction and scalability: how many biocomputing gates can be interconnected in a circuit? *J Phys Chem B*. 2008;**112**:11777–11784.

11. Joachim C, Martrou D, Rezeq M, Troadec C, Jie D, Chandrasekhar N, Gautier S. Multiple atomic scale solid surface interconnects for atomic circuits and molecule logic gates. *J Phys: Condens Matter*. 2010;**22**:084025.
12. George D, Jaros B. *The HTM Learning Algorithms*. Numenta Inc Report. 2007.
13. Hawkins J, George D. *Hierarchical Temporal Memory. Concepts, Theory, and Terminology*. Numenta Inc Report. 2006.
14. Warneke B, Last M, Liebowitz B, Pister KSJ. Smart dust: Communicating with a cubic-millimeter computer. *Computer*. 1997;**34**:2–9.
15. Link JR, Sailor MJ. Smart dust: Self-assembling, self-orienting photonic crystals of porous Si. *Proc Natl Acad Sci*. 2003;**100**:10607–10610.
16. Sailor MJ, Link JR. Smart dust: nanostructured devices on a grain of sand. *Chem Commun*. 2005:1375–1383.
17. Katz E, Willner I. Integrated nanoparticle-biomolecule hybrid systems: synthesis, properties and applications. *Angew Chem Int Ed*. 2004;**43**:6042–6108.
18. Katz E, Willner I, Wang J. Electroanalytical and bioelectroanalytical systems based on metal and semiconductor nanoparticles. *Electroanalysis*. 2004;**16**: 19–43.
19. Burns A, Ow H, Wiesner U. Fluorescent core-shell silica nanoparticles: towards 'Lab-on-a-particle' architectures for nanobiotechnology. *Chem Soc Rev*. 2006;**35**: 1028–1042.
20. Burns A, Sengupta P, Zedayko T, Baird B, Wiesner U. Core/shell fluorescent silica nanoparticles for chemical sensing: towards single-particle laboratories. *Small*. 2006;**2**:723–726.
21. Gill R, Zayats M, Willner I. Semiconductor quantum dots for bioanalysis. *Angew Chem Int Ed*. 2008;**47**:7602–7625.
22. Uchiyama S, McClean GD, Iwai K, de Silva AP. Membrane media create small nanospaces for molecular computation. *J Am Chem Soc*. 2005;**127**:8920–8921.
23. de Silva AP, Dobbin CM, Vance TP, Wannalerse B. Multiply reconfigurable 'plug and play' molecular logic via self-assembly. *Chem Commun*. 2009:1386–1388.
24. Pallavicini P, Diaz-Fernandez YA, Pasotti L. Micelles as nanosized containers for the self-assembly of multicomponent fluorescent sensors. *Coord Chem Rev*. 2009;**253**:2226–2240.
25. Dahl JA, Maddux BLS, Hutchison JE. Towards greener nanosynthesis. *Chem Rev*. 2007;**107**:2228–2269.
26. Zabet-Khadousi A, Dhirani A. Charge transport in nanoparticle assemblies. *Chem Rev*. 2008;**108**:4072–4124.
27. Talapin DV, Lee J-S, Kovalenko MV, Shevchenko EV. Prospects of colloidal nanocrystals for electronic and optoelectronic applications. *Chem Rev*. 2010; **110**:389–458.
28. Li J, Zhang JZ. Optical properties and applications of hybrid semiconductor nanomaterials. *Coord Chem Rev*. 2009;**253**:3015–3041.
29. Smith AM, Nie S. Semiconductor nanocrystals: structure, properties and band gap engineering. *Acc Chem Res*. 2010;**43**:190–200.
30. Tao F, Bernasek SL, Xu G-Q. Electronic and structural factors in modification and functionalization of clean and passivated semiconductor surfaces with aromatic systems. *Chem Rev*. 2009;**109**:3991–4024.
31. Ashkenasy C, Cahen D, Cohen R, Shanzer A, Vilan A. Molecular engineering of semiconductor surfaces and devices. *Acc Chem Res*. 2002;**35**:121–128.

32. Vilan A, Cahen D. How organic molecules can control electronic devices. *Trends Biotechnol*. 2002;**20**:22–29.
33. Lebedev MV. Surface modification of III–V semiconductors: chemical processes and electronic properties. *Progr Surf Sci*. 2002;**70**:153–186.
34. Heimel G, Romener L, Zojer E, Bredas J-L. The interface energetics of self-assembled monolayers on metals. *Acc Chem Res*. 2008;**41**:721–729.
35. Cooper DR, Nadeau JL. Nanotechnology for *in vivo* neuroscience. *Nanoscale*. 2009;**1**:183–200.
36. Kotov NA, Winter JO, Clements IP, Jan E, Timko BP, Campidelli S, Pathak S, Mazzatenta A, Lieber CM, Prato M, Bellamkonda RV, Silva GA, Kam NWS, Patolsky F, Ballerini L. Nanomaterials for neural interfaces. *Adv Mater*. 2009;**21**:3970–4004.
37. Kumar S, Nann T. Shape control of II–VI semiconductor materials. *Small*. 2006;**2**:316–329.
38. Pileni MP. Control of the size and shape of inorganic nanocrystals at various scales from nano to macrodomains. *J Phys Chem C*. 2007;**111**:9019–9038.
39. Viswanath B, Kundu P, Halder A, Ravishankar N. Mechanistic aspects of shape selection and symmetry breaking during nanostructure growth by wet chemical methods. *J Phys Chem C*. 2009;**113**:16866–16883.
40. Siegfried MJ, Choi K-S. Directing the architecture of cuprous oxide crystals during electrochemical growth. *Angew Chem Int Ed*. 2005;**44**:3218–3223.
41. Choi K-S. Shape control of inorganic materials via electrodeposition. *Dalton Trans*. 2008:5432–5438.
42. Grzybowski BA. *Chemistry in Motion: Reaction–Diffusion Systems for Micro- and Nanotechnology*. Chichester: John Wiley & Sons Ltd. 2009.
43. de Mello Donega C. Synthesis and properties of colloidal heteronanocrystals. *Chem Soc Rev*. 2011;**40**:1512–1546.
44. Alivisatos AP. Perspectives on the physical chemistry of semiconductor nanocrystals. *J Phys Chem*. 1996;**100**:13226–13239.
45. Burda C, Chen X, Narayanan R, El-Sayed MA. Chemistry and properties of nanocrystals of different shapes. *Chem Rev*. 2005;**105**:1025–1102.
46. Kaprov SV, Mikushev SV. Electron-hole excitation in CdSe quantum dots under strong and intermediate confinement conditions. *Phys Solid State*. 2010;**52**(8): 1750–1756.
47. Laheld UEH, Pedersen FB, Hemmer PC. Excitons in type-II quantum dots: finite offsets. *Phys Rev B*. 1995;**52**(4):2697.
48. Ivanov S, Piryatinski A, Nanda J, Tretiak S, Zavadil K, Wallace W, Werder D, Klimov V. Type-II core/shell CdSe/ZnSe nanocrystals: synthesis, electronic structures and spectroscopic properties. *J Am Chem Soc*. 2007;**129**:11708–11719.
49. Bishop KJM, Wilmer CE, Soh S, Grzybowski BA. Nanoscale forces and their use in self-assembly. *Small*. 2009;**5**:1600–1630.
50. Grzybowski BA, Wilmer CE, Kim J, Browne KP, Bishop KJM. Self-assembly: from crystals to cells. *Soft Matter*. 2009;**5**:1110–1128.
51. Gracias DH, Tien J, Breen TL, Hsu C, Whitesides GM. Forming electrical networks in three dimensions by self-assembly. *Science*. 2000;**289**:1170–1172.
52. Gawęda S, Podborska A, Macyk W, Szaciłowski K. Nanoscale optoelectronic switches and logic devices. *Nanoscale*. 2009;**1**:299–316.
53. Li SS. *Semiconductor Physical Electronics*. New York: Springer Science+Business Media LLC 2006.

54. Chen W, Qi D, Gao X, Tye Shen Wee A. Surface transfer doping of semiconductors. *Progr Surf Sci*. 2009;**84**:279–321.
55. Hebda M, Stochel G, Szaciłowski K, Macyk W. Optoelectronic switches based on wide bandgap semiconductors. *J Phys Chem B*. 2006;**110**:15275–15283.
56. Szaciłowski K, Macyk W, Hebda M, Stochel G. redox-controlled photosensitization of nanocrystalline titanium dioxide. *Chem Phys Chem*. 2006;**7**:2384–2391.
57. Macyk W, Stochel G, Szaciłowski K. Photosensitization and photocurrent switching effect in nanocrystalline titanium dioxide functionalized with iron(II) complexes: a comparative study. *Chem Eur J*. 2007;**13**:5676–5687.
58. Szaciłowski K, Macyk W. Photoelectrochemical photocurrent switching effect: a new platform for molecular logic devices. *Chimia*. 2007;**61**:831–834.
59. Di Iorio Y, Rodríguez HB, San Román E, Grela MA. Photoelectrochemical behavior of alizarin modified TiO_2 films. *J Phys Chem C*. 2010;**114**:11515–11521.
60. Gawęda S, Stochel G, Szaciłowski K. Bioinspired nanodevice based on the folic acid/titanium dioxide system. *Chem Asian J*. 2007;**2**:580–590.
61. Gawęda S, Stochel G, Szaciłowski K. Photosensitization and photocurrent switching in carminic acid/titanium dioxide hybrid material. *J Phys Chem C*. 2008;**112**:19131–19141.
62. Podborska A, Gaweł B, Pietrzak Ł, Szymańska IB, Jeszka JK, Łasocha W, Szaciłowski K. Anomalous photocathodic behavior of CdS within the Urbach tail region. *J Phys Chem C*. 2009;**113**:6774–6784.
63. Szaciłowski K, Macyk W. Chemical switches and logic gates based on surface modified semiconductors. *Comp Rend Chimie*. 2006;**9**:315–324.
64. Balzani V, Venturi M, Credi A. *Molecular Devices and Machines – A Journey into Nanoworld*. Weinheim: Wiley-VCH 2003.
65. Gawęda S, Kowalik R, Kwolek P, Macyk W, Mech J, Oszajca MF, Podborska A, Szaciłowski K. Nanoscale digital devices based on the photoelectrochemical photocurrent switching effect: preparation, properties and applications. *Isr J Chem*. 2011;**51**:36–55.
66. Szaciłowski K, Macyk W, Stochel G. Light-driven OR and XOR programmable chemical logic gates. *J Am Chem Soc*. 2006;**128**:4550–4551.
67. Podborska A, Szaciłowski K. Towards 'Computer-on-a-particle' devices: optoelectronic 1:2 demultiplexer based on nanostructured cadmium sulfide. *Aust J Chem*. 2010;**63**:165–168.
68. Mech J, Kowalik R, Podborska A, Kwolek P, Szaciłowski K. Arithmetic device based on multiple Schottky-like junctions. *Aust J Chem*. 2010;**63**:1330–1333.
69. Choi J-W, Nam YS, Oh SY, Lee WH, Fujihira M. 'AND' logic function of molecular photodiode consisting of GFP/TCNQ hetero-film. *Mol Cryst Liq Cryst*. 2002;**377**:249–252.
70. Choi J-W, Nam Y-S, Lee WH. 'OR' logic function of molecular photodiode consisting of gfp/viologen/cytochrome c hetero-film. *Mol Cryst Liq Cryst*. 2003;**407**:485–492.
71. Matsui J, Mitsuishi M, Aoki A, Miyashita T. Optical logic operation based on polymer Langmuir-Blodgett film assembly. *Angew Chem Int Ed Engl*. 2003;**42**: 2272–2275.
72. Matsui J, Mitsuishi M, Aoki A, Miyashita T. Molecular optical gating devices based on polymer nanosheets assemblies. *J Am Chem Soc*. 2004;**126**:3708–3709.

73. Yasutomi S, Morita T, Imanishi Y, Kimura S. A molecular photodiode system that can switch photocurrent direction. *Science*. 2004;**304**:1944.
74. Yasutomi S, Morita T, Kimura S. pH-controlled switching of photocurrent direction by self-assembled monolayer of helical peptides. *J Am Chem Soc*. 2005; **127**:14564–14565.
75. Sakamoto A, Matsuo Y, Matsuo K, Nakamura E. Efficient bidirectional photocurrent generation by self-assembled monolayer of penta(aryl)[60]fullerene phosphonic acid. *Chem Asian J*. 2009;**4**:1208–1212.
76. Nitahara S, Akiyama T, Inoue S, Yamada S. A photoelectronic switching device using a mixed self-assembled monolayer. *J Phys Chem B*. 2005;**109**:3944–3948.
77. Nitahara S, Tasaki N, Akiyama T, Yamada S. Molecular logic devices using mixed self-assembled monolayers. *Thin Solid Films*. 2006;**499**:354–358.
78. Szaciłowski K, Macyk W. Working prototype of an optoelectronic XOR/OR/YES reconfigurable logic device based on nanocrystalline semiconductors. *Solid State Electron*. 2006;**50**:1649–1655.
79. Gao GG, Deng Y, Kispert LD. Photovoltaic response of carotenoid-sensitized electrode in aqueous solution: ITO coated with a mixture of TiO_2 nanoparticles, carotenoid, and polyvinylcarbazole. *J Chem Soc, Perkin Trans 2*. 1999:1225–1230.
80. Meng H, Yang Y, Chen Y, Zhou Y, Liu Y, Chen X, Ma H, Tang Z, Liu D, Jiang L. Photoelectric conversion switch based on quantum dots with i-motif DNA scaffolds. *Chem Commun*. 2009;**17**:2293–2295.
81. Gill R, Patolsky F, Katz E, Willner I. Electrochemical control of the photocurrent direction in intercalated DNA/CdS nanoparticle systems. *Angew Chem Int Ed*. 2005;**44**:4554–4557.
82. Sheeney-Haj-Ichia L, Cheglakov Z, Willner I. Electroswitchable Photoelectrochemistry by Cu^{2+}–polyacrylic acid/CdS-nanoparticle assemblies. *J Phys Chem B*. 2004;**108**(1):11–15.
83. Gissler W. Photoelectrochemical investigation on trigonal selenium film electrodes. *J Electrochem Soc*. 1980;**127**:1713–1716.
84. Agostinelli G, Dunlop ED. Local inversion of photocurrent in cadmium telluride solar cells. *Thin Solid Films*. 2003;**431–432**:448–452.
85. Zhang P, Cheng C, Jiao P, Li Y, He Z, Zhang H. Well improved photoswitching characteristic of CdSe nanorods via CdS nanoparticle-decoration. *Mater Lett*. 2008;**62**:1151–1154.
86. Minoura H, Tsuiki M. Cathodic photo-effect at CdS electrode in aqueous polysulfide solution. *Chem Lett*. 1978:205–208.
87. Mueller N, Hodes G, Vainas B. Cathodic current photoenhancement at mechanically damaged CdS electrodes. *J Electroanal Chem*. 1984;**172**:155–165.
88. Vainas B, Hodes G. A photocathodic effect at the CdS–electrolyte interface. *J Electroanal Chem*. 1981;**130**:391–394.
89. Ogawa S, Hu K, Fan F, Bard AJ. Photoelectrochemistry of films of quantum size lead sulfide particles incorporated in self-assembled monolayers on gold. *J Phys Chem B*. 1997;**101**:5707–5711.
90. Harakeh M, Alawieh L, Saouma S, Halaoui L. Charge separation and photocurrent polarity-switching at CdS quantum dots assembly in polyelectrolyte interfaced with hole scavengers. *Phys Chem Chem Phys*. 2009;**11**:5962–5973.

91. Matsui J, Abe K, Mitsuishi M, Aoki A, Miyashita T. Quasi-solid-state optical logic devices based on redox polymer nanosheet assembly. *Langmuir*. 2009;**25**:11061–11066.
92. Oszajca MF, McCall KL, Robertson N, Szaciłowski K. Photocurrent switching effects in TiO_2 modified with ruthenium polypyridine complexes. *J Phys Chem C*. 2011;**115**:12187–12195.

24

Photorefractive (PR) polymers and their recent applications

J. THOMAS, University of Central Florida, USA

DOI:10.533/9780857098764.3.676

Abstract: Photorefractive (PR) polymers offer the high potential of developing a large area platform for recording dynamic holograms. They are made by carefully manipulating the energy levels and concentrations of the components required for charge generation, transport and trapping, and modulation of refractive index. The fact that dynamic holograms can be recorded in photorefractive polymers without the necessity of any special processing techniques makes it very attractive for many applications including refreshable 3D displays. This chapter describes the basic theory used to develop photorefractive polymers, molecular energy level requirements and the components used to develop highly efficient polymer composites. In addition, it describes some of the recent applications of photorefractive polymers.

Key words: photorefractive (PR) polymers, nonlinear optical properties, dynamic holograms, photoconducting polymers, refreshable 3D display, photoconducting polymers.

24.1 Introduction

Significant advances have been accomplished in the development of photorefractive (PR) polymers since their discovery in 1991, and exhibit high efficiencies and fast response times. [1–4] The PR effect, originally discovered in inorganic crystals more than 40 years ago,[5,6] initially drew attention as an apparent detriment to non-linear applications. However, development was pursued because of some unique properties relevant to some other perceived applications. Primarily, the PR process was reversible though also can be fixable, allowing both read/write and read-only applications, as opposed to standard photographic films which could only be written once.[7] Moreover, the signature non-local nature of the process allowed coupling and energy transfer to occur between two coherent beams.[8] The inherent advantage of organic polymer materials to manipulate the properties that suit to a given application is very attractive in making optical and photonic devices at low cost. The structural constraints were also relaxed, allowing polymers to be custom made into different geometries, such as optical devices and displays unlike typical inorganic crystals. The smaller dielectric constant reduces the electric field screening of trapped charges increasing the quality factor. The drawbacks of polymer materials for optical and display applications is that they are more dispersive in nature, both optically and electronically, and the theory behind their operation is significantly more complicated than regular crystals, though much work has been done in these areas.[9,10] Currently, PR polymers outperform their inorganic counterparts in properties like diffraction efficiency, two-beam coupling gain, and sensitivity.[11,12]

Owing to this remarkable progress many applications have appeared, including optical communication, correlation, and imaging through scattering media, all with different material challenges that can be met by these highly versatile polymers.[13–18] Recently, they have been shown to function in dynamic holographic displays, which can be of use in medical imaging, industrial design, among other emerging areas such as telepresence.[19] Unlike other permanent media for recording holograms, PR polymers are reversible and require no post-processing. They demonstrate fast response time and high diffraction efficiency, which are necessary material properties for such an application. However, progress in other areas has not been as rapid, particularly in the area of sensitivity. In the visible, the sensitivity is still orders of magnitude smaller than that of permanent films used for recording static holograms. There are also few routes to extend operation into the infrared (IR), and only recently have these begun to bear fruit. Some of these include using organic photosensitizers with one photon absorption in the near IR, [20,21] using semiconductor nanocrystals where the absorption band is tunable,[22] and using two-photon absorption to charge sensitizers that are otherwise transparent.[23.24]

This chapter describes the fundamentals of photorefractivity, material concepts behind fabrication of organic PR polymer composites, including the functional components, their respective roles in devices, and the basic physical mechanisms that must be taken into account when designing devices. Recent progress in these areas is also discussed, including new hole-transporting polymers for reduced glass transition temperature (T_g) and high mobility. Many novel sensitizers are also reviewed, which is a very dynamic area of research. New composites with excellent sensitivity in the near IR wavelengths have extended the range of high performing polymers beyond the visible. Finally, some material considerations necessary for specific applications are also taken into account, such as pulsed writing for high speed operation of many devices, and updatable holographic displays. Optimized materials have been shown to exhibit good performance even under single pulse nanosecond writing times, enabling operation at 100 Hz or more, which is faster than continuous wave (CW) recording schemes. The materials for holographic displays will be briefly discussed and extensions to reflection geometry and video-rate response times will be examined, since still higher sensitivities and trap densities are needed to accomplish these goals.

24.2 Fundamentals of photorefractivity

In the PR effect, an index of modulation is accomplished by the generation of space charge field by redistribution of photogenerated charges based on non-uniform irradiation and electro-optic effect resulting from nonlinear chromophore. Therefore, the necessary requirements for photorefractivity are photoconductivity and electro-optic effect. Ideally, a PR material should be an insulator in the dark and a good conductor under illumination. *Charge generation*, *transport*, *trapping* and *electro-optic activity* are represented in Fig 24.1. A spatially modulated intensity pattern obtained from two coherent interfering beams can be written as:

$$I(x) = I_0[1 + m\cos(2\pi/\Lambda)], \qquad [24.1]$$

where $I_0 = I_1 + I_2$ is the total incident intensity, i.e., the sum of the two beam intensities; the fringe visibility, $m = 2(I_1I_2)^{1/2}/(I_1 + I_2)$ and the periodicity or fringe visibility is Λ.[25,26] In tilted transmission geometry Λ is given by:

$$\Lambda = \frac{\lambda}{2n\sin[(\alpha_2 - \alpha_1)/2]}, \qquad [24.2]$$

where n is the refractive index of the material, λ the optical wavelength in vacuum, and α_1 and α_2 are the incident internal angles of the two writing beams with respect to the sample normal. Periodicity, λ represents the sinusoidal light distribution which depends on the writing beam

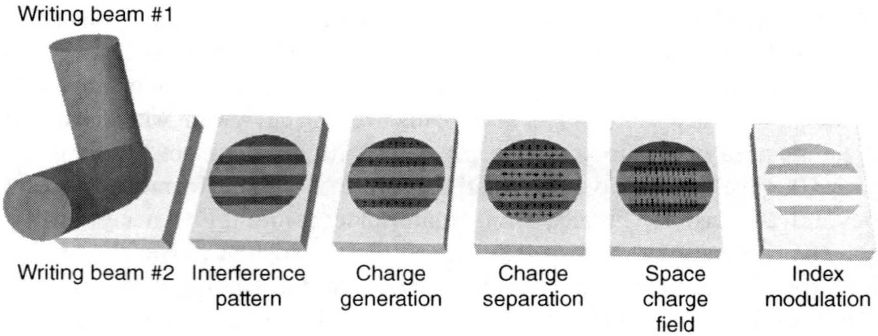

24.1 Cartoon showing the light interference, charge generation, separation, space charge field build up and refractive index modulation in photorefractive polymers.

angle and can vary from a sub-micron to a few tens of microns. The two writing beams yield an interference pattern within the sample and charge carriers are generated in the high light intensity areas. Since the mobility of the majority carriers (usually holes in organic materials) are higher, they move from the bright areas leaving behind the carriers of opposite charge. This movement is diffusion controlled due to concentration gradient or a drift due to an externally applied electric filed. The traps present in the low-intensity (dark) region limit the migration process. An inhomogeneous space charge distribution results from the charges trapped in low intensity regions and those opposite charges left behind in high intensity regions. The space charge distribution $\rho(x)$ in a material with a dielectric constant ε induces an internal space charge field $E_{sc}(x)$ under a one-dimensional light distribution is given by Poisson's equation is $dE_{sc}/dx = 4\pi\rho/\varepsilon$. The phase shift between the light distribution due to the interfering beams and the generated space charge field will be $\pi/2$ if the transport is driven by diffusion alone. However, it can be different if carrier drift due to an external field is also involved. The spatial derivative appearing in Poisson's equation explains the non-local response in PR materials. The life time of the space charge field is governed by the type of charge traps present in the sample. The space charge field formation is followed by the electro-optic modulation of the refractive index. The signature of photorefractivity is the phase-shift between the initial light distribution due to interfering beams and the modulated refractive index.

The Kukhtarev model[27–29] provides the first order component of the steady-state space charge field generated by a sinusoidal light distribution in inorganic crystals. The amplitude of the space charge field (E_{sc}) in this model is described by:

$$E_{sc} = m\left(\frac{(E_0^2 + E_D^2)}{(1+E_D/E_q)^2 + (E_0/E_q)^2}\right)^{1/2}, \qquad [24.3]$$

where E_0 is the component of the applied field along the grating vector. The diffusion field E_D is defined as:

$$E_D = \frac{Kk_BT}{e}, \qquad [24.4]$$

where $K = 2\pi/\Lambda$ is the grating vector, k_B the Boltzmann constant, T the temperature and e the elementary charge of an electron. The trap limited field E_q is represented by:

$$E_q = \frac{eN_{eff}}{K\varepsilon\varepsilon_0}, \qquad [24.5]$$

where ε_0 is the permittivity and N_{eff} the effective density of traps in the photorefractive sample. The phase-shift between the space charge field and the light intensity pattern is given by:

$$\Theta = \arctan\left[\frac{E_D}{E_0}\left(1+\frac{E_D}{E_q}+\frac{E_0^2}{E_D E_q}\right)\right], \quad \text{for} \quad E_0 \neq 0 \qquad [24.6]$$

and $\Theta = \pi/2$ for $E_0 = 0$.

The magnitude of the index modulation $\Delta n(x)$ is related to the magnitude of the space charge field is given by:

$$\Delta n(x) = -\frac{1}{2} n^3 r_{eff} E_{sc}(x), \qquad [24.7]$$

r_{eff} is the effective electro-optic coefficient. The electro-optic coefficient depends on the symmetry and orientation of the sample and the polarization of the light beams. This model successfully provides a framework for the photorefractivity in inorganic crystals where the linear electro-optic effect is the prevailing factor. However, many phenomena in polymer PR materials like field dependence of charge generation, transport, trapping and recombination processes cannot be explained by this model alone. In addition, the high diffraction efficiency and two bean coupling gain accomplished in PR polymers are much larger than that expected simply from the electro-optic effect of the nonlinear component. These enhanced effects observed in PR polymers are explained by a mechanism called 'orientational enhancement' (OE). This effect will be discussed below.

24.2.1 Electro-optic PR polymers

An organic PR sample is made up of different components, and each serves a specific purpose. An oriented gas model was developed to provide a reasonable approximation for the description of poled polymers.[30,31]

Oriented gas model

Under a given applied field and molecular density, macroscopic optical properties of materials can be obtained from the orientational distribution of the molecules present. The degree of orientation of chromophore molecules can be achieved by applying statistical mechanics to the molecular population. A charge transfer push–pull molecule with an electron accepting group connected to the electron donating group through a π-conjugated bridge is considered in this approximation (Fig. 24.2). These molecules are highly polarizable with a cylindrical symmetry exhibiting strong nonlinear optical response.

24.2 A push–pull molecular concept with donor group connected to the acceptor through a π-bridge.

Considering second-order effects, the molecular polarization expressed in the molecular reference frame $\{x, y, z\}$ can be given as

$$p_i = \varepsilon_0(F_0\mu_i + F_1\alpha_{ij}E_j + D^{(2)}F_2\beta_{ijk}E_jE_k + \ldots), \tag{24.8}$$

where μ_i is the permanent ground state dipole moment, α_{ij} the linear molecular polarizability tensor, β_{ijk} the first molecular hyperpolarizability tensor, and $D^{(2)}$ is a degeneracy factor that accounts for the number of distinct permutations of the interacting applied field frequency components. The field correction factors F_0, F_1 and F_2 gives a proper estimation of the screening effects of the surrounding dielectric on the applied fields interacting with the chromophore molecules. The value of these correction parameters depends on the frequency of the fields inducing the polarization and they are usually included into the values of the polarizabilities and molecular dipole moments. For example, in the electro-optic effect, the 'dressed' values are given as

$$\mu_i^* = f_0\mu_i, \quad \alpha_{ij}^* = f_\omega\alpha_{ij}, \quad \beta_{ijk}^* = f_0 f_\omega^2\beta_{ijk}, \tag{24.9}$$

where f_0 and f_ω are the Onsager and Lorentz–Lorenz field correction factors, respectively, defined as

$$f_0 = \frac{\varepsilon_{\mathrm{dc}}[\varepsilon(\omega)+2]}{[2\varepsilon_{\mathrm{dc}}+\varepsilon(\omega)]}; \quad f_\omega = \frac{\varepsilon(\omega)+2}{3}, \tag{24.10}$$

where $\varepsilon(\omega)$ is the dielectric constant at optical frequencies (i.e. $n^2 = \varepsilon(\omega)$) and $\varepsilon_{\mathrm{dc}}$ the zero frequency or direct current (DC) dielectric constant.

It is possible to relate molecular polarizability and macroscopic polarization if an appropriate statistical average is taken using the dressed values from the framework provided by the oriented gas model. In order to explain electric field (applied along the laboratory Z-direction) induced electro-optic effect, the model assumes (i) push–pull chromophore molecules are used; (ii) the only nonvanishing first hyperpolarizability tensor element is β_{zzz}, and third order effects are negligible; (iii) chromophores can freely reorient under an applied field without any interactions with the surroundings and (iv) any other chromophore interactions are negligible.

The electric field-induced macroscopic polarization calculated using the above approximations is

$$P_I = \frac{1}{V}\left(\sum p_i\right)_I, \tag{24.11}$$

where V is the volume and the subscripts I and i refer to the laboratory and molecular frames, respectively. This summation has very little practical value since a large number of chromophore molecules are present in a unit volume. But (24.11) can be replaced by a thermodynamic average for an arbitrary function $g(\Omega)$ defined as

$$\langle g(\Omega)\rangle = \frac{\int g(\Omega) f(\Omega) \mathrm{d}\Omega}{\int f(\Omega) \mathrm{d}\Omega}, \tag{24.12}$$

where $f(\Omega)$ is the statistical orientational distribution which follows the Maxwell–Boltzmann distribution under steady-state conditions. For a set of polarizable dipoles the distribution is defined as

$$f(\Omega)\mathrm{d}\Omega = \exp\left[-\frac{U(\theta)}{k_{\mathrm{B}}T}\right]\sin\theta \mathrm{d}\theta, \tag{24.13}$$

where $U(\theta)$ is the interaction energy between the DC poling field E_{p} and the molecular dipole moment μ, which is written as

$$\begin{aligned} U(\theta) &= -\mu^* \cdot E_{\mathrm{p}} \\ &= -|\mu^*||E_{\mathrm{p}}|\cos\theta \\ &= -\xi k_{\mathrm{B}} T \cos\theta, \end{aligned} \tag{24.14}$$

where $\xi = \dfrac{\mu^* E_{\mathrm{p}}}{k_{\mathrm{B}}T}$.

It is interesting to note that the interaction energy between the poling field and the electronic contribution to the polarization has been neglected in this analysis. The justification of this approximation is based on the assumption that the chromophores have large dipole moments. Hence (24.12) becomes a Langevin function if $g(\Omega) = \cos^m\theta$, where

$$\langle\cos^m\theta\rangle = \frac{\int_0^\pi \cos^m\theta \exp(\xi\cos\theta)\sin\theta \mathrm{d}\theta}{\int_0^\pi \exp(\xi\cos\theta)\sin\theta \mathrm{d}\theta} \equiv L_m(\xi). \tag{24.15}$$

If the poling field is small, the Langevin functions can be simplified to make $\xi \ll 1$. Applying this approximation the first three Langevin functions can be written as

$$L_1(\xi) \approx \frac{\xi}{3}; \quad L_2(\xi) \approx \frac{1}{3} - \frac{2\xi^2}{45}; \quad L_3(\xi) \approx \frac{\xi}{5}. \tag{24.16}$$

Oriented gas model-based equations (24.12 to 24.15) provide the basic outline to relate microscopic and bulk properties in polymer composite poled by an electric field.

Refractive index modulation

A correlation between the microscopic and the macroscopic polarization can be achieved from (24.8) as

$$P_{\{X,Y,Z\}} = N\langle \mathrm{A}p_{\{x,y,z\}}\rangle, \qquad [24.17]$$

where N is the total number of chromophore molecules present in a sample volume of V and A is a transformation tensor from the microscopic to the macroscopic frame. However, it is essential to derive expressions that permit the macroscopic susceptibility to be correlated with the microscopic polarizability because the index of refraction is directly linked to the susceptibility tensor. The various contributions to (24.17) and the subsequent modulation of the refractive index are mentioned in the proceeding sections.

Pockels electro-optic effect

Index modulation due to the Pockels electro-optic effect is accomplished by applying both optical and electrical field to a non-centrosymmetric material. The macroscopic polarization component induced along the Z axis when both fields are applied along the same axis is described by

$$P_Z = N\varepsilon_0 \langle a_{zZ}a_{zZ}a_{zZ}\rangle \beta^*_{ZZZ} E_Z(0)E_Z(\omega), \qquad [24.18]$$

where a_{ij} are the direction cosines of the transformation tensor which can be written as

$$a_{iJ} \equiv \hat{i}\cdot\hat{J};\quad \text{i.e.}\; a_{zZ} \equiv \hat{z}\cdot\hat{Z} = \cos\theta \quad \text{and} \quad a_{zX} \equiv \hat{z}\cdot\hat{X} = \sin\theta\cos\phi, \qquad [24.19]$$

where angles θ and ϕ are given in Fig. 24.3.

Based on (24.15) and (24.19), the second order susceptibility tensor can be described in terms of the microscopic first hyperpolarizability as

$$\begin{aligned} \chi^{(2)}_{ZZZ} &= N\langle a_{zZ}a_{zZ}a_{zZ}\rangle \beta^*_{ZZZ}, \\ &= N\langle \cos^3\theta\rangle \beta^*_{ZZZ} \\ &= NL_3(\xi)\beta^*_{ZZZ} \end{aligned} \qquad [24.20]$$

If an optical field in the X-direction is applied, other elements of $\tilde{\chi}^{(2)}$ are obtained as shown below

$$\begin{aligned}\chi^{(2)}_{ZXX} &= N\langle a_{zZ}a_{zX}a_{zX}\rangle\beta^{*}_{ZZZ}\\ &= N\langle\cos\theta\sin^2\theta\cos^2\phi\rangle\beta^{*}_{ZZZ}\\ &= \frac{N}{2}[(L_1(\xi)-L_3(\xi))\beta^{*}_{ZZZ}],\end{aligned} \qquad [24.21]$$

Therefore (24.7) which is the space charge field induced refractive index modulation and the electro-optic tensor can be derived from (24.20) and (24.21)

$$\begin{aligned}\Delta n &= -\frac{1}{2}n^3 r_{\text{eff}} E_{\text{SC}}\\ r_{\text{eff}} &= -\frac{2}{n^4}\chi^{(2)}_{\text{eff}}\end{aligned} \qquad [24.22]$$

and

$$\chi^{(2)}_{\text{eff}} = \hat{e}^{*}_{\text{d}}\cdot\tilde{\varepsilon}(\tilde{\chi}^{(2)}:\hat{k}\cdot\tilde{\varepsilon}\cdot\hat{e}_{\text{i}})$$

The unit polarization vectors for the diffracted $\hat{e}_{\text{d}}$ and incident beam $\hat{e}_{\text{i}}$, the dielectric tensor $\tilde{\varepsilon}$ and the unitary grating vector $\hat{k}$ are the quantities defined inside the tensorial product.

Orientational birefringence

The index of modulation occurs in inorganic ferroelectric crystals like lithium niobate is by a linear electro-optic or Pockels effect. These non-centrosymmetric crystals are first subjected to poling to align their crystal domain. In high T_g PR polymers, a similar poling procedure is followed to align the nonlinear chromophore at a high temperature (above T_g). The polymer is cooled to room temperature with the field applied. However, in low T_g photorefractive polymers, the poling is carried out at room temperature during the writing process itself. This is due to the orientational freedom of the chromophores in the polymer composite. Hence the chromophores can give rise to index modulation known as orientational birefringence. Therefore, in photorefractive polymers, orientational birefringence dominates the electro-optic response, but it does not prevent other effects from adding to the total electro-optic response.

In inorganic crystals, the index modulation is provided by the linear electro-optic or Pockels effect, which leads to an electric-field induced change in the refractive index. The Pockels effect occurs only in crystals that are non-centrosymmetric. In ferroelectric crystals like lithium niobate or barium titanate, the crystal domains must first be aligned by applying a poling field. It has been believed that electro-optic effect is entirely responsible for the index of modulation in photorefractive polymers. This is true in the case of high T_g polymers in which the poling of the nonlinear chromophores is carried out at high temperature (at or above T_g) using a high applied electric field and cooling down to room temperature (with the field applied). However, the poling is carried out at room temperature during the writing process if the polymer T_g is at or near room temperature. In low T_g polymers, an index of modulation can be achieved without heating the sample due to the orientational freedom of the chromophore molecules. Hence the chromophores can give rise to index modulation known as orientational birefringence.

Considering the polarizability in the molecular x–y plane is identical and non-negligible compared to the input along the z-axis, the linear polarizability tensor can be given as

$$\tilde{\alpha}^* = \begin{pmatrix} \alpha_{xx}^* & 0 & 0 \\ 0 & \alpha_{yy}^* & 0 \\ 0 & 0 & \alpha_{zz}^* \end{pmatrix} = \begin{pmatrix} \alpha_\perp^* & 0 & 0 \\ 0 & \alpha_\perp^* & 0 \\ 0 & 0 & \alpha_\parallel^* \end{pmatrix} \quad [24.23]$$

Since it is presumed that the chromophore molecule is a rod-like dipole, it has different linear polarizability in directions along and perpendicular to the molecular axis. The macroscopic polarization in the presence of a poling field and an optical field along the Z-axis can be presented as

$$\begin{aligned} &P_Z = \varepsilon_0 \chi_{ZZ}^{(1)} E_Z(\omega) \\ &\chi_{ZZ}^{(1)} = N\langle \alpha_\perp^* + \Delta\alpha^* \cos^2\theta \rangle \quad and \quad \Delta\alpha^* = \alpha_\parallel^* - \alpha_\perp^* \end{aligned} \quad [24.24]$$

In the absence of a poling electric field, the chromophores are randomly aligned, the interaction energy in the Boltzmann distribution disappears and the distribution function is defined as $f_0(\Omega)d\Omega = \sin\theta d\Omega$. This distribution is beneficial in determining the polarization of unpoled material in the presence of an optical field. The purpose of the poling field is to change the distribution function defined by (24.13). The change in the linear susceptibility under the influence of a poling field is written as

$$\Delta\chi_{ZZ}^{(1)} = N\left(\langle \alpha_\perp^* + \Delta\alpha^* \cos^2\theta \rangle - \langle \alpha_\perp^* + \Delta\alpha^* \cos^2\theta \rangle_0\right), \quad [24.25]$$

By using appropriate integrals it can be demonstrated that $\langle \cos^2\theta \rangle_0 = 1/3$ thus (24.25) becomes

$$\Delta\chi_{ZZ}^{(1)} = N\Delta\alpha^* \left[\langle \cos^2\theta \rangle - \frac{1}{3} \right]. \quad [24.26]$$

The change in susceptibility leads to a change in refractive index given by

$$\Delta n_{ZZ} = \frac{\varepsilon_0}{2n} \Delta\chi_{ZZ}^{(1)}. \quad [24.27]$$

If the field is polarized along either the X or Y axis instead of the Z axis, a similar route provides the following relation between the changes in susceptibility induced in each case

$$\Delta\chi_{XX}^{(1)} = \Delta\chi_{YY}^{(1)} = -\frac{1}{2}\Delta\chi_{ZZ}^{(1)}. \qquad [24.28]$$

The optimization of chromophores for photorefractive applications can be performed based on (24.21) and (24.26) since they directly relate to the macroscopic index modulation. Additionally, it is possible to draw a relationship between molecular polarizability and macroscopic susceptibility using (24.14) and (24.16). We find:

$$\Delta\chi_{ZZ}^{(1)} \approx -N\Delta\alpha^{*}\frac{2}{45}\left(\frac{\mu^{*}}{k_{\mathrm{B}}T}\right)^{2}E_{\mathrm{p}}^{2}, \qquad [24.29]$$

and

$$\chi_{ZZZ}^{(2)} \approx N\frac{\mu^{*}\beta_{ZZZ}^{*}}{5k_{\mathrm{B}}T}E_{\mathrm{p}} \qquad [24.30]$$

$$\chi_{ZXX}^{(2)} \approx N\frac{\mu^{*}\beta_{ZZZ}^{*}}{15k_{\mathrm{B}}T}E_{\mathrm{p}}. \qquad [24.31]$$

The above equations establish that molecular dipole moment, polarizability anisotropy and first hyperpolarizability of the nonlinear chromophore are the factors determining the modulation of refractive index. In tilted transmission geometry for writing gratings, the total field is the superposition of modulated space charge field and external field (Fig. 24.3). The chromopores present in the PR polymer align based on the total field. Therefore, refractive index modulation has contributions from birefringence as well as electro-optic effect unlike inorganic crystals where the contribution is only from electro-optic effect. The birefringence is due to

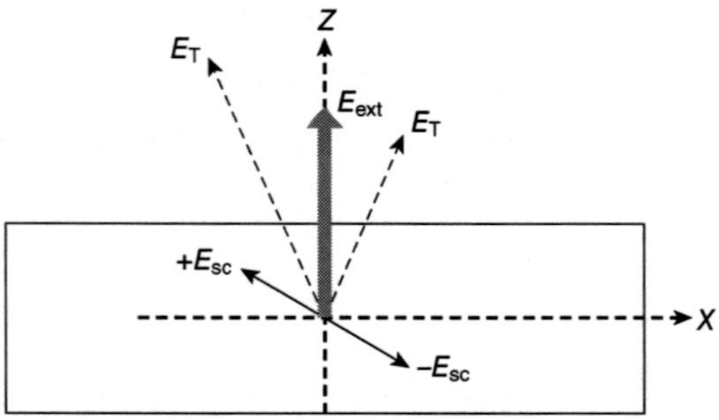

24.3 The total poling field in a low glass transition temperature polymer for standard tilted transmission geometry.

the anisotropy of the linear polarizability whereas electro-optic effect is due to the second-order nonlinearity. The high diffraction efficiency in PR polymers is due to the combined effect of these two contributions.

The total poling field is:

$$E_{\mathrm{T}}(r) = [E_{\mathrm{sc}}(\vec{r})\sin\varphi]\hat{X} + [E_{\mathrm{ext}} + E_{\mathrm{sc}}(\vec{r})\cos\varphi]\hat{Z} \qquad [24.32]$$

where the spatially modulating space-charge field is given by $E_{\mathrm{sc}}(\vec{r})=E_{\mathrm{sc}}\exp(i\vec{K}\vec{r})$, φ is the angle between the external field and grating vector and E_{ext} is the external field. The periodically changing angle ξ with respect to the laboratory Z axis under the total poling field is given by:

$$\xi = \arctan\left[\frac{E_{\mathrm{sc}}(\vec{r})\sin\varphi}{E_{\mathrm{ext}} + E_{\mathrm{sc}}(\vec{r})\cos\varphi}\right]. \qquad [24.33]$$

The contribution of the birefringence to the refractive index modulated at the spatial frequency K for a given slanted four-wave mixing transmission geometry can be written as:

$$\Delta n_{\mathrm{eff}}^{(1)} = \frac{2\pi}{n}\hat{e}_{\mathrm{d}}^{*}.T(r).\Delta\tilde{\chi}^{(1)}T^{-1}(r).\hat{e}_{\mathrm{i}}, \qquad [24.34]$$

where $T(r)$ is the transformation matrix between laboratory and local coordinates and $\Delta\tilde{\chi}^{(1)}$ is a diagonal matrix of $\Delta\chi_{XX}^{(1)}, \Delta\chi_{YY}^{(1)}$ and $\Delta\chi_{ZZ}^{(1)}$ given in equations (24.28) and (24.29).

The different contributions to the modulation of refractive index for each polarization of the probe beam is detailed as

$$\Delta n_{K,S}^{(1)} = -\frac{2\pi}{n}BE_{\mathrm{ext}}E_{\mathrm{sc}}\cos\varphi, \qquad [24.35]$$

$$\Delta n_{K,P}^{(1)} = \frac{2\pi}{n}BE_{\mathrm{ext}}E_{\mathrm{sc}}\left[2\sin\alpha_1\sin\alpha_2\cos\varphi - \cos\alpha_1\cos\alpha_2\cos\varphi + \frac{3}{2}\sin(\alpha_1+\alpha_2)\sin\varphi\right] \qquad [24.36]$$

where,

$$B = \frac{2}{45}Nf_{\infty}(a_{\parallel} - a_{\perp})\left(\frac{\mu^{*}}{k_{\mathrm{B}}T}\right)^{2}. \qquad [24.37]$$

The birefringence contribution is doubled when a p-polarized reading beam is used in the tilted transmission geometry.

Likewise, the refractive index modulation due to the electro-optic effect is given by the expression

$$\Delta n_{\text{eff}}^{(2)} = \frac{4\pi}{n}\hat{e}_{\text{d}}^{*}.T(r).\Delta\tilde{\chi}^{(2)}T^{-1}(r).\hat{e}_{\text{i}}. \quad [24.38]$$

The second order expression for s and p polarizations when the second order susceptibility is approximated to a diagonal matrix with diagonal components $\chi_{ZXX}^{(2)}$ and $\chi_{ZZZ}^{(2)}$ can be expressed as:

$$\Delta n_{K,S}^{(2)} = \frac{8\pi}{n}CE_{\text{ext}}E_{\text{sc}}\cos\varphi \quad [24.39]$$

$$\Delta n_{K,P}^{(2)} = \frac{8\pi}{n}CE_{\text{ext}}E_{\text{sc}}[\cos\alpha_1\cos\alpha_2\cos\varphi + 3\sin\alpha_1\sin\alpha_2\cos\varphi + \sin(\alpha_1+\alpha_2)\sin\varphi] \quad [24.40]$$

where,

$$C = \frac{Nf_0f_\infty\beta\mu^*}{15k_\text{B}T}. \quad [24.41]$$

The contribution from the electro-optic effect is enhanced by a factor of three when a p-polarized reading beam is used in slanted transmission geometry. However, when low T_g polymers are used, it is improved by only a factor of two.

The summation of the first-order (birefringence) and second-order (electro-optic) contributions give the total modulation of the refractive index. Based on this summation, one can define the following figure of merit (FOM) for chromophores[32] as

$$\text{FOM} = \frac{2}{9k_\text{B}T}\mu^2\Delta\alpha + \mu\beta \quad [24.42]$$

In order to accomplish a high orientational birefringence, the polarizability anisotropy of the chromophore $\Delta\alpha = (\alpha_{\text{II}} - \alpha_{\perp})$ should be high. A high hyperpolarizability β is desirable for a considerable contribution from the electro-optic effect along with high dipole moment μ for better refractive index modulation.

24.3 Functions of photorefractive (PR) components

A typical PR composite consists of charge transporting polymer matrix, photosensitizer, nonlinear chromophore molecules, and a plasticizer. There are two approaches to develop a PR sample: either mixing all the components together in a common solvent and develop into a solid composite or chemically attach different components to a polymer backbone. In a PR polymer, the photosensitizer is a photo-reducible molecule that can either absorb light or form a charge-transfer complex with the hole-transport polymer. When it is excited with a suitable wavelength the photo-sensitizer injects a hole into the transport system. The NLO chromophore is responsible for the field-dependent refractive index. Since birefringence is the major factor in generating high index modulations, a T_g close to room temperature is a desired property. Plasticizer molecules are added to the mixture to lower the T_g.

Significant advances have been achieved in the field of organic PR since their discovery. Several PR samples have been developed with high diffraction efficiency, fast response time and large two beam coupling gain. This includes functionalized polymers, amorphous glass and polymer-dispersed liquid crystals. Guest–host composites are the best studied PR samples because of its high degree of customizability of composition. That is, in order to accomplish a specific property, its constituents can be easily engineered. This versatility sometimes comes at the cost of potential phase separation due to the mixing of components with immiscible characteristics. Hence, material manipulations are extremely important in developing high-quality PR polymer composites.

24.3.1 Charge transport polymer (CTP)

The charge transport polymer (CTP) is usually the host polymer material and will be an efficient charge conductor enabling the formation of space charge field. In PR samples, holes are the majority carriers and the hole transport has been extensively studied. The CTP are so selected that they are highly conjugated with delocalized π-electrons. It should also be an electron donor capable of accepting a hole from the sensitizer molecule (in the case of hole transport). The latter condition requires the highest occupied molecular orbital (HOMO) level of CTP to be above (lower in energy) that of the sensitizer to energetically facilitate charge transfer. Hereafter in this chapter, all HOMO energy levels discussed will be with reference to that of the CTP. For the most commonly used CTPs, the HOMO is typically between –5.5 and –6.0 eV (referenced to vacuum level). A component with a HOMO energy level higher than that of the CTP means that its ionization potential is lower and a HOMO energy level lower than that of CTP means that its ionization potential is higher. The loading of CTP in the PR composite will be high enough so that transport can occur by hopping.[33] In PR polymers, an electric field is necessary for transport since the mobility is highly field dependent.[34]

A few of the commonly used transporting agents are shown in Fig. 24.4. Carbazole-containing polymers are among the thoroughly studied materials. Among them, poly(vinyl carbazole) (PVK) is the first high-performing composite utilized for making PR devices. A few other conjugate polymers have drawn considerable attention as well, due to the generally higher drift mobilities and reduced polarity leading to more stable mixing.

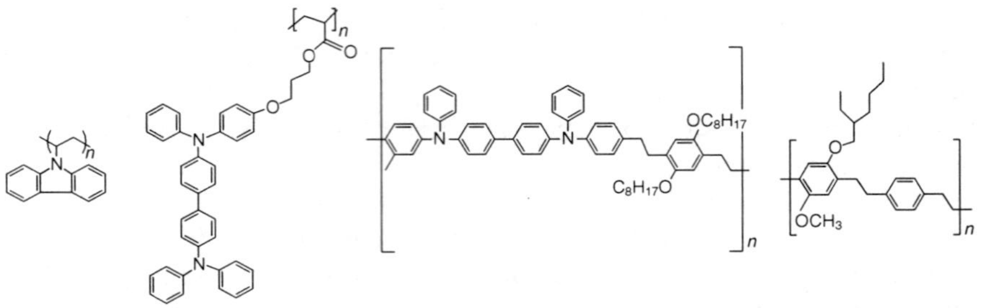

(a) PVK (b) PATPD (c) TPD-PPV (d) *p*-MEH-PPV

24.4 Commonly used charge transporting polymers.

Triarylamine-containing side chain polymers, such as poly(acrylic tetraphenyldiaminobiphenyl) (PATPD) have been as successful as PVK-based samples, and the response time is not dependent on the history of illumination,[35] as discussed below. Other polymers like poly(phenylene vinylene) (PPV), showed better steady-state characteristics compared to PVK systems.[36,37]

24.3.2 Nonlinear molecules

The nonlinear chromophore molecules provide the refractive index modulation necessary for PR effect in response to the development of the space-charge field. Orientational birefringence or the linear Pockels electro-optic effect is instrumental in achieving high refractive index modulation. Thus, the molecule should exhibit either a large linear polarizability anisotropy (birefringence) or first hyperpolarizability (electro-optic), and in both cases the requirement is a large ground state dipole moment. An extensively accepted expression for quantifying the chromophore quality and optimizing these two contributions to the index modulation is given by equation 24.42.[38]Maximization of hyperpolarizability for specific donor and acceptor strengths for a given conjugated bridge is predicted by a two-state four-orbital model assuming non-interacting electrons.[39] These predictions agree well with the experimentally observed values for β obtained from electric field-induced second harmonic generation measurements (EFISH), which determines the projection of the second-order polarizability tensor along the direction of the molecular dipole moment. As a guide for optimizing the projection by adjusting the strengths of the acceptor and donor molecules, and the conjugated π-bridge length, a bond order alternation model has been proposed.[40] It is noted that the contribution from the linear polarizability anisotropy is about an order of magnitude higher than that from the first hyperpolarizability in most of the commonly used chromophores in PR composites.[38] This is the molecular manifestation of the orientational enhancement, and states that the orientational birefringence is mostly responsible for the enhanced PR performance in polymer composites. It needs to be noted that this expression for the FOM arises from the oriented gas model, which foresees the macroscopic properties of the material from the orientational distribution, density of molecules, and microscopic non-linear properties. However, in an intimate mixture of components in a PR polymer composite, inter molecular interactions can happen, which are not accounted for in the model and can influence the performance, such as the interaction between the polar chromophore and the transport manifold, as discussed below.

An important parameter in optimizing the chromophores' properties is the dipole moment, as it can affect not only the refractive index modulation, but also the molecular aggregation, and charge transport and trapping in PR composite. Strong donor–acceptor termination groups are used to create electron separation across the π-conjugate bridge to accomplish a permanent dipole moment. In order to accomplish the second order non-linear properties, the delocalization along a π-conjugated bridge permits rapid electronic redistribution in the presence of an electric field. These molecules referred to as push–pull molecules,[41] the dipole moment can also be engineered by altering the length of the bridge, similar in principle to the quantum mechanical particle-in-a-box. However, longer bridge lengths may lead to reduced orientational freedom and increased optical absorption. Even though, larger dipole moments lead to high FOM, phase instability of highly polar chromophores in presence of other non-polar molecules can cause a major problem in PR composites. Hence an optimum chromophore density is to be determined even though large loading enhances modulation efficiency. Moreover, chromophores with high dipole moment can reduce the mobility of charges due to the energetic disorder introduced in the polymer matrix.[42]

It is to be noted that the chromophores can also act as a sensitizer[43] if the energetics is appropriate at the operational wavelength. For this to be effective, the HOMO level must be deeper than that of the CTP. If instead a chromophore molecule with a HOMO level shallower than CTP is used, it may act as a deep trap for holes already injected into the transport manifold from other sensitizers.[44]

Thus, the ionization potential is an important parameter in determining the magnitude of the SC field and the response time of PR polymers.[45] The concentration of the chromophore is also an important parameter. If the HOMO levels are aligned properly, at low concentrations the chromophore can be used as traps. As the concentration increases the chromophore may also begin to participate in charge transport via hopping,[46] which will also lead to a tradeoff between a large SC field and fast response as chromophore also participate in charge transport. In general, care must be taken

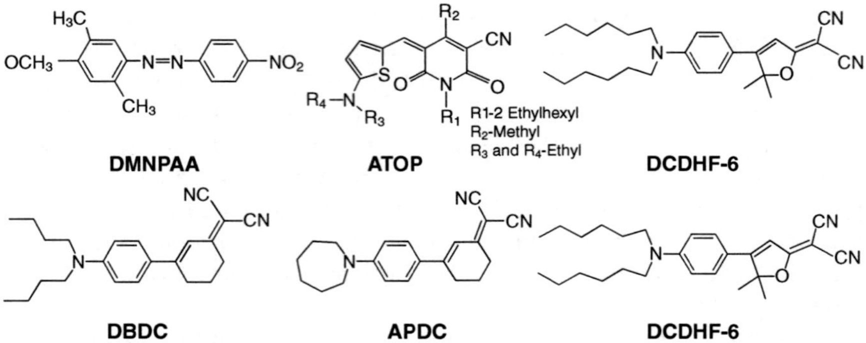

24.5 **Chemical structures of nonlinear optical chromophores used in photorefractive composites.**

to select a chromophore with proper optical and electronic properties for the anticipated application, as this can affect not only the index modulation but also the charge generation, transport, and trapping.

Several high-performing chromophores have been synthesized for use in PR composites. Some of the most common ones are dicyanostyrenes (DCST),[47] azo-dye derivatives (DMNPAA),[48] and oxypyridine dyes (ATOP).[49] There are also some recent reports about tricyano molecules (DCDHF), which shows high photoconduction and two-beam coupling (TBC) gain.[50,51] A number of reviews are available on the structure and design of chromophore for PR composites.[12,49,50] The structures for some of these chromophores are shown in Fig. 24.5.

The design of nonlinear optical chromophores is also an active area, since this can affect nearly every functional property of the composites. There is no shortage of high performance composites. For example, PVK/DCDHF composites have exhibited TBC gains of 200 cm^{-1} with an absorption coefficient of 13 cm^{-1} at 647 nm and at only 30 V/μm.[47] PSX-based composites with the 2-piperidino-5-thienylmalononitrile (P-TH-DC) chromophore have recently shown near 100% diffraction efficiency at moderate fields of 50–70 V/μm and response times of about 100 ms.[52] At 532 nm, which is of interest for visual applications, PATPD and FDCST composites without C_{60} sensitization also have high diffraction efficiency.[19] In this case, the sensitization comes from the chromophore itself, and the ionization potential can be adjusted based on the position and degree of fluorination.[53] Liquid chromophores have also been used to achieve the PR effect without a plasticizer, demonstrating a response time of 46 ms.[54] Arylimine derivatives have exhibited near 100% diffraction efficiency and over-modulation at low fields (30–50 V/μm) with relatively low concentrations (25%),[55] though the gain coefficient is well below 100 cm^{-1}.

24.3.3 Photosensitizers

Charge generation at the wavelength of interest in PR samples is accomplished by photosensitizing molecules. The photosensitizers may form a charge transfer complex with the CTP, facilitating the charges to be efficiently transferred. In the case of primarily hole conduction in PR polymer composites, the sensitizer after photexciting an electron from its HOMO to LUMO will inject a hole into the CTP by accepting an electron, becoming reduced. For the PR effect to be reversible, it should also be oxidizable to allow it to return to the original state. In order to act as a sensitizer, its HOMO level must be well below that of the CTP so that efficient hole generation can be possible. According to Marcus's theory, the difference between the ionization potential of the donor and acceptor should be large to accomplish large photogeneration efficiency.[56,57]

Photosensitizers such as fullerenes may also serve as traps after being reduced, as there is a close relationship observed between the anion density and the trap density deduced from near-IR spectroscopic and PR characterization.[52,58] Moreover, the chromophore molecules can participate in this process as well, as it can provide the compensating sites for the holes to balance the anion density. These sites can help in reducing the recombination of charges. Thus, the ionization potential of the PR component is highly important in establishing the trap density.[53,59]

Some of the most common sensitizers for visible wavelengths are C_{60}, TNF, and TNFDM (Fig. 24.6) as they will form a CT complex with donor molecules. A highly soluble fullerene derivative, PCBM, is sometimes used in place of C_{60}, and is also used for the wavelength sensitivity in the visible region. A recently developed DBM has been used for sensitivity at 1550 nm via two-photon absorption.[14] A multitude of wavelength sensitivity was also achieved by inorganic crystals like CdSe and PbS. This approach has the benefit of tuning the absorption band by changing the size of the particles.[60]

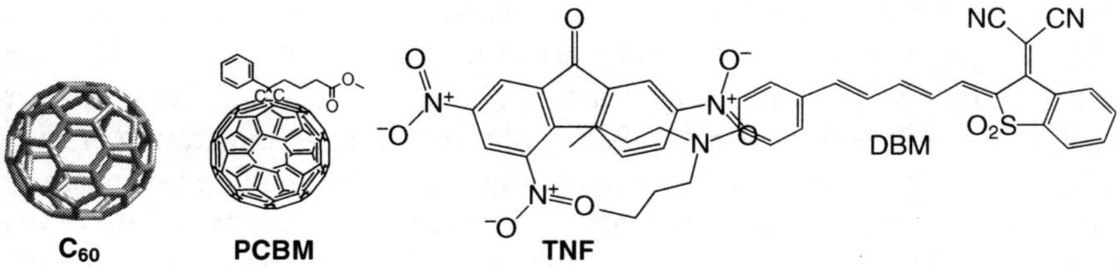

24.6 Photorefractive sensitizer molecules.

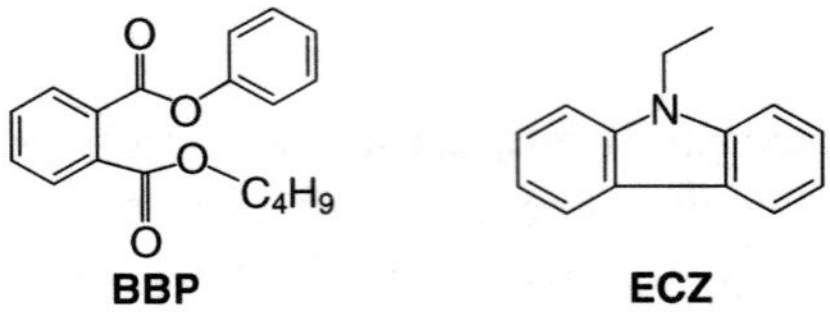

24.7 Benzyl butyl phthalate (BBP) and 9-ethyl carbazole plasticizers (ECZ).

24.3.4 Plasticizers in PR samples

In the majority of cases, the T_g of the PR composites is much higher than room temperature. As discussed previously, since the largest contribution to the index modulation is achieved by the orientational birefringence, the T_g of the material must be at or near the operating (room) temperature. In order to bring the T_g to room temperature, plasticizers may be added to the composite. They are typically inert with respect to charge generation and trapping, though they do reduce the functional volume by diluting the charge transport matrix.

The commonly used plasticizer in PVK-based PR composite is benzyl butyl phthalate (BBP) at a loading level of 10–15 wt%. An alternate approach is to make the plasticizer less inert by using the hole transporting monomer unit as part of the design. The monomer, 9-ethyl carbazole (ECZ), has been highly successful as a plasticizer even though its contribution to charge transport and trapping has never been carefully scrutinized. These two plasticizers are shown in Fig. 24.7.

24.4 Photorefractive (PR) characterization techniques

24.4.1 Steady-state four-wave mixing (SSFWM)

The fundamental characterization of PR material involves steady-state four-wave mixing (SSFWM). In SSFWM, the intensity of the diffracted beam is monitored as a function of the applied field.[61] The schematic diagram of a four-wave mixing experiment is shown in Fig. 24.8. Beams (1) and (2) are two interfering beams which create the gratings and a reading beam (3) is diffracted from the grating written inside the sample. The sample surface is usually tilted 60° (ψ) with respect to the sample bisector. The writing beam incident angles are α_1 and α_2. The grating vector is given by K and d is the sample thickness. In order to minimize the grating erasure a low-intensity beam is used as the writing beam. The reading beam can be at the same or a different wavelength from the writing beams and therefore requires a tilt angle in its direction to satisfy the Bragg condition. A well-known configuration is degenerate four-wave mixing where all beams have

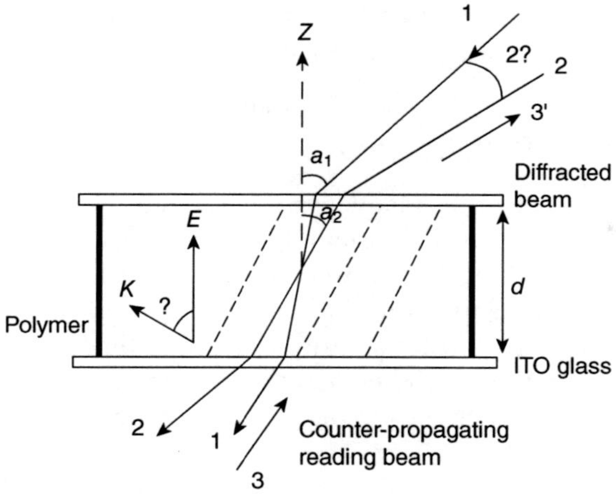

24.8 Photorefractive four-wave mixing geometry.

the same wavelength and the reading beam (3) partially diffracts from the grating written by the writing beams to produce a fourth beam (3′). In this configuration, the diffracted beam (3′) is counterpropagating with respect to the writing beam (1) and the Bragg phase-matching condition is automatically fulfilled. Beam 4 is the transmitted beam counterpropagating with respect to the second writing beam (2). A non-degenerate four wave mixing is where a different readout wavelength than the writing beam wavelength is used. A cautious selection of the polarizations and relative powers of the writing and reading beams maximizes the visibility of the phase grating and minimizes the interaction between the beams. The diffraction efficiency provides the strength of the gratings which can either be internal or external. The internal diffraction efficiency is the ratio of the diffracted beam intensity to that of the transmitted beam intensity in the absence of a grating ($\eta_{int} = I_{diff}/I_{tran}$) whereas the external diffraction efficiency is the ratio of the diffracted beam intensity to the incident intensity ($\eta_{ext} = I_{diff}/I_{inci}$).

The functional form of the index modulation can be extracted from the field-dependent diffraction efficiency. The experimental data is normally fit with a simplified form of Kogelnik's equation given by

$$\eta = A\sin^2(BE^P) \qquad [24.43]$$

The index modulation can be calculated by relating the argument of the sine in (24.43) with that of the full Kogelnik relation once the fitting parameters A, B and p have been determined as given below:

$$BE^P = \Delta n\left(\frac{\pi d}{\lambda\sqrt{\cos\alpha_1\cos\alpha_2}}\hat{e}_1\cdot\hat{e}_2\right). \qquad [24.44]$$

where d is the thickness of the sample, λ is the wavelength, and the dot product is between the polarization vectors of the writing beams.

24.4.2 Transient PR characterization

The dynamic response of a PR samples is determined by transient four-wave mixing experiments. In a transient PR measurement, the reading beam and one of the writing beams were turned on using an electrical shutter at a constant field applied to the sample. A Pockels cell turns the second writing beam on and off for a desired period of time. The electrical shutter remains open with the reading and the writing beams illuminating the sample until the end of the experiment, at which point the shutter is closed.

A simplified version of Kogelnik's equation is used to fit the time-dependent diffracted power. The temporal evolution of the index modulation is described by the weighted biexponential function[52]

$$\eta = A\sin^2[B(1-me^{-t/t_1}+(1-m)e^{-t/t_2})], \qquad [24.45]$$

where A, B, m, are amplitudes and t_1 and t_2 are the time constant fitting parameters. This functional form provides an excellent fit to experimental results. The weighed biexponential function has been commonly used for organic samples. However, some materials have shown severe disagreement with (24.45) such that other functional forms have also been considered.

The performance of a holographic device is defined in terms of its sensitivity.[62] A widely used expression for sensitivity is

$$S = \frac{\sqrt{\eta_{\text{ext}}(t_{\text{exp}})}}{I_{\text{WB,ext}}t_{\text{exp}}}, \qquad [24.46]$$

where $\eta_{\text{ext}}(t_{\text{exp}})$ is the total external diffraction efficiency achieved after a short exposure time t_{exp} when writing with a total external writing beam intensity $I_{\text{WB,ext}}$. Sensitivity enhancement can be achieved by (i) accomplishing high space charge field (E_{sc}), (ii) optimizing the EO chromophore structures and enhancing their orientational mobility in the polymer matrix, (iii) decreasing absorption losses, and (iv) reducing the sample exposure time by increasing the recording speed.

24.4.3 Two-beam coupling (TBC)

An identical experimental geometry as FWM is used in TBC characterizations except that no reading beams are used and two writing beams are p-polarized. Two optical beams, probe and pump, intersect inside the PR polymer creating a nonlocal diffraction grating and subsequently, the beams diffract from the gratings they have just created. This results in the amplification of one of the beams and the attenuation of the other. This phenomenon is unique to PR gratings due to their non-local character, leading to the nonzero phase shift between the optical interference pattern and the grating.

A reversal of the direction of energy transfer is achieved by reversal of the applied field due to the physical nature of the phase shift. This is because in low T_g organic materials, the energy transfer direction is determined by the drift of the majority carrier in the electric field, and therefore a change in polarity of the electric field can reverse the direction of energy transfer. This phenomenon of asymmetric two-beam coupling is a fingerprint of the PR nature of the photoinduced grating in any material; the energy exchange between the beams is generally monitored as a function of the applied field.

24.4.4 High-performing PR polymers

All high-performance PR polymers developed in the early ages of its discovery utilized PVK, which has since become one of the standards, with fast speed and high efficiency. However, there are a number of properties that limit the usefulness of PVK in applications, such as component aggregation and deterioration of the response time upon illumination.[53,58] PR composites with polysiloxane derivatives (PSX) with a siloxane chain, which are similar to PVK in having a carbazole group have also been developed. PSX composites exhibit a lower T_g compared to PVK-based composites with similarly high performance,[63,64] and difficult to phase separate.[65] A variant class of the transporting polymers uses carbazole as a pendant group on a rigid backbone, such as poly (*p*-phenylene terephthalate) (PPT).[66] These materials have a tendency to self-organize into layered microstructures, and have a glass transition below room temperature without the addition of a plasticizer. TBC gains over 200 cm^{-1} and diffraction efficiencies close to 100% have been reported on these samples. They show superior performance over PVK in reflection geometry where the small grating spacing often limits PR characteristics.[67,68]

Recently, a different type of polymer called PATPD has been developed, which attains hole transport without the degradation of the PR properties.[35] This was accomplished by manipulating the HOMO level of PATPD, which is above that of most chromophores. The energy levels of some typical constituents in PATPD and PVK composites are shown in Fig. 24.9. With PVK as the CTP, the mobile holes can be trapped in the chromophore HOMO, enhancing the ionized acceptor density that is associated with deep trapping. With PATPD polymer composite, it is energetically unfavorable

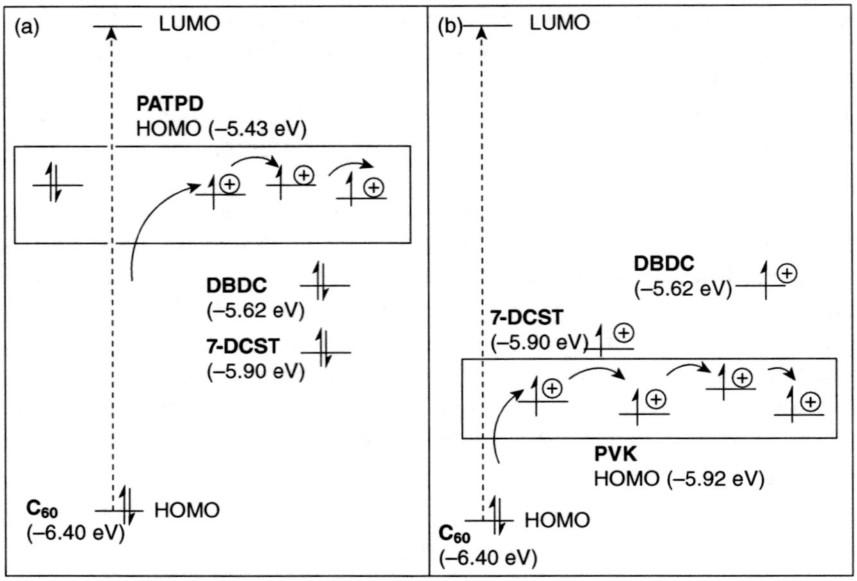

24.9 Intrinsic energy states and charge generation, transport and trapping process in (a) PATPD and (b) PVK with various chromophores. Reprinted with permission from Ref. 35, Copyright Wiley-VCH (2004).

for the mobile holes to be injected into the chromophore HOMO. Thus, the possibility of the chromophore acting as a compensating trap, leading to the formation of C_{60} anions in the composite, is reduced. The accumulation of traps degrades the photoconductivity and the response time of the grating as the material is exposed, and with PATPD, this is avoided because of the HOMO energy level. Indeed, in PATPD composites video-rate response times were maintained over a large range of exposures, whereas the response times of similar PVK-based composites deteriorated by about an order of magnitude, as shown in Fig. 24.9. Both the composites studied were sensitized with sensitizer C_{60} and contained either 7-DCST, DBDC, or a combination of chromophores. Response times were measured under 633 nm illumination with a total fluence of 1.1 W/cm^2.

The tetraphenyldiaminobiphenyl (TPD) is attached to a polyacrylic backbone as a pendent group through a flexible alkoxy linker. This alkoxy linker reduces the T_g of the composite by providing structural flexibility and orientational freedom to the pendent TPD groups. Thus, it is expected that the photorefractive traps are shallow, and conformational or structural in nature. Attempts have been made to engineer the chromophore HOMO levels to mimic the situation in PVK-based composites, but this was achieved only at the expense of a dramatic reduction in the dynamic range of the device.[68]

The phase stability of the composite was also enhanced, allowing chromophore loadings of 35–40%, and external diffraction efficiencies approaching 70% were accomplished at fields of about 50 V/μm with net TBC gain coefficients well above 100 cm^{-1}. The hole mobility in TPD is reported to be about two orders of magnitude higher than in PVK,[67,69] which may help to increase the photogeneration efficiency; this high mobility is one of the reasons why TPD-based polymers have also found use in organic light-emitting diodes.[54,55,68,69]

24.5 Photorefractive (PR) polymer composites for applications

The advent of PR polymers which can be made into large area devices led to the laboratory demonstration of many applications, for which all the previous material developments assisted to various degrees. Some of the important applications proposed are tissue imaging,[18] beam cleanup,[70] data storage,[71] and dynamic displays.[72] In this section, some of the specific material aspects required to realize a few of these applications will be reviewed, namely pulsed writing required for many high-speed devices, and holographic displays.

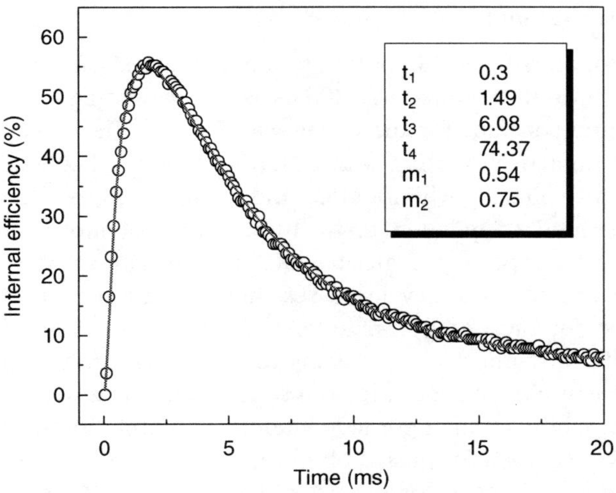

24.10 Plot demonstrating the simultaneous rise and decay of the diffraction efficiency in a photorefractive sample under single pulse illumination. Pulse width was 1 ns and energy density was 4 mJ/cm^2, applied field was 95 V/μm. (Reproduced with permission from Ref. 73. Copyright 2006 American Institute of Physics.)

24.5.1 Pulsed PR recording

One of the primary advantages of recording a grating with pulsed writing beams is that sufficient writing energy can be delivered in much shorter period of time than in CW. Pulsed recording highly improves the writing time as well as make the system vibration insensitive. Pulsed recoding requires materials that can respond to brief impulses.

In one of the previous pulsed recording PR studies conducted,[73] a composite consisting of PATPD/7-DCST/ECZ/C_{60} (54.5/25/20/0.5 wt.%), 105 μm thick, was illuminated with two 532 nm writing beams about 1 ns in pulse duration (total fluence of 4 mJ/cm^2). Under single pulse writing, a maximum diffraction efficiency of 56% was observed in 1.8 ms after illumination, as charge transport, trapping, and chromophore orientation continued after each pulse. An applied field of 95 V/μm was used in a standard geometry with a 60° slant and an inter-beam angle of 20°. In CW recording, the same sample shows near 100% efficiency with a response time of 4 ms under a similar fluence.

Quantitative temporal characteristics were obtained by fitting the curve to a modified exponential function. Since grating build-up and decay are occurring simultaneously, each of which is typically characterized by two time constants, a total of four time constants were used:

$$\begin{aligned} \Delta n &\propto [1 - m_1 \exp(-t/t_1) - (1 - m_1)\exp(-t/t_2)] \\ &\quad \times [m_2 \exp(-t/t_3) + (1 - m_2)\exp(-t/t_4)] \\ \eta &\propto \sin^2(B\Delta n^2) \end{aligned} \qquad [24.47]$$

All the fit parameters for single pulse illumination are shown in Fig. 24.10. The fast time constant obtained was 300 μs, with a weighting factor of 0.54. The slowest time constant for the decay was 74.4 ms. The effect of pulse energy was examined by writing with different fluences. The fact that the peak power is also altering will not affect the PR process, since the number of charges generated is dependent on the total number of photons absorbed. The results of various pulse energies from 0.2 to 3 mJ/cm^2 are shown in Fig. 24.11. The maximum efficiency increases more or less linearly with the energy, at least for the energy range used. However, the recording time does not significantly improve, suggesting the dynamics are not limited on these timescales by charge generation but by the charge transport and alignment of chromophores, which are independent of writing intensity.

The devices developed for pulsed photorefractivity needs to be operated at the fastest possible speed for the desired application, at least as fast as the pulse repetition rate, which can be around hundreds of Hz. To accomplish this goal, substantial grating formation and decay should be within 10 ms so that new information can be recorded. The grating builds within 1–2 ms, and the decay can be accelerated by uniform illumination. Gated

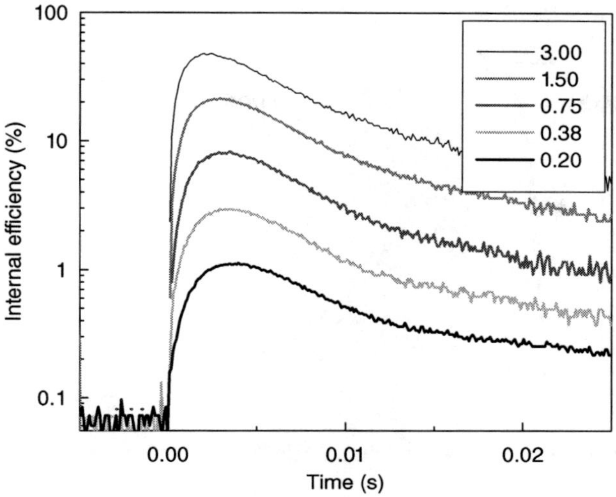

24.11 Time response of diffraction efficiency under single pulse exposure with energy densities varying from 0.20 to 3.00 mJ/cm^2. (Reproduced with permission from Ref. 73. Copyright 2006 American Institute of Physics.)

illumination for 5 ms with a CW 532 nm beam was used. The dynamics of this process are shown in Fig. 24.12. Since it cannot be present during the grating formation, a variable time delay is used for the gating. At an energy density of 6 mJ/cm^2, the entire write–read–erase process was completed within 10 ms. The delay as well as the intensity and nature (cw or pulsed) of the erasing can be adjusted to optimize the signal-to-noise ratio during the reading process.

While pulse writing constitutes a significant increase in the dynamic response, some display and storage applications require even faster writing times. Recently, Nau *et al.*[74] has demonstrated a grating written with 130 fs pulses at 800 nm in PVK samples sensitized with TNFM. With an average intensity of 3.5 mW/cm^2, the diffraction efficiency reported is 2%, lower than for the previously reported PATPD samples under ns exposure, but much higher repetition rates are achievable (80 MHz) with these lasers, thus requiring materials with response times on the order of 10 ns. However, as faster lasers are used, the spectral bandwidth increases, which may reduce contrast from unintentional spectral multiplexing.

24.5.2 Updatable 3D PR displays

Three-dimensional perception is fundamental to the human interaction with the world, since our brains are accustomed to processing more than

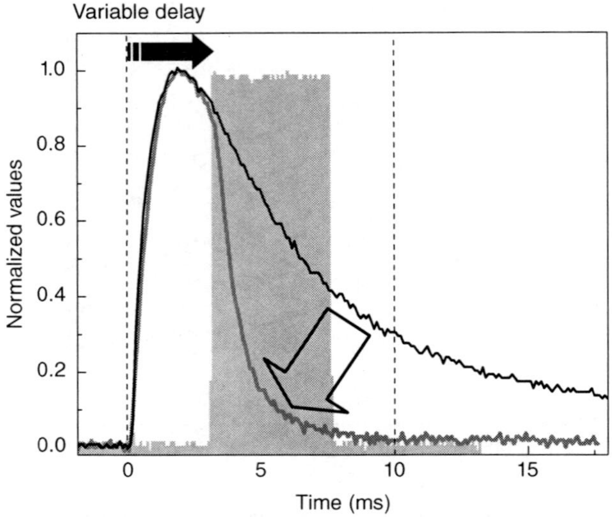

24.12 Diffraction efficiency decay as a function of time with and without a separate uniform illumination to accelerate the decay. The shaded bar shows the duration of the CW erasing beam. The grating has completely decayed in 10 ms, showing the potential for operation at 100 Hz. (Reproduced with permission from Ref. 73. Copyright 2006 American Institute of Physics.)

just a single perspective to understand our surroundings. A 3D image has the ability to make people better understand: for example, a 3D viewing would be highly advantageous in medical imaging, industrial design, and terrain mapping.

Currently, 3D imaging techniques rely on stereoscopic rendering that requires special eye-glasses, at the expense of user fatigue and lack of the number of available perceptions. Holographic displays can reproduce very high-quality 3D images without these drawbacks, but to date have been made with either static photopolymers[75] or dynamic media with small-size and low resolution.[76,77]

The attractive properties of PR polymers like large diffraction efficiencies, fast writing times, large area, phase stability, and reversible recording process are attractive for holographic display applications. The first updatable holographic 3D display based on PR polymers has been recently developed utilizing all these properties.[19] A copolymer approach has been used to reduce the phase separation, allowing higher loading of the chromophores. The copolymer consists of a polyacrylic backbone with pendant group TPD-type and carbaldehyde aniline (CAAN) attached through an alkoxy linker (PATPD-CAAN) in a ratio of 10 : 1. The function of CAAN

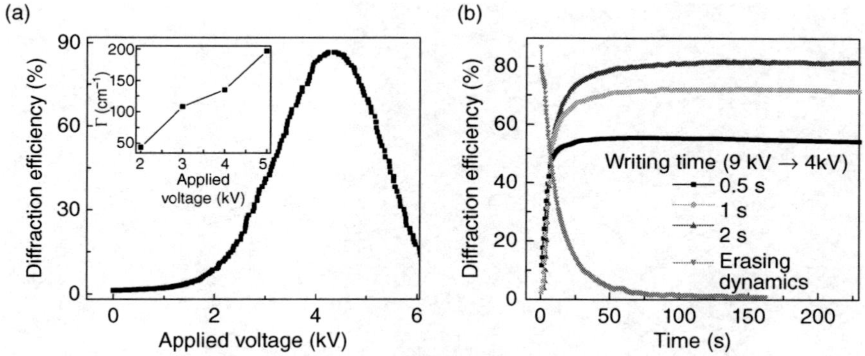

24.13 (a) Diffraction efficiency (90%) exhibited by a 105 μm thick sample at 4 kV in the standard slanted transmission geometry at 532 nm. (b) Transient plots demonstrating writing and erasing of photorefractive gratings. (Reproduced with permission from Ref. 19. Copyright 2008 Nature Publishing Group.)

group was to provide phase stability to the composite. A fluorinated dicyanostyrene (FDCST) NLO chromophore was added to provide sufficient refractive index modulation and charge generation at the wavelength of interest (532 nm). ECZ was used as a plasticizer to bring the glass -transition temperature near to room temperature. The ratios of PATPD : CAAN/FDCST/ECZ were 50/30/20 wt.%. The sample for holographic display was 105 μm thick with an active area of 4 inches × 4 inches (100 mm × 100 mm).

The high diffraction efficiency (90%) exhibited by a 105 μm thick sample at 4 kV in the standard slanted transmission geometry at 532 nm is shown in Fig. 24.13a. The transient plots of a105 μm thick sample in the standard slanted transmission geometry at 532 nm shown in Fig. 24.13b. It is to be noted that this sample is slower than the typical fast PR composites, though at 9 kV an intermediate peak is reached after 1 s of writing. This allows writing the display at this voltage and sets the timescale of the recording process and is utilized in a new approach called voltage kick-off technique. This technique provided high-speed writing and long persistency for PR samples.[78]

An integral image holography was used to make the hologram for the 3D display. The technique involves the processing of dozens of 2D perspectives of an object on a computer which is followed by optically multiplexing these perspectives onto the recording medium in a manner such that when reconstructed, the sensation of depth is created via parallax. The two writing beams interfere inside the sample in Fourier transform geometry. The hologram was recorded by successively writing approximately 120 holographic pixels, or 'hogels' side by side. Taking into account the geometry of the human vision, horizontal parallax only was used, so the hogels

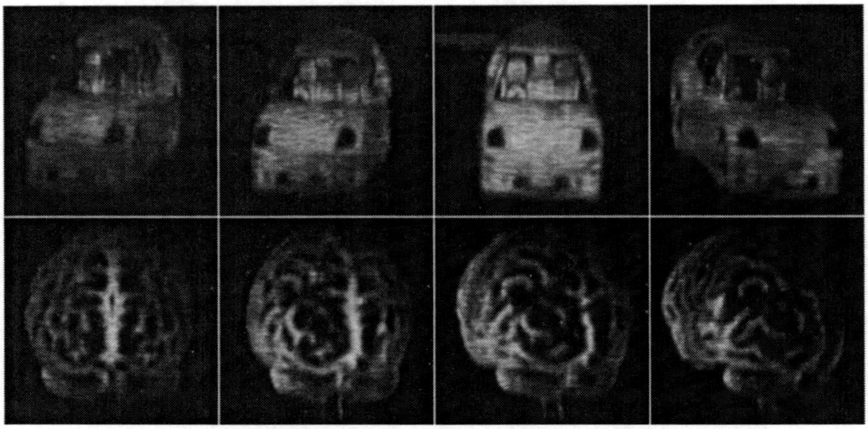

24.14 2D images of the 3D holographic images produced from updatable display setup. Images are taken at different camera locations to demonstrate occlusion and parallax. The right-most images are one frame from the computer model used to generate the holograms. (Reproduced with permission from American Chemical Society.)

are vertical stripes. For recording, 532 nm light was used with a total writing intensity of 0.1 W/cm^2. Given the transients, each hogel is written for 1 s, so the overall writing time is about 3–4 minutes for a 4 inch × 4 inch sample. After writing, the sample was translated to a new position and the hologram was read using a 650 nm light-emitting diode (LED) lamp. The image was visible for a period of about 3 hours, but erasing can be done at any time by illuminating the sample with a homogeneous beam at 532 nm. Some images produced by this process are shown in Fig. 24.14.

Even though this is a significant progress in the area of PR polymers and 3D holographic displays, many applications will require even faster samples for use in video-rate displays, and practicality dictates the viewability in standard white light room conditions.

In reflection recording configuration, the grating spacing is 0.2 μm, which is considerably (6.5 times) smaller than in transmission geometry. In order to achieve high diffraction efficiency, higher trap densities are required. For this, PCBM was added at 0.5 wt% to the 3D display composition both as a sensitizer and a charge trap.[79] This approach would also improve the speed of sample for use in video-rate displays, as discussed above. The results of FWM measurements in reflection geometry for this sample are also included, we observe 40% diffraction efficiency at 9 kV and an offset of 3°. The increased trap density due to PCBM leads directly to a larger SC field, but likely reduces the charge recombination, both of which explain the increase in efficiency.

24.6 Conclusion and future trends

Photorefractive polymers have transformed to such an extent that they can be used to display refreshable 3D holograms. Currently, they outperform their inorganic counterparts in diffraction efficiency, sensitivity and two beam coupling gain. The sensitivity of the devices can be tuned from the visible to the infrared, and highly customizable fabrication enables to make devices of various shapes and sizes with relative ease. Thus, the current focus of the research is on developing materials for specific applications and assessing their performance in real-world schemes. In particular, holographic 3D displays will have potential applications in medical technologies, military planning, and the entertainment industry.

Despite the rapid progress, there is still plenty of room for improvement. One aspect which needs particular enhancement is the sensitivity of PR polymers. It is much less than photographic films at least for holographic applications, though the updatability is a significant advantage. Many of the complex processes like charge transport and trapping are still not completely understood and modeling efforts will need to be implemented to better understand the mechanism and guide developments. Moreover, the electrical and chemical stability is still a concern in some high-performing samples. Currently emerged applications like updatable 3D display will continue to drive the scientific community in the development of new materials to meet these challenges.

24.7 References

1. Ducharme, S.; Scott, J. C.; Moerner, W. E. *Phys. Rev. Lett.* **1991**, 66, 1846.
2. Meerholz, K.; Volodin, B. L.; Sandalphon; Kippelen, B.; Peyghambarian, N. *Nature* **1994**, 371, 497.
3. Ostroverkhova, O.; Moerner, W. E. *Chem. Rev.* **2004**, 104, 3267.
4. Thomas, J.; Norwood, R. A.; Peyghambarian, N. in *New Directions in Holography and Speckle*; Caulfield, H. J., Vikram, C. S., Eds.; American Scientific Publishers: California, 2008.
5. Ashkin, A.; Boyd, G. D.; Dziedzic, J. M.; Smith, R. G.; Ballman, A. A.; Levinste, J. J., Nassau, K. *Appl. Phys. Lett.* **1966**, 9, 72.
6. Chen, F. S. *J. Appl. Phys.* **1967**, 38, 3418.
7. Staebler, D. L. *Appl. Phys. Lett.* **1975**, 26, 182.
8. Staebler, D. L.; Amodei, J. J. *J. Appl. Phys.* **1972**, 43, 1042.
9. Schildkraut, J. S.; Cui, Y. *J. Appl. Phys.* **1992**, 72, 5055.
10. Ostroverkhova, O.; Singer, K. D. *J. Appl. Phys.* **2002**, 92, 1727.
11. Moerner, W. E.; Grunnet-Jepsen, A.; Thompson, C. L. *Annu. Rev. Mater. Sci.* **1997**, 27, 585.
12. Marder, S. R.; Kippelen, B.; Jen, A. K.-Y.; Peyghambarian, N. *Nature* **1997**, 388, 845.
13. Günter, P.; Huignard, J.-P. *Photorefractive Materials and Their Applications I, II and III*; Springer-Verlag: Berlin, 1988.

14. Tay, S.; Thomas, J.; Eralp, M.; Li, G.; Kippelen, B.; Marder, S. R.; Meredith, G.; Schülzgen, A.; Peyghambarian, N. *Appl. Phys. Lett.* **2004**, 85, 4561.
15. Volodin, B. L.; Kippelen, B.; Meerholz, K.; Peyghambarian, N.; Javidi, B. A. *Nature* **1996**, 383, 58.
16. Kippelen, B.; Marder, S. R.; Hendrickx, E.; Maldonado, J. L.; Guillemet, G.; Volodin, B. L. Steele, D. D.; Enami, Y.; Sandalphon; Yao, Y. J.; Wang, J. F.; Röckel, H.; Erskine, L.; Peyghambarian, N. *Science* **1998**, 279, 54.
17. Winiarz, J. G.; Ghebremichael, F.; Thomas, J.; Meredith, G.; Peyghambarian, N. *Opt. Express* **2004**, 12, 2517.
18. Salvador, M.; Prauzner, J.; Köber, S.; Meerholz, K.; Turek, J. J.; Jeong, K.; Nolte, D. D. *Opt. Express* **2009**, 14, 11834.
19. Tay, S.; Blanche, P.-A.; Voorakaranam, R.; Tunç, A. V.; Lin, W.; Rokutanda, S.; Gu, T.; Flores, D.; Wang, P.; Li, G.; St. Hilaire, P.; Thomas, J.; Norwood, R. A.; Yamamoto, M.; Peyghambarian, N. *Nature* **2008**, 451, 694.
20. Eralp, M.; Thomas, J.; Tay, S.; Li, G.; Meredith, G.; Schulzgen, A.; Peyghambarian, N., Walker, G. A.; Barlow, S.; Marder, S. R. *Appl. Phys. Lett.* **2004**, 85, 1095.
21. Köber, S.; Prauzner, J.; Salvador, M.; Kooistra, F.; Hummelen, J.; Meerholz, K. *Adv. Mater.* **2010**, 22, 1383.
22. Choudhury, K. R.; Sahoo, Y.; Jang, S. J.; Prasad, P. N. *Adv. Funct. Mater.* **2005**, 15, 751.
23. Tay, S.; Thomas, J.; Eralp, M.; Li, G.; Norwood, R. A.; Schülzgen, A.; Yamamoto, M.; Barlow, S.; Walker, G. A.; Marder, S. R.; Peyghambarian, N. *Appl. Phys. Lett.* **2005**, 87, 171105.
24. Blanche P.-A.; Kippelen, B.; Schulzgen, A.; Fuentes-Hernandez, C.; Ramos-Ortiz, G.; Wang, J. F.; Hendrickx, E.; Peyghambarian, N.; Marder, S. R. *Opt. Lett.* **2002**, 27, 19.
25. Kippelen, B.; Meerholz, K.; Peyghambarian, N. 'An introduction to photorefractive polymers,' in *Nonlinear Optics of Organic Molecules and Polymers*, H. S. Nalwa and S. Miyata, eds, CRC Press, Boca Raton, (1997), Page 465.
26. Eichler, H. J.; Günter, P.; Pohl, D. W. *Laser-induced Dynamic Gratings*, Springer-Verlag, Berlin (1986).
27. Kuktharev, N. V. *Pis'ma Zh. Tekh. Fiz.* **1976**, 2, 114, *Sov. Tec. Phys. Lett.* **1976**, 2, 438.
28. Kukhtarev, N. V.; Markov, V. B.; Odulov, S. G. *Opt. Commun.* **1977**, 23, 338.
29. Kukhtarev, N. V.; Markov, V. B.; Odulov, S. G.; Soskin, M. S.; Vinetskii, V. L. *Ferroelectrics* **1979**, 22, 949.
30. Williams, D. J. in *Nonlinear Optical Properties of Organic Molecules and Crystals*, Chemla, D. S. and Zyss, J. Academic Press, New York (1987).
31. Singer, K. D.; Kuzyk, M. G.; Sohn, J. E. J. *Opt. Soc. Am.* **1987**, 4, 968.
32. Ostroverkhova, O.; Moerner, W. E. *Chem. Rev.* **2004**, 104, 3267.
33. Bässler H. *Phys. Stat Sol. B* **1993**, 175, 15.
34. Pautmeier L.; Richert R.; Bässler, H. *Synth. Met.* **1990**, 37, 271.
35. Thomas, J.; Fuentes-Hernandez, C.; Yamamoto, M.; Cammack, K; Matsumoto, K.; Walker, G. A.; Barlow, S.; Kippelen, B.; Meredith, G.; Marder, S. R.; Peyghambarian, N. *Adv. Mater.* **2004**, 16, 2032.
36. Mecher, E. H.; Brauchle, C.; Horhold, H. H.; Hummelen, J. C.; Meerholz, K. *Phys. Chem. Chem. Phys.* **1999**, 1, 1749.
37. Horhold, H. H.; Tillmann, H.; Raabe, D.; Helbig, M.; Elflein, W.; Brauer, A.; Holzer, W.; Penzkofer, A. *Proc. SPIE* **2001**, 4105, 431

38. Wortmann, R.; Poga, C.; Twieg, R. J.; Geletneky, C.; Moylan, C. R.; Lundquist, P. M.; DeVoe, R. G.; Cotts, P. M.; Horn, H.; Rice, J. E.; Burland, D. M. *J. Chem. Phys.* **1996**, 105, 10637.
39. Marder S. R.; Beratan D. N.; Cheng L. T. *Science* **1991**, 252, 103.
40. Meyers, F.; Marder, S. R.; Pierce, B. M.; Bredas, J. L. *J. Am. Chem. Soc.* **1994**, 116, 10703.
41. Kippelen, B.; Meerholz, K.; Peyghambarian, N. 'An introduction to photorefractive polymers,' in *Nonlinear Optics of Organic Molecules and Polymers*, Nalwa, H. S. and Miyata, S. (eds.), CRC Press, Boca Raton (1997).
42. Goonesekera, A.; Ducharme, S. *J. Appl. Phys.* **1999**, 85, 6506.
43. Quintana J. A.; Boj P. G.; Villalvilla J. M.; Ortiz J.; Fernandez-Lazaro F.; Sastre-Santos A.; Diaz-Garcia M. A. *Appl. Phys. Lett.* **2005**, 87, 261111.
44. Daubler T. K.; Bittner R.; Meerholz, K.; Cimrova, V.; Neher, D. *Phys. Rev. B* **2000**, 61, 13515.
45. Van Steenwinckel D.; Hendrickx, E.; Persoons, A.; Van den Broeck, K.; Samyn, C. *J. Chem. Phys.* **2000**, 112, 11030.
46. Oh, J.; Lee, C.; Kim, N. *J. Appl. Phys.* **2008**, 104, 073709.
47. Wright, D.; Diaz-Garcia, M. A.; Casperson, J. D.; DeClue, M.; Moerner, W. E.; Twieg, R. J.; *Appl. Phys. Lett.* **1998**, 73, 1490.
48. Bauml, G.; Schloter, S.; Hofmann, U.; Haarer, D. *Synth. Met.* **1998**, 97, 165.
49. Meerholz, K.; De Nardin, Y.; Bittner, R.; Wortmann, R.; Wurthner, F. *Appl. Phys. Lett.* **1998**, 73, 4.
50. Ostroverkhova, O.; Gubler, U.; Wright, D.; Moerner, W. E.; He, M.; Twieg, R. *Adv. Funct. Mat.* **2002**, 12, 621.
51. Hou, Z.; You, W.; Yu, L. P. *Appl. Phys. Lett.* **2003**, 82, 3385.
52. Meerholz, K.; De Nardin, Y.; Bittner, R.; Wortmann, R.; Würthner, F. *Appl. Phys. Lett.* **1998**, 73, 4.
53. Herlocker J. A.; Fuentes-Hernandez, C.; Ferrio, K. B.; Hendrickx, E.; Blanche, P.-A.; Peyghambarian, N.; Kippelen, B.; Zhang, Y.; Wang, J. F.; Marder, S. R. *Appl. Phys. Lett.* **2000**, 77, 2292.
54. Miller, R. D.; Lee, V. Y.; Twieg, R. J. *Chem. Commun.* **1995**, 245.
55. Yamamori, A.; Adachi, C.; Koyama, T.; Taniguchi, Y. J. *J. Appl. Phys.* **1999**, 86, 4369.
56. Marcus, R. A. *Rev. Mod. Phys.* **1993**, 65, 599.
57. Hendrickx, E.; Van Steenwinckel, D.; Persoons, A.; Samyn, C.; Beljonne D.; Bredas J. L. *J. Chem. Phys.* **2000**, 113, 5439.
58. Grunnet-Jepsen, A.; Wright, D.; Smith, B.; Bratcher, M. S.; DeClue, M. S.; Siegel, J. S.; Moerner W. E. *Chem. Phys. Lett.* **1998**, 291, 553.
59. Hendrickx, E.; Zhang, Y. D.; Ferrio, K. B.; Herlocker, J. A.; Anderson, J.; Armstrong, N. R.; Mash, E. A.; Persoons, A. P.; Peyghambarian, N.; Kippelen, B. *J. Mat. Chem.* **1999**, 9, 2251.
60. Li, X.; Chon, J. W. M.; Gu, M. *Aust. J. Chem.* **2008**, 61, 317.
61. Thomas, J.; Norwood, R.; Peyghambarian, N. *J. Mater. Chem.* **2009**, 19, 7476.
62. Mecher, E.; Gallego-Gomez, F.; Meerholz, K.; Tillmann, H.; Hörhold, H. H.; Hummelen, J. C. *ChemPhysChem* **2004**, 5, 277.
63. Zobel, O.; Eckl, M.; Strohriegl, P.; Haarer, D. *Adv. Mat.* **1995**, 7, 911.
64. Choi, C.; Nguyen, Q. V.; Kim, N. *Macromol. Res.* **2010**, 18, 279.
65. Moon, I. K.; Choi, C.; Kim, N. *Opt. Mater.* **2009**, 31, 1017.
66. Kwon, O. P.; Lee, S. H.; Montemezzani, G.; Gunter, P. *Adv. Funct. Mater.* **2003**, 13, 434.

67. Kwon, O. P.; Montemezzani, G.; Gunter, P.; Lee, S. H. *Appl. Phys. Lett.* **2004**, 84, 43.
68. Kwon, O. P.; Kwon, S. J.; Jazbinsek, M.; Lee, S. H.; Gunter, P. *Polymer* **2005**, 46, 10301.
69. Ogino, K.; Nomura, T.; Shichi, T.; Park, S.-H.; Sato, H.; Aoyama, T.; Wada, T. *Chem. Mater.* **1997**, 9, 2768.
70. Winiarz, J. G.; Ghebremichael, F.; Thomas, J..; Meredith, G.; Peyghambarian, N. *Opt. Express* **2004**, 12, 2517.
71. Nau, D.; Christ, A.; Giessen, H.; Wagner, A.; Euteneuer, A.; Salvador, M.; Mecher, E.; Meerholz, K. *Appl. Phys. B.* **2009**, 95, 31.
72. Blanche, P.; Tay, S.; Voorakaranam, R.; Saint-Hilaire, P.; Christenson, C.; Gu, T.; Lin, W.; Flores, D.; Wang, P.; Yamamoto, M.; Thomas, J.; Norwood, R. A.; Peyghambarian, N. *J. Disp. Technol.* **2008**, 4, 424.
73. Eralp, M.; Thomas, J.; Tay, S.; Schulzgen, A.; Norwood, R. A.; Yamamoto, M.; Peyghambarian, N. *Appl. Phys. Lett.* **2006**, 89, 114015.
74. Nau, D.; Christ, A.; Giessen, H.; Wagner, A.; Euteneuer, A.; Salvador, M.; Mecher, E.; Meerholz, K. *Appl. Phys. B.* **2009**, 95, 31.
75. Zebra Imaging, Inc.; Klug, M. A.; Newswanger, C.; Huang, Q.; Holzbach, M. E. *Active digital hologram display*; US Patent 7, 227,674; June 5, 2007.
76. Hilaire, P. S.; Lucente, M. A.; Benton, S. A. *J. Opt. Soc. Amer. A* **1992**, 9, 1969.
77. Huebschman, M. L.; Munjuluri, B.; Garner, H. R. *Opt. Express* **2003**, 11, 437.
78. Blanche, P.; Tay, S.; Voorakaranam, R.; Saint-Hilaire, P.; Christenson, C.; Gu, T.; Lin, W.; Flores, D.; Wang, P.; Yamamoto, M.; Thomas, J.; Norwood, R. A.; Peyghambarian, N. *J. Disp. Technol.* **2008**, 4, 424.
79. Christenson, C. W.; Blanche, P.-A.; Tay, S.; Voorakaranam, R.; Gu, T.; Lin, W.; Wang, P.; Yamamoto, M.; Thomas, J.; Norwood, R. A.; Peyghambarian, N. *J. Disp. Technol.* **2010**, 6, 1.

25

Organic waveguides, ultra-low loss demultiplexers and electro-optic (EO) polymer devices

R. A. NORWOOD, University of Arizona, USA, C. T. DEROSE, Sandia National Laboratories, USA, C. GREENLEE, Lightwave Logic Corp., USA and A. YENIAY, Photon-X, LLC, USA

DOI: 10.533/9780857098764.3.709

Abstract: The focus of this chapter is to introduce important background information on the physical properties of polymers and then to discuss the significant progress that has occurred in polymers for passive waveguide operations, such as high-performance filters for dense wavelength division multiplexing (DWDM) systems, and in both waveguide and free-space electro-optic (EO) polymer modulators, which have applications in both high-speed communications and radio-frequency (RF) photonics.

Key words: optical polymers, electro-optic (EO) polymer, arrayed waveguide grating, perfluoropolymer, spatial light modulator, radio-frequency (RF) photonics.

25.1 Introduction and motivation for using polymer (opto)electronic components

The glassy state has long been recognized to be practically ideal for light transmission, owing to the isotropic and transparent nature of glassy media. Inorganic glass science has developed over thousands of years, reaching its zenith in the realization of optical fiber with less than 0.2 decibels/kilometer (dB/km) propagation loss, a tremendous feat that enabled the creation of the Internet as we know it. Glass science is based, fundamentally, on working with atoms, where certain atoms are known to be effective at forming glass networks, while others are better at breaking them up and also frustrating crystallization.

It has been only one hundred years since glassy synthetic polymers were first made. Polymer science progressed dramatically during the 20th century, as it was realized that manipulating molecules, not atoms, was more efficient and less costly. Polymers now dominate much of our material world owing to their low-cost (and low energy) manufacture, light weight, flexibility, and ability to be continually tailored to address the application at hand.

Polymers have played an important role in the development of information technology with epoxy composites emerging as the primary printed circuit board material, polymer-based photoresist driving the lithographic patterning on ever-denser integrated circuits, and optical polymers, which have enabled the production of compact disks, DVDs, polymer optical fiber, and claddings for glass optical fibers.

More recently, advanced optical polymers have been developed possessing exceptional properties, and that have enabled ultra-low loss optical waveguides [1], high-performance variable optical attenuators [2], 100 GHz electro-optic (EO) modulators [3], polymeric light emitting diodes [4], polymer photovoltaics [5], polymers for optical limiting [6, 7], and 3D holographic display polymers [8, 9]. The focus of this chapter is to introduce important background information on the physical properties of polymers and then to discuss the significant progress that has occurred in polymers for passive waveguide operations, such as high-performance filters for dense wavelength division multiplexing (DWDM) systems, and in both waveguide and free-space electro-optic polymer modulators, which have applications in both high-speed communications and radio-frequency (RF) photonics.

25.2 General polymer science

The extraordinary richness of organic chemistry provides the basis for polymer science, even though inorganic polymer chemistry has made significant strides in recent years. In this work we will focus our discussion on organic polymers and their optical properties.

Polymers are built up from individual units called monomers, where the monomers can comprise a large range of organic compounds that are able to react in such a way as to create a high molecular weight 'molecule' from the single molecule present at the outset. These high molecular weight species, the polymer chains, generally have a distribution of chain lengths, and possess dramatically different properties from the starting monomers, which are typically in either liquid or crystalline powder form at room temperature. These properties generally include excellent film-forming ability, moldability, flexibility, and compatibility with a large range of substrates.

25.2.1 Basic polymer chemistry

The field of polymer chemistry is very active with new approaches to polymer synthesis emerging at a steady pace. For our purposes we can learn a lot about polymer chemistry by considering two of the most popular optical polymers, poly (methylmethacrylate) (PMMA) and polycarbonate, both of which have a long and successful history in a broad range of optical applications.

PMMA is formed from the methylmethacrylate (MMA) monomer (Fig 25.1a and b). This transformation from monomer to polymer is generally accomplished by a free radical polymerization. In this approach, the monomers are combined with small amounts of a molecule called the initiator. In the case of PMMA, a common initiator is azobisisobutyronitrile (AIBN), which is typically incorporated at a fraction of a mole percent. At elevated temperatures, AIBN initiates the formation of a free radical on the MMA monomer as shown in Fig. 25.2. The resulting MMA free radical is highly reactive and, in particular, can undergo a reaction that links the free radical species to another MMA monomer, which then carries the free radical end group. This 'dimer' can then react with another MMA monomer, and so forth, leading to the formation of a long polymer chain, with the reaction conditions being principally controlled by AIBN concentration, temperature, and time.

In order to describe the polymer that is then formed we introduce the concepts of *number-averaged* and *weight-averaged* molecular weight. To determine the number-averaged molecular weight, we note that the polymer molecule weight, M, must be a multiple of the monomer molecular weight, M_0. For free radical polymerizations there are no atoms removed during

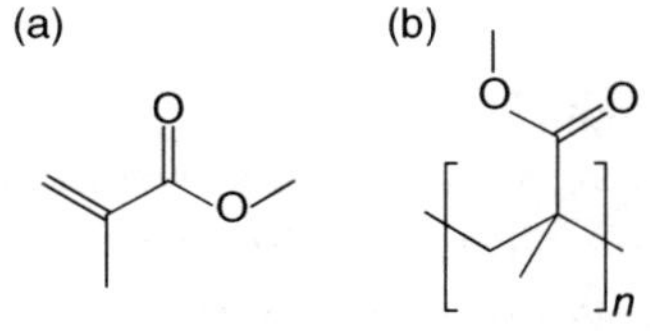

25.1 Chemical structures of (a) methyl methacrylate and (b) polymethylmethacrylate.

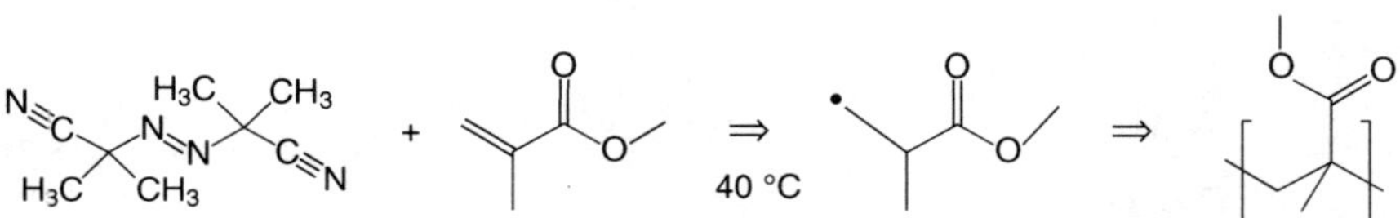

25.2 Free radical polymerization of MMA by AIBN.

polymerization, so this definition works nicely. Thus, the possible values of M can be written as M_i, where $M_i = iM_0$. The total weight is given by

$$\sum_{i=1}^{\infty} N_i M_i, \qquad [25.1]$$

where N_i is the number of polymer chains with molecular weight, M_i, while the total number of molecules is

$$\sum_{i=1}^{\infty} N_i, \qquad [25.2]$$

thus we write

$$\bar{M}_N = \frac{\sum_{i=1}^{\infty} N_i M_i}{\sum_{i=1}^{\infty} N_i}. \qquad [25.3]$$

Note that the number fraction of polymers with molecular weight M_i is

$$X_i = \frac{N_i}{\sum_{i=1}^{\infty} N_i}, \qquad [25.4]$$

whence we may write

$$\bar{M}_N = \sum_{i=1}^{\infty} X_i M_i, \qquad [25.5]$$

for the number-averaged molecular weight.

Often the properties of polymers depend on the size or weight of a given polymer molecule. We merely replace N_i with N_iM_i, the weight of chains that are i monomers long, in our previous expression to obtain

$$\bar{M}_W = \frac{\sum_{i=1}^{\infty} N_i M_i^2}{\sum_{i=1}^{\infty} N_i M_i}. \qquad [25.6]$$

Since

$$\frac{N_i M_i}{\sum_{i=1}^{\infty} N_i M_i}, \qquad [25.7]$$

is the weight fraction of polymer with molecular weight M_i we can write

$$\bar{M}_W = \sum_{i=1}^{\infty} w_i M_i. \qquad [25.8]$$

The ratio of the number-averaged molecular weight to the weight-averaged molecular weight is known as the polydispersity index (PDI) given by

$$\mathrm{PDI} = \frac{\bar{M}_W}{\bar{M}_N}, \qquad [25.9]$$

where large values of PDI indicate a broad molecular weight distribution, while low values (~1) indicate a narrow distribution.

It is important to know $\bar{M}_W$, $\bar{M}_N$, and PDI for polymer samples of interest, since these values profoundly affect important polymer properties such as solubility, mechanical strength, and viscosity in the melt.

25.2.2 Polymer physics

The physics of polymers is dominated by the fact that most polymers are in an amorphous or semicrystalline state, the nature of which dictates their mechanical, electrical, and optical properties among other aspects. In this chapter our discussion centers almost exclusively on amorphous polymers, which are generally required for integrated optics, since semicrystalline materials suffer from extensive light scattering owing to the presence of boundaries between the amorphous and crystalline regions.

Amorphous polymers as a class share some common features such as

- *low light scattering*, which results in high transparency in regions where the polymer's monomeric constituents do not have light absorption;
- *excellent dielectric properties*, owing to the difficulty of charge transport in an amorphous polymer matrix;
- *mechanical properties* that depend largely on the glass transition temperature, T_g, and the degree of crosslinking that exists between the polymer chains.

Amorphous polymers are usually weaker than their semicrystalline counterparts, which can be a challenge for optical applications in which the polymer is expected to provide structural strength, such as windows. Furthermore, polymer properties can often change in the course of repeated exposure to moisture and ultraviolet radiation leading to changes of mechanical properties in use that can be quite significant.

The glass transition temperature plays a crucial role in the physics of amorphous polymers. A glass transition is a second order phase transition, which means that while the specific heat is continuous at T_g, its derivative is not. Figure 25.3 illustrates this behavior for an idealized glass transition. Other mechanical properties, such as the coefficient of thermal expansion, are discontinuous at T_g, since the specific volume is continuous, but its derivative is not.

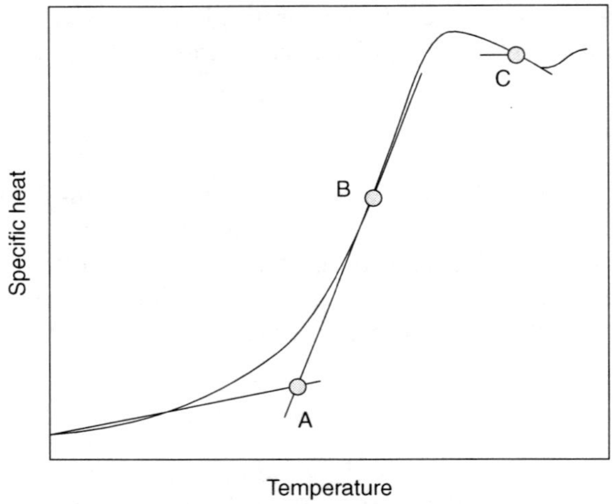

25.3 Specific heat as a function of temperature for a polymer undergoing a glass transition – the glass transition occurs at A.

For applications, it is important that the glass transition be far from the operating range. Ordinarily, we think of this as being *above* the operating range, but in many cases, such as heavily crosslinked polymers, it is also acceptable to have the T_g be well *below* the operating range. This is indeed the case for one of the most successful and rugged polymers, rubber.

The amorphous nature of optical polymers also results in a number of related physical properties. Optical polymers are generally good dielectrics, with high resistivity, low dielectric constant, and low dielectric loss. Owing to their amorphous nature, they generally have large amounts of free volume, leading to rapid diffusion of both gases and vapors into polymer films. This same free volume lends these materials high flexibility especially for films thinner than 50 μm or so.

25.3 Polymer processing

25.3.1 Solution processing

Thin polymer films, such as those used in integrated optics, are usually processed in solution and deposited by casting, dip coating or spin coating. In casting the solution is dropped onto the substrate and then allowed to dry, perhaps with heating to accelerate the evaporation of the solvent. Dip coating is usually performed using a process such as that shown in Fig. 25.4. The polymer solution is held in a container into which the substrate is lowered and then removed at a constant speed. The film thickness is a function of the concentration of the polymer in solution and the speed at which

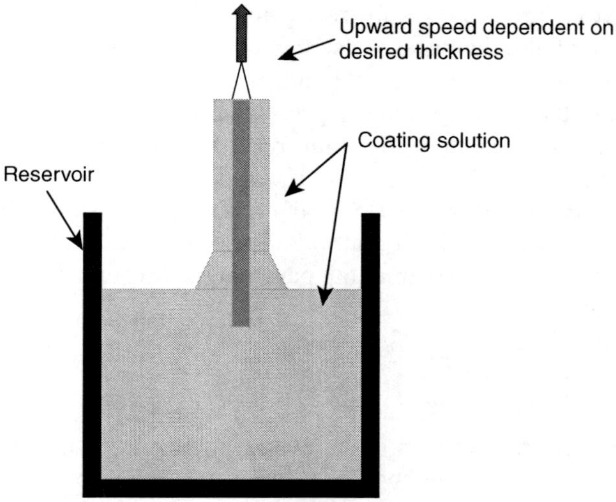

25.4 Schematic of dip-coating process for depositing polymer films.

the substrate is withdrawn. Dip coating can be used to deposit a wide range of film thicknesses (100 to 10 μm) with fairly high uniformity and reasonably good control over the absolute value of the polymer thickness. The thickness of a dip-coated film is related to the experimental parameters such as through the Landau–Levich expression [10]

$$h = \frac{0.94(\eta \cdot v)^{2/3}}{\gamma_{LV}^{1/6}(\rho \cdot g)^{1/2}} \qquad [25.10]$$

where h is the film thickness, v is the speed at which the substrate is withdrawn, g is the acceleration due to gravity, ρ is the density of the solution, and η is the viscosity. It is straightforward to set up a dip-coating apparatus since the primary requirement is a variable speed automated translation stage that can be controlled with practically arbitrary precision by using a computer for control.

For ultra-high uniformity films requiring nanometer level control over the film thickness, the preferred method for deposition of a dissolved polymer is spin coating. In spin coating, a controlled volume of the polymer solution is dispensed onto the intended substrate, often silicon owing to its superb flatness, the presence of a passivating native oxide, and control over the substrate thickness. The substrate is held, usually by vacuum, to a flat metal plate or chuck. The substrate/polymer solution are then spun at rates from 500 rpm up to approximately 5000 rpm, with the resulting polymer thickness being a function of spin speed, polymer concentration in solution and solution viscosity. Often a low-speed 'spread' cycle is used to spread the solution evenly around the substrate prior to high-speed spinning used to achieve the final film thickness. After spin coating or dip-coating, samples are dried at elevated temperatures (>100 °C) either in ovens or on hot plates. Hot plate baking is frequently preferred, since the residual solvent is then naturally forced out of the film, resulting in dense films with no trapped solvent. Oven baking works well if it is necessary to heat the film for an extended period of time, such as an annealing step, so heating cycles are often a combination of both methods. For spin coating, the film thickness can be related to experimental parameters through the following expressions such as [11]

$$h = \left(1 - \frac{\rho_A}{\rho_{A_0}}\right)\left(\frac{3\eta \cdot m}{2\rho_{A_0} \cdot \omega^2}\right)^{1/3} \qquad [25.11]$$

where h is the film thickness, m is the evaporation rate of the solvent (in length per unit time), ω is the angular velocity, η is the viscosity, ρ_{A_o} is the initial mass of volatile solvent per unit volume, and ρ_A is the final mass of volatile solvent per unit volume (after spinning). We see that concentration, spin speed, and, to a lesser extent, viscosity are the most important factors to control for spin coating.

25.3.2 Melt processing

By melt processing, we mean any process where we make use of the fact that an amorphous polymer is generally a glass and, as such, softens considerably at temperatures above T_g, becoming effectively molten at temperatures approximately 100 °C above T_g in most cases. One of the challenges of melt processing of polymers is that at the temperatures required many polymers will have an increased tendency to suffer from thermo-oxidative decomposition, and this is especially the case when dye molecules are incorporated as in the case of EO polymers. Once the polymer is well-above T_g, it can easily be stretched and formed into practically arbitrary shapes.

Melt processing is extensively used in the production of polymer optical fibers, transparent rolls of polymer film, molded lenses, and compact disks/DVDs, among many other examples. Polymer optical fibers can either be drawn from a preform, as in the case of glass optical fiber, or coextruded, a process that is used by Mitsubishi Rayon, the main producer of polymer optical fiber [12]. Figure 25.5 shows a polymer optical fiber being drawn using a conventional draw tower, which can be much shorter than glass fiber draw towers, since the processing temperatures are 200–300 °C as opposed to ≥ 1000 °C for most glasses.

Recently, there has been a renewed effort devoted to melt processing for the formation of sub-wavelength features that are useful for polarizers and

25.5 Fiber draw tower for polymer optical fiber.

waveplates, as well as optical waveguides. Figure 25.6 shows an scanning electron micrograph (SEM) of a waveguide that we have molded in an EO polymer that is single-mode [13]. As the quality of optical polymers improves and mass consumption of photonics components commences, fueled by the ever-increasing bandwidth demands of the Internet and cloud computing, there will be an urgent need for molding-based technology that can cheaply and efficiently deliver polymer photonic components.

25.4 Ultra-low loss polymer waveguide devices: materials science

25.4.1 Perfluoropolymer platform

In this section we will review ultra-loss loss single mode perfluoropolymer waveguides in terms of overall material properties, fabrication processes, devices and their applications in optical telecommunications and RF

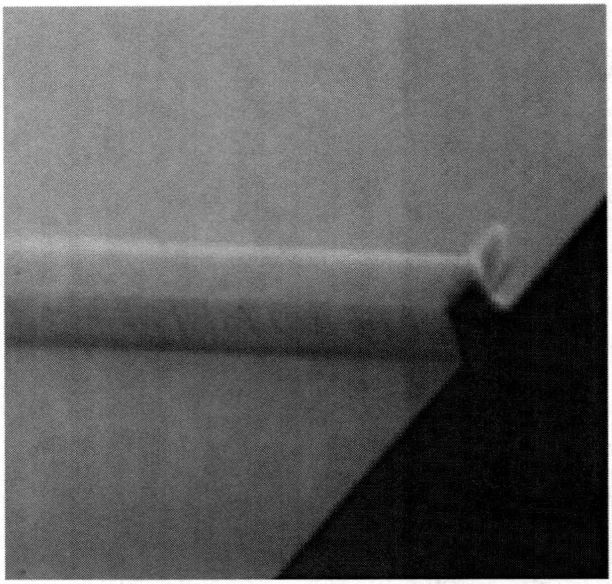

25.6 SEM of molded EO polymer waveguide – waveguide is approximately 5 × 5 μm².

photonics. Perfluoropolymers have the lowest absorption limited loss of available optical polymers, owing to the absence of carbon–hydrogen bonds, which contribute overtone infrared absorption in regions of interest for communications. Particularly, we will discuss single mode perfluoropolymer waveguide structures exhibiting polarization independent ultra-low loss of <0.05 dB/cm at 1310 nm and 1550 nm bands with a high Δn (i.e. 1.6%) as well as their optical power-handling reliability properties, which are important for both telecommunications and avionics applications. Moreover, porous structures that arise during the waveguide fabrication process will be discussed by considering implications to the propagation loss based on the Rayleigh–Mie scattering loss mechanism. We will show that the porosity needs to be reduced to below 40 nm to realize waveguide structures with ultra-low propagation losses (i.e. <0.1 dB/cm) based on the fabrication and measurement of morphologies with various degrees of porosity. Further, bending loss and respective polarization-dependent propagation loss behavior will be briefly discussed as a means to determine a minimum bending radius for realizing compact devices with ultra-low loss and polarization independent features.

Based on these insights into perfluoropolymer low-loss structures, we demonstrated long delay lines (i.e. 1 m) with compact features and highly efficient athermal arrayed waveguide grating (AWG) multiplexers and demultiplexers will be demonstrated as key building blocks for various applications. In an attempt to further improve cross-talk (xtalk) and insertion loss features of AWGs with high channel counts and high resolutions, various techniques and their implications will also be investigated, i.e. shadow mask fabrication, array tapering, etc. Specifically, RF photonics applications that benefit from the combination of AWG filtering and long delay lines (i.e. selective true time delays (TTD) and wide-band photonic channelizer modules) will be discussed as promising, potentially ground-breaking applications of this platform.

Planar waveguide technologies offer unique integrated photonic platforms for various devices with increased functionality and reduced size and cost for optical communication and RF photonics applications [14]. Numerous single-mode devices have already been realized on silica-on-silicon, ion-exchanged glass, InP and polymer waveguide platforms. The polymer waveguide platform presents several advantages that include design of the optical properties, ease in processing and device fabrication, high-density device integration, cost effectiveness, and high-volume production [15]. However, material and process-induced losses need to be addressed carefully to minimize propagation losses. Earlier reported polymer waveguides were usually fabricated with structures that contain hydrocarbon bonds (C–H), which limit the waveguide intrinsic absorption loss within the 1300–1650 nm communication window to be no less than 0.2 dB/cm [15, 16]. An effective way to reduce the intrinsic absorption loss is to design polymers in which the highly absorbing C–H groups are replaced with very low absorption loss fluorocarbon groups (C–F) [17, 18]. Various material systems such as partially fluorinated acrylates, polyimides, and silicone resins have been investigated to reduce the absorption loss; however, in order to achieve polymer waveguides with propagation loss comparable to silica waveguides, a fully fluorinated polymer material system with little, or no, C–H content is required.

Besides material absorption loss, scattering and process-induced losses also contribute to the propagation loss in polymer waveguides [19, 20]. The origin of scattering loss in polymer waveguides can be both extrinsic and intrinsic. In the simplest case, extrinsic scattering loss may result from dust, dissolved bubbles, unreacted monomer, voids and cracks in the polymer waveguide, for example. If present these features are generally large inclusions (i.e. >1 μm) with largely wavelength independent scattering. In order to avoid these potential extrinsic sources, it is routinely necessary to perform all material process and device fabrication steps under traditional clean-room conditions. On the other hand, intrinsic scattering can result, for example, from compositional inhomogeneities, or static and dynamic density fluctuations, both of which can occur on short length scales (i.e. <0.1 μm) [21, 22]. In these cases, the scattering intensity is isotropic and can be highly wavelength dependent, i.e., Rayleigh scattering. However, under ordinary conditions, loss values as small as <0.1 dB/cm are achievable in typical, high-quality, optical polymer waveguides with only these intrinsic scattering loss sources. Moreover, polymer materials generally tend to form porous structures upon spin coating and annealing due to spinodal decomposition [23]. These porous structures form additional Rayleigh–Mie scattering loss centers, where these structures must be controlled and reduced to below 40 nm size to realize waveguide structures with ultra-low propagation losses (i.e. <0.1 dB/cm).

In addition to propagation loss, polarization dependent loss (PDL) is also a critical factor. PDL usually originates from material birefringence, asymmetric waveguide cross-sections, or induced stresses, for example. Stress usually arises during waveguide fabrication steps, especially when the materials of the waveguide structure and supporting substrate have different coefficients of thermal expansion (CTE) [24]. Polarization independent performance can be obtained by utilizing very large-scale integration (VLSI) processing methods in fabricating symmetric waveguides together with a CTE matched polymer substrate [25]. In general, ultra-low loss and PDL independent performance must be complemented with a high Δn (e.g. >1.5%) in order to make it feasible to realize integrated functionalities with compact layouts; 1m long delay lines need to be bent with a maximum of 4 mm radius in order to place 8 delay line chips on a 4″ (100 mm) wafer. Thus, high Δn is another key parameter in the waveguide platform enabling long propagation lengths and high-density device integration as well as further increasing manufacturing efficiency and bringing down material and fabrication cost.

The perfluoropolymer waveguide platform is ideal for a variety of integrated optic devices that would benefit from ultra-low loss, high Δn and polarization independent loss properties, for example, delay lines, thermo-optic (TO) switches, splitters, AWGs, tunable filters, etc. In principle, devices like AWGs exhibit high temperature dependence due to their phase-sensitive nature, e. g., wavelength shifts of around 11 pm/°C for conventional silica-on-silicon based AWGs. A few different approaches have been investigated to achieve athermal AWGs. One approach involves canceling out the TO coefficient of the waveguide core material with either a reversed TO coefficient cladding material, or replacing a section of the waveguide with a reversed thermal optic coefficient material [24]. Another approach utilizes a moving fiber at the input slab waveguide so that the AWG structure is effectively dynamically changing with temperature to reduce temperature effects [25]. For the perfluoropolymer platform, one can achieve athermality by choosing a substrate with a specific CTE that matches the TO coefficient of the waveguide, where all polymer configurations may provide the ideal approach owing to polymers' intrinsically positive CTE and negative TO coefficient properties [26]. Further, the all-polymer design may be configured with a superstrate overlay approach that sandwiches the polymer waveguides in between a polymer superstrate and a polymer substrate. By tuning the superstrate thickness and therefore the effective thermal expansion coefficient, the AWG temperature dependence may be tuned to achieve athermal performance [27]. Using this technique, the temperature shift of a perfluoropolymer AWG was shown to be around –0.5 pm/°C. In an attempt to further improve xtalk and insertion loss features of AWGs with high channel counts and high resolution, various techniques and their implications are discussed such as shadow mask fabrication and array tapering, etc.

Besides known applications such as cost-effective photonic devices for wavelength division multiplexing (WDM) networks (e.g. splitters, TO switches, AWGs), the perfluoropolymer waveguide platform offers unique benefits for RF photonics applications, especially the integration of AWG filtering with long delay lines for selective TTD and high-resolution wide band photonic channelizer modules. The optical TTD technique is an important element for both RF photonics and optical telecommunication applications [28, 29]. On the RF photonics side, besides many advantages of optical signal processing techniques that include low distribution loss, reduced size, higher frequency operation and increased immunity to electromagnetic interference (EMI), optical TTD is a key enabling technique particularly for phased array antenna (PAA) systems where electronic techniques are inadequate. On the optical telecommunications side, TTD is also a promising technique for providing dynamic optical buffering elements for optical routing in all-optical networks as well as encoder/decoder elements for wavelength code division multiple access (CDMA) networks to increase the correlation properties of codes and network security [30]. Both the optical telecommunication and RF photonics applications require similar specifications for TTD modules such as low insertion loss, low coupling loss to single mode fiber (SMF) and low polarization dependence, all of which can be met by the perfluoropolymer waveguide platform.

The channelizer module is another promising application for RF photonics as a means to enable a large portion of the RF spectrum (i.e. 2–20 GHz) to be simultaneously monitored and analyzed. A channelizer is basically an AWG with extremely high resolution that can be realized with large length differences among the array grating lengths, for example 1 GHz resolution requires ~20 cm length differences between the shortest and longest grating arms [31]. The channelizer requires innovative AWG designs with precise fabrication processes and active phase control of individual grating arms. Although there are still fabrication challenges in achieving a wide band channelizer, the initial results are promising, where interlayer coupling and coarse filtering have been achieved [32]. The photonic channelizer module offers an order of magnitude reduction in the power, weight, and cost as compared to an electronic system, yet several research objectives (i.e. active phase control via TO, multilayer coupling control, etc.) must be met to realize such unique devices on the perfluoropolymer platform.

25.4.2 Perfluoropolymers

Perfluoropolymer material systems offer superior properties for optical waveguides in terms of transparency and stability owing to their hydrocarbon free and stable molecular structures. There are several commercially available perfluoropolymer materials with various grades, for example poly[2,3-(perfluoroalkenyl) perfluorotetrahydrofuran] (CYTOP™, from Asahi Glass), poly[2,2-bistrifluoromethyl-4,5-difluoro-1,3-dioxole-co-tetrafluoroethylene] (Teflon AF™ from Dupont Chemicals) and poly[2,2,4-trifluoro-5-trifluoromethoxy-1,3-dioxole-co-tetrafluoroethylene] (HYFLON AD™, from Solvay Solexis). Figure 25.7 illustrates the corresponding chemical structures for (a) Cytop™, (b) Teflon AF™, and (c) Hyflon AD™. A combination of these systems with each other or with various dopants can be utilized for refractive index tailoring to desired Δn values. These systems generally need to go through purification processes in a cleanroom environment before waveguide fabrication processes.

Table 25.1 shows the vibrational overtones with corresponding central peak wavelengths, normalized intensities and absorption losses. Polymer waveguide material absorption loss in the 1300–1650 nm range predominately comes from the overtones of C–H bond vibrations. The strong bond and high effective mass of the C and F bonded atoms result in low vibrational energies (~10 μm) and, correspondingly, relatively low associated absorption coefficients. As seen from the Table 25.1, with hydrogen being the smallest atom, the C–H bond has a vibrational absorption peak at 3390 nm, and, thus, second and third overtone modes, extending well into the 1300–1650 nm range accompanied by high absorption. As an example, acrylate polymers containing C–H bonds exhibit an absorption loss of about

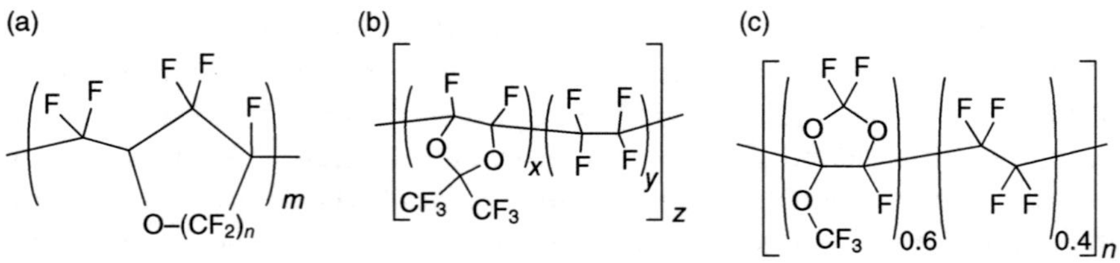

25.7 Chemical structures of several perfluoropolymers of interest for low-loss optical waveguides: (a) Cytop™, (b) Teflon AF™, and (c) Hyflon AD™.

Table 25.1 Wavelengths, intensities and absorption losses of some important vibrational bond structures

Bond	Wavelength (nm)	Intensity (a.u.)	Absorption (dB/cm)
C–H	$n_1 = 3390$	1	500
C–H	$n_2 = 1729$	7.2×10^{-2}	36
C–H	$n_3 = 1176$	6.8×10^{-3}	3.4
C–D	$n_3 = 1541$	1.6×10^{-3}	0.8
C–D	$n_4 = 1174$	1.3×10^{-4}	0.065
C–F	$n_5 = 1626$	6.4×10^{-6}	0.0032
C–F	$n_6 = 1361$	1.9×10^{-7}	0.000095
C–F	$n_7 = 1171$	6.4×10^{-9}	0.0000032
C=O	$n_3 = 1836$	1.2×10^{-2}	6.0
C=O	$n_4 = 1382$	4.3×10^{-4}	0.22
C=O	$n_5 = 1113$	1.8×10^{-5}	0.009
O–H	$n_2 = 1438$	6.0×10^{-2}	30
O–H	$n_2 = 979$	5.2×10^{-3}	2.6

0.75 dB/cm at 1550 nm. As a comparison, C–F bond absorption only starts to extend into the 1300–1650 nm range with its fifth, sixth, and seventh overtones with associated relative absorption intensities several orders of magnitude lower than that of the C–H bond. For perfluoropolymers with no C–H bonds, waveguide losses of <0.03 and <0.07 dB/cm were realized for low and high Δn waveguide structures, respectively, in the 1300–1650 nm range [17, 18]. The absorption limited loss of such perfluoropolymer waveguide is well below 0.001 dB/cm.

25.5 Ultra-low loss polymer waveguide fabrication and process-induced losses

As is well known in addition to absorption loss, scattering and process-induced effects can also contribute to the propagation loss in single mode polymer waveguides [13–22]. Under ordinary conditions, values as negligibly small as <0.01 dB/cm are achievable in typical, high-quality, optical polymer waveguides from intrinsic scattering loss sources. Various conditions affecting these general contributions to loss in perfluoropolymer waveguide structures remain under continued study.

Process-induced scattering losses in polymer waveguides can be encountered from surface roughness and stress-induced scattering. Polymer thin films are typically formed by spin coating process steps that include depositing from solution and subsequently drying in an oven. The resultant films are generally uniform but may have roughness features that contribute to scattering losses. Single mode waveguides that are then fabricated from these thin films can also contribute additional roughness features, especially by way of lithographic process steps in which dry or wet etching define the waveguide structure. Thus, these spin coating and etch process steps must be performed with sufficient care to keep the overall surface roughness to a minimum, for example, <40 nm RMS is required for <0.03 dB/cm loss. Another process-induced loss contribution is stress-induced scattering. A general observation is that in the deposition of a thin film on a solid substrate, a crucial aspect of the process is managing the naturally occurring stress that is primarily induced from the film–substrate materials mismatch. In planar glass waveguide fabrication, for example, the effects of stress are even more pronounced because of the high temperature deposition (e.g. >1000 °C) involved, often necessitating follow-on annealing operations. In polymers, thin films for waveguide fabrication are usually deposited from liquid solution at relatively low temperatures (e.g., ~*RT*) with solvent bakes and annealing steps occurring at moderate temperatures (e.g., 100°C). It is well known that these temperature excursions, when coupled with the mismatch in the CTE between the waveguide and substrate, may yield high stress values and consequent large birefringence and polarization dependence. In recognition of these potential problems, polymer waveguide and polymer substrate materials have been arranged with appropriate CTEs to avoid large scattering losses [17].

Typical fabrication process steps for perfluoropolymer waveguides are illustrated in Fig. 25.8. We fabricated buried channel waveguides with our

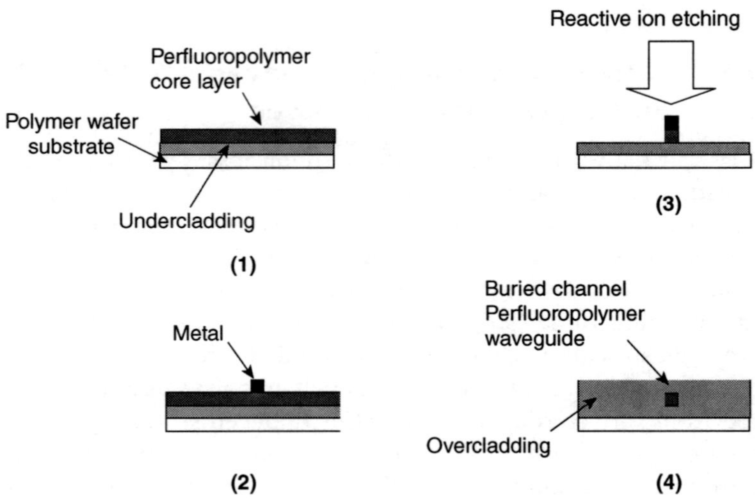

25.8 Process steps that were used for perfluoropolymer waveguide fabrication.

perfluoropolymer materials utilizing standard silicon VLSI fabrication steps, such as spin coating of undercladding and core layers (1), photolithography (2), reactive ion etching (3) and a final spin coating of overcladding (4). Figure 25.9(a) and (b) show an SEM micrograph of a channel waveguide after core etching processing, and a micrograph of the cross-section of our waveguides, respectively. The waveguide is symmetric (4 μm × 4 μm) with a core-cladding refractive index difference of about 1.6%. We measured the propagation loss spectrum of our waveguide to high precision by fabricating a 100 cm long waveguide chip with a footprint of about 3 cm × 4 cm. The layout of the chip follows a spiral pattern, as depicted in Fig. 25.10, to ensure minimal bending loss. Coupling and propagation losses are distinguished by measuring the loss of a 5 cm long reference channel on the same chip. There are also four channels following the spiral and references in order to ensure consistency and waveguide yield. The measured propagation loss spectra over the 1310, 1440, and 1550 nm bands are shown in Fig. 25.10. The waveguides were pigtailed with a high numerical aperture (NA) fiber with a mode field diameter matched to our waveguide; coupling loss was measured to be 0.4 dB per interface. As seen in Fig. 25.10, the waveguide structure exhibits a propagation loss of <0.05 dB/cm at 1310 nm and 1550 nm while <0.1 dB/cm at the 1440 nm bands for all four channels. Furthermore, we measured the overall PDL of the waveguides to be random and less than 0.004 dB/cm owing to the use of a CTE matched polymer as the substrate. This CTE matching enabled low-stress loading during the waveguide fabrication process, resulting in polarization independent performance. Note that insensitivity to wavelength is an important feature for

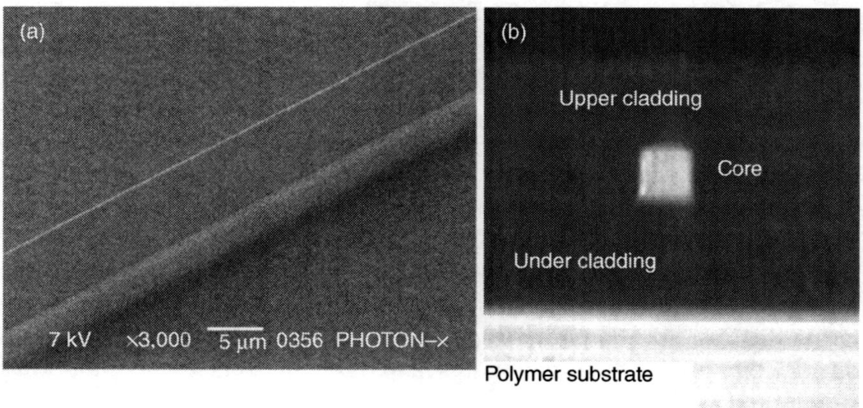

25.9 (a) SEM micrograph of a channel waveguide after core etching, (b) optical micrograph of a perfluoropolymer waveguide interface with light emitting from a 4 μm × 4 μm symmetric waveguide.

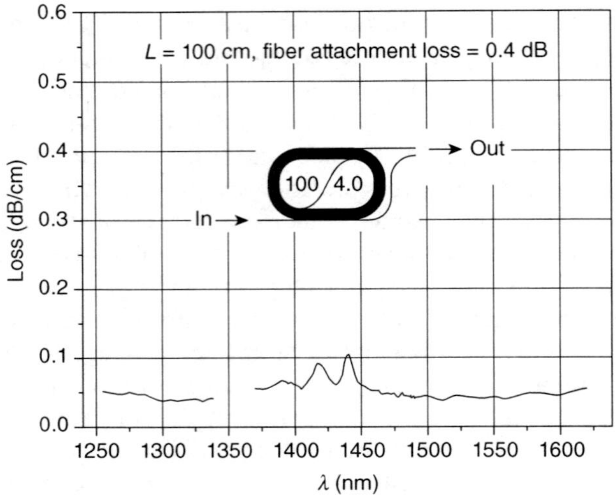

25.10 Measured loss spectra of the perfluoropolymer waveguides.

devices operating over a broadband. Although there is still a minimal O–H absorption peak at 1440 nm due to residual solvent or polymer end groups, the overall loss spectrum is comparable to that of commercially available silica optical waveguides.

25.5.1 Nanoporosity effect

As mentioned above, the total optical loss for an optical waveguide consists of absorption and scattering losses and should be kept to a minimum. Perfluoropolymer waveguides exhibit minimal absorption losses in the visible and near-infrared ranges. However, there are contributions to the optical scattering loss consisting of volume scattering due to discontinuities in the refractive index from density or composition variations, and surface scattering, arising from waveguide wall roughness. The volume scattering contribution is intrinsic and associated with the well-known classic $1/\lambda^4$ dependence of Rayleigh scattering that is present in the limit where the spatial dimensions of the refractive index discontinuities are much smaller than the vacuum propagation wavelength λ of the light guided inside the waveguide. There is an important additional contribution to volume scattering that can arise in perfluoropolymer thin films, namely a contribution due to the presence of nanopores and their clusters into nanoporous structures.

The formation of these nanoscopic aero-structures is presumed to occur as follows. Under standard spin-coating conditions of the perfluoropolymer solution, there is a natural interplay between various material properties, such as surface energy, solubility, polymer glass transition temperature, entropy, and processing conditions that include temperature, pressure, and ambient atmosphere, which can result in amorphous perfluoropolymer thin films exhibiting extensive nanopore and nanoporous structures. The behavior of the polymer solution is described by treating the solution as a binary system of perfluoropolymer and miscible solvent that is unstable to temperature or concentration fluctuations. Under standard thin film fabrication procedures, the polymer solution is first prepared under critical miscibility conditions such that, at room temperature, the polymer is completely miscible with the solvent, and the solution is thermodynamically stable at a critical concentration C_c and critical miscibility temperature T_c. During spin coating, rapid solvent evaporation takes place causing the binary system to cool to temperatures below T_c into a thermodynamically unstable regime, where the solution becomes a metastable liquid–liquid dispersion. Upon further cooling and solution concentration, the metastable dispersion undergoes a highly nonlinear, evaporation-induced phase separation, or equivalently polymer precipitation, driven by spinodal decomposition. The major result is the formation of a perfluoropolymer thin film containing large regions of nanopores and arrays of nanopores clustered into extended nanoporous structures.

The morphologies of these nanoporous structures have been observed to demonstrate critical dependence on the thermal annealing profile for our waveguide material following the spin-coating process. Temperature ramping rates, especially near the T_g, and total annealing duration at T_g can be configured to obtain a wide range of nanoporosity morphologies. We experimentally optimized the annealing profile for the minimum pore size and volume that would yield the lowest scattering losses. The wafer annealing process usually starts from room temperature, from which the temperature is slowly ramped (e.g. 1 °C/min) up to the T_g of the core material. Then, the temperature is kept at T_g for approximately one hour to release the stress originated by film formation and the difference in CTE of core and cladding materials. Next, another slow temperature ramp is applied up to 20 °C above T_g, where the wafer remains for 24 h. Finally, a reverse temperature ramp profile with the identical rates is applied to cool down the wafer. Figures. 25.11(a) and (b) show SEM images for two processed thin films with different temperature annealing processes. The perfluoropolymer nanoporous structures are clearly evident directly under SEM. The structure, shown in Fig. 25.11(a), was annealed at 20 °C above T_g for only several minutes with the same ramping rate for both baking and cooling processes whereas the structure, shown in Fig. 25.5(b), followed the optimized process as described above. Numerous experiments also indicated that annealing at 20 °C above T_g for at least 24 h results in both reducing the existing pore sizes and pore numbers. Moreover, annealing at lower temperatures with

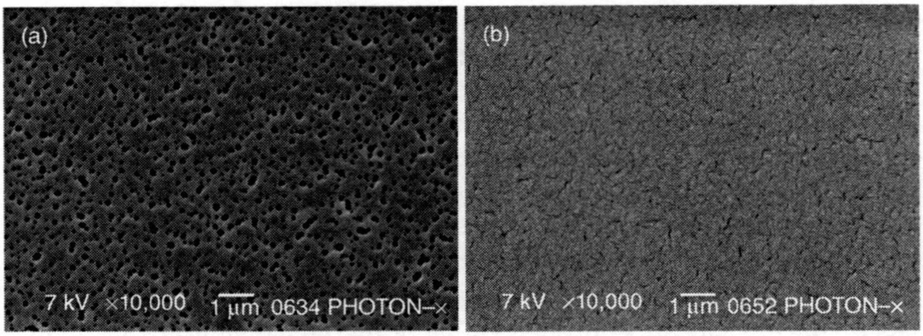

25.11 SEM micrographs of waveguides after core etching with (a) nanopore sizes of 100 nm and (b) nanopore sizes of 10 nm.

the same duration generally resulted in larger pore sizes. We further observed that higher temperature (i.e. >20 °C above T_g) annealing caused permanent waveguide shape deformations although the porous structure still remained similar to that of Fig. 25.11(b). With regard to implications on the induced loss, this extrinsic scattering contribution arising from the presence of the nanopores is described by the well-known Rayleigh–Mie theory [20, 21]. The nanopores are here viewed as nonabsorbing spheres of refractive index n_{por} evenly distributed in the perfluoropolymer host matrix. The associated scattering loss α, in dB per unit length, is dependent on the nanopore diameter d, which is assumed to be smaller than the vacuum propagation wavelength λ of the light guided inside the waveguide, the refractive index ratio of the nanopores and the surrounding host material $m = n_{por}/n_{sur}$, and the volume fraction of the nanopores in the host V_p. To fabricate a certain optical component with a nanopore-induced scattering loss budget of α, the pore diameter d should satisfy the following relationship:

$$d < \left[\alpha \frac{1}{1.692 \times 10^3} \left(\frac{m^2+2}{m^2-1} \right)^2 \frac{\lambda^4}{V_p} \right]^{1/3} \qquad [25.12]$$

It is challenging to distinguish each process-induced loss contribution accurately, since these mechanisms are somewhat interdependent of each other. For example, high porosity induces larger wall roughness, material nonuniformity and deformation, each of which separately contributes to the overall propagation loss value in addition to scattering induced propagation loss. In addition, the refractive index also tends to reduce with larger porosity, which may also lead to a slight reduction in the Δn of the waveguide, resulting in relaxed light confinement and larger bending loss values, compared to those of structures with a smaller porosity. Yet, useful prediction can be obtained from (25.12). As an example, for a structure with $n_{sur} = 1.34$

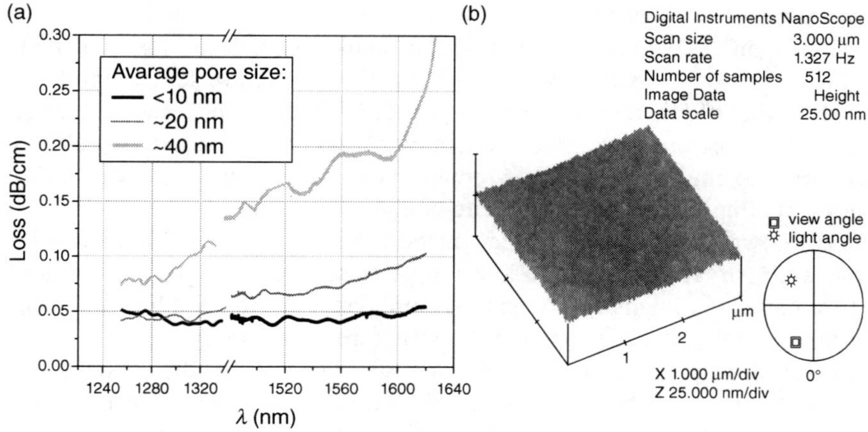

25.12 (a) Measured loss spectra of the perfluoropolymer waveguides with various degrees of nanoporosities and (b) atomic force microscopy (AFM) image of a channel waveguide with <10 nm pore sizes after core etching.

and $V_p = 1\%$ at $\lambda = 1550$ nm, the nanopore diameter d must be smaller than 50 nm for a Rayleigh–Mie scattering induced loss budget of $\alpha < 0.1$ dB/cm.

Figure 25.12(a) shows measured propagation loss spectra over the 1310 and 1550 nm bands for three structures exhibiting porous morphology with approximately <10, ~20, and ~40 nm pore sizes. The structures with different pore sizes were fabricated by simply varying the overall annealing time at 20 °C above T_g, i.e. ~24, 4, and 1 h, respectively. The waveguide structure with <10 nm pore sizes exhibits a uniform propagation loss of 0.05 dB/cm at 1550 nm while the structures with ~20 and ~40 nm yield approximately 0.07 and 0.18 dB/cm propagation losses at 1550 nm, respectively. Note that wavelength dependence becomes predominant as the pore sizes increase for both structures, a clear indication of Rayleigh–Mie scattering. It is also seen that the loss spectrum of the structure with the ~40 nm pore sizes produce several peaks that may also result from the remaining solvent due to the short annealing time. An atomic force microscopy (AFM) image of a surface of the optimized waveguide (i.e. <10 nm pore sizes) is shown in Fig. 25.12(b), where the waveguide shows a superior surface with a smoothness of <0.4 nm in the absence of porosity.

25.5.2 Bending loss

Bending loss is another key property for optical waveguides, where the minimum lossless bend radius eventually determines the size and component density that can be integrated on a single chip. Here, the main underlying loss mechanism is the field mismatch at the transition between the straight and curved waveguide that leads to radiation loss and additional scattering loss from wall roughness [33]. There are several modeling and computational methods that include full-vectorial finite-element, effective index, eigenmode expansion, and planar wave boundary methods reported to analyze the induced radiation loss, mode coupling, and polarization coupling mechanism due to the bending in optical waveguides [34–36]. It is known that bending losses can be reduced either by increasing the radius of curvature or by utilizing a large refractive index profile with high Δn waveguide designs. Here, experimental results on the bending loss properties of our particular waveguide structures are briefly presented. The measured propagation loss spectra for two different bending radii (i.e., 2 and 4 mm) over 1550 nm band are shown in Fig. 25.13(a). The measured waveguides are 100 cm long with a core cross-section of 4 μm × 4 μm and Δn of 1.6%. Fig. 25.13(b) illustrates 50 cm and 100 cm long fabricated delay lines with 4 mm bending radius, where the layout of the chip follows a spiral pattern to ensure minimal bending loss. As can be seen in Fig. 25.13(a), the waveguide structure with 4 mm bending radius exhibits a propagation loss of <0.05 dB/cm while a smaller bending radius of 2.5 mm gives rise to <0.075 dB/cm propagation loss at 1550 nm, as well as demonstrates a typical bending loss induced wavelength dependence. Moreover, bending loss breaks the symmetry between polarization modes and causes transverse electric (TE) and transverse magnetic (TM) polarized modes to exchange power in the bent waveguides, leading to polarization – xtalk [37]. Consist-

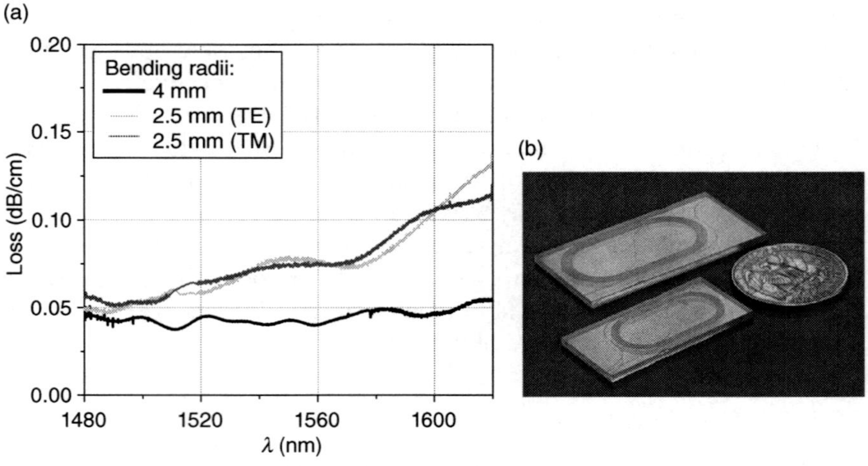

25.13 (a) The measured propagation loss spectra for two different bending radii over 1550 nm band and (b) 50 and 100 cm long fabricated waveguide chips with 4 mm bending radius.

ent polarization dependence was also observed by introducing bending loss for waveguide structures with a reduced bending radius of 2.5 mm, while the structure with a bending radius of 4 mm exhibited a polarization independent loss spectrum. In addition to the optimized annealing process, CTE matching between waveguide and substrate materials enabled low stress loading during the waveguide fabrication process, resulting in negligible polarization dependent propagation loss.

25.5.3 Power-handling reliability

Earlier generation polymer waveguides had severe optical power-handling reliability issues that prevented the realization of practical devices, primarily due to infrared absorption induced photodegradation. In particular, the oxygen molecule has strong absorption in the near infrared and in the excited state it can convert to the highly active singlet form which can react with even mildly conjugated polymers, causing photo-induced local variations of refractive index, with the resulting loss of optical confinement causing the degradation in performance. These problems are absent from perfluoropolymer waveguides with an ultra-low absorption of <0.001 dB/cm in the near IR telecom windows and no active sites for singlet oxygen attack. Figure 25.14 shows the power-handling properties of perfluoropolymer

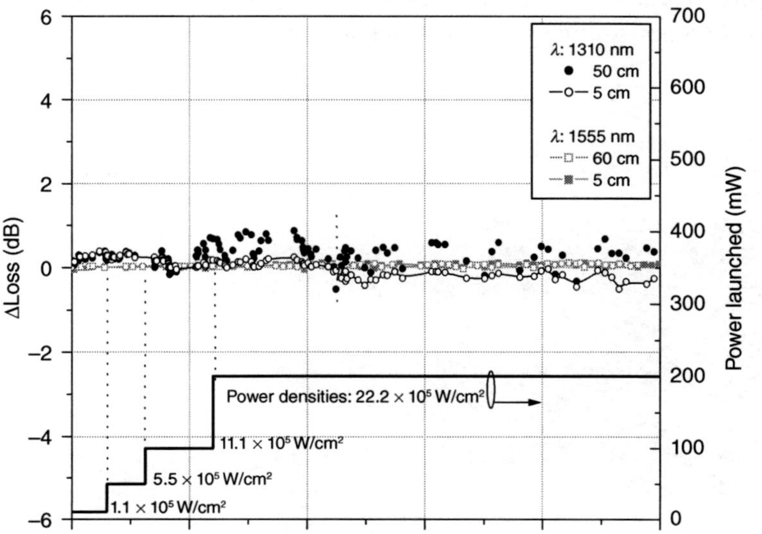

25.14 **Perfluoropolymer waveguide insertion loss as a function of time at various 1550 and 1310 nm input signal power levels.**

waveguides with 0.05 dB/cm and 0.07 dB/cm propagation losses for 1310 and 1550 nm, respectively, where the insertion loss is plotted as a function of time for input power at 10, 50, 100, and 200 mW for 1310 and 1550 nm wavelengths. Here, 5 and 50 cm waveguides are used to identify any potential failure in fiber coupling or in waveguide loss [17]. The results indicate that both the 5 and the 50 cm long waveguides have an insertion loss variation less than 0.1 and 0.9 dB for 1550 and 1310 nm wavelengths, respectively, after 2000 h of exposure. The slightly higher loss at 1310 nm may come from oxygen molecule absorption, which is higher near 1310 nm.

25.5.4 Athermal phase sensitive waveguide devices

AWG multi/demultiplexers are key components for high channel-count wavelength division multiplexing systems because of their use in multiplexers, demultiplexers, add-drop multiplexing, and routing. AWG multi/demultiplexers have already been realized using silica-on-silicon, ion-exchange glass, InP, Si, and polymers. In comparison with other passive waveguide approaches, perfluoropolymers offer an attractive platform owing to their ultra-low loss and potential low cost fabrication on various kinds of substrates. Figure 25.15(a) shows an SEM micrograph of the waveguide structures at the output of the perfluoropolymer AWG, fabricated with VLSI fabrication steps as described above for a 4 μm × 4 μm waveguide with a Δn of 1.6%. The AWG operates in the 66th order, free spectral range (FSR) of 23.5 nm, grating waveguide number of 126 and waveguide path differ-

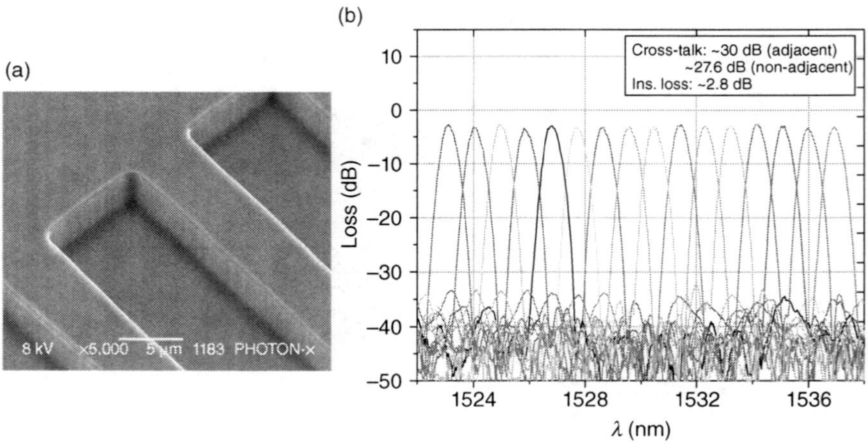

25.15 (a) SEM micrograph of perfluoropolymer AWG output region and (b) measured fiber to fiber transmission spectra of the perfluoropolymer AWG.

ence of 77 μm [17]. The fiber-to-fiber spectral transmission characteristics of the AWG were measured with a tunable laser and an optical spectrum analyzer. The measured data in Fig. 25.15(b) shows adjacent xtalk levels of about –30 ± 2 dB, non-adjacent xtalk of –28 ± 2 dB with an insertion loss of 2.8 ± 0.3 dB. The polarization dependent shift is less than 0.1 nm.

In addition, owing to their phase-sensitive nature, AWG output channel wavelengths usually exhibit highly sensitive temperature dependence as mentioned above. A few different approaches have been developed to achieve athermal AWGs so as to eliminate temperature controllers and their associated issues such as higher cost, increased power consumption, reduced reliability, and greater device size [24, 25]. In contrast, all polymer configurations offer athermal operation once waveguide and substrate material properties are carefully adjusted [26, 27].

The wavelength shift versus temperature is given by $d\lambda/dT = (\lambda/n)(dn/dT + n\cdot\alpha)$ where λ is the wavelength, T is temperature, n is effective index of the waveguide, α is CTE, and dn/dT is the TO coefficient experienced by the waveguide array section within the AWG device. The athermal condition is therefore: $dn/dT = -n\cdot\alpha$. In the usual case where the thin waveguide layers are placed on top of a much thicker substrate, the substrate CTE dictates the value of α. As depicted in Fig. 25.16(a), the temperature dependence of this configuration for the perfluoropolymer AWG on a polymer substrate is 12 ± 0.4 pm/°C, which is essentially the same as the silica-on-silicon value. It is also found that by attaching a polymer superstrate with a lower thermal expansion coefficient and adjusting its thickness, the value

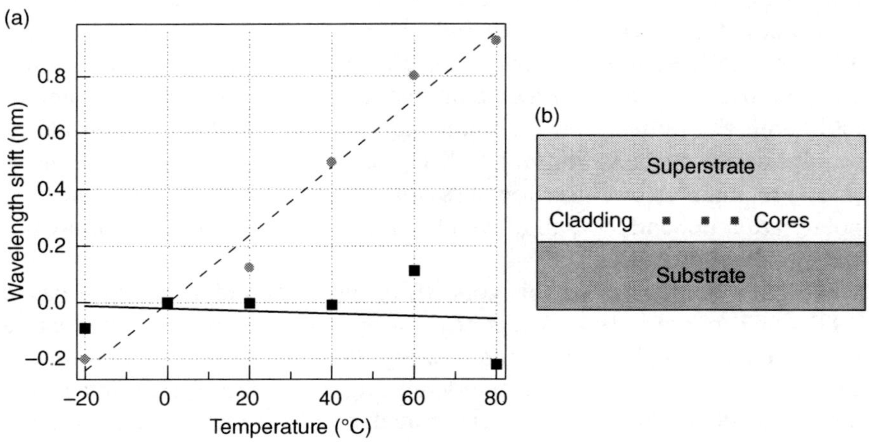

25.16 Measured temperature dependence of AWG channel wavelength shift (a) without superstrate (-------) and with superstrate (____); (b) schematic of superstrate structure.

of α can be fine-tuned so that the athermal condition is satisfied as shown in Fig. 25.16(b). With superstrate tuning, the temperature dependence of the channel wavelength shift can be reduced to –0.5 ± 0.4 pm/°C over a wide range from –20 °C to 80 °C as shown in Fig. 25.16(a). This technique has the immediate advantage of tunability after the device is fabricated. Further, the superstrate can be chosen among a large variety of materials since it need not meet the requirements of the waveguide substrate, such as surface smoothness and stability under device fabrication conditions.

25.6 Perfluoropolymer-based true time delay (TTD) modules

The perfluoropolymer waveguide platform offers unique benefits for RF photonics applications, especially the integration of AWG filtering with long delay lines for selective TTD and high-resolution wide band photonic channelizer modules. The optical TTD functionality is important for both RF photonics and optical telecommunication applications [28, 29]. On the RF photonics side, optical signal processing techniques present various advantages that include low distribution loss, reduced size, higher frequency operation, and increased immunity to EMI. More importantly, TTD configurations can be implemented via photonic techniques that would revolutionize PAA systems for squint-free, agile, wide-band and frequency independent operations where electronic techniques are inadequate. TTD modules also have applications in wavelength hopping optical CDMA, which is a practical network control strategy because it overcomes the limitations of switching time associated with optoelectronic devices and the processing delay associated with network protocols; in this respect optical TTD is ideally suited for performing the function of an optical CDMA encoder–decoder, encoding a data bit in both wavelength and time domains [30]. Both the optical telecommunication and RF photonics applications require similar specifications for TTD modules. These overlapping specifications are mainly low insertion loss, low coupling loss to SMF and low polarization dependence, all of which can be met by the perfluoropolymer waveguide platform.

We have demonstrated this concept using an all-polymer AWG-based TTD in a symmetric feedback configuration, also known as a recirculating photonic filter (RPF), as illustrated in Fig. 25.17(a). The modulated optical carrier was steered by the waveguide grating to the appropriate integrated delay line, depending on the carrier wavelength. The signal is delayed and fed back into the symmetric input port via optical fiber delay lines. The grating then focuses the delayed beam into the common output port. This TTD can be considered topologically equivalent to two AWGs in series with

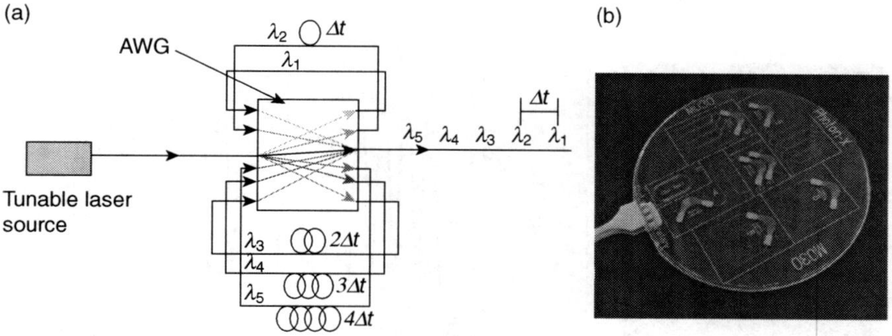

25.17 (a) Schematic of proposed wavelength selective TTD system; (b) assembly of monolithically integrated AWG and delay lines for 4-bit operation.

a delay element in between. Thus, three sequential operations are performed in a single AWG: wavelength demultiplexing, time delay, and wavelength multiplexing. It is important to note that the RPF has no $1/N$ loss; all the power at a given wavelength is diffracted into the output port. Thus, high resolution (e.g., 8 bits) can be obtained with low loss in a compact integrated device. Furthermore, digital control of the TTD module can be effected at a tunable laser module with up to 8-bit configurations, while monolithically integrated AWG and delay lines provide all passive processes, resulting in more simplicity in the system. Figure 25.17(b) shows a 4″ (100 mm) polymer wafer with several fabricated perfluoropolymer 4-bit TTD chips with various delay increments. VLSI fabrication of monolithically integrated AWG and delay lines enable superior control of the resolution of the delay lines (~1 μm), thus providing sub-picosecond time delay control. Figure 25.18(a) shows a mask design of a TTD module in the RPF configuration. In this design, all wavelength channels will loop back into the AWG and focus into a single channel waveguide at the output end. This design provides 4-bit operation with a maximum relative time delay of 465.5 ps with 10 ps spacing. If longer delay is required, the delay lines in the looped back region can be designed with spiral turns to generate more delay time without a significant increase in chip size. Figure 25.18(b) shows the output characteristic of a 16-channel TTD device with 100 GHz channel spacing. The fiber-to-fiber spectral transmission characteristics of the TTD device were measured in high resolution by synchronizing the wavelength sweeping function of an external cavity laser (ECL) with an optical spectrum analyzer (OSA) via external triggering. In this technique, the measurement resolution bandwidth is determined by the narrow linewidth of the ECL (i.e. 200 kHz). All channels go through the AWG twice and respective delay lines in between. The insertion loss of the module varies from 3.3 to

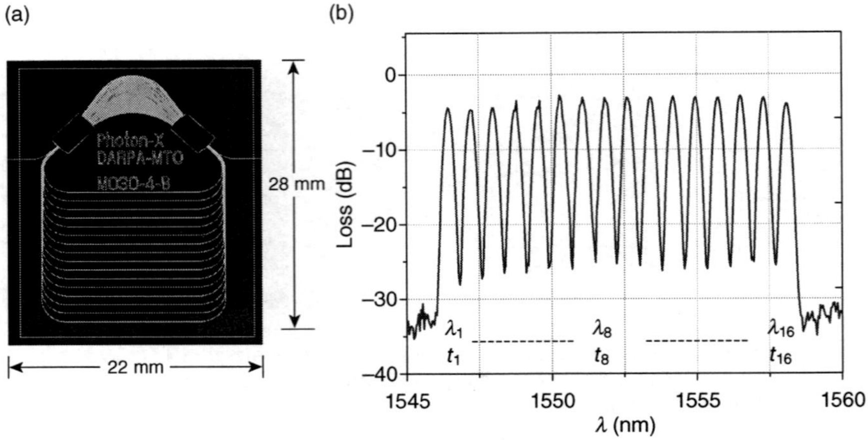

25.18 (a) Low-loss 4-bit module mask design; (b) transmission spectra of the TTD channels with 10 ps increments.

4.8 dB due to additional loss acquired from delay line changes and AWG transmission characteristics, where the side channels suffer slightly higher loss compared to those at the center. TTD minimum and maximum delay times were arranged to be 315.5 ps (Channel#16) and 465.5ps (Channel#1), respectively. The overall TTD chip size is 22 mm × 28 mm. Furthermore, the delay length of each individual line can be set at a desired length for different applications, and the resolution is within a micron since the features are directly transferred from the photomask with high fidelity. The unique athermal perfluoropolymer platform assures that the delay time of this TTD module will not change (i.e. $\Delta\lambda \leq 0.1$ nm) within the operational temperature range from 0 to 65 °C.

In considering further performance improvement (i.e. insertion loss and xtalk reduction), of the perfluoropolymer-based AWGs, we can see that the total loss of an AWG has two main origins; one is waveguide propagation loss, and the other is mode mismatch loss that occurs at the junction region between the slab and arrayed waveguides. The AWG loss from mode mismatch occurring between the slab modes and arrayed waveguide modes in the junction region is due to the presence of physical gaps between the arrayed waveguides at the junction. The gaps are formed because of the finite resolution of photolithography together with isotropic etching taking place during reactive ion etching (RIE) processing steps. There are several methods to reduce the effect of the presence of these gaps. These include (a) reducing the spacing between the gaps (i.e., gap reduction) [38]; (b) introducing island-like structures in the immediate vicinity of the gaps (i.e., island structures) [39]; or (c) gradually smoothing, or tapering, out the mode transition region (i.e., vertically tapered structures) [40]. The individual methods can also be combined to achieve even greater improvement.

Upon looking at all three of these methods singly and in various combinations, one finds that gap reduction will result in a direct reduction in loss; however, it reaches a limit where new lithography techniques and etching methods have to be introduced and applied. Further reduction of the gap size will introduce unstable and irregular features in the photolithographic mask that will transfer into the patterned AWG as scattering centers, resulting in increased loss and xtalk.

Island structures are significantly more effective when the designed gap size is relatively large. The island functions as a local wavefront velocity modifier so that when the wavefront reaches the arrayed waveguides, the mode shape is similar for each channel waveguide cross-section. The presence of the islands theoretically can reduce the loss in a manner similar to gap reduction. However, island structures also have a resolution limit associated with the slightly rounded edges and corners of each individual island that greatly reduce the efficiency and even increases loss.

Vertically tapered structures cause the field in the transition region to change gradually. The height of the 3D taper is the same as the height of the slab and decreases with distance from the junction, finally reaching zero at a distance of a few hundred microns from the slab-arrayed waveguide junction. With this approach, light propagating from the slab propagates smoothly into the arrayed waveguide region via the 3D tapered waveguides. Fabrication of the tapered structures can be implemented by the auxiliary mask technique, routinely done in VLSI processing for semiconductor device manufacturing. Thus, it is expected that further implementations of gap reduction via higher-resolution lithography and tapering via the auxiliary mask technique will improve the above AWG performances beyond that of silica-based counterparts in addition to the other discussed advantages of the perfluoropolymer platform.

25.7 Wide band channelizer with high-resolution arrayed waveguide grating (AWG)

The channelizer module is another promising application for RF photonics as a means to enable a large portion of the RF spectrum (i.e., 2 to 20 GHz) to be simultaneously monitored and analyzed. A channelizer is basically an AWG with extremely high resolution (e.g., <5 GHz) that can be realized with large length differences among the array grating lengths; for example, 1 GHz resolution requires ~20 cm length difference between the shortest and longest grating arms. The channelizer requires innovative AWG designs with a precise fabrication process and active phase control of individual grating arms. Although there are still fabrication challenges in achieving a wide band channelizer, the initial results are promising, with interlayer coupling and coarse filtering having been achieved [31, 32]. Figure 25.19 shows an SEM picture of an intersection of the fabricated perfluoropolymer two-layer waveguide structure for interlayer coupling and wide band channelizer applications. Multilayer designs provide much longer delay lengths and individual tapping to slab regions. An exemplary design layout is shown in Fig. 25.20(a), where a single long delay line and hundreds of taps are arranged in a two-layer configuration as required for high-resolution RF processing. This potentially allows meters of path length difference to be achieved in an area of a few square centimeters based on the all polymer platform.

In principle, optical power is tapped off the long delay using evanescent taps of varying coupling strength. The tap optical power is coupled through feeder waveguides into the slab waveguide area. The slab and feeder waveguides are physically located in a second waveguiding layer parallel to the waveguiding layer containing the spiral delay line. Figure 25.20(b) shows the preliminary performance of such a design with approximately maximum 0.6 m length differences between the grating arms for spectrally separating the signal into <1 GHz channels. Despite the fabrication challenges, the initial results (i.e. 12 dB peak at 400 MHz) are encouraging [32]. The photonic channelizer module is promising to offer orders of magnitude reduction in the power, weight, and cost as compared with an electronic system, however several additional research objectives (i.e. active phase control via TO, multilayer coupling control, etc.) must be met to realize such unique devices on the perfluoropolymer platform.

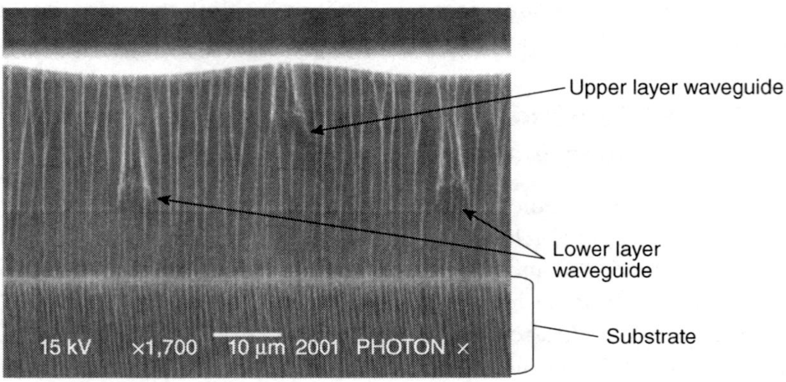

25.19 SEM micrograph of an intersection of the fabricated perfluoropolymer two-layer waveguide structure for interlayer coupling in a wide band channelizer.

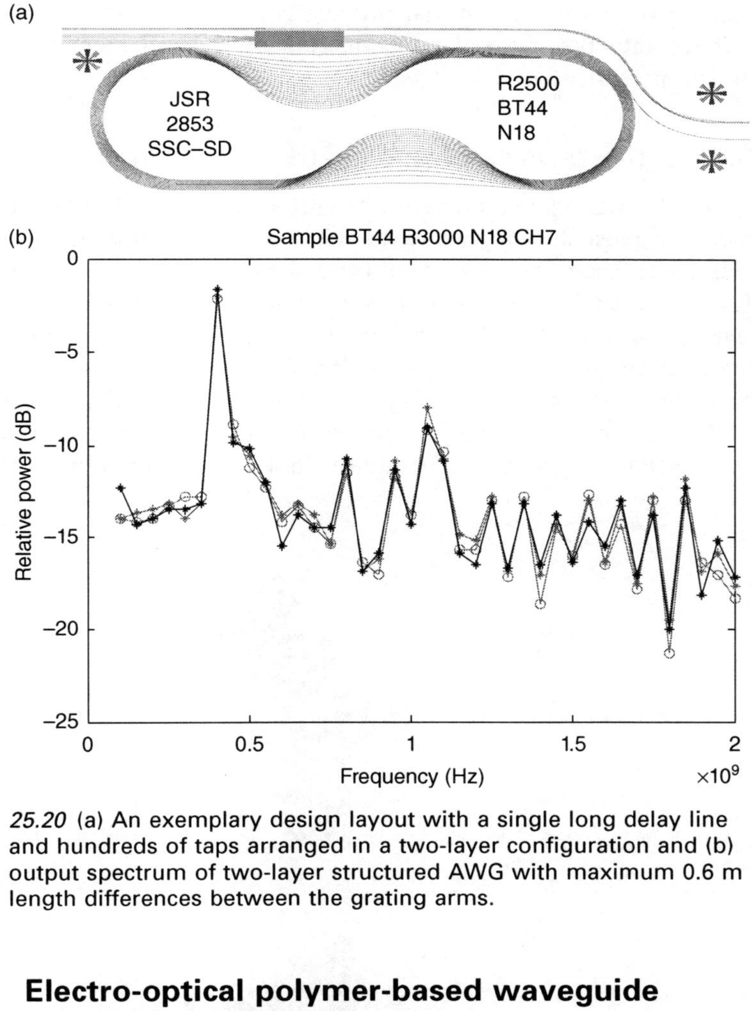

25.20 (a) An exemplary design layout with a single long delay line and hundreds of taps arranged in a two-layer configuration and (b) output spectrum of two-layer structured AWG with maximum 0.6 m length differences between the grating arms.

25.8 Electro-optical polymer-based waveguide devices: materials science

The Pockels coefficient of EO polymers has been rapidly increasing over the last decade, reaching more than 300 pm/V [41, 42]. This ultra-high second order nonlinear activity makes EO polymers an attractive alternative to the industry standard EO material lithium niobate. Along with the ability to heterogeneously integrate EO polymers on any substrate [43–47], their optical and electro-optical properties can be tailored with engineering at the molecular level [48–50], a unique advantage over crystalline nonlinear optical materials. The purpose of this section is two-fold: (1) to introduce the reader to the basic material theory and physics of guest–host type EO polymers and (2) to demonstrate that through system level considerations potential trade-offs in material properties can be determined that allow feedback into the molecular engineering of EO polymers for attaining ultimate performance.

25.8.1 The physics of guest–host EO polymers

A guest–host EO polymer, here after referred to as EO polymer, is composed of a guest chromophore loaded into a host polymer matrix, as shown in Fig. 25.21. The EO coefficient of EO polymers has rapidly increased from a few tens of pm/V in the 1980s to more than 300 pm/V today [41, 42, 48–67]. Devices utilizing EO polymers have been fabricated with high operating speeds, up to greater than 100 GHz [68–72], low optical loss [73] and with operating voltages that are below 1 V [44, 74–77]. Much of the gain in EO activity and overall device performance can be attributed to an improved understanding of the requisite molecular characteristics of organic

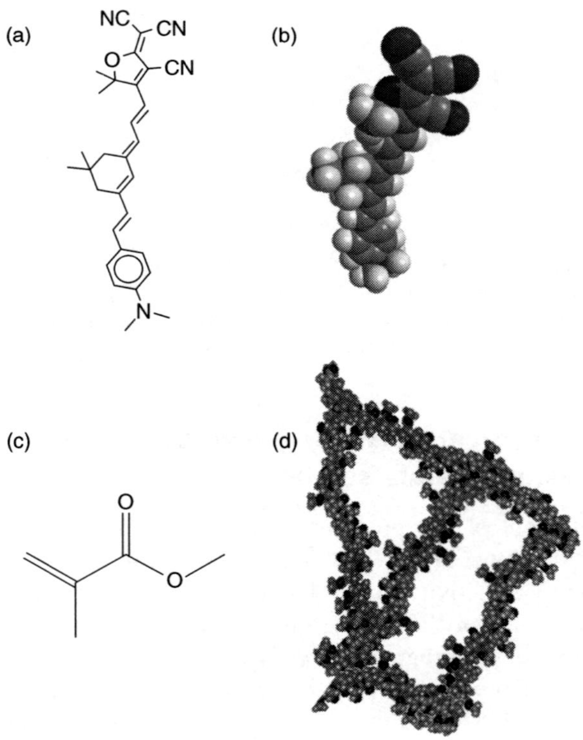

25.21 (a) The chemical structure of a *guest* EO chromophore (CLD1 [61]) (b) 3 D representation of CLD1 (c) chemical structure of the MMA and (d) 3 D representation of the common host polymer PMMA.

nonlinear optical materials. However, in order to better understand the desired properties and molecular structure of chromophores a basic understanding of the macroscopic phenomena they are to provide is required.

25.8.2 Introduction to nonlinear optics

In linear optics the optical properties of a material are independent of the interaction of said material with electromagnetic fields, unlike nonlinear optics where the presence of electromagnetic fields can modify the optical properties of a material. The EO effect is a second order nonlinear optical effect where the presence of a low-frequency electric field can induce a change in a material's refractive index. Hence, in this section the basic classical theory of nonlinear optics is outlined. For a more thorough discussion of nonlinear optics the interested reader is referred to [63, 78–82].

As with most electromagnetic phenomena the classical theory of nonlinear optics begins with Maxwell's equations, which, in SI units are given by

$$\nabla \cdot D = \rho \qquad [25.13]$$

$$\nabla \cdot B = 0 \qquad [25.14]$$

$$\nabla \times E = -\frac{\partial B}{\partial t} \qquad [25.15]$$

$$\nabla \times H = J + \frac{\partial D}{\partial t} \qquad [25.16]$$

where D is the electric displacement field in units of C/m^2, ρ is the space charge density in units of C/m^3, B is the magnetic field in units of T, E is the electric field in units of V/m, H is the auxiliary magnetic field in units of A/m and J is the current density in units of A/m^2. Maxwell's equations are supplemented by constitutive relationships

$$D = \varepsilon_0 \varepsilon E \qquad [25.17]$$

$$B = \mu_0 \mu H \qquad [25.18]$$

$$J = \sigma E \qquad [25.19]$$

where ε_0 is the permittivity of free space in units of F/m, ε is the dimensionless electric permittivity tensor, μ_0 is the permeability of free space in units of H/m, μ is the dimensionless magnetic permeability tensor, and σ is the conductivity in units of S/m. It is necessary to mention that ε_0, and σ all have a dependence on the frequency of the applied electromagnetic fields.

To better understand the interaction of light and matter (25.17) can be rewritten as a physically meaningful additive relation

$$D = \varepsilon_0 E + P \qquad [25.20]$$

with

$$P = \varepsilon_0 \chi E \tag{25.21}$$

where P is the material polarization in units of C/m^2 that can be shown to represent the dipole moment density [83], and χ is the electric susceptibility tensor, a dimensionless quantity that quantifies how easily a material polarizes in response to an applied electric field. Assuming we are in a nonmagnetic, non-conductive material, that is $\mu = 1$ and $\sigma = 0$, a reasonable assumption for EO polymers, we can take the curl of (25.15), interchange the order of the space and time derivatives and replace $\nabla \times B$ with $\mu_0(\partial D/\partial t)$ and with (25.17), (25.19) and (25.20), we have

$$\nabla \times \nabla \times E = -\mu_0 \frac{\partial^2 D}{\partial t^2}. \tag{25.22}$$

Replacing D with the right hand side of (25.20), using the vector identity $\nabla \times \nabla \times A = \nabla(\nabla \cdot A) - \nabla^2 A$ and realizing that $\varepsilon_0\mu_0 = 1/c^2$, where c is the speed of light in vacuum results in

$$\nabla^2 E - \frac{1}{c^2}\frac{\partial^2 E}{\partial t^2} - \nabla(\nabla \cdot E) = \frac{1}{\varepsilon_0 c^2}\frac{\partial^2 P}{\partial t^2} \tag{25.23}$$

which is the electromagnetic wave equation. Unlike for linear dielectric isotropic homogeneous media, in general for nonlinear media the term $\nabla(\nabla \cdot E)$ is not identically zero. However, for transverse electromagnetic plane waves this term does vanish and in many cases of interest it can be shown to be small compared with the other terms of the wave equation, therefore it will be assumed to be zero for the rest of this chapter.

Following from the discrete time Fourier transform, without loss of generality the E field can be written as the sum of discrete frequency components

$$E = \sum_{n=-N}^{N} E_n(r)\mathrm{e}^{-i\omega_n t} \tag{25.24}$$

with ω_n denoting the nth frequency component, $E_n(r)$ is the spatially varying field of the nth frequency component, r is the spatial position vector, N is the total number of frequency components in the field and $E_{-n} = E_n^*$, where * denotes the complex conjugate. (25.24) can be rewritten in a shortened notation as

$$E = \sum_{n=-N}^{N} E(\omega_n) \tag{25.25}$$

(25.21) holds true for linear materials, however, for nonlinear optical materials the material polarization responds nonlinearly to the applied electric field and so the material polarization can be represented as a power series in E as

$$P = \varepsilon_0(\chi^{(1)}E + \chi^{(2)}E^2 + \chi^{(3)}E^3 + \ldots) \qquad [25.26]$$

where $\chi^{(n)}$ is a rank n+1 tensor. Then replacing E in (25.26) with (25.24) we have

$$P_i = \varepsilon_0 \sum_j \sum_n \chi_{ij}^{(1)}(\omega_n)E_j(\omega_n) + \varepsilon_0 \sum_{jk} \sum_{nm} \chi_{ijk}^{(2)}(\omega_n + \omega_m)E_j(\omega_n)E_k(\omega_m) + \ldots \qquad [25.27]$$

where the subscripts ijk represent the Cartesian coordinate unit vector directions. From (25.27) it is clear that the polarization is a combination of linear and nonlinear terms, $P = P^{(1)} + P^{(NL)}$, where the nonlinear terms act as source terms for electric fields with frequencies of, for example, $\omega_n + \omega_m$. Since the polarization involves mixed frequency terms, separate wave equations can be found for each frequency.

Assuming a material with a nonzero $\chi^{(2)}$, by combining (25.23) and (25.27), a nonlinear wave equation can be written in the frequency domain as

$$\nabla^2 E(\omega_n + \omega_m) + \frac{(\omega_n + \omega_m)^2}{c^2}\varepsilon^{(1)}(\omega_n + \omega_m)E(\omega_n + \omega_m) = -\frac{2(\omega_n + \omega_m)^2}{c^2}\sum_{jk}\chi_{ijk}^{(2)}(\omega_n + \omega_m)E_j(\omega_n)E_k(\omega_m) \qquad [25.28]$$

where $\varepsilon^{(1)} = n^2$ and n is the linear refractive index. This equation leads to many interesting effects such as second harmonic generation, parametric oscillation, difference frequency generation, optical rectification, and the EO effect. By letting one frequency component of the electromagnetic field be an applied radio frequency signal and the other an optical field, $\omega_m = \omega_{RF}$, where $\omega_{RF} << \omega_n$, so that the assumption $\omega_{RF} + \omega_n \approx \omega_n$ is valid then, if for simplicity we assume that E is a scalar (25.28) becomes

$$\nabla^2 E(\omega_n) + \frac{\omega_n^2}{c^2}\varepsilon^{(1)}(\omega_n)E(\omega_n) = -\frac{2\omega_n^2}{c^2}\chi^{(2)}(\omega_n)E(\omega_n)E(\omega_{RF}) \qquad [25.29]$$

which can be rearranged to give

$$\nabla^2 E(\omega_n) + \frac{\omega_n^2}{c^2}\{(\varepsilon^{(1)} + 2\chi^{(2)}(\omega_n)E(\omega_{RF})\}E(\omega_n) = 0 \qquad [25.30]$$

Assuming that $\chi^{(2)}E(\omega_{RF}) << 1$, (25.30) looks just like the linear wave equation for a material with a refractive index $n = n_0 + \chi^{(2)}E(\omega_{RF})$, where $n_0 = \sqrt{\varepsilon^{(1)}}$. It is clear then, that by applying an RF wave to a material with a nonzero $\chi^{(2)}$, the refractive index can be modified in proportion to $E(\omega_{RF})$. This is commonly known as the EO or Pockels effect.

The derivation of the EO effect required a material with a nonzero $\chi^{(2)}$; however, a key observation is that in order for a material to have a nonzero $\chi^{(2)}$ it must be non-centrosymmetric. This can be easily proved as follows: if a material is centrosymmetric than it must possess inversion symmetry in other words, the material polarization for an applied electric field E, $P(E)$, must change sign if the sign of E changes, that is to say $P(-E) = -P(E)$. However for a material with a nonzero $\chi^{(2)}$, $P^{(2)}(E) = \chi^{(2)}E^2$, so $P^{(2)}(-E) = \chi^{(2)}(-\mathrm{E})^2$, which is equal to $-\chi^{(2)}(E)^2$ if and only if $\chi^{(2)} = 0$, proving that in order for a material to possess a nonzero $\chi^{(2)}$ it must be non-centrosymmetric. This proof can be generalized for all even order material nonlinearities. Interestingly this also shows that no special material symmetries are required for odd order nonlinearities, therefore all materials (even air) possess odd order nonlinearities. These facts will prove to be critical in understanding the design and characterization of chromophores for EO polymers.

25.8.3 EO formalism

Since the Pockels effect was discovered more than 50 years prior to the development of the field of nonlinear optics, the formalism is significantly different than that used for the rest of the field. This section will provide a brief discussion of the EO formalism and how the EO effect modifies the refractive index of an anisotropic material.

The index ellipsoid

In order for a material to be electro-optic it must generally be anisotropic and so D is related to E by a tensor relationship

$$D = \varepsilon_0 \begin{pmatrix} \varepsilon_{xx} & \varepsilon_{xy} & \varepsilon_{xz} \\ \varepsilon_{yx} & \varepsilon_{yy} & \varepsilon_{yz} \\ \varepsilon_{zx} & \varepsilon_{zy} & \varepsilon_{zz} \end{pmatrix} E \qquad [25.31]$$

For optical wavelengths that are far from the material absorption band edge and absorption resonances, this relationship can be simplified by rotation into the principle axis frame as

$$D = \varepsilon_0 \begin{pmatrix} \varepsilon_{xx} & 0 & 0 \\ 0 & \varepsilon_{yy} & 0 \\ 0 & 0 & \varepsilon_{zz} \end{pmatrix} E \qquad [25.32]$$

In the principal axis coordinate system (25.32) can be shown to represent a material whose refractive index is described by an ellipsoid as [79, 83]

$$\frac{x^2}{n_x^2}+\frac{y^2}{n_y^2}+\frac{z^2}{n_z^2}=1 \qquad [25.33]$$

with $n_i^2 = \varepsilon_{ii}$. Poled EO polymers are uniaxial and so (25.33) simplifies even further to

$$\frac{x^2}{n_o^2}+\frac{y^2}{n_o^2}+\frac{z^2}{n_e^2}=1 \qquad [25.34]$$

where n_o and n_e represent the ordinary and extraordinary refractive indices, respectively. (25.34) can be used to find the refractive indices for an electromagnetic wave of a given polarization by finding the intersection of the index ellipsoid with the plane perpendicular to the direction of propagation that passes through the origin. Since poled EO polymers are uniaxial the ordinary wave will always have refractive index n_o, and the extraordinary wave refractive index will change depending on the angle of the propagation vector with respect to the material's principal axes. This principle is illustrated in Fig. 25.22.

The Pockels coefficient

The relationship between D and E can also be represented in terms of the impermeability tensor η, where $\eta = \varepsilon^{-1}$

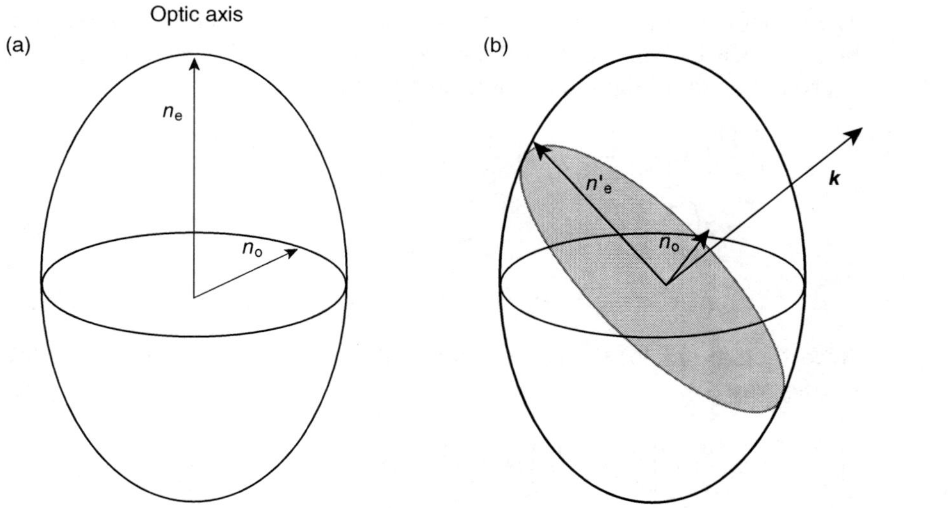

25.22 (a) Light propagating along the principal axis in a uniaxial material will not observe birefringence. (b) When the light is propagating in a direction $\boldsymbol{k}$ other than the optic axis birefringence will be observed with an ordinary refractive index n_o and an extraordinary refractive index n_e which depends on the projection of $\boldsymbol{k}$ onto the optic axis.

$$E = \frac{1}{\varepsilon_0}\begin{pmatrix} \eta_{xx} & \eta_{xy} & \eta_{xz} \\ \eta_{yx} & \eta_{yy} & \eta_{yz} \\ \eta_{zx} & \eta_{zy} & \eta_{zz} \end{pmatrix} D \qquad [25.35]$$

similarly to the description of the material polarization as a power series in the E field, the impermeability elements η_{ij} can be represented as a power series in E

$$\eta_{ij} = \eta_{ij}^{(0)} + \sum_k r_{ijk} E_k + \dots \qquad [25.36]$$

For electromagnetic fields with frequencies that are far from absorption resonances the impermeability tensor must be symmetric which is to say $\eta_{ij} = \eta_{ji}$, and so a shortened notation for the r-coefficients can be used, $r_{ijk} = r_{hk}$, where

$$h = \begin{cases} 1 & ij = 11 \\ 2 & ij = 22 \\ 3 & ij = 33 \\ 4 & ij = 23 \quad or \quad 32 \\ 5 & ij = 13 \quad or \quad 31 \\ 6 & ij = 12 \quad or \quad 21. \end{cases}$$

The r-coefficients are also known as the Pockels coefficients so named after Friedrich Carl Alwin Pockels, the discoverer of the EO effect. In a general coordinate system the index ellipsoid has the form:

$$\left(\frac{1}{n^2}\right)_1 x^2 + \left(\frac{1}{n^2}\right)_2 y^2 + \left(\frac{1}{n^2}\right)_3 z^2 + 2\left(\frac{1}{n^2}\right)_4 yz + 2\left(\frac{1}{n^2}\right)_5 xz + 2\left(\frac{1}{n^2}\right)_6 xy = 1 \qquad [25.37]$$

$$= \eta_{11}x^2 + \eta_{22}y^2 + \eta_{33}z^2 + 2\eta_{23}yz + 2\eta_{13}xz + 2\eta_{12}xy \qquad [25.38]$$

According to (25.36) the EO effect modifies the index ellipsoid in the following way

$$\left[\left(\frac{1}{n^2}\right)_1 + \Delta\left(\frac{1}{n^2}\right)_1\right]x^2 + \left[\left(\frac{1}{n^2}\right)_2 + \Delta\left(\frac{1}{n^2}\right)_2\right]y^2 + \left[\left(\frac{1}{n^2}\right)_3 + \Delta\left(\frac{1}{n^2}\right)_3\right]z^2 + 2\left[\left(\frac{1}{n^2}\right)_4 + \Delta\left(\frac{1}{n^2}\right)_4\right]yz + 2\left[\left(\frac{1}{n^2}\right)_5 + \Delta\left(\frac{1}{n^2}\right)_5\right]xz + 2\left[\left(\frac{1}{n^2}\right)_6 + \Delta\left(\frac{1}{n^2}\right)_6\right]xy = 1 \qquad [25.39]$$

where

$$\begin{pmatrix} \Delta\left(\frac{1}{n^2}\right)_1 \\ \Delta\left(\frac{1}{n^2}\right)_2 \\ \Delta\left(\frac{1}{n^2}\right)_3 \\ \Delta\left(\frac{1}{n^2}\right)_4 \\ \Delta\left(\frac{1}{n^2}\right)_5 \\ \Delta\left(\frac{1}{n^2}\right)_6 \end{pmatrix} = \begin{pmatrix} r_{11} & r_{12} & r_{13} \\ r_{21} & r_{22} & r_{23} \\ r_{31} & r_{32} & r_{33} \\ r_{41} & r_{42} & r_{43} \\ r_{51} & r_{52} & r_{53} \\ r_{61} & r_{62} & r_{63} \end{pmatrix} \begin{pmatrix} E_x \\ E_y \\ E_z \end{pmatrix}. \qquad [25.40]$$

Oftentimes symmetry arguments can be used to simplify the 18 element $\boldsymbol{r}$ tensor to just a few nonzero elements. EO polymers have C_{2v} symmetry [84] and so their $\boldsymbol{r}$ tensor can be shown to reduce to

$$r = \begin{pmatrix} 0 & 0 & r_{13} \\ 0 & 0 & r_{13} \\ 0 & 0 & r_{33} \\ 0 & r_{13} & 0 \\ r_{13} & 0 & 0 \\ 0 & 0 & 0 \end{pmatrix}. \qquad [25.41]$$

So, for an EO polymer in the principal axis coordinate system, the index ellipsoid is modified as follows for an applied direct current (DC) electric field polarized in the z-direction

$$\left(\frac{1}{n_o^2} + r_{13}E_z\right)x^2 + \left(\frac{1}{n_o^2} + r_{13}E_z\right)y^2 + \left(\frac{1}{n_e^2} + r_{33}E_z\right)z^2 = 1. \qquad [25.42]$$

Interestingly, the principal axes remain the same in this case, which is not generally true. The new values of the ordinary and extraordinary refractive indices are given by the well-known expressions

$$n_o' = n_o - \frac{1}{2}n_o^3 r_{13} E_z \qquad [25.43]$$

$$n_e' = n_e - \frac{1}{2}n_e^3 r_{33} E_z, \qquad [25.44]$$

assuming $n^2 r E_z < <1$, so $n \cdot (1 + n^2 r E_z)^{-1/2}$ can be approximated as the first two terms of its Taylor series, a reasonable assumption for moderate applied

E-fields. Furthermore these expressions yield a useful figure of merit for EO polymers, n^3E_z.

25.9 Molecular theory of electro-optic (EO) polymers

The previous section introduced the macroscopic theory of nonlinear optics and discussed the EO effect. This section will describe the molecular theory of electro-optically active chromophore molecules, and how these molecules can be turned into macroscopic EO polymer films. Since the material polarization P represents the dipole moment density of a material, assuming that local field factors can be ignored, justifiable for moderate densities of molecules, the material polarization of a collection of molecules can be written as

$$P = \frac{1}{V}\sum_{i=1}^{N} d_i \qquad [25.45]$$

where d_i is the induced dipole moment of the ith molecule, N is the total number of molecules and V is the volume over which the dipoles exist. This expression will prove valuable in translating the properties of individual molecules to those of a bulk material.

25.9.1 The Lorentz oscillator model

While a full description of the interaction of light with matter requires quantum electrodynamics, intuition for the light matter interaction can be gained by considering a purely classical description, the Lorentz oscillator model. In this model the *linear* molecule–light interaction is described by considering the force between the nucleus and electron(s) of a molecule to be due to a spring binding the two together, with an applied electromagnetic field represented as an external driving force on the spring system. The overall material response can then be determined by realizing that the displacement of the electron from its equilibrium position induces a dipole moment that can be simply related to the material polarization by (25.45). For simplicity the discussion of the electric fields and forces will be limited to one dimension, so vector notation will not be used. The model is pictorially described in Fig. 25.23. The Lorentz force F_L exerted on a charge $-q$ by an electric field E is given by

$$F_L = -qE \qquad [25.46]$$

The effect of the magnetic field is omitted from the Lorentz force since for non-relativistic electron velocities, the term $v \times B$ is negligible. According to Hooke's law the restoring force, F_R, of the bond between the electron and the nucleus can be represented as

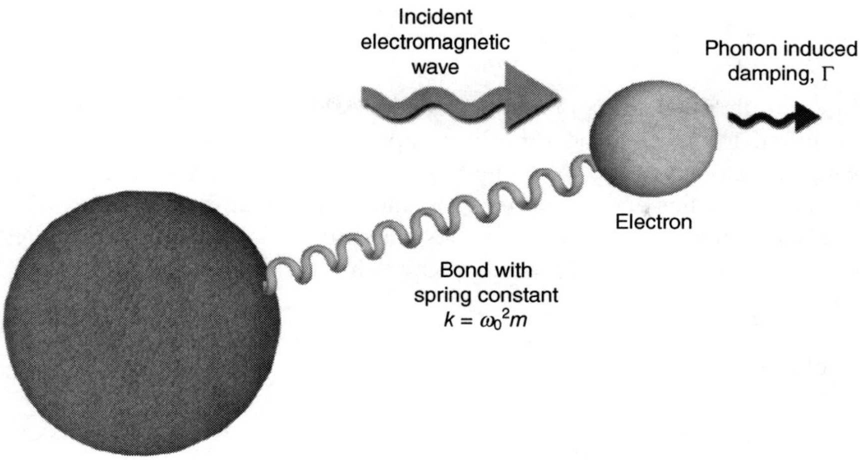

25.23 In the Lorentz oscillator model the bond between a valence electron and the nucleus is represented by a spring, the incident electromagnetic field acts as a driving term whose energy is damped by electron–phonon interaction and other non-radiative processes.

$$F_R = -m\omega_0^2 x \qquad [25.47]$$

where m is the mass of the electron, ω_0 is the natural frequency of the oscillator and x is the displacement from equilibrium position. In this description we have assumed that the mass of the nucleus is much larger than the mass of the electron(s), so only the motion of the electron is significant. Finally, a damping force, F_D, proportional to the electron velocity exists

$$F_D = -2m\Gamma \frac{dx}{dt} \qquad [25.48]$$

The origin of the damping force is optical absorption, and it is related to electron–phonon interactions and other non-radiative electronic transitions and therefore requires quantum mechanics to specify. However, it should be clear that some damping force exists, otherwise an oscillator set in motion would continue perpetually. By inserting these forces into Newton's equation $F = ma$, where m is the mass of electron(s) attached to the nucleus, we have

$$-qE - m\omega_0^2 x - 2\Gamma \frac{dx}{dt} = m\frac{d^2x}{dt^2} \qquad [25.49]$$

By rearranging (25.49), the familiar driven, damped harmonic oscillator equation emerges as

$$\frac{d^2x}{dt^2} + 2\Gamma\frac{dx}{dt} + m\omega_0^2 x = \frac{-q}{m}E \qquad [25.50]$$

In quantum mechanics an analogous expression for the Hamiltonian can be found for the hydrogen atom by applying the dipole approximation, which is valid when the wavelength of the driving field is much larger than the atomic radius and is true for most molecules of interest at optical frequencies, showing the validity of this classical model. Letting the applied electric field be a monochromatic plane wave of the form

$$E = E_0 e^{i(\omega t - kx)} + E_0 e^{-i(\omega t - kx)} \qquad [25.51]$$

a solution of the form

$$x = \frac{-q}{m}\frac{1}{\omega_0^2 - 2i\Gamma\omega - \omega^2}(E_0 e^{i\omega t} + E_0 e^{-i\omega t}) \qquad [25.52]$$

can be found. The dipole moment induced by the applied electric field is given by $-qx$, then the induced material polarization is given by $P = -Nqx$, where N is the number density of molecules in units of m^{-3}, whence it follows that

$$P = \frac{Nq^2}{m}\frac{1}{\omega_0^2 - 2i\Gamma\omega - \omega^2}(E_0 e^{i\omega t} + E_0 e^{-i\omega t}) \qquad [25.53]$$

Using the relationships $P = \varepsilon_0\chi E$, $\varepsilon_r = 1 + \chi$, and $\sqrt{\varepsilon_r} = n + i\kappa$, assuming that N is such that $|\chi| < 1$ so that $(1+\chi)^{1/2} \sim 1 + \chi/2$, a valid assumption given that local field effects were assumed to be negligible in this definition of P, which is valid only for small N, we have

$$n = 1 + \frac{Nq^2}{2\varepsilon_0 m}\frac{\omega_0^2 - \omega^2}{(\omega_0^2 - \omega^2)^2 + 4(\Gamma\omega)^2} \qquad [25.54]$$

$$\kappa = \frac{Nq^2}{\varepsilon_0 m}\frac{\Gamma\omega}{(\omega_0^2 - \omega^2)^2 + 4(\Gamma\omega)^2} \qquad [25.55]$$

Plots of n and k can be seen in Fig. 25.24. It is interesting to observe that in the region near ω_0, anomalous dispersion of the refractive index n occurs. It is also worth noting that the addition of the damping term (25.48) to (25.49) leads directly to the existence of an optical absorption coefficient α, where $\alpha = 4\pi\kappa/\lambda$, and λ is the wavelength of the incident electromagnetic field, supporting the introduction of a phenomenological damping term.

25.9.2 The anharmonic oscillator

It is relatively straightforward to extend the Lorentz oscillator model to the nonlinear regime. The fundamental physics of nonlinear optical effects arise

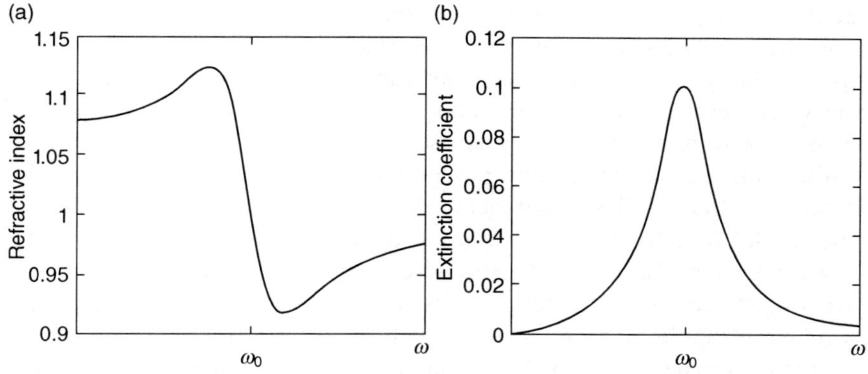

25.24 (a) The refractive index and (b) extinction coefficient near the resonance of a molecule as predicted by the Lorentz model.

from the deviation of the electronic displacement from harmonic behavior. One way to break the harmonicity of a driven oscillator is with the addition of a term of the form ax^2 to (25.50). This leads to an ordinary nonlinear differential equation

$$\frac{\mathrm{d}^2x}{\mathrm{d}t^2} + 2\Gamma\frac{\mathrm{d}x}{\mathrm{d}t} + m\omega_0^2 x + ax^2 = \frac{-q}{m}E. \qquad [25.56]$$

When this equation is solved using perturbation theory, one finds for an applied E field at frequency ω terms of the induced molecular dipole moment that are proportional to E^2, which oscillate at a frequency 2ω (for a full discussion on solving this equation using perturbation theory the interested reader is referred to [80]). Even higher-order harmonic terms can be found by adding terms to the restoring force such as bx^3 and cx^4. By analogy with the material polarization, P, it is reasonable to write the induced molecular dipole moment, μ as a power series in E

$$\mu = \alpha E + \beta E^2 + \gamma E^3 + \ldots \qquad [25.57]$$

where α is the molecular polarizability tensor, β is the second-order molecular hyperpolarizability tensor, and γ is the third-order molecular hyperpolarizability tensor.

25.9.3 The internal field model

Owing to their delocalized π-electron cloud, conjugated molecules and polymers have been found to have large third-order molecular hyperpolarizabilities [63, 79, 81, 82]. Simple conjugated molecules are centrosymmetric and therefore do not possess a β coefficient; however, by perturbing a conjugated molecule with an electronegative substituent on one end and an electropositive substituent on the other, the symmetry is broken and a nonzero β emerges. This is why second order nonlinear chromophores are often called push–pull molecules and can be simply described as D-π-A, which represents an electron-donor, D, connected with a conjugated element, π, to an electron-acceptor, A.

The equivalent internal field model is a simple approach that provides intuition for the physical origin of the EO effect in nonlinear chromophores. This model, first described by Oudar and Chemla [85], states that if a centrosymmetric conjugated molecule is perturbed by adding an electronegative or electropositive substituent to one of the hydrogen positions a permanent dipole moment is created which forms a field, E_{int}, internal to the molecule. According to (25.57) the induced dipole moment of the non-perturbed molecule can be written as

$$\mu = \alpha E + \gamma E^3 \quad [25.58]$$

After perturbing the molecule with an electronegative or electropositive substituent, the induced dipole moment responds to the total field $E_{\text{int}} + E(\omega)$, and so in a scalar approximation

$$\begin{aligned} \mu &= \alpha[E_{\text{int}} + E(\omega)] + \gamma[E_{\text{int}} + E(\omega)]^3 \\ &= \alpha E_{\text{int}} + \gamma E_{\text{int}}^3 + (\alpha + 3\gamma E_{\text{int}}^2)E(\omega) + 3\gamma E_{\text{int}} E^2(\omega) + \gamma E^3(\omega). \end{aligned} \quad [25.59]$$

From (25.59) we can reason that

$$\beta = 3\gamma E_{\text{int}} \quad [25.60]$$

From (25.59) it can be seen that the magnitude of the permanent dipole moment created by perturbing the conjugated molecule is $\Delta\mu = \alpha E_{\text{int}} + \gamma E_{\text{int}}^3 \sim \alpha E_{\text{int}}$ so, $\text{E}_{\text{int}} = \Delta\mu/\alpha$ and

$$\beta = \frac{3\gamma\Delta\mu}{\alpha} \quad [25.61]$$

So, according to the internal field model, highly electronegative and electropositive substituents are desired to produce a large dipole moment, $\Delta\mu$, and thus a large second order hyperpolarizability, β. Although in order to accurately calculate the hyperpolarizability coefficients for real molecules sophisticated *ab initio* quantum chemistry methods are required [64, 86], this is an excellent starting point for understanding the physics of highly nonlinear chromophores.

25.9.4 The two-level model

A slightly more complex model for second-order molecular hyperpolarizability proposed by Oudar and Chemla [87] is known as the two-level model, which was developed when much higher molecular second-order hyperpolarizablities than were predicted by the internal field model were experimentally observed. According to the theory, the molecular second-order hyperpolarizability can be decomposed into two parts

$$\beta = \beta_{add} + \beta_{ct} \qquad [25.62]$$

where β_{add} is an additive portion arising from the interaction of the individual substituents and the π-electron system (which can be described by the internal field model) and β_{ct} is the contribution arising from the charge transfer interaction between acceptor and donor groups. In this model, the charge transfer properties of the molecule are characterized by only two states, a ground state and a low-lying charge transfer excited state. With this assumption, the expression for β_{ct} can be derived to be

$$\beta_{ct} = \frac{3e^2}{2m} \frac{\hbar\omega_{gn} f_{gn} \Delta\mu_{gn}}{[(\hbar\omega_{gn})^2 - (2\hbar\omega)^2][(\hbar\omega_{gn})^2 - (\hbar\omega)^2]} \qquad [25.63]$$

where e is the fundamental charge, m is the electron mass, g and n represent the ground and excited states respectively, f_{gn} is the oscillator strength of the ground to excited state transition, $h\omega_{gn}/2\pi$ is the energy difference between the ground and excited state, $\Delta\mu_{gn}$ is the difference in dipole moments between ground and excited states, and ω is the frequency of the optical wave. This expression shows that there is in fact dispersion in the EO coefficient. As the optical frequency nears the absorption resonance, β is enhanced.

25.10 Electric-field assisted poling in polymer films

We have seen the molecular origin of the EO activity of push–pull chromophores; however, in order to translate this molecular phenomenon to a macroscopic polymer film, electric-field assisted poling is required. When a guest–host polymer film is first deposited, the chromophores are randomly oriented and therefore the bulk material is centrosymmetric. This symmetry can be broken by applying a strong electric field and heating the polymer to its, T_g. As discussed above, the glass transition of a polymer is a second order phase transition. Above T_g molecular chains are able to slide past each other when a force is applied, allowing the chromophore molecule to rotate so that its permanent dipole moment partially aligns to the applied electric field. After the chromophores have aligned to the applied DC poling field, the film is cooled to room temperature with the poling field left on, locking the order into place and breaking the centrosymmetry. A pictorial description of the poling process can be seen in Fig. 25.25.

The physics of the poling process has been described in an oriented gas model [88, 89]. Within this model the degree of alignment of the

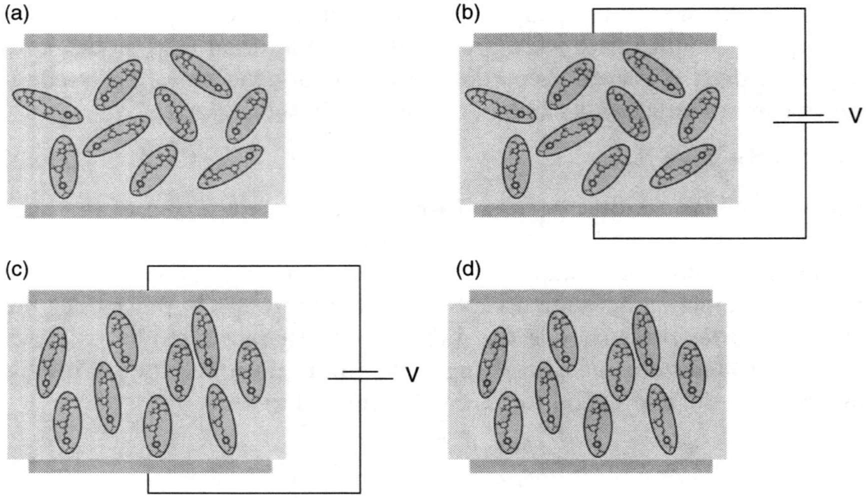

25.25 The poling process: (a) initially the chromophores are randomly oriented, (b) the EO polymer is heated to its glass transition temperature with an applied electric field, (c) after reaching the glass transition temperature the chromophores align to the applied electric field, (d) the polymer is cooled to room temperature with the poling field applied after which the field is removed and the order is locked into place. Note that this is a cartoon description of the poling process and the degree to which the chromophores align to the applied poling field depends strongly on the strength of the applied poling field.

chromophore molecules to the applied poling field can be analyzed using thermodynamic arguments. In the oriented gas picture of poling, assuming the chromophore molecules are far apart so there are no local field effects and no interchromophore interaction, for a chromophore with a single large hyperpolarizability tensor component, β_{zzz}, it can be shown that

$$\chi_{33}^{(2)} \approx \frac{N\mu E_{\mathrm{p}}}{5kT}\beta_{zzz} \tag{25.64}$$

$$\chi_{31}^{(2)} \approx \frac{N\mu E_{\mathrm{p}}}{15kT}\beta_{zzz} \tag{25.65}$$

where N is the number density of chromophores, μ is the permanent dipole moment of the chromophore, β_{zzz} is the molecular second order hyperpolarizability tensor element, E_{p} is the applied poling field strength, k is Boltzmann's constant, and T is the poling temperature. Therefore, the EO tensor for a poled EO polymer is reduced to

$$r = \begin{pmatrix} 0 & 0 & \frac{1}{3}r_{33} \\ 0 & 0 & \frac{1}{3}r_{33} \\ 0 & 0 & r_{33} \\ 0 & \frac{1}{3}r_{33} & 0 \\ \frac{1}{3}r_{33} & 0 & 0 \\ 0 & 0 & 0 \end{pmatrix}, \qquad (25.66)$$

a fact which is often used in the measurement of the Pockels coefficient of poled polymer films [90].

The oriented gas model works well to first order; however, when chromophores are loaded in a polymer matrix at large densities the effect of inter-chromophore interactions, which are assumed to be small in the oriented gas model, are no longer negligible. At large loading levels chromophores tend to form dimers oriented in a head-to-tail fashion, this dimer formation causes a rolling off of the *r*-coefficient at high loading densities [91], as well as optical scattering from larger aggregates which form at high loading densities.

While poling single layer EO polymer films is relatively simple, poling in multi-layer stacks, which is typically required for devices and applications, can be quite challenging. If great care is not taken, excess optical loss can be generated by inhomogeneities created by the poling process [92]. However, with a careful choice of cladding material low poling induced loss can be achieved [70]; furthermore, an enhancement in the r_{33} coefficient can be achieved by increasing the dielectric breakdown strength of the polymer with a properly chosen cladding material [93–97].

25.11 Device and system level analysis for electro-optical polymer waveguides

When designing a guest–host EO polymer it is useful to define a material figure of merit in order to quantify material trade-offs. A naive approach would be to simply maximize the material r_{33} at all cost. However, in this section it will be shown that from device and system level considerations, a material figure of merit can be defined which truly optimizes material performance, namely n^3r_{33}/α. The system under consideration for this analysis will be an RF photonic link.

In an RF photonic link a radio-frequency electrical signal is encoded onto an optical carrier, with the optical signal being propagated some distance after which it is converted back to the electrical domain with a photodetector. There are several advantages to using an optical carrier to transmit the RF signal, one being that optical waves can be sent several kilometers through optical fibers with very little attenuation whereas RF electrical signals transmitted over copper lines are severely attenuated at these distances. A block diagram of an RF photonic link is shown in Fig. 25.26.

The RF signal is encoded onto the optical carrier using an optical modulator, and it is in this component that EO polymers can make a large impact. The type of modulator that will be considered for this analysis is the commonly used waveguide Mach–Zehnder modulator (MZM). Assuming the electrical bandwidth of the modulator is sufficiently high for the RF frequencies of interest, the optical transfer function, T, for an EO polymer waveguide MZM is given by

$$T = 0.5 \cdot P_0 \cdot T_{\text{FF}} \cdot \left[1 + \cos\left(\frac{2\pi}{\lambda}\Gamma n^3 r_{33}\frac{V}{d}\cdot L + \varphi\right)\right] \qquad (25.67)$$

where P_0 is the optical power input to the device, T_{FF} is the fiber-to-fiber transmission of the modulator, λ is the wavelength of the optical carrier, n is the refractive index of the EO polymer, Γ is the overlap integral of the waveguide mode with the EO polymer, r_{33} is the Pockels coefficient of the EO polymer, V is the voltage applied to the modulator, d is the separation between the electrodes, L is the length of the waveguide, and ϕ is the optical phase difference between the two arms of the Mach-Zehnder. The fiber-to-fiber transmission can be given by

$$T_{\text{FF}} = \alpha_{\text{coupling}}^2 \cdot e^{-\alpha \cdot L} \qquad (25.68)$$

where α_{coupling} is the fraction of light coupled from an optical fiber to the input/ output waveguide mode of the modulator, α is the propagation loss of the EO polymer in cm^{-1} and L is the total length of the waveguide modulator in *cm*.

There are several important system parameters in RF photonic links, two critical ones being the link gain and noise figure (NF). The link gain is

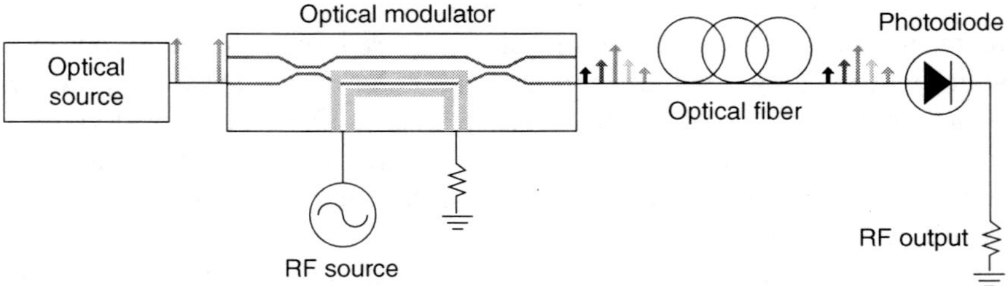

25.26 Block diagram of an RF photonic link.

defined as the ratio of the RF power output of the link to that of the input RF power and to first order is given by:

$$g = \left(\left. \frac{\partial T}{\partial V} \right|_{V=0} \cdot V \cdot \boldsymbol{R} \right)^2 \cdot R_{\text{load}} \frac{R_{\text{mod}}}{V^2}$$
$$= \left[\frac{\pi \cdot P_0 \cdot \alpha^2_{\text{coupling}} \cdot e^{-\alpha \cdot L} \cdot \Gamma \cdot n^3 \cdot r_{33} \cdot L}{\lambda \cdot d} \cdot \sin(\varphi) \cdot \boldsymbol{R} \right]^2 R_{\text{mod}} \cdot R_{\text{load}} \quad (25.69)$$

where $\boldsymbol{R}$ is the responsivity of the photodetector, R_{mod} is the characteristic impedance of the modulator and R_{load} is the load impedance. Typically, systems are built with all impedances matched to 50 Ω. From (25.69) it is clear that having a phase offset ϕ of $\pi/2$ maximizes the link gain, which biases the modulator at quadrature. An additional benefit of biasing the modulator at quadrature is that it forces all even order derivatives of the transfer function to zero, which means that all even order harmonic distortion terms are also zero.

Since (25.69) is dependent on L, it is interesting to consider whether there is a waveguide MZM length, L_{opt}, which maximizes the link gain. According to (25.69) if the function $\exp(-\alpha L) \cdot L$ is maximized, the link gain will also be maximized with respect to the waveguide length. Simple calculus shows that $L_{\text{opt}} = 1/\alpha$. Then, assuming $R_{\text{mod}} = R_{\text{load}} = R$, for the optimal waveguide length the gain can be written:

$$g_{\text{opt}} = \left(\frac{\pi \cdot P_0 \cdot \alpha^2_{\text{coupling}} \cdot \Gamma \cdot \dfrac{n^3 \cdot r_{33}}{\alpha}}{\lambda \cdot d \cdot e} \cdot \sin(\varphi) \cdot R \cdot \boldsymbol{R} \right)^2 . \quad (25.70)$$

An interesting aspect of (25.70) is that the only parameter that depends on the EO polymer material properties is the ratio $n^3 r_{33}/\alpha$, which thereby emerges as the material figure of merit for EO polymers.

Let's look at another system parameter, the NF, and see if this figure of merit is still pertinent. The link NF is defined as the ratio of the input signal to noise ratio (SNR) to the SNR out of the link, and is given in dB. There are three prominent sources of noise that degrade the link NF; Johnson noise

$$N_{\text{Johnson}} = kT_0 \cdot (1 + g) \quad (25.71)$$

where k is the Boltzmann constant and T_0 is the temperature of operation, the 1 in the $(1 + g)$ factor represents Johnson noise in the load resistor, and the g represents the amplified input Johnson noise; next we consider shot noise

$$N_{\text{shot}} = 2qI_{\text{det}}R \quad (25.72)$$

where I_{det} is the average current produced at the photodetector and q is the fundamental electric charge; and finally relative intensity noise (RIN)

$$N_{RIN} = 10^{\frac{RIN}{10}} I_{det}^2 R \tag{25.73}$$

where RIN is a factor measured in dB/Hz and depends on the laser source. Therefore the link NF is given by:

$$NF = 10\log_{10}\left(\frac{N_{Johnson} + N_{shot} + N_{RIN}}{gkT_0}\right) \tag{25.74}$$

Using (25.67) and (25.69), (25.74) can be rewritten as

$$NF = 10\log_{10}\left\{1 + \left[\frac{\lambda d}{\pi P_0 R\boldsymbol{R}\Gamma n^3 r_{33}\alpha_{coupling}^2 e^{-\alpha L} L\sin(\varphi)}\right]^2 \cdot \left[1 + \frac{P_0 q R\boldsymbol{R}(1+\cos(\varphi))\alpha_{coupling} e^{-\alpha L}}{kT_0} + \frac{P_0^2 R\boldsymbol{R}^2 10^{\frac{RIN}{10}}(1+\cos(\varphi))^2 \alpha_{coupling}^2 e^{-2\alpha L}}{4kT_0}\right]\right\} \tag{25.75}$$

It is interesting to note that according to (25.75) the NF is always greater than zero and therefore an improvement in the SNR can never be gained in an RF photonic link. Depending on the optical power input to the modulator and the RIN factor the NF can be Johnson, shot or RIN limited.

The following analysis will consider each of the three noise limiting cases and determine if there is a waveguide length that optimizes the link NF analogously to the link gain case. For the analysis we will assume an optical wavelength of 1550 nm, an EO polymer with a refractive index of 1.6, a propagation loss of 1 cm^{-1}, an r_{33} of 200 pm/V, a photodetector with a responsivity of 0.9 A/W, a waveguide MZM biased at quadrature with coupling losses of 1 dB and an overlap integral, Γ, a system impedance of 50 Ω, a laser source with a RIN of –170 dB/Hz and a background temperature of 300 K. A significant advantage of biasing the modulator at quadrature is that MZMs with differential outputs are available and therefore balanced detection can be used which reduces the RIN noise by the common mode rejection ratio, and although lasers with ultra-low RINs of –170 dB/Hz are commercially available, even sources with significantly worse RINs could be used and the link would not suffer a penalty. Examples of all three noise limits can be seen in Fig. 25.27.

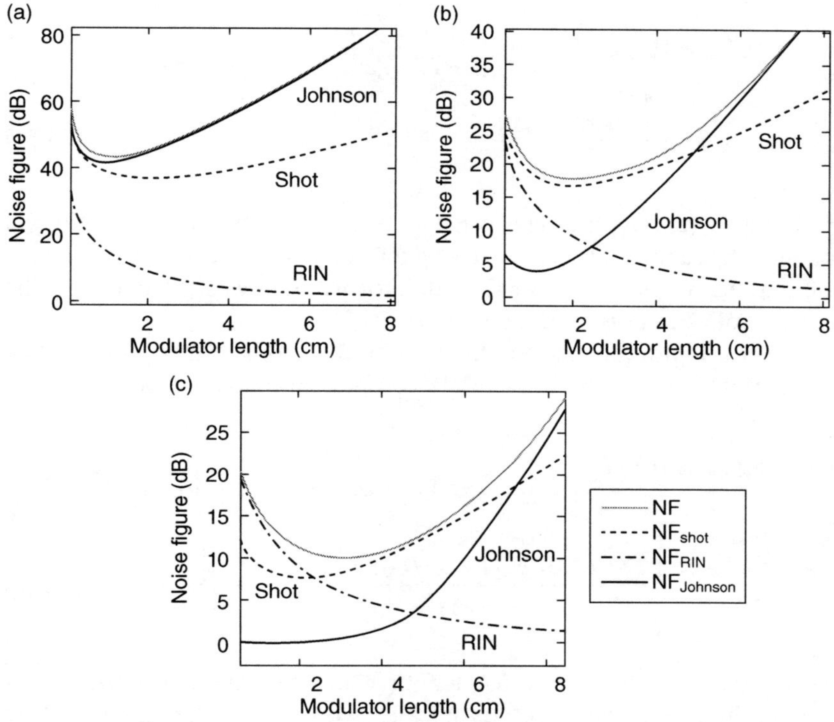

25.27 For the example system described in the text (a) an optical input power of 1 mW creates a Johnson noise limited link (b) input optical power of 100 mW creates a shot noise limited link, and (c) and an input optical power of 1 W causes a shift in the optimal modulator length for shot noise limited performance due to RIN limiting.

For input powers below about 1 mW the example system is Johnson noise limited. In the Johnson noise limit the NF is approximately given by:

$$NF_{\text{Johnson}} = 10\log_{10}\left\{1+\left[\frac{\lambda d}{\pi P_0 R \boldsymbol{R} \Gamma n^3 r_{33} \alpha_{\text{coupling}}^2 \mathrm{e}^{-\alpha L} L \sin(\varphi)}\right]^2\right\} \tag{25.76}$$

In this limit it is clear that the device length that will minimize the NF will be the one which maximizes the function $\exp(-\alpha L)\cdot L$, which occurs for a length given by $L_{\text{opt}} = 1/\alpha$, exactly the device length which maximizes the link gain. Therefore in the low-power limit, a device can be designed which maximizes link gain and minimizes the link NF, an intriguing result.

For intermediate powers between approximately 10 and 100 mW the example link is shot noise limited. In the shot noise limit the NF can be approximated as:

$$\mathrm{NF_{shot}} = 10\log_{10}\left\{1+\left[\frac{\lambda d}{\pi P_0 R\boldsymbol{R}\Gamma n^3 r_{33}\alpha_{\mathrm{coupling}}^2 \mathrm{e}^{-\alpha L} L\sin(\varphi)}\right]^2 \cdot\left[\frac{P_0 q R\boldsymbol{R}(1+\cos(\varphi))\alpha_{\mathrm{coupling}}\mathrm{e}^{-\alpha L}}{kT_0}\right]\right\} \tag{25.77}$$

According to (25.77), the device length which will minimize the NF will maximize the function $L^2 \exp(-\alpha L)$, which is $L_{\mathrm{opt}} = 2/\alpha$. Therefore in a shot noise limited system the optimal modulator will be twice the length of the one for a Johnson noise limited system.

Finally for high input powers above 1 W the link is RIN limited. When RIN dominates the noise, the link NF can be approximated as:

$$\mathrm{NF_{RIN}} = 10\log_{10}\left\{1+\left[\frac{\lambda d}{\pi P_0 R\boldsymbol{R}\Gamma n^3 r_{33}\alpha_{\mathrm{coupling}}^2 \mathrm{e}^{-\alpha L} L\sin(\varphi)}\right]^2 \cdot\left[\frac{P_0^2 R\boldsymbol{R}^2 10^{\frac{RIN}{10}}(1+\cos(\varphi))^2\alpha_{\mathrm{coupling}}^2\mathrm{e}^{-2\alpha L}}{4kT_0}\right]\right\} \tag{25.78}$$

According to (25.78) the device length which will minimize the NF in the RIN limit decreases monotonically with device length, therefore the optimal modulator length will be given by the length after which the shot noise begins to dominate over RIN noise. This length is given approximately by

$$L_{\mathrm{opt}} = \frac{W\left(\dfrac{P_0\boldsymbol{R}10^{-\mathrm{RIN}/10}\alpha_{\mathrm{coupling}}^2}{2qe^2}\right)+2}{\alpha} \tag{25.79}$$

where $W(x)$ is the Lambert W function.

It is interesting to note that in any noise limit the optimal modulator length is simply inversely proportional to α. Therefore at the optimum waveguide length in any noise limit, the dimensionless link noise factor, $F = 10^{\mathrm{NF}/10}$, is simply inversely proportional to n^3r_{33}/α and since the noise factor is a quantity which is desired to be minimized, then the EO polymer material figure of merit chosen to optimize link gain also optimizes the link noise figure and noise factor.

By choosing an example system and EO polymer-based device it is possible to derive an EO polymer material figure of merit that is not simply the r_{33} coefficient. For analog RF photonic links it has been shown that the proper figure of merit to optimize when designing EO polymers at the molecular level is n^3r_{33}/α.

25.12 Electro-optic (EO) polymer spatial light modulators: theory

In this section we discuss recent progress in ultrafast spatial light modulators (SLM) employing electro-optic polymers. Existing commercial SLM technologies are based either on liquid crystals or digital micromirror devices, both of which are limited to speeds in the tens of microseconds to milliseconds. In the section below we describe the first comprehensive effort to demonstrate a spatial light modulator based on EO polymers, which in principle should be capable of operating in the nanosecond and even sub-nanosecond regimes.

25.12.1 Voltage-induced resonant wavelength shift

The light transmitted through a tilted Fabry–Perot interferometer (FPI) is represented by

$$T_{FP} = \frac{T_a T_b}{\left(1-\sqrt{R_a R_b}\right)^2} \frac{1}{1+F\sin^2\delta}, \tag{25.80}$$

where the subscripts a and b denote a particular mirror, T is the transmittance of each mirror, R is the reflectance of each mirror, F is the finesse of the cavity, $\delta = (\phi_a + \phi_b)/2 - \delta'$, $\delta' = 2\pi nd \cos\theta_p/\lambda$, φ is the mirror reflectance phase, n is the polymer refractive index, d is the thickness of the polymer, θ_p is the refracted angle in the polymer, and λ is the wavelength of the incident light. Fig. 25.28 illustrates how light propagates through a tilted FPI as a voltage is applied to the electrodes/mirror introducing a birefringence in the EO polymer layer.

In order to find the resonant wavelength shift due to an applied electric field we need to find the FPI phase that gives maximum light transmission. Applying a voltage to the tilted FPI can be represented by taking the derivative of (25.80), which yields the following:

$$\frac{\partial T_{FP}}{\partial V} = -2\frac{T_a T_b}{\left(1-\sqrt{R_a R_b}\right)^2} \frac{F\sin\delta\cos\delta}{1+F\sin^2\delta}\left(\frac{\partial\delta}{\partial V}\right) = 0 \tag{25.81}$$

Solving (25.81) for $\partial\delta/\partial V$ yields:

$$\frac{\partial\delta}{\partial V} = \left(\frac{\partial\phi_a}{\partial V} + \frac{\partial\phi_b}{\partial V}\right)\Big/2 - \frac{\partial\delta'}{\partial V} = 0 \tag{25.82}$$

By including the phase dispersion with respect to wavelength, polymer refractive index, and tilt angle, the voltage-induced change in mirror reflectance phase of (25.82) becomes:

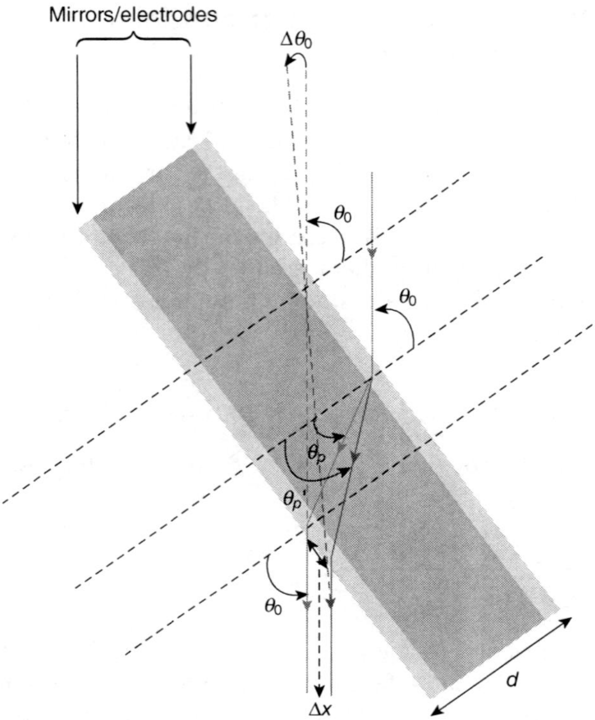

25.28 Simple diagram of a tilted electro-optic Fabry–Perot interferometer that shows the change in light path when a voltage is applied.

$$\left(\frac{\partial \phi_a}{\partial V}+\frac{\partial \phi_b}{\partial V}\right)=\frac{\partial n}{\partial V}\left(\frac{\partial \phi_a}{\partial n}+\frac{\partial \phi_b}{\partial n}\right)+\frac{\partial \lambda}{\partial V}\left(\frac{\partial \phi_a}{\partial \lambda}+\frac{\partial \phi_b}{\partial \lambda}\right)+\frac{\partial \theta_0}{\partial V}\left(\frac{\partial \phi_a}{\partial \theta_0}+\frac{\partial \phi_b}{\partial \theta_0}\right) \quad (25.83)$$

The voltage-induced change in phase of the polymer spacer layer of (25.82) simplifies to:

$$\frac{\partial \delta'}{\partial V}=\frac{2\pi}{\lambda}\left(d\cos\theta_P\frac{\partial n}{\partial V}+n\cos\theta_P\frac{\partial d}{\partial V}-nd\sin\theta_P\frac{\partial \theta_P}{\partial V}-\frac{nd\cos\theta_P}{\lambda}\frac{\partial \lambda}{\partial V}\right) \quad (25.84)$$

From Snell's law the refracted angle inside the polymer layer is given by

$$\theta_P=\sin^{-1}\left(\frac{n_0}{n}\sin\theta_0\right), \quad (25.85)$$

where θ_0 is the light incident angle and n_0 is the refractive index of the outside medium. Taking the derivative of (25.85) with respect to voltage gives voltage-induced change in refracted angle inside the polymer and is given as

$$\frac{\partial \theta_P}{\partial V} = \frac{\partial}{\partial V}\left[\sin^{-1}\left(\frac{n_0}{n}\sin\theta_0\right)\right] \tag{25.86}$$

Simplifying (25.86) yields:

$$\frac{\partial \theta_P}{\partial V} = \left(\frac{n_0}{n}\cos\theta_0\frac{\partial \theta_0}{\partial V} - \frac{n_0}{n^2}\sin\theta_0\frac{\partial n}{\partial V}\right)\Bigg/\left(1-\frac{n_0^2\sin^2\theta_0}{n^2}\right)^{\frac{1}{2}} \tag{25.87}$$

Simplifying (25.87) yields:

$$\frac{\partial \theta_P}{\partial V} = \left(\frac{n_0}{n}\cos\theta_0\frac{\partial \theta_0}{\partial V} - \frac{n_0}{n^2}\sin\theta_0\frac{\partial n}{\partial V}\right)\Bigg/(1-\sin^2\theta_P)^{\frac{1}{2}} \tag{25.88}$$

Simplifying (25.88) yields:

$$\frac{\partial \theta_P}{\partial V} = \left(\frac{n_0}{n}\cos\theta_0\frac{\partial \theta_0}{\partial V} - \frac{n_0}{n^2}\sin\theta_0\frac{\partial n}{\partial V}\right)\Bigg/\cos\theta_P \tag{25.89}$$

Substituting (25.83) and (25.84) into (25.82) yields:

$$\begin{aligned}\frac{\partial \delta}{\partial V} = \frac{1}{2}\Bigg[&\frac{\partial n}{\partial V}\left(\frac{\partial \phi_a}{\partial n}+\frac{\partial \phi_b}{\partial n}\right)+\frac{\partial \lambda}{\partial V}\left(\frac{\partial \phi_a}{\partial \lambda}+\frac{\partial \phi_b}{\partial \lambda}\right)+\frac{\partial \theta_0}{\partial V}\left(\frac{\partial \phi_a}{\partial \theta_0}+\frac{\partial \phi_b}{\partial \theta_0}\right)\\ &-\frac{2\pi}{\lambda}\Bigg(d\cos\theta_P\frac{\partial n}{\partial V}+n\cos\theta_P\frac{\partial d}{\partial V}-nd\sin\theta_P\frac{\partial \theta_P}{\partial V}\\ &-\frac{nd\cos\theta_P}{\lambda}\frac{\partial \lambda}{\partial V}\Bigg)\Bigg]\end{aligned} \tag{25.90}$$

Substituting (25.89) into (25.90) and setting the equation equal to zero we arrive at the following approximation:

$$\begin{aligned}0 = \frac{\lambda}{2\pi}\Bigg[&\Delta n\left(\frac{\partial \phi_a}{\partial n}+\frac{\partial \phi_b}{\partial n}\right)+\Delta\lambda\left(\frac{\partial \phi_a}{\partial \lambda}+\frac{\partial \phi_b}{\partial \lambda}\right)+\Delta\theta_0\left(\frac{\partial \phi_a}{\partial \theta_0}+\frac{\partial \phi_b}{\partial \theta_0}\right)\\ &-\frac{2nd\cos\theta_P}{\lambda}\Delta\lambda+2d\cos\theta_P\Delta n-2n\cos\theta_P\Delta d\\ &+2nd\sin\theta_P\left(\frac{\frac{n_0}{n}\cos\theta_0\Delta\theta_0-\frac{n_0}{n^2}\sin\theta_0\Delta n}{\cos\theta_P}\right)\Bigg],\end{aligned} \tag{25.91}$$

where Δn is the change in refractive index due to the Pockel's effect, $\Delta\lambda$ is the resonant wavelength shift, $\Delta\theta_0$ is the apparent change in refracted angle at the output of the FPI, and Δd is the change in thickness due to the piezoelectric and electrostrictive effects of the material. Simplifying and reorganizing (25.91) yields:

$$0=\frac{\lambda}{2\pi}\left[\Delta n\left(\frac{\partial\phi_a}{\partial n}+\frac{\partial\phi_b}{\partial n}\right)+\Delta\lambda\left(\frac{\partial\phi_a}{\partial\lambda}+\frac{\partial\phi_b}{\partial\lambda}\right)+\Delta\theta_0\left(\frac{\partial\phi_a}{\partial\theta_0}+\frac{\partial\phi_b}{\partial\theta_0}\right)\right.$$
$$-\frac{2nd\cos\theta_P}{\lambda}\Delta\lambda+2d\cos\theta_P\Delta n-2n\cos\theta_P\Delta d$$
$$\left.+2n_0 d\tan\theta_\mathrm{P}\cos\theta_0\Delta\theta_0-2\frac{n_0}{n}d\tan\theta_P\sin\theta_0\Delta n\right] \quad (25.92)$$

Simplifying and reorganizing (25.92) yields:

$$\left[\frac{\lambda}{2\pi}\left(\frac{\partial\phi_a}{\partial\lambda}+\frac{\partial\phi_b}{\partial\lambda}\right)+\frac{2nd\cos\theta_P}{\lambda}\right]\Delta\lambda$$
$$=-\left[\frac{\lambda}{2\pi}\left(\frac{\partial\phi_a}{\partial n}+\frac{\partial\phi_b}{\partial n}\right)-2d\cos\theta_P-2\frac{n_0}{n}d\tan\theta_P\sin\theta_0\right]\Delta n$$
$$-\left[\frac{\lambda}{2\pi}\left(\frac{\partial\phi_a}{\partial\theta_0}+\frac{\partial\phi_b}{\partial\theta_0}\right)+2n_0 d\tan\theta_P\cos\theta_0\right]\Delta\theta_0+2n\cos\theta_\mathrm{P}\Delta d \quad (25.93)$$

Several additional simplification and reorganization steps yield:

$$\left[\frac{\lambda}{2\pi}\left(\frac{\partial\phi_a}{\partial\lambda}+\frac{\partial\phi_b}{\partial\lambda}\right)+\frac{2nd\cos\theta_P}{\lambda}\right]\Delta\lambda$$
$$=-\left[\frac{\lambda}{2\pi}\left(\frac{\partial\phi_a}{\partial\theta_0}+\frac{\partial\phi_b}{\partial\theta_0}\right)-2n_0 d\tan\theta_P\cos\theta_0\right]\Delta\theta_0$$
$$-\left[\frac{\lambda}{2\pi}\left(\frac{\partial\phi_a}{\partial n}+\frac{\partial\phi_b}{\partial n}\right)-\frac{2d}{\cos\theta_P}\right]\Delta n+2n\cos\theta_\mathrm{P}\Delta d \quad (25.94)$$

Solving (25.94) for $\Delta\lambda$ yields:

$$\Delta\lambda=\frac{\left[\frac{2d}{\cos\theta_\mathrm{P}}-\frac{\lambda}{2\pi}\left(\frac{\partial\phi_a}{\partial n}+\frac{\partial\phi_b}{\partial n}\right)\right]\Delta n-\left[\frac{\lambda}{2\pi}\left(\frac{\partial\phi_a}{\partial\theta_0}+\frac{\partial\phi_b}{\partial\theta_0}\right)+2n_0 d\tan\theta_P\cos\theta_0\right]\Delta\theta_0+2n\cos\theta_\mathrm{P}\Delta d}{\frac{2nd\cos\theta_\mathrm{P}}{\lambda}+\frac{\lambda}{2\pi}\left(\frac{\partial\phi_a}{\partial\lambda}+\frac{\partial\phi_b}{\partial\lambda}\right)} \quad (25.95)$$

If light hits the FPI at normal incidence then (25.95) simplifies to the following familiar equation:

$$\Delta\lambda=\frac{\left[2d-\frac{\lambda}{2\pi}\left(\frac{\partial\phi_a}{\partial n}+\frac{\partial\phi_b}{\partial n}\right)\right]\Delta+2n\Delta d}{\frac{2nd}{\lambda}+\frac{\lambda}{2\pi}\left(\frac{\partial\phi_a}{\partial\lambda}+\frac{\partial\phi_b}{\partial\lambda}\right)} \quad (25.96)$$

For a tilted FPI the output beam experiences a lateral displacement which could be useful as part of beam steering device design. This lateral offset Δx is illustrated in Fig. 25.29. The relationship between Δx and $\Delta\theta_0$ as shown in Fig. 25.29 is represented by the following:

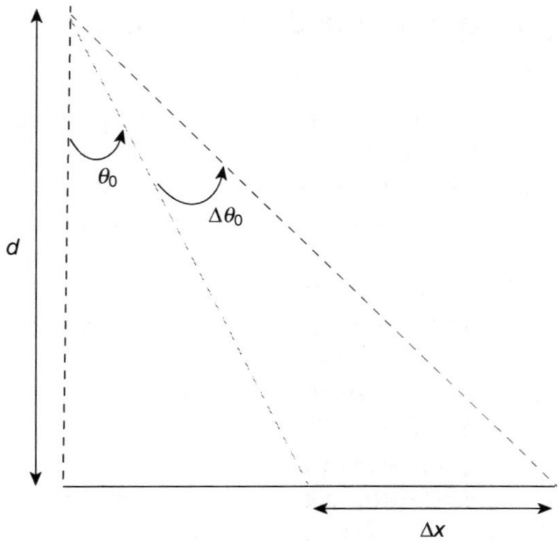

25.29 Geometry that illustrates the FPI output beam lateral displacement caused by the apparent change in output angle due to Pockels effect.

$$\frac{\Delta x}{d} = \tan\theta_0 - \tan(\theta_0 - \Delta\theta_0) \tag{25.97}$$

By rearranging terms and assuming that $\Delta\theta_0$ is small for (25.97) we can perform a Taylor series approximation that yields:

$$\tan\theta_0 - \frac{\Delta x}{d} \approx \tan\theta_0 - (1 + \tan^2\theta_0)\Delta\theta_0 \tag{25.98}$$

Solving (25.98) for $\Delta\theta_0$ and simplifying yields:

$$\Delta\theta_0 \approx \frac{\Delta x}{d}\cos^2\theta_0 \tag{25.99}$$

Substituting (25.99) into (25.95) gives the voltage-induced resonant wavelength shift as a function of output lateral beam displacement as follows:

$$\Delta\lambda = \frac{\begin{array}{l}\left[\dfrac{2d}{\cos\theta_P} - \dfrac{\lambda}{2\pi}\left(\dfrac{\partial\phi_a}{\partial n} + \dfrac{\partial\phi_b}{\partial n}\right)\right]\Delta n \\ -\left[\dfrac{\lambda}{2\pi}\left(\dfrac{\partial\phi_a}{\partial\theta_0} + \dfrac{\partial\phi_b}{\partial\theta_0}\right) + 2n_0 d\tan\theta_P\cos\theta_0\right] \\ \dfrac{\Delta x}{d}\cos^2\theta_0 + 2n\cos\theta_P\Delta d\end{array}}{\dfrac{2nd\cos\theta_P}{\lambda} + \dfrac{\lambda}{2\pi}\left(\dfrac{\partial\phi_a}{\partial\lambda} + \dfrac{\partial\phi_b}{\partial\lambda}\right)} \tag{25.100}$$

25.13 Spacial light modulator device design and fabrication

The general design for a thin film FPI has been described previously in the literature [98–101]. The fabricated device consists of an 8 × 8 array of individually addressable pixels which were designed to operate at a resonant wavelength of 1550 nm. This array is divided into four quadrants each containing 16 identical pixels with pixel sizes of 800, 692, 565, and 400 μm. The quadrant with 565 μm pixels was chosen for analysis due to this region having the least amount of surface roughness in the EO polymer layer. The structural layout of this region is illustrated in Fig. 25.30.

A high reflectance distributed Bragg reflector (DBR) mirror consisting of 10 periods of alternating SiO_2 and Ta_2O_5 quarter-wave layers serves as one mirror in the Fabry–Perot cavity. The other mirror is a thin film of Au that also serves as the top electrode. The Au is coated with a quarter-wave layer of SiO_2 which serves as both an anti-reflection coating and encapsulaton layer to protect the Au surface from damage during the poling process. The pixels are defined by etched 2% Al-doped ZnO (AZO) which also serve as the bottom electrodes. A layer of Ge is patterned in between the pixels for the purpose of drastically reducing the amount of light transmitted due to the high absorption of Ge at 1550 nm coupled with the multiple reflections inside the cavity. This has the effect of increasing the image contrast of the SLM significantly. The spacer layer consists of the SEO100 EO polymer from Soluxra. Simulation of this multi-layer thin film design was performed using Essential Macleod software.

A step by step overview of the fabrication process is illustrated in Fig. 25.31. Initially a high reflectance DBR mirror was sputtered onto a clean 2 in (50 mm) diameter, 2 mm thick fused silica substrate. Afterwards, a 50 nm layer of AZO was deposited onto the surface by pulsed laser deposition (PLD) then annealed at 400°C. Next, a bi-layer lift-off process to deposit

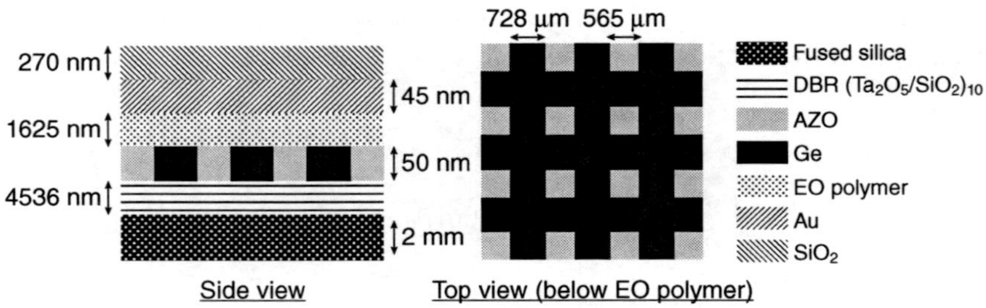

25.30 Layout of fabricated SLM device test region at a design wavelength of 1550 nm. Reproduced with permission from the Optical Society of America.

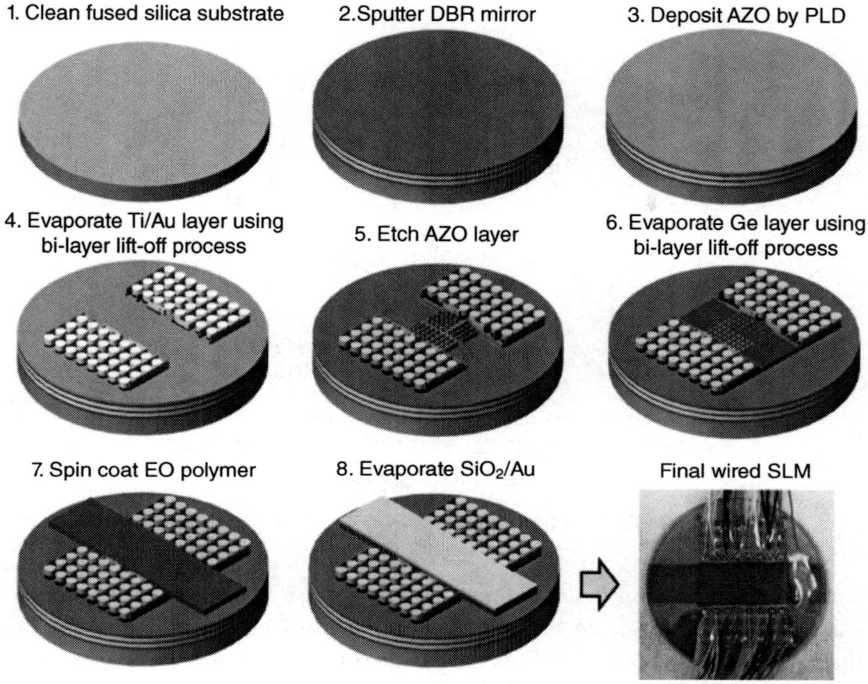

25.31 Fabrication process procedure for creating the spatial light modulator. Reproduced with permission from the Optical Society of America.

contact electrodes with connecting traces consisting of layers of Ti/Au was performed as follows: A 400 nm layer of MicroChem 3A lift-off resist (LOR) was spin-cast onto the substrate then softbaked on a hotplate for 5 min at 175°C. A 1.3 μm layer of Shipley 1813 positive photo-resist was spin-cast on top of the lift-off resist then softbaked for an additional 2 min at 115°C. The sample was then patterned by UV lithography using a dark field mask and the positive resist was developed using Shipley CD-26 metal ion free developer. The exposure time was determined by measuring the power of the UV lamp and determining the required expsoure dose from the interference curve obtained from the photoresist data sheet and illustrated in Fig. 25.32.

The sample was baked on a hotplate for an additional 5 min at 125°C to harden the positive resist in prepartion for the undercutting step. The sample was then submerged in CD-26 developer long enough to undercut the lift-off resist sufficiently to ensure good lift-off of the evaporated material and photo-resist. A 30 nm layer of Ti, which serves as an adhesion layer between the Au and substrate and a 250 nm layer of Au were then

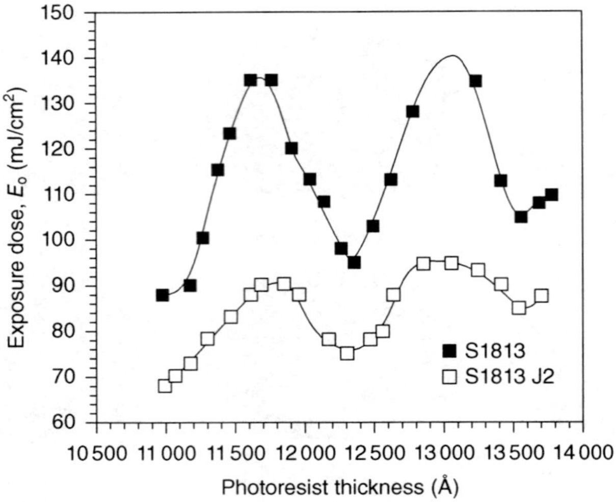

25.32 Shipley 1813 positive photo-resist exposure dosage curve.

deposited by electron-beam evaporation. The masked regions of the sample were removed by placing the sample in a bath of Microposit Remover PG that was heated to 80°C and placed inside a ultrasonicator to ensure complete removal of material. The detailed lift-off procedure is illustrated in Fig. 25.33(a).

Next, an etching procedure to pattern the AZO was performed as follows: a 1.3 μm layer of Shipley 1813 positive photo-resist was spin-cast onto the substrate then softbaked on a hotplate for 2 min at 115°C. The sample was patterned by UV lithography using a clear field mask, then developed using Shipley 352 developer. Etching of the unmasked regions of AZO was done using a dilute etchant consisting of HCl and H_2O in a ratio of 1:800 which gives an etch rate of approximately 5 nm/s. The resist was then stripped to reveal patterned pixels with connecting traces that are aligned directly underneath the Ti/Au traces created in the previous lift-off step. The detailed etching procedure is illustrated in Fig. 25.33(b).

A Ge contrast enhancment layer was then deposited using the previously described lift-off procedure. The higher the cavity finesse is, the lower the thickness of Ge required to block essentially all light in between pixels. A thickness of 50 nm was sufficient for this application. The EO polymer used for this device is the Soluxra produced guest–host polymer SEO100 which is a blended polymer and chromophore mixture that is dissolved in dibromomethane to create a polymer solution that is 6 wt%. The EO polymer was spin-cast to create a 1625 nm thick film and dried under vacuum at 80°C overnight. A 45 nm layer of Au and a 270 nm layer of SiO_2 were deposited on top of the polymer by electron-beam evaporation. Kynar

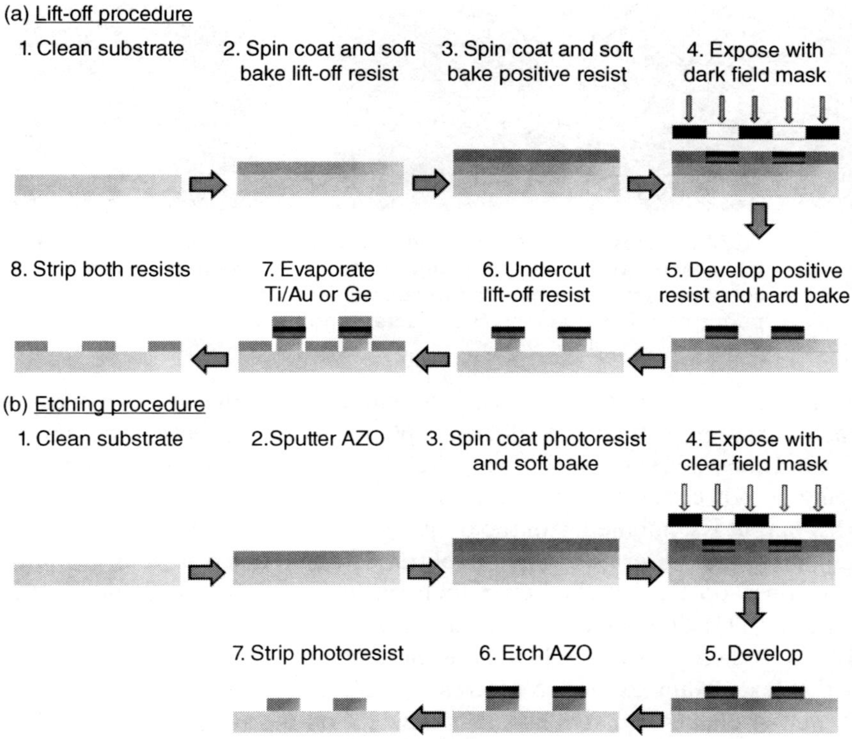

25.33 Detailed description of UV lithography: (a) lift-off procedure and (b) etching procedure. Reproduced with permission from the Optical Society of America.

30-gauge wires were soldered to the bottom Au electrodes. Since the top Au electrode/mirror is very thin it was not possible to solder a wire directly to it. Instead a two-part high operating temperature silver conductive epoxy was used to attach a 30-gauge wire. All wires were covered in two-part insulating epoxy to provide additional mechanical stability. The device was poled using a field of 75 V/μm on a hot plate that was initially at room temperature while increasing the temperature as fast as possible until it reached 133 °C. The poled device was then cooled to ambient room temperature and the applied field was removed.

25.14 Spacial light modulator device characterization

The detailed setup created for characterizing SLM device performance is illustrated in Figs. 25.34 and 25.35. The setup is structured such that the upper portion is used for characterizing the throughput of the SLM

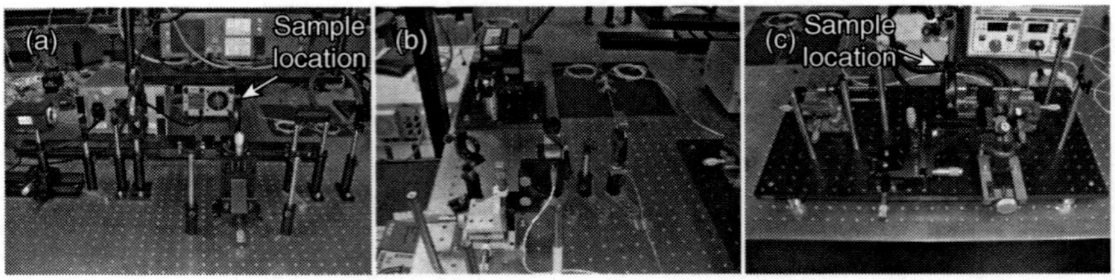

25.34 Pictures of characterization setup for SLM: (a) portion of setup for measuring image throughput of SLM; (b) portion of setup for coupling lasers and broadband source; (c) portion of setup for performing individual pixel characterization.

while the lower portion is used to characterize the performance of each pixel individually. A fiber coupled C-band tunable laser and a C-band amplified spontaneous emission (ASE) source are combined further using a 2-1 coupler with the output connected to a fiber collimator. This output is combined with the output from a spatially filtered free-space C-band tunable laser using a beam splitter. The light that is traveling towards the upper portion of the setup then passes through a variable neutral density (ND) filter and is folded using a plane mirror oriented at 45 °C. The light then goes through an adjustable polarizer and is expanded using a Galilean beam expander to create a collimated beam approximately 30 mm in diameter that is normally incident on the sample. The sample is attached to a rigid kinematic mount that provides two-axis translation. A magnification system consisting of two plano-convex lenses is used after the sample to completely fill the CCD array of the camera (Goodrich SU320M) with an iris being used to remove any stray background light. The reason for coupling the two C-band tunable lasers is to help correct the non-uniformity in the device throughput caused by thickness variations in the spin-cast EO polymer films. The divided light that travels towards the lower portion of the setup passes through a reversed Keplerian beam expander that decreases the size of the beam to a collimated spot size of approximately 500 μm that is normally incident on the sample. The sample is translated in the transverse directions to couple light through each pixel. An objective is used to focus the light that passes through the pixel onto a high-speed detector (InGaAs) or a multimode fiber that is attached to a precision 5-axis stage and connected to an optical spectrum analyzer (OSA).

25.14.1 DC performance

After poling, pixel-by-pixel characterization was performed on the SLM. Using a C-band ASE source the optical spectrum of each pixel is measured

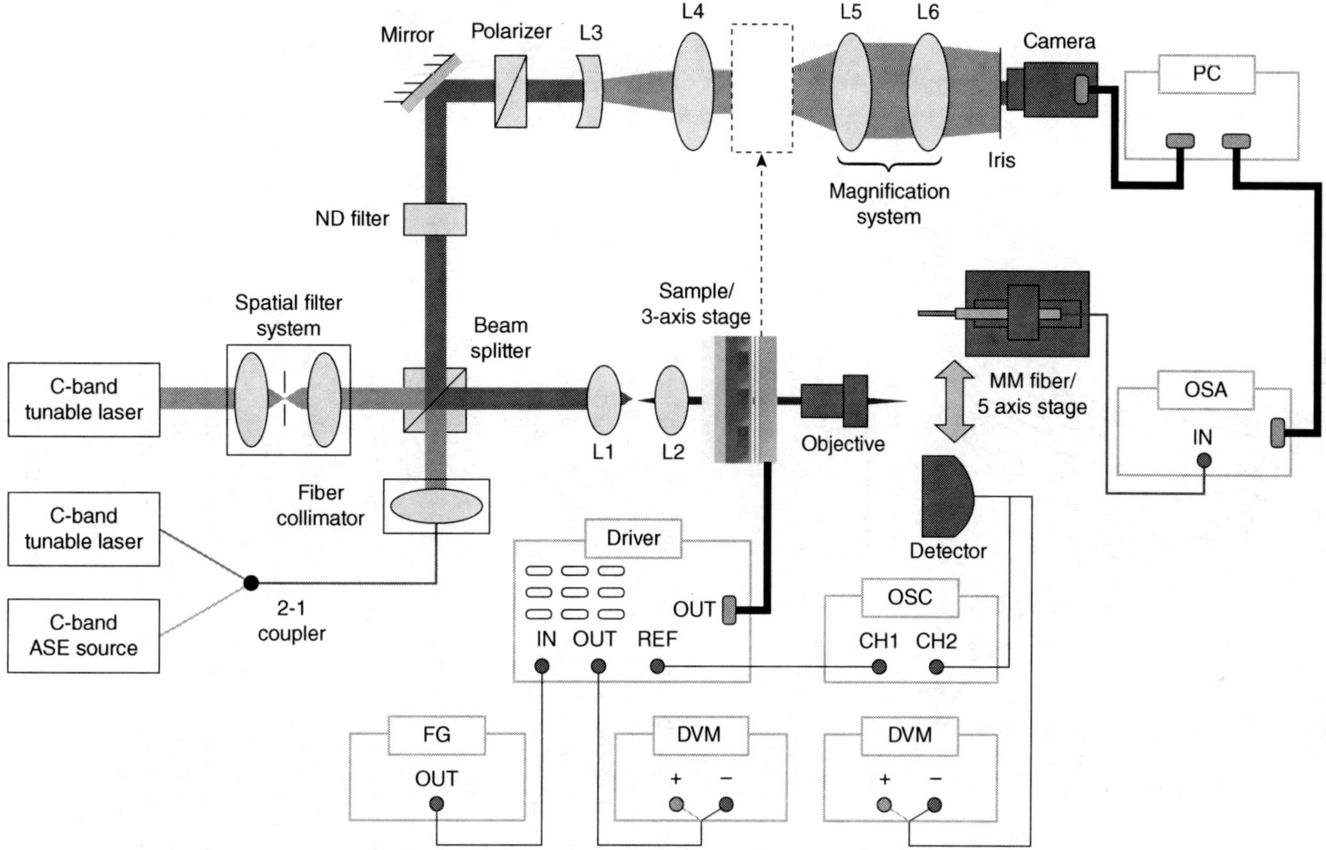

25.35 Schematic of characterization setup for SLM. FG: function generator, DVM: digital voltage meter, OSC: oscilloscope, OSA: optical spectrum analyzer, PC: personal computer, and L: lens. Reproduced with permission from the Optical Society of America.

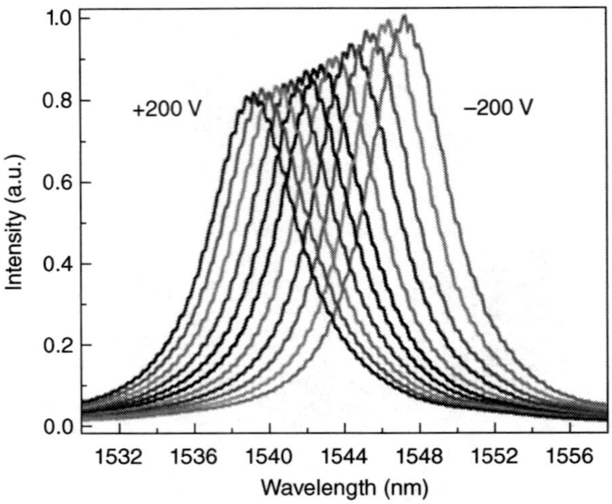

25.36 **The spectral shift for one pixel as applied DC voltage is varied from –-200 V to 200 V. Reproduced with permission from the Optical Society of America.**

by coupling light through the multimode fiber that is connected to the OSA. By taking into account the phase dispersion at the active medium interfaces of the SLM the average finesse is estimated to be 52. To measure the spectrum shift due to an applied electric field, a DC voltage is applied and varied from –200 V to 200 V with the results shown for an arbitrary pixel in Fig. 25.36.

The noticeable change in peak intensity as the applied voltage is varied is due to the unbalanced mirror reflectance of the Fabry–Perot cavity. The average tunability is calculated to be 20 pm/V and the average resonant wavelength shift is illustrated in Fig. 25.37. The ASE source is then replaced with a C-band tunable laser and tuned to the resonant wavelength of each pixel. The average insertion loss is 22 dB and determined by measuring the ratio of the power throughput of each pixel over the power after the sample is removed. The applied DC voltage is varied from –200 V to 200 V while the pixel output is monitored using a detector. From this data the average normalized intensity and the average isolation ratio, which is the average normalized intensity in dB, are calculated with the results being displayed in Fig. 25.38.

25.14.2 Modulation performance

To measure the modulation performance of the SLM a sinusoidal voltage is applied simultaneously to all pixels while performing optical measure-

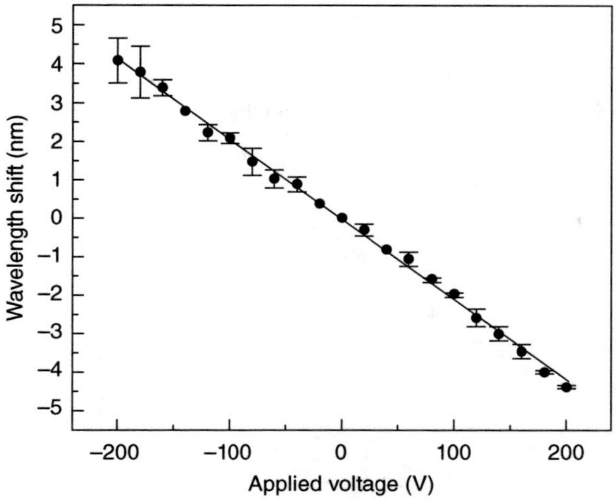

25.37 The average peak wavelength shift as applied DC voltage is varied from −200 V to 200 V. A linear fit to the data is performed. Reproduced with permission from the Optical Society of America.

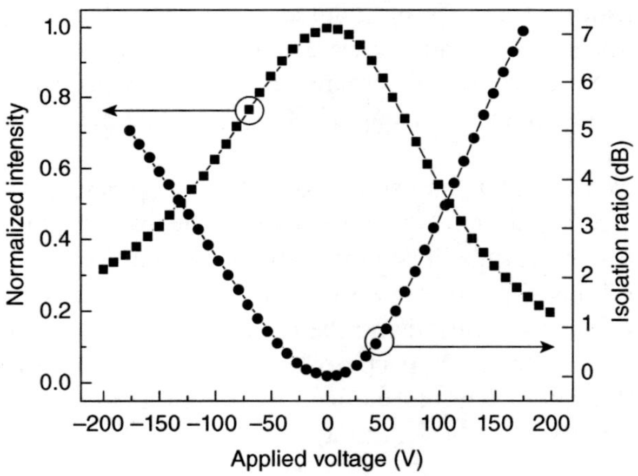

25.38 The average normalized intensity and isolation ratio as applied DC voltage is varied from −200 V to 200 V. Reproduced with permission from the Optical Society of America.

ments on each pixel individually. The reason for applying voltages to all pixels at the same time was to ensure that high-speed performance could be observed with the device in full operation. A C-band tunable laser is tuned to the resonant wavelength and the applied voltage is varied from 10 V_{rms} to 100 V_{rms} at a frequency of 100 kHz to measure the average modulation contrast for all pixels with the results illustrated in Fig. 25.39.

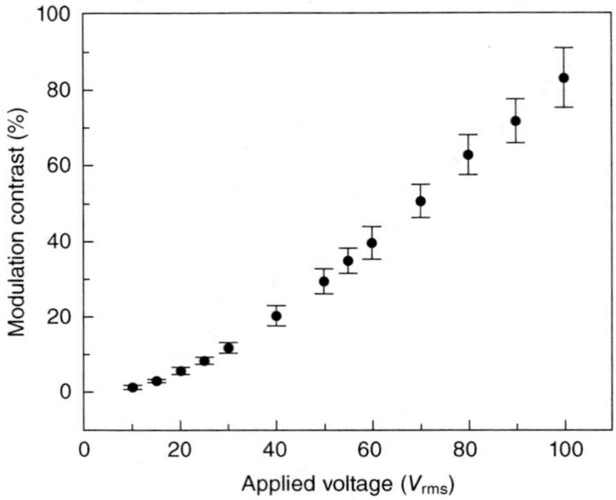

25.39 The relationship between average modulation contrast and applied RMS voltage at a drive frequency of 100 kHz.

Since the transmitted light is modulated regardless of the sign of the applied voltage, the signal has a response frequency that is twice the modulation frequency. An oscilloscope screenshot showing a pixel modulation contrast of 28% and a drive signal of 50 V_{rms} at 500 kHz is shown in Fig. 25.40.

The Pockel's coefficient r_{13} is estimated to be 22.3 pm/V based on the device performance, which is in close agreement to the measurement results obtained from the reflection ellipsometry [90] and Mach–Zehnder interferometry [102] methods. To determine the broadband performance of each pixel the drive frequency is varied and the modulation contrast is recorded. Due to the structure of the SLM each pixel can be approximately modeled as a first order RC circuit. The normalized modulation contrast frequency response for select pixels is shown in Fig. 25.41.

25.14.3 Device throughput

Next the device throughput of the SLM was examined. The C-band ASE source was used to show the uniformity of the SLM throughput over a broad range of wavelengths with the resulting image shown in Fig. 25.42(a). To demonstrate the spatial operation of the SLM array a single wavelength source is needed while a DC voltage is applied. Due to the surface of the polymer being rough we chose to use two coupled C-band lasers tuned to wavelengths of 1545 nm and 1555 nm to increase image uniformity. A 200 V_{DC} signal was applied to selected pixels to demonstrate the contrast

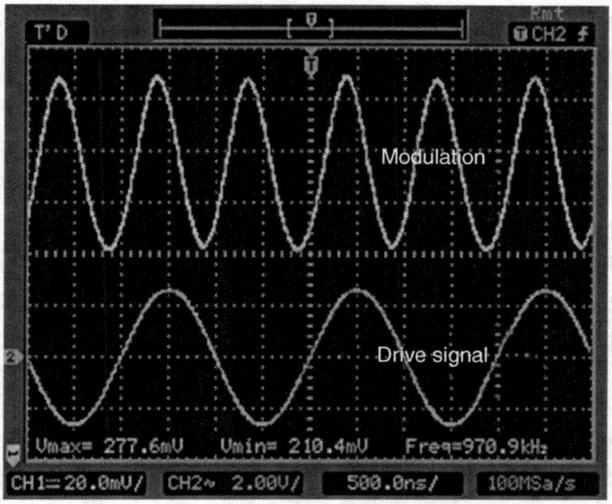

25.40 Oscilloscope screenshot that shows pixel modulation depth of approximately 28% for a drive signal of 50 V_{rms} at 500 kHz. Reproduced with permission from the Optical Society of America.

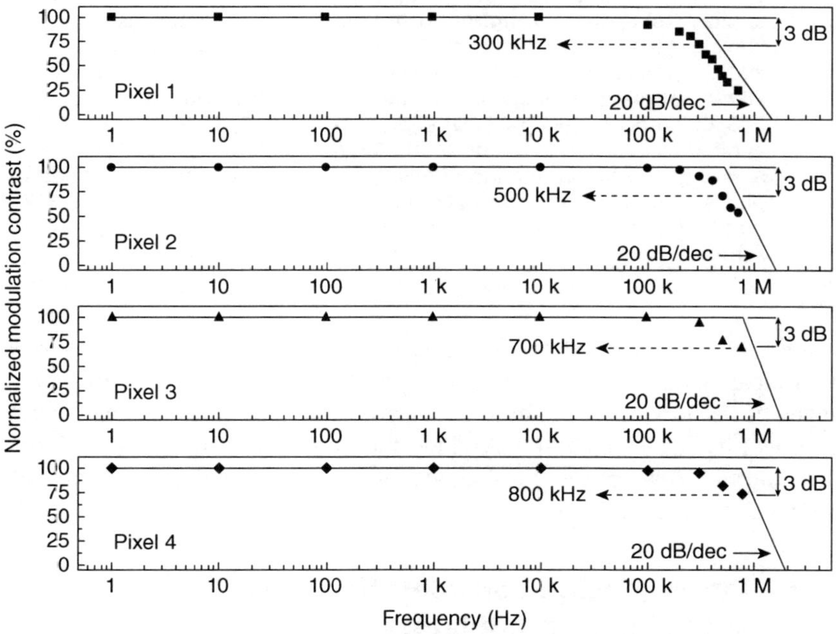

25.41 The normalized modulation contrast frequency response for select pixels. Reproduced with permission from the Optical Society of America.

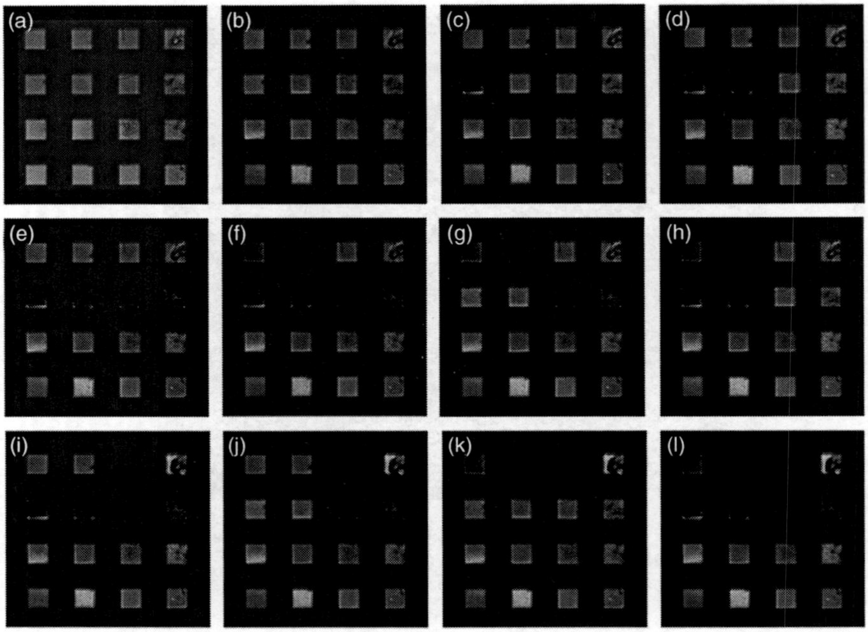

25.42 (a) Device throughput using C-band ASE source with no voltage applied; (b) device throughput using two C-band tunable lasers at wavelengths of 1545 nm and 1555 nm with no voltage applied; (c) 200 V_{DC} applied to pixel 5; (d) 200 V_{DC} applied to pixels 5 and 6; (e) 200 V_{DC} applied to pixels 5, 6, 7, and 8; (f) 200 V_{DC} applied to pixels 1, 2, 5, 6, 7, and 8; (g) 200 V_{DC} applied to pixels 1, 2, 7, and 8; (h) 200 V_{DC} applied to pixels 1, 2, 5, and 6; (i) 200 V_{DC} applied to pixels 3, 4, 5, 6, 7 and 8; (j) 200 V_{DC} applied to pixels 3, 4, 7, and 8; (k) 200 V_{DC} applied to pixels 1, 2, 3, and 4; and (l) 200 V_{DC} applied to pixels 1, 2, 3, 4, 5, 6, 7, and 8. Reproduced with permission from the Optical Society of America.

between the on and off states. The contrast of the camera was adjusted so that a clear distinction between the on and off states could be observed. Images of the device throughput as a voltage is applied to selected pixels are shown in Fig. 25.42(b)–(l). A video file demonstrating the modulation of the top two rows of the 4×4 array with a drive signal 150 V_{rms} at 2 Hz is shown in reference [103].

25.15 Future design considerations for spatial light modulators

Owing to the difficulty in controlling both the amplitude and phase of the light transmitted through the SLM, this device is best suited for applications that require operation in binary mode. Dynamic holography [8, 9] is a current application that needs both wavelength stability and high-speed modulation which current SLM technology is not able to provide. Even though our current SLM design shows potential to be suitable for holography applications there are a few design improvements that need to be made in our next generation of devices to make this a reality.

In order to provide the high resolution and pixel count that holography requires, a set of photomasks with much smaller pixel sizes and an electronic driver with a much larger number of channels would need to be designed. Driving an SLM using a large drive voltage at high frequencies in large densely packed pixel arrays may prove to be difficult, therefore the device voltage requirements need to be addressed. The modulation voltage can be dramatically reduced by increasing the finesse of the device or using alternative poling methods such as coplanar electrode poling to utilize the larger Pockels coefficient, r_{33}. The major source of loss in the SLM is caused by the high absorption of the Au electrode/mirror and the absorption of the AZO layer due to the multiple reflections inside the cavity. The insertion loss can be reduced dramatically by replacing the Au layer with the combination of a DBR mirror and an AZO or other transparent conductive oxide (TCO) layer to serve as the top electrode. If both bottom and top electrode TCO layers are placed outside of the cavity the insertion loss can be further reduced to less than 1 dB but the tradeoff will be an increase in both modulation voltage and RC time constant due to the voltage drop across the DBR mirrors. The speed is currently limited because of the high resistance of the traces that consist of an AZO region overlapped with a Ti/Au region. The variation in speed for each pixel is due to the different AZO trace length for each pixel. Speeds in the microwave regime can be achieved simply by making the traces entirely out of Ti/Au and using microwave engineering design software to properly design the transmission lines and impedance match the device to the drive electronics.

25.16 Conclusion

Photonic polymers have advanced dramatically over the last decade, with passive polymers reaching unprecedented standards of optical loss and uniformity, while EO polymers have achieved exceptionally high EO coefficients, passing Telcordia testing and making their way into real products. In this chapter we have reviewed some fundamental aspects of polymers, presented sophisticated passive polymer devices such as arrayed waveguide gratings and photonic true time delay, reviewed EO polymers, developed a figure of merit for EO polymer modulators, and discussed a new application for high EO coefficient EO polymers, namely ultrafast spatial light modulators. It is clear that photonic polymers are poised to play a significant role in telecommunications and data center photonics in the near future.

25.17 References

1. L. Eldada, R. Blomquist, M. Maxfield, D. Pant, G. Boudoughian, C. Poga and R. A. Norwood 'Thermo-optic planar polymer Bragg grating OADMs with broad tuning range,' *IEEE Phot. Tech. Lett.* **11**, 448 (1999).
2. M.-C. Oh, S.-H. Cho, Y.-O. Noh, H.-J. Lee, J.-J. Joo, and M.-H. Lee, 'Variable optical attenuator based on large-core single-mode polymer waveguide,' *IEEE Phot. Tech. Lett.* **17**, 1890 (2005).
3. R. A. Norwood, C. DeRose, Y. Enami, H. Gan, C. Greenlee, R. Himmelhuber, O. Kropachev, C. Loychik, D. Mathine, Y. Merzylak, M. Fallahi, and N. Peyghambarian, 'Hybrid sol–gel electro-optic polymer modulators: Beating the drive voltage/loss tradeoff,' *J. Nonlinear Optical Phys. & Mater.* **16**, 217 (2007).
4. J. Roncali, 'Synthetic principles for bandgap control in linear π-conjugated systems,' *Chem. Rev.* **97**, 173 (1997).
5. G. Dennler, M. C. Scharber, and C. J. Brabec, 'Polymer-fullerene bulk-heterojunction solar cells,' *Adv. Mat.* **21**, 1323 (2009).
6. X. Zhu, J. Wang, D. Nguyen, J. Thomas, R. A. Norwood, and N. Peyghambarian, 'Linear and nonlinear optical properties of Co_3O_4 nanoparticle-doped polyvinyl-alcohol thin films,' *Opt. Mat. Express* **2**, 103 (2012).
7. X. Zhu, J. Wang, P. Lau, D. Nguyen, R. A. Norwood, and N. Peyghambarian, 'Nonlinear optical performance of periodic structures made from composites of polymers and Co_3O_4 nanoparticles,' *Appl. Phys. Lett.* **97**, 093503 (2010).
8. S. Tay, P. A. Blanche, R. Voorakaranam, A. V. Tunc, W. Lin, S. Rokutanda, T. Gu, D. Flores, P. Wang, P. St. Hilaire, G. Li, J. Thomas, R. A. Norwood, M. Yamamoto and N. Peyghambarian, 'An updateable 3D holographic display,' *Nature* **451**, 694 (2008).
9. P. A. Blanche, A. Bablumian, R. Voorakaranam, C. Christenson, W. Lin, T. Gu, D. Flores, P. Wang, W.-Y. Hsieh, M. Kathaperumal, B. Rachwal, O. Siddiqui, J. Thomas, R. A. Norwood, M. Yamamoto, and N. Peyghambarian, 'Holographic three-dimensional telepresence using large-area photorefractive polymer,' *Nature* **468**, 80 (2010).
10. L. D. Landau, B. G. Levich, 'Dragging of a liquid by a moving plate', *Acta Physiochim* **17**, 42 (1942).
11. D.Meyerhofer, 'Charateristics of resist films produced by spinning', *J. Appl. Phys.* **49**, 3993 (1978).
12. Y. Koike, T. Ishigure, and E. Nehei, 'High bandwidth graded index polymer optical fiber,' *IEEE J. Lightwave Tech.* **13**, 1475 (1995).
13. J. Thomas, P. Gangopadhyay, E. Araci, R. A. Norwood, and N. Peyghambarian, 'Nanoimprinting by melt processing: An easy technique to fabricate versatile nanostructures,' *Adv. Mat.* **23**, 4782 (2011).
14. K. Okamoto, *Fundamentals of Optical Waveguides*, Academic Press; 2nd Edition, (Burlington, MA 2005).
15. T. Kaino, *Polymers for Lightwave and Integrated Optics*, edited by L. A. Hornak, Chap. 1 (Marcel Dekker, New York 1992).

16. G. Fischbeck, R. Moosburger, C. Kostrzewa, A. Achen, and K. Petermann, 'Single mode optical waveguides using high temperature stable polymer with low losses in the 1555 nm range,' *Elec. Lett.* **33**, 518, (1997).
17. A. Yeniay, R. Y. Gao, K. Takayama, R. F. Gao, and A. F. Garito, 'Ultra low loss polymer waveguides' *IEEE J. Lightwave Tech.* **22**, 154 (2004).
18. Y. Kuwana, S. Takenobu, K. Takayama, S. Yokotsuka, and S. Kodama, 'Low loss and highly reliable polymer optical waveguides with perfluorinated dopant-free core,' in *OFC and NFOEC*, Technical Digest, Paper OWF2, Optical Society of America (2006).
19. R. A. Norwood, R. Gao, J. Sharma, and C.C. Teng, 'Sources of loss in single-mode polymer optical waveguides,' *Proc. SPIE* **4439**, 19–28 (2001).
20. C. Pitois, A. Hult, and D. Wiesmann 'Absorption and scattering in low-loss polymer optical waveguides,' *J. Opt. Soc. Am.* **B19** (2001).
21. H. C. van de Hulst, *Light Scattering by Small Particles*, Chapters 9, 114–128, (Dover Publications, New York, 1981).
22. R. M. Almeida, P. J. Morais, and C. Vasconcelos, 'Optical loss mechanism in nanocomposite sol-gel planar waveguides,' *Proc. SPIE* **3136**, 296 (1997).
23. F. S. Bates and P.Wiltzius, 'Spinodal decomposition of a symmetric critical mixture of deuterated and protonated polymer', *J. Chem. Phys.* **91**, 3258 (1989).
24. Y. Kokubun, M. Takizawa, and S. Taga, 'Three-dimensional athermal waveguides for temperature independent lightwave devices,' *Elec. Lett.* **30**, 1223 (1994).
25. A. Kaneko, S. Kamei, Y. Inoue, H. Takahashi, and A. Sugita, 'Athermal silica-based arrayed-waveguide grating (AWG) multi/demultiplexers with new low loss groove design,' *Elec. Lett.* **36** (2000).
26. N. Keil, H. H. Yao, C. Zawadski, J. Bauer, C. Dreyer, and J. Shneider, 'Athermal all-polymer arrayed waveguide grating multiplexer,' *Elec. Lett.* **31**, 2192 (1995).
27. R. Y. Gao, R. Gao, A. Yeniay, K. Takayama, and A. F. Garito, 'Athermal arrayed waveguide grating multi/demultiplexer based on perfluorinated polymers,' *ECOC 2002*, Munich, Germany (2002).
28. H. Zmuda, R. A. Soref, P. Payson, S. Johns, and E. N. Toughlian, 'Photonic beamformer for phased array antennas using a fiber grating prism,' *IEEE Phot. Tech. Lett.* **9**, 241 (1997).
29. P. F. McManamon, T. A. Dorschner, D. L. Corkum, L. J. Friedman, D. S. Hobbs, M. Holz, S. Liberman, H. Q. Nguyen, D. P. Resler, R. C. Sharp, and E. A. Watson, 'Optical phased array technology,' *Proc. IEEE* **84**, 268 (1996).
30. D. K. Hunter *et al.*, 'Buffering in optical packet switches,' *IEEE J. Lightwave Tech.* **16**, 2081 (1998).
31. E. W. Jacobs, R. B. Olsen, J. S. Rodgers, D.C. Evans, T. E. Weiner, and C. Lin, 'RF-over-fiber and optical processing for Navy applications,' *OFC and NFOEC 2007*, Paper OWU1, pp. 1–3, Anaheim, CA (2007).
32. J. S. Rodgers, 'Compact photonic wide band spectrum analyzer,' *SPIE Optics and Photonics Symposium Photonics Technology for Space Environments III*, San Diego, CA (2007).
33. M. Rajarajan, S. S. A. Obayya, B. M. A. Rahman, K. T. V. Grattan, and H. A. El-Mikathi, 'Characterization of low-loss waveguide bends with offset optimization for compact photonic integrated circuits,' *IEE Proc.-Optoelectron.* **147**, 382 (2000).

34. P. Bienstman, E. Six, M. Roelens, M. Vanwolleghem, and R. Baets, 'Calculation of bending losses in dielectric waveguides using eigenmode expansion and perfectly matched layers,' *IEEE Phot. Tech. Lett.* **14**, 164 (2002).
35. S. Kim and A. Gopinath, 'Vector analysis of optical dielectric waveguide bends using finite-difference method,' *IEEE J. Lightwave Tech.* **14**, 2085 (1996).
36. A. Nesterov and U. Troppenz, 'A plane-wave boundary method for analysis of bent optical waveguides,' *IEEE J. Lightwave Tech.* **21**, 2434 (2003).
37. C. Yeung, T. Rozzi, and G. Cerri, 'Crosspolarisation coupling in curved dielectric rib waveguides,' *IEE Proc.-Optoelectron.* **135**, 281 (1988).
38. A. Sugita, A. Kaneko, K. Okamoto, M. Itoh, A. Himeno, and Y Ohmori, 'Very low insertion loss arrayed-waveguide grating with verticallytapered waveguides,' *IEEE Phot. Tech. Lett.* **12**, 1180 (2000).
39. J. Yamauchi *et al.*, 'Novel AWG demultiplexer composed of slabs with islands', *OFC 2002*, **4**, 1, pp. 662–663, (2002).
40. Y. Hibino, 'Recent advances in high-density and large-scale AWG multi/demultiplexers with higher index-contrast silica-based PLCs,' *IEEE J. Sel. Top. Quant. Elec.* **8** (2002).
41. L. Dalton, B. Robinson, A. Jen, P. Ried, B. Eichinger, P. Sullivan, A. Akelaitis, D. Bale, M. Haller, Jingdong Luo, Sen Liu, Yi Liao, K. Firestone, N. Bhatambrekar, S. Bhattacharjee, J. Sinness, S. Hammond, N. Buker, R. Snoeberger, M. Lingwood, H. Rommel, J. Amend, Sei-Hum Jang, Antao Chen, and W. Steier, 'Electro-optic coefficients of 500 pm/V and beyond for organic materials,' *Proc. SPIE* **5935**, 593502-1–593502-12 (2005).
42. T. D. Kim, J. W. Kang, J. D. Luo, S. H. Jang, J. W. Ka, N. Tucker, J. B. Benedict, L. R. Dalton, T. Gray, R. M. Overney, D. H. Park, W. N. Herman, and A. K. Y. Jen, 'Ultralarge and thermally stable electro-optic activities from supramolecular self-assembled molecular glasses,' *J. Am. Chem. Soc.* **129**, 488 (2007).
43. T. Baehr-Jones, M. Hochberg, G. X. Wang, R. Lawson, Y. Liao, P. A. Sullivan, L. Dalton, A. K. Y. Jen, and A. Scherer, 'Optical modulation and detection in slotted silicon waveguides,' *Optics Express* **13**, 5216 (2005).
44. T. Baehr-Jones, B. Penkov, J. Q. Huang, P. Sullivan, J. Davies, J. Takayesu, J. D. Luo, T. D. Kim, L. Dalton, A. Jen, M. Hochberg, and A. Scherer, 'Nonlinear polymer-clad silicon slot waveguide modulator with a half wave voltage of 0.25 V', *Appl. Phys. Lett.* **92**, 163303 (2008).
45. B. A. Block, T. R. Younkin, P. S. Davids, M. R. Reshotko, P. Chang, B. M. Polishak, S. Huang, J. Luo, and A. K. Y. Jen, 'Electro-optic polymer cladding ring resonator modulators,' *Optics Express* **16**, 18326 (2008).
46. L. Dalton, A. Harper, A. Ren, F. Wang, G. Todorova, J. Chen, C. Zhang, and M. Lee, 'Polymeric electro-optic modulators: From chromophore design to integration with semiconductor very large scale integration electronics and silica fiber optics,' *Indu. Eng. Chem. Res.* **38**, 8 (1999).
47. M. Hochberg, T. Baehr-Jones, G. X. Wang, M. Shearn, K. Harvard, J. D. Luo, B. Q. Chen, Z. W. Shi, R. Lawson, P. Sullivan, A. K. Y. Jen, L. Dalton, and A. Scherer, 'Terahertz all-optical modulation in a silicon-polymer hybrid system,' *Nature Materials* **5**, 703 (2006).
48. L. Dalton, 'Rational design of organic electro-optic materials,' *J. Phy. Condensed Matter* **15**, R897–R934 (2003).

49. S. R. Marder, B. Kippelen, A. K. Y. Jen, and N. Peyghambarian, 'Design and synthesis of chromophores and polymers for electro-optic and photorefractive applications,' *Nature* **388**, 845 (1997).
50. P. A. Sullivan, H. Rommel, Y. Liao, B. C. Olbricht, A. J. P. Akelaitis, K. A. Firestone, J. W. Kang, J. D. Luo, J. A. Davies, D. H. Choi, B. E. Eichinger, P. J. Reid, A. T. Chen, A. K. Y. Jen, B. H. Robinson, and L. R. Dalton, 'Theory-guided design and synthesis of multichromophore dendrimers: An analysis of the electro-optic effect,' *J. Am. Chem. Soc.* **129**, 7523 (2007).
51. M. Ahlheim, M. Barzoukas, P. V. Bedworth, M. BlanchardDesce, A. Fort, Z. Y. Hu, S. R. Marder, J. W. Perry, C. Runser, M. Staehelin, and B. Zysset, 'Chromophores with strong heterocyclic acceptors: A poled polymer with a large electro-optic coefficient,' *Science* **271**, 335 (1996).
52. D. M. Burland, R. D. Miller, and C. A. Walsh, 'Second-order nonlinearity in poled-polymer systems,' *Chem. Rev.* **94**, 31 (1994).
53. Y. J. Cheng, J. D. Luo, S. Hau, D. H. Bale, T. D. Kim, Z. W. Shi, D. B. Lao, N. M. Tucker, Y. Q. Tian, L. R. Dalton, P. J. Reid, and A. K. Y. Jen, 'Large electro-optic activity and enhanced thermal stability from diarylaminophenyl-containing high-beta nonlinear optical chromophores,' *Chem. Mat.* **19**, 1154 (2007).
54. Y.-J. Cheng, J. Luo, S. Huang, X. Zhou, Z. Shi, T.-D.Kim, D. H. Bale, S. Takahashi, A. Yick, B. M. Polishak, S.-H. Jang, L. R. Dalton, P. J. Reid, W. H. Steier, and A. K.-Y. Jen, 'Donor–acceptor thiolated polyenic chromophores exhibiting large optical nonlinearity and excellent photostability,' *Chem. Mat.* **20**, 5047 (2008).
55. L. Dalton, 'Polymeric electro-optic materials: optimization of electro-optic activity, minimization of optical loss, and fine-tuning of device performance,' *Opt. Eng.* **39**, 589 (2000).
56. L. Dalton, 'Nonlinear optical polymeric materials: From chromophore design to commercial applications,' *Polym. Photonics Applications* I **158**, 1 (2002).
57. L. Dalton, 'Organic electro-optic materials,' *Pure Appl. Chem.* **76**, 1421 (2004).
58. L. R. Dalton, W. H. Steier, B. H. Robinson, C. Zhang, A. Ren, S. Garner, A. T. Chen, T. Londergan, L. Irwin, B. Carlson, L. Fifield, G. Phelan, C. Kincaid, J. Amend, and A. Jen, 'From molecules to opto-chips: organic electro-optic materials,' *J. Mat. Chem.* **9**, 1905 (1999).
59. S. Gilmour, S. R. Marder, J. W. Perry, and L. T. Cheng, 'Large 2nd-order optical nonlinearities and enhanced thermal stabilities in extended thiophene-containing compounds,' *Adv. Mat.* **6**, 494 (1994).
60. T. D. Kim, J. D. Luo, J. W. Ka, S. Hau, Y. Q. Tian, Z. W. Shi, N. M. Tucker, S. H. Jang, J. W. Kang, and A. K. Y. Jen, 'Ultralarge and thermally stable electro-optic activities from Diels-Alder crosslinkable polymers containing binary chromophore systems,' *Adv. Mater.* **18**, 3038 (2006).
61. H. Ma, B. Q. Chen, T. Sassa, L. R. Dalton, and A. K. Y. Jen, 'Highly efficient and thermally stable nonlinear optical dendrimer for electrooptics,' *J. Am. Chem. Soc.* **123**, 986 (2001).
62. S. R. Marder and J. W. Perry, 'Molecular materials for 2nd-order nonlinear-optical applications,' *Adv. Mater.* **5**, 804 (1993).
63. S. R. Marder, J. E. Sohn, and G. D. Stucky, *Materials for Nonlinear Optics: Chemical Perspectives* (American Chemical Society, 1991).

64. B. H. Robinson, L. R. Dalton, A. W. Harper, A. Ren, F. Wang, C. Zhang, G. Todorova, M. Lee, R. Aniszfeld, S. Garner, A. Chen, W. H. Steier, S. Houbrecht, A. Persoons, I. Ledoux, J. Zyss, and A. K. Y. Jen, 'The molecular and supramolecular engineering of polymeric electro-optic materials,' *Chem. Phys.* **245**, 35 (1999).
65. Z. Shi, S. Hau, J. Luo, T.-D. Kim, N. M. Tucker, J.-W. Ka, H. Sun, A. Pyajt, L. Dalton, A. Chen, and A. K. -Y. Jen, 'Highly efficient Diels-Alder crosslinkable electro-optic dendrimers for electric-field sensors,' *Adv. Func. Mat.* **17** (2007), 2557 (2007).
66. C. Zhang, C. G. Wang, L. R. Dalton, H. Zhang, and W. H. Steier, 'Progress toward device-quality second-order nonlinear optical materials. 4. A trilink high mu beta NLO chromophore in thermoset polyurethane: A "guest–host" approach to larger electrooptic coefficients,' *Macromol.* **34**, 253 (2001).
67. X.-H. Zhou, J. Luo, S. Huang, T.-D. Kim, Z. Shi, Y.-J. Cheng, S.-H. Jang, D. B. Knorr Jr., M. Overney, and A. K.-Y. Jen, 'Supramolecular self-assembled dendritic nonlinear optical chromophores: Fine-tuning of arene-perfluoroarene interactions for ultralarge electro-optic activity and enhanced thermal stability,' *Adv. Mat.* **21**, 1976 (2009).
68. B. Bortnik, Y. C. Hung, H. Tazawa, B. J. Seo, J. D. Luo, A. K. Y. Jen, W. H. Steier, and H. R. Fetterman, 'Electrooptic polymer ring resonator modulation up to 165 GHz', *IEEE J. Sel. Top. Quant. Elec.* **13**, 104 (2007).
69. D. T. Chen, H. R. Fetterman, A. T. Chen, W. H. Steier, L. R. Dalton, W. S. Wang, and Y. Q. Shi, 'Demonstration of 110 GHz electro-optic polymer modulators,' *Appl. Phys. Lett.* **70**, 3335 (1997).
70. H. Chen, B. Chen, D. Huang, D. Jin, J. D. Luo, A. K. Y. Jen, and R. Dinu, 'Broadband electro-optic polymer modulators with high electro-optic activity and low poling induced optical loss,' *Appl. Phys. Lett.* **93**, 043507 (2008).
71. R. J. Michalak, Y. H. Kuo, F. D. Nash, A. Szep, J. R. Caffey, P. M. Payson, F. Haas, B. F. McKeon, P. R. Cook, G. A. Brost, J. D. Luo, A. K. Y. Jen, L. R. Dalton, and W. H. Steier, 'High-speed AJL8/APC polymer modulator,' *IEEE Phot. Tech. Lett.* **18**, 1207 (2006).
72. C. C. Teng, 'Traveling-wave polymeric optical-intensity modulator with more than 40 GHz of 3-dB electrical bandwidth,' *Appl. Phys. Lett.* **60**, 1540 (1992).
73. C. T. Derose, R. Himmelhuber, D. Mathine, R. A. Norwood, J. Luo, A. K.-Y. Jen, and N. Peyghambarian, 'High Δn strip-loaded electro-optic polymer waveguide modulator with low insertion loss,' *Opt. Express* **17**, 3316 (2009).
74. Y. Enami, C. T. DeRose, C. Loychik, D. Mathine, R. A. Norwood, J. Luo, A. K. Y. Jen, and N. Peyghambarian, 'Low half-wave voltage and high electro-optic effect in hybrid polymer/sol-gel waveguide modulators,' *Appl. Phys. Lett.* **89**, 143506 (2006).
75. Y. Enami, C. T. Derose, D. Mathine, C. Loychik, C. Greenlee, R. A. Norwood, T. D. Kim, J. Luo, Y. Tian, A. K. Y. Jen, and N. Peyghambarian, 'Hybrid polymer/sol-gel waveguide modulators with exceptionally large electro-optic coefficients,' *Nature Photonics* **1**, 180 (2007).
76. Y. Enami, D. Mathine, C. T. DeRose, R. A. Norwood, J. Luo, A. K. Y. Jen, and N. Peyghambarian, 'Hybrid cross-linkable polymer/sol-gel waveguide modulators with 0.65 V half wave voltage at 1550 nm,' *Appl. Phys. Lett.* **91**, 093505 (2007).

77. Y. Q. Shi, C. Zhang, H. Zhang, J. H. Bechtel, L. R. Dalton, B. H. Robinson, and W. H. Steier, 'Low (sub-1-volt) halfwave voltage polymeric electro-optic modulators achieved by controlling chromophore shape,' *Science* **288**, 119 (2000).
78. N. Bloembergen, *Nonlinear Optics*, 4th edition (World Scientific, 1996).
79. R. W. Boyd, *Nonlinear Optics*, 3rd edition (Academic Press, 2008).
80. C. Flytzanis and J. L. Oudar, *Nonlinear Optics: Materials and Devices* (Springer, Berlin, 1985).
81. P. N. Prasad and D. J. WIlliams, *Introduction to Nonlinear Optical Effects in Molecules and Polymers* (Wiley Interscience, 1991).
82. D. J. Williams, *Nonlinear Optical Properties of Organic and Polymeric Materials* (American Chemical Society, 1983).
83. M. Born and E. Wolf, *Principles of Optics*, 7th edition (Cambridge University Press, 1999).
84. H. R. Cho, M. J. Shin, S. H. Han, and J. W. Wu, 'Mach-Zehnder interferometer measurement of the Pockels effect in a poled polymer film with a coplanar electrode structure,' *Appl. Phys. Lett.* **69**, 3788 (1996).
85. J. L. Oudar and D. S. Chemla, 'Theory of second-order optical susceptibilities of benzene substitutes,' *Opt. Comm.* **13**, 164 (1975).
86. B. H. Robinson and L. R. Dalton, 'Monte Carlo statistical mechanical simulations of the competition of intermolecular electrostatic and poling-field interactions in defining macroscopic electro-optic activity for organic chromophore/polymer materials,' *J. Phys. Chem.* **A104**, 4785 (2000).
87. J. L. Oudar and D. S. Chemla, 'Hyperpolarizabilities of the nitroanilines and their relations to the excited state dipole moment,' *J. Chem. Phys.* **66**, 2664 (1977).
88. K. D. Singer, M. G. Kuzyk, and J. E. Sohn, 'Second-order nonlinear-optical processes in orientationally ordered materials: Relationship between molecular and macroscopic properties,' *J. Opt. Soc. Am. B* **4**, 968 (1987).
89. D. J. Williams, 'Organic polymeric and non-polymeric materials with large optical nonlinearities,' *Angew. Chem. Inter. Ed.* **23**, 690 (1984).
90. C. C. Teng and H. T. Man, 'Simple reflection technique for measuring the electrooptic coefficient of poled polymers,' *Appl. Phys. Lett.* **56**, 1734 (1990).
91. A. W. Harper, S. Sun, L. R. Dalton, S. M. Garner, A. Chen, S. Kalluri, W. H. Steier, and B. H. Robinson, 'Translating microscopic optical nonlinearity into macroscopic optical nonlinearity: the role of chromophore chromophore electrostatic interactions,' *J. Opt. Soc. Am.* **B15**, 329 (1998).
92. C. C. Teng, M. A. MortazaviI, and G. K. Boudoughian, 'Origin of the poling-induced optical loss in a nonlinear-optical polymeric wave-guide,' *Appl. Phys. Lett.* **66**, 667 (1995).
93. R. Blum, M. Sprave, J. Sablotny, and M. Eich, 'High-electric-field poling of nonlinear optical polymers,' *J. Opt. Soc. Am.* **B15**, 318 (1998).
94. C. T. DeRose, Y. Enami, C. Loychik, R. A. Norwood, D. Mathine, M. Fallahi, N. Peyghambarian, J. D. Luo, A. K. Y. Jen, M. Kathaperumal, and M. Yamamoto, 'Pockels coefficient enhancement of poled electro-optic polymers with a hybrid organic–inorganic sol-gel cladding layer,' *Appl. Phys. Lett.* **89**, 131102 (2006).

95. J. P. Drummond, S. J. Clarson, J. S. Zetts, F. K. Hopkins, and S. J. Caracci, 'Enhanced electro-optic poling in guest–host systems using conductive polymer-based cladding layers,' *Appl. Phys. Lett.* **74**, 368 (1999).
96. S. Huang, T. D. Kim, J. D. Luo, S. K. Hau, Z. W. Shi, X. H. Zhou, H. L. Yip, and A. K. Y. Jen, 'Highly efficient electro-optic polymers through improved poling using a thin TiO_2-modified transparent electrode,' *Appl. Phys. Lett.* **96**, 243311 (2010).
97. M. Sprave, R. Blum, and R. Eich, 'High electric field conduction mechanisms in electrode poling of electro-optic polymers,' *Appl. Phys. Lett.* **69**, 2962 (1996).
98. H. Gan, H. Zhang, C. T. DeRose, R. A. Norwood, N. Peyghambarian, M. Fallahi, J. Luo, B. Chen, and A. K.-Y. Jen, 'Low drive voltage Fabry–Perot etalon device tunable filters using poled hybrid sol–gel materials,' *Appl. Phys. Lett.* **89**, 041127 (2006).
99. H. Gan, H. Zhang, C. T. DeRose, R. A. Norwood, M. Fallahi, J. Luo, A. K.-Y. Jen, B. Liu, S.-T. Ho, and N. Peyghambarian, 'Hybrid Fabry–Perot etalon using an electro-optic polymer for optical modulation,' *Appl. Phys. Lett.* **89**, 141113 (2006).
100. H. Gan, C. Greenlee, C. Sheng, R. A. Norwood, M. Fallahi, S. Wang, W. Lin, M. Yamamoto, K. Mohanalingam, and N. Peyghambarian, 'Near-resonance electro-optic activity enhancement and improved modulation performance for polymer based Fabry-Perot interferometers,' *Appl. Phys. Lett.* **92**, 203302 (2008).
101. E. Garmire, 'Theory of quarter-wave-stack dielectric mirrors used in a thin Fabry-Perot filter,' *Appl. Opt.* **42**, 5442 (2003).
102. C. Greenlee, A. Guilmo, A. Opadeyi, R. Himmelhuber, R. A. Norwood, M. Fallahi, J. Luo, S. Huang, X.-H. Zhou, A. K.-Y. Jen, and N. Peyghambarian, 'Mach–Zehnder interferometry method for decoupling electro-optic and piezoelectric effects in poled polymer films,' *Appl. Phys. Lett.* **97**, 041109 (2010).
103. C. Greenlee, J. Luo, K. Leedy, B. Bayraktaroglu, R. A. Norwood, M. Fallahi, A. K.-Y. Jen, and N. Peyghambarian, 'Electro-optic polymer spatial light modulator based on a Fabry–Perot interferometer configuration,' *Opt. Express* **19**, 12750 (2011).

25.18 Appendix: acronyms

AFM – atomic force microscopy
AIBN – azoisobutyronitrile
ASE – amplified spontaneous emission
AWG – arrayed waveguide grating
AZO – aluminum doped zinc oxide
CDMA – code division multiple access
CTE – coefficient of thermal expansion
DBR – distributed Bragg reflection
DWDM – dense warelength division multiplexing
DC – direct current
ECL – external cavity laser

EMI – electromagnetic interference
EO – electro-optic
FPI – Fabry–Perot interferometer
LOR – lift-off resist
MMA – methyl methacrylate
MZM – Mach–Zehder modulator
NA – numerical aperture
NF – noise figure
OSA – optical spectrum analyzer
PAA – phased array antenna
PDI – polydispersity index
PDL – polarization dependent loss
PLD – pulsed laser deposition
PMMA – poly (methylmethacrylate)
RF – radio frequency
RIN – relative intensity noise
RPF – recirculating photonic filter
rpm – revolutions per minute
SEM – scanning electron micrograph
SLM – spatial light modulator
SMF – single mode fiber
SNR – signal to noise ratio
TCO – transparent conducting oxide
TE – transverse electric
TM – transverse magnetic
TO – thermo-optic
TTD – true time delay
VLSI – very large scale integration

Index

Lightning Source UK Ltd.
Milton Keynes UK
UKOW03n0609280713

214406UK00001B/5/P

9 780857 092656